Vertebrates

Comparative Anatomy, Function, Evolution

Seventh edition

Seventh edition

Vertebrates

Comparative Anatomy, Function, Evolution

Kenneth V. Kardong, Ph.D.

Washington State University

VERTEBRATES: COMPARATIVE ANATOMY, FUNCTION, EVOLUTION, SEVENTH EDITION

2 3 4 5 6 7 8 9 0 QVS/QVS 1 0 9 8 7 6 5 4

ISBN 978–0–07–802302–6
MHID 0–07–802302-5

Senior Vice President, Products & Markets: *Kurt L. Strand*
Vice President, General Manager, Products & Markets: *Marty Lange*
Vice President, Content Production & Technology Services: *Kimberly Meriwether David*
Managing Director: *Michael Hackett*
Marketing Manager: *Patrick Reidy*
Director, Content Production: *Terri Schiesl*
Senior Content Project Manager: *Melissa M. Leick*
Buyer: *Susan K. Culbertson*
Compositor: *Laserwords Private Limited*
Typeface: *10/12 Adobe Caslon Pro*
Printer: *Quad/Graphics*

All credits appearing on page or at the end of the book are considered to be an extension of the copyright page.

Library of Congress Cataloging-in-Publication Data

Kardong, Kenneth V.
    Vertebrates : comparative anatomy, function, evolution / Kenneth V. Kardong, Ph.D.—Seventh edition.
        pages cm
    ISBN 978–0–07–802302–6 (alk. paper)
 1. Vertebrates—Anatomy.   2. Vertebrates—Physiology.   3. Anatomy, Comparative.
 4. Vertebrates—Evolution.   I. Title.
    QL805.K35 2015
    596—dc23
                                                                    2013042281

www.mhhe.com

Dedicated with pleasure and gratitude to
T. H. Frazzetta
who, like me, remembers fondly
Richard C. Snyder

Brief Contents

CHAPTER **one**
INTRODUCTION   1

CHAPTER **two**
ORIGIN OF CHORDATES   48

CHAPTER **three**
THE VERTEBRATE STORY   82

CHAPTER **four**
BIOLOGICAL DESIGN   128

CHAPTER **five**
LIFE HISTORY   161

CHAPTER **six**
INTEGUMENT   212

CHAPTER **seven**
SKELETAL SYSTEM: THE SKULL   241

CHAPTER **eight**
SKELETAL SYSTEM: THE AXIAL SKELETON   294

CHAPTER **nine**
SKELETAL SYSTEM: THE APPENDICULAR SKELETON   325

CHAPTER **ten**
THE MUSCULAR SYSTEM   372

CHAPTER **eleven**
THE RESPIRATORY SYSTEM   413

CHAPTER **twelve**
THE CIRCULATORY SYSTEM   451

CHAPTER **thirteen**
THE DIGESTIVE SYSTEM   503

CHAPTER **fourteen**
THE UROGENITAL SYSTEM   545

CHAPTER **fifteen**
THE ENDOCRINE SYSTEM   592

CHAPTER **sixteen**
THE NERVOUS SYSTEM   625

CHAPTER **seventeen**
SENSORY ORGANS   671

CHAPTER **eighteen**
CONCLUSIONS   714

Contents

Preface   xv

CHAPTER one

INTRODUCTION   1

COMPARATIVE VERTEBRATE MORPHOLOGY   1
- Designs of Students   2
- Vertebrate Design—Form and Function   3
- Grand Design   4

HISTORICAL PREDECESSORS—EVOLUTION   4
- The Process behind the Change   5
- Linnaeus   5
- Naturalists   6
- J-B. de Lamarck   6
  - *Acquired Characteristics   7*
  - *Upward to Perfection   7*
- Natural Selection   8
- A. R. Wallace   8
- Charles Darwin   9
- Critics and Controversy   10

HISTORICAL PREDECESSORS—MORPHOLOGY   11
- Georges Cuvier   11
- Richard Owen   12

WHY ARE THERE NO FLYING ELEPHANTS?   14

MORPHOLOGICAL CONCEPTS   15
- Similarities   15
- Symmetry   16
- Segmentation   16

EVOLUTIONARY MORPHOLOGY   19
- Function and Biological Role   19
- Preadaptation   20
- Evolution as Remodeling   21

PHYLOGENY   21
- Of Bean Stalks and Bushes   21
- Simplification   23
- Patterns of Phylogeny   23
- Grades and Clades   24

PALEONTOLOGY   29
- Fossilization and Fossils   30
- Recovery and Restoration   32
- From Animal to Fossil   35
- Dating Fossils   37
  - *Stratigraphy   37*
  - *Index Fossils   38*
  - *Radiometric Dating   39*
  - *Geological Ages   39*

TOOLS OF THE TRADE   41
- The Question   41
- The Function   42
- The Biological Role   46

OVERVIEW   47

CHAPTER two

ORIGIN OF CHORDATES   48

CHORDATE PHYLOGENY   48

CHORDATE CHARACTERISTICS   50
- Notochord   51
- Pharyngeal Slits   52
- Endostyle or Thyroid Gland   53
- Dorsal and Tubular Nerve Cord   54
- Postanal Tail   54
- Chordate Body Plan   54

PROTOCHORDATES   54
- General Protochordate Features   55
- Hemichordata   56
  - *Enteropneusta—"Acorn Worms"   57*
  - *Pterobranchia   60*
  - *Hemichordate Phylogenetic Affinities to Chordates   60*
  - *Hemichordate Phylogenetic Affinities to Echinoderms   61*
- Cephalochordata   61
- Urochordata   66
  - *Ascidiacea—"Sea Squirts"   68*

Larvacea (Appendicularia)   70
Thaliacea   73
Overview of Protochordates   73

CHORDATE ORIGINS   74
Chordates from Annelids and Arthropods   75
Chordates from Echinoderms   76
Auricularian Hypothesis   76
Larval Echinoderm to Chordate Tadpole   77
Chordate Origins and Phylogeny   78
Dorsoventral Inversion   78

OVERVIEW   80

CHAPTER three

THE VERTEBRATE STORY   82

INTRODUCTION   82
Innovations   83
Vertebral Column   83
Head   84
Origin of Vertebrates   84
Step 1: Prevertebrate   85
Step 2: Agnathan   85
Step 3: Gnathostome   86
Vertebrate Classification   86

AGNATHANS   86
Living Agnathans   86
Myxinoidea   86
Petromyzontiformes   88
Early Vertebrate Fossils   89
Conodonts   89
Ostracoderms   90
Pteraspidomorphi   93
Other Ostracoderms (Osteostracans,
Anaspids, Thelodonts)   93
Overview of Agnathan Evolution   93

GNATHOSTOMES   94
Placodermi   94
Chondrichthyes   95
Elasmobranchii—Sharks and Rays   96
Holocephali—Chimaeras   97

TELEOSTOMI   97
Acanthodii   97
Osteichthyes   98
Actinopterygii   99
Sarcopterygii   101
Overview of Fish Phylogeny   104

TETRAPODS   104
Primitive Tetrapods   104
Labyrinthodonts   104
Lissamphibia—Modern Amphibians   106
Urodela (Caudata)-Salamanders   107

Salientia (Anura)–Frogs   107
Gymnophiona (Apoda)–Caecilians   108
Lepospondyls   108

AMNIOTES   108
Stem-Amniotes   110
Sauropsids   110
Mesosaurs   110
Reptilia   111
Synapsida   119
Pelycosauria   119
Therapsida   120
Mammalia   121

OVERVIEW   126

CHAPTER four

BIOLOGICAL DESIGN   128

INTRODUCTION: SIZE AND SHAPE   128

SIZE   129
Relationships Among Length, Area, and Volume   129
Surface Area   133
Volume and Mass   133

SHAPE   134
Allometry   134
Transformation Grids   134

ON THE CONSEQUENCES OF BEING
THE RIGHT SIZE   137

BIOMECHANICS   137
Fundamental Principles   138
Basic Quantities—Length, Time, and Mass   138
Units   139
Derived Quantities—Velocity, Acceleration,
Force, and Relatives   139
Reference Systems   140
Center of Mass   140
Vectors   140
Basic Force Laws   141
Free Bodies and Forces   142
Torques and Levers   142
Land and Fluid   144
Life on Land: Gravity   144
Life in Fluids   145
Machines   148
Strength of Materials   148
Loads   149
Biological Design and Biological Failure   149
Tissue Response to Mechanical Stress   151
Responsiveness of Bone   152

BIOPHYSICS AND OTHER PHYSICAL PROCESSES   155
Diffusion and Exchange   156
Pressures and Partial Pressures   156

    Countercurrent, Concurrent,
    and Crosscurrent Exchange   *156*
  Optics   158
    Depth Perception   *158*
    Accommodation   *159*

OVERVIEW   159

CHAPTER five
LIFE HISTORY   161

INTRODUCTION   161

EARLY EMBRYOLOGY   163
  Fertilization   163
  Cleavage   164
    Amphioxus   *164*
    Fishes   *165*
    Amphibians   *165*
    Reptiles and Birds   *165*
    Mammals   *166*
    Overview of Cleavage   *167*
  Gastrulation and Neurulation   167
    Amphioxus   *169*
    Fishes   *169*
    Amphibians   *171*
    Reptiles and Birds   *171*
    Mammals   *173*

ORGANOGENESIS   176
  Histogenesis   177
  Epithelium   177
    Covering and Lining Epithelium   *179*
    Glandular Epithelium   *180*
  Connective Tissues   180
    General Connective Tissues   *181*
    Special Connective Tissues   *181*
  Bone Development and Growth   182
    Endochondral Bone Development   *183*
    Intramembranous Bone Development   *184*
    Comparative Bone Histology   *186*
    Bone Remodeling and Repair   *186*
    Joints   *187*
  Neural Crest and Ectodermal Placodes   189

EXTRAEMBRYONIC MEMBRANES   190
  Reptiles and Birds   190
  Mammals   191
    Eutherian Placenta   *192*
    Other Placentae   *192*

OVERVIEW OF EARLY EMBRYONIC
DEVELOPMENT   193

DEVELOPMENT OF THE COELOM
AND ITS COMPARTMENTS   196

MATURATION   197
  Metamorphosis   197
  Heterochrony   198
    Peramorphosis   *199*
    Paedomorphosis   *199*

ONTOGENY AND PHYLOGENY   201
  Biogenetic Law   201
  Von Baer's Law   202
  Overview of the Biogenetic Laws   202
  Hox Genes and Their Kingdoms   204
    Egg to Adult   *204*
    Shaping Up: Positions and Parts   *204*
    Evolutionary Significance   *204*
  Epigenomics   205
    Induction   *206*
    Phylogeny   *206*

OVERVIEW   210

CHAPTER six
INTEGUMENT   212

EMBRYONIC ORIGIN   213

GENERAL FEATURES OF THE INTEGUMENT   213
  Dermis   213
  Epidermis   215

PHYLOGENY   216
  Integument of Fishes   216
    Primitive Fishes   *217*
    Chondrichthyes   *217*
    Bony Fishes   *218*
  Integument of Tetrapods   219
    Amphibians   *219*
    Reptiles   *220*
    Birds   *221*
    Mammals   *228*

SPECIALIZATIONS OF THE INTEGUMENT   233
  Nails, Claws, Hooves   233
  Horns and Antlers   234
  Baleen   235
  Scales   236
  Dermal Armor   236
  Mucus   237
  Color   238

OVERVIEW   239

CHAPTER seven
SKELETAL SYSTEM: THE SKULL   241

INTRODUCTION   242

Chondrocranium 242
    Embryology 242

Splanchnocranium 244
    Embryology 244
    Origin of Jaws 246
    Types of Jaw Attachments 247

Dermatocranium 248
    Parts of the Dermatocranium 249
        Dermal Bone Series 249

Overview of Skull Morphology 250
    Braincase 250
    Jaws 253
    Hyoid Apparatus 253

Cranial Kinesis 253

Phylogeny of the Skull 255
    Agnathans 255
        Early Vertebrates 255
        Ostracoderms 255
        Cyclostomes 255
    Gnathostomes 255
        Fishes 256
        Early Tetrapods 263
        Primitive Amniotes 264
        Modern Reptiles 268
        Birds 272
        Synapsids 274

Overview of Skull Function and Design 284
    Prey Capture 285
        Feeding in Water 285
        Feeding in Air 287
    Swallowing 288

Overview 288
    Cranial Neural Crest 288
    Emergence of Mammals 289
    Evolutionary Modifications of Immature Forms: Akinesis in Mammals 291
    Composite Skull 292

CHAPTER **eight**

SKELETAL SYSTEM: THE AXIAL SKELETON 294

Introduction 294

Basic Components 295
    Vertebrae 295
    Regions of the Vertebral Column 295
    Centra 295
    Ribs 297

Sternum 299
Gastralia 299

Embryonic Development 301
    Fishes 301
    Tetrapods 302

Phylogeny 304
    Fishes 304
        Agnathans 304
        Gnathostomes 304
    Tetrapods 309
        Early Tetrapods 309
        Amniotes 312

Form and Function 314
    Fluid Environment 314
    Terrestrial Environment 315
    Design of Vertebrae 317
        Direction of the Neural Spine 317
        Height of the Neural Spine 317
    Regionalization of the Vertebral Column 318

Overview 321

CHAPTER **nine**

SKELETAL SYSTEM: THE APPENDICULAR SKELETON 325

Introduction 325

Basic Components 326
    Fins 326
    Limbs 326

Origin of Paired Fins 327
    Gill-Arch Theory 327
    Fin-Fold Theory 328
    Embryonic Development of Tetrapod Limbs 330

Phylogeny 331
    Fishes 331
        Agnathans 331
        Placoderms 331
        Chondrichthyans 333
        Acanthodians 334
        Bony Fishes 334
    Tetrapods 336
        Pectoral Girdle 336
        Pelvic Girdle 339
        Manus and Pes 339

Evolution of the Appendicular System 346
    Dual Origin of the Pectoral Girdle 346
    Adaptive Advantage of Lobe Fins 347
    Onto the Land 347

FORM AND FUNCTION   348
    Swimming   349
    Terrestrial Locomotion   350
      Early Gaits   350
      Early Modes of Locomotion   350
      Cursorial Locomotion   353
      Aerial Locomotion   358
    Origin of Bird Flight   367
    Fossorial Locomotion   369

OVERVIEW   370

CHAPTER **ten**
THE MUSCULAR SYSTEM   372

INTRODUCTION   372

ORGANIZATION OF MUSCLES   373
    Classification of Muscles   373
    Skeletal Muscle   374
    Cardiac Muscle   375
    Smooth Muscle   375
    Structure of Skeletal Muscles   375
    Tendons   376
    Basis of Muscle Contraction   376
      Resting and Active Muscle   376
      Molecular Mechanisms of Contraction   376

MUSCLE FUNCTION   377
    Muscle Fibers   377
    Tension-Length Curves for a Single Muscle Fiber   377
    Properties of Muscle Fibers   377
    Muscle Organs and Fibers   379
      Whole Muscle Force Generation   379
      Tension-Length Curves for a Whole Muscle   380
      Graded Force   380
      Cross-Sectional Area   383
      Fiber Orientation   383
      Velocity of Shortening   385
      Distance of Shortening   385
    Bone–Muscle Lever Systems   385
    Sequencing of Muscle Actions   387
    Overview of Muscle Mechanics   388
    Muscle Actions   388
    Muscle Homologies   390

EMBRYONIC ORIGIN OF MUSCLES   391
    Postcranial Musculature   392
      Appendicular Musculature   392
      Axial Musculature   393
    Cranial Musculature   393
      Jaw and Pharyngeal Musculature   393
      Extrinsic Eye Muscles   393

COMPARATIVE ANATOMY   394
    Postcranial Musculature   394

      Axial Musculature   394
      Appendicular Musculature   397
    Cranial Musculature   405
      Branchiomeric Musculature   405
      Hypobranchial Musculature   408

OVERVIEW   411

CHAPTER **eleven**
THE RESPIRATORY SYSTEM   413

INTRODUCTION   413

RESPIRATORY ORGANS   416
    Gills   416
    Gas Bladders   416
      Lungs   416
      Swim Bladders   416
    Cutaneous Respiratory Organs   417
    Accessory Air-breathing Organs   417
    Breathing and Embryos   418

VENTILATORY MECHANISMS   421
    Cilia   421
    Muscular Mechanisms   421
      Water Ventilation: Dual Pump   421
      Air Ventilation: Buccal Pump   422
      Air Ventilation: Aspiration Pump   423

PHYLOGENY   424
    Agnathans   424
    Elasmobranchs   426
    Bony Fishes   427
    Overview of Fish Respiration   427
      Gills   427
      Lungs and Swim Bladders   428
    Amphibians   430
      Amphibian Larvae   430
      Amphibian Adults   432
    Reptiles   433
    Mammals   435
      Ventilation   435
      Gas Exchange   437
    Birds   437

FORM AND FUNCTION   440
    Patterns of Gas Transfer   440
    Rates of Gas Transfer   443
    Breathing in Water   444
    Breathing in Air   444

EVOLUTION OF RESPIRATORY ORGANS   444
    Acid-Base Regulation   444
    Ventilation   446
      Ciliary Pumps   446
      Muscular Pumps   446

Water-to-Land Transition   446
   Air-breathing Organs   446
   Advantages of Movement to Land   447
   Air-breathing Mechanisms   448
Bird Lungs and Air Sacs   449

OVERVIEW   450

CHAPTER **twelve**

THE CIRCULATORY SYSTEM   451

INTRODUCTION   451

CARDIOVASCULAR SYSTEM   452
Blood   452
Arteries, Veins, and Capillaries   452
   Arteries   453
   Hemodynamics of Circulation   453
   Veins   454
   Microcirculation   454
Single and Double Circulation   455
Embryonic Development of the Cardiovascular
System   456
Phylogeny of the Cardiovascular System   457
   Arterial Vessels   461
   Venous Vessels   466
Hearts   473
   Basic Vertebrate Heart   473
   Fishes   475
   Amphibians   478
   Reptiles   480
   Birds and Mammals   489
Cardiovascular System: Matching Design
to Environmental Demands   490
   Accessory Air-breathing Organs   491
   Diving Birds and Mammals   491
   Heart Flow   492
   Ontogeny of Cardiovascular Function   492
   Fetal Circulation in Placental Mammals   492
   Changes at Birth   492
   Heat Transfer   494

LYMPHATIC SYSTEM   496
Lymphatic Vessels   496
Lymphatic Tissue   498
Form and Function   498

OVERVIEW   499

CHAPTER **thirteen**

THE DIGESTIVE SYSTEM   503

INTRODUCTION   503
Preview   503

COMPONENTS OF THE DIGESTIVE SYSTEM   504
Buccal Cavity   504
   Boundaries   504
   Palate   505
   Teeth   506
   Tongue   516
Pharynx   517
Alimentary Canal   520
   Esophagus   521
   Stomach   521
   Intestines   523
   Cloaca   525
   Specializations of the Alimentary Canal   525
   Vascularization of the Gastrointestinal Tract   527
   Fishes   527
   Tetrapods   528
Associated Glands of Digestion   531
   Oral Glands   531
   Liver   533
   Pancreas   533

FUNCTION AND EVOLUTION
OF THE DIGESTIVE SYSTEM   535
Absorption   535
Feces   535
Mechanical Breakdown of Food   536
   Mastication   536
   Gizzards   536
Chemical Breakdown of Food   536
   Foregut Fermentation   537
   Hindgut Fermentation   540
Foregut versus Hindgut Fermenters   540
   Size and Fermentation   541
Digesting Toxins   542
Feeding and Fasting   542

OVERVIEW   543

CHAPTER **fourteen**

THE UROGENITAL SYSTEM   545

INTRODUCTION   545

URINARY SYSTEM   545
Embryonic Development   546
   Nephrotome to Nephric Tubules   546
   Tripartite Concept of Kidney Organization   547
Kidney Phylogeny   550
   Fishes   550
   Tetrapods   551
Kidney Function and Structure   553
   *Excretion: Removing the Products
   of Nitrogen Metabolism   553*
   Osmoregulation: Regulating Water and Salt Balance   554

Evolution 562
Preadaptation 562
Origin of Vertebrates 562

REPRODUCTIVE SYSTEM 563
Structure of the Mammalian Reproductive System 563
Embryonic Development 563
Gonads and Gametes 563
Reproductive Tracts 565
Overview 565
Female Reproductive System 567
Ovary 567
Genital Ducts 567
Oviduct 569
Uterus 570
Male Reproductive System 572
Testis 572
Genital Ducts 572
Copulatory Organs 576
Cloaca 582
Urinary Bladder 586
Function and Evolution 588
Potency and Fertility 588
External and Internal Fertilization 588
Delays in Gestation 589

OVERVIEW 589

CHAPTER **fifteen**

THE ENDOCRINE SYSTEM 592

SURVEY OF ENDOCRINE GLANDS 592
Thyroid Gland 592
Structure and Phylogeny 592
Function 594
Ultimobranchial Body and Parathyroid Gland 596
Ultimobranchial Body 596
Parathyroid Gland 597
Form and Function 598
Adrenal Gland 598
Structure and Phylogeny 598
Function 601
Pancreatic Islets 602
Structure and Phylogeny 602
Function 602
Pituitary Gland 604
Structure 604
Phylogeny 606
Function 607
Gonads 610
Pineal Gland 610
Secondary Endocrine Glands 611
Gastrointestinal Tract 611
Kidneys 612

ENDOCRINE COORDINATION 613
Mammalian Reproduction 613
Male 613
Female 613
Metamorphosis in Frogs 616
Fundamentals of Hormonal Control 620
Functional and Structural Linkage 620
Target Tissue Responses 621
The Endocrine System and the Environment 621

EVOLUTION 622

OVERVIEW 624

CHAPTER **sixteen**

THE NERVOUS SYSTEM 625

INTRODUCTION 625
Types of Cells within the Nervous System 625
Neuroglia 625
Neurons 625
Transmission of Information 626
Neurosecretory Cells 628

PERIPHERAL NERVOUS SYSTEM 628
Spinal Nerves 629
Cranial Nerves 630
Evolution 637
Functions of the Peripheral Nervous System 638
Spinal Reflexes 638
The Autonomic Nervous System 641

CENTRAL NERVOUS SYSTEM 646
Embryology 646
Spinal Cord 649
Spinal Reflexes 649
Spinal Tracts 650
Brain 651
Phylogeny 651
Form and Function 654
*Functional Associations of Parts
of the Central Nervous System 666*
Limbic System 666

OVERVIEW 669

CHAPTER **seventeen**

SENSORY ORGANS 671

INTRODUCTION 671

COMPONENTS OF A SENSORY ORGAN 672

GENERAL SENSORY ORGANS 672
    Free Sensory Receptors 672
    Encapsulated Sensory Receptors 673
    Associated Sensory Receptors 673
      Proprioception 673
    Mechanisms of Perceiving Stimuli
    from General Sensory Receptors 674

SPECIAL SENSORY ORGANS 674
    Chemoreceptors 674
      Nasal Passages 675
      Vomeronasal Area 677
      Mouth 678
    Radiation Receptors 679
      Photoreceptors 681
      Infrared Receptors 691
    Mechanoreceptors 693
      Lateral Line System 694
      Vestibular Apparatus 695
      Auditory System 696
      Functions of the Ear 701
    Electroreceptors 709
      Structure and Phylogeny 709
      Form and Function 709
    Additional Special Sensory Organs 712

OVERVIEW 712

CHAPTER **eighteen**
CONCLUSIONS 714

INTRODUCTION 714

STRUCTURAL ANALYSIS 717

FUNCTIONAL ANALYSIS 718
    How Does It Work? 718
    Functional Coupling, Functional Compromise 719
    Multiple Functions 720
    Performance 721

ECOLOGICAL ANALYSIS 722

EVOLUTIONARY ANALYSIS 722
    Historical Constraints 722
    Primitive and Advanced 722
    Diversity of Type/Unity of Pattern 723
    Mosaic Evolution 725
    Morphology and Modules 725

MODE AND TEMPO OF EVOLUTION 727
    Remodeling 728
    Embryonic Changes 729
    Hox Genes 729
    Evolutionary Significance 730

THE PROMISE OF VERTEBRATE MORPHOLOGY 730

APPENDIX **A**
VECTOR ALGEBRA 731

APPENDIX **B**
INTERNATIONAL UNITS (SI) 733

APPENDIX **C**
COMMON GREEK AND LATIN COMBINING FORMS 736

APPENDIX **D**
CLASSIFICATION OF CHORDATES LINNAEAN 740
CLASSIFICATION OF CHORDATES CLADISTIC 743

GLOSSARY 744
CREDITS 758
INDEX 764

Preface

If you are a student coming to the study of vertebrates for the first time, several introductory remarks may be helpful, especially on how this textbook will support your work. First, the discipline of vertebrate biology is diverse and inclusive. It brings together themes from molecular biology, genes and genomes, evolution and embryology, biomechanics and experimental physiology, and it incorporates continuing and astonishing new fossils into the vertebrate story. Much of what you have met in earlier courses you will meet again here in an integrated way.

Second, to unify these themes, I have again written and revised this seventh edition within the unifying framework of form, function, and evolution. The first few chapters set this up, and the subsequent chapters treat vertebrates system by system. You may notice that each of these subsequent chapters begins with a discussion of morphology, followed by a discussion of function and evolution. Each chapter is therefore self-contained—form, function, evolution.

Third, as a student you likely enter this course after some background in the sciences, perhaps expecting to equip yourself with practical knowledge useful later in professional schools or in health-related careers. Certainly this course, in part, delivers such practical information. But because vertebrate morphology is an integrative discipline, it brings together physiology, embryology, behavior, and ecology and also deploys modern methods of systematics and new finds in paleontology. Consequently, you will move beyond memorizing facts in isolation or as an end in themselves, and instead begin to meet and understand larger concepts to which the morphology testifies. What may come as a surprise is that many theories, especially evolutionary theories within vertebrate biology, are still unsettled and unresolved, inviting a new idea or fresh approach open to anyone. This is one of the reasons I have included various controversies, and support your efforts to become engaged in the thinking and scientific process.

For faculty who have used this textbook before, you will find it retains a familiar and inviting organization with the science updated and the student support enhanced. For those coming to this textbook for the first time, you will notice that the morphology receives generous treatment within a phylogenetic context. But, today we expect our students to develop academic and professional skills beyond just facility with anatomical terminology. In general, we expect our students to develop skills in critical thinking and a facility with scientific concepts. Each of us will find our own way of composing a course in vertebrate morphology that serves such course objectives. This textbook was written to support such course objectives as individual instructors build their courses. It is flexible. You can mix and match, change order to suit your course, and give emphasis to those systems that most suit the organization of your course. Because each chapter integrates form, function, and evolution pertinent to that system, each chapter is coherent within itself. Where information or concepts are treated in greater detail outside a particular chapter, they are cross referenced to help guide the student and clarify the discussion. Although discussed in earlier editions, let me repeat the specific strategy built into this textbook to improve student success and to help them develop skills in critical thinking and conceptual understanding.

For the Student

A number of strategic features within the textbook enhance its usefulness for students. It is richly **illustrated** with figures that include new information and provide fresh perspectives. Each chapter opens with an **outline**. Important **concepts** and major anatomical terms are boldfaced. **Cross references** direct students to other areas of the text where they can refresh their understanding or clarify an unfamiliar subject. Each chapter concludes with a **chapter overview,** which draws attention to some of the concepts developed within the chapter. **Box Essays** are included along the way in most chapters. Their purpose is to present subjects or historical events that students should find interesting and, perhaps from time to time, even fun. A **glossary** of definitions is included at the end of the book.

In addition to its practical features, the textbook also uses selected topics within vertebrate structure to develop student skills in critical thinking and mastery of concepts within a coherent framework.

Critical Thinking

Within the sciences, critical thinking is the ability to marshal factual information into a logical, reasoned argument. Especially if accompanied by a laboratory, a course in vertebrate morphology delivers hands-on experience with the anatomy of representative animals. Students can be directly engaged in the discovery of vertebrate form. But they can be encouraged to go beyond this. Instructors can lead students into larger issues—How does it function? How did it evolve? For example, early on in the textbook, students are introduced to "Tools of the Trade," methods by which we empirically examine how parts work and how we can place organisms within a phylogenetic context. After a discussion of basic morphology, each chapter discusses how these systems work and how they evolved.

I have deliberately included new, neglected, or competing views on function and evolution. Many of these ideas come from Europe, where they have been known for a long time. Personally, I find many of these ideas compelling, even elegant. Others strike me, frankly, as thin and unconvincing. Despite my own skepticism, a few contrary ideas are included. My purpose is to get students to think about issues of form, function, and evolution.

Several theories on the evolution of jaws are discussed, as are several theories of the origin of paired fins. Often students expect that today we have the final answers. Students implore, "Just tell me the answer." The debate about dinosaur physiology is a wonderful opportunity to show students the ongoing process of scientific investigation. Most have seen the Hollywood films and expect the issue settled. But we know that science is an ongoing process of refinement, challenge, and sometimes revolutionary change. One Box Essay sets forth the early case for dinosaur endothermy. That debate spawned further investigation that now returns to challenge such a view of dinosaurs as "hot-blooded" beasts. The second Box Essay on dinosaur endothermy presents this newer and contrary evidence, and thereby showcases how, even in extinct animals, it is possible to test hypotheses about their physiology, morphology, and lifestyles.

Concepts

Vertebrate morphology also helps develop an appreciation and understanding of the scientific concepts that unite biology and reflect on "how" science works. As John A. Moore put it, science is a "way of knowing" (Moore, *American Zoologist*, 1988). Comparative morphology throws into clear relief differences and similarities between organisms. The concepts of homology, analogy, and homoplasy help us understand the basis of these comparative features. Many of the concepts were birthed in the nineteenth century and have grown into the guiding themes of biology today. Evolution, defined as descent with modification through time, is one of the foundation concepts in biology. Vertebrate morphology provides a showcase of adaptive change on the basic vertebrate body plan. But evolution is change in a highly integrated organism, a connected system of parts and their functions. This too was recognized within the nineteenth century, suggesting constraints on evolutionary modification. Vertebrate morphology provides compelling examples of how an integrated organism might evolve. For example, a remarkable fossil record documents an undeniable change in jaw articulation within synapsids, seeing the two participating bones (articular, quadrate) of basal synapsids replaced by two different bones in derived groups, including mammals. Fossil intermediates between the two conditions mark the anatomical changes, but they also suggest how functional changes, which must accompany evolving systems, also change without disrupting performance.

Within many vertebrate systems, the close coupling of form and function with lifestyle is illustrated. Built on a basic vertebrate plan, the tetrapod locomotor system illustrates the close relationship between limbs and axial skeleton, and the type of locomotion—flight, cursorial, burrowing. The cardiovascular system, especially in organisms that exploit water and air, illustrates the close relationship between vascular morphology and the physiological flexibility that permits. The basic concepts of form, function, and adaptive evolution parade before us as we move from system to system in vertebrate morphology.

Evolution proceeds most often by remodeling, modification of a basic underlying plan, not by all new construction. This is illustrated in the skeletal system, as well as within the cardiovascular (aortic arches) system.

Organizational Strategy and Rationale

I have written this book within the unifying framework of form, function, and evolution. These are common themes that run throughout. The vertebrate groups are organized phylogenetically, and their systems discussed within such a context. Morphology is foremost, but I have developed and integrated an understanding of function and evolution into the discussion of anatomy of the various systems. The first five chapters prepare the way.

Chapter 1 introduces the discipline, evaluates the intellectual predecessors to modern morphology, defines central concepts, and alerts students to misunderstandings they may unknowingly bring with them to the study of evolutionary processes. Chordates and their origins are covered in chapter 2. Considerable attention is given to the neglected protochordates and their evolution. This sets the stage for an extended discussion of the cast of characters in the vertebrate radiation, which occupies us for the remainder of the book, beginning next in chapter 3. Here we discuss vertebrates, their origins, and basic taxonomic relationships. Chapter 4 introduces basic concepts of biomechanics and biophysics, preparing for their use later in understanding aspects of vertebrate design and function. Chapter 5 includes a summary of descriptive embryology and concludes with a discussion of the role embryonic processes play in vertebrate evolutionary events.

The remaining chapters develop each major system. Besides carrying overall themes, each chapter internally follows a consistent organization. Each begins with a basic introduction to the morphology, and then proceeds to discuss function and evolution. This way, the overall themes are repeated in each chapter, bringing consistency of presentation to each chapter and coherence throughout.

New and Expanded in the Seventh Edition

New fossil finds, modern experimental research, and new phylogenies continue to enrich vertebrate biology, sometimes solving old questions or surprising us with a new understanding of how vertebrates function and how they evolved. Much of this is added to this new edition.

Feathered Dinosaurs. New and remarkable fossil finds of dinosaurs, especially in China, continue to be discovered, some showing evidence of surface feathers. In other words, feathers evolved before birds. This means that these skin specializations addressed biological roles before they addressed flight. This is discussed in the chapter on integument (chapter 6) with new supportive illustrations.

Shark Skin. In addition to favorably affecting fluid flow across their surface, shark placoid scales also bristle, lift up, when boundary layer separation begins to occur to reduce its effects. This newly discovered feature of shark skin is now also included in chapter 6.

Evo-Devo. I have built on the genetic section on evolution and development (chapter 5) introduced in earlier editions. Examples throughout show how master control genes (*Hox* genes) and developmental genes preside over the construction of the vertebrate body and its various systems. For example, in chapter 8 wonderful experimental work in mice using "knockout" techniques has shown how various *Hox* genes control differentiation of the mammalian axial column. In the concluding chapter, I emphasize how these special evo-devo gene sets provide the basis for understanting the genetic mechanisms underlying major evolutionary changes.

Phylogenetic Relationships. Thanks to continuing use of improved genetic and morphological data sets, phylogenetic relationships are becoming better resolved, and natural groups are emerging from this analysis with better clarity. This is the basis for revision in chapter 3, but these updated phylogenies are carried forward throughout the book or new phylogenies added, as for example with the phylogeny of the axial column (chapter 8).

Lungs and the Rise of Archosaurs. The especially efficient lungs of birds are well known with air sacs and one-way flow of air. But new experimental evidence (chapter 11) identifies similar one-way air flow, even without air sacs, in crocodiles. If true of archosaurs in general, it may represent a respiratory adaptation to low oxygen levels in the early Mesozoic and account for the rise of Archosaurs.

Turning over Chordates. New developmental genetics, discussed in the previous editions, informs us that the immediate chordate ancestors flipped over, reversing dorsal and ventral surfaces. That view seems to hold still and therefore remains the surprising basis of the chordate body plan today.

Updated and Revised. Countless changes and revisions throughout this new edition have been made, some major, some small. These changes have corrected misinformation, updated information, and often better clarified an explanation. For this I am indebted to students, reviewers, and colleagues for bringing these suggestions to my attention.

Serving the Student. Features of the textbook have been further expanded to make its presentation more clear and inviting. The use of **color** brightens these sections of the book. Color has also been used to better correlate and compare structures between figures in these chapters. Where feasible, within color signatures, for example, I have added more color to the illustrations. Many **illustrations** are new, revised, or relabeled to improve clarity. For example, besides those illustrations mentioned earlier, new/revised figures illustrate a feathered dinosaur, clarify embryonic development of the urogenital system; and various changes have been made in figures elsewhere. Scientific references are available to the students, online, if they would like to follow up or read more about a particular subject. The accompanying laboratory dissection guide (authored with E. J. Zalisko) is closely cross-referenced to this textbook. In addition, selective **functional laboratories** are available, online, to provide students with firsthand experience of working between the anatomy and its functional and evolutionary significance.

Serving Instructors. This seventh edition—new, revised, updated—can serve as reference and resource support for the course you put together on vertebrates. In addition to this, resources are available to you online. The functional laboratories may be downloaded and used as they supplement your course. **PowerPoint images,** chapter by chapter, are available online along with additional images from McGraw-Hill that can be used to compose lectures and laboratory presentations.

Supplements

Comparative Vertebrate Anatomy: A Laboratory Dissection Guide

Newly revised, *Comparative Vertebrate Anatomy: A Laboratory Dissection Guide,* Seventh Edition, by Kenneth V. Kardong and Edward J. Zalisko, is now available. At the end of this dissection guide, the authors include a Student Art Notebook. This notebook, promoted by students, is a reprinted collection of the most important and commonly used dissection figures in the current edition of the laboratory manual. It addresses a frustration inherent in most dissection guides, especially when comparing homologous systems between representative animals, of having to flip between text and distantly placed illustrations. This laboratory manual weaves the functional and evolutionary concepts from this textbook, *Vertebrates: Comparative Anatomy, Function, Evolution,* into the morphological details of the laboratory exercises. Using icons, the laboratory manual identifies cross references to this textbook, so students can quickly move from the dissection guide to this textbook to consult the expanded treatment of function and evolution. Each chapter of the dissection guide first introduces the system, makes comparisons, and demonstrates common themes in the animal systems. Then the written text carefully guides students through dissections, which are richly illustrated. Anatomical terms are boldfaced and concepts italicized. The dissection guide is written so that instructors have the flexibility to tailor-make the laboratory to suit their needs.

Website for *Vertebrates: Comparative Anatomy, Function, Evolution,* Seventh Edition

A website for this textbook, available at *www.mhhe.com/ kardong7e,* includes further useful information upon which instructors can depend and students can consult. Here can be found the **functional laboratories,** helpful in a linked laboratory if available, or helpful selectively in lecture. End-of-chapter **selected references,** giving students a start into the literature, are located here.

Biology Digitized Video Clips

McGraw-Hill is pleased to offer digitized biology video clips on DVD. Licensed from some of the highest-quality science video producers in the world, these brief segments range from about five seconds to just under three minutes in length and cover all areas of general biology, from cells to ecosystems. Engaging and informative, McGraw-Hill's digitized biology videos will help capture students' interest while illustrating key biological concepts and processes. Includes video clips on mitosis, Darwin's finches, amoeba locomotion, tarantula defense, nematodes, bird/water buffalo mutualism, echinoderms, and much more! ISBN: 978-0-07-312155-0 (MHID: 0-07-312155-X)

Electronic Textbook

CourseSmart is a new way for faculty to find and review eTextbooks. It's also a great option for students who are interested in accessing their course materials digitally and saving money. CourseSmart offers thousands of the most commonly adopted textbooks across hundreds of courses from a wide variety of higher education publishers. It is the only place for faculty to review and compare the full text of a textbook online, providing immediate access without the environmental impact of requesting a print copy. At CourseSmart, students can save up to 50% off the cost of a print book, reduce their impact on the environment, and gain access to powerful web tools for learning—including full text search, notes and highlighting, and email tools for sharing notes between classmates. www.CourseSmart.com

Art and Artists

Please indulge me a final moment of lament. For illustrations, modern textbooks have substituted photographs and figures made of computer generated plastic figures, especially when rending molecular events. Nothing wrong with this but what has been eliminated is involvement of the human touch, directly, namely the artist. Preparing this textbook has given me the chance, at no expense to students by the way, of engaging some of the best artists of our day. They bring a sharp eye and traditional talent to rendering of enlightening pieces of art. Many have contributed, but one is L. Laszlo Meszoly (Harvard University), who has contributed special figures to this and earlier editions. Another is Kathleen M. Bodley, whose remarkable ability to render soft tissue in particular has enriched our dissection guide and is quite stunning. Her work also graces the cover of this textbook. My thanks to these two in particular. This is a wonderful tradition of scientific illustration they carry forward.

Acknowledgments

I am indebted to reviewers, students, and colleagues who have generously shared with me their suggestions to improve this edition of the textbook. My hope is that these colleagues will see, if not their point of view, at least their influence within this edition, and accept my sincere thanks for their thoughtful suggestions and criticisms. For their special help in this edition and earlier editions I recognize:

Daniel Blackburn
Trinity College

Richard W. Blob
Clemson University

Carol Britson
University of Mississippi

Stephen Burnett
Clayton State University

George Cline
Jacksonville State University

C. G. Farmer
University of Utah

T. H. Frazetta
University of Illinois

Nick Geist
Sonoma State University

Ira F. Greenbaum
Texas A & M University

Maria Laura Habegger
University of South Florida

Christine M. Janis
Brown University

Amy W. Lang
University of Alabama

Jon M. Mallatt
Washington State University

Sue Ann Miller
Hamilton College

Philip J. Motta
University of South Florida

Barbara Pleasants
Iowa State University

Calvin A. Porter
Xavier University of Louisiana

Tamara L. Smith
Westridge School

Jeffrey Thomas
Queens University Charlotte

David Varricchio
Montana State University

Mindy Walker
Rockhurst University

Andrea Ward
Adelphi University

Jeanette Wyneken
Florida Atlantic University

It has been a special pleasure for me to work previously with several especially supportive and helpful colleagues. In particular, I note the extensive help of Christine M. Janis in several difficult chapters, as well as the patient and especially informative education I received on regenerating bird feathers from P. F. A. Maderson and W. J. Hillenius. I also want to note especially the help from Ira R. Greenbaum for engaging in a pleasurable and productive exchange on all things vertebrate.

For answering my queries, supplying me with their critical thoughts, and/or for earlier participation in this and previous editions, I gratefully recognize the following: Neil F. Anderson, Alejandra Arreola, Miriam A. Ashley-Ross, Ann Campbell Burke, Walter Bock, Warren W. Burggren, Anindo Choudhury, Michael Collins, Mason Dean, Ken P. Dial, Alan Feduccia, Adrian Grimes, Maria Laura Habegger, Linda Holland, Marge Kemp, Amy W. Lang, William T. Maple, Jessie Maisano, David N. M. Mbora, Philip Motta, David O. Norris, R. Glenn Northcutt, Kevin Padian, Kathryn Sloan Ponnock, Michael K. Richardson, Timothy Rowe, John Ruben, J. Matthias Starck, James R. Stewart, Billie J. Swalla, Steven Vogel, Alan Walker, and Bruce A. Young.

I am indebted to the patient, able, and supportive people at McGraw-Hill who were so important in bringing this revised seventh edition along. I thank again the McGraw-Hill field staff who link the summary effort of all who helped in this revision to faculty and students who use it. In turn, these field reps return your comments of what you do and do not like, and thereby aid in the improvement of this textbook, making it a shared work in progress. Lori Bradshow was indispensible as Developmental Editor. It was a pleasure to work with Ligo Alex and her talented team of copyeditors at SPi-Global.

To friends and family I remain grateful and thank them for their support during various editions of this textbook.

CHAPTER 1

Introduction

COMPARATIVE VERTEBRATE MORPHOLOGY
Designs of Students
Vertebrate Design—Form and Function
Grand Design

HISTORICAL PREDECESSORS—EVOLUTION
The Process behind the Change
Linnaeus
Naturalists
J-B. de Lamarck
Acquired Characteristics
Upward to Perfection
Natural Selection
A. R. Wallace
Charles Darwin
Critics and Controversy

HISTORICAL PREDECESSORS—MORPHOLOGY
Georges Cuvier
Richard Owen

WHY ARE THERE NO FLYING ELEPHANTS?

MORPHOLOGICAL CONCEPTS
Similarities
Symmetry
Segmentation

EVOLUTIONARY MORPHOLOGY
Function and Biological Role
Preadaptation
Evolution as Remodeling

PHYLOGENY
Of Bean Stalks and Bushes
Simplification
Patterns of Phylogeny
Grades and Clades

PALEONTOLOGY
Fossilization and Fossils
Recovery and Restoration
From Animal to Fossil
Dating Fossils
Stratigraphy
Index Fossils
Radiometric Dating
Geological Ages

TOOLS OF THE TRADE
The Question
The Function
The Biological Role

OVERVIEW

Comparative Vertebrate Morphology

Comparative morphology deals with anatomy and its significance. We focus on animals, in particular vertebrate animals, and the significance these organisms and their structure may hold. The use of "comparison" in comparative morphology is not just a convenience. It is a tool. Comparison of structures throws similarities and differences into better relief. Comparison emphasizes the functional and evolutionary themes vertebrates carry within their structures. Comparison also helps formulate the questions we might ask of structure.

For example, different fishes have different tail shapes. In the **homocercal** tail, both lobes are equal in size, making the tail symmetrical (figure 1.1a). In the **heterocercal** tail, found in sharks and a few other groups, the upper lobe is elongated (figure 1.1b). Why this difference? The homocercal tail is found in teleost fishes—salmon, tuna, trout, and the like. These fishes have a swim bladder, an air-filled sac that gives their dense bodies neutral buoyancy. They neither sink to the bottom nor bob to the surface, so they need

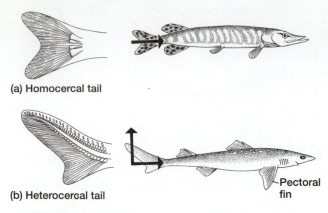

(a) Homocercal tail

(b) Heterocercal tail

Pectoral fin

FIGURE 1.1 Homocercal and heterocercal fish tails. Form differs because function differs. (a) Sweeping, side-to-side movements of the homocercal tail, common in fishes with neutral buoyancy, drive the body forward. (b) Swimming strokes of the heterocercal tail propel the fish forward, and motion of the long extended upper lobe imparts an upward lift to the posterior end of the fish. Sharks, which are a good deal denser than water, need the upward forces provided by the extended lobe of the tail to counteract a tendency to sink.

not struggle to keep their vertical position in the water. Sharks, however, lack swim bladders, and so tend to sink. The extended lobe of their heterocercal tail provides lift during swimming to help counteract this sinking tendency. So, the differences in structure, homocercal versus heterocercal, are related to differences in function. Why an animal is constructed in a particular way is related to the functional requirements the part serves. Form and function are coupled. Comparison of parts highlights these differences and helps us pose a question. Functional analysis helps answer our question and gives us a better understanding of animal design. **Functional morphology** is the discipline that relates a structure to its function.

Comparative analysis thus deploys various methods to address different biological questions. Generally, comparative analysis is used either in a historical or a nonhistorical context. When we address historical questions, we examine evolutionary events to work out the history of life. For example, on the basis of the comparison of characters, we may attempt to construct classifications of organisms and the evolutionary phylogeny of the group. Often such historical comparisons are not restricted to classification alone but center on the process of evolution behind morphological units, such as jaws, limbs, or eyes.

When we make nonhistorical comparisons, as is frequently the case, we look outside an evolutionary context, with no intention of concluding with a classification or elucidation of an evolutionary process. Nonhistorical comparisons are usually extrapolative. For example, by testing a few vertebrate muscles, we may demonstrate that they produce a force of 15 N (newtons) per square centimeter of muscle fiber cross section. Rather than testing all vertebrate muscles, a time-consuming process, we usually assume that other muscles of similar cross section produce a similar force (other things being equal). The discovery of force production in some muscles is extrapolated to others. In medicine, the comparative effects of drugs on rabbits or mice are extrapolated to tentative use in humans. Of course, the assumed similarities upon which an extrapolation is based often do not hold in our analysis. Insight into the human female reproductive cycle is best obtained if we compare the human cycle with those in higher primates because primate reproductive cycles, including the human one, differ significantly from those of other mammals.

Extrapolation allows us to make testable predictions. Where tests do not support an extrapolation, science is well served because this forces us to reflect on the assumptions behind the comparison, perhaps to re-examine the initial analysis of structures and to return with improved hypotheses about the animals or systems of interest. Comparison itself is not just a quick and easy device. The point to emphasize is this: Comparison is a tool of insight that guides our analysis and helps us set up hypotheses about the basis of animal design.

Designs of Students

Such philosophical niceties, however, usually do not entice students into their first course in morphology. Most students first venture into a course in vertebrate morphology on their way into some other profession. Customarily, morphology courses prepare students headed into technical fields such as human medicine, dentistry, or veterinary medicine. Vertebrate form and function will be the foundation for these medical fields. As will, for example, diagnostic medicine that benefits from the development of anatomically and functionally correct prosthetic devices to replace injured body parts lost to disease and trauma.

In addition, morphology is important to taxonomists who use the structure of animals to define characters. In turn, these characters are used as the basis for establishing relationships between species.

Morphology is central to evolutionary biology as well. Many scientists, in fact, would like to see a discipline devoted to the combined subject, namely, **evolutionary morphology.** Evidence of past evolutionary changes is inscribed in animal structure. Within the amphibian limb are the structural reminders of its fish-fin ancestry; within the wing of a bird are the evidences of its derivation from the reptilian forelimb. Each modern group living today carries forward mementos of the evolutionary course traveled by its ancestors. For many biologists, a study of the morphological products of the past gives insight into the processes that produced them, insight into the natural forces that drove evolutionary changes, and insight into the limitations of evolutionary change.

The Scientific Method—What They Tell You and What They Don't

Formally, the scientific method includes formulation of a hypothesis, design of a test, carrying out of an experiment, analysis of results, corroboration or falsification of the hypothesis, and formulation of a new hypothesis. In practice, science does not follow such a stately and linear sequence. Broken equipment, uncooperative animals, paperwork, and committee meetings all conspire against the well-laid plans of mice, men, and women. It is more than the "expected unexpected" that affects experiments and tests one's blood pressure. The intellectual questions themselves do not always find satisfying answers. Accidents, chance, and even dreams are part of the creative process.

Otto Loewi shared the 1936 Nobel Prize in medicine with Henry Dale for demonstrating that nerve impulses pass from one nerve cell to the next in series across the space between them, the synapse, by a chemical transmitter. Early in the twentieth century, opinion had been divided between those physiologists who felt that this neuron-to-neuron transmission was chemical and those who felt that it was electrical. A definitive experiment settling the issue was needed. One night, when he was deep in sleep, the definitive experiment came to Loewi and woke him. Relieved and satisfied, he went back to sleep looking forward to the next day. When awaking the next morning, he remembered dreaming the experiment but had forgotten what it was. Several frustrating weeks passed until, once again deep in sleep, Loewi dreamt the same dream, and the experimental design came back. Leaving nothing to chance this time, he got up, dressed, and in the middle of the night went to his laboratory to begin the experiment that would settle the issue

of transmission and years later win him a share of the Nobel Prize.

Loewi's experiment was as simple as it was elegant. He removed the heart and associated vagus nerve from the body of a frog and isolated them in a beaker of saline. Next he stimulated the free vagus nerve, causing the heart rate to slow. Loewi then took this saline and poured it over another isolated frog heart from which the vagus had been removed. The rate of this heart also slowed, providing clear evidence that a chemical produced by the stimulated vagus nerve controlled heart rate. Transmission between nerve (vagus) and organ (heart) was brought about by chemical agents, not by electrical currents.

As a young cell biologist, Herbert Eastlick began a series of experiments to pursue his interest in embryonic development of young muscle. He transplanted the still formative hindlimbs of a chick to the side of a host chick while the host was still developing in its egg. The transplanted hindlimbs were usually received and grew well enough on the side of the host chick to allow study. One day, when a local supplier was temporarily out of the white leghorn eggs Eastlick had used, he substituted brown leghorns, a breed with brown feathers. After three days of incubation, one egg was opened and both leg-forming areas of a brown leghorn were transplanted to a white leghorn host. Results were puzzling. The right transplanted leg from the brown leghorn developed brown feathers, the left transplanted leg from the same brown leghorn developed white feathers. What caused these contrary results?

Eastlick checked his notes, repeated his experiments, and used great care in performing more transplants. Still some

transplanted legs were brown and some were white. It then dawned on him that the stump of the transplanted limb might in some instances include nearby neural crest cells, but not in all instances. Neural crest cells form first on top of the nerve tube and then normally disperse about the embryo. He tried limbs with and without accompanying neural crest cells. That was it. Those brown leghorn limbs with neural crest cells produced brown feathers. Those without lacked pigment cells and were white. Eastlick, who started out working on muscles, confirmed what a few had guessed at the time, namely, that one derivative of neural crest cells is pigment cells that give feathers their color.

Alexander Fleming (1881–1955), while studying bacteria, noticed that when molds occasionally contaminated cultures, the bacteria next to the molds failed to grow. Hundreds of students and fellow bacteriologists before Fleming had seen molds and likely noticed the stunted growth of bacteria. But it was Fleming's curiosity that precipitated the serious question, "What causes this reaction?" In answering it, he discovered that molds produced penicillin, a bacterial inhibitor. Fleming's question opened the way for development of a new branch of pharmacology and a new industry. His answer established the basis of disease control through antibiotics.

Testing of a well-crafted hypothesis forms the center of the scientific method. But where the next hypothesis comes from cannot always be predicted. A thought in the middle of the night, an experiment gone wrong, a close observation of the ordinary, these too may inspire a new scientific hypothesis and are part of the method of science.

Vertebrate Design—Form and Function

Morphology offers more than charitable assistance to other disciplines. The study of morphology provides its own pleasure. It raises unique questions about structure and offers a method to address these questions. In brief, vertebrate morphology seeks to explain vertebrate design by elucidating the reasons for and processes that produce the basic structural plan of an organism. For most scientists today, evolutionary processes explain form and function. We might hear it said

that the wings of birds, tails of fishes, or hair of mammals arose for the adaptive advantages each structure provided, and so they were favored by natural selection. Certainly this is true, but it is only a partial explanation for the presence of these respective features in bird, fish, and mammal designs. The external environment in which an animal design must serve certainly brings to bear evolutionary pressures on its survival, and thus on those anatomical features of its design that convey adaptive benefits.

Internal structure itself also affects the kinds of designs that do or do not appear in animals. No terrestrial vertebrate rolls along on wheels. No aerial vertebrate flies through the air powered by a rotary propeller. Natural selection alone cannot explain the absence of wheels in vertebrates. It is quite possible to imagine that wheels, were they to appear in certain terrestrial vertebrates, would provide considerable adaptive advantages and be strongly favored by natural selection. In part, the explanation lies in the internal limitations of the structure itself. Rotating wheels could not be nourished through blood vessels nor innervated with nerves without quickly twisting these cords into knots. Wheels and propellers fall outside the range of structural possibility in vertebrates. Structure itself contributes to design by the possibilities it creates; evolution contributes to design by the favored structures it preserves. We must consult both structure and evolution to understand overall design. That is why we turn to the discipline of morphology. It is one of the few modern sciences that addresses the natural unity of both structure (form and function) and evolution (adaptation and natural selection). By wrapping these together in an integrated approach, morphology contributes a holistic analysis of the larger issues before contemporary biology. Morphology is concerned centrally with the emergent properties of organisms that make them much more than the reduced molecules of their parts.

Grand Design

Vertebrate design is complex, often elegant, and sometimes remarkably precise. To many early-day morphologists, this complexity, this elegance, and this precision implied the direct intervention of a divine hand in guiding the production of such sophisticated designs. However, not everyone was convinced. Spectacular mountain ranges do not require recourse to divine intervention to explain them. Plate tectonics offers a natural explanation. Under pressure from colliding tectonic plates, the Earth's crust crumples to produce these ranges. With knowledge, scientific explanations uncover the mysteries that shroud geological events.

Similarly, biology has found satisfying natural explanations to replace what were once assumed to be direct divine causes. Modern principles of evolution and structural biology offer a fresh approach to vertebrate design and an insight into the processes responsible for producing that design. Just as processes of plate tectonics help geologists understand the origin of the Earth's surface features, structural and evolutionary processes help biologists understand the origin of plant and animal life. Life on Earth is a product of these natural processes. Humans are not exempt nor are we given special dispensation from these processes. Like our fellow vertebrates, humans too are products of our evolutionary past and basic structural plan. The study of morphology, therefore, brings us an understanding of the integrated processes that forged us. To understand the processes behind our design is to understand the product, namely, humans themselves, both what we are and what we can become.

But, I am getting ahead of the story. We have not had an easy intellectual journey in reaching the clarity of morphological concepts we seem to enjoy at the moment. The principles were not always so obvious, the evidence not always so clear. In fact, some issues prevalent over 100 years ago remain unresolved. The significance of underlying structure to the evolution of design, central to much of biology early in the nineteenth century, is only recently being re-examined for its potential contribution to modern morphology. Morphology has often been internally beset by unhappy contentions between those scientists centered on structure and those centered on evolution. To some extent, the fundamental principles of both structure and evolution have grown from different intellectual sources and different intellectual outlooks. To understand this, we need to examine the historical development of morphology. Later in this chapter, we examine the intellectual roots of theories about structure. But first, let's look to the intellectual roots of theories about evolution.

Historical Predecessors—Evolution

The concept of evolution is tied to the name Charles Darwin (figure 1.2). Yet most persons are surprised to learn that Darwin was not first, nor was he ever foremost, in proposing that organisms evolve. In fact, the idea of change through time in animals and plants dates back to ancient schools of Greek philosophy. Over 2,500 years ago, Anaximander developed ideas about the course of change from fishlike and scaly animals to land forms. Empedocles saw original creatures come together in oddly assembled ways—humans

FIGURE 1.2 **Charles Darwin** (1809–1882), about 30 years old and three years back from his voyage aboard H.M.S. **Beagle**. Although *The Origin of Species* was still just a few notebooks in length and several decades away from publication, Darwin had several accomplishments behind him, including his account of *The Voyage of the Beagle,* a collection of scientific observations. At this time, he was also engaged to his cousin Emma Wedgwood, with whom he would live a happy married life.

with heads of cattle, animals with branches like trees. He argued that most perished, but only those creatures who came together in practical ways survived. Even at their best, these armchair views are more poetic than scientific, so it would be an exaggeration to characterize this Greek philosophical thought as a practical predecessor of modern evolutionary science. Nevertheless, the idea of evolution existed long before Darwin, thanks to these Greek philosophers.

The Process behind the Change

What the Englishman Charles Darwin contributed was not the idea that species evolve. Rather, Darwin proposed the conditions for and mechanism of this evolutionary change. He proposed three conditions:

First, if left unchecked, members of any species increase naturally in number because all possess a *high reproductive potential.* Even slow-breeding elephants, Darwin pointed out, could increase from a pair to many millions in a few hundred years. We are not up to our rooftops in elephants, however, because as numbers increase, resources are consumed at an accelerating rate and become scarce. This brings about condition two, *competition* for the declining resources. In turn, competition leads to condition three, *survival of the few.* Darwin termed the mechanism now determining which organisms survive and which do not **natural selection,** nature's way of weeding out the less fit. In this struggle for existence, those with superior adaptations would, on average, fare better and survive to pass on their successful adaptations. Thus, descent with modification resulted from the preservation by natural selection of favorable characteristics.

As simple as this sounds today, Darwin's insight was profound. He performed no decisive experiment, mixed no chemicals in test tubes, ground no tissue in a blender. Rather, Darwin's insight arose from observation and reflection. The controversy over evolutionary processes emerges at one of three levels—fact, course, mechanism—and asks a different question at each level. The first level addresses the *fact of evolution* and asks if organisms change through time. Did evolution occur? The fact that evolution has occurred is today well established by many lines of evidence, from gene changes to the fossil record. But this does not mean that all controversies over evolution are comfortably settled. At the next level, we might ask: What *course* did evolution then take? For example, anthropologists who study human evolution usually agree on the fact that humans did evolve, but they often disagree, sometimes violently, over the course of that evolution. Finally we can ask: What *mechanism* produced this evolution? At this third level in the evolutionary debate, Darwin made his major contribution. For Darwin, natural selection was the mechanism of evolutionary change.

Verbal scuffles over the fact, course, and mechanism of evolution often become prolonged and steamy because opponents ask questions at different levels and end up arguing at cross-purposes. Each of these questions had to be settled historically as well to bring us to an understanding of the evolutionary process. Historians have taken much notice of the violent public reaction to Darwin's ideas on evolution, a reaction spurred by their challenge to religious convention. But what of the scientific climate at that time? Even in scientific circles, opinion was strongly divided on the issue of "transmutation" of species, as evolution was termed then. The issue initially centered around the fact of evolution. Do species change?

Linnaeus

Foremost among the scientists who felt that species were fixed and unchangeable was Carl von Linné (1707–1778), a Swedish biologist who followed the custom of the day by latinizing his name to Carolus Linnaeus, by which he is most recognized today (figure 1.3). Linnaeus devised a system for naming plants and animals, which is still the basis of modern taxonomy. Philosophically he argued that species were unchangeable, created originally as we find them today. For several thousand years, Western thought had kept company with the biblical view, namely, that all species resulted from a single and special act of divine creation, as described in Genesis, and thereafter species remained unchanged.

FIGURE 1.3 Carolus Linnaeus (1707–1778). This Swedish biologist devised a system still used today for naming organisms. He also firmly abided by and promoted the view that species do not change.

Although most scientists during the 1700s sought to avoid strictly religious explanations, the biblical view of creation was a strong presence in Western intellectual circles because it was conveniently at hand and meshed comfortably with the philosophical arguments put forth by Linnaeus and those who argued that species were immutable (unchanging). However, it was more than just the compatibility of Genesis with secular philosophy that made the idea of immutable species so appealing. At the time, evidence for evolution was not assembled easily, and the evidence available was ambiguous in that it could be interpreted both ways, for or against evolution.

Naturalists

Today we understand the perfected adaptations of animals—the trunks of elephants, the long necks of giraffes, the wings of birds—as natural products of evolutionary change. Diversity of species results. To scientists of an earlier time, however, species adaptations reflected the care exercised by the Creator. Diversity of plant and animal species was proof of God's almighty power. Animated by this conviction, many sought to learn about the Creator by turning to the study of what He had created. One of the earliest to do so was the Reverend John Ray (1627–1705), who summed up his beliefs along with his natural history in a book entitled *The Wisdom of God Manifested in the Works of the Creation* (1691). He tackled the tricky question of why the Divine made obnoxious creatures. To paraphrase Ray, consider lice: They harbor and breed in clothes, "an effect of divine providence, designed to deter men and women from sluttishness and sordidness, and to provoke them to cleanliness and neatness." William Paley (1743–1805), archdeacon of Carlisle, also articulated the common belief of his day in his book *Natural Theology; or Evidences of the Existence and Attributes of the Deity Collected from the Appearances of Nature* (1802). Louis Agassiz (1807–1873), curator of the Museum of Comparative Zoology at Harvard University, found much public support for his successful work to build and stock a museum that collected the remarkable creatures that were this world's manifestations of the divine mind (figure 1.4). For most scientists, philosophers, and laypeople, there was, in the biological world of species, no change, thus no evolution. Even in secular circles of the mid-nineteenth century, intellectual obstacles to the idea of evolution were formidable.

J-B. de Lamarck

Among those taking the side of evolution, few were as uneven in their reputation as Jean-Baptiste de Lamarck (figure 1.5a). Most of his life, Lamarck lived on the border of poverty. He did not even hold the equivalent of a professorship at the Jardin du Roi in Paris (later the Muséum National d'Histoire Naturelle; figure 1.5b). Abrupt speech, inclination to argument, and strong views did little to endear Lamarck to his colleagues. Yet his *Philosophie Zoologique*, generally dismissed when published in 1809 as

FIGURE 1.4 Louis Agassiz (1807–1873) was born in Switzerland but came to his second and permanent home in the United States when he was 39. He studied fossil fishes and was first to recognize evidence of the worldwide ice ages, episodes of glaciation in Earth's history. He founded the Museum of Comparative Zoology at Harvard University. Although brilliant and entertaining in public and in anatomical research, Agassiz remained unconvinced of Darwinian evolution to the end of his life.

the amusing ruminations of a "poet," eventually established the theory of evolutionary descent as a respectable scientific generalization.

Lamarck's ideas spoke to the three issues of evolution—fact, course, and mechanism. As to the fact of evolution, Lamarck argued that species changed through time. Curiously, he thought that the simplest forms of life arose by spontaneous generation; that is, they sprang ready-made in muck from inanimate matter but thereafter evolved onward and upward into higher forms. As to the course of evolution, he proposed a progressive change in species along an ascending scale, from the lowest on one end to the most complex and "perfect" (meaning humans) on the other. As to the mechanism of evolution, Lamarck proposed that need itself produced heritable evolutionary change. When environments or behaviors changed, an animal developed new needs to meet the demands the environment placed upon it. Needs altered metabolism, changed the internal physiology of the organism, and triggered the appearance of a new part to address these needs. Continued use of a part tended to develop that part further; disuse led to its withering. As environments changed, a need arose, metabolism adjusted, and new organs were created. Once acquired, these new characteristics were passed on to offspring. This, in summary, was Lamarck's view. It has been called evolution by means of the *inheritance of acquired characteristics*. Characters were "acquired" to meet new needs and then "inherited" by future generations.

(a)

(b)

FIGURE 1.5 (a) J-B. de Lamarck (1744–1829) worked most of his scientific life at the Muséum National d'Histoire Naturelle (b). His academic position gave him a chance to promote the idea that species change.

While a debt is owed Lamarck for championing evolutionary change and so easing the route to Darwin, he also created obstacles. Central to his philosophy was an inadvertent confusion between physiology and evolution. Any person who begins and stays with a weight-lifting program on a regular basis can expect to see strength increase and muscles enlarge. With added weight, use (need) increases; therefore, big muscles appear. This physiological response is limited to the exercising individual because big muscles are not passed genetically to offspring. Charles Atlas, Arnold Schwarzenegger, and other bodybuilders do not pass newly acquired muscle tissue to their children. If their children seek large muscles, they too must start from scratch with their own training program. Somatic characteristics acquired through use cannot be inherited. Lamarck, however, would have thought otherwise.

Unlike such physiological responses, evolutionary responses involve changes in an organism that are inherited from one generation to the next. We know today that such characteristics are genetically based. They arise from gene mutation, not from somatic alterations due to exercise or metabolic need.

Acquired Characteristics

Lamarck's proposed mechanism of inheritance of acquired characteristics failed because it confused immediate physiological response with long-term evolutionary change. Yet most laypeople today still inadvertently think in Lamarckian terms. They mistakenly view somatic parts arising to meet immediate needs. Recently, a moderator of a nature program on giraffes unknowingly resorted to a Lamarckian explanation when he informed us that the origin of the long neck helped giraffes meet the "needs" of reaching treetop vegetation. But, environmental demands do not reach into genetic material and directly produce appropriate heritable improvements to address new needs or new opportunities. Bodybuilding changes muscles, not DNA. That route of inheritable modification does not exist in any organism's physiology.

The other side of the Lamarckian coin is disuse, loss of a part following loss of a need. Some fishes and salamanders live in deep caves not reached by daylight. These species lack eyes. Even if they return to the light, eyes do not form. Evolutionarily, the eyes are lost. It is tempting to attribute this evolutionary loss of eyes to disuse in a dark environment. That, of course, would be invoking a Lamarckian mechanism. Contrary to Lamarck's theory, somatic traits are not inherited.

Because it comes easily, it is difficult to purge a Lamarckian explanation from our own reasoning. We fall automatically and too comfortably into the convenient habit of thinking of parts as rising to meet "needs," one creating the other. For Darwin, and for students coming to evolution fresh today, Lamarck's theory of acquired characteristics impedes clear reasoning. Unfortunately, Lamarck helped popularize an erroneous outlook that current culture perpetuates.

Upward to Perfection

The proposed course of evolution championed by Lamarck also remains an intellectual distraction. The concept of the "scale of nature" (Latin, *scala naturae*) goes back to Aristotle and is stated in various ways by various philosophers. Its central theme holds that evolving life has a direction beginning with the lowest organisms and evolving to the highest, progressively upward toward perfection. Evolutionists, like Lamarck, viewed life metaphorically as ascending a ladder one rung at a time, up toward the complex and the perfected. After a spontaneous origin, organisms progressed up this metaphorical ladder or scale of nature through the course of many generations.

The concept of a ladder of progress was misleading because it viewed animal evolution as internally driven in a particular direction from the early, imperfect, soft-bodied forms up toward perfected humans. As water runs naturally downhill, descent of animals was expected to run naturally to the perfected. Simple animals were not seen as adapted in their own right but rather as springboards to a better future. The scale of nature concept encouraged scientists to view animals as progressive improvements driven by anticipation of a better tomorrow. Unfortunately, remnants of this idea still linger in modern society. Certainly humans are perfected in the sense of being designed to meet demands, but no more so than any other organism. Moles and mosquitoes, bats and birds, earthworms and anteaters all achieve an equally perfect match of parts-to-performance-to-environmental demands. It is not the benefits of a distant future that drive evolutionary change. Instead, the immediate demands of the current environment shape animal design.

The idea of perfection rooted in Western culture is perpetuated by continued technological improvements. We bring it unnoticed, like excess intellectual baggage, into biology where it clutters our interpretation of evolutionary change. When we use the terms *lower* and *higher*, we risk perpetuating this discredited idea of perfection. Lower animals and higher animals are not poorly designed and better designed, respectively. Lower and higher refer only to order of evolutionary appearance. Lower animals evolved first; higher animals arose after them. Thus, to avoid any suggestion of increasing perfection, many scientists prefer to replace the terms *lower* and *higher* with the terms **primitive** and **derived** to emphasize only evolutionary sequence of appearance, early and later, respectively.

To Lamarck and other evolutionists of his day, nature got better and animals improved as they evolved "up" the evolutionary scale. Thus, Lamarck's historical contribution to evolutionary concepts was double sided. On the one hand, his ideas presented intellectual obstacles. His proposed mechanism of change—inheritance of acquired characteristics—confused physiological response with evolutionary adaptation. By championing a flawed scale of nature, he diverted attention to what supposedly drove animals to a better future rather than to what actually shaped them in their present environment. On the other hand, Lamarck vigorously defended the view that animals evolved. For many years, textbooks have been harsh in their treatment of Lamarck, probably to ensure that his mistakes are not acquired by modern students. However, it is also important to give him his place in the history of evolutionary ideas. By arguing for change in species, Lamarck helped blunt the sharp antievolutionary dissent of contemporaries like Linnaeus, gave respectability to the idea of evolution, and helped prepare the intellectual environment for those who would solve the question of the origin of species.

Natural Selection

The mechanism of evolution by means of natural selection was unveiled publically by two persons in 1858, although it was conceived independently by both. One was Charles Darwin; the other was Alfred Wallace. Both were part of the respected naturalist tradition in Victorian England that encouraged physicians, clergymen, and persons of leisure to devote time to observations of plants and animals in the countryside. Such interests were not seen as a way to pass idle time in harmless pursuits. On the contrary, observation of nature was respectable because it encouraged intercourse with the Creator's handiwork. Despite the reason, the result was thoughtful attention to the natural world.

A. R. Wallace

Alfred Russel Wallace, born in 1823, was 14 years younger than Darwin (figure 1.6). Although following the life of a naturalist, Wallace lacked the comfortable economic circumstances of most gentlemen of his day; therefore, he turned to a trade for a livelihood. First he surveyed land for railroads in his native England, and eventually, following his interest in nature, he took up the collection of biological specimens in foreign lands to sell to museums back home. His search for rare plants and animals in exotic lands took him to the Amazon jungles and later to the Malay Archipelago in the Far East. We know from his diaries that he was impressed by the great variety and number of species to which his travels introduced him. In early 1858, Wallace fell ill while on one of the Spice Islands (Moluccas) between New Guinea and Borneo. During a fitful night of fever, his mind recalled a book he had read earlier by the Reverend Thomas Malthus entitled *An Essay on the Principle of Population, as It Affects the Future Improvement of Society*. Malthus, writing of human populations, observed that unchecked breeding causes populations to grow geometrically, whereas the supply of food grows more slowly. The simple, if cruel, result is that people increase faster than food. If there is not enough food to go

FIGURE 1.6 Alfred Russel Wallace (1823–1913) in his thirties.

around, some people survive but most die. The idea flashed to Wallace that the same principle applied to all species. In his own words written some years later:

> It occurred to me to ask the question, Why do some die and some live? And the answer was clearly, that on the whole the best fitted lived. From the effects of disease the most healthy escaped; from enemies, the strongest, the swiftest, or the most cunning; from famine, the best hunters or those with the best digestion; and so on.
>
> Then I at once saw, that the ever present variability of all living things would furnish the material from which, by the mere weeding out of those less adapted to the actual conditions, the fittest alone would continue the race.
>
> There suddenly flashed upon me the idea of the survival of the fittest.
>
> The more I thought over it, the more I became convinced that I had at length found the long-sought-for law of nature that solved the problem of the Origin of Species.

(Wallace, 1905)

Wallace began writing that same evening and within two days had his idea sketched out in a paper. Knowing that Darwin was interested in the subject, but unaware of how far Darwin's own thinking had progressed, he mailed the manuscript to Darwin for an opinion. The post was slow, so the journey took four months. When Wallace's paper arrived out of the blue with its stunning coincidence to his own ideas, Darwin was taken by complete surprise.

Charles Darwin

Unlike Wallace, Charles Darwin (1809–1882) was born into economic security. His father was a successful physician, and his mother part of the Wedgwood (pottery) fortune. He tried medicine at Edinburgh but became squeamish during operations. Fearing creeping idleness, Darwin's father redirected him to Cambridge and a career in the church, but Darwin proved uninterested. At formal education, he seemed a mediocre student. While at Cambridge, however, his long-standing interest in natural history was encouraged by John Henslow, a professor of botany. Darwin was invited on geological excursions and collected biological specimens. Upon graduation, he joined as *de facto* naturalist of the government's H.M.S. *Beagle* over the objections of his father, who wished him to get on with a more conventional career in the ministry.

He spent nearly five years on the ship and explored the coastal lands it visited. The experience intellectually transformed him. Darwin's belief in the special creation of species, with which he began the voyage, was shaken by the vast array of species and adaptations the voyage introduced to him. The issue came especially to focus on the Galápagos Islands off the west coast of South America. Each island contained its own assortment of species, some found only on that particular island. Local experts could tell at sight from which of the several islands a particular tortoise came. The same was true of many of the bird and plant species that Darwin collected.

Darwin arrived back in England in October 1836 and set to work sorting his collection, obviously impressed by the diversity he had seen but still wedded to misconceptions about the Galápagos collection in particular. He had, for instance, thought that the Galápagos tortoise was introduced from other areas by mariners stashing reptilian livestock on islands to harvest during a later visit. Apparently Darwin dismissed reports of differences among the tortoises of each island, attributing these differences to changes that attended the animals' recent introductions to new and dissimilar habitats. However, in March of 1837, almost a year and a half after departing the Galápagos, Darwin met in London with John Gould, respected specialist in ornithology. Gould insisted that the mockingbirds Darwin had collected on the three different Galápagos Islands were actually distinct species. In fact, Gould emphasized that the birds were endemic to the Galápagos—distinct species, not just varieties—although clearly each was related to species on the South American mainland. It seemed to have suddenly dawned on Darwin that not only birds, but plant and tortoise varieties, were distinct as well. These tortoises geographically isolated on the Galápagos were not only derivatives of ancestral stocks but now distinct island species.

Here then was the issue. Was each of these species of tortoise or bird or plant an act of special creation? Although distinct, each species also was clearly related to those on the other islands and to those on the nearby South American mainland. To account for these species, Darwin had two serious choices. Either they were products of a special creation, one act for each species, or they were the natural result of evolutionary adaptation to the different islands. If these related species were acts of special divine creation, then each of the many hundreds of species would represent a distinct act of creation. But if this were so, it seemed odd that they would all be similar to each other, the tortoises to other tortoises, the birds to other birds, and the plants to other plants on the various islands, almost as if the Creator ran out of new ideas. If, however, these species were the natural result of evolutionary processes, then similarity and diversity would be expected. The first animal or plant washed or blown to these oceanic islands would constitute the common stock from which similar but eventually distinct species evolved. Darwin sided with a natural evolution.

But Darwin needed a mechanism by which such evolutionary diversification might proceed, and at first he had none to suggest. Not until his return to England did Darwin's experiences from the Galápagos Islands and throughout his voyage crystallize. Two years after his return, and while in the midst of writing up his results of other studies from the *Beagle*, Darwin read for amusement the essay on population by Malthus, the same essay Wallace would discover years later.

The significance struck Darwin immediately. If animals, like humans, outstripped food resources, then competition for scarce resources would result. Those with favorable adaptations would fare best, and new species incorporating these favored adaptations would arise. "Here then I had at last got a theory by which to work" wrote Darwin. In a moment of insight, he had solved the species problem. That was 1838, and you would think the excitement would have set him to work on papers and lecturing. Nothing of the sort happened. In fact, four years lapsed before he wrote a first draft, which consisted of 35 pages in pencil. Two years later, he expanded the draft to over 200 pages in ink, but he shoved it quietly into a drawer with a sum of money and a sealed letter instructing his wife to have it published if he met an untimely death. A few close friends knew what he had proposed but most did not, including his wife with whom he otherwise enjoyed a close and loving marriage. This was Victorian England. Science and religion fit hand and glove.

Darwin's delay testifies to how profoundly he understood the larger significance of what he had discovered. He wanted more time to gather evidence and write the volumes he thought it would take to make a compelling case. Then in June 1858, 20 years after he had first come upon the mechanism of evolution, Wallace's manuscript arrived. Darwin was dumbfounded. By coincidence, Wallace had even hit upon some of the same terminology, specifically, natural selection. Mutual friends intervened, and much to the credit of both Wallace and Darwin, a joint paper was read in the absence of both before the Linnaean Society in London the following month, July 1858. Wallace was, as Darwin described him, "generous and noble." Wallace, in "deep admiration," later dedicated his book on the Malay Archipelago to Darwin as a token of "personal esteem and friendship." Oddly, this joint paper made no stir. But Darwin's hand was now forced.

Critics and Controversy

Darwin still intended a thick discourse on the subject of natural selection but agreed to a shorter version of "only" 500 pages. This was *The Origin of Species,* published at the end of 1859. By then word was out, and the first edition sold out as soon as it appeared.

Largely because he produced the expanded case for evolution in *The Origin of Species,* and because of a continued series of related work, Darwin is remembered more than Wallace for formulating the basic concept. Darwin brought a scientific consistency and cohesiveness to the concept of evolution, and that is why it bears the name Darwinism.

Science and religion, especially in England, had been tightly coupled. For centuries, a ready answer was at hand for the question of life's origin, a divine explanation, as described in Genesis. Darwinism challenged with a natural explanation. Controversy was immediate, and in some remnant backwaters, it still lingers today. Darwin himself retired from the fray, leaving to others the task of public defense of the ideas of evolution.

Sides quickly formed. Speaking before the English Parliament, the future prime minister Benjamin Disraeli safely chose his friends: "The question is this—Is man an ape or an angel? My lord, I am on the side of the angels."

Despite the sometimes misguided reactions, two criticisms stuck and Darwin knew it. One was the question of variation, the other the question of time. As to time, there seemed not to be enough. If the evolutionary events Darwin envisioned were to unfold, then the Earth must be very old to allow time for life to diversify. In the seventeenth century, James Ussher, Archbishop of Armagh and Primate of All Ireland, made an honorable effort to calculate the age of the Earth. From his biblical studies of who begot whom and from historical dates available at the time, Ussher determined that the first day of Creation began in 4004 B.C. on Saturday, October 22, at nightfall. A contemporary, Dr. John Lightfoot, vice-chancellor at Cambridge University, estimated further that humans were created five days later, at 9:00 in the morning, presumably Greenwich mean time. Many took this date as literally accurate, or at least as indicative of the recent origin of humans, leaving no time for evolution from apes or angels. A more scientific effort to age the Earth was made by Lord Kelvin, who used temperatures taken in deep mine shafts. Reasoning that the Earth would cool from its primitive molten state to present temperatures at a constant rate, Kelvin extrapolated backward to calculate that the Earth was no more than 24 million years old. He did not know that natural radioactivity in the Earth's crust keeps the surface hot. This fact deceptively makes it seem close in temperature and thus in age to its molten temperature at first formation. The true age of the Earth is actually several billion years, but unfortunately for Darwin, this was not known until long after his death.

Critics also pointed to inheritance of variation as a weak spot in his theory of evolution. The basis of heredity was unknown in Darwin's day. The popular view held that inheritance was blending. Like mixing two paints, offspring received a blend of characteristics from both parents. This view, although mistaken, was taken seriously by many. It created two problems for Darwin. From where did variation come? How was it passed from generation to generation? If natural selection favored individuals with superior characteristics, what ensured that these superior characteristics were not blended and diluted out of existence in the offspring? If favored characters were blended, they would effectively be lost from view and natural selection would not work. Darwin could see this criticism coming and devoted much space in *The Origin of Species* to discussing sources of variation.

Today we know the answers to this paradox. Mutations in genes produce new variations. Genes carry characteristics unaltered and without dilution from generation to generation. This mechanism of inheritance was unknown and unavailable to Darwin and Wallace when they first sought answers to the origin of species. It was probably no coincidence that the intellectual breakthroughs of both were fostered by voyages of separation from the conventional scientific climate of their day. Certainly, study of nature was encouraged, but a ready interpretation of the diversity and order they observed awaited such naturalists. Although the biblical story of creation in Genesis was conveniently at hand and taken literally by some to supply explanations for the presence of species, there were scientific obstacles as well. Confusion between physiological and evolutionary adaptation (Lamarck), the notion of a scale of nature, the idea of fixity of species (Linnaeus and others), the young age of Earth (Kelvin), and the mistaken views of variation and heredity (blending inheritance) all differed from predictions of evolutionary events or confused the picture. It is testimony to their intellectual insight that Darwin and Wallace could see through the obstacles that defeated others.

Historical Predecessors—Morphology

We might expect that the study of structure and the study of evolution historically shared a cozy relationship, each supporting the other. After all, the story of evolution is written in the anatomy of its products, in the plants and animals that tangibly represent the unfolding of successive changes through time. For the most part, direct evidence of past life and its history can be read in the morphology of fossils. By degrees, living animals preserve evidence of their phylogenetic background. It might seem then that animal anatomy would have fostered early evolutionary concepts. For some nineteenth-century anatomists, this was true. T. H. Huxley (1825–1895), remembered for many scientific contributions including monographs on comparative anatomy, remarked upon first hearing Darwin's ideas of natural selection words to the effect, "How truthfully simple. I should have thought of it." Huxley was won over (figure 1.7). Although Darwin

retired from public controversy following the publication of *The Origin of Species*, Huxley pitched in with great vigor, becoming "Darwin's Bulldog" to friend and foe alike.

Not all anatomists joined the evolutionary bandwagon so easily, however. Some simply misread morphology as giving evidence of only stasis, not change. On the other hand, many raised solid objections to Darwinian evolution, some of which still have not been addressed even today by evolutionary biologists. To understand the contribution of morphology to intellectual thought, we need to backtrack a bit to the anatomists who preceded Darwin. Foremost among these was the French comparative anatomist, Georges Cuvier.

Georges Cuvier

Georges Cuvier (1769–1832) brought attention to the function that parts performed (figure 1.8). Because parts and the function they served were tightly coupled, Cuvier argued that organisms must be understood as functional wholes. Parts had dominant and subordinate ranking as well as compatibility with each other. Certain parts necessarily went together, but others were mutually exclusive. Possible combinations were thus limited to parts that meshed harmoniously and met necessary conditions for existence; therefore, the number of ways parts could be assembled into a workable organism was predictable. Given one part of an organism, Cuvier once boasted, he could deduce the rest of the organism. Parts of organisms, like parts of a machine, serve some purpose. Consequently, for the entire organism (or machine) to perform properly, the parts must harmonize. Sharp carnivore teeth would be necessarily set in jaws suited for biting, into a skull that buttressed the jaw, on a body with claws for snaring prey, with a digestive tract for digesting meat, and

FIGURE 1.7 **Thomas H. Huxley** (1825–1895) at age 32.

FIGURE 1.8 **Georges Cuvier** (1769–1832). His life spanned the French Revolution, which at first won his sympathies, but as lawlessness and bloodshed became more of its character, he grew increasingly dismayed by its excesses. His life also overlapped with Napoleon's rule. Cuvier came to Paris in 1795 to take a post at the Muséum National d'Histoire Naturelle, where he pursued administrative duties and studies in paleontology, geology, and morphology for most of his remaining life.

so forth (figure 1.9). Alter one part, and the structurally and functionally integrated machinery of the organism would fail. If one part is altered, function of connected parts is disrupted, and performance fails. Evolution could not happen. If an animal were altered, harmony among the parts would be destroyed, and the animal would no longer be viable. Change (evolution) would cease before it began. Cuvier's functional morphology put him in intellectual company with Linnaeus but in opposition to Lamarck's evolutionary ideas.

Cuvier took comfort as well from the known fossil record of his day. Gaps existed between major groups, as would be expected if species were immutable and evolution did not occur. During his time, ancient Egyptian mummies of humans and animals were being pilfered by Napoleon's armies and sent to European museums. Dissection proved that these ancient animal mummies were structurally identical to modern species. Again, this was evidence of no change, at least to Cuvier. Today, with a more complete fossil record at our disposal and a realization that evolution occurred over millions of years, not just within the few millennia since the time of the pharoahs, we could enlighten Cuvier. In his day, however, the mummies were for Cuvier sweet pieces of evidence confirming what his view of morphology required.

Parts were adapted to perform specific functions. If a part was changed, function failed and an animal perished. Thus, there was no change and no evolution of species.

Richard Owen

English anatomist Richard Owen (1804–1892) believed like Cuvier that species were immutable, but unlike Cuvier, he felt that the correspondence between parts (homologies) could not be left without explanation (figure 1.10a). Virtually the same bones and pattern are present in the flipper of a dugong, the forelimb of a mole, and the wing of a bat (figure 1.10b). Each possesses the same bones. Why?

From our twentieth-century perspective, the answer is clear. Out of a common ancestry, evolution passes along similar structures to perform new adaptive functions. But Owen, opposed to evolutionary ideas, was determined to find an alternative explanation. His answer centered around **archetypes.** An archetype was a kind of biological blueprint, a supposed underlying plan upon which an organism was built. All parts arose from it. Members of each major animal group were constructed from the same essential, basic plan. All vertebrates, for instance, were thought to share the same archetype, which explained why all possessed the same

FIGURE 1.9 Irreducible design. Cuvier recognized that organisms were complex functional wholes. Certain parts necessarily fit together. Remove a part and the whole organism fails. Consequently, Cuvier boasted that given one part, he could deduce the rest. Start with a carnivore's tooth and it necessarily fit into a strong jaw, part of a robust skull, aided by clawed limbs to snare prey, set into a predator's body, and so forth.

(a)

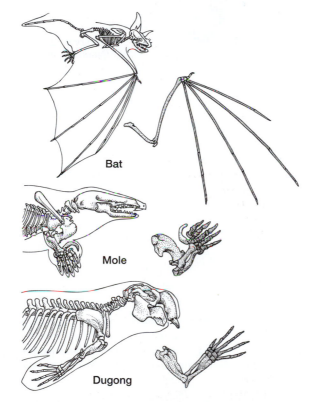

Bat

Mole

Dugong

(b)

FIGURE 1.10 **Richard Owen** (1804–1892). (a) Although admired for his anatomical research, Owen was a difficult man from the accounts of those who worked or tangled with him. He agreed with Cuvier's emphasis on adaptation; however, he felt some explanation for homologies was required and, therefore, introduced the idea of archetypes. (b) Forelimbs of bat, mole, and dugong. Owen noted that each limb performs a different function—flight, digging, and swimming, respectively—and each is superficially different, but he could trace all three to an underlying common plan he called the archetype. Today we recognize that common ancestry accounts for these underlying similarities, although we would join Owen in crediting adaptation for the superficial differences among these homologous parts.

(b) From R. Owen.

fundamental parts. Specific differences were forced on this underlying plan by particular functional needs. Owen was fuzzy about why he ruled out an evolutionary explanation, but he was vigorous in promoting his idea of archetypes.

He even carried this idea to repeated parts within the same individual (figure 1.11a). For example, he envisioned that the vertebrate skeleton consisted of a series of idealized segments he termed vertebrae (figure 1.11b). Not all available parts of these serially repeated vertebrae were expressed at each segment, but all were available if demanded. Taken together, this idealized series of vertebrae constituted the archetype of the vertebrate skeleton. Johann Wolfgang von Goethe (1749–1832), although perhaps best remembered as a German poet, also dabbled in morphology and was the first to suggest that the vertebrate skull was created from modified and fused vertebrae. His idea was expanded by others, such as Lorenz Oken (1779–1851), so by Owen's time, the concept was well known. Owen considered the skull to be formed of vertebrae extended forward into the head. He held that all four vertebrae contributed, and even went so far as to derive human hands and arms from parts of the fourth contributing vertebra, "the occipital segment of the skull."

T. H. Huxley, in a public lecture (published in 1857–1859), took to task the "vertebral theory of the skull," as it had become known. Bone by bone, he traced homologies and developmental appearances of each skull component. He reached two major conclusions. First, all vertebrate skulls are constructed on the same plan. Second, this developmental plan is *not* identical to the developmental pattern of the vertebrae that follow. The skull is not an extension of vertebrae, at least according to Huxley. Ostensibly, the subject of Huxley's public lecture was the skull, but his target was Owen and the archetype. The archetype is, wrote Huxley, "fundamentally opposed to the spirit of modern science."

Certainly Owen was the leader of those morphologists who idealized structure and pushed the vertebral theory of the skull too far and too literally. On the other hand, Huxley succeeded too well in discrediting the concept of archetypes. The two men clashed over archetypes and came down on opposite sides of evolution as well (Huxley for, Owen against). With the eventual triumph of Darwinian evolution in the twentieth century, the issues raised by morphologists such as Owen and Cuvier also tended to be forgotten. In a sense, the baby got thrown out with the bath water; that is, serious morphological issues were forgotten as evolutionary concepts triumphed.

The rise of molecular biology in recent times has further contributed to the displacement of morphology. Molecular biology has won a deserved place in modern science, with its successes in medicine and insights into the molecular machinery of the cell. Unfortunately, in some circles, all significant biological issues that humans face have been reduced to the chemical laws that govern molecules. In its extreme, such a reductionist view sees an organism as nothing more than the simple sum of its parts—know the molecules to know the person.

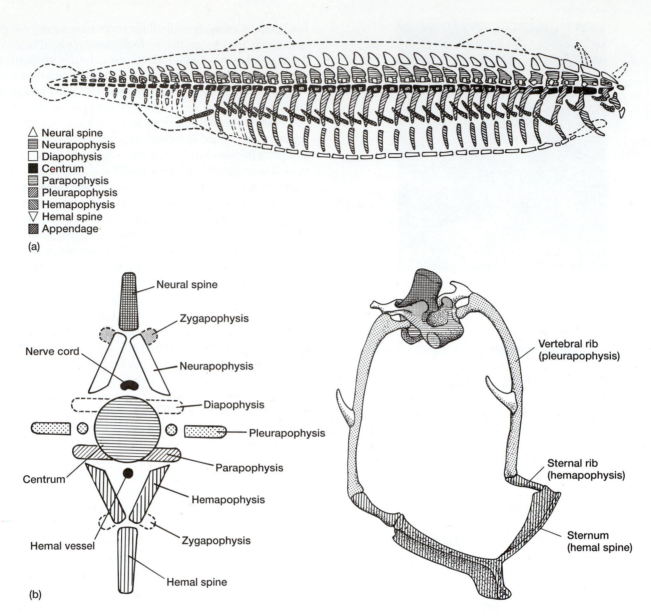

Neural spine
Neurapophysis
Diapophysis
Centrum
Parapophysis
Pleurapophysis
Hemapophysis
Hemal spine
Appendage

(a)

Neural spine
Zygapophysis
Nerve cord
Neurapophysis
Diapophysis
Pleurapophysis
Parapophysis
Centrum
Hemapophysis
Hemal vessel
Zygapophysis
Hemal spine

(b)

Vertebral rib
(pleurapophysis)

Sternal rib
(hemapophysis)

Sternum
(hemal spine)

FIGURE 1.11   Vertebrate archetype. Richard Owen saw the underlying pattern of the vertebrate body as a repeating series of vertebral units, collectively the vertebrate archetype (a). Owen supported the view that these vertebral units, carried forward into the head, even produced the basic elements of the skull. (b) Ideal vertebra. Each vertebra potentially included numerous elements, although not all were expressed in each segment. An actual section from a bird's skeleton indicates how this underlying plan might be realized.

From R. Owen.

Certainly this is naive. A long distance separates the molecules of DNA from the final product we recognize as a fish or a bird or a human. Furthermore, as obvious as it might sound, the action of DNA does not reach upward to affect the agency of natural selection, but rather natural selection acts downward on DNA to affect the genetic structure of populations. A great deal of what we need to understand about ourselves comes from the world around us, not just from the DNA within.

Practitioners of morphology have begun to take these issues that occupied Cuvier and Owen a century and a half ago and bring them forward in a modern context. Cuvier's

emphasis on adaptation has been given new life because of the clarity it brings to our appreciation of biological design. The idea of a pattern underlying the process of design has also been revisited. The result of this has been quite surprising. To explain biological design, we need more than Darwinism. Morphology, too, must be seen as a cause of design.

Why Are There No Flying Elephants?

Not all animal designs are equally likely. Some imaginable animal concoctions simply do not work mechanically, so they never arise. Their bulk is too great or their design unwieldy.

An elephant with wings would literally never fly; that is obvious. Yet many modern evolutionary biologists tend to forget about physical limitations when discussing animal design. Most resort solely to evolutionary explanations. It is tempting to be satisfied with such comfortable explanations of animal design—the long necks of giraffes give them reach to treetop vegetation, the hair of mammals insulates their warm-blooded bodies, the fins of fishes control their swimming, the venom of vipers improves their hunting success.

These and other examples of animal design were favored by natural selection, presumably for the adaptive advantages each conferred. This is reasonable, as far as it goes, but it is only half an explanation. Figuratively, natural selection is an external architect that chooses designs to fit current purposes. But the raw materials or morphology of each animal is itself a factor in design. To build a house with doors, walls, and roof, the architect lays out a scheme, but the materials available affect the character of the house. Use of brick, wood, or straw will place limits or constraints on the design of the house. Straw cannot bear several stories of weight like bricks, but it can be bent into rounded shapes. Wood makes for economical construction but is susceptible to rot. Opportunities and limitations for design lie in each material.

To explain form and design, we must certainly consider the environment in which an animal resides. Among bird groups, there are no truly burrowing species that are counterparts to mammalian moles. So-called burrowing owls exist, but these are hardly equal to moles in exploiting a subterranean existence. Most modern amphibians occur near water because of their moisture requirements. Gliding fishes exist, but truly flying forms with strong wings do not. Elephants are large and ponderous in construction, which precludes a flying form on the elephant plan no matter how strongly natural selection favors it.

To understand form and to explain design, we must evaluate both external and internal factors. The external environment assaults an organism with a wrath of predators, challenges of climate, and competition from others. Natural selection is a manifestation of these factors. Internal factors play a part as well. Parts are integrated into a functionally whole individual. If design changes, it must do so without serious disruption of the organism. Because parts are interlocked into a coherent whole, there exist limits to change before the organism's machinery will fail. The internal construction of an organism sets boundaries to allowable change. It establishes possibilities engendered by natural selection. As new species appear, further possibilities open. But natural selection does not initiate evolutionary changes in design. Like a jury, natural selection acts only on the possibilities brought before it. If natural selection is strong and possibilities are few, then extinction occurs or diversification along that particular evolutionary course is curtailed. As a result, the avian design for delicacy of flight offers few possibilities for evolution of robust design and powerful forelimbs for digging. On the other hand, the avian design allows for the further evolution of airborne vertebrate species. Not all evolutionary changes are equally probable, in large part because not all morphologies (combinations of parts) are equally available to natural selection.

Morphology embraces the study of form and function, of how a structure and its function become an integrated part of an interconnected design (the organism), and of how this design itself becomes a factor in the evolution of new forms. The term **morphology** is not just a synonym for the word **anatomy.** It has always meant much more; for Cuvier, it meant the study of structure with function; for Owen, it meant the study of archetypes behind the structure; and for Huxley, it meant a study of structural change over time (evolution). Today, diverse schools of morphology in North America, Europe, and Asia all generally share an interest in the structural integration of parts, the significance of this for the functioning of the organism, and the resulting limitations and possibilities for evolutionary processes. Morphology does not reduce explanations of biological design to molecules alone. Morphological analysis focuses on higher levels of biological organization—at the level of the organism, its parts, and its position within the ecological community.

Morphological Concepts

To analyze design, concepts of form, function, and evolution have developed. Some of the most useful of these address similarity, symmetry, and segmentation.

Similarities

In different organisms, corresponding parts may be considered similar to each other by three criteria—ancestry, function, and appearance. The term **homology** applies to two or more features that share a common ancestry, the term **analogy** to features with a similar function, and the term **homoplasy** to features that simply look alike (figure 1.12).

FIGURE 1.12 Similarities. Parts may be similar in ancestry, function, and/or appearance. Respectively, these are defined as homology, analogy, or homoplasy. None of these types of similarities is mutually exclusive. Parts may simultaneously be homologous and analogous and homoplastic.

These terms date back to the nineteenth century but gained their current meanings after Darwin established the theory of common descent.

More formally, features in two or more species are homologous when they can be traced back in time to the same feature in a common ancestor. The bird's wing and the mole's arm are homologous forelimbs, tracing their common ancestry to reptiles. Homology recognizes similarity based upon common origin. A special case of homology is **serial homology,** which means similarity between successively repeated parts in the *same* organism. The chain of vertebrae in the backbone, the several gill arches, or the successive muscle segments along the body are examples.

Analogous structures perform similar functions, but they may or may not have similar ancestry. Wings of bats and bees function in flight, but neither structure can be traced to a similar part in a common ancestor. On the other hand, turtle and dolphin forelimbs function as paddles (analogy) and can be traced historically back to a common source (homology). Analogy recognizes similarity based upon similar function.

Homoplastic structures look alike and may or may not be homologous or analogous. In addition to sharing a common origin (homology) and function (analogy), turtle and dolphin flippers also look superficially similar; they are homoplastic. The most obvious examples of homoplasy come from mimicry or camouflage, where an organism is in part designed to conceal its presence by resembling something unattractive. Some insects have wings shaped and sculptured like leaves. Such wings function in flight, not in photosynthesis (they are not analogous to leaves), and certainly such parts share no common ancestor (they are not homologous to leaves), but outwardly they have a similar appearance to leaves; they are homoplastic.

Such simple definitions of similarities have not been won easily. Historically, morphology has struggled to clarify the basis of structural similarities. Before Darwin, biology was under the influence of idealistic morphology, the view that each organism and each part of an organism outwardly expressed an underlying plan. Morphologists looked for the essence or ideal type behind the structure. The explanation offered for this ideal was the unity of plan. Owen proposed that archetypes were the underlying source for an animal's features. Homology for Owen meant comparison to the archetype, not to other adjacent body parts and not to common ancestors. Serial homology meant something different too, based again on this invisible archetype. But Darwinian evolution changed this by bringing an explanation for similarities, namely common descent.

Analogy, homology, and homoplasy are each separate contributors to biological design. Dolphins and bats live quite different lives, yet within their designs we can find fundamental likenesses—hair (at least some), mammary glands, similarities of teeth and skeleton. These features are shared by both because both are mammals with a distinct but common ancestry. Dolphins and ichthyosaurs belong to quite different vertebrate ancestries, yet they share certain likenesses—flippers in place of arms and legs and stream-lined bodies. These features appear in both because both are designed to meet the common hydrodynamic demands of life in open marine waters. In this example, convergence of design to meet common environmental demands helps account for likenesses of some locomotor features (figure 1.13). On the other hand, the webbed hindfeet of gliding frogs and penguins have little to do with common ancestry (they are not closely related) or with common environmental demands (the frog glides in air, the penguin swims in water). Thus, structural similarity can arise in several ways. Similar function in similar habitats can produce convergence of form (analogy); common historical ancestry can carry forward shared and similar structure to descendants (homology); occasionally, accidents or incidental events can lead to parts that simply look alike (homoplasy). In explaining design, we can invoke one, two, or all three factors in combination. To understand design, we need to recognize the possible contribution of each factor separately.

Symmetry

Symmetry describes the way in which an animal's body meets the surrounding environment. **Radial symmetry** refers to a body that is laid out equally from a central axis, so that any of several planes passing through the center divides the animal into equal or mirrored halves (figure 1.14a). Invertebrates such as jellyfishes, sea urchins, and sea anemones provide examples. With **bilateral symmetry,** only the **midsagittal plane** divides the body into two mirrored images, left and right (figure 1.14b).

Body regions are described by several terms (figure 1.14c). **Anterior** refers to the head end **(cranial), posterior** to the tail **(caudal), dorsal** to the back, and **ventral** to the belly or front. The midline of the body is **medial;** the sides are **lateral.** An attached appendage has a region **distal** (farthest) and **proximal** (closest) to the body. The **pectoral region** or chest supports the forelimbs; the **pelvic region** refers to hips supporting the hindlimbs. A **frontal plane (cononal plane)** divides a bilateral body into dorsal and ventral sections, a **sagittal plane** splits it into left and right portions, and a **transverse plane** separates it into anterior and posterior portions.

Because humans carry the body upright and walk with the belly forward, the terms **superior** and **inferior** generally replace the terms *anterior* and *posterior,* respectively, in medical anatomy. Like many terms used only in the descriptive anatomy of humans, *superior* and *inferior* are poor ones to employ in general comparative research because few animals other than humans walk upright. If you venture into the study of human anatomy, you can expect to meet such specialized terms.

Segmentation

A body or structure built of repeating or duplicated sections is segmented. Each repeated section is referred to as a **segment** (or **metamere**), and the process that divides

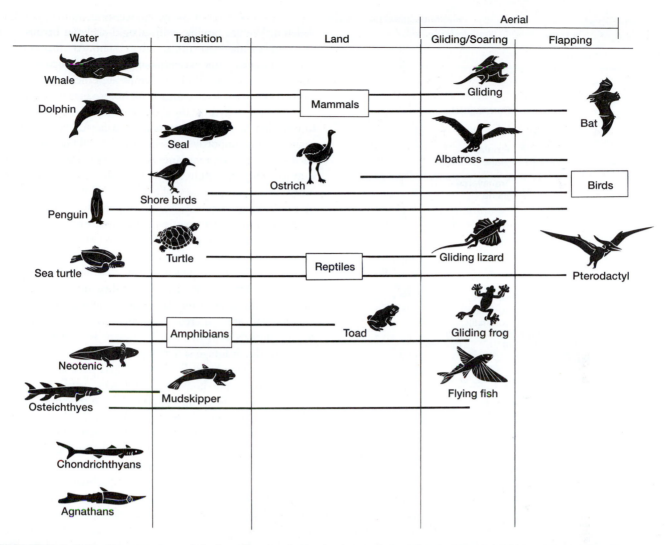

FIGURE 1.13 **Convergence of design.** Groups of animals often evolve in habitats that differ from those of most other members of their group. Most birds fly, but some, such as ostriches, cannot and live exclusively on land; others, such as penguins, live much of their lives in water. Many, perhaps most, mammals are terrestrial, but some fly (bats) and others live exclusively in water (whales, dolphins). "Flying" fishes take to the air. As species from different groups enter similar habitats, they experience similar biological demands. Convergence to similar habitats in part accounts for the sleek bodies and fins or flippers of tuna and dolphins because similar functions (analogy) are served by similar parts under similar conditions. Yet tuna and dolphins come from different ancestries and are still fish and mammal, respectively. Common function alone is insufficient to explain all aspects of design. Each design carries historical differences that persist despite similar habitat.

a body into duplicated sections is called **segmentation** (or **metamerism**). The backbone, composed of repeating vertebrae, is a segmental structure; so is the lateral body musculature of fish that is built from repeating sections of muscle.

Not all body segmentation is the same. To understand design based upon segmentation, we need to turn our attention to invertebrates. Among some invertebrates, segmentation is the basis for amplifying reproductive output. In tapeworms, for example, the body begins with a head (the scolex) followed by duplicated sections called proglottids (figure 1.15). Each section is a self-contained reproductive "factory" housing complete male and female reproductive organs. The more sections, the more factories, and the more

eggs and sperm produced. Some overall body unity is established by simple but continuous nerve cords and excretory canals that run from segment to segment. Other than this, each segment is semiautonomous, a way to replicate sex organs and boost overall reproductive output, which is quite unlike segmentation found in other animals.

Annelids, such as earthworms and leeches, have segmented bodies that provide support and locomotion rather than reproduction. Annelid segmentation differs from that of tapeworms because the annelid body coelom is fluid filled and forms a hydrostatic skeleton. The hydrostatic skeleton is one of two basic types of supportive systems found in animals.

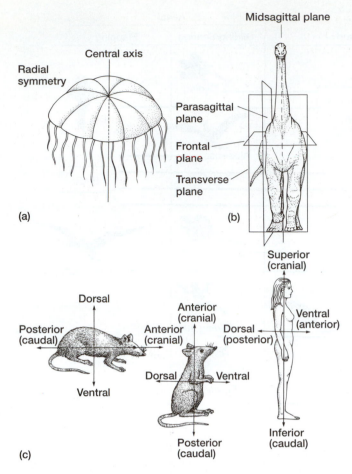

FIGURE 1.14 **Body symmetry.** Radial and bilateral are the two most common body symmetries. (a) Radially symmetrical bodies are laid out regularly around a central axis. (b) Bilaterally symmetrical bodies can be divided into mirror images only through the midsagittal plane. (c) Dorsal and ventral refer to back and belly, respectively, and anterior and posterior to cranial and caudal ends, respectively. In animals that move in an upright position (e.g., humans), superior and inferior apply to cranial and caudal ends, and ventral and dorsal apply to anterior and posterior sides, respectively.

The other supportive system we see in animals is a rigid skeleton. We are familiar with a rigid skeleton because our bones and cartilage constitute such a system. Another example is the chitinous outer skeletons of arthropods, such as crabs, lobsters, and insects. Rigid skeletons are efficient systems of levers that allow selective muscle use to produce movement.

Although hydrostatic skeletons are perhaps less familiar to you, they are common among animals. As the term *hydro* suggests, this supportive system includes a fluid-filled cavity enclosed within a membrane. A hydrostatic skeleton usually is further encased within a muscular coat. At its simplest, the muscular coat is composed of circular and longitudinal bands of muscle fibers (figure 1.16). Movement is accomplished by controlled muscle deformation of the hydrostatic skeleton. In burrowing or crawling animals, movement is usually based on peristaltic waves produced in the body wall. Swimming motions are based on sinusoidal waves of the body.

The advantage of a hydrostatic skeleton is the relatively simple coordination. Only two sets of muscles, circular and longitudinal, are required. Consequently, the nervous system of animals with hydrostatic systems is usually simple as well. The disadvantage is that any local movement necessarily involves the entire body. Because the fluid-filled cavity extends through the entire body, muscle forces developed in one region are transmitted through the fluid to the entire animal. Thus, even when movement is localized, muscles throughout the body must be deployed to control the hydrostatic skeleton.

In truly segmented animals, **septa** sequentially subdivide the hydrostatic skeleton into a series of internal compartments. As a consequence of compartmentalization, the body musculature is also segmented, and in turn the nerve and blood supply to the musculature are segmentally arranged as well. The locomotor advantage is that such segmentation allows for more localized muscle control and localized changes in shape (figure 1.17). For instance, the segmented body of an earthworm is capable of localized movement.

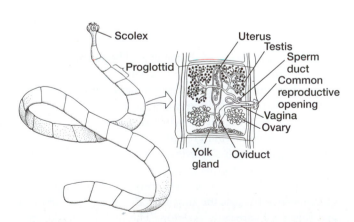

FIGURE 1.15 **Segmented tapeworm.** Each section, or proglottid, is a reproductive factory producing eggs and sperm.

FIGURE 1.16 **Hydrostatic skeleton.** At its simplest, changes in shape and movement involve two mechanical units, the muscle layers of the body wall (longitudinal and circular) and the fluid-filled body coelom within. Contraction of the circular muscles lengthens the shape; contraction of longitudinal muscles shortens the body. The fluid within is incompressible so that muscular forces are spread throughout the body to bring about changes in shape.

FIGURE 1.17 **Locomotion of a segmented worm.**
Fluid within the body cavity flows into selected compartments, filling and expanding each. This ballooning of the body is controlled selectively by each body segment and coordinated overall by the worm's nervous system. As the fluid passes backward from one compartment to the next, each expanded segment pushes against the surrounding soil in turn and establishes a firm hold on the walls of the worm's tunnel-shaped body. Extension of the anterior body pushes the head forward in order for the worm to make progress through the soil.

After Gray and Lissmann.

Segmentation among vertebrates is less extensive than segmentation among invertebrates. Lateral body musculature is laid out in segmental blocks, and nerves and blood vessels supplying it follow this segmental pattern. But segmentation goes no deeper. The viscera are not repeated units, and the body cavity is not serially compartmentalized. Locomotion is provided by a rigid skeleton, and the vertebral column (or notochord) is served by segmental body musculature; however, segmentation of the outer body musculature does not extend inward to the coelom and viscera.

Although the vertebrate body is not composed of a hydrostatic skeleton, selected organs are based on the principle of hydrostatic support. The notochord, for instance, contains a core of fluid-engorged cells tightly wrapped in a sheath of fibrous connective tissue. This incompressible but flexible rod is a hydrostatic organ that functions to keep the body at a constant length. The penis is another example of a hydrostatic organ. When properly stimulated, cavities within it fill firmly with fluid, in this case with blood, to give the penis an erect rigidity of some functional significance.

Evolutionary Morphology

As mentioned previously, evolution and morphology have not always been happy companions. On the brighter side, the more recent cooperation between scientists in both disciplines has clarified our understanding of animal design. With this cooperation, concepts of design and change in design have come into better relief.

Function and Biological Role

For most of us, the concept of function is rather broad and used loosely to cover both how a part works in an organism and how it serves adaptively in the environment. The cheek muscles in some small mice act to close their jaws and chew food. In so doing, these muscles perform the adaptive role

of processing food. The same structure works both within an organism (chewing) and in the role of meeting environmental demands (resource processing). To recognize both services, two terms are employed. The term **function** is restricted to mean the action or property of a part as it works *in an organism*. The term **biological role** (or just **role**) refers to how the part is used *in the environment* during the course of the organism's life history.

In this context, the cheek muscles of mice function to close the jaws and serve the biological role of food processing. Notice that a part may have several biological roles. Not only do jaws serve a role in food processing, but they might also serve the biological role of protection or defense if used to bite an attacking predator. One part may also serve several functions. The quadrate bone in reptiles functions to attach the lower jaw to the skull. It also functions to transmit sound waves to the ear. This means that the quadrate participates in at least two biological roles: feeding (food procurement) and hearing (detection of enemies or prey). Body feathers in birds provide another example (figure 1.18a–c). In most birds, feathers function to cover

(a)   Thermoregulation

(b)   Contour

(c)   Courtship

FIGURE 1.18 **Biological roles.** The same structure may serve several biological roles. For example, in addition to producing lift for flight, feathers play a part in (a) thermoregulation (insulation), to prevent heat loss to a cold environment; (b) aerodynamic contouring (flight), to streamline the body; and (c) reproduction (courtship), to display colors to rivals or mates.

the body. In the environment, the biological roles of feathers include insulation (thermoregulation), aerodynamic contouring of body shape (flight), and in some, display during courtship (reproduction).

Functions of a part are determined largely in laboratory studies; biological roles are observed in field studies. Inferring biological roles only from laboratory studies can be misleading. For example, some harmless snakes produce oral secretions in which laboratory biologists discovered toxic properties. Many leaped to the conclusion that the biological role of such toxic oral secretions must be to kill prey rapidly, but field studies proved that this was not the case. Humans also produce a saliva that is mildly toxic (function), but certainly we do not use it to envenomate prey (biological role). Saliva serves the biological role of processing food by initiating digestion and lubrication of food. Toxicity is an inadvertent by-product of human saliva, without any adaptive role in the environment.

Preadaptation

For many scientists, the word **preadaptation** is chilling because it seems to invite a misunderstanding. Alternative terms have been proposed (protoadaptation, exaptation), but these really do not help and only congest the literature with redundant jargon. If we keep in mind what preadaptation does not mean as well as what it signifies, then the term should present no special difficulty. Preadaptation means that a structure or behavior possesses the necessary form and function *before* (hence pre-) the biological role arises that it eventually serves. In other words, a preadapted part can do the job before the job arrives. The concept of preadaptation does not imply that a trait arises in anticipation of filling a biological role sometime in the future. Adaptive traits serve roles of the moment. If there is no immediate role, selection eliminates the trait.

For example, feathers likely evolved initially in birds (or in their immediate ancestors) as insulation to conserve body heat. Like hair in mammals, feathers formed a surface barrier to retard the loss of body heat. For warm-blooded birds, feathers were an indispensable energy-conserving feature. Today, feathers still play a role in thermoregulation; however, for modern birds, flight is the most conspicuous role of feathers. Flight came later in avian evolution. Immediate ancestors to birds were ground- or tree-dwelling, reptilelike animals. As flight became a more important lifestyle in this evolving group, feathers already present for insulation became adapted into aerodynamic surfaces in order to serve flight. In this example, we can say that insulating feathers were a preadaptation for flight. They were ready to serve as aerodynamic surfaces before that biological role actually arose.

Similarly, the wings of diving birds are preadapted as paddles. In pelicans and auks, they are used to swim while the bird is submerged. If, as now seems likely, primitive lungs for respiration arose early in fishes, then they were preadapted to become swim bladders, buoyancy devices of later fishes. Fish fins were preadapted to become tetrapod limbs.

One hypothetical scheme of preadaptation traces the origin of birds from reptiles through a series of five preflight stages (figure 1.19). Beginning with reptiles that lived in or frequented trees, the sequence shows that some leaped from branch to branch in order to escape pursuing predators or get to adjacent trees without making a long journey down one tree and back up the other. Such behavior established the animal's practice of taking to the air temporarily. Next came parachuting, in which the animal spread its limbs and flattened its body to increase resistance and slow descent during the vertical drop, softening the impact on landing. Gliding was next. The animal deflected from the line of fall, so horizontal travel increased. Flailing, an early stage of active flight, further increased the horizontal distance. Flapping flight gave access to habitats unavailable to terrestrial species. In fact, a new mode of life was achieved and modern birds are the result.

Such a view, although hypothetical, presents a plausible sequence by which flight in birds might have arisen. It helps address several criticisms leveled at morphological processes of evolutionary change. One long-standing complaint against the concept of evolutionary change is that many structures, such as large, complicated wings and feathers, could not possibly have had any selective value when they first appeared. Such **incipient structures** would be small and formative when they first made their evolutionary debut. The argument goes like this: Incipient structures would not enjoy selective favor until they were large and elaborate enough to perform the role that brought an adaptive advantage, such as flapping flight. However, this example shows that large, complicated structures need not have evolved all at once in one large evolutionary binge. In the hypothesized five-stage evolution of bird flight, no preceding stage anticipated the next. There was no drive in the stages themselves propelling them necessarily to the next stage. Each stage

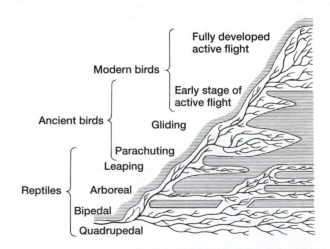

FIGURE 1.19 Evolution of bird flight modeled as a series of successive steps, each preadapted to the next, that trace the evolution of birds from reptiles. Each step is adaptive in its own right, but after having been achieved, each sets the stage for the next.

After Bock.

was adaptive in its own right, for the immediate advantages enjoyed. If conditions changed, organisms may have evolved further, but there were no guarantees.

Some mammals, like "flying" squirrels, are still gliders. They are well adapted to conifer forests. Others, such as bats, are full-fledged, powered fliers. In an evolutionary sense, gliding squirrels are not necessarily "on their way" to becoming powered fliers like bats. Gliding is sufficient to meet demands the squirrels face when moving through the canopy of northern conifer forests. Gliding in these squirrels serves the environmental demands of the present. It does not anticipate powered flight in the distant future.

The example of bird flight also reminds us that a new biological role usually precedes the emergence of a new structure. With a shift in roles, the organism experiences new selective pressures in a slightly new niche. The shift from leaping to parachuting, or from parachuting to gliding, or from gliding to early flailing flight initially placed old structures in the service of new biological roles. This initial shift in roles exposed the structure to new selection pressures favoring those mutations that solidify a structure in its new role. First comes the new behavior, and then the new biological role follows. Finally a change in structure becomes established to serve the new activity.

Evolution as Remodeling

The scheme that traces the evolution of bird flight also tells us that evolutionary change usually involves renovation, not new construction. Old parts are altered, but seldom are brand new parts added. Almost always, a new structure is just an old part made over for present purposes. In fact, if a complete novelty made a sudden appearance, it would probably disrupt the organism's smooth, functional harmony and would be selected against.

Because evolution proceeds largely through the process of remodeling, descendant organisms bear the traces of ancestral structures. Preadaptation does not cause change but is only an interpretation of evolutionary outcomes after they occur. Preadaptation is hindsight, a look backward to see out of what ancestral parts present structures arose. In hindsight, we might see that leaping preceded parachuting, parachuting preceded gliding, and gliding preceded flailing. Each preceding step preadapted to the next. The conceptual mistake would be to interpret these steps as internally driven inevitably from grounded reptiles to flying birds. Nothing of the sort is intended. We do not know ahead of time the future course of evolution, so we cannot tell which structures are preadapted until after they have evolved into new roles.

Phylogeny

The course of evolution, known as **phylogeny,** can be summarized in graphic schemes, or **dendrograms,** that depict treelike, branched connections between groups. Ideally the representation is a faithful expression of the relationships between groups. But the choice of dendrogram is based on intellectual bent and practical outcome. Dendrograms summarize evolution's course. This brevity gives them their attractiveness. All have risks, all flirt with oversimplification, and all take shortcuts to make a point. Let us look at the advantages and disadvantages of several types of dendrograms.

Of Bean Stalks and Bushes

In 1896, Ernst Haeckel wrote *The Evolution of Man,* in which he depicted the human pedigree or human phylogeny (figure 1.20). The book is a useful summary of his thoughts on the subject. Some today might wish to correct points in Haeckel's explicit phylogeny, but what does not stick out so readily is the assumption behind his dendrogram, namely,

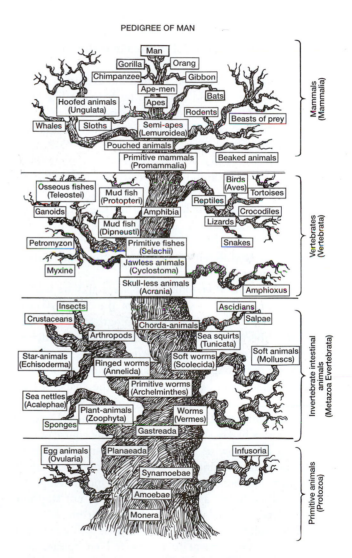

FIGURE 1.20 Haeckel's phylogeny. Like a tree, this phylogeny displays the proposed branching of species. Although many lines of evolution are shown, Haeckel chose to call it the "Pedigree of Man," subtle evidence of the common view that humans represent the culmination of evolution's efforts.

From Ernst Haeckel.

that humans are the pinnacle of evolution. Neither then (nineteenth century) nor now (twenty-first century) was Haeckel alone in assuming that nature climbed from one species to the next like rungs on a ladder, from primitive to perfected, from lower forms to humans at the top of the scale of nature. What such a dendrogram subtly promotes is the mistaken view that humans stand alone as the sole possessor of the top rung of the evolutionary ladder.

In reality, the human species is just one of thousands of recent evolutionary products. Evolution does not proceed up a single ladder, but bushes outward along several simultaneous courses. Although mammals continued to prosper largely on land, birds evolved concurrently and teleost fishes diversified in all waters of the world. Birds, mammals, fishes, and all species surviving today represent the current, and still evolving, species within their groups. No single species is a Mount Everest among the rest. Humans share the current evolutionary moment with millions of other species, all with long histories of their own. All adapted in their own ways to their own environments.

To reflect this diverse pattern of evolution faithfully, dendrograms should look like bushes, not like bean stalks or ladders (figure 1.21a, b). After birds evolved from reptiles, reptiles not only persisted but actually diversified and continued to evolve and prosper. The same holds for amphibian ancestors that gave rise to reptiles and for fishes that gave rise to these amphibian ancestors. Certainly modern amphibians have carried forward primitive features from their early ancestors; however, they have also continued to evolve independently of reptiles since the two lineages parted company over 300 million years ago. Frogs are structurally quite different, for instance, from the earliest amphibian ancestors.

Dendrograms that look like bean stalks or ladders are quick, uncomplicated summaries of the course of evolution (figure 1.21a). This is their strength. But they can also mislead because they imply that the most significant achievement of an earlier group is to serve as the source for a derivative group—fish for amphibians, amphibians for reptiles, and so on. Dendrograms in the shape of ladders warp our view in that more recent groups are somehow depicted as better perfected than earlier groups. Dendrograms that look like bushes not only track the course of new groups but also show us that after one group gives rise to another, both may continue to evolve concurrently and adapt to their own environments (figure 1.21b). Once a new group is produced, evolution among ancestors does not stop nor does a derived group necessarily replace its ancestors.

The evolution of life is a continuous and connected process from one moment to the next. New species may evolve gradually or suddenly, but there is no point of discontinuity, no break in the lineage. If a break occurs in the evolving lineage, the consequence is extinction, a finality not redeemed. When taxonomists study current living species, they examine an evolutionary cross section of time in that they view only the most recent but continuing species

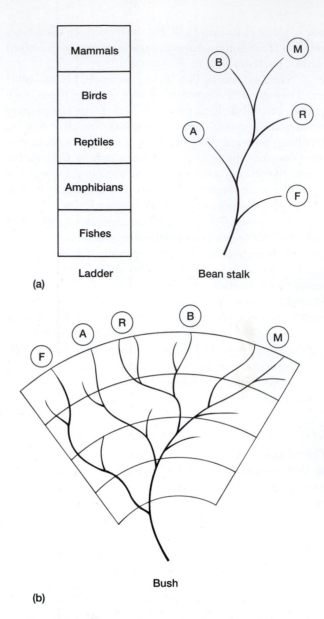

FIGURE 1.21 **Bean stalks and bushes.** (a) The "ladder of creation" is a misleading metaphor. Evolution proceeds not in a stately fashion up a ladder of species, one to the next, but along parallel lines that branch outward. Dendrograms shaped like bean stalks illustrate the order in which a group appeared, but nurture the misleading view that species evolved in linear sequence up to the present time. (b) The diversity of unfolding evolution is better represented by a dendrogram shaped like a bush.

with a long diverging history behind them. The apparent discreteness of species or groups at the current moment is partly due to their previous divergence. When followed back into their past, the connectedness of species can be determined. A dendrogram showing lineages in three dimensions (figure 1.22) emphasizes this continuity. If reduced to a two-dimensional branching dendrogram, the relationships stand out better but imply an instant distinctiveness of species at

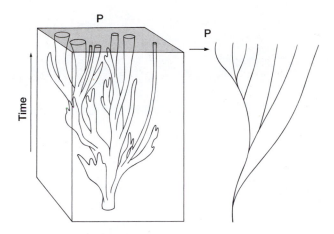

FIGURE 1.22 **Evolution of dendrograms.** The course of evolution, with some branches becoming extinct, is depicted by the left dendrogram. We stand at the time horizon plane (P) to observe the lineages that have persisted to the present. The illustration on the right is one possible two-dimensional dendrogram that represents only the major surviving lines of descent.

branch points. The sudden branches are a taxonomic convention but may not faithfully represent the gradual separation and divergence of species and new groups.

Simplification

Most dendrograms intend to make a point and are simplified accordingly. For example, the evolution of vertebrates is depicted in figure 1.23a to make a point about steps along the way. Although this representation is considerably simplified, it is a convenient summary; but if taken literally, the dendrogram is quite implausible. The first four species are living, so they are unlikely direct ancestral species in the steps. A more plausible representation of their evolution is shown in figure 1.23b. Species at each division point lived millions of years ago and are certainly extinct by now. Only distantly related derivative ancestors survive to the present and are used to represent steps in the origin of vertebrates.

A more complicated dendrogram of birds is shown in figure 1.24. Many groups are included, their likely evolution traced, and the relationships between them proposed. Thus, their phylogeny is more faithfully represented, although the complexity of the diagram makes major trends less apparent. Notice how the more complete detail makes the dendrogram hard to read and, hence, less useful in identifying major trends. In choosing a dendrogram, we should strike some compromise between simple (but perhaps misleading) and complex (but perhaps overwhelming).

Patterns of Phylogeny

Dendrograms can be used to express relative abundance and diversity. The swollen and narrowed shapes of the "balloons" in figure 1.25 roughly represent the relative numbers of vertebrates that existed in each group during various

FIGURE 1.23 **Steps in vertebrate evolution.**
(a) Examples of a hemichordate, a cephalochordate, a urochordate larva, a lamprey, and a salamander (from left to right). All are living species, so they are not likely the immediate ancestors of each succeeding group, as this scheme mistakenly implies.
(b) Their actual ancestors (from A to D, respectively) lived millions of years ago and are now extinct. Modified descendants that represent these species today carried forward some of the primitive traits of their extinct ancestors, but they also evolved additional modifications.

geological times. The first mammals and birds arise within the Mesozoic but do not become abundant and prominent components in terrestrial faunas until much later—in fact, not until after the decline of the contemporaneous reptiles at the end of the Cretaceous. Shapes of branches within a dendrogram convey this additional information.

Rates at which new species appear can also be represented by the sharpness of branching within a dendrogram. One dendrogram is sharply angular, which implies rapid change and relatively sudden appearance of new species (figure 1.26a). The other shows smooth branches, implying the gradual appearance of new species (figure 1.26b).

FIGURE 1.24 Phylogeny of birds. This dendrogram attempts to detail the relationships and the time of origin of each group of modern birds. Although it expresses the hypotheses of these relationships in detail, the diagram is too complex and difficult to view easily. General trends are less evident as well.

© J. Fisher, "Fossil Birds and Their Adaptive Radiation," in The Fossil Record, The Geological Society of London, 1967. Reprinted with permission of the Geological Society of London.

Behind these two types of dendrograms stand different sets of assumptions about the process of evolution. One sees evolution working gradually to produce new species. The other sees the process as an event in which species persist for long stretches with relatively little change followed by a rather abrupt appearance of a new species. In the 1940s, G. G. Simpson termed such long intervals of unchanged evolution occasionally interrupted by short bouts of rapid change as **quantum evolution.** Efforts to celebrate this in dendrograms have recently found favor again, termed **punctuated equilibrium** by those sharing Simpson's view.

Grades and Clades

Living vertebrates derive from a succession of distant ancestors and differ considerably from them. Modern vertebrates carry forward the collective results of these changes upon changes—thousands of them. Taken together, these collective changes produce the modern groups as we meet them today. To reconstruct this history, we may examine particular characters, using them to track the history of these changes. Formally, the earlier (or ancestral) state of a character is its **primitive condition,** referred to as a **plesiomorphic trait;** its later (or descendant) state after transformation is its **derived condition,** referred to as a **synaphomorphic trait.** A **taxon** is simply a named group of organisms. A taxon may be a **natural taxon,** one that accurately depicts a group that exists in nature resulting from evolutionary events. Or, a taxon may be an **artificial taxon,** one that does not correspond to an actual unit of evolution. A **sister group** is the taxon most closely related to the group we are studying. Using transformed characters as our guide, we inspect the pattern of vertebrate evolution and assign names for taxa accordingly, but we may do so with different goals in mind.

If a group of organisms carries a large number of distinctive derived characteristics, we might wish to recognize this by suggesting that the group has reached a new stage, step, or grade in its organization. In a traditional sense, a **grade** was meant to be an expression of the degree of change or level of adaptation reached by an evolving group. Treating groups as grades has been used in the past in some taxonomic schemes. For example, the fused and distinctive shell of turtles might be seen as a drastic reorganization of the skeleton requiring taxonomic recognition. This could be done by elevating turtles to a distinctive taxonomic rank coequal with birds. In this sense of grade, evolving groups collect such a large number of derived characteristics that they pass an imagined threshold that earns them a high taxonomic rank. By such a view, mammals could be considered a taxonomic grade; so could birds. Although sometimes useful

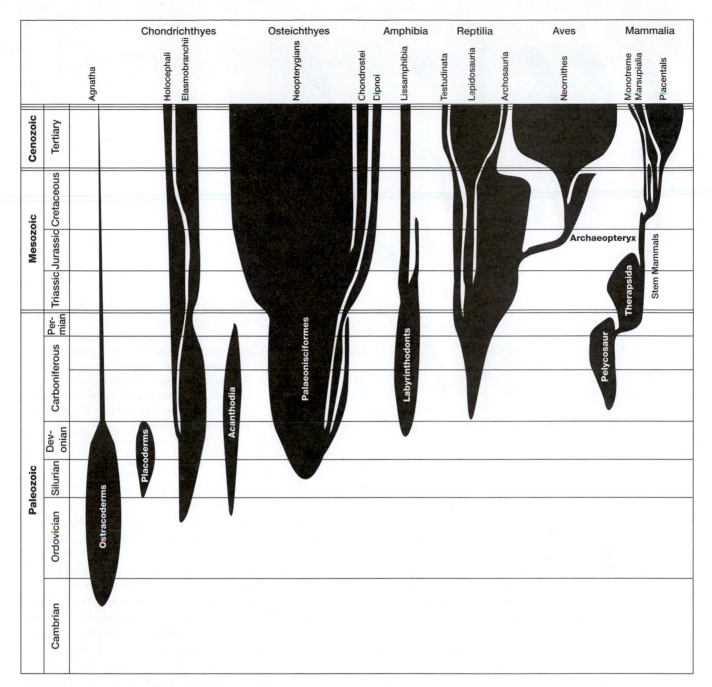

FIGURE 1.25 Abundance phylogeny. This dendrogram attempts to represent the first time each vertebrate group appeared and the relative abundance of each group (depicted by the size of each balloon).

as a way of recognizing the degree of anatomical divergence between groups, grades can be misleading. The group Reptilia traditionally includes members with scales and a shelled egg (cleidoic egg). But such a grade does not represent a single evolving group. Instead, the reptilian grade has been reached independently, once within the line to modern reptiles and once early within the line to mammals. Conversely, current groups may not seem to look alike—crocodiles and birds, for example. But these are survivors of a common

lineage that places them more closely related to each other than either is to modern reptiles. Therefore, we may prefer to recognize groups based on their genealogy rather than on a subjective judgment of the degree of change alone.

If members of a group of organisms share a unique common ancestor, we can recognize this by naming the lineage itself. A **clade** is a lineage—all organisms in a lineage plus the ancestor they have in common. **Traditional systematics** places together organisms with similar or homologous

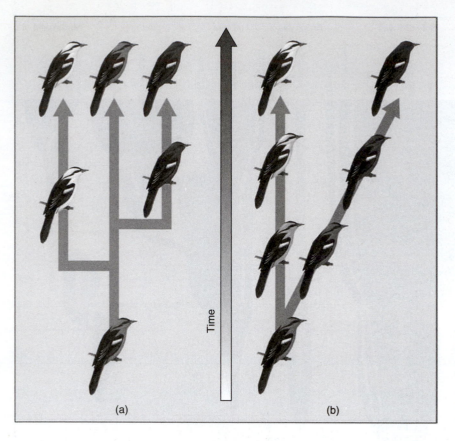

FIGURE 1.26 Patterns of evolution.
A dendrogram may be intended to represent the abrupt (a) or gradual (b) appearance of new species represented by a new branch. Although the two dendrograms agree on the relationships of species, they depict two different processes behind their evolution, namely, a rapid evolutionary process (a) or a gradual process of evolution (b).

characteristics. The newer **phylogenetic systematics** places together organisms belonging to the same clade, and hence is also called **cladistics.** Within cladistics, the taxon name refers to the clade—to the genealogy itself—not necessarily to characters per se. Clades are recognized without concern for the amount of anatomical variation within the taxon. Consequently, some clades might include members very homogeneous in their basic morphology (e.g., birds, snakes, frogs) or quite heterogeneous (e.g., actinopterygian fishes). Genealogy, not within-group variation, is the basis for recognizing a clade. The dendrogram depicting this genealogy is a **cladogram,** a hypothesis about the lineages and their evolutionary relationships. The advantage of such cladograms is the clarity and ease with which they may be critiqued. A practical disadvantage is the swiftness by which a cladogram may be replaced with a newer cladogram, leaving us with abandoned taxon names replaced with newer names for more recent hypotheses of relationship. Character transformations play a central part in producing cladograms. In particular, derived characteristics are most important.

Relationships between groups are recognized on the basis of derived characteristics. The more derived characteristics shared by two groups, the more likely it is they are closely related. The assortment of taxa we are interested in examining is our **ingroup;** the **outgroup** is close to but not part of this assortment and is used as a reference. In particular, the outgroup helps us make decisions about which

character state represents the derived condition. The sister group is the first outgroup we might consult because it is the most closely related. But, we might also successively make comparisons to more distantly related second or third outgroups. Often, at this point, fossils may play an important reference role so that we can better decide primitive and derived states of a character. Once the degree of shared, derived characteristics is determined, we can represent associations in a Venn diagram (figure 1.27a). Because evolution proceeds by descent with modification, as Darwin helped establish, we expect those groups most closely related to be part of a common lineage. Therefore, from this diagram, we produce our hypothesis of genealogy, the cladogram (figure 1.27b), based on the characters we have examined. The layers of brackets above the cladogram represent the levels of inclusiveness of our groups within clades. As we name each clade, we would be producing our classification of this ingroup. In our cladogram, we could mark the sites at which particular character transformations occur. We could thereby use the cladogram to summarize important points of character transformation in the evolution of the groups and identify the distinctive derived characters that are associated with each clade.

Cladistics demands that we staunchly follow the practice of naming clades that recognize genealogy (figure 1.28). A clade is **monophyletic** in that it includes an ancestor and *all* its descendants—but *only* its descendants. Groups formed

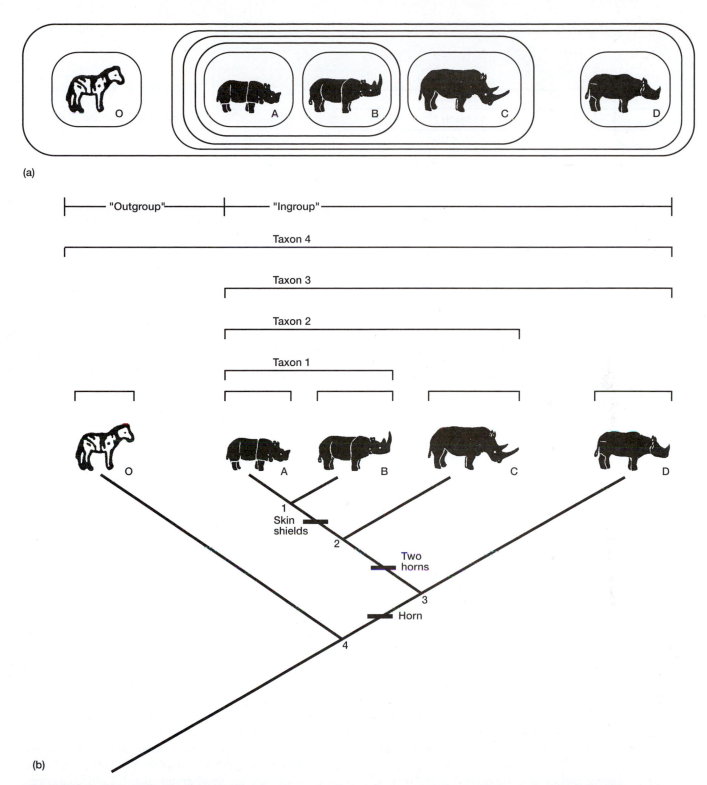

(a)

(b)

FIGURE 1.27 Classification. (a) Venn diagrams sort individuals into successive boxes of relatedness. Individuals of the same species are most closely related and put together in the smallest group—A, B, C, D, and O. If species A and B share more unique, derived features in common than with any others, then we would place them in a common group, and so on, expanding our diagram to include those more distantly related. (b) The genealogy of these species can be expressed in the branching diagram, with the brackets representing successive clades of common descent. The point of branching is the node, the distance between nodes the internode. Taxon 1 includes Species A and B, together with their common ancestor 1 at the node. Taxon 2 includes the clade of Species A, B, and C plus their common ancestor 2, represented at the node, and so on. To make the genealogy more familiar, each taxon would be named. For example, Taxon 3 might be named "Rhinocerotidae." To make the genealogy even more useful, we could identify at the internodes some of the many character transformations that occurred. For example, a horn first arises between nodes 4 and 3; a second horn between nodes 3 and 2; thick skin shields between nodes 2 and 1.

(b) Modified from Classification, British Museum (Natural History).

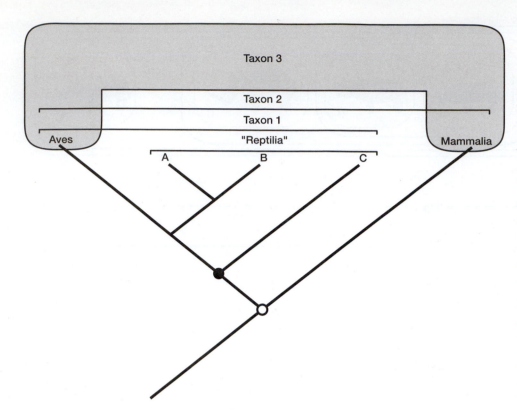

FIGURE 1.28 Cladistic concepts. Monophyletic groups include an ancestor and all of the descendant groups. Taxon 1 is monophyletic because it includes the common ancestor (solid circle at node) plus all descendants— groups A, B, C, and Aves. However, "Reptilia" is paraphyletic, an artificial grouping that deletes Aves, one of the descendants of the same ancestor groups A, B, and C share. Taxon 3 is polyphyletic, also an artificial group, because it places Aves (birds) and Mammalia together on the mistaken view that their endothermy is a homologous feature. Taxon 2 (Amniota) is also monophyletic because it unites all groups descended from the same common ancestor (open circle at the node).

on the basis of nonhomologous characters are **polyphyletic.** If we combined birds and mammals together because we mistook their endothermic physiology (warm-blooded) as the result of common descent, we would be forming an artificial, polyphyletic group. Groups that include a common ancestor and some, but not all, of its descendants are **paraphyletic.** This can happen with some traditional definitions of Reptilia. Modern reptiles and birds derive from a common ancestor. If birds were left out of the clade that represented this common lineage, then what remained would be a paraphyletic group. If paraphyletic groups are used for convenience, the names are usually placed in quotation marks to signal the unnatural composition of the group. Both polyphyletic and paraphyletic taxa are artificial taxa. They do not reflect the actual, complete course of evolution within a common lineage. Further, within cladistics we discover a second meaning for the term *grade.* Here, grade is a synonym for a paraphyletic group. When we visit specific vertebrate groups in chapter 3, we shall meet these issues directly.

By producing explicit and uncluttered hypotheses of relationship, cladograms have become part of the modern language of evolutionary analysis. But the starkness of these straightened cladograms should not obscure the bushiness of the evolutionary pattern they represent. If for reasons of convenience or incompleteness fossils are excluded, then a cladogram based only on living taxa can be rather barren (figure 1.29a). This does not suggest that modern birds

evolved from crocodiles (or crocodiles from birds), only that among recent taxa birds are more closely related to crocodiles than they are to any other living group. Adding only a few of the fossil taxa (figure 1.29b) should make it clear that the cladogram could be enlarged to better reflect the richness and actual diversity of evolution within these vertebrate groups. Adding fossil groups also helps us understand the transitional steps between living groups. In this phylogeny (figure 1.29a), only representatives of the living groups are represented. If we had only these living groups to reconstruct the steps in early vertebrate evolution, then a great deal of connecting information would be unavailable. However, a rich series of fossil groups provides concrete intermediate steps, giving us greater confidence in our interpretation of this phylogeny.

Formally, the **crown group** is the smallest clade that includes all living members of a group, and any fossils nested within. The **stem group** is the set of extinct taxa that are not in the crown group but are more closely related to the crown group than to any other. Together the crown and stem groups constitute the **total group** (figure 1.30). For example, in figure 1.29, the crown groups are the boxed groups; the stem groups are unboxed.

Students should recognize dendrograms as summaries of information about the course of vertebrate evolution. But you should realize that dendrograms also contain, even if inadvertently, hidden expressions of intellectual preference and personal bias. Dendrograms are practical devices

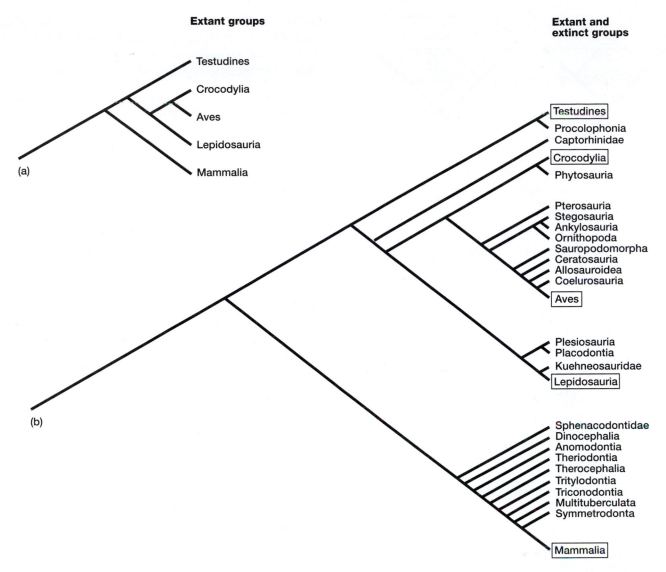

Extant groups

- Testudines
- Crocodylia
- Aves
- Lepidosauria
- Mammalia

(a)

Extant and extinct groups

- Testudines
- Procolophonia
- Captorhinidae
- Crocodylia
- Phytosauria
- Pterosauria
- Stegosauria
- Ankylosauria
- Ornithopoda
- Sauropodomorpha
- Ceratosauria
- Allosauroidea
- Coelurosauria
- Aves
- Plesiosauria
- Placodontia
- Kuehneosauridae
- Lepidosauria
- Sphenacodontidae
- Dinocephalia
- Anomodontia
- Theriodontia
- Therocephalia
- Tritylodontia
- Triconodontia
- Multituberculata
- Symmetrodonta
- Mammalia

(b)

FIGURE 1.29 **Extant and extinct groups.** (a) The cladogram of living groups of amniotes clearly shows the close relationship between birds (Aves) and crocodiles (Crocodylia), but should not suggest that modern groups derive directly from each other. (b) Adding extinct groups illustrates the richness of the historical associations back through which modern groups (boxed) trace their evolution to a common ancestor. Fossils, when added to the analysis, also help to determine the primitive and derived states of characters, and thereby help improve our ability to sort groups on the basis of shared, derived features.

Modified from A. B. Smith.

designed to illustrate a point. Sometimes this requires complex sketches, and other times just a few simple branches on a phylogenetic tree serve our purposes.

Paleontology

The late paleontologist Alfred Romer once poetically referred to the grandeur and sweep of vertebrate evolution as the "vertebrate story." And in a sense it is exactly that, a story with twists and turns that could not have been known beforehand—the debut of new groups, the loss of old ones, the mysteries of sudden disappearances, the evolutionary tales told by the parade of characters. Like a good story, when we finish it we will know the characters better, and because we ourselves are part of this story, we will come to know ourselves a little better as well. The vertebrate story unfolds over a span of half a billion years, a depth of time almost unimaginable (figure 1.31). To help us fathom this vastness of time, we consult paleontology, the discipline devoted to events of the distant past.

The vertebrate story is a narrative spoken partially from the grave, because of all species ever to exist most are now extinct. The evolutionary biologist and paleontologist G. G. Simpson once estimated that of all animal species ever to

FIGURE 1.30 Extant and extinct in phylogenies. The cladogram shows the relationship between extant (black lines) and extinct (white lines) groups. The stem groups include all the intermediate but now extinct fossil groups. Together, crown plus stem groups constitute the total group, the monophyletic clade.

evolve, roughly 99.9% are extinct today. So in this story of life on Earth, most of the cast of characters are dead. What survives are their remnants, the fossils and the sketchy vignettes these fossils tell of the structure and early history of vertebrates.

Fossilization and Fossils

When we think of fossil vertebrates, we probably picture bones and teeth, the hard parts of a body that more readily resist the destructive processes following death and burial. Certainly most fossil vertebrates are known from their skeletons and dentition. In fact, some extinct species of mammals are named on the basis of a few distinct teeth, the only remnants to survive. The calcium phosphate compound composing bones and teeth is a mineral usually preserved indefinitely, with little change in structure or composition. If groundwater seeps through bones lying in soil or rock, over time other minerals such as calcite or silica may soak into the tiny spaces of bone to add further minerals and harden it.

FIGURE 1.31 Geologic time. The gathering of cosmic gases under gravity's pull created Earth some 4.6 billion years ago. Yet life became neither abundant nor complicated until the Cambrian Period, about 542 million years ago, when the first vertebrates appeared.

Source: After U.S. Geological Survey publication, Geologic Time.

Thomas Jefferson, when vice-president of the United States, reported before a scientific society and subsequently published in 1797 a paper on *Megalonyx*, a fossil ground sloth, whose bones had been discovered in Virginia (later named *Megalonyx jeffersonii*). He also knew of large bones of mastodons and other great fossil animals from the eastern United States. When president, he set up the Lewis and Clark expedition to lay a claim to the land, gather scientific information, and find a northwest passage. Part of their goal was to see if mastodons or any other animals discovered as fossils still existed alive in the vastness of the western continent. In 1806, the expedition found a giant leg bone near Billings, Montana (USA) that was certainly a dinosaur. Unfortunately, no living mastodons were found. We now know they disappeared from North America at least 8,000 years earlier.

Prior to the American Revolution, the French naturalist, George Louis LeClerc de Buffon, proposed that compared to his rich European environment, the North American environment was impoverished, unable to support any animals robust in character. His patriotic pride stung, Jefferson countered using the mastodon as an example of such an animal that had thrived in the New World.

Fossils are more than bones and teeth, however. Occasionally products of vertebrates, such as eggs, will fossilize. If tiny young bones are preserved inside, we can identify them and the group to which they belong (figure 1.32). This tells us more than just the structure of this species; it also tells us something about its reproductive biology. The discovery in Montana of fossilized clumps of eggs belonging to duck-billed dinosaurs testified to the reproductive style of this species, but there was accompanying circumstantial evidence to imply even more. The clumps or clutches of eggs were near each other, about two adult body lengths apart, suggesting that the area was a breeding colony. Analysis of the rock sediments in which they were found indicates that the colony was on an island in the middle of a runoff stream from the nearby Rocky Mountains. At the same site, bones from duck-billed dinosaurs of different sizes, and thus different ages, were present. This could happen only if young stayed around the nest until they were fully grown. Perhaps the parents even gathered food and brought it back to nourish the newly hatched young. For this species of duck-billed dinosaurs, the emerging picture is not one of a dispassionate reptile that laid its eggs and departed. Instead, this reptile appears to have had sophisticated parental care and supportive social behavior. Gathering of food, protecting and teaching of young, and bonding of pairs are implied by the fossils.

A marine fossil of an ichthyosaur, a dolphinlike reptile, was recovered from limestone rocks dating to 175 million years ago (figure 1.33). This adult specimen appears to be a female fossilized in the act of giving birth. Several small (young) skeletons remain with her body, one apparently emerging through the birth canal and another already born

FIGURE 1.32 Fossil eggs. Examination of the fetal bones within these eggs reveals that these are of *Protoceratops*, a Cretaceous dinosaur that lived in what is today Mongolia.

FIGURE 1.33 Fossil ichthyosaur. Small skeletons are seen within the adult's body and next to it. This may be a fossilized birth, with one young already born (outside), one in the birth canal, and several more still in the uterus. Such special preservations suggest the reproductive pattern and live birth process in this species.

(a)

(b)

FIGURE 1.34 **Archaeopteryx.** The original feathers have long since disintegrated, but their impressions left in the surrounding rock confirm that the associated bones are those of a bird.

FIGURE 1.35 **Mummification.** (a) Fossil mummified carcass of the duck-billed dinosaur *Anatosaurus*. (b) Detail shows the surface texture of the skin.

lying beside her (figure 1.33). If this represents a "fossilized birth," then unlike most reptiles, ichthyosaurs bore live young who were fully functional, like young dolphins today.

Occasionally, fossils preserve more than just their hard parts. If a full animal skeleton is discovered, microscopic analysis of the region occupied in life by the stomach might reveal the types of foods eaten shortly before its death. Dung is sometimes fossilized. Although we might not know which animal dropped it, we can gain some notion about the types of foods eaten. Soft parts usually decay quickly after death and seldom fossilize. A dramatic exception to this has been the discovery of woolly mammoths, distant relatives to elephants, frozen whole and preserved in the arctic deep freeze of Alaska and Siberia. When thawed, these mammoths yielded hair, muscles, viscera, and digested food, exceptional finds indeed. Rarely are paleontologists so lucky. Occasionally soft parts leave an impression in the terrain in which they are buried. Impressions of feathers in the rock around the skeleton of *Archaeopteryx* demonstrate that this animal was a bird (figure 1.34). Similar impressions of skin tell us about the surface textures of other animals—scaly or smooth, plated or fine beaded (figure 1.35a, b).

The past behavior of now extinct animals is sometimes implied by their fossilized skeletons. Nearly complete skeletons of fossilized snakes have been found in lifelike positions in rocks dating to 32 million years ago. These natural aggregations seem to represent, as in many modern species of temperate snakes, a social event to prepare for hibernation during the cold winter season. Other vertebrate behaviors, or at least their locomotor patterns, are implied in fossilized footprints (figure 1.36). Size and shape of footprints, together with our knowledge of animal assemblages of the time, give us a good idea of who made them. With dinosaur tracks, it has been possible to estimate the velocity of the animal at the time the tracks were made.

Three-and-a-half-million-year-old volcanic ash, now hardened to stone, holds the footprints of ancestral humans. Discovered in present-day Tanzania by Mary Leakey, the sets of footprints are those of a large individual, a smaller individual, and a still smaller individual walking in the steps of the first. These human footprints confirm what had been deciphered from skeletons, namely, that our ancestors of over 3 million years ago walked upright on two hindlegs.

Recovery and Restoration

Paleontologist and artist combine talents to re-create the extinct animal as it might have looked in life. Remnants of long-dead animals provide source material from which basic anatomy is reassembled. After such a length of time in the ground, even mineral-impregnated bone becomes brittle. If the original silty sediments around bone have hardened to stone, they must be chipped or cut away to expose the fossilized bone encased within. Picks and chisels help to partially expose the upper surface and sides of the bone, which are wrapped in protective plaster and allowed to harden (figure 1.37).

(a)   (b)

FIGURE 1.36 **Dinosaur tracks.** (a) Tracks from the Late Jurassic were made in soft sand that later hardened to form rock. Two sets are present: the large tracks of (b) a sauropod and the three-toed tracks of a smaller carnosaur, a bipedal carnivorous dinosaur.

(a)   (b)

FIGURE 1.37 **Fossil dig in Wyoming.** (a) Partially exposed dinosaur bones. The work crew prepares the site and notes the location of each excavated part. (b) This *Triceratops* femur is wrapped in a plaster jacket to prevent disintegration or damage during transport back to the museum.

Photos courtesy of Dr. David Taylor, Executive Director, NW Museum of Natural History, Portland, Oregon.

Following this procedure, the remainder of the bone is exposed and the plaster wrap extended to encase it completely. The brittle bones are shipped to laboratories within their plaster support. Once specimens reach the lab, the plaster, along with any further rock, is removed. Tiny needles were once used to pick away the rock. Today, a stream of fine sand from a pencil-sized nozzle is used to sandblast or carve away rock to free the fossil.

Confidence in a restored version of a fossil rests largely on direct fossil evidence and knowledge of modern, living counterparts, which indirectly supply the likely biology of the fossil (figure 1.38). Size and body proportions are readily determined from the skeleton. Muscle scars on bones help determine how muscles might have run. When added to the skeleton, these give us an idea of body shape. General feeding type—herbivore or carnivore—is implied by the type of teeth; and lifestyle—aquatic, terrestrial, or aerial—is determined by the presence of specialized features such as claws, hooves, wings, or fins. The type of rock from which the fossil was recovered—marine or terrestrial deposits, swamp or dryland—further testifies to its lifestyle. Comparison with related and similarly structured living vertebrates helps fill in locomotor style and environmental requirements (figure 1.38a–c).

The presence or absence of ears, proboscis (trunk), nose, hair, and other soft parts must be guessed at. Living relatives help in this process. For instance, all living rodents have vibrissae, long hairs on the snout, so these might be included in restorations of extinct rodents. Except for some burrowing or armored forms, most mammals have a coat of fur, so it is fair to cover a restored mammal with hair. All living birds have feathers, and reptiles have scales, both of which can logically be added to restored avian or reptilian fossils, although the length or size must be guessed. Surface colors or patterns, such as stripes or spots, are never preserved directly in an extinct vertebrate. In living animals, colored patterns camouflage appearance or emphasize courtship and territorial behaviors. Reasonably, surface patterns had similar functions among extinct animals, but specific colors and patterns chosen for a restoration must usually be produced from the artist's imagination.

However, recovered genetic material can sometimes help. The dark and light hairy coats preserved with frozen woolly mammoths suggested color variations, but it was initially not known if this represented natural color variation or if it was an artifact of preservation. Now genetic material isolated from the leg bone of a 43,000-year-old mammoth includes a gene that in at least two living forms, mice and people, produces shades of light and dark hair. Restoring mammoths as blondes or brunettes is now more reasonable. In a nice piece of forensic paleontology, scientists have also extracted enough DNA from extinct Neandertal skulls to isolate a gene for hair color. In modern humans, this same gene causes red hair. If it acted the same in Neandertals, then at least some were red-heads.

(a)

(b)

(c)

FIGURE 1.38 Reconstruction of an extinct animal. (a) The skeleton of the extinct short-faced bear, *Arctodus simus,* is positioned in its likely posture in life. (b) Scars on the bones from muscular attachments and knowledge of general muscle anatomy from living bears allow paleontologists to restore muscles and hence to create the basic body shape. (c) Hair added to the surface completes the picture and gives us an idea of what this bear might have looked like in its Alaskan habitat 20,000 years ago.

Some dinosaurs and fossil birds retain melanosomes, packets of pigment granules in their feathers. The shapes of these melanosomes imply colors. In modern birds, oblong-shaped melanosomes make black-gray colors; spherical-shaped make brown to red tones. The presences of these shapes in dinosaurs argue that dinosaurs were covered, at least in part, by plumage from blacks, grays, and brown to red.

A dynamic museum mural showing dinosaurs at battle or Neandertals on the hunt may satisfy our curiosity for what they might have looked like in life. However, in any such restorations human interpretation stands between the actual bones and the fully colored reconstruction.

New fossil finds, especially of more complete skeletons, improve the evidence upon which we build a view of extinct vertebrates. Often, however, new insights into old bones arise from an inspired reassessment of the assumptions upon which original restorations were based. Such is largely the case with recent reassessment of dinosaurs. Their structures, size, and success now seem to make them warm-blooded, active vertebrates living a lifestyle less like reptilian lizards and turtles of today and more like mammals or birds. New fossil discoveries got us thinking, but the major change in the way artists and paleontologists restore dinosaurs today reflects new courage in interpreting them as predominantly active land vertebrates of the Mesozoic.

Reconstruction of human fossils has followed fashion as well as new discoveries. When first unearthed in the late nineteenth century, Neandertal bones were thought to be those of a single individual, a Cossack soldier from the Napoleonic wars fought a few decades earlier. In the beginning of the twentieth century, this view gave way to a stoop-shouldered, beetle-browed, and dim-witted image. Neandertals were reassessed to be a breed apart from modern *Homo sapiens,* and restoration reflected this demoted image. Today Neandertals are classified again as a human species, *Homo sapiens neanderthalensis.* Shaven and suited, the claim goes, a Neandertal could walk the streets of New York without drawing a second glance or a raised eyebrow. In New York, perhaps, but this "new" elevation of Neandertal to modern status has been inspired by current artists' restorations that make the species look human.

The point is not to smirk at those who err or follow fashion but to recognize that any restoration of a fossil is several steps of interpretation away from the direct evidence of the bones themselves. Reconstructing the history of life on Earth improves with new fossil discoveries as well as with improved knowledge of basic animal biology. The better we understand the function and physiology of animals, the better our assumptions will be when we restore life to the bones of dead fossils. It is worth the risks and pitfalls to re-create the creatures of the past because in so doing we recover the unfolding story they have to tell us about life on Earth.

From Animal to Fossil

The chance is extremely remote that an animal, upon death, will eventually fossilize. Too many carrion eaters await within the food chain (figure 1.39). Disease or age or hunger may weaken an animal, but a harsh winter or successful predator is often the immediate instrument of death. Its flesh is consumed by carnivores and its bones broken and picked over by marauding scavengers that follow. On a smaller scale, insect larvae and then bacteria feed on what remains. By stages, the

FIGURE 1.39 Almost fossils. Upon death, few animals escape the keen eyes of scavengers looking for a meal. Bacteria and bugs descend upon the flesh that is left. Small animals seeking calcium chew up bones. Little, if anything, is left to fossilize.

deceased animal is broken down to its chemical components, which reenter and recycle through the food chain. In a small forest, hundreds of animals die each year, yet as any hiker or hunter can attest, it is rare to find an animal that has been dead for any length of time. Scavengers and decomposers go quickly to work. Even rodents, whose customary food is seeds or foliage, will gnaw on bones of dead animals to obtain calcium. To escape this onslaught, something unusual must intervene before all trace of the dead animal is literally eaten up.

Animals living in water or near the shore are more likely to be covered by mud or sand when they die (figure 1.40). Upland animals die on ground exposed to scavengers and decay; thus, most fossil-bearing rocks (i.e., sedimentary rock) are formed in water. Even if successfully buried, bones are still in peril. Under pressure and heat, silt turns to rock. Shifting and churning and settling of rock layers can pulverize fossils within. The longer a fossil lies buried, the greater the chance these tectonic events will obliterate it. This is why older rock is less likely to harbor fossils. Finally, the fossil must be discovered. Theoretically, one could begin to dig straight down anywhere through the Earth's crust at any site and eventually hit fossil rocks. Excavations for roads or buildings occasionally unearth fossils in the process. Usually such a freelance approach to fossil discovery is too chancy and expensive. Instead, paleontologists visit natural **exposures** where sheets of crustal rock have fractured and slipped apart or been cut through by rivers, revealing the edges of rock layers perhaps for the first time in millions of years. In these layers, or **strata,** the search begins for surviving fossils.

Taphonomy is the study of how decay and tissue disintegration affect fossilization. Certainly organisms may be lost to such destructive processes, but even if eventually fossilized, the preceding decay may produce a misleading fossil. For example, the earliest chordates are known only from soft-bodied organisms, predating the evolution of hard parts

FIGURE 1.40 Making fossils. Extinct animal remains that persist have escaped the appetites of scavengers, decomposers, and later tectonic shifting of the Earth's crustal plates in which they reside. Usually water covers a dead animal so that it escapes the notice of marauding scavengers. As more and more silt is deposited over time, the fossil becomes even more deeply buried in soil compacted into hardened rock. For the fossil held in the rock to be exposed, the Earth must open either by fracture or by the knifing action of a river.

such as the vertebrate skeleton. Taphonomic studies of modern counterparts discovered that characteristics that diagnose derived organisms decay before primitive characters associated with earlier ancestors. The consequence is to produce a carcass artificially simplified from its natural derived condition before fossilization. At least in chordates, the loss to decay first of synaphomorphies then second to the decay-resistant plesiomorphic characters produces a bias. Depending on the race between decay and fossilization, this simplification can be significant and the resulting interpretation wrongly places the fossil lower within a phylogenetic tree.

Dating Fossils

To discover a fossil is not enough. Its position in time with regard to other species must be determined as well, because this will help place its morphology in an evolutionary sequence. Techniques for dating fossils vary, and preferably several are used to verify age.

Stratigraphy

One such technique is **stratigraphy,** a method of placing fossils in a relative sequence to each other. It occurred to Giovanni Arduino as early as 1760 that rocks could be arranged from oldest (deepest) to youngest (surface). By the time the British geologist Charles Lyell published his great three-volume classic, *Principles of Geology,* during 1830–1833, a system of relative dating of rock layers was well established. The principle is simple. Similar strata, layered one on top of another, are built in chronological order (figure 1.41). As in construction of a tower, the oldest rocks are at the bottom, with later rocks in ascending sequence to the top where the most recent rocks reside. Each layer of rock is called a **time horizon** because it contains the remains of organisms from one slice in time. Any fossils contained within separate layers can be ordered from the oldest to the most recent, bottom to top. Although this gives no absolute age, it does produce a chronological sequence of fossil species relative to

FIGURE 1.41 Stratigraphy. Sediment settling out of water collects at the bottoms of lakes. As more sediment collects, the deeper layers are compacted by the ones above until they harden and become rock. Animal remains become embedded in these various layers. Deeper rock forms first and is older than rock near the surface. Logically, fossils in deeper rock are older than those above, and their position within these rock layers gives them a chronological age relative to older (deeper) or younger (surface) fossils.

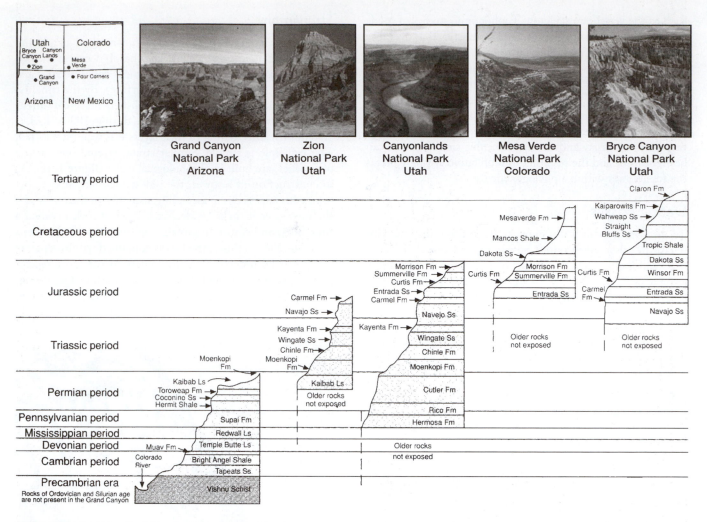

FIGURE 1.42 **Building a chronology of fossils.** Each exposure of rocks can be of a different age from other exposures. To build up an overall sequence of fossils, various exposures can be matched where they share similar sedimentary layers (same ages). From five sites in the southwest United States, overlapping time intervals allow paleontologists to build a chronology of fossils greater than that at any single site.

each other. By placing fossils in their stratigraphic sequence, we can determine which arose first and which later, relative to other fossils in the same overall rock exposure.

Index Fossils

By matching rock strata in one location to comparable rocks in another exposed location, we can build up an overlapping chronological sequence longer than that represented at any single location by itself (figure 1.42). The actual correlation of rock strata between two distantly located sites is done by comparison of mineral content and structure. Index fossils are distinctive markers that can facilitate matching of rock strata. These are species of animals, usually hard-shelled invertebrates, that we know from previous work occur only within one specific time horizon. Thus, the presence of an index fossil confirms that the stratigraphic layer is equivalent in age to a similar layer containing the same fossil species elsewhere (figure 1.43).

Locality 1   Locality 2   Locality 3

FIGURE 1.43 **Index fossils.** After careful study at many well-dated sites, paleontologists can confirm that certain fossils occur only at restricted time horizons (specific rock layers). These distinctive index fossils are diagnostic fossil species used to date rocks in new exposures. In this example, the absence of index fossils confirms that layer B does not exist at the third location. Perhaps rock-forming processes never reached the area during this time period, or the layer was eroded away before layer C formed.

After Longwell and Flint.

Radiometric Dating

Relative stratigraphic position is useful, but to assign an age to a fossil, a different technique is used. This is radiometric dating, a technique that takes advantage of the natural transformation of an unstable elemental isotope to a more stable form over time (figure 1.44a). Such radioactive decay of an element from one isotope state to another occurs at a constant rate, expressed as the characteristic half-life of an isotope. The half-life is the length of time that must pass before half the atoms in the original sample transform into product atoms (figure 1.44b). Common examples include "decay" of uranium-235 to lead-207 (half-life of 713 million years) and potassium-40 to argon-40 (half-life of 1.3 billion years). When rocks form, these radioactive isotopes are often incorporated. If we compare the ratios of product to original and if we know the rate at which this transformation occurs, then the age of the rock and, hence, the age of fossils it holds can be calculated. If, for instance, our sample of rock showed lots of argon relative to potassium, then the rock would be quite old and our estimated age quite high (figure 1.44c). Most of the potassium would have decayed to argon, its product. Conversely, if there were little argon compared with potassium, then only a little time would have passed and our calculated age would be young.

Some natural processes help to purify the sampling. Zircon crystals, a mixture of elements, form in underground chambers as molten rock cools. When the tight molecular structure of these crystals solidifies, zircon incorporates uranium atoms but excludes lead atoms. As the captured uranium decays to lead through time, only the lead derived from the decay accumulates in the crystals. By measuring the ratio of remaining uranium to lead, the age of the zircon crystals can be calculated.

Because of the sometimes capricious uptake of isotopes when rocks form, not all rocks can be dated by radiometric techniques. But when available and cross-checked, radiometric dating yields the absolute ages of rocks and the fossils these rocks contain.

Geological Ages

Geological time is divided and subdivided in turn into eons, eras, periods, and epochs (figure 1.45). The oldest rocks on Earth, with an age of 3.8 billion years, are found in Canada. However, radiometric dates of meteorite fragments fallen to Earth give age estimates of 4.6 billion years. Since astronomers assume that our solar system and everything within it—planets, sun, comets, meteors—formed at about the same time, most geologists take this figure as the

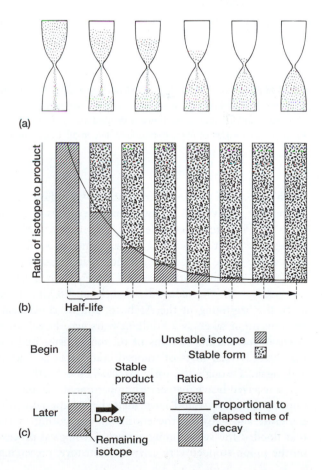

FIGURE 1.44 Radiometric dating. (a) Sand flows regularly from one state (upper portion) to another (lower portion) in an hourglass. The more sand in the bottom, the more time has passed. By comparing the amount of sand in the bottom with that remaining in the top and by knowing the rate of flow, we can calculate the amount of time that has elapsed since the flow in an hourglass was initiated. Similarly, knowing the rate of transformation and the ratios of product to original isotope, we can calculate the time that has passed for the radioactive material in rock to be transformed into its more stable product. (b) Half-life. It is convenient to visualize the rate of radioactive decay in terms of half-life, the amount of time it takes an unstable isotope to lose half its original material. Shown in this graph are successive half-lives. The amount remaining in each interval is half the amount present during the preceding interval. (c) A radioactive material undergoes decay, or loss of mass, at a regular rate that is unaffected by most external influences, such as heat and pressure. When new rock is formed, traces of radioactive materials are captured within the new rock and held along with the product into which it is transformed over the subsequent course of time. By measuring the ratio of product to remaining isotope, paleontologists can date the rock and thus date the fossils they contain.

FIGURE 1.45 Geological time intervals. The Earth's history, from its beginnings 4.6 billion years ago, is divided into major eons, the Precambrian (Cryptozoic) and the Phanerozoic. These eons are divided into eras of unequal length—such as Paleozoic, Mesozoic, and Cenozoic. Each era is divided into periods, and periods into epochs. Only epochs of the Cenozoic are listed in this figure. Arrows mark times of mass extinctions, five in all. In parentheses are the relative magnitudes of these catastrophic drops in diversity.

After Raup and Sepkoski, 1982.

Earth's age. The span of Earth history, 4.6 billion years to the present, is divided into four unequal eons back through time: the Phanerozoic (visible life), Proterozoic (early life), Archean (ancient rocks), and Hadean (molten rocks). The earliest eon is the Hadean, when most water existed in gaseous form and the Earth was still largely molten, leaving no rock record. The oldest dated rocks at 3.8 billion years mark the beginning of the Archean, and its conclusion is by convention taken as 2.5 billion years ago. Fossils of the Archean include impressions of microorganisms and stromatolites, layered mats of trapped cyanobacteria, bacteria, and algae. Through the early Archean, the Earth and its moon received heavy meteorite bombardment. Around each impact, the crust would have melted, perhaps puncturing the crust and allowing the enormous outpouring of lavas that flooded the surrounding surface. Geological processes on the moon stopped very early in its history, preserving a

glimpse of the cratered Archean landscape. The heavy meteorite bombardment reworked much of the early Earth's crust, leaving it cratered as well. But continuing geological processes on Earth, formation of new and remelting of old continental crusts, obliterated much of these early rocks and cratered continents.

From the Archean into the Proterozoic, the fossil record changes little. Stromatolites and microfossils are still present. Microorganisms, termed eukaryotes, with a nucleus and ability for sexual reproduction, rather than just dividing, appear late in the Proterozoic. This was also a time when the world's continents were joined into one or perhaps two large continental blocks. This later part of the Proterozoic experienced a long, severe ice age. The ice cap developed on all continents, extending almost to the equator. Together, the first three eons are sometimes termed the Precambrian.

Understandably, rocks from these early eons are rare, and those surviving rocks contain traces of only microscopic organisms, the first primeval forms to appear as life on Earth gained momentum. At 542 million years ago, or as we now know, slightly earlier, complicated multicellular organisms made a sudden appearance, which is why we start the Phanerozoic eon here.

The Phanerozoic divides into three eras: Paleozoic (old animal life), Mesozoic (middle animal life), and Cenozoic (recent animal life). Invertebrates predominated during the Paleozoic era, as they still do today. But, among the vertebrates, fishes were then most conspicuous and diverse, so that the Paleozoic might be termed the Age of Fishes. The first tetrapods appear in the Paleozoic, and by late in this era an extensive radiation was well underway. But, the extraordinary diversity of reptiles in the Mesozoic took them into nearly every conceivable environment. So extensive was this radiation that the Mesozoic is often termed the Age of Reptiles. The following era, the Cenozoic, is often called the Age of Mammals. Until then, mammals included species small in size and few in numbers. The vast extinctions at the end of the Mesozoic, which were to see the demise of the dinosaurs and many allied groups of reptiles, seem to have opened evolutionary opportunities for mammals, who then enjoyed a period of their own expansive radiation into the ensuing Cenozoic. Yet, this radiation must be kept in perspective. If the Cenozoic were to be named for the vertebrate group with the most species, it would properly be termed the Age of Teleost Fishes, or secondly, the Age of Birds, or thirdly still, the Age of Reptiles. Despite the previous Mesozoic extinctions that depleted their ranks, reptiles today still outnumber mammals in terms of numbers of species. However, in the Cenozoic, mammals displayed for the first time a radiation unequaled in their history, and they occupied dominant positions within most terrestrial ecosystems. Because we, of course, are mammals and it is our taxonomic class that is on the rise, the Cenozoic to most is the Age of Mammals.

Eras divide into periods, whose names originated in Europe. The Cambrian, Ordovician, and Silurian were named by British geologists working in Wales. Respectively, Cambria was the Roman name for Wales, and the Ordovices and Silures were names for Celtic tribes who existed there before the Roman conquest. Devonian was named for rocks near Devonshire, also on British soil. The Carboniferous ("coal-bearing") period similarly celebrates the British coal beds upon which so much of Great Britain's participation in the Industrial Revolution depended. In North America, coal-bearing rocks of this age match with the Lower and Upper Carboniferous; American geologists sometimes refer to these divisions of the Carboniferous as the Mississippian and Pennsylvanian periods, after rocks in the Mississippi Valley and the state of Pennsylvania. The Permian, although named by a Scotsman, is based on rocks in the province of Perm in western Siberia. The Triassic takes its name from rocks in Germany; the Jurassic from the Jura Mountains between France and Switzerland; and the Cretaceous from the Latin word for chalk (*creta*), which refers to the white chalk cliffs along the English Channel.

It was once thought that geologic eras could be divided into four parts—Primary, Secondary, Tertiary, and Quaternary—oldest to youngest, respectively. This proved untenable for the eras, but two names, Tertiary and Quaternary, survive in U.S. usage as the two periods of the Cenozoic. Internationally, however, these terms are replaced with Paleogene and Neogene.

Throughout the geological timescale, periods divide into epochs usually named after a characteristic geographic site of that age. Sometimes boundaries between epochs are marked by changes in characteristic fauna. For example, in North America the late part of the Pliocene epoch is recognized by the presence of particular species of fossil deer, voles, and gophers. The early part of the succeeding Pleistocene is recognized by the appearance of mammoths. The boundary or transitional time between both of these epochs is defined by a fauna that includes extinct species of jackrabbits and muskrats, but not mammoths. Most names of epochs are not in general use and will not be referred to in this book.

The character and pattern of life around us today owe as much to what has become extinct as to what new species have arisen. Had the dinosaurs not become extinct at the end of the Mesozoic, mammals may never have experienced the opportunity to radiate as they did during the Cenozoic. The world would be different. Looking back on it, all species are but passing actors on the stage of life. Some go out with a bang, some with a whimper. Some members of a taxonomic group are carried out in **uniform,** or **background, extinctions** characterized by the gradual loss of species over long time periods. In **catastrophic,** or **mass, extinctions,** the loss of species occurs in many different groups, taking out large numbers of species, and occurring abruptly over a relatively short period of geological time. At least five such episodes of mass extinctions are known from the Phanerozoic (figure 1.45). Note that the Cretaceous extinctions, including the dinosaurs, are dwarfed by the extinctions at the Permo-Triassic boundary, wherein perhaps as many as 96% of marine invertebrates became extinct. Although more extensive, the large, charismatic land vertebrates had not yet evolved in numbers, so that the Permo-Triassic extinctions mostly affected small marine invertebrates.

Tools of the Trade

Analysis of vertebrate design proceeds in three general steps, each enhancing the other.

The Question

A specific question about design is formulated first in any analysis. This is not so trivial or simple as it may sound. A well-formed question focuses thought, suggests the appropriate experiment or line of research to pursue, and promises a productive answer. Physicists of the late nineteenth century believed that space contained a kind of fixed, invisible substance called "ether," which accounted for how light traveled through space. Like sound in air, light in ether was thought to propagate by setting it in motion. As planets circled the sun, they sped through this ether's wind like a person sitting in the open bed of a truck rushing through air. Physicists asked the question, "How might light be affected as it passes with or against the ether wind?" After a series of experiments with light, they found no effect of the ether. For a time, they and other scientists were stumped. As it turned out, they had asked the wrong question. Ether, as an invisible occupant of space, does not exist. No ether, no wind. They should have asked first if ether existed! Our opinion of these physicists should not be harsh, however, because even mistakes inspire better questions and an eventual more sound answer.

In morphology, several practical tools can be used to help define the question. One is dissection, the careful anatomical description of an animal's structural design. Supporting dissection is the newer technique of high-resolution digital tomography, a derivative of the CAT (computer-aided tomography) scan used in human and animal medicine. It is based on X-ray images made sequentially along an animal, then assembled into a 3-D overall image of the subject (figure 1.46a). It can resolve details as small as a few tens of microns in size, even when imaging objects made of high-density materials. Another tool is taxonomy, the proposed relationships of the animal (and its parts) to other species. From these techniques, we gain insight into the morphological design and can place this design in a comparative relationship with other organisms. The specific questions we then formulate about the structure of the organism might be about its function or evolution.

The Function

To determine how a structure performs within an organism, various techniques are used to inspect the functioning organism or its parts directly. Radiography, X-ray analysis,

(a)

(b)

FIGURE 1.46 **Egyptian cobra.** (a) When threatened, this cobra elevates its ribs displaying a "hood," expansion of skin, which may sport an image thought to intimidate along with this defiant, mouth-open defensive upright pose. Modern methods of imaging permit morphologists to examine details of the underlying anatomy, here the elevated ribs, which advertise a threatening retaliatory strike by this highly-venomous cobra. (b) Cobra displaying partial hooding, mouth closed.

(a) From Bruce A. Young (University of Massachusetts) and Kenneth V. Kardong (Washington State University), "Naja haje" (On-line), Digital Morphology, at http://digimorph.org/specimens/Naja_haje. Take a look. (b) Kindly supplied by Bruce A. Young.

allows direct inspection of hard parts or marked parts during performance (figure 1.47). Expense or accessibility, however, often makes radiography of a living organism impractical. High-speed videotape or cinematographic film can sometimes be used instead. The event, feeding or running, for example, is filmed with the camera set at a high rate of speed so that the event unfolds in slow motion when played back at normal projection speed. The tape or film preserves a record of the event, and slow-motion playback permits careful inspection of motions at a speed where sudden displacements are obvious. Natural markers—for instance, bulging muscles or visible hard parts such as teeth or hooves—allow inferences about the functioning of attached or underlying bones and muscles. Inferences can also be made from gentle manipulation of parts in a relaxed or anesthetized animal. Thus, with radiography or high-speed tape/film, displacement of individual points on the animal can be followed, measured, and plotted frame by frame. From this careful record of displacements, velocity and acceleration of parts can be calculated to describe the motion of parts quantitatively. Together with information on simultaneous muscle activity, this produces a description of the part and an explanation of how its bone and muscle components achieve a characteristic level of performance.

Visceral functions can be addressed in other ways. Thin tubes (cannulae) inserted into blood vessels and connected to calibrated and responsive instruments (transducers) allow us to study an animal's circulatory system (figure 1.48). Similar approaches to kidney and gland function have been used. Radiopaque fluids, those visible on radiographs, can be fed to animals, allowing us to follow mechanical events of the digestive tract. Muscles, when active, generate low levels of stray electric charges. Electrodes inserted into muscles can detect these on monitors, allowing the investigator to determine when a particular muscle is active during performance of some event. This activity can be compared with activity of other muscles. This is the technique of electromyography (EMG); the electric record of the muscle is an **electromyogram.**

Figure 1.49 depicts an experimental setup combining several techniques simultaneously to analyze the feeding strike of a venomous snake. With the snake under anesthesia and with proper surgical technique, four pairs of bipolar, insulated, fine-wire electrodes are inserted into four lateral jaw muscles to record electromyograms of each during the strike. A strain gauge is affixed with glue to a suitable location on top of the snake's head where it can detect motion of underlying skull bones. The wires, termed *leads,* from the four bipolar electrodes and strain gauge are sutured to its skin, carefully bound into a cable, and connected to preamplifiers that boost the very low signals from the jaw muscles. Interference with these signals from stray electrical "noise" in the room can be reduced if the snake and apparatus are placed in an electrically shielded cage, a Faraday cage (not shown in figure 1.49). From preamplifiers, each circuit, called a channel, is run next to an amplifier. The strain gauge enters the amplifier at this junction as well (channel 5); special electrical balancing of its signal may be necessary.

The snake is centered on a force platform that records forces produced in the three planes of space (forward/back, up/down, left/right), and the leads enter the last three channels to fill this eight-channel system. A permanent record of the rapid strike is made by a high-speed camera or video

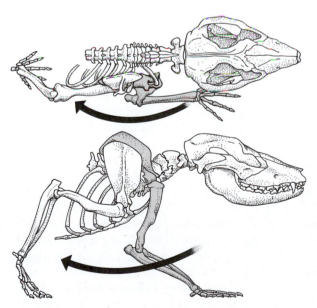

FIGURE 1.47 **Walking stride of a opossum.** The propulsive phase is depicted in these tracings of motion radiographs from overhead and side views. Change in the position of the shoulder blade (scapula) is evident.

Based on the research of F. A. Jenkins and W. A. Weijs.

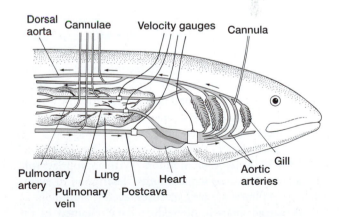

FIGURE 1.48 **Analysis of viscera in a lungfish.** To monitor blood pressure, cannulae (small tubes) are inserted into blood vessels. To monitor rate of blood flow, velocity gauges are placed around selected vessels. From such information, it is possible to determine changes in the rate of blood flow in this lungfish when it breathes in water or gulps air into its lung.

After K. Johansen.

FIGURE 1.49 **Experimental analysis of function.** Careful surgery allows insertion of bipolar electrodes into four selected jaw muscles on the right side of the snake. A strain gauge is fixed over a movable point in the snake's skull. Leads from these electrodes are connected to preamplifiers and then to amplifiers to boost and filter the signals. Channels from the force platform join these four electrodes and carry responses in the three planes of space. The electrical output is displayed on monitors and saved on the computer. The snake strike is filmed by a high-speed camera or video that is pulse-synchronized with the other electrical outputs. Voice comments may be added. Electrical "noise" in the room can be reduced by placing the snake (but not the recording instruments) in a shielded Faraday cage (not shown). Later, slow playback from storage to the monitors permits manual analysis of data, or playback can be directed into a computer for analysis. Comparison of separate events is easier if all are recorded simultaneously, but parts can be done in separate runs and then later matched.

system. The camera produces a pulsed output that is simultaneously combined with the rest of the electrical outputs to permit matching of film events with EMG, strain gauge, and force platform data. A background mirror tilted at 45° allows a carefully placed camera to record dorsal and lateral views of the strike simultaneously. Notes on temperature, time, and other environmental data are recorded by hand.

Outputs are displayed on a monitor for immediate viewing, and they are saved on computer storage as a permanent record. Later, the stored data can be played back slowly, and redisplayed on monitors. With appropriate software, a computer allows quantitative description of events, matching of film/video with electrical events, and so on.

A partial analysis of feeding data obtained in this way is illustrated in figure 1.50a–c. Three instants during the snake's strike are shown—just before, at the beginning of, and during venom injection. Its head positions at these three points are traced from the film record, and below each position are the outputs from the first five channels (electromyograms 1–4, strain gauge 5). The snake's instantaneous movement unfolds at the beginning (left) of each

record and travels across each trace from left to right. From prior dissections, structural components hypothesized to be important in strike performance are set forth in a proposed morphological model, to which these functional data are now added. Before onset of the strike, all muscle channels are silent because no contraction is occurring, and the strain gauge trace indicates that the snake's mouth is closed (figure 1.50a). As the strike begins, the lower jaw starts to open. This is initiated by contraction of muscle 1 and indicated by activity on the electrical trace for the first time (figure 1.50b). The initial rotation of the fang is detected by the strain gauge. At the third point in the strike, the snake closes its jaws firmly on the prey, and all the jaw-closing muscles, including the first, show high levels of activity (figure 1.50c). The strain gauge indicates changes in the jaw positions during this bite, from fully open at first to jaw closure on the prey. Thus, the first muscle opens the lower jaw, but its high electrical activity slightly later during the bite indicates that it continues to play a role. The other three muscles are powerful jaw-closing muscles, adductors, and act primarily during the bite.

Taken literally, a "living fossil" is a contradiction in terms because, of course, fossils are dead. But, occasionally a species survives up to the present having changed little in external appearance since the inception of its lineage. In these living fossils, evolution is arrested. Because they retain in their bodies ancient characteristics, and because they are living, they carry forward the physiology and behavior missing in preserved fossils. All living animals, not just a privileged few, retain at least a smattering of characteristics that are throwbacks to an earlier time in their evolution. The duckbill platypus, a furry mammal of Australia, still lays eggs, a holdover from its reptilian ancestors. Even humans retain ancient features. We have hair, for example, that comes down from the most ancient of mammals. I suppose we could even count our backbone as a retained feature of fishes!

However, what most scientists mean by a living fossil is an unspecialized species, alive today, that is built from the same ancient features that first appeared in the early days of the lineage. In terms of head and body shape, crocodiles have been labeled as living fossils, as have sturgeons and *Amia,* the bowfin. Along the coasts of New Zealand persists a lizardlike reptile, *Sphenodon.* Four-legged and scaled, it looks like a squat but otherwise average lizard. Under the skin, however, the skeletal system, especially the skull, is quite ancient. One of the most surprising living fossils is the surviving sarcopterygian, *Latimeria,* a coelacanth. This fish is a distant relative of the group giving rise to the first tetrapods. And until 1939, *Latimeria* was thought to have been extinct for millions of years.

Latimeria retains many ancient sarcopterygian creations: well-developed notochord, unique snout, fleshy appendages, divided tail. Its discovery excited great interest because the last members of this line had apparently expired 75 million years ago. In 1938, Goosen, a commercial fishing captain working the marine waters off the southern tip of Africa, decided, on an impulse, to fish the waters near the mouth of the Chalumna River. He was about 5 km offshore, over the submarine shelf, when he lowered his trawling nets into 40 fathoms (240 ft, about 73 m) of water. An hour or so later,

the nets were retrieved and opened to spill onto the deck a ton and a half of edible fish, two tons of sharks, and one coelacanth. None of these old salts had ever seen such a fish, and they had little idea about what it was except to recognize its uniqueness. As was the custom, the crew saved the fish for the curator of the tiny museum in East London, Africa, their port city. (Although this was in South Africa, a British heritage inspired local names, hence East London for this museum situated in Africa.)

The curator was Ms. M. Courtenay-Latimer (box figure 1). The museum's budget was thin, to say the least, so to build local enthusiasm and support she had emphasized exhibits representing local sea life. She encouraged crews of fishing trawlers to watch for unusual specimens. If any were caught, they were included in the pile of inedible rubbish fish at the end of the day, and Courtenay-Latimer was called to come pick what specimens she could use. On this particular day while sorting through fish, she spotted the heavy-scaled, blue coelacanth with fins like arms. It was 1.6 m in length and weighed 60 kg. When caught it had snapped at the fishermen, but it was now dead and beginning to decompose in the hot sun. By training, Courtenay-Latimer was not an ichthyologist nor was she blessed with a staff of experts. Besides curator, she was also treasurer and secretary of the museum. Although she did not recognize the coelacanth for exactly what it was, she was keen enough to realize that it was special and convinced a reluctant taxi driver to deliver her, her assistant, and the rather smelly fish back to the museum. Thin budgets again plagued her as there were no freezers or equipment to preserve such a large fish. It was then taken to a taxidermist

who was instructed to save even the parts not needed for the job. But, after three days in the hot weather and no return word from the nearest fish expert whom Courtenay-Latimer contacted, the taxidermist discarded the soft parts. When she told the chairman of the museum's board of trustees what she suspected, he scoffed, suggesting that "all her geese were swans." Apparently he entertained the idea of discarding it but eventually relented and authorized the stuffing and mounting of the fish.

Unfortunately, her letter to the closest fish expert took 11 days to reach him because East London was still a rather remote area of South Africa and it was the holiday season. The expert whom she contacted was J. L. B. Smith, an instructor in chemistry by profession, an ichthyologist by determination. The letter included a description and rough sketch of the fish, which was enough to tell Smith that this could be the scientific find of the decade. As anxious as he was to see and confirm the fish, however, he could not leave to make the 560-km (350-mile) journey to East London. He had examinations to administer and score. Eventually, his excitement and hopes were realized when he finally did visit the museum and peered on the fish for the first time. It was a coelacanth until then known to science only from Mesozoic fossils. In honor of the person (Courtenay-Latimer) and the place (Chalumna River), Smith named it *Latimeria chalumna.*

Since then, other *Latimeria* have been discovered off the coast of eastern Africa and in Indonesia. They seem to be predators living at depths of 40 to 80 fathoms. Thanks largely to a captain, a curator, and a chemist, *Latimeria* is a living fossil again today.

BOX FIGURE 1
M. Courtenay-Latimer, while curator of the East London Museum in South Africa. Her quick sketch and notes of the coelacanth sent to J. L. B. Smith for his opinion are shown next to her.

FIGURE 1.50 Initial analysis of morphological and functional data. Three points in the feeding strike of a venomous snake are illustrated: (a) just before the strike, (b) at the onset of the strike, and (c) during the bite. Electrical traces from the four muscles (channels 1–4) and strain gauge (channel 5) are shown below each. The biomechanical models (right) of the snake's skull during each stage are based on prior anatomical analysis. (a) No myograms are evident prior to the strike, and no bone or fang displacement occurs. (b) The muscle opening the snake's jaw (channel 1) and the strain gauge records (channel 5) are first to show changes on the myograms. The model incorporates these changes by showing the start of fang erection. (c) The snake's jaws close firmly to embed its fully erected fang within its prey. Electromyograms show that all jaw muscles are active, and the strain gauge indicates that the snake's mouth is closing on the prey. These events are incorporated into the model (right), where solid arrows represent the onset and direction of contraction vectors.

This form-function analysis is far from complete. Many more muscles are involved, and events on both sides of the animal need to be followed. Presentations of different sizes of prey might result in modifications of jaw function and so on. Anatomical analysis produces a knowledge of basic structure. From this, a set of testable questions about design can be formulated. Which structural elements are critical to performance? How do they function? Functional data address these questions.

It is best if motion and muscle events are recorded simultaneously, to make comparisons between them easier. Often this is not feasible, however. Equipment may be unavailable or the animal uncooperative. Thus, it is not uncommon and certainly acceptable to perform parts of the functional analysis separately, then later match up bone displacements and muscle activity. It is becoming common now to include analysis of the nervous system along with simultaneous muscle and bone events. This produces a more complete explanation of performance. Not only is the immediate basis of motion described, but the basis for neural control of these displacements and for initiating muscle activity are described as well. Activity of muscles at appropriate moments can be seen also.

The Biological Role

To discover the adaptive role of a part, scientists eventually venture into the field to document how the animal actually deploys the morphological design in the environment. Careful observation of the organism in its environment must be incorporated with techniques of population biology to assess overall ecological performance of a part's form and function. **Ecomorphology** is the term that has been coined to recognize the importance of ecological analysis in the examination of a morphological system.

By this late point in an analysis, one usually has a good idea of how a structure might be used under natural conditions. Occasionally there are surprises. For example, unlike other finches, the "woodpecker" finch of the Galápagos uses its beak to break off a sharp needle or twig, and uses this "tool" as a spear or probe to jab insect grubs hidden under the bark of trees. Deer mice chew rough seeds and grasses but also grab an occasional insect to eat as well. The jaws of deer mice consequently function as more than just a grinding mill of tough seeds. The pronghorn, a deer-like animal of the North American plains, can attain speeds in excess of 96 km/hr, but no natural predator today or in the past existed with a comparable ability. High speed itself, therefore, is not just an adaptation for escape from predators. Instead, pronghorns cruise at 30 to 50 km/hr in order to move between scattered resources. This, not just escape from predators, seems to be the most important aspect of pronghorn speed and design.

Thus, laboratory studies determine the form and function of a design. Field studies assess the biological role of the feature; that is, how the form and function of the feature serve the animal under natural conditions. A feature's biological role, in turn, suggests the kinds of selection pressures brought to bear on the organism and how the feature might be an adaptation that addresses these evolutionary forces. Carrying this a step further, comparison of homologous features from one group to another, or from one class to another, provides insight into how change in animal design might reflect changes in selection pressures.

The story of vertebrate evolution is the story of transition and adaptive change—transition from water to land (from fish to tetrapod), from land to air (from reptiles to birds), and in some cases, the reinvasion of water (dolphins, whales) or return to a terrestrial mode of life (e.g., ostriches). In the study of vertebrate evolution, it is useful to think of how a particular design adapts the organism to the particular demands of its present environment, and how structure itself places limitations on or opens opportunities for the kinds of adaptations that might eventually arise.

Overview

Anatomy and its significance are the province of comparative morphology. Our task is to understand how organisms work and how they evolved. Although today form, function, and evolution together provide this understanding, reaching this harmonious union coursed a difficult and contentious history. To morphology, Darwin added and united issues of biological design into a common context: descent with modification. Morphology enjoyed its own independent intellectual history, recognizing the tight coupling of form and function, along with the basic underlying anatomical patterns upon which organisms were built. From this came the recognition of the separate influences of history (homology), function (analogy), and simple similarity (homoplasy) upon vertebrate design. Comparison is one of our techniques, as is the experimental evaluation of functions, and the representation of evolutionary events in dendrograms. Dendrograms both summarize phylogenetic patterns and suggest the process producing change through time. The major steps of evolution can be summarized simply (figure 1.23), but this can underestimate their complexity. The complexity can be summarized (figure 1.24), but this may produce a bewildering dendrogram without indication of abundance. Abundance can be summarized (figure 1.25), but this loses some of the detailed genealogy. The genealogy is summarized in cladograms (figure 1.29), but this gives primacy to lineage alone and oversimplifies evolutionary events, especially if fossils are not included.

Most species ever to live are today extinct. Consequently, we turn to the fossil record, where we recover the larger cast of characters in the vertebrate story. Hard bones and teeth most likely survive the rough and violent process of fossilization. Occasionally, footprints, impressions, and soft parts survive to disclose further insights into the life of organisms of the past. Reconstructions from the fossil materials bring animals of the past back to life. Reconstructions are hypotheses, susceptible to fashion, but also improved by new facts, sounder phylogenies, and a better biology. In morphological studies, a better biology emerges through new techniques of functional analysis—high-speed analysis of motion and careful monitoring of physiological processes. As we bring our understanding of vertebrate form and function into the environment where the animal lives, we bring comparative morphology to bear upon the adaptive role of an organism's particular features. The adaptive basis of an organism's survival cannot be reduced to its genome. It is the whole organism, integrated and dynamic, not its genes, that directly meets the environment. Survival depends upon form and function matched adaptively to the selection forces met in the environment where the feature serves. We embark then upon a discovery of this remarkable vertebrate story, seeking to explain how vertebrate design works and how it has evolved.

WEB LINK

Visit the text website at *www.mhhe.com/kardong7e* for additional study aids including selected references, functional laboratories, and web links to chapter topics.

CHAPTER 2

Origin of Chordates

CHORDATE PHYLOGENY

CHORDATE CHARACTERISTICS
  Notochord
  Pharyngeal Slits
  Endostyle or Thyroid Gland
  Dorsal and Tubular Nerve Cord
  Postanal Tail
  Chordate Body Plan

PROTOCHORDATES
  General Protochordate Features
  Hemichordata
    Enteropneusta—"Acorn Worms"
    Pterobranchia
    Hemichordate Phylogenetic Affinities to Chordates

Hemichordate Phylogenetic Affinities to Echinoderms
Cephalochordata
Urochordata
    Ascidiacea—"Sea Squirts"
    Larvacea (Appendicularia)
    Thaliacea
Overview of Protochordates

CHORDATE ORIGINS
  Chordates from Annelids and Arthropods
  Chordates from Echinoderms
    Auricularian Hypothesis
    Larval Echinoderm to Chordate Tadpole
  Chordate Origins and Phylogeny

OVERVIEW

Chordates are neither the most diverse nor the largest of the animal phyla, although in terms of the number of species, they come in a respectable fourth behind arthropods, nematodes, and molluscs (figure 2.1). Living chordates consist of three groups of unequal size: cephalochordates (amphioxi or lancets), urochordates (tunicates or "sea squirts"), and the largest group, the vertebrates (fishes, amphibians, reptiles, and mammals). Tucked away within this phylum is a small family, the hominids, that includes humans. In part, our interest in chordates derives from the fact that humans belong to this phylum, so studying chordates brings topics concerning us close to home. But we have more than just a vested interest in chordates. Many chordates are constructed of hard parts that survive to yield a respectable history in the fossil record, which has made them especially useful in defining ideas about evolutionary processes. Advanced chordates are also some of the most intricate animals ever to appear. They therefore introduce us to questions about the complexity of biological organization and about the special mechanisms important in evolution.

Chordate Phylogeny

Chordates have a fluid-filled internal body cavity termed a **coelom.** They are part of a major radiation within the Bilateria, animals built upon a bilateral, symmetrical body plan. Within the Bilateria, two apparently distinct and independent evolutionary lines are present. One line is the **protostomes,** which includes molluscs, annelids, arthropods, and many smaller groups. The protostome lineage itself divides into Lophotrochozoa and Ecdysozoa (figure 2.2). The other bilaterian line is the **deuterostomes,** which includes ambulacraria (echinoderms, hemichordates) and chordates (figure 2.2). The distinction between protostomes and deuterostomes was originally recognized on the basis of certain embryological characteristics (table 2.1). Recently, molecular studies have confirmed and clarified these two lines of evolution within the bilaterians. More will be said later about embryonic development, but here some general introductory features can help clarify the differences between protostomes and deuterostomes.

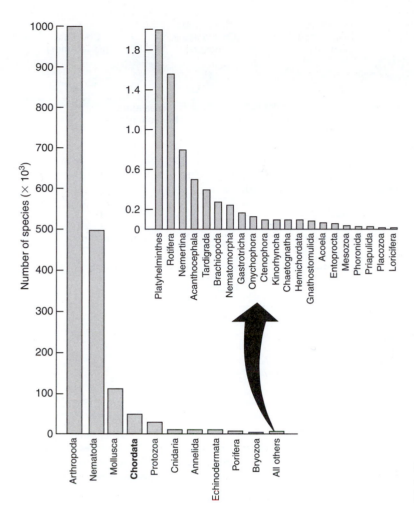

FIGURE 2.1 **Relative abundance of species within the animal phyla.** When finally counted, Nematoda may outnumber Arthropoda.

Embryonic development; details of early cleavage (p. 164)

In both bilaterian groups, the egg begins to divide repeatedly after fertilization, a process termed **cleavage,** until the very young embryo is made up of many cells formed from the original single-celled egg (figure 2.3). In some animals, dividing cells of the embryo are offset from each other, a pattern known as **spiral cleavage.** In others, the dividing cells are aligned, a pattern termed **radial cleavage.** At this point, the embryo is little more than a clump of dividing cells that soon become arranged into a round, hollow ball, with cells forming the outer wall around a fluid-filled cavity within. One wall of this ball of cells begins to indent and grow inward, a process called **gastrulation.** The opening into this indentation is the **blastopore,** and the indented cells themselves are destined to become the gut of the adult. Indentation continues until cells reach the opposite wall, where they usually break through, forming a second opening into the primitive gut (the blastopore being the first).

The now multicellular embryo is composed of three basic tissue layers: an outer **ectoderm,** an inner **endoderm** that forms the lining of the gut, and a **mesoderm** that forms the layer between the two. If a solid mass of mesodermal cells splits to form the body cavity within them, the result is a **schizocoelom** (figure 2.3a). If, instead, the mesoderm arises as outpocketings of the gut that pinch off to form the body cavity, the result is an **enterocoelom** (figure 2.3b).

Protostomes, literally meaning "first mouth," are animals in which the mouth arises from or near the blastopore. Additionally, they tend to have spiral cleavage, a schizocoelom, and a skeleton derived from the surface layer of cells (figure 2.3a). Deuterostomes, literally meaning "second mouth," are animals in which the mouth arises not from the blastopore but secondarily at the opposite end of the gut as the blastopore itself becomes the anus (figure 2.3b). Additionally, embryonic development of deuterostomes includes radial cleavage, an enterocoelom, and a calcified skeleton, when present, derived generally from mesodermal tissues. These embryological characteristics shared by

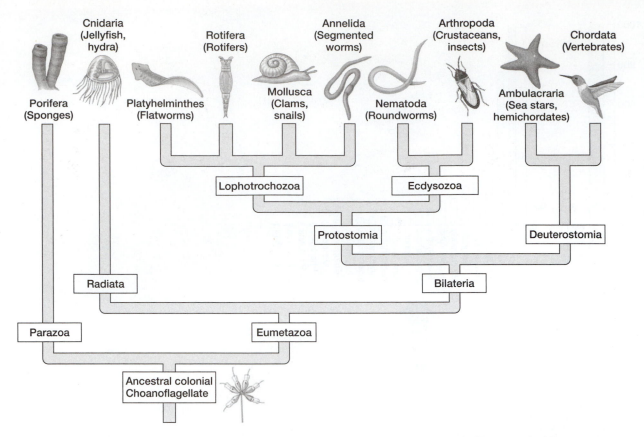

FIGURE 2.2 **Animal evolution.** After sponges (Parazoa) diverge, leaving all other animals (Eumetazoa), differences in symmetry testify to two groups (Radiata, Bilateria). Within Bilateria embryonic differences diagnose the Protostomia and Deuterostomia. Note that chordates are deuterostomes along with the Ambulacraria (echinoderms and hemichordates). Common members of each taxon in parentheses.

deuterostomes testify that they are more closely related to each other in an evolutionary sense than to any of the protostomes. Embryological characteristics, modern molecular phylogenies, and the fossil record all imply that there was an ancient and fundamental divergence between the protostomes and deuterostomes.

Chordates evolved within the deuterostomes. Their mouth forms opposite to the blastopore, their cleavage generally is radial, their coelom is an enterocoelom, and their skeleton arises from mesodermal tissues of the embryo. But we should be clear about the character of the chordates themselves. It is easy to forget that two of the three chordate

taxa are technically invertebrates—the Cephalochordata and the Urochordata. Strictly speaking, the invertebrates include all animals except members of the vertebrates.

The earliest chordate fossils appear in the Cambrian period, about 530 million years ago. Although later chordates evolved hard bones and well-preserved teeth that left a substantial fossil testimony to their existence, ancestors to the first chordates likely had soft bodies and left cues of the evolutionary pathway taken from prechordate to chordate. Thus, to decipher chordate origins, we derive evidence from anatomical and molecular (codes of gene sequences) clues carried in the bodies of living chordates. In order to evaluate the success of our attempts at tracing chordate origins, we first need to decide what defines a chordate. We will then attempt to discover the animal groups that are the most likely evolutionary precursors of the chordates.

Chordate Characteristics

At first glance, the differences among the three chordate taxa are more apparent than the similarities that unite them. Most vertebrates have an endoskeleton, a system of rigid internal elements of bone or cartilage beneath the skin.

TABLE 2.1 Fundamental Patterns in Bilateria Development	
Protostomes	**Deuterostomes**
Blastopore (mouth)	Blastopore (anus)
Spiral cleavage	Radial cleavage
Schizocoelic coelom	Enterocoelic coelom
Ectodermal skeleton	Mesodermal skeleton

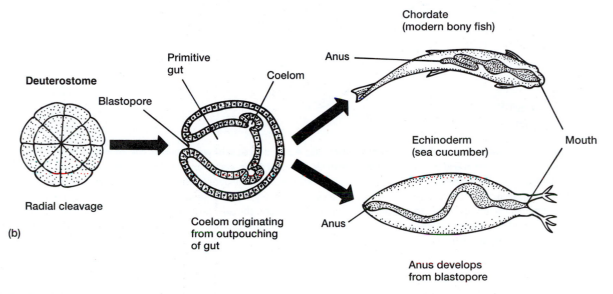

FIGURE 2.3 **Protostomes and deuterostomes.** Bilateria are divided into two major groups on the basis of embryonic characteristics. (a) Protostomes usually show spiral cleavage, coelom formation by splitting of the mesoderm, and derivation of the mouth from the blastopore. (b) Deuterostomes often exhibit radial cleavage, coelom formation by outpocketing of the gut, and derivation of the anus from or in the vicinity of the blastopore.

The endoskeleton participates in locomotion, support, and protection of delicate organs. Some vertebrates are terrestrial, and most use jaws to feed on big food particles. But cephalochordates and urochordates are all marine animals, none are terrestrial, and all lack a bony or cartilaginous skeleton. However, their support system may involve rods of collagenous material. Cephalochordates and urochordates are suspension feeders, having a sticky sheet of mucus that strains small food particles from streams of water passing over a filtering apparatus.

All three taxa, despite these superficial differences, share a common body design similar in at least five fundamental features: **notochord, pharyngeal slits, endostyle** or **thyroid gland, dorsal hollow nerve cord** forming the simple central nervous system, and **postanal tail** (figure 2.4a–c). These five features diagnose the chordates, and taken together, distinguish them from all other taxa. We look next at each characteristic separately.

Notochord

The notochord is a slender rod that develops from the mesoderm in all chordates. It lies dorsal to the coelom but beneath and parallel to the central nervous system (dorsal nerve cord). The phylum takes the name Chordata from this structure. Typically, the notochord is composed of a core of cells and fluid encased in a tough sheath of fibrous tissue (figure 2.5a). Sometimes the fluid is held within swollen cells called vacuolated cells; other times it resides between core cells of the notochord. The notochord has the mechanical properties of an elastic rod, so it can be flexed laterally from side to side (figure 2.5c), but cannot be collapsed along its length like a telescope (figure 2.5b). This mechanical property results from the cooperative action of the outer fibrous sheath and the fluid core it encloses. If the fluid were drained, like letting air from a balloon, the outer sheath would collapse and form no useful mechanical

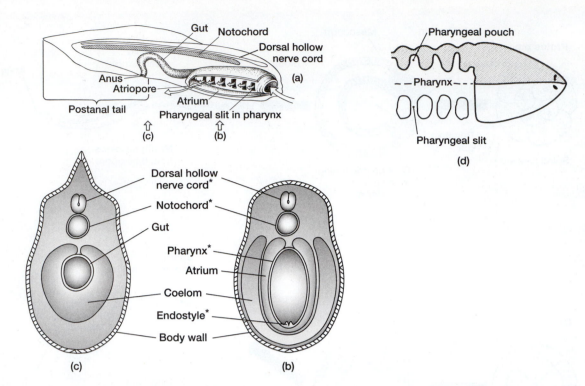

FIGURE 2.4 **Generalized chordate characteristics.** (a) A single stream of water enters the chordate mouth, flows into the pharynx, and then exits through several pharyngeal slits. In many lower chordates, water exiting through the slits enters the atrium, a common enclosing chamber, before returning to the environment via the single atriopore. The endostyle is a glandular-groove that runs along the floor of the pharynx. (b) Cross section through the pharynx showing the tube (pharynx) within a tube (body wall) organization. (c) Cross section through region posterior to the pharynx. (d) Frontal section through pharynx of generalized chordate embryo, showing (top) early formation of pharyngeal pouches which later (bottom) break through the walls to delineate the pharyngeal slits. Asterisks indicate chordate synapomorphic characters.

device. The fluid that normally fills the notochord remains static and does not flow. Such mechanical structures, in which the outer wall encloses a fluid core, are called **hydro-static organs.** The notochord is a hydrostatic organ with elastic properties that resist axial compression. It lies along the body axis to allow lateral flexion but prevents collapse of the body during locomotion (figure 2.5d).

To understand the notochord's mechanics, imagine what would occur if one block of muscle contracted on one side of an animal without a notochord. As the muscle shortens, it shortens the body wall of which it is part and telescopes the body. In a body with a notochord, the longitudinally incompressible cord resists the tendency of a contracting muscle to shorten the body. Instead of shortening the body, the contraction of the muscle sweeps the tail to the side. Thus, upon contraction, the body's segmentally arranged musculature acts upon the notochord to initiate swimming motions that produce lateral pressure against the surrounding water medium. Upon muscle relaxation, the springy notochord straightens the body. Thus, the notochord prevents the collapse or telescoping of the body and acts as the muscle's antagonist in order to straighten the body. As a result, alternating side-to-side muscle contractions in partnership with the notochord generate lateral

waves of body undulation. This form of locomotion may have been the initial condition that first favored the evolution of the notochord.

The notochord continues to be an important functional member throughout most groups of chordates. Only in later forms, such as in bony fishes and terrestrial vertebrates, is it largely replaced by an alternative functional member, the vertebral column. Even when replaced by the vertebral column, the notochord still appears as an embryonic structure, inducing the neural tube to develop above it into the brain and spinal cord and serving as a scaffold for the growing embryonic body. In adult mammals with a full vertebral column, the notochord is reduced to a remnant, the **nucleus pulposus.** This is a small core of gel-like material within each intervertebral disk that forms a spherical pad lying between successive vertebrae.

Structure and embryonic development
of the notochord (pp. 164, 284)

Pharyngeal Slits

Although debuting before the chordates in the hemichordates, **pharyngeal slits** were passed to and incorporated into the chordate body plan (figure 2.4). The **pharynx** is a part

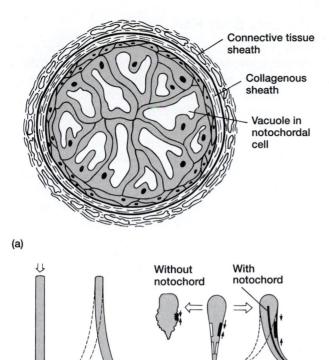

Connective tissue sheath

Collagenous sheath

Vacuole in notochordal cell

(a)

Without notochord   With notochord

(b)   (c)   (d)

FIGURE 2.5   Notochord. (a) Cross section of the notochord of a frog tadpole. (b) The notochord lies above the body cavity and is axially incompressible; that is, it resists shortening in length. (c) The notochord is flexible laterally, however. (d) As seen from above, the consequences of muscle contraction in a body with and without a notochord. Without a notochord, lateral muscle contraction telescopes the body uselessly. A notochord prevents collapse of the body, and muscle contractions on alternating sides efficiently flex the body in swimming strokes.

of the digestive tract located immediately posterior to the mouth. During some point in the lifetime of all chordates, the walls of the embryonic pharynx push out into a series of bays, the pharyngeal pouches (figure 2.4d). These may later nearly pierce, or in aquatic chordates actually pierce, the walls to form a longitudinal series of openings, the pharyngeal slits (also called pharyngotremy, literally meaning "pharyngeal holes"). The term *gill slits* is often used in place of pharyngeal slits for each of these openings, but a "gill" proper is a specialized derived structure of fish and larval amphibians composed of tiny plates or folds that harbor capillary beds for respiration in water. In such vertebrates, gills form adjacent to these pharyngeal slits. The slits are openings only, often with no significant role in respiration. In many primitive chordates, these openings serve primarily in feeding, but in embryos they play no respiratory role; therefore *gill slits* is a misleading term.

When pharyngeal slits first evolved, they likely aided in feeding. As openings in the pharynx, they allowed the one-way flow of a water current—in at the mouth and out through the pharyngeal slits (figure 2.4). Secondarily, when

the walls defining the slits became lined with gills, the passing stream of water also participated in respiratory exchange with the blood circulating through the capillary beds of these gills. Water entering the mouth could bring suspended food and oxygen to the animal. As it passed across the vascularized gills and then exited through the slits, carbon dioxide was given up to the departing water and carried away. Therefore, the current of water passing through pharyngeal slits can simultaneously support feeding and respiratory activities.

In gill-less primitive chordates, the pharynx itself is often expanded into a **pharyngeal** or **branchial basket,** and the slits on its walls are multiplied in number, increasing the surface area exposed to the passing current of water. Sticky mucus lining the pharynx snatches food particles from suspension. Sets of cilia, also lining the pharynx, produce the water current. Other cilia gather the food-laden mucus and pass it into the esophagus. This mucus and cilia system is especially efficient in small, **suspension-feeding** organisms, those that extract food floating in water. Such a feeding system is prevalent in primitive chordates and in groups that preceded them.

In the earliest vertebrates that depended upon gill respiration to support an active lifestyle, mucus and cilia served less well. Cilia are weak pumps, ineffective against gill resistance. In such vertebrates, a pharyngeal pump worked by muscles takes the place of cilia to now move the water that ventilates the gills. The muscular pump, in place of mucus and cilia, also becomes the basis for procurement and processing of large food items. Slits still serve as convenient exit portals for excess or spent water, while adjacent gill structures function in respiration. In fishes and aquatic amphibians, the pharyngeal slits that appear during embryonic development usually persist into the adult and form the exit channel through which water associated with feeding and respiration flows. Except for parts of the ear, for vertebrates that reside on land, the embryonic pharyngeal pouches slits normally never open and thus do not give rise directly to any adult derivative.

Why cilia are replaced by muscles as body size increases (p. 131)

Endostyle or Thyroid Gland

The endostyle is a glandular groove in the floor of the pharynx. It is involved in filter feeding. The thyroid gland is an endocrine gland that produces two major hormones. The thyroid gland, like the endostyle, arises embryologically from the floor of the pharynx. And the thyroid gland, like the endostyle, is involved in iodine metabolism, further suggesting a homology between the two, with the endostyle, being the phylogenetic predecessor of the thyroid. Supporting this, the jawless fish called lampreys have a true endostyle when they are young larvae, that becomes a true thyroid when they become adults. Thus, all chordates have endostyles (urochordates, cephalochordates, larval lamprey) or thyroids (adult lamprey, all other vertebrates).

Thyroid gland (p. 592)

Dorsal and Tubular Nerve Cord

A third chordate characteristic is a dorsal hollow nerve cord derived from ectoderm (figure 2.6b). The central nervous system of all animals is ectodermal in embryonic origin, but only in chordates does the nerve tube typically form by a distinctive embryonic process, namely, by **invagination.** Initially, the surface ectoderm of the back thickens into a plate. This **neural plate** of cells folds or rolls up and sinks inward from the surface (invaginates) as a tube to take up residence dorsally within the embryo, just above the notochord. In most nonchordate embryos, by contrast, the ectodermal cells destined to form the central nervous system do not amass as thickened surface plates (placodes); instead, surface cells individually move inward to assemble into the basic nervous system. More importantly, the major nerve cord in most nonchordates is ventral in position, below the gut, and solid. In chordates, however, the nerve cord lies above the gut and is hollow along its entire length; or more accurately, it surrounds the **neurocoel,** a fluid-filled central canal (figure 2.6b). The advantage, if any, of a tubular rather than a solid nerve cord is poorly understood, but this distinctive feature is found only among chordates.

Nerve tube formation (p. 167)

Postanal Tail

Fourth, chordates possess a postanal tail that represents a posterior elongation of the body extending beyond the anus. The chordate tail is primarily an extension of the chordate locomotor apparatus, the segmental musculature and notochord. In contrast, the anus in nonchordates is terminal, at the extreme posterior end of the body. More will be said later about the role of this postanal tail in swimming.

Swimming in fishes (pp. 306–308)

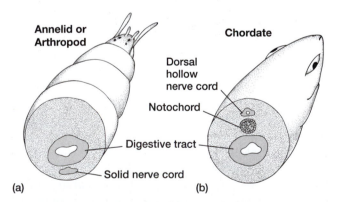

FIGURE 2.6 Dorsal hollow nerve cord. (a) Basic body plan of an annelid or arthropod. In such animals, a definitive nerve cord, when present, is ventral in position, solid, and lies below the digestive tract. (b) Basic chordate body plan. The nerve cord of chordates lies in a dorsal position above the digestive tract and notochord. Its core is hollow, or more correctly, it has a fluid-filled central canal, the neurocoel, indicated as the white spot in the dorsal hollow nerve cord.

Chordate Body Plan

What is common to all chordates are these five primary features: notochord, pharyngeal slits, endostyle or thyroid, dorsal hollow nerve cord, and postanal tail. These characteristics may be present only briefly during embryonic development, or they may persist into the adult stage, but all chordates exhibit them at some point during their lifetimes. Taken together, they are a suite of characters found only among chordates. Chordates also show segmentation. Blocks of muscle, or **myomeres,** are arranged sequentially along the adult body and tail as part of the outer body wall (see, for example, figure 2.16). The myomeres are straight (tetrapods); Σ-shaped (fishes); or $>$-shaped (cephalochordates).

Now that we have an idea about the basic and secondary characteristics of chordates, let us turn our attention to the evolutionary origin of this group. Biologists interested in such questions often consult an assortment of primitive chordates and their immediate ancestors whose structure and design inform us about how and why the early chordate body plan arose. These animals are the protochordates.

Protochordates

The protochordates are an informal assemblage of animals including a prechordate (hemichordates) and two primitive chordates (cephalochordates, urochordates) (figure 2.7). The member taxa include some of the earliest or "first"; hence, "proto-," chordates. The protochordates are not a proper taxonomic group, but a collection of convenience where members share some or all five features of the fundamental chordate body plan. Because the fossil record reveals little about chordate ancestors, living protochordates have been scrutinized for clues to chordate origins. Living protochordates are themselves, of course, products of a long evolutionary history independent of other taxa. Their anatomy is simple, and their phylogenetic position ancient.

Our reasons for giving our attention specifically to them is because their morphologies and lifestyles provide tantalizing clues to the first appearance and advantages of the various characters that comprise the chordate body plan. Molecular data, used to decipher phylogenetic relationships, have both confirmed and surprised our previous understanding of evolutionary events based on morphology, especially on larval morphology. For many years, scientists thought the first chordates resembled either baglike urochordates or wormlike enteropneusts, which then gave rise to streamlined, fish-shaped cephalochordates and from there, to true fishes (vertebrates). However, it was long suspected that echinoderms and hemichordates were more closely related to each other than to other deuterostomes. That is corroborated by molecular data and these two are now placed in the Ambulacraria (figure 2.7). Then more recent molecular and anatomical evidence prompted a more radical change, wherein cephalochordates are now seen to be basal chordates and

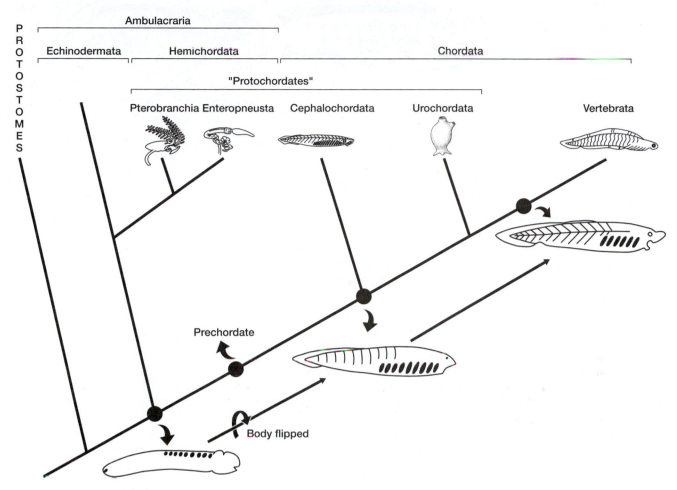

FIGURE 2.7 **Phylogenetic relationships within the "protochordates."** Protochordates are compared to echinoderms and more distantly, to protostomes. A few living protochordates are shown along the top; three hypothetical ancestors are shown below.

urochordates occupy a more derived position close to vertebrates (figure 2.7). This implies that cephalochordates may be a good model for the first chordates and, in fact, resemble the ancestors to the chordates. But, there is even more.

Specific sets of major genes working through the signal proteins they manufacture act to determine which part of the embryo becomes dorsal (back) and which ventral (belly). Specifying general regions of an embryo is termed **patterning,** and this particular type determining the body axis is dorsoventral patterning. The trade-off between the gene set for dorsal and the opposing gene set for ventral eventually establishes the dorsoventral axis. Molecular investigations have discovered that in chordates, the actions of these gene sets is the reverse of all other animals including hemichordates. Ventral gene action in non-chordates is dorsal in chordates. This means that between hemichordates and chordates the body plan became flipped over or inverted (figure 2.7)!

More on body inversion later (p. 79)

The hope is that within living members of the protochordates we will not only discover the steps from prechordate to early chordate, but also come to understand why and how features of the chordate body plan evolved in the first place and the surprises along the way. Before embarking on a quest to understand this challenging, complex, and astonishing history of chordate origins, let's first meet the participants.

General Protochordate Features

All protochordates are marine animals that feed by means of cilia and mucus. But they often live quite different lives as young larvae than they do as adults. As larvae, they may be **pelagic,** residing in open water between the surface and the bottom. Although unattached, most free-floating larvae have limited locomotor capability and are therefore **planktonic,** riding from place to place primarily in currents and tides rather than by their own efforts of long-distance swimming. As adults, they are usually **benthic,** living on or within a bottom marine substrate. Some **burrow** into the

FIGURE 2.8 **Hemichordate, generalized acorn worm.** Proboscis, collar, and trunk regions are shown in partial cutaway view, revealing the coelom in each region and the associated internal anatomy of the worm. Within the proboscis is the stomochord, an extension of the digestive tract. The food-laden cord of mucus (spiral arrow at right) enters the mouth together with water. Food is directed through the pharynx into the gut. Excess water exits via the pharyngeal slits. Several slits open into each branchial pouch, a common compartment with a branchial pore that opens to the outside environment.

Modified from Gutmann.

substrate or are **sessile** and attached to it. Some adults are **solitary,** living alone; others are **colonial** and live together in associated groups. Some are **dioecious** (literally, two houses), with male and female gonads in separate individuals; others are **monoecious** (one house), with both male and female gonads in one individual.

This informal category of convenience, the protochordates, usually includes three groups: hemichordates, cephalochordates, and urochordates. We look next at each.

Hemichordata

Members of the hemichordates are marine "worms" with apparent links to chordates on the one hand and to echinoderms on the other. They share with chordates unmistakable pharyngeal slits (figure 2.8). Most of their nervous system is a network in the skin epidermis. But in the collar region, the epidermis and dorsal nerve cord are invaginated into a deeper **collar cord** (figure 2.10). This method of formation, its dorsal position, and the fact that it may be hollow in parts resembles the chordate dorsal, hollow nerve tube, suggesting homology between them. However, if the chordate body is inverted, then this collar cord is in the wrong position, suggesting instead that it is a unique feature of hemichordates alone, and that hemichordates lack a dorsal, hollow nerve cord, even in part. Some hemichordates have a postanal appendage, a larval structure or, as adults, a device helping to hold them in a burrow or tunnel. But, this appendage, when present, is not a derivative of the locomotor system, and hence, hemichordates lack a true postanal tail. They also lack a notochord. Although in possession of pharyngeal slits, overall hemichordates lack other homologous equivalents of other major chordate features; hence, the name *hemi-* or *half-*chordates.

As larvae, some of these worms pass through a small planktonic stage called the **tornaria larva** (figure 2.9). This planktonic larva is equipped with ciliated bands on its surface and a simple gut. In its ciliated structure, simple digestive system, and planktonic lifestyle, the tornaria larva resembles the **auricularia larva** of echinoderms. Such morphological similarities testify to a close phylogenetic link between hemichordates (tornaria larva) and echinoderms (auricularia

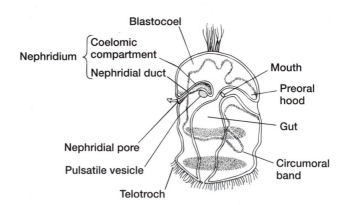

FIGURE 2.9 **Hemichordate, generalized tornaria larva.** The simple gut begins at the mouth under a preoral hood and passes through the body of the larva. On the surface, a meandering circumoral band of cilia runs along each side of the larva. A tuft of cilia projects from the anterior end, and the telotroch, an apron of cilia, runs along the posterior end. The excretory organ is a nephridium consisting of a coelomic compartment lined by podocytes that extends toward the exterior via a ciliated nephridial duct and opens through a nephridial pore.

Based on Ruppert and Balser.

larva). This close relationship is confirmed by recent phylogenetic analyses based on molecular (gene expression) studies, which unite them in the taxon ambulacraria (figure 2.7).

Hemichordates, like both echinoderms and chordates, are deuterostomes. Their mouth forms opposite to the embryonic blastopore, and they exhibit the characteristic deuterostome patterns of embryonic cleavage and coelom formation. The similarities of hemichordates as larvae to echinoderms, on the one hand, and as adults to chordates, on the other, are tantalizing. Perhaps they stand close to the evolutionary route taken by both pre-chordates and pre-echinoderms, and still hold clues to the origin of the chordate body plan. But remember that living hemichordates are themselves millions of years departed from the actual ancestors they might share with early prechordates. Their own evolution has dealt them specialized structures serving their sedentary habits. Within the hemichordates are two taxonomic groups, the **enteropneusts,** burrowing forms, and the **pterobranchs,** usually sessile forms.

Enteropneusta—"Acorn Worms"

The enteropneusts, or acorn worms, are marine animals of both deep and shallow waters. Some species reach over a meter in length, but most are shorter than this. Most live in mucus-lined burrows and have a body with three regions—**proboscis, collar, trunk**—each with its own coelom (figures 2.8 and 2.10a–c). The proboscis, used in both locomotion and feeding, includes a muscular outer wall that encloses a fluid-filled coelomic space. Muscular control over the shape of the proboscis gives the animal a useful probe to shape a tunnel or inflate itself against the walls of the burrow

FIGURE 2.10 **Hemichordata, Enteropneusta.** The hemichordates depicted in this figure are enteropneusts, known informally as acorn worms. (a) External features and body regions of an adult worm. (b) Acorn worm *Balanoglossus* in burrow. (c) Nervous system of the acorn worm *Saccoglossus.* The nervous system is organized into dorsal and ventral nerve cords on the body surface from which nerve nets spread to all parts of the body.

(a, b) After Stiasny; (c) after Knight-Jones.

to anchor its body in place (figure 2.10b). Tucked away in their burrows, many species ingest loosened sediment, extract the organic material it contains, pass the spent sediment through their simple gut, and deposit a casting (fecal waste) on the surface of the substrate where changing tides flush it away. Some wide-bodied, deep-sea enteropneusts crawl and glide along the abyssal ocean bottom.

Other species are suspension feeders, extracting tiny bits of organic material and plankton directly from the water. In these forms, the synchronous beating of cilia on the outer surface of the proboscis sets up water currents that flow across the animal's mucous surface (figure 2.11). Suspended materials adhering to the mucus on the proboscis are swept along ciliary tracks to the mouth. The muscular lip of the collar can be drawn over the mouth to reject or sort larger food particles.

Excess water that enters the mouth exits through numerous pharyngeal slits located along the lateral walls of the pharynx. Sets of adjacent slits open into a common chamber, the dorsally placed **branchial pouch,** that in turn pierces the outer body wall to form the **branchial pore,** an undivided opening to the outside environment (figure 2.8). Excess water departing from the pharynx thus passes first through a slit, then through one of the several branchial pouches, and finally exits through the branchial pore to the outside (figure 2.12c).

A ciliated **hypobranchial ridge** (ventral) and a ciliated **epibranchial groove** (dorsal) run along the midline of the pharynx. These, and the walls of the pharynx, secrete mucus and move the captured food particles. Particle movement is from dorsal to ventral, and then posteriorly to the gut. If the body plan of chordates is inverted relative to hemichordates, then the epibranchial ridge may be homologous to the endostyle, the ciliated food-groove, that is placed ventrally in other protochordates. However, in hemichordates the

binding of iodine and the secretion of mucous sheets occur generally throughout its pharynx, and are not centered on a single groove. The later endostyle of other protochordates, where iodine concentrates and mucous sheets are secreted, may not represent a homologous structure. Instead, this local "endostyle" may simply represent only a specialized region of this more general iodine-binding ability throughout the pharynx in hemichordates.

During ontogeny, perforations developing in the lateral walls of the pharynx form the original pharyngeal slits (figure 2.12a). However, each such slit next becomes partially subdivided by the **tongue bar,** a downward growth from the top rim of the opening (figure 2.12b). The fleshy bars between the original slits are referred to as the **primary pharyngeal bars** (or septa) and the tongue bars that come to divide them are the **secondary pharyngeal bars.** The **lateral**

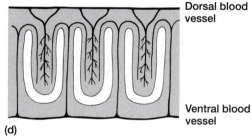

FIGURE 2.12   Hemichordate pharynx. Lateral view of tongue bar formation (a) to (b). During development, slits appear in the pharynx (a). This is followed by the partial subdivision of each slit by the downward growth of a process, the tongue bar. M-shaped skeletal rods appear within the primary and secondary bars (b). (c) Cross section through branchial bars. Cilia lining these bars move water from the pharynx past the edges of each tongue bar, past each primary bar, into the common branchial pouch and then out through a branchial pore. (d) Vascular supply to the tongue bars. Branches from the dorsal and ventral blood vessels supply each tongue bar, suggesting that respiratory exchange also occurs in the pharyngeal slits of the hemichordate.

FIGURE 2.11   Suspension mucous feeding. Direction and movement of food and mucus are indicated by arrows. Food material, carried along in the water current generated by surface cilia, travels across the proboscis and into the mouth where it is captured in mucus and swallowed. Rejected food material collects in a band around the collar and is shed.

After Burdon-Jones.

cilia covering the edges of both primary and secondary pharyngeal bars move water currents through the pharynx. The **frontal cilia** move mucus and occur in mucus-secreting epithelium along the medial edges of tongue bars and elsewhere within the lining of the pharynx (figure 2.12c). A network of afferent and efferent branchial vessels supplies the tongue bars, possibly participating in respiratory exchange with the passing stream of departing water (figure 2.12d).

The **stomochord** (figure 2.8) arises in the embryo as an outpocketing from the roof of the embryonic gut anterior to the pharynx. In the adult, the stomochord retains a narrow connection to what becomes the buccal cavity, but it usually enlarges as it projects forward into the cavity of the proboscis to form a preoral diverticulum. The surface of the stomochord is associated with components of the vascular and excretory systems. Its walls consist of epithelial cells, like those of the buccal cavity, as well as ciliated and glandular cells. Its hollow interior communicates with the buccal cavity.

Excretion in acorn worms probably occurs partly through the skin, but they also possess a **glomerulus** (figure 2.13), a dense network of blood vessels within the proboscis. Vascular fluid entering the glomerulus from the dorsal blood vessel is presumably filtered, yielding "urine" that is released into the proboscis coelom and eventually eliminated through the proboscis pore. Within the collar, a pair of ciliated collar ducts that extend from the collar coelom to the exterior via the first pharyngeal pore are also thought to be excretory in function.

The circulatory system is represented by two principal vessels, a **dorsal** and a **ventral blood vessel** (figure 2.12d). The blood, which contains few cells and lacks pigment, is propelled by muscular pulsations in these major vessels. From the dorsal vessel, blood passes forward into a **central blood sinus** at the base of the proboscis. Riding on top of this sinus is the **heart vesicle** (figure 2.13), which exhibits muscular pulsations and provides additional motive force to drive blood from the blood sinus forward into the glomerulus. From the glomerulus, blood flows to the ventral blood vessel and posteriorly beneath the digestive tract, which the ventral vessel supplies.

The nervous system in acorn worms consists mainly of a diffuse network of nerve fibers at the base of the epidermis of the skin (figure 2.10c). Dorsally and ventrally, the nerve network is consolidated into longitudinal nerve cords joined by nerve interconnections. This is largely different from the internalized central nervous systems of chordates, but as mentioned above in some species, the section of dorsal nerve cord in the collar invaginates from the surface ectoderm, sinks downward, and pinches itself off from the ectoderm to form a **collar cord.**

Enteropneust gonads are housed in the trunk, the sexes are dioecious, and fertilization is external. Early cleavage is radial, and formation of the body cavities is usually enterocoelic. In some species, development proceeds directly from egg to young adult. In most, however, there is a tricoelomic tornaria larval stage in that the three body cavities include an anterior **protocoel,** a middle **mesocoel,** and a posterior **metacoel,** which become the coelom of the proboscis, collar, and trunk, respectively (figure 2.13). The tornaria feeds and may remain a planktonic larva for several months before undergoing metamorphosis into the benthic adult.

The adult body is covered by a ciliated epithelium of varying prominence, interspersed with glandular cells that produce a mucous coating. The musculature varies between regions and between species. It lacks segmentation, but instead muscle fibers are laid out generally in circular and longitudinal orientations. This body musculature is best developed dorsally, as is the coelom. The wall of the straight digestive tract is mostly devoid of intrinsic musculature, although localized bands of circular fibers may occur in the branchial and esophageal regions.

The tornaria has a nephridium (figure 2.9), an excretory organ through which the larva regulates its internal ionic environment and rids itself of metabolic wastes. It consists of a blind-ended tube within the anterior region of the larva. During metamorphosis, the nephridium enlarges into the proboscis coelom (protocoel) of the adult, but in the larva, the ciliated nephridial duct (pore canal) conveys waste to the surface and opens to the outside via the **nephridial pore** (hydropore, proboscis pore; figure 2.9). In addition to ciliated cells, the walls of the nephridium are lined by

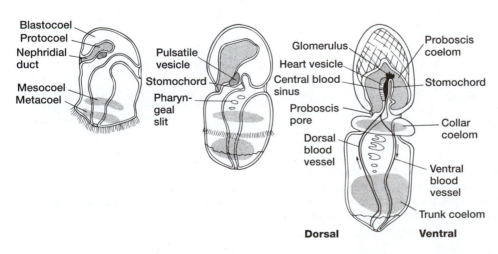

FIGURE 2.13
Metamorphosis of hemichordate larva. Transformation of larva into juvenile, from left to right. The three coeloms of the larva—protocoel, mesocoel, metacoel—give rise to the three respective body cavities of the adult—proboscis, collar, and trunk.

podocytes, specialized excretory cells that form a porous boundary between the lumen of the nephridium and the blastocoel, the larval cavity in which it resides. It is thought that the beat of cilia draws excess fluid from the blastocoel across the porous layer of podocytes, into the lumen of the nephridium, and out the nephridial pore. A small contractile **pulsatile vesicle** lies next to the nephridium. All of these structures persist and are functional within the proboscis of the adult, with claims that they have cellular homologues to the kidney tubules of vertebrates.

Vertebrate kidney (p. 551)

Upon the larva's metamorphosis, the nephridium expands to become the proboscis coelom, its canal becomes the proboscis duct, and much of its lining becomes the muscle and connective tissue of the proboscis. The pulsatile vesicle becomes the contractile tissue, or heart vesicle, that settles on top of the forming central venous sinus. Podocytes become associated with the specialized blood vessels, the glomerulus (figure 2.13).

Pterobranchia

Pterobranchs evolved from acorn worms. Most pterobranchs, of which there are only two genera, live in secreted tubes in oceanic waters (figure 2.14a). These species are small and colonial. Because individual identity is often lost, each contributing individual to the colony is commonly referred to as a **zooid.** Proboscis, collar, and trunk are present in each zooid, although they may be quite modified. The collar, for instance, is drawn out into two or more elaborate tentacles, part of the animal's suspension-feeding apparatus (figure 2.14a). The trunk is U-shaped, with the anus bending back to open at the top of the rigid tube in which the animal resides. An extension of the body, the stalk, attaches it to its tube and jerks the animal safely inside when it is disturbed (figure 2.14b).

Excretory organs of pterobranchs include a glomerulus in the proboscis and perhaps a ciliated pair of collar ducts. A stomochord is usually present. The nervous system is even simpler than that of acorn worms. A tubular nerve cord is absent. The **collar ganglion,** which is the closest a pterobranch comes to possessing a central nervous system, lies next to the epidermis in the dorsal region of the collar. Nerve branches emanate forward from the collar ganglion to the tentacles and posteriorly into the trunk. A few pharyngeal slits are present in most species.

Hemichordate Phylogenetic Affinities to Chordates

With links to chordates on the one hand and echinoderms on the other, hemichordates hold out among living groups the best promise of connecting chordates to their ancestral

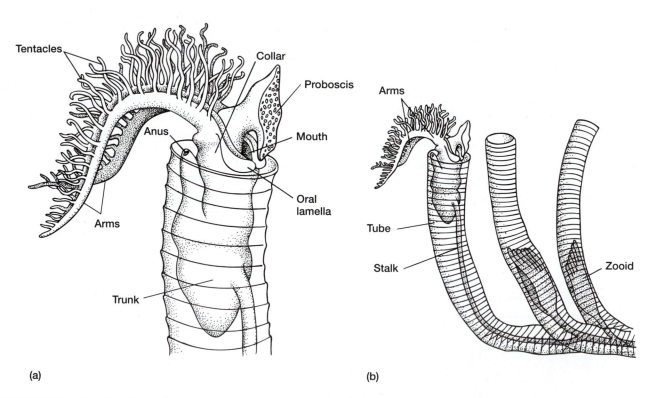

(a)

(b)

FIGURE 2.14 Hemichordata, Pterobranchia. (a) The sessile pterobranch *Rhabdopleura.* Notice that this pterobranch has the same body plan as an acorn worm—proboscis, collar, trunk—but these three features are modified and the whole animal lives in a tube. (b) Pterobranchs in tubes. When disturbed, the stalk shortens to pull a pterobranch to safety inside the tube. Because they live merged into a colony, each contributing pterobranch individual is often called a zooid.

After Dawydoff.

source among the invertebrates. Some recognized this early in the twentieth century, but enthusiasm, perhaps overenthusiasm, led to overinterpretation of hemichordate structure into the chordate camp. The stomochord within the proboscis was originally deemed to be a notochord and championed as a further structural link with chordates. But, such a claim is unfounded. Unlike a true notochord, the hemichordate stomochord is hollow, originates anterior to the pharynx from endoderm, and lacks the fibrous sheath necessary to give it the structural integrity of a rigid notochord. Current studies using gene expression similarly fail to find homology between stomochord (hemichordate) and notochord (chordate). Although the pharyngeal similarities provide a convincing link to chordates, the hemichordate body plan, composed of proboscis, collar, and trunk, is quite unlike the body plan of any other protochordate. And we cannot neglect the larval and molecular evidence that places hemichordates closer to echinoderms, even though echinoderms have obviously undergone extensive modifications (evolution of calcium carbonate surface plates) and radical makeover of their adult body plan (five-rayed symmetry in living forms).

Hemichordate Phylogenetic Affinities to Echinoderms

One of the most familiar echinoderms is the sea star (figure 2.15a). Characteristic of the group, its unsegmented adult body is based on a pentaradial (five-rayed or armed) symmetry, departing from the bilateral symmetry of most other coelomate groups. An endoskeleton of calcium carbonate is formed anatomically of distinctive ossicles produced by genes unique to echinoderms. These ossicles may form a solid case (sea urchins, sand dollars) or be reduced to isolated ossicles in a thick skin (sea cucumbers). Each ossicle is a single crystal of calcium carbonate common to and a diagnostic character uniting all echinoderms. A unique,

internal, fluid-filled plumbing system, the water-vascular system, works the tube feet used by some in locomotion and others in food gathering. Some echinoderms move by arm swings (brittle stars) whereas others are attached by a stalk to the substrate (sea lilies). No head or brain is present, and the nervous system consists of radial nerves that depart from a central nerve ring and reach into the arms and other parts of the body. All are marine, and the group is old (like chordates, it is at least Early Cambrian, maybe earlier).

Adults of modern echinoderms and hemichordates do not look much alike, but the **stylophora,** a fossil group of echinoderms, carry some potentially intermediate characters supportive of the view that echinoderms had a hemichordate-like ancestor. The stylophora, sometimes divided into solutes, cornutes, and mitrates, are known only from marine rocks dated 505 to 325 million years ago (figure 2.15b). Although they lack pentaradial symmetry and lack a water-vascular system, their body walls are formed of articulated plates of calcium carbonate, testifying that they are echinoderms. They interest us because they, or at least some, possessed a pharynx with pharyngeal slits, and roughly bilateral symmetry like hemichordates and chordates. Today the echinoderms and hemichordates are united in the ambulacrarians (figure 2.7). In retrospect, ambulacrarians document, with the presence of pharyngeal slits, an evolutionary start to the building of chordates.

Cephalochordata

Cephalochordates resemble the earliest chordates, at least based on their current phylogenetic position (figure 2.7). If current gene-expression studies are correct, then their basic body plan is dorsoventrally inverted compared to that of earlier deuterostomes including hemichordates. This is quite a jump with no fossil intermediates to help. Let's see what we do have in hand.

(a)   (b)

FIGURE 2.15 Echinoderms. (a) Sea star. This echinoderm illustrates the basic five-armed (pentaradial) body symmetry and its "spiny" surface built from underlying calcium carbonate ossicles. (b) Stylophora. An early echinoderm, known only from fossils. Small imbricated plates, also of calcium carbonate, cover the non-pentaradial body from which extends a whiplike stalk.

(a) From Bryan Wilbur, Ashley Gosselin-Ildari, 2001, "Piaster sp." Digital morphology at http://digimorph.org/specimens/Piaster_sp/. (b) After Jefferies.

Living cephalochordates occur worldwide in warm temperate and tropical seas. They are built upon the characteristic chordate pattern that includes pharyngeal slits, tubular nerve cord, notochord, and postanal tail (figure 2.16b–d). These animals are anatomically simple, with an approach to food gathering we have seen in other protochordates, namely, suspension feeding based on a pharyngeal filtering apparatus surrounded by an atrium. Their diet consists of microorganisms and phytoplankton. Slits open the walls of the extensive pharynx to allow exit of a one-way feeding current driven by cilia. Supporting edges of each slit constitute the primary pharyngeal bars (figure 2.17). During embryonic development, a tongue bar grows downward from the upper rim of each slit and joins the ventral rim, thereby completely dividing each original pharyngeal slit into two. This dividing support, derived from the tongue bar, constitutes a secondary pharyngeal bar. The primary bar includes an extension of the coelom; the secondary bar does not. **Supportive rods** of primitive cartilage support all the pharyngeal bars internally. Short, connecting rods, **synapticules,** cross-link these pharyngeal bars.

Major ciliated food corridors line the pharynx. The ventral channel is the endostyle, the dorsal channel is the **epibranchial groove,** and the inside edges of the primary and secondary pharyngeal bars carry ciliary tracts. An **oral hood** encloses the anterior entrance to the pharynx and supports an assortment of food-processing equipment. **Buccal cirri** which prevent entrance of large particles, project from the free edge of the oral hood, like a sieve. The inside walls of the oral hood hold ciliated tracts that sweep food particles into the mouth. The coordinated motion of these cilia gives the impression of rotation and inspired the name **wheel organ** for these tracts (figure 2.16d). One of these dorsal tracts, usually located below the right side of the notochord, bears a ciliated invagination that secretes mucus to help collect food particles and is known as **Hatschek's pit** or **groove.** Hatschek's pit occurs in the roof of the buccal cavity, a similarity shared with the vertebrate pituitary gland, part of which also forms by invagination from the roof of the buccal cavity. This has led some to propose that Hatschek's pit has an endocrine (hormone-secreting) function.

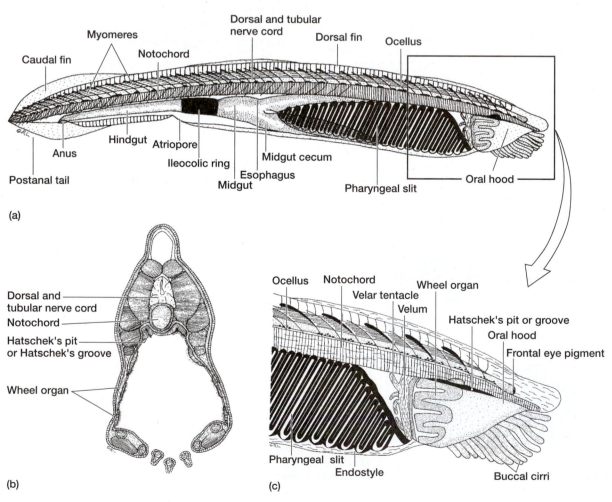

FIGURE 2.16 **Cephalochordate.** (a) *Branchiostoma lanceolatum,* a living cephalochordate known as amphioxus, shown in lateral view, (b) cross section through the oral hood, and (c) enlargement of anterior end.

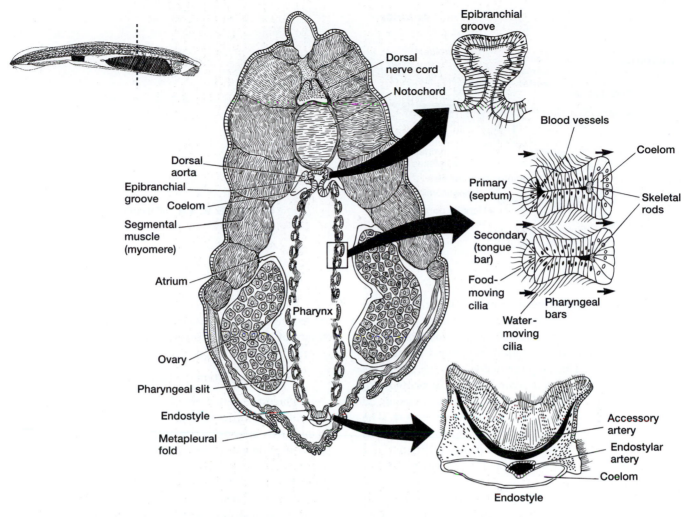

FIGURE 2.17 **Cross section of amphioxus.** The slanted pharyngeal bars encircle the pharynx. On the right, two of the pharyngeal bars are enlarged. Notice that they are cut transversely at right angles to their long axis. The coelom continues into the primary branchial bars but is absent from the secondary branchial bars that form as downgrowths to subdivide each pharyngeal slit. The cross section is drawn at about the point indicated in the inset, upper left.

After Smith; Møller and Philpott; Baskin and Detmers.

The posterior wall of the oral hood is defined by the **velum,** a partial diaphragm that supports short, sensory **velar tentacles.** Suspended material faces a gauntlet of testing, sifting, and sorting devices before passing through the central opening in the velum and entering the pharynx. Mucus, secreted by the endostyle and secretory cells of the pharyngeal bars, is driven up the walls of the pharynx by cilia. Food particles adhere and then are gathered dorsally into a thread in the epibranchial groove, from which they are conveyed back to the gut. The filtered water passes out through pharyngeal slits to the atrium and finally departs posteriorly via the single **atriopore.**

Parts of the cephalochordate digestive system may be precursors of vertebrate organs. For instance, the endostyle of amphioxus collects iodine as does the thyroid gland, the pharyngeal endocrine gland of vertebrates. The midgut cecum, a forward extension of the gut, is thought by some to

be a forerunner of the liver (because of its position and blood supply) and of the pancreas (because cells in its walls secrete digestive enzymes). Whatever their phylogenetic fate, these and other parts of amphioxus are a reflection of the specialized demands of suspension feeding.

The blood of amphioxus is a colorless plasma lacking blood cells and lacking oxygen-carrying pigments. Paired **anterior** and **posterior cardinal veins** return blood from the body, joining in paired common cardinal veins (ducts of Cuvier) (figure 2.18). The paired common cardinal veins and the single **hepatic vein** meet ventrally in the swollen **sinus venosus.** Blood flows anteriorly from the sinus venosus into the **endostylar artery** (ventral aorta). Below each primary pharyngeal bar, the endostylar artery branches into a set of ascending vessels to supply the primary bar (figure 2.18). At their departure from the endostylar artery, some of these vessels form swellings termed bulbilli. The secondary

Anterior

Posterior

Paired dorsal aortae

Glomerulus

To head
To head

Dorsal aorta

Cardinal vein

Intestine

Bulbillus

Endostylar artery (ventral aorta)

Pharyngeal complexes

Common cardinal vein

Hepatic vein

Hepatic portal vein

Subintestinal vein

Sinus venosus

Midgut cecum

FIGURE 2.18   Circulatory system of amphioxus.

After Alexander.

pharyngeal bars are not directly supplied from the endostylar artery. Instead, blood flows from the primary to the secondary bars through small vessels in the cross-connecting synapticules. Within the secondary bars, blood travels dorsally in the vessels. Dorsal to the pharyngeal slits, some vessels from the primary and secondary bars anastomose to form saclike renal glomeruli. Dorsal to these glomeruli, all pharyngeal-bar vessels join the paired **dorsal aortae.** The anterior end of amphioxus is supplied by forward extensions of the dorsal aortae. Posterior to the pharynx, left and right aortae fuse into an unpaired aorta that supplies the rest of the body (figure 2.18).

Thus, blood circulation in amphioxus is laid out on the same general pattern as that in vertebrates. Blood courses forward into the ventral aorta (endostylar artery), upward into the dorsal aorta, then posteriorly from there in the dorsal aorta. Afferent and efferent vessels move blood to and from the midgut cecum, respectively. Blood flow does not reverse in an ebb-and-flow pattern. As in vertebrates, capillary-like vascular networks in the major organs of amphioxus connect afferent and efferent vessels. However, amphioxus has no heart. The sinus venosus is positioned like a vertebrate heart, but it lacks heartlike pulsations. Instead, the job of contraction is distributed among other vessels: the hepatic vein, ventral aorta (endostylar artery), bulbilli, and others. These pump blood. Smooth or striated muscles are absent from their walls, but specialized, contractile myoepithelial cells are present. These cells are the presumed source of the pumping forces that move the blood.

In the pharynx, two parallel vessels travel uninterrupted through each pharyngeal arch, rather than the single aortic arch typical of vertebrates. Within secondary bars these two vessels form loops connected at their bends, with adjacent vessels in the primary branchial bar. The two blood vessels are referred to collectively as a **pharyngeal arch complex** (figure 2.18). Although structurally distinct, a pharyngeal arch complex is perhaps analogous to the afferent and efferent arteries of vertebrate gills.

The pharynx and its branchial bars serve feeding, and are less significant in respiration. Instead, the whole body surface of amphioxus through simple diffusion is a major contributor to respiration.

Circulatory system (p. 461)

The excretory system of amphioxus consists of paired nephridia opening into the atrium via a nephridial tubule and an unpaired nephridium opening into the buccal cavity. The paired nephridia arise from mesodermal cells, unlike those in most coelomate invertebrates, where they are derived from ectodermal cells. A nephridium consists of clusters of **podocytes** (figure 2.19a, b). Each podocyte is a single cell with cytoplasmic **pedicels,** projections that contact the nearby glomerulus, connected to the dorsal aorta. From the other side of the podocyte, a long circular strand of microvilli, with a single long flagellum down the center, projects across the coelomic space to enter the **nephridial tubule.** Each nephridial tubule receives a cluster of podocytes and in turn opens into the atrium. Podocytes occur commonly among invertebrates. The podocytes of amphioxus, with pedicels that embrace nearby glomerular blood vessels, are very similar to the foot processes of podocytes, cells found in the vertebrate kidney. The exact excretory function podocytes play in amphioxus is unclear, but their arrangement between blood vessel and atrium suggests a role in eliminating metabolic wastes removed from the blood and flushed away by the stream of water passing through the atrium.

Vertebrate kidney (p. 551)

The cephalochordate larval stage is planktonic, lasting from 75 to over 200 days. The young larval amphioxus is markedly asymmetrical in its head and pharynx (figure 2.20). For example, the first pair of coelomic pouches gives rise to two different structures: the left to Hatschek's pit, the right to the lining of the head coelom. Left and right series of pharyngeal slits appear at different times as well. The left series of slits appears first near the ventral midline

From their early discovery, cephalochordates seemed destined to be a lesson in taxonomic etiquette. Demeaned in 1774 by the first attempt to classify them, they were thought to be slugs and dubbed *Limax lanceolatus* by the German zoologist P. S. Pallas (although to be fair, he had only a scruffy, ill-preserved specimen from which to work).

In 1836 William Yarrell recognized the special nature of these animals and named them *Amphioxus* (meaning pointed at both ends) *lanceolatus*. Alas, this name stuck too well because much later it was discovered that O. G. Costa, actually two years before Yarrell, had christened them *Branchiostoma* meaning "grill" mouth for the buccal cirri,

and by rules of taxonomic priority the species should, and now does, carry this official generic name. Amphioxus, however, is a familiar name ingrained in common usage. It is not quite the mouthful of *Branchiostoma*, so we shall keep amphioxus (without italics or capitalization) as one common name, with "lancelet" as another.

---

and proliferates to perhaps as many as 14. The last slits in this series degenerate, leaving eight slits on the left side. The larval mouth forms on the left side of the body. Some think that the resulting asymmetry of the head might be related to larval amphioxus's spiral body movements during feeding. Next, the remaining slits on the left migrate up the left side

FIGURE 2.19   Nephridium of amphioxus. (a) Dorsal region of the pharynx showing the relationship of podocytes to the vascular glomerulus on the one end and the atrium on the other. (b) Nephridium structure enlarged. Podocytes embrace the walls of the glomerulus through cytoplasmic pedicels and reach the nephridial tubule through microvilli that have a central flagellum.

(b) After Brandenburg and Kümmel.

of the pharynx to their final lateral position. At the same time, the right pharyngeal slits make their first appearance, symmetrically positioned with those on the left. More slits are now added on both sides, together with the appearance of tongue bars that divide them as they form.

The larva lacks an atrium. During metamorphosis, the atrium is added from metapleural folds. These ventrolateral folds appear on either side, grow down over the pharyngeal slits, meet at the ventral midline beneath the pharynx, and fuse to complete the surrounding atrium (figure 2.17). The velum is now added to the mouth, and the buccal cirri and wheel organ are then added in front of the mouth. During this metamorphosis, the larva sinks out of the plankton to a substrate in which it will take up a burrowing residence as an adult.

Although adults are good swimmers, they usually live buried in coarse sediments with their oral hood protruding into the overlying water. Amphioxus prefers coastal waters and lagoons well aerated by tides but not churned by heavy wave action. Its locomotor system, based on myomeric segmental muscles of the body wall and a hydrostatic notochord, serves amphioxus in such habitats. Likely a reflection of amphioxus's buried lifestyle, its stiff notochord goes to the rostral tip of the body, leading to the name "cephalochordate" (head, notochord). The cephalochordate notochord originates from the roof of the gastrocoel during development, as it does in most other chordates. However, unlike notochords of other chordates, the notochord of amphioxus consists of a series of striated muscle cells arranged transversely and this sets it apart from all other protochordate and vertebrate notochords (figure 2.21). Fluid-filled spaces separate muscle cells, and both cells and spaces are enclosed in a dense connective tissue sheath. In the larva, a single row of tightly packed, highly vacuolated cells forms the notochord, whereas in the adult, most cellular vacuoles disappear and extracellular fluid-filled spaces emerge between these cells. Notochordal muscles receive their innervation by connections to the dorsal nerve cord through cytoplasmic extensions that run dorsally through the connective tissue sheath to the surface of the spinal cord, at which point they meet nerve endings within the cord.

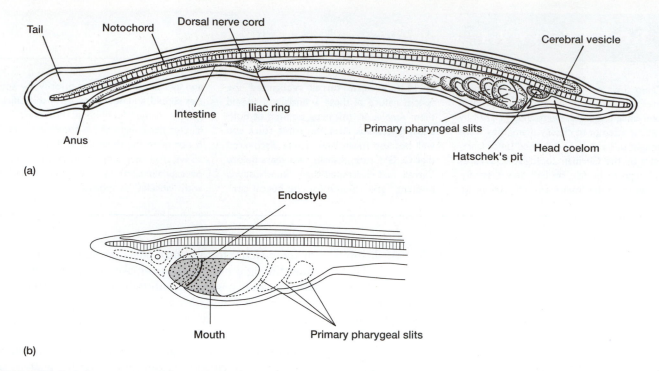

(a)

(b)

FIGURE 2.20 **Larval amphioxus.** (a) Pharyngeal slits appear only on the left side of the body during this early stage of development, but the basic chordate pattern is evident from the notochord, dorsal nerve cord, and short postanal tail. The atrium around the pharynx will not appear until metamorphosis. (b) Left side of early amphioxus larva showing asymmetrically placed mouth, which at metamorphosis will assume a midventral position. These primary pharyngeal slits, along with others that appear posteriorly, will become divided by tongue bars.

(a) After Lehman; (b) after Wiley.

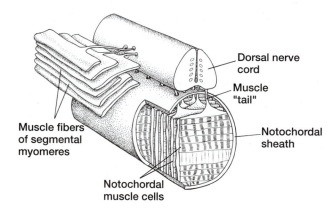

FIGURE 2.21 **Specialized notochord of amphioxus.** Like a row of poker chips, plates of slowly contracting muscles are packed within the notochordal sheath. Each plate is a single or sometimes double muscle cell containing contractile fibrils arranged transversely. Cytoplasmic extensions of these plates, called "tails," pass upward through holes in the notochordal sheath and synapse with the surface of the dorsal nerve cord. Fluid-filled spaces occur between these muscle cells, although a few vacuoles lie within these specialized cells. Muscle cells of the myomeres also send "tails" to the surface of the adjacent nerve cord where they synapse. The nerve cord directly stimulates these muscle cells through these synapses.

After Flood; Flood, Guthrie, and Banks.

When these muscle cells contract, the tough notochordal sheath prevents ballooning, internal pressure rises, and the notochord stiffens. Stiffening may strengthen burrowing or increase the intrinsic vibration rate of amphioxus to aid it in swimming fast.

The tubular nerve cord of amphioxus does not enlarge anteriorly into a differentiated brain. That is it does not show swellings indicative of forebrain, midbrain, and hindbrain. However, microscopic evidence suggests cellular homologies to parts of the vertebrate brain, in particular, similarities to hindbrain and the diencephalon part of the forebrain.

Muscle cells in the myomeres make contact with the spinal cord not by dedicated motor nerves reaching out peripherally to the muscles, but by thin processes of the muscles themselves of each myomere that, through cytoplasmic extensions, reach centrally to the surface of the spinal cord (figure 2.21).

Urochordata

Urochordates, also called **tunicates**, have a baglike, simplified body, especially as adults (figure 2.7). But, at some point in their life histories, they show all five shared derived chordate characteristics: notochord, pharyngeal slits, endostyle, tubular nerve cord, and postanal tail (figure 2.22a). Consequently, they are proper chordates placed within the Chordata.

(a)

(b)

(c)

FIGURE 2.22 **Urochordata, Ascidiacea larva.** (a) Larva of the ascidian *Distaplia occidentalis*. (b) Enlarged view of the anterior larval nervous system of *Diplosoma* (mostly its "brain"); see rectangle in (a). (c) Cross section of the larval tail of *Diplosoma*. During development, the tail twists the dorsal fin to the left side of the body, but in this figure, the tail is rotated 90° and drawn upright. Notice that the ventral and dorsal fins are formed from the outer layer of the tunic and that the central notochord is surrounded by sheets of muscle. The dorsal nerve cord is composed of ependymal cells around a neurocoel cavity, with axons of motor nerves coursing along its side.

From Cloney and Torrence; Torrence; Torrence and Cloney. (a) R. A. Cloney, "Ascidian larvae and the events of metamorphosis," American Zoologist, 22:817–826, 1982. American Society of Zoologists. Reprinted by permission of the Society for Integrative and Comparative Biology (Oxford University Press).

Urochordates are specialists at feeding on suspended matter, especially very tiny particulate plankton. In most, the pharynx is expanded into a complex straining apparatus, the branchial basket. In a few species, however, the filtering apparatus is secreted by the epidermis and surrounds the animal. All species are marine. Urochordates are divided into several major taxonomic classes. Ascidiacea are sessile as adults, but have swimming larvae, whereas the taxa Larvacea and Thaliacea are permanently pelagic and drift in the plankton, unattached to any fixed substrate.

Urochordate literally means "tail backstring," a reference to the notochord. The familiar name, tunicates, is inspired by the characteristic flexible outer body cover, the **tunic.** It is secreted by the underlying epidermis with contributions from scattered cells within the tunic itself. This tunic, sometimes referred to by the more general name **test,** characterizes the urochordates.

Ascidiacea—"Sea Squirts"

Ascidians, or sea squirts, are marine animals that are often brightly colored. Some species are solitary, others colonial. Adults are sessile, but larvae are planktonic.

Larva   The larva, sometimes called the **ascidian tadpole,** does not feed during its short sojourn of a few days as a free-living member of the plankton, but it disperses to and selects the site at which it will take up permanent residence as an adult. Only the larval stage exhibits all five chordate characteristics simultaneously. The small pharynx bears slits in the tadpoles of colonial species. The tubular nerve cord extends into a tail supported internally by a turgid notochord. Vacuolated cells are absent from the ascidian notochord. Instead, in most solitary and colonial species, the notochord has a cell-less interior and is therefore tubular. Its walls are composed of a single layer of epithelial cells covered externally by a circumferential sheath of collagen fibers. The epithelial layer encloses an extracellular gel- or fluid-filled lumen (figure 2.22c). Therefore, the ascidian notochord is a turgid, tubular rod closed at both ends.

In solitary ascidian species, the gut does not fully differentiate in the nonfeeding larva, so an anus is not present to mark the point beyond which the tail continues. In many colonial species, however, the larval gut may be fully differentiated, including an anus that opens into the atrial chamber, and feeding may begin within 30 minutes after settlement. The "postanal" tail is present, although sometimes twisted or rotated about 90° to the body. Individual striated muscle cells lie in chains or sheets along the sides of the tail, but do not form segmental blocks of myomeres. Special **myomuscular** and gap junctions join these muscle cells together so that all cells on one side act as a unit, contracting together to bend the tail. Like the adult, the ascidian larva is covered by a tunic. The surface of the tunic is covered by thin inner and outer cuticular layers. The outer cuticular layer forms the larval tail fins, but is cast off at metamorphosis. The inner cuticular layer remains after metamorphosis to form the outermost surface of the juvenile. Beneath the tunic, the epidermis at the anterior end of the body forms **adhesive papillae** that serve to attach the larva to a substrate at the end of its planktonic existence.

The central nervous system forms dorsally in typical chordate fashion, from an embryonic neural plate that rolls up. It has three subdivisions: (1) sensory vesicle and (2) visceral ganglion, both of which form a rudimentary brain, and (3) the dorsal, hollow nerve cord extending into the tail. The **sensory vesicle** (figure 2.22a) located next to the rudimentary pharynx contains navigational equipment thought to be involved in orientating the larva during its planktonic existence. Within the sensory vesicle is a light-sensitive **ocellus** ("little eye") and a gravity-sensitive **otolith** (figure 2.22b). A

rudimentary **cerebral ganglion,** functional only after metamorphosis, and a **visceral ganglion** are nearby and send nerves to various parts of the body. The nerve cord includes ciliated **ependymal cells** around the neurocoel and **nerve tracts** that arise from the visceral ganglion and pass lateral to the ependymal cells in order to supply the tail muscles (figure 2.22c). Sensory nerves return from the tail and adhesive papillae to the visceral ganglion.

In the vertebrates, the neural crest is a special group of early embryonic cells that depart from the neural tube, and migrate through defined pathways, and differentiate into a vast array of cell types. In fact, such multipotential, migratory cells may be a unique feature, elaborated by vertebrates. Recently, in ascidians, migratory cells have been identified as well and are proposed to be forerunners of neural crest cells. However, these ascidian cells migrate as single cells (rather than streams of cells), and give rise only to pigment cells of the body wall and siphon (rather than being multipotential). Therefore, migratory neural crestlike cells debut in urochordates, but it is in the vertebrates we find neural crest cells with additional functions and an expanded repertoire of structures to which they contribute.

Neural Crest and Ectodermal Placodes (p. 189)

Circulating blood cells and a rudimentary heart are present (figure 2.22a). In a few colonial species, the blood cells become mature and the heart beats. Like the adult hearts, such larval hearts periodically reverse the direction of pumping.

Metamorphosis   At the end of its short planktonic stage, the ascidian larva makes contact with the substrate of choice, usually in a dark or shaded location, adhesive papillae take hold to attach it, and metamorphosis to a young adult begins almost immediately (figure 2.23). Within a few minutes of attachment, contraction of either notochordal or epidermal cells draws the **axial complex** (tail and all its contents) into the body. The notochord cells separate from each other, the extracellular fluid leaks from the central lumen, and the notochord becomes limp. The axial complex is then resorbed over the next several days and its constituents redistributed to support the young, growing juvenile. Lost too are the outer tunic layer, sensory vesicle, and visceral ganglion; however, the pharynx enlarges, slits in its walls increase in number, and the attached individual begins feeding for the first time.

Most of the chordate features that made their debut in the larva, namely, notochord, tail, and dorsal nerve tube, disappear in the forming adult. Although the pharynx persists and even expands, it becomes highly modified. The slits in its walls proliferate and each subdivides repeatedly, producing smaller openings called **stigmata.** This remodeled pharynx forms the barrel-shaped branchial basket (expanded pharynx plus numerous stigmata) of the adult sea squirt (figure 2.24a).

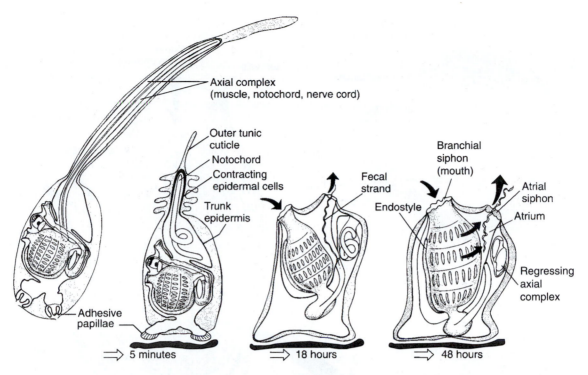

FIGURE 2.23 **Metamorphosis of the ascidian larva** *Distaplia,* **proceeding from left to right.** The planktonic nonfeeding larva settles and attaches to a substrate. Adhesive papillae hold the larva in place, contraction of tail epidermis pulls the axial complex into the body, and the larva sheds its outer cuticle following attachment. By 18 hours, the branchial basket rotates to reposition the siphons, and the appearance of a fecal strand testifies that active feeding has begun. By 48 hours, most of the axial complex is resorbed, rotation is complete, and attachment to the substrate is firm. At this point, the juvenile is clearly differentiated.

Based on the research of R. A. Cloney.

Adult The tunic, composed of a unique protein, **tunicin,** and a polysaccharide similar to plant cellulose, forms the body wall of an ascidian adult. The branchial basket, a large atrial cavity around this basket, and the viscera are all enclosed within the walls formed by the tunic (figure 2.24). The tunic attaches the base of the animal to a secure substrate (figure 2.24a). Incurrent (branchial) and excurrent (atrial) siphons form entrance and exit portals for the stream of water that circulates through the body of the tunicate. Tiny, fingerlike sensory tentacles encircle the incurrent siphon to examine the entering water and perhaps exclude excessively large particles before water enters the branchial basket. The complex pharyngeal slits, the stigmata, sieve the passing water before it flows from the branchial basket into the **atrium,** the space between basket and tunic (figure 2.24a). From here, the current exits via the excurrent siphon.

Rows of cilia line the branchial basket. The mucus-producing **endostyle,** a mid-ventral food-groove like that of amphioxus, is connected by continuous ciliated bands or tracks around the inside to the **dorsal lamina.** Particulate matter is extracted from the passing stream of water by a netlike sheet of mucus lining the branchial basket. The rows of cilia collect the food-laden mucus to move it from ventral to dorsal, delivering it to the dorsal lamina, which in turn conveys it posteriorly to the gut.

The sea squirt's heart, located near the pharynx, is tubular, with a single layer of muscle-like striated myoepithelial cells forming its wall (figure 2.24d). The surrounding **pericardial cavity** is the only remnant of the coelom. Contraction of the heart pushes blood out to the organs and tunic. After a few minutes, the flow reverses to return blood along the same vessels to the heart. Unlike the vertebrate circulatory system, there is no continuity between the heart myoepithelium and the blood vessels. Ascidian blood vessels are not lined by an endothelium. Instead, they are true hemocoels—that is, connective tissue spaces. The blood contains a fluid plasma with many kinds of specialized cells, including **amoebocytes** that resemble vertebrate lymphocytes. They are phagocytic, and some accumulate waste materials. No specialized excretory organ has been found in tunicates.

The adult nervous system consists of the brainlike **cerebral ganglion** located between the siphons (compare figure 2.22 and figure 2.24b). Nerves that pass to the siphons, branchial basket, and visceral organs arise from each end of the ganglion. Beneath the ganglion lies the **subneural gland,** a structure of unknown function that is left over from the larva and joined to the branchial basket via a **ciliated funnel.**

Smooth muscle bands run the length of the body and encircle the siphons their contraction bring about changes in the shape and size of the adult. When the ascidian is

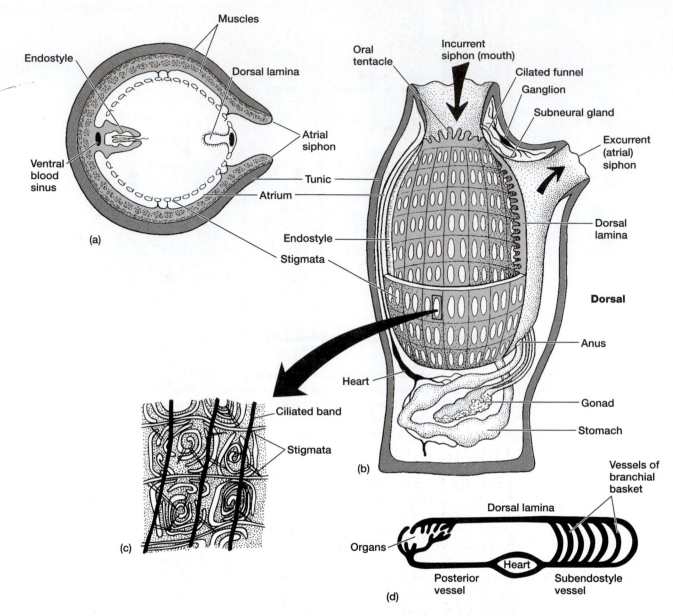

FIGURE 2.24 Adult solitary ascidian. (a) Schematic cross section of the body at the level of the atrial siphon, with dorsal to the right. Food captured in the lining mucus is moved dorsally, collected in the dorsal lamina, and passed to the stomach. (b) Whole animal, with left side of the body and part of the branchial basket cut away. Oral tentacles exclude large particles entering with the stream of water via the incurrent siphon. The water passes from incurrect siphon into branchial basket, through the pharyngeal slits (stigmata), into the surrounding atrium, and exits via the excurrent siphon. (c) The structures of several highly subdivided pharyngeal slits, the stigmatae, are depicted. (d) Diagram of urochordate circulation. Blood flows in one direction and then reverses itself rather than maintaining a single direction of flow.

threatened or wave buffeted, especially when exposed during low tides, these muscles quickly contract, reducing body size, and water is jetted out via the siphons; hence, the common name, "sea squirts."

All ascidians are hermaphrodites; both sexes occur in the same individual (monoecious), although self-fertilization is rare. Solitary ascidians reproduce only sexually, while colonial ascidians reproduce sexually and asexually (figure 2.25). Asexual reproduction involves **budding.** Rootlike **stolons** at the base of the body may fragment into pieces that produce more individuals, or buds may arise along blood vessels or

viscera. In colonial species, buds even appear in the larva before metamorphosis. Such budding gives the tunicate a way to propagate rapidly when conditions improve, thereby avoiding risky, planktonic dispersal of vulnerable larvae. In some species, buds seem especially hardy and are adept at surviving temporary adversity.

Larvacea (Appendicularia)

Members of the worldwide class Larvacea are tiny marine animals that reach only a few millimeters in length and reside within the planktonic community. Larvacea received

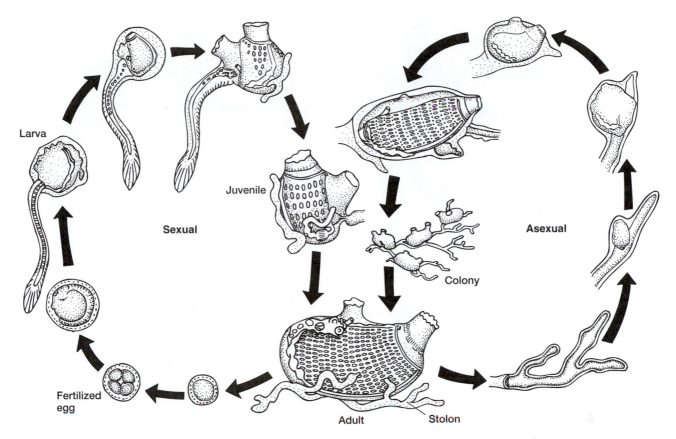

FIGURE 2.25 **Urochordate-ascidian life cycle.** The life cycle of colonial ascidians includes a sexual (left) and an asexual (right) phase. In the sexual phase, the tunicate larva develops from a fertilized egg. This larva is planktonic and persists for a few hours or a few days at most. It soon settles on a solid substrate and undergoes metamorphosis into a sessile juvenile that grows into adulthood. The asexual phase begins with external buds from the rootlike stolon or internal buds from organs within the body, depending on the species. These buds grow and differentiate into adults, often forming a colony of tunicates.

After Plough.

their name because the adults retain larval characteristics similar in some ways to the ascidian tadpole with its tail and trunk (figure 2.26a–c). The implication was that adult larvaceans derived from the larval stages of ascidians. In fact, more recent phylogenetic analyses now suggest otherwise—larvaceans and ascidians are equally ancient. Larvaceans became so highly modified for a rapid life cycle that their immediate ancestor is difficult to imagine.

Larvaceans secrete a most remarkable feeding apparatus that consists of three components: **screens, filters,** and expanded **gelatinous matrix.** This feeding apparatus is outside the animal, and not part of its pharynx as in other urochordates. Because the larvacean lives within the gelatinous matrix it constructs, this matrix is termed a "house." This house also holds the feeding screens and filters, and forms the channels through which streams of water carry suspended food particles. Houses and feeding styles differ among the various species, but generally the undulating tail of the larvacean creates a feeding current that draws water into the house. Incoming water passes first through the meshwork of screens that exclude large particles; consequently, the screens serve as an initial sorting device. This

water continues its flow through internal passages and then up the sides and through the mucous feeding-filters, where tiny suspended food particles are removed. When the current of water is emptied of its suspended material, it leaves the house through an excurrent opening. The larvacean takes advantage of its convenient central position at the base of the feeding-filters to gather all intercepted food particles. By means of ciliary action, the animal sucks trapped particles from the filters into its pharynx every few seconds. There, mucus, secreted by an endostyle, gathers the food. Excess water exits from the pharynx via a pair of pharyngeal slits and joins the current departing through the excurrent opening.

If the filters become clogged with food, a reverse flow may clear them. If that fails, the house is abandoned and a new one secreted (figure 2.27). Actively feeding larvaceans might abandon and build new houses every few hours. Disturbance of captured larvaceans, perhaps simulating predator attack, can prompt an even more frequent cycle of abandon and build.

The rudiment of a new house (tunic), which is secreted by the epithelium, is already present while the animal still occupies its old one. Some houses split to release the larvacean;

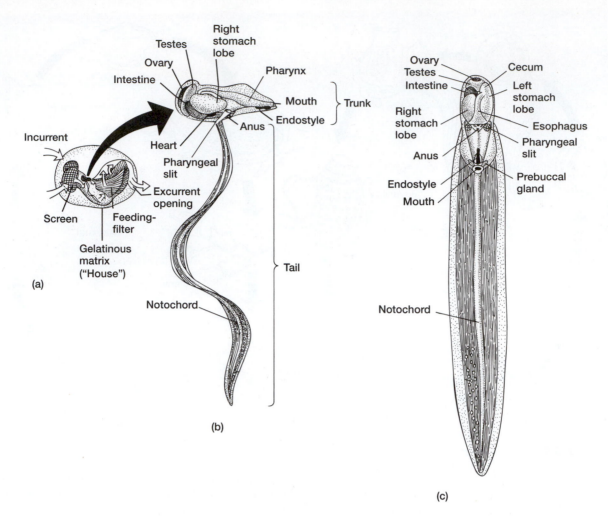

FIGURE 2.26 **Urochordata, Larvacea (Appendicularia),** *Oikopleura albicans.* (a) Tiny *Oikopleura* is shown within its larger gelatinous house. The animal's feeding-filter obtains food from the incurrent of circulating water (small arrows). This larvacean resides at the base of the screen where it sucks food off these screens. (b and c) Enlarged lateral and dorsal views, respectively, of the isolated larvacean. The undulating tail, supported by a notochord, is active in producing the current of food-bearing water that moves through the internal channels of the house and through the feeding-filter.

(a) After Flood; (b, c) after Alldredge.

others have special escape hatches. Almost immediately upon exiting its old house, the animal initiates a vigorous series of motions that enlarge the new house rudiment to a size it can enter. Once inside, expansion of the new house continues with the addition of feeding screens and filters. Sometimes within the space of only a few minutes, the new house is complete and the larvacean is once again feeding actively.

All species, except one, are monoecious, and most of these are **protandrous;** that is, sperm and eggs are produced by the same gonad (of the same individual) but at different times during its life. Maturation is so rapid that within 24 to 48 hours of fertilization, miniature larvaceans secrete a house and are set up for feeding.

Their rapid reproduction and special feeding apparatus give larvaceans a competitive advantage over other aquatic suspension feeders. Larvaceans are especially adept at gathering ultraplankton, very minute, bacteria-sized organisms. Collectively, ultraplankton are the major producers in most open oceans, but they are generally too small to be captured by the filtering traps of most suspension feeders. These tiny organisms that escape the clutches of other suspension feeders fall prey to the efficient filtering gear of larvaceans. Larvacea are able to sift through large volumes of water, ingest a wide range of plankton sizes, including the very tiniest, and proliferate rapidly in response to local blooms in food supply.

The trunk of the larvacean holds its major body organs, although which organs are present varies among the three families of Larvacea. Members of the smallest family, the Kowalevskiidae, lack endostyle and heart. In the Fritillaridae, the stomach consists of only a few cells. In the Oikopleuridae, the family best studied, the digestive system includes a U-shaped digestive tube, a pharynx with a

FIGURE 2.27 House building by the appendicularian *Oikopleura*. Clogged filters apparently trigger an appendicularian to abandon its house (1). Vigorous movements enlarge the rudiment of a new house (2 and 3) until there is room enough for the animal to enter (4). Thereafter, the house is further enlarged, filters are secreted, and feeding begins again (12).

After Alldredge.

pair of pharyngeal slits, and an endostyle that manufactures mucus. The blood of larvaceans, which is mostly devoid of cells, circulates through a system of simple sinuses driven by the pumping action of a single heart and the movement of the tail.

The tail is thin and flat. Within it, muscle bands act on a notochord to produce movement. A tubular nerve cord is present.

Thaliacea

Like larvaceans, the thaliaceans are free-living, pelagic urochordates, but unlike larvaceans, thaliaceans are apparently derivatives of the adult ascidian rather than of the tadpole morph (figure 2.28a–c). A few pharyngeal slits are present. Details of feeding are unresolved, although cilia, mucus, and a branchial basket certainly participate.

Some species of thaliaceans are built like colonial ascidians, except incurrent and excurrent siphons lie at opposite ends of the body (figure 2.28c). The outer body, or test, encloses a water-filled chamber. Most thaliaceans possess encircling (circumferential) bands of muscle within the wall of the test. Slow contraction of these muscle bands constricts the test and squeezes the water in the chamber out through the posterior aperture. When muscles relax, the elastic test

expands, drawing in water through the anterior aperture to refill the chamber. Repeated cycles of muscle contraction and test expansion produce a one-way flow of water through the thaliacean, creating a jet propulsion system for locomotion.

Overview of Protochordates

Protochordates possess some of (hemichordates) or all (cephalochordates, urochordates) five characteristics that define a chordate—notochord, pharyngeal slits, endostyle or thyroid gland, dorsal hollow nerve tube, postanal tail—although these characteristics may be present at one stage in the life history and not at another. Always marine, their adults are usually benthic and their larval stage planktonic. Consequently, larva and adult live quite different lifestyles and are structurally quite different in design. Their food consists of suspended particles extracted from a stream of water propelled by cilia. Food particles are collected on sheets of mucus and directed to the gut. Water flowing in with food is diverted outside through lateral pharyngeal slits in order to prevent turbulence that might disrupt the carefully gathered mucous cords laden with food. When present, the notochord, along with tail muscles, is part of the locomotor apparatus, giving the animal more mobility than afforded by cilia alone.

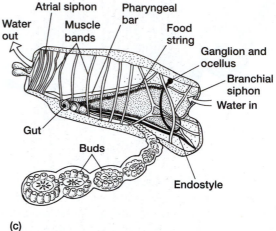

FIGURE 2.28 **Urochordata, Thaliacea.** (a) Colony of thaliaceans. (b) Isolated zooid. Longitudinal section of the body of this individual member is removed from its "house." Small arrows indicate direction of water flow. (c) An order of thaliacean known as salps. Branchial and atrial siphons are at opposite ends, turning the feeding current of water into a modest propulsive jet. Asexual buds are produced.

(a, b) After Brien; (c) after Berrill.

Protochordates have a phylogenetic history that precedes the vertebrates. They have enjoyed a long and independent evolution of their own dating back over 520 million years. Their relationships to each other and the sequence of their evolutionary emergence have received attention from biologists for over a century. With this introductory knowledge of protochordates, let us turn to the question of their evolutionary origins.

Chordate Origins

Fossils relevant to chordate origins are scant, and most living invertebrates are highly derived. Most living invertebrate groups diverged from each other over half a billion years ago and have since gone their own separate evolutionary ways. Whichever invertebrate group we favor as an immediate ancestor to chordates, it could not be converted directly from its modern form into a chordate without drastic reorganization. Although living groups are consulted for possible clues they retain of ancestral relationships, biologists realize that the actual ancestors to chordates are long extinct.

Faced with these intrinsic obstacles and with little evidence from the fossil record to help, it is hardly surprising that disagreement over the origin of chordates has been common. At one time or another, almost every group of invertebrates has been cited as the immediate evolutionary source of chordates. Although preposterous, even protozoa have been suggested as more or less direct ancestors of chordates! Less extreme but also tenuous are chordate origins among the nemertines, or ribbon worms, with the eversible proboscis of these worms said to give rise to the chordate notochord, the pharynx to the branchial region, and so forth.

Coming to our aid are modern methods of phylogenetic reconstruction, especially those employing molecular probes of taxonomic relationships. These provide the chronology of chordate character appearance, primitive to derived, and the hypothesized relationships among chordates and their immediate ancestors (figure 2.7). Unfortunately, even the best of these phylogenetic reconstructions are only descriptive hypotheses. They do not speak directly to the causes of evolutionary change, namely *how* these changes occurred and *why* they occurred. There is no inevitability for chordates. Their rise must be based on plausible morphological remodelings and on the survival advantages that gave the emerging chordate features favorable adaptive benefits.

This effort to understand chordate origins goes back to the beginning of the nineteenth century. One such view traces chordate origins back to annelids and arthropods.

Chordates from Annelids and Arthropods

The first person to propose that the chordate body plan was derived from a flipped-over version of an arthropod's was Geoffroy Saint-Hillaire, French zoologist. In 1822 he put forth this theory, perhaps inspired by his dissections of lobsters, but also as part of his larger view that all animals shared a common, underlying body plan that nature molded into dramatic variations. Georges Cuvier led those opposing this theory, and in 1830 publicly debated Saint-Hillaire. Cuvier ticked off a long list of differences that surpassed the similarities, and temporarily quieted the issue for most scientists.

Separate theories put forth later in the nineteenth century revived this view that annelids or arthropods might be chordate ancestors. In the early twentieth century, biologist W. H. Gaskell and, shortly thereafter, William Patten revived the issue and presented a closely argued case in support of annelid and/or arthropod ancestry for chordates.

Their reasoning, taken together, went as follows. Annelids and arthropods share with chordates similarities of basic body design. All three groups are segmented. All exhibit similarities in gross brain regionalization, with forebrain and hindbrain. Finally, the basic chordate body plan is present in annelids and arthropods, although upside down (figure 2.29a, b). In annelids and arthropods, the nerve cord occupies a ventral position below the gut along with a major blood vessel. If an annelid or an arthropod is flipped over, this brings the nerve cord into a dorsal position, along with the major blood vessel, which becomes the dorsal aorta. In this reversed position, the inverted annelid or arthropod body becomes the fundamental chordate body.

This argument has since been embellished by the imaginative work of others, but it suffers from some major weaknesses. For example, many of the supposed linking similarities between chordates and annelids or arthropods result from homoplasy rather than homology. The segmentation and jointed appendages that are part of an arthropod body are quite unlike the chordate myotomal segmentation. The main nerve cord of annelids and arthropods is solid, not hollow as in chordates, and develops embryologically in a fundamentally different way. Furthermore, the usual positions

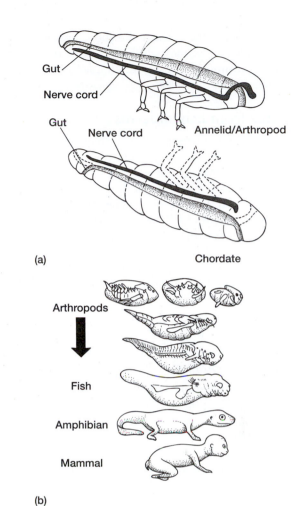

FIGURE 2.29 Proposed evolution from annelid/ arthropod to chordate. (a) If details are ignored, the basic annelid/arthropod body turned on its back produces the basic chordate body with the nerve cord now dorsally positioned above, rather than below, the gut. (b) Elaboration of this upside-down theory begins with the nauplius larva of crustaceans and other arthropods that swim with their legs up and back down. Through imagined transitional forms, these changes supposedly gave rise to an "inverted" vertebrate.

Source: W. Patten, The Evolution of the Vertebrates and Their Kin, 1912. Philadelphia: P. Blakiston's Son & Co.

for a chordate's mouth and anus are ventral, whereas an annelid or an arthropod rolled on its back would turn both mouth and anus up, pointing skyward. Inverting an annelid or arthropod to produce a chordate body plan would require migration of the mouth and anus back ventrally or formation of new ones ventrally. Unfortunately for this theory, the embryology of chordates preserves no hint of such an event.

The embryonic history of chordates also is fundamentally different in method of coelom formation, derivation of mesoderm, and in basic pattern of early cleavage. Even the body axis is different. In protostomes, such as annelids and arthropods, the anterior end forms on the side with the embryonic blastopore. In deuterostomes, such as chordates, anterior is pointed in the opposite direction, away from the blastopore. Collectively, these difficulties with theories of an annelid or arthropod ancestry for chordates encouraged alternative proposals.

Chordates from Echinoderms

Echinoderms, like chordates, are deuterostomes, as embryonic similarities testify. It is perhaps just this underlying similarity that inspired W. Garstang, a biologist of the late nineteenth and early twentieth centuries, to put forth an alternative theory outlining the origin of chordates. Garstang reasoned that because of these embryonic affinities, echinoderms or a group very similar to echinoderms were the likely chordate ancestors.

At first this seemed farfetched. Adult echinoderms, such as starfish, sea urchins, sea cucumbers, and crinoids, offer little to suggest a phylogenetic affinity with chordates. They have tube feet, calcium carbonate plates in their skin, and pentaradial (five-armed) body symmetry. Let's examine Garstang's hypothesis more closely.

Auricularian Hypothesis

Both echinoderms and chordates are deuterostomes that share embryonic similarities of cleavage and mesodermal and coelomic formation. Echinoderm larvae, like chordates generally, are bilaterally symmetrical.

This echinoderm-to-chordate hypothesis is often called the **auricularian hypothesis,** in reference to a particular type of echinoderm larva, the auricularia larva found in holothurians (sea cucumbers). The auricularian hypothesis begins with a dipleurula larva, an idealized version of this auricularian larva, thought to represent the simplified ancestor of all echinoderm larvae. Garstang proposed that, in fact, chordate characteristics first debuted in this dipleurula larva

(figure 2.30). This dipleurula larva had bilateral symmetry and possessed a simple one-way gut. Near the mouth was an **adoral band** of feeding cilia; across its lateral body surface meandered a long row of cilia, the **circumoral band,** by which the larva was propelled. In the path to chordate, Garstang envisioned that the larval body elongated, becoming increasingly muscular, and formed a tail that, together with a notochord, could generate lateral undulations as a means of aquatic locomotion. Body elongation drew out the circumoral ciliated band and brought its left and right halves dorsally, where they met at the midline together with the underlying nerve tract, the antecedent of the nerve tube (figure 2.30). Garstang pointed to the rolling up of the neural tube during vertebrate embryology as an embryonic remnant of this phylogenetic event. The segmental body musculature with notochord evolved simultaneously with the neural tube that controlled it. On the other hand, lengthening of the adoral band near the mouth and into the pharynx provided the beginnings of an endostyle. Pharyngeal slits then appeared to complete its transformation into a full-fledged chordate.

But starting from an echinoderm larva, evolution does not look hopefully ahead to the distant advantages of a chordate lifestyle (see chapter 3). Changes in the echinoderm larva must have been driven by some immediate adaptive advantage at the time of their first appearance. What could these be?

Suppose, for instance, that the larva of this ancestral echinoderm spent more and more time in its planktonic stage, fed, and thereby grew in size. Larger size is an advantage in escaping predation and in becoming established on a substrate once metamorphosis begins. If the larva's size increased for these or other reasons, this change in size alone would require compensating changes in two systems, locomotion and feeding, for the same reason.

FIGURE 2.30 Garstang's theory of the origin of the chordate body plan. The proposed common ancestor of chordates (left) was bilaterally symmetrical and had the external appearance of a young echinoderm larva. The ancestor's circumoral ciliated bands and their associated underlying nerve tracts moved dorsally to meet and fuse at the dorsal midline, forming a dorsal nerve cord in the chordate body plan. The adoral ciliated band gave rise to the endostyle and ciliated tracts within the pharynx of the chordate. Scientists other than Garstang noted that the appearance of pharyngeal slits improved efficiency by providing a one-way flow for the food-bearing stream of water. A notochord appeared later, and with the swimming musculature, is a locomotor advantage in the larger organism.

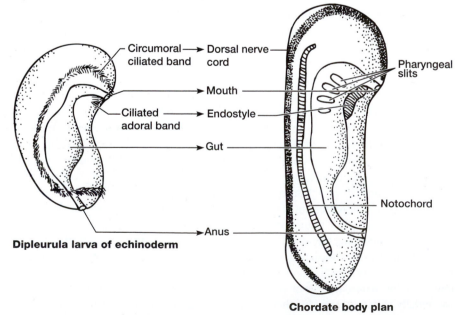

Circumoral ciliated band → Dorsal nerve cord

Ciliated adoral band → Mouth

Endostyle

Gut

Pharyngeal slits

Notochord

Anus

Dipleurula larva of echinoderm

Chordate body plan

The reason is geometry. As an object gets larger, surface and mass increase unevenly relative to each other. Body mass increases proportionately to the *cube* of the linear dimensions, but surface area increases only by the *square* of linear dimensions. In a larva that increased in size, surface cilia propelling the larva would not increase fast enough to keep up with the expanding mass. The locomotor surface would fall behind as the larva got bigger. As a result, there would be relatively fewer surface cilia to move a relatively greater bulk. This, the argument goes, favored the development of an alternative locomotor system. Segmental swimming musculature, elongated body, and stiffened bar (notochord) are the supposed solutions, first supplementing and then replacing the faltering ciliary system.

Similarly, the mode of feeding had to change, and for the same reason, namely, a geometric mismatch between surface area bearing cilia and body mass requiring nutritional support. The surface around the mouth supported feeding cilia that sweep suspended particles into the mouth. But as larval size increases, body mass outstrips the ability of these surface cilia to meet nutritional needs. An adoral ciliary band expanded into an endostyle would improve food transport. Perforations (slits) in the pharynx would allow one-way flow of a feeding current. Both changes would increase the efficiency of the feeding mechanism. These feeding structures may have been favored by just such selective pressures.

Consequences of size on surface and volume ratios (p. 132)

Larval Echinoderm to Chordate Tadpole

However, the problem of larval metamorphosis into an adult echinoderm still remains. Sooner or later, the planktonic echinoderm had to transform into a benthic adult. But how might this echinoderm larva, now endowed with chordate characters, achieve a separate evolutionary destiny from the echinoderm adult it is fated to become at metamorphosis?

Garstang's answer to this was clever again. He suggested that the adult stage was eliminated and the larval stage enhanced. The pelagic larva is adapted to a free-living lifestyle, the adult to its benthic lifestyle. If the modified larva enjoyed success, and the adult did not, then time spent in this larval stage might be extended at the expense of time spent as an adult. If the larva became sexually mature while still a larva, it could reproduce, an adult function, and thus escape from a life cycle tied to a benthic adult. This process is termed paedomorphosis.

More on Paedomorphosis (pp. 197–201)

A paedomorphic echinoderm larva equipped with chordate features might enjoy the adaptive advantages of greater pelagic mobility and, now reproductive, depart along an independent evolutionary course. Those scientists favoring an echinoderm ancestry for chordates have been quick to invoke paedomorphosis within phylogenetic schemes. Garstang, for instance, suggested that vertebrates might have evolved from echinoderms first through hemichordate-like, and then through urochordate-like, ancestors via paedomorphosis (figure 2.31).

FIGURE 2.31 Summary of Garstang's view of vertebrate origins. Beginning with an echinoderm larva, Garstang proposed a series of literal evolutionary steps through the larval stages that involved paedomorphosis (✱) and eventually produced chordates. (Notice that Garstang's phylogeny is mistaken about Urochordate and Cephalochordate placement.)

Chordate Origins and Phylogeny

From the early to late twentieth century, most vertebrate biologists adopted Garstang's auricularian hypothesis, especially the part deriving cephalochordates and vertebrates from urochordate larvae via paedomorphosis. However, this hypothesis has recently fallen on hard times. Unlike many other earlier suggestions for chordate origins, the auricularian hypothesis has the advantage that it makes specific scientific predictions that can be, and have been, tested. In fact, it is more of a scenario, a set of specific predictions or hypotheses together composing a larger integrated view of chordate origins.

Perhaps because it has many parts it has become an easy target. For example, many structures simply appear, as if from thin air (e.g., notochord, segmental muscles). Most problematic is newer information from gene expressions and phylogenies based on gene sequences. The difficulty is to reconcile the auricularian scenario with this more recent molecular phylogeny of chordate origins.

Perhaps most challenging to the aricularian hypothesis is reconciling it with modern developmental mapping. In chordates, genes involved in patterning the body axis are expressed the whole length of the nerve cord. But, in dipleurula-type larvae, the expression of these same genes is mainly restricted to the head region only. This implies that the dipleurula larva is a head without the rest of the body! Obviously this is a problem for the auricularian hypothesis, which assumes it had an equivalent to the chordate trunk and tail. Perhaps the expression territory (domain) over which these dipleurula genes preside could have enlarged to produce a full chordate pattern, and thereby these dipleurula genes could have given rise to the entire chordate body plan. But, this is piling speculation upon speculation, and new evidence of body inversion makes the auricularian hypothesis even less applicable to chordate origins.

Recall that earlier in this chapter, I mentioned that chordates appear to be dorsoventrally inverted relative to all other bilaterians, including echinoderms and hemichordates. Further, cephalochordates, not urochordates, are the most basal group of living chordates. These ideas are summarized in figure 2.32, and expanded in figure 2.33.

Dorsoventral Inversion

Chordate Origins and Phylogeny In bilaterians, two sets of genes, working through proteins they produce, specify the dorsoventral axis of the body (figure 2.34). The conventional and ancestral dorsal surface is determined during embryonic development by the signaling protein BMP (bone morphogenetic protein), and the ventral surface by the protein chordin. However, in chordates this is reversed—BMP is expressed ventrally, and chordin is expressed dorsally. This could only happen if ancestral ventral became later chordate dorsal (and vice versa). Anatomical landmarks help us track these changes. Notice that in hemichordates, pharyngeal slits open dorsally; but in chordates, they open ventrally (figure 2.34). BMP and chordin have a gradient and antagonistic effects on dorsoventral patterning. Curiously, the mouth is an exception, opening ventrally under both conditions before and after invention. It may have migrated during the inversion or a new mouth formed ventrally in chordates.

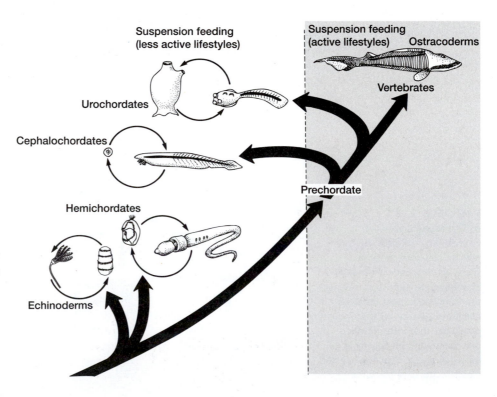

FIGURE 2.32 **Lifestyles, prechordate to chordates.** Pharyngeal slits are present early within protochordates. Upon reaching the prechordate, other chordate characteristics would be present—notochord, postanal tail, dorsal hollow nerve cord—all serving a more active lifestyle. One view, shown here, is that this prechordate was a suspension-feeder, although an active one, thereby accounting for the presence of its basic chordate features. The other view is that this prechordate was an incipient predator; these predatory features were enhanced into vertebrates, but reversed in cephalochordates and urochordates secondarily returning them to suspension-feeding habits.

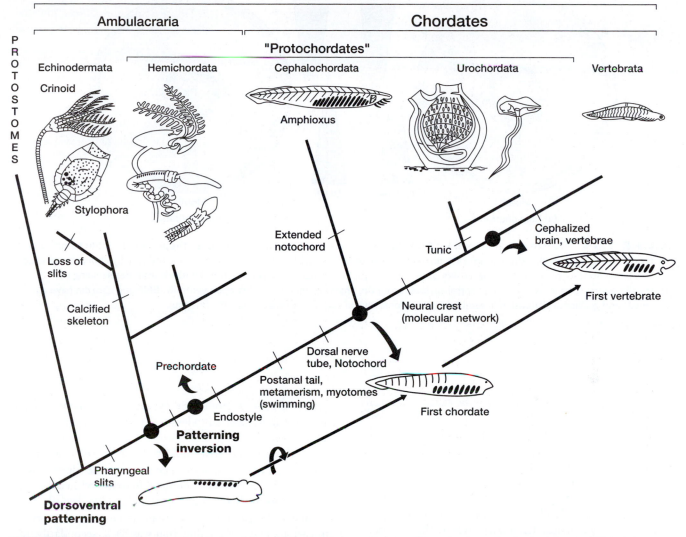

FIGURE 2.33 **Phylogenetic relationships within the deuterostomes.** Note that between the Ambulacraria (Echinodermata + Hemichordata) and Cephalochordata a body inversion occurs, reversing the dorsoventral axis. Other major changes in character states are shown along the way.

Based on Mallatt, 2009.

Although reminiscent of Saint-Hillaire's nineteenth-century idea of deriving chordates from an inverted protostome (annelid/arthropod), these molecular data do not support such a direct derivation. Instead, the inversion occurred within the deuterostomes.

This inversion helps clear up some curiosities and anomalies of chordate features that are just the inverse of hemichordate features. For example, in hemichordates (enteropneusts), pharynx particles trapped in mucus are transported from dorsal to ventral, then to the gut; in chordates by contrast, food-laden mucous sheets are transported from ventral to dorsal pharynx, then to the gut. Blood flows forward in the dorsal vessel, backward in the ventral vessel (hemichordates); but in chordates, blood flows backward in the dorsal vessel, forward in the main ventral vessel. The body musculature is best developed dorsally in hemichordates, but

best developed ventrally in chordates. Inverting the body, at the point shown in figure 2.33, reconciles these differences and helps to explain why they exist. Further, note that the mouth of the young amphioxus resides asymmetrically on the left side (figure 2.20b), as if this early larval stage captures this ancient phylogenetic inversion in progress.

Why prechordates flipped over is not known. One view is that the prechordate worms lived vertically in burrows, making dorsoventral orientation less important. A second view is that these prechordate ancestors, when they began swimming in the water column, swam comfortably on their backs, belly-up, much as fairy shrimps and some other animals do today. A third view is that these prechordate worms did not differ much in their dorsal and ventral surfaces, so it was of little consequence how they lay on the ocean floor, whether on bellies, backs, or sides.

FIGURE 2.34 **Dorsoventral inversion: Chordate body is a hemichordate body plan flipped over.** (a) In hemichordates (and in protostomes generally), the anatomically dorsal side is determined by BMP expression and the ventral side by chordin expression. From the hemichordate pharynx, pharyngeal slits open dorsally. (b) In chordates, the reverse holds with chordin determining the anatomically dorsal surface and BMP the ventral surface. From the pharynx, pharyngeal slits open ventrally. BMP and chordin have an antagonistic and gradient relationship in establishing the dorsoventral axis. P, Pharyngeal slits; Px, Pharynx.

After Lowe et al., 2006.

Dorsoventral inversion must still be taken tentatively. The developmental evidence comes from only a few species of arthropods and chordates, one annelid, and one hemichordate. The picture may become much more complicated when more species are investigated (the reason my colleague Billie Swalla warns me to be more cautious).

Whatever the story on the inversion, we can take advantage of the current molecular phylogeny to now map the steps in the origin of chordates. Chordates arose within the deuterostome radiation (figure 2.33) before the Cambrian Period. These early prechordates were likely mobile, bottom-dwelling worms similar perhaps to enteropneust hemichordates. Pharyngeal slits debuted here as aids in the ciliary and mucous system of feeding, taking advantage of thick mats of bacteria and other microorganisms covering the ocean bottoms characteristic of the time. Other chordate characteristics were to follow likely in service to locomotion that was becoming more active. An increase in body size or availability of larger food items may have favored such a change in lifestyle and locomotion. Accompanying more active locomotion was muscle segmentation (myomeres), along with an elastic but anticompressive rod (notochord) to prevent telescoping of the body, and the tail extension to aid propulsion. The nervous tissue serving the segmental myomeres became consolidated for more effective control of contraction (dorsal, hollow nerve cord). The summary consequence of these collective changes was to produce a true chordate (postanal tail), roughly resembling amphioxus in body form. Although subsequent evolution to vertebrates continued to emphasize a progressively more

active lifestyle, many surviving today, such as amphioxus and ascidians, reverted to less active or even sessile lifestyles as adults.

Two general views hold today as to the character of this first chordate. On the one hand, some view it as an incipient predator with a better differentiated head, a simple pharynx with few slits, eyes, and a large mouth. Such an active, predaceous prechordate is envisioned to evolve in two directions. One direction was back, secondarily, to a less active filter-feeding system (e.g., cephalochordates, urochordates), the other direction was onward to vertebrates, enhancing predaceous traits.

On the other hand, the more likely first chordate was not an incipient predator but rather a filter feeder, although an active one (figure 2.32). Its active lifestyle would account for the conditions that similarly favored the evolution of basic chordate characteristics for active locomotion, while also explaining the almost identical filter-feeding equipment of cephalochordates and urochordates. Subsequent modifications of the pharynx in particular, which we will examine in chapter 3, represent a later stage, producing predatory vertebrates from such a suspension feeding, first chordate.

Overview

Ongoing phylogenetic research and the availability of new molecular methods provide an improved, although certainly unfinished, view of protochordate evolution (figure 2.33). Vertebrates arise within the deuterostome radiation, part of the chordate clade. The other deuterostome clade includes the echinoderms plus the hemichordates, which are more closely related to each other than to chordates based on shared similarities in larval morphology and molecular features (gene sequences and expression). Some fossil echinoderms preserved a bilateral symmetry, but most, including all living groups, diverged dramatically, becoming pentaradial, losing pharyngeal slits and a distinct neurulated nerve cord. Hemichordates are monophyletic, with pterobranchs arising within the enteropneusts, and exhibit an early chordate characteristic (pharyngeal slits).

Cephalochordates have long been considered close to vertebrates, but molecular phylogeny argues otherwise and now places them further away as basal chordates. However, this gives them a new status among living animals as representing approximate ancestral chordates. Unfortunately, it is complicated to derive an amphioxus-like body from an inverted hemichordate worm. The bodies and nervous systems are very different. The absence of fossils makes the intermediate stages even more difficult to envision.

Urochordates are monophyletic, now the sister group to vertebrates. The neural crest, which will be discussed often in these pages, is an important synapomorphy uniting these groups. However, the larval and adult stages of urochordates are very simplified, having jettisoned segmental muscles (metamerism), nephridia, and some major gene complexes. This loss of a significant part of the urochordate genome means a significant loss of informative molecular data by which they might be compared to and placed within the protochordates and keep this phylogeny unsettled for a while longer. These lost genes are present earlier in cephalochordates and present later in vertebrates, where they are needed to form the basic body plan of vertebrates. Thus, larval tunicates, or any other urochordate, are not themselves likely immediate ancestors to vertebrates. Instead, both urochordates and vertebrates more probably arose from a common ancestor, simplified in urochordates but elaborated in vertebrates.

This phylogenetic view (figure 2.33) suggests that a wormlike ancestor, perhaps similar to an enteropneust worm, evolved into the hemichordates/echinoderms on one side of the deuterostomes and into a chordate on the other. Strictly speaking, this means that chordates did not evolve *from* echinoderms (cf. Garstang) and certainly not from annelids/arthropods (cf. Saint-Hillaire, Patten). Although Garstang worked with a misleading phylogeny, he brought to the study of chordate origins ideas about the mechanisms of change and the adaptive basis of evolutionary change.

Although controversial in its specifics, the origin of chordates lies certainly somewhere among the invertebrates, a transition occurring in remote Proterozoic times. Within the chordates arose the vertebrates, a group of vast diversity that includes some of the most remarkable species of animals ever to grace the land, air, and waters of Earth. Within the early chordates the basic body plan was established: namely, pharyngeal slits, notochord, endostyle or thyroid, dorsal hollow nerve cord, and postanal tail. Feeding depended upon the separation of suspended food particles from the water and involved the pharynx, a specialized area of the gut with walls lined by cilia to conduct the flow of food-bearing water. Mucus coated the pharynx walls to snatch suspended particles. Pharyngeal slits allowed a one-way flow of water. Locomotor equipment included a notochord and segmentally arranged muscles extending from the body into a postanal tail.

Feeding and locomotion were activities that favored these novel and specialized structures in early chordates. Subsequent evolutionary modifications would center around feeding and locomotion and continue to characterize the wealth of adaptations found within the later vertebrates.

WEB LINK

Visit the text website at *www.mhhe.com/kardong7e* for additional study aids including selected references, functional laboratories, and web links to chapter topics.

CHAPTER 3

The Vertebrate Story

INTRODUCTION
Innovations
Vertebral Column
Head
Origin of Vertebrates
Step 1: Prevertebrate
Step 2: Agnathan
Step 3: Gnathostome
Vertebrate Classification

AGNATHANS
Living Agnathans
Myxinoidea
Petromyzoniformes
Early Vertebrate Fossils
Conodonts
Ostracoderms
Pteraspidomorpha
Other Ostracoderms (Osteostracans, Anaspids,
*   Thelodonts)*
Overview of Agnathan Evolution

GNATHOSTOMES
Placodermi
Chondrichthyes
Elasmobranchii—Sharks and Rays
Holocephali—Chimaeras

TELEOSTOMI
Acanthodii
Osteichthyes
Actinopterygii
Sarcopterygii
Overview of Fish Phylogeny

TETRAPODS
Primitive Tetrapods
Labyrinthodonts
Lissamphibia—Modern Amphibians
Urodela (Caudata)
Salientia (Anura)
Gymnophiona (Apoda)
Lepospondyls

AMNIOTES
Stem-Amniotes
Sauropsids
Mesosaurs
Reptilia
Synapsida
Pelycosauria
Therapsida
Mammalia

OVERVIEW

Introduction

The vertebrate story unfolds over a span of more than half a billion years, an unimaginable depth of time (figure 3.1). During this time, some of the largest and most complex animals ever known evolved among the vertebrates. Vertebrates occupy marine, freshwater, terrestrial, and aerial environments, and exhibit a vast array of lifestyles. Like amphioxus and tunicates, vertebrates are proper chordates and possess at some time during their lives all five defining chordate characteristics: notochord, pharyngeal slits, tubular and dorsal nerve tube, postanal tail, and endostyle. The diversity vertebrates enjoy might be attributed to opportunity. They arose at a time when few large predators existed. Their success may be due to their great variety of innovations as well. Two of these innovations—the vertebral column and the cranium—provide names for this major taxon.

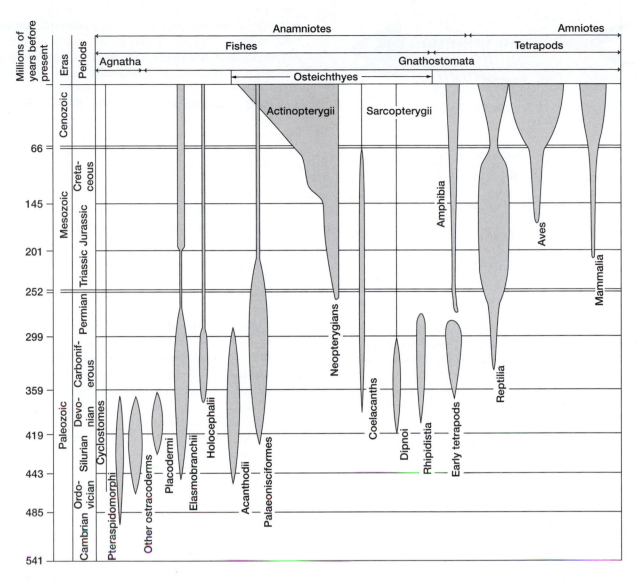

FIGURE 3.1 **Vertebrate diversity.** The vertical scale on the left represents geological time in millions of years before the present. Names of geological eras and periods are listed in conjunction with geological time. Each gray column of the graph begins with the first known fossil traces of the specific group. The varied widths of each column express subjective estimates of the relative abundance and diversity of that particular group through time. The agnathans are the oldest. The Chondrichthyes are represented by two taxa: the elasmobranchs and the holocephalans. The Osteichthyes are represented by two subgroups: the actinopterygians (Palaeonisciformes, neopterygians) and the sarcopterygians (dipnoans, Rhipidistia). The traditional broad groups of vertebrates, indicated across the top of the graph, include Agnatha and Gnathostomata, fishes and tetrapods, and anamniotes and amniotes, which encompass the taxa below them. Conodonts are not indicated, but their presence in the fossil record begins in about the Middle Cambrian and lasts through the Triassic.

Innovations

Vertebral Column

The **vertebral column** inspires the name *vertebrates* and is composed of **vertebrae,** a series of separate bones or cartilage blocks firmly joined as a backbone that defines the major body axis. Squeezed between successive vertebrae are thin compression pads, the **intervertebral disks** or **bodies.** A typical vertebra (figure 3.2) consists of a solid cylindrical body, or **centrum,** that often encloses the notochord, a dorsal **neural arch** enclosing the spinal cord, and a ventral **hemal arch** enclosing blood vessels. Extensions of these arches are **neural** and **hemal spines,** respectively. The earliest vertebrates (*Haikouella, Haikouichthys*) relied upon a strengthened notochord to meet mechanical demands of body support and locomotion. But, these earliest vertebrates apparently also possessed rudimentary vertebrae. In these and other early fishes, the vertebral elements rode

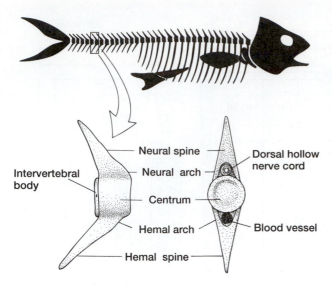

FIGURE 3.2 Basic vertebra. Vertebrae replace the notochord as the predominant means of body support in derived fishes and tetrapods. A typical vertebra usually consists of a single centrum, with a neural arch and a neural spine dorsally and a hemal arch and a hemal spine ventrally. The notochord may become enclosed in the centrum, or more usually is lost. Intervertebral bodies are cartilaginous or fibrous pads that separate vertebrae. In adult mammals, these bodies are called intervertebral disks, which retain gel-like cores that are remnants of the embryonic notochord.

upon or surrounded a notochord that continued to serve as the major structural component of the animal's body. In later fishes and terrestrial vertebrates, successive vertebrae take over the adult functions of body support and movement. As the role of the vertebral column enlarged, that of the notochord in the adult declined. In adults of most advanced vertebrates, the embryonic notochord disappears, although in mammals, it persists only as a small, springy, gel-like core within each intervertebral disk called the **nucleus pulposus.**

Head

The other major innovation that evolved in vertebrates is the **cranium,** or skull, which inspires the other name, *craniates,* for the vertebrates or a vast subgroup of them. The cranium is a composite structure of bone and/or cartilage that supports sensory organs in the head and encases or partially encases the brain. The term **cephalization** refers to the anterior clustering of specialized sensory organs such as paired eyes, ears, nose, and other sensory receptors. The anterior part of the neural tube, which services these sense organs, enlarges to form a distinct brain with bulges, called forebrain, midbrain, and hindbrain. The cranium, including the cephalized nervous tissue, constitutes the head.

Vertebrate evolution has been characterized by a fresh and vast array of head structures. However, there are currently no better diagnostic vertebrate features than the presence of **neural crest cells** and **epidermal placodes,** both embryonic features found so far only in vertebrates. Although these embryonic structures cannot be directly observed in fossils, their adult derivatives can. They give rise to most sensory organs of the head, some parts of the cranium, and to distinctive kinds of teeth. Because these cells are embryonic, they are transient and seldom spring to mind when we think of vertebrate characteristics. But, these special neural crest cells and placodes are the source of most adult structures that distinguish vertebrates from the other chordates.

Neural crest cells (p. 189); formation of the vertebrate head (p. 244)

Origin of Vertebrates

The origin and early evolution of vertebrates took place in marine waters; however, at one time, fossil and physiological evidence seemed to point to a freshwater origin. Many early vertebrate fossils were recovered from what appeared to be freshwater or delta deposits (Ordovician). Some of these earliest fish fossils consisted of fragments of bony armor worn smooth, as if upon death the bodies were washed and tumbled down freshwater streams, eventually coming to rest in the silt and sand that collect in deltas at the mouths of rivers. In the 1930s, the physiologist Homer Smith argued that the vertebrate kidney worked well to rid the body of any osmotic influx of excess water, a problem among freshwater but not among marine animals. However, the discovery of still older fish fossils (Cambrian) confirmed the earliest vertebrates in marine waters. It was shown from this find that the vertebrate kidney, while good at maintaining water balance, need not be interpreted as an innovation of freshwater forms. The kidneys of lobsters and squid work in similar ways, yet these invertebrates and their ancestors have always been marine. Further, the Ordovician sediments first thought to be from fresh water, instead proved to be from shallow, near-shore parts of the sea. Today, few scientists insist that the very first vertebrates were products of freshwater environments.

Kidney physiology and early vertebrate evolution (p. 562)

Evolution of early vertebrates was characterized by increasingly active lifestyles hypothesized to proceed in three major steps. Step 1 comprised a suspension-feeding *prevertebrate* resembling amphioxus. The prevertebrate deployed only cilia to produce the food-bearing current that entered the pharynx. Step 2 comprised an *agnathan,* an early vertebrate lacking jaws but possessing a muscular pump to produce a food-bearing water current. Step 3 comprised a *gnathostome,* a vertebrate with jaws, whose food collection was targeted. This gnathostome fed on larger food items with a muscularized mouth and jaws that rapidly snatched selected prey from the water. These three steps possibly unfolded as follows.

Step 1: Prevertebrate

This prevertebrate arose within the protochordates (see chapter 2). Recall that currently there are two contrasting views of what features characterized this prevertebrate (see figure 2.32). One view is that the prevertebrate was an incipient predator; the other that it was a suspension feeder. Suspension feeding based on ciliary pumps is common to hemichordates, urochordates, and cephalochordates. Although tentative, I shall consequently take the view that the first prevertebrate deployed a similar method of suspension feeding (figure 3.3). As we have just discussed, this would have been a marine form, perhaps very similar to amphioxus, but a nonburrowing, free-swimmer better able to tolerate estuary environments in which rivers enter and mix with the sea. The shift from such a prevertebrate to a vertebrate condition involved two mechanical changes in the pharynx that together produced a muscular pump.

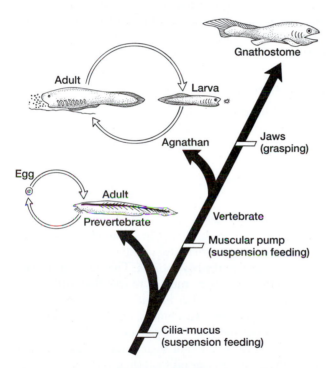

FIGURE 3.3 Origin of vertebrates. A more active, predaceous lifestyle characterized vertebrate evolution, leading away from the suspension feeding that typified vertebrate ancestors. Prevertebrates are envisioned as suspension feeders, perhaps something like amphioxus, but they changed and came to depend on a muscularized pharynx to produce feeding and respiratory currents of water. Following prevertebrates, an agnathan stage developed in which adults might have been benthic feeders, but larvae continued the trend toward a more active lifestyle. Selection and capture of specific prey may have next led to the jawed gnathostomes. Thus, the early trend in vertebrate evolution was from ciliary to muscular mechanisms of moving feeding currents, and then to jaws that directly snatched prey from water.

First, the pharynx developed an encircling band of muscles. Second, strong and springy cartilage replaced the collagen in the pharyngeal bars. Contraction of the muscle bands constricted the pharynx, squeezing water out the pharyngeal slits. Upon muscle relaxation, the cartilaginous supports sprang back to expand the pharynx, restore its original shape, and draw in new water. Initially, this new muscular pump merely supplemented the existing ciliary pumps in moving water through the pharynx. But in larger animals, surface ciliary pumps became less effective in supplying their greater body mass. Increased body size favored prominence of the muscular pump and loss of the ciliary mechanism for moving water. The appearance of an active muscular (and cartilaginous bar) pump now removed the limits to size imposed by a ciliary pump.

Along with its contribution to more-efficient suspension feeding, the muscular pump also addressed the demands of another evolving key vertebrate innovation, gills. Protochordates have pharyngeal slits, but not gills. **Gills** are complex, folded respiratory organs on the pharyngeal pouches, whose folds (lamellae) contain complex blood capillary beds. Gills are bathed by water, laden with suspended food and high in oxygen content that is pumped through the pharynx. Placed in this current of water, gills necessarily increase the resistance to fluid flow through the pharynx. Therefore, besides serving feeding, the strong muscular pump also helped push water across the newly evolving gills, thereby supporting the increased respiratory demands in this active prevertebrate.

By the end of this step, the first vertebrates would have become quite good at actively finding food and swimming away from predators. They would have a cephalized nervous system with eyes, nose, balance organ (ears), and a distinct brain in support.

The New Mouth (p. 509)

Step 2: Agnathan

Appearance of a muscular pharyngeal pump brought early vertebrate evolution to the agnathan stage. The ensuing diversification of these jawless fishes was extensive in its own right and exploited an expanded pharyngeal pump. These agnathans are envisioned to be suspension feeders on an unusually thick soup of particles, or deposit feeders, mud grubbers that pushed their mouths into loose organic or silty mud and drew in sediment rich in organic particles and microorganisms. Although cilia and mucus of the branchial basket still served to collect these passing particles from suspension and transport them to the esophagus, the new muscularized pharynx, not cilia, forced the stream of rich organic material through the mouth. Some fossil agnathans (ostracoderms) possibly created thick suspensions of food by using roughened structures around their mouth to scrape growths of algae from rock surfaces into free suspension that could then be sucked into the mouth by action of the same muscular pharynx.

Step 3: Gnathostome

Prevertebrates, with their ciliary pump, and probably early agnathan vertebrates with their muscular pump, fed on suspended particles. Food-bearing currents carried enough organic material past mucus, where some collided with it and was gathered up. The transition from agnathan to gnathostome involved a switch in feeding method taking advantage of larger particles with more food mass. Transitional species became raptorial feeders that plucked individual food particles selectively from suspension or off surfaces. Some chosen food items would have been wary zooplankton ready to dash off when approached. Other, still larger items would have been particles with significant inertia, such as small worms, that required forceful suction effort to be ingested. Raptorial and suction feeding favored a sudden and forceful expansion of the pharyngeal pump followed by firm mouth closure to prevent escape of captured food. Elastic recoil of the springy cartilaginous bars allowed early jawless vertebrates to produce some suction, drawing food into the mouth; but, this system was too weak to allow forceful capture and ingestion. With the advent of jaws powered by quick muscle action, pharyngeal expansion and suction became strong and active. Muscles serving the anterior pharyngeal bar (near the mouth) became especially large, to open the mouth quickly and close it with a strong bite, securing the "inhaled" prey. The anterior pharyngeal bar enlarged, becoming the grasping jaws, so that limits to prey size were also removed so that even large, squirming prey could be secured. Active predation on large prey became a common lifestyle in subsequent vertebrate radiation.

Vertebrate Classification

The traditional taxonomy divides vertebrates into classes. These classes can be merged into convenient vertebrate groups that share distinctive features. Amphibians, reptiles, birds, and mammals are collectively termed **tetrapods.** Tetrapod literally means four-footed, but the group is understood to include descendants of four-footed ancestors, such as snakes, legless lizards, legless amphibians, flippered marine mammals, and birds, as well as proper **quadrupeds** (four-footed) vertebrates. All other vertebrates are **fishes.** Vertebrates with jaws are **gnathostomes** (meaning "jaw" and "mouth"); fishes without jaws are **agnathans** (meaning "without jaws") that lack such rigid hinged elements supporting the borders of the mouth. Embryos of reptiles birds, and mammals possess a delicate, transparent, saclike membrane, the **amnion,** that encases the embryo in a protective water compartment. Vertebrates producing embryos wrapped in such an amnion are **amniotes**; those without an amnion are **anamniotes** (fishes and amphibians).

Some of these groups are paraphyletic, but still retain informal usefulness. We shall use formal methods to identify natural groups, but students should become comfortable with informal names of simple convenience as well.

Embryonic amnion (p. 190); grades and clades (p. 24)

Agnathans

The vertebrate story begins with agnathans. A mouth is of course present, but these "jawless" fishes lack jaws, a biting apparatus derived from pharyngeal bars (5 branchial arches). Vertebrates have a deep past, debuting within the Early Cambrian explosion of animal types half a billion years ago. Hagfishes and lampreys carry this history of jawless vertebrates into the present. Together, these two living groups are known as **cyclostomes** (meaning "round" and "mouth"). They are often treated as proxies for the most primitive of vertebrates, but they are highly modified, adapted to specialized lifestyles, and therefore depart in many ways from the general ancestral state. From the Late Cambrian, bony shards of carapace attest to the presence of vertebrates and to a bony body. These animals were the **ostracoderms** (meaning "shell" and "skin"), ancient vertebrate fishes encased in bony armor. Finally, we can add remarkable fossil impressions and carbon remains of soft-bodied vertebrates that come from the very dawn of vertebrate origins. Relationships are still controversial and changing frequently with new phylogenetic analyses, but we can certainly summarize analyses now in progress and the extraordinary story unfolding (figure 3.4). Let us begin with the living agnathans.

Living Agnathans

The fossil history of hagfishes and lampreys reaches to the Late Devonian, but most likely living agnathans arose much before this. All living agnathans lack bone and possess a single nostril.

Myxinoidea

The **hagfishes**—deep sea, mud-burrowing, eel-like scavengers—feed on dead or dying invertebrates and other fishes, or prey on invertebrates in the mud (figure 3.5a, b), and are included in the Myxinoidea. They use teethlike processes on their muscular "tongue" to rasp flesh from prey or reel in worms. Slime glands beneath the skin release mucus through surface pores. This mucus, or "slime," may serve to slip them from the grip of a predator or clog its gills. In addition, hagfishes can knot their bodies to escape capture or give them force to tear off food (figure 3.5e).

Ovaries and testes occur in the same individual, but only one kind of gonad is functional; so hagfishes are not practicing hermaphrodites. The eggs are large and yolky, with up to 30 per individual. Development from yolk-filled eggs is direct; that is, without a larval stage or metamorphosis.

Hagfishes possess a single, wide nostril placed terminally, at the anterior end of the head. Water enters this single nasal opening, passes down a nasohypophyseal duct, and over an unpaired nasal sac on its way to the pharynx and gills. The vestibular apparatus or ear is a balancing organ, and includes a single semicircular canal. Adult hagfishes have no trace of vertebrae on or around their notochord but some larval hagfishes do. Vertebra-like elements

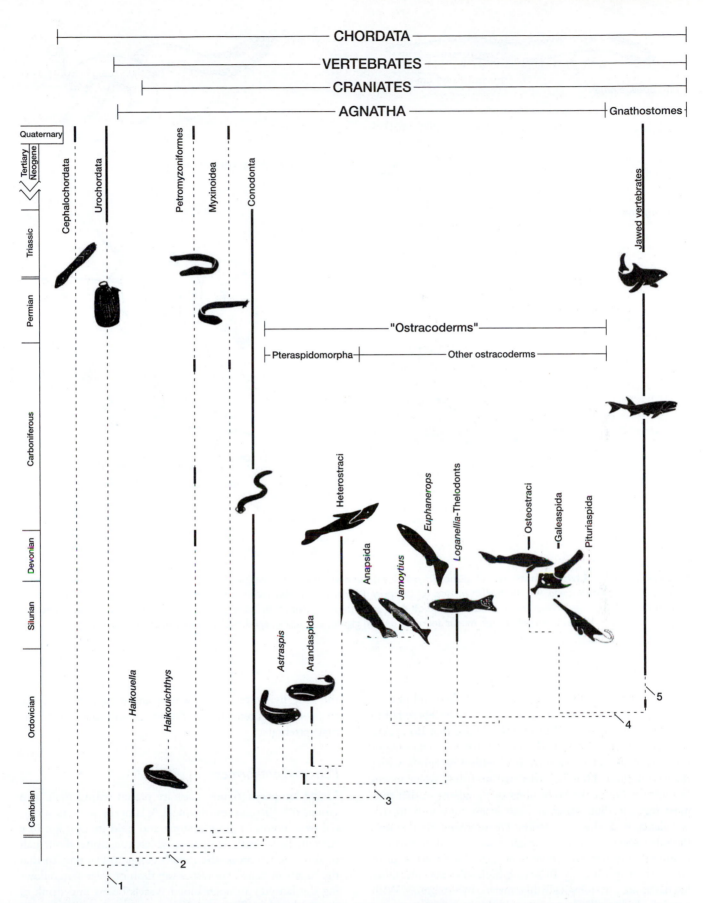

FIGURE 3.4 **Agnathan phylogenetic relationships.** Dotted lines indicate the likely phylogenetic relationships and inferred geologic range. Solid lines show stratigraphic ranges. Dermal bone fragments from the Late Cambrian imply early presence of ostracoderms, probably an unnamed member of the Pteraspidomorpha. Major synapomorphies at nodes: (1) Notochord, tubular and dorsal nerve cord, pharyngeal slits, postanal tail, endostyle (thyroid). (2) Cephalized brain, vertebrae. (3) Extensive dermal skeleton, lateral line system in grooves. (4) Pectoral fins. (5) Jaws, pelvic fins. *Modified from Donoghue, Fore, and Aldridge, with additions based on Janvier and on Mallatt.*

(a) *Bdellostoma*

1 cm

(b) *Myxine*

1 cm

(c) *Petromyzon*

1 cm

(d) *Ammocoetes*

1 cm

(e) "Knotting" behavior in feeding

(f) "Knotting" behavior in escape

FIGURE 3.5   **Living agnathans.** Hagfishes (Myxinoidea), lampreys (Petromyzontiformes), and lamprey larvae (=Ammocoetes). (a) The slime hag *Bdellostoma*. (b) The hagfish *Myxine*. (c) Lamprey, *Petromyzon*. (d) *Ammocoetes*, lamprey larva. (e) "Knotting" behavior. Hagfishes are scavengers. When pulling pieces of food off dead prey, they can twist their bodies into a "knot" that slips forward to help tear pieces free. (f) Knotting, together with mucus secreted by skin glands, also helps hagfishes slip free of an unfriendly grip.

(c) After Dean; (e, f) after Jensen, 1966.

form in the postanal tail of embryos just below and in contact with the notochord. Apparently, actual vertebrae fail to develop in the truck and in the tail. Those that do appear are lost in the adult. Body fluid of hagfishes is also unique. In other vertebrates, seawater is roughly two-thirds saltier than body fluid. Thus, in other marine fishes, water moves osmotically out of the body along this gradient so that they must regulate their water and salt levels constantly to stay in balance with the surrounding environment. By contrast, the salt concentrations in hagfish tissues are similar to surrounding seawater, and there is no net flow of water in or out of the hagfish body. In having high salt concentrations, hagfishes are physiologically like marine invertebrates. With physiological similarity to invertebrates and distinctiveness from other vertebrates, hagfishes have been considered to be

the only living vertebrate whose ancestors never lived in fresh water but remained in salt water since the time of the first vertebrates.

Petromyzontiformes

Living **lampreys** (figure 3.5c) are placed within their own group, the Petromyzontiformes. A lamprey uses its suctorial oval mouth to grasp a stone and hold its position in a current. In parasitic lampreys, which constitute about half of all living lampreys, the mouth clings to live prey so that the rough "tongue" can rasp away flesh or clear skin, allowing the lamprey to open blood vessels below and drink of the fluid within. Some species are marine, but all spawn in fresh water. The marine forms often migrate long distances to reach spawning grounds upriver. During spawning, fertilized eggs are deposited in a prepared nest in loose pebbles. An **ammocoete larva** hatches from an egg (figure 3.5d). Unlike its parents, the ammocoete is a suspension feeder that lies buried in loose sediment on the bottom of freshwater streams with its hooded mouth protruded. Upon metamorphosis, the ammocoete transforms into a an adult. In some species, the larval stage may last up to seven years, at which time metamorphosis yields a nonfeeding adult that reproduces and soon dies.

Medial fins are present, but paired fins and limbs are absent. Vertebrae are represented by individual blocks of cartilage that ride atop the lamprey's prominent notochord. The ear, or vestibular apparatus, includes two semicircular canals. The single medial nasohypophyseal opening, not involved in respiration, is suggestive of a relationship with some ostracoderm groups, as are similarities of the brain and cranial nerves. However, a broader-based analysis of morphological characters establishes their distinctiveness.

Like hagfishes, lampreys lack bone or surface scales. Under experimental manipulation, lampreys exhibit the latent capacity to calcify their endoskeleton. However, hagfishes normally do not produce, nor can they be experimentally manipulated to produce, a mineralized skeleton externally, within the dermis of the skin. Both cyclostomes are assumed to primitively lack bone. We will consider this shortly in our overview of agnathan evolution. But first, let's complete our survey of early vertebrates.

Early Vertebrate Fossils

The earliest vertebrates lacked mineralized tissues. Consequently, early vertebrate history can be documented only by fossils forming under very unusual and favorable conditions where the carbon stamp of soft tissues is preserved. All come from the early Cambrian of China (Chengjiang).

Haikouella (figure 3.6a), and a very similar animal called *Yunnanozoon*, lived near the beginning of the Cambrian period, part of the main burst of the Cambrian explosion, and may lie at the very base of the vertebrate radiation. It has an amphioxus-like atrium and atriopore associated with water flow through the pharynx, along with all distinctive chordate features. In addition, it has characteristics associated with vertebrates—vertebrae ("protovertebrae" by some paleontologists), pharyngeal bars with attached gill filaments, a dorsal nerve cord with a relatively large brain, a head with possible lateral eyes, and a ventrally situated buccal cavity with short tentacles around the mouth. Its characteristics (upper lip like ammocoetes, gills, cephalization of anterior sensory organs), and early position (Cambrian) also satisfy the predictions of vertebrate origins (figure 3.3). It lacked a skull and ear capsule, and myomeres were straight, rather than V-shaped.

Two other amazing finds from China are clearly early vertebrates. One is *Myllokunmingia,* the other *Haikouichthys,* both from the early Cambrian. Both lacked bone, but

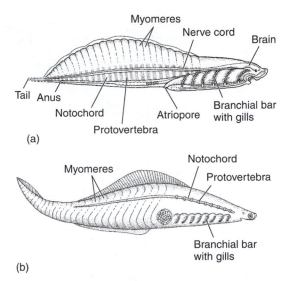

FIGURE 3.6 **Early vertebrate fossils.** (a) *Haikouella* from the early Cambrian. Note the gills, brain, and other vertebrate features; 2.75 cm in length. (b) *Haikouichthys* from early Cambrian, 2.75 cm in length.

(a) After reconstruction kindly supplied by Mallatt and Chen; (b) after reconstruction kindly supplied by Shu.

simple skull elements are present such as ear, nose (and perhaps eye) capsules. However, both were equipped with gill bars, typical V-shaped myomeres, a heart, a head, and possible vertebrae, as well as the distinctive chordate features (notochord, pharyngeal slits, postanal tail). If not the same, *Myllokunmingia* and *Haikouichthys* were at least very close taxonomically. Compared to *Haikouella*, both had more derived features such as large eyes, V-shaped myomeres, and an ear evident in at least one fossil (figure 3.6b).

Conodonts

For almost a century and a half, toothlike microfossils known as conodont elements have been important index fossils in many geological studies. Although extremely common in rocks from the late Cambrian to the end of the Triassic, the organism owning this assortment of pointed and comblike phosphatic elements was not known, leading to speculation that it might be a mollusc, or chordate, or even a part of aquatic plants. This mystery was solved in the early 1980s with the discovery of fossils of a laterally compressed, slender, soft-bodied animal bearing a complete set of conodont elements in its pharynx. But there was much more than this. These fossils bore evidence that conodonts were in fact vertebrates. The trunk exhibited evidence of a series of V-shaped myomeres, a notochord down the midline, and caudal fin rays on what could be interpreted to be a postanal tail. Above the notochord was a streak consistent with the interpretation of a dorsal nerve cord (figure 3.7). Some favorable fossils show evidence of big eyes and an otic capsule. One shows pharyngeal slits. Histological examination

FIGURE 3.7 **Conodont.** (a) Whole, restored animal. (b) Cross section through pharynx showing position of P elements. (c) Isolated conodont apparatus, showing P, S, and M elements.

After Alldridge and Purnell, 1996.

of conodont elements suggested the presence of mineralized dental tissues known from vertebrates, such as cellular bone, calcium phosphate crystals, calcified cartilage, enamel, and dentin. Dentin is laid down by odontoblast cells, embryonic derivatives of ectomesoderm, thereby providing indirect evidence for the presence of neural crest, a typical vertebrate tissue.

The conodont apparatus was thought by some to be a filter-feeding system serving animals with relatively small bodies, 3–10 cm long, although some reached perhaps 30 cm or more. However, evidence of wear on some of the elements suggests instead that the bladelike posterior (P) elements were used to slice and crush food, quite unlike the endostyle-based, filter-feeding system of protochordates. A recent fossil, wherein these elements were preserved in their natural positions within the pharyngeal floor, implies that the anterior S and M elements (figure 3.7) were attached to tonguelike or cartilaginous plates that moved out and in of the mouth catching and delivering, respectively, impaled food. This unusual feeding apparatus, locomotor system (notochord, myotomes), and relatively large eyes moved by extrinsic muscles suggest further that conodonts selected and fed on larger food particles—prey, not suspended material—within the marine waters where they lived and swam.

In some ways, conodonts differ significantly from vertebrates and remain puzzling. Dental elements of the conodont apparatus show evidence of replacement, but worn or broken dental elements also exhibit evidence of growth by bouts of renewed layering (=appositional growth). The conodont apparatus was a very specialized structure (figure 3.7). If carried on a tonguelike structure, it would function similar to the lingual feeding mechanism of hagfish.

Ostracoderms

Following the conodonts, other groups of agnathans appeared in the very Late Cambrian and enjoyed their greatest radiation in the Silurian and Early Devonian (figure 3.8). Like the conodonts, they had complex eye muscles and dentin-like tissues. A few possessed paired appendages. They were

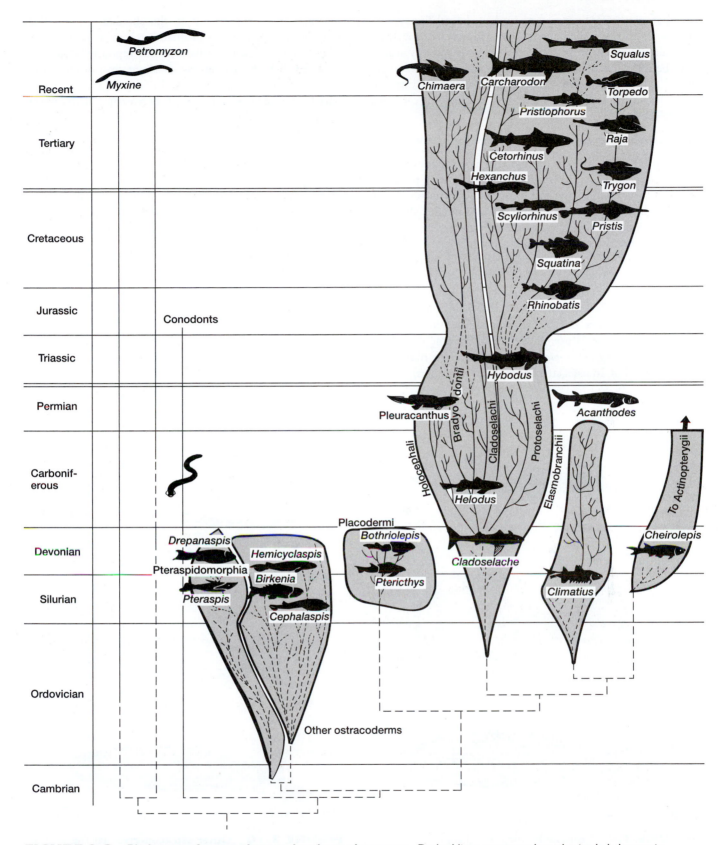

FIGURE 3.8 **Phylogeny of euagnathans and early gnathostomes.** Dashed lines represent hypothesized phylogenetic associations where fossil intermediates are unknown.

After J. Z. Young.

the first vertebrates to possess an intricate lateral-line system and bone, although the bone is located almost exclusively in the outer exoskeleton that encases the body in bony armor just beneath the epidermis, in the dermis (hence, dermal bone). The endoskeleton of these later agnathans was not well developed and, when present, was usually of cartilage inside the body.

Comparison of exoskeleton and endoskeleton (chapter 7)

Most ostracoderms (figures 3.9 and 3.10) were minnow-sized, not more than a few centimeters in length. Bony plates of the head were often large and fused into a composite **head shield.** Plates on the trunk were typically smaller, allowing lateral flexibility for swimming. Beneath the superficial bony plates, an endoskeleton of bone is seldom evident in fossils, which suggests that the vertebral column, if present, was cartilaginous or that axial support was provided by a notochord. Spines and lobes projecting from the armored bodies of many ostracoderms probably offered them some protection from predators and perhaps contributed to their stability as they moved through the water. In some thelodont ostracoderms (e.g., *Phlebolepis*), a lateral pair of ribbonlike fins was present. In osteostracans (e.g., *Hemicyclaspis*), paired muscular fins were evident in the shoulder region. They resemble the paired fins of gnathostomes in position and vasculature, but their internal anatomy is poorly known.

Small body size, absent or slight fins, heavy dermal armor, dorsoventral flattening, and, of course, absence of jaws, have led to the view that most of these early agnathans were slow, poor swimmers, and bottom dwellers that may have extracted suspended food from organic sediments. Natural groupings of ostracoderms are still being worked out. One group in particular, the thelodonts, remains especially difficult to place. This taxonomic uncertainty is largely due to the scarcity of characters from fossil material, especially internal characters. Some of the controversy also results from paraphyletic assemblages, convenient but unnatural assortments of similar-looking fossils. Generally, most ostracoderms fall into several distinct clades (pteraspidomorphs, osteostracans, anaspids) and into a scattered assemblage (thelodonts).

(a) Galeaspid

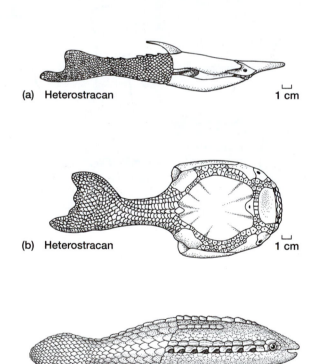

(a)   Heterostracan

(b)   Heterostracan

(c)   Astraspian

(b) Osteostracan

(c) Anaspid

(d) Thelodont

FIGURE 3.9 **Pteraspidomorphs.** All are extinct fishes of the early Paleozoic, with plates of bony armor that developed in the head. (a) The heterostracan *Pteraspis.* (b) The heterostracan *Drepanaspis.* (c) *Astraspis* from North America.

(a) After Gross; (b) after White; (c) after Elliott.

FIGURE 3.10 **Other ostracoderms.** (a) The galeaspid *Yunnanogaleaspis,* for which only the head shield is known. (b) The osteostracan *Hemicyclaspis.* (c) The anaspid *Pharyngolepis.* (d) The thelodont *Phlebolepis.*

(a) After Pan and Wang; (b, c) after Moy-Thomas and Miles; (d) after Ritchie.

Phylogenetic relationships are based on comparisons of shared derived traits. But different sets of traits can produce contradictory hypotheses of relationship. For example, the relationships among living hagfishes, lampreys, and jawed vertebrates are still debated because phylogenies built on morphological (and physiological) traits contradict those built on molecular traits—DNA and RNA sequences. Morphological data suggest that cyclostomes are not a monophyletic group, but instead that hagfishes are more basal and lampreys alone are the sister group of jawed vertebrates (e.g., Gess et al., 2006). However, extensive gene-sequence data suggest otherwise, namely that hagfishes and lampreys form a natural monophyletic group, which is the living sister group to jawed vertebrates (e.g., Mallatt and Winchell, 2007). Sometimes combining data sets (morphological plus molecular) can help, but other times, as with cyclostomes and gnathostomes, a large data set (e.g., molecular) can dramatically swamp the effect of otherwise revealing information in a small data set (e.g., morphological) thereby artificially altering the phylogenetic interpretation (e.g., Near, 2009). Sometimes fossils can help resolve such conflicts by revealing ancient suites of traits that favor one view over another. Unfortunately, when the earliest lampreys turn up in the Devonian, they are already very similar to modern lampreys. That is not surprising because cyclostomes likely diverged from other vertebrates much earlier.

Pteraspidomorphi

The pteraspidomorphs appear in the Ordovician (possibly Late Cambrian), although they are represented at first only by splinters of primitive bone lacking true bone cells (acellular bone). These bone fragments have been recovered from benthic sediments associated with marine invertebrates. The group extends into the Late Devonian, where more complete fossils have been found. Although some species are incomplete, a vestibular apparatus with two semicircular canals and the presence of paired nasal openings seem to characterize most pteraspidomorphs.

Most pteraspidomorphs had head shields formed by fusion of several large bony plates (figure 3.9a–c). Behind the head shield, the exoskeleton was composed of small plates and scales. Occasionally, lateral and dorsal spines projected from this shield. No pteraspidomorph is known to have paired fins.

Other Ostracoderms (Osteostracans, Anaspids, Thelodonts)

Body shape of these ostracoderms is quite varied, suggesting varied lifestyles (figure 3.10a–d). In osteostracans and anaspids, the single nasal opening merges with a single opening of the hypophysis (pituitary gland) on top of the head to form the common nasohypophyseal opening, often keyhole-shaped. The fossil record of this group extends from the Late Ordovician to the Late Devonian.

Nasal sacs (p. 676)

One distinct group, the osteostracans, possessed a vestibular apparatus consisting of two semicircular canals. They were heavily armored with bony plates (figure 3.10a, b) that formed a head shield and smaller scales that covered the rest of the body. Bodies were either **fusiform** (spindle-shaped) or flattened. In some, lobes projected back from the edges of the head shield now thought to be homologous to the paired pectoral (shoulder) fins of gnathostomes and, like them, may have conferred some degree of stability during active swimming (figure 3.10b). Anaspids, another ostracoderm group appearing late in the Silurian, showed a small head shield, increased flexibility of the body armor, and a tail that is hypocercal (extended ventral lobe), all of which suggest a trend toward more open-water swimming (figure 3.10c). Some anaspids and anaspid-like forms resembled lampreys in important ways, as we will see later. Thelodonts are perhaps their own ostracoderm clade or a scattered ensemble of smaller, primitive groups. Their exoskeleton is composed entirely of tiny scales, eyes are small, gill slits are ventrally placed, and the large tail is fork-shaped (figure 3.10d). They lived in shallow-marine environments.

Fish tail types (p. 307)

Overview of Agnathan Evolution

The detailed evolutionary history of agnathans is still contentious, but the recovery of extraordinary new fossils has at least clarified early events and narrowed controversies. Discovery of remarkably preserved *Haikouella* and *Haikouichthys*, plus closely related forms, furnishes critical missing information about the very earliest of the vertebrates. The soft-bodied Cambrian fossils, *Haikouella* and *Haikouichthys*, occupy basal positions within vertebrates (figure 3.4). These recent fossil finds of soft-bodied vertebrates place vertebrate origins in the Cambrian explosion over a half-billion years ago. Conodonts, known only from toothlike microfossils a few decades ago, now are restored on the basis of soft-tissue fossil impressions, which supply a remarkably detailed picture of these tiny, eel-like animals. New species of ostracoderms have been recovered as well.

Based upon current molecular studies, living hagfishes and lampreys group together into the cyclostomes. But where cyclostomes stand within the rest of the agnathans is less well decided. Usually, cyclostomes are placed basally, or nearly so, within the agnathans, as we have done here (figures 3.4 and 3.8). With such a phylogenetic placement, absence of mineralized skeletal tissues in hagfishes and lampreys would be primary, and such tissues evolved after them in conodonts. Note that conodonts, largely because of their mineralized feeding apparatus, are more derived than hagfishes or lampreys. But, the absence of such mineralized tissues and the possession of other simple (reduced?) morphological features in hagfishes and lampreys is what also drives them low on the agnathan phylogeny. Consequently, some biologists hold out the possibility of cyclostomes arising later, perhaps derivatives of ostracoderms, such as anaspids or *Jamoytius*. If derived from ostracoderms, then the absence of mineralized tissues in cyclostomes is a secondary trait.

Whichever of these placements of cyclostomes proves correct, bone nevertheless first evolved in ostracoderms, a paraphyletic group, basal to the gnathostomes. The ostracoderms may prove to be grades of diversification.

Gnathostomes

One of the most significant changes in early vertebrate evolution was the development of jaws in primitive fishes. These were prey-grasping and biting devices derived from anterior pharyngeal arches. Two early groups of jawed fishes are known. The acanthodians appeared first in the Early Silurian, although fragmentary evidence may date to the middle Ordovician about 30–70 million years after the appearance of the first ostracoderms. A second group, the placoderms, is known first from the Early Silurian (figure 3.11). Jaws could

grasp, bite, or crush prey, allowing these fishes to capture and process larger food. This adaptation opened up an expanded predatory way of life.

Early gnathostomes also possessed two sets of **paired fins.** One set, **pectoral fins,** was placed anteriorly; the other set, **pelvic fins,** was located posteriorly. Both pairs articulated with supportive bony or cartilaginous girdles within the body wall. Supported on girdles and controlled by specialized musculature, paired fins conferred stability and control, allowing a swimming animal to maneuver within and prowl its marine environment actively. Compared to the ostracoderms that preceded them, early gnathostomes probably enjoyed more active lives, venturing into new habitats in search of food, breeding sites, retreats, and unexploited resources.

Overall, this radiation of gnathostomes proceeded along two major lines of evolution—one produced the **Chondrichthyes,** the other the **Teleostomi** (figure 3.11). The chondrichthyans include the sharks and their allies. The teleostomians include the bony fishes, the most diverse of any vertebrate group, and within them emerge the tetrapods. Phylogenetically basal to these two major lines of gnathostomes are the placoderms, with which we will begin our survey.

Placodermi

Fossil **placoderms** (meaning "plate" and "skin") date from the Early Silurian, but they flourished in the Devonian. Primitive placoderms were similar to earlier ostracoderms in some ways. Most were encased in a heavy bony armor, the tail was small, and the head shield was composed of large, fused plates of dermal bone (figure 3.12a–g). The dermal bone, forming the external encasing skeleton, lacked dental

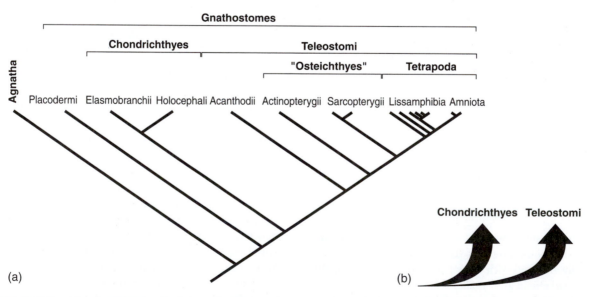

(a)

(b)

FIGURE 3.11 **Gnathostomes, phylogenetic relationships.** (a) Phylogeny of major gnathostome groups. (b) Note that gnathostomes, above the placoderms, evolved along two major lines—the Chondrichthyes and the Teleostomi. Quotes on Osteichthyes reminds that here the term is paraphyletic.

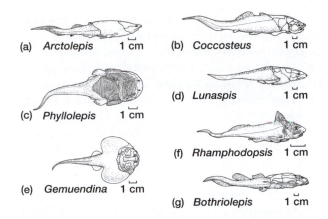

(a) *Arctolepis*   1 cm

(b) *Coccosteus*   1 cm

(c) *Phyllolepis*   1 cm

(d) *Lunaspis*   1 cm

(e) *Gemuendina*   1 cm

(f) *Rhamphodopsis*   1 cm

(g) *Bothriolepis*   1 cm

FIGURE 3.12 **Placoderms.** Most placoderms possessed a dermal armor composed of bony plates on the head and thorax that were broken up into small scales on the midbody and tail. Many placoderms were large and most were active predators. (a) The arthrodire *Arctolepis*. (b) The arthrodire *Coccosteus*. (c) The phyllolepid *Phyllolepis*. (d) The petalichthyid *Lunaspis*. (e) The rhenanid *Gemuendina*. (f) The ptyctodontid *Rhamphodopsis*. (g) The antiarch *Bothriolepis*.

After Stensiö, 1969.

tissues such as denticles found in the skin of chondrichthyan fishes. Unlike ostracoderms, all placoderms had jaws carrying bony projections and rimmed in some advanced placoderms with teeth. Consequently, placoderms were not limited to a diet of suspended organic particles, but now with predaceous jaws and teeth they were able to exploit larger food or bite big chunks out of unwary prey. A bony thorax shield articulated with a bony head shield helps distinguish placoderms as a clade. Paired pectoral and pelvic fins were present. A prominent notochord that supplied longitudinal support to the body was often accompanied by ossified neural and hemal arches. Although true centra were absent, neural and hemal arches (in the region just behind the head) were often fused into a sturdy composite bone called the synarcual. This provided a fulcrum with which the braincase articulated and may have facilitated raising of the head. Other than this synarcual, endochondral bone was generally absent.

Placoderms are a diverse assemblage of generally heavily armored fish, perhaps even polyphyletic in origin. Some were hand-sized, other species were 30 feet (10 meters) in length, and as a group had a nearly worldwide distribution. Without any living counterparts, it is difficult to interpret the lifestyle of such fish encased in bony shields. They are usually depicted as bottom feeders. Most had a flattened body form. Together with heavy armor and slight paired fins, such a body form suggests a benthic life. Although most placoderms were benthic, some had reduced and lightened bony armor along the body. In addition, large size, strong jaws, sleek bodies, and strengthened axial column suggest that some placoderms had an active and predaceous lifestyle.

Placoderms radiated along several lines. Some were adapted to the open ocean, whereas others spread from marine environments, in which they arose, to fresh water. Some were specialized bottom dwellers such as the rhenanids, which were skate- or raylike forms. The more robust arthrodires enjoyed pelagic lives, cruising in pursuit of food. Some tapered forms, such as the ptyctodontids, resembled modern chimaeras (ratfishes). Ptyctodontid males usually possessed a set of **pelvic claspers,** which were specialized pelvic fins associated with the practice of internal fertilization. Further evidence of reproductive habits comes from a small (25 cm) female ptyctodont fossilized giving birth to a live young with umbilical cord intact, testifying to internal fertilization and to live birth (vivipary)

Their varied lifestyles produced varied forms, resulting in a diversity that often makes it difficult to place placoderms in a settled phylogenetic sequence, and even questioning whether they constitute a unified group. They appear at about the time when intermediates between ostracoderms and modern gnathostomes might be expected, but placoderms are too specialized to be such direct intermediates. Placoderms dominated the Devonian seas but died out abruptly at its end to be replaced in the Early Carboniferous by the ascending Chondrichthyes (cartilaginous fishes) and Osteichthyes (bony fishes). No living fishes carry extensive plates of external bony armor similar to that of placoderms, so it is difficult even to understand the mechanical or physiological advantages such armored bodies might have enjoyed. Today, most view placoderms as a more or less natural but specialized group that underwent early, extensive diversification, then vanished. They are without any living descendants and are not even closely related to the cartilaginous or bony fishes that replaced them. This gives placoderms the distinction of being the only major group of jawed vertebrates to ever become entirely extinct without descendants.

Chondrichthyes

The modern chondrichthyans consist of two groups, the sharks and rays (elasmobranchs) and the chimaeras (holocephalans) (figure 3.13a, b). Some systematists suggest that each group arose independently, but anatomical and molecular evidence argues otherwise. For example, both groups have similar fin structures, cartilaginous skeleton (especially prismatic), and pelvic claspers (in males); primitive members show similarities in serial replacement of teeth.

The placoid scales of chondrichthyans are distinctive in that they are usually tiny, pointed, or cone shaped, and show no signs of growth. Initially they form beneath the skin and erupt to the surface. Such scales first appeared in the Middle Ordovician, thus possibly placing the first Chondrichthyes in this period. However, the first of two major episodes of chondrichthyan radiation began later, in the Early Devonian, and extended throughout the rest of the Paleozoic when they were, for a time, more common than

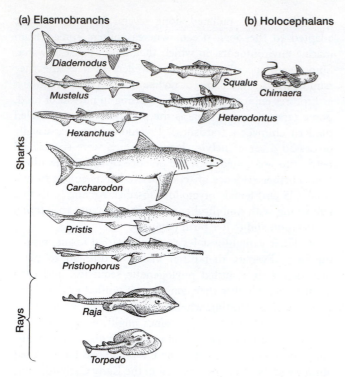

(a) Elasmobranchs   (b) Holocephalans

Diademodus

Squalus

Chimaera

Mustelus

Heterodontus

Hexanchus

Sharks

Carcharodon

Pristis

Pristiophorus

Rays

Raja

Torpedo

FIGURE 3.13 **Chondrichthyans.** (a) Elasmobranchs, including various sharks and rays. (b) Holocephalans.

After J. Z. Young.

bony fishes. The earliest remains, mainly teeth, are found in marine waters, and the group has remained predominantly marine ever since, although one shark order (Xenacanthimorpha) of the Paleozoic was almost exclusively freshwater. Even today sharks can tolerate limited time in fresh water streams and a few modern species make such waters their home. Most chondrichthyans had sleek, fusiform bodies, suggesting that they were active swimmers. The second major episode of radiation began in the Jurassic and extends to the present.

As the name Chondrichthyes (meaning "cartilage" and "fish") suggests, members of this group have skeletons composed predominantly of cartilage impregnated with calcium. However, as we have seen, bone was present in the very earliest agnathan vertebrates; therefore, its near absence in later chondrichthyans must represent a secondary loss. Such a view is supported by traces of bone found in the placoid scales and in the teeth. Bone is also found as a thin veneer on the vertebrae of some modern sharks. A fossil shark from the Permian even had a thick layer of bone around its lower jaw.

Like most fishes, chondrichthyans are denser than water, so they tend to sink. For bottom-dwelling rays, this is not a problem, but open-water swimmers must expend extra effort to counteract this tendency. Large livers holding buoyant oils, an anterior body that acts like a hydrofoil, and heterocercal tails supply lift to help chondrichthyans maintain their place within the vertical water column.

Unlike most bony fishes, cartilaginous fishes produce relatively small numbers of young. Some females lay eggs usually enclosed in a tough, leathery case; others retain their young in the reproductive tract until they are fully developed. Gestation may last a long time. In dogfish sharks, embryos are carried within the uterus for nearly two years, and young are nourished directly from the yolk. However, for some shark embryos retained in the reproductive tract, yolk is supplemented by a nutrient-rich material secreted from the walls of the uterus. In others, a placenta-like association, complete with an umbilical cord, develops between embryo and mother. The pelvic fin of males is often modified into a **pelvic claspers** used to engage the female and aid internal fertilization.

The braincase of chondrichthyans is usually extensive, but without sutures between elements. In the earliest species, the notochord reigned as the major structural member of the axial skeleton, although a few cartilaginous neural spines were strung in a series along its dorsal surface. Modern chondrichthyans have a vertebral column composed mostly of cartilage that largely replaces the notochord as the functional support of the body. The first gill slit is generally reduced and may close before birth; in elasmobranchs it remains open as a small, rounded opening termed a **spiracle.**

Elasmobranchii—Sharks and Rays

Among the cartilaginous fishes, *sharks* occupy the spotlight, likely because most are formidable carnivores, with great whites and bull sharks being extreme examples. Most modern sharks occur in the oceans of the world. Some species frequent great depths along deep oceanic trenches. The sleeper shark has been photographed with remote cameras at depths of over 1,600 meters. Gill slits, usually five to seven, open directly to the exterior. In most sharks, the mouth is armed with serrated, pointed teeth. The functional teeth are backed by rows of replacement teeth, each ready to rotate into position to take the place of a broken or lost functional tooth. This turnover can be rapid. In young sharks, each forward tooth can be replaced weekly.

Tooth replacement (p. 507)

Cladoselache, a 2-m chondrichthyan, was an early shark of the Devonian. As in its modern counterparts, tooth replacement in *Cladoselache* was continuous. Fins were supported by paired girdles, but these girdle halves were not yet united as a single bar reaching across the midline. The dogfish shark, basking shark, and whale shark are examples of modern elasmobranchs. The dogfish shark, a delicacy in restaurants when fresh, and when preserved a frequent companion of many biology students in comparative anatomy classes, seldom exceeds 1 m in length. The basking shark and whale shark reach 10 m and 20 m, respectively, making them, after the baleen whales, the largest living vertebrates. Neither of these sharks, however, is a slashing predator. Instead, both strain food from the water each summer.

The basking shark feeds by swimming forward with the mouth agape. In this way, it strains up to hundreds of pounds of zooplankton, mainly copepods, daily from the water. During winter months, plankton stocks decline in the subpolar and temperate waters. The basking shark is thought to recline on the bottom in deep water during this slow season. Whale sharks feed on plankton all year with rakers on gill bars that are modified into great sieves. When feeding, they approach plankton, usually a school of shrimp-like krill, from below and sweep rapidly upward, engulfing both krill and water at once. Excess water exits through their gill slits, and krill are strained out and swallowed.

All *rays* belong to the Batoidea. Modern rays are bottom-dwelling specialists with a fossil record from the Early Jurassic. The pectoral fins are greatly enlarged and fused to the head to give the flattened body an overall disk-shaped appearance. The tail is reduced, and flapping of the pectoral fins provides propulsion. The teeth are designed to crush prey, mostly molluscs, crustaceans, and small fishes discovered buried in sand. On their whip-shaped tail, sting-rays carry a jagged-edged, sharp spine that they can lash at attackers. Electric rays can even administer severe shocks, generated by modified blocks of muscle, to thwart enemies or stun prey. Manta rays and devil rays, some of the largest members of this group, measure up to 7 m across from fin tip to fin tip. They are often pelagic and gracefully cruise tropical waters in search of plankton that they strain by means of modified gill bars.

Electric organs (p. 373)

Rays have a round spiracle located dorsally and behind the eyes. The spiracle is the primary means by which some rays get water into the mouth and across the gills. Rays rest on their flattened ventral surface, "belly," and have full gill slits ventrally placed and eyes dorsally placed on the body. Rays should not be confused with flatfishes (flounders, sole, halibut). [Flatfishes are bony fish that rest on one side against the substrate, and have gill slits *and* eyes twisted around to the "top" side of the body.]

The terms *skate* and *ray* are used loosely and even interchangeably because there is no natural biological or tax-onomic difference between them. Generally, skates have a **rostrum,** a pointed, noselike extension of the braincase, and produce eggs encased in a leathery case. They are placed in one family, the Rajidae. Most rays lack a rostrum and give birth to live young, but they are placed in several different families. Taxonomically, however, skates are a type of ray and members of the Batoidea.

Holocephali—Chimaeras

Chimaeras (or ratfishes) are modern representatives of the holocephalans and are exclusively marine. The large pec-toral fins primarily handle the job of swimming through strong, sweeping strokes. This apparently accounts for why the tail does not end in an enlarged propulsive caudal fin;

instead, it is long and tapered to a point, inspiring the name *ratfish.* Fossil holocephalans are known from the Late Devonian onward.

Chimaeras differ from sharks in many ways. The upper jaws of chimaeras are firmly fused to the braincase. Their gill openings are not exposed to the surface, but covered exte-riorly by an **operculum.** However, their operculum is an extended flap of skin, not a bony plate as in bony fishes. In adult chimaeras, the small, circular spiracle, derived from the first gill slit, is absent and appears only as a transitory embry-onic structure. Their diet includes seaweed and molluscs that the grinding or crushing plates of their teeth can accommo-date. Scales are absent. In addition to pelvic claspers, males sport a single median hook, the **cephalic clasper,** on their head, which is thought to clench the female during mating.

Today, there are only about 25 species of chimaeras. They spend most of their time in deep waters over 80 m, and they have no commercial value. Because these factors have discouraged study of this group, chimaeras remain poorly known.

Teleostomi

The Teleostomi is a large group embracing the acanthodi-ans (sister group of the bony fishes), the bony fishes, and their tetrapod derivatives (see figure 3.15). Arising within these teleostomes are the teleosts (Teleostei), which today comprise most of the living fishes.

Acanthodii

Acanthodians are represented by spines in the Early Silurian, with some debated evidence that they were pres-ent very late in the Ordovician. They reached peak diversity during the Devonian and persisted well into the Permian,

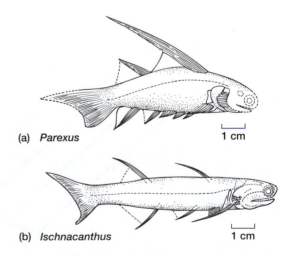

(a) *Parexus*   1 cm

(b) *Ischnacanthus*   1 cm

FIGURE 3.14 **Acanthodians.** Note the spines along the body of each that in life supported a web of skin. (a) *Parexus*, Lower Devonian. (b) *Ischnacanthus*, Lower-Middle Devonian.

After Watson.

long after the placoderms had become extinct. The largest acanthodian was over 2 m in length, but most were minnow sized (under 20 cm) with streamlined bodies. Early acanthodians were marine, but later ones tended to occupy fresh water.

Acanthodii means "spiny forms," a reference to rows of spines along the top and sides of the body. Each fin, except the caudal fin, was defined on its leading edge by a prominent, fixed spine that probably supported a thin web of skin (figure 3.14a, b). In many species, intermediate spines ran between the paired pectoral and pelvic girdles, and in some reduced but unmistakable true bony fin elements were tucked away at the base of at least the pectoral spine. Their vertebral column incorporated an ossified series of neural and hemal arches along a prominent notochord, which extended well into the long dorsal lobe of the tail and served as the major mechanical support for the body. Compared with ostracoderms, the dermal armor was considerably reduced and replaced by many small scales across the surface of the body. Dermal armor occurred on the head, but these bony plates were small and formed no composite unit such as a head shield. In some, gill slits opened separately, as in chondrichthyans, but in others the gill slits were covered externally, at least partially, by a bony operculum.

Acanthodians have been bounced around within gnathostome taxonomy, a reflection of their still uncertain relationship to other primitive jawed fishes. Their early fossil debut and partial exoskeleton invite their comparison with placoderms. On the other hand, a relationship to chondrichthyes is suggested by their subterminal mouth below the snout (in contrast with the terminal mouth of bony fishes), caudal fin with projecting dorsal lobe, nonoverlapping scales, and basic jaw structure. In addition, the sleek shape and partially ossified internal skeleton of acanthodians point to a relationship with osteichthyes, the advanced bony fishes. They are usually placed between cartilaginous and bony fishes, but may be united with bony fishes in the Teleostomi (figure 3.15) in recognition of similarities in the skull.

Osteichthyes

Most living vertebrates are bony fishes, members of the Osteichthyes (figures 3.11, 3.15). Small overlapping scales from the Late Silurian are the first fossil remains known of this group. Osteichthyans are not the only fishes to contain bone in their skeletons, but the taxonomic term *Osteichthyes* (meaning "bone" and "fish") recognizes the pervasive presence of bone, especially throughout the endoskeleton, among members of this class. In early bony fishes, much of the internal skeleton was ossified and surface scales rested upon a foundation of dermal bone. In most later descendants, ossification persisted or progressed in the internal skeleton, although the skull and scales tended toward reduced ossification. This trend toward more complete ossification of the internal skeleton reversed

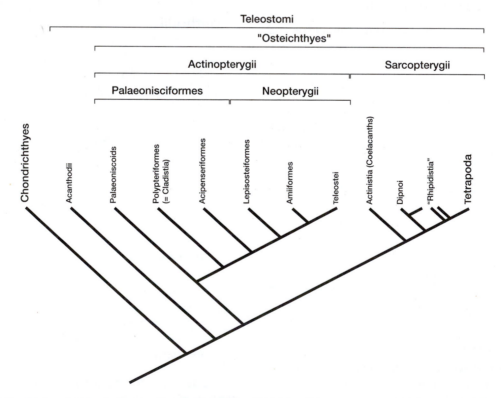

FIGURE 3.15 **Teleostomi, phylogenetic relationships.** "Rhipidistia" in quotes to notice that, as presently constructed, it may be paraphyletic with one lineage related to Dipnoans and another to derivates from other lineages. "Osteichthyes" in quotes to also notice possible paraphyletic association.

only in a few groups, such as sturgeons, paddlefishes, and some later lungfishes, in which endoskeletons are primarily cartilaginous. Whereas cartilaginous fishes address problems of buoyancy with oily livers and hydrofoil fins, most bony fishes possess an adjustable, gas-filled **swim bladder** that provides neutral buoyancy, so they need not struggle to keep from sinking or bobbing to the surface.

Swim bladder and its distribution within fishes (p. 416)
Bony fish scales (p. 218)

No single feature alone distinguishes them from other fishes. Rather, bony fishes have a suite of characteristics, including a bony operculum, a swim bladder, possibly modified from lungs, and an extensive ossification of the endoskeleton. Dermal bones may cover the body, especially in primitive groups, but these are never large and platelike as in ostracoderms or placoderms. Instead, the body is usually covered by overlapping scales. The mouth is terminal, meaning that it resides at the very anterior tip of the body, rather than subterminal as in some other fishes such as sharks. A bony operculum covers the external series of gill slits. Fins are often strengthened by **lepidotrichia,** slender bony rods or "rays" that provide a fanlike internal support.

Bony fishes consist of two groups of quite unequal size. The **actinopterygians** compose the vast majority of bony fishes and have been the dominant group of fishes since the mid-Paleozoic (figure 3.16). The other group of bony fishes is the **sarcopterygians.** Although small in numbers today, this group is important to the vertebrate story because it gave rise to the tetrapods, all land vertebrates, and their descendants.

Actinopterygii

Actinopterygians are called "ray-finned" fishes because of their distinctive fins, which are internally supported by numerous slender, endoskeletal lepidotrichia (rays). Muscles that control fin movements are located *within* the body wall, in contrast to the muscles of sarcopterygians that are located outside the body wall along the projecting fin.

Some fish biologists divide actinopterygians into **chondrosteans, holosteans,** and **teleosts,** each intended to represent primitive, intermediate, and advanced groups of ray-finned fishes, respectively, representing an increase in ossification (figure 3.17). As mentioned earlier, the internal skeleton sees increasing ossification in many groups, but the

FIGURE 3.16 Actinopterygian phylogeny.

After J. Z. Young.

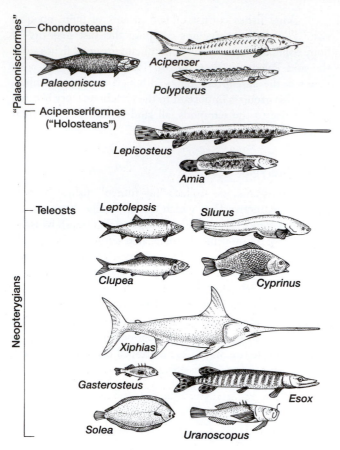

FIGURE 3.17 Representative actinopterygians.

After J. Z. Young.

and acanthodians similar to that of palaeoniscoids. This may reflect a phylogenetic relationship or an early convergence of a successful feeding style based on quick snatching of prey. Palaeoniscoids occupied marine as well as freshwater habitats. They reached their greatest diversity during the Late Paleozoic but were replaced in the early Mesozoic by neopterygians.

Scale types in fishes (p. 218)

Surviving Palaeonisciformes include acipenserids, sturgeons and paddlefish, placed in the Acipenseriformes (= chondrostean), and bichirs placed in the Polypteriformes (= *Cladistia*). In most acipenserids, the first gill slit is reduced to a spiracle; longitudinal support of the body comes from a prominent notochord. In a departure from palaeoniscoids and other primitive bony fishes, acipenserids usually lack ganoid scales except for a few enlarged scales arranged in separate rows along the sides of the body. Reversing a trend toward ossification, the skeleton is almost entirely cartilaginous. Paddlefishes occur in fresh waters in North America and China. They are open-water filter feeders of plankton. Sturgeons, the largest species of freshwater fishes, can reach up to 8 m and 1,400 kg. Some migrate between fresh and marine waters, making treks of over 2,500 kilometers. These toothless bottom feeders eat buried invertebrates, dead fishes, and young fingerlings of other fish species. Some may live to be 100 and do not reach sexual maturity until they are almost 20 years old. Their **roe** (eggs) are sold commercially as Russian caviar. Although once considered a nuisance species, sturgeons are a favored food today, especially smoked. Over 50,000 are harvested annually from the Columbia River in North America alone.

Bichirs (pronounced "beechers") share rhomboidal ganoid scales, similar patterns of skull bones, and a spiracle with other primitive chondrosteans. They inhabit swamps and streams of Africa and include the living genera *Polypterus* and *Erpetoichthys*. They possess a swim bladder that is more like a paired, ventral lung. Species of *Polypterus* will drown if they cannot occasionally inhale fresh gulps of air to replenish the air in their lungs. Their pectoral fins are "fleshy" as well. Because of their paired lungs and fleshy fins, bichirs were formerly classified with lungfishes as sarcopterygians. But today, most see the fleshy pectoral fins as a distinctive feature that evolved independently of the fleshy (lobe) fins of sarcopterygians. Some taxonomists classify bichirs within the bony fishes as *Cladistia* a sister group to Acipensiformes plus Neopterygii.

Neopterygii In the early Mesozoic, neopterygians replaced Palaeonisciformes as the most dominant group of fishes and have flourished ever since. They display a great range of morphologies and have adapted to a variety of habitats in all parts of the world. In the course of neopterygian evolution, changes in the skull accommodated increased jaw mobility during feeding and offered attachment sites for associated feeding musculature. Scales became rounder and thinner.

skull and scales undergo reduced ossification, and in some groups the endoskeleton is even cartilaginous. Teleost is still serviceable, but chondrostean is now used synonymously with acipenseriformes, which we meet on the next page, and holostean may be a paraphyletic group. In our classification scheme, we use two divisions current at the moment: the **Palaeonisciformes,** encompassing primitive ray-finned fishes, and the **Neopterygii,** encompassing derived ones. These two groups are further divided into lower categories (see Appendix D).

Palaeonisciformes The extinct palaeoniscoids are the best known of the primitive Palaeonisciformes and probably the earliest bony fishes. One species reached half a meter in length, but most were smaller. The notochord provided axial support, although ossified neural and hemal arches accompanied it, as the notochord reached well into the extended tail. The fusiform palaeoniscoid body, suggesting an active life, was covered by small, overlapping rhomboidal scales arranged in parallel rows set closely to one another. The base of each scale consisted of bone, the middle was composed of dentin, and the surface was covered with ganoine, an enamel-like substance that gave them their name of **ganoid scales.** Many find the head shape of early sharks

Lungfishes: Dealing with Drought

Many lungfishes live in swamps that dry out on an annual basis. As the water level begins to fall, the lungfish burrows into the still soft mud, forming a bottle-shaped hole into which it curls up (box figure 1a). When the mud dries, mucus secreted by the skin hardens to form the cocoon, a thin lining that resists further water loss within the burrow holding the lungfish (box figure 1b). Usually, the fish's metabolic rate drops as well, thereby curtailing its caloric and oxygen needs. Such a reduced physiological state in response to heat or drought is termed *estivation* (box figure 1c). As long as there is standing water above the burrow, the lungfish occasionally comes to the surface to breathe air through the neck of the burrow. After the surface dries completely, the neck of the burrow remains open to allow direct breathing of air.

Estivation enjoys a long history. Burrows of lungfish from the early Permian and Carboniferous have been discovered. The African lungfish normally estivates for four to six months, the length of the dry summer season, but it can sustain longer

(a)

Water level

(b)   **(c)**

BOX FIGURE 1   **African lungfish during estivation within its burrow.** Reduced metabolism requires only infrequent breathing. The lungfish draws in fresh air through the neck of the burrow that maintains continuity with the environment above. (a) While declining water still covers the swamp, the lungfish burrows into the soft mud, establishes the basic U-shaped burrow, and reaches to the surface to breathe. (b) As the water level drops further, the lungfish moves into a cocoon lined with mucus and maintains contact with the air through breathing holes. (c) In the cocoon, the rolled up lungfish enters an estivative stage during which its metabolic rate drops and its respiratory requirements decrease.

After Grasse.

periods of estivation if forced to do so. The South American lungfish estivates as well, but it does not form a mucous cocoon nor does it fall into such a deep metabolic torpor. Although the Australian lungfish does not estivate, it can use its lungs to breathe air when oxygen levels drop in the water it frequents.

In contrast the thick, overlapping scales of palaeoniscoids afforded protection but restricted flexibility. Reduction of surface scalation probably accompanied development of more active swimming. The notochord was replaced by increasingly ossified vertebrae that also promoted efficient swimming. The asymmetrical heterocercal tail of palaeoniscoids was generally replaced by a symmetrical homocercal tail.

Although primitive neopterygians (previously called "holosteans") had a homocercal tail, internal vestiges remain of a heterocercal ancestry, a spiracle is absent, and scales are reduced. These primitive living neopterygians include gars (Lepisosteiformes), which still retain large, rhomboidal ganoid scales, and bowfins (Amiiformes). Both have more flexible jaws than palaeoniscoids, but less flexible than advanced neopterygians.

The most recent group of ray-finned fishes is the derived neopterygians, or Teleostei (meaning "terminal" and "bony fish"). This very diverse group encompasses close to 20,000 living species that enjoy extensive geographic distribution. Representatives occur from pole to pole and at elevations ranging from alpine lakes to deep-ocean trenches. Teleosts have a long history dating back over 225 million years to the Late Triassic. Nevertheless, it seems to be a monophyletic group. Generally, teleosts share a suite of characteristics, including homocercal tail, circular scales without ganoine, ossified vertebrae, swim bladder to control buoyancy, and skull with complex jaw mobility allowing for rapid capture and manipulation of food.

Some of the more familiar groups of living teleosts include the clupeomorphs (herrings, eels), salmonids (salmon, trout, whitefishes, pikes, smelts), percomorphs (perches, basses, seahorses, sticklebacks, sculpins, halibut), cyprinids (minnows, carp, suckers, pikeminnows), siluroids (catfishes), and atherinomorphs (flying fishes, silversides, grunion).

Sarcopterygii

Sarcopterygians are the second group of bony fishes. Unlike in the ray-finned actinopterygians, the paired fins of sarcopterygians rest at the ends of short, projecting appendages with internal bony elements and soft muscles, hence the name "fleshy-finned fish." Although sarcopterygians never were a diverse group, they are significant because they gave rise to the very first terrestrial vertebrates. The tetrapod limbs evolved from sarcopterygian fins; however, these fins do not support the sarcopterygian body, nor do they serve the fish on land. Instead, fleshy fins are aquatic devices that sarcopterygians seem to use for pivoting or maneuvering in shallow waters or for working bottom habitats in deeper waters.

Sarcopterygians were common in fresh water during most of the Paleozoic, but today the only surviving sarcopterygians are three genera of lungfishes (dipnoans) living in tropical streams and rare coelacanths, confined to the deep waters of the Indian Ocean. To these living groups are added fossils, many newly discovered, that provide a rich picture of this fish group within which evolved the tetrapods. A variety of historical names have attempted to keep pace with our changing characterization of this group. To some, the sarcopterygians were known once as Choanichthyes, in recognition of external nostrils opening internally to the mouth through holes termed **choanae.** Differences in embryonic development have raised doubts about the homology of choanae among fishes, however, and now dampen enthusiasm for this alternative name. Once sarcopterygians were divided into two subgroups, **dipnoans,** and all others combined into the **crossopterygians** (tassle-finned fish). Dipnoi are a monophyletic group, but crossopterygians are now considered to be paraphyletic, including coelacanths (Actinistea) and rhipidistians, which we meet next.

Choanae or internal nares (p. 262)

Other than fleshy fins, primitive sarcopterygians differ from other bony fishes in having scales covered with **cosmine.** These **cosmoid scales,** initially rhomboidal in shape, tend to be reduced to thin, circular disks without cosmine in later sarcopterygians. Early species usually had two dorsal fins and heterocercal tails (figure 3.18a, b). In later species, the dorsal fins usually were reduced, and the caudal fin became symmetrical and **diphycercal,** with the vertebral column extending straight to the end of the tail with equal areas of fin above and below it (figure 3.18c).

Scale types of bony fishes (p. 219)
Fish tail types (p. 307)

Actinistia (Coelacanths) Coelacanths first appeared in the Middle Devonian and survived into the Late Mesozoic, when they were thought to become extinct. The chance recovery of one in the 1930s in the marine waters off the coast of southern Africa provided science with a "living fossil." (See Box Essay, p. 46.) This African fish was *Latimeria,* which inhabits deep oceanic shelves of 100–400 meters. Other populations have been found off the coast of Tanzania, on the east coast of Africa. A second species was discovered in waters around Indonesia, also at depth.

Throughout this group, the braincase is divided by a hingelike joint transversely across the top of the skull. The vertebral centra are tiny but the notochord is especially prominent. Most coelacanths were marine. In living coelacanths, the swim bladder does not serve in respiration but is filled with fat. During the day, they typically rest in small

(a) *Dipterus*　　　　1 cm

(b) *Osteolepis*　　　　1 cm

(c) *Latimeria*　　　　1 cm

FIGURE 3.18 **Sarcopterygians.** (a) *Dipterus,* fossil lungfish of the Devonian. Note the heterocercal tail. (b) *Osteolepis,* a rhipidistian of the Devonian that also had a heterocercal tail. (c) *Latimeria* is a living sarcopterygian (coelacanthiformes) exhibiting a diphycercal tail.

(a, b) After Traquair; (c) after Millot.

groups in volcanic caves along steep slopes. The lobefins help hold and position the fish within currents. A newly discovered and photographed coelacanth occurs in marine waters in Indonesia. It has not been studied in detail, but may represent a new species of coelacanth.

Discovery of living coelacanth (p. 46)
Coelacanth cranial kinesis (p. 264)

Dipnoi The fossil record of lungfishes extends back to the Devonian. *Styloichthys* (Early Devonian), the earliest known lungfish, shared some characteristics with rhipidistians as well, suggesting that it might be a transitional species between rhipidistians and modern lungfishes. All Devonian lungfishes were marine but modern forms occupy fresh water. Three surviving genera occur in continental streams and swamps (figure 3.19a–c). With paired lungs, dipnoans can breathe during periods when oxygen levels in the water fall or when pools of water evaporate during dry seasons. Modern lungfishes lack cosmine, have a skeleton composed mostly of cartilage, and exhibit a prominent notochord.

Rhipidistians date back to the Early Devonian. Although the notochord is still predominant, it is now accompanied by ossified neural and hemal arches as well as by concentric centra that tend to constrict it and supplement its function. During the Late Paleozoic, rhipidistians were the dominant freshwater predators among bony fishes. The rhipidistian braincase had a hingelike joint running transversely across its middle so that the front of the braincase swiveled on the back of the braincase. This ability, together with modifications in skull bones and jaw musculature, represents design changes accompanying a specialized feeding style thought to involve a powerful bite. "Rhipidistians" is a paraphyletic group lumping some lungfish, such as **porolepiforms**, with other sarcopterygians that are stem groups to tetrapods, such as **osteolepiforms** and **panderichthyids**. These stem groups had jaws that carried **labyrinthodont teeth** characterized by complex infolding of a tooth wall around a central pulp cavity. Rhipidistians gave rise to tetrapods during the Devonian but themselves became extinct early in the Permian.

Cranial kinesis (p. 263); labyrinthodont teeth (p. 512)

Erik Jarvik, a Swedish paleontologist, provided important descriptions of *Eusthenopteron*, an osteolepiform (figure 3.20a). Its lobefins and skull place it close to the ancestry of tetrapods. *Panderichthys* (figure 3.20b), known from the Late Devonian (or slightly earlier), possesses lobefins, braincase structure, and an intracranial joint like that of *Eusthenopteron*. But the skull roof of *Panderichthys* is flattened, the parietal bone paired, and eyes moved upward and posteriorly, resembling the condition in the first tetrapods.

Perhaps most remarkable of these transitional sarcopterygians between fishes and tetrapods is the newly described Late Devonian fish *Tiktaalik* about 3 million years younger than *Panderichthys* (figure 3.20c). *Tiktaalik* is an intermediate link between fishes and land vertebrates, exhibiting a body covered in rhombus-shaped, cosmoid bony scales, absence of bony links of the skull to the pectoral girdle as in later terrestrial vertebrates, a flattened skull likely advantageous for quick snaps at prey in shallow water, and loss of the bony gill cover suggesting a change in gill ventilation toward supplementary use of a lung. Its presence in river channel sediments indicates a freshwater lifestyle. In shallow waters, tipping up to breathe air might have been difficult. Instead, when basking at the surface, *Tiktaalik* may simply have used its dorsally positioned spiracle, an opening to the buccal cavity, to draw in air. Large ribs hint at better support

FIGURE 3.19 Sarcopterygians—Living lungfishes. (a) Australian lungfish, *Neoceratodus*. (b) African lungfish, *Protopterus*. (c) South American lungfish, *Lepidosiren*.

FIGURE 3.20 Sarcopterygians— "Rhipidistians." These Devonian fishes are closely related to tetrapods. (a) The osteolepiform *Eusthenopteron* has lobed pectoral and pelvic fins with internal bony supports. (b) The panderichthyid *Panderichthys*, also equipped with lobed pectoral and pelvic fins, has a flattened body, eyes on top of the head, and lacks dorsal and anal fins. (c) The tetrapod-like fish *Tiktaalik* is a remarkable intermediate between other fossil sarcopterygians on the one side and early tetrapods on the other. Note that elements of the shoulder girdle and appendage are present, as in tetrapods, but appendages end not in digits but in fin rays as in other sarcopterygian fish. Covering scales removed; about 1 m in overall length.

(a) After Carroll; (b) after Vorobyeva and Schultz; (c) after Daeschler, Shubin, and Jenkins.

(a)

(b)

(c)

when taking sorties onto land. Its pectoral fins are almost but not quite forelimbs in that they include a robust internal skeleton, but terminate not in digits but in fin rays, like other sarcopterygian fishes.

Fish scale (p. 219)

Overview of Fish Phylogeny

All modern fishes except cyclostomes belong to either the Chondrichthyes or the Osteichthyes. Fishes are diverse in morphology and worldwide in distribution. They outnumber all other vertebrates combined and are one of the most successful groups of animals.

In the earliest fishes, the ostracoderms, bone was already a major part of their external design. In many later groups, there was a tendency for ossification to extend to the internal skeleton, but bone was secondarily reduced or lost in chondrichthyans and some bony fishes, such as Acipenseriformes and lungfishes. Overall two trends characterize the gnathostomes: On the one hand, chondrichthyans lose perichondral bone, bone around cartilage, replaced by prismatic calcified cartilage; on the other hand, osteichthyans tend to gain bone as part of their endoskeleton.

Fishes are major players in the vertebrate story. Within the fish group, jaws and paired fins first appeared. Ray-finned fishes have been the dominant aquatic vertebrates since the mid-Paleozoic. Lobe-finned fishes gave rise to land vertebrates, the tetrapods. In a sense then, the story of tetrapods is a continuation of what began with fishes. We recognize this common lineage within the Teleostomi. Tetrapods inherited paired appendages, jaws, backbones (vertebrae), and lungs from fishes. We celebrate this close relationship by placing land vertebrates as a subgroup of the sarcopterygians (figure 3.15). The demands of terrestrial life and the new opportunities available led to a rather extensive remodeling of the fish design as tetrapods diversified into terrestrial and eventually aerial modes of life. Tetrapod design is the part of the vertebrate story to which we next turn.

Tetrapods

Vertebrates stepped tentatively onto land during the Late Paleozoic after the great, single supercontinent of Pangaea had formed. These first tetrapods still lived mostly in water, but could use their formative limbs to navigate the shallow fresh waters where they lived and perhaps make an occasional sortie onto land. From these beginnings, tetrapods subsequently underwent an extensive radiation, so that today included in tetrapods are fully terrestrial vertebrates as well as many amphibious, aquatic, and flying groups. Literally, tetrapod means four-footed, although it includes some derived groups with secondary loss of limbs, such as snakes. Formally, Tetrapoda is characterized by a **chiridium,** a muscular limb with well-defined joints and digits (fingers and toes). There was a move afoot (no pun intended) to replace

the term *Tetrapoda* with an alternative esoteric terminology. But this has failed in part because it would leave many groups with hands and feet out of the tetrapods, which means "four feet." So here, we shall use the term "Tetrapoda" in its established way, diagnosed by the presence of a chiridium, and phylogenetically by the clade of species more closely related to each other than they are to rhipidistians (figure 3.21).

Struggles with tetrapod taxonomy are not new, and reflect serious efforts to recognize natural groupings and evolutionary events. Once the anatomy of the backbone was used to track tetrapod lineages, but such single-character analysis proved too limited and susceptible to functional convergence (analogy) rather than reliably signaling common ancestry (homology). The phylogeny based only on vertebrae is now abandoned, but taxon names inspired by their vertebral structure survive—temnospondyls, embolomeres, lepospondyls. The **labyrinthodonts,** named originally for their complex tooth structure, were once thought to be an early tetrapod clade. Now recognized as a paraphyletic stem group (figure 3.21), "labyrinthodont" still supplies us with a term of convenience for early tetrapods. The labyrinthodonts document the amazing transition between their fish ancestors within the sarcopterygians ("rhipidistia") on the one hand and later terrestrial tetrapods on the other.

Early tetrapods are known only from fossils. Thus, molecular techniques, which depend on living representatives, cannot complement morphological taxonomic studies. Further, there are two major gaps in the amphibian fossil record, one in the Late Paleozoic lasting almost 100 million years between living taxa and their oldest known fossil ancestors; the other for the first 30 million years of the Early Carboniferous, "Romer's gap" named after the noted paleontologist, Alfred Romer, during which all major later lineages arose. But hundreds of preserved footprints and trackways occur from early to throughout the Paleozoic. Although not connected to particular species, these could have been made in the then soft mud only by early tetrapods walking on land. Still, this history of early tetrapods and their subsequent establishmenton land is remarkable.

Primitive Tetrapods

Labyrinthodonts

Ancient tetrapods retained bony scales, although these were generally restricted to the abdominal region. Many were surprisingly large in body length, with proportionately large skulls as well. *Eogyrinus,* a Carboniferous species, reached 5 m in length (figure 9.18a). Grooves etched in the skulls of some juveniles carried the **lateral line system,** a strictly aquatic sensory system found in fossils of young, but absent in adults of the same species. At metamorphosis, living terrestrial amphibians also lose the lateral line system of their aquatic larvae. Thus, many ancient tetrapods, like modern amphibians, were probably aquatic as juveniles and terrestrial as adults.

FIGURE 3.21 **Tetrapod phylogenetic relationships.** Paraphyletic groups in quotes.

Based on Coates, Ruta, and Friedman.

The earliest groups of labyrinthodonts date from the Late Devonian. One was *Acanthostega*, which could be described aptly as a "four-footed fish" because of its close similarity to the rhipidistian fishes from which it evolved (figure 3.22). It is an intriguing and suggestive group in many ways. In addition to inheriting the distinctive rhipidistian vertebra with unconstricted notochord, *Acanthostega*, like its fish ancestors, also possessed radial fin rays supporting a tail fin, a lateral line system, and labyrinthodont teeth. As in rhipidistians, an intracranial joint was present. Yet, *Acanthostega* was clearly a tetrapod with a more characteristic tetrapod pattern of dermal skull bones, limbs with digits, and weight-bearing girdles. Although its ear region possessed a stapes derived from part of the second gill arch (hyomandibula), *Acanthostega* lacked an auditory system specialized for detection of airborne sounds. Its stapes served primarily as a mechanical brace at the back of the skull. Alternatively, the stapes is hypothesized as being used to control passage of respiratory air streams to and from the lungs through the spiracle. This is not too surprising because *Acanthostega*, like most early tetrapods, was still a predominantly aquatic animal. In fact, *Acanthostega* retained "fish" branchial arches supporting internal gills, implying that *Acanthostega* lived exclusively in water. If representative of early tetrapods, then this suggests that digits arose first in an aquatic environment, later to serve on land. Further, the five-toed, five-fingered pattern (pentadactyl chiridium) that became standard in

post-Devonian tetrapods was not yet fixed in these early tetrapods that had more than five (polydactyl chiridium). *Acanthostega* had eight fingers and eight toes; *Ichthyostega* had seven toes; and *Tulerpeton*, another Devonian tetrapod, had six fingers (other digits were not preserved in the fossils).

Another early tetrapod was *Ichthyostega* (figures 3.23a and 9.17). A large, unconstricted notochord extended into the braincase. Unlike *Acanthostega* and most other early tetrapods, the vertebral column of *Ichthyostega* was specialized for some sort of dorsoventral flexion, of unknown function (see figure 9.17). It possessed radial fin rays supporting a tail fin, a lateral line system, labyrinthodont teeth, and internal gills.

Other fossil groups were present as well, although their placement phylogenetically continues to be especially uncertain, largely because of the extraordinary diversity within each group. The Temnospondyli (figure 9.18b, c) are typically robust bodied, with completely roofed but flat skulls. Numerous groups were fully aquatic as adults. They were the only labyrinthodonts to survive the Paleozoic, producing Mesozoic forms that were flattened, fully aquatic predators, with some groups actually invading the sea. Anthracosaur skulls tended to be deep, and they also tended to be terrestrial as adults.

Unlike their rhipidistian ancestors, early tetrapods were soon adapting to sojourns onto land. The limbs and supportive girdles were generally more ossified and stronger, and the vertebral column tended to increase in prominence.

FIGURE 3.22 *Acanthostega,* **early tetrapod.** A tetrapod from the Devonian showing transitional features from fish to tetrapod. Note the polydactyl feet. About 60 cm overall length.

Based on Coates, 1996.

(a) *Ichthyostega*   1 cm

(b) *Seymouria*   1 cm

FIGURE 3.23 **Labyrinthodont tetrapods.**
(a) *Ichthyostega,* from the Late Devonian, is a member of the ichthyostegid group. The animal was about 1 m long. (b) Skeleton of *Seymouria,* a later highly terrestrial anthracosaur from the early Permian. About 50 cm long.

(b) After Willston.

In early tetrapods, as in a few rhipidistian fishes before them (e.g., *Tiktaalik*), the connection of the shoulder girdle with the skull was absent and a mobile neck region developed, allowing the head to move in all directions relative to the body. The opercular bones were lost along with the internal gills they protected. Primitive tetrapods probably inherited lungs as well as their aquatic mode of reproduction from their rhipidistian ancestors. Fertilization was likely external, with large numbers of small eggs laid in water. As in modern salamanders, fossil larval stages of Paleozoic tetrapods exhibit external gills. Utilization of land was almost certainly an adult occupation following metamorphosis from an aquatic larval stage. Most frequented freshwater environments, although some fossils were recovered in sediments from estuarine or even coastal marine environments.

Lissamphibia—Modern Amphibians

The Lissamphibia arose within the labyrinthodont radiation, specifically from temnospondyls (figure 3.21), although many labyrinthodont features, such as infolded labyrinthine teeth, have been lost in or by the time lissamphibians debut. The lissamphibia includes fossil and living forms. The term **amphibian** was once applied to all early tetrapods, but recent taxonomic analysis makes this too encompassing. Today some would apply it as an equivalent to lissamphibia. But here we restrict it to a subgroup of lissamphibia, namely to all living forms—**salamanders, frogs,** and **caecilians**— which date back over 200 million years to the Jurassic and today include almost 4,000 species displaying a wide range of life histories (figure 3.24a–c). Except for an absence from some oceanic islands, they occur throughout the tropical and temperate regions of the world. Amphibian eggs, which lack shells and amniotic membranes, are laid in water or moist locations. External fertilization characterizes frogs, whereas internal fertilization characterizes most salamanders and probably all caecilians. Typically, paired lungs are present,

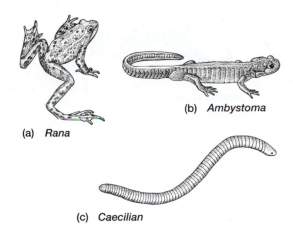

(a) *Rana*

(b) *Ambystoma*

(c) *Caecilian*

FIGURE 3.24 **Lissamphibia.** (a) Frog (*Rana*).
(b) Salamander (*Ambystoma*). (c) Gymnophiona (*Caecilian*).

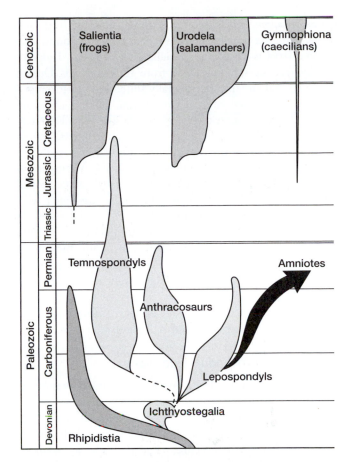

FIGURE 3.25 **Times of appearances of modern amphibian groups.** The three orders of labyrinthodonts (ichthystegalians, temnospondyls, anthracosaurs) and the three orders of Lissamphibia (Salientia, Urodela, Gymnophiona) are shown separately. Rhipidistian fishes, from which ancient tetrapods arose, are included as well.

although they may be reduced or even absent entirely in some families of salamanders. Mucous glands of the skin keep amphibians moist, and granular (poison) skin glands produce chemicals unpleasant or toxic to predators.

Modern amphibians in some ways stand between fishes and later tetrapods; therefore, they supply us with approximate living intermediates in the vertebrate transition from water to land. In their own right, however, living amphibians are specialized and represent a considerable departure in morphology, ecology, and behavior from the ancient Paleozoic tetrapods (figure 3.25). Many bones of the ancient skull and pectoral girdle are lost. Scales are absent, except in caecilians, which allows respiration to occur through the moist skin. Living amphibians are small. The fossil record preserves no intermediate common ancestor that definitely connects them with either the lepospondyls or the labyrinthodonts. Salamanders appeared first in the Upper Jurassic. When frogs first appeared in the Triassic, they were essentially modern in their skeletal design, already exhibiting the highly derived saltatory (jumping) locomotor system.

Living amphibians share some common characteristics. Most modern forms are small, respire through their skin, have unique **pedicellate** teeth with a suture dividing the tooth base from the tip, and possess an extra bone associated with the ear, the **auricular operculum.** Living amphibians characteristically undergo metamorphosis from larva to adult, a remodeling of the larval form that may be subtle as in salamanders or dramatic as in tadpole to adult frog. Currently, most taxonomists treat all living amphibians as members of their own group, the Lissamphibia.

Urodela (Caudata)—Salamanders

Urodela, or Caudata, contains the salamanders. Informally, "newts" are aquatic salamanders belonging to the family Salamandridae. In general body form, salamanders resemble Paleozoic tetrapods, having paired limbs and a long

tail. Terrestrial salamanders usually protrude their tongue to feed, but aquatic forms part their jaws rapidly to create a suction that gulps in the food. Compared to the ancestral tetrapod skull, the urodele skull is broader and more open, with many bones lost or fused. Salamanders have no "eardrum," or tympanum, nor do they have a **temporal notch,** an indentation at the back of the skull. Among primitive salamanders, fertilization is external, but in advanced groups, the male produces the **spermatophore,** a package of sperm perched on a pedicle that is placed on the ground before a female who collects it into her reproductive tract, thereby facilitating sperm transfer. After the female collects all or some of the sperm into her cloaca, eggs are fertilized internally within the reproductive tract.

Salientia (Anura)—Frogs

Frogs and toads make up the Salientia. Adult frogs are without a tail, hence the name *anurans* ("no tail"). Their long hindlegs are part of their leaping equipment, inspiring their alternative name of *salientians* (jumpers). Except in the genus *Ascaphus,* fertilization is external in most frogs and toads. Eggs are usually laid in water or moist locations. The tadpole larva is a striking specialization of frogs. Tadpoles usually feed by scraping algae from rock surfaces. During this stage, salientians are especially suited to exploit temporary food resources, such as spring algae blooms in drying ponds. Typically, after a brief existence, the tadpole undergoes a rapid and radical change, or **metamorphosis,** into an adult with quite a different design. The adult has a stout body and usually protrudes its tongue to feed. A **tympanum** (eardrum), is usually present and is especially well developed in males where it picks up vocalizations associated with courtship and territorial defense.

The terms *frog* and *toad* are imprecise. In a strict sense, toads are frogs belonging to the family Bufonidae. More informally, the term "toad" is used for any frog having "warty" skin and **parotoid glands**—large, raised glandular masses behind the eyes. "Warts" consist of clumps of skin glands scattered across the body surface. Other frogs have smooth skin without warts and lack parotoid glands.

Gymnophiona (Apoda)–Caecilians

The wormlike gymnophionans, or caecilians, show no trace of limbs or girdles; hence, they are sometimes called *apodans* ("no feet"). All are restricted to damp tropical habitats, where they live a burrowing lifestyle. Unlike the open skull of frogs and salamanders, the caecilian skull is solid and compact. Although their life histories are not well known, males possess a copulatory organ; thus, fertilization is internal. Primitive caecilians lay eggs that hatch into aquatic larvae; more advanced species produce live terrestrial young. I treat them here as arising within the temnospondyls but some argue for an independent origin within the lepospondyls.

Lepospondyls

The lepospondyls (figures 3.21, 3.26) can be distinguished from labyrinthodonts by many skeletal features, especially features associated with a generally large size, absence of labyrinthodont teeth, and reduction in many dermal skull bones. What unites them, and further distinguishes them from labyrinthodonts, is the presence of a solid vertebra in which all three elements—neural spine, two centra—are fused onto a single, spool-shaped centrum. The lepospondyls appeared quite early in the Carboniferous, were never as abundant as labyrinthodonts, and became extinct by the mid Permian, much earlier than the labyrinthodonts.

Vertebrae types (p. 310)

The distinctive nectridean lepospondyls were apparently entirely aquatic, reversing a trend in most other early tetrapods. Their paired limbs were small and ossification was reduced, but the tail of some species was quite long. Skulls of "horned" nectrideans of the Early Permian were flattened and drawn out into distinctive long, winglike processes (figure 3.26a, b).

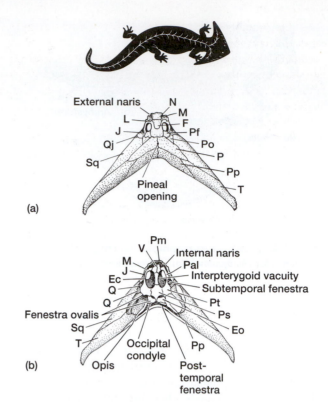

FIGURE 3.26 *Diploceraspis,* a lepospondyl, was a "horned" nectridean of the early Permian. Overall body length was about 60 cm. (a) Dorsal and (b) ventral views of the skull. The various bones of the skull are ectopterygoid (Ec), exoccipital (Eo), frontal (F), jugal (J), lacrimal (L), maxilla (M), nasal (N), parietal (P), palatine (Pal), postfrontal (Pf), premaxilla (Pm), postorbital (Po), postparietal (Pp), parasphenoid (Ps), pterygoid (Pt), quadrate (Q), quadratojugal (Qj), squamosal (Sq), tabular (T), and vomer (V).

After Beerbower.

Microsaurs (meaning "small" and "lizard") were not lizards, but lepospondyls, despite their misleading name. Most were small, around 10 cm, but varied in design. The group was primarily terrestrial, although several families were secondarily aquatic exhibiting lateral line grooves on the face and a few were burrowing forms.

Amniotes

Amniotes bear embryos enveloped in extraembryonic membranes. The embryo, together with these membranes, is usually packaged in a calcareous or leathery shelled egg. In living forms, reproduction by amniotic eggs can be observed directly. Because of their close affinities to living amniotes, it is inferred that many Mesozoic vertebrates laid such shelled eggs. But fossil animals, especially in basal groups, seldom leave direct evidence of their reproductive style to diagnose their taxonomic position. Instead, phylogenetic studies, using many characters, place groups in relationship to each other, helping to at least delineate the amniote lineages.

The amniote radiation is composed of two major lineages, the **Sauropsida** and **Synapsida** (figure 3.27). As fossils document, they diverged very early, certainly by the Carboniferous and perhaps earlier. The sauropsids include birds, dinosaurs, modern reptiles, and many of the diverse assemblages of the Mesozoic. The sauropsids diversified along two major lineages, the **Parareptilia** and the **Eureptilia.** The Synapsida is a monophyletic lineage producing many various forms, including therapsids and modern mammals.

Skull Fenestration Traditionally, relationships within these amniote groups were based on characteristics of the temporal region of the skull, the area behind each eye. This region seemed to be a reliable indicator of evolutionary lineages, and to a large extent, its use has proven to be justified. As a consequence of the attention given to the temporal region, a formal terminology grew up to describe the amniote skull.

The temporal region in amniotes varies in two ways: in the number of openings, termed **temporal fenestrae,** and in the position of the **temporal arches,** or **bars,** made up of defining skull bones. From these two criteria, up to four primary skull types were recognized. In primitive amniotes, as well as in their non-amniote ancestors, the temporal region is covered completely by bone that is not pierced by temporal openings (figure 3.28a). This **anapsid skull**

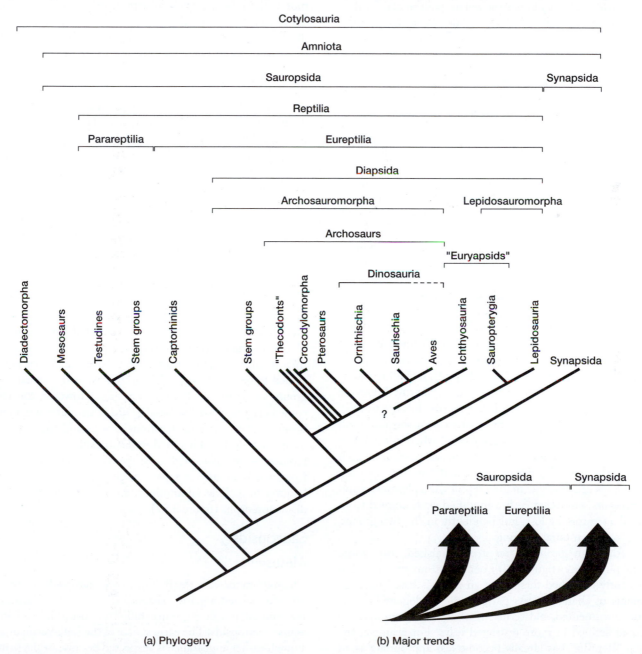

(a) Phylogeny (b) Major trends

FIGURE 3.27 Amniotes, phylogenetic relationships. (a) Phylogeny of major groups. Note the major trends within amniotes, as summarized in (b). Paraphyletic groups in quotes.

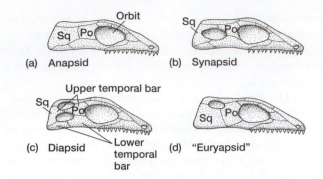

(a) Anapsid   (b) Synapsid

(c) Diapsid   (d) "Euryapsid"

FIGURE 3.28 Amniote skull types. Differences among the skulls occur in the temporal region behind the orbit. Two, one, or no fenestrae may be present, and the position of the arch formed by postorbital (Po) and squamosal (Sq) bones varies. (a) The anapsid skull has no temporal fenestrae. (b) The synapsid skull has a bar above its single temporal fenestra. (c) The diapsid skull has a bar between its two temporal fenestrae. (d) The "euryapsid" skull has a bar below its single temporal fenestra. Rather than being a separate skull type, the euryapsid skull is thought to be derived from a diapsid skull that lost its lower temporal bar and opening.

is characteristic of the first amniotes and the later turtles and their allies. The **synapsid skull** found in mammalian ancestors represents an early divergence from the anapsid. The synapsid skull has a single pair of temporal openings bordered above by a temporal bar formed by squamosal and postorbital bones (figure 3.28b). In another group that diverged from anapsids, we recognize a **diapsid skull** characterized by two pairs of temporal openings separated by this temporal bar. As points of formal anatomical reference, this squamosal-postorbital bar is designated as the **upper temporal bar.** The **lower temporal bar,** formed by jugal and quadratojugal bones, defines the lower rim of the lower temporal fenestra (figure 3.28c). Diapsids, including pterosaurs and dinosaurs, were predominant during the Mesozoic and gave rise to birds and all living reptiles (except turtles). The "euryapsid" skull, once thought to be a separate skull type, actually is a modified diapsid skull wherein the lower temporal bar is lost, leaving the squamosal-postorbital arch to form the lower border of the paired fenestral opening. Two groups of Mesozoic marine reptiles, the plesiosaurs and ichthyosaurs, possessed such a modified skull derived from diapsid ancestors by loss, independently in the two groups, of the lower temporal opening.

Although probably far from finished, this more recent phylogeny (figure 3.27) benefits from the inclusion of recently described fossils and an analysis based on large numbers of characters. As parts of the phylogeny become better documented, old terms will likely be abandoned or become defined in more restricted ways. For example, the group "Reptilia" has already become less appropriate as an embracing taxon for all of these early amniotes. The reptile *Sphenodon*, living on islands near New Zealand, may be

unfamiliar, but most of us know of snakes, lizards, turtles, and crocodiles. From these living forms, we have some composite image of what constitutes a "reptile." Living reptiles have scales (but no hair or feathers) composed partly of surface epidermis. They usually reach preferred body temperature by absorbing heat from the environment. Respiration is primarily through the lungs, with very little occurring through the skin. Thus, we may find it odd that taxonomists still quibble over what constitutes a reptile. However, reptiles, as traditionally understood, turn out to be a taxonomic grade, with specializations associated with different diets, patterns of locomotion, and body size. Among modern groups, for example, crocodiles have more features in common with birds than they do with lizards, snakes, or turtles. To reflect these natural groupings, we need to restrict traditional names and in some cases abandon them for a more accurate evolutionary phylogeny.

Stem-Amniotes

At present, the most likely sister group to the amniotes is the diadectomorphs, such as the hulking and pig-sized *Diadectes*, which, unlike most of its contemporaries, exhibits some evidence of herbivory. Members of this primitive group arose in the Late Carboniferous and, together with lissamphibia, show affinities to the anthracosaurs (figure 3.21). Certainly they are an important transitional group between non-amniote and amniote tetrapods. With relationships of early amniotes still being resolved, we might provisionally place the diadectomorphs within the anthracosaurs. Alternatively, we might resuscitate an old term, Cotylosauria, and place them within that taxon (figure 3.27).

These terms permeate past literature and may serve in the future, so a brief introduction is needed. The anthracosaurs have, at one time or another, included Seymouromorpha and various other late non-amniote groups. Whatever their membership, the anthracosaurs have usually been envisioned to be closely related to amniotes, or the actual basal group, ancestors to all later amniotes. However, the term coined originally for this basal group of amniotes was **Cotylosauria.** It has included various groups and the term has been used loosely. The cotylosaurs, meaning literally "stem reptiles," were envisioned to be the basal group of amniotes from which all later groups issue. Here we shall use Cotylosauria to include all amniotes and their sister group, the diadectomorphs (figure 3.27).

Sauropsids

Mesosaurs

The mesosaurs were the first of many sauropsids to embark on a specialized aquatic existence (figure 3.29). Fossils are few, and this puzzling group exhibits no close affinity to other aquatic sauropsids. The earliest date to the Late Permian, but a much earlier appearance is suspected because of the primitive skull that lacked temporal fenestrae. The elongate snout bristled with sharp, long teeth, which may have formed a

FIGURE 3.29 Mesosaur. This aquatic amniote lived in the Middle Permian. The long tail was used for swimming, and the limbs were probably paddle shaped. Overall length, about 1 meter.

After McGregor; von Huene.

filter-feeding device suited for sifting crustaceans or an effective snare for fish. Like many sauropsids that are secondarily aquatic, mesosaurs had paddle-shaped feet, a laterally compressed tail, and long neck. Neural arches of the trunk were expanded and slightly overlapping, thereby resisting torsion, but favoring lateral bending. No other sauropsids are known from Late Permian ages in South America and southern Africa, except for mesosaurs that frequented both sides of the Atlantic Ocean Basin. Such a distribution suggests a much closer position of both continents to each other and became the first biological evidence for continental drift.

Reptilia

The taxon Reptilia, in a restricted sense, applies to the Parareptilia and Eureptilia, which together share similarities of the braincase that distinguish them from the mesosaurs. The taxon Anapsida was once used for this first group and Diapsida for the second. Anapsida were reptiles diagnosed by skulls without temporal fenestrae; Diapsida by skulls with two temporal fenestrae. However, analysis of multi-characters, compared to just skull temporal region, reveals a slightly different evolutionary history. Some reptiles with anapsid skulls do not occur within Anapsida; captorhinids have anapsid skulls, but do fall within the old "diapsids." Today, anapsida applies to a skull type, but it is used less as a taxon name. Taxonomically, Diapsida is now used in a more restricted sense for a monophyletic lineage within the Eureptilia (figure 3.27). Note that birds get scooped up in this cladistic grouping as well. This simply recognizes that birds are a natural but specialized derivative of earlier reptiles. We return to birds later, but first we complete our review of sauropsids.

Parareptilia Testudines (turtles) and an assortment of fossil groups (e.g., *Pareiasaurus*) are included in the Parareptilia. They have a distinctive ear region wherein the eardrum is supported by the squamosal (rather than by the quadrate) and by the retroarticular process, a backward projection of the lower jaw. Further, the foot is unique in the way the digits articulate with the ankle bones.

The only surviving members are turtles. When turtles first appeared in the Late Triassic, they already possessed a distinctive shell made up of a dorsal **carapace** of expanded ribs and surface skin plates (scutes) and a connected ventral **plastron** of fused bony pieces (figure 3.30). An oddity of turtles, found in them and nowhere else, is the incorporation of the limbs and articulated girdles from splayed positions outside the body to inside the embracing and protective bony shell (carapace plus plastron). This seems to be an abrupt re-design of the amniote body plan into the specialized turtle body plan. The most primitive turtles for which there is a fossil record (Late Triassic) already possessed this shell housing limbs within. Where fossils fail to clarify, modern molecular genetics suggests the underlying mechanism—the fundamental change in a few turtle *Hox* genes—is the apparent basis for the radical transformation of the amniote body plan into the unique design of turtles.

Turtle axial skeleton (p. 315)

(a) *Proganochelys* (b) (c) *Pareiasaurus*

FIGURE 3.30 Parareptilia, Testudinata. (a) *Proganochelys,* a turtle of the Triassic exhibiting a pattern of skin scutes that overlay the carapace. (b) Fossil *Proganochelys* skull showing absence of temporal fenestrae. Overall length, 2 meters. (c) Pareiasaurus from the Late Permian. Bones of the skull (b) include the angular (An), articular (Ar), dentary (D), frontal (F), jugal (J), lacrimal (L), maxilla (M), nasal (N), parietal (P), postorbital (Po), prefrontal (Prf), premaxilla (Pm), quadratojugal (Qj), surangular (Sa), and squamosal (Sq).

(a, b) After Jaekel; (c) after Gregory.

Modern turtles belong either to the Pleurodira or to the Cryptodira, depending on the method they employ to retract their head into their shell. Pleurodires flex their neck laterally to retract the head, whereas cryptodires flex their neck vertically. These two groups seem to share a common ancestor, *Proganochelys*, from the Late Triassic. The term *tortoise* is sometimes applied to turtles restricted to land, but no formal taxonomic distinction is made between turtle and tortoise.

Traditionally, turtles are viewed as the only living representatives of the earliest reptiles, a sister group to diapsids, as I treat them here (figure 3.27). However, I should concede that some recent morphological evidence, that includes fossils, favors placement of turtles within the diapsid radiation, not primitively within reptiles generally. Although currently unresolved, the outcome of this debate is important because if turtles turn out to be diapsids, then a major revision would be in order in comparative studies. First, turtles (if diapsids) could no longer be assumed to represent the ancestral reptilian condition against which other derived groups could be compared. Second, the absence of temporal fenestrae in turtles (again, if diapsids) would imply a secondary condition in which the two fenestrae close up again, not an ancestral condition. Third, this would leave the Parareptilia composed of only fossil forms, with no living representative. We shall have to wait and see.

Eureptilia Within the Eureptilia, the Diapsida is diagnosed by two temporal fenestrae, together with a palatine fenestra within the roof of the mouth. Based on these skull characteristics, the oldest diapsid has been considered to be *Petrolacosaurus*, an araeoscelidian reptile from the Late Carboniferous in what is today Kansas. The body was about 20 cm long, with slightly elongated neck and limbs, and the tail added another 20 cm to the overall length. The skull was typically diapsid, with a pair of temporal openings defined by complete temporal bars. Other primitive diapsid species became quite specialized. *Coelurosauravus* had greatly elongated ribs that in life likely supported a gliding membrane. *Askeptosaurus* was about 2 m in length, slender, and probably aquatic in habits.

The most basal eureptilian is not an araeoscelidian, but a member of the Captorhinidae, also known from the Carboniferous. The captorhinids lack temporal fenestrae and so represent the stage just before appearance of the diapsid condition. However, captorhinids share with other eureptilians characteristic long and slender limbs, similar digit and ankle articulations, and similarities of bone structure within the skull. Captorhinid reptiles were small, about 20 cm in length, and generally similar to modern lizards in that they had a well-ossified skeleton (figure 3.31). Rows of tiny, sharp teeth along the margins of the jaws and across the roof of the mouth as well as an agile body suggest that insects might have been a major part of their diet, as they are in the similarly designed small, modern lizards. Captorhinids are

Eocaptorhinus   1 cm

FIGURE 3.31   Captorhinomorph. Skeleton of the reptile *Eocaptorhinus,* from the Permian. A North American member of the captorhinid family.

After Heaton and Reisz.

broadly similar to anthracosaurs, but captorhinids possess reptilian features, such as strong jaw musculature, and reptilian structural details in their skull, limbs, and vertebral column. The first captorhinid occupied tree stumps away from standing bodies of water, providing additional testimony that they exploited the land further, more characteristic of reptiles than of their early tetrapod ancestors.

Eureptilian Radiation Within the eureptilian radiation, there were three major lineages, all built on a diapsid design (figure 3.27). One is the **Lepidosauromorpha,** including fossil forms as well as snakes, lizards, and allies. The second is the **Archosauromorpha,** including dinosaurs, birds, and related groups. The third, Euryapsida, includes Mesozoic marine reptiles, ichthyosaurs and sauropterygians. Perhaps because of their highly modified aquatic specializations, informally the "euryapsids," these marine forms of the Mesozoic continue to frustrate the best efforts of phylogenetic analyses to place them taxonomically. We place sauropterygia allied to the ichthyosaurs but in the **Lepidosauromorpha** (figure 3.27).

Ichthyopterygia During the Mesozoic, several major diapsid lineages became specialized for aquatic existence. Among them were the **ichthyosaurs,** although their exact placement within the diapsid radiation is not settled (figure 3.32a). From deposits of the Early Triassic, the first ichthyosaurs appear already to have been aquatic specialists. Advanced ichthyosaurs had a porpoise-like body design, but their tail swept from side to side to provide propulsion, unlike the porpoise tail that moves in a dorsoventral direction. Sleek bodies, paddlelike limbs, and teeth around the rim of a beaklike mouth testify to an active predaceous lifestyle. Preserved in fine-grained sediments, ichthyosaur stomach contents include prodigious amounts of belemites (squidlike mollusks), fish, and in some, hatchling turtles. Relatively huge eyes gave them light-gathering power for dim waters and also the ability to fine-focus on small, quick prey. Fossil remains of a pregnant female show fully-formed young ready for birth or in the process, evidence for live birth (not egg laying) in these marine reptiles (see figure 1.33). One of the largest ichthyosaurs rivaled or exceeded the size of a modern sperm whale.

Sauropterygia Sauropterygians were, along with ichthyosaurs, the other Mesozoic lineage of diapsids specialized to an aquatic mode of life. Their current placement, tentatively, is within the lepidosauromorpha. This group includes the early **nothosaurs** (Triassic) and the later **plesiosaurs** (Jurassic-Cretaceous) that evolved from them. The plesiosaur body was heavy, the neck often long, and the limbs were modified into paddles that acted like oars or hydrofoils to propel the animal in water (figure 3.32b).

Lepidosauria Modern snakes, lizards, *Sphenodon*, and their ancestors constitute the lepidosaurs. A Late Permian/Early Triassic group of lepidosaurs, the **Eosuchia,** are most likely the ancestors of all modern lepidosaurs. *Sphenodon*, the tuatara, is the sole survivor of a once widespread Mesozoic reptilian group called rhynchosaurs. It survives today only on parts of New Zealand and nearby islands (figure 3.33a). This genus carries forward the primitive eosuchian skull with complete temporal bars defining upper and lower temporal openings. In lizards, the lower temporal bar is absent. Snakes lack both upper and lower temporal bars. Because these connecting constraints are deleted in the skulls of lizards and snakes, both of these vertebrate groups, especially snakes, have increased jaw mobility that enhances their ability to capture and swallow prey.

Functional consequences of the loss of temporal arches (p. 268)

The **squamates** include snakes, lizards, and a group of tropical or subtropical reptiles, the amphisbaenids. Some taxonomists place amphisbaenids with lizards; others treat them as a distinct group. All amphisbaenids are burrowers; most are limbless and prey upon arthropods (figure 3.33b).

FIGURE 3.32 **Marine reptiles of the Mesozoic.**
(a) Ichthyosaur, a porpoise-like reptile about 1 m long.
(b) Sauropterygian, a plesiosaur, about 7 m in length.

(a) After Romer; (b) after Andrews.

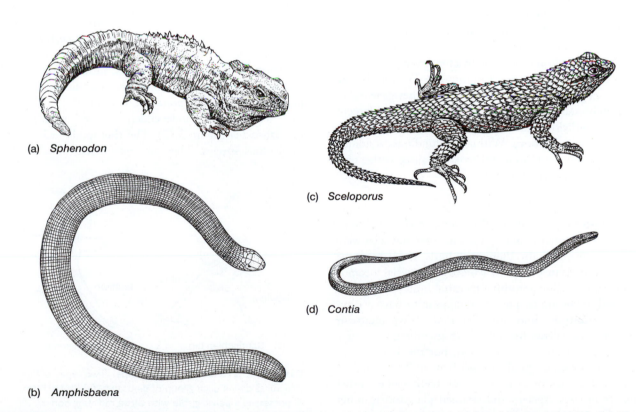

(a) *Sphenodon*

(b) *Amphisbaena*

(c) *Sceloporus*

(d) *Contia*

FIGURE 3.33 **Lepidosaurs.** (a) *Sphenodon*. (b) Amphisbaenian, a burrowing lepidosaur. (c) Lizard (*Sceloporus*). (d) Snake (*Contia*).

The majority of living squamates are lizards or snakes (figure 3.33c, d). Many people are surprised to learn that some species of lizards (other than amphisbaenids) are limbless, like snakes; therefore, the presence or absence of limbs alone does not distinguish snakes from true lizards. Instead, differences in internal skeletal anatomy, especially in the skull, are used to diagnose the two groups. Further, lizards have movable eyelids and most have an external auditory meatus (opening). Snakes lack both structures.

Archosauromorpha Encompassed within the archosauromorphs are several stem groups, small assemblages of diapsids known from fossils, and a very large group, the **archosaurs**, which includes familiar forms such as crocodiles, dinosaurs, and birds. Archosaurs display a trend toward increasing **bipedalism**, or two-footed locomotion. The forelimbs tend to be reduced, whereas the hindlimbs are drawn under the body to become the major weight-bearing and locomotor appendages. The skull is diapsid, but an additional fenestra opens on the face between maxilla and lacrimal bones, the **antorbital fenestra**, as well as a mandibular fenestra in the lower jaw.

The term *archosaur*, meaning "ruling reptile," recognizes the extraordinary radiation and preeminence of this group during the Mesozoic. Formally then, archosaurs include "thecodonts," the most primitive of the group, crocodiles, birds, pterosaurs, and two large groups, the Saurischia and Ornithischia. Taken together, Saurischia and Ornithischia constitute what laypeople informally think of as the "dinosaurs." However, birds are descendants. They evolved within the dinosaur radiation (figure 3.27) so, formally, they should be included.

The **thecodonts**, a paraphyletic group, arose late in the Permian and prospered during the Triassic. Before becoming extinct by the end of the Triassic, they gave rise to all later archosaurs. Thecodonts take their name from teeth set in deep, individual sockets (thecodont condition) rather than in a common groove. Within the hindlimb, a unique ankle design appeared in some thecodonts along with a tendency to bipedal, upright posture.

Ankle types (p. 347)

The **pterosaurs**, often called pterodactyls after members of one subgroup, could glide and soar but also were capable of powered flight. Pterosaurs, birds, and bats are the only three vertebrate groups to achieve active aerial locomotion. Because of their antorbital fenestra, limb posture, and specialized ankle joints, pterosaurs appear to have a phylogenetic affinity to dinosaurs. The first known pterosaur was already specialized for flight with membranous wings. Many were sparrow- to hawk-sized, but the Late Cretaceous *Quetzalcoatlus*, found in fossil beds in Texas, had an estimated wingspan of 12 m. Pterosaur teeth suggest a diet of insects in some species and strained plankton in other species. Fossilized stomach contents confirm that one species ate fish.

(a) *Rhamphorhynchus*   (b) *Pteranodon*

FIGURE 3.34   Pterosaurs. The lengthened forelimb of pterosaurs supported a membrane derived from skin to form the wing. (a) *Rhamphorhynchus*. Wingspan was about 1.5 m. (b) Skeleton of *Pteranodon*. Wingspan was about 8 m.

(a) After Williston; (b) after Eaton.

The early **rhamphorhynchoids** are pterosaurs distinguished by long tails and teeth (figure 3.34a). The later pterodactyloids lacked a tail and teeth and often had a projecting crest at the back of the head (figure 3.34b).

Dinosaurs include two groups of archosaurs: the Saurischia and Ornithischia. The two dinosaur groups differ in the pelvic structure. In saurischians, the three bones of the pelvis—ilium, ischium, pubis—radiate outward from the center of the pelvis (figure 3.35a). In ornithischians, the ischium and part of the pubis lie parallel and project backward toward the tail (figure 3.35b). All dinosaurs have either a saurischian or an ornithischian type of pelvis. As we shall see, birds are part of the saurischian clade, and therefore, some include them within the dinosaurs. Consequently, authors wishing to distinguish birds from other dinosaurs may refer to saurischians plus ornithischians as "nonavian dinosaurs."

There are two independent lines of evolution within the Saurischia (figure 3.27). The **theropods** include mostly carnivorous species. They are comfortable bipeds, adapted

Ilium

Pubis

Ischium

(a) Saurischian hip

Ilium

Ischium

Pubis

(b) Ornithischian hip

FIGURE 3.35   Dinosaur hips. Two types of hip structures define each group of dinosaurs. (a) Saurischians all possessed a pelvic girdle with three radiating bones. (b) Ornithischians had a hip with pubis and ischium bones lying parallel and next to each other.

to upright locomotion on two hindlegs. The theropods include *Velociraptor*, *Tyrannosaurus* and *Allosaurus*, and within this radiation, birds evolve. The mostly herbivorous **sauropodomorphs** constitute the other saurischian line. They arose in the Triassic, and by its end, split into distinctive groups, **prosauropods** and **sauropods**. Familiar sauropodomorphs include *Apatosaurus* (formerly *Brontosaurus*), *Diplodocus,* and *Brachiosaurus.*

There are several evolutionary lines within the exclusively herbivorous ornithischians (figures 3.36, 3.37). One includes **stegosaurs, ankylosaurs,** and their allies; others include **ornithopods** (e.g., duck-billed dinosaurs), **pachycephalosaurs** (bipedal, head-butting dinosaurs), and **ceratopsians** (e.g., *Triceratops*). Plant material was clipped by a horny bill, then sliced and ground by cheek-teeth. Ornithischians were rare in the Triassic, but are more common in the Jurassic (figure 3.37).

Along with birds, the crocodiles, alligators, and their close allies **(gavials, caimans)** are the only reptilian members of the archosaurs to survive the Mesozoic and live in modern times. In many features, especially the skull and ankle joint, alligators and crocodiles are not far removed from primitive thecodonts. Modern crocodilian families are known from the Late Cretaceous.

Aves Birds outnumber all vertebrates except fishes. They can be found virtually everywhere, from the edge of the polar ice to tropical forests. They are derived diapsids. Among extant amniotes, birds are most closely related to crocodiles, and share many of the same basic features, despite their superficial differences. Both lay eggs encased in shells and have similar bone and muscle structures. Over a century ago, these features led T. H. Huxley to call birds "glorified reptiles." Taking fossils into account, most place the origin of birds within the Saurischia, as we have done here (figure 3.37), specifically within the theropods. Therefore, birds are part of the dinosaur radiation. Evidence for this close association with dinosaurs comes especially from similarities in hip, wrist, and wishbone.

Within this theropod radiation arose the coelurosaurs (figure 3.37), which share features even more closely related to birds such as a wishbone (furcula) and fused sternum (breast bone). Of particular surprise and interest has been the discovery of feathers within some members of this group. Some filamentous feathers were fluffy and downlike, a few species had vaned feathers—flat and symmetrical feathers on both sides of a central shaft. These would have been ill-suited for powered flight, leading some to suggest that feathers arose initially perhaps as surface insulation, aids in thermoregulation of body temperature. However, these coelurosaurian dinosaurs and earliest birds lacked nasal tubinates, a characteristic diagnostic of warm-blooded physiology. Whatever their initial biological roles, feathers evolved before birds.

Feathers (p. 222)
Dinosaurs: Hot to Cold—The Sequel (p. 123)
Turbinates (pp. 276, 495)

Flight.—Only birds, bats, and pterosaurs evolved the capacity for powered flight, but not all bird flight is the same. Some birds soar, some hover, some are sprint fliers, others long-distance fliers, and some don't fly at all. The wings of the flightless penguins serve as flippers. Ostriches have lost use of their wings altogether and depend entirely on running for locomotion. In fact, some of the largest birds ever to evolve were flightless. *Gastornis* (*Diatryma*), a 2-meter-tall flightless bird, cruised the forests of Europe and North America 55 million years ago (figure 3.38b).

Phorusrhacus, a similar flightless bird, lived in South America 30 million years ago (figure 3.38c). Both, although unrelated, were large terrestrial predators. Although they left no descendants, other large flightless birds evolved and even survived into recent times along with primitive humans. Examples are the elephant bird (*Aepyornis*) of Madagascar and the 3-m tall moas (*Dinornis*) of New Zealand. Moas belonged to a family of large ground birds that were plant eaters in New Zealand when no native land mammals resided there. Unfortunately for moas and modern scientists, the Polynesians who arrived in about A.D. 1300 hunted

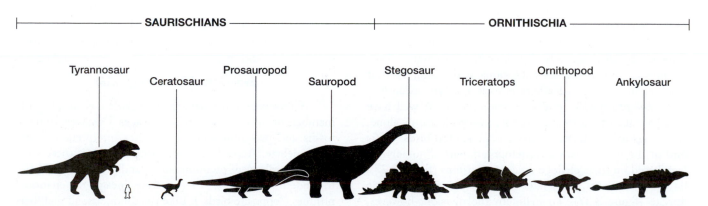

FIGURE 3.36 Sizes of selected dinosaurs. Relative sizes of adults are shown. A 2-m human in light outline is diminutive by comparison.

FIGURE 3.37 **Phylogeny of Dinosauria.** Dinosaurs are composed of two lineages, the ornithischia (left) and saurischia (right). Within the ornithischia are several clades: (1) Thyreophora, including ankylosaurs and stegosaurs; (2) Ornithopods, including duck-billed dinosaurs; (3) Pachycephalosaurs; and (4) Ceratopsia. Within the saurischia are two major clades: (5) Sauropodomorpha and (6) Theropods, which encompass allosaurs, various other carnivorous dinosaurs, (7) coelurosauria, (8) maniraptora, and (9) birds (Aves).

Modified from Paul C. Sereno, "The Evolution of Dinosaurs," Science, 25 June 1999, 284: 2137–2147. Copyright © 1999 American Association for the Advancement of Science.

moas for food and colorful feathers. By the time Western explorers visited, moas were extinct, with only fossil remains to tell their story.

Feathers and Origins.—Size, flight, and anatomy alone do not distinguish birds from other vertebrates. Birds are dressed in feathers, specializations of the skin. Were it not for impressions of feathers in rock, the early avian fossil *Archaeopteryx* (150 million years ago) might well have been mistaken for a reptile from its skeletal anatomy alone. It belongs to the Archaeornithes, or "ancestral birds." This bird of the Jurassic was a contemporary with dinosaurs. In fact, based on their phylogenetic relationships (figure 3.27), birds are dinosaurs, just a later branch in this monophyletic lineage (figure 3.37). To maintain the division, however, some speak of non-avian dinosaurs to distinguish this vast group from birds (aves).

As to feathers, new fossil finds, especially in China, show non-avian dinosaurs with feathers. Amongst these is *Anchiornis* (160 million years ago), a pre-*Archaeopteryx* theropod with flight feathers on its forearms, fuzzy coating over its body, and strangely, long feathers on its hindlegs.

Evolution of Feathers (p. 226)

Cretaceous birds have been regarded as simply early members of current and familiar lineages. Discovery of more fossils and taxa now suggests another interpretation, still under debate. According to this interpretation, these Cretaceous fossils were predominantly landbirds belonging not to modern lineages, but to a separate lineage, the enantiornithines ("opposite birds"). Like the dinosaurs, all members of this taxon became extinct at the end of the Cretaceous. Other Cretaceous birds belonged to the smaller ornithurine,

Dinosaurs: Heresies and Hearsay— The Heated Debate

The claim is that dinosaurs were warm-blooded, like birds and mammals, not cold and sluggish like lizards and snakes. To be specific, the issue is not really whether the blood of dinosaurs ran hot or cold. After all, on a hot day with the sun beating down, even a so-called cold-blooded lizard can bask, heat its body, and, strictly speaking, have warm blood circulating in its arteries and veins. The issue is not blood temperature, hot or cold, but whether the source of the heat is internal or external. To clarify this issue, two useful terms need to be defined, *ectotherm* and *endotherm*. Animals that depend largely upon sunlight or radiation from the surrounding environment to heat their bodies are cold-blooded, or more accurately, ectotherms ("heat from outside"). Turtles, lizards, and snakes are examples. "Warm-blooded" animals produce heat inside their bodies by metabolizing proteins, fats, and carbohydrates. To be more accurate, warm-blooded animals are endotherms ("heat from within"). Birds and mammals are obvious examples.

Were dinosaurs ectotherms or endotherms? The source of their body heat is in dispute, not their blood temperature.

Heat for *ectotherms* is cheaply won. They need only bask in the sun. The trouble with such a lifestyle is that the sun is not available at night nor is it always available in cold temperate climates. By contrast, heat for *endotherms* is expensive. A digested meal, often caught with great effort, produces fats, proteins, and carbohydrates necessarily spent in part to generate heat to keep the endotherm body warm. Where endotherms have an advantage is that their activity need not be tied to heat available from the environment. These different physiologies are accompanied by different lifestyles. Ectotherms bask; on cold nights they become sluggish; and in freezing winters they hibernate. Endotherms remain metabolically active throughout each day and each season, despite the cold or inclement weather. Certainly there are exceptions—bears and some small mammals hibernate—but endothermy requires continuous activity in most cases. Thus, the issue of warm-bloodedness in dinosaurs is not just about physiology but about the type of accompanying lifestyle they enjoyed.

Because dinosaurs have traditionally been classified as reptiles, they were for many years envisioned to be ectotherms just like their living counterparts—lizards, snakes, turtles, and crocodiles. Initially, the case for endothermic dinosaurs was built around four principal lines of evidence. Let's look at the arguments.

Insulation. First, some mid- to late-Mesozoic reptiles had surface insulation, or at least they seemed to. For ectotherms, a surface insulation would only block absorption of the sun's rays through the skin and interfere with efficient basking. But, for endotherms, a surface layer holding in their internally manufactured heat might be an expected adaptation. Unfortunately, soft insulation is rarely preserved, but in a few fossils of the Mesozoic, impressions in the surrounding rock indicate the presence of an insulating layer of feathers (*Archaeopteryx*). In fact, feathers likely first arose as thermal insulation and only later evolved into aerodynamic surfaces. Apparently, then, some Mesozoic reptiles had insulation like endotherms rather than bare skin like ectotherms.

Large and Temperate. Second, large Mesozoic reptiles are found in temperate regions. Today, large reptiles such as great land tortoises and crocodiles do not occur in temperate regions. They live in warm tropical or subtropical climates. The only modern reptilian inhabitants of temperate regions are small or slender lizards and snakes. The reason is easy to understand. When winter arrives in temperate regions and freezing cold settles in, these small ectothermic reptiles squeeze themselves into deep crevices where they safely hibernate until spring and escape the freezing temperatures of winter. On the other hand, for a large and bulky animal, there are no suitably sized cracks or crannies into which they can retreat to avoid the winter cold. Large animals must be endothermic to survive in temperate climates. Even though the world of the Mesozoic was warmer than today, with no polar ice caps, winters in northern temperate regions would have been cool and days short. Thus, the presence of large reptiles in temperate climates of the Mesozoic suggests that they were warm-blooded. Like wolves, coyotes, elk, deer, moose, bison, and other large temperate mammals today, the large Mesozoic reptiles depended on heat produced physiologically to see them through.

Predator-to-Prey Ratios. Third, the ratio of predators to prey argues for endothermic dinosaurs. Endotherms, in a sense, have their metabolic furnaces turned up all the time, day in and day out, to maintain a high body temperature. A single endothermic predator, therefore, requires more "fuel," in the form of prey, to keep the metabolic furnaces stoked than an ectothermic predator of similar size. Paleontologist Robert Bakker thus reasoned that there should be few predators but lots of prey (lots of fuel to feed the few predators) in ecosystems dominated by endothermic reptiles. But, if ectothermic reptiles dominated, then proportionately more predators should be present. By selecting strata that stepped through the rise of dinosaurs, Bakker compiled the ratios. If Mesozoic archosaurs were becoming endothermic, then the ratio of predator to prey should drop. That happens. As this ratio was followed from early reptiles, to pre-dinosaurs, and to dinosaurs, it dropped. There were proportionately fewer predators and more prey.

Bone Histology. Fourth, the microarchitecture of dinosaur bone is similar to that of endothermic mammals, not to that of ectothermic reptiles. Bones of ectothermic reptiles show growth rings, like those of trees, and for much the same reason as the rings in trees, they grow in seasonal spurts. Endothermic mammals, with constant body temperature year round, lack such growth rings in their bones. When various groups of dinosaurs were examined, the microarchitecture of their bones told a clear story—no growth rings.

Dinosaurs then became active animals. They romped and played, chased prey, and dashed for cover. Endothermically, they were formidable. They even made it to the silver screen, snorting warm air from hot bodies as they made meals of mammals—people—in the movie *Jurassic Park*.

The important point to keep before us is that dinosaurs were in their own right an extraordinary group. These active animals occupied almost every conceivable terrestrial habitat. Their social systems were complex, and the adults of some species were enormous. If dinosaurs were endotherms, their complete demise at the end of the Mesozoic can only be more mysterious and the loss of the awesome splendor of this group all the more intriguing.

Although dinosaurs died out, the debate over what kind of reptile they were continues to evolve (see Box Essay 3.5).

or modern-type birds. Not until the Cretaceous extinctions did these birds, like their eutherian mammal counterparts, begin their radiations. Specifically, it was through transitional "shorebirds" within the ornithurines that two great groups of modern birds arose, the Paleognathae and the Neognathae. The paleognaths, or "ratites," include ostriches, rheas, emus, cassowary, kiwi, tinamous, and several extinct groups such as moas and elephant birds. The neognaths include all other living groups of birds.

Feather types, development, and function (p. 222)

Diversity.—The basic avian design has proved highly adaptable, and birds have undergone extensive diversification. For instance, the Cretaceous ornithurines included a primitive flamingo and *Hesperornis*, a toothed diving bird with such small wings that it was certainly flightless. *Ichthyornis* was a small, ternlike seabird recovered from Cretaceous rock of Kansas (figure 3.38a). By the Late Mesozoic, water birds had already diverged widely.

Birds have continued to be successful in exploiting aquatic resources (figure 3.39). Some species dive deep beneath the surface and use their wings to propel themselves in pursuit of fishes. Others are specialized for plunging and use their diving speed to carry them to fishes below. Many species feed at the water's surface, either skimming it from the air or dipping for resources as they float. A few species prowl the air above the water, surprising other birds and pirating their catch.

Raptors are birds with **talons,** specialized feet used to stun or grasp prey. Hawks, eagles, and owls are examples. Many hunt prey on the ground. Other raptors, such as the prairie falcon, strike their quarry, usually a dove or a slow migrating duck, in the air and then follow it to the ground to dispatch the injured prey (figure 3.40).

Feet and wings reflect functions performed. Paddling birds have webbed feet, and raptors have talons. Feet of running species are robust, and those of waddlers are broad. Birds that soar on strong winds usually have long and narrow wings like the wings of glider planes. High-speed or migratory birds have narrow, often swept-back wings. Pheasants and other birds that deploy short bursts of flight in enclosed bushy or forest habitats have broad elliptical wings for maneuverability. Slotted high-lift wings are seen in birds that soar on warm air updrafts over inland areas.

Aerodynamics and wing designs serving flight (p. 363)

(a) *Ichthyornis*

(b) *Diatryma*   (c) *Phorusrhacus*

FIGURE 3.38 **Extinct birds.** (a) *Ichthyornis* was pigeon-sized, likely sought fish for food, and lived in North America about 100 million years ago. (b) *Diatryma* lived 55–50 million years ago. It was a flightless bird that stood over 2 m tall and likely ran down small prey much as the diminutive roadrunner does today. (c) *Phorusrhacus*, another flightless predator, lived in South America some 30 million years ago.

After Peterson.

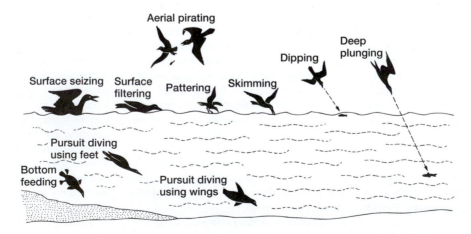

FIGURE 3.39 **Lifestyles of waterbirds.**

The discovery of *Archaeopteryx* was especially timely. In 1861 the fossil was quarried from a site in Bavaria in what is present-day Germany. Only two years before, Charles Darwin had published *The Origin of Species,* which immediately ignited public debate. These were the early days of paleontology, with relatively few recovered fossils and even fewer serious scientists to dig them up. Critics of Darwin were quick to point out the absence of fossil intermediates between groups, which his theory of evolution anticipated. If one group gave rise to another, as Darwin's ideas suggested, then transitional forms should occur. *Archaeopteryx* helped address this objection. It was such an intermediate fossil because it possessed features of both birds (feathers) and reptiles (skeleton, teeth).

The discovery of *Archaeopteryx* prompted interest in the possibility of other ancient avian fossils that might further narrow the gap between reptiles and birds. Reptiles have teeth but modern birds do not. Somewhere between the two, evolutionary intermediates developed a bill and lost teeth. Thus, uncovering a fossil bird with reptilian teeth would be of considerable significance and help supply details about this evolutionary transition. O. C. Marsh, an American

BOX FIGURE 1   **Bird with teeth.** Hesperornis lived 100 million years ago in inland seas of North America. Although larger (almost 1 m overall) in shape, its features and probable lifestyle resembled the modern loon. This bird also retained teeth, a characteristic held over from its reptilian ancestors.

paleontologist of the mid-1800s, discovered just such birds with teeth, although they were later than *Archaeopteryx* (box figure 1).

Despite the significance of Marsh's discoveries, enemies of evolution in the U.S. Congress protested the use of taxpayer's money to search out fossils with bird teeth, which everyone knew did not exist (until Marsh discovered them, of course). Today, as in the nineteenth century, science is a predominant feature of our culture. Most politicians who govern today have no better training in biology or in any science than did politicians in Marsh's time. Law schools and businesses still supply most of our public figures. A background heavy in lawyers and light on scientists gives only lopsided preparation to persons who guide the destiny of science in society.

Archaeopteryx occasionally still makes the news. Recently, a well-known astronomer dabbling in paleontology claimed that the Bavarian fossils of *Archaeopteryx* were forgeries. Fossil forgeries have occasionally turned up, but *Archaeopteryx* is decidedly not one of them. Regrettably, this astronomer's cavalier opinion cast undeserved doubt upon these fossils. Although the popular media picked up and spread gleefully the premature rumors of a forgery, they failed to report equally the results of an extensive reinvestigation that showed these forgery claims to be completely groundless. Charitably said, this astronomer could have saved everyone lots of wasted time had he simply made an effort to bring his naive ideas before someone familiar with the pitfalls into which he stumbled.

Stay tuned. *Archaeopteryx* seems to have a public life of its own.

FIGURE 3.40   **Falcon attack.** The falcon's midair blow delivered with the talons is a "stoop" intended to stun and knock the prey from the air. The prey is finally controlled and killed on the ground.

Synapsida

Synapsids arose late in the Paleozoic, about 300 million years ago. During the Late Carboniferous and through the ensuing Permian, synapsids were the most abundant terrestrial vertebrates, diversifying into small to large carnivores and herbivores. Within synapsids, three major radiations occurred: **pelycosaures, therapsids,** and **mammals** (figure 3.41).

Synapsids are amniotes with a single temporal fenestra bounded above by the upper temporal bar (squamosal-postorbital bones: figure 3.28). They exhibit early some characteristics of body posture and tooth formation that became elaborated within the later mammals. Anticipating this, paleontologists once referred to some early synapsids (pelycosaurs + therapsids) as "mammal-like reptiles," an unfortunate designation, because they are neither reptiles nor mammals, and this tempts one to glance over these early synapsids in order to emphasize mammals alone. Yet synapsids exhibit a remarkable fossil record with a great variety of distinctive forms. On display within synapsids is the transition from ectothermic to endothermic amniotes (figure 3.42).

Pelycosauria

Pelycosaurs are a paraphyletic group, an assemblage of early synapsids whose relationships are still being sorted out. Pelycosaurs arose in the Late Carboniferous from primitive

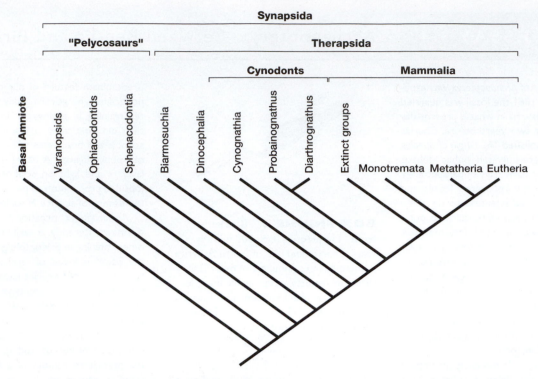

FIGURE 3.41 Synapsida, phylogenetic relationships. "Pelycosaurs" in quotes to remind that it is paraphyletic.

amniotes and soon enjoyed an extensive radiation through the Early Permian, coming to constitute about half of the known amniote genera of their time. Some, like *Edaphosaurus*, were herbivorous. Most, however, were carnivores and preyed on fish and aquatic amphibians. Different species of pelycosaurs differed in size, but they were not very diverse in design, perhaps because of their specialized lifestyle. The most notable specialization in some species was a broad "sail" along the back consisting of an extensive flap of skin supported internally by a row of fixed neural spines projecting from successive vertebrae (figure 3.43a, b). If the sail was brightly colored in life, it might have been deployed in courtship or in bluff displays with rivals, as elaborate ornamentations of birds are today. The sail might also have been a solar collector. When turned broadside to the sun, blood circulating through the sail was warmed and then carried to the rest of the body.

Rather suddenly, pelycosaurs declined in numbers and were extinct by the end of the Permian. Therapsids evolved from them and largely replaced them for a time as the dominant terrestrial vertebrates.

Therapsida

Therapsids appeared in the Early Permian, and prospered during the Triassic. However, therapsids are basically all gone by the end of the Triassiac, with only a few straggling species persisting just into the Early Cretaceous.

The end of the Permian was a violent period in Earth history. Volcanic activity intensified, introducing acid rain into the atmosphere and producing ash clouds encircling the globe. Climate cooled with formation of polar ice caps, ushering in an Ice Age. Not surprising, with such stress on ecosystems, a huge and defining extinction event struck at the Permo-Triassic boundary, taking over half of the marine animals, and included extinction of many of the therapsid species on land. A few surviving therapsid groups (cynodonts and dicynodonts) reradiated in the Triassic but soon declined and became extinct in the Early Cretaceous. Milder, warm climates and reduced volcanic activity returned in the Triassic. Reradiating therapsids apparently exploited terrestrial habitats more extensively than the pelycosaurs before them; consequently, they exhibit greater diversity of body design. Some trends in therapsids were conservative, however. Their stance was quadrupedal and their feet had five digits (figure 3.44a, b). Teeth were differentiated into distinct types, perhaps with specialized functions. The skull, especially the lower jaw, became simplified. Some herbivorous therapsids became specialized for rooting or grubbing, some for digging, some for browsing, and one even arboreal (tree-dwelling). The overall selection for more active terrestrial locomotion and for feeding specializations resulted in great diversity within therapsids (figure 3.42). There is even some evidence from bone histology and latitudinal distribution that therapsids were becoming endothermic beginning in the Triassic.

Cynodonts One especially successful group of therapsids was the cynodonts. Some were herbivores, but most were carnivores. They arose in the Late Permian, and became the dominant land carnivores in the early part of the Triassic,

FIGURE 3.42 **Relative diversity of terrestrial vertebrates.** Geological time is represented on the vertical axis, vertebrate diversity on the horizontal. Synapsids and the mammals that arose within the synapsids are darkly shaded. Note the large, early diversity of therapsids that abruptly gave way to the diapsid reptiles during the mid-Mesozoic.

until largely replaced by terrestrial sauropsids in the Late Triassic. Cynodonts had teeth specialized for slicing. Temporal bones and muscles of the skull changed substantially during their evolution as well, giving them modified jaw mechanics. Further, extensive turbinates were present in the nose. These are thin, scrolled, and folded plates of bone that warm and humidify the incoming air (as well as hold the olfactory epithelium). In the roof of the mouth, a secondary palate divided the food passage from the air passage through the nose. As breath is expelled through the nose, turbinates recapture much of this heat and moisture, reducing heat and water loss. These characteristics suggest that cynodonts were experimenting with active lifestyles based on an endothermic metabolism.

During their evolution, cynodonts underwent a marked decline in body size, from that of a large dog to many only slightly larger than a weasel. However, by the Late Triassic, most cynodonts went into a precipitous decline, except for one group that remained and eventually prospered after the great dinosaur extinctions at the end of the Cretaceous. This surviving cynodont group is the mammals (figure 3.42).

Mammalia

The mammals arose within the therapsid radiation in the Late Triassic, initially small and shrewlike. These Mesozoic mammals contended with a terrestrial fauna then dominated by dinosaurs, especially the saurischians generally.

(a) *Dimetrodon*

(b) *Edaphosaurus*

FIGURE 3.43   **Pelycosaurs.** (a) *Dimetrodon*, a predator, reached 3 m in length (Lower Permian of Texas). (b) *Edaphosaurus*, a herbivore (Late Carboniferous and Early Permian), was about 3 m long.

After Romer.

(a) *Moschops*

(b) *Titanophoneus*

FIGURE 3.44   **Therapsids.** (a) *Moschops*, about 5 m in length. (b) *Titanophoneus*, about 2 m.

(a) After Gregory; (b) after Orlov.

Most Mesozoic mammals were shrew-sized and the largest not much bigger than a cat up until the mass extinctions closed the Mesozoic. The radiation of modern mammal groups began early in the Cenozoic, especially among the eutherian mammals. Now, more diverse and many larger forms appeared, perhaps related to the breakup of the earlier, large landmasses into the smaller continents we recognize today. About 20,000 years ago, as the climate began to warm, most of the large mammals, megafauna, began to disappear either because of this climate change or because

human societies, based on hunting, were spreading, especially in the northern hemisphere. Living forms include the **monotremes** (platypus and spiny anteaters) and the **Theria,** made of **metatherians** (pouched marsupials such as kangaroos and opossums) and **eutherians** (placental mammals).

Characteristics of Mammals   The two primary characteristics that define living mammals are hair and mammary glands. In general, mammals are endothermic furry animals nourished from birth with milk secreted by their mothers. All have hair, although in whales, armadillos, and some other mammals, it is reduced considerably. A thick coat of hair, the **pelage,** primarily insulates the mammalian body to hold in heat. Hair also has a sensory function as a recorder of fine touch. The bases of sensory hairs stimulate associated nerves when the hair is moved. "Whiskers," especially evident in the faces of carnivores and rodents, are specialized long hairs called sensory **vibrissae.**

Sebaceous glands of the mammalian skin are associated with hair. Their products condition the skin and allow evaporative loss of excess body heat. The embryonic similarity between skin glands and mammary glands suggests that milk glands were derived from these specialized skin glands. Few mammals have numerous sweat glands, humans being one exception and, hence, the likely reason these glands receive such disproportional attention. Most cooling in mammals is done by panting (e.g., dogs) or special circulatory mechanisms. In addition, mammalian red blood cells that transport oxygen lose their nuclei and most other cell organelles when they mature and enter the general circulation.

Hair, mammary glands, sebaceous glands, and red blood cells without nuclei (anucleate) are unique to living mammals. Other characteristics that are not necessarily restricted to this class include large brain in relation to body size, maintenance of high body temperature (except in some young and during resting periods of torpor), and modifications of the circulatory system from that of other amniotes.

Hair and mammary glands rarely are preserved in fossils, so they are of little practical value in tracking the early evolution of mammals. Alternatively, fossil mammals usually exhibit three distinct skeletal characteristics. The first is a chain of three tiny bones, confined to the middle ear, that conduct sound from the tympanum to the sensory apparatus of the inner ear. Reptiles have only one primary middle ear bone, but never three. Second, the lower jaw of mammals is composed only of the dentary, a single bone, whereas several bones make up the lower jaw of reptiles. The third skeletal feature is a joint between the dentary and squamosal bones of the jaws. In reptiles, other bones form the jaw joint. Even these three features are not always preserved in fossils, so paleontologists often resort to other backup features such as tooth structure. For instance, most teeth in mammals are replaced just once in a lifetime, not continuously, and occlusion of teeth is more precisely controlled than in reptiles.

Mammalian teeth, their development, and functions (p. 508)

Seldom does the first announcement of new evidence meet with instant scientific acceptance. We should be professional skeptics until the evidence is evaluated, independently examined, and checked again. When scientists do this, the result is often the appearance of a new perspective, different from any of the theories that guided us in the first place. Dinosaurs, hot or cold, may be an example.

Insulation. Evidence of surface hair in therapsids is at best ambiguous. But even if hair was present, therapsids are synapsids, standing well outside the dinosaur radiation. As to pterosaurs, in fact, they lacked hair. Recently described pterosaur fossils, extraordinarily well preserved, show that the flight membranes stretched across their forearms were internally supported by an exquisite network of ordered connective tissue. Superficially, this produced a fine-lined pattern on the skin, which had been mistaken for "hair." This internal webbing, reacting to air pressure, permitted the wing membrane to shape itself into an aerodynamic surface to meet demands while in flight.

Bone histology. Although some dinosaurs seem to lack growth rings typical of ectotherms, and therefore meet one prediction of endothermy, some early birds do have growth rings. Bones from Cretaceous birds, enanthiornithines, show evidence of annual growth rings, like those of ectothermic animals. If these birds were ectotherms, then their immediate known relative, *Archaeopteryx,* likely had rings as well, as did the primitive saurischian dinosaurs from which *Archaeopteryx* presumably evolved. This histology fits with conclusions from a comparison of the respiratory physiology of living birds and mammals, which suggests that flight may have come before endothermy in Cretaceous birds. Finally, recent examination of bone from an early sauropod, *Massospondylus,* revealed faint growth rings at least in this particular dinosaur.

Noses. Turbinates are folds of bone in the nose, across which air is directed when entering and departing from the lungs. The turbinates support membranes that warm and humidify entering air, and dehumidify air departing, thereby recovering water otherwise lost. Where breathing rate is high to support endothermy, turbinates are present in the nasal passage. Mammals and birds have them. Dinosaurs apparently do not. CAT (computer-aided tomography) scans of dinosaur fossils showed no evidence of these respiratory turbinates.

Despite the evidence against dinosaur endothermy, their growth rates were apparently high, much like endotherms. And they seemed to be built for active lives. The debate has not yet cooled; dinosaur heresies may heat up again.

Note that the characteristics we most associate with mammals (hair and mammary glands) are unavailable to paleontologists. Wherever the mammalian grade is reached, it will be somewhat arbitrary. We cannot be certain that a boundary fossil with a mammalian skull or tooth pattern also possessed hair and milk glands in life. Therefore, using the lineage (clade) provides a more objective basis for assigning taxonomic status, and more accurately represents the evolutionary history of mammals. Reasonably then, mammals are a type of cynodont, cynodonts a type of therapsid, and therapsids part of the synapsid lineage (figure 3.41).

Extinct Mammals Mammals are Late Triassic therapsids, making early mammals contemporaries of Mesozoic reptiles, such as pterosaurs, crocodiles, turtles, and dinosaurs. Extinct mammals include several groups with elaborate names: kuehneotherids, haramiyoids, sinoconodontids, multituberculates, and morganucodonts, to mention a few. Generally, these early mammals were the size of a shrew. They were probably nocturnal and endothermic, and most had sharp, pointy teeth. Brain size was larger, for a given body size, than in their reptilian contemporaries. Teeth in primitive mammals did more than just snag prey or clip vegetation. Teeth were **heterodont,** meaning different in general appearance throughout the tooth row—incisors at the front of the mouth, canines, premolars, and molars along the sides of the mouth. This permits division of labor, allowing some teeth to tear or clip food, others to break it up mechanically and prepare it for rapid digestion. Muscular cheeks kept the food between tooth rows that chewed the food. Specialized tooth function implies, but does not prove, that primitive mammals were endothermic. If they were, they probably had a coat of insulating fur. Early mammals presumably hatched from eggs and nursed from mammary glands like the monotremes, the most primitive mammals living today.

Living Mammals Marsupial and eutherian mammals trace their ancestry to a common group in the Early Cretaceous. They share several derived features including live birth, not egg laying, and so are placed together within the Theria. Monotremes most likely arose in Australia, diverged early from the Theria probably in the Lower Jurassic, and have been much on their own course ever since. The three species of living monotremes include the platypus that inhabits Australia and the adjacent island of Tasmania, and the two species of spiny anteaters that inhabit Australia and New Guinea. Like therian mammals, monotremes have hair, suckle their young, and are endotherms. However, unlike other mammals, monotremes lack nipples, lack external ears, and embryos develop in shelled eggs, primitive features retained from the generalized amniote condition.

Monotreme embryology (p. 166)

Today, a substantial radiation of marsupials remains in South America and Australia, which, together with surrounding regions, holds a diversity of marsupials (Australian kangaroos are a familiar example). Tiny kangaroo young are born at an early developmental stage, pull themselves into their mother's pouch, and suckle there until they grow considerably larger. No male marsupial has a pouch. This is a female feature, although some female marsupials are also pouchless. Specialized forms still present in Australia, such as a burrowing marsupial (marsupial "mole") and a species that glides through the air (marsupial "flying squirrel"), suggest that marsupials once enjoyed great diversity. The large mammals in Australia are built predominantly upon a marsupial design, but placental mammals in the form of rodents also reached Australia about 4 million years ago, and there radiated into many smaller endemic species.

Marsupials (chapter 5—embryology, chapter 14—reproductive organs, and chapter 15—hormonal control of the breeding cycle)

Eutherian mammals are today by far the most numerous and widespread of any mammalian group (figure 3.45). The nutritional and respiratory needs of the young are provided through a **placenta,** a vascular organ connecting the fetus and the female uterus. Such a vascular association between fetus and mother is not unique to eutherian mammals. A temporary "placenta" forms between the early embryo and the female uterus in some marsupials. In fact, nutritional and respiratory support of the embryo is found to varying degrees today in some reptiles, fishes, and even a few amphibians. What distinguishes eutherian mammals is that reproduction in *all* species is based on a placenta.

Vertebrate placentae (p. 192)

Diversification and adaptive radiation of eutherian mammals evolved against the backdrop of environmental changes produced by alteration in terrestrial floras, mass extinctions at the end of the Cretaceous, continental fragmentation and drift, and in climate change. The taxonomy (figure 3.45) captures this in four natural groups—Afrotheria, Xenathra, Euarchontoglires, and Laurasiatheria. Although the Afrotheria currently seem to promise to be the eutherian root, more recent analysis places the

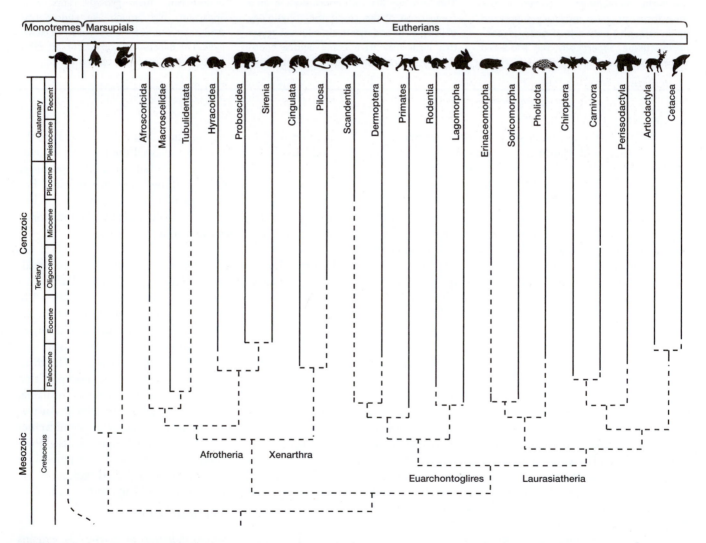

FIGURE 3.45   Living mammals. Monotremes, marsupials, and eutherians are the three groups of mammals living today, the placentals being the largest group.

Cingulata (anteaters) as the most primitive group of living eutherians. Bats **(Chiroptera)** are the only mammals with powered flight, although gliding placental mammals arose three times—as flying lemurs **(Dermoptera)** in Asia and within the rodents (flying squirrels and scaly-tailed squirrels). Two eutherian groups are fully aquatic: the **Cetacea,** which includes toothed whales **(odontocetes)** and the baleen whales **(mysticetes),** and the **Sirenia,** which includes the manatees.

The term *ungulate* is a descriptive term of convenience and refers to hoofed animals, comprising about a third of all living and extinct mammalian genera. Ungulates include the **Perissodactyla** (horses, rhinoceroses, tapirs) and **Artiodactyla** (pigs, camels, cattle, deer, etc.), **Cetacea** (whales and porpoises), together with what are often called the subungulate. (subungulates = paenungulates). Subungulates loosely include the **Proboscidea** (elephants), **Sirenia** (sea cows), **Tubulidentata** (aardvark), and **Hyracoidea** (hyraxes). A rumen is present in most artiodactyls. This is a specialized part of the digestive tract, from which **ruminants** derive their common name. Giraffes, deer, cattle, bison, sheep, goats, antelopes, and their allies in the Artiodactyla, with few exceptions such as pigs, are all ruminants.

Within the **Carnivora,** the term *fissiped* is used informally for land carnivores (cats, dogs, bear, skunks), and the term *pinniped* refers to semiaquatic carnivores (seals and walruses).

Rodentia is the largest of the eutherian groups and is often divided informally into the **sciuromorphs** (squirrel-like), the **myomorphs** (mouselike), and the **hystricomorphs** (porcupinelike). **Primates** are arboreal or had ancestors who were, and they possess grasping fingers and toes tipped by nails. The **lower primates,** or prosimians (strepsirrhini), include lemurs and lorises. The **higher primates,** or anthropoids (haplorrhini), encompass the **catarrhine** (Old World) monkeys, which lack prehensile tails, and the **platyrrhine** (New World) monkeys, some of which use a prehensile tail. The word *monkey* is a general term that has no formal taxonomic definition. The term *apes* refers to the paraphyletic *pongids* (orangutans, gorillas, chimps), and the term *hominids* to humans and their immediate ancestors of the **Hominidae.**

The place of origin and routes of dispersal of therian mammals are still debated, although the known fossil record indicates that the earliest marsupial and eutherian species arose in the Early Cretaceous of China, although the subsequent Cretaceous radiation of marsupials was in North America and eutherians were there a bit afterward in the Late Cretaceous. Continental drift was then beginning to break up the few large continents of the Mesozoic into smaller landmasses, separating them from each other by open ocean. The Atlantic Ocean was growing but still small, and most continents were still in contact. The Late Cretaceous climate, even in polar regions, was mild. During this time, marsupials dispersed to Asia, Antarctica, and Australia, while eutherians migrated into Africa and to the New World (figure 3.46). As the continents fragmented further during the Cenozoic, these stocks of mammals were carried into semi-isolation and served as the founding stocks for the distinctive mammalian groups that subsequently evolved on the separating continents.

Early in the Mesozoic, all continents were joined into one large supercontinent, Pangaea. But in an active Earth, this supercontinent began to rift in two so that by the Late

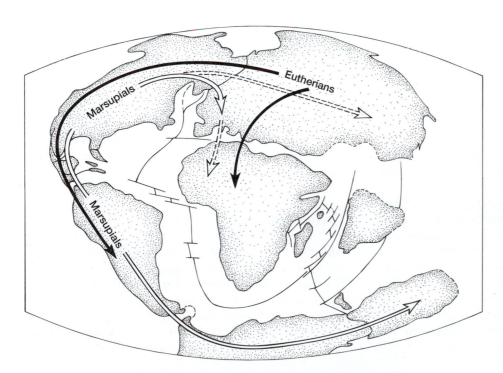

FIGURE 3.46 Therian radiation. Position of the continents during the late Mesozoic is shown. Although today most marsupials live in Australia, their center of origin was apparently the New World (North America) of the late Cretaceous. From there they spread in two directions. One direction during the Eocene was to Europe and North Africa, although they subsequently became extinct on both those continents (dashed arrows). The other direction in which marsupials spread was through South America and Antarctica to Australia before these continents separated. Eutherians originated in the Old World and spread to the New World via land connections that existed between the continents during the Mesozoic.

After Marshall.

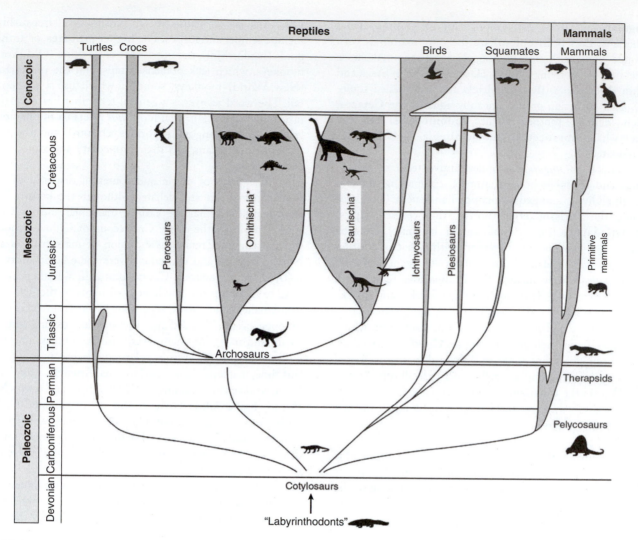

FIGURE 3.47 **Mesozoic Diversity and Extinctions.** The widths of each group subjectively express the estimates of relative abundance. Note the extinctions of not just the dinosaurs but other Mesozoic groups as well; and note that birds and mammals are contemporaries of the dinosaurs but become prominent only after their extinctions.

* Ornithischia plus Saurischia equal the dinosaurs. "Labyrinthodonts" in quotes to remind that it is a term of convenience and paraphyletic.

Mesozoic Pangea had split into two regions introducing a North/South geographical division in land masses. In turn, during the Cenozoic, these regions continued to further fragment and rotate into the recognizable continents we are familiar with today. This fragmentation affected evolution of mammals during the Late Mesozoic and through the Cenozoic. The newer molecular phylogeny, upon which figure 3.45 is based, also detects several major groupings within eutherian mammals. Many of these major groupings (Xenarthra, Afrotheria) are endemic to southern continents, which has led to the view that eutheria as a whole had a southern origin. However, the fossil evidence speaks otherwise, namely that therians originated in northern regions of Laurasia during the Early Jurassic, and subsequently spread to other continents while land connections still existed. The

South American–Antarctica–Australian connection persisted into the Early Cenozoic affording migrations along this route. This time, Early Cenozoic, was probably when species from northern continents entered Africa also across land bridges.

Overview

Agnatha The vertebral column consists of a chain of vertebrae, a segmental series of cartilage or bone blocks, and characterizes the vertebrates. The earliest vertebrates were soft bodied, from the Cambrian—*Haikouella* and *Hiakouichthys*. Later ostracoderms were encased in protective shells of dermal bone. Today the only living representatives of these agnathans are the boneless cyclostomes—hagfishes and lampreys. Jawless, these first vertebrates were likely limited in lifestyle until the invention of jaws.

Gnathostomes The evolution of jaws gave the earliest gnathostomes equipment to bite or crush prey, and included the early acanthodians and placoderms. These early gnathostomes also possessed two sets of paired fins (or spines). Lifestyles were more active and varied. In general, the radiation of gnathostomes proceeded along two major lines of evolution—one producing the Chondrichthyes (sharks and their allies), and the other the Osteichthyes (bony fishes)—the actinopterygians and sarcopterygians. During the Late Paleozoic, tetrapods arose within the sacrcopterygians and vertebrates moved to land for the first time. These first tetrapods are known only from fossils. The first to survive to the present were members of the lissamphibia, we know as living amphibians (frogs, salamanders, caecilians). Amniotes arose within this early radiation of tetrapods, producing sauropsids on the one hand and synapsids on the other. Sauropsids would produce within their radiation the familiar turtles, lizards, snakes, crocodiles, and birds, and during the Mesozoic (figure 3.47) one of the most remarkable groups to ever grace the Earth, the dinosaurs in two subgroups—Ornithisichia and Saurischia. The synapsids independently underwent their own special radiation, eventually giving us therapsids and modern mammals—monotremes, marsupials, and eutherian (placental) mammals.

WEB LINK

Visit the text website at *www.mhhe.com/kardong7e* for additional study aids including selected references, functional laboratories, and web links to chapter topics.

CHAPTER 4

Biological Design

INTRODUCTION: SIZE AND SHAPE

SIZE
  Relationships Among Length, Area, and Volume
  Surface Area
  Volume and Mass

SHAPE
  Allometry
  Transformation Grids

ON THE CONSEQUENCES OF BEING
  THE RIGHT SIZE

BIOMECHANICS
  Fundamental Principles
    Basic Quantities—Length, Time, and Mass
    Units
    Derived Quantities—Velocity, Acceleration, Force,
      and Relatives
    Reference Systems
    Center of Mass
    Vectors

Basic Force Laws
Free Bodies and Forces
Torques and Levers
Land and Fluid
  Life on Land: Gravity
  Life in Fluids
Machines
Strength of Materials
  Loads
  Biological Design and Biological Failure
Tissue Response to Mechanical Stress
  Responsiveness of Bone

BIOPHYSICS AND OTHER PHYSICAL PROCESSES
  Diffusion and Exchange
    Pressures and Partial Pressures
    Countercurrent, Concurrent, and Crosscurrent
      Exchange
  Optics
    Depth Perception
    Accommodation

OVERVIEW

Introduction: Size and Shape

Bodies, like buildings, obey laws of physics. Gravity will bring down an ill-designed dinosaur just as certainly as it will fell a faulty drawbridge. Animals must be equipped to address biological demands. The long neck of a giraffe gives it access to treetop vegetation; the claws of cats hook prey; a thick coat of fur gives the bison protection from the cold of winter. In order for animals to catch food, flee from enemies, or endure harsh climates, structures have evolved that serve animals against these challenges to survival. But there is more to an animal's environment than predators and prey, climate and cold. An animal's design must address physical demands. Gravity acts on all structures within its reach. Heavy terrestrial vertebrates must exert much effort to move

a massive body from one place to another. Bones and cartilage must be strong enough to bear the weight. If these skeletal structures fail, so does the organism, and its survival is at risk. Animals at rest or in motion experience forces that their structural systems must withstand. As the British biologist J. B. S. Haldane put it,

> "It is easy to show that a hare could not be as large as a hippopotamus, or a whale as small as a herring. For every type of animal there is a most convenient size, and a large change in size inevitably carries with it a change in form."
>
> (Haldane, 1956, p. 952)

In this chapter we examine how structures built by humans and those evolved by natural selection have design features that incorporate and address common problems posed by basic physical forces. For example, living organisms come in a great variety of sizes (figure 4.1); however, not all designs work equally well for all sizes (figure 4.2).

A grasshopper can jump a hundred or more times its own body length. From time to time, this feat has tempted some people to proclaim that if we were grasshoppers we could leap tall buildings in a single bound. The implication is that grasshoppers possess special leaping devices absent in humans. Certainly grasshoppers have suitably long legs that launch them great distances. But the more important reason why grasshoppers and humans differ in their relative jumping abilities is a matter of size, not a matter of long legs. If a grasshopper were enlarged to the size of a human, it too would be unable to leap a hundred times its new body length, despite its long legs. Differences in size necessarily bring differences in performance and in design.

To illustrate this point, let us look at two examples, one from music and one from architecture. A small violin, although shaped generally like a bass, encloses a smaller resonance chamber; therefore, its frequency range is higher (figure 4.3). The larger bass encloses a larger resonance chamber and consequently has a lower frequency range. A Gothic cathedral, because it is large, encloses relatively more space than a small, brick-and-mortar neighborhood church. Large cathedrals include devices to increase surfaces through which light may pass in order to illuminate the congregation within (figure 4.4). The end and sidewalls of cathedrals are designed with outpocketings that architects call apses and transepts. The side walls are pierced by slotted openings, clerestories, and tall windows. Together, apses, transepts, clerestories, and windows allow more light to enter, so they compensate for the proportionately larger volume enclosed within. Later in this chapter, we will see that this principle applies to animal bodies as well.

Shipbuilders often resort to a scale model to test ideas for hull design. But the model, because it is many times smaller than the ship it represents, responds differently to the wave action in a testing tank. Thus, a model alone may not reliably mimic the performance of a larger ship. To compensate, shipbuilders minimize the size discrepancy with a trick—they use slower speeds for smaller models to keep the ship-to-wave interactions about the same as those that large vessels meet on open seas.

Size and shape are functionally linked whether we look inside or outside of biology. The study of size and its consequences is known as **scaling**. Mammals, from shrews to elephants, fundamentally share the same skeletal architecture, organs, biochemical pathways, and body temperature. But an elephant is not just a very large shrew. Scaling requires more than just making parts larger or smaller. As body size changes, the demands on various body parts change disproportionately. Even metabolism scales with size. Oxygen consumption per kilogram of body mass is much higher in smaller bodies. Size and shape are necessarily linked, and the consequences affect everything from metabolism to body design. To understand why, we look first to matters of size.

Size

Because they differ in size, the world of an ant or a water strider and the world of a human or an elephant offer quite different physical challenges (figure 4.5a, b). A human coming out of his or her bath easily breaks the water's surface tension and, dripping wet, probably carries without much inconvenience 250 g (about half a pound) of water clinging to the skin. However, if a person slips in the bath, he or she has to contend with the force of gravity and risks breaking a bone. For an ant, surface tension in even a drop of water could hold the insect prisoner if it were not for properties of its chitinous exoskeleton that make it water repellent. On the other hand, gravity poses little danger. An ant can lift 10 times its own weight, scamper upside down effortlessly across the ceiling, or fall long distances without injury. Generally, the larger an animal, the greater the significance of gravity. The smaller an animal, the more it is ruled by surface forces. The reason for this has little to do with biology. Instead, the consequences of size arise from geometry and the relationships among length, surface, and volume. Let us consider these.

Relationships Among Length, Area, and Volume

If shape remains constant but body size changes, the relationships among length, surface area, volume, and mass change. A cube, for instance, that is doubled in length and then doubled again is accompanied by larger proportional changes in surface and volume (figure 4.6a). Thus, as its length doubles and redoubles, its edge length increases by first 2 and then 4 cm, or factors of 2 and 4. However, the total surface area of its faces increases by factors of 4 and 16. The cube's volume increases in even faster steps, by factors of 8 and 64, for the doubling and redoubling. The shape of the cube stays constant, but because, and *only* because, it is larger, the biggest cube encloses relatively more volume per unit of surface area than does the smallest cube. In other words, the biggest cube has relatively less surface area per unit of volume than the smallest cube (figure 4.6b).

It is certainly no surprise that a large cube has, in *absolute* terms, more total surface area and more total volume than a smaller cube. But notice the emphasis on *relative* changes between volume and surface area, and between surface area and length. These are a direct consequence of changes in size. These relative changes in surface area in relation to volume have profound consequences for the design of bodies or buildings. Because of them, a change in size inevitably requires a change in design to maintain overall performance.

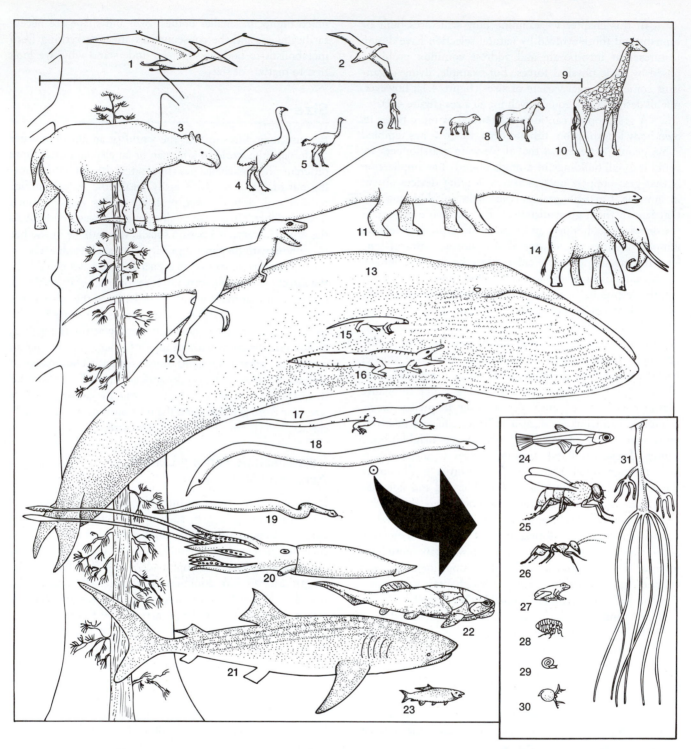

FIGURE 4.1 **Animal sizes range over many orders of magnitude.** The largest animal is the blue whale, the smallest adult vertebrate a tropical frog. All organisms are drawn to the same scale and are numbered as follows: (1) the pterosaur *Quetzalcoatlus* is the largest extinct aerial reptile; (2) the albatross is today the largest flying bird; a fossil bird (not shown) from South America had an estimated wing span of 20 feet; (3) *Baluchitherium* is the largest extinct land mammal; (4) *Aepyornis* is the largest extinct bird; (5) ostrich; (6) a human figure represented by this scale is 6 feet tall; (7) sheep; (8) horse; (9) this line designates the length of the largest tapeworm found in humans; (10) the giraffe is the tallest living land animal; (11) *Diplodocus* (extinct); (12) *Tyrannosaurus* (extinct); (13) the blue whale is the largest known living animal; (14) African elephant; (15) the Komodo dragon is the largest living lizard; (16) the saltwater crocodile is the largest living reptile; (17) the largest terrestrial lizard (extinct); (18) *Titanoboa* at 43 feet is the longest extinct snake; (19) the reticulated python is the longest living snake; (20) *Architeuthis*, a deep-water squid, is the largest living mollusc; (21) the whale shark is the largest fish; (22) an arthrodire is the largest placoderm (extinct); (23) large tarpon; (24) female *Paedocypris progenetica* from peat swamps of Sumatra; (25) housefly; (26) medium-sized ant; (27) this tropical frog is the smallest tetrapod; (28) cheese mite; (29) smallest land snail; (30) *Daphnia* is a common water flea; (31) a common brown hydra. The lower section of a giant sequoia is shown in the background on the left of the figure with a 100-foot larch superimposed.

After H. G. Wells, J. S. Huxley, and G. P. Wells.

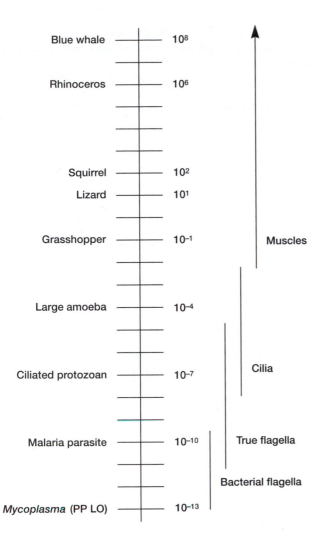

FIGURE 4.2 Body size and locomotion. The masses of living organisms are given on a logarithmic scale. The blue whale tops the scale. *Mycoplasma,* a prokaryotic, bacterium-like organism, is at the bottom. The locomotor mechanism ranges from bacterial flagella to muscle as size increases. Size imposes constraints. Cilia and flagella that move a small mass will become less suitable for locomotion of larger masses. Bigger animals require muscles to drive locomotion.

After McMahon and Bonner.

FIGURE 4.3 Influence of size on performance. The four members of the violin family are similarly shaped but they differ in size. Size differences alone produce different resonances and account for differences in performance. The bass is low, the violin high, and the middle-sized cello and viola produce intermediate frequencies.

After McMahon and Bonner.

Church of Little Tey   10 m

FIGURE 4.4 Influence of size on design. The floor plans of a small medieval church (top) and a large Gothic cathedral (bottom) both in England are drawn to about the same length. The medieval church is about 16 m in length, the Gothic cathedral about 139 m. Because the Gothic cathedral is larger in life, however, it encloses relatively greater space. Transept, chapels, and slotted windows of the sidewalls of the cathedral must let in more light to compensate for the larger volume and to brighten the interior.

For an extended account of the consequences of size on design, see Gould, 1977.

Norwich Cathedral   10 m

(a)   (b)

FIGURE 4.5 Consequences of being large or small. Gravity exerts an important force on a large mass. Surface tension is more important for smaller masses. (a) The large elephant has stout, robust legs to support its great weight. (b) The small water strider is less bothered by gravity. In its diminutive world, surface forces become more significant as it stands on water supported by surface tension.

More formally stated, the surface area (*S*) of an object increases in proportion to ($\propto$) the square of its linear dimensions (*l*):

$$S \propto l^2$$

But volume (*V*) increases even faster in proportion to the cube of its linear dimensions (*l*):

$$V \propto l^3$$

This proportional relationship holds for any geometric shape expanded (or reduced) in size. If we enlarge a sphere, for example, from marble size to soccer ball size, its diameter increases 10 times, its surface increases 10^2 or 100 times, and its volume increases 10^3 or 1,000 times. Any object obeys these relative relationships imposed by its own geometry. A tenfold increase in the length of an organism, as can occur during growth, would bring a 100-fold increase in surface area and a 1,000-fold increase in volume if its shape did not change in the

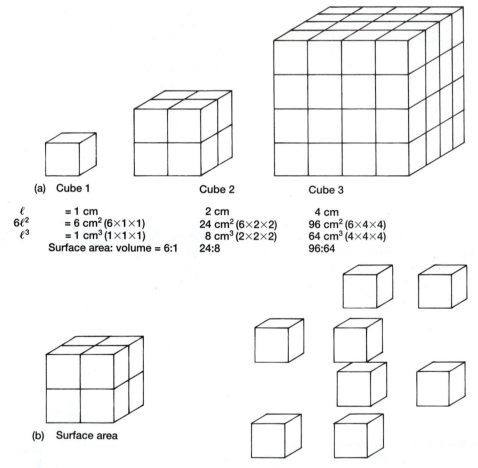

(a)   Cube 1   Cube 2   Cube 3

l	= 1 cm	2 cm	4 cm
$6l^2$	= 6 cm² (6×1×1)	24 cm² (6×2×2)	96 cm² (6×4×4)
l^3	= 1 cm³ (1×1×1)	8 cm³ (2×2×2)	64 cm³ (4×4×4)
	Surface area: volume = 6:1	24:8	96:64

(b)   Surface area

FIGURE 4.6 Length, surface, and volume. (a) Even if shape remains the same, a size increase alone changes the proportions among length, surface, and volume. The length of each edge of the cube quadruples from the smallest to the largest size shown. Cubes 1, 2, and 3 are 1, 2, and 4 cm in length (*l*) on a side, respectively. The length (*l*) of a side increases by a factor of 2 as we go from cube 1 to cube 2 and from cube 2 to cube 3. The surface area jumps by a factor of 4 (2^2) with each doubling of length, and the volume increases by a factor of 8 (2^3). A large object has relatively more volume per unit of surface than a smaller object of the same shape. (b) Surface area. By dividing an object into separate parts, the exposed surface area increases. The cube shown on the left has a surface area of 24 cm², but when it is broken into its constituents, the surface area increases to 48 cm² (8 × 6 cm²). Similarly, chewing food breaks it into many pieces and so exposes more surface area to the action of digestive enzymes in the digestive tract.

process. So to maintain performance, an organism would have to be designed differently when enlarged simply to accommodate an increase in its volume. Consequently, the same organism is necessarily different when large and must accordingly be designed differently to accommodate different relationships among its length, surface, and volume. With this in mind, let us next turn to surface area and volume as factors in design.

Surface Area

To start a fire, a single log is splintered into many small pieces of kindling. Because the surface area is increased, the fire can start more easily. Similarly, many bodily processes and functions depend on relative surface area. Chewing food breaks it into smaller pieces and increases the surface area available for digestion. The efficient exchange of gases, oxygen and carbon dioxide, for instance, depends in part upon available surface area as well. In gills or lungs, large blood vessels branch into many thousands of tiny vessels, the capillaries, thereby increasing surface area and facilitating gas exchange with the blood. Folds in the lining of the digestive tract increase surface area available for absorption. Bone strength and muscle force are proportional to the cross-sectional areas of parts that particular bones and muscles support or move. Vast numbers of bodily processes and functions depend on relative surface area. These examples show that some designs maximize surface area, while others minimize it. Structures (lungs, gills, intestines, capillaries) that are adapted to promote exchange of materials typically have large surface areas.

Because, as we have seen, surface and volume scale differently with changing size, processes based on relative surface area must change with increasing size. For example, in a tiny aquatic organism, surface cilia stroke in coordinated beats to propel the animal. As the animal gets larger, surface cilia have to move proportionately more volume, so they become a less effective means of locomotion. It is no surprise that large aquatic organisms depend more on muscle power than on ciliary power to meet their locomotor needs. The circulatory, respiratory, and digestive systems rely particularly on surfaces to support the metabolic needs required by the mass of an animal. Large animals must have large digestive areas to ensure adequate surface for assimilation of food in order to sustain the bulk of the organism. Large animals can compensate and maintain adequate rates of absorption if the digestive tract increases in length and develops folds and convolutions. Rate of oxygen uptake by lungs or gills, diffusion of oxygen from blood to tissues, and gain or loss of body heat are all physiological processes that rely on surface area. As J. B. S. Haldane once said: "Comparative anatomy is largely the story of the struggle to increase surface in proportion to volume" (Haldane, 1956, p. 954). We will not be surprised then, when in later chapters we discover that organs and whole bodies are designed to address the relative needs of volume in relation to surface area.

As body size increases, oxygen consumption per unit of body mass decreases (figure 4.7). In absolute terms, a large animal, of course, takes in more total food per day than

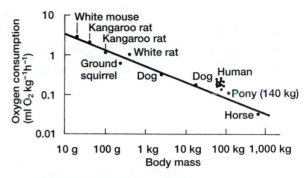

FIGURE 4.7 **Relationship between metabolism and body size.** Physiological processes, like anatomical parts, scale with size. The graph shows how oxygen consumption decreases per unit of mass as size increases. This is a log-log plot showing body mass along the horizontal scale and oxygen consumption along the vertical. Oxygen consumption is expressed as the volume (ml) of oxygen (O_2) per unit of body mass (kg) during one hour (h).

After Schmidt-Nielsen.

does a small animal to meet its metabolic needs. Certainly, an elephant eats more each day than a mouse does. A cougar may consume several kilograms of food per day, a shrew only several grams. But in relative terms, metabolism per gram is less for the larger animal. The several grams the shrew consumes each day may represent an amount equivalent to several times its body weight; the cougar's daily food intake is a small part of its body mass. Small animals operate at higher metabolic rates; therefore, they must consume more oxygen to meet their energy demands and maintain necessary levels of body temperature. This is partly due to the fact that heat loss is proportional to surface area, whereas heat generation is proportional to volume. A small animal has more surface area in relation to its volume than a larger animal does. If a shrew were forced to slow its weight-specific metabolic rate to that of a human, it would need an insulation of fur at least 25 cm thick to keep warm.

Volume and Mass

When a solid object increases in volume, its mass increases proportionately. Because body mass is directly proportional to volume, mass (like volume) increases in proportion to the cube of a body's linear dimensions.

In terrestrial vertebrates, the mass of the body is borne by the limbs, and the strength of the limbs is proportional to their cross-sectional area. Change in body size, however, sets up a potential mismatch between body mass and cross-sectional limb area. As we learned earlier in this section, mass increases faster than surface area when size increases. A tenfold increase in diameter produces a 1,000-fold increase in mass but only a 100-fold increase in cross-sectional area of the supporting limbs. If shape is unchanged without compensatory adjustments, weight-bearing bones fall behind the

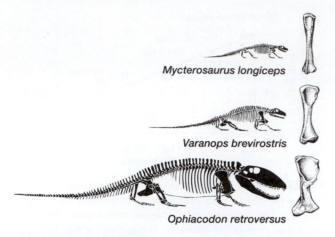

Mycterosaurus longiceps

Varanops brevirostris

Ophiacodon retroversus

FIGURE 4.8 **Body size and limb design in pelyco-saurs.** Relative sizes of three pelycosaur species are illustrated. The femurs of each, drawn to the same length, are shown to the right of each species. The larger pelycosaur carries a relatively larger mass, and its more robust femur reflects this supportive demand.

| Years | 0.42 | 0.75 | 2.75 | 6.75 | 12.75 | 25.75 |

FIGURE 4.9 **Allometry in human development.** During growth, a person changes shape as well as size. As an infant grows, its head makes up less of its overall height and its trunk and limbs make up more. Ages, in years, are indicated beneath each figure.

From McMahon and Bonner; modified from Medawar.

mass they must carry. For this reason, bones of large animals are relatively more massive and robust than the bones of small animals (figure 4.8). This disproportionate increase in mass compared with surface area is the reason why gravity is more significant for large animals than for small ones.

Whether we look at violins, Gothic cathedrals, or animals, the consequences of geometry reign when it comes to size. Objects of similar shape but different size must differ in performance.

Shape

To remain functionally balanced, an animal must have a design that can be altered as its length and area and mass grow at different rates. As a result, an organism must have different shapes at different ages (sizes).

Allometry

As a young animal grows, its proportions may also change. Young children, too, change in proportion as they grow; children are not simply miniature adults. Relative to adult proportions, the young child has a large head and short arms and legs. This change in shape in correlation with a change in size is called **allometry** (figure 4.9).

Detection of allometric scaling rests on comparisons, usually of different parts as an animal grows. For instance, during growth, the bill of the godwit, a shorebird, increases in length faster than its head. The bill becomes relatively long compared to the skull (figure 4.10). Generally, the relative sizes of two parts, *x* and *y*, can be expressed mathematically in the allometric equation

$$y = bx^a$$

where *b* and *a* are constants. When the equation is graphed on log-log paper, a straight line results (figure 4.11a, b).

Allometric relationships describe changes in shape that accompany changes in size. Size changes do not occur only during ontogeny. Occasionally, a phylogenetic trend within a group of organisms includes a relative change in size and proportion through time. Allometric plots describe these trends as well. Titanotheres are an extinct group of mammals that comprise 18 known genera from the Early Cenozoic. A plot of skull length versus horn height for each species shows an allometric relationship (figure 4.12). In this example, we track evolutionary changes in the relationship between parts through several species.

Compared with a reference part, the growing feature may exhibit positive or negative allometry, depending on whether it grows faster than (positive) or slower than (negative) the reference part. For example, compared with skull length, the bill of the godwit shows positive allometry. The term **isometry** describes growth in which the proportions remain constant, and neither positive nor negative allometry occurs. The cubes shown in figure 4.6 exemplify isometry, as do the salamanders illustrated in figure 4.13.

Transformation Grids

D'Arcy Thompson popularized a system of transformation grids that express overall changes in shape. The technique compares a reference structure to a derived structure. For instance, if the skull of a human fetus is taken as a reference structure, a rectilinear transformation grid can be used to define reference points at the intersections of the horizontal and vertical grid lines (figure 4.14). These reference points on the fetal skull are then relocated on the adult skull. Next,

FIGURE 4.10 Allometry in the head of a black-tailed godwit. Differences in relative growth between skull length (lines A and B) and bill length (lines B and C) are compared. Notice that for each increase in skull length, the bill grows in length as well, but at a faster rate. As a result, the bill is shorter than the skull in the chick (top) but longer than the skull in the adult (bottom).

Specimen	Skull dimensions (mm)	
	x	y
A	1	3.5
B	2	11.8
C	3	23.9
D	4	39.6
E	5	58.5
F	6	80.5
G	7	105.4
H	8	133.0
I	9	163.7
J	10	196.8
K	20	662.0
L	30	1,345.9
M	40	2,226.8
N	50	3,290.5
O	60	4,527.2
P	70	5,929.1
Q	80	7,489.9
R	90	9,204.3

(a)

(b)

FIGURE 4.11 Graphing allometric growth. (a) If we organize a range of skulls from the same species in order of size (A–R), we can measure two homologous parts on each skull and collect these data points in a table. (b) If we plot one skull dimension (y) against the other (x) on log-log paper, a line connecting these points describes the allometric relationship between the points during growth in size of the members of this species. This can be expressed with the general allometric equation, $y = bx^a$, wherein y and x are the pair of measurements and b and a are constants, b being the y-intercept and a being the slope of the line. In this example, the slope of the line (a) is 1.75. The y-intercept (b) is 3.5, observed on the graph or calculated by placing the value of x equal to 1 and solving for y. The equation describing the data is $y = 3.5x^{1.75}$.

Adapted with permission from On Size and Life by T. McMahon and J. T. Bonner, Copyright 1983 by Thomas McMahon and John Tyler Bonner. Reprinted by permission of Henry Holt and Company, LLC.

FIGURE 4.12 Allometric trends in phylogeny. The skull and horn lengths of titanotheres, an extinct family of mammals, are plotted. The horn length increases allometrically with increasing size of the skull of each species.

After McMahon and Bonner.

Fetal skull

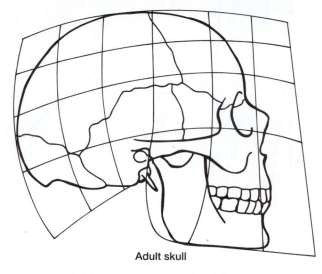

Adult skull

FIGURE 4.14 Transformation grids in ontogeny. Shape changes in the human skull can be visualized more easily with correlated transformation grids. Horizontal and vertical lines spaced at regular intervals can be laid over a fetal skull. The intersections of these lines define points of reference on the fetal skull that can be relocated on the adult skull (bottom) and used to redraw the grid. Because the adult skull has a different shape, the reference points from the fetal skull must be reoriented. A reconstructed grid helps to emphasize this shape change.

After McMahon and Bonner, based on Kummer.

FIGURE 4.13 Isometry. These six species of salamanders differ in size; yet, the smallest is almost the same shape as the largest because body proportions within this genus (*Desmognathus*) remain almost constant from species to species.

Kindly supplied by Samuel S. Sweet, UCSB.

these reference points are connected again to redraw the grid, but because the shape of the skull has changed with growth, the redrawn grid too is differently shaped. Thus, the grid graphically depicts shape changes. Similarly, transformation grids can be used to emphasize, graphically, phylogenetic differences in shape between species, such as the fishes shown in figure 4.15.

Transformation grids and allometric equations do not explain changes in shape; they only describe them. However, in describing changes in proportions, they focus our attention on how tightly shape couples with size.

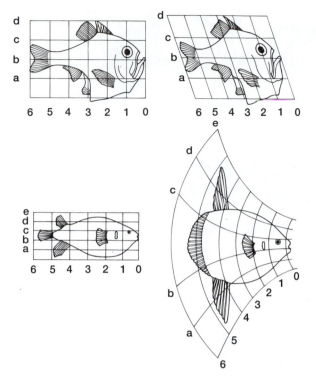

FIGURE 4.15 Transformation grids in phylogeny. Changes in shape between two or more species, usually closely related, can also be visualized with transformation grids. One species is taken as the reference (left), and the reference points are relocated in the derived species (right) to reconstruct the transformed grid.

Modified from Thompson.

On the Consequences of Being the Right Size

Animals, large or small, enjoy different advantages because of their sizes. The larger an animal, the fewer are the predators that pose a serious threat. Adult rhinoceroses and elephants are simply too big for most would-be predators to handle. Large size is also advantageous in species in which physical aggression between competing males is part of reproductive behavior. On the other hand, small size has its advantages as well. In fluctuating environments struck by temporary drought, the sparse grass or seeds that remain may sustain a few small rodents. Because they are small, they require only a few handfuls of food to see them through. When the drought slackens and food resources return, the surviving rodents, with their short reproductive and generation times, respond in short order and their population recovers. In contrast, a large animal needs large quantities of food on a regular basis. During a drought, a large animal must migrate or perish. Typically, large animals also have long generation times and prolonged juvenile periods. Thus, populations of large animals may take years to recover after a severe drought or other devastating environmental trauma.

The larger an animal, the more its design must be modified to carry its relatively greater weight, a consequence of the increasing effects of gravity. It is no coincidence that the blue whale, the largest animal on Earth today, evolved in an aquatic environment in which its great weight received support from the buoyancy of the surrounding water. For terrestrial vertebrates, an upper size limit occurs when supportive limbs become so massive that locomotion becomes impractical. The movie creators of Godzilla were certainly unaware of the impracticality of their design as this great beast crashed about stomping buildings. For lots of reasons, not the least of which is his size, Godzilla is an impossibility.

Body parts used for display or defense often show allometry, as the adult ram horns in figure 4.16 illustrate. As a male lobster grows, its defensive claw grows too, but much more rapidly than the rest of its body. When the lobster attains a respectable size, its claw has grown into a formidable weapon (figure 4.17). The claw exhibits **geometric growth;** that is, its length is *multiplied* by a constant in each time interval. The rest of the body shows **arithmetic growth** because a constant is *added* to its length in each time interval. To be effective in defense, the claw must be large, but a young lobster cannot yet wield so heavy a weapon because of its small size. Only after attaining substantial body size can such a claw be effectively deployed in defense. The accelerated growth of the claw brings it in later life up to fighting size. Before that, the small lobster's major defensive tactic is to dash for cover under a rock.

This example shows that size and shape are sometimes linked because of biological function, as with the lobster and its claw. More often, however, design is concerned with the consequences of geometry. Changes in the relationship among length, surface, and volume as an object increases in size (figure 4.6) are the major reason why change in size is necessarily accompanied by change in shape. As we see time and again throughout the book, size itself is a factor in vertebrate design and performance.

Biomechanics

Physical forces are a permanent part of an animal's environment. Much of the design of an animal serves to catch prey, elude predators, process food, and meet up with mates. But biological design must also address the physical demands placed upon the organism. In part, analysis of biological design requires an understanding of the physical forces an animal experiences. Those in the field of **bioengineering** or **biomechanics** borrow concepts from engineering mechanics to address these questions.

Mechanics is the oldest of the physical sciences, with a successful history dating back at least 5,000 years to the ancient pyramid builders of Egypt. It continues up to the present with the engineers who send spaceships to the planets. Through the course of its history, engineers of this discipline have developed principles that describe the physical

FIGURE 4.16 **Changes in horn shape.** These species and subspecies of Asiatic sheep show changes in horn shape across their geographic distribution (top). The first in the lineup is the Barbary sheep (*Ammotragus*) from North Africa. The others belong to species or subspecies of *Ovis,* Asiatic sheep of the argali group that extend into central Asia. The last sheep on the right is the Siberian argali (*Ovis ammon ammon*). As a young male bighorn grows in size (bottom illustration), its horns change shape as well. In the adult ram, these horns are used in social displays and in combat with male rivals.

Modified from Geist, 1971.

FIGURE 4.17 **Lobster allometry.** Although the defensive claw is small at first, it grows geometrically while body length increases only arithmetically. Because of this, when the body is large enough to use the claw, the claw has increased dramatically in size to become an effective weapon. The dashed line indicates the size the claw would reach if it did not show allometry; that is, if it grew arithmetically instead of geometrically.

properties of objects, from bodies to buildings. Ironically, engineers and biologists usually work in reverse directions. An engineer starts with a problem, for instance, a river to span, and then designs a product, a bridge, to solve the problem. A biologist, however, starts with the product, for instance, a bird wing, and works back to the physical problem it solves, namely, flight. Nonetheless, reducing animals to engineering analogies simplifies our task of understanding animal designs.

Fundamental Principles

Animals certainly are more than just machines. But the perspective of biomechanics gives a clarity to biological design that we might not otherwise expect. An introduction to a few basic biomechanical principles follows.

Basic Quantities—Length, Time, and Mass

Most of the physical concepts we deal with in biomechanics are familiar. **Length** is a concept of distance, **time** is a concept of the flow of events, and **mass** is a concept of inertia.

Length and time come to us easily. But when it comes to the concept of **mass,** however, our intuition not only fails but actually interferes, because what most people call "weight" is not equivalent to "mass." Mass is a property of matter, weight a measure of force. One way to think of the difference is to consider two objects in outer space, say, a pen and a refrigerator. Both would be weightless and neither would exert a force on a scale. However, both still have mass, although the mass of each is different. To toss the pen to a companion astronaut would require little effort, but to move the massive but weightless refrigerator would require a mighty heave even in the weightlessness of space. Contrary to intuition, therefore, weight and mass are not the same concepts.

Units

Units are not concepts but conventions. They are standards of measurement that, when attached to length, time, and mass, give them concrete values. A photograph of a building alone gives no necessary indication of its size (figure 4.18); therefore, a friend is often pressed into service to stand in the picture to give a sense of scale to the building. Similarly, units serve as a familiar scale. But different systems of units have grown up in engineering, so a choice must be made.

In a few English-speaking countries, mainly the United States, the "English system" of measurement—pounds, feet, seconds—has been preferred. Initially, these units grew up from familiar objects such as body parts. The "inch" was originally associated with the thumb's width, the "palm" was the breadth of the hand, about 3 inches, the foot equaled 4 palms, and so on.

FIGURE 4.18 **Units of reference.** We use familiar objects as references of size. If denied familiar references, such as fellow humans (top), we can easily overestimate the true size of the cathedral (bottom). Units of measurement such as inches, feet, and pounds are conventions attached to quantities in order for us to set standard references for expressing distances and weight.

Although poetic, the English system can be cumbersome when converting units. For instance, to change miles to yards requires multiplying by 1,760. To convert yards to feet, we must multiply by 3, and to convert feet to inches, we multiply by 12. During the French Revolution, a simpler system based on the meter was introduced. Changing kilometers to meters, meters to centimeters, or centimeters to millimeters requires only moving the decimal point. The **Système Internationale,** or **SI,** is an extended version of the older metric system. Primary units of the SI include meter (m), kilogram (kg), and second (s) for dimensions of length, mass, and time, respectively. In this book, as throughout physics and biology, SI units are used. Table 4.1 lists the common units of measurement in both the English and the SI systems.

Derived Quantities—Velocity, Acceleration, Force, and Relatives

Velocity and **acceleration** describe the motion of bodies. Velocity is the rate of change in an object's position, and acceleration in turn is the rate of change in its velocity. In part, our intuition helps our understanding of these two concepts. When traveling east by car on an interstate highway, we may change our position at the rate of 88 km per hour (velocity) (about 55 mph if you are still thinking in the English system). Step on the gas and we accelerate; hit the brake and we decelerate or, better stated, we experience negative acceleration. With mathematical calculations, negative acceleration is a better term to use than deceleration because we can keep positive and negative signs in a more straightforward way. The sensation of acceleration is familiar to most, but in common conversation, units are seldom mentioned. When they are properly applied, units may sound strange. For instance, suddenly braking a car may produce a negative acceleration of -290 km h^{-2}

English System	**Physical Quantity**	**Système Internationale (SI)**
Slug or pound mass	Mass	Kilogram (kg)
Foot (ft)	Length	Meter (m)
Second (s)	Time	Second (s)
Feet/second (fps)	Velocity	Meters/second (m s^{-1})
Feet/second^2 (ft sec^2)	Acceleration	Meters/second^2 (m s^{-2})
Pound (lb)	Force	Newtons (N or kg m s^{-2})
Foot-pound (ft-lb)	Moment (torque)	Newton meters (Nm)

TABLE 4.1 Common Fundamental Units of Measurement

(about -180 mph h^{-1}). The units may be unfamiliar, but the experience of acceleration, like that of velocity, is an everyday event.

Force describes the effects of one body acting on another through their respective mass and acceleration. **Density** is mass divided by volume. Water has a density of 1,000 kilograms per cubic meter (kgm m^{-3}). **Pressure** is force divided by the area over which it acts—pounds ft^{-2} or N m^{-2} for instance. **Work** is the force applied to an object times the distance the object moves in the direction of the force with a joule (after James Joule, 1818–1889) as the unit. Oddly, if the object does not move, much force may be applied but no work is accomplished. A chain holding a chandelier exerts a force keeping the object in place, but if the chandelier remains in position, no displacement occurs so no work occurs. **Power** is the rate at which work gets done, therefore power equals work divided by the time it takes. The unit is the watt (after James Watt, 1736–1819), and one watt is a joule per sec (J s^{-1}).

Common conversation has allowed these terms to drift in meaning, which we need to avoid when we use them in a physical sense. I have already mentioned the misuse of weight (a result of gravity) and of mass (a result of the object's own properties independent of gravity). We might speak of a strong arm squeeze as a lot of force, when our discomfort may in fact result from force per concentrated area—pressure. We might express admiration for a person lifting a weight by saying he or she exerts a lot of power, when in fact we are not talking about the rate of doing work but the force generated to lift the weight. In physical terms, we might speak ambiguously. If we say something is heavy we could mean either it is massive or that it is very dense. Even units have slipped. The calorie listed on food packaging is actually a kilocalorie, but calorie sounds leaner.

Reference Systems

When preparing to record events, a conventional frame of reference is selected that can be overlaid on an animal and its range of activity. But be prepared. A reference system can be defined relative to the task at hand. For instance, when you walk back to the restroom in the tail section of an airplane, you use the plane for reference and ignore the fact that you are really walking forward, with respect to the Earth below. A bird can't get its tail feathers ruffled by flying with a tailwind—it just goes that much faster with respect to the Earth below. For our purposes, and for most engineering applications as well, the coordinate system is usually defined relative to the surface of the Earth.

For reference systems, there are several choices, including the polar and cylindrical systems. The most common, however, is the **rectangular Cartesian reference system** (figure 4.19). For an animal moving in three-dimensional space, its position at any moment can be described exactly on three axes at right angles to each other. The horizontal axis is x, the vertical axis is y, and the axis at right angles to these is z.

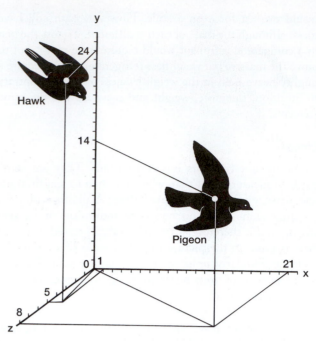

FIGURE 4.19 **A three-axis Cartesian coordinate reference system defines the position of any object.** Customarily the horizontal, vertical, and axis at right angle to these two are identified as x, y, and z, respectively. The three intersect at the origin (0). The direct projection line of an object to each axis defines its position at that instant along each axis. Thus, the three projections fix an object's position in space— 1, 24, 5 for the hawk and 21, 14, 8 for the pigeon. The white dot graphically represents the center of mass of each bird.

Once defined, the orientation of these reference systems cannot be changed, at least not during the episode during which we are taking a series of measurements.

Center of Mass

If we are interested in the motion of a whole organism rather than the separate motion of its parts, we can think of the mass of the animal as being concentrated at a single point called its **center of mass.** The center of mass, in laypersons' terms the center of gravity, is the point about which an animal is evenly balanced. As a moving animal changes the configuration of its parts, the position of its center of mass changes from one instant to the next (figure 4.20).

Vectors

Vectors describe measurements of variables with a magnitude and a direction. Force and velocity are examples of such variables because they have magnitude (N in the SI, mph in the English system) and direction (e.g., northwesterly direction). A measurement with only magnitude and no direction is a **scalar quantity.** Time duration and temperature have magnitude but no direction, so they are scalar, not vector, quantities. A force applied to an object can also be

FIGURE 4.20 Center of mass. The single point at which the mass of a body can be thought to be concentrated is the center of mass. As the configuration of this jumper's body parts changes from takeoff (left) to midair (right), the instantaneous location of the center of mass (red dot) changes as well. In fact, note that the center of mass lies momentarily outside the body. A high-jumper or pole-vaulter can pass over the bar even though his or her center of gravity moves under the bar.

represented along a rectangular Cartesian reference system. When we use such a reference system, trigonometry helps us to calculate vector values. For example, we can measure the force applied to a dragged object (F in figure 4.21), but the portion of that force acting horizontally against surface friction (F_x) is more difficult to measure directly. However, given the force (F) and angle (θ), we can calculate both horizontal and vertical components (F_x and F_y). And, of course, conversely, if we know the component forces (F_x and F_y), we can calculate the combined resultant force (F).

Basic Force Laws

Much of engineering is based on laws that were formulated by Isaac Newton (1642–1727). Three of his laws are fundamental:

1. *First law of inertia.* Because of its inertia, every body continues in a state of rest or in a uniform path of motion until a new force acts on it to set

it in motion or change its direction. **Inertia** is the tendency of a body to resist a change in its state of motion. If the body is at rest, it will resist being moved, and if it is in motion, it will resist being diverted or stopped.

2. *Second law of motion.* Simply stated, the change in an object's motion is proportional to the force acting on it (figure 4.22). Or, a force (F) is equal to the mass (m) of an object times its experienced acceleration (a):

$$F = ma$$

Units of this force in newtons (N), kg m s^{-2}, are the force needed to accelerate 1 kg mass at 1 meter per second^2.

3. *Third law of action, reaction.* Between two objects in contact, there is for each action an opposite and equal reaction. Applying a force automatically generates an equal and opposite force—push on the ground and it pushes back on you.

Albert Einstein's (1879–1954) theories of relativity placed limits on these Newtonian laws. But these limitations become mathematically significant only when the speed of an object approaches the speed of light (186,000 miles/s). Newtonian laws serve space travel well enough to get vehicles to the moon and back, and so they will serve us here on Earth as well.

FIGURE 4.21 Vectors. When dragging the seal, the polar bear produces a resultant force (**F**) that can be represented by two small component forces acting vertically (F_y) and horizontally (F_x). The horizontal force acts against surface friction. If we know the resultant force (**F**) and its angle (θ) with the surface, we can calculate the component forces using graphic or trigonometric techniques.

FIGURE 4.22 Forces of motion. (a) The force a frog produces at liftoff is the result of its mass and acceleration at that instant ($F = ma$). (b) Forces produced collectively by both feet of a hefty frog and the ground are opposite but equal. The vector parallelograms represent the components of each force. If a frog of 50 g (.05 kg) accelerates 100 m s^{-2}, a force of 5 N = (100 × .05) is generated along the line of travel. By using trigonometric relations, we can calculate the component forces. If liftoff is at 37°, then these component forces are 4 N = (cos 37° × 5 N) and 3 N = (sin 37° × 5 N).

In biomechanics, Newton's second law, or its modifications, are most often used because the separate quantities can be measured directly. In addition, knowing the forces experienced by an animal often gives us the best understanding of its particular design.

Free Bodies and Forces

To calculate forces, it often helps to isolate each part from the rest in order to look at the forces acting on that part. A **free-body diagram** graphically depicts the isolated part with its forces (figure 4.23a, b).

When you walk across a floor, you exert a force upon it. The floor gives ever so slightly and imperceptibly until it returns a force equal to yours, which exemplifies the action and reaction principle described by Newton's third law. If the floor did not push back equally, you would fall through. Think of a diver perched at the end of a diving board. The board bends until it pushes back with a force equal to the force exerted on it by the diver. Diver and board are separated in the free-body diagram, and the forces on each are shown in figure 4.23a. If both forces are equal and opposite, they cancel and the two are in equilibrium. If not, motion is produced (figure 4.23b).

As a practical matter, mechanics is divided between these two conditions. Where all forces acting on an object balance, we are dealing with that part of mechanics known as **statics.** Where acting forces are unbalanced, we are dealing with **dynamics.**

Torques and Levers

In the vertebrates, muscles generate forces and skeletal elements apply these forces. There are several ways to represent this mechanically. Perhaps the most intuitive representation is with torques and levers. The mechanics of torques and levers are familiar because most persons have firsthand experience with a simple lever system, the teeter-totter or seesaw of childhood. Action of the seesaw depends on the opposing weights seated on opposite ends and on the distances of these weights from the pivot point, or **fulcrum.** This distance from weight to fulcrum is the **lever arm.** The lever arm is measured as the perpendicular distance from force to fulcrum. Shorten the lever arm and more weight must be added to keep the board in balance (figure 4.24a). Lengthen it sufficiently, and a little sister can keep several big brothers balanced on the opposite end.

A force acting at a distance (the lever arm) from the fulcrum tends to turn the seesaw about this point of rotation; or more formally, it is said to produce **torque.** When levers are used to perform a task, we also recognize an **in-torque** and **out-torque.** If more output force is required, shortening the "out" and lengthening the "in" lever arms increases the out-torque. Conversely, if out-torque speed (= velocity) or travel distance is required, then lengthening the out and shortening the in lever arms favor greater velocity and distance in the out-torque (figure 4.24c). Of course,

FIGURE 4.23 **Free-body diagrams.** (a) Two physical bodies, the board and the diver, each exert a force on the other. If the forces of the two bodies are equal, opposite, and in line with each other, then no linear or rotational motion results. Although the forces are present, the two bodies are in equilibrium (left). To depict these forces (right), the two bodies are separated in free-body diagrams, and the forces acting on each body are represented by vectors (arrows). (b) If forces are unequal, motion is imparted. The diver has, by sudden impact, pushed the board down farther than it would go under his weight alone. So it then pushes upward with a force greater than the weight of the diver, and the diver is accelerated upward.

this increased speed and distance are achieved at the expense of force in the out-torque. In engineering terms, torque is more commonly described as the **moment** about a point and the lever arm as the **moment arm.**

The mechanics of levers mean that output force and output speed are inversely related. Long output lever arms favor speed, whereas short ones favor force. Regardless of how desirable it would be to have both in the design of, say, an animal's limb, simple mechanics do not permit it. Similarly, long output lever arms sweep through a greater distance, whereas short ones move through a shorter distance. For a given input, both output force and output speed cannot be maximized. Compromises and trade-offs in design must be made.

Consider the forelimbs of two mammals, one a runner specialized for speed, the other a digger specialized to generate large output forces. In figure 4.25a, the relatively long

Lever arm (*l*)   Lever arm (*l*)

(a)

(b)

PAINT

(c)

FIGURE 4.24 **Principles of lever systems.** (a) The balance of forces about a point of pivot (fulcrum) depends on the forces times their distances to the point of pivot, their lever arms (*l*). (b) To get more output force, the point of pivot is moved closer to the output and farther from the input force. In this diagram, the short output lever arm (l_o) and long input lever arm (l_i) work in favor of more output force. (c) To produce high output speed, the pivot point is moved closer to the input force (l_i). Other things being equal, speed is achieved at the expense of output force.

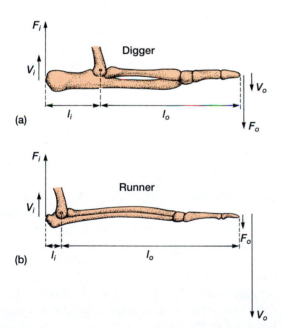

(a)

(b)

FIGURE 4.25 **Strength versus speed.** The forearms of a digger (a) and a runner (b) are drawn to the same overall length. Input forces (F_i) and input velocities (V_i) are the same, but output forces (F_o) and velocities (V_o) differ. The differences result from the differences between the lever arm ratios of the two forearms. Output force is greater in the digger than in the runner, but output velocity of the digger is less. Formally, these differences can be expressed as differences in mechanical advantages and in velocity ratios.

elbow process and short forearm of the digger favor large force output. In the runner (figure 4.25b), the elbow is short, the forearm long. Lever arms are less favorable to force output in the forearm of the runner but more favorable to speed. The speed of the elbow is magnified by the relatively greater output lever arm, but this is accomplished at the expense of output force.

More formally, we can express the mechanics of input and output forces at different velocities with simple ratios. The ratio of F_o/F_i, the output to input force, is the **mechanical advantage** (or **force advantage**). The ratio of the output to input lever arms, l_o/l_i, is the **velocity ratio** (or **speed** or **distance advantage**).

As we might expect, diggers enjoy a greater mechanical advantage in their forearm, but runners enjoy a greater velocity advantage in their forearm. There are, of course, other ways of producing output force or speed. Increased size and, hence, force, of input muscles and emphasis of fast-contracting muscle cells both affect output. The lever systems of an animal, in turn, set the relationships between force and speed (or distance).

Artiodactyls, such as deer, have limbs designed to produce both high forces, as during acceleration, and high speed, as when velocity of escape is required (figure 4.26). Two muscles, the medial gluteus and the semimembranosus, with different mechanical advantages, make different contributions to force or to speed output. The medial gluteus enjoys a higher velocity ratio (l_o/l_i = 44, compared

Medial gluteus

l_i

l_i

Semimembranosus

l_o

FIGURE 4.26 High and low gear muscles. Both the medial gluteus and the semimembranosus muscles turn the limb in the same direction, but they possess different mechanical advantages in doing so. A muscle's lever arm is the perpendicular distance to the point of rotation or pivot point (black dot) from the line of muscle action (dotted line). The velocity ratio is higher in the medial gluteus, which can move the limb faster. But the semimembranosus moves the limb with greater output force because of its longer lever arm. Lever arms in both the muscles (l_i) and the common lever arm out (l_o) are indicated.

After Hildebrand.

with $l_o/l_i = 11$ for the semimembranosus), a leverage that favors speed. If we compare these muscles with the gears of a car, the medial gluteus would be a "high" gear muscle.

On the other hand, the semimembranosus has a mechanical advantage favoring force and would be a "low" gear muscle. During rapid locomotion, both are active, but the low gear muscle is most effective mechanically during acceleration, and the high gear muscle is more effective in sustaining the velocity of the limb.

The two muscles of the deer limb swing it in the same direction. But each muscle acts with a different lever advantage. One is specialized for large forces, the other for speed. This represents one way biological design may incorporate the mechanics of torques and levers to provide the limb of a running animal with some degree of both force and speed output. Just as a seesaw does not have a single fulcrum that can maximize output force and output speed simultaneously, similarly one muscle cannot maximize both. A single muscle has leverage that can maximize either its force output or its speed output, but not both, a limitation that arises from the nature of mechanics, not from any necessity of biology. To work around this, two or more muscles may divide the various mechanical chores amongst them and impart to the limb favorable force, speed, or distance during limb rotation. Biological design must abide by the laws and limits of mechanics when mechanical problems of animal function arise.

Land and Fluid

For terrestrial vertebrates, most external forces they experience arise ultimately from the effects of gravity. Vertebrates in fluids, such as fishes in water or birds in flight, experience additional forces from the water or air around them. Because the forces are different, the designs that address them differ as well.

Life on Land: Gravity

Gravity acts on an object to accelerate it. On the surface of the Earth, the average acceleration of gravity is about 9.81 m s^{-2} acting toward the Earth's center. Newton's second law ($F = ma$) tells us that an animal with a mass of 90 kg produces a total force of 882.9 N (90 kg $\times$ 9.81 m s^{-2}) against the Earth upon which it stands. An object held in your hand exerts a force against your hand, which results from the object's mass and gravity's pull. Release the object, and the acceleration from gravity's effects becomes apparent as the object picks up speed as it falls to Earth (figure 4.27). Gravity's persistent attempt to accelerate a terrestrial animal downward constitutes the animal's weight. In tetrapods, this is resisted by the limbs.

The weight of a quadrupedal animal is distributed among its four legs. The force borne by fore- and hindlimbs depends on the distance of each from the center of the animal's mass. Thus, a large *Diplodocus* might have distributed its 18 metric tons (39,600 pounds) with a ratio of 4 tons to its forelimbs and 14 to its hindlimbs (figure 4.28).

When we explored the consequences of size and mass at the beginning of this chapter, we noted that large animals have relatively more mass to contend with than small

FIGURE 4.27 Gravity. The clam released by the seagull accelerates under gravity's pull and picks up speed as it falls to the rocks. With equal intervals of time designated by each of the six arrows, note the accelerating positions of the clam.

animals. A small lizard scampers safely across tree limbs and vertical walls; a large lizard is earthbound. Gravity, like other forces, is a part of an animal's environment and affects performance in proportion to body size. Size is also a factor for animals that live in fluids, although forces other than gravity tend to be predominant.

Life in Fluids

Dynamic Fluids   Water and air are fluids. Air is certainly thinner and less viscous than water, but it is a fluid nonetheless. The physical phenomena that act on fishes in water generally apply to birds in air. Air and water differ in viscosity, but they place similar physical demands on animal designs. When a body moves through a fluid, the fluid exerts a resisting force in the opposite direction to the body's motion. This resisting force, termed **drag,** may arise from various

FIGURE 4.28 Weight distribution. (a) The estimated center of mass of this dinosaur lies closer to the hindlimbs than to its forelimbs, so the hindlimbs bear most of the animal's weight. For *Diplodocus,* its 18 metric tons (39,600 pounds) might have been carried by a ratio of 4 tons to its forelimbs and 14 tons to its hindlimbs. (b) If *Diplodocus* lifted its head and forefeet up to reach high vegetation, then all 18 tons would have been carried by hindlimbs and tail, which, forming three points of support, give each limb and the tail 6 tons to carry.

physical phenomena, but forces caused by **friction drag** (or skin friction) and by **pressure drag** (see below) are usually the most important. As an animal moves through a fluid, the fluid flows along the sides of its body. As fluid and body surface move past each other, the fluid exerts a resisting force (drag) on the surface of the animal where they make contact. This force creates friction drag and depends, among other things, on the viscosity of the fluid, the area of the surface, the surface texture, and the relative speed of fluid and surface.

Individual particles in a fluid traveling in a flow describe individual paths. If the average direction of these particles is plotted and points connected along the line of overall flow, nonoverlapping streamlines that represent the general layered pattern of fluid flow are produced. The derived **streamlines** therefore express the statistical summary of layered flows slipping smoothly across one another within a moving fluid. Special and often complex events occur within the **boundary layer,** the thin, fluid layer closest to the surface of the body. Generally it is a thin gradient slowing from the velocity of the general flow down to zero on the surface of the object across which the fluid flows. In your car traveling at 60 mph (96 kmh), the air velocity drops from 60 mph down to zero in the boundary layer, which is one reason clinging insects on your windshield in this boundary layer can hold on. This layer can be very thin. In a Boeing 747, the boundary layer is about 1 in. thick at the trailing edge of the wing. Natural instabilities in the boundary layer may cause the fluid to become chaotic and the flow is spoken of as turbulent. This increases drag dramatically. Where the flow is smooth and nonchaotic, it is described as laminar.

If the particles in the boundary layer passing around an object are unable to make the sharp turn smoothly behind the object, then the layers within the flow tend to part; this is termed **flow separation** (figure 4.29a). The fluid behind the object moves faster and pressure drops leading to pressure drag, which may be seen as a wake of disturbed fluid behind

(a) Fast

Separation

(b)

FIGURE 4.29 Streamlining. (a) Particles in the boundary layer are unable to make the sharp change in direction and velocity to negotiate the turn around the cylinder-shaped object, and flow separation occurs behind the object. (b) Extension and tapering of the object into the area of disturbance helps prevent separation and results in a streamline shape.

a boat. Physically, the flow separation results from a substantial pressure differential (pressure drag) between the front and the back of the animal. An extended, tapering body fills in the area of potential separation, encourages streamlines to close smoothly behind it, and thereby reduces pressure drag (figure 4.29b). The result is a streamlined shape common to all bodies that must pass rapidly and efficiently through a fluid. An active fish, a fast-flying bird, and a supersonic aircraft are all streamlined for much the same reason—to reduce pressure drag (figure 4.30a–d).

A golf ball in flight meets the same problems but is engineered differently to address drag forces. The dimpled surface of a golf ball helps hold the boundary layer longer, smoothes the streamlines, reduces the size of the disturbed wake, and, hence, reduces pressure drag. As a result, a golf ball with dimples travels, other things being equal, about twice as far as one with a smooth surface.

Together, friction and pressure drag contribute to **profile drag,** which is related to the profile or shape an object presents to the moving fluid. If you place your cupped hand out the window of a fast-moving car, you can feel the difference when presented edge-on or palm-on to the onrushing air. A change in profile changes the drag. A thin, broad wing of a bird meeting the air edge-on presents a small profile. But as the wing tips up, changing the angle of attack, the broad profile of the wing meets the air, increasing drag. Fish fins or seal flippers, when used to make sharp turns, are moved with the broadest profile to the water, much like the power stroke of a boat oar, taking advantage of profile drag to help generate cornering forces.

Engineers examine the physical problems associated with motion through fluids within the disciplines of hydrodynamics (water) or aerodynamics (air). Applied to animal designs that move through fluids, these disciplines reveal how size and shape affect the way the physical forces of a fluid act on a moving body.

In general, four physical characteristics affect how the fluid and body dynamically interact. One of these is the *density,* or mass per unit volume of the fluid. A second is the *size* and *shape* of the body as it meets the fluid. The resistance a rowboat oar experiences when the blade is pulled broadside-on is, of course, quite different than when it is pulled edge-on. The third physical characteristic of a fluid is its *velocity.* Finally, the *viscosity* of a fluid refers to its resistance to flow. These four characteristics are brought together in a ratio known as the Reynolds number:

$$Re = \frac{\rho l U}{\mu}$$

where ρ is the density of the fluid and μ is a measure of its viscosity; l is an expression of the body's characteristic shape and size; and U is its velocity through the fluid.

(a)

(b)

Alula

(c)

(d)

FIGURE 4.30 Life in fluids. (a) The airplane wing, shown in cross section, encourages smooth flow of the passing airstream. (b) As the angle of the wing increases relative to air flow, separation of flow from the upper wing surface suddenly forms and lift is lost. (c) Birds, such as this falcon, have a small tuft of feathers (alula) that can be lifted to smooth the airflow at high angles of attack. (d) When separation begins, this small airfoil can be lifted to form a slot that accelerates air over the top of the wing, preventing separation and thus stalling—a sudden drop in lift.

Modified from McMahon and Bonner.

Perhaps because we ourselves are large land vertebrates, we have some intuition about the importance of gravity but no special feeling for all that the Reynolds number has to tell us about life in fluids. The units of all variables of the ratio cancel each other, leaving the Reynolds number without units—no feet per second, no kilograms per meter, nothing. It is dimensionless, a further factor obscuring its message; yet, it is one of the most important expressions that summarizes the physical demands placed upon a body in a fluid. The Reynolds number was developed during the nineteenth century to describe the nature of fluid flow, in particular, how different circumstances might result in fluid flows that are dynamically similar. The Reynolds number tells us how properties of an animal affect fluid flow around it. In general, at low Reynolds numbers, skin friction is of great importance; at high Reynolds numbers, pressure drag might predominate. Perhaps most importantly, at least for a biologist, the Reynolds number tells us how changes in size and shape might affect the physical performance of an animal traveling in a fluid. It draws our attention to the features of the fluid (viscosity) and the features of the body (size, shape, velocity) that are most likely to affect performance.

For scientists performing experiments, the Reynolds number helps them to build a scale model that is dynamically similar to the original. For example, several biologists wished to examine air ventilation through prairie dog burrows but lacked the convenient space to build a life-sized tunnel system in the laboratory. Instead, they built a tunnel system ten times smaller but compensated by running winds ten times faster through it. The biologists were confident that the scale model duplicated conditions in the full-sized original because a similar Reynolds number for each tunnel verified that they were dynamically similar even though their sizes differed.

Static Fluids   Fluids, even thin, low-density fluids such as air, exert a pressure on objects within them. The unit of pressure, Pascal (Pa), is equivalent to 1 newton acting over 1 square meter. The expression "as light as air" betrays the common misconception that air has almost no physical presence. In fact, air exerts a pressure in all directions of about 101,000 Pa (14.7 psi, or pounds per square inch) at sea level, which is equivalent to 1 **atmosphere** (atm) of pressure. The envelope of air surrounding Earth extends up to several hundred kilometers. Although not dense, the column of air above the surface of Earth is quite high, so the additive weight at its base produces a substantial pressure at Earth's surface. We and other terrestrial animals are unaware of this pressure since it comes from all directions and is counterbalanced by an equal outward pressure from our bodies. Thus, all forces on our bodies balance, inside with out. Respiratory systems need only produce relatively small changes in pressure to move air in and out of the lungs.

If we drive from low elevation to high elevation in a short period of time, we might notice the unbalanced pressure that builds uncomfortably in our ears until a yawn or stretch of our jaw "pops" and equilibrates the inside and outside pressures to relieve the mismatch. Most of us have experienced increasing pressure as we dive deeper in water. At a given depth, the pressure surrounding an animal in water is the same from all sides. The deeper the animal, the greater is the pressure. In fresh water, with each meter of depth, atmospheric pressure increases by about 9.8×10^3 Pa. At 5 meters, atmospheric pressure would be about 49×10^3 Pa. Scaled for a human, that would be like trying to breathe with a 90-kg slab placed upon the chest. A fully submerged sauropod would experience 49×10^3 N on each square meter of its entire chest (figure 4.31). It is not likely that even the massive chest muscles of this dinosaur could overcome so much pressure when it drew in a breath. Therefore, *Brachiosaurus* and other long-necked animals probably did not live aquatic lives with their bodies deeply submerged and their heads reaching far above to the surface to snorkel air. Snorkeling works only for small creatures close to the water's surface such as mosquito larvae or a blow-hole-equipped cetacean breaching the surface.

Buoyancy describes the tendency of a submerged object in a fluid to sink or to rise. Long ago, Archimedes (287–212 B.C.E.) figured out that buoyancy was related to the *volume* an object displaces compared to its own weight.

FIGURE 4.31   Water pressure. Water pressure increases with depth, but at any given depth, pressure is equal from all directions. For each meter below the surface, the pressure in fresh water increases by about 9,800 Pa. A large sauropod submerged up to its chin would experience water pressure of about 49,000 Pa (5 m × 9,800 Pa) around its chest, too much pressure to allow chest expansion against this force. Breathing would be impossible. Sauropods such as *Brachiosaurus* were probably not so completely aquatic as shown here, and certainly they would not have used a tall neck to snorkel air well above their submerged chest.

If the density of the submerged object is less than that of water, then buoyancy will be a positive upward force; if its density is greater than water, then the buoyancy is negative and it is forced down. Since density is related to volume, any change in volume will affect the tendency of the object to rise or sink. Many bony fishes possess a flexible gas bladder (swim bladder) that can be filled with various gases. As the fish dives deeper, pressure increases, compressing the air, reducing volume, and thereby effectively making the fish denser. The negative buoyancy now pushes the fish down, and it starts to sink. As we will see in chapter 11, such a fish can add more gas into the gas bladder to increase its volume and return it, overall, to neutral buoyancy.

Machines

When we are interested in the motions of parts of the same animal, it is customary to represent each movable part with a link. A joined series of links is a **kinematic chain** representing the main elements of an animal. If these linkages are floppy and without control, then the chain is said to be unconstrained. A kinematic chain restricted in motion is constrained and formally constitutes a **mechanism.** The motion of one link will impart a definite and predictable motion in all other links of the same mechanism (figure 4.32a).

A kinematic mechanism simulates the relative motions of the parts of the animal it represents, so it helps identify the role of each element. For example, several bony elements on both sides of a lizard's skull are involved when it lifts its snout during feeding. These elements can be represented by a kinematic chain that constitutes the jaw mechanism of the lizard (figure 4.32b, c).

Often we are interested in more than just the motion of a mechanism. We might want to know something about the transfer of actual forces. Such devices that transfer forces are **machines.** Formally defined, a machine is a mechanism for transferring or applying forces. In a car's engine, the pistons transfer the explosive forces of gasoline combustion to the connecting rod, the rod to the crankshaft, and the crankshaft in turn to the gears, axles, and wheels. Pistons to wheels collectively form a "machine" that transfers energy from the ignited gasoline to the road. Levers that transfer forces qualify as machines too. The input force brought into a machine by a lever arm is applied elsewhere as an output force by the opposite lever arm. In this engineering sense, the jaws of a herbivore are a machine whereby the input force produced by the jaw muscles is transferred along the mandible as an output force to the crushing molar teeth (figure 4.33).

Strength of Materials

A weight-bearing structure carries or resists the forces applied to it. These forces, termed a **load,** can be experienced in three general ways. Forces pressing down on an object to compact it are **compressive forces,** those that stretch it are **tensile forces,** and those that slide its sections are **shear forces** (figure 4.34a–c). Surprisingly, the same structure is

FIGURE 4.32 Kinematic chain. (a) This four-linkage mechanism is joined by pin connections so that the motion of link 3 imparts a specific motion to the three other links. (b) The four-linkage chain of a lizard skull (ignoring the lower jaw) is constrained. (c) Again, motion of link 3 imparts a specific motion to each of the other links. After T. H. Frazzetta

not able to withstand the three types of force applications equally. For any structure, the maximum force a structure sustains in compression before breaking is its **compressive strength;** in tension, is its **tensile strength;** and in shear, is its **shear strength.** Internal forces, termed **stress,** are the reaction to these external forces loaded on the structure.

Table 4.2 lists the strengths of several materials when they are exposed to compressive, tensile, and shear forces. Notice from this table that most materials are strongest in resisting compressive forces and weakest in their ability to withstand tension or shear. This is very significant in design. Ordinarily, supportive columns of buildings bear the load in a compressive fashion, their strongest weight-bearing orientation. However, if the column bends slightly, tensile forces, to which they are more susceptible, appear.

When any object bends, compressive forces build up on the inside of the bend and tensile forces on the outside. Opposite sides experience different force applications. The column may be strong enough to withstand compressive forces, but the appearance of tensile forces introduces forces it is intrinsically weaker in resisting. If bending persists, breaks may originate on the side experiencing tension,

FIGURE 4.33 **Jaws as machines.** A machine transfers forces. Here the lower jaw of a herbivore transfers the force of the temporalis muscle (open arrow) to the tooth row (solid arrow) where food is chewed. Rotation occurs about the condyle.

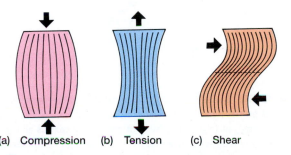

(a) Compression   (b) Tension   (c) Shear

FIGURE 4.34 **Direction of force application.** The susceptibility of a material to breaking depends on the direction in which the force is applied (arrows). Most materials withstand compression best (a) and are weaker when placed in tension (b) or shear (c).

TABLE 4.2 Strength of Different Materials Exposed to Compressive, Tensile, and Shear Forces			
Material	**Compressive Strength (Pa)**	**Tensile Strength (Pa)**	**Shear Strength (Pa)**
Bone	165×10^6	110×10^6	65×10^6
Cartilage	27.6×10^6	3.0×10^6	0.26×10^6
Concrete	24.1×10^6	4.0×10^6	1.6×10^6
Cast iron	620.5×10^6	310.2×10^6	379.2×10^6
Granite	103×10^6	10×10^6	13.8×10^6

Ultimate strengths shown.

Source: Adapted from J. E. Gordon, 1978. Structures, or why things don't fall down, DaCapo Press, NY. Other sources have also been used.

propagate through the material, and cause the column to fail. Flying buttresses, side braces on the main support piers (columns) of Gothic cathedrals, were used to prevent the piers from bending, thus keeping them in compression and allowing them to better carry the weight of the cathedral's arched roof (figure 4.35).

Loads

How a load is positioned upon a supportive column affects its tendency to bend (figure 4.36a–c). When the load is arranged evenly above the column's main axis, no buckling is induced and the column primarily experiences the load as compressive force (figure 4.36b). The same load placed asymmetrically off center causes the column to bend (figure 4.36c). Tensile (and shear) forces now appear. Tensile and compressive forces are greatest at the surfaces of the column, least at its center. Development of surface tensile forces is especially ominous because of the intrinsic susceptibility of supportive elements to such forces—what we notice as cracks.

Biological Design and Biological Failure

Fatigue Fracture With prolonged or heavy use, bones, like machines, can fatigue and break. When initially designed, the working parts of a machine are built with materials strong enough to withstand the calculated stresses they will experience. However, with use over time, these parts often fail, a

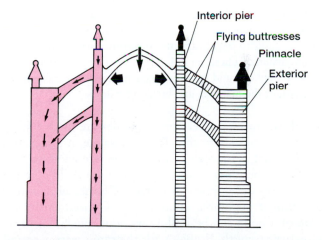

FIGURE 4.35 **Gothic cathedral.** The right side of the cathedral shows its structural elements. The left side illustrates how these structures carry the thrust lines of forces. At its simplest, the cathedral includes the exterior pier topped by a pinnacle, the main interior pier, and the flying buttresses between the interior and exterior piers. The weight of the vault (roof) produces an oblique thrust against the interior piers. Wind pressure or snow load accentuates this lateral pressure, which tends to bend the main interior piers. Flying buttresses act in an opposite direction to resist this bending and help carry lateral thrust from the roof to the ground (small arrows).

After Gordon.

(a)   (b)   (c)

FIGURE 4.36   **Loading.** (a) When a material bends under a load, compressive forces (solid arrows) develop along the concave side, tensile forces (open arrows) along the convex side. (b) When a supportive column is loaded symmetrically (with the weight centered), the only type of force experienced is compressive force. The distribution of the 100-kg mass within a representative section is depicted graphically. The lengths of the down arrows show equal distribution of compressive forces within this representative section. (c) Asymmetrical loading of the same mass causes the column to bend. The column experiences compressive forces (down arrows) and tensile forces (up arrows). Both compressive and tensile forces are greatest near the surface and least toward the center of the column.

condition known to engineers as **fatigue fracture.** Not long after the Industrial Revolution, engineers noticed that moving parts of machinery occasionally broke at loads within safe limits. Axles of trains, in use for some time, suddenly broke for no apparent reason. Cranks or cams that had withstood peak loads many times before sometimes suddenly broke under routine operation. Eventually engineers appreciated that one of the factors leading to these failures was fatigue fracture. Although a moving part might be strong enough initially to bear up easily to peak loads, over time tiny microfractures form in the material. These are insignificant individually, but cumulatively they can add up to a major fracture that exceeds the strength of the material, and breakage follows.

Load Fracture   In vertebrates, bones are loaded symmetrically or, where that is not possible, muscles and tendons act as braces to reduce the tendency for a load to induce bending of a bone (figure 4.37). The greatest stresses develop at the surface of bone, while forces are almost negligible at its center. Consequently, the core of a bone can be hollow without much loss of its effective strength. Probably for the same reason, cattail reeds, bamboos, bicycle frames, and fishing poles are hollow as well. This economizes on material without a great loss in strength.

Most fractures likely begin on the side of the bone experiencing tensile forces. To start it, a fracture requires energy as intermolecular bonds begin to break, but as it propagates, more energy is released than is consumed so the fracture tends to grow easily and quickly. Think for instance of tearing a piece of paper—the tear starts with some effort but rips (fractures) more easily once underway. In bone, a fracture propagates through the matrix, causing failure.

Bone, however, is a composite material consisting of several substances with different mechanical properties. Together these substances resist the propagation of a fracture better than any one constituent alone (figure 4.38a–c). This same principle of composites gives fiberglass its resistance to breaking. Fiberglass consists of glass fibers embedded in a plastic resin. Glass is brittle, resin weak, but together they are strong because they blunt small cracks and prevent their spread. As a crack approaches the boundary between the two fiberglass materials, the resin gives slightly. That spreads out

FIGURE 4.37   **Braces.** The weight of the upper body is carried on the heads of the femurs (left). This means that during the striding gait (right), the head of one femur carries all the upper body weight. Consequently, the long shaft of the femur is loaded asymmetrically, increasing its tendency to bend. The iliotibial tract, the long tendon of the tensor fasciae latae muscle that runs laterally across the femur, partially counteracts this tendency to bend and thereby reduces the tensile forces that would otherwise develop within the femur.

(a)   (b)   (c)

FIGURE 4.38   **Fracture propagation.** (a) Failure of a structure begins with the appearance of a microfracture that spreads rapidly. (b) In composite materials, such as bone, the advancing fracture is preceded by stress waves that may cause the concentrated force to spread at the boundary between the composite materials, where they give slightly. (c) As the fracture line meets this boundary, its sharp tip is blunted and its progress curtailed.

the force and thus reduces the stress at the tip of the crack that makes it advance. A space in the material can do the same job; thus hard foams resist cracking. Bone makes use of both stress relief and small voids or spaces to reduce crack propagation.

Collagen fibers and hydroxyapatite crystals are the main materials of bone matrix. They are thought to act in a manner analogous to the glass and resin of fiberglass to blunt small fractures. Further, the orientation of collagen fibers alternates in successive layers so that they better receive tensile and compressive forces.

Teeth, too, seem to be built to stop cracks. The outer part of a tooth is enamel, the inner part dentin. Enamel is almost pure ceramic, a calcium phosphate mineral called hydroxyapatite; but dentin, in addition to hydroxyapatite, also includes about 40% of the protein collagen. The result—enamel and dentin have different physical properties. When a microcrack propagates through the enamel toward dentin at the tooth interior, it stops at the boundary with the dentin. At this enamel-dentin interface, the surface is scalloped, which deflects the trajectory of the arriving crack, decreasing its full force, and thereby serves to stop the crack's spread.

Bone structure (p. 182); tooth anatomy (p. 507)

Tissue Response to Mechanical Stress

Tissues can change in response to mechanical stress. If living tissue is unstressed, it tends to decrease in prominence, a condition termed **atrophy** (figure 4.39a). If it experiences increased stress, tissue tends to increase in prominence, a condition termed **hypertrophy** (figure 4.39b). Cell division and proliferation under stress are termed **hyperplasia.** Thus, in response to exercise, the muscles of an athlete will increase in size. This overall increase is primarily due to an increase in the size of existing muscle cells, not to an increase in cell

FIGURE 4.39 Loss (atrophy) and increase (hypertrophy) of bone. (a) Cross section of a normal foot bone from a dog is illustrated on the left. Cross section of the same bone from the opposite foot (right) that was immobilized in a cast for 40 weeks reveals significant atrophy. (b) Cross section of a normal femur from a pig is depicted on the left. Cross section of a femur from a pig that had been vigorously exercised on a regular basis for over a year shows increased bone mass (right). Hypertrophy is evident from the thickening and greater density of the bone cortex.

number (hypertrophy but not much hyperplasia). During pregnancy, smooth muscles of the uterus increase both in size and number (hypertrophy and hyperplasia).

Muscle response to chronic exercise (p. 382)

Tissues can, under some circumstances, change from one type to another, a transformation called **metaplasia.** Metaplastic transformations are often pathological. For example, the normal ciliated pseudostratified columnar epithelium of the trachea may become stratified squamous epithelium in tobacco smokers. But some metaplastic changes seem to be part of normal growth and repair processes as well. For example, reptiles exhibit metaplastic bone formation during growth of long bones. Chondrocytes become osteoblasts and cartilaginous matrix becomes osseous as cartilage undergoes direct transformation into ossified bone. During bone repair in reptiles, amphibians, and fishes, the cartilaginous callus appears to arise from connective tissue through metaplasia.

Tissue types (p. 177)

All tissues retain some physiological ability to adjust to new demands, even after embryonic development is complete. Weight training causes an athlete's existing muscles to increase and his or her tendons to strengthen. Regular long distance running enhances circulation, increases blood volume, improves oxygen delivery to tissues, and metabolizes stored lipids more efficiently. Although the number of nerve cells does not usually increase in response to the physiological stress of exercise, coordination of muscle performance often does. Tissues continue to adapt physiologically to changes in demand throughout the life of the individual. One of the best examples is bone because it illustrates the complexity of tissue response.

Responsiveness of Bone

While performing in a protective or supportive role, bone cannot significantly deform or change shape. Leg bones that telescope or bend like reeds would certainly be ineffective as supports for the body. Bones must be firm. But because living bone is dynamic and responsive, it gradually changes during the life of an individual. The genetic program of a person sets forth the basic form a bone takes, but immediate environmental factors also contribute to ultimate bone form. Some peoples of the New World developed the practice of wrapping a baby's head against a cradle board (figure 4.40a, left). As a result, the normal skull shape of the baby was altered so that the side pressed to the board was flattened. In parts of Africa and Peru, prolonged bandaging of the back of the skull caused elongation of the cranium (figure 4.40a, right). Until recent times, young Chinese girls who looked forward to a leisured life had their feet permanently folded and tightly bound to produce tiny feet in adulthood. The toes were crowded and the arch exaggerated (figure 4.40b, right). The normal and, by comparison, large foot was considered ugly in women (figure 4.40b, left). Because foot-binding

(a)

(b) Normal foot   Bound foot

(c)

FIGURE 4.40 Responsiveness of bone to mechanical stress. (a) The continuous mechanical pressure of a cradle board flattened the back of the Navajo Indian skull (left), and wrapping of the Peruvian native skull (right) caused it to elongate. (b) Historically, many Chinese followed the practice of binding the feet of young girls with tight wrappings. The small, deformed foot shown on the right was considered socially "attractive." (c) A nutritional calcium deficiency during infancy led to rickets, which weakened this woman's skeleton, shown here at age 70. Her bones bent under the normal load of her body.

After Halsted and Middleton.

When bone sustains prolonged stress, microdamage occurs in the form of micro-fractures. The response of bone is to physiologically adapt by mending these microfractures. The forelimb of tetrapods includes two bones (ulna, radius) we will meet in detail later, but for now we need only recognize that each mechanically buttresses the other. Remove one of these two bones and the other now, without its partner, experiences increased stress up to four times greater. However, after a few months these stresses decline and return to normal. What happens? The bone physiologically adjusts by laying down new bone to address the new stresses. The overload causes an increase in microdamage, which stimulates an increase in repair rate of new bone formation, remodeling it. This in turn reduces surface stress, which returns the bone's mechanical challenges to where it was before being damaged. At a genetic level, the changes in mechanical load stimulate important genes involved in bone formation. Think of it. Mechanical events reach into the bone cells and activate genes that produce not just new bone, but bone in the right places to address the new mechanical stresses. This is a wonderful insight. Consider the comments of one scientist. In older humans who fall and experience a hip fracture, over half may never live independently again and 20% are dead in half a year. Grim statistics. But now that the linkages from mechanical events to gene action are understood, there is promise.

impaired biomechanical performance, this also had what was considered the proper social consequence of keeping women literally "in their place."

Environmental Influences Four types of environmental influences alter or enhance the basic shape of bone set down by the genetic program. One is infectious disease. A pathogenic organism can act directly to alter the pattern of bone deposition and change its overall appearance. Or, the pathogen can physically destroy regions of a bone. A second environmental influence is nutrition. With adequate diet, normal bone formation is usually taken for granted. If the diet is deficient, bones can suffer considerable abnormalities. Rickets, for example, caused by a deficiency of calcium in humans, results in buckling of weight-bearing bones (figure 4.40c). Ultraviolet radiation transforms dehydrocholesterol into vitamin D, which the human body needs to incorporate calcium into bones. Sunshine and supplements of milk are usually enough to prevent rickets. Hormones are the third factor that can affect bone form. Bone is a calcium reservoir, which is perhaps its oldest function. When demanded, some calcium is removed from bone matrix. Calcium drains occur during lactation when the female produces calcium-rich milk, during pregnancy when the fetal skeleton begins to ossify, during egg laying when hard shell is added, and during antler growth when the bony rack of antlers is developing.

Endocrine control of bone calcium (p. 596)

The fourth environmental influence on bone form is mechanical stress (figure 4.40b, c). Each weight-bearing bone experiences gravity, and muscles tug on most bones. Forces produced by gravity and muscle contraction place bone in an environment of stresses that determine the final shape of bone. Throughout an individual's life, these stresses

upon bone change. As a young animal gains its footing and daring, it becomes more active. As an adult, it might migrate, battle for territory, or increase its foraging to support offspring of its own. As the animal grows bigger, scaling becomes a factor. Geometric increase in the mass of a growing animal places greater mechanical demands upon the supportive elements of its body. Human athletes on a continuous training program intentionally increase their loads on bones and muscles to stimulate physiological adaptation to the heightened activity. Conversely, age or inclination can lead to declining activity and reduced stress on bones. Teeth might decay, and this changes the stress pattern experienced by the jaws. An injury can lead to favoring one limb over another. For a variety of reasons, then, the forces experienced by bones change.

Atrophy and Hypertrophy The response of bone to mechanical stresses depends upon force duration. If bone experiences continuous pressure, bone tissue is lost and atrophy occurs. Continuous pressure against bone arises occasionally with abnormal growths, such as brain tumors that bulge from the surface of the brain and press on the underside of the bony skull. If this continuous pressure is prolonged, the bone erodes, forming a shallow depression along the surface of contact. Aneurysms, ballooning of blood vessels at weak spots in the vessel wall, can exert continuous pressure against nearby bone and cause it to atrophy. Orthodontic braces cinched to teeth by a dentist force teeth up against the sides of the bony sockets in which they sit. Resorption of continuously stressed bone opens the way for teeth to migrate slowly but steadily into new and presumably better positions within the jaws.

Thus, bone that experiences a continuous force atrophies; however, so does bone that experiences no force. When forces are absent, bone density actually thins. People restricted to prolonged bed rest without exercise show signs of osteoporosis. This has been studied experimentally in dogs on whom a cast has been applied to one leg. The immobilizing cast eliminates or considerably reduces the normal loads carried on a leg bone. Bones so immobilized exhibit significant signs of resorption, which can occur rather quickly. Experiments with immobilized wings of roosters show that within a few weeks wing bones become extensively osteoporetic. Rarification of bone matrix occurs in astronauts during extended periods of weightlessness. Calcium salts leave bones, circulate in the blood, and this excess is actually excreted. When astronauts return to Earth's gravitational forces, their skeletons gradually recover their former density. Even over long voyages, the ossified skeleton is not likely to disappear altogether, but it may fall to a genetically determined minimum. And, of course, muscle contractions maintain some regime of forces on bone. But during deep space travel lasting many months, bone atrophy can progress far enough to make return to Earth's gravity hazardous. Prevention of bone atrophy in space travel remains an unsolved problem.

Between continuously stressed and unstressed bone is the third type of force application, *intermittent* stress. Intermittent stress stimulates bone deposition, or hypertrophy. The importance of intermittent forces on bone growth and form has long been suspected from the fact that bone atrophies when intermittent forces are removed. Conversely, when rabbit bones were intermittently stressed by a special mechanical apparatus, hypertrophy occurred. More recently, bones of the rooster wing were stressed once daily with compressive loads but otherwise left immobilized. After a month, the artificially stressed bones did not exhibit osteoporosis but did show growth of new bone, clearly an appropriate physiological response to the artificially induced intermittent stresses.

Internal Design   The overall shape of a bone reflects its role as part of the skeletal system. Internal bone tissue consists of areas of **compact bone** and **spongy bone.** Distribution of compact and spongy bone is also thought to be directed by mechanical factors, although there is little hard evidence to support this correlation. According to an engineering theory called the trajectorial theory, when a load is placed on an object, the material within the object carries the resulting internal stress along stress trajectories or paths that pass these forces from molecule to molecule within the object (figure 4.41a). A beam embedded at its base in the wall will bend under its own weight. The lower surface of the beam experiences compressive forces as the material is pushed together, and the upper surface of the beam experiences tension as material here is pulled apart. The resulting compressive and tensile stresses are carried along stress trajectories that cross at right angles to each other and bunch under the beam's surface.

Culmann and Meyer, nineteenth-century engineers, applied this trajectorial engineering theory to the internal architecture of bone. Because the femur carries the load or weight of the upper body, they reasoned that similar stress trajectories must arise within this bone. In order for the body to build a strong structure and yet be economical with material, bony tissue should be laid down along these stress trajectories, the lines along which the load is actually carried. After looking at sections of bone, Culmann suggested that nature arranged bone spicules (**trabeculae**) into a lattice of spongy bone at the ends of long bones (figure 4.41b). Because these lines of stress move to the surface near the middle of the bone, the trabeculae follow suit, and the overall result is a tubular bone. If the trabeculae of bone follow internal lines of stress, these trabeculae might be expected to form a lattice of spongy bone after birth when functional loads are first experienced. This is borne out. Trabeculae of young fetuses display random honeycomb architecture. Only later do they become arranged along presumed lines of internal stress.

Wolff's Law   As applied mechanical forces change, bone responds dynamically to adapt physiologically to changing stresses. Wolff's Law, named for a nineteenth-century scientist who emphasized the relationship between bone form and function, states that remodeling of bone occurs in proportion to the mechanical demands placed upon it.

When bone experiences new loads, the result is often a greater tendency to buckle. When buckling occurs, tensile forces appear. Bones are less able to withstand tensile forces than they can compressive forces. In order to compensate, bone undergoes a physiological remodeling to better adapt to the new load (figure 4.42a–c). Initially, adaptive remodeling entails thickening along the wall experiencing compression. Eventually overall remodeling restores the even, tubular shape of the bone. How are cells along the compressive side selectively stimulated to deposit new bone? Nerves penetrate throughout bone, so they might be one way of promoting and coordinating the physiological response of osteocytes to changes in loading. However, bones to which nerves have been cut still abide by Wolff's Law and adjust to changes in mechanical demand.

Muscles pulling on bone affect the shape of vascular channels near their points of attachment to bone, which alters the blood pressure in vessels supplying bone cells. Increased muscle activity accompanying increased load might, via such blood pressure changes, stimulate bone cells to remodel. However, muscle action on bone, even if sufficient to change blood pressure, seems too global a mechanism to lead to the specific remodeling responses actually observed in bone.

Bone cells occupy small lacunae, spaces within the calcium matrix of a bone. Slight configurational changes in the lacunae occupied by bone cells offer a more promising mechanism. Under compression, lacunae tend to flatten;

FIGURE 4.41 **Stress trajectories.** (a) A beam projecting from a wall tends to bend under its own weight, placing internal stresses on the material from which it is made. Engineers visualize these internal stresses as being carried along lines called stress trajectories. Compressive forces become concentrated along the bottom of the beam, tensile forces along the top. Both forces are greatest at the surface of the beam. (b) Stress trajectories in living bone. When this theory is applied to living bone, the matrix of bone appears to be arranged along the lines of internal stress. The result is an economical latticework of bone, with material concentrated at the surface of a tubular bone. A cross section through the proximal end of a femur reveals the lattice of bone spicules within the head that become concentrated and compacted along the wall within the shaft of the femur.

Latticework model kindly supplied by P. Dullemeijer, after Kummer.

(a)

(b)

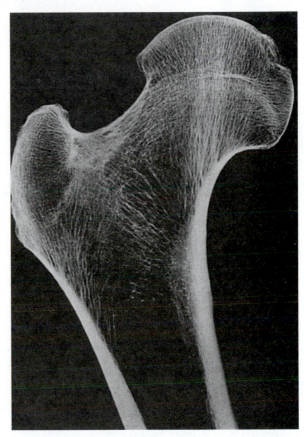

under tension they tend to become round. If these configurational changes produced under load could be read by the bone cells occupying the lacunae, then bone cells might initiate a remodeling matched to the type of stress experienced.

Another mechanism might involve **piezoelectricity,** or low-level electric charges. These are surface charges that arise within any crystalline material under stress—negative charges appear on the surface under compression, and positive charges on the surface in tension. Bone, with its structure of hydroxyapatite crystals, experiences piezoelectric charges when it is loaded. It can be easily imagined that under a new load, a new environment of piezoelectric charges would appear within the tissue of stressed bone. If individual bone cells could key off these localized piezoelectric charges, then a specific remodeling response might follow.

Although promising, each of these proposed mechanisms by itself seems insufficient to account for the physiologically adaptive remodeling that occurs during bone response to the functional demands placed upon it. This is a challenging area for further research.

Biophysics and Other Physical Processes

Biophysics is concerned with principles of energy exchange and the significance of these principles for living organisms. The use of light, the exchange of heat, and the diffusion of molecules are fundamental to the survival of an organism. Biological design and its limits are determined by the

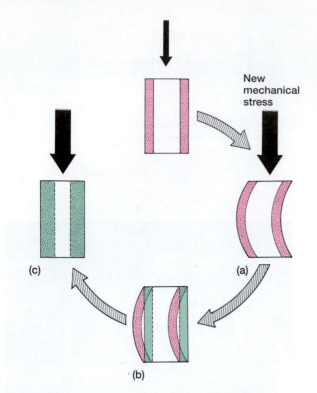

FIGURE 4.42 Bone remodeling. When tubular bone experiences a new and more distorting stress (a), it undergoes a physiological response that both thickens and straightens it. New bone forms along the concave surface (b), remodeling the bone, and the straight shape is restored (c). Further remodeling returns the bone to its original shape (top), although the walls would now be thicker to withstand the new, increased load.

physical principles governing energy exchange between an organism and its environment, and internally between active tissues within the organism. One of the most important of these physical principles applies to the exchange of gases.

Diffusion and Exchange

Pressures and Partial Pressures

Air pressure varies slightly with weather conditions, such as low- and high-pressure fronts, and with temperature. When animals ascend in altitude, air pressure drops significantly as the air thins (becomes less dense) and breathing becomes more labored. This drop in pressure of the gases, especially oxygen, creates the difficulty. Air is a mixture of nitrogen (about 78% by volume), oxygen (about 21% by volume), carbon dioxide, and trace elements. Each gas in air acts independently to produce its own pressure irrespective of the other gases in the mixture. Of the total 101,000 Pa (pressure of air) at sea level, oxygen contributes 21,210 Pa (101,000 Pa × 21%) to the total, nitrogen 78,780 Pa (101,000 Pa × 78%), and the remaining gases 1,010 Pa. Because each gas contributes only a part of the total pressure, its contribution is its **partial pressure.** The rate at which oxygen can be inhaled depends on its partial pressure. At 5,300 m (18,000 ft), air pressure drops to about 0.5 atm (atmosphere), or 50,500 Pa. Oxygen still composes about 21% of the air, but because the air is thinner, there is less total oxygen present. Its partial pressure falls to 10,605 Pa (50,500 Pa × 21%). With a drop in the partial pressure of oxygen, the respiratory system picks up less and breathing becomes more labored. Animals living in the high mountains, and especially high-flying birds, must be designed to address this change in atmospheric pressure.

Because water weighs much more than air per unit of volume, an animal descending through water experiences pressure changes much more quickly than one descending through air. With each descent of about 10.3 m (33.8 ft), water pressure increases by about 1 atm. Thus, a seal at a depth of 20.6 m experiences almost two additional atmospheres of pressure more than it experiences when basking on the beach. The effect of this pressure change on body fluids and solids is probably inconsequential, but gas in the lungs or in the gas bladders of fishes is compressed significantly. Each 1-meter descent in water adds 9,800 Pa of pressure, or about 1.5 lb of pressure per square inch of chest wall. Compressing the lungs or the gas bladder reduces their volume and thus affects buoyancy. The movement of gases into and out of the bloodstream is affected by the difference in the partial pressure of oxygen breathed in at the surface and its partial pressure when it is diffused into the blood once the animal is submerged. We look specifically at these properties of gases and the way in which the vertebrate body is designed to accommodate them when we examine the respiratory and circulatory systems in chapters 11 and 12, respectively.

Countercurrent, Concurrent, and Crosscurrent Exchange

Exchange is a large part of life. Oxygen and carbon dioxide pass from the environment into the organism or from the organism into the environment. Chilled animals bask to pick up heat from their surroundings; large, active animals lose heat to their surroundings to prevent overheating. Ions are exchanged between the organism and its environment. This process of exchange, whether it involves gases, or heat, or ions is sometimes supplemented by air or water currents passing one another. Efficiency of exchange depends on whether the currents pass in opposite or equivalent directions.

Imagine two parallel, but separate, identical tubes through which streams of water flow at the same speed. Water entering one tube is hot, and water entering the other is cold. If the tubes are made of conducting material and contact each other, heat will pass from one to the other (figure 4.43a, b). Water flow may be in the same direction, as in **concurrent exchange,** or in opposite directions, as in **countercurrent exchange.** The efficiency of heat exchange between the tubes is affected by the directions of flow.

If the streams are concurrent, as the two tubes come in contact, the temperature difference will be at its maximum but will drop as heat is transferred from the hotter to

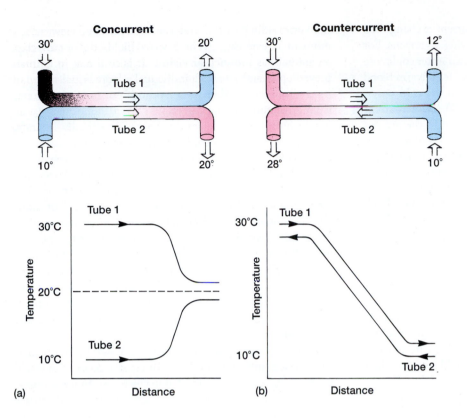

Concurrent

30° 20°
Tube 1
Tube 2
10° 20°

Countercurrent

30° 12°
Tube 1
Tube 2
28° 10°

(a)

Temperature
30°C — Tube 1
20°C
10°C — Tube 2
Distance

(b)

Temperature
30°C — Tube 1
10°C — Tube 2
Distance

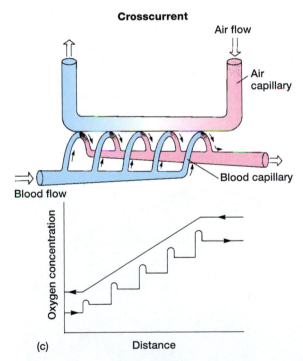

Crosscurrent

Air flow
Air capillary
Blood capillary
Blood flow

Oxygen concentration
Distance

(c)

FIGURE 4.43 **Systems of exchange.** Direction and design of exchange tubes affect the efficiency of transfer regardless of whether the exchange involves heat, gases, ions, or other substances. The first two examples (a and b) illustrate heat transfer. The third example (c) depicts gas exchange. (a) Concurrent exchange describes the condition in which separated fluids flow in the same direction. Because the temperature gradient between the fluids is high when they enter the tubes and rather low when they exit, the average difference in heat exchanged between the two fluids is relatively high. The fluid in tube 2 is at 10° when it enters and at 20° when it exits. (b) In countercurrent exchange, the fluids pass in opposite directions within the two tubes so that the temperature difference between them remains relatively low all along their lengths. The fluid in tube 2 is at 10° when it enters and at 28° when it exits. Thus, more heat is transferred with countercurrent exchange than with concurrent exchange. (c) In a stepwise crosscurrent exchanger, each blood capillary branch passes across an air capillary at about right angles to it and picks up oxygen. The levels of oxygen rise serially in the departing blood. Arrows indicate the direction of flow.

the colder tube. The cold stream of water will warm, the hot stream will cool; so at their point of departure, both streams of water approach the average of their two initial temperatures (figure 4.43a). If we take the same tubes and same starting temperatures but run the currents in opposite directions, we have a countercurrent exchange; heat transfer becomes much more efficient than if both currents flowed in the same direction (figure 4.43b). A countercurrent

flow keeps a differential between the two passing streams throughout their entire course, not just at the initial point of contact. The result is a much more complete transfer of heat from the hot stream to the cold stream. When the tubes are separated, the cold stream is nearly as warm as the adjacent hot stream. Conversely, the hot stream gives up most of its heat in this countercurrent exchange, so that its temperature has fallen to almost that of the entering cold stream.

This physical principle of countercurrent exchange can be incorporated into the design of many living organisms. For example, endothermic birds that wade in cold water could lose much of their critical body heat to the icy water if warm blood circulated through their feet was exposed to the cold water. Replacing this lost heat could be expensive. A countercurrent heat exchange between outgoing warm blood in the arteries supplying the feet and returning cold blood in the veins prevents heat loss in wading birds. In the upper legs of such birds, small arteries come in contact with small veins, forming a **rete,** a network of intertwining vessels. Because arterial blood in these vessels passes in opposite directions to venous blood, a countercurrent system of heat exchange is established. By the time the blood in the arteries reaches the feet, it has given up almost all of its heat to the blood in the veins returning to the body. Thus, there is little heat lost through the foot into the cold water. The countercurrent system of the rete forms a **heat block,** preventing the loss of body heat to the surroundings. Estimates indicate that the rete is so efficient in heat transfer that if boiling water were poured through a wading bird's arteries at one end and ice water through its veins at the other end, blood vessels in the feet would lose less than 1/10,000 of a degree in temperature.

Respiration in many fishes is characterized by a countercurrent exchange also. Water high in oxygen flows across the gills, which contain blood capillaries low in oxygen flowing in the opposite direction. Because water and blood pass in opposite directions, gas exchange between the two fluids is very efficient.

In bird lungs, and perhaps in other animals as well, gas exchange is based on another type of flow, a stepwise crosscurrent exchange between blood and air capillaries. Because blood capillaries cross at nearly right angles to the air capillaries in which gas exchange occurs, a crosscurrent is created (figure 4.43c). Blood capillaries run sequentially from an arteriole to supply each air capillary. When blood capillaries cross an air capillary, oxygen passes into the bloodstream and CO_2 is given up to the air. Each blood capillary contributes stepwise to the rising level of oxygen in the venule it joins. Partial pressures vary along the length of an air capillary, but the additive effect of these blood capillaries in series is to build up efficient levels of oxygen in venous blood as it leaves the lungs.

Optics

Light carries information about the environment. Color, brightness, and direction all arrive coded in light. Decoding this information is the business of light-sensitive organs. However, the ability to take advantage of this information is affected by whether the animal sees in water or in air, and it is affected by how much the two eyes share overlapping fields of view.

Depth Perception

The position of the eyes on the head represents a trade-off between panoramic vision and depth perception. If the eyes are positioned laterally, each scans separate halves of the surrounding world, and the total field of view at any moment is extensive. Where visual fields do not overlap, an animal has **monocular vision.** It is common in animals preyed upon and gives the individual a large visual sweep of its environment to detect the approach of potential threats from most surrounding directions. Strict monocular vision, in which the visual fields of the two eyes are totally separate, is relatively rare. Hagfish, lampreys, some sharks, salamanders, penguins, and whales have strict monocular vision.

Where visual fields overlap, vision is **binocular vision.** Extensive overlap of visual fields characterizes humans. We have as much as 140° of binocular vision, with 30° of monocular vision on a side. Binocular vision is important in birds (up to 70°), reptiles (up to 45°), and some fishes (as much as 40°). Within the area of overlap, the two visual fields merge into a single **stereoscopic image** (figure 4.44). The advantage of stereoscopic vision is that it gives a sense of depth perception. Closing one eye and maneuvering about a room demonstrates how much sense of depth is lost when the visual field of only one eye is used.

Depth perception results from how the brain processes visual information. With binocular vision, the visual field seen by each eye is divided in the brain. In most mammals, half goes to the same side, and the other half crosses via the **optic chiasma** to the opposite side of the brain. For a given part of the visual field, inputs from both eyes are brought together on the same side of the brain. Within the brain, the **parallax** of the two images is compared. Parallax is the slightly different views one gets of a distant object when it is viewed from two different points. Look at a distant lamppost from one position, and then step a few feet laterally and look at it again from this new position.

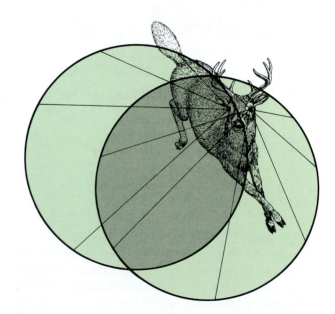

FIGURE 4.44 **Stereoscopic vision.** Where the conical visual fields of the deer overlap, they produce stereoscopic vision (shaded area).

Slightly more of one side of the lamppost can be seen, less of the opposite side, and the position of the post relative to background reference points changes as well. The nervous system takes advantage of parallax resulting from eye position. Each visual image gathered by each eye is slightly offset from the other because of the distance between the eyes. Although this distance is slight, it is enough for the nervous system to produce a sense of depth resulting from the differences in parallax.

Depth perception and stereoscopic vision (p. 688)

Accommodation

Sharp focusing of a visual image upon the retina is termed **accommodation** (figure 4.45a). Light rays from a distant object strike the eye at a slightly different angle than rays from a nearby object. As a vertebrate alters its gaze from close to distant objects of interest, the eye must adjust, or accommodate, to keep the image focused. If the image falls behind the retina, **hyperopia,** or farsightedness, results. An image focused in front of the retina produces **myopia,** or nearsightedness (figure 4.45b, c).

The lens and the cornea are especially important in focusing entering light. Their job is considerably affected by the **refractive index** of the media through which light passes, a measure of the bending effects on light passing from one medium to another. The refractive index of water is similar to the refractive index of the cornea; therefore, when light passes through water to the cornea in aquatic vertebrates, there is little change in the amount it bends as it converges on the retina. But when light passes through air to the liquid medium of the cornea in terrestrial vertebrates, it bends considerably. Similarly, aquatic animals viewing an object in air must compensate for the distortion produced by differences in the refractive indexes of air and water (figure 4.46). As a consequence of these basic optical differences, eyes are designed to work either in water or in air. Underwater vision is not necessarily out of focus. It just looks that way to our air-adapted eyes when we jump into a clear stream and attempt to focus our eyes. If we place a pocket of air in front of our eyes (e.g., a diving face mask), the refractive index our eyes are designed to accommodate returns and things become clear.

Accommodation can be accomplished by mechanisms that change the lens or the cornea. Hagfish and lampreys have a corneal muscle that changes the shape of the cornea to focus entering light. In elasmobranchs, a special protractor muscle changes the position of the lens within the eye. The elasmobranch eye is focused for distant vision. For near objects, the protractor muscle moves the lens forward. In most amniotes, the curvature of the lens changes to accommodate the eye's focus on objects near or far. Ciliary muscles act on the lens to change its shape and thus alter its ability to focus passing light (which, by the way, have nothing to do with microscopic cilia).

Eyes and mechanisms of accommodation (chapter 17)

Overview

Size matters; so does shape. Like any other characteristic of an organism, size and shape have survival consequences. Large organisms have fewer serious enemies. Small organisms find strength in numbers. But size and shape have physical consequences, in and of themselves. For a small

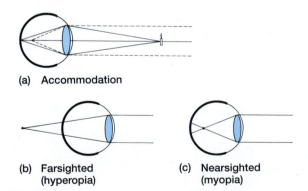

(a) Accommodation

(b) Farsighted (hyperopia)

(c) Nearsighted (myopia)

FIGURE 4.45 **Accommodation.** (a) Normal vision in which the image, solid lines, is in sharp focus on the retina of the eye. (b) Farsighted condition (hyperopia) in which the lens brings the light rays to focus behind the retina. (c) Nearsighted condition (myopia) in which the sharpest focus falls in front of the retina.

FIGURE 4.46 **Refraction.** Differences in the refractive indexes of water and air bend light rays that enter the water from the insect. The result makes the insect seem to be in a different position than it really is, indicated by the dashed lines. The archer fish must compensate in order to shoot a squirt of water accurately and hit the real, not the imaginary, image.

animal, gravity presents almost no dangers. Small lizards might scamper up walls and across ceilings. But for large animals, gravity may be more of a threat than predators. As J. B. S. Haldane reminds us, because of differences in scaling, a change in size inevitably requires a change in form. This is not for reasons of biology, but is instead a necessary consequence of geometry. Surface area increases rapidly with increase in size, scaling in proportion to the square of the linear dimensions; volume (mass) is even more affected, increasing by the cube of linear dimensions. Inevitably, larger organisms have relatively greater mass with which to contend and, consequently, the supportive and locomotor systems must be built differently and stronger to meet the accompanying physical demands.

Shape changes in proportion to size, termed *allometry*, are common during the growth of a young organism into the larger adult. These can be illustrated with graphs or transformation grids. The result, relative to body size, is often to accelerate the development of a body part, bring it to full size later in life when the adult is large enough and mature enough to use it. Shape is important for animals that move at significant speeds through fluids. A thin shape, presented to the fluid flow, helps reduce drag that would otherwise retard progress. Turned broadside, a fin or flipper uses profile drag to generate forces. A favorable shape, such as streamlining, encourages smooth, nonseparating flow. The Reynolds number tells us how changes in size and shape might affect performance of an animal in fluid, and emphasizes the importance of both in meeting physical demands of the fluid environment.

Forces, produced by muscles, are conveyed through levers, the skeletal system. The Newtonian laws of motion identify the physical forces an animal meets arising from inertia, motion, and action/reaction. When initiating motion, the bone-muscle system overcomes inertia, accelerates limbs or body into motion, and the contacted fluid or ground returns reaction forces. Muscles put a force into a lever system, and the lever system outputs that force as part of a task. The output to input ratio represents the mechanical advantage, a way of expressing whether a muscle has a leverage that increases either force output or speed output. Linked chains of jointed bones work as machines to transfer input forces from one part of the mechanism to another.

When conveying or receiving forces, the bone-muscle system itself is exposed to stresses that may be experienced as compression, tension, or shear forces. Failure level under each is different, with bones generally strongest in compression and most susceptible to breakage in shear. Further, the resulting stresses are carried unevenly within the skeletal element. Wolff's Law notices that bone remodels internally in proportion to the level and distribution of these stresses.

We also meet the fundamentals for gas diffusion and optics, which we will apply more fully in several of the later chapters.

In this chapter, we recognize that organisms face physical demands that endanger their survival. Consequently, we turn to the discipline that studies such a physical relationship between design and demands, namely, engineering. From this, we usefully apply its insights from biomechanics and biophysics to understand more of the adaptive basis of animal architecture.

WEB LINK

Visit the text website at *www.mhhe.com/kardong7e* for additional study aids including selected references, functional laboratories, and web links to chapter topics.

CHAPTER 5

Life History

INTRODUCTION

EARLY EMBRYOLOGY
    Fertilization
    Cleavage
        Amphioxus
        Fishes
        Amphibians
        Reptiles and Birds
        Mammals
        Overview of Cleavage
    Gastrulation and Neurulation
        Amphioxus
        Fishes
        Amphibians
        Birds and Reptiles
        Mammals

ORGANOGENESIS
    Histogenesis
    Epithelium
        Covering and Lining Epithelium
        Glandular Epithelium
    Connective Tissues
        General Connective Tissues
        Special Connective Tissues
    Bone Development and Growth
        Endochondral Bone Development
        Intramembranous Bone Development
        Comparative Bone Histology
        Bone Remodeling and Repair
        Joints
    Neural Crest and Ectodermal Placodes

EXTRAEMBRYONIC MEMBRANES
    Reptiles and Birds
    Mammals
        Eutherian Placenta
        Other Placentae

OVERVIEW OF EARLY EMBRYONIC DEVELOPMENT

DEVELOPMENT OF THE COELOM AND ITS COMPARTMENTS

MATURATION
    Metamorphosis
    Heterochrony
        Peramorphosis
        Paedomorphosis

ONTOGENY AND PHYLOGENY
    Biogenetic Law
    Von Baer's Law
    Overview of the Biogenetic Laws
    Hox Genes and Their Kingdoms
        Egg to Adult
        Shaping Up: Positions and Parts
        Evolutionary Significance
    Epigenomics
        Induction
        Phylogeny

OVERVIEW

Introduction

The English politician Benjamin Disraeli put it as follows, "Youth is a blunder; manhood a struggle; old age a regret!" The unfolding course of normal events from embryo to death constitutes an individual's life history. Whether it falls into blunder, struggle, and regret, as Disraeli proposed, is a matter for poets to debate. For biologists, life history begins with fertilization, followed by embryonic development, maturation, and in some cases senescence, each stage being a

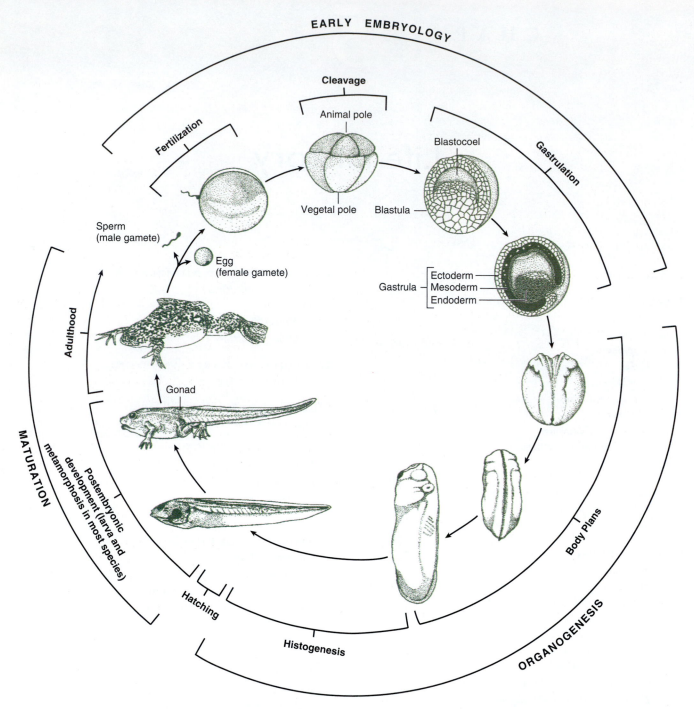

FIGURE 5.1 **From single cell to millions of cells—life cycle of a frog.** A sperm fertilizes the single-celled egg, and cell division (cleavage) begins, leading to a multicellular blastula with a fluid-filled core (blastocoel). Major rearrangements (gastrulation) of formative cellular layers (ectoderm, mesoderm, endoderm) lead next to an embryonic stage wherein these formative embryonic cells become arranged into organs (organogenesis) and specific tissues (histogenesis). Upon hatching, the larva feeds and grows further, eventually undergoing a major anatomical change (metamorphosis), becoming a juvenile and then an adult frog, which reproduces to repeat the cycle.

prelude to the next. Embryonic **development,** or **ontogeny,** extends from fertilization to birth or hatching. During this time, a single cell, the egg, is fertilized and divides into millions of cells from which the basic structural organization of the individual takes shape. Maturation includes the time from birth to the point of sexual maturity. Maturation usually involves growth in size and acquisition of learned skills

as well as appearance of anatomical features that distinguish the reproductive-ready adult. Prereproductive individuals are called juveniles or immatures. If juvenile and adult are strikingly different in form and the change from the one to the other occurs abruptly, the transformation is termed **metamorphosis.** A familiar example of metamorphosis is the transformation of a tadpole into a frog (figure 5.1).

Loss of physical vigor and reproductive ability accompany **senescence,** or **aging.** This phenomenon is apparent in humans but rare in wild animals. In fact, senescent animals usually provide an unwilling but easy meal for ready predators. Most examples among animals other than humans come from zoos, because zoo animals are spared the natural fate of their free-ranging colleagues. Only a few examples of senescence in the wild are known. Some species of salmon senesce quickly after spawning and die within a few hours. Aging individuals occasionally survive in social species, such as canids and higher primates. But humans are unusual among vertebrates in that aging individuals commonly enjoy an extended postreproductive life. Even before life-extending medicines and health care, senior citizens characterized ancient human societies. The value of the elderly to human societies does not stem from their services as warriors or hunters or tillers of the soil, because physical vigor has declined, nor can it be found in their procreative capacity. Perhaps aging humans were valued because of the child care service they could perform. Or perhaps aging individuals were living libraries, repositories of knowledge gained from a lifetime of experiences. Whatever the reasons, most human societies are unusual in protecting senescent individuals within the safety of society and not casting them to the wolves.

Early Embryology

At the other end of an individual's life history are the events of early embryology. This is a complex, fascinating study in its own right (figure 5.1). Embryonic development has profoundly contributed to evolutionary biology and to morphology. Early in development, the cells of the embryo become sorted into three primary germ layers: **ectoderm, endoderm,** and **mesoderm.** In turn, each layer gives rise to specific regions that form body organs. Structures of two species that pass through closely similar steps of embryonic development can be taken as evidence of homology between these structures. Close homology testifies to the phylogenetic relationship of both species.

Although embryonic development is an unbroken, continuous process, we recognize stages in this progression in order to follow events and compare the developmental processes between groups. The youngest stage of the embryo is the fertilized egg, or **zygote,** which develops subsequently through the **morula, blastula, gastrula,** and **neurula** stages. During these early stages, the **embryonic area** becomes delineated from the **extraembryonic area** that supports the embryo or delivers nutrients but does not become a part of the embryo itself. The delineated embryo first becomes organized into basic germ layers and then passes through **organogenesis** (meaning "organ" and "formation") during which the well-established germ layers differentiate into specific organs.

Fertilization

Union of two mature sex cells, or **gametes,** constitutes fertilization. The male gamete is the **sperm** and the female the **ovum,** or egg. The sperm and egg carry genetic material from each parent. Both are **haploid** at maturity, with each containing half the chromosomes of each parent. The sperm's passage through the outer layers of the ovum sets in motion, or **activates,** embryonic development.

Although an egg can be very large, as is a chicken egg, it is but a single cell with a nucleus, cytoplasm, and cell membrane, or **plasma membrane.** While still in the ovary, the ovum accumulates **vitellogenin,** a transport form of yolk formed in the liver of the female and carried in her blood. Once in the ovum, vitellogenin is transformed into **yolk platelets** consisting of storage packets of nutrients that help support the growing needs of the developing embryo. The quantity of yolk that collects in the ovum is specific to each species. Eggs with slight, moderate, or enormous amounts of yolk are **microlecithal, mesolecithal,** or **macrolecithal,** respectively (table 5.1). Further, the yolk can be evenly distributed **(isolecithal)** or concentrated at one pole **(telolecithal)** of the spherical ovum. When yolk and other constituents are unevenly arranged, the ovum shows a **polarity** defined by a **vegetal pole,** where most yolk resides, and an opposite **animal pole,** where the prominent haploid nucleus resides.

The region immediately beneath the plasma membrane of the ovum is referred to as the **cortex** of the ovum. It often contains specialized **cortical granules** activated at fertilization. Outside the plasma membrane, three envelopes surround the ovum. The first, the **primary egg envelope,** lies between the plasma membrane and the surrounding cells of the ovary. The most consistent component of this primary layer is the **vitelline membrane,** a transparent jacket of fibrous protein. In mammals, the homologous structure is called the **zona pellucida.** When the zona pellucida is viewed with a light microscope, a thin, striated line that was once called the "zona radiata" seems to constitute another discrete component of this primary layer. However, the high-resolution electron microscope reveals that the zona radiata is not a separate layer but an effect produced by a dense stand of microvilli projecting from the surface of the ovum. These microvilli intermingle with microvilli reaching

TABLE 5.1 Comparison of Cleavage Patterns and Yolk Accumulation in Representative Vertebrates		
Cleavage Pattern	**Yolk Accumulation**	**Representative Animals**
Holoblastic	Microlecithal	Amphioxus
	Mesolecithal	Lampreys, bowfins, gars, amphibians
Meroblastic	Macrolecithal	Elasmobranchs, teleost fishes
Discoidal[a]	Macrolecithal	Reptiles, birds, monotremes

[a]Discoidal cleavage is an extreme case of meroblastic cleavage.

inward from the surrounding cells of the ovary. This stand of microvilli increases surface contact between the ovum and its environment within the ovary. Often after fertilization, a **perivitelline space** opens between the vitelline membrane and the plasma membrane.

The **secondary egg envelope** is composed of **ovarian,** or **follicle, cells** that immediately surround and help transfer nutrients to the ovum. In most vertebrates, follicle cells fall away from the ovum as it departs from the ovary. However, in eutherian mammals, some follicle cells cling to the ovum, becoming the **corona radiata** that accompanies the ovum on its journey to the uterus. The successful sperm must penetrate all three layers—follicle cells (in eutherian mammals), vitelline membrane, and plasma membrane.

The **tertiary egg envelope,** the exterior wrapping around the egg, forms in the oviducts. In some sharks, it consists of an egg case. In birds, reptiles, and monotremes, it includes the shell, shell membranes, and albumen enveloping the ovum. The tertiary layer is added after fertilization when the ovum travels down the uterine tubes. Vertebrates laying eggs encapsulated in such shells or other tertiary egg envelopes are **oviparous** (meaning "egg" and "birth"). If parents nestle over the eggs to add warmth, the eggs are **incubated.** Those vertebrates giving birth to embryos without such shells are **viviparous** (meaning "live" and "birth"). The **gestation** period includes the time the embryo develops within the female.

Viviparity has independently evolved over a hundred times in vertebrates. Many of these occasions occur in fishes, but most are in squamates. Oddly, no instance of viviparity is known in turtles, crocodiles, or birds, perhaps because they use the eggshell as a calcium reservoir on which the embryo draws when its own skeleton is undergoing ossification. In squamates, calcium is stored in the embryonic yolk, so that evolutionary loss of the eggshell creates no loss of access to calcium stores. Viviparity has evolved repeatedly in squamates.

In some species, shelled eggs are retained within the oviducts of the female until they hatch or until the shells fall away. Shortly thereafter, the young are released to the world from the oviduct. Such reproductive patterns make it clear we must distinguish between the act of giving birth and the mode of supplying the fetus with nutrition. Specifically, **parturition** is the act of giving birth via viviparity, and **oviposition** is the act of laying eggs. The general term **parition** includes parturition and oviposition. Two general terms describe patterns of fetal nutrition. Embryos that draw nutrients from the yolk of the ovum are **lecithotrophic.** Lecithotrophic nutrition occurs through direct transfer of yolk to the connecting part of the digestive tract, as in some fishes, or through the vitelline arteries and veins that provide a vascular connection between the embryo and its yolk reserves. If nutrients are drawn from alternative sources, the embryos are **matrotrophic.** Vascular placentae or secretions of the oviduct that deliver nutrients to embryos are examples of matrotrophy. If offspring are provisioned with nutrients after birth or hatching, matrotrophy can be continued. In mammals, nutrient delivery changes from preparitive matrotrophy (placenta) to postparitive matrotrophy (lactation).

Release of the ovum from the ovary is **ovulation.** Fertilization usually takes place soon thereafter. With the fusion of egg and sperm, the **diploid** chromosome number is restored. Activation of development, initiated by sperm penetration, ushers in the next process, **cleavage.**

Cleavage

Repeated mitotic **cell division** of the zygote occurs during **cleavage.** The embryo experiences little or no growth in size, but the zygote is transformed from a single cell into a solid mass of cells called the morula. Eventually the multicelled and hollow blastula forms (figure 5.2a–c). The **blastomeres** are the cells resulting from these early cleavage divisions of the ovum.

The first cleavage furrows appear at the animal pole and progress toward the vegetal pole. Where yolk is sparse, as in the microlecithal eggs of amphioxus and eutherian mammals, cleavage is **holoblastic**—mitotic furrows pass successfully through the entire zygote from animal to vegetal pole. After the first few furrows pass from the animal to the vegetal pole, subsequent furrows perpendicular to these develop until a hollow ball of cells forms around an internal fluid-filled cavity. Structurally, the blastula is the hollow ball of cells around the internal **blastocoel** cavity. In embryos where yolk is plentiful, cell division is impeded, mitotic furrowing is slowed, only a portion of the cytoplasm is cleaved, and cleavage is said to be **meroblastic.** In extreme cases, such as in the eggs of many fishes, reptiles, birds, and monotremes, meroblastic cleavage becomes **discoidal** because extensive yolk material at the vegetal pole remains undivided by mitotic furrows and cleavage is restricted to a cap of dividing cells at the animal pole.

In all chordate groups, cleavage converts a single-celled zygote into a multicellular, hollow blastula. Variations in the fundamental cleavage process result from characteristic differences in the amount of accumulated yolk reserves. The simplest pattern occurs in amphioxus, where little yolk is present. Eggs of amphibians possess substantially more yolk than those of amphioxus. In most fishes, reptiles, birds, and monotremes, great stores of yolk are packed into the egg. Eutherian mammals have little yolk present (figure 5.2a–d; table 5.1).

Amphioxus

Eggs of amphioxus are microlecithal. The first cleavage plane passes from animal to vegetal pole, forming two blastomeres. The second cleavage plane is at right angles to the first and also passes from animal to vegetal poles, producing an embryo of four cells that resembles an orange with

Zygote Morula Blastula Sagittal section of blastula

(a) Amphioxus

Blastoderm

(b) Amphibians

Blastoderm
Subgerminal space

(c) Reptiles/birds

Inner cell mass

(d) Mammals
     (eutherian)

FIGURE 5.2 **Cleavage stages
in five chordate groups.** Relative
sizes are not to scale. (a) Amphioxus.
(b) Amphibian. (c) Reptiles and birds.
(d) Eutherian mammal.

four wedges. The third cleavage plane is at right angles to
the first two and lies between poles just above the equator,
producing the eight-celled morula stage (figure 5.2a). Subsequent divisions of the blastomeres, now less and less in
synchrony with each other, yield the 32-celled blastula surrounding the fluid-filled blastocoel.

Fishes

In gars and bowfins, cleavage is holoblastic, although cleavage furrows of the vegetal pole are slowed. Most cell division is restricted to the animal pole (figure 5.3). Blastomeres
in the vegetal pole are relatively large and hold most of the
yolk reserves; those in the animal pole are relatively small
and form the **blastoderm,** a cap of cells arched over a small
blastocoel. The blastula produced is very much like that of
amphibians.

In hagfishes, chondrichthyans, and most teleosts,
cleavage is strongly discoidal, leaving most of the yolky cytoplasm of the vegetal pole undivided (figure 5.4a–d). Cleavage
in teleosts produces two cell populations in the blastula. One
of these is the blastoderm, also called the **blastodisc** because
it is a discrete patch of embryonic tissue, or **embryonic disc**
because it is destined to form the embryonic body. The other
cell population formed is the **periblast,** a syncytial layer
closely adhering to the uncleaved yolk. The periblast helps
to mobilize this yolk so that it can be used by the growing
embryo (see figure 5.9).

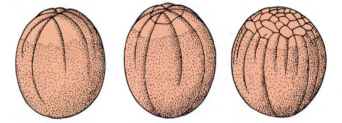

FIGURE 5.3 Holoblastic cleavage in the bowfin,
Amia.

After Korschelt.

Amphibians

As in the gars and bowfins, blastomeres of the animal pole
divide more often than those of the vegetal pole, in which
cell division is presumably slowed by abundant yolk platelets. Consequently, cells of the vegetal pole, having undergone fewer divisions, are larger than those of the more active
animal pole. When the blastula stage is reached, the small
blastomeres of the animal pole constitute the blastoderm
and form a roof over the emerging blastocoel.

Reptiles and Birds

In reptiles and birds, the fertilized egg does not directly
produce an embryo. Rather, the early cleaving cells form
a blastoderm that eventually separates to form an epiblast

Blastoderm
Blastocoel
Vitelline membrane
Periblast
Yolk mass

(a)   (b)   (c)   (d)

FIGURE 5.4 **Discoidal cleavage in a teleost (zebra fish).** (a) Cleavage begins with the appearance of the first mitotic furrow. After successive mitotic divisions (b), the blastula (c) results. (d) Cross section of the blastula. A cap of blastoderm rests on the uncleaved yolk mass, and a vitelline membrane is still present around the entire blastula.

After Beams and Kessel in Gilbert.

(future embryo) and a hypoblast (supportive structures). Yolk is so prevalent within the vegetal pole that cleavage furrows do not pass through it at all; thus, cleavage is discoidal. Blastomeres resulting from successful cleavage clump at the animal pole, forming the blastoderm (descriptively termed a *blastodisc* in reptiles and birds) that rests atop the undivided yolk (figure 5.2c). The term **subgerminal space** applies to the fluid-filled cavity between blastoderm and yolk at this point in development.

The blastoderm becomes **bilaminar** (two-layered). Cells at its border migrate forward, beneath the blastoderm, toward the future anterior end of the embryo. Along the way, these cells are joined by cells dropping from the blastoderm, an event known as **ingression.** Migrating cells together with ingressing cells form the new **hypoblast.** Cells remaining in the depleted blastoderm now properly constitute the **epiblast.** The space between the newly formed hypoblast and epiblast is the compressed blastocoel.

Mammals

In mammals, the blastula stage is termed a **blastocyst.** The three living groups of mammals differ in their modes of reproduction. The most primitive living mammals, the monotremes, retain the reptilian mode of reproduction and lay shelled eggs. Marsupials are viviparous, but the neonate is born at a very early stage in its development. Eutherian mammals retain the embryo within the uterus until a later stage in development and supply most of its nutritional and respiratory needs through the specialized placenta. Because of such differences, embryonic development in these three groups will be treated separately.

Monotremes In monotremes, yolk platelets collect in the ovum to produce a macrolecithal egg. When the ovum is released from the ovary, the follicle cells are left behind. Fertilization occurs in the oviduct. The walls of the oviduct secrete first an "albumenlike" layer and then a leathery shell before the egg is laid. Cleavage, which is discoidal, begins during this passage of the embryo down the oviduct and gives rise to the blastoderm, a cap of cells that rests atop the undivided

yolk. The blastoderm grows around the sides of the yolk and envelops it almost completely (figure 5.5a).

Marsupials In marsupials, the ovum accumulates only modest amounts of yolk. Upon ovulation, it is surrounded by a zona pellucida but lacks follicle cells (lacks the corona radiata; figure 5.5b). Once the ovum is fertilized, the oviduct adds a **mucoid coat** and then an outer, thin **shell membrane.** The shell membrane is not calcified, but it is generally similar in mode of formation, chemical composition, and structure to the shell of monotremes and some oviparous reptiles. It is an acellular layer secreted by the luminal epithelium of the fallopian tube and uterus. It remains around the embryo throughout cleavage and blastocyst formation, perhaps serving to give support to the developing embryo. At the end of gestation, it is finally shed.

Early cleavage in marsupials does not result in formation of a morula. Instead, blastomeres spread around the inner surface of the zona pellucida, forming a single-layered **protoderm** around a fluid-filled core. Initially the blastocyst is a **unilaminar** (single-layered) protoderm around a blastocoel. Through uptake of uterine fluids, the blastocyst and its enveloping mucoid and shell membranes expand in size. Cells at one pole of the blastocyst give rise to the embryo and to its amnion, whereas the remaining cells give rise to a **trophoblast.** Trophoblastic cells help establish the embryo during its brief residence within the uterus, after which they participate in physiological exchange between maternal and fetal tissues, contribute to extraembryonic membranes, and possibly protect against the female's premature immunological rejection of the embryo before birth.

Eutherians In eutherians, the ovum contains very little yolk when it is released from the ovary. It is surrounded by the zona pellucida and clinging follicle cells, which form the corona radiata. After fertilization, cleavage results in the morula, a compact ball of blastomeres still within the zona pellucida and with an added exterior mucoid coat. The appearance of fluid-filled cavities within the morula ushers in the blastocoel. Cells organize around the blastocoel to form the blastocyst. The zona pellucida prevents the blastocyst from prematurely attaching to the oviduct until it reaches the uterus. Upon arrival in the uterus, the blastocyst lyses a small hole in the zona pellucida and squeezes out. At this point, the blastocyst consists of an outer sphere of trophoblastic cells and an inner cell mass clumped against one wall (figure 5.5c). The trophoblast contributes to the extraembryonic membranes that will establish a nutritive and respiratory association with the uterine wall. The inner cell mass contributes additional membranes around the embryo and eventually forms the body of the embryo itself.

Doubt has recently been raised about the homology of the trophoblastic layers in eutherians and marsupials. Marsupials lack a morula and an inner cell mass and differ from eutherians in other aspects of cleavage as well.

FIGURE 5.5 **Cleavage in three groups of living mammals.** (a) Monotremes exhibit discoidal cleavage, with a blastocyst composed of a cap of blastoderm atop uncleaved yolk. (b) In marsupials, cleavage does not result in a distinct morula stage composed of a mass of solid cells. Instead, cells produced during cleavage spread along the inside of the zona pellucida and directly become the protoderm. The oviduct produces a mucoid coat and thin shell membrane. (c) Eutherian mammals pass from a morula to a blastocyst, where cells are set aside as an inner cell mass and an outer wall (the trophoblast). Very little yolk is present. A mucoid coat and zona pellucida are present around the morula but are not shown here.

Monotreme based on Flynn and Hill.

The terms *choriovitelline membrane* and *chorioallantoic membrane* have been suggested as replacements for the term *trophoblast*. The issue is not just a struggle over names. If the trophoblast proves to be unique to eutherians, this implies that it arose as a new embryonic structure in the Cretaceous when eutherian mammals emerged. This new trophoblast would have been a vital component in the emerging reproductive style of eutherians, allowing prolonged intrauterine exchange between fetal and maternal tissues. But the trophoblast of marsupials accomplishes most of the same functions as the trophoblast of eutherians. Until the evidence is more persuasive, we will follow the conventional view of a homologous trophoblast in both eutherian and marsupial mammals.

Overview of mammalian evolution (p. 121)

Overview of Cleavage

During cleavage, repeated cell divisions produce a multicellular blastula, each cell of which is a parcel containing within its walls some of the original cytoplasm of the egg. Because ingredients within the original polarized ovum were unevenly distributed, each cell holds a slightly different cytoplasmic composition that it carries during migration to new positions within the embryo. In some species, the blastula imbibes uterine fluids to swell in size, but it does not grow by incorporating new cells. During gastrulation, the stage following blastula formation, most cells arrive at their final destinations. Some of the initial ability of these cells to differentiate along many pathways has been narrowed, however, so that most cells at this stage are fated to contribute to just one part of the embryo. During subsequent embryonic stages, cell fate narrows further until each cell eventually differentiates into a terminal cell type.

Gastrulation and Neurulation

Cells of the blastula undergo major rearrangements within the embryo to reach the gastrula and neurula stages. **Gastrulation** (meaning "gut" and "formation") is the process by which the embryo forms a distinct endodermal tube that constitutes the early gut. The space enclosed within the gut is the **gastrocoel**, or **archenteron**. **Neurulation** (meaning "nerve" and "formation") is the process of forming an ectodermal tube, the **neural tube**. This tube is a forerunner of the central nervous system and encloses the **neurocoel**. Gastrulation and neurulation occur simultaneously in some species and include other embryonic events with far-reaching consequences. During this time, the three germ layers come to occupy their characteristic starting positions: ectoderm on the outside, endoderm lining the primitive gut, and mesoderm between the other two (figure 5.6a). Sheets of mesoderm become tubular, and the resulting body cavity enclosed within the mesoderm is the **coelom** (figure 5.6b).

Cleavage is characterized by cell division; gastrulation is characterized by major rearrangements of cells. By the end of gastrulation, large populations of cells, originally on the surface of the blastula, divide and spread toward the inside of the embryo, a process that is much more than simple cell shuffling. As a result of this reorganization, tissue layers and cell associations are established strategically within

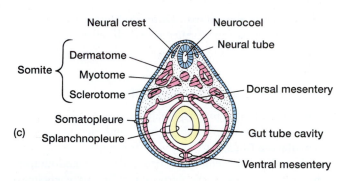

FIGURE 5.6 **General steps in successive differentiation of the mesoderm and neural tube.** Mesoderm initially comes to lie between the other two germ layers (a), and differentiates into three major regions: the epimere, the mesomere, and the hypomere (b). Each of these gives rise to specific layers and groups of mesodermally derived cell populations (c). Neurulation begins with a dorsal thickening of the ectoderm into a neural plate (a). This plate folds (b), and its folds fuse into a hollow neural tube (c). Note the formation and separation of the neural crest (c) from the edges of the original neural plate.

the embryo. How they are positioned will largely determine their subsequent interactions with each other. Tissue-to-tissue interactions, or inductive interactions, are one of the major determinants of later organ formation.

Although the pattern of gastrulation varies considerably among chordate groups, it is usually based on a few methods of cell movement in various combinations. Cells may spread across the outer surface as a unit **(epiboly)**; cells

may turn inward and then spread over the internal surface **(involution)**; a wall of cells may indent or simply fold inward **(invagination)**; sheets of cells may split into parallel layers **(delamination)**; or individual surface cells may migrate to the interior of the embryo **(ingression)**.

By whatever method, cells moving to the interior leave behind the sheet of surface cells that constitute the ectoderm. The most common method of neurulation is **primary neurulation** wherein the neural tube is formed through folding of the dorsal ectoderm. Specifically, the surface ectoderm thickens into a strip of tissue that forms the **neural plate** along what is to be the dorsal side and anterior-posterior axis of the embryo (figure 5.6a). In tetrapods, sharks, lungfishes, and some protochordates, the margins of the neural plate next grow upward into parallel ridges that constitute the **neural folds** (figure 5.6b). The neural folds eventually meet and fuse at the midline, forming the neural tube that encloses the neurocoel (figure 5.6c). The neural tube is destined to differentiate into the brain and spinal cord (the central nervous system). Just before or just as the neural folds fuse, some cells within these ectodermal folds separate out and establish a distinct population of **neural crest cells.** In the embryo's trunk, these neural crest cells are organized initially into cords, but in the head, they usually form into sheets. From their initial position next to the forming neural tube, neural crest cells migrate out along defined routes to contribute to various organs. Such cells are unique to vertebrates and are discussed at length later in this chapter.

In lampreys and teleost fishes (and in the tail region of tetrapods), the neural plate does not directly form a tubular nerve cord via folding. Instead, the neural tube forms by a process of **secondary neurulation** wherein the neurocoel appears via cavitation within a previously solid cord. Specifically, the thickened neural plate sinks inward from the surface along the dorsal midline, forming directly a *solid* rod of ectodermal cells, the **neural keel.** Later, a neurocoel appears by cavitation within the core of the formerly solid neural keel to produce the characteristic dorsal tubular nerve tube. Because there is no infolding of the neural plate to form a tube, strictly speaking, there is no "crest" in fishes to yield a neural "crest." In lampreys and teleost fishes, neural crest cells segregate from the dorsolateral edges of the neural keel. Thereafter, their behavior and contributions to tissues are similar to those in tetrapods. They become organized into distinct populations of crest cells before (trunk) or after (cranial) they begin ventral migration along distinct pathways to locations of eventual differentiation.

The endoderm is derived from cells moving inward from the outer surface of the blastula. At first the endoderm forms the walls of a simple gut extending from anterior to posterior within the embryo. But as development proceeds, outpocketings from the gut and its interactions with other germ layers produce associated glands and their derivatives.

The mesoderm also is derived from cells entering from the outer surface of the blastula. Mesodermal cells proliferate as they expand into a tissue sheet around the insides of the body between outer ectoderm and inner endoderm. Occasionally, rather than forming a sheet, mesodermal cells become dispersed to produce a network of loosely connected cells called **mesenchyme.** (The term **ectomesenchyme** applies to the loose confederation of cells derived from the neural crest.) The notochord arises from the dorsal midline between lateral sheets of mesoderm. Each lateral sheet of mesoderm becomes differentiated into three regions: a dorsal **epimere** or **paraxial mesoderm,** a middle **mesomere** or **intermediate mesoderm,** and a ventral **hypomere** or **lateral plate mesoderm** (figure 5.6b). The central cavity within the mesoderm is the paired **primary** or **embryonic coelom.** Parts of the embryonic coelom often become enclosed in the mesoderm, forming a **myocoel** within the epimere, a **nephrocoel** within the mesomere, and simple **coelom** (body cavity) within the lateral plate mesoderm.

Two processes can produce these cavities within the mesoderm. In **enterocoely,** the most primitive method of coelom formation among chordates, the interior cavity is contained within the mesoderm when it first pinches off from other tissue layers. In **schizocoely,** the mesoderm forms first as a solid sheet and splits later to open the cavity within. If you remember that vertebrates are deuterostomes, which are characterized by enterocoely, you may be surprised to learn that schizocoely predominates in this group as a whole. In fact, cephalochordates and lampreys are the only chordates in which the coelom is formed by strict enterocoely. This has led many to conclude that the method of coelom formation is not a useful criterion for characterizing superphyletic groups. Others hold that the absence of enterocoely in most vertebrates is likely a secondary condition derived from enterocoelous ancestors. Coelom formation via mesodermal splitting can be attributed to developmental modifications, perhaps to accommodate enlarged yolk stores. In this view, schizocoely evolved independently in vertebrates and protostomes. Until this dilemma is resolved, we will follow the view that schizocoelic coelom formation in vertebrates is derived from enterocoelous ancestors.

Phylogenetic significance of coelom formation (p. 50)

The paraxial mesoderm (epimere) forms as a pair of cylindrical condensations adjacent to and parallel with the notochord. The paraxial mesoderm becomes organized into connected clusters of loosely whorled mesenchymal cells, termed **somitomeres.** Beginning at about the neck and progressing posteriorly, clefts form between somitomeres to delineate anatomically separate condensed clumps of mesoderm, **somites.** The somitomeres in the head remain connected, and may number seven in amniotes and teleosts, and four in amphibians and sharks. They give rise to striated muscles of the face, jaws, and throat, with the connective tissue component derived from the neural crest. The somites, in series with the somitomeres, vary in number with species. Somites in turn split into three separate mesodermal populations. These populations of somite cells contribute to the skin

musculature **(dermatome),** the body musculature **(myotome),** and the vertebrae **(sclerotome).** The mesomere gives rise to portions of the kidney. As the coelom expands within the hypomere, inner and outer mesodermal sheets of cells are defined. The inner wall of the hypomere is the **splanchnic mesoderm,** and the outer wall the **somatic mesoderm** (figure 5.6b). These sheets of mesoderm come into association with endoderm and ectoderm, with which they interact later to produce specific organs. Collectively, the paired sheet of splanchnic mesoderm and the adjacent sheet of endoderm form the **splanchnopleure;** the somatic mesoderm and the adjacent ectoderm form the **somatopleure** (figure 5.6c).

Amphioxus

Gastrulation in amphioxus occurs by invagination of the vegetal wall (figure 5.7a). As vegetal cells grow inward, they obliterate the blastocoel. Cells on the inside next separate into endoderm and mesoderm. Some investigators prefer to emphasize the potential of this single layer, calling it future endoderm and mesoderm. Others refer to it as **endomesoderm** in recognition of its present unity. The endomesoderm eventually moves up against the inside wall of the ectoderm and forms the primitive gut. The gastrocoel communicates to the exterior through the **blastopore** (figure 5.7a). The embryo is consequently transformed during early gastrulation from a single layer of blastomeres to a double layer of cell sheets consisting of the ectoderm and the endomesoderm. Each layer will give rise to specific adult tissues and organs.

Delineation of the mesoderm occurs during neurulation in the amphioxus embryo. A series of paired outpocketings form and pinch off from the mesoderm. These cavities merge to become the coelom (figure 5.7b). As the paired mesodermal outpocketings take shape, the mesoderm at the dorsal midline between them differentiates into the **chordamesoderm.** In addition to giving rise to the notochord, the chordamesoderm stimulates differentiation of the overlying ectoderm into the central nervous system. The epidermis lateral to the early neural plate detaches and moves across the neural plate. Only after the two sides meet and form a continuous sheet of epidermis does the neural tube below round up (figure 5.7b). The mesoderm then becomes delineated into epimere, mesomere, and hypomere.

Fishes

Gastrulation, like cleavage, is modified in proportion to the amount of yolk present. The amount of yolk varies considerably from one group of fishes to the next, so the patterns of gastrulation are quite varied in fishes as well.

In lampreys and primitive bony fishes, the onset of gastrulation is marked by the appearance of an indentation, the dorsal edge of which is the **dorsal lip of the blastopore** (figure 5.8a–j). The dorsal lip of the blastopore is an important organizing site within the embryo. Surface cells flow to the blastopore by epiboly, slip over its lip and turn inward,

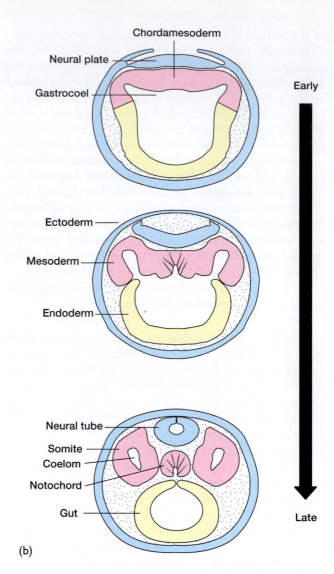

FIGURE 5.7 **Amphioxus gastrulation and neurulation.** (a) At the left, invagination at the vegetal pole pushes cells into the interior of the blastula. The blastocoel is eventually obliterated, and the new space these ingrowing cells define becomes the gastrocoel. The arrow indicates the anteroposterior axis of the embryo. (b) Successively older cross sections taken along the plane (P) defined in the illustration on the left (a). As development proceeds, mesodermal outpocketings appear and pinch off to form somites, leaving the endoderm to form the lining of the gut.

and then begin to spread along the internal ceiling of the embryo. These entering surface cells constitute the endomesoderm, the name reminding us again of the two germ layers (endoderm and mesoderm) into which it will separate. The endomesoderm surrounds a gastrocoel and obliterates the blastocoel as it grows.

During gastrulation in sharks and teleost fishes, the blastoderm grows over the surface of the yolk, eventually engulfing it completely to form the extraembryonic **yolk sac.** While this is occurring, the endomesoderm arises under and at the edges of the spreading blastoderm (figure 5.9a, b). The endomesoderm is continuous with the surface layer of the blastoderm, but its source is disputed. Some claim that it is formed by cells flowing around the edge of the blastoderm and inward. Others claim that deep cells already in place become rearranged to produce the endomesoderm. Whatever its embryonic source, the endomesoderm tends to be thickest at the posterior edge of the blastoderm where it becomes concentrated into the **embryonic shield** that produces the body of the embryo (figure 5.10a–d).

Separation of the endomesoderm into endoderm and mesoderm occurs next. When finally separate, the endoderm is a flat sheet of cells stretched over the adjacent yolk, but it does not grow around the entire yolk mass. A recognizable gastrocoel has not yet appeared. The mesoderm forms chordamesoderm at the midline. Chordamesodermal cells give rise to the notochord and lateral plates of mesoderm that grow around the yolk. Thus, the yolk is eventually enclosed by a membrane consisting of periblast, mesoderm, and ectoderm, but no endoderm.

Gastrulation is based on several noteworthy differences within different groups of fishes. In lampreys, the coelom is enterocoelic, forming as the mesoderm pinches off from the rest of the endomesoderm. This is similar to the enterocoelic process in amphioxus and suggests that enterocoely represents the primitive method of coelom formation. We do not know how coelom formation occurs in hagfishes, but in all other vertebrates, the coelom forms by schizocoely in which the solid sheet of mesoderm splits to open spaces that become the body cavity.

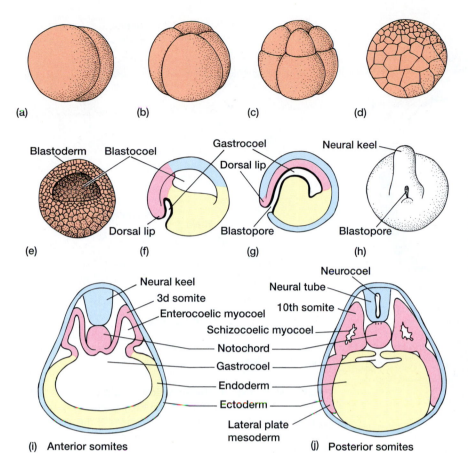

(a) (b) (c) (d)

Blastoderm Blastocoel

Gastrocoel

Dorsal lip

Neural keel

Dorsal lip

Dorsal lip

Blastopore

Blastopore

(e) (f) (g) (h)

Neurocoel

Neural keel
3d somite
Enterocoelic myocoel

Neural tube
10th somite
Schizocoelic myocoel

Notochord

Gastrocoel

Endoderm

Ectoderm

Lateral plate
mesoderm

(i) Anterior somites

(j) Posterior somites

FIGURE 5.8 Lamprey early embryonic development. (a–d) Cleavage stages leading to a blastula. (e) Cross section of the blastula. (f, g) Cross section of successive stages in gastrulation. (h) Exterior view of the entire gastrula. Formation of the myocoel within the somites is different in the anterior (i) compared with the posterior (j) region. In the anterior region, the myocoel is enterocoelous; posteriorly, it is schizocoelous. No open neural plate is formed. Instead, a solid cord of ectodermal cells sinks to the interior from the dorsal midline, forming the solid neural keel. This solid cord of cells becomes secondarily hollowed out to form the characteristic dorsal tubular nerve cord.

After Lehman.

Amphibians

In amphibians, a superficial indentation marks the beginning of gastrulation and establishes the dorsal lip of the blastopore. Three major and simultaneous cell movements occur. First, the movement of surface cells by epiboly creates a stream of cells flowing toward the blastopore from all directions (figure 5.11a). Second, these cells involute over the lips of the blastopore. Third, entering cells move to and take up specific sites of residence within the embryo. Cells entering by such migratory routes become part of the endomesoderm surrounding the gastrocoel. The chordamesoderm, forerunner to the notochord, arises middorsally within the endomesoderm. Separation of the endomesoderm into distinct germ layers begins with the appearance of paired projections of tissue growing out from the endomesoderm's lateral inner wall and upward to meet beneath the forming notochord. These paired projections of tissue, together with the ventral region of yolk-laden endomesoderm, separate into the endoderm proper. The rest of the endomesoderm becomes the mesoderm proper. From a solid sheet of cells, the mesoderm grows downward between the newly delineated endoderm and outer ectoderm. Distinct epimere, mesomere, and hypomere become evident in the mesoderm, and by schizocoely the solid mesodermal layer splits to produce the coelom within (figure 5.11b).

Gastrulation establishes the ectoderm and endoderm, obliterates the blastocoel, forms the new gastrocoel, and leaves a blastopore that is partially plugged by yolk-laden cells not completely drawn into the interior of the embryo.

Neurulation in amphibians usually begins before the endomesoderm has separated into its distinct germ layers. As in all tetrapods, neurulation proceeds by thickening of the neural plate that rolls up into the hollow neural tube (figure 5.11b). An external view of anuran development from fertilization to growth of operculum and forelimbs is illustrated in figure 5.12.

Reptiles and Birds

In reptiles and birds, enormous accumulations of yolk alter embryonic processes. The flattened blastula includes the superficial epiblast, the hypoblast beneath, and the blastocoel between them.

Reptiles Early development processes in reptiles are less well studied than those in birds, but they are assumed to be basically the same. Based on the morphology of the early embryos, the cleavage stages of reptiles and birds are similar, as is neurulation. One notable difference is gastrulation. Reptiles retain a discrete blastopore associated with movement of cells during gastrulation; in birds, the blastopore is lost, and

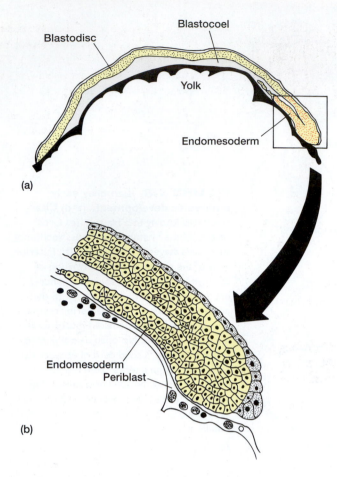

FIGURE 5.9 Gastrulation at an early stage in a teleost fish (trout). (a) Cross section of the blastoderm arched over a compressed blastocoel. (b) Enlarged view of the posterior region of the blastoderm when the second layer, the endomesoderm, first appears.

Figure from Boris I. Balinsky. Introduction to Embryology, Fifth Edition, fig. 155. © 1981 Brooks/Cole, a part of Cengage Learning, Inc. Reproduced by permission. www.cengage.com/permissions.

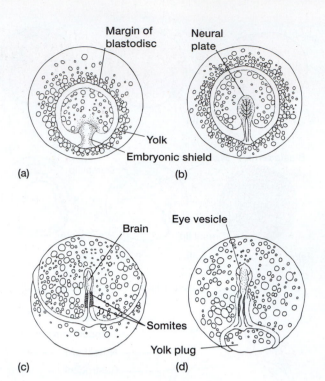

FIGURE 5.10 Differentiation within the embryonic shield of a teleost fish (trout). (a) Early gastrulation. (b) Later gastrulation. (c) Formation of anterior regions of the embryo within the embryonic shield. (d) Blastoderm has nearly completed its overgrowth of the yolk.

Figure from Boris I. Balinsky. Introduction to Embryology, Fifth Edition, fig. 156. © 1981 Brooks/Cole, a part of Cengage Learning, Inc. Reproduced by permission. www.cengage.com/permissions.

gastrulation involves cell streaming through the primitive streak (see below, next).

Birds   The onset of gastrulation is marked in the epiblast by the appearance of a thickened area at what will eventually be the posterior region of the embryo. This thickened area constitutes the **primitive streak** (figure 5.13a, d) and originates as a raised clump of cells called the **primitive node** (Hensen's node). The **primitive groove** is a narrow gully that runs down the middle of the primitive streak. Cells spread across the surface of the epiblast through epiboly and reach the primitive streak, where they involute at the edges of the streak and enter the embryo. Cells entering via the primitive streak either contribute to the mesoderm by spreading between epiblast and hypoblast or form the endoderm by sinking farther to the level of the hypoblast. At this level, they displace hypoblastic cells by pushing them to the periphery (figure 5.13b).

By the end of gastrulation, many surface cells originally belonging to the epiblast have migrated to new positions within the embryo. Cells remaining on the surface now constitute a proper ectoderm. Involuting cells have pushed cells of the hypoblast to the extraembryonic area. In their place over the yolk are the newly arrived cells of the embryonic endoderm. Between ectoderm and endoderm is the mesoderm, also composed of cells arriving via involution through the primitive streak. Along the midline, a notochord differentiates within the mesoderm (figure 5.13c).

Neurulation involves formation of a neural tube from a neural plate precursor. At the onset of neurulation, the three germ layers have already been delineated (figure 5.14a), and reorganization of the lateral mesoderm begins. Initially, the mesoderm is a plate of solid tissue that lies lateral to the notochord with recognizable epimere, mesomere, and hypomere. The hypomere splits, forming the splanchnic and somatic layers of mesoderm and the schizocoelic coelom between. Association of these mesodermal layers with adjacent endoderm and skin ectoderm produces the composite splanchnopleure and somatopleure (figure 5.14c). Although the primitive streak does not have an opening like a blastopore, it functions like a blastopore as the site through which superficial cells enter the embryo.

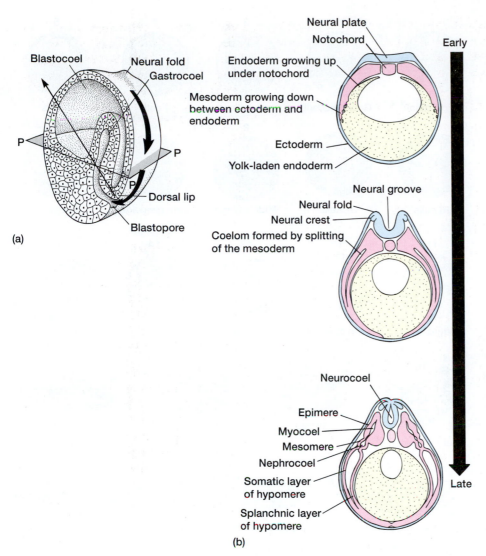

Blastocoel
Neural fold
Gastrocoel

P
P

P

Dorsal lip

Blastopore

(a)

Neural plate
Notochord

Endoderm growing up under notochord

Mesoderm growing down between ectoderm and endoderm

Ectoderm

Yolk-laden endoderm

Early

Neural groove
Neural fold
Neural crest
Coelom formed by splitting of the mesoderm

Neurocoel
Epimere
Myocoel
Mesomere
Nephrocoel
Somatic layer of hypomere
Splanchnic layer of hypomere

Late

(b)

FIGURE 5.11 **Amphibian gastrulation and neurulation.** (a) Sagittal section of an amphibian gastrula. Cells move along the surface (epiboly) and turn inward at the blastopore to form the enlarging gastrocoel. Solid arrows indicate the surface movements of cells. The long arrow indicates the anteroposterior axis of the embryo. (b) Successively older cross sections taken through the plane (P) illustrated in the sagittal section (a). As development proceeds, wings of endoderm grow, fuse, and become distinct from the mesoderm. The mesoderm grows downward and differentiates into various body regions. Notice that the coelom forms within the mesoderm by a splitting of this mesodermal layer.

Mammals

Monotremes Gastrulation, like cleavage, is quite different in the three living groups of mammals. In monotremes, as in reptiles, gastrulation involves a blastodisc atop a large yolk mass. At the end of cleavage, the blastocyst is unilaminar. The blastoderm is five to seven cells thick at its center but thinned at its margins. This sheet of blastoderm grows through mitotic cell division and spreads around the yolk. During pregastrulation, the monotreme blastocyst becomes bilaminar. As the blastoderm grows around the yolk, it sheds cells inward, forming distinct endodermal and ectodermal layers. The endoderm is formed from these inward-moving cells; the ectoderm is composed of cells that remain behind on the surface. These two layers grow into the vegetal pole so that the yolk becomes completely enclosed within the embryo.

As in birds, gastrulation in monotremes begins with the appearance of the primitive streak. From its initial development as a thickened area in the ectoderm to which surface cells converge, the primitive streak becomes a major elongate axis around which the embryonic body is organized. The neural plate (= medullary plate of older literature) forms as an early thickening of ectodermal cells before a distinctive primitive streak can be distinguished. Details of cell movements are not known, but presumably epiboly and involution bring surface cells around the primitive streak into the interior of the embryo.

The second event of gastrulation is the appearance of a mesodermal sheet (figure 5.15a). It probably arises from cells entering via the primitive streak and becomes interposed in its customary position between existing ectoderm and endoderm.

In echidnas (monotremes), the embryo within the uterus increases in size because it takes up fluid secreted by the uterus before it is enveloped in a leathery outer shell. This absorbed fluid is thought to provide nutrition for embryonic growth during the last days of gestation and the ten-day incubation period.

Neurulation in monotremes appears to involve the rolling up of a neural plate into a hollow neural tube.

1	2	3	4	5
Fertilization	Gray crescent	2 cells	4 cells	8 cells
6	7	8	9	10
16 cells	32 cells	Midcleavage	Late cleavage	Dorsal lip
11	12	13	14	15
Midgastrula	Late gastrula	Neural plate	Neural folds	Rotation
16	17	18	19	
Neural tube	Tail bud	Muscular response	Heartbeat	
20		24		37
Gill circulation		Operculum development		Limb bud

FIGURE 5.12 **External view of anuran development.** Beginning with fertilization (1), morula (6–8), blastula (9 and 10), gastrula (11 and 12), and neurula (13–16) stages follow successively. In later development, the tail bud forms (17), muscular twitches begin (18), heartbeat commences (19), functional external gills develop (20), and blood circulation occurs through the caudal fin. Subsequent events include the formation of an operculum (24), a flap of head skin that grows over and covers the gills. The hindlimbs develop first and then the forelimbs. Eventually the embryo undergoes metamorphosis into a juvenile frog. Stages 21–23 and 25–36 are not illustrated.

After Duellman and Trueb.

Marsupials In marsupials, the blastocyst is composed of a single layer of protodermal cells spread around the inside wall of the zona pellucida. The marsupial blastocyst is distinct among mammals, forming neither a blastodisc like monotremes nor an inner cell mass like eutherians. Strictly speaking, it is unilaminar at the end of cleavage. During pregastrulation, this unilaminar blastocyst is transformed into a bilaminar embryo with an ectoderm and an endoderm. Cells of the protoderm proliferate near the animal pole and migrate around their own inner surface, forming a deeper endodermal layer (figure 5.15b). The superficial protoderm layer at this point is now called the ectoderm. As the two germ layers become delineated, the primitive streak appears in the ectoderm, marking the beginning of gastrulation.

Surface cells stream to the primitive streak and involute to the interior of the embryo. Once inside, they contribute to the mesoderm that spreads between outer ectoderm and deeper endoderm (figure 5.15b).

As in other vertebrates, the neural plate rolls into a neural tube during neurulation.

Eutherians In **eutherians,** the blastocyst is composed of two distinct populations of cells at the end of cleavage, an outer trophoblast and an inner cell mass. During pregastrulation, reorganization of the inner cell mass produces a bilaminar embryonic disc composed of epiblast (future ectoderm and mesoderm) and hypoblast (future extraembryonic tissue; figure 5.15c). This occurs when some cells depart from the inner cell mass and migrate into and around the periphery of the blastocoel, forming a thin hypoblastic layer, which is sometimes referred to as the endoderm at this point. The remaining population of cells of the depleted inner cell mass is the epiblast. The now flattened and circular epiblast, together with the adjacent and underlying cells of the hypoblast, constitute the embryonic disc. At this point, the epiblast contains all cells that will produce the actual embryo. The **exocoelomic membrane** is a term occasionally applied to endodermal cells outside the embryonic disc; this is based on the unproven theory that endodermal cells arise from the trophoblast rather than from the inner cell mass as hypoblastic cells do.

(a)

(b)

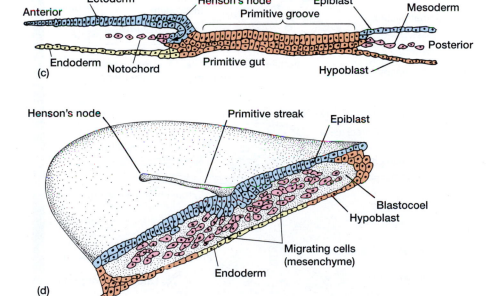

FIGURE 5.13 Bird gastrulation. (a) Dorsal view of the primitive streak. Arrows indicate the direction of major cell movements from the surface through the primitive streak to the interior. (b) A cross section through the embryo illustrates the inward flow of cells. Some of these cells contribute to the mesoderm; others displace the hypoblast to form the endoderm. (c) A longitudinal medial section through the embryo shows the forward migration of a separate stream of cells that produce the notochord. (d) Three-dimensional view of the primitive streak during early gastrulation.

(a, b) After Carlson; (c) after Balinsky; (d) after Duband and Thiery.

In eutherians, as in birds, appearance of a primitive streak marks the beginning of gastrulation (figure 5.16a). Surface cells of the epiblast stream toward the primitive streak (epiboly) and over its edges (involution) to reach the inside. As in reptilian and avian embryos, some entering cells move deep into the embryo, displacing the hypoblast to the periphery where its cells contribute to extraembryonic tissues. Other entering cells become organized into a middle mesoderm. These mesodermal cells grow outward between the deep hypoblast (now more correctly termed the endoderm) and the superficial epiblast (now termed the ectoderm) that has been depleted of cells. The notochord arises from these entering cells (figure 5.16b). The laterally placed mesoderm is at first a solid sheet of tissue but subsequently becomes differentiated into epimere, mesomere, and hypomere. Splitting of the solid mesodermal layer produces the coelom by schizocoely and defines somatic and splanchnic mesodermal sheets.

As regionalization of the mesoderm takes place, neurulation results in development of a tubular nerve tube from a neural plate (figure 5.16b).

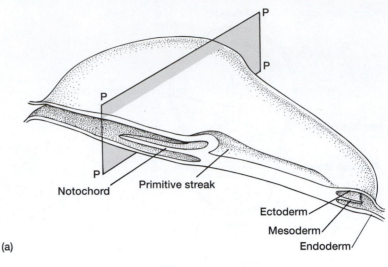

Notochord   Primitive streak

Ectoderm
Mesoderm
Endoderm

(a)

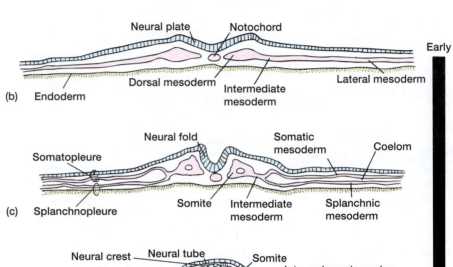

Neural plate   Notochord

Early

(b)   Endoderm

Dorsal mesoderm   Intermediate mesoderm   Lateral mesoderm

Neural fold   Somatic mesoderm   Coelom

Somatopleure

(c)   Splanchnopleure

Somite   Intermediate mesoderm   Splanchnic mesoderm

Neural crest   Neural tube   Somite
Dorsal aorta   Intraembryonic coelom
Lateral body fold

(d)

Primitive gut   Extraembryonic coelom

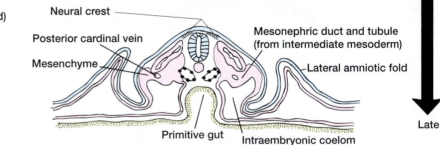

Neural crest

Posterior cardinal vein   Mesonephric duct and tubule (from intermediate mesoderm)

Mesenchyme   Lateral amniotic fold

Late

Primitive gut   Intraembryonic coelom

(e)

FIGURE 5.14 Bird gastrulation and neurulation.
(a) Sagittal section of the embryonic disc showing the primitive streak and the extent of the three primary germ layers. (b–e) Successively older cross sections through the plane (P) indicated in the top figure (a). As gastrulation proceeds, cells entering through the primitive streak form the mesoderm and the endoderm. The mesoderm becomes further differentiated into specific regions, and the endoderm displaces the former hypoblast to the periphery. Successive cross sections show neurulation proceeding from neural plate to neural folds to hollow nerve tube. Note also the regionalization of the mesoderm and the appearance of extraembryonic membranes (lateral amniotic fold).

Organogenesis

By the end of neurulation, several major reorganizations of the embryo have been accomplished. First, polarity based on the animal-vegetal pole axis of the egg has been superseded by bilateral symmetry based on an anterior-posterior axis of the emerging embryonic body. Second, the three primary germ layers have been delineated: ectoderm, endoderm, mesoderm. In all vertebrates, ectoderm gives rise to nervous tissue and epidermis; endoderm to the lining of the digestive and respiratory tubes; mesoderm to skeletal, muscular, and circulatory systems and to connective tissues (figure 5.17). There are exceptions, but generally across vertebrate groups, the same major adult tissue has as its source the same specific

FIGURE 5.15 **Gastrulation in mammals.** In all three mammalian groups, a primitive streak is formed through which cells enter to contribute to the mesoderm. (a) Monotreme. (b) Marsupial. (c) Eutherian.

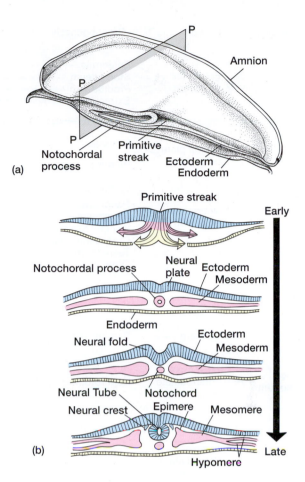

FIGURE 5.16 **Eutherian mammal gastrulation and neurulation.** (a) Sagittal section of the embryonic disc. (b) Successively older cross sections through the plane (P) indicated in the top illustration (a). As gastrulation proceeds, cells entering through the primitive streak form the mesoderm that differentiates into various body regions (bottom cross section).

germ layer of the embryo. Third, the three germ layers become strategically positioned next to one another so that they can mutually interact during **organogenesis,** the differentiation of organs from tissues. Two germ layers frequently and subsequently combine to form a single organ. For example, the alimentary canal is derived from both endoderm (lining, secretory layer) and mesoderm (smooth muscle and investing outer layers). Smooth muscle (mesoderm) is added to the lining (endoderm) of the respiratory tree. The integument includes the combined epidermis (ectoderm) and dermis (mesoderm). Mesoderm is especially important in organogenesis because of its cooperative associations with both ectoderm and endoderm. It is partially supported in its own differentiation by the other two layers, but in turn it stimulates or induces the other layers to form parts of organs.

Histogenesis

The environment immediately around the cell is the **extracellular matrix,** meaning "outside the cell," or the **interstitial space (interstitium),** meaning "around the cell." But separate cells functioning in isolation are seldom found within the body. Instead, similar cells are usually associated into sheets or confederations of cells. Where these aggregations of similar cells are specialized to perform a common function,

they constitute a **tissue.** One early accomplishment of development is to place cells produced during cleavage into one of the cellular germ layers: ectoderm, mesoderm, endoderm. In turn, these formative germ layers differentiate into proper tissues through the process of **histogenesis** (meaning "tissue formation"). There are four primary categories of adult tissues: **epithelium, connective tissue, muscle tissue,** and **nervous tissue.** Muscle and nervous tissues are discussed more fully in chapters 10 and 16, respectively. Because we meet epithelia and connective tissues repeatedly, they are introduced next, and aspects of their embryonic development are discussed.

Epithelium

Epithelial tissues are formed of closely adjoined cells with very little extracellular matrix between them. Usually one side of the epithelium rests upon a **basal layer.** For many years, **basement membrane** was the term used to describe this basal layer, but the electron microscope revealed that the basement membrane is a blend of two structures with separate origins,

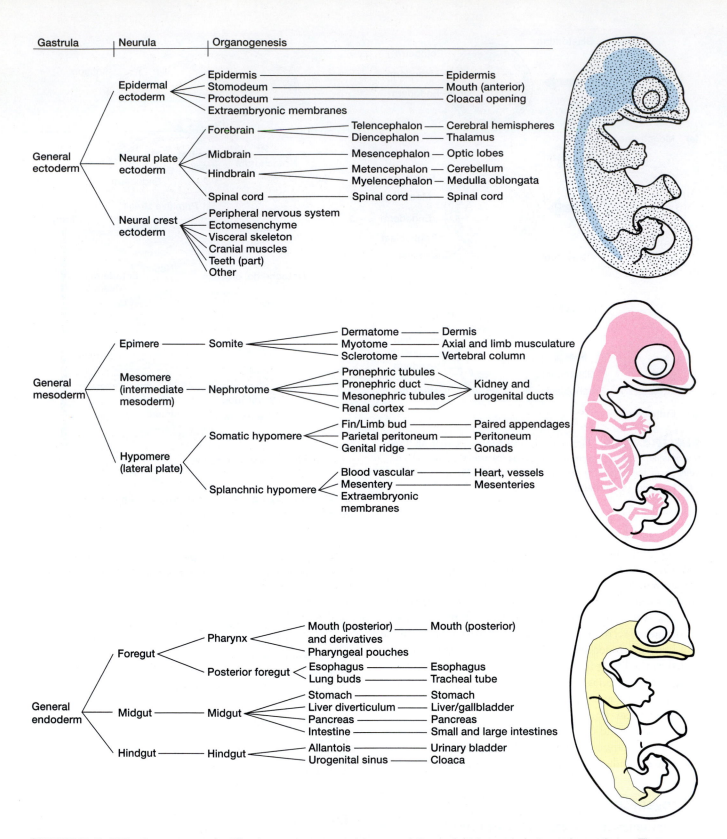

FIGURE 5.17 Organogenesis. The three primary germ layers are delineated during gastrulation and neurulation. Thereafter, they become differentiated into various body regions, and these regions produce the major organs of the vertebrate body. The embryonic origin of each organ or part of an organ can be traced back to these specific germ layers. In general, ectoderm produces the skin and nervous system; mesoderm the skeleton, muscle, and circulatory system; and endoderm the digestive tract and its visceral derivatives.

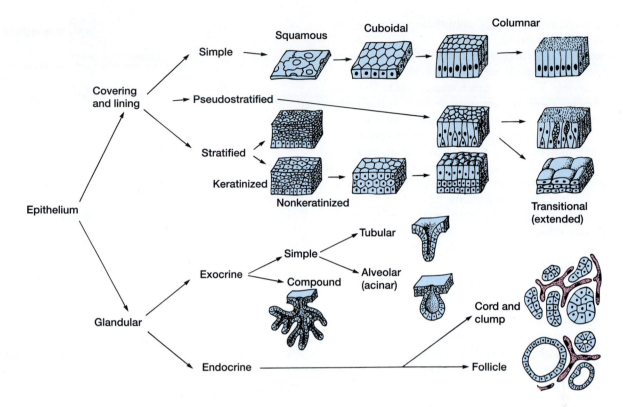

FIGURE 5.18 Classification of epithelia. Epithelia fall into one of two groups: (1) membranes that line or cover cavities or (2) glands that secrete products that act elsewhere in the body. Membranes are single-layered (simple) or multilayered (stratified) sheets of cells. Cells in the sheets can be squamous, cuboidal, or columnar in shape. Exocrine glands release their products (secretions) into ducts that are single (simple) or branched (compound). Endocrine glands release their products into blood vessels; they are arranged into clusters (cord and clump) or into tiny balls (follicles).

After Leeson and Leeson.

the **basal lamina** (derived from epithelium) and the **reticular lamina** (derived from connective tissue). By convention, the choice of terms depends on what can be resolved by the microscope, a basement membrane (light microscope) or basal and reticular laminae (electron microscope). The basement membrane anchors cells into sheets, acts as a selective barrier to the passage of metabolites, and regulates cell behavior through cell signaling—cell to cell communication. Opposite to the basal layer is the **free surface,** or **apical surface,** which faces a **lumen** (cavity) or the exterior environment. With one end resting on the basement membrane and the opposite facing the lumen, epithelial cells have a distinct polarity. The free surface is the usual site at which secretory products are released from the cell **(exocytosis)** or materials are taken into the cell **(endocytosis).** The free surface is most likely to form tiny fingerlike processes, such as microvilli and cilia. The **stereocilia** are very long microvilli. Epithelia are divided into two categories, sheets (membranes) and glands (secretory; figure 5.18).

Covering and Lining Epithelium

Epithelial membranes cover surfaces or line body cavities, ducts, and lumina of vessels. Arranged in sheets, epithelia can be either (1) **simple,** composed of a single layer of cells or (2) **stratified,** composed of more than one layer of cells. The cells themselves can be **squamous** (flat), **cuboidal** (cube shaped), or **columnar** (tall) in form. Epithelial names take advantage of these features of arrangement and cell shape. For example, simple squamous epithelium is made up of a single layer (hence, simple) of flat cells (hence, squamous). Simple squamous epithelium most commonly lines body cavities and vessels. The tissue lining blood and lymph vessels is called **endothelium,** and that lining body cavities is **mesothelium.** Simple cuboidal epithelium appears in many ducts. Simple columnar epithelium lines the digestive tract and some other tubular structures (figure 5.18).

In stratified squamous epithelium, characteristic of the skin, mouth, and esophagus, cells occur stacked in layers (stratified), and surface cells are flat (squamous). Stratified cuboidal and stratified columnar epithelia are rare. In mammals, cells of the male urethra and cells of the Graafian follicles of the ovary are examples.

In addition to simple and stratified epithelia, the third type of lining epithelium is **pseudostratified epithelium** found in the trachea. Cells appear stacked when first inspected under a microscope, but a more careful look reveals that they are falsely layered. The staggered arrangement of cell nuclei is responsible for this false ("pseudo-")

stratification. Actually, all cells, even those at the top, rest on the basement membrane.

Transitional epithelium is a special kind of pseudostratified epithelium found only in the bladder and ducts of the urinary system. The cells stretch when the bladder is distended, allowing them to accommodate changes in bladder size. When relaxed, transitional cells become bunched and deceptively appear to constitute a multilayered epithelium. Recent study indicates that, even when relaxed, each cell touches the basement membrane, so properly the tissue is a pseudostratified epithelium. The name *transitional epithelium* is a misnomer held over from the time when this tissue was erroneously thought to be intermediate (hence, transitional) between other types of epithelia.

Glandular Epithelium

Cells specialized to secrete a product are called **glands.** Glands with ducts that collect and carry away the product are **exocrine glands;** if the product is carried away by the circulatory system, the glands are **endocrine glands.** Glands usually arise from **glandular epithelium.** The ectoderm and endoderm of the early embryo are lining epithelia; therefore, adult organs derived from them are epithelial organs. Epithelial glands arise as tubes or solid cords through invagination and outgrowths from these two epithelial germ layers. Strictly speaking, however, not all cells that produce secretions are epithelial glands derived from ectoderm or endoderm. Some connective tissue cells derived from mesenchyme secrete products that are carried away by ducts or blood vessels; or their products simply collect in the extracellular matrix around the secreting cell. Thus most, but not all, glands of the vertebrate body are epithelial in origin.

A **multicellular gland** is composed of many secretory cells in aggregation, and a **unicellular gland** has only a single secretory cell. Exocrine glands can be **tubular** (cylindrical) or **alveolar** (acinar; rounded in shape). Glands can be **simple,** drained by a single duct; or **compound,** drained by multiple branching ducts. The **myoepithelial cells** are derived from ectoderm (hence, they are epithelial), but they possess contractile properties (hence, myo-). They are associated with the basal regions of secretory cells and mechanically assist with release of products from exocrine glands. Endocrine glands are composed of cells aggregated into **cords** and **clumps** (sheets and solid masses) or **follicles** (tiny, hollow spheres; figure 5.18).

Connective Tissues

Connective tissues generally include bone, cartilage, fibrous connective tissue, adipose tissue, and blood (figure 5.19). At first glance, connective tissues seem to be the misfits of histology—the leftovers after all other tissues have been categorized. Connective tissues have a variety of functions and occur in diverse contexts. Adipose tissue stores lipids; bone and cartilage support the body; blood transports respiratory gases; dense connective tissue packs organs. Bone cells reside in a hard casing of calcium phosphate (hydroxyapatite); blood cells

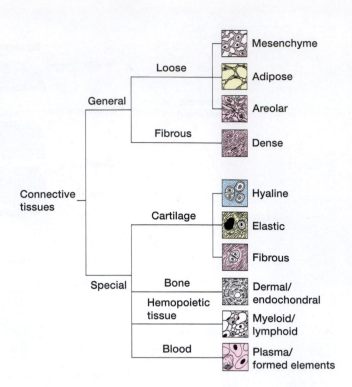

FIGURE 5.19 **Categories of connective tissue.** Bone, cartilage, fibrous tissue, adipose tissue, and blood are some of the body's connective tissues. Each type of connective tissue includes a distinctive cell type surrounded by an extracellular matrix.

occur in liquid plasma. To complicate matters, schemes for classifying connective tissues vary among different textbooks. Elegant but futile efforts have been made to find a common denominator for all connective tissues. Some physiologists define them functionally on the basis of their mechanical role in support. Bone, cartilage, and perhaps fibrous connective tissues qualify as supportive tissue, but blood certainly does not. Others define connective tissues as developing from mesenchyme. Certainly, many connective tissues arise from mesoderm, but there are exceptions to this as well. For instance, connective tissues of jaw muscles and some cranial bones arise from neural crest cells, not from mesoderm.

Rather than search for an overly restrictive definition, perhaps it is best to view connective tissues as bringing a convenient order to what otherwise would be a jumble of tissue types. Generally, each type of connective tissue includes a distinctive *cell type* that is isolated from other cells and surrounded by or embedded in a relatively abundant *extracellular matrix.* Of course, adipose tissue is the exception, because almost no matrix surrounds individual adipose cells.

The consistency of the extracellular matrix surrounding connective tissues determines the physical properties of the tissue and, hence, its functional role. In bone, the matrix is hard; in loose connective tissue, it is gel-like; in blood, it is fluid. The matrix is made up of protein fibers and a surrounding **ground substance.** Consistency of the ground substance varies from liquid to solid, depending on the tissue type.

Connective tissues can be categorized as general or special as well.

General Connective Tissues

General connective tissues are dispersed widely throughout the body. The most common is fibrous connective tissue that forms tendons and ligaments as well as much of the dermis of the skin and the outer capsules of organs. The distinctive cell is a **fibroblast,** and the extracellular matrix secreted by fibroblasts is principally a network of protein fibers in a ground substance of polysaccharide gel. Mesenchyme, which we met earlier in the chapter as an embryonic tissue, should not be confused with an embryonic germ layer (e.g., ectoderm, mesoderm, endoderm). Unlike epithelial cells, mesenchymal cells are not polarized; they are not coupled together by prominent junctional complexes, nor do they rest on a basement membrane. Usually there is intercellular space between mesenchymal cells, which may persist as a source of formative cells that differentiate into and replenish damaged cells in the adult.

Special Connective Tissues

Examples of special connective tissues are bone, cartilage, blood, and hemopoietic tissues. The two types of **hemopoietic tissues** form blood cells: **Myeloid tissue** is located inside cavities of bone, and **lymphoid tissue** occurs in the spleen, lymph nodes, and elsewhere. It was once thought that myeloid and lymphoid tissues each produced only one type of circulating blood cell, **myelocytes** and **lymphocytes,** respectively. Today we realize that both types of hemopoietic tissue are capable of manufacturing either of these blood cell types.

Some specialized connective tissues may undergo **mineralization,** a general process wherein various inorganic ions (e.g., iron, magnesium, calcium) are deposited in the organic matrix of tissues to harden them. Dentin, enamel, and the ganoine of ganoid fish scales are a few examples. **Calcification** is a specialized type of mineralization wherein calcium carbonate (invertebrates) or calcium phosphate (vertebrates) is deposited in the organic matrix. Calcification occurs early during some types of bone development, during bone repair, and in some fishes such as sharks. In chondrichthyans, three calcification types may be recognized: areolar calcification of densely calcified tissue formed in concentric rings used to age these fishes; globular calcification formed of fused spherules of calcium phosphate; and dense prismatic calcification producing refraction of light: hence prismatic. Prismatic calcification is a chondrichthyan synapomorphy. **Ossification** is a specialized type of calcification, unique to vertebrates, involving the deposition of hydroxyapatite (calcium phosphate) in the organic matrix leading to bone formation.

Cartilage and bone are mineralized connective tissues in which inorganic salts and protein fibers have been deposited in the matrix. They differ in cell type (chondrocytes in cartilage, osteocytes in bone), in composition of the matrix (chondroitin sulfate in cartilage, calcium phosphate in bone), and in vascularization (cartilage is avascular, bone is typically vascular). They also differ in their microarchitecture: Bone can be highly ordered into osteons, and cartilage is usually less organized. On their surfaces, both are covered by a similar coat of fibrous connective tissue. Although virtually identical, these fibrous sheaths are logically termed the **perichondrium** around cartilage and the **periosteum** around bone.

FIGURE 5.20 **Types of cartilage.** The cartilage cell, or chondrocyte, is surrounded by a matrix composed of a ground substance and protein fibers. (a) Fibers are not apparent in the matrix of hyaline cartilage when it is viewed with a light microscope. (b) Collagen fibers are abundant in fibrocartilage, giving it mechanical resistance to tensile forces. (c) Elastin, the predominant protein fiber in elastic cartilage, makes the cartilage springy and flexible.

Cartilage   Cartilage is a firm but flexible special connective tissue. The matrix primarily consists of chondroitin sulfate (ground substance) and collagenous or elastic proteins (fibers). The cartilage of living agnathans lacks collagen, suggesting that collagen became the predominant structural protein of cartilage later with the origin of gnathostomes. Spaces within the matrix called **lacunae** house cartilage cells, or **chondrocytes.** The physical properties of cartilage and, hence, its functional roles are determined largely by the type and abundance of protein fibers in the matrix. There are three types of cartilaginous tissue.

The most widespread is **hyaline cartilage.** In the embryo, hyaline cartilage makes up many bones before they undergo **ossification** (bone formation). In the adult, hyaline cartilage persists at the articular ends of long bones, at the tips of ribs, in tracheal rings, and in many parts of the skull. Collagen fibrils are present in the matrix but not in sufficient abundance to be easily seen with a light microscope. The name hyaline, meaning "glassy," refers to the homogenous appearance of the matrix, which resembles pieces of frosted glass (figure 5.20a).

Where cartilage is subjected to tensile or to warping loads, the ground substance is liberally reinforced with collagen fibers, which are obvious under microscopic examination. Such cartilage is **fibrocartilage** (figure 5.20b). The solid ground substance is especially effective in resisting compressive forces, whereas the embedded collagen fibers are better at addressing tensile forces. Fibrocartilage occurs in intervertebral disks, the pubic symphysis, disks within the knee, and selectively in other sites.

As the name suggests, **elastic cartilage** is flexible and springy, a property due to the presence of elastic fibers in the matrix (figure 5.20c). The internal support for your ear and epiglottis is a good example of elastic cartilage.

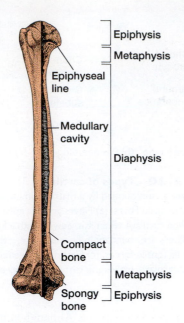

Epiphysis

Metaphysis

Epiphyseal line

Medullary cavity

Diaphysis

Compact bone

Metaphysis

Spongy bone — Epiphysis

FIGURE 5.21 **Regions of a long bone.** The middle section of bone is the diaphysis (shaft), containing the medullary cavity. In mammals and a few other groups, the ends of the bone develop secondary centers of ossification, or epiphyses, although this term is sometimes used loosely to mean simply the end of a bone. Between the diaphysis and the epiphysis is the metaphysis, the actively growing region of bone. Compact bone is dense; spongy (or cancellous) bone is porous. The medullary cavity and all spaces in spongy bone are filled with hemopoietic (blood-forming) tissues.

Cartilage does not receive its blood supply directly. Blood vessels reside only within the perichondrium on its surface. Thus, nutrients and gases must pass between blood and chondrocytes by long-range diffusion through the intervening matrix. Similarly, no nerves directly penetrate cartilage. Cartilage may be heavily invested with calcium salts as, for example, in the skeletons of chondrichthyan fishes, but cartilage is never as highly organized as bone.

Bone   Bone is a specialized connective tissue in which calcium phosphate and other organic salts are deposited in the matrix. Bone cells are identified on the basis of their activity: **osteoblasts** engage in osteogenesis (i.e., produce new bone); **osteoclasts** remove existing bone; and osteocytes maintain fully formed bone.

There are several criteria by which we classify bone. From its visual appearance, we see two types of bone: **cancellous** or **spongy bone,** which is porous, and **compact bone,** which appears dense to the naked eye (figure 5.21). From its position, we recognize **cortical bone** in the outer boundary or cortex of a bone and **medullary bone,** which lines the core of bone.

The presence or absence of bone cells gives us **cellular** and **acellular** bone, respectively. Bone may be described as **vascular** or **avascular** if many or few blood (= vascular) channels traverse it, respectively. The organization of the bone, especially the orientation of collagen and the ordered placement of bone cells within the matrix, is one of the most used criteria to classify bone. Because such a criterion has been widely applied to various extant and extinct vertebrate groups, a varied terminology has grown up. For our purposes, we shall recognize two general categories of bone based on descriptive criteria—lamellar and nonlamellar bone. **Nonlamellar bone** (= fibrolamellar bone; woven bone) is characterized by the disorderly, irregular arrangement of collagen within the matrix (figure 5.22a). It is typical of fast-growing bone. **Lamellar bone** is characterized by the orderly, regular arrangement of collagenous fibers within the matrix, usually accompanied by the regular orientation of bone cells (figure 5.22b). It is typical of slow-growing bone. A layer of bone matrix with its closely packed collagen fibers aligned in parallel is termed a **lamella.** Successive layers of lamellae may have their alignment of collagen oriented at different angles to adjacent lamellae. This produces a "plywood-like" structure, adding strength to the bone. A special type of lamellar bone is **Haversian bone** (figure 5.22c). Organic salts are arranged in a regular and highly ordered unit known as **osteon** (figure 5.22c, d; figure 5.23). Each osteon is a series of concentric rings made up of bone cells and layers of bone matrix around a central canal through which blood vessels, lymphatic vessels, and nerves travel. Volkmann's canals, running diagonally through this system, interconnect blood vessels between osteons.

Many bones exhibit **lines of arrested growth (LAGs)** during which growth ceases, or there may even be slight erosion or absorption of previously deposited bone (figure 5.22). LAGs form as a result of seasonal activity, nutritional abundance or quality of food, differences in rates of growth at various ages, or as interruptions in pulses of growth due to environmental stress (e.g., cold weather). Typically, ectothermic animals exhibit LAGs on an annual or seasonal basis, but a few endotherms, such as various species of Eurasian grouse, Old World field mice, the common dolphin, various species of voles, minks, Sika deer, and gibbons, also produce LAGs. Bone exhibiting periods of sustained deposition interrupted by LAGs or rings is termed **zonal bone** (figure 5.22).

One of the most important criteria to classify bone is the pattern of embryonic development, of which there are two basic types: **endochondral** and **intramembranous bone.** In the following section and subsections on bone, we trace these two types of bone development.

Bone Development and Growth

In both endochondral and intramembranous development, the first formative bone to appear is nonlamellar, sometimes said to be **immature bone** or **woven bone,** wherein lots of cells are interspersed amongst irregularly strewn bundles of collagen. As development proceeds to a more orderly arrangement of the matrix, the bone becomes lamellar, or sometimes called **mature bone.** Both developmental patterns begin with local aggregations of loosely arranged mesenchymal cells. Thereafter, the two processes differ. In intramembranous development, bone is formed directly

(a) Nonlamellar bone   (b) Lamellar bone   (c) Haversian bone

FIGURE 5.22 **General types of bone.** (a) Nonlamellar (= fibrolamellar) bone, based on a young American alligator. (b) Lamellar bone, based upon several living and extinct turtles. (c) Haversian bone, a specialized form of lamellar bone. (d) Enlarged section of Haversian bone. Note lines of arrested growth (LAGs), which may appear in all types of bone, here delineating zonal regions between them. Bone interrupted by LAGs is sometimes called zonal bone.

(a) After Reid, 1997; (b) after de Ricqlès, 1976 and others; (c) after Krstić. Artist: L. Laszlo Meszoly.

with no cartilage intermediate; in endochondral development, cartilage is formed initially and subsequently replaced by bone. It is impossible to tell from the gross visual appearance of mature bone whether it was produced by endochondral or intramembranous development. We look at these two types of development in more detail.

Endochondral Bone Development

Endochondral means within or from cartilage, and bones resulting from this developmental process are sometimes referred to as **cartilage** or **replacement bones.** During endochondral development, we can recognize in some bones up to three regions. The middle shaft is the **diaphysis;** each end is an **epiphysis;** and the region between is the **metaphysis** or **epiphyseal plate** (figure 5.21). Endochondral bone development involves the formation of a cartilage model of the future bone from mesenchymal tissue and the subsequent replacement of this cartilage model by bone tissue. Replacement of cartilage continues throughout most of an individual's early life.

The steps of this process are illustrated in figure 5.24a–g. First, loose collections of mesenchymal cells condense to form a hyaline cartilage model surrounded by a perichondrium (figure 5.24a). Second, the periosteal bone collar forms in the region of the diaphysis (figure 5.24b). Cells on the inner surface of the diaphyseal perichondrium become osteoblasts and deposit the bone collar. As the bone collar is being formed, inorganic calcium salts accumulate in the matrix to calcify the cartilage in the core of the diaphysis (figure 5.24c).

Calcium salts also seal off chondrocytes from nutritional and gas exchange with blood vessels on the surface of the cartilage. The entombed chondrocytes die as calcification proceeds. Next, the vascular system invades this calcified cartilage. These proliferating blood vessels erode away the cartilage debris to form the initial spaces of the marrow cavity.

Finally, osteoblasts appear within the core of the bone, and the primary center of ossification is established (figure 5.24d). Within this center of ossification, old bits of calcified cartilage become overlaid by new bone. Spike-like **trabeculae** are transitional composites of new bone and resorbing calcified cartilage. Later, when an ossified matrix predominates, trabeculae are called **bone spicules.** Additional osteoblasts circulating in blood are brought in by the invading vascular tissue. At about the same time, osteoclasts appear as well, signaling the active nature of bone remodeling through matrix deposition (osteoblasts) and removal (osteoclasts). Cartilage replacement, begun in the diaphysis, continues in the metaphysis. The epiphyseal plate is the active area of cartilage growth, calcification, cartilage removal, and new bone deposition. As the ossification process approaches, the chondrocytes proliferate and hypertrophy while the surrounding matrix calcifies (figure 5.24, bottom inset). Blood vessels invade and erode the calcified cartilage. Ossification is the last process to overtake a region and finally replace the cartilage remnants.

Proliferation of cartilage in the epiphyses lengthens bone. Continued deposition of bone under the diaphyseal periosteum

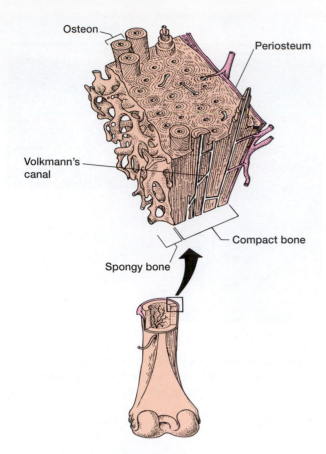

FIGURE 5.23 **Bone architecture.** Osteons make up compact bone. Each osteon is a series of concentric rings of osteocytes and their matrix. Nerves and blood vessels pass through a central canal within each osteon. Diagonal connections, known as Volkmann's canals, allow blood vessels to interconnect between osteons. As new osteons form, they usually override existing, older osteons as part of the ongoing dynamic process of bone remodeling.

After Krstić.

contributes to increased growth in bone girth. Bones in fishes, amphibians, and reptiles grow throughout their lifetimes, although growth slows in later life. Thus, some fishes, turtles, and lizards can reach quite large sizes. In birds and mammals, however, bone growth ceases when adult size is attained.

In mammals and in the bones of some lizards and birds, secondary centers of ossification arise in the epiphyses (figure 5.24e, f). The events that occur are similar to those that occurred during primary ossification in the bone shaft; namely, cartilage calcifies, blood vessels invade the epiphyses, osteoblasts appear, and new bone is deposited. In humans, these secondary centers of ossification appear at two to three years of age.

At or shortly after mammals reach sexual maturity, the epiphyseal plates and the metaphyseal regions they occupy ossify completely (figure 5.24g). Stated another way, the zone of ossification overtakes cartilage proliferation. At this point, the mammal's major growth phase is over. However, active remodeling and reorganization of the bone matrix continue throughout the remainder of life.

Intramembranous Bone Development

In intramembranous bone development, bone forms directly from mesenchyme without a cartilage precursor (figure 5.23). Initially the mesenchyme is compacted into sheets or membranes; hence, the resulting bones occasionally are referred to as "membrane bones."

As mesenchymal cells condense, they quickly become richly supplied with blood vessels. Between these compacted cells there appears a gel-like ground substance. Dense bars of bone matrix are deposited within this ground substance, and osteoblasts simultaneously become evident for the first time. The dense bars of matrix become more numerous, eventually replacing the gel-like ground substance. Subsequent growth proceeds by application of successive layers of new bone to the surface of these existing bone matrix bars (figure 5.25a–c). There are three types of specialized intramembranous bone development: dermal bone, sesamoid bone, and perichondral bone.

Dermal bones form directly through ossification of mesenchyme. They are called dermal bones because the mesenchymal source of these bones lies within the dermis of the skin. Many bones of the skull, pectoral girdle, and integument are examples. Occasionally, dermal bones structurally and functionally replace endochondral elements. The human mandible, as in many derived groups, starts as a cartilage element but then becomes ensheathed in dermal bone bearing teeth.

Sesamoid bones form directly within tendons, which are themselves derived from connective tissue. The patella of the knee and the pisiform bone of the wrist are examples. Sesamoid growth seems to be a response of tendons to mechanical stresses.

Perichondral and **periosteal bone** are formed from the deep cell layer of the fibrous connective tissue covering cartilage (perichondrium) or bone (periosteum). This type of bone develops early and retains the ability to form bone directly in the adult. Osteoblasts differentiate within this inner layer of the perichondrium or periosteum to produce bone without a cartilage precursor. Such direct, surface bone formation is termed **appositional growth.**

After ossification occurs and bone is formed, breakage or trauma to this bone may be followed by the appearance of cartilage. Because this cartilage forms after initial bone formation, it is called **secondary cartilage.** Following breakage, cartilage holds the ends of broken bone together and is replaced soon after through endochondral bone ossification. Repair of a fracture involving cartilage is widespread throughout vertebrates. Some embryologists prefer a more restrictive definition, recognizing as secondary cartilage only the cartilage that arises on the margins of intramembranous bones from periosteal cells in response to mechanical stresses. Once formed, this cartilage may ossify or remain as cartilage throughout life. In this restricted sense, secondary cartilage is known only in birds and mammals.

Periosteum

Marrow cavity

Osteocyte

(a)

(b)

(c)

(d)

(e)

(f)

(g)

Zone of hyaline cartilage

Hyaline cartilage

Zone of proliferation

Zone of hypertrophy

Calcified cartilage matrix

Zone of calcification

Blood vessels invade

Zone of ossification

Bone matrix

Blood vessel

Osteoblast (on surface)

Osteocyte (in bone matrix)

Myeloid

Osteoclast (multinucleate)

Marrow cavity

Calcified cartilage matrix

FIGURE 5.24 **Steps in endochondral bone growth.** (a) Hyaline cartilage model. (b) Appearance of a bone collar. (c) Calcification of cartilage in the diaphysis followed by invasion of blood vessels. (d) Onset of ossification. (d, e, f) Appearance of secondary centers of ossification (epiphyses). (g) At maturity, the growth center (metaphysis) disappears. The top inset illustrates a portion of the wall of the diaphysis in which perichondral bone appears under the periosteum. The bottom inset is a section through the metaphysis showing successive proliferation of new cartilage, calcification, and replacement by the advancing line of ossification.

Bone is found only in vertebrates. Why it should make an evolutionary debut here and not in some other animal group is not known. One theory holds that bone arose first not as a supportive tissue but as a stored form of calcium or phosphate. Because salts of calcium and phosphate and other minerals occur in greater concentration in seawater than in the tissues of marine organisms, they tend to invade an animal's body, seeking an equilibrium. Excess salts and minerals can be excreted by the kidney or deposited out of the way, perhaps in the skin. Calcium and phosphate ions participate usefully in cellular metabolic pathways, so if they were stored rather than excreted, they would be easily accessible during times of increased metabolic demand. Large calcium and phosphate stores, if located superficially, would also form a hard surface, protecting vertebrates from physical assault by predators. This secondary protective role might then favor development of the more extensive bony armor characteristic of early fishes. The appearance of a bony internal skeleton came still later, under selection for enhanced mechanical support.

As plausible as this hypothesis might be, it does not account for the particular form calcium takes in vertebrate bone. The hard, inorganic fraction of vertebrate bone is calcium phosphate in the crystalline form of hydroxyapatite, rather than calcium carbonate in the crystalline form of calcite or aragonite that characterizes most invertebrate skeletons. Perhaps, as recently suggested, calcium in vertebrate bone is more stable under conditions of physiological stress associated with active lifestyles. In contrast to most invertebrates, vertebrates show an ancient and unusual lifestyle characterized by intense bursts of activity. Bursts of activity lead to lactic acid formation followed by marked fluctuations of blood pH accompanied by prolonged acidosis (more acidic) before resting pH levels return. Under conditions of acidosis, calcium carbonate of invertebrates tends to literally dissolve, whereas calcium phosphate of vertebrates is more stable. A skeleton that tended to dissolve following extended activity would obviously weaken. This would also flood the circulating blood with excess calcium, perhaps further complicating normal metabolism of internal organs.

Thus, a skeleton of calcium would afford some mechanical protection, but one of calcium phosphate in particular (but not of calcium carbonate) would make bone matrix more stable. It would also reduce the physiological disadvantages bone dissolution otherwise might create for an animal that depended on bursts of activity. This hypothesis for the evolution of vertebrate bone also fits well with the views of those who see early vertebrates or prevertebrates as animals abandoning sedentary lifestyles of their ancestors in favor of more active ones (see also Ruben and Bennett, 1987).

Early chordate evolution (p. 74)

Comparative Bone Histology

Bone composed of osteons is found throughout gnathostomes, but it is not the only histological pattern of bone nor is it even the most common. In many teleost fishes, bone is acellular, entirely lacking osteocytes within the calcium phosphate matrix. During growth, osteoblasts at the surface secrete new matrix. However, these cells remain at the surface of the bone and do not become encased in their own secretions; so the bone they produce is acellular. Ostracoderms as well as some other groups of fishes have both acellular and cellular bone. In amphibians and reptiles, bone is often lamellar and cellular, with osteocytes present. Occasionally, osteons are present as well, being formed secondarily during continued growth and remodeling. More frequently, however, new bone is formed on a seasonal basis, producing growth rings in the cortex.

The view of bone as composed of an extensive osteon system comes from human bone and may generally apply to higher primates, but even among mammals this pattern shows differences. In many nonprimate mammals, large areas of acellular and even nonvascular bone may be found within the same individual. Bone from rats exhibits few osteons. In many marsupials, insectivores, artiodactyls, and carnivores, osteons may be absent from bone or from large regions of bone.

Bone Remodeling and Repair

Microfractures accumulate in the mineralized matrix of bone over time. If left unattended, these microfractures might coalesce into a major fracture and the bone failing at a critical moment. To repair damage before it weakens bone significantly, new bone must replace older bone on a regular basis. An advancing front of osteoclasts in partnership erodes channels through existing bone. In the wake of these osteoclasts, a large population of osteoblasts gathers to line the newly eroded channel and deposit new bone in characteristic concentric rings, forming a new osteon that often overrides lamellae of older osteons (figure 5.26).

This bone repair process is not only an important part of preventative maintenance, it is also a continuous remodeling process through which bone adapts to new functional demands throughout an individual's lifetime. In spite of preventative maintenance, however, an unexpected blow or twist might break a bone.

A break initiates a four-step repair process. First, a blood clot forms between the broken ends of bone (figure 5.27a). Smooth muscle contraction and normal clotting seal the severed ends of blood vessels that run through the bone. Second, a callus develops between the ends of the break, mostly from the activity of cells within the periosteum (figure 5.27b). The callus is

(a)

(b)

(c)

FIGURE 5.25 **Intramembranous bone formation.** (a) Mesenchymal cells converge and produce osteoid tissue, a precursor of bone matrix. (b) Blood vessels invade, osteoblasts appear, and initial osteoid tissue becomes enriched with calcium, forming the matrix of immature bone. (c) After more and denser matrix forms, the cells within are more properly called osteocytes. Those on the surface still actively produce more bone matrix and so are osteoblasts.

After Krstić.

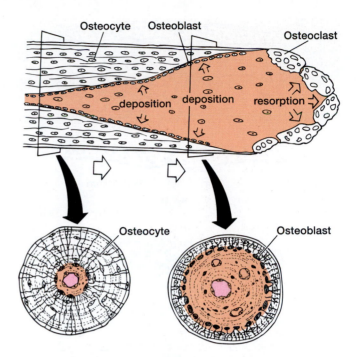

FIGURE 5.26 **Formation of a new osteon.** An advancing line of osteoclasts removes bone cells by eroding through existing bone matrix to open a channel. Osteoblasts appear along the perimeter of the channel and immediately begin to form concentric rings of new matrix organized around a central blood vessel (bottom). As these osteoblasts themselves become surrounded by the matrix, they become proper osteocytes.

After Lanyon and Rubin.

composed of hyaline and fibrocartilage, often with remaining bits of the blood clot. A few new bony spicules appear at this time as well. Third, the cartilaginous callus is replaced by bone, largely through a process reminiscent of endochondral bone formation. Cartilage calcifies, chondrocytes die, vascular tissue invades, osteoblasts and osteoclasts arrive, and bone matrix appears (figure 5.27c). After the cartilage is replaced, the two broken bone ends are usually knitted together by irregular spicules of bone (figure 5.27d). Finally, osteoclasts and osteoblasts participate in the remodeling of this rough mend to finish the repair process. This final remodeling step can continue for months. If the original break was severe, the area of repair can remain rough and uneven for many years (figure 5.27e).

In 1843, Dr. David Livingstone (of "Dr. Livingstone, I presume"), the famous Scot who explored Africa in the early nineteenth century, was badly mauled by a lion. He sustained a severe fracture of his upper arm but survived to carry on a prolonged missionary campaign. After his death 30 years later, his remains were returned to England and positively identified, in part by the distinctive fracture callus still evident.

Joints

Where separate bone or cartilage elements make contact, **joints** or **articulations** are formed. Joints may be defined functionally depending on whether they are movable or not. If a joint permits considerable movement, it is said to be a **synovial joint** or a **diarthrosis.** If a joint is restrictive or permits no relative movement between articulated elements, it is termed a **synarthrosis.** Joints may be defined structurally depending on the type of connective tissue joining them across the articulation. Structurally, a synovial joint (diarthrosis) is defined by a **synovial (or joint) capsule** whose walls consist of dense fibrous connective tissue lined by a **synovial membrane,** which secretes a lubricating **synovial fluid** into the confined space; the ends of contacting bones are capped with **articular cartilage.** Synarthroses lack synovial structures (capsule, membrane, fluid) and, in this, are structurally distinguished from diarthroses. Within synarthroses, if the connection between elements is of bone, it is a **synostosis;** if it consists of cartilage, a **synchondrosis;** if of fibrous connective tissue, a **syndesmosis.** Where a synostosis represents the fusion of formerly separate bones,

(a) Clot   Periosteum

Cartilage

(b) New bone (spicules)

(c) Woven bone

(d)

(e)

FIGURE 5.27 Repair of breaks in bone. (a) When a fracture occurs, a callus of clotted blood and debris initially forms between the ends of the broken bone (b), but it is soon replaced by cartilage. The cartilage becomes calcified, blood vessels invade, osteoblasts and osteoclasts appear, and new bone matrix is laid down. (c) The spicules of woven bone hold the broken ends of the fracture together and through remodeling (d) come to replace the broken section of bone. (e) A healed fracture. Most breaks in bone heal, and the bone's nearly normal shape returns after a period of remodeling, but not always. If the break is severe and "setting" the bone in proper realignment is poorly done, then repair may be imperfect. This humerus, from Dr. David Livingstone, shows the site of a fracture sustained during a lion's attack 30 years earlier.

(a–d) After Krstić; Ham; (e) after Halsted and Middleton.

In the vertebrates, neural crest arises at the border of the neural plate, and expresses a diverse array of cell types (Table 5.2) as streams of these cells migrate to final locations. Because of their significant contribution to the vertebrate head and body, their evolutionary debut has been proposed to be a key event fostering the transition from filter feeding to active predation. However, the origin of neural crest among the protochordates remains controversial as homologous cell types have not been conclusively identified. Recent manipulation of genetic and regulatory systems in urochordates (ascidians) finds evidence of a pigment cell lineage that likely represents a molecular neural crest precursor. This suggests that if not the cell type, then the molecular regulatory network is found in ascidians, thereby predating vertebrates. This molecular machinery was probably co-opted later into the lateral plate ectoderm of vertebrates to produce their distinctive cellular neural crest derivatives.

---

the firm union is regarded as **ankylosed.** Most synchondroses, especially if formed at the midline, are termed **symphyses.** The mandibular and pubic symphyses would be examples. Most **sutures** are syndesmoses.

Such dual criteria for defining joints—one functional, one structural—are based largely on articulations found in mammals. While these terms apply generally to other vertebrates, exceptions occur. In snakes, the "mandibular symphysis" permits considerable relative movement of jaw rami and may lack cartilage all together. Syndesmoses between lateral cranial elements introduce considerable degrees of freedom, not restrictions; freely movable articulating surfaces may be capped with articular cartilages, but lack a complete synovial capsule. In the dolphin flipper, the usual mammalian synovial joints between individual phalanges have secondarily been replaced by firm syndesmoses, which make the flipper stiff but strong in its role as a hydrodynamic device. In birds, some cranial bones form syndesmoses. However, the articulated bones may be thinned, thereby permitting significant flexion or bending through the joint as part of the bird's system of cranial kinesis. As a result of such variation, joint function (movability) cannot always be predicted from joint structure (connective tissue type) alone, or vice versa.

Neural Crest and Ectodermal Placodes

Neural crest cells, ectodermal placodes, and their many derivatives have been known since the nineteenth century, but their extraordinary significance to vertebrate evolution has only recently received the attention deserved. In vertebrates, migratory neural crest cells and ectodermal placodes contribute to a great variety of adult structures, are set aside early in vertebrate development.

Before complete closure of the neural folds, neural crest cells break loose from the crest of these folds and adjacent surface epithelium to assemble temporarily into distinct cords above the forming neural tube. This is a staging area from which they subsequently migrate out along defined routes within the embryo to permanent sites at which they differentiate into a great variety of structures, including ganglia of spinal and cranial nerves, Schwann cells that form the insulating sheath around peripheral nerves, chromaffin cells

of the adrenal medulla, pigment cells of the body (except in retina and central nervous system), and several types of widely dispersed hormone-producing cells. In the head, neural crest cells give rise to most cartilage and bone of the lower jaw and to most connective tissue of the voluntary muscles. Within the cores of teeth, **odontoblasts** that secrete the inner layer of dentin also arise from neural crest cells. The derivatives of neural crest cells are summarized in table 5.2.

Ectodermal placodes are anatomically distinct from neural crest cells, although both may arise by common developmental interactions. In teleosts, nasal and otic placodes behave like the neural keel; they form as solid buds, which then secondarily cavitate. In other vertebrates, all placodes are thickenings of the surface ectoderm that sink inward to form specific sensory receptors (figure 5.28). Sensory fibers of the spinal nerves departing from along the length of the spinal cord arise embryologically from neural crest cells. Cranial nerves arise from neural crest cells and ectodermal placodes in the embryo. In fishes and amphibians, placodes contributing to cranial nerves are located in two rows within the head.

TABLE 5.2 Neural Crest Derivatives

Peripheral nervous system
  Including dorsal root (spinal) and autonomic ganglia, and their postganglionic neurons

Hormone-producing cells
  Chromaffin cells of the adrenal medulla
  Calcitonin cells

Schwann cells

Parts of meninges

Branchial cartilage cells

Chromatophore cells (except in retina and central nervous system)

Odontoblasts

Dermis of facial region

Vasoreceptors

Sensory capsules and parts of neurocranium

Cephalic armor and derivatives (implied)

Heart
  Connective tissue
  Smooth muscle of outflow septa

FIGURE 5.28 Ectodermal placodes in a representative vertebrate. There are two paired sets of ectodermal placodes, the dorsolateral and the epibranchial placodes, as well as the olfactory and the optic placodes. All form sensory organs or receptors. Not shown: adenohypophseal placode which pinches off from the mouth lining, joins the neurophypopysis, and together form the pituitary gland.

The upper row of **dorsolateral placodes** and the lower row of **epibranchial placodes** lie sequentially just above the gill slits. Some cells of the dorsolateral placodes also contribute to other sensory systems as well. They migrate to positions over the head and along the body where they differentiate into receptor cells and associated sensory nerves of the lateral line sensory system. The **otic placode,** an especially prominent member of the dorsolateral series of placodes, sinks inward from the surface as a unit to form the vestibular apparatus concerned with balance and hearing.

> Cranial and spinal nerves (chapter 16); sensory organs
> derived from placodes (chapter 17)

The paired **olfactory placodes** form at the tip of the head and differentiate into sensory receptors of smell that grow to and connect with the brain. Between the olfactory placodes, and perhaps sharing a common phylogenetic origin, is the unpaired **adenohypophyseal placode,** a medial evagination of the ectoderm that enlarges into the hypophyseal pouch (Rathke's pouch), contributing the adenohypophysis to the pituitary. The paired **optic placodes** form laterally to produce the lens of the eye. Placodes can interact with the neural crest but do not arise from it. All vertebrate placodes, except the optic placode, differentiate into sensory nerves. The derivatives of ectodermal placodes are summarized in table 5.3.

The vertebrate body, especially the head, is in large measure a collection of structures of neural crest or placode origin. Although integrated harmoniously in the adult, these unique derivatives distinguish vertebrates from all other chordates.

TABLE 5.3 Placodes and Their Derivatives

Placode	Derivative
Dorsolateral	
Lateral line	Lateral line mechanoreceptors and electroreceptors
Otic	Vestibular apparatus
Cranial nerve	Sensory nerve ganglia
Epibranchial	
Cranial nerve	Sensory nerve ganglia, VII, IX, X
Olfactory	Sensory epithelium
Optic	Lens of eye

Extraembryonic Membranes

While the embryo is in the ovary (teleosts) or during its passage down the oviduct (most vertebrates), it gains extrinsic secondary and tertiary egg envelopes. Intrinsic membranes should not be confused with these wrappings added by the oviducts. Intrinsic membranes that arise from the embryonic germ layers and grow to surround the developing embryo are **extraembryonic membranes** (figure 5.29a–d). They function in sequestering waste products, transporting nutrients, and exchanging respiratory gases. They create a tiny aquatic environment, enveloping the embryo in a self-contained, fluid-filled capsule. Once the extraembryonic membranes form, the embryo effectively floats in an almost weightless environment, with gravity having only a slight effect upon its delicate and growing tissues. Extraembryonic membranes also protect the young embryo within its own moist environment so that an external body of water is not needed.

Vertebrates whose embryos possess extraembryonic membranes are **amniotes,** the **amnion** being one of the several extraembryonic membranes. Amniotes include reptiles, birds, and mammals. **Anamniotes,** meaning without an amnion, include fishes and amphibians. Fishes lay their eggs in water, and amphibians seek moist spots or return to water to deposit their eggs. Embryos of fishes and amphibians lack most of the extraembryonic membranes of amniotes, but have yolk sacs.

Extraembryonic membranes appear early and continue to enlarge throughout development, keeping pace with the enlarging metabolic needs of the growing embryo. At birth or hatching, the young individual breaks free of these membranes and must depend on its own internal organs to meet its nutritional (digestive tract) and respiratory (lungs) needs. The four extraembryonic membranes and their origins in reptiles, birds, and mammals are summarized in table 5.4 and discussed in detail in the following subsections.

Reptiles and Birds

In birds and generally in reptiles, the extraembryonic membranes form soon after the basic germ layers are established. The germ layers that contribute to the extraembryonic membrane are continuous with the germ layers from the body of the embryo, but they spread outward, extending away from

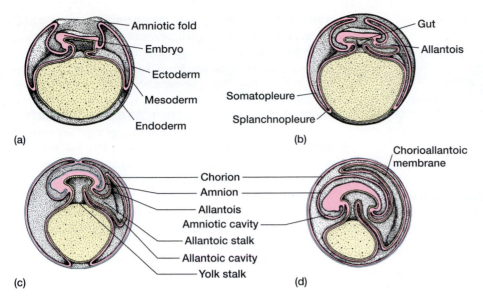

(a)

(b)

(c)

(d)

Amniotic fold
Embryo
Ectoderm
Mesoderm
Endoderm

Gut
Allantois
Somatopleure
Splanchnopleure

Chorioallantoic membrane

Chorion
Amnion
Allantois
Amniotic cavity
Allantoic stalk
Allantoic cavity
Yolk stalk

FIGURE 5.29 **Extraembryonic membrane formation in a bird (sagittal sections).** Somatopleure lifts upward (a), forming amniotic folds that join (b) and fuse (c) above the embryo to produce the chorion and the amnion. The expanding allantois comes into association with the chorion to produce the chorioallantoic membrane (d). An extensive vascular network forms within the mesoderm and serves as a site of respiratory exchange for gases passing through the porous shell (not shown).

After Arey.

the embryo. The bilaminar splanchnopleure of endoderm and splanchnic mesoderm form one membrane sheet that spreads around the yolk, eventually enclosing it as the **yolk sac.** Blood vessels develop in the mesodermal component of the spreading splanchnopleure and form a network of **vitelline vessels.** This vascularization network is important in mobilizing the energy and nutrients of the yolk during embryonic growth. The somatopleure of surface ectoderm and the somatic mesoderm form the other bilaminar sheet that spreads outward from the embryonic body (figure 5.30). The somatopleural sheet grows upward over the embryo as **amniotic folds,** which eventually meet and fuse at the midline. Two membranes are produced from the amniotic folds. One is the amnion that immediately surrounds the embryo and encloses it in a fluid-filled **amniotic cavity.** The other is the more peripheral **chorion** (figure 5.29c).

As the amniotic folds develop, the **allantois,** a diverticulum of the hindgut endoderm, grows outward, carrying splanchnic mesoderm with it. The endoderm and

splanchnic mesoderm of the allantois continue to expand, slipping between the amnion and the chorion; and between the yolk sac and chorion. Eventually, the outer allantois and the chorion fuse to form a single composite membrane, the **chorioallantoic membrane** (figure 5.29c, d). The mesoderm sandwiched within this membrane forms an extensive network of **allantoic vessels** that function in respiratory exchange through the porous shell. The **allantoic cavity** bounded by the allantois becomes a repository for the embryo's excretory wastes.

Mammals

Structures homologous to the four extraembryonic membranes of reptiles and birds appear in mammals: amnion, chorion, yolk sac, and allantois. In monotremes, the extraembryonic membranes are formed in much the same way they are formed in reptiles and birds (table 5.4). The vascular allantois retains a respiratory function both before and after the egg is laid. The

TABLE 5.4	Sources of the Four Extraembryonic Membranes in Most Reptiles, Birds, and Mammals				
	EXTRAEMBRYONIC MEMBRANE				
Vertebrate Group	Amnion	Chorion	Allantois	Yolk Sac	Respiratory Membrane
	GERM LAYER SOURCES				
Birds	Ectoderm, somatic mesoderm	Ectoderm, somatic mesoderm	Endoderm, splanchnic mesoderm	Endoderm, splanchnic mesoderm	Chorion, allantois
Reptiles	Ectoderm, somatic mesoderm	Ectoderm, somatic mesoderm	Endoderm, splanchnic mesoderm	Endoderm, splanchnic mesoderm	Chorion, allantois
Monotremes	Ectoderm, somatic mesoderm	Ectoderm, somatic mesoderm	Endoderm, splanchnic mesoderm	Endoderm, splanchnic mesoderm	Chorion, allantois
Marsupials	Ectoderm, somatic mesoderm	Ectoderm, somatic mesoderm	Endoderm, splanchnic mesoderm	Endoderm, splanchnic mesoderm	Chorion, splanchno-pleure
Eutherian mammals	Ectoderm (trophoblast), somatic mesoderm	Ectoderm (trophoblast), somatic mesoderm	Endoderm, splanchnic mesoderm	Endoderm, splanchnic mesoderm	Chorion, allantois

Allantoic cavity
Allantois
Amnion
Amniotic cavity
Shell
Embryo
Chorion
Yolk sac
Albumen
Extraembryonic
coelom
Splanchnopleure
Somatopleure

FIGURE 5.30 **Cross section of a bird embryo within the shelled egg after about eight hours of incubation.** Note the early formation of the allantois and the amnion.

After Patten.

vascular yolk sac may be applied to the uterine wall, absorbing nutrients. After the shell membranes are added and the egg is laid, this vascular yolk sac continues to mobilize stored nutrients but from the yolk now bundled within the shelled egg. In marsupials and some eutherian mammals, such as dogs and pigs, the amnion forms from amniotic folds in the somatopleure as it does in reptiles, birds, and monotremes. In other eutherian mammals, such as humans, fluid-filled spaces appear within the inner cell mass prior to the establishment of germ layers. These spaces coalesce to form the initial amniotic cavity.

In therian mammals, a structure homologous to the yolk sac is present, but it contains only a few (marsupials) or no (eutherians) yolk platelets. Instead, it is filled with fluid. The embryonic disc is suspended between the amniotic cavity and the yolk sac. As in other amniotes, the allantois begins as an outgrowth of the hindgut that expands outward, becoming surrounded by a layer of mesoderm as it grows. The chorion of eutherian mammals is bilaminar, as in reptiles and birds. It forms from the trophoblast and the adjacent mesodermal layer. The expanding allantois grows into contact and fuses with much of the internal wall of the chorion, producing the chorioallantoic membrane. The allantoic vessels, or the **umbilical vessels,** as they are more often called in mammals, develop within the mesodermal core of the chorioallantoic membrane. These vessels function in respiration and nutritional exchange with the uterus of the mother.

Eutherian Placenta

A **placenta** is a composite structure formed in part from tissues of the fetus and in part from tissues of the mother wherein both establish intimate vascular contact (figure 5.31). In eutherian mammals, two extraembryonic membranes separately or together may produce a placenta, depending upon species. One membrane is the chorioallantoic membrane. The chorioallantoic placenta is often called the **allantoic placenta** because the allantois of the fetus supplies the blood vessels. The other extraembryonic structure is the yolk sac, which, if it supplies the blood vessels, produces a **yolk sac placenta.** The expanded yolk sac makes contact with the chorion to form the composite choriovitelline

membrane, which invades the uterine walls to form a placenta. Part of the choriovitelline membrane may be vascular, other parts avascular, thereby forming, respectively, vascular and avascular yolk sac placentae. In some eutherian mammals, such as dogs, the yolk sac placenta is transitory, while in others, such as raccoons and mice, it remains functional until birth. In placenta, blood from the mother does not pass into the fetus. Rather, the placenta brings capillary beds of both fetus and female into close association, but not into direct union, to allow transfer of nutrients and oxygen from the mother to the fetus and nitrogenous wastes and carbon dioxide from the fetus to the mother.

Eutherian mammals are also called **placental mammals** because eutherian reproduction is characterized by a placenta. The placenta of eutherian mammals begins to form when the blastocyst first makes contact with the wall of the readied uterus. In humans, implantation of the blastocyst results in its taking up residence in the uterine wall about six days after ovulation (figure 5.32a–d). In some species, implantation is postponed for weeks or months as further development of the blastocyst is temporarily arrested. This postponement, termed **delayed implantation,** extends the length of gestation so as to prevent an inopportune birth of a new individual while the female is still nursing young of a previous litter or while seasonal resources are slight. Badgers, bears, seals, some deer, and camels have delayed implantation.

Upon implantation, cells of the trophoblast proliferate to form two recognizable layers. Cells of the outer **syntrophoblast (syncytiotrophoblast)** layer lose their boundaries to form a multinucleated syncytium. The syntrophoblast helps the embryo enter the uterine wall and establish an association with maternal blood vessels. The second derivative layer of the trophoblast is the **cytotrophoblast,** the cells of which retain their boundaries and contribute to the extraembryonic mesoderm (figure 5.32c).

In summary, the placenta is formed of fetal and maternal tissues. Blood vessels of the fetus grow out into the syntrophoblast, where they establish a close association with maternal blood vessels. The placenta supports respiratory and nutritional functions of the fetus. Hormones produced by the placenta stimulate other endocrine organs of the mother and help maintain the uterine wall with which the embryo is associated.

Placental blood circulation (p. 492)

Other Placentae

Most people are surprised to learn that placentae develop in marsupials and even in some fishes, amphibians, and reptiles. In fact, birds are the only major vertebrate taxon in which no members possess a placenta. Like eutherian mammals, both marsupials and reptiles have allantoic placentae and yolk sac placentae. One of the most widespread placental types among marsupials is the yolk sac placenta, which provides gas and nutrient exchange between fetal and uterine tissues (figure 5.33d). In some marsupials, such as koalas and bandicoots, both a yolk sac placenta and

FIGURE 5.31 **Uterus during pregnancy.** (a–e) Primate embryo and its membranes are shown at successive stages of development. The decidua is the inner lining of the uterus, the myometrium is the outer muscular wall. That part of the decidua associated with the fetal chorion is the decidua basalis. Together the maternal decidua basalis and fetal chorion form the placenta. The decidua parietalis and the decidua capsularis compose the remainder of the decidua. Once the placenta is formed, the umbilical cord carries the paired umbilical arteries, unpaired umbilical vein, and stalk of the yolk sac from the placenta to the embryo. The amniotic cavity continues to grow with the embryo until term, at which time it contains liquid (the so-called water bag).

After Patten and Carlson.

an allantoic placenta are present (figure 5.33c). Implantation in bandicoots is similar to that in eutherian mammals (figure 5.33f) in that the chorion invades the uterus, bringing fetal and maternal capillary beds into close association.

Most reptiles, like birds, lay eggs, but many lizards and snakes give birth to live young. These live-bearing reptiles have both yolk sac placentae and allantoic placentae. In some reptiles, such as South American **Mabuya lizards,** the chorioallantoic membrane interdigitates with the uterine epithelium to form a **placentome** (figure 5.33b), which is specialized chorioallantoic placenta *over* the embryo for exchange of nutrients and gases.

Overview of Early Embryonic Development

Yolk stores affect the pattern of cleavage and subsequent gastrulation. When yolk accumulates in the ovum in large quantities, it mechanically interferes with the formation of mitotic furrows and restricts cleavage to the relatively yolk-free area at the animal pole. In extreme cases, such as in teleost fishes, reptiles, birds, and monotremes, cleavage is discoidal, with the blastodisc confined to a cap of cells on top of the yolk. Subsequent gastrulation involves rearrangement of surface cells that move through an embryonic shield or primitive streak. Like blastopores, both embryonic shields and primitive streaks function as embryonic organizing areas. Both may be homologous to blastopores but flattened to accommodate the large amount of yolk.

Discoidal cleavage evolved independently in teleost fishes on the one hand and in reptiles, birds, and monotremes on the other. Of course, we do not know what cleavage pattern characterized the early lissamphibia. Modern amphibians have mesolecithal eggs and holoblastic cleavage. If lissamphibia had the same cleavage pattern as modern amphibian descendants, then the discoidal cleavage seen in modern reptiles, birds, and monotremes must represent a derived condition that evolved independently of the discoidal cleavage of teleosts.

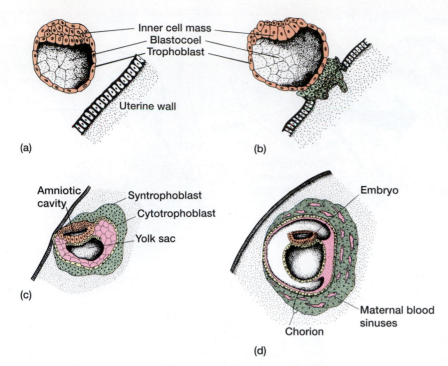

Inner cell mass
Blastocoel
Trophoblast

Uterine wall

(a)

(b)

Amniotic cavity

Syntrophoblast

Cytotrophoblast

Yolk sac

(c)

Embryo

Maternal blood sinuses

Chorion

(d)

FIGURE 5.32 Implantation of a mammalian embryo (human) within the wall of the uterus. (a) The blastocyst is not yet attached to the wall at about five days, but note that the inner cell mass, the trophoblast, and the blastocoel are already present and the zona pellucida has been shed. (b) Initial contact of the blastocyst with the uterine wall. (c) Deeper penetration of the blastocyst into the wall of the uterus. The trophoblast gives rise to an outer syntrophoblast, which is a syncytium, and the inner cytotrophoblast. The amniotic cavity forms by cavitation within the inner cell mass. (d) Blood sinuses of the maternal circulation course through the syntrophoblast to provide nutritional support and respiratory exchange for the embryo.

After McLaren in Austin and Short.

In eutherian mammals, the yolk sac is almost entirely devoid of yolk, yet cleavage is discoidal and gastrulation occurs via a primitive streak just as if large quantities of yolk were present and cells had to move around such an obstruction. This cleavage process likely represents the retention of features inherited from ancestors with yolk-laden eggs. Without reference to the phylogenetic background of eutherian mammals, such a pattern of early embryonic development would be difficult to explain.

Division of vertebrates into amniotes and anamniotes reflects a fundamental difference in mechanism of embryonic support. Appearance of the amnion along with other extraembryonic membranes in the reptiles represents an adaptation to an increasingly important terrestrial mode of life that took advantage of many new possibilities. Most reptiles, birds, and monotremes have shelled, or **cleidoic, eggs.** Once the cleidoic egg evolved, females no longer needed to trek long distances to bodies of water to lay their eggs in safety. The cleidoic egg is a self-contained little world. The yolk sac holds nutrients to support the developing embryo, the allantois serves as a repository into which nitrogenous waste products can be sequestered safely away from the embryo, and the amnion floats the embryo in a water jacket to prevent desiccation and lessen mechanical shocks. Either the yolk sac or the allantois becomes vascularized to serve a respiratory function.

Among mammals, we see a range of compromises in the pattern of embryonic development. With less yolk, the embryo correspondingly increases its dependence on oviducts and uterus for nutrients. This is true in monotremes where embryos do retain a store of yolk, but the relative volume is considerably less than in reptiles. Before being shelled and laid, the monotreme embryo may employ a vascular yolk sac to absorb nutrients from the uterine wall. This vascular yolk sac continues to marshal yolk sac nutrients while the embryo develops later in the shelled egg. In marsupials, early gestation is relatively slow. Once the outer shell membrane breaks down, a modest vascular placenta establishes association with the uterus and organogenesis becomes more rapid. However, prolonged development within the female poses additional problems. As the embryo gets larger, respiratory demands increase, and oxygen delivery must be improved. In eutherian mammals, a well-developed placenta develops to exchange gases with the maternal blood and address this problem. However, another potential problem arises for the embryo because the placenta keeps it in such close association with maternal tissues. At least half of the embryo is immunologically foreign because half its proteins are produced by the male's genetic contribution. If recognized as foreign, the mother's immune system will try to reject the embryo.

In marsupials, the embryo spends a relatively brief time within the uterus and is born at an early stage in development. Adult kangaroos can reach 70 kg, but the young weigh less than 1 g when they are born. A short gestation period addresses, in part, possible immunological rejection and provides evidence for why a fetus is born early. Further, the marsupial blastocyst is protected initially from immunological recognition by an inert eggshell membrane of strictly maternal origin that is retained throughout most of the brief gestation period. In eutherian mammals, the outer layer of the trophoblast is thought to promote implantation and prevent rejection of the embryo during its prolonged gestation.

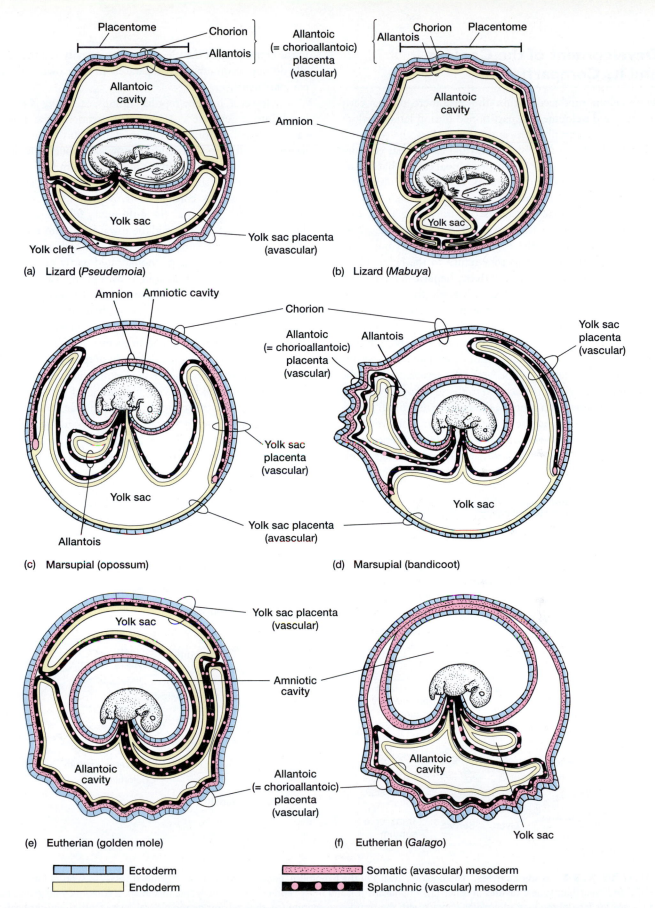

FIGURE 5.33 **Fetal extraembryonic membranes.** Convergence characterizes the evolution of extraembryonic membranes in amniotes. (a) Yolk sac and allantoic placentae, *Pseudemoia* (Australian grass skink). (b) Allantoic placenta, *Mabuya* (South American Mabuya skink). (c) Yolk sac placenta (vascular and avascular regions) of the opossum, *Didelphis* (marsupial). (d) Yolk sac and allantoic placentae, bandicoot (marsupial). (e) Yolk sac and allantoic placentae, golden mole (eutherian). (f) Allantoic placenta, a primate bush baby (eutherian). Cells that help digest the yolk invade the yolk cleft, which is present in many squamates.

Some, after Dawson; with special thanks to James R. Stewart.

Development of the Coelom and Its Compartments

The coelom produced within the hypomere during early embryonic development is partitioned during later development. In fishes, amphibians, and most reptiles, the coelom is subdivided into an anterior **pericardial cavity** that contains the heart and a posterior **pleuroperitoneal cavity** that houses most other viscera (figure 5.34a–c). The name pleuroperitoneal cavity originated in tetrapods, but is applied even to sharks and other fishes that lack lungs (pleuro-). The **transverse septum** is a complicated fibrous partition that separates these two compartments of the coelom. Large embryonic veins pass through this septum as they return to the heart. These veins eventually make contact with the **hepatic diverticulum** from the gut, which is destined to become the liver. As the hepatic diverticulum grows into the mesenchymal core of the septum, it meets these large embryonic veins that subdivide into the vascular sinusoids of the liver. As growth continues, the liver bulges from the confines of the transverse septum. The septum's posterior wall becomes the **serosa** covering the liver, and a constricted connection to the septum becomes the **coronary ligament.** In reptiles, the transverse septum generally lies oblique within the body rather than dorsoventrally. This results from its posterior shift to a position beneath the pleuroperitoneal cavity that is situated dorsally. Lungs reside in the cranial end of the pleuroperitoneal cavity, but they usually do not become housed separately in their own coelomic compartments.

However, in some reptilian groups, each lung is sequestered into a separate coelomic compartment, the **pleural cavity.** Pleural cavities form in crocodiles, turtles, and some lizards as well as in birds and mammals, although the developmental pattern is different in mammals from that in other groups. In reptiles that have a pleural cavity and in all birds, the pleural cavity is cordoned off by a thin, nonmuscular, oblique septum known as the **pulmonary fold** (figure 5.35). This fold grows from the midline toward and into the serosa of the liver. Growth continues until the pulmonary fold joins the body wall. Thus, the pulmonary fold partly suspends the liver and sequesters each lung in its own pleural cavity.

In mammals, a **coelomic fold (pleuroperitoneal membrane)** originating in the dorsal body wall grows ventrally, eventually to meet and fuse with the transverse septum. This fusion confines each lung in its own pleural cavity. The coelomic fold becomes a muscularized **diaphragm,** so that its contractions directly influence lung ventilation after hatching or birth (figure 5.34d, e). Muscularization of the diaphragm is complex. Some muscle cells populating the outer rim of the diaphragm arise in thoracic myotomes in the adjacent body wall and are innervated by respective thoracic spinal nerves. Further, mesenchyme associated with the foregut at

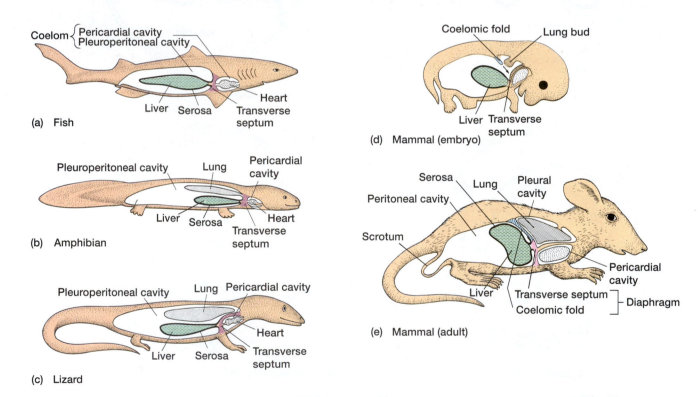

(a) Fish

(b) Amphibian

(c) Lizard

(d) Mammal (embryo)

(e) Mammal (adult)

FIGURE 5.34 Body cavities. The coelom, arising in the hypomere, becomes divided by a fibrous transverse septum into pericardial and pleuroperitoneal cavities in fishes (a), amphibians (b), and most reptiles (c). In embryonic mammals, a coelomic fold grows past the posterior face of the lung and makes contact with the transverse septum (d), thus separating the pleural cavity from the peritoneal cavity. This fold subsequently becomes invested with muscle primordia and together with the transverse septum becomes the muscular prehepatic diaphragm of the adult (e). In the males of some species, a posterior extension of the coelom through the body wall produces the scrotal pouch (scrotum) that receives the testes.

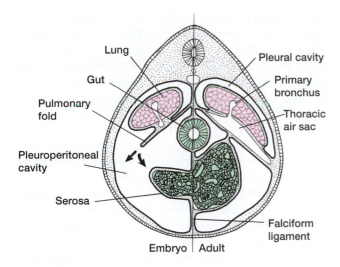

FIGURE 5.35 Avian body cavities. Cross section of a bird illustrating the embryonic (left) and adult (right) cavities. In the embryo, the pulmonary fold grows obliquely to establish contact with the liver and the body wall. This confines the lung to its pleural cavity.

the level of the lumbar vertebrae condenses to form bands of diaphragm muscle, collectively the asymmetrical left and right **crura** (sing., *crus*). These muscle bands originate on the vertebral column and insert into the dorsomedial diaphragm. However, most muscle cells arise in cervical myotomes far anterior to the diaphragm. These cervical muscle primordia enter the coelomic fold when it is opposite the cervical region. The differential growth of the embryo causes a gradual caudal displacement of the coelomic fold, thereby carrying these muscle primordia posteriorly within the body. The ventral transverse septum remains relatively unmuscularized and forms the **central tendon** of the dome-shaped diaphragm. The **phrenic nerve,** a collection of several cranial nerves, develops in the neck region adjacent to the cervical myotomes. As these myotomes are carried posteriorly, the phrenic nerve accompanies them, serving to innervate most of the diaphragm. The diaphragm's position anterior to the liver makes it a **prehepatic diaphragm** (figure 5.34d). Only mammals have such a prehepatic diaphragm; however, many vertebrates possess an analogous sheet or sheets of striated muscle located posterior to the liver. These sheets function in lung ventilation and are called **posthepatic diaphragms.** In crocodiles, for instance, **diaphragmatic muscles** function collectively as a posthepatic diaphragm to pull the liver posteriorly, using it as a plunger to help inflate the lungs.

Vertebrate diaphragms and lung ventilation (p. 435)

Thin cellular sheets of mesothelium, a special category of epithelium that forms from lateral plate mesoderm, line the coelom and its subdivisions. Mesothelium secures the integrity of cavities, defines spaces in which active organs operate more freely, and helps sequester organs with conflicting activities. For example, the pericardial cavity separates the heart from other viscera to allow the transient buildup of favorable

pressure around this organ at critical stages in its pumping cycle so that its chambers can be refilled. The pleuroperitoneal cavity accommodates the intestine through which peristaltic waves move food during digestion. The cavity gives the intestine freedom of movement during digestive episodes, yet the digestive tract activity remains controlled by the mesenteries that suspend it. Division of the coelom into compartments also allows more localized control of internal organs. For example, within the pleural cavities, lungs are placed directly under the control of muscles that ventilate them. Some mammals possess a scrotum, a coelomic pouch that protrudes outside the body cavity into which the testes descend, and there find a cooler environment favorable to sperm production and storage (figure 5.34e). Mesothelium from opposite sides of the body meets, wraps around internal organs, and forms a connecting stalk that suspends these organs within the cavity and connects adjacent organs to each other. This connecting stalk is a **mesentery** consisting of two layers of mesothelium with connective tissue, blood vessels, and nervous tissue in between.

Maturation

Metamorphosis

As the events of early development come to a close, the embryo takes shape. If this emerging individual is free-living and fundamentally unlike the adult, it is termed a larva and will eventually undergo metamorphosis, a radical and abrupt postembryonic change in structure to become an adult. Even in vertebrates lacking a distinct metamorphosis, the newborn still undergoes a period of maturation during which it develops from a juvenile to an adult. Strictly speaking, the overall process of **ontogeny** (development) is ongoing throughout the life of the individual and does not end at hatching or birth.

It is not uncommon for larva and adult, or juvenile and adult, to live different lives in quite different environments. Among marine chordates such as tunicates, larvae are unattached and mobile or freely carried by currents to new locations. Such larvae are dispersal stages. Less restricted than sessile, bottom-bound adults, the tunicate larvae select the specific location that will be their permanent residence as adults. The adult tunicate is a feeding and reproductive stage. In frogs, the young larva, or tadpole, is typically a feeding stage through which the individual takes advantage of fleeting resources in a drying puddle or seasonal algal bloom. The sexually mature adult stage is less confined to bodies of water. If larva and adult live in different environments, they necessarily will have dissimilar designs.

If conditions experienced by the larva are more hospitable than those endured by the adult, the balance of time an individual spends as a larva compared with its stretch as an adult might change adaptively as well. For example, in some species of lampreys, the individual may persist in larval form for several years, metamorphosing into the brief adult form only long enough for a few weeks of breeding before dying. The sole function of the adult is reproduction (figure 5.36a).

In some species of salamanders, the ancestral adult form fails to appear during the life cycle. Instead, the larval form becomes sexually mature and breeds. In lowland populations of the northwestern salamander, *Ambystoma gracile,* individuals remain as aquatic larvae for several years and then metamorphose into sexually mature terrestrial adults that breed. In high montane populations of this same species, many individuals stop short of metamorphosis (figure 5.36b). Their larval forms become sexually mature and breed. For these montane individuals, forgoing metamorphosis means

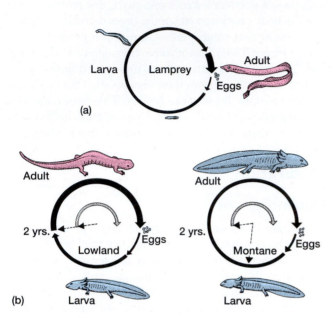

FIGURE 5.36 **Heterochrony.** (a) Life cycle of a lamprey. In many species, the larval stage lasts the longest, perhaps several years, and metamorphosis produces an adult that breeds during a brief few weeks and then dies. (b) The northwestern salamander, *Ambystoma gracile,* ranges from lowlands to high mountainous regions of the Pacific Northwest. The *Ambystoma* larva is primarily aquatic, the adult more terrestrial. In lowland populations, the larva undergoes metamorphosis at about two years of age, becomes an adult, and reproduces. In montane populations, this species usually does not metamorphose, although individuals become sexually mature and reproduce at about two years of age. Thus, in montane populations, two-year-old individuals are anatomically larval in appearance and habits, but they are capable of reproduction. Paedomorphosis describes an individual larval in anatomy, but one that is sexually mature. Neoteny is a special case of paedomorphosis wherein sexual maturity occurs but somatic development slows, allowing juvenile characteristics to persist. Like hands on a clock, both sexual maturity (small hand) and somatic development (large hand) arrive during metamorphosis at about the same time (two years of age) in lowland populations of the northwestern salamander. In montane populations, somatic development lags and metamorphosis does not occur, but sexual maturity arrives at more or less the right time. This is indicated by a slowing of the big hand of the clock (somatic development) relative to the small hand (reproductive development). These are neotenic forms—adults (sexually mature) but in juvenile garb (anatomically larval). The outer series of arrows (solid black) follow somatic changes. The inner arrow (light) shows the onset and extent of sexual maturity during a salamander's life history.

that they avoid becoming a terrestrial form exposed to harsh alpine winters. By remaining larval, they retain their aquatic lifestyle in which they can safely overwinter in the unfrozen depths of ponds. Theoretically, the transformed adult could scamper safely back into ponds with the onset of winter, but the larval form already possesses external gills and feeding jaws that are better suited to pond life. Occasionally, it is the larval stage that confronts formidable risks. Some frog larvae hatch into an aquatic environment teeming with predators. Apparently such challenges favor **direct development,** wherein the young embryo develops directly into the young froglet, bypassing the larval stage. The mostly arboreal species of Puerto Rican tree frogs, denizens of Central and South America as well as the most southern parts of Florida, lack a larval stage in their life cycles. Their eggs, laid in water trapped in tree branches, hatch directly into young froglets, tiny replicas of the adults, bypassing a larval (tadpole) stage.

Heterochrony

All vertebrates pass from embryo through a larval and/or juvenile stage on their way to the adult stage. Evolutionary modification of adults often begins here first, in the early stages of ontogeny, by relative chances in timing of developmental events. The term **heterochrony** describes such a phyletic change, wherein there is an ontogenetic shift in the onset or timing of a feature's appearance in a descendant species compared to its ancestor. For example, the oral disk of lampreys enlarges early in their larval ontogeny. However, in a few parasitic forms, it remains small until later at sexual maturity, when it enlarges. Note that heterochrony is determined on a relative basis—descendant compared to ancestor, or more formally, ingroup compared to outgroup. It is an ontogenetic change with phylogenetic consequences. Because heterochrony links ontogeny with phylogeny, it has often been the center of debates over the processes behind evolutionary change. New and sometimes dramatic evolutionary novelties of adults may be built out of changes in the timing of developmental events. Because these changes result from shifts in existing ontogenies, new morphologies can arise rapidly on an evolutionary timescale, producing new adaptive possibilities. Unfortunately, terms have proliferated, been misused, and changed in meaning. We will look at just a few that might be helpful and can be rescued from the muddle of previous usage.

Heterochrony includes several ontogenetic processes that affect the *rate* of growth of a part, the *onset* of its debut during embryology, or the *offset* of its growth period. Depending upon the trade-offs of rate, onset, and offset during ontogeny, the consequence is to produce two major phylogenetic results: paedomorphosis and peramorphosis. In **paedomorphosis** (meaning "child" and "form"), embryonic or *juvenile* characteristics of ancestors appear in the *adults* of descendants. In **peramorphosis** (meaning "beyond" and "form"), the adult characters of ancestors, exaggerated or extended in shape, appear in *adults* of descendants. For example, in humans, our flat face, rather than a muzzle, seems to be a paedomorphic feature, the retention of the juvenile characteristic of young primates. However, our relatively long limbs, compared to those of primates, seem to be a peramorphic feature, the result of prolonged growth in hindlimb length after sexual maturity.

Peramorphosis

Building new adult morphologies by exaggeration or extension of ancestral adult morphologies occurs through several processes, wherein a feature grows longer, grows faster, or begins development relatively earlier. Respectively, these processes are hypermorphosis, acceleration, and predisplacement. In **hypermorphosis,** ontogeny is longer and offset is late, so that allometric growth of parts continues beyond the normal endpoint of ancestors. In **acceleration,** the character grows faster during ontogeny compared to its pace in the ancestor. In **predisplacement,** onset is earlier and the character starts growth earlier, so it is more advanced in its development than its ancestor at a similar stage. Starting earlier, the character has a head start and is relatively ahead of other developing tissues. By these three processes, a character or feature becomes exaggerated in the finished adult, compared to its development in the ancestor (figure 5.37).

Most vertebrate examples of peramorphosis to date have been theoretical or difficult to support. Phylogenetic increase in horn size in brontotheres was thought to represent an example of peramorphosis (see figure 4.12), specifically by the process of hypermorphosis. However, reinterpretation of the data questions the simple allometric relationship on technical grounds and suggests that several different ontogenetic processes may have been involved. The Irish "elk," an extinct artiodactyl of the Pleistocene, possessed an enormous rack of antlers. The Irish "elk" was actually a deer, the largest ever to live. Within this deer family, measurements of antler size to skull size (or to body size) in adults show a strong, positive allometric relationship. But the Irish elk falls out appropriately on this allometric relationship as simply the largest of the deer. Apparently, growth of the Irish elk continued beyond the normal endpoint of ancestors (hypermorphosis), producing an exaggerated deer with large antlers to accompany a large body size.

Paedomorphosis

Adults are paedomorphic if they resemble juveniles of their ancestors. Stated slightly differently, paedomorphosis results when the larval form becomes reproductively mature. Adaptively, it may represent a trade-off between the advantages or disadvantages of larval versus adult morphologies. Paedomorphosis occurs through several processes wherein adult morphologies are built of juvenile features because the adult features exhibit early offset, grow slower, or begin development relatively late. Respectively, these are progenesis, neoteny, and postdisplacement (figure 5.37).

In **progenesis,** there is an early cessation of somatic development. Growth stops at a relatively earlier age; the individual becomes sexually mature at that earlier age and therefore possesses, as an adult, juvenile features. Relative to somatic development, sexual maturity is accelerated. Progenesis is found in some amphibian and insect lineages. For example, during early development in members of the tropical salamander genus *Bolitoglossa*, hands and feet are webbed and paddlelike. Only late in their development do distinct digits finally become delineated. Unlike other members of this tropical genus, *Bolitoglossa occidentalis* lives in trees. It has webbed feet and a small body, both adaptations to arboreal life. The flat, paddlelike feet give it grip on slippery leaves, and the small body reduces the risks of gravity's downward pull. Because growth ceases at a still small juvenile size, a small body results. As a consequence of this early cessation of development, other developmental processes in *B. occidentalis* are also arrested early. Limb development stops before digits become delineated, leaving the animal with webbed, paddle-shaped hands and feet. Other stunted characteristics occur as well, again truncated by the early cessation of development. Not all changes correlated with small body size necessarily have adaptive significance; but small size and webbed feet seem to have overriding advantages. Sexual maturity in *B. occidentalis,* when compared with other closely related species, occurs earlier relative to somatic development, giving us an example of paedomorphosis that results from progenesis.

In **neoteny,** features grow at a slower rate, compared to their rate in an ancestor. Normal sexual maturity overtakes slowed somatic development, producing a paedomorphic adult (figure 5.36b). The mudpuppy *Necturus maculosus* is permanently neotenic. It lives on the bottoms of lakes and retains its gills throughout life. However, populations of the tiger salamander *Ambystoma tigrinum* exhibit neoteny in response to immediate environmental conditions. In western North America, some populations are neotenic and reproduce as aquatic, gill-breathing forms; others lose their gills, develop lungs, and metamorphose into sexually mature adults. As mentioned previously, some populations of the northwestern salamander also exhibit neoteny (figure 5.36b).

In **postdisplacement,** a feature appears late in development, relative to its time of appearance in an ancestor. Starting late, the feature does not reach adult form by the end of maturation, keeps its juvenile quality, and becomes a paedomorphic feature in the adult.

Each stage in ontogeny is adaptive in its own right. To be a successful adult, the individual must first be a successful infant or juvenile. Larval and juvenile characteristics function not just as predecessors to adult structures to come, but most serve the individual in the environment it currently occupies. The entire ontogeny of an individual is the sum total of adaptive responses to different environments and selective pressures during its entire lifetime. Change in emphasis between larval and adult morphologies reflects this adaptive change in the time an individual spends within each stage of its life history.

The term "recapitulation" is an old and troubled term, related to heterochrony. In **recapitulation,** descendant species as *embryos* or juveniles resemble *adult* stages of ancestors.

HETEROCHRONY

Paedomorphosis

Progenesis — Neoteny — Postdisplacement

(early offset) — (slow rate) — (late onset)

Peramorphosis

Hypermorphosis — Acceleration — Predisplacement

(late offset) — (fast rate) — (early onset)

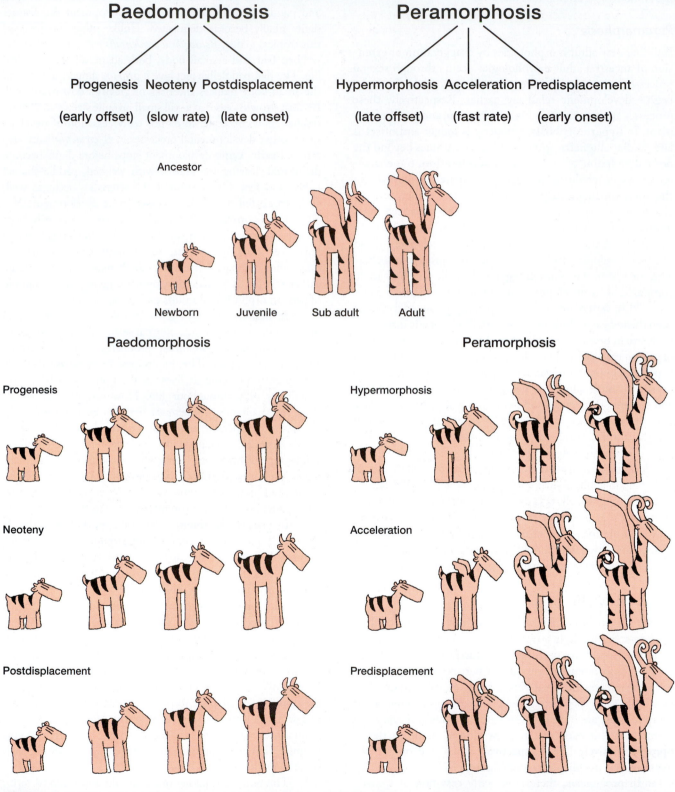

FIGURE 5.37 **Heterochrony.** Heterochrony results in an adult that retains juvenile features, paedomorphosis, or exhibits extended features, peramorphosis. In progenesis, there is early offset of somatic growth, and juvenile characters, relative to the ancestor, characterize the adult. In hypermorphosis, there is late offset, producing continued growth of horns, tail, and wings. In neoteny, growth rate of horns and tail is slow; in acceleration, these grow fast. In postdisplacement, onset of horn and tail growth is late; in predisplacement, onset is early.

Modified from Kenneth J. McNamara, Shapes of Time; new drawings by Sarah Long.

It is a troubled term because it has been used in various ways by different scientists, and it has been the center of debate over the "biogenetic law." Recent attempts to recycle it employ it to describe the phylogenetic consequences of heterochronic processes. For example, as juvenile Irish elks grew up, they likely passed through the adult size stages of smaller ancestors, "recapitulating" those stages, until they reached their exaggerated and derived adult sizes. Some characterize paedomorphosis as "reverse recapitulation," meaning that the ancestor is the reverse of the derived members of the phylogenetic sequence. Although not a view shared by all evolutionary biologists, it may be time to retire the term "recapitulation," both because of its historical misuse and because more useful terms are available. To understand this, we next review its history, going back to the nineteenth century.

Ontogeny and Phylogeny

Biogenetic Law

It has long been supposed that ontogeny, especially early events of embryonic development, retains current clues to distant evolutionary events. Ernst Haeckel, a nineteenth-century German biologist, stated this boldly in 1866 in what became

known as the **biogenetic law.** Pharyngeal slits, numerous branchial arches, and other fish characteristics even appear in the early embryos of reptiles, birds, and mammals, but they are lost as these tetrapod embryos proceed to term (figure 5.38). Although lost as tetrapod development unfolds, these and many similar structures are remnants of fish features from the evolutionary past. Haeckel argued that from ovum to complete body, the individual passes through a series of developmental stages that are brief, condensed repetitions of stages through which its successive ancestors evolved. The biogenetic law states that ontogeny in abbreviated form recapitulates (repeats) phylogeny.

Haeckel certainly recognized that recapitulation was approximate. Comparing it to an alphabet, he suggested that the ancestry behind each organism might be a sequence of stages: A, B, C, D, E, . . . Z, whereas the embryology of a descendant individual might pass through an apparently defective series: A, B, D, F, H, K, M, and so on. In this example, several evolutionary stages have fallen out of the developmental series. Although the ancestry of an organism might include an entire series of steps, Haeckel did not believe that all of these would necessarily appear in the ontogeny of a later individual. Evolutionary stages could disappear from the developmental series. Nevertheless, he

FIGURE 5.38 Embryology and evolution. Six species are shown across the figure. The youngest development stage of each is at the top of the figure followed by two successively older stages below. Note that "gill slits" (= pharyngeal slits) are present in fishes and amphibians yielding adult derivatives. But in tetrapods dependent on lungs, and without adult gills, these same embryonic gill slits still debut in young embryos. Even in humans, a tail is present early.

Photos kindly supplied by M. Richardson. See also Richardson et al. 1997, 1998.

felt that the basic series of major ancestral stages remained the same and, thus, the biogenetic law applied.

Development certainly exhibits a conservatism wherein ancient features persist like heirlooms in modern groups. Ontogeny, however, is not so literally a repeat of phylogeny as Haeckel supposed. A contemporary of his, Karl Ernst von Baer (1792–1876), cited examples from embryos of descendant animals that did not conform to the biogenetic law: Chick embryos lack the scales, swim bladders, fin rays, and so forth of adult fishes that evolutionarily preceded them. Furthermore, the order of appearance of ancestral structures is sometimes altered in descendant embryos. Haeckel allowed for exceptions; von Baer did not. Von Baer said that these exceptions and "thousands" more were too much. He proposed alternative laws of development.

Von Baer's Law

Von Baer proposed that development proceeds from the *general* to the *specific*, later codified as von Baer's **law of the general to the specific** (figure 5.39a). Development begins with undifferentiated cells of the blastula that become germ layers, then tissues, and finally organs. Young embryos are undifferentiated (general), but as development proceeds, distinguishing features (specific) of the species appear—horns, hooves, feathers, shells. Each embryo, instead of passing through stages of distant ancestors, departs more and more from them. Thus, the *embryo* of a descendant is never like the *adult* of an ancestor and only generally like the ancestral embryo. Other scientists since von Baer have also dissented from strict application of the biogenetic law. What can be made of all this?

Overview of the Biogenetic Laws

First, the biogenetic law, as proposed by Haeckel, does not hold as a broad description of ontogeny and phylogeny. No general correspondence occurs between descendant *embryos* and ancestor *adults*. As von Baer pointed out, what we observe at best is a correspondence between the *embryos* of descendants and the *embryos* of their ancestors.

Second, embryonic similarity followed by increasing differentiation toward adult stages does occur commonly as von Baer proposed, from general to specific (figure 5.39a).

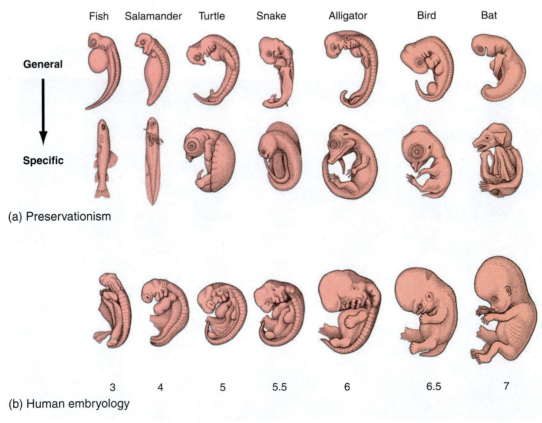

(a) Preservationism

(b) Human embryology

FIGURE 5.39 **Principles of embryology.** (a) Preservationism. Across the early embryos, *general* features are preserved such as gill slits, tail, early limbs. But as embryonic development proceeds, it proceeds to the *specific* where the particular features of the adult to be are now established. Note, for instance, the changes in the snake and the bat. (b) Human embryology. Note that the embryo does not become first a tiny fish, followed by an amphibian, reptile (or bird), before becoming human. There is no embryonic recapitulation of adult ancestors during human development. Approximate embryo age in weeks is given beneath each.

Source: Fish, snake, bat based on Richardson; amphibian after Harrison; turtle after Miller; alligator after Ferguson; bird after Patten.

General features appear first. You can tell that an early embryo is a vertebrate rather than an arthropod; a tetrapod rather than a fish; a bird rather than a reptile; a raptor rather than a duck. As a fish embryo approaches hatching, its "limb" buds become fins, a bird's become wings, a mammal's become paws or hooves or hand, and so forth. For example, in humans, one of our early embryonic stages possesses gill slits, a tail, and other general vertebrate structures, but as development proceeds human embryos do not successively become slippery fish, slimy amphibians, scaly reptiles (or feathered birds), before becoming hairy mammals (figure 5.39b). Our embryology is not an abbreviated rerun of evolution from fish to mammal. Instead our embryonic development proceeds from the general to the specific, from a generalized vertebrate embryo to a recognizable, particular human. There is, however, an element of deep conservatism in ontogeny, even if it is not an exact telescoping of evolutionary events. After all, the young embryos of mammals, birds, and reptiles do develop gill slits that never become functional breathing devices. Is this recapitulation? No. It is better to think of this as **preservationism,** for reasons not too difficult to imagine.

Each adult part is the developmental product of prior embryonic preparation. The zygote divides to form the blastula; gastrulation brings germ layers to their proper positions; mesoderm interacts with endoderm to form organ rudiments; tissues within organ rudiments differentiate into adult organs. Skip a step, and the whole cascade of ensuing developmental events may fail to unfold properly.

In mammals, the notochord of the embryo is replaced almost entirely in the adult by the solid vertebral column (figure 5.40). For the young embryo, the notochord provides an initial axis, a scaffolding along which the delicate body of the embryo is laid out. The notochord also stimulates development of the overlying nerve tube. If the notochord is removed, the nervous system does not develop. The adult supportive role is taken over by the vertebral column, but the notochord performs a vital *embryonic* role before disappearing; namely, it serves the young embryo as a central element of embryonic organization. A notochord that persists in the mammalian embryo should not be interpreted as a sentimental memento of a distant phylogenetic history. Instead, it should be seen as a functioning component of early embryonic development.

Another reason for developmental conservatism is pleiotropy, wherein a single gene may have multiple effects on many different and even unrelated traits. Where a gene or genes control a group of traits, simple one-to-one modification becomes difficult. Modification of one gene linked into multiple traits is therefore likely to unfavorably disrupt a whole suite of characters within its cascade of effects.

Structures, genes, and developmental processes intertwine to produce the conservatism evident in development. They are not easily eliminated without a broad disruption of ensuing events. Anatomical innovations, new structures brought into service in the adult, are usually added at the end of developmental processes, not at the beginning. A new structure inserted early into the developmental process would require many simultaneous replacements of many disrupted developmental processes thereafter. Evolutionary innovations thus usually arise by remodeling rather than by entirely new construction.

The forelimbs of ancestors that supported the body and allowed the organism to romp over the surface of the land are renovated into the wings that carry bats and birds aloft. We need look no further than our own human bodies to find similar examples of evolutionary remodeling. The backbone and legs that carried our distant ancestors comfortably on all fours hold us upright in a bipedal stance. The arms and hands that can control the delicate strokes of a paintbrush or the writing of a novelist come refashioned

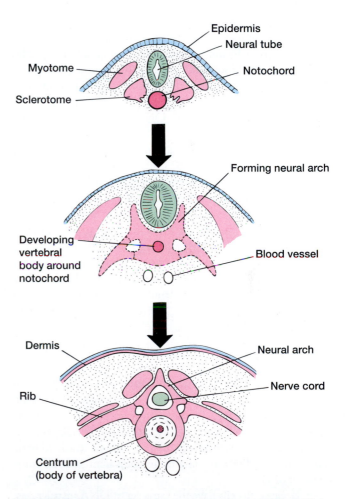

FIGURE 5.40 **Vertebrae replace the notochord in mammalian embryos.** The sclerotomes are segmented clusters of cells that gather around the notochord and differentiate into the segmentally arranged vertebrae, collectively known as the vertebral column. Vertebrae protect the nerve cord and provide sites of attachment for muscles. The vertebral column functionally replaces the notochord that persists only as a small core of the intervertebral disks between successive vertebrae.

from ancient forelegs that carried a hefty trunk and helped our ancestors dash from predators. The past is hard to erase. When parts are already available, renovation is easier than new construction.

Hox Genes and Their Kingdoms

We owe the term *homeotic* to William Bateson (1861–1926) and to his interest in biological variation. He noted that normal body parts of animals and plants were often switched, transforming a part into the likeness of another, producing odd varieties. For example, he observed, on occasion, the stamens of a flower transformed into petals. In 1894, he called such varieties **homeotic** (*homeo-*, same; *-otic*, condition) **mutants.** A more recent example comes from fruit flies. The repeating body segments of a normal fly are clumped into three body regions—head, thorax, and abdomen. The *head* includes eyes, mouthparts, and sensory antennae; the *thorax* has wings, legs, and haltere (balancing organ); the *abdomen* holds most of the body organs but lacks legs, wings, antennae, or other appendages. Occasionally, in one generation, an abrupt, transforming mutation occurs. Close up, the homeotic mutant looks like it stepped out of a science fiction movie. A leg replaces the antennae on the head, or a second wing-bearing segment is added to the thorax, giving the mutant two pairs of wings. One body part is replaced by another.

Today we know that such major changes are due to **homeotic genes**—master gene switches bringing under their command legions of secondary genes responsible in turn for the formation of body parts. Although first worked out in arthropods, fruit flies in particular, similar homeotic genes have been found throughout the animal kingdom and even in plants and fungi (yeast). Although sometimes restricted to vertebrates, the term **Hox genes** is now more commonly used to embrace all of these homeotic genes wherever they occur. Before looking at the details of *Hox* gene action and their evolutionary significance, we first need to understand the context in which they act.

Egg to Adult

The egg is one cell, the adult is millions of cells. To get from egg to adult, repeated cell division must occur, beginning with fertilization. Initially, division is restricted to cleaving the egg, but eventually proliferation of dividing cells contributes as well to growth in size of the embryo. Each somatic cell formed by division contains an equivalent and full complement of DNA.

Because all cells have the same set of DNA instructions, any particular cell anywhere in the embryo could form muscle, or nerve, or contribute to an arm or leg. But these cells and parts cannot appear randomly, or the embryo will be a scramble of bits and pieces in odd places. Arms must develop in the front, hindlimbs at the back; eyes must be on the head and, in fact, the head must be on the front end, and so on. Placement and appearance of body parts must sprout in the embryo in

the right positions. Organization is required. This organization begins by establishing basic body symmetry—front to back, top to bottom. Formally, a body **polarity** is established in the young embryo wherein anterior and posterior ends (front and back) and dorsal and ventral (top and bottom) regions are delineated. Usually this is done through chemical gradients, where distinct chemicals are concentrated at one region and decrease toward the other, as, for example, from front to back. Such chemical gradients, along with other chemical information, provide **positional information** within the embryo. The chemicals act as guideposts directing the subsequent positioning and placement of parts. By setting up this axis early, it is in place as a blueprint or chemical scaffolding to guide ensuing placement and building of body parts. In some animals, *Hox* genes actually turn on to set up body polarity; in others, polarity is established in the unfertilized egg. By whichever means, positional information is set up early, ready to direct placement of subsequent embryonic body parts and events.

Shaping Up: Positions and Parts

With the body polarity in place, the embryo can now be built, and most of the *Hox* genes work in this embryonic environment. Positional information within the embryo and environmental cues working through chemical intermediaries activate *Hox* genes, and in turn they activate large banks of structural genes. *Hox* genes are *regulatory genes* that manage parts of the genetic program controlling structural genes; *structural genes* actually make products involved in building the phenotype. Particular *Hox* genes determine where paired wings form or where legs develop. *Hox* genes are called master control genes because they may regulate 100 or more structural genes. Consequently, even a small change in one *Hox* gene can magnify into huge effects through the downstream structural genes over which it presides. There is amazing molecular similarity in *Hox* genes throughout the animal kingdom, further testimony here at the molecular level to the underlying evolutionary continuity between groups.

Hox genes are found in clusters with their loci lined up on chromosomes. The order of *Hox* genes in the cluster is in the same front-to-back order as the body part it affects (figure 5.41). A small change in one *Hox* gene in a cluster can produce large changes in the body region over which it presides, adding segments, or legs, or wings, or removing them.

Evolutionary Significance

Research continues. Many answers await research outcomes. But some promising correlations between *Hox* gene changes and major evolutionary events are apparent (figure 5.42). Major changes between major animal phyla are correlated with duplications in *Hox* genes or an increase in the number of *Hox* genes (figure 5.42a). The number of body regions over which a *Hox* gene presides may expand, thereby adding

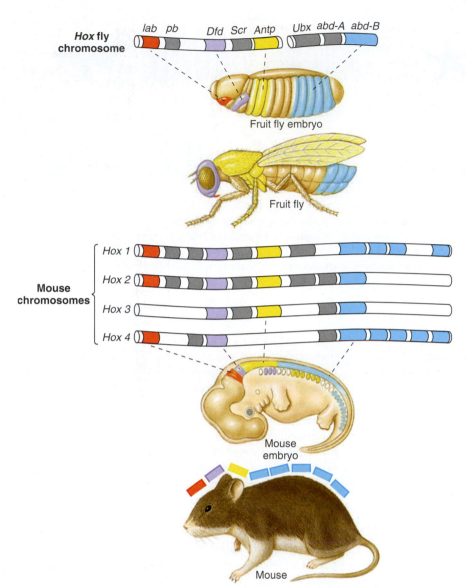

Hox fly chromosome

lab *pb* *Dfd* *Scr* *Antp* *Ubx* *abd-A* *abd-B*

Fruit fly embryo

Fruit fly

Mouse chromosomes

Hox 1
Hox 2
Hox 3
Hox 4

Mouse embryo

Mouse

FIGURE 5.41 *Hox* **genes.** In the fruit fly (*Drosophila melanogaster*), *Hox* genes are located in clusters on a single chromosome, the *HOX* fly chromosome. In the mouse (*Mus musculus*), similar genes are located on four chromosomes. In the fly and mouse, these genes control the development of front-to-back parts of the body.

segments, or may change the character of typical segments (figure 5.42b, c). Through mutations that change downstream gene action, parts on segments are added or eliminated (figure 5.42d).

Hox genes are elegant and complex. They are highly conserved anatomically (nucleotide sequences) and uniform in their expression (regulatory genes). What seems to have evolved is how they are activated and how downstream target genes respond in turn. Research is turning up a more complex story. Apparently, some *Hox* genes are turned on and off repeatedly during embryonic development, responding to the changing chemical and anatomical conditions within the developing embryo. Not only do *Hox* genes simultaneously turn on legions of structural genes, but some can directly and selectively control single, individual downstream genes as well. *Hox* genes turned on at one stage in embryonic development may be turned on again later but

produce a different effect. *Hox* genes and their triggers may remain more or less the same, but downstream tissues respond differently. Within flies, the pair of halteres, riding on the thoracic segment behind the single pair of wings, is apparently a modification of the wings that occupied that position in ancestors (figure 5.42d). As we meet them, we will examine examples of *Hox* genes at work within various vertebrate systems, contributing the genetic basis of rapid evolutionary change.

Epigenomics

Organisms are more than just the products of genes. Strictly speaking, genes make only varieties of RNA. Thereafter, some varieties of RNA assemble different amino acids into proteins. Proteins construct parts of cells, which cooperate in making whole cells, which join to form tissues, and so on until an organism is finally assembled. As these events move

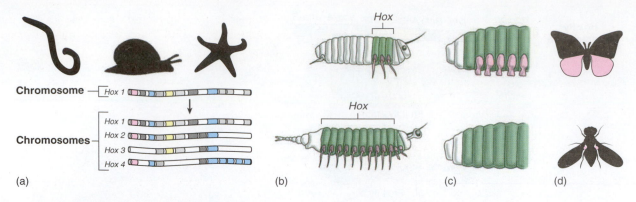

FIGURE 5.42 **Evolutionary Changes via *Hox* Genes.** Several major changes are thought to be based on changes in *Hox* genes and in their pathways of control of structural genes. These include changes in the number of *Hox* genes producing phyla-level changes (a), broad changes of *Hox* expression over body regions (b), local changes of *Hox* expression (c), and changes in regulation of downstream genes or in function, here changing the second-segment wings of a moth or butterfly into the haltere of flies (d).

After Gellon and McGinnis, 1998.

further from the genes, the genes have less and less direct hold on how the organism is eventually formed. The mutual associations established among cells and tissues play a large role in the eventual developmental outcome. These events are **epigenomic** (= epigenetic), literally above the genes or genome. Each level of organization—proteins, cells, tissues, organs, and so on—comes under the jurisdiction of additional constraints by which further development proceeds. An example might help.

Induction

During early development, the predecessors to the chain of vertebrae composing the vertebral column appear as a series of paired blocks or segments of tissue, the sclerotomes, nestled along either side of the neural tube (figure 5.40). If development proceeds normally, the sclerotomes give rise to cartilages that eventually ossify into vertebrae, and the neural tube gives rise to the spinal cord. If a section of the neural tube is removed experimentally at this early stage, then of course the affected stretch of the spinal cord fails to develop. Surprisingly, however, the adjacent vertebral column fails as well, even though the sclerotomes are not directly affected. That is because the neural tube, in addition to providing the foundation for the spinal cord, also stimulates proper development of the neighboring sclerotomes. This stimulatory effect between developing tissues of the embryo is known as **induction.** Developmental events are coupled and locked into step with each other. In the adult, the vertebral column comes to protect the spinal cord by surrounding it. Nerves that reach out from the cord squeeze between successive vertebrae. To create well-fit structures in the adult, nerves and vertebrae must match and grow together. Induction between neural tube and sclerotomes ensures that they are paced with each other so

that neither races ahead prematurely. Tissue interactions, not genes, are the most immediate developmental events to promote and shape the outcome.

Between neural tube and sclerotomes, induction is a one-way street—neural tube to sclerotomes. The reverse experiment, removal of the sclerotome, leads to little interruption in the growth of the neural tube. Reciprocal induction between tissues is common, however. The growing embryonic tetrapod limb provides an example. Two pairs of limb buds sprout along the sides of the body, being the first evidence of the future fore- and hindlimbs. As each limb bud lengthens like a sprouting branch, proximal, middle, and distal parts take shape in that order. Within the early limb bud, there is a recognizable mesodermal core and a surface thickening of ectoderm at the tip, the **apical ectodermal ridge (AER).** Both mesoderm and AER must interact to produce limb development. If the AER is removed, limb development ceases immediately. AER promotes outward growth of the limb bud. The core of mesoderm determines whether the limb produced is a forelimb or a hindlimb. Exchange of mesodermal cores between anterior and posterior limbs in birds results in a reversal of the arrangement of wings and legs. The AER stimulates growth of the mesoderm, but it in turn is maintained by the underlying core of mesoderm.

Phylogeny

The tight coupling of AER and mesoderm arises from interactions between the tissues themselves rather than from distant dictates of genes. The course of developmental events arises predominantly out of these mutual inductions between tissues. A tiny alteration of one tissue can have a profound effect on the adult structures produced. Such epigenetic interactions have been central not only in development but in evolution. Legless lizards exemplify this.

Most lizards possess four legs used to great advantage. But in some species in which habitats favor sleekness, legless forms have evolved. These forms use their whole body like snakes to slip through crowded terrain. As in other vertebrates, limbs of lizards also grow as lateral body sprouts along the sides of the young embryo. Additionally, nearby somites, clumps of mesoderm, grow downward to contribute cells to the core of mesoderm in the limb bud and establish an interaction with the AER (figure 5.43a). In limbless lizards, these somites fail to grow downward completely, AER regresses, and limbs do not appear (figure 5.43b). In these specialized lizards, a major adaptive change to limblessness has occurred by simple modification of an early developmental pattern. In this case, the evolution of limblessness did not require the accumulation of hundreds of mutations, each closing down one tiny anatomical part of the limb, one for the thumb, one for the second finger, and so on. Rather, a few changes in limb bud growth during early

developmental stage apparently gave rise to the limbless condition that found adaptive favor in the specialized habitat frequented by these lizards.

Much the same developmental pattern occurs in pythons, which are primitive snakes with rudimentary hindlimbs. Hindlimb buds appear in the embryo, but the AER does not materialize or become active. (Forelimb buds make no appearance, having been lost earlier in evolution.) The genetic basis of this developmental pattern has been discovered. Master control genes, *Hox* genes, regulate the expression of AER but fail to activate the AER, which in turn fails to initiate limb bud outgrowth. The result is a limbless snake. Specifically, it is hypothesized that *Hox* genes controlling the expression of the thorax or chest region in ancestors expanded their domain posteriorly, taking over responsibility for the development of the rest of the body in early snakes. Stated another way, the body of a snake, neck to cloaca, is an expanded chest. Limbs in snake ancestors appear in front and behind the thorax, but not within the thorax region itself. Consequently, the progressive caudal expansion of the *Hox* gene thorax domains would be accompanied by no limb expression. In fact, as the thorax domain expanded, it would actually suppress any local limb bud growth. This accounts for the loss of forelimbs and hindlimbs via accompanying AER suppression. Such large-scale changes in morphology, initiated by the relatively few but important master control genes, could be the basis for rapid evolutionary changes.

Limbless snakes evolved from legged lizards. Some primitive snakes, such as pythons, still retain hindlimb vestiges as did a fossil snake that retained small but well-differentiated hindlimbs. This suggests that one of the first stages in snake evolution was loss of the forelimbs. This can be explained by *Hox* gene expression. In vertebrates, such as the bird, forelimbs develop just anterior to the most anterior expression of the domain of the *Hox* gene, *Hoxc6*. Posterior to this point, *Hoxc6* and *Hoxc8* overlap and together specify thoracic vertebrae bearing ribs characteristic of the chest. However, in pythons, the domains of these two *Hox* genes are extended together throughout most of the body, so that forelimb development does not occur but telling the vertebrae within their domains to form thoracic vertebrae (figure 5.44). The body of the snake is essentially an expanded chest. Loss of hindlimbs seems to have occurred by a different mechanism. *Sonic hedgehog*, a gene important in maintaining the apical epidermal ridge, is not expressed, apparently leading to failure of hindlimb development.

A similar alteration of a developmental pattern seems to have been the basis for evolution of the specialized foot of modern birds. In reptiles, the tibia and fibula bones of the lower leg are about equal in length and articulate with several small bones of the ankle (figure 5.45a). In *Archaeopteryx*, this feature began to change. Although both tibia and fibula are equal in length, the ankle bones were

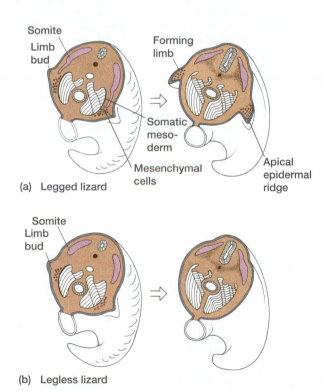

FIGURE 5.43 **Limb formation in lizards.** Cross sections through the posterior end of the embryo are depicted. (a) Mesenchymal cells normally depart from the somatic mesoderm, enter the forming limb bud, and become the core of the growing limb. Ventral processes from the local somites reach into this area of migrating mesenchymal cells. (b) In legless lizards, early rudimentary limb buds form, but somites do not grow into this vicinity. This apparently denies an inductive influence over events. The apical epidermal ridge regresses, the limb bud recedes, and no limbs develop.

FIGURE 5.44 **Limb loss in snakes.** In the chick, the domain of *Hoxc6* extends forward to promote development of the forelimbs. Where its domain overlaps with *Hoxc8,* the two together specify thoracic vertebrae with ribs. In the python, the expression of both genes extends forward, telling vertebrae to form ribs. *Hoxc6* serves no region alone, as in the chick, but overlaps throughout with *Hoxc8* such that forelimbs do not form.

Based on Cohn and Tickle.

FIGURE 5.45 **Hindlimbs of a reptile (a), a primitive bird (*Archaeopteryx*) (b), and a modern bird (c, d).** The several ankle bones of reptiles and the pair of shank bones, fibula and tibia, are reduced through this series to modern birds. Although Müller did nothing to alter the genotype of experimental chicks, his mechanical barrier separating differentiating regions produced an embryonic (d) and adult (e) hindlimb similar to those of *Archaeopteryx* and especially similar to reptile limbs (a). Apparently, the underlying developmental program in modern birds was altered very little in the course of evolution. Müller was able to re-create much of the ancestral condition in the foot by making only modest changes in the developmental pattern in modern birds.

After Müller and Alberch.

reduced to two in *Archaeopteryx* (figure 5.45b). In modern birds, the fibula is short and tiny, but the tibia has enlarged to engulf the two ankle bones and form a single composite bone (figure 5.45c).

In an attempt to clarify the evolution of the bird leg, embryologist Armand Hampé performed experiments in which he either separated tibia and fibula or provided additional mesenchyme to the ankle during early development of the limb bud. The limb produced in both cases bore a remarkable resemblance to the limb of *Archaeopteryx*. Tibia and fibula were of equal length; separate ankle bones were again present (figure 5.45d). These experiments were extended by Gerd Müller, who used inert barriers inserted into early chick hindlimbs to separate regions differentiating into tibia and fibula. The resulting limb suggested to Müller similarities to reptile limbs in that tibia and fibula were of equal length and were not closely adjacent (figure 5.45e). Further, the musculature of the experimental chick hindlimb reverted to a characteristic reptilian pattern of insertion. Such experimental manipulations could not have affected the genome because only the developmental pattern was altered. Could it be that Hampé and Müller had experimentally run evolution in reverse and discovered the simple method by which profound changes were initially brought about in birds? In ancient archosaurs, a few mutations affecting the supply of cells to or interaction with the fibula could have had a cascading effect on ankle development and resulted in an extensively altered adult design.

It is tempting to interpret other specialized structures in a similar light. Among modern horses, only a single toe (the middle or third toe) persists on each leg to form the functional digit (figure 5.46a). However, ancestral horses, such as *Protorohippus,* had four toes on the front foot and three on the back. Occasionally, modern horses develop vestiges of these old second and fourth toes (figure 5.46b–d). When this occurs, we get a glimpse

of the underlying developmental pattern that produces the foot. The reduction of toes in horses enjoyed adaptive favor because it contributed to locomotor performance. Literally hundreds of structural changes occurred in bones, muscles, ligaments, nerves, and blood vessels from the four- or five-toed ancestors to the single-toed modern horses. If this, like the evolution of limblessness in some lizards, was based on a narrowing of the underlying developmental pattern, then these hundreds of changes could be accomplished with relatively few gene mutations.

Evolutionary alteration of developmental patterns offers a simple way to produce profound anatomical changes. But we must keep in mind that we see only the successes in retrospect, not the failures. If appropriate gene mutations fail to appear at a timely moment, there is nothing the organism can do to summon them up to produce a desired part. Needs do not usher in the desired genetic improvements. For horses, birds, legless lizards, and lampreys, the fortuitous but timely appearance of new genes affecting developmental patterns produced renovated adult structures that found adaptive favor at the time. For every structure that succeeded and persisted, many have failed and perished.

FIGURE 5.46 **Atavisms, extra toes in modern horses.** (a) Modern horses have only one enlarged digit on each foot, a single toe. The one toe evolved from ancestors with three or four toes. During the course of their evolution, the peripheral toes IV, II, and I were lost and the central toe (III) emphasized. (b, c) On rare occasions, however, these "lost" toes or their remnants reappear, testifying to the lingering presence of the underlying ancestral developmental pattern. (d) On rare occasions, modern horses, such as the one illustrated, exhibit additional toes. Such toe remnants in modern horses apparently represent the partial reemergence of an ancient ancestral pattern.

For more information on extra toes in modern horses, see Gould, S. J. 1983. Hen's teeth and horse's toes. Further reflections in natural history. New York: W. W. Norton.

BOX ESSAY 5.3   Panda's Thumb

The giant panda from the forests of China is related to bears. Unlike bears, which are omnivores and eat almost anything, the giant panda feeds almost exclusively on bamboo shoots for about 15 hours per day. Pandas strip leaves by passing the stalks between their thumb and adjacent fingers. In addition to diet, pandas are unique among bears in apparently possessing six digits on the forelimb instead of the customary five. The extra digit is the "thumb," which is actually not a thumb at all but an elongated wrist bone controlled by muscles that work it against the other five digits to strip leaves from bamboo stalks. The true thumb is committed to another function, so it is unavailable to act in opposition to the other fingers. The radial sesamoid bone of the wrist has become remodeled and pressed into service as the effective "thumb" (box figure 1).

BOX FIGURE 1 **The panda's thumb.** The panda has five digits like most mammals; however, opposing these is another digit, a "thumb," which is actually not a thumb at all but an elaborated wrist bone.

The availability of parts narrows or enlarges evolutionary opportunities. Had the wrist bone been locked irretrievably into another function, as the original thumb was, the doors for evolution of bamboo feeding might have been closed and this endearing bear might never have evolved (see also Davis, 1964, and Gould, 1980).

Overview

During its lifetime, an organism begins as a fertilized egg; undergoes embryonic development; is birthed or hatched, perhaps passing through subsequent maturation as a larva or juvenile; reaches sexual maturity; and may undergo senescence before dying. This is its life history. It is not just the adult vertebrate that evolves, but its entire life cycle, as we were reminded in salamanders and lampreys (figure 5.36).

During embryonic development, different types of cells become delineated. This increase of cellular diversity is termed **differentiation.** As cells differentiate, they also undergo major displacements within the embryo, eventually taking up positions where they form the basic organs and body configuration of the embryo, which will be the shape of the basic adult built on this embryonic plan. These movements and reorganization of the cells as part of tissue layers is **morphogenesis** (body + form).

Before fertilization, constituents within the ovum may be organized unevenly during its time in the ovary, thereby defining an animal and vegetal pole. Fertilization activates the ovum into development and restores the diploid complement of chromosomes. Rapid mitotic division characterizes cleavage, producing from the single-celled zygote a multicellular, hollow blastula (blastocyst). The pattern is very diverse among embryos, largely due to the quantity and distribution of yolk. During gastrulation and neurulation, the basic germ layers—ectoderm, mesoderm, endoderm—become delineated through specific morphogenetic processes such as surface spreading (epiboly), inward spreading (involution), infolding (invagination), and/or splitting (delamination) of sheets of cells. Primary neurulation proceeds by the rolling up of neural folds. Secondary neurulation proceeds by cavitation in a solid neural keel (medullary cord), and is found in teleosts throughout their body and in the tail region of all tetrapods. Although derived from ectoderm, the neural crest might be considered a fourth germ layer because of its importance within vertebrates, giving rise to ganglia and their fibers of sensory nerves, medullar cells of the adrenal gland, melanophores of the skin, skeletal and connective tissue components of the branchial arches, and other cell types. During gastrulation, a bilateral symmetry replaces the animal-vegetal pole symmetry of the zygote. Four basic tissue types, derived from these basic germ layers, differentiate—epithelium, connective, muscle, and nervous tissues. These tissues interact during organogenesis to form organs.

In amniotes, the fertilized egg does not directly result in an embryo. Instead, the embryo emerges as a distinct population of cells within the cleaving egg. The remainder of the cells produced contribute to the extraembryonic membranes. These membranes support the embryo and its nutritional and respiratory needs, and in amniotes, hold the embryo in a water environment. Hatching or birth may bring the young vertebrate into an environment where it spends its larval or juvenile life. Maturation gradually, or by metamorphosis abruptly, brings it to sexual maturity, which is sometimes accompanied by a change in environment.

The earlier developmental stages of a life cycle are not just steps to the adult but adaptive designs in their own right, as for example heterochronic (neotonic) larval stages of salamanders that adapt them to the aquatic environment in which they live at that stage. Heterochrony is a phylogenetic change based on an embryonic change in relative timing of developmental events. Changes in relative timing, compared to an ancestor, may occur in up to three ways: onset, rate, or offset of a part's growth relative to the rest of the embryo. The result is peramorphosis or paedomorphosis, which may affect a part of an organism or the whole organism. Evolutionary alteration of developmental patterns offers a simple way to produce profound anatomical changes. But we must keep in mind that we see only the successes in retrospect, not the failures. If appropriate gene mutations fail to appear at a timely moment, there is nothing the organism can do to summon them up to produce a desired part. Needs do not usher in the desired genetic improvements. The fortuitous appearance of new genes affecting developmental patterns produces renovated adult structures that may or may not find adaptive favor at the time. For every structure that succeeded and persisted, many have failed and perished.

The variation in vertebrate embryos, especially during gastrulation and neurulation, is often attributed to accommodation to the relative amounts of stored yolk around which morphogenetic processes build the early embryonic body. But there is also some conservatism within these events, some preservation of embryonic structures and processes, even across different vertebrate groups. The embryo of eutherian mammals holds very little yolk yet its developmental processes progress *as if* much yolk were present—a primitive streak forms and gastrulation proceeds around its borders, much as happens in other vertebrate groups where the egg is packed with yolk. This conservatism results from the continued importance of primitive functions (e.g., notochord induction), linked effects of pleiotropic genetic networks, and profound adaptive adult changes that may require only modest adjustments in the embryo.

Common genetic and tissue processes now are seen to underlie fundamental developmental events. Early embryonic events result in **patterning** of the embryo, establishing first the basic body regions—dorsal, ventral, anterior, and posterior. [We met this in chapter 2 when noticing there-patterning in the evolution of chordates.] Cell interaction during organogenesis involves **cell signaling,** wherein cells communicate either through direct contact of via molecules released by one cell, and carried to another, which may govern its cellular activity and coordinate its related actions. Sections of DNA, such as *Hox* genes, act as master control "switches"—genetic switches. They do not encode any proteins but regulate when and where other genes are turned on or off.

These patterns of development and evolution teach that no part is an island. All parts are linked and integrated with the rest of the organism. Consequently, there is no one-to-one correspondence between genes and body parts. Some genes affect many parts; some parts are affected by many genes. Evolution does not necessarily proceed gene by gene, each bringing a tiny change that over millions of years eventually adds up to a new structure. With its influence dispersed, one small genetic change especially in gene switches, can produce large, integrated structural modifications that are the basis for major and rapid evolutionary changes in design.

WEB LINK

Visit the text website at *www.mhhe.com/kardong7e* for additional study aids including selected references, functional laboratories, and web links to chapter topics.

CHAPTER 6

Integument

EMBRYONIC ORIGIN

GENERAL FEATURES OF THE INTEGUMENT
  Dermis
  Epidermis

PHYLOGENY
  Integument of Fishes
    Primitive Fishes
    Chondrichthyes
    Bony Fishes
  Integument of Tetrapods
    Amphibians
    Reptiles
    Birds
    Mammals

SPECIALIZATIONS OF THE INTEGUMENT
  Nails, Claws, Hooves
  Horns and Antlers
  Baleen
  Scales
  Dermal Armor
  Mucus
  Color

OVERVIEW

The integument (or skin) is a composite organ. On the surface is the **epidermis,** below it is the **dermis,** and between them lies the **basement membrane** (basal lamina and reticular lamina). The epidermis is derived from the ectoderm and produces the basal lamina (figure 6.1a). The dermis develops from mesoderm and mesenchyme and produces the reticular lamina. Between the integument and deep body musculature is a transitional subcutaneous region made up of very loose connective and adipose tissues. In microscopic examination, this region is termed the **hypodermis.** In gross anatomical dissection, the hypodermis is referred to as the **superficial fascia** (figure 6.1b).

The integument is one of the largest organs of the body, making up some 15% of the human body weight. Epidermis and dermis together form some of the most varied structures found within vertebrates. The epidermis produces hair, feathers, baleen, claws, nails, horns, beaks, and some types of scales. The dermis gives rise to dermal bones and osteoderms of reptiles. Collectively, epidermis and dermis form teeth, denticles, and scales of fish. In fact, the developmental destinies of dermis and epidermis are so closely linked across the basement membrane that in the absence of one, the other by itself is incapable of or inhibited from producing these specialized structures. In terms of embryonic development, then, epidermis and dermis are tightly coupled and mutually necessary.

As the critical border between the organism and its exterior environment, the integument has a variety of specialized functions. It forms part of the exoskeleton and thickens to resist mechanical injury. The barrier it establishes prevents the entrance of pathogens. Skin swabs from healthy human volunteers identified over 200 different genera of resident bacteria, including many that, if given the chance through a breach in the skin, would produce serious staph infections, acne, and eczema, among other pathologies. The integument helps hold the shape of an organism as well. Osmotic regulation and movement of gases and ions to and from the circulation are aided by the integument in conjunction with other systems. Skin gathers needed heat, or radiates the excess, and houses sensory receptors. It specializes in feathers for locomotion, hair for insulation, and horns for defense. Skin pigments block harmful sunlight and display bright colors during courtship. The list of functions can easily be extended.

The remarkable variety of skin structures and roles makes it difficult to briefly summarize the forms and functions of the integument. Let us begin by examining the embryonic origin and development of the skin.

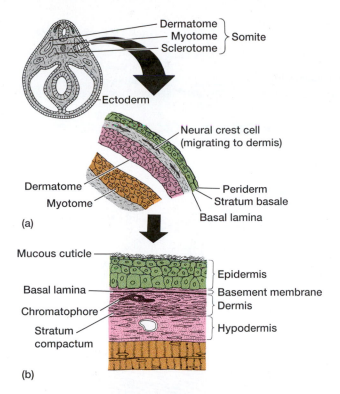

(a)

(b)

FIGURE 6.1 **Embryonic development of the skin.**
(a) Cross section of a representative vertebrate embryo. The ectoderm initially differentiates into a deep stratum basale, which replenishes the outer periderm. The dermatome settles in under the epidermis to differentiate into the connective tissue layer of dermis. As migrating neural crest cells pass between dermis and epidermis, some settle between these layers to become the chromatophores. (b) The epidermis further differentiates into a stratified layer that often has a mucous coat or cuticle on the surface. Within the dermis, collagen forms distinctive plies (layers) that constitute the stratum compactum. The basement membrane lies between the epidermis and the dermis. Beneath the dermis and the deeper layer of musculature is the hypodermis, a collection of loose connective and adipose tissues.

Embryonic Origin

By the end of neurulation in the embryo, most skin precursors are delineated. The single-layered surface ectoderm proliferates to give rise to the multilayered epidermis. The deep layer of the epidermis, the **stratum basale (stratum germinativum),** rests upon the basement membrane. Through active cell division, the stratum germinativum replenishes the single layer of outer cells called the **periderm** (figure 6.1a). Additional skin layers are derived from these two as differentiation proceeds.

The dermis arises from several sources, principally from the dermatome. The segmental **epimeres** (somites) divide, producing the **sclerotome** medially, the embryonic source of the vertebrae, and the **dermomyotome** laterally. Inner cells of the dermomyotome become rearranged into the **myotome,** the major source of skeletal muscle. The outer wall of the dermomyotome spreads out under the ectoderm as a more or less distinct **dermatome** that differentiates

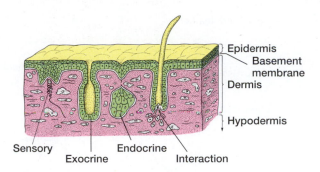

FIGURE 6.2 **Specializations of the integument.**
Sensory receptors reside in the skin. Exocrine glands with ducts and ductless endocrine glands form from invaginations of the epidermis. Through a dermal-epidermal interaction, specialized skin structures such as hair, feathers, and teeth arise.

into the connective tissue component of the dermis. Connective tissue within the skin is usually diffuse and irregular, although in some species collagen bundles are arranged into a distinct, ordered layer within the dermis. This layer is called the **stratum compactum** (figure 6.1b). Cells of neural crest origin migrate into the region between dermis and epidermis, contributing to bony armor and to skin pigment cells called **chromatophores** (meaning "color" and "bearing"). Usually chromatophores reside in the dermis, although in some species they may send pseudopods into the epidermis or take up residence there themselves. Often, chromatophores are scattered within the hypodermis. Nerves and blood vessels invade the integument to round out its structural composition.

Fundamentally, the integument is composed of two layers, epidermis and dermis, separated by the basement membrane. Vascularization and innervation are added, along with contributions from the neural crest. From such simple structural ingredients, a great variety of integumentary derivatives arises. The integument houses sensory organs that detect arriving stimuli from the external environment. Invagination of the surface epidermis forms skin glands: exocrine if they retain ducts, and endocrine if they separate from the surface and release products directly into blood vessels (figure 6.2). Interaction between epidermis and dermis stimulates specializations such as teeth, feathers, hair, and scales of several varieties (figure 6.3a–i).

General Features of the Integument

Dermis

The dermis of many vertebrates produces plates of bone directly through intramembranous ossification. Because of their embryonic source and initial position within the dermis, these bones are called **dermal bones.** They are prominent in ostracoderm fishes but appear secondarily even in derived groups, such as in some species of mammals.

Dermal (intramembranous) bone development (p. 184)

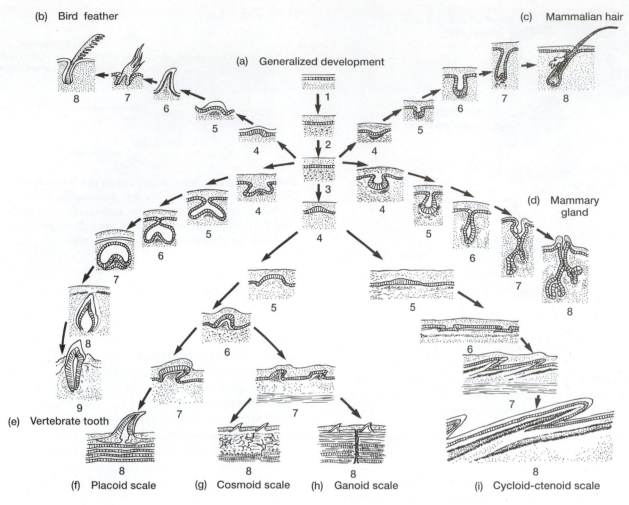

FIGURE 6.3 Skin derivatives. (a) Out of the simple arrangement of epidermis and dermis, with a basement membrane between them, a great variety of vertebrate integuments develops. Interaction of epidermis and dermis gives rise to feathers in birds (b), hair and mammary glands in mammals (c and d), teeth in vertebrates (e), placoid scales in chondrichthyans (f), and cosmoid, ganoid, and cycloid-ctenoid scales in bony fishes (g–i).

Based on the research of Richard J. Krejsa in Wake.

The most conspicuous component of the dermis is the fibrous connective tissue composed mostly of collagen fibers. Collagen fibers may be woven into distinct layers called **plies.** The dermis of the protochordate amphioxus exhibits an especially ordered arrangement of collagen within each ply (figure 6.4). In turn, plies are laminated together in very regular, but alternating, orientation. These alternating layers act like warp and weft threads of cloth fabric, giving some shape to the skin and preventing it from sagging. In aquatic vertebrates, such as sharks, the bundles of collagen lie at angles to each other, giving the skin a **bias,** like cloth; that is, the skin stretches when it is pulled at an angle oblique to the direction of the bundles. For example, if you take a piece of cloth, like a handkerchief, and pull it along either warp or weft threads, the cloth extends very little under this parallel tension. But if you pull at opposite corners, tension is applied obliquely at a 45° angle to the threads, and the cloth stretches considerably (figure 6.5a, b). This principle seems to govern the tightly woven collagen of shark skin. Its flexible skin bias accommodates lateral bending of the body but simultaneously resists distortions in body shape. As a result, the skin stretches without wrinkling. Because it does not wrinkle, water flows smoothly and without turbulence across the surface of the body (figure 6.5d).

In fishes and aquatic vertebrates, including cetaceans and aquatic squamates, collagen fibers of the dermis are usually arranged in orderly plies that form a recognizable stratum compactum. In terrestrial vertebrates, the stratum compactum is less obvious because locomotion on land depends more on the limbs and less on the trunk. And, of course, any wrinkling of the skin is less disruptive to a terrestrial vertebrate moving through air. Consequently, collagen fibers are present, even abundant, in the skin of terrestrial vertebrates, but they are much less regularly ordered and usually do not form distinct plies.

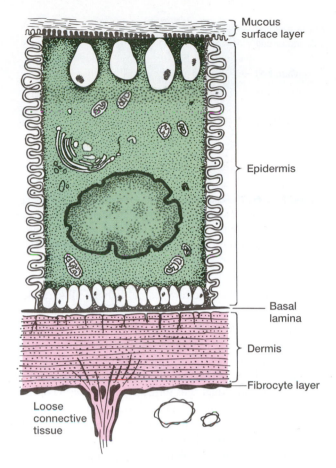

FIGURE 6.4 Protochordate, skin of amphioxus. The epidermis is a single layer of cuboidal or columnar cells that secrete a mucus that coats the surface and rests upon a basal lamina. The dermis consists of very highly ordered collagen fibers arranged in alternating plies (layers) to form a "fabric" that brings structural support as well as flexibility to the outer body wall. Pigment is secreted by the epidermal cells themselves.

After Olsson.

Epidermis

The epidermis of many vertebrates produces mucus to moisten the surface of the skin. In fishes, mucus seems to afford some protection from bacterial infection and helps ensure the laminar flow of water across the body surface. In amphibians, mucus probably serves similar functions and additionally keeps the skin from drying during the animal's sojourns on land.

In terrestrial vertebrates, the epidermis covering the body often forms an outer **keratinized** or **cornified** layer, the **stratum corneum.** It is one of the tetrapod innovations that help them address life in a drying and abrasive terrestrial environment.

All cells in the stratum corrneum are dead cells. New epidermal cells are formed by mitotic division, primarily in the deep stratum basale. These new epidermal cells push more superficial ones toward the surface, where they tend to self-destruct in an orderly fashion. During their demise, various protein products accumulate and collectively form **keratin** in a process called **keratinization.** The resulting superficial stratum

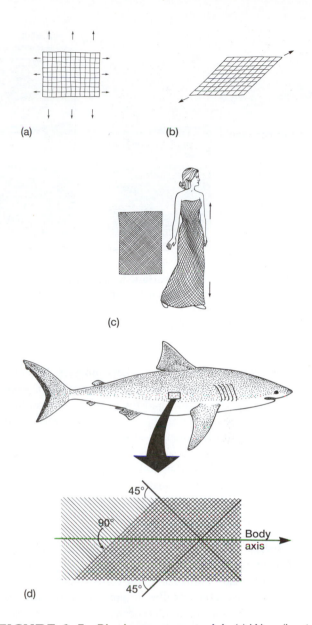

FIGURE 6.5 Bias in woven material. (a) Warp (longitudinal threads) and weft (cross threads) compose the fibers of fabric. If the tensile force is parallel with the threads (indicated by the arrows), little distortion of the fabric occurs. (b) However, tension along the bias at 45° to the threads results in a substantial change in shape. (c) Fashion designers take advantage of these features of fabric when designing clothes. In the loose bias direction, the fabric falls into folds and pleats but can hold its shape along the warp-and-weft threads. (d) Plies of collagen of the stratum compactum of fish skin act in a similar way. The flexible bias of the skin is oriented at 45° to the body length, thus accommodating lateral bending during swimming. This arrangement keeps the skin flexible but tight so that surface wrinkling does not occur and turbulence is not induced in the streamlines passing over the body as the fish swims.

After Gordon.

corneum is a nonliving layer that serves to reduce water loss through the skin in dry terrestrial environments. Keratin is a class of proteins produced during keratinization, and the specific

FIGURE 6.6 **Keratinization.** In places where mechanical friction increases, the integument responds by increasing production of a protective keratinized callus, and the stratum corneum thickens as a result. Key: E, epidermis; D, dermis.

FIGURE 6.7 **Lamprey skin.** Among the numerous epidermal cells are separate unicellular glands, the granular cells and the club cells. Note the absence of keratinization. The dermis consists of regularly arranged collagen and chromatophores.

epidermal cells that participate are **keratinocytes.** In sauropsids, the epidermis produces two types of keratinocytes—one type containing alpha (soft) and the other beta (hard) forms of keratin. Alpha-keratins are present in most flexible epidermal layers where shape changes occur. Beta-keratins are more common in specializations such as hard scales, claws, beak, and feathers. In synapsids, only alpha-keratins are present.

Keratinization and formation of a stratum corneum also occur where friction or direct mechanical abrasion insult the epithelium. For example, the epidermis in the oral cavity of aquatic and terrestrial vertebrates often exhibits a keratinized layer, especially if the food eaten is unusually sharp or abrasive. In areas of the body where friction is common, such as the soles of the feet or palms of the hands, the cornified layer may form a thick protective layer, or **callus,** to prevent mechanical damage (figure 6.6). The stratum corneum may be differentiated into hair, hooves, horn sheathes, or other specialized cornified structures. The term *keratinizing system* refers to the elaborate interaction of epidermis and dermis that produces the orderly transformation of keratinocytes into such cornified structures.

Finally, scales form within the integument of many aquatic and terrestrial vertebrates. Scales are basically folds in the integument. If dermal contributions predominate, especially in the form of ossified dermal bone, the fold is termed a **dermal scale.** An epidermal fold, especially in the form of a thickened keratinized layer, produces an **epidermal scale.**

Phylogeny

Integument of Fishes

With few exceptions, the skin of most living fishes is nonkeratinized and covered instead by mucus. Exceptions include keratinized specializations in a few groups: The "teeth" lining the oral disk of lampreys, the jaw coverings

of some herbivorous minnows, and the friction surface on the belly skin of some semiterrestrial fish are all keratinized derivatives. However, in most living fishes, the epidermis is alive and active on the body surface, and there is no prominent superficial layer of dead, keratinized cells. Surface cells are often patterned with tiny **microridges** that perhaps hold the surface layer of mucus. The mucous layer is formed from various individual cells in the epidermis with contributions from multicellular glands. This mucous coat, termed a **mucous cuticle,** resists penetration by infectious bacteria, probably contributes to laminar flow of water across the surface, makes the fish slippery to predators, and often includes chemicals that are repugnant, alarming, or toxic to enemies.

Two types of cells occur within the epidermis of fishes: **epidermal cells** and specialized **unicellular glands.** In living fishes, including cyclostomes, prevalent epidermal cells make up the stratified epidermis. Superficial epidermal cells are tightly connected through cell junctions and contain numerous secretory vesicles that are released to the surface where they contribute to the mucous cuticle. Epidermal cells of the basal layer are cuboidal or columnar. Mitotic activity is present in but not restricted to the basal layer.

Unicellular glands are single, specialized, and interspersed among the epidermal cell population. There are several types of unicellular glands. The **club cell** is an elongate, sometimes binucleate, unicellular gland (figure 6.7). Some chemicals within club cells excite alarm or fear. They are thought to be released by observant individuals to warn others of imminent danger. The **granular cell** is a diverse cell found in the skin of lampreys and other fishes (figure 6.7). Both granular and club cells contribute to the mucous cuticle, but their other functions are not fully understood. The **goblet cell** is a type of unicellular gland that is absent from lamprey skin but is usually found in other bony and cartilaginous fishes. It too contributes to the mucous cuticle and is recognized by its "goblet" shape, namely, a narrowed basal stem and wide apical end holding secretions. The electron microscope has helped distinguish an additional type of unicellular gland in the epidermis, the **sacciform cell.** It holds a large, membrane-bound secretory product that seems to function as a repellent or toxin against enemies once it is released. As increased attention is given to the study of fish skin, other cell types are being recognized. This growing list of specialized cells within the epidermis reveals a complexity and variety of functions that were not previously appreciated.

Collagen within the stratum compactum is regularly organized into plies that spiral around the body of the fish, allowing the skin to bend without wrinkling. In some fishes, the dermis has elastic properties. When a swimming fish bends its body, the skin on the stretched side stores energy that helps unbend the body and sweeps the tail in the opposite direction.

The fish dermis often gives rise to dermal bone, and dermal bone gives rise to dermal scales. In addition, the surface of fish scales is sometimes coated with a hard, acellular **enamel** of epidermal origin and a deeper **dentin** layer of dermal origin. Until recently, both enamel and dentin were recognized on the basis of appearance, not on their chemical composition. As the surface appearance of scales changed between fish groups, so did the terminology. Enamel was thought to give way phylogenetically to "ganoin" and dentin to "cosmine." These terms were inspired by the superficial appearance of scales, not by their chemical composition nor even by their histological organization. Perhaps it is best to think of ganoin as a different morphological expression of enamel, and cosmine as a different morphological expression of dentin and to be prepared for subtle chemical differences as we meet them.

Primitive Fishes

In ostracoderms and placoderms, the integument produced prominent bony plates of dermal armor that encased their bodies in an exoskeleton. Dermal bones of the cranial region were large, forming the head shields; but more posteriorly along the body, the dermal bones tended to be broken up into smaller pieces, the dermal scales. The surface of these scales was often ornamented with tiny, mushroom-shaped tubercles. These tubercles consisted of a surface layer of enamel or an enamel-like substance over an inner layer of dentin (figure 6.8). One or several radiating pulp cavities resided within each tubercle. The dermal bone supporting these tubercles was **lamellar,** organized in a layered pattern.

The skin of living hagfishes and lampreys departs considerably from that of primitive fossil fishes. Dermal bone is absent, and the skin surface is smooth and without scales. The epidermis is composed of stacked layers of numerous living epidermal cells throughout. Interspersed among them are unicellular glands, namely, the large granular cells and elongate club cells. In addition, the skin of hagfishes includes **thread cells** that discharge thick cords of mucus to the skin surface when the fish is irritated. The dermis is highly organized into regular layers of fibrous connective tissue. Pigment cells occur throughout the dermis. The hypodermis includes adipose tissue. Within the dermis, hagfishes also possess multicellular **slime glands** that release their products via ducts to the surface.

FIGURE 6.8 Section through an enlarged ostracoderm scale. The surface consists of raised tubercles capped with dentin and enamel enclosing a pulp cavity within. These tubercles rest upon a foundation of dermal bone, part of the dermal armor covering the body.

After Kiaer.

Chondrichthyes

In cartilaginous fishes, dermal bone is absent, but surface denticles, termed **placoid scales,** persist. These scales are what give the rough feel to the surface of the skin (figure 6.9a). Numerous secretory cells are present in the epidermis as well as stratified epidermal cells. The dermis is composed of fibrous connective tissue, especially elastic and collagen fibers, whose regular arrangement forms a fabriclike warp and weft in the dermis (figure 6.5d). This gives the skin strength and prevents wrinkling during swimming.

The placoid scale itself develops in the dermis but projects through the epidermis to reach the surface. A cap of enamel forms the tip, dentin lies beneath, and a pulp cavity resides within (figure 6.9a, b). Chromatophores occur in the lower part of the epidermis and upper regions of the dermis.

In sharks, separation of fluid flow from the surface of the body can significantly increase drag. The skin of many sharks has at least two mechanisms that control this boundary layer separation and thereby reduce drag and both mechanisms involve the placoid scales. First, the surface of

FIGURE 6.9 Shark skin. (a) Surface view of the skin showing regular arrangement of projecting placoid scales. (b) Section through a placoid scale of a shark. The projecting scale consists of enamel and dentin around a pulp cavity.

The skin of most amphibians contains glands that secrete products that are distasteful or even toxic to predators. Often these skin toxins are not manufactured by the frogs themselves, but instead are co-opted from toxins in their prey, such as from ants or other arthropods. Some ant specialists suggest that the insects, in turn, may get the toxins or toxin precursors from their diet. In tropical regions of the New World lives a group of frogs, the poison dart frogs, with especially toxic skin secretions (box figure 1). Native peoples of the region often gather these frogs, hold them on sticks over a fire to stimulate release of these secretions, and then collect the secretions on the tips of

BOX FIGURE 1   Poison dart frog. Its bright colors advertise toxic skin secretions that are poisonous to most predators.

their darts. Game shot with these toxin-laced darts is quickly tranquilized or killed. Cooking denatures the toxins, making the game safe for humans to eat. If confirmed by future research, the toxins these native peoples use would travel the ecological web back, through frogs, to insects, to their diet.

---

the scale spine is sculptured into riblets, parallel keels that control boundary layer separation (figure 6.10a). Second, in regions of the shark body particularly susceptible to separation, such as the flanks, the placoid scales are passively erected, bristled, by the adverse pressure gradient. At maximum girth upstream from the flanks, streamlines necessarily speed up, reducing pressure there and so induce flow close to the surface to reverse towards this forward point of low pressure. This flow reversal, close to the body, lifts the placoid scales. So bristled, they reduce flow separation and thereby minimize pressure drag (figure 6.10b).

Dynamic Fluids (boundary layer) (p. 145)

Bony Fishes

The dermis of bony fishes is subdivided into a superficial layer of loose connective tissue and a deeper layer of dense fibrous connective tissue. Chromatophores are found within the dermis. The most important structural product of the dermis is the scale. In bony fishes, dermal scales do not actually pierce the epidermis, but they are so close to the surface they give the impression that the skin is hard (figure 6.11a, b). The epidermal covering includes a basal layer of cells. Above this layer are stratified epidermal cells. As they move toward the surface, epidermal cells undergo cytoplasmic transformation, but they do not become keratinized. Within these layered epidermal cells occur single unicellular glands, the secretory and club cells. These unicellular glands, along with epidermal cells, are the source of the mucous cuticle, or surface "slime."

On the basis of their appearance, several types of scales are recognized among bony fishes. The **cosmoid scale,** seen in primitive sarcopterygians, resides upon a double layer of bone, one layer of which is vascular and the other lamellar.

On the outer surface of this bone is a layer that is now generally recognized as dentin, and spread superficially on the dentin is a layer now recognized as enamel. The unusual appearance of these enamel and dentin coats inspired, in the older literature, the respective names of "ganoin" and "cosmine," on the mistaken belief that ganoin was fundamentally a different mineral from enamel and cosmine different from dentin. Although the chemical nature of these layers is now clear, the earlier names have stuck to give us the terms for distinctive scale types. In the cosmoid scale, there is a thick, well-developed layer of dentin (cosmine) beneath a thin layer of enamel (figure 6.12a).

The **ganoid scale** is characterized by the prevalence of a thick surface coat of enamel (ganoin), without an underlying layer of dentin (figure 6.12b). Dermal bone forms the foundation of the ganoid scale, appearing as a double layer of vascular and lamellar bone (in palaeoniscoid fishes) or a single layer of lamellar bone (in other ancestral actinopterygians). Ganoid scales are shiny (because of the enamel), overlapping, and interlocking. Living polypteriforms and gars retain ganoid scales. However, in most other lines of bony fishes, ganoid scales are reduced through the loss of the vascular layer of bone and loss of the enamel surface. This produces, in teleosts, a rather distinctive scale.

The **teleost scale** lacks enamel, dentin, and a vascular bone layer. Only lamellar bone remains, which is acellular and mostly noncalcified (figure 6.12c). Two kinds of teleost scales are recognized. One is the **cycloid scale,** composed of concentric rings, or **circuli.** The other is the **ctenoid scale,** with a fringe of projections along its posterior margin (figure 6.12d). New circuli are laid down, like rings in a tree, as a teleost fish grows. Annual cycles are evident in the groupings of these circuli, and from this pattern in the scales we can determine the age of individual fish.

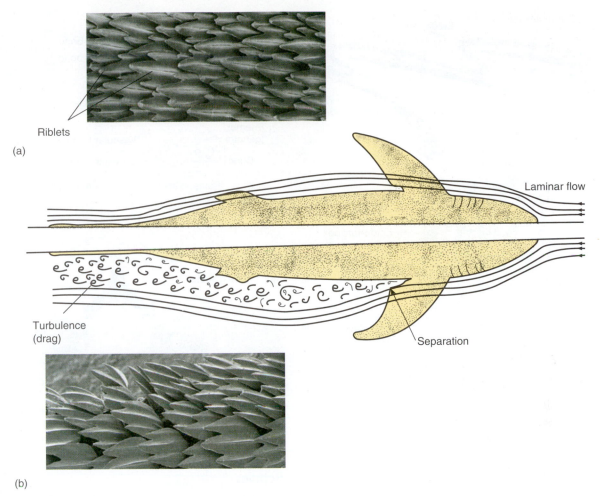

FIGURE 6.10 **Shark skin drag reduction—mako shark.** (a) Laminar flow is established across the surface of the shark and skin scales are relaxed (top inset). (b) When separation of the boundary layer occurs it results in turbulence and therefore an increase in pressure drag. But, adverse surface flow reversal acts on the scales to passively actuate or erect them, bristle them (lower inset). This produces a favorable effect on the flow separation. Scales manually erected in (b) to illustrate bristling.

Based on the research of Philip Motta with Amy W. Lang and Maria Laura Habegger, who also kindly supplied the photos through the University of South Florida.

Integument of Tetrapods

Although keratinization occurs in fishes, among terrestrial vertebrates it becomes a major feature of the integument. Extensive keratinization produces a prominent outer cornified layer, the stratum corneum, that resists mechanical abrasion. Lipids are often added during the process of keratinization or spread across the surface from specialized glands. The cornified layer along with these lipids increases the resistance of the tetrapod skin to desiccation.

Multicellular glands are more common in the skin of tetrapods than in the skin of fishes. In fishes, the mucous cuticle and secretions of the unicellular glands at or near the surface of the skin coat it. In contrast, among tetrapods, multicellular glands usually reside in the dermis and reach the surface through common ducts that pierce the cornified layer. Thus, the stratum corneum that protects the skin and prevents desiccation also controls the release of secretions directly to the surface. If it were not for these openings in the stratum corneum, the surface of the skin could not be coated or lubricated by these secretions.

Amphibians

Amphibians are of special interest because during their lives they usually metamorphose from an aquatic form to a terrestrial form. In most modern amphibians, the skin is also specialized as a respiratory surface across which gas exchange occurs, with the capillary beds in the lower epidermis and deeper dermis. In fact, some salamanders lack lungs and depend entirely on **cutaneous respiration** through the skin to meet their metabolic needs.

Cutaneous respiration (p. 417)

The most primitive tetrapods had scales like the fishes from which they arose. Among living amphibians, dermal scales are present only as vestiges in some species of tropical caecilians (Apoda). Frogs and salamanders lack all traces of dermal scales (figure 6.13a). In salamanders, the skin of the aquatic larvae includes a dermis of fibrous connective tissue, consisting of superficial loose tissue over a compact deep layer. Within the epidermis are deep basal cells and surface apical cells. Scattered throughout are large **Leydig cells** thought to secrete substances

FIGURE 6.11 **Bony-fish skin.** (a) Arrangement of dermal scales within the skin of a teleost fish (arrows indicate direction of scale growth). (b) Enlargement of epidermis. Note epidermal cells and club cells.

(a) After Spearman.

FIGURE 6.12 **Scale types of bony fishes.** Cross section of a cosmoid scale (a), a ganoid scale (b), and a teleost scale (c). Surface views of the two types of the teleost scale, cycloid and ctenoid scales (d).

that resist entry of bacteria or viruses (figure 6.13b). In terrestrial adults, the dermis is similarly composed of fibrous connective tissue. In the epidermis, Leydig cells are now absent, but distinct regions can be recognized, such as the strata basale, spinosum, granulosum, and corneum. Presence of a thin stratum corneum affords some protection from mechanical abrasion and retards loss of moisture from the body without unduly shutting off cutaneous gas exchange. During the breeding season, **nuptial pads** may form on digits or limbs of male salamanders or frogs. Nuptial pads are raised calluses of cornified epidermis that help the male hold the female during mating.

Generally, the skin of frogs and salamanders includes two types of multicellular glands: mucous and poison glands. Both are located in the dermis and open to the surface through connecting ducts (figure 6.13b). The **mucous glands** tend to be smaller, each being made up of a little cluster of cells that release their product into a common duct. The **poison glands** (granular glands) tend to be larger and often contain stored secretions within the lumen of each gland. Secretions of poison glands tend to be distasteful or even toxic to predators. However, few persons handling amphibians are bothered by this secretion, nor need they be concerned, because it is potentially harmful only if eaten or injected into the bloodstream.

Chromatophores may occasionally be found in the amphibian epidermis, but most reside in the dermis. Capillary beds, restricted to the dermis in most vertebrates, reach into the lower part of the epidermis in amphibians, a feature serving cutaneous respiration.

Reptiles

The skin of reptiles reflects their greater commitment to a terrestrial existence. Keratinization is much more extensive, and skin glands are fewer than in amphibians. Scales are present, but these are fundamentally different from the dermal scales of fishes, which are built around bone of dermal origin. The reptilian scale usually lacks the bony undersupport or any significant structural contribution from the dermis. Instead, it is a fold in the surface epidermis, hence, an epidermal scale. The junction between adjacent epidermal scales is the flexible **hinge** (figure 6.14a). If the epidermal scale is large and platelike, it is sometimes termed a **scute.** Additionally, epidermal scales may be modified into crests, spines, or hornlike processes.

Although not usually associated with scales, dermal bone is present in many reptiles. The **gastralia,** a collection of bones in the abdominal area, are examples. Where dermal bones support the epidermis, they are called **osteoderms,** plates of dermal bone located under the epidermal scales. Osteoderms are found in crocodilians, some lizards, and some extinct reptiles. Some bones of the turtle shell are probably modified osteoderms.

The dermis of reptilian skin is composed of fibrous connective tissue. The epidermis is generally delineated into three regions: stratum basale, stratum granulosum, and stratum corneum. However, this changes prior to molting in those reptiles that slough large pieces of the cornified skin layer. In turtles and crocodiles, sloughing of skin is modest, comparable to birds and mammals, in whom small flakes fall off at irregular

(a)

(b)

FIGURE 6.13 Amphibian skin. (a) Section through an adult frog skin. A basal stratum basale and a thin, superficial stratum corneum are present. The transitional layer between them includes a stratum spinosum and a stratum granulosum. (b) Diagrammatic view of amphibian skin showing mucous and poison glands that empty their secretions through short ducts to the surface of the epidermis.

intervals. But in lizards, and especially in snakes, shedding of the cornified layer, termed **molting** or **ecdysis,** results in removal of extensive sections of superficial epidermis. As molting begins, the stratum basale, which has given rise to the strata granulosum (inner) and corneum (outer), duplicates the deeper layers of granulosum and corneum, pushing up under the old layers. White blood cells invade the **stratum intermedium,** a temporary layer between old and new skin (figure 6.14b). These white blood cells are thought to promote the separation and loss of the old superficial layer of the skin.

Integumental glands of reptiles are usually restricted to certain areas of the body. Many lizards possess rows of **femoral glands** along the underside of the hindlimb in the thigh region. Crocodiles and some turtles have **scent glands.** In alligators of both sexes, one pair of scent glands opens into the cloaca, another pair opens on the margins of the lower jaw. In some turtles, scent glands can produce quite pungent odors, especially when the animal is alarmed by handling. Most integumental glands of reptiles are thought to play a role in reproductive behavior or to discourage predators, but the glands and their social roles are not well understood.

Birds

Basic Structure The feathers of birds have been called nothing more than elaborate reptilian scales. This oversimplifies the homology. Certainly the presence of epidermal scales along the legs and feet (figure 6.15a) of birds testifies to their debt to reptiles. If not a direct remodeled reptilian scale, then the feather is as example of yet another more fundamental homology of the underlying interaction of the epidermal-dermal layers producing such a skin specialization (see figure 6.3).

The dermis of bird skin, especially near the feather follicles, is richly supplied with blood vessels, sensory nerve endings, and smooth muscles. During the brooding season, the dermis in the breast of some birds becomes increasingly vascularized, forming a **brood patch** in which warm blood can come into close association with incubated eggs.

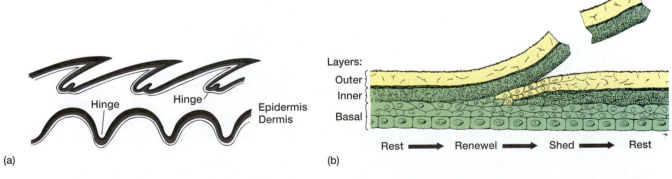

(a)   (b)

FIGURE 6.14 Reptile skin. (a) Epidermal scales. Extent of projection and overlap of epidermal scales varies among reptiles and even along the body of the same individual. Snake body scales (top) and tubercular scales of many lizards (bottom) are illustrated. Between scales is a thinned area of epidermis, a "hinge" allowing skin flexibility. (b) Skin shedding. Just before the old outer layer of epidermis is shed, the basal cells produce an inner epidermal generation. White blood cells collect in the splitting zone to promote separation of new from old outer epidermis.

(a) After Maderson; (b) after Landmann.

The Asian snake, *Rhabdophis tigrinus,* provisions its defenses with toxins garnered from the natural prey it consumes, a poisonous toad. This toad possesses skin glands toxic to most vertebrates, but this snake, the tiger keelback snake, can tolerate these toxins. In fact, upon digestion of the toad, the snake harvests these toxins to be redeployed in special nuchal (neck) glands. When the snake is bitten by its own predator, these glands on the snake's neck burst, releasing the toxic contents that produce a burning distaste or even blindness if squirted into the eyes, thereby discouraging or deterring the attacker. There is even some evidence that female snakes can pass the toxins to their young embryos, equipping the youngsters with a ready defensive chemical arsenal when born. Sequestering toxins in invertebrates is well known, but this discovery in *R. tigrinus* may lead to discovery of similar systems in other snakes feeding on amphibians with poisonous skin glands.

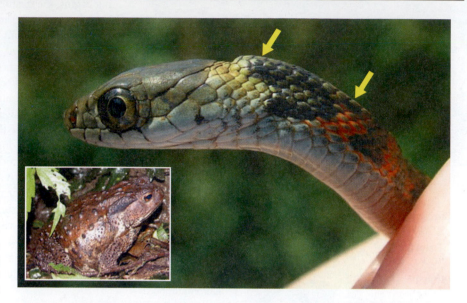

BOX FIGURE I   **Tiger keelback snake and the Japanese toad, its toxic prey.** Note the raised keel (arrows) in the neck of this brightly colored snake, where are stored the toxins gathered from the poisonous toad *Bufo japonicus* (inset), which it has eaten and digested.

Photos kindly supplied by Deborah A. Hutchinson and Alan H. Savitzky, part of the research team including A. Mori, J. Meinwald, F. C. Schroeder, and G. M. Burghardt.

The epidermis comprises the stratum basale and the stratum corneum. Between them is the transitional layer of cells transformed into the keratinized surface of the corneum (figure 6.15b).

Bird skin has few glands. The **uropygial gland,** located at the base of the tail (figure 6.16a), secretes a lipid and protein product that birds collect on the sides of their beak and then smear on their feathers. Preening coats the feathers with this secretion, making them water repellent, and probably conditions the keratin of which they are composed. Following a molt, preening also helps the newly regenerated feather unfurl and assume its functional shape. The other gland, located on the heads of some birds, is the **salt gland,**

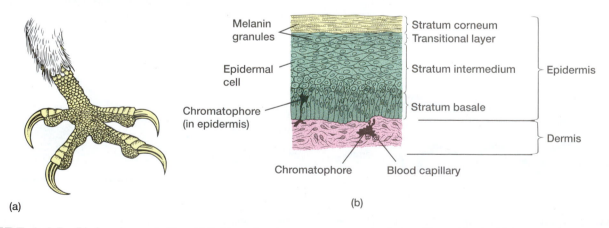

(a)

(b)

FIGURE 6.15   **Bird scales and skin.** (a) Epidermal scales are present on the feet and legs of birds. (b) Section of skin showing the stratum basale and the keratinized surface layer, the stratum corneum. Cells moving out of the basal layer move through first the stratum intermedium and transitional layer before reaching the surface. These middle layers are equivalent to the spinosum and granulosum layers of mammals.

(a) After Smith; (b) after Lucas and Stettenheim.

which is well developed in marine birds. Salt glands excrete excess salt obtained when these birds ingest marine foods and seawater.

Salt excretion (p. 560)

Feathers distinguish birds from all other living vertebrates. Feathers can be structurally elaborate and come in a variety of forms. Yet feathers are nonvascular and nonnervous products of the skin, principally of the epidermis and the keratinizing system. They are laid out along distinctive tracts, termed **pterylae,** on the surface of the body (figure 6.16a). Via one or several molts, they are replaced each year.

Generally the modern bird feather is built from a tubular central shaft, the **rachis,** which carries on either side a **vane,** a series of **barbs** with interlocking connections termed **barbules** (hooklets). The rachis and attached vanes constitute the **spathe** (figure 6.16b). The rachis continues proximal as the barbless **calamus,** or quill, which anchors the feather to the body and often is moved by attached dermal muscles. In modern birds, feathers are of many types serving different functions (figure 6.16c). Flight feathers are long and the vanes asymmetrical about the stiffening rachis; those flight feathers on the wings are *remiges* (sing, *remix*) and those on the tail *rectrices* (sing, *rectrix*). Contour feathers, or **pennaceous feathers,** cover the body and usually have symmetrical vanes about a rachis.

(a)

(b) Feather types

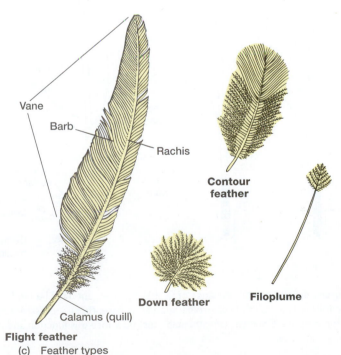

(c) Feather types

FIGURE 6.16 **Feather tracts and feather morphology.** (a) Feathers arise along specific pterylae or feather tracts. (b) General morphology of contour and down feathers. (c) Feather types. Flight feathers constitute the major locomotor surfaces. Contour feathers on the body aerodynamically shape the surface of a bird. Filoplumes are often specialized for display. Down feathers lie close to the skin as thermal insulation

(a, c) After Smith; (b) after Spearman.

Down feathers, or **plumulaceous feathers,** lack a distinctive rachis and noninterlocking barbs extend out from the calamus as a fluffy feather important in insulation (figure 6.16b).

Feathers develop embryologically from **feather follicles,** invaginations of the epidermis that dip into the underlying dermis. The root of the feather follicle, in association with the dermal pulp cavity, begins to form the feather.

The old feather is shed (molt), and the beginning of a new feather, the feather filament (or pin or blood feather), soon grows out of the follicle as a consequence of cell proliferation at the follicular base (figure 6.17a). The new epidermal cells form three distinct tissues: a supportive but later disposable sheath around the growing feather; the main feather tissues themselves that later unfurl to assume their final, functional shape; and pulp caps that temporarily protect the delicate dermal core. As the growing spathe begins to unfurl (figure 6.17b), new protective caps form one below the other as older pulp caps are shed together with the upper portions of the sheath sloughed during preening. Further growth lengthens the feather as its spathe continues to deploy (figure 6.17c). When spathe development is complete, the development of the calamus begins within the

same sheath at the base of the feather. The fully formed feather, embraced about its base by the feather follicle, is now in place (figure 6.17d).

In a sense, a feather is a sheet of mature, dead keratinocytes that is full of slits. This is accomplished by the remarkable patterning zone that determines number, shape, and spacing between cells and cell populations that form the feather primordium. As barb ridges of the young spathe are delineated by the patterning zone, so are the future slits and spaces that will appear between them. Calamus formation differs from that of the spathe in that no slits appear, and a tubular calamus is produced.

These regeneration events are summarized in figure 6.18, which shows a highly schematic, telescoped abstract of feather development. During feather regeneration, inductive interaction between the dermal papilla and base of the follicle establishes a zone of cell proliferation, where new keratinocytes are produced, and a zone of patterning just above it that will generate the morphogenetic signals presiding over the fates of these keratinocytes. In the follicle, rings of outer cells (beta-keratin), the sheath and feather itself, form more or less concentrically around the inner stratum cylindricum and pulp caps

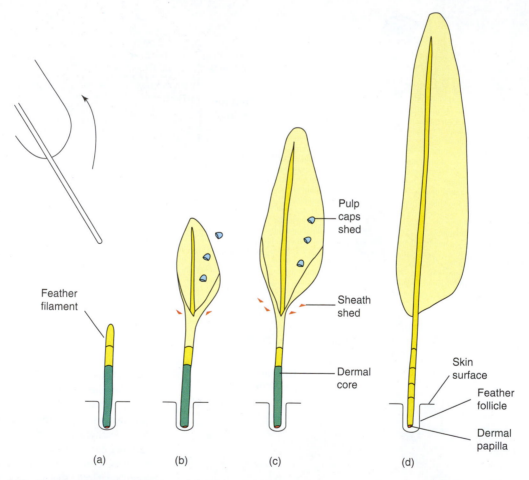

FIGURE 6.17 Feather growth. Molting and developmental sequence of feather replacement. (a) The old feather is shed (molt), and a new feather filament soon grows out of the follicle as a result of cell proliferation at its base. (b, c) Successive stages in spathe development. Note that some tissues necessary for initial development (sheath, pulp caps) now lose this function and are sloughed off as the mature feather emerges. (d) Mature, new feather in place.

Based on the research of P. F. A. Maderson and W. J. Hillenius.

(alpha-keratin), and dermal core (figure 6.18, cross sections). The feather filament continues to grow out from the follicle accompanied by the highly vascularized dermal core, which extends through the follicle mouth above the surrounding integument. Core tissues are protected from desiccation and trauma by a succession of pulp caps derived from the stratum cylindricum. The protective sheath, important initially as a scaffold to the developing feather, is eventually lost to preening once the differentiated feather is mature and ready to unfurl. The spathe is the first part of the young feather to differentiate beneath the sheath. As the tip of the spathe unfurls, the base of the spathe is still under construction. When the spathe completes its differentiation, the calamus is next formed, also in the same region beneath the sheath. As calamus formation proceeds, pulp caps continue to form within its hollow core as the dermal core regresses within the follicle. Dermal muscles, connected in a network of muscles, act to erect mobile feathers.

The patterning process is complex. New keratinocytes formed in the proliferation zone move up in the follicle but their fates are determined in the patterning zone by morphogenetic signals emanating from the patterning zone. Here cells become programmed to form sheath, pulp caps, barbs, barbules, or rachis. Cells moving through the patterning zone receive different signals than cells that precede or follow them, leading to the highly specific differentiation of the emerging feather. As the spathe is being differentiated, the patterning process sets aside populations of keratinocytes, for example, tissues of the future barb, barbules, and rachis. Additionally, other signals also establish precisely patterned fates where cells will lose their connections to one another and form the future spaces and slits between barbs and barbules. Thus the patterning zone not only sets cell fates forming structures of the spathe, but also presides over the ultimate spacing between feather parts. During deployment, this spacing allows adjacent barbs and barbules to separate as they unfurl. Preening of the emerging spathe encourages the overlap and interlocking of barbules as the mature feather takes final shape. When this completes, calamus formation begins. The patterning process now specifies a different outcome, namely, uninterrupted adhesion of keratinocytes and no spaces, thereby forming this tubular base of the feather. Thus the patterning zone not only sets cell fates but also presides over the spacing between feather parts, and also programs cells destined to form the sheath, pulp caps, and possibly the stratum cylindricum as well as the feather primordium itself. The rachis is not formed by the fusion of several barbs but also by this patterning process.

Functions   There are several types of feathers (figure 6.16c). Contour feathers aerodynamically shape the surface of the bird. Down feathers lie close to the skin as thermal insulation. Filoplumes are often specialized for display, and flight feathers constitute the major aerodynamic surfaces. Flight feathers of the wings are a type of contour feather. They are characterized by a long rachis and prominent vanes (figure 6.16). These feathers have some value as insulation, but their primary function is locomotion. Most feathers receive sensory stimuli and carry colors for display or courtship. Chromatophores occur within the epidermis, and their pigments are carried into the feathers to give them color. But light refraction on the feather barbs and barbules also creates some of the iridescent colors that feathers display.

Evolution of Feathers   When we think of feathers, we think of their roles in flight, but they likely had other functions when they first arose. One view is that feathers, or their scaly predecessors, played a role in surface insulation. Surface insulation, of course, holds heat in or shields the body from taking up excess heat. Either may have been the initial advantage of feathers. Surface insulation would have interfered with the absorption of environmental heat, a disadvantage if the ancestors of birds were ectothermic. However, many species of ectothermic lizards have enlarged surface scales. Once the basking lizard is warmed, it turns so that the scales act like many tiny parasols to shade the skin surface and block further uptake of solar radiation (figure 6.19). Once enlarged and shaped for heat exclusion, these protofeathers would be preadapted for heat retention or for flight.

Others argue that the ancestors of birds were endothermic. In this view, protofeathers initially functioned to conserve internally produced body heat. The evolution of aerodynamic devices serving flight came later.

Whether beginning with an ectotherm or an endotherm, many still suggest that feathers played a role in surface insulation when they first appeared and became secondarily co-opted for a role in flight. However, early birds and their immediate dinosaurian ancestors lacked nasal turbinates, a character diagnostic of warm-blooded physiology. If serving as insulation, these first feathers would have had a more complicated role than previously thought.

Dinosaurs: Hot to Cold—The Sequel (p. 123); Turbinates (pp. 276, 495)

A different view entirely stems from the argument that feathers evolved initially as aids to gliding and then to flight. Feathers were selected because of their favorable effect on the airstream passing over the body or limbs of a gliding animal. If the protoavian limb were not streamlined, then pressure drag would result and turbulence would have reduced aerodynamic efficiency. However, surface scales projecting from the trailing edge of the limb would have streamlined the limb, reduced drag, and thus been favored by selection.

Aerodynamic principles (pp. 141, 361)

Regardless of whether they evolved first for gliding or for insulation, feathers were modified from reptilian scales or at least from the common inductive interaction between dermis and epidermis. In modern birds, feathers that serve flight are highly modified. Interlocking barbs and barbules give some structural integrity to the flexible flight feather. In flight feathers, the rachis is offset, making the vane asymmetrical (figure 6.20a). This design affects the action of the flight feather during wing beats. On the downstroke, the pressure on the underside of each

FIGURE 6.18 **Feather regeneration, summary.** The highly schematic, compressed summary of feather development is shown. Cross sections of the regenerating feather are at the right to show the arrangement of concentric layers involved. At the base of the feather follicle, morphogenetic signaling between the dermal papilla and epidermal wall of the follicle establishes a proliferation zone and patterning zone. The new feather, first the spathe and later the calamus, develops between the sheath and stratum cylindricum, which together are wrapped around the dermal core. The dermal core is highly vascularized and has both supportive and nutritive functions. S. cylindricum = Stratum cylindricum. (a) and (b) Arrows indicate approximately where the cross sections are taken.

Based on the research of P. F. A. Maderson and W. J. Hillenius.

Beginning in the late twentieth century, remarkable new dinosaur finds were reported from China based on exquisitely preserved fossils. But the most unexpected was that some of these dinosaurs were covered in a thin coat of fluffy feathers (box figure 1a)! The scientific sensation was immense and immediate. So was the controversy. Box figure 1b is a summary phylogeny of current claims of feathers on dinosaurs, specifically within the theropod radiation. Some had a fluffy, down-like surface coating (plumulaceous); others also possessed shafted feathers with a symmetrical vane (pennaceous); others still had asymmetrical vanes implying flight; in some forearms carried wing feathers. The controversy surrounds two features of these fossils. The first is a challenge to the interpretation of these fossil "feather" impressions, which some claim are not feathers, at least not in early theropods, but degraded collagen skin fibers. The second challenge is to the phylogeny, which seems to be in considerable flux. That is certainly true especially as new fossils are added to the data base, but that is also true of most phylogenies that are part of ongoing research. Currently, node 1 is where we might identify "Avis"; node 2 is deinonychosauria; node 3 is unnamed. But as more fossils are discovered, our definition of Avis might change to node 2 or even 3.

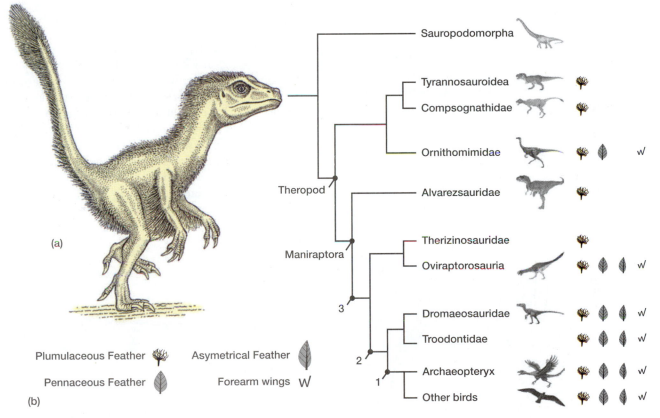

BOX FIGURE 1 **Feathered dinosaurs.** (a) A possible reconstruction of an early, feathered therapod. (b) Summary phylogeny showing current claims for the occurrence of feathers in therapods. In some taxa, only a few species show evidence of feathers. Earliest to appear are down feathers (plumulaceous); then contour feathers (pennaceous); asymmetrical feathers (flight); feathers on the forerms (W).

feather acts along its anatomical midline, the **center of pressure.** But because the rachis is offset, the result is to twist the feather slightly about its point of attachment to the limb, forcing feathers of the wing together into a broad surface that presses against the air and drives the bird forward (figure 6.20b). On the upstroke, the center of pressure is now across the topside of the asymmetrical feather and forces it to twist in the opposite direction, opening a channel between the feathers (figure 6.20c). This reduces their resistance to the airstream and allows the wing to recover and prepare for the next power downstroke.

Bird flight (p. 359)

This controlled twisting of flight feathers passively responding to wing beat depends on the asymmetrical design of the feather and hence on the action of air pressure

FIGURE 6.19 **Hypothetical scale, intermediate stage between an enlarged reptile scale and an early bird feather.** Some living lizards use enlarged scales to reflect away excess solar radiation. Subdivision of the scale provides the flexibility required for free movement in an active animal.

After Regal.

against it during powered flight. A close look at the wing feathers of *Archaeopteryx* also reveals an offset rachis and an asymmetrical vane (figure 6.21). This suggests that by the time of *Archaeopteryx* powered flight had already evolved.

Mammals

As in other vertebrates, the two main layers of the mammalian skin are epidermis and dermis, which join and interface through the basement membrane. Beneath lies the hypodermis, or superficial fascia, composed of connective tissue and fat.

Epidermis The epidermis may be locally specialized as hair, nails, or glands. Epithelial cells of the epidermis are keratinocytes and belong to the keratinizing system that forms the dead, superficial cornified layer of the skin. The surface keratinized cells are continually exfoliated and replaced by cells arising primarily from the deepest layer of the epidermis, the stratum basale. Cells within the basale divide mitotically, producing some that remain to maintain the population of stem cells and others that are pushed outward. As they are displaced to higher levels, they pass through keratinization stages exhibited as distinct, successive layers toward the surface: **stratum spinosum, stratum granulosum,** often a **stratum lucidum,** and a **stratum corneum** (figure 6.22). The process of keratinization is most distinct in regions of the body where the skin is thickest, as on the soles of the feet. Elsewhere, these layers, especially the lucidum, may be less apparent.

Keratinocytes are the most prominent cell type of the epidermis. Other types are recognized, although their functions are less clearly known. The **Langerhans cells** are stellate cells dispersed singly throughout the upper parts of the stratum spinosum. Current evidence suggests that they play a role in cell-mediated actions of the immune system. The **Merkel cells,** originating from the neural crest and associated with nearby sensory nerves, are thought to respond to tactile stimulation (mechanoreceptors).

In addition to these epithelial cell types, the other prominent cell type that becomes secondarily associated with the epidermis is the chromatophore. Chromatophores arise from

(a)

(b) Downstroke

(c) Upstroke

FIGURE 6.20 **Flight function of asymmetrical feather vane.** (a) Wing is extended, as it might appear during the middle of the power downstroke. One of the primary feathers (highlighted) is removed to show the axis of rotation about its calamus, where it attaches to the limb. (b, c) Cross sections through three flight feathers during the downstroke (b) and upstroke (c). During the power downstroke, air pressure against the underside of the wing would be experienced by each feather along its center of pressure down the anatomical midline of the feather. Because the rachis is offset, however, this center of pressure forces the feather to rotate about its axis, and the primary feathers temporarily form a closed, uniform surface. During the recovery upstroke, air pressure against the back of the wing forces rotation in the opposite direction, spaces open between feathers, and air slips between the resulting slots, thus reducing resistance to wing recovery.

FIGURE 6.21 *Archaeopteryx* **feather.** This feather from the wing of *Archaeopteryx* shows the asymmetrical design of the vane, suggesting that it may have been used during powered flight as in modern birds.

Based on Ostrom.

embryonic neural crest cells and may be found almost anywhere within the body. Those that reach the skin occupy sites within the deeper parts of the epidermis itself. They secrete granules of the pigment melanin, which are passed directly to epithelial cells and eventually carried into the stratum corneum or into the shafts of hair. Skin color results from a combination of the yellow stratum corneum, the red underlying blood vessels, and the dark pigment granules secreted by chromatophores.

Dermis The mammalian dermis is double layered. The outer **papillary layer** pushes fingerlike projections, termed **dermal papillae,** into the overlying epidermis. The deeper **reticular layer** includes irregularly arranged fibrous connective

tissue and anchors the dermis to the underlying fascia. Blood vessels, nerves, and smooth muscle occupy the dermis but do not reach the epidermis. The mammalian dermis produces dermal bones, but these contribute to the skull and pectoral girdle and only rarely form dermal scales in the skin. One exception is *Glyptodon*, a fossil mammal whose epidermis was underlaid by dermal bone. A similar situation exists in the living armadillo. These species represent secondary developments of dermal bone in the mammalian integument.

Blood vessels and nerves enter the dermis. Hair follicles and glands project inward from the epidermis (figure 6.21). The dermis is usually composed of irregularly arranged fibrous connective tissue that is often impregnated with elastic fibers to give it some stretch but return it to its original shape. As a person ages, this elasticity is lost, and the skin sags.

Hair Hairs are slender, keratinous filaments. The base of a hair is the **root.** Its remaining length constitutes the non-living **shaft.** The outer surface of the shaft often forms a scaly **cuticle.** Beneath this is the **hair cortex,** and at its core is the **hair medulla** (figure 6.22).

The hair shaft projects above the surface of the skin, but it is produced within an epidermal **hair follicle** rooted in the dermis. The surface of the epidermis dips down

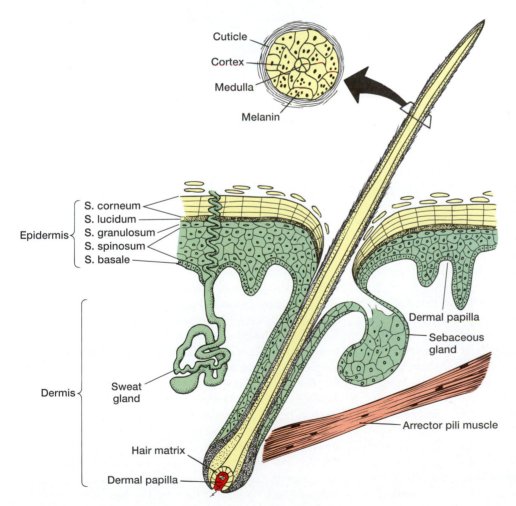

FIGURE 6.22 Mammalian skin. The epidermis is differentiated into distinct layers. As in other vertebrates, the deepest is the stratum basale, which through mitotic division produces cells that as they age become successively part of the stratum spinosum, the stratum granulosum, often the stratum lucidum, and finally the surface stratum corneum. The dermis pokes up dermal papillae that give the overlying epidermis an undulating appearance. Sweat glands, hair follicles, and sensory receptors lie within the dermis. Notice that the sweat ducts pass through the overlying epidermis to release their watery secretions on the surface of the skin. Key: S., Stratum.

The hooded pitohui is a brightly colored songbird, one of perhaps half a dozen related species endemic to the forests of New Guinea. This species is the first documented toxic bird. Its skin and feathers are laced with a potent neurotoxin, which if touched, causes numbness and tingling. This distributed neurotoxin is thought to provide the pitohui with some protection from ectoparasites. Apparently, the bird's poison also works to repel snake and hawk attacks, which might also account for the bright coloration announcing its toxicity to predators. The neurotoxin itself is not manufactured by the pitohui, but is acquired from a beetle that the bird can safely eat and commandeer its toxins for the pitohui's own defense. This same neurotoxin, formally batrachotoxin, is also found in some poison dart frogs, likely obtained from the same or similar insect source.

BOX FIGURE I   The hooded pitohui, black headed and bright orange breast, held in a protected hand. The *choresine* beetle, inset and slightly enlarged proportionately, is the source of the neurotoxin that forms the bird's chemical defense from ectoparasites and natural predators.

Photos kindly supplied by and based on the research of John P. Dumbacher.

---

into the dermis to form the hair follicle. At its expanded base, the follicle receives a small tuft of the dermis, the **hair papilla.** This papilla seems to be involved in stimulating activity of the **matrix cells** of the epidermis, but does not itself directly contribute to the hair shaft. The tiny clump of living matrix cells, like the rest of the stratum basale, is the germinal region that starts the process of keratinization to produce hair within the follicle. Unlike keratinization within the epidermis, which is general and continuous, keratinization within the hair follicle is localized and intermittent.

The hair shaft grows out of the living hair follicle, which goes through a cycle of activity with three stages—growth, degeneration, rest. During growth, there is active proliferation of cells in the hair papilla at the base of the hair, producing successive addition to the hair shaft, which emerges from and continues to lengthen from the skin surface. At the end of the growth stage, the hair-producing cells become inactive and die, entering the degeneration stage. Thereafter the follicle enters the resting stage, which may last for several weeks or months. Eventually, stem cells in the papilla produce a new follicle and the growth stage begins again. At about this time the old hair shaft falls out to be replaced by the new growing shaft. The cycle is intrinsic and cutting the hair seems not to accelerate hair growth.

Chromatophores in the follicle contribute pigment granules to the hair shaft to give it further color. The **arrector pili** muscle, a thin band of smooth muscle anchored in the dermis, is attached to the follicle and makes the hair stand erect in response to cold, fear, or anger. As humans (and many other mammals) age their hair becomes gray, no matter what their original youthful color. This occurs because special stem cells responsible for hair color within the hair follicle begin to die. In early life, these stem cells differentiate into specialized melanocytes that manufacture pigments of hair and skin. But as an individual grows older, these cells die off, eliminating a source of pigments, and adding no color to the hair shaft, leaving only its intrinsic gray or silver color.

A thick covering of hair, **fur,** or **pelage,** is generally composed of guard hairs and underfur. The **guard hairs,** the larger, coarse hairs, are the most apparent on the outer surface of the fur. The **underfur** is stationed beneath the guard hairs and is usually much finer and shorter. Both guard hairs and underfur function largely as insulators. In most marine mammals, the underfur is reduced or lost entirely, and only a few guard hairs are evident. Hair has grain; that is, it is laid out in a particular direction. Strokes against the grain are resisted (figure 6.23). An exception occurs in moles, which lack turnaround space and must back up in their tunnels. The hair of moles can be combed forward or backward without much difference in grain.

FIGURE 6.23 **Hair tracts.** Hair grows with a grain, a particular direction in which it slants. Notice the various growth directions (arrows) in which the hairs of the marsupial bandicoot lie.

After Lyne.

Some hairs are specialized. Sensitive nerves are associated with the roots of **vibrissae,** or "whiskers," around the snouts of many mammals. Such a sensory function may have been the first function of hair in the form of vibrissae, appearing before an insulative pelage evolved. Not surprisingly, vibrissae are common today in nocturnal mammals and in mammals that live in burrows with limited light. The **quills** of porcupines are stiff, coarse hairs specialized for defense.

Evolution of Hair Fossil skin impressions from the Middle Jurassic give evidence for the presence then of hair, and presumably of mammals that owned it. But the initial adaptive value of hair remains speculative. One view holds that hair arose initially as surface insulation, retaining body heat in primitive mammalian endotherms. The presence of nasal turbinates in synapsids, earlier from the Late Permian, suggests an early endothermy and, hence, a role for insulating hair. An alternative view is that hair evolved first as tiny projecting rods in the hinges between scales and served as tactile devices. These "protohairs" could help monitor surface sensory data when an animal was hiding from an enemy or retreating from the weather. If such a role increased in importance, it would have favored longer shafts and perhaps the evolution of structures resembling vibrissae. This sensory protohair might then have evolved secondarily into an insulative pelage as mammals became endothermic. Although insulative in modern mammals, hair still retains a sensory function.

Being soft and decomposable, hair does not leave a reliable trace in the fossil record. Some therapsids, ancestors to mammals, have tiny pits in the facial region of their skulls. These pits resemble pits on skulls associated with sensory vibrissae in modern mammals. Some have interpreted these pits as indirect evidence of hair in therapsids. But the skulls of some modern lizards with scales have similar pits and, of course, lizards have no hair. Thus, such pits are not conclusive evidence of the presence of hair. Further, one especially well-preserved skin impression of *Estemmenosuchus,* a therapsid from the Upper Permian, shows no evidence of hair. The epidermis was smooth, without scales, and undifferentiated, although it was supplied with glands. In life, its skin was probably soft and pliable. Thus, we still do not know when hair first arose in primitive mammals or in their therapsid ancestors.

Glands Principally there are three main types of integumental glands in mammals: sebaceous, eccrine, and apocrine. Scent, sweat, and mammary glands are derived from them. Sebaceous glands are globular or saclike in shape; eccrine and apocrine glands are long, coiled invaginations of the epidermis that reach deep with the dermis but maintain continuity through the surface of the skin even through the cornified stratum corneum.

The **sebaceous glands** produce an oily secretion, **sebum,** that is released into hair follicles in order to condition and help waterproof fur. Sebaceous glands are absent from the palms of hands and soles of feet, but they are present, without associated hair, at the angle of the mouth, on the penis, near the vagina, and next to the mammary nipples. At these sites, their secretion lubricates the skin surface. The **wax glands** of the outer ear canal, which secrete earwax, and **Meibomian glands** of the eyelid, which secrete an oily film over the surface of the eyeball, probably are derived from sebaceous glands.

The **eccrine glands** produce thin, watery fluids, are not associated with hair follicles, begin to function before puberty, and are innervated mainly by cholinergic nerves. In most mammals these glands are associated with the soles of the feet and hands, prehensile tails, and other sites in contact with abrasive surfaces. Chimpanzees and humans have the greatest number of eccrine glands, including some on the palms and soles. In the platypus, these glands are limited to the snout. In mice, rats, and cats, they are present on the paws, and they appear around the lips in rabbits. In elephants, eccrine glands and sebaceous glands are absent entirely.

The **apocrine glands** produce a viscous, lipid-containing fluid, are associated with hair follicles, begin to function at puberty, and are innervated mainly by adrenergic nerves. Their secretions primarily function in chemical signaling.

Surface evaporation of the products of these glands helps to dissipate heat, wherein they are termed "sweat glands." However, sweat glands are not found in all mammals, and in fact humans are exceptional in our perfuse use of evaporative cooling to thermoregulate. Our sweat glands are derived from eccrine glands but those of horses derive from apocrine glands, an example of convergent evolution.

Sweat also contains waste products; therefore, the integument represents one avenue for elimination of metabolic by-products.

Cholinergic and adrenergic innervation (p. 642)

The **scent glands** are derived from apocrine glands and produce secretions that play a part in social communication. These glands may be located almost anywhere on the body, as on the chin (some deer, rabbits), face (deer, antelope, bats), temporal region (elephants), chest and arms (many carnivores), anal region (rodents, dogs, cats, mustelids), belly (musk deer), back (kangaroo rats, peccaries, camels, ground squirrels), or legs and feet (many ungulates). Secretions of these glands are used to mark territory, identify the individual, and communicate during courtship.

The **mammary glands** produce **milk,** a watery mixture of fats, carbohydrates, and proteins that nourishes the young. Ectodermal mammary ridges, within which mammary glands form, are located along the ventrolateral side of the embryo. The number of mammary glands varies among species. Release of milk to a suckling is **lactation.**

Lactation has been reported in males of a species of Malaysian fruit bat, opening the possibility that these fruit bat males may actually suckle the young they sire. Outside of this fruit bat, lactation has been reported only in domesticated male animals, the likely result of abnormal inbreeding and in pathological conditions. With these exceptions, mammary glands become functional only in females. Mammary glands consist of numerous **lobules.** Each lobule is a cluster of secretory alveoli in which milk is produced. The alveoli can open into a common duct that, in turn, can open directly to the surface through a raised epidermal papilla, or **nipple.** The nipple is usually surrounded by a circular pigmented area of skin called the **areola.** Alveolar ducts also can open into a common chamber, or **cistern,** within a long collar of epidermis called the **teat.** The teat forms a secondary duct carrying milk from the cistern to the surface (figure 6.24a–c). Adipose tissue can build up beneath the mammary glands to produce **breasts.**

In monotremes, nipples and teats are absent, and breasts do not form. Milk is released from ducts onto the flattened milk patch, or areola, on the surface of the skin (figure 6.24a). The front of the infant's snout is shaped to fit the surface, permitting vigorous suckling. In short 20- to 30-minute bursts of suckling, a young echidna can take in milk equivalent to about 10% of its body weight. In marsupials and eutherian mammals, either teats or nipples are present (figure 6.24b, c). At sexual maturity, adipose tissue builds up under the mammary gland to produce the breast. Enlargement of the mammary glands occurs under hormonal stimulation shortly before the birth of suckling young. Suckling stimulates a neural response to the nervous system that results in release of **oxytocin,** the hormone that stimulates contraction of myoepithelial cells enveloping the alveoli, and hence milk is released. In common language, this active release of milk is termed *letdown.*

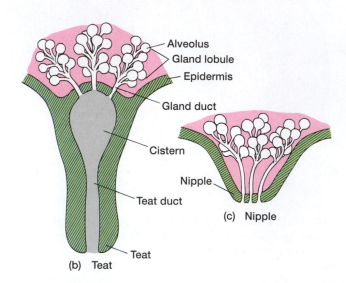

FIGURE 6.24 **Mammary glands.** Glandular mammary tissue derived from the integument lies in the dermis, but ducts reach through the epidermis to the surface. Mammary glands are arranged in lobules, each lobule being a collection of alveoli and their immediate ducts. (a) In monotremes, the mammary glands open directly to the unspecialized skin surface, and the young press their shaped snout to the patch of skin where these glands open. (b) In some marsupials and in many placental mammals, the mammary ducts open through specializations of the integument. The teat is a tubular specialization of the epidermis that is expanded at its base into the cistern, a chamber that receives milk from the mammary glands before passing it along the common teat duct to the suckling infant. (c) The nipple is a raised epidermal papilla around which the supple lips of the infant fit directly to drink the released milk.

The origin of lactation in mammals remains a complex issue. The earliest mammals, with perhaps only one exception (*Sinoconodon*), had diphyodont dentition ("milk" teeth and permanent teeth) as in modern suckling mammals, compared to polyphyodont dentition (continuous tooth replacement) in most other vertebrates. Diphyodonty implies the presence of mammary glands and suckling because suckling for nutrition, rather than foraging of hard food, allows substantial skull growth before teeth are required to process hard food.

Teeth (p. 508)

The detailed similarities of mammary glands in living monotremes, marsupials, and eutherians argue for a monophyletic origin of these glands, perhaps by the combination of parts of preexisting sebaceous and apocrine glands. This has led Daniel Blackburn to hypothesize a series of steps in the evolution of mammalian lactation, beginning in ancestors who incubated eggs or hatchlings. Skin gland secretions of the females that had antimicrobial properties would protect the surface of the young from bacteria, fungi, or other pathogens. Ingestion of these secretions in small quantities by hatchlings would reduce pathological affects and microorganisms in the digestive tract. If this secretion included maternal immunoglobulins, this would also directly deliver immunity to the offspring. Any nutritional value included with these secretions of ancestral mammary glands would be of further adaptive significance. By such a scenario, maternal skin glands of an incubation patch first supply protection from pathogens and then, secondly, become involved in nutrition of the young. Subsequent evolution included enhancement of the secretion into a copious flow that was much more nutritious (i.e., milk). Increase in the milk-nutrient content replaced the large yolk supply as the basis for support of the embryo. This was further favored by increased efficiency in infant feeding (suckling), anatomical specializations (teats in therians), and physiological control of milk production and release (hormones).

Milk release (p. 608)

Specializations of the Integument

Nails, Claws, Hooves

Nails are plates of tightly compacted, cornified epithelial cells on the surface of fingers and toes; thus, they are products of the keratinizing system of the skin. The **nail matrix** forms new nail at the nail base by pushing the existing nail forward to replace that which is worn or broken at the free edge. Nails protect the tips of digits from inadvertent mechanical injury. They also help stabilize the skin at the tips of the fingers and toes, so that on the opposite side the skin can establish a secure friction grip on objects grasped.

Only primates have nails (figure 6.25a). In other vertebrates, the keratinizing system at the terminus of each digit produces claws or hooves (figure 6.25b, c). **Claws,** or **talons,** are curved, laterally compressed keratinized projections from the tips of digits. They are seen in some amphibians and in most birds, reptiles, and mammals. **Hooves** are enlarged keratinized plates on the tips of the ungulate digits.

The horse hoof consists of the hoof wall, sole, and the frog (figure 6.26). The **hoof wall** is U-shaped and open at the heel, a derivative of the integument. It consists of a keratinized **stratum externa** (= tectorium), a thin, shiny surface layer; the **stratum medium,** thicker and also keratinized and permeated with coiled, tubular channels; and an inner **stratum internum** (= lamellatum), a highly and regularly laminated, infolded layer that interdigitates with the dermis (= corium) beneath. The hoof wall grows out from its base, the germinal region (matrix cells), not from the underlying dermis, at about 6 mm per month, taking 9 to 12 months overall for the toe to renew.

(a) Nail

(b) Claw

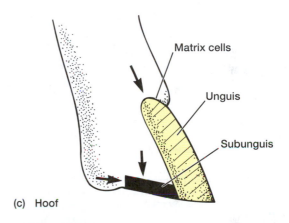

(c) Hoof

FIGURE 6.25 Epidermal derivatives. The unguis (nail, claw, or hoof) is a plate of cornified epithelium growing outward from a bed of proliferating matrix cells at its base and from a softer subunguis nearer its tip.

After Spearman.

The bottom of the hoof, the ground surface, includes the wedge-shaped **frog,** a mostly keratinized derivative of the integument that fills the opening in the heel of the hoof wall. The **sole** fills the ground surface space between the wall and triangular frog (figure 6.26b). It consists of epidermis and thickened dermis, the corium of the sole. Deep to the sole is the fatty **digital cushion,** or **pad,** a derivative of the hypodermis.

The hoof wall transfers energy of impact across the laminations to the digit. Much of the energy of impact is also absorbed by the slight flaring of the U-shaped hoof wall. The forces from the sole, frog, and digital cushion, and the outward movement of the lateral cartilages, create a hydraulic component of shock absorption. Brief surges in venous plexuses in the hoof reach high values during contact of 600–800 mm Hg.

FIGURE 6.26 Horse hoof. (a) Horse forefoot showing internal bones and hoof wall. Enlarged cross section shows layers from outer wall to inner bone. (b) Longitudinal section through horse forefoot.

From William J. Banks, Applied Veterinary Histology, *2nd Edition. 1986. Lippincott Williams & Wilkins.*

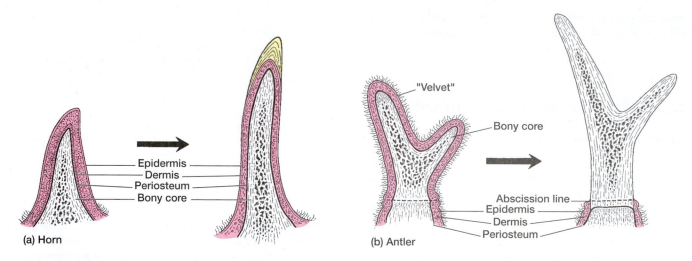

FIGURE 6.27 Horns and antlers. (a) Horns appear as outgrowths of the skull beneath the integument, which forms a keratinized sheath. Horns occur in bovids of both sexes and are usually retained year-round. (b) Antlers also appear as outgrowths of the skull beneath the overlying integument, which is referred to as "velvet" because of its appearance. Eventually this overlying velvet dries and falls away, leaving the bony antlers. Antlers are restricted to members of the deer family and, except for caribou (reindeer), they are present only in males. Antlers are shed and replaced annually.

After Modell.

Horns and Antlers

"Horned" lizards have processes extending from behind the head that look like horns but are specialized, pointed epidermal scales. Mammals, dinosaurs, and extinct turtles are the only vertebrates with true horns or antlers.

The skin, together with the underlying bone, contributes to both true horns and antlers. As these structures take shape, the underlying bone rises up, carrying the overlying integument with it. In **horns,** the associated integument produces a tough, cornified sheath that fits over the bony core that is never branched (figure 6.27a). In **antlers,** the overlying living skin (called "velvet") apparently shapes and provides

vascular supply to the growing bone. Eventually the velvet falls away to unsheath the bare bone, the actual material of the finished antlers that is branched (figure 6.27b).

True antlers occur only in members of the Cervidae (e.g., deer, elk, moose). Typically, only males have antlers, which are branched and shed annually. There are notable exceptions. Among caribou, both sexes have seasonal antlers. In deer, the antler usually consists of a **main beam,** from which branch shorter **tines,** or **points.** In yearling bucks, antlers are usually no more than prongs or spikes that may be forked. The number of tines tends to increase with age, although not exactly. In old age, antlers may even be deformed. In caribou and especially in moose, the main antler beam is compressed and **palmate,** or shovel-like, with a number of points projecting from the rim.

The annual cycle of antler growth and loss in the white-tail deer, for example, is under hormonal control. In the spring, increasing length of daylight stimulates the pituitary gland at the base of the brain to release hormones that stimulate antlers to sprout from sites on the skull bones. By late spring, the growing antlers are covered by velvet. By fall, hormones produced by the testes inhibit the pituitary, and the velvet dries. By thrashing and rubbing, the deer wipes the velvet off to expose the fully formed, but now dead, bone of the antlers (figure 6.28a–e). Males use their antlers during clashes with other males to maintain access to reproductively receptive females. Following this brief mating season, further hormonal changes lead to a weakening of the antler at its base where it attaches to the living bone of the skull. The antlers break off, and for a short time during winter, deer are without antlers.

Among mammals, **true horns** are found among members of the family Bovidae (e.g., cattle, antelope, sheep, goats, bison, wildebeests). Commonly, horns occur in both males and females, are retained year-round, and continue to grow throughout the life of the individual. The horn is unbranched and formed of a bony core and a keratinized sheath (figure 6.29). Those of the males are designed to withstand the forces encountered during head-butting combat. In large species, females usually have horns as well, although they are not as large and curved as in males. In small species, females are often hornless.

(a) (b) (c) (d) (e)

FIGURE 6.28 Annual growth of elk antlers. (a, b) New antlers begin to grow in April. (c) By May, antlers are nearly fully formed even though they are still covered by the living integument (velvet). (d) By late summer, the velvet has begun to dry and peel off. (e) Fully formed bony antlers are in place.

After Modell.

FIGURE 6.29 True horns of the mountain sheep (Bovidae). The cornified covering of the horn of the mountain sheep is removed on the right side of the skull to reveal the bony core.

Unlike the true horns of bovids, horns of the pronghorn, family Antilocapridae, are forked in adult males. The old outer cornified sheath, but not the bony core, is shed annually in early winter (figure 6.30a). The new sheath beneath, already in place, becomes fully grown and forked by summer. Female pronghorns also have horns whose keratinized sheath is replaced annually, but these are usually much smaller and only slightly forked. The horns of giraffes are different still. They develop from separate, cartilaginous processes that ossify, fuse to the top of the skull, and remain covered with living, noncornified skin (figure 6.30b). The rhinoceros horn does not include a bony core, so it is exclusively a product of the integument. It forms from compacted keratinous fibers (figure 6.30c).

Baleen

The integument within the mouths of mysticete whales forms plates of **baleen** that act as strainers to extract krill from water gulped in the distended mouth. Although it is sometimes referred to as "whalebone," baleen contains no bone. It is a series of keratinized plates that arise from the integument. During its formation, groups of dermal papillae extend and lengthen outward, carrying the overlying epidermis. The epidermis forms a cornified layer over the surface of these projecting papillae. Collectively, these papillae and their covering of epidermis constitute the plates of frilled baleen (figure 6.31).

Feeding by baleen whales (p. 288)

Scales

Scales have many functions. Both epidermal and dermal scales are hard, so when they receive mechanical insult and surface abrasion, they prevent damage to soft tissues beneath. The density of scales also makes them a barrier against invasion of foreign pathogens, and they retard water loss from the body. In sharks and other fishes, scales dampen the boundary layer turbulence to increase swimming efficiency. Some reptiles regulate the amount of

FIGURE 6.30 Other types of horns. (a) In pronghorns, the bony core of the horns is unbranched, but the cornified sheath is branched. (b) Giraffe horns are small, ossified knobs covered by the integument. (c) Rhinoceroses have several horns that rest on a low knob on the skull, but these horns have no inner core of bone. As outgrowths of the epidermis alone, they are mainly composed of compacted keratinized fibers.

After Modell.

(a) Pronghorn

(b) Giraffe

(c) Rhinoceros

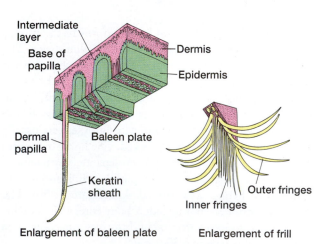

Enlargement of baleen plate   Enlargement of frill

FIGURE 6.31 Baleen from a whale. The lining to the mouth includes an epithelium with the ability to form keratinized structures. Groups of outgrowing epithelium become keratinized and frilly to form the baleen. Longitudinal sections of the keratinized layer and the frill are enlarged below the diagram of the baleen.

After Pivorunas.

surface heat they absorb by turning their bodies toward or away from the sun. This determines whether the sun rays are deflected off the full face of the scale or shine under the lifted posterior edge of the scale to reach the thin epidermis beneath.

Epidermal scales are the major component of the skin of reptiles. They are also present in birds along their legs, and in some mammals, such as the beaver, they cover the tail (figure 6.32a).

Dermal Armor

Dermal bone forms the armor of ostracoderm and placoderm fishes. Being a product of the dermis, dermal bone finds its way into alliances with a great variety of structures. Dermal bone supports the scales of bony fishes but tends to be lost in tetrapods. It is absent in the skin of birds and most mammals. Exceptions have been noted earlier, namely, in the fossil mammal *Glyptodon* (figure 6.32b) and in the skin of the living armadillo. However, selected dermal bones take up residence in the fish skull and pectoral girdle and have persisted into modern groups of vertebrates. Most dermal bones of the skull and shoulder girdle all began phylogenetically in the skin and later sank inward to become parts of the skeleton. This sharing of available parts between systems reveals again the remodeling character of evolution.

The shell of turtles is a composite structure. The dorsal half of the shell is the **carapace,** formed by fusion of dermal bone with expanded ribs and vertebrae

In humans, the conversion of dehydrocholesterol to vitamin D, which is necessary for normal bone metabolism, requires small amounts of ultraviolet radiation. If vitamin D is insufficient, bones become soft and deformed. On the other hand, too much ultraviolet radiation can be very damaging to deep living tissues. Skin alone is not especially effective in reflecting or safely absorbing these wavelengths of solar radiation. This task falls to the chromatophores and the pigment they produce.

Only a few minutes of exposure to sunlight each day is necessary to convert enough precursor (dehydrocholesterol) into vitamin D to meet an individual's metabolic needs. In tropical regions near the equator, sunlight passes directly through the absorbing layers of the atmosphere to strike the surface of the Earth. Terrestrial vertebrates covered with hair, feathers, or scales have some external protection against sun exposure. Humans, who essentially lack a thick coat of hair, do not. Too much ultraviolet radiation can produce harmful amounts of vitamin D, sunburn, and a higher incidence of skin cancer. The evolution of increased numbers of chromatophores in the skin of people in tropical regions protects against too much ultraviolet radiation. In temperate regions away from the equator, the angle of incidence of sunlight is low, passing more diagonally through more atmosphere and thus filtering out much of the ultraviolet radiation. Fewer chromatophores in the skin compensate for decreased availability of ultraviolet radiation, apparently allowing just enough radiation to convert dehydrocholesterol to a sufficient amount of vitamin D. Differences in skin color among the human populations are a result of these adaptive compromises.

Thus, the number of chromatophores in the skin is an evolutionary adaptation to the level of exposure to ultraviolet radiation. In addition, the production of pigment granules can change in response to short-term changes in exposure to sunlight. If exposure to sunlight is reduced, the chromatophores decrease their level of synthesis of pigment granules, and the skin lightens. If exposure is increased, pigment granule production increases, and the skin darkens. Such tanning occurs in all humans, but it is most conspicuous in light-skinned Caucasians. Sudden exposure to high levels of sunlight may result in sunburn, or radiation damage to the integument. As it does with a burn from a hot stove, the skin repairs itself and sheds the damaged layers. This is why the skin "peels" several days following a sunburn.

(b) *Glyptodon*

(a) Beaver tail

(c) Turtle shell

FIGURE 6.32 Epidermal and dermal derivatives. (a) Epidermal scales occur in some mammalian structures, such as this tail of a beaver (dorsal view). (b) Although dermal scales are rare, they are present and fused into the armor of *Glyptodon*, a fossil mammal. (c) Turtle shell. The encasing shell of a turtle is derived from three sources—the ribs and the vertebrae of the endoskeleton (stippled) and the dermal bone arising in the integument (white). The surface of this bony shell is covered by large, thin epidermal scales (not shown).

After Smith.

(figure 6.32c). Ventrally, the **plastron** represents fused dermal bones along the belly. On the surface of both carapace and plastron, keratinized plates of epidermis cover this underlying bone.

Turtle shell (p. 315)

Mucus

Mucus produced by the skin serves several functions. In aquatic vertebrates, it inhibits entrance of pathogens and may even have some slight antibacterial action. In terrestrial amphibians, mucus keeps the integument moist, allowing it to function in gas exchange. Although cutaneous respiration is prominent in amphibians, it occurs in other vertebrates as well. For example, many turtles rely on cutaneous gas exchange as they hibernate submerged in ice-covered ponds during the winter. Their shells are too thick, of course, to allow significant gas exchange, but exposed areas of skin around the cloaca offer a suitable opportunity. Sea snakes may depend on cutaneous respiration for up to 30% of their oxygen uptake. Similarly, fishes such as the plaice, European eel, and mudskipper may depend on some cutaneous gas exchange to meet their metabolic requirements.

Cutaneous respiration (p. 417)

Mucus is also involved in aquatic locomotion. As a surface coat, it smoothes the irregularities and rough surface features on the epidermis to reduce the friction met by a vertebrate swimming through relatively viscous water.

Color

Skin color results from complex interactions among physical, chemical, and structural properties of the integument. Changes in blood supply can redden the skin, as in blushing. The **differential scattering** of light, referred to as Tyndall scattering, is the basis for much color in nature. This is the phenomenon that makes the clear-day sky appear blue. In birds, air-filled cavities within feather barbs take advantage of this scattering phenomenon to produce the blue feathers of kingfishers, blue jays, bluebirds, and indigo buntings. Many black, brown, red, orange, and yellow colors result from pigments that produce color by selective light reflection. Interference phenomena are responsible for **iridescent colors.** As light is reflected from materials with different refractive indices, interference between different wavelengths of light produces iridescent colors. In many birds, iridescent colors result from interference of light reflected off the tiny barbs and barbules of the feathers.

Many of the pigments producing colors by this variety of physical phenomena are synthesized by and held in specialized chromatophores. Because these are cells, the suffix *-cyte* instead of *-phore* might seem more logical; however, the tradition of using the suffix *-phore* (meaning "bearer of") for chromatophores and for all the various types of chromatophores is an entrenched convention, especially applied to pigment cells of ectothermic vertebrates and all invertebrates with chromatophores. We follow the widespread practice in this text. Most chromatophores arise from embryonic neural crest and can take up residence almost anywhere within the body. It is not uncommon to find them associated with the walls of the digestive tract, within the mesenteries, or around the reproductive organs. Their function at these remote sites is not resolved, but they are thought to protect deep cell layers from penetrating solar radiation.

On the basis of form, composition, and function, four groups of chromatophores are currently recognized. The most well known of these is the **melanophore** that contains the pigment melanin. Cellular organelles called **melanosomes** house these melanin granules that intercept sunlight striking the surface of an animal to prevent penetration of harmful radiation. They, of course, also add color to the integument that may camouflage an animal, making it less detectable, or brighten a part that contributes to a behavioral display. There are two types of melanophores. The **dermal melanophore** is a broad, flat cell that changes color rapidly and is found only in ectotherms. The **epidermal melanophore** is a thin, elongated cell prominent in endotherms but present in all vertebrates. By contributing melanosomes, it adds color to keratinocytes, hair, and feathers.

The **iridophore,** which contains light-reflecting, crystalline guanine platelets, is a second type of chromatophore. It is found in ectothermic vertebrates and in the iris of the eye of some birds. Two other types of chromatophores are the xanthophore, containing yellow pigments, and the **erythrophore,** so called because of its red pigments. In addition, a few chromatophores contain several of these

pigments but are not classified. For example, in the iris of the Mexican ground dove, chromatophores contain both reflecting platelets (as expected in iridophores) and melanin (as in melanophores). This suggests that differentiation of chromatophores from neural crest stem cells must be responsive to a variety of developmental cues that produce pigment cells with intermediate properties.

Sunlight can influence physiological changes in chromatophore activity. Increased exposure stimulates increased production of pigment granules, resulting in darker skin over a period of days. In some vertebrates, the response is more immediate. Some fishes and lizards can change their colors almost instantly. The true chameleon, for example, can change colors to match its environment, at least if the background is light brown to dark green. Some fishes, such as the flounder, can change not only their color but also their color pattern to resemble the background (figure 6.33a). This physiological adjustment of color to background is mediated by the endocrine system and involves redistribution of pigment granules within the chromatophores. It was once thought that chromatophores themselves changed shape, sending out cytoplasmic pseudopods. Now it appears that color changes are not based on changes in cell shape. Instead, chromatophores assume a relatively fixed shape, and in response to hormonal stimulation, their pigment granules are either shuttled out into the previously positioned pseudopods or returned to become concentrated centrally within the cell (figure 6.32b).

Endocrine control of melanophores (p. 610)

(a)

(b)

FIGURE 6.33 Color changes. (a) The flounder changes its surface color as texture and pattern of the substrate changes. (b) Mediated by the endocrine system, chromatophores in the integument change the position of pigment granules within their cellular processes to change the hue and pattern of skin color.

Overview

Our skin is a dynamic organ. About every two weeks we shed it or, rather, we replace it. Our epidermis turns over that fast. Stem cells replace transforming cells. Vertebrate skin is also the phylogenetic source of surface specializations.

In a general way, it is easy to homologize integumentary structures (figure 6.34). Hair, feathers, and reptilian scales all are products of the epidermis, so they are all broadly homologous. But taken separately, controversies persist about homologies. For instance, some would claim that hair is a transformed reptilian scale, originally protective in function. Others argue that hair is a derivative of epidermal bristles, originally sensory in function. Some authors point to the similarity in structure between placoid scales and shark teeth to support the view that vertebrate teeth arose from shark scales. Others dissent, pointing out that teeth were present in early fishes before sharks evolved, so shark scales could not be the forerunners of vertebrate teeth.

When approaching controversies surrounding skin evolution, we need to remember that the skin consists of two layers, an epidermis and a dermis, not a single evolving structure. Interactions between these two layers play a part in their evolution. The dermis helps maintain, regulate, and specify the types and proliferation of epidermal cells. This has been explored with experimental embryology. For example, the epidermis from the leg of a chick embryo destined to form leg scales can be peeled away from its underlying dermis and kept alive in isolation with sufficient nutrients. Cells of such living but isolated epidermis cease to proliferate. If recombined in vitro with embryonic dermis, the epidermal cells resume proliferation and scales form. We know that the stimulus is within the dermis because if any other tissue, such as cartilage or muscle, is substituted, the epidermis fails to respond.

In a few instances, the epidermis acts autonomously from the dermis. When exposed to air, isolated chick epidermis shows the intrinsic capacity to transform itself into a keratinized layer without contact with an underlying dermis. Although not well understood, this degree of epidermal autonomy seems dependent on the ability of the epidermis to reconstruct the basement membrane or its chemical equivalent.

Despite its occasional independence, activity of the epidermis is largely influenced by the underlying dermis. Its direction of differentiation is also set by the dermis. For example, in the chick embryo, dermis from the leg promotes overlying epidermis to form keratinized scales, and dermis from the trunk induces overlying epidermis to produce feathers. If trunk dermis is experimentally replaced by leg dermis, the overlying trunk epidermis that customarily would produce feathers instead produces scalelike thickenings characteristic of the transplanted dermis. In the guinea pig, if dermis from the trunk, ear, or sole of the foot is transplanted beneath the epidermis elsewhere in the body, the epidermis responds by producing epidermal derivatives characteristic of the trunk, ear, or sole of the foot, respectively. In some lizards, the skin

(a)

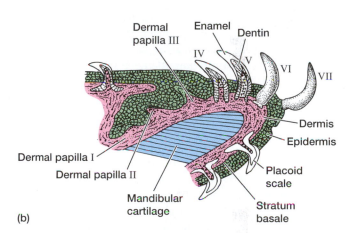

(b)

FIGURE 6.34 Interaction between epidermis and dermis. (a) Interactions between dermis and epidermis produce a variety of structures, such as teeth and placoid scales. (b) Shark teeth are derivatives of the oral epithelium. The basic similarities of composition (enamel, dentin) and method of formation (epidermis-dermis) suggest that teeth and placoid scales are homologous as integumentary structures. Successive stages in tooth development are indicated by Roman numerals.

(b) After Smith.

forms two types of epidermal scales, one tiny and tubercular in shape, the other large and overlapping. Scale type is determined by the underlying dermis. If embryonic dermis is switched between the two developing scale types, the overlying epidermis differentiates in accordance with its transplanted dermis. In mouse embryos, dermis specifies the type of hair as well as the general pattern of hairs produced. Upper lip dermis promotes the formation of vibrissae; trunk dermis promotes the formation of guard hairs.

In large measure, then, response of the epidermis is specific to the type of underlying dermis. To some extent, the age of an experimentally transplanted dermis or epidermis influences this response. Results tend to vary if the source of one transplant is from a young embryo and the source of the other transplant is from an older embryo or adult. Nevertheless, under normal circumstances, the dermis seems to bring a necessary physical substrate and organization along with nutrient supply to the epidermis. This stimulatory effect of dermis upon epidermis is **embryonic induction.** Although the dermis does not directly contribute cells of its own to the epidermal derivative (hair, feather, scale), it induces the type of epidermal specialization. The epidermis responds by altering activity of its germinal layer to produce the specified structure.

This epidermal-dermal interaction is evident even in tissue transplants between species from different classes. However, the dermis usually cannot induce the epidermis to form a specialization that is not typical of its class. Lizard epidermis can be paired with chick or mouse dermis. Likewise, chick epidermis can be paired with lizard or mouse dermis, and mouse epidermis can be paired with lizard or chick dermis. In these reciprocal transplants between reptile (lizard) and bird (chick) and between reptile (lizard) and mammal (mouse), the type of skin specialization induced (scale, feather, or hair) conforms to the origin of the epidermis, not to the transplanted dermis. Thus, the lizard epidermis is induced to form a reptilian scale, the chick epidermis to form a feather, the mouse epidermis to form hair, irrespective of the origin of the dermis with which it is paired. Interestingly, if the transplanted dermis does not come from a region producing a skin specialization, it seems to lack the necessary ability to induce the cross-class epidermis to form a specialization. Further, specializations induced by these cross-class transplants of dermis do not develop fully. Lizard scales, bird feathers, and mammal hair form but cease to grow after a certain stage. Apparently, the foreign dermis is sufficient to stimulate epidermal proliferation, but it cannot specify the type of epidermal derivative.

The evolution of the skin, in particular its specializations, has apparently involved changes in the ability of the dermis to induce and the epidermis to respond, as well as in the interactions between them. From experimental embryology on living forms, we realize that if we speak only of the evolution of epidermal structures, we are neglecting the role of the dermis in this process. Although the dermis may not actually contribute cells to specialized skin derivatives, it is indispensable for their normal formation. Remove the dermis from the pulp cavity of a placoid scale, and the enamel and dentin fail to form normally. If the dermis is missing beneath the pulp cavity of a forming tooth, the tooth enamel forms incompletely. And the reverse holds. Remove the epidermis, and the dermis alone is unable to form a placoid scale or vertebrate tooth properly. Interaction of epidermis and dermis is necessary to produce a normal skin derivative.

Experimental embryology has extended this insight into evolutionary events. Modern birds, of course, lack teeth. The young chick breaking out of the eggshell uses what is termed an "egg tooth." In reality, this is not a tooth at all but a projection on the cornified bill. That is why recent research has been surprising. Koller and Fisher took tooth-inducing dermis from a mouse jaw, placed it under a bird's beak, and allowed the pair to differentiate. In several successful experiments, rudimentary teeth appeared. Chick epidermis had been induced by mouse dermis to form teeth! Although teeth do not form in modern birds, the epidermis of birds has not entirely lost its tooth-forming potential. This latent potential in birds is not expressed because the inductive interaction between bird dermis and epidermis has been lost. The epidermis is present and the dermis is present, but in birds their interaction has changed.

Perhaps the focus of evolutionary events in the integument has been as much on this interaction as on the layers themselves. Obviously, interactions do not fossilize and they are hard to characterize structurally. It is little wonder that controversies about homology exist. If we think of the epidermis, the dermis, and their interaction as an evolving unit, then their specialized products (hair, feathers, and reptilian scales) are broadly homologous. Shark scales, vertebrate teeth, and bony-fish scales can be seen as products of this interacting epidermal-dermal system; therefore, they are homologous integumentary structures.

WEB LINK

Visit the text website at *www.mhhe.com/kardong7e* for additional study aids including selected references, functional laboratories, and web links to chapter topics.

CHAPTER 7

Skeletal System: The Skull

INTRODUCTION

CHONDROCRANIUM
Embryology

SPLANCHNOCRANIUM
Embryology
Origin of Jaws
Types of Jaw Attachments

DERMATOCRANIUM
Parts of the Dermatocranium
Dermal Bone Series

OVERVIEW OF SKULL MORPHOLOGY
Braincase
Jaws
Hyoid Apparatus

CRANIAL KINESIS

PHYLOGENY OF THE SKULL
Agnathans
Early Vertebrates
Ostracoderms
Cyclostomes

Gnathostomes
Fishes
Early Tetrapods
Primitive Amniotes
Modern Reptiles
Birds
Synapsids

OVERVIEW OF SKULL FUNCTION
AND DESIGN
Prey Capture
Feeding in Water
Feeding in Air
Swallowing

OVERVIEW
Cranial Neural Crest
Emergence of Mammals
Evolutionary Modifications of Immature Forms:
Akinesis in Mammals
Composite Skull

The skeleton gives the vertebrate body shape, supports its weight, offers a system of levers that together with muscles produces movement, and protects soft parts such as nerves, blood vessels, and other viscera. Because it is hard, bits of the skeleton often survive fossilization better than does soft tissue anatomy, so our most direct contact with long-extinct animals is often through their skeletons. The story of vertebrate function and evolution is written in the architecture of the skeleton.

The skeletal system is composed of an exoskeleton and an endoskeleton (figure 7.1a). The **exoskeleton** is formed from or within the integument, the dermis giving rise to bone and the epidermis to keratin. The **endoskeleton** forms deep within the body from mesoderm and other sources, not directly from the integument. Tissues contributing to the endoskeleton include fibrous connective tissue, bone, and cartilage.

During the course of vertebrate evolution, most bones of the exoskeleton stay within the integument and protect surface structures. Dermal armor of ostracoderms and bony scales of fishes are examples. Other bones have sunk inward, merging with deeper bones and cartilaginous elements of the endoskeleton to form composite structures. As a practical matter, this makes it difficult to examine the exoskeleton and the endoskeleton separately. Parts of one are often found in company with the other. Instead, we select composite structural units and follow their evolution. This way of dividing the skeleton for study gives us two units: the skull, or **cranial skeleton,** and the **postcranial skeleton** (figure 7.1b). The postcranial skeleton includes the vertebral column, limbs, girdles, and associated structures, such as ribs and shells. In chapters 8 and 9, we examine the postcranial skeleton. Our discussion of the skeleton begins with the skull.

FIGURE 7.1 **Organization of skeletal tissues in vertebrates.** Components of the skeletal system function together as a unit but, as a convenience, they can be divided into manageable parts for closer analysis. (a) As a protective and supportive system, the skeleton can be divided into structures on the outside (exoskeleton) and inside (endoskeleton) of the body. (b) On the basis of position, the skeleton can be treated as two separate components, the cranial skeleton (skull) and the postcranial skeleton. The postcranial skeleton includes the axial and appendicular skeletons.

Introduction

Although merged into a harmonious unit, the vertebrate skull, or **cranium,** is actually a composite structure formed of three distinct parts. Each part of the skull arises from a separate phylogenetic source. The most ancient part is the **splanchnocranium (visceral cranium),** which first arose to support pharyngeal slits in protochordates (figure 7.2a). The second part, the **chondrocranium,** underlies and supports the brain and is formed of endochondral bone or of cartilage, or both (figure 7.2b). The third part of the skull is the **dermatocranium,** a contribution that in later vertebrates forms most of the outer casing of the skull. As its name suggests, the dermatocranium is composed of dermal bones (figure 7.2c).

Endochondral and dermal bone (p. 182)

In addition to these formal components, two general terms apply to parts of the cranium. The **braincase** is a collective term that refers to the fused cranial components immediately surrounding and encasing the brain. Structures of the dermatocranium, the chondrocranium, and even the splanchnocranium can make up the braincase, depending on the species. The **neurocranium** is used as an equivalent term for the chondrocranium by some morphologists. Others expand the term to include the chondrocranium along with fused or attached sensory capsules—the supportive nasal, optic, and otic capsules. Still others consider the neurocranium to be only the ossified parts of the chondrocranium. Be prepared for slightly different meanings in the literature. Although we use the term *neurocranium* sparingly, neurocranium is understood to include the braincase (ossified or not) plus associated sensory capsules.

Chondrocranium

Elements of the chondrocranium appear to lie in series with the bases of the vertebrae. This arrangement inspired several morphologists of the nineteenth century to propose that the primitive vertebral column initially extended into the head to produce the skull. By selective enlargement and fusion, these intruding vertebral elements were seen as the evolutionary source of the chondrocranium. Consequently, the idea grew that the head was organized on a segmental plan like the vertebral column that produced it. Today this view is not held as confidently, although many allow that the occipital arch forming the back wall of the skull may represent several ancient vertebral segments that now contribute to the posterior wall of the chondrocranium (table 7.1).

In elasmobranchs, the expanded and enveloping chondrocranium supports and protects the brain within. However, in most vertebrates, the chondrocranium is primarily an embryonic structure serving as a scaffold for the developing brain and as a support for the sensory capsules.

Embryology

Although the embryonic formation of the chondrocranium is understood, details may differ considerably, especially as neural crest contributions may vary widely among species. Further, differences in research tools (gene markers, vital dyes, tissue transplants) can also lead to different interpretations. Generally, condensations of head mesenchyme form elongate cartilages next to the notochord. The anterior pair are the **trabeculae,** the posterior pair the **parachordals,** and in some vertebrates, a pair of **polar cartilages** lies between them (figure 7.3a). Behind the parachordals, several **occipital cartilages** usually appear as well. In addition to these cartilages, the sensory capsules associated with the nose, eyes, and ears develop supporting cartilages: **nasal, optic,** and **otic capsules,** respectively. Two types of embryonic cells differentiate to form the chondrocranium. Neural crest cells contribute to the nasal capsule, trabeculae (possibly only the anterior part), and perhaps to part of the otic capsule (figure 7.4a). Mesenchyme of mesodermal origin contributes to the rest of the chondrocranium (figure 7.4b). As development proceeds, these cartilages fuse. The region between the nasal capsules formed by the fusion of the anterior tips of the trabeculae is the **ethmoid plate.** The parachordals grow together across the midline to form the **basal plate** between the otic capsules. The occipitals grow upward and around the nerve cord to form the **occipital arch** (figure 7.3b). Collectively, all of these expanded and fused cartilages constitute the chondrocranium.

In elasmobranchs, the chondrocranium does not ossify. Instead the cartilage grows still farther upward and

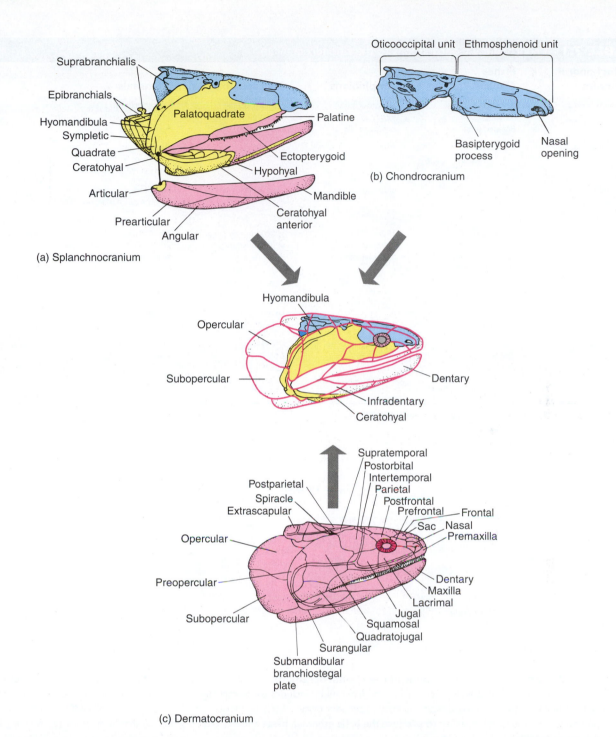

FIGURE 7.2 Composite skull. The skull is a mosaic composed of three primary contributing parts: the chondrocranium, the splanchnocranium, and the dermatocranium. Each has a separate evolutionary background. The skull of *Eusthenopteron,* a Devonian rhipidistian fish, illustrates how parts of all three phylogenetic sources contribute to the unit. (a) The splanchnocranium (yellow) arose first and is shown in association with the chondrocranium (blue) and parts of the dermatocranium (red). The right mandible is lowered from its point of articulation better to reveal deeper bones. (b) The chondrocranium in *Eusthenopteron* is formed by the union between the anterior ethmosphenoid and the posterior oticooccipital units. (c) The superficial wall of bones composes the dermatocranium. The central figure depicts the relative position of each contributing set of bones brought together in the composite skull. (Sac: nasal series)

TABLE 7.1

TABLE 7.1   Endochondral Contributions to the Chondrocranium

Endochondral Structure	Fishes (Teleost)	Amphibians	Reptiles/Birds	Mammals	
Occipital bones	Supraoccipital Exoccipital Basioccipital	Supraoccipital Exoccipital Basioccipital	Supraoccipital Exoccipital Basioccipital	Supraoccipital Exoccipital Basioccipital	} Occipital bone
Mesethmoid bone	Mesethmoid[a] (internasal)	Absent	Absent	Mesethmoid (absent in primitive mammals, ungulates)	} Ethmoid
Ethmoid region	Ossified	Unossified	Unossified	Turbinals (ethmo-, naso-, maxillo-)	
Sphenoid bones *Sphenethmoid* *Orbitosphenoid* *Basisphenoid* *Pleurosphenoid*	*Sphenethmoid* *Orbitosphenoid* *[Basisphenoid]*[b] *Pleurosphenoid*	*Sphenethmoid* *Orbitosphenoid* *Basisphenoid* ?	*Sphenethmoid* *Orbitosphenoid* *Basisphenoid* *Pleurosphenoid* (crocodilians, amphisbaenians)	Presphenoid *Orbitosphenoid* *Basisphenoid* Absent	} Sphenoid[c]
Laterosphenoid			Laterosphenoid (snakes)	Absent	
Otic capsule *Periotic*	{ Prootic *Epiotic* *Sphenotic*	Prootic *Opisthotic*	Prootic *Opisthotic* *Epiotic* (absent in birds)	Petrosal with *mastoid process*	

[a]*This bone is of dermal origin, so it is not strictly homologous to tetrapod mesethmoid.*

[b]*This bone is usually absent or reduced in some fishes.*

[c]*Alisphenoid from the splanchnocranium contributes.*

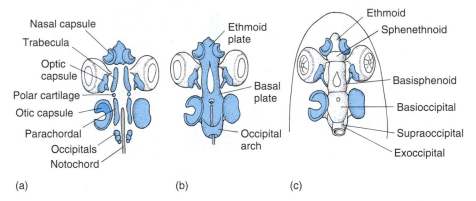

FIGURE 7.3   Embryonic development of the chondrocranium. Cartilage (blue) appears first but in most vertebrates is replaced by bone (white) later in development. The chondrocranium includes these cartilaginous elements that form the base and back of the skull together with the supportive capsules around sensory organs. Early condensation of mesenchymal cells differentiates into cartilage (a) that grows and fuses together to produce the basic ethmoid, basal, and occipital regions (b) that later ossify (c), forming basic bones and sensory capsules.

After deBeer.

over the brain to complete the protective walls and roof of the braincase. In most other vertebrates, the chondrocranium becomes partly or entirely ossified (figure 7.3c).

Splanchnocranium

The splanchnocranium is an ancient chordate structure. In amphioxus, the splanchnocranium, or at least its forerunner, is associated with the filter-feeding surfaces.

Among vertebrates, the splanchnocranium generally supports the gills and offers attachment for the respiratory muscles. Elements of the splanchnocranium contribute to the jaws and hyoid apparatus of gnathostomes.

Embryology

Within vertebrates, the splanchnocranium arises embryologically from neural crest cells, *not* from lateral plate mesoderm like the smooth muscle in the walls of the

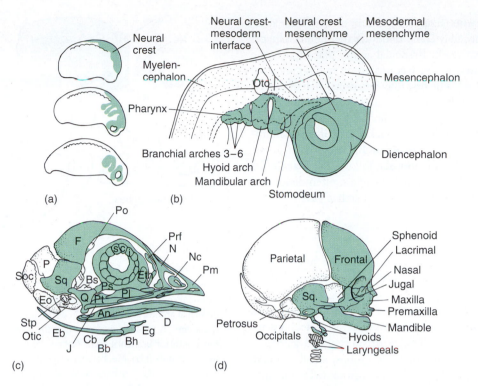

FIGURE 7.4 **Neural crest contributions to the skull.** (a) Salamander embryo illustrating the sequential spread of neural crest cells. During early embryonic development, neural crest cells contribute to the head mesenchyme, which is called the ectomesoderm because of its neural crest origin. (b) Also contributing to the head mesenchyme are cells of mesodermal origin, the mesodermal mesenchyme. The position of the mesodermal (stippled) and the neural crest (shaded) mesenchyme, and the approximate interface between them, are indicated in the chick embryo. Skull of a chick (c) and a human fetus (d) show bones or portions of bones derived from neural crest cells (shaded). Abbreviations: angular (An), basibranchial (Bb), basihyal (Bh), basisphenoid (Bs), ceratobranchial (Cb), dentary (D), epibranchial (Eb), entoglossum (Eg), exoccipital (Eo), ethmoid (Eth), frontal (F), jugal (J), nasal (N), cartilage nasal capsule (Nc), parietal (P), palatine (Pl), premaxilla (Pm), postorbital (Po), prefrontal (Prf), parasphenoid (Ps), pterygoid (Pt), quadrate (Q), scleral ossicle (Sci), supraoccipital (Soc), squamosal (Sq), stapes (Stp).

After Noden, Couly et al; LeDourain and Kalcheim.

digestive tract. This common embryonic origin unites the elements of the splanchnocranium into a community of elements. In protochordates, neural crest cells debut in urochordates, where they migrate from the neural tube to the body wall, and there differentiate into pigment cells, their only currently known derivative. Pharyngeal bars in protochordates arise from mesoderm and form the unjointed branchial basket, the phylogenetic predecessor of the vertebrate splanchnocranium. However, in vertebrates the neural crest gives rise to a great variety of adult structures, including the jaws and gill arches. Cells of the neural crest depart from the sides of the neural tube and move into the walls of the pharynx between successive pharyngeal slits to differentiate into the respective pharyngeal arches. Pharyngeal arches of aquatic vertebrates usually are associated with their respiratory gill system. Because of this association, they are referred to as **branchial arches,** or **gill arches.**

Each arch can be composed of a series of up to five articulated elements per side, beginning with the **pharyngobranchial** element dorsally and then, in descending order, the **epibranchial, ceratobranchial, hypobranchial,** and

FIGURE 7.5 **Primitive splanchnocranium.** Seven arches are shown. Up to five elements compose an arch on each side, beginning with the pharyngobranchial dorsally and in sequence to the basibranchials most ventrally. The first two complete arches are named: mandibular arch for the first and hyoid arch for the second that supports it. The characteristic five-arch elements are reduced to just two in the mandibular arch: the palatoquadrate and Meckel's cartilage. The large hyomandibula, derived from an epibranchial element, is the most prominent component of the next arch, the hyoid arch. Behind the hyoid arch are variable numbers of branchial arches I, II, and so on. Labial cartilages are not included.

basibranchial elements (figure 7.5). One or more of these anterior branchial arches may come to border the mouth, support soft tissue, and bear teeth. Branchial arches that support the mouth are called **jaws,** and each contributing arch is numbered sequentially or named. The first fully functional arch of the jaw is the **mandibular arch,** the largest and most anterior of the modified series of arches. The mandibular arch is composed of the **palatoquadrate** dorsally and **Meckel's cartilage** (mandibular cartilage) ventrally. The **hyoid arch,** whose most prominent element is the **hyomandibula,** follows the mandibular arch. A varying number of branchial arches, often designated with Roman numerals, follow the hyoid arch (figure 7.5).

Origin of Jaws

In agnathans, the mouth is neither defined nor supported by jaws. Instead, the splanchnocranium supports the roof of the pharynx and lateral pharyngeal slits. Lacking jaws, ostracoderms would have been restricted to a diet of small, particulate food. The ciliary-mucous feeding surfaces of protochordates probably continued to play a large part in the food-gathering technique of ostracoderms. In some groups, small teethlike structures, derived from surface scales, surrounded the mouth. Perhaps ostracoderms used these rough "teeth" to scrape rock surfaces and dislodge encrusted algae or other organisms. As these food particles became suspended in water, ostracoderms drew them into their mouths with the incurrent flow of water. The mucus-lined walls of the pharynx collected these dislodged food particles from the passing stream.

Jaws appear first in acanthodian and placoderm fishes that used them as food traps to grab whole prey or take bites from large prey. Within some groups, jaws also served as crushing or chewing devices to process food in the mouth. With the advent of jaws, these fishes became more free-ranging predators of open waters.

Jaws arose from one of the anterior pair of gill arches. Evidence supporting this comes from several sources. First, the embryology of sharks suggests that jaws and branchial arches develop similarly in series (figure 7.6) and both arise from neural crest. The spiracle appears to have once been a full-sized gill slit, but in modern sharks it is crowded and much reduced by the enlarged hyoid arch next in series. Furthermore, nerves and blood vessels are distributed in a pattern similar to branchial arches and jaws. Finally, the musculature of the jaws appears to be transformed and modified from branchial arch musculature.

So it seems reasonable to conclude that branchial arches phylogenetically gave rise to jaws. But the specifics remain controversial. For example, we are not sure whether jaws represent derivatives of the first, second, third, or even fourth branchial arches of primitive ancestors. Derivation of the mandibular arch also excites some controversy. The **serial theory** is the simplest view and holds that the first or perhaps second ancient branchial arch gave rise exclusively to the mandibular arch, the next branchial arch exclusively to the hyoid arch, and the rest of the arches to the branchial arches of gnathostomes (figure 7.7a).

FIGURE 7.6 **Shark embryo, the dogfish *Scyllium*** Jaws appear to be in series with the branchial arches. The mandibular arch is first, followed by the hyoid and then several branchial arches. Such a position of the jaws, in series with the arches, is taken as evidence that the jaws derive from the most anterior branchial arch. Abbreviations: ceratohyal (Ch), hyomandibula (Hy), Meckel's cartilage (Mk), neural arch (Ne), occipital arch (Oa), orbital cartilage (Oc), polar cartilage (Pc), palatoquadrate (Pq), trabecula (Tr). Labial cartilages are not included.

After deBeer.

Erik Jarvik, a Swedish paleontologist, proposed the **composite theory,** a more complex view based on his examination of fossil fish skulls and embryology of living forms (figure 7.7b). He hypothesized that ten branchial arches were present in primitive species, the first and following arches being named terminal, premandibular, mandibular, hyoid, and six branchial arches. Rather than the "one arch, one mandible" view, he envisioned a complex series of losses or fusions between selective parts of several arches that came together to produce the single composite mandible. According to his theory, the mandibular arch of gnathostomes is formed by fusion of parts of the premandibular arch and parts of the mandibular arch of jawless ancestors. The palatoquadrate forms from the fusion of the epibranchial of the premandibular arch with the epibranchial and one pharyngobranchial of the mandibular arch. Meckel's cartilage arises from the expanded ceratobranchial element. Next, the hyoid arch arises phylogenetically from the epibranchial, ceratobranchial, and hypobranchial elements of the third primitive gill arch. The remaining branchial arches persist in serial order. The other elements of the primitive arches are lost or fused to the neurocranium.

Descriptive embryology provides much of the evidence put forth in these theories. However, descriptive embryology alone cannot trace arch components from embryo to adult structures with complete confidence. We can look forward to the use of more modern techniques to help settle this. For example, populations of cells can be marked with chemical or cellular markers early in embryonic development and followed to eventual sites of residence in the adult. These markers

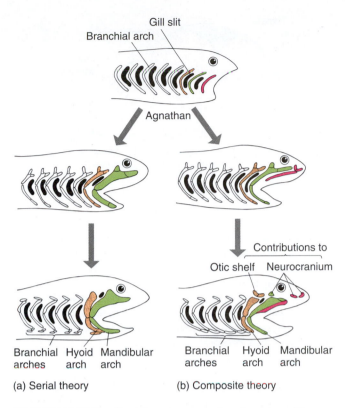

Gill slit

Branchial arch

Agnathan

Contributions to

Otic shelf   Neurocranium

Branchial   Hyoid   Mandibular
arches   arch   arch

Branchial   Hyoid   Mandibular
arches   arch   arch

(a) Serial theory

(b) Composite theory

FIGURE 7.7   Serial and composite theories of jaw development. (a) The serial theory holds that jaws arise completely from one of the anterior branchial arches. Elements may be lost within it, but other elements from other arches do not contribute. (b) In the composite theory, the mandibular arch is formed from elements of several adjacent arches that also contribute to the neurocranium.

would permit us to detect the contributions of gill arches to jaws or chondrocranium.

Such work is at the moment under way using molecular and genetic probes, but these so far have produced unsettled and somewhat contradictory results. For example, in the early work on a species of jawless lamprey, the first results identified a *Hox* gene expressed in the mandibular arch of lamprey embryos, but this same *Hox* gene was not expressed in the mandibular arch of jawed vertebrates. This suggested that this *Hox* gene suppressed jaw development (lamprey) and its absence (gnathostomes) removed this inhibition to facilitate evolution of jaws. However, more recent work on other species of lampreys found no such *Hox* expression. This suggested instead that this *Hox* gene is not a key component of jaw evolution.

More consistent results have been found with other important genes. The emerging, but still tentative, view of jaw evolution based on this molecular evidence is that the mandibular arch of jawless fishes became divided—the dorsal part contributing to the neurocranium and the remainder evolving into the mandible proper of gnathostomes. This is closest to the composite theory but differs in detail, which remains to be settled. Nevertheless, all these views share the same basic consensus: namely, that in general the vertebrate jaws are derivatives of ancient gill arches (table 7.2).

Types of Jaw Attachments

Because of the mandible's prominence, evolution of the jaws is often traced through how the mandible is attached (i.e., its **suspensorium**) to the skull (figure 7.8). Agnathans represent the earliest **paleostylic** stage in which none of the arches

Arch	Sharks	Teleosts	Amphibians	Reptiles/Birds	Mammals
I	Meckel's cartilage	Articular[a]	Articular	Articular	Malleus[b]
	Palatoquadrate	Quadrate Epipterygoid	Quadrate Epipterygoid	Quadrate Epipterygoid	Incus[b] Alisphenoid
II	Hyomandibula	Hyomandibula Symplectic Interhyal	Stapes Extracolumella	Stapes Extracolumella	Stapes[b]
	Ceratohyal	Ceratohyal Hypohyal	Ceratohyal Hypohyal	Ceratohyal	Anterior horn hyoid
	Basihyal	Basihyal		Body of hyoid	Body of hyoid
III	Pharyngobranchial Epibranchial Ceratobranchial Hypobranchial	Pharyngobranchial Epibranchial Ceratobranchial Hypobranchial	Body of hyoid	Second horn of hyoid	Second horn of hyoid
IV	Branchial arch		Last horn and body of hyoid Laryngeal cartilages (?)	Last horn and body of hyoid Laryngeal cartilages (?)	Thyroid cartilages (?)
V	Branchial arch	Branchial arch	Laryngeal cartilages (?)	Laryngeal cartilages (?)	Laryngeal cartilages
VI	Branchial arch	Branchial arch	Not present	Not present	Not present
VII	Branchial arch	Branchial arch			

TABLE 7.2   Derivatives of Branchial Arches in Sharks, Teleosts, and Tetrapods

[a]Sometimes dermal bone contributes.

[b]See figure 7.53 and related text for discussion of middle ear evolution.

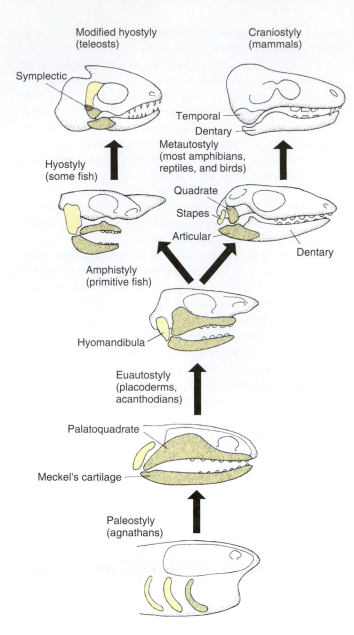

FIGURE 7.8 **Jaw suspension.** The points at which the jaws attach to the rest of the skull define the type of jaw suspension. Note the mandibular arches (yellow, crosshatched areas) and hyoid arches (yellow areas). The dermal bone (white areas) of the lower jaw is the dentary.

FIGURE 7.9 **Skull of armadillo embryo.** During embryonic formation of the three middle ear ossicles (incus, stapes, malleus), the incus and stapes arise from the mandibular arch, testifying to the phylogenetic derivation of these bones from this arch. The dermal dentary is cut away to reveal Meckel's cartilage, which ossifies at its posterior end to form the malleus. (Blue, chondrocranium contribution; yellow, splanchnocranium contribution; red, dermatocranium.)

After Goodrich.

attaches themselves directly to the skull. The earliest jawed condition is **euautostylic,** found in placoderms and acanthodians. The mandibular arch is suspended from the skull by itself (hence, "auto"), without help from the hyoid arch. In early sharks, some osteichthyans, and rhipistians, jaw suspension is **amphistylic;** that is, the jaws are attached to the braincase through two primary articulations, anteriorly by a ligament connecting the palatoquadrate to the skull and posteriorly by the hyomandibula. Many, perhaps most, modern sharks exhibit a variation of amphistylic jaw suspension. In most modern bony fishes, jaw suspension is **hyostylic** because the mandibular arch is attached to the braincase primarily through

the hyomandibula. Often a new element, the **symplectic bone,** aids in jaw suspension. The visceral cranium remains cartilaginous in elasmobranchs, but within bony fishes and later tetrapods, ossification centers appear, forming distinctive bony contributions to the skull. In most amphibians, reptiles, and birds, jaw suspension is **metautostylic.** Jaws are attached to the braincase directly through the quadrate, a bone formed in the posterior part of the palatoquadrate (figure 7.8). The hyomandibula plays no part in supporting the jaws; instead, it gives rise to the slender **columella** or **stapes,** involved in hearing. Other elements of the second arch and parts of the third contribute to the **hyoid** or **hyoid apparatus** that supports the tongue and the floor of the mouth. In mammals, jaw suspension is **craniostylic.** The entire upper jaw is incorporated into the braincase, but the lower jaw is suspended from the dermal **squamosal** bone of the braincase. The lower jaw of mammals consists entirely of the **dentary** bone, which is also of dermal origin. The palatoquadrate and Meckel's cartilages still develop, but they remain cartilaginous except at their posterior ends, which give rise to the **incus** and **malleus** of the middle ear, respectively (figure 7.9). Thus, in mammals, the splanchnocranium does not contribute to the adult jaws or to their suspension. Instead, the splanchnocranium forms the hyoid apparatus, styloid, and three middle ear bones: malleus, incus, and stapes. Through Meckel's cartilage, the splanchnocranium contributes the scaffolding around which the dentary bone forms.

Dermatocranium

Dermal bones that contribute to the skull belong to the dermatocranium. Phylogenetically, these bones arise from the bony armor of the integument of early fishes and sink inward to become applied to the chondrocranium and splanchnocranium. Bony elements of the armor also become associated with the endochondral elements of the pectoral girdle to give rise to the dermal components of this girdle.

Dermal girdle (p. 336)

Dermal bones first become associated with the skull in ostracoderms. In later groups, additional dermal bones of the overlying integument also contribute. The dermatocranium forms the sides and roof of the skull to complete the protective bony case around the brain; it forms most of the bony lining of the roof of the mouth, and encases much of the splanchnocranium. Teeth that arise within the mouth usually rest on dermal bones.

As the name suggests, bones of the dermatocranium arise directly from mesenchymal and ectomesenchymal tissues of the dermis. Through the process of intramembranous ossification, these tissues form dermatocranial bones.

Parts of the Dermatocranium

Dermal elements in modern fishes and living amphibians have tended to be lost or fused so that the number of bones present is reduced and the skull simplified in comparison to their ancestors. In amniotes, bones of the dermatocranium predominate, forming most of the braincase and lower jaw. The dermal skull may contain a considerable series of bones joined firmly at sutures in order to box in the brain and other skull elements. As a convenience, we can group these series and recognize the most common bones in each (figure 7.10; table 7.3).

Dermal Bone Series

Facial Series   The facial series encircles the external naris and collectively forms the snout. The maxilla and premaxilla (incisive) define the margins of the snout and usually bear teeth. The nasal lies medial to the naris. The septomaxilla is a small dermal bone of the facial series that is often absent. When present, it is usually sunken below the surface bones and aids in forming the nasal cavity.

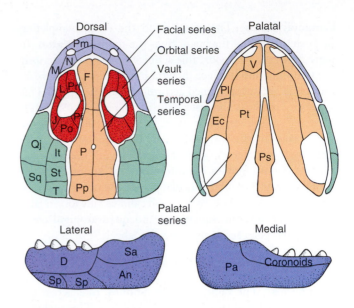

FIGURE 7.10   Major bones of the dermatocranium. Sets of dermal bones form the facial series surrounding the nostril. The orbital series encircles the eye, and the temporal series composes the lateral wall behind the eye. The vault series, the roofing bones, runs across the top of the skull above the brain. Covering the top of the mouth is the palatal series of bones. Meckel's cartilage (not shown) is encased in the mandibular series of the lower jaw. Abbreviations: angular (An), dentary (D), ectopterygoid (Ec), frontal (F), intertemporal (It), jugal (J), lacrimal (L), maxilla (M), nasal (N), parietal (P), prearticular (Pa), palatine (Pl), premaxilla (Pm), postorbital (Po), postparietal (Pp), prefrontal (Prf), parasphenoid (Ps), pterygoid (Pt), quadratojugal (Qj), surangular (Sa), splenial (Sp), squamosal (Sq), supratemporal (St), tabular (T), vomer (V).

Orbital Series   The dermal bones encircle the eye to define the orbit superficially. The **lacrimal** takes its name from the nasolacrimal (tear) duct of tetrapods that passes through or near this bone. The **prefrontal, postfrontal,** and **postorbital** continue the ring of bones above and behind the orbit. The **jugal** usually completes the lower rim of the orbit. Not to be confused with these dermal bones are the **scleral ossicles** of neural crest origin that, when present, reside within the orbit defined by the ring of dermal bones.

TABLE 7.3	Major Dermal Bones of the Skull				
BRAINCASE					**MANDIBLE**
Facial Series	**Orbital Series**	**Temporal Series**	**Vault Series**	**Palatal Series**	**Mandibular Series**
Premaxilla	Lacrimal	Intertemporal	Frontal	Vomer	Lateral bones:
Maxilla	Prefrontal	Supratemporal	Parietal	Palatine	Dentary (teeth)
Nasals (septomaxilla)	Postfrontal Postorbital	Tabular	Postparietal	Ectopterygoid	Splenials (2)
	Jugal	Squamosal Quadratojugal		Pterygoid Parasphenoid (unpaired)	Angular Surangular
					Medial bones: Prearticular Coronoids

Temporal Series The area behind the orbit, completing the posterior wall of the braincase is the **temporal region.** In many primitive tetrapods, this series is indented posteriorly by a **temporal notch.** Once thought in life to suspend an eardrum, this notch was named accordingly an otic notch. This now seems unlikely, and instead the notch perhaps accommodated a spiracle, a respiratory tube. Openings called **fenestrae** (sing., *fenestra*) arise within this temporal region of the outer braincase in many tetrapods in association with the jaw musculature. A row of bones, the **intertemporal, supratemporal,** and **tabular,** makes up the medial part of the temporal series. This row is reduced in early tetrapods and usually lost in later species. Laterally, the **squamosal** and **quadratojugal** complete the temporal series and form the "cheek."

Temporal bone (p. 280)

Vault Series The vault, or **roofing bones,** runs across the top of the skull and covers the brain beneath. This includes the **frontal** anteriorly and the **postparietal** (interparietal) posteriorly. Between them is the large **parietal,** occupying the center of the roof and defining the small **parietal foramen** if it is present. The parietal foramen is a tiny skylight in the skull roof that exposes the pineal gland, an endocrine gland, to direct sunlight.

Palatal Series The dermal bones of the **primary palate** cover much of the roof of the mouth. The largest and most medial is the **pterygoid.** Lateral to it are the vomer, palatine, and **ectopterygoid.** Teeth may be present on any or all four of these palatal bones. In fishes and lower tetrapods, there also is an unpaired medial dermal bone, the **parasphenoid.**

Mandibular Series Meckel's cartilage is usually encased in dermal bones of the mandibular series. Laterally, the wall of this series includes the tooth-bearing **dentary** and one or two **splenials,** the **angular** at the posterior corner of the mandible and the **surangular** above. Many of these bones wrap around the medial side of the mandible and meet the **prearticular** and one or several **coronoids** to complete the medial mandibular wall. Left and right mandibles usually meet anteriorly at the midline in a **mandibular symphysis.** If firm, the mandibular symphysis unites them into an arched unit. Most notably in snakes, the mandibular symphysis is composed of soft tissues, permitting independent movement of each mandible.

Overview of Skull Morphology

Braincase

In chondrichthyan fishes, the braincase is an elaborate cartilaginous case around the brain. The dermatocranium is absent, reflecting the elimination of almost all bone from the skeleton. However, in most bony fishes and tetrapods, the braincase is extensively ossified with contributions from several sources. For descriptive purposes, it is useful to think of the braincase as a box with a platform of endoskeletal

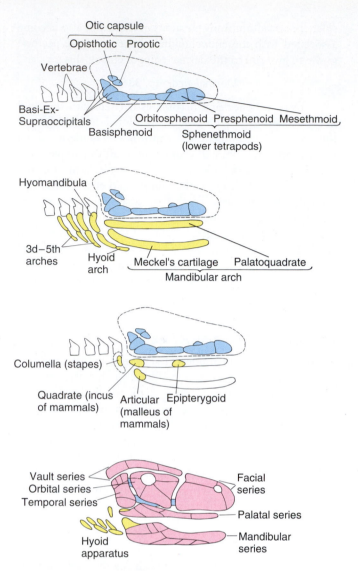

FIGURE 7.11 **Contributions to the skull.** The chondrocranium (blue) establishes a supportive platform that is joined by contributions from the splanchnocranium (yellow), in particular the epipterygoid. Other parts of the splanchnocranium give rise to the articular, quadrate, and hyomandibula, as well as to the hyoid apparatus. The dermatocranium (pink) encases most of the chondrocranium together with contributions from the splanchnocranium.

elements supporting the brain, all encased in exoskeletal bones (figure 7.11). The endoskeletal platform is assembled from a series of **sphenoid** bones. The **occipital** bones, which apparently are derived from anterior vertebrae, form the end of this sphenoid platform. These occipital bones, up to four in number **(basioccipital, supraoccipital,** and paired **exoccipitals),** close the posterior wall of the braincase except for a large hole they define, the **foramen magnum,** through which the spinal cord runs. Articulation of the skull with the vertebral column is established through the **occipital condyle,** a single or double surface produced primarily within the basioccipital but with contributions from the exoccipitals in some species.

The otic capsule rests on the posterior part of the endoskeletal platform and encloses the sensory organs of the ear. The

The idea that the skull is derived from serial compacted vertebrae dates to the eighteenth century. The German naturalist and poet W. Goethe (1749–1832) was apparently the first to think of but not the first to publish this idea. Goethe gave us the word *morphology,* which meant to him the search for underlying meaning in organic design or form. Among his discoveries was the observation that plant flowers are modified stem petals compacted together. His venture into vertebrates and vertebrate skulls in particular occurred in 1790 while he was strolling in an old cemetery in Venice. He spied a dried ram's skull disintegrated at its bony sutures but held in sequence by the soil. The separated bones of the ram's skull seemed to be the foreshortened anterior vertebrae of the backbone, but Goethe did not publish this idea until about 1817. Public credit for this idea and for elaborating it goes to another German naturalist, L. Oken (1779–1851). In 1806, Oken was strolling in a forest and came upon a dried sheep skull. He was similarly struck by its serial homology with the vertebrae, and shortly thereafter published the idea (box figure 1a).

Next, the vertebral theory of skull origin fell into the hands of Richard Owen and became part of his much embellished theoretical view on animal archetypes (box figure 1b). Because of Owen's prominence in early nineteenth-century science, the idea of skull from vertebrae became a central issue within European scientific communities. One of the most persuasive dissenters from this view of a vertebral source for the skull was T. H. Huxley, who based his critique upon a detailed comparative study of vertebrate skulls and their development. This came to a head (no pun intended) in an invited lecture, the Croonian lecture of 1858, in which Huxley argued that the development of the skull showed that it was not composed of vertebrae. He suggested that the "skull was no more derived from vertebrae, than vertebrae are derived from

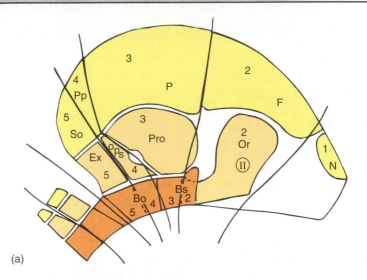

(a)

BOX FIGURE I   **Getting a head.** Derivation of the head from anterior vertebrae was proposed separately by Goethe and Oken. Richard Owen (nineteenth century) expanded on their ideas. (a) Ram's skull, showing how its presumed segmental pattern might be interpreted as being derived from parts of anterior vertebrae that expanded. (b) Richard Owen's elaborated view of head segmentation from vertebrae. Owen proposed that anterior vertebrae within the body moved forward to contribute to skeletal elements to the head. Therefore, Owen believed, the bony elements of the head could be homologized to the parts of a fundamental vertebral pattern. (c) Taking several vertebrates, he indicated how named parts of the skull might represent respective parts of this underlying vertebral pattern from which they derive. (d) T. H. Huxley proposed alternatively that, rather than being derived from vertebrae moved forward into the head, the components of the head were derived from a basic segmentation unrelated to the vertebral segmentation behind the skull. These basic segments (Roman numerals) are laid out across a generalized vertebrate skull to show the respective contributions to specific parts. Abbreviations: basioccipital (Bo), basisphenoid (Bs), exoccipital (Ex), frontal (F), nasal (N), opisthotic (Ops), orbitosphenoid (Or), parietal (P), postparietal (Pp), prootic (Pro), supraoccipital (So).

(a) After Jollie.

the skull." The skull, Huxley argued, arose in much the same way in most vertebrates, by fusing into a unit, not as a jointed series. Skull ossification showed no similarity with ossification of the following serial vertebrae. Although Huxley was probably right about this for most of the skull, the occipital region does ossify in a manner similar to vertebrae.

While disposing of the vertebral theory, Huxley substituted a segmental theory, tracing the segmentation to somites, not to vertebrae (box figure 1c). He took the otic capsule housing the ear as a "fixed" landmark and envisioned four somites (preotic) in front and five somites (postotic) behind it as segmental sources for segmental adult derivatives of the head.

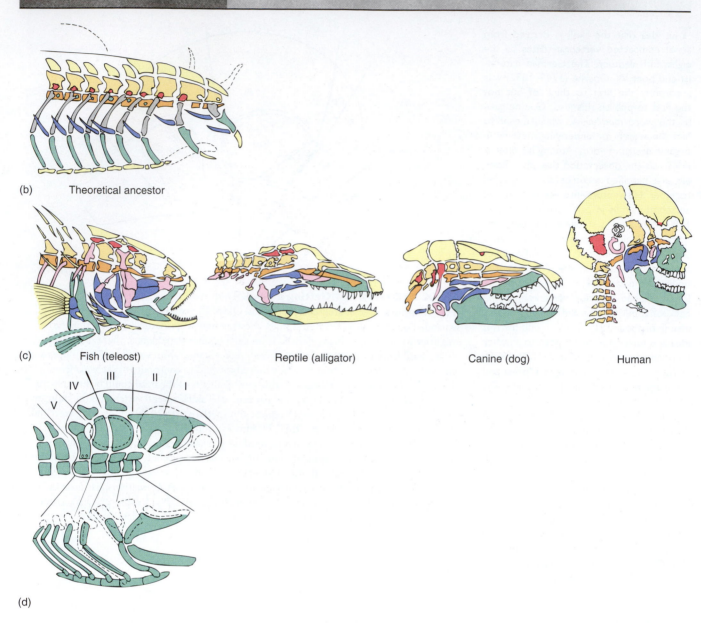

(b)   Theoretical ancestor

(c)   Fish (teleost)   Reptile (alligator)   Canine (dog)   Human

(d)

(b–c) After Reader; (d) after Jollie.

Today, some would argue that the head is a unique developmental system without any tie to the segmental somites (somitomeres). The neural crest cells that also contribute to parts of the skull show no segmental pattern in the head. However, at least in fishes, the branchial arches are segmental, as is the head paraxial mesoderm (somitomeres), and segmentation apparently can be carried into the accompanying neurocranium. Matched shading in the vertebrate series (box figure 1c) shows derivatives from parts of a theoretical ancestor (box figure 1b).

splanchnocranium contributes the **epipterygoid** (**alisphenoid** of mammals) to the endoskeletal platform and gives rise to one (columella/stapes) or more (malleus and incus of mammals) of the middle ear bones housed in the otic capsule.

In most vertebrates, these endoskeletal elements, along with the brain and sensory organs they support, are enclosed by the exoskeletal elements, derivatives of the dermis, to complete the braincase.

Jaws

The **upper jaw** consists of the endoskeletal palatoquadrate in primitive vertebrates. The palatoquadrate is fully functional in the jaws of chondrichthyans and primitive fishes, but in bony fishes and tetrapods, the palatoquadrate usually makes limited contributions to the skull through its two derivatives: the epipterygoid, which fuses to the neurocranium, and the **quadrate,** which suspends the lower jaw except in mammals. The dermal maxilla and premaxilla replace the palatoquadrate as the upper jaw.

The **lower jaw,** or **mandible,** consists only of Meckel's cartilage in chondrichthyans. In most fishes and tetrapods, Meckel's cartilage persists but is enclosed in exoskeletal bones of the dermatocranium, which also support teeth. Meckel's cartilage, encased in dermal bone, usually remains unossified, except in some tetrapods where its anterior end ossifies as the **mental** bone. In most fishes and tetrapods (except mammals), the posterior end of Meckel's cartilage can protrude from the exoskeletal case as an ossified **articular** bone.

In mammals, the lower jaw consists of a single bone, the dermal dentary. The anterior tooth-bearing part of the dentary is its **ramus.** Jaw-closing muscles are inserted on the **coronoid process,** an upward extension of the dentary. Posteriorly, the dentary forms the transversely expanded **mandibular condyle,** a rounded process that articulates with the **glenoid fossa,** a depression within the temporal bone of the braincase. Thus, in mammals, the mandibular condyle of the dentary replaces the articular bone as the surface of the lower jaw through which is established mandibular articulation with the braincase.

Hyoid Apparatus

The hyoid, or hyoid apparatus, is a ventral derivative of the splanchnocranium behind the jaws. In fishes, it supports the floor of the mouth. Elements of the hyoid apparatus are derived from the ventral parts of the hyoid arch and from parts of the first few branchial arches. In larval and paedomorphic amphibians, the branchial bars persist but form a reduced hyoid apparatus that supports the floor of the mouth and functional gills. In adults, the gills and the associated part of the hyoid apparatus are lost, although elements persist within the floor of the mouth usually to support the tongue. Typically, the hyoid apparatus includes a main body, the **corpus,** and extensions, the **cornua** ("horns"). In many mammals, including humans, the distal end of the hyoid horn fuses with the otic region of the braincase to form the **styloid process.**

Cranial Kinesis

Kinesis means movement. Cranial kinesis refers literally, then, to movement within the skull. But if left this general, the definition becomes too broad to provide a useful context in which to discuss skull function. Some authors restrict the term to skulls with a transverse, hingelike joint across the skull roof and a transverse, sliding basal joint in the roof of the mouth. But this restricted definition precludes most teleost fishes, despite their highly mobile skull elements. Here, we use cranial kinesis to mean movement between the upper jaw and the braincase about joints between them (figure 7.12a). Such **kinetic skulls** characterize most vertebrates. They are found in ancient fishes (rhipidistians and probably palaeoniscoids), bony fishes (especially teleosts), very early tetrapods, most reptiles (including most Mesozoic forms), birds, and early therapsid ancestors to mammals. Kinetic skulls are not present in modern amphibians, turtles, crocodiles, and mammals (with the possible exception

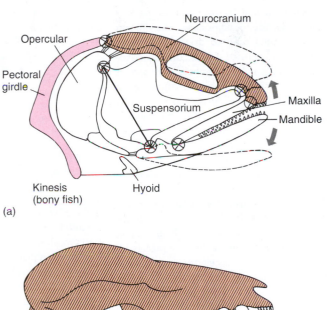

(a)

(b)

FIGURE 7.12 Mobility of skull bones. (a) The fish skull is kinetic. The upper jaw and other lateral skull bones rotate upon each other in a linked series, resulting in displacements of these bones (dashed outline) during feeding. Circles represent points of relative rotation between articulated elements. (b) The mammal skull is akinetic because no relative movement occurs between the upper jaw and the braincase. In fact, the upper jaw is incorporated into and fused with the braincase. There are no hinge joints through the braincase nor any movable linkages of lateral skull bones.

In hares, or "jackrabbits" (but not in distantly related pikas or in their fossil ancestors), a suture between regions of the fetal braincase remains open in the adult, forming an intracranial joint (box figure I). This intracranial joint runs along the sides and base of the adult braincase and hinges across the top via the postparietal. The joint permits relative motion between anterior and posterior parts of the braincase. It has been hypothesized that this joint helps absorb the impact forces sustained as the forelimbs strike the ground when a rabbit runs. Upon impact, mechanical deformation of the joint would absorb some kinetic energy as the hinge is strained. This deformation and absorption would reduce the shock sustained by the anterior part of the braincase. Additionally, the impact forces would tend to drive blood from intracranial sinuses into a complex association of venous channels and spaces within the skull. This would help dissipate these kinetic forces further as they acted against resistance offered by the walls of the blood vascular system.

The external ears (pinnae) of hares radiate heat generated during strenuous activity, but apparently only after locomotor exercise ceases. During locomotion, the ears are usually held erect by strong muscles at their bases. It has been hypothesized that these erect ears help reopen the intracranial joint as the hare pushes off on another leap to accelerate again, thus in a sense "resetting" this cranial mechanism and preparing it to act as a shock-absorbing device when the forelimbs again strike the ground (box figure Ic).

The functional significance of the intracranial joint is still debated. However, if such hypotheses are confirmed, this specialized joint in hares, together with their projecting ears, might also serve

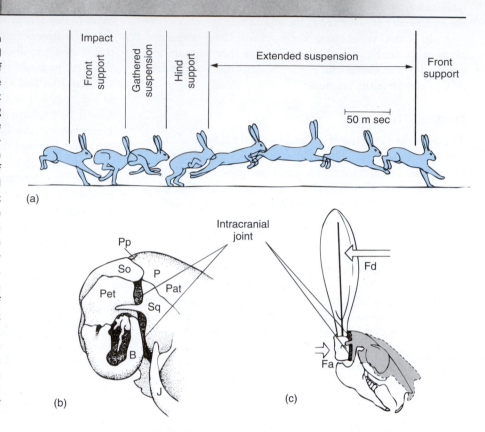

BOX FIGURE I   Possible cranial kinesis in hares. (a) Phases during a running stride are illustrated. Note that the forelimbs receive the initial impact upon landing. (b) Posterior regions of the skull of the jackrabbit *Lepus*. The intracranial joint extends along the sides of the skull between squamosal (Sq) and otic regions and then along the base of the skull. The interparietal bone forms the hinge across the top of the skull. (c) External ears held erect and attached to the posterior part of the skull may help to reposition the posterior part of the skull relative to the anterior part during the extended suspension phase of running. The presumed motion (slightly exaggerated) of the anterior braincase relative to the posterior braincase is indicated. Fa is the force vector due to acceleration resulting from thrust, and Fd is the force vector due to drag of the ears in the oncoming wind. Abbreviations: bulla (B), postparietal (Pp), jugal (J), parietal (P), petrosal (Pet), supraoccipital (So), squamosal (Sq). (Based on the research of D. Bramble.)

to reduce jarring of the eyes carried in the anterior braincase. Among mammals, rabbit kinesis represents an independent and apparently unique condition that did not evolve from therapsid kinesis. Further, it evolved not for its advantages during feeding but rather for its advantages during rapid locomotion.

---

of rabbits, see Box Essay 7.2). The widespread presence of cranial kinesis among vertebrates, but its essential absence among mammals, seems to create a problem for humans. Because we, like most other mammals, have **akinetic skulls** with no such movement between upper jaw and braincase, we tend to underestimate its importance (figure 7.12b).

Kinesis and akinesis each have advantages. Cranial kinesis provides a way to change the size and configuration of the mouth rapidly. In fishes and other vertebrates that feed in water, rapid kinesis creates a sudden reduction of pressure in the buccal cavity so that the animal can suck in a surprised prey. This method of prey capture, which takes advantage of a sudden vacuum to gulp in water carrying the intended food, is known as **suction feeding.** Cranial kinesis also allows tooth-bearing bones to move quickly into strategic positions during rapid feeding. Some teleost fishes, for instance, swing their anterior tooth-bearing bones forward at the last moment to reach out quickly at the intended prey. In many venomous snakes, linked bones along the sides of the skull can rotate forward. The venomous viper erects the maxillary bone bearing the fang and swings it from a folded position along its upper lip to the front of the mouth, where it can more easily deliver venom into prey. In many fishes and reptiles with kinetic skulls, teeth on the upper jaw can be reoriented with respect to the prey in order to assume a more favorable position during prey capture or to align crushing surfaces better during swallowing. Here, cranial kinesis brings near simultaneous contact and closure of both upper and lower jaws on the prey. Without this, the first jaw to make contact singly would tend to knock prey away, foiling capture. On the other hand, loss of kinesis in mammals leaves them with an akinetic skull, which allows infants to suckle easily. Juvenile and adult mammals can chew firmly with sets of specialized teeth that work accurately from a secure, akinetic skull.

Tooth structure and occlusion (p. 507)

Phylogeny of the Skull

The skull is a composite structure derived from the splanchnocranium, dermatocranium, and chondrocranium. Each component of the skull comes from a separate phylogenetic source. The subsequent course of skull evolution is complex, reflecting complex feeding styles. With a general view of skull structure now in mind, we can turn to a more specific look at the course of this evolution.

Agnathans

Early Vertebrates

The earliest vertebrates, *Haikouella* and *Haikouichthys*, are known from soft tissue impressions only, as mineralized tissues are absent. Each is reputed by some biologists to have protovertebrae or forerunners to vertebrae. However, any formed elements of a skull are lacking.

Ostracoderms

Osteostracans were one of the more common groups of ostracoderms. They possessed a head shield formed from a single piece of arched dermal bone, two close-set eyes dorsally placed with a single pineal opening between them, and a median nostril in front of the pineal opening. Along the sides of the head shield ran what are believed to be sensory fields, perhaps electrical field receptors or an early lateral line system sensitive to currents of water.

The broad, flattened head shield lowered the profile of ostracoderms, perhaps allowing them to hug the bottom surface, and their slight body suggest that they were benthic-dwelling fishes. The head shield formed the roof over the pharynx and held the sequential branchial arches that stretched like beams across the roof of the pharynx. Paired **gill lamellae** supported on **interbranchial septa** were stationed between these bars. Reconstructions of the head of *Hemicyclaspis*, a cephalospidomorph, indicate that a throat plate, presumably of cartilage, stretched across the floor of the pharynx (figure 7.13a). Muscle action is thought to have raised and lowered this plate to draw a stream of water first into the mouth, then over the gills, and finally out the branchial pores along the ventral side of the head. Suspended particles held in the stream of water could be captured within the pharynx before the water was expelled (figure 7.13b).

Anaspids were another group of early ostracoderms. Instead of a single bony shield, many small bony scales covered the head (figure 7.14a–c). The eyes were lateral, with a pineal opening between them and a single nostril in front. The body was streamlined, suggesting a slightly more active life than other ostracoderms enjoyed.

Heterostracans had flat to bullet-shaped heads composed of several fused bony plates (figure 7.15a). Their eyes were small and laterally placed, with a median pineal opening but no median nostril. Presumably, water flowed through the mouth, across the gill slits of the large pharynx, into a common tunnel, and out a single exit pore. The mouth of some heterostracans was rimmed with sharp, pointed oral scales that could have been used to dislodge food from rocks, allowing it to join the stream of water that entered the mouth (figure 7.15b).

Some scientists think that a few ostracoderms were predaceous, using the buccal cavity to gather up large prey, but because ostracoderms lacked strong jaws, feeding could not be based on powerful biting or crushing. The heavily plated heads and slight bodies of most ostracoderms argue for a relatively inactive lifestyle spent feeding on detritus and organic debris stirred up and drawn into the pharynx.

Cyclostomes

Hagfishes and lampreys are the only surviving agnathans. However, subsequent specializations have left these cyclostomes with anatomies quite unlike those of the early ostracoderms. Cyclostomes lack bone entirely and are specialized for parasitic or scavenging lives that depend on a rasping tongue to scrape up tissue for a meal. Lampreys have a single medial nostril and a pineal opening, together entering into the nasohypophyseal opening. Branchial pouches are present. The braincase is cartilaginous. Branchial arches, although present, form an unjointed branchial basket. Hagfishes have a median nostril but no external pineal opening.

Gnathostomes

All vertebrates, except agnathans, have jaws and form the embracing group gnathostomes ("jaw mouth"). Some biologists mark the advent of vertebrate jaws as one of the most important transitions in their evolution. Powerful closing muscles, derivatives of the branchial arch musculature, make the jaws strong biting or grasping devices. It is not surprising,

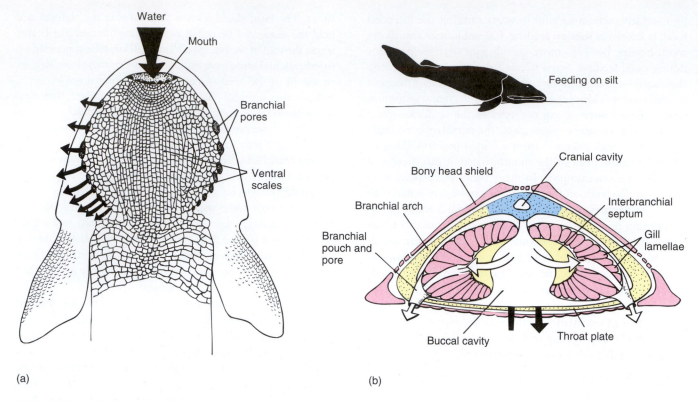

FIGURE 7.13 Ostracoderm *Hemicyclaspis* a cephalaspidomorph. (a) Ventral view showing branchial pores, the presumed sites of exit for water moving through the pharynx. (b) Cross section through the pharynx illustrating respiratory gill lamellae and supporting branchial arch. Presumably, the floor of the pharynx could be raised and lowered to actively draw water into the mouth and drive it out through the several branchial pores. The current crossed the respiratory gills before exiting. Suspended food may have been collected in the pharynx and then passed to the esophagus.

After Jarvik.

then, that with the advent of jaws, gnathostomes experience a dietary shift away from suspension feeding of the ostracoderms to larger food items. With a change in diet also comes a more active lifestyle.

Fishes

Placoderms As much as a third to a half of the anterior placoderm body was composed of heavy plates of dermal bone that also enclosed the pharynx and braincase. The rest of the body was covered with small bony scales. The dermal plates of the head were thick and tightly joined into a unit termed the **cranial shield** (figure 7.16a, b). Although the pattern of these dermal plates has been compared to scales of bony fishes, their arrangement was sufficiently different that it seems best to follow the convention of using different names until some agreement is reached on their homologies. The braincase was heavily ossified, and the upper jaws attached to it. In most, a well-defined joint existed between the braincase and the first vertebra. A spiracle was apparently absent. Water departing from the mouth exited posteriorly at the open junction between cranial and trunk shields. Most placoderms were about 1 m in length, although one species possessing strong jaws reached nearly 6 m overall.

Acanthodians The gnathostomes with one of the earliest surviving fossil record are the acanthodians. Most were small, several centimeters in length, with streamlined bodies, suggesting an active swimming lifestyle. Their bodies were covered with nonoverlapping, diamond-shaped, dermal bony scales. The bony scales of the head region were enlarged into small plates. The pattern of cranial dermal scales resembled bony fishes, but as with placoderms, these are usually given their own names. Some species had an **operculum,** a bony flap that covered the exit gill slits. Eyes were large, suggesting that visual information was especially important to these fishes. *Acanthodes* (early Permian) possessed a **lateral cranial fissure,** a gap that partially divided the posterior braincase. This fissure is an important fixture in actinopterygian fishes, where it allows exit of the tenth cranial nerve. The mandibular arch that formed the jaws was much like that of sharks and bony fishes. Three centers of ossification appear within the palatoquadrate: The metapterygoid and **autopalatine** both articulated with parts of the braincase, and the posterior quadrate articulated with the ossified Meckel's cartilage (figure 7.17a). A dermal bone, the **mandibular,** reinforced the ventral edge of the lower jaw. A hyoid arch and five successive branchial arches were present in *Acanthodes* (figure 7.17b).

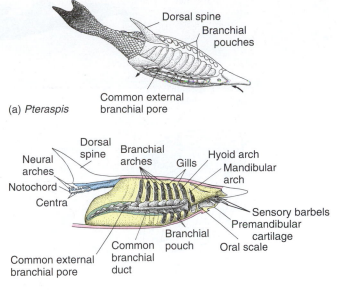

(a) *Pteraspis*

(b) *Poraspis*

FIGURE 7.15 **Ostracoderm feeding.** (a) Lateral view of *Pteraspis*, a heterostracan. Water flowed through the mouth, over the gills, suspended in branchial pouches, and into a common chamber before finally exiting via the branchial pore. Large, fused bony plates formed the head shield. Throughout the tail, the bony scales were small to accommodate the lateral bending of the tail. (b) Schematic reconstruction of the head of a heterostracan. Pointed, rough oral scales rimmed the mouth and might have been used to scrape or dislodge food from rock surfaces. This reconstruction of a heterostracan is based primarily on *Poraspis*.

After Stensiö.

FIGURE 7.14 **Ostracoderm *Pterolepis*, an anaspid.** (a) Exposed skull. The splanchnocranium included a few elements around the mouth, and the chondrocranium held the eye. A notochord was present and vertebral elements rested on it. (b, c) Restoration of muscles and some of the surface scales. The throat cartilages supported the floor of the buccal cavity, which might have been part of a pump to draw water into the mouth and then force it across the gills and out through the external branchial pores.

After Stensiö.

Chondrichthyans Cartilaginous fishes possess almost no bone. Denticles are present, vestiges of scales made up of the minerals enamel and dentin. A dermatocranium is absent. Instead, the chondrocranium has been expanded upward and over the top of the head to form the braincase. As a consequence, the chondrocranium is a much more prominent component of the skull than it is in most other vertebrates. The **ethmoid** and **orbital** anterior regions and posterior **oticooccipital** region are merged into an undivided braincase. The splanchnocranium is present. In primitive chondrichthyans, six gill arches trailed the mandibles (figure 7.18a, b). The upper jaw (palatoquadrate) of primitive sharks was supported by the braincase and probably by the hyomandibula.

Modern sharks usually lack a strong, direct attachment between hyomandibula and palatoquadrate. Instead, the jaws are suspended at two other sites, by the ceratohyal and Meckel's cartilage and by a strong, ligamentous connection running from the base of the nasal capsule to the orbital process of the palatoquadrate. As the ceratohyal, and to some extent the hyomandibula, have moved in to aid in supporting the jaws, the gill slit in front has become crowded, leaving only a small opening, the **spiracle.** In some sharks (great whites, makos, hammerheads) and in most living bony fishes, the spiracle has vanished altogether. In chondrichthyans, such as holocephalans, the jaws mechanically crush hard shells of prey, but in active chondrichthyans, such as predaceous sharks, the jaws capture prey.

Sharks may use suction to draw small prey toward or into the mouth, but more commonly, they attack prey directly, approaching it head-on. As sharks raise their head, the lower jaw descends (figure 7.19a). Upper and lower jaws articulate with each other, and both in turn are suspended like a pendulum from the hyoid arch. The hyoid arch swings about its attachment to the braincase, which permits the

(a)

Orbital cavity

IX

X

Pineal opening

Fenestra endonarina

Nasal capsule

Branchial arch

Hyomandibula

Meckel's cartilage

Palatoquadrate

(b)

Lateral line

Paranuchal plate

Nuchal plate

Postpineal

Sclerotic rim

Articular process for pectoral appendage

Premedian

Lateral plate

Anterior ventrolateral plate

Extra-lateral

Lateral plate

Prelateral plate

Lateral plate

FIGURE 7.16 **Placoderm skull.** *Bothriolepis* was about 15 cm long and lived in the middle Devonian. (a) Lateral view of splanchnocranium and chondrocranium. (b) Skull with overlying dermatocranium in place. Note the dermal plates.

After Stensiö, 1969.

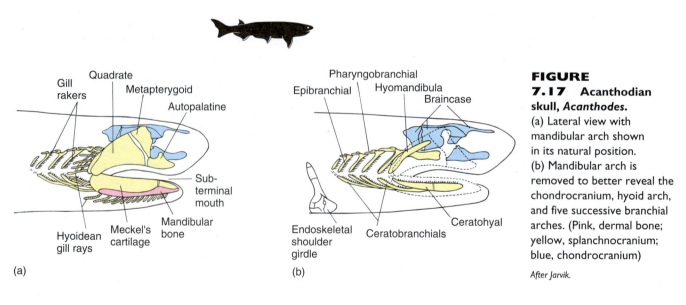

(a)

Gill rakers

Quadrate

Metapterygoid

Autopalatine

Sub-terminal mouth

Hyoidean gill rays

Meckel's cartilage

Mandibular bone

(b)

Pharyngobranchial

Epibranchial

Hyomandibula

Braincase

Endoskeletal shoulder girdle

Ceratobranchials

Ceratohyal

FIGURE 7.17 **Acanthodian skull, *Acanthodes*.** (a) Lateral view with mandibular arch shown in its natural position. (b) Mandibular arch is removed to better reveal the chondrocranium, hyoid arch, and five successive branchial arches. (Pink, dermal bone; yellow, splanchnocranium; blue, chondrocranium)

After Jarvik.

jaws to descend and shift downward and forward over the prey (figure 7.19b). Teeth along the upper (palatoquadrate) and lower (Meckel's cartilage) jaws are often oriented with their points in an erect position to engage the surface of the prey. Occasionally a nictitating membrane, a movable flap of opaque skin, is drawn protectively across each eye.

Jaw protrusion may also assist the synchronized meeting of upper and lower jaws on the prey. If the lower jaw alone is responsible for closing the mouth, it might prematurely strike the prey before the upper jaw is suitably positioned to assist. Protracting the mandibles away from the head allows

the jaws to assume a more favorable geometric configuration so that they meet the prey simultaneously and avoid deflecting it when they close. As the jaws clamp on the prey, the mandibular arch often is protracted near the end of closure. If the prey is large, the shark may violently shake its head to cut free a section of the prey and swallow it.

When protracted, the jaws disrupt the streamlined body silhouette characteristic of an active, open-water fish. Retraction of the jaws following feeding restores the hydrodynamic, streamlined shape of the fish and tucks the jaws back up against the chondrocranium.

(a)   (b)

FIGURE 7.18 Shark skull. (a) Primitive shark *Cladoselache,* a Late Devonian shark that reached perhaps 55 cm in length. Mandibles were followed by a complete hyoid arch and five branchial arches. Full gill slits were present between each arch. (b) Modern shark *Squalus,* the dogfish shark. The hyoid arch, second in series, is modified to support the back of the mandibular arch. As the hyoid moves forward to help suspend the jaw, the gill slit in front is crowded and reduced to the small spiracle. Although fused into one unit, the three basic regions of the chondrocranium are ethmoid, orbital, and oticooccipital. Abbreviations: basihyal (Bh), ceratohyal (Ch), hyomandibula (Hy), Meckel's cartilage (Mk), palatoquadrate (Pq).

(a) After Zangerl.

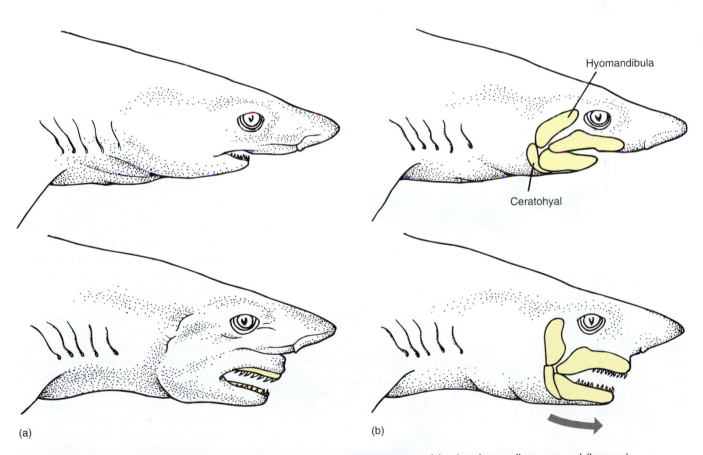

(a)   (b)

FIGURE 7.19   Feeding in sharks. (a) Sketches of shark with jaws retracted (top) and manually protracted (bottom). (b) Interpreted positional changes in the mandibular arch as it rides forward on its suspension from the ceratohyal. Position depicted is near the completion of jaw closure on the prey. Arrow indicates ventral and forward shift of the jaws.

Based on, and simplified from, the research of T. H. Frazzetta.

Actinopterygians Early actinopterygians had relatively large eyes and small nasal capsules. The jaws were long, extending to the front of the head. The jaws carried numerous teeth, and an operculum covered the gill arches. The hyoid arch increased its support of the mandibles. Homologies of dermal bones in some groups have been difficult to assign, partly because of the proliferation of extra bones, especially facial bones. Around the external naris, there may be many tiny bones variously ascribed by position to nasals, rostral, antorbitals, and others. One common scheme is shown in figure 7.20a, b, but several varieties occur as well. Notice in particular the set of **opercular bones** covering the gills and the set of **extrascapulars** at the dorsal, posterior rim of the skull. These are major dermal bones in actinopterygians that are lost in tetrapods (figure 7.21a, b).

Within actinopterygians, an extraordinary radiation occurred that continues to the present. It is difficult to generalize about trends within the skull because so many varied specializations of modern bony fishes are part of this radiation. If a common trend exists, it is for increased liberation of bony elements to serve diversified functions in food procurement.

Most living actinopterygians employ rapid suction feeding, with prey capture completed within 1/40 of a second. The almost explosive expansion of the buccal cavity creates a vacuum to accomplish swift capture. Negative pressure, relative to ambient pressure, sucks a pulse of water carrying the prey into the mouth. Once captured, teeth hold the prey. Compression of the buccal cavity expels excess water posteriorly out the gill slits. Fishes that feed by suction take in larger chunks of food than suspension feeders. Larger food particles have more inertia and require a stronger feeding device. Suction feeders consequently possess a well-muscularized buccal cavity and powerful, kinetic jaws.

(a)   Dorsal   (b)   Palatal

FIGURE 7.20   Major skull bones of an actinopterygian fish. (a) Dorsal view. (b) Palatal (ventral) views. Opercular bones are represented by dashed lines. Abbreviations: ectopterygoid (Ec), extrascapulars (Es), intertemporal (It), jugal (J), lacrimal (L), maxilla (M), nasal (N), parietal (P), palatine (Pal), premaxilla (Pm), postorbital (Po), postparietal (Pp), parasphenoid (Ps), pterygoid (Pt), quadrate (Q), rostral (R), supratemporal (St), tabular (T), vomer (V).

In primitive actinopterygians, such as the fossil *Cheirolepis* and living *Amia* (figures 7.21a, b and 7.22a, b), the feeding apparatus includes several units. One is the neurocranium, to which the premaxilla and maxilla are usually fused. The posterior part of the neurocranium articulates with and is free to rotate on the anterior vertebra. The opercular bones form a unit along the side of the head. The **suspensorium** is formed

FIGURE 7.21   Skull of the primitive palaeoniscoid fish *Cheirolepis*, from the Late Devonian. Overall length of the fish was about 24 cm. (a, b) Dorsal and lateral views of the skull, respectively. Bones of the pectoral girdle (pink) are tightly connected to the posterior wall of the skull. Abbreviations: angular (Ang), branchiostegals (Br), clavicle (Cl), cleithrum (Ct), dentary (D), dermohyal (Dhy), external naris (En), lateral extrascapular (Esl), intertemporal (It), jugal (J), lacrimal (L), maxilla (M), nasal (N), opercular (O), accessory opercular (Opa), parietal (P), premaxilla (Pm), postorbital (Po), preopercular (Pop), posttemporal (Pot), postparietal (Pp), preorbital (Pro), quadratojugal (Qj), rostral (R), sclerotic ring (Sc), supracleithrum (Sct), subopercular (Sop), supraorbital (Spo), supratemporal (St).

After Carroll.

(a)   (b)

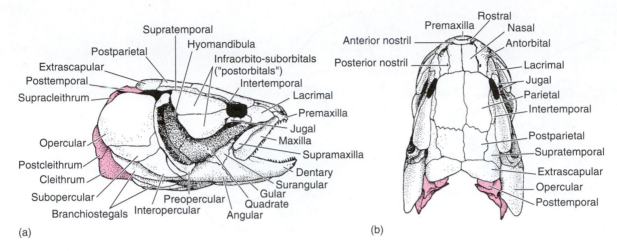

FIGURE 7.22 **Skull of the bowfin, *Amia*, a chondrostean.** Lateral (a) and dorsal (b) views. Pectoral girdle (pink).

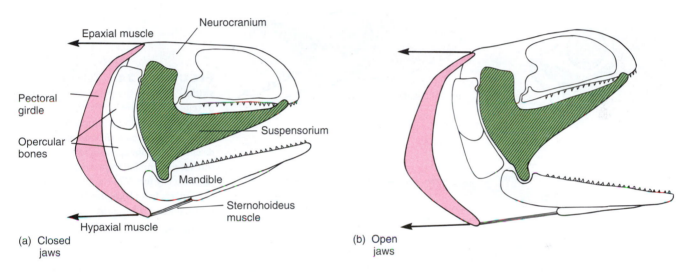

FIGURE 7.23 **Jaw opening in a primitive actinopterygian fish.** (a) Jaws are closed. (b) Jaws are open. The mandible rotates on its articulation with the suspensorium, which in turn is articulated with the opercular bones. The pectoral girdle remains relatively fixed in position, but the neurocranium rotates on it to lift the head. Lines of action of major muscles are shown by arrows. Pectoral girdle (pink).

After Lauder.

from the fusion of various bones in different species but usually includes the hyomandibula, various pterygoids, and quadrate. The suspensorium is shaped like an inverted triangle, its two upper corners articulating with the snout and braincase, its third lower corner articulating with the mandible. During jaw opening, epaxial muscles of the trunk raise the neurocranium and the attached upper jaw. Sternohyoideus muscles in the throat move the hyoid apparatus to lower the mandible (figure 7.23a, b). Strong adductor muscles of the jaws run from the suspensorium directly to the mandible to close the lower jaw.

In advanced actinopterygians, the teleosts, there is usually even greater freedom of skull bone movement (figure 7.24a–e). The premaxilla and maxilla are now usually freely articulated with each other and with the neurocranium

(figure 7.25). During jaw opening, the neurocranium is raised, and the mandible is lowered. In addition, the geometric arrangement of the jaws allows it to move forward. The hyoid apparatus forms struts within the floor of the buccal cavity. When pulled backward by the throat musculature, these hyoid struts help push the lateral walls of the buccal cavity apart and so contribute to its sudden enlargement and creation of suction within.

Sarcopterygians In early lungfishes, the upper jaw (palatoquadrate) was fused to the ossified braincase, which was a single unit with teeth flattened into plates. This suggests that the earliest lungfishes fed on hard foods, like their living counterparts that have similar tooth plates and jaws

Premakilla

Opercular

Maxilla

(a) Smallmouth bass

(b) Sheepshead fish

(c) Butterfly fish

(d) Cutlass fish

(e) Needlefish

FIGURE 7.24 Teleost skulls. Despite the great diversification of teleosts in many habits, the basic pattern of skull bones is preserved. (a) Smallmouth bass (*Micropterus dolomieu*). (b) Sheepshead fish (*Archosargus probatocephalus*). (c) Butterfly fish (*Chaetodon ocellatus*). (d) Cutlass fish (*Trichiurus lepturus*). (e) Needlefish (*Tylosurus marinus*).

After Radinsky.

for feeding on shellfishes, snails, and crustaceans. Another group of sarcopterygians, the rhipidistians, had strong jaws with small, pointed teeth. However, in contrast to the teeth of other fishes, the walls of rhipidistian teeth were extensively infolded, producing distinct **labyrinthodont teeth.** Large teeth were carried on the dentary of the lower jaw and along the lateral bones of the palate—vomer, palatine, ectopterygoid. Bones of the dermatocranium resembled those of actinopterygians, and like actinopterygians, the palatoquadrate articulated anteriorly with the nasal capsule and laterally with the maxilla. Unlike actinopterygians and extant lungfishes, the braincase of rhipidistians typically ossified into two articulated units: an anterior **ethmoid unit** (ethmosphenoid unit) and a posterior oticooccipital unit, with a flexible joint between them. In the dermal roofing bones above this joint, a hinge formed between the parietal and postparietal. Consequently, the snout could rotate upward about the rest of the skull, a displacement thought to be important during feeding (figure 7.26). The functional notochord also extended well forward into the head, passing

through a tunnel in the oticooccipital segment, eventually abutting the back of the ethmoid unit and perhaps bringing added support into this region of the skull.

Labyrinthodont teeth (p. 512)

Nasal Capsules   From fishes to tetrapods, the nasal capsules have had a complex history. The nasal capsules hold the olfactory epithelium in the form of a paired **nasal sac** (figure 7.27a). In actinopterygians, the nasal sac typically does not open directly into the mouth. Instead, its anterior (incurrent) and posterior (excurrent) narial openings establish a route for one-way water flow across the olfactory epithelium, delivering to it fresh chemical odors. By contrast, each nasal sac of tetrapods opens directly into the mouth via an internal naris, or **choana** (figure 7.27b). Each nasal sac also opens to the exterior by way of an **external naris** (nostril), thus establishing a respiratory route for airflow in and out of the lungs. In addition to internal and external nares, a third opening within the nasal sac begins as a tube,

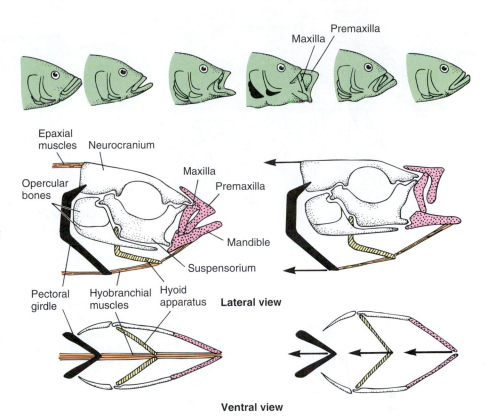

FIGURE 7.25 **Suction feeding of a teleost fish.** Top series are traces from a high-speed film of jaw opening (food not shown). Note changes in position of the jaws. Lateral and ventral views, respectively, of the major kinetic bones of the skull are shown when jaws are closed (left) and when they are open (right). Note the forward movement of the jaws (stippled areas) and outward expansion of the buccal cavity. Lines of muscle action are shown by arrows.

After Liem.

the **nasolacrimal duct,** that runs toward the orbit in order to drain away excess secretions of the adjoining lacrimal gland after helping to moisten the surface of the eye.

Olfactory organs (p. 680)

Among sarcopterygians, the nasal capsules of rhipidistians are similar to those of tetrapods. In rhipidistians, the nasolacrimal duct is an adaptation that benefits surface fishes that poke their eyes and nostrils out of the water. The lacrimal gland moistens exposed sensory organs that are subjected to drying. The nasolacrimal duct is probably an extension of the nasal sac of actinopterygian fishes. Rhipidistians (but not coelacanths) also possess internal nares, apparently representing a new derivative of the nasal sac connecting it with the mouth. However, lungfishes probably lack internal nares, although this is still debated. In lungfishes, the posterior (excurrent) naris opens near the margin of the mouth but does not pierce the palatal series of dermal bones as does the true internal naris of rhipidistians and tetrapods.

Early Tetrapods

The earliest tetrapods arose from rhipidistian ancestors and retained many of their skull features, including most of the bones of the dermatocranium. Numerous bones in the snout were reduced, leaving a distinct nasal bone occupying a position medial to the external naris (figure 7.28a, b). Beginning in tetrapods, the hyomandibula ceases to be involved in jaw suspension and instead becomes dedicated eventually to hearing as the stapes (or columella) within the middle ear. As mentioned earlier, the hyomandibula/stapes in rhipidistian fishes may have presided over the intake and exhalation of air to and from the lungs via the spiracle. But as terrestrial life evolved in early tetrapods, it became dedicated to hearing in air. The opercular series of bones covering the gills are typically lost. Extrascapulars across the back of the fish skull also disappear in primitive tetrapods. Along with this, the pectoral girdle loses its attachment to the back of the skull. Roofing bones and chondrocranium become more tightly associated, reducing the neurocranial mobility of the snout in comparison with rhipidistians.

The lateral line system, an aquatic sensory system, is evident in skulls of the earliest tetrapods, at least among the juveniles that were presumably aquatic stages (figure 7.29a, c). The skull is flattened, and in some a temporal notch at the back of the skull is present. The stapes conveys sound vibrations to the inner ear. But the stapes in early tetrapods is still a robust bone that also seems to be a buttress between the braincase and the palatoquadrate. Teeth were conical in labyrinthodonts, with the enamel folded into complex patterns. Teeth of lepospondyls lacked the highly folded enamel, and the otic notch was absent.

The skulls of modern amphibians are greatly simplified compared with those of their fossil ancestors, with many of the dermal bones being lost or fused into composite bones. Caecilian skulls are compact and firmly ossified,

(a)

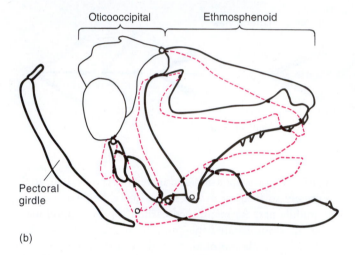

(b)

FIGURE 7.26 Cranial kinesis of a coelacanth, *Latimeria.* (a) Lateral view of the skull. (b) Biomechanical model of major functional elements showing displacement pattern during jaw opening (solid lines) compared with closed position (dashed lines). Pterygoid complex includes entopterygoid, ectopterygoid, and epipterygoid.

(a) After Millot, Anthony, and Robineau; (b) based on Lauder.

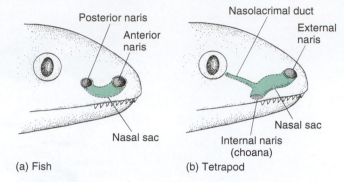

(a) Fish   (b) Tetrapod

FIGURE 7.27 Openings of the nasal sac. (a) In an actinopterygian fish, the nasal sac typically has an anterior naris through which water enters and a posterior naris through which water exits, but the nasal sac does not open into the mouth. (b) In a tetrapod, the nasal sac has an external naris (homologous with the anterior naris of the fish) and a nasolacrimal duct to the orbit (an extension of the nasal sac). In addition to these, a third extension of the nasal sac, the internal naris (homologous with the posterior naris of fish), opens into the buccal cavity through the roof of the mouth and now, in this internal position, is called a choana.

The splanchnocranium, a major component of the fish skull, is reduced in amphibians. In modern amphibians, the hyomandibula plays no role in jaw suspension. This task is taken over exclusively by the articular and quadrate bones through which the mandible articulates with the skull. The branchial arches composing the hyobranchial apparatus support external respiratory gills in the larvae, but when the larvae metamorphose into the adult, these arches are reduced to the hyoid apparatus that supports the action of the tongue.

Salamanders commonly use suction feeding in water. The floor of the throat is rapidly expanded and the jaws parted enough so that the pulse of water carrying the intended prey enters (figure 7.32). Excess water gulped in with the prey exits at the back of the mouth through the gill slits. In salamanders, as in fishes, there is a **unidirectional** flow of food and water into the mouth and out the gill slits. In metamorphosed salamanders and in adult frogs, gill slits are absent, so excess water entering the mouth during feeding must reverse its flow to exit via the mouth. Such flow is said to be **bidirectional.** On land, amphibians commonly use a sticky, projectile tongue. At close range, muscles catapult the tongue over the parted mandibles and into contact with the prey. At longer range, muscle action works in cooperation with fluid-filled spaces within the tongue to accelerate it along the hyoid apparatus. Retraction of the tongue returns the attached food to the mouth, and teeth close on it to control the struggling prey.

Primitive Amniotes

The first amniotes were small and would probably remind us of lizards in general appearance. The skull roof, like that of early tetrapods, was formed from the dermatocranium, with

although the pattern of dermal bones can be quite varied. In salamanders, the chondrocranium consists primarily of **orbitosphenoid** and **prootic** bones, with exoccipitals closing the posterior wall of the braincase (figure 7.30). Nasal bones are usually present. Up to four pairs of roofing bones contribute to the skull: Frontals and parietals are present in all, but prefrontals and lacrimals vary among groups. In anurans (figure 7.31), ossification of the chondrocranium is highly variable, usually with just five bones present, a single sphenethmoid and paired prootics and exoccipitals. A nasal bone is present, but only a paired, composite **frontoparietal** remains of the roofing bones. In both frogs and salamanders, the single parasphenoid has expanded to form a large plate that has crowded other palatal bones.

(a) Dorsal

| Rhipidistian | Early tetrapod | Modern amphibian (salamander) |

(b) Ventral

FIGURE 7.28 Diagrammatic views of skull modifications from rhipidistian to early tetrapod to modern amphibian (salamander). (a) Dorsal views. (b) Ventral (palatal) views. Skull bones lost in the derived group are shaded in the skull of the preceding group. Abbreviations: basioccipital (Bo), basisphenoid (Bs), ectopterygoid (Ec), exoccipital (Eo), frontal (F), intertemporal (It), jugal (J), lacrimal (L), maxilla (M), nasal (N), opercular (Op), opisthotic (Ops), parietal (P), palatine (Pal), postfrontal (Pf), premaxilla (Pm), postorbital (Po), prefrontal (Prf), prootic (Pro), postparietal (Pp), parasphenoid (Ps), pterygoid (Pt), quadrate (Q), quadratojugal (Qj), supratemporal (St), squamosal (Sq), tabular (T), vomer (V).

openings for eyes, pineal organ, and nostrils (figure 7.33a–d). Robust attachment flanges and processes are evidence of strong jaw-closing muscles. The palatoquadrate of the mandibular arch was reduced to the small epipterygoid and separate quadrate. The hyoid arch produced a stapes, a stout bone that braced the back of the dermatocranium against the chondrocranium. These early tetrapods lacked a temporal notch. Sound transmission to the inner ear may have occurred along bones of the lower jaw.

Skull Fenestrae As mentioned earlier, the temporal region of the dermatocranium contains features particularly revealing of amniote lineages (figure 7.34). Fenestrae are openings in the outer dermatocranium. The anapsid skull lacks temporal

fenestrae. In recent turtles, **emarginations** often encroach upon the posterior margin of the skull roof. These emarginations are large notches that function like fenestrae, but they are independent phylogenetic derivatives. The diapsid skull includes two temporal fenestrae, a condition carried forward in *Sphenodon* and in crocodiles and their allies. However, the lower and/or upper temporal bars are often lost in other modern forms. This gives us several contemporary varieties of a modified diapsid skull in which the diapsid condition is substantially altered, such as birds, lizards, and especially snakes.

The synapsid skull of pelycosaurs, therapsids, and modern mammals contains a single temporal opening. Loss of the postorbital bone in modern mammals allows merging of the temporal fenestra with the orbit.

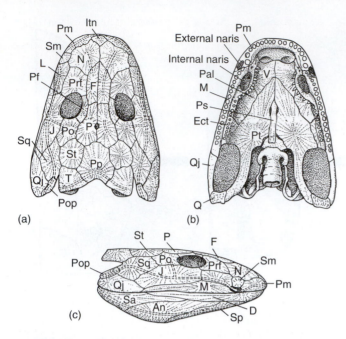

(a)

(b)

(c)

FIGURE 7.29 **Skull of a primitive tetrapod of the late Devonian.** Dorsal (a), ventral (b), and lateral (c) views. Parallel tracks of dashed lines indicate course of the aquatic lateral line system on the skull bones. Abbreviations: angular (An), dentary (D), ectopterygoid (Ect), frontal (F), internasal (Itn), jugal (J), lacrimal (L), maxilla (M), nasal (N), parietal (P), palatine (Pal), postfrontal (Pf), premaxilla (Pm), postorbital (Po), preopercular (Pop), postparietal (Pp), prefrontal (Prf), parasphenoid (Ps), pterygoid (Pt), quadrate (Q), quadratojugal (Qj), surangular (Sa), septomaxilla (Sm), splenial (Sp), supratemporal (St), squamosal (Sq), tabular (T), vomer (V).

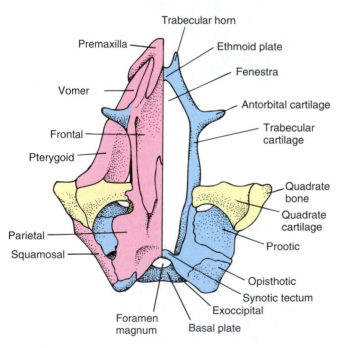

FIGURE 7.30 **Skull of *Necturus*, a modern amphibian.** Superficial skull bones are indicated on the left. These bones have been removed to reveal the chondrocranium and derivatives of the splanchnocranium on the right.

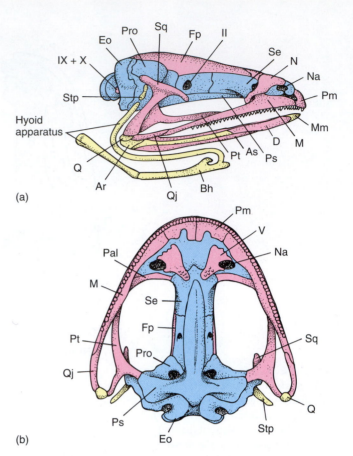

(a)

(b)

FIGURE 7.31 **Frog skull.** Lateral (a) and ventral (b) views. Abbreviations: articular (Ar), angulosphenoid (As), basihyal (Bh), dentary (D), exoccipital (Eo), frontoparietal (Fp), maxilla (M), mento-Meckelian (Mm), nasal (N), naris (Na), palatine (Pal), premaxilla (Pm), prootic (Pro), parasphenoid (Ps), pterygoid (Pt), quadrate (Q), quadratojugal (Qj), sphenethmoid (Se), squamosal (Sq), stapes (Stp), vomer (V). Roman numerals indicate foramina serving specific cranial nerves.

After Marshall.

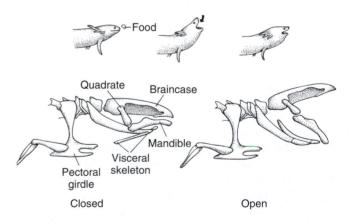

Closed

Open

FIGURE 7.32 **Suction feeding by an aquatic salamander.** Before, during, and after suction feeding traced from a high-speed-film feeding sequence (top series). Note the interpreted positions of the skull elements when the jaws are closed (bottom left) and open (bottom right).

After Lauder.

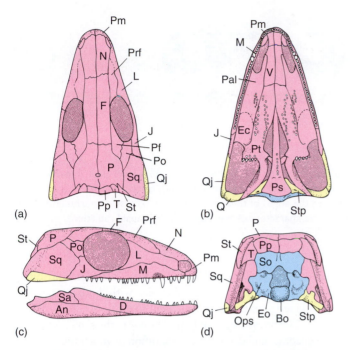

(a)

(b)

(c)

(d)

FIGURE 7.33 **Skull of an early amniote from the Carboniferous.** Dorsal (a), ventral (b), lateral (c), and posterior (d) views. Abbreviations: angular (An), basioccipital (Bo), dentary (D), ectopterygoid (Ec), exoccipital (Eo), frontal (F), jugal (J), lacrimal (L), maxilla (M), nasal (N), opisthootic (Ops), parietal (P), palatine (Pal), postfrontal (Pf), premaxilla (Pm), postorbital (Po), postparietal (Pp), prefrontal (Prf), parasphenoid (Ps), pterygoid (Pt), quadrate (Q), quadratojugal (Qj), surangular (Sa), supraoccipital (So), supratemporal (St), squamosal (Sq), stapes (Stp), tabular (T), vomer (V).

After Carroll.

Although used by taxonomists to delineate phylogenetic lineages within tetrapods, the functional significance of fenestrae is not clear. With few exceptions, most notably lepospondyls, fenestrae are absent in early tetrapods and primitive amniotes. Because fenestrae are associated with strong jaw adductor muscles, it has been suggested that they open space in the skull for these muscles to bulge during contraction (figure 7.35a–c). But it is difficult to see how such a function could have afforded some initial advantage favoring their evolution. Initially fenestrae would have been too small to provide space for bulging muscles that presumably favored their appearance. Alternatively, some have suggested that unstressed bone of the dermatocranium might have little selective value if it did not contribute to muscle attachment. Its loss would have been expected, leading to the initial appearance of fenestrae in these areas. More positively, it has been proposed that the rims of open fenestrae offer a more secure attachment site for muscles than does a flat surface. Muscle tendons merge with the periosteum, spread the tensile forces around the rim, and distribute them across the extended surface of the bone. This might render the attachment site less susceptible to being torn loose from the bone.

Whatever the function of fenestrae, their presence would be possible only if holes would not unduly weaken the ability of the skull to withstand stresses. Their absence in labyrinthodonts and primitive amniotes, their presence in later amniotes, and the appearance of emarginations by a different route in turtles imply a complex and not fully understood interaction between function and design in early tetrapods.

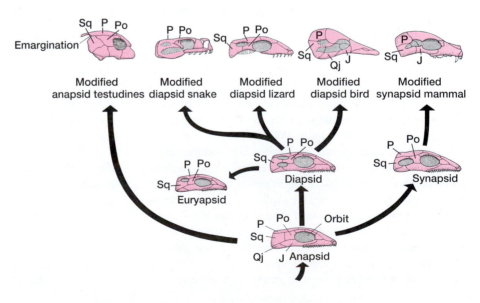

FIGURE 7.34 **Major lineages of dermatocranium evolution within amniotes.** The anapsid skull occurs in primitive amniotes and their modern descendants, turtles and tortoises. Two major groups, the diapsids and synapsids, independently evolved from the anapsids. *Sphenodon* and crocodilians retain the primitive diapsid skull, but it has been modified in diapsid derivatives such as snakes, lizards, and birds. Shading indicates positions of temporal fenestrae and orbit. Abbreviations: jugal (J), parietal (P), postorbital (Po), quadratojugal (Qj), squamosal (Sq).

FIGURE 7.35 **Temporal fenestrae.** The shift in jaw muscle attachment to the skull is shown. (a) Anapsid skull. In early amniotes, temporal muscles run from the neurocranium to the lower jaw. Such a skull is retained in modern turtles. (b) Perforation in the dermatocranium opens fenestrae, and attachment of jaw muscles expands to the edges of these openings. (c) Extensive attachment of jaw muscles to the surface of the dermatocranium. Such development of fenestrae characterizes the diapsid and synapsid radiations.

After Smith.

FIGURE 7.36 **Cranial kinesis in squamates.**
(a) There are three types of cranial kinesis based largely on the position at which the hinge (X) lies across the top of the skull. The hinge may run across the back of the skull roof (metakinesis), behind the orbit (mesokinesis), or in front of the orbit where the snout articulates (prokinesis). (b) The ability of the quadrate to rotate about its dorsal end is called streptostyly.

Cranial Kinesis in Reptiles Skull elements of reptiles exhibit varying degrees of mobility. The most extensive motions are found in the skulls of lizards and especially snakes. In these two groups, a transverse hinge extends across the skull roof, a **transcranial joint.** Depending on the position of this hinge, three names apply. Where a hinge passes across the back of the skull, permitting rotation between the neurocranium and outer dermatocranium, the skull is said to exhibit **metakinesis** (figure 7.36a). If a joint passes through the dermatocranium behind the eye, the skull exhibits **mesokinesis.** If a joint in the dermatocranium passes in front of the orbits, the skull exhibits **prokinesis.** Depending on the number of hinges, the skull may be **monokinetic,** having one joint, or **dikinetic** (amphikinetic), having two joints. Although rare, mesokinesis is possibly present in amphisbaenians and some burrowing lizards. Prokinesis is typical in snakes and birds. Most modern lizards are dikinetic, with both meta- and mesokinetic joints across their skull roofs.

The term **streptostyly** applies not to the skull roof but to the quadrate and describes the condition in which the quadrate is free to undergo some degree of independent rotation about its dorsal connection with the braincase (figure 7.36b). Most lizards, snakes, and birds are streptostylic.

Modern Reptiles

Modern turtles have anapsid skulls, but emarginations that develop from the posterior region forward often result in the opening of large regions within outer bones of the dermatocranium (figure 7.37a–e). Large jaw-closing muscles occupy this space. Although turtles lack teeth, the opposing surfaces of upper and lower jaws are usually covered with keratinized "tooth" plates that deliver powerful biting forces to food.

Several modern reptiles are surviving members of the diapsid radiation. In *Sphenodon,* complete upper and lower temporal bars firmly join the front and the back of the lateral skull wall (figure 7.38a–d). A mobile transcranial joint and movable palate are absent. Consequently, no significant mobility is permitted within the dermatocranium. However, the mandible slides back and forth on the fixed quadrate from which it is suspended. The single row of teeth of the mandible moves between a double row of teeth on the upper jaw, an action that seems to be important in slicing through some types of prey.

Loss of the lower temporal bar produces the modified diapsid skull of lizards (figure 7.39). Loss of this lower bony strut laterally liberates the posterior part of the skull from the snout, thereby permitting streptostyly, and hence promoting the mesokinetic part of lizard dikinesis. Lizard ancestors, the younginiformes, apparently possessed a single metakinetic joint across the back of the skull. A second kinetic joint, the mesokinetic joint, has been added to this in most modern lizards, making the skulls of most lizards dikinetic.

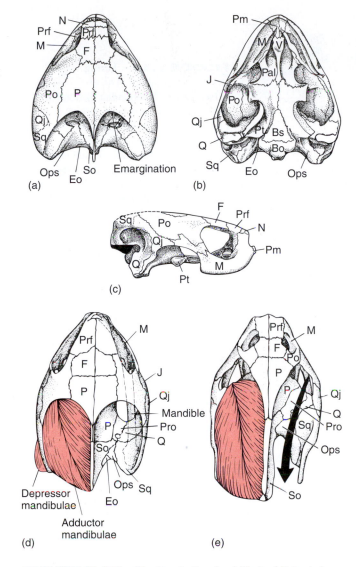

FIGURE 7.37 **Turtle skulls.** (a–c) Skull of *Pleisochelys,* from the Late Jurassic. *Pleisochelys* is the earliest known member of the cryptodires. Note the absence of any temporal fenestrae but the presence of emarginations etched in the dorsal, posterior rim of the skull. Dorsal (a), ventral (b), and lateral (c) views. (d) European pond turtle *Emys,* showing site of residence of jaw opening (depressor mandibulae) and closing (adductor mandibulae) muscles in relation to emargination. (e) Modern softshell turtle *Trionyx,* showing line of action of adductor mandibulae, solid arrow, from lower jaw to skull within enlarged emargination. Abbreviations: basioccipital (Bo), basisphenoid (Bs), exoccipital (Eo), frontal (F), jugal (J), maxilla (M), nasal (N), opisthotic (Ops), parietal (P), palatine (Pal), prefrontal (Prf), premaxilla (Pm), prootic (Pro), postorbital (Po), pterygoid (Pt), quadrate (Q), quadratojugal (Qj), supraoccipital (So), squamosal (Sq), vomer (V).

(a–d) After Carroll; (e) after Romer.

Although skulls of some specialized lizards, such as burrowers, anteaters, and some herbivores, seem monokinetic, this is likely a secondary condition. This kinetic machinery of lizard jaws has been modeled as a four-bar linkage system (figure 7.40a, b). One unit is the triangular-shaped snout. Its posterior wall forms one of the four linkages. The dorsal corner of the snout participates in the mesokinetic joint and forms a second mechanical link with the dorsal end of the quadrate through the top of the skull. The quadrate represents the third link. The fourth mechanical link connects the quadrate's lower end (where it meets the pterygoid) forward to the posterior lower corner of the snout to complete and close the four-bar kinematic chain.

Biomechanical mechanisms (p. 147)

Without such a kinematic series of linkages, jaw closure would be scissorslike, and jaw-closing forces on the prey would have a forward component that might deflect or squirt the prey out of the mouth, increasing the chance of prey loss (figure 7.40c). However, in the skull of many lizards, rotation of the four linkages permits changes in geometric configuration. As a consequence, these lizards can alter the angle of the tooth row borne by the snout as it closes on the prey. Upper and lower jaws close and meet the prey nearly simultaneously, delivering forces directed at the prey; thus the lizard is less likely to experience prey loss.

The metakinetic joint is not directly part of this linkage train of bones, although its transverse axis is coincident with the quadrate-parietal joint of the four-bar linkage mechanism. The metakinetic joint permits the dermatocranium, to which the linkage chain is joined, to move relative to the deeper neurocranium. The axis of the metakinetic joint is almost coincident with the superficial joint between the dorsal end of the quadrate and the braincase, but it is not part of this outer set of linkages. Thus, rotation about this metakinetic joint lifts the whole dermatocranium along with the entire set of linkages relative to the neurocranium.

Some lizards, like many terrestrial salamanders, project their tongues during feeding. When the tongue is prominently deployed, a lizard engages in **lingual feeding** (figure 7.41a). The jaws part, and the sticky tongue is projected at the prey. In chameleons, a circular **accelerator muscle** wraps around the **lingual (= entoglossal) process** of the hyoid apparatus (figure 7.41b, c). Upon contraction, the accelerator muscle squeezes the lingual process, picks up speed as it slides down the tapered process, perhaps like squeezing a slippery bar of soap, and carries along the glandular tip of the tongue (figure 7.41d). The gathered momentum of the tip of the tongue launches it out of the mouth toward the prey. Upon impact, the fleshy glandular tip of the tongue flattens against the target, establishing firm adhesion. Retraction of the tongue back into the mouth retrieves the prey. The jaws then close to hold the captured prey.

FIGURE 7.38 **Living rhynchocephalian.** The two temporal fenestrae are still bounded by bone in *Sphenodon*, a living diapsid. Dorsal (a), ventral (b), lateral (c), and posterior (d) views. Abbreviations: angular (An), articular (Ar), basioccipital (Bo), basisphenoid (Bs), coronoid (C), dentary (D), ectopterygoid (Ec), exoccipital (Eo), epipterygoid (Ept), frontal (F), jugal (J), maxilla (M), nasal (N), opisthotic (Ops), parietal (P), prearticular (Pa), palatine (Pal), postfrontal (Pf), premaxilla (Pm), postorbital (Po), prefrontal (Prf), prootic (Pro), parasphenoid (Ps), pterygoid (Pt), quadrate (Q), quadratojugal (Qj), surangular (Sa), supraoccipital (Soc), squamosal (Sq), stapes (Stp), vomer (V).

After Carroll.

FIGURE 7.39 **Lizard skull.** Lizards are modified diapsids. Two fenestrae are present, but the ventral bony border of the lower fenestra is absent, a result of changes serving increased cranial kinesis. Abbreviations: angular (An), articular (Ar), basioccipital (Bo), basisphenoid (Bs), coronoid (C), dentary (D), ectopterygoid (Ec), exoccipital (Eo), epipterygoid (Ept), frontal (F), jugal (J), lacrimal (L), maxilla (M), nasal (N), opisthotic (Ops), orbitosphenoid (Osp), parietal (P), prearticular (Pa), postfrontal (Pf), premaxilla (Pm), postorbital (Po), prefrontal (Prf), parasphenoid (Ps), prootic (Pro), pterygoid (Pt), quadrate (Q), surangular (Sa), septomaxilla (Sm), splenial (Sp), supraoccipital (So), squamosal (Sq), supratemporal (St), stapes (Stp), vomer (V).

After Jollie.

FIGURE 7.40 Kinesis of a lizard skull. (a) Joints within the skull allow the snout to lift upward or bend downward about its mesokinetic articulation with the rest of the braincase. This results in a change in the angle of closure of the teeth when the animal grabs its prey. (b) These movable units of the lizard skull can be represented as a kinematic mechanism by linkages (heavy lines) and points of rotation (circles). Compared with the rest position of these linkages (a), geometric changes are shown during opening (middle) and closing (bottom) on the prey. (c) The functional significance of cranial kinesis in lizards is related to the resulting change in angle of tooth rows. Kinesis bends the snout so that both rows close directly on the prey (bottom). Were this not the case (middle and top), jaw closure would be more of a scissors action, tending to squirt the prey back out of the mouth.

Based on the research of T. H. Frazzetta.

In snakes, the frontal and parietal roofing bones have grown down around the sides of the skull to form most of the walls of the braincase as well (figure 7.42). Their enlargement results in crowding or loss of many of the other dermal bones. Snake skulls are prokinetic. A joint across the skull forms in front of the orbit between frontal and nasal regions. However, most of the extensive mobility of the snake jaw results from changes in skull design in the lateral bones. Both upper and lower temporal bars are lost, thus removing struts that in other diapsid skulls form restrictive braces across the temporal region. The kinematic machinery of the snake skull includes more elements than the linkage system of lizards (figure 7.43a–c). The quadrate, as in lizards, is streptostylic but more loosely articulated with the pterygoid. Muscle forces imparted directly to the pterygoid are transmitted to the tooth-bearing maxilla via the linking

ectopterygoid. The maxilla rotates upon the prefrontal, from which it is suspended from the braincase. In many snakes, especially in advanced venomous snakes such as vipers, the prefrontal and supratemporal also enjoy some degree of rotation upon the braincase. Thus, the kinetic system can be modeled on a linkage chain with up to six links (supratemporal, quadrate, pterygoid, ectopterygoid, maxilla, prefrontal) suspended at either end from a seventh link, the braincase (figure 7.43d).

The mandible of snakes, suspended from the quadrate, includes a tooth-bearing dentary that articulates with a posterior **compound bone** derived from the fused surangular, prearticular, and articular. A tiny splenial is usually present on the medial side. Both halves of the lower jaw are joined at the mandibular symphysis, not by bony fusion but by flexible soft tissues that tether

FIGURE 7.41 **Lingual feeding in lizards.** (a) Jackson's chameleon uses its projectile tongue to "shoot" long distances at prey. (b) Hyoid apparatus of the chameleon includes an elongated lingual process (Lp) along which the tongue slides during launch. (c) Hyoid apparatus of a lizard without a projectile tongue. (d) Mechanical basis of tongue projection. The accelerator muscle, a circular band around the lingual process, contracts to squeeze the lingual process. The squeeze of accelerator muscles causes the muscle to slide rapidly toward the tip of the lingual process, carrying with it the glandular surface of the tongue. With gathered momentum, the tongue is launched from the lingual process toward the prey. The folded glossohyal muscle attached to the tongue's tip is carried out as well and eventually is responsible for retrieving the tongue and the adhering prey. Abbreviations: anterior process (Ap), basihyal (Bh), ceratobranchials I and II (CbI and CbII), ceratohyal (Ch), lingual process (Lp).

(b, c) After Bramble and Wake.

together the tips of the mandibles. Mandibular tips thus enjoy generous independent movement. Because there are no bony cross connections between chains of movable bones on left and right sides, each kinematic set of linkages can spread and move independently of the other on the opposite side. This is particularly important during swallowing when alternating left and right sets of bones are walked over the prey (see figure 7.63 and Box Essay 7.3). It is a mistaken view that snakes "unhinge" their jaws when they swallow. Instead, the great freedom of rotation between elements of the kinematic chains, the independent movement of each, and the ability of flaring the flexible jaws outward to accommodate bulky prey all account for the suppleness of snake jaws. These processes, not disarticulation, permit snakes to swallow (although slowly) relatively large, whole prey.

Crocodilians, together with *Sphenodon* and squamates (lizards and snakes), represent the surviving reptiles with a diapsid skull. The crocodilian skull is a composite of chondro-, dermato-, and splanchnocranial elements, although the dermatocranium tends to predominate (figure 7.44). Both temporal bars are present, and the skull is firm, without any evidence of cranial kinesis. However, crocodile ancestors possessed kinetic skulls, suggesting that modern forms have lost this feature. Further, modern crocodiles possess a secondary palate, a further departure from diapsid ancestors. Within the roof of the mouth, marginal bones (premaxilla, maxilla, palatine) grow inward to meet at the midline beneath the sphenoid region. Together with the pterygoid, these marginal bones produce the bony secondary palate that separates the nasal passageway from the mouth.

Birds

Birds also arise from a diapsid ancestry, but like squamates, they show considerable modification of this skull pattern (figure 7.45). The braincase is much inflated and ossified in birds, accommodating a relatively expanded brain within.

FIGURE 7.42

Diagrammatic comparison of a derived modern snake skull with an amniote skull. Bones lost in the modern snake are indicated by shading in the primitive captorhinomorph. Abbreviations: exoccipital (Eo), frontal (F), jugal (J), lacrimal (L), maxilla (M), nasal (N), parietal (P), postfrontal (Pf), premaxilla (Pm), postorbital (Po), postparietal (Pp), prefrontal (Prf), pterygoid (Pt), quadrate (Q), quadratojugal (Qj), supraoccipital (So), squamosal (Sq), supratemporal (St), tabular (T).

Primitive amniote

Derived reptile (viperid snake)

(a)

(b)

(c)

(d)

FIGURE 7.43 Kinematic model of movable skull bones in a venomous snake, the water moccasin.

Whole head (a) with successive removal of skin and muscles (b) reveals bones of the skull. (c) Linkage bones movable relative to the braincase are in color; lower jaw is crosshatched. (d) Biomechanical model of movable bones rotatable about pin connections. Movable bones include the ectopterygoid (Ec), maxilla (M), pterygoid (Pt), prefrontal (Prf), quadrate (Q), supratemporal (St). Location of the main venom gland (Vg) is shown as well.

After Kardong.

The jaws of snakes are highly kinetic, with great freedom of motion. Skull bones that in other reptiles are fixed to the braincase or have restricted movement are joined in snakes into linked chains with extensive motion relative to the braincase. Further, the series of linked bones on left and right sides are not joined directly; so they experience independent displacement, a feature allowing alternating left and right reciprocating motion of jaw bones over

the prey being swallowed. This independent motion and outward spreading of the jaws (not "unhinging" of the jaws) allows most snakes to swallow large prey. Little by little, the distended jaws are walked in alternating steps over the prey until it is completely engulfed.

During the rattlesnake strike, the forward swing of these linked bones quickly erects the maxilla and the fang it carries into position to inject venom into the prey.

Snake fangs are modified teeth with hollow cores so that venom flows from their base into the prey. The fangs of most venomous snakes are longer than other teeth in the mouth, and the fangs of vipers and pit vipers are especially long. Extensive rotation of the fangbearing maxilla in such snakes allows this long fang to be folded up and out of the way along the upper lip when it is not in use.

Sutures between bones are usually overgrown in the adult so that boundaries are not easily delineated. The palatal bones are quite varied, but generally all show some degree of reduction and lightening. Vomers and ectopterygoids are small, pterygoids are short struts articulating with the quadrate, and epipterygoids are usually lost (figure 7.46a–d).

Like turtles and some dinosaurs, birds are toothless, and their jaws are covered by keratinized sheaths. Birds that feed on slippery prey, such as fish-eating shorebirds, have beaks whose keratinized margins are serrated to improve friction grip. The jaws are drawn out into a **beak.** The upper temporal bar is absent, and the lower temporal bar is a slender rod called the jugal bar (quadratojugal-jugal bar), which extends from the beak posteriorly to the side of the movable (streptostylic) quadrate. The skull is prokinetic. A strong **postorbital ligament** extends from behind the eye to the lower jaw. In neognathous birds, the palate is functionally divided at the pterygo-palatine joint (figure 7.47a, b). The converging paired pterygoids meet or nearly meet at the midline to couple there with the palatines (figure 7.47b). At this coupling, these bones form a joint that slides along the ventral edge of the orbital septum. The nasofrontal and palato-maxilla joints are thin, flexible, bony joints, not synovial, but they can be represented as hinges about which rotation occurs. The mechanically important bones are modeled as a linkage system on each side (figure 7.47c). When inserting muscles pull the quadrate and palate forward, the divided palate slides along the septum pushing the palatines forward, which in turn push against the base of the beak, rotating it about the nasofrontal joint, and raising the beak. The coupled pair of gliding linkages is a **slider-crank mechanism.** (Muscles acting directly on the lower jaw activate its opening.) Jaw-closing muscles act in the opposite fashion to move the beak downward to grasp food, and return the kinetic machinery to the rest position. The thin jugal bar usually bows outward during jaw opening, but does not contribute significantly to the mechanism producing upper jaw elevation.

Many birds use their beak like a probe to reach buried grubs or insects embedded in tree bark or soft soil. Such birds often use a form of rhynchokinesis, the lifting of the beak tips about points of rotation within the jaws (figure 7.47e). The jaws need not be parted far to seize the food. Other birds have beaks that open tough seeds and short, stout jaws that concentrate closing forces at the base of their beaks.

In paleognathous birds such as emus and ostriches, the pterygoids do not meet at the midline, but slide on projecting struts, the basipterygoid processes (figure 7.47d). So structurally distinct is the palate that it has been used to argue that all paleognaths (ratites and tinamous) represent a primitive condition and a monophyletic group.

Synapsids

Primitive Synapsids   The sail-backed pelycosaur *Dimetrodon* represents a primitive synapsid. Therapsids continue the synapsid line and exhibit considerable diversity (figure 7.48). For a time in the Permian and Early Triassic, they were fairly abundant. Some were herbivores; most were carnivores. Most skull bones of early amniotes persist, but characteristic of synapsids, the temporal region develops a single opening bound horizontally along its lower border by a bony connection between jugal and squamosal bones bowing outward in the cheek region. This bony squamosal-jugal bar is now commonly called the **zygomatic arch.** Throughout synapsid evolution there is a tendency for the temporal opening to enlarge, likely related to increasing mass and specialization of nearby jaw musculature. In advanced therapsids and primitive mammals, the vertical bar dividing the orbit from the single temporal fenestra is lost.

Mammals   The skull of mammals represents a highly modified synapsid pattern. Various dermal elements are lost in therian mammals, including the prefrontal, postorbital, postfrontal, quadratojugal, and supratemporal (figure 7.49). The postparietals, typically paired in reptiles, fuse into a single, medial **interparietal** in therapsids, which in mammals

FIGURE 7.44 Alligator skull. A composite skull design characteristic of vertebrates. The skull is a combination of elements receiving contributions from the chondrocranium (blue), the splanchnocranium (yellow), and the dermatocranium (pink).

may incorporate the tabular and fuse with the occipital bones. Monotremes retain several early synapsid skull features, including prefrontal, postfrontal, and pleurosphenoid bones together with unfused occipitals. Monotremes are somewhat specialized as well. The lacrimal of therians is absent in them, and the jugal bones are small (figure 7.50a–d). A tympanic ring encircles the middle ear bones of monotremes and occasionally of marsupials, but in most eutherians, other bones expand into a large swollen capsule, the auditory bulla, that houses the middle ear ossicle (figure 7.51a–c).

Eutherian Mammals Fusions between separate centers of ossification produce composite bones in the skull of placental mammals. The single occipital bone represents the

fusion of basioccipital, paired exoccipitals, supraoccipital, and interparietal (and perhaps tabular) (figure 7.52a). The occipital bone defines the foramen magnum and closes the posterior wall of the braincase. As in monotremes and marsupials, there is a ventrally located, bilobed occipital condyle that articulates with the **atlas,** the first vertebra of the cervical region. Dorsally, a raised **nuchal crest** may form across the back of the occipital region, offering a secure attachment site for neck muscles and ligaments that support the head.

Several embryonic centers contribute to the sphenoid bone, representing the orbitosphenoid, presphenoid, basisphenoid, and a large alisphenoid (the epipterygoid of lower vertebrates) (figure 7.52b).

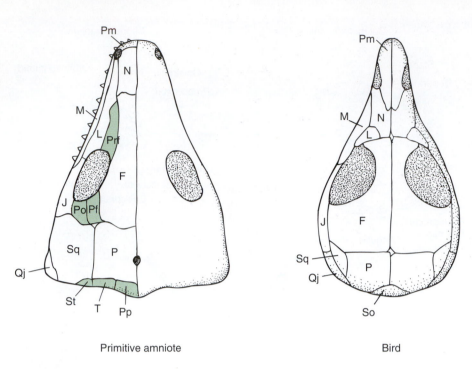

FIGURE 7.45 Diagrammatic comparison of a derived bird skull with a primitive amniote. Bones lost in birds are shaded in the primitive reptile. Abbreviations: frontal (F), jugal (J), lacrimal (L), maxilla (M), nasal (N), parietal (P), postfrontal (Pf), premaxilla (Pm), postorbital (Po), postparietal (Pp), prefrontal (Prf), quadratojugal (Qj), supraoccipital (So), squamosal (Sq), supratemporal (St), tabular (T).

Primitive amniote   Bird

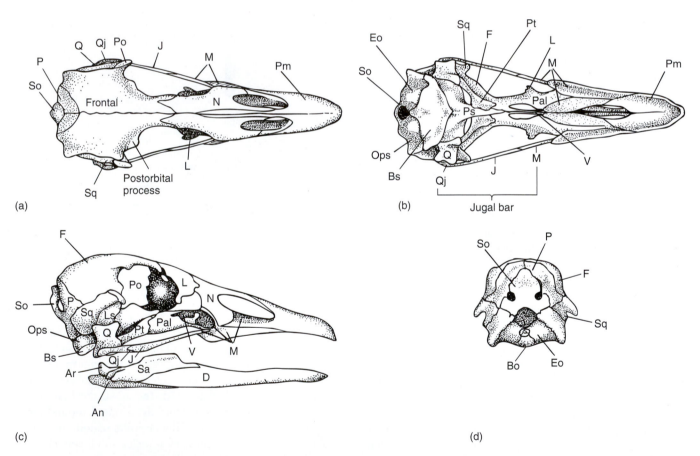

FIGURE 7.46 Bird skull. In the adult bird, sutures between skull bones fuse to obliterate identifiable borders. Dorsal (a), ventral (b), lateral (c), and posterior (d) views of the skull of a young gosling (*Anser*) before bones fuse. Abbreviations: angular (An), articular (Ar), basioccipital (Bo), basisphenoid (Bs), dentary (D), exoccipital (Eo), frontal (F), jugal (J), lacrimal (L), laterosphenoid (Ls), maxilla (M), nasal (N), opisthotic (Ops), parietal (P), palatine (Pal), premaxilla (Pm), parasphenoid (Ps), pterygoid (Pt), quadrate (Q), quadratojugal (Qj), surangular (Sa), supraoccipital (So), squamosal (Sq), vomer (V).

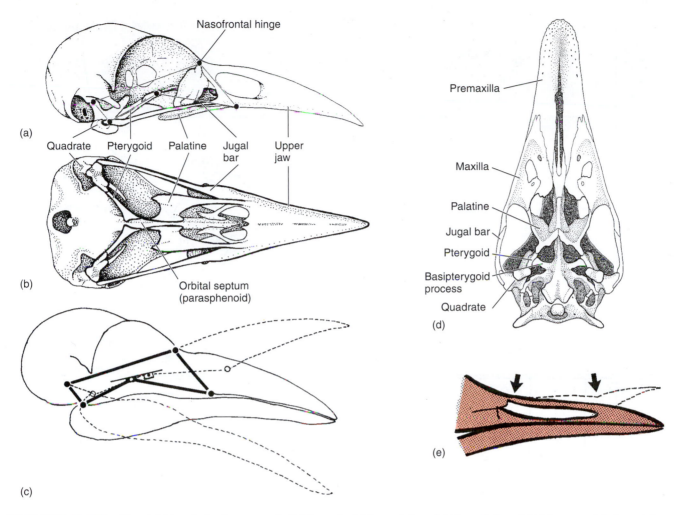

FIGURE 7.47 Cranial kinesis in the crow skull (*Corvus*). (a) Lateral view. (b) Ventral view. (c) Linkage model of cranial kinesis, slider-crank mechanism. From the rest position (solid lines), the point of coupling between pterygoids and palatines slides forward along the orbital septum to a new position (dashed lines), which lifts the upper jaw about the prokinetic joint (nasofrontal hinge). (d) Paleognath. Ostrich palate. (e) Rhynchokinesis. Flexions within the beak permit the tips of the upper and lower jaws to part without opening of the entire mouth.

On the side of the braincase behind the orbit, a large **temporal** bone is formed by the fusion of contributions from all three parts of the skull (figures 7.52c and 7.53). The dermatocranium contributes the squamosal and the **tympanic bulla** (a derivative of the angular) in many mammals. The chondrocranium contributes the **petrosal,** itself a derivative of prootic and opisthotic bones (figure 7.52c). The petrosal often bears a ventrally directed projection, the **mastoid process.** The splanchnocranium contributes three tiny middle ear bones (malleus, incus, stapes) and the styloid (figure 7.54).

In most tetrapods, the nasal capsule remains unossified. However, in mammals, the ethmoid portion ossifies to form scroll-like **turbinates** (turbinals, conchae). There are usually three sets of turbinates attached to respective neighboring bones: the **nasoturbinate, maxilloturbinate,** and **ethmoturbinate.** The coiled walls of the turbinates support the mucous membrane within the nasal passage. Air entering these passages is warmed and moistened before reaching the lungs, functions that are especially important in endotherms. Absent in ungulates but present in most other orders, such as rodents, carnivores, and primates, is another region of the nasal capsule, the mammal **mesethmoid.** This element forms the septum between the nasal capsules and usually remains cartilaginous. Between the nasal area and cranial cavity stands the transverse and finely perforated **cribriform plate** (figure 7.54). Olfactory nerves originating in the olfactory epithelium of the nasal capsule pass through this plate to reach the olfactory bulb of the brain.

Middle Ear Bones   Two profound changes in the lower jaw mark the transition from therapsid to mammal (figure 7.55). Both changes go hand in glove. They result in such an alteration in skull design that some anatomists doubted them until the surprisingly good fossil record made the evolutionary transition undeniable. One of these changes is the loss of the postdentary bones of the lower jaw. The other is the presence of three middle ear bones.

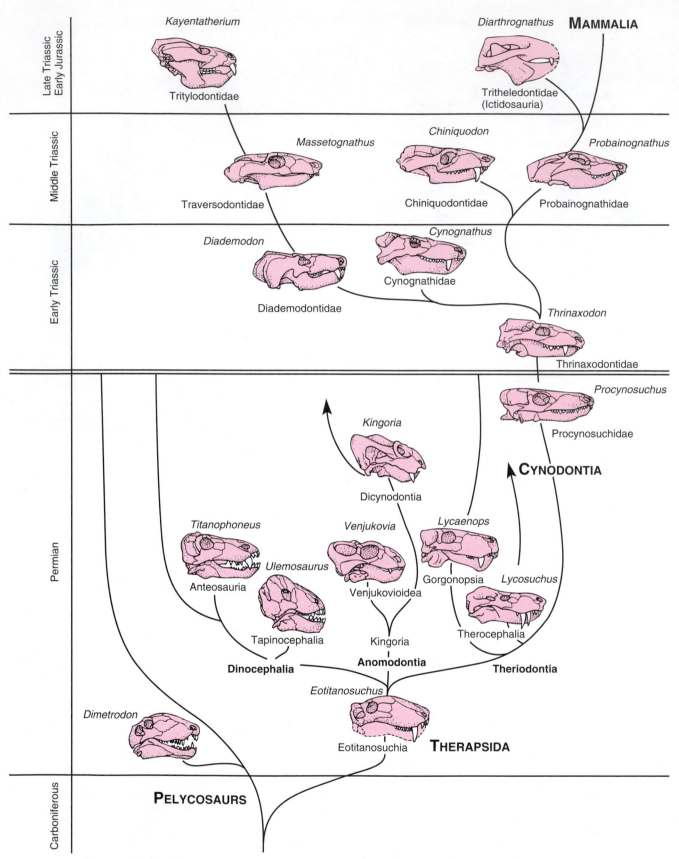

FIGURE 7.48 **Early radiation of therapsids.** Therapsids continue the synapsid lineage and exhibit considerable diversity. For a time in the Permian and Early Triassic, they were fairly abundant. Some were herbivores; most were carnivores. Anomodontia continued into the Cretaceaous; Therocephalians continued to the mid Triassic.

Based on the research of James A. Hopson.

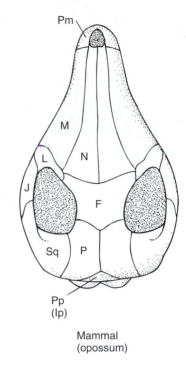

FIGURE 7.49 **Diagrammatic comparison of a derived mammal skull with a primitive amniote skull.** Bones lost in the derived mammal are shaded in the primitive amniote. In mammals, orbital and temporal openings merge. Abbreviations: frontal (F), jugal (J), interparietal (Ip), lacrimal (L), maxilla (M), nasal (N), parietal (P), postfrontal (Pf), premaxilla (Pm), postorbital (Po), postparietal (Pp), prefrontal (Prf), quadratojugal (Qj), squamosal (Sq), supratemporal (St), tabular (T).

Primitive amniote

Mammal (opossum)

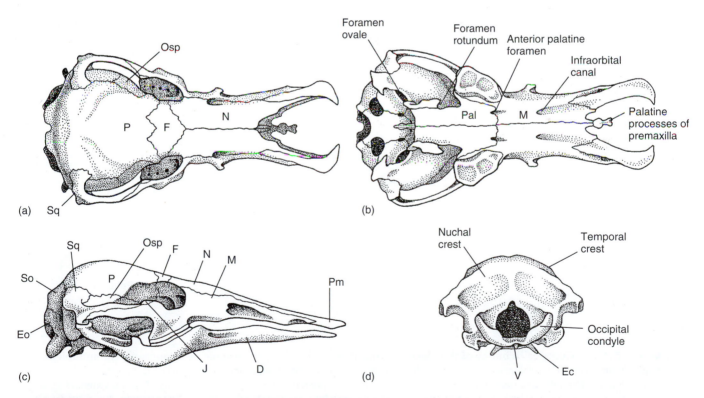

FIGURE 7.50 **Monotreme, skull of the platypus *Ornithorhynchus*.** Dorsal (a), ventral (b), lateral (c), and posterior (d) views. Abbreviations: dentary (D), ectopterygoid (Ec), exoccipital (Eo), frontal (F), jugal (J), maxilla (M), nasal (N), orbitosphenoid (Osp), parietal (P), palatine (Pal), premaxilla (Pm), supraoccipital (So), squamosal (Sq), vomer (V).

(a)

(b)

(c)

FIGURE 7.51 Marsupial, skull of the opossum *Didelphis*. Dorsal (a), palatal (b), and lateral (c) views. Abbreviations: basioccipital (Bo), basisphenoid (Bs), dentary (D), frontal (F), jugal (J), lacrimal (L), maxilla (M), nasal (N), parietal (P), palatine (Pal), premaxilla (Pm), presphenoid (Prs), pterygoid (Pt), squamosal (Sq).

After Carroll.

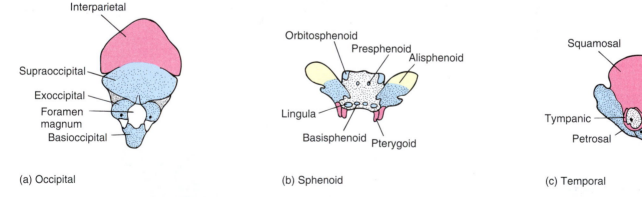

(a) Occipital

(b) Sphenoid

(c) Temporal

FIGURE 7.52 Composite bones of the placental mammal skull during embryonic development, *Homo sapiens*. (a) Occipital bone has centers of ossification that include the interparietal (postparietal), supraoccipital, paired exoccipital, and the basioccipital. (b) The sphenoid bone is a fusion of the orbitosphenoid, presphenoid, basisphenoid, pterygoid, and alisphenoid (epipterygoid). In many mammals, these fused bones are joined by parts of the pterygoid and lingula. (c) The temporal bone results primarily from the merger of the squamosal, tympanic, and petrosal (prootic plus opisthotic).

After Hyman.

Canis dirus

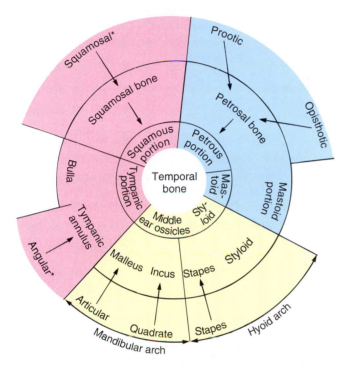

FIGURE 7.53 Mammalian temporal bone. The temporal bone forms phylogenetically from the dermatocranium (angular, squamosal), and chondrocranium (prootic, opisthotic), into which is set contributions from the splanchnocranium (articular, quadrate, stapes, styloid). The separate bony elements in early amniotes (outer circle) contribute to the composite temporal bone of mammals (middle and inner circle). Some of these contributions are dermal bones (*). The otic capsule is buried beneath the surface of the skull, leaving the exposed and often elongated mastoid process. The bulla or auditory bulla forms, at least in part, from the tympanic annulus, itself a phylogenetic derivative of the angular bone. The exposed squamous portion of the temporal bone is illustrated in color on the skull of the Pleistocene wolf *Canis dirus*. Abbreviations: auditory bulla (B), dentary (D), frontal (F), infraorbital foramen (Fio), jugal (J), lacrimal (L), maxilla (M), occipital (Oc), parietal (P), premaxilla (Pm), squamosal (Sq).

In vertebrates, the inner ear is embedded deep within the otic capsule and holds the sensory apparatus responsive to sounds. The hyomandibula or its derivatives deliver sound vibrations to the sensitive inner ear. In all tetrapods, the hyomandibula tends to become reduced to a slender, light bone called the stapes (= columella). Sometimes there is a second hyomandibula-derived bone, the **extracolumella.** The stapes is usually suspended in the middle ear cavity where damping by restrictive attachments is minimized. As sounds set the tympanum into motion, these vibrations are imparted to the small, responsive stapes. Its opposite end often expands to reach the sensitive inner ear apparatus that responds to the vibrations the stapes delivers.

In mammals, two tiny, additional bones join the stapes in the middle ear. Together these bones transmit sound to the inner ear. Specifically, these three middle ear bones are malleus (derived from the articular), incus (derived from the quadrate), and stapes (derived from the hyomandibula). So distinctive is the presence of three middle ear bones that many anatomists mark the fossil transition to mammals at the point of their acquisition.

Anatomy and function of the ear (p. 705)

Coupled with the derivation of the three middle ear bones are changes in the posterior bones of the mandible. In early synapsids (pelycosaurs), the lower jaw includes the tooth-bearing dentary in addition to several postdentary bones (angular, articular, coronoid, prearticular, splenial, surangular) (figure 7.56). In derived synapsids (mammals), this set of postdentary bones has been entirely lost from the lower jaw, and the dentary has enlarged to assume the exclusive role of lower jaw function. From pelycosaur to therapsid to mammal, the anatomical details of these changes are well documented in an ordered time sequence by the fossil record. In pelycosaurs, the articular (future malleus) resides at the back of the mandible and establishes lower jaw articulation with the quadrate (future incus). In early to later therapsids, these two bones become reduced, along with the postdentary bones, eventually moving out of the lower jaw and taking up a position in the middle ear. The functional reason for these changes is thought to be related to improved hearing, especially to a wider range of sounds. The phylogenetic reduction in size of these bones would reduce their mass and thus increase their oscillatory responsiveness to airborne vibrations. Their removal from the jaw joint permits their more specialized role in transmitting sound to the inner ear. Alternatively, or along with such changes related to hearing, some morphologists have proposed that changes in feeding style led to changes in the preferred site of insertion of the jaw-closing muscles, specifically a shift forward on the dentary and closer to the teeth. Larger jaw muscles acting close to the tooth row lessen the stresses at the back of the jaw where it articulates with the skull. Loss of postdentary bones then might reflect this shift in forces forward to the tooth row and away from the joint these bones formed.

FIGURE 7.54 **Diagram of a dog's skull.** Sources of the various bones are outlined: dermatocranium (pink), chondrocranium (blue), and splanchnocranium (yellow).

After Evans.

These changes in the lower jaw were accompanied by changes in the method of food preparation prior to swallowing. Most reptiles bolt their food, swallowing it whole or in large chunks. Mammals typically chew their food before swallowing it, a process termed **mastication.** Mastication also occurs in a few groups of fishes and lizards. But, it is within mammals that feeding strategy is based on mastication of food. If mastication became a more characteristic part of food preparation, then changes in jaw-closing muscles might be expected, with greater emphasis shifting to the dentary.

Secondary Palate and Akinesis   In addition to changes in the mammalian lower jaw, the presence of a secondary palate is also related to mastication. The secondary palate includes a **hard palate** of bone and a posterior continuation of fleshy tissue, the **soft palate** (figure 7.57a, b). The hard palate is formed from the inward growth of bony processes of the premaxilla, maxilla, and palatine that meet at the midline as a bony platform (figure 7.58a–c). This hard palate and its fleshy continuation effectively separate the food chamber below from the respiratory passage above. Some turtles and also crocodilians have a secondary palate, and they benefit from the advantages of separation of routes for food and air. But chewing requires that food stays in the mouth for an extended period of time in mammals; therefore, separation of respiratory and oral passages is especially important. Mastication can proceed without impeding regular breathing. Similarly, the secondary palate completes the firm roof of the food chamber, so that the pumping action of the throat of a suckling infant creates effective negative pressure within the mouth without interfering with the respiratory passage.

Mastication (p. 536)

Mastication in mammals has been accompanied by very precise tooth occlusion to serve the mechanical breakdown of food. Precise, strong occlusion requires a firm skull, so mammals have lost cranial kinesis, leaving them an akinetic skull. The mammalian mandibular condyle fits into a very precise articulation with the squamosal bone. When jaws close about this joint, upper and lower rows of teeth are placed in very precise alignment. This allows specialized teeth to function properly. As a further consequence of precise occlusion, the pattern of tooth eruption in mammals differs from that of most other vertebrates. In lower vertebrates, teeth wear and are replaced continuously **(polyphyodonty);** therefore, the tooth row is always changing. If teeth function primarily to snag prey, this causes little difficulty. However, continuous replacement means that at some location in the jaws, worn teeth are missing or new ones are moving into position. To avoid disruption of occlusion, teeth in most mammals are not continuously replaced. Mammals exhibit **diphyodonty.** Only two sets of teeth erupt during the lifetime of a mammal, the "milk teeth" of the young and the "permanent" teeth of the adults.

Tooth types and their development (p. 507)

The chain of events leading from mastication to akinesis and to diphyodonty should not be viewed as inevitable. Some fishes chew their food but retain kinetic skulls and polyphyodonty. But evolutionary events that produced the mammalian skull underscore the importance of examining anatomical changes in partnership with functional changes that must accompany phylogenetic modification of vertebrate design. Form and function necessarily go together, an issue to which we next turn.

FIGURE 7.55 Evolution of the mammalian middle ear bones. Left column, a medial view of the left mandibular ramus; right column, lateral view of right mandibular ramus and quadrate. No teeth are shown to make comparisons clear. From primitive pelycosaurs, to therapsids, to the first mammals, changes in the postdentary bones are indicated along with incorporation of the quadrate (incus) and articular (malleus) into the middle ear. The fossil species used to follow these changes are shown in relationship to their occurrence in the geological record. Abbreviations: angular (An), articular (Ar), coronoid (C), anterior coronoid (Ca), condyle of dentary (Co), coronoid process (Cp), dentary (D), incus (I), malleus (Ma), prearticular (Pa), quadrate (Q), quadratojugal (Qj), stapes (S), surangular (Sa), splenial (Sp), tympanic annulus (Ty).

Based on the research of James A. Hopson and Edgar F. Allin.

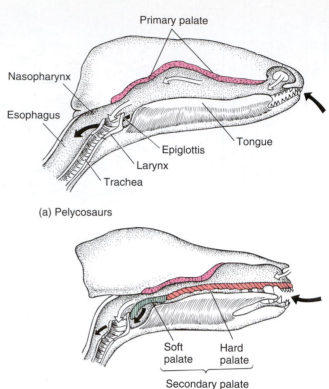

FIGURE 7.57 **Secondary palate.** (a) In most reptiles, air entering the nostrils and food entering the front of the mouth share a common passage, although each exits differently at the back. Air moves down the trachea, and food travels down the esophagus. (b) In mammals, the routes of air and food are separated from the beginning of their travel by the secondary palate, a structure of bone (hard palate) and soft tissue (soft palate). White arrows indicate the path of air; dark arrows indicate the path of food.

FIGURE 7.56 **Changes in jaw articulation during transition from early (pelycosaurs) to late (mammals) synapsids.** In mammals, the postdentary bones of the lower jaw are mostly lost and the dentary enlarges. Bones involved in jaw articulation in pelycosaurs, the articular and the quadrate, become reduced and move in to contribute to the inner ear ossicles of mammals. Jaw articulation in mammals is taken over by the dentary and squamosal. The stapes is not shown.

Overview of Skull Function and Design

The skull performs a variety of functions. It protects and supports the brain and its sensory receptors. It may house cooling equipment to cool the brain during sustained activity or during a rise in environmental temperature. In many active terrestrial mammals, the nasal epithelium lining the nasal passages dissipates excess heat by evaporation as air moves across this moist lining. A similar function has been proposed for the elaborate air passageway in some groups of hadrosaurs, the duck-billed dinosaurs (figure 7.59). Air

entering their nostrils would have coursed through intricate passageways formed within premaxilla and nasal bones to provide evaporative cooling. The skull of many animals also supports the voice box and occasionally serves as a sound resonator to deepen or amplify an animal's call. The Weddell seal takes advantage of its jaws to open and maintain its breathing holes in surface ice (figure 7.60).

These examples remind us that the skull is a multipurpose "tool" involved in a great variety of functions. Its design reflects and incorporates these multiple roles. Generalizations about skull design can be misleading if we ignore its multiple functions. However, if we are cautious, we can understand how skull design reflects fundamental functional problems. The skull primarily functions as part of the feeding system of vertebrates. How it addresses problems of feeding depends largely upon whether feeding occurs in air or in water. Each medium presents different limitations and opportunities. The viscosity of water and the buoyancy of tiny organisms within it mean that water, much more than thin air, holds a richer community of floating planktonic organisms. Suspension feeding and harvesting of these tiny organisms

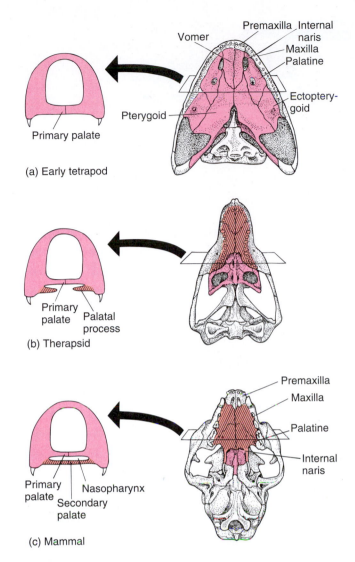

FIGURE 7.58 Evolution of the secondary palate.
(a) Early tetrapod with a primary palate in cross section (left) and ventral (right) views. (b) Therapsid with a partial secondary palate formed by the medial extension of the premaxilla and maxilla. (c) Mammal with a secondary palate that, in addition to extensions of the premaxilla and maxilla, includes part of the palatine bone.

After Smith.

becomes economical, and filter-feeding devices enjoy some adaptive favor. Generally, feeding proceeds in two steps, food capture and then swallowing. We look at each of these in turn.

Prey Capture

Feeding in Water

The first step in feeding is food capture, which depends generally on the medium in which feeding takes place. The higher viscosity of water presents both problems and opportunities for the animal feeding in water. Feeding in water poses a disadvantage in that water easily carries shock or pressure waves ("bow waves") immediately in front of the predator approaching its food. These pressure waves can

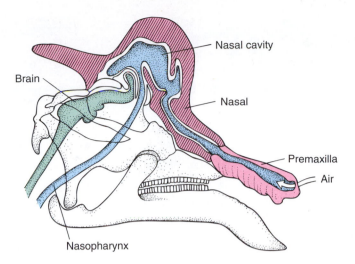

FIGURE 7.59 Air passage of duck-billed dinosaurs. The air passageway is formed by the premaxillae and nasal bones of the hadrosaur. Air that flowed through the nasal cavity on its way to the lungs cooled the nasal epithelial lining and hence the blood flow through it. Although the vascular system of the hadrosaur is not known, if it were similar to some mammals, then this cooled blood might have circulated in such a fashion as to precool the blood flowing to the brain. In this way, the brain was protected from elevated temperatures. Alternatively or additionally, such an expanded air passage might have been a resonating chamber to amplify vocalizations.

After Wheeler.

arrive an instant before the advancing predator and alert or deflect the intended prey. On the other hand, when a vertebrate quickly gulps water into its mouth, the viscosity of the water drags along the prey as well. This viscosity makes **suction feeding,** used with relatively large prey, possible.

To capture small foods, aquatic animals use **suspension feeding.** The density of water gives it the viscosity to retard the fall of particulate material out of suspension. Compared to air, water holds a floating bounty of tiny organic particles and microorganisms, a rich potential nutritional resource for an organism with the equipment to harvest it. Cilia move and control currents of water (and transport captured food), and sticky mucus snatches suspended food from the current of water as it glides by.

Suspension Feeding Suspension feeding is a feeding strategy confined largely, perhaps exclusively, to animals living in water. Some, making an esoteric point, argue that bats "filter" insects "suspended" in air, but this misses the point. Air is too thin to hold suspended food for long. Bats catch or grasp prey, but they do not really use a filtering apparatus nor face the same mechanical problems as aquatic organisms in a viscous medium; so they are not suspension feeders. Most suspension feeders are benthic (bottom-dwelling) organisms or are associated with a herbivorous/detritus feeding style. Respiration and feeding are tightly coupled. Often the same water currents support both activities.

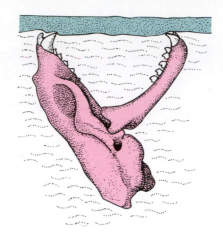

FIGURE 7.60 **Weddell seal.** In addition to feeding, the jaws of this seal are used to ream away ice to open or reopen a frozen breathing hole.

After Kooyman.

Suspension feeders use several methods to intercept and gather nutrients traveling within streams of water. Captured particles are usually smaller than the pores of the filter. They may collide directly with the filter (figure 7.61a), or because of their inertia, they deviate from the streamlines to collide with the mucus-covered surface of the filter (figure 7.61b). Upon impact, the particles cling to the sticky mucus and are rolled up in mucous cords, and then are passed by cilia into the digestive tract.

Less commonly, a sieve can be used to strain suspended particles larger than the pores of the sieve. As the stream of water passes through the sieve, the particles are held back and then collected from the face of the selective filter (figure 7.61c). This method is rare among animals, perhaps because the relatively large particles filtered tend to plug and foul the sieve. The buccal cirri of amphioxus intercept large particles, apparently to prevent them from entering the pharynx and clogging the suspension-feeding system. The gill rakers of bony fishes also remove particulate matter. When the filter becomes clogged, these fishes can clear the material by a kind of cough or quick expansion of the gill arches. Larvaceans (urochordates) abandon their filter when it becomes clogged, secrete a new filter, and continue straining microorganisms from the circulating current of water.

In some invertebrates, the mucus is electrically charged. Mild attraction pulls particles out of suspension and into contact with the walls of the filtering device. However, such mechanisms of suspension feeding are unknown in vertebrates and protochordates.

In amphioxus, the endostyle and lining of the pharynx secrete mucus, which is swept upward by the action of cilia also lining the pharynx. The main current, driven by cilia, passes the cirri upon entry into the mouth and pharynx, through the pharyngeal slits, into the atrium, and via the atriopore exits to the outside environment once again. Small suspended particles in the current pass the pharyngeal bars. Some deviate from the stream of water to collide with and become entrapped in the mucous layer. The mucus and its captured particles are gathered dorsally in the epibranchial groove where they are formed into a mucous cord that other cilia sweep into the digestive tract.

In the ammocoetes larvae of lampreys, suspension feeding is similar to that of amphioxus except that a pair of muscular velar flaps, rather than cilia, beat rhythmically to create the current that flows into the pharynx. Mucus, secreted along the sides of the pharynx, is driven upward by cilia into the epibranchial groove. A row of cilia at the base of this groove forms mucus and captured food into a cord that is passed into the digestive tract. The ventral endostyle of ammocoete adds digestive enzymes to the forming mucous food cord but does not secrete mucus.

Although some envision the ostracoderms as employing new modes of feeding, the absence of strong jaws would have made this unlikely. They seem to have lacked even the muscular tongue of cyclostomes to break up food and place it into suspension. Thus, ostracoderms likely carried forward a suspension-feeding style similar to that of the protochordates before them. Not until gnathostomes do we see a significant tendency away from suspension feeding.

In gnathostomes, suspension feeding is less common. Some actinopterygians use gill rakers like a sieve to filter larger particles from the stream of passing water. The larvae of anurans employ a buccal pump. They draw in a stream of water containing food particles or scrape rock surfaces to enrich the entering stream with these dislodged materials.

The success and efficiency of suspension feeding depends on the size and speed of passing particles. It is most effective with small food particles that neither foul the filter nor escape the sticky mucous lining. To take advantage of large food items, another feeding style evolved, namely, suction feeding.

Suction Feeding Like most fishes, amphibians living in water typically use suction feeding (see figures 7.25 and 7.32). The buccal cavity expands rapidly, pressure

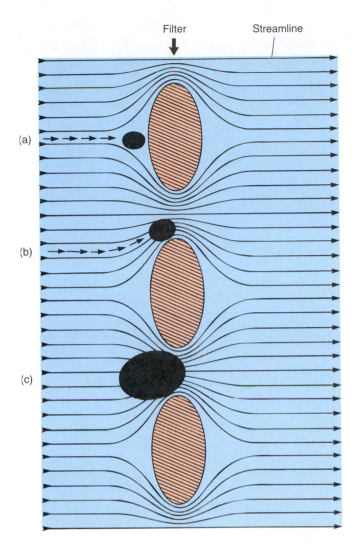

FIGURE 7.61 **Suspension feeding: Interception of suspended food particles.** (a) Direct interception of food particles occurs when particles strike the filtering device. Food is carried in streamlines flowing around bars through openings in the feeding filter. (b) Small, dense particles flow along the streamlines until the fluid is sharply diverted. Particle inertia causes food particles to deviate from the streamlines, collide with the filtering device, and adhere to the mucous coat of the filter. (c) The filtering apparatus can function as a sieve by holding back large particles that fail to pass through the small pores. Cilia drive the food-laden mucus to the digestive tract.

FIGURE 7.62 **Terrestrial feeding by a salamander.** In this filmed sequence, the salamander's jaws open (top). Its tongue starts to project (middle), approaches (bottom), and then makes contact with the prey.

Based on the research of J. H. Larsen.

drops, and food is aspirated into the mouth. Geometry and enlargement of the buccal cavity are controlled by the muscularized visceral skeleton. Excess water, gulped in with the food, is accommodated in several ways. In salamanders prior to metamorphosis and in fishes, gill slits at the back of the mouth offer an exit for excess water. Flow is unidirectional. In salamanders after metamorphosis, in frogs, and in all other aquatic vertebrates, gill slits are absent, so excess water entering the mouth reverses its direction of flow to exit via the same route. Flow is bidirectional. Turtles possess

an expansive esophagus that receives and temporarily holds this excess water until it can be slowly expelled without losing the captured prey.

The early stages in vertebrate evolution took place in water, mostly in marine waters, occasionally in fresh water. Adaptations for feeding and respiration took advantage of these conditions. Adaptations for suspension and suction feeding are present in early vertebrates. With the transition of vertebrates to land and air, neither suspension nor suction feeding provided efficient ways to procure or process food. The jaws became specialized for grasping.

Feeding in Air

Terrestrial feeding in most amphibians and many lizards requires a projectile tongue. The term **lingual feeding** recognizes the use of a rapid, projected, and sticky tongue to capture prey (figures 7.62 and 7.41a–d). However, in many other

---

Skeletal System: The Skull   287

(a)

(b)

FIGURE 7.63 **Swallowing by a rat snake, *Elaphe,* as seen in dorsal view.** (a) Outline of snake's head during successive swallowing motions, left to right. Previous head position is indicated by the dotted outline. With alternating left and right advances, the jaws walk over the prey along a line of progress, the axis of swallowing, until the jaws pass over the entire prey. These jaw-walking displacements place the prey at the back of the throat where contractions of neck muscles move the prey along to the stomach. (b) Movable bones of the skull (shaded) on one side swing outward from the prey and advance farther forward where they come to rest momentarily on the surface of the prey at a new position. Movable bones of the opposite side now take their turn. By such reciprocating motion, jaws walk along the prey. In addition to jaw displacement, the skull itself swings outward from the swallowing axis (arrow) through an angle (β) in the direction of the advancing bones to place them farther along the prey. Abbreviations: ectopterygoid (Ec), maxilla (M), mandible (Md), palatine (Pal), pterygoid (Pt), quadrate (Q), supratemporal (St).

After Kardong.

animals, prey is captured by **prehension,** a method by which the animal rapidly grasps the prey with its jaws. In such animals, the jaws are prey traps, designed to snare the unwary.

As a strategy to capture prey, prehension does not always involve the jaws. Birds of prey snatch quarry with their talons, and mammalian predators often use claws to catch and then control intended prey. Jaws are used secondarily to help hold the struggling victim or to deliver a killing bite.

Swallowing

Once an animal has captured and dispatched its prey, it must swallow the prey in order to digest it. In suspension feeders, the food-laden cords of mucus are swept by synchronized ciliary action into the esophagus. Other animals usually swallow prey whole or in large pieces. Suction feeders rapidly expand the buccal cavity repeatedly to work the captured prey backward into the esophagus. Terrestrial vertebrates use the tongue to reposition the food bolus and work it toward the back of the mouth. The highly kinetic skull of snakes allows great freedom of jaw movement. A snake swallows a relatively large animal by stepping the tooth-bearing bones over the dispatched prey (figure 7.63a, b).

Swallowing mechanisms of terrestrial vertebrates (p. 517)

In many vertebrates, swallowing involves mastication (the chewing of food). Mastication occurs in a few groups of fishes and lizards. Within mammals, mastication has had a profound influence upon skull design, producing an akinetic skull with precise tooth occlusion and only two replacement sets of teeth, a secondary palate, large jaw-closing musculature, and changes in lower jaw structure.

Overview

Cranial Neural Crest

The cranial neural crest cells leave their initial sites next to the neural tube and form streams of cells contributing to the mesenchyme, which differentiates into bone, cartilage, cranial nerve cells, and various other structures in the head. In particular, the hindbrain is segmented into compartments called rhombomeres. Neural crest cells derived from particular rhombomeres migrate into and come to populate particular pharyngeal arches, which in turn give rise to particular cranial structures (figure 7.64). Generally, in tetrapods, neural crest cells from rhombomeres 1 and 2 (and in some taxa, neural crest from forebrain and midbrain) migrate into the first pharyngeal (mandibular) arch, producing the palatoquadrate and Meckel's cartilage, some of the facial bones, and, in mammals, the incus and malleus. Cells of rhombomere 4 enter the second pharyngeal (hyoid) arch, producing the stapes, styloid process, and contribute to the hyoid. Rhombomere 6 releases neural crest cells into pharyngeal arches 3 and 4, which contribute to the hyoid and to the thyroid cartilages. Rhombomere 7 also contributes to pharyngeal arch 4. Cells from rhombomeres 3 and 5 do not migrate through their adjacent mesenchyme, but instead enter the streams of neural crest cells on either side of them.

The largest animal ever is the blue whale. It makes its living as a filter feeder. The filtering device is the baleen, a brushlike specialization of keratin (not enamel) from the oral epithelium occupying the site where teeth might otherwise be expected in the upper jaw. "Whalebone" is a misnomer for baleen. The term is inaccurate because no bone occurs within the baleen. Because baleen is frayed, it acts like a strainer to hold back food from the stream of water passed through it. Food preference depends a little on species, but most baleen whales strain small fishes or shrimplike crustaceans called "krill," which school or gather in dense swarms. Food collected in the baleen is licked free by the tongue and swallowed.

Blue and humpback whales represent one subgroup of baleen whales called fin whales, or rorquals. Right whales are the other subgroup. In both groups, the teeth are absent, the baleen is present, and the skull is long and arched to hold the filter-feeding equipment.

To feed, right whales part their jaws slightly and swim through swarms of krill. The stream of water enters the front of the mouth and passes out the lateral suspended wall of baleen. Here the krill become entangled in the frilly baleen and are licked up and swallowed (box figure 1a). The blue whale feeds differently. As it approaches a school of fishes or krill, it opens its mouth wide to swim over and engulf the concentrated prey and accompanying water. Pleated furrows along its neck and belly allow the throat to inflate like a pouch and fill with this huge mass of water (box figure 1b).

BOX FIGURE 1   Whales feeding. (a) The right whale has long plates of baleen suspended from its upper jaw, and it feeds by swimming through plankton with its mouth parted. Water enters, passes along the sides of the tongue, and then departs through the curtain of baleen, leaving the plankton entangled in the baleen. (b) As a fin whale approaches a concentration of planktonic organisms, usually krill, it opens its mouth and engulfs these organisms together with the water in which they reside. Its pleated throat allows for considerable expansion of the mouth in order to accommodate the plankton-filled water. The whale lifts its throat to force water out through its baleen, which holds back the food but allows excess water to filter out. Its tongue licks this food from the baleen and swallows it.

Somewhat surprisingly, in at least tetrapods, cranial neural crest cells that migrate into the first pharyngeal arch form separate populations, one dorsal and one ventral. The ventral population forms both elements of the mandibular arch, the palatoquadrate and Meckel's cartilage. The dorsal population does not contribute to the jaws, as once thought, but instead contributes to parts of the chondrocranium.

Hox genes expressed in various combinations and regions of the neural crest (figure 7.64) instruct the cells quite early as to which tissues they will form. Although details are still forthcoming, many major evolutionary changes in vertebrate jaws, teeth, and facial bones seem based on changes in the placement or instructed fate of

neural crest cells. For example, a *Hox* gene is expressed in the mandibular arch of a species of lamprey, but not in gnathostomes. This apparent loss of *Hox* gene expression may have been the genetic basis for facilitating the evolution of jaws in gnathostomes.

Neural crest (pp. 244, 290)

Emergence of Mammals

Mammals brought many innovations to the vertebrate design; several involved the skull. One change already noted is in the lower jaw. In early amniotes, as in gnathostomes

Up to 70 tons of water are temporarily held in the expanded throat. The whale then contracts the swollen pouch, forcing water through the baleen, where the food is strained, collected up by the tongue, and swallowed.

Humpback whales have been observed to release air bubbles while circling a school of prey swimming above them. As the air bubbles rise, they form a "bubble cloud" that may corral or drive the school up to the surface ahead of the whale. The bubble cloud may also immobilize or confuse schools of prey, causing them to clump together, or it may disguise the whale as it surges upward with its mouth open through the center of the bubble cloud. Some humpbacks begin feeding at the surface by lopping or slapping their tail against the water as they dive. Just as the tail flukes are about to reenter the water, a whale flexes its tail so that the flukes hit the water, leaving a bubbling effervescence at the surface. This is thought to startle prey and stimulate them to clump tightly together in a school. The whale then releases a bubble cloud as it dives, which it follows by a feeding lunge back up through the bubble cloud to collect the prey in its mouth.

The oldest fossil whales come from the Oligocene and bear unmistakable resemblances to primitive terrestrial mammals. Distinct incisor, canine, premolar, and molar teeth were present. From these earliest whales, two major modern lines soon arose. One is the baleen whales, formally called the mysticetes. The other major line of whales is the toothed whales, or the odontocetes, including sperm, killer, and other whales with teeth.

In both baleen and toothed whales, the skull is telescoped. Some bones are pushed together and even overlap, yet a long snout persists (box figure 1c). In the odontocetes, the backward lengthening of the facial bones creates the snout. In the mysticetes, the occipital bones are pushed forward. Although achieved differently, the result is the same—to reposition the nostrils to a more central and dorsal position. When a whale surfaces to breathe, this position of the nostrils allows easy venting of the lungs and drawing in of fresh air without the whale having to tip its entire head out of water.

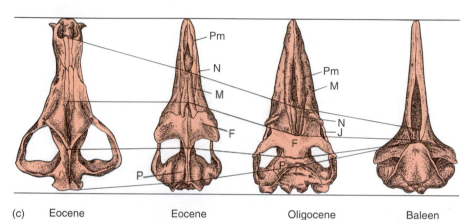

(c)   Eocene   Eocene   Oligocene   Baleen
        carnivore   cetacean   whale   whale
        (*Andrewsarchus*)   (*Prozeuglodon*)   (*Actiocetus*)

(c) Skulls of whales have been highly modified during their evolution, especially the design of the face and the position of the nostrils. *Andrewsarchus*, a terrestrial carnivorous ungulate from the Eocene, may have belonged to a group from which early cetaceans may have evolved. For comparison, Eocene (*Prozeuglodon*), Oligocene (*Actiocetus*), and a modern baleen whale are illustrated. Although not on a direct evolutionary line with each other, these comparisons show the changes in cetacean skull design, especially in the facial region. Abbreviations: frontal (F), jugal (J), maxilla (M), nasal (N), parietal (P), premaxilla (Pm).

(a, b) After Pivorunas; (c) after Olsen; Romer.

generally, the jaws articulate with the braincase via the articular-quadrate joint. In mammals, this is quite different. The jaws articulate via the dentary-temporal joint. Several postdentary bones become lost during this transition to mammals; the quadrate and articular move to the middle ear. The dentary expands posteriorly to form a new articulation with the skull, namely, via the dentary-temporal joint. Although the factors favoring these changes are disputed, the reality of these changes is not. Bones located at the back of the primitive amniote lower jaw were either lost or altered in function from jaw articulation to hearing. But this fact raises a new problem. How could bones involved in jaw suspension change function without disrupting the intermediate species? If postdentary bones moved to the middle ear, how could they abandon jaw suspension without producing an individual with no method of supporting the jaw against the skull? G. Cuvier, nineteenth-century French anatomist, would have denied such a possibility. He argued that evolution could not occur for just such a reason, because a change in structure would disrupt function and stop evolution in its tracks before it had begun.

Diarthrognathus, a late cynodont close to primitive mammals, suggests an answer. Its name means two (di-) sites of articulation (arthro-) of the jaw (-gnathus). In addition to the articular-quadrate joint inherited from reptiles, a dentary-squamosal joint was apparently present. We do

(a) (b) (c)

FIGURE 7.64 Cranial neural crest migration and *Hox* genes, generalized tetrapod. (a) Various cranial structures derived from particular pharyngeal arches. (b) In turn, these pharyngeal arches are themselves populated by neural crest migrating (arrows) out of the rhombomeres of the hindbrain. (c) Embryo showing location of pharyngeal arches and hindbrain. *Hox* expression patterns in the neural crest show the boundaries of these gene domains. Key to abbreviations: r2–r8, rhombomeres 2–8.

(a, b) After McGinnis and Krumlauf; (c) after Carlson.

not know the feeding style of *Diarthrognathus*, so we cannot be sure of the biological role played by this second articulation.

What do living vertebrates suggest? Some birds, such as the skimmer, for example, feed by holding their lower jaw just below the water's surface and flying swiftly along until they strike a fish. Then the jaws snap shut to snatch the fish. A secondary articulation seems to strengthen the lower jaw and help prevent its dislocation as it collides with the fish. *Diarthrognathus* did not feed on fish, but it may have wrestled with struggling prey or fought with competitors. A second jaw articulation would make the jaw stronger. Whatever its advantages, a dentary-squamosal joint was established before the postdentary bones departed from the lower jaw; therefore, when the quadrate and articular bones departed, an alternative method of lower jaw-skull articulation was already in place. This is significant because loss or movement of these bones to support hearing did not disrupt the function they abandoned, jaw suspension. The existing dentary-squamosal articulation was in a sense "ready to serve," preadapted for a new or expanded function.

Preadaptation (p. 18)

Probainognathus, another late cynodont, like *Diarthrognathus* exhibits a posterior extension of the dentary to establish a secondary point of jaw articulation with the skull (figure 7.65). *Probainognathus*, *Diarthrognathus*, and several other late cynodonts with similar transitional double jaw articulations suggest how a harmonious transition in form and function might have occurred. They remind us again that a series of anatomical changes alone is an incomplete statement about evolutionary events. The anatomical changes must be coupled with hypotheses about the accompanying functional series of changes. Form and function go together, and both must receive attention if we are to bring some understanding to the process of evolutionary change.

Evolutionary Modifications of Immature Forms: Akinesis in Mammals

We sometimes forget that an evolutionary modification can debut in an embryonic or infant stage and later become incorporated or expanded in the adult. Such may have been the case with akinesis in mammals. In all mammals, infants suckle milk from their mothers. Suckling requires a pump and a seal. Fleshy lips provide the

FIGURE 7.65 **Double jaw articulation.** Skull of *Probainognathus*, a late cynodont (therapsid). Shown in lateral (a) and ventral (b) views, double jaw articulation occurs between the quadrate and articular (solid arrow), the primitive condition, and another articulation occurs between dentary and temporal (squamosal) (open arrow) that came to predominate in later mammals. (c) Posterior view, with window cut to show the stapes and its contact with the reduced quadrate. The new articulation, dentary-temporal (via squamosal), open arrow, as well as the primitive jaw articulation, quadrate-articular, are present. Abbreviations: angular (An), articular (Ar), basioccipital (Bo), basisphenoid (Bs), dentary (D), epipterygoid (Ept), jugal (J), lacrimal (L), maxilla (M), nasal (N), parietal (P), prefrontal (Prf), premaxilla (Pm), prootic (Pro), postorbital (Po), pterygoid (Pt), quadrate (Q), surangular (Sa), septomaxilla (Sm), squamosal (Sq), vomer (V).

After Carroll; after Romer.

seal around the teat of the mammary gland, the mouth is the chamber that receives the milk, and the up-and-down action of the tongue pumps the milk from the mother to the infant's mouth and esophagus. If respiration and feeding shared a common chamber, as in most reptiles, the infant would have to interrupt nursing and release its attachment to the nipple in order to breathe. A secondary palate makes this inefficient interruption in feeding unnecessary. It separates feeding from breathing by separating the mouth from the nasal chambers. But a secondary palate that separates the mouth from the nasal passages also fuses left and right halves of the skull, thus preventing any movement within or across the braincase. The result is an akinetic skull.

Further changes in the adult evolved later. With loss of kinesis, the skull is firm and ready to serve strong jaw-closing muscles. Mastication, development of specialized teeth to serve chewing (accurately occlusal tooth rows), and a muscular tongue (to move food into position between tooth rows) might then find adaptive favor. Certainly there are other ways to chew food. Some fishes with kinetic skulls and teeth that are continuously replaced chew their food. In mammals, the conditions seem especially favorable for mastication, and we find this adaptation in almost all mammalian species, including its appearance in some late therapsids. Analysis of evolutionary events often centers on adult stages, yet understanding of these events must come from a knowledge of the entire life history of species.

Composite Skull

Dermatocranium, chondrocranium, and visceral cranium contribute to the skull. Although their phylogenetic backgrounds are different, parts of each combine into a functional unit, the skull. If species were stamped out one at a time, each being a unique creation, then there would be little reason to expect a composite skull. Yet, clearly the skull is a combination from different phylogenetic sources. In Darwin's day, as now, we can point to this as evidence that supports evolution, not special creation. Evolution is behind the origin of new structures and the species that display them.

The skull also illustrates a point made earlier. Evolution proceeds by remodeling, seldom by new construction. Consider the splanchnocranium. From fishes to mammals, it is variously modified to serve the adaptive demands of the organism of the moment, first as support for gill slits, and then as the source of jaws, as support for the tongue, and in mammals as part of the ear (figure 7.66). In this series of transformations seen in retrospect, we have no reason to believe that the splanchnocranium of the first fish anticipates its eventual contribution to the hearing devices of mammals. Evolution does not look ahead. The sequences or trends we see are apparent only after the fact, after evolution has taken place. There is no evolutionary arrow anticipating the future. We have seen that changes in the splanchnocranium, chondrocranium, and dermatocranium serve biological roles of the moment. They are not predestined changes preparing for the future.

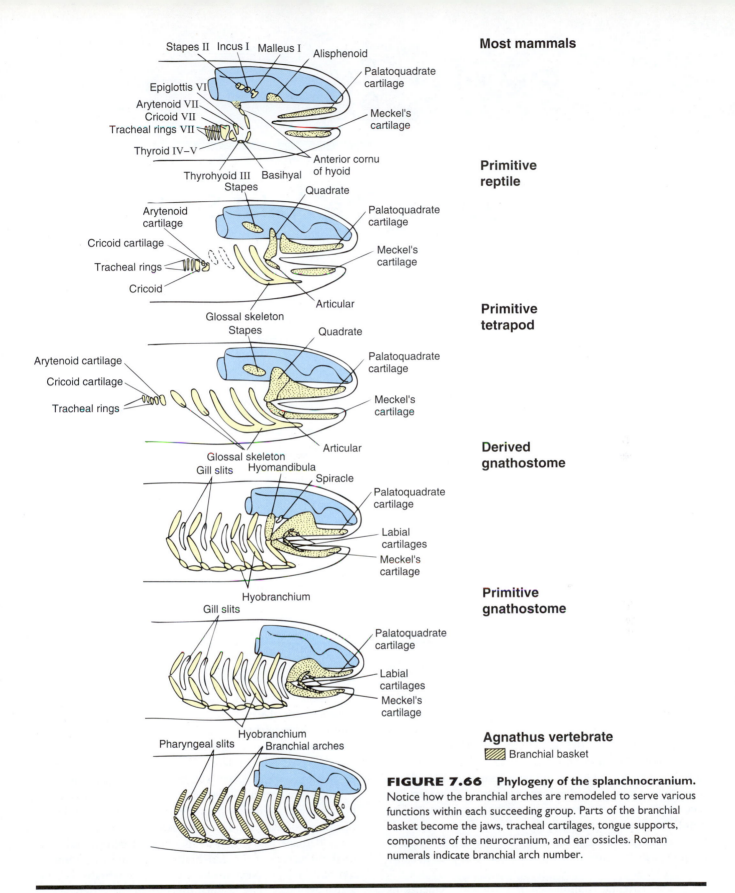

Most mammals

Stapes II Incus I Malleus I Alisphenoid
Palatoquadrate cartilage
Epiglottis VI
Arytenoid VII
Cricoid VII
Tracheal rings VII
Meckel's cartilage
Thyroid IV–V
Anterior cornu of hyoid
Thyrohyoid III Basihyal

Primitive reptile

Stapes Quadrate
Arytenoid cartilage
Palatoquadrate cartilage
Cricoid cartilage
Meckel's cartilage
Tracheal rings
Cricoid
Articular

Primitive tetrapod

Glossal skeleton
Stapes Quadrate
Arytenoid cartilage
Palatoquadrate cartilage
Cricoid cartilage
Tracheal rings
Meckel's cartilage
Articular

Derived gnathostome

Glossal skeleton
Gill slits Hyomandibula
Spiracle
Palatoquadrate cartilage
Labial cartilages
Meckel's cartilage
Hyobranchium

Primitive gnathostome

Gill slits
Palatoquadrate cartilage
Labial cartilages
Meckel's cartilage
Hyobranchium

Agnathus vertebrate

Pharyngeal slits Branchial arches

▨ Branchial basket

FIGURE 7.66 **Phylogeny of the splanchnocranium.**
Notice how the branchial arches are remodeled to serve various
functions within each succeeding group. Parts of the branchial
basket become the jaws, tracheal cartilages, tongue supports,
components of the neurocranium, and ear ossicles. Roman
numerals indicate branchial arch number.

WEB LINK

Visit the text website at *www.mhhe.com/kardong7e* for
additional study aids including selected references, functional
laboratories, and web links to chapter topics.

Skeletal System: The Axial Skeleton

INTRODUCTION

BASIC COMPONENTS
Vertebrae
Regions of the Vertebral Column
Centra
Ribs
Sternum
Gastralia

EMBRYONIC DEVELOPMENT
Fishes
Tetrapods

PHYLOGENY
Fishes
Agnathans
Gnathostomes

Tetrapods
Early Tetrapods
Amniotes

FORM AND FUNCTION
Fluid Environment
Terrestrial Environment
Design of Vertebrae
Direction of the Neural Spine
Height of the Neural Spine
Regionalization of the Vertebral Column

OVERVIEW

Introduction

Two structural components combine to define the long axis of the vertebrate body, offer sites for muscle attachment, prevent telescoping of the body, and support much of the weight. One is the notochord, the other, the vertebral column. The **notochord** is a long, continuous rod of fibrous connective tissue wrapping a core of fluid or fluid-filled cells. The **vertebral column** consists of a discrete but repeating series of cartilaginous or bony elements. The notochord is phylogenetically the oldest of the two structural components, but it tends to give way to the vertebral column, which assumes the role of body support in most later vertebrates.

The vertebral column debuts early with evidence of protovertebrae, segmental blocks along a notochord in the fossils *Haikouella* and *Haikouichthys*. Living hagfishes possess a notochord but adults lack vertebrae. In lampreys, a few small cartilaginous elements, neural arches and spines,

ride upon a prominent notochord, but the main bodies of vertebrae are absent. The evolution of the vertebral column in fishes and tetrapods is complicated, partly because some components became enlarged, others were lost, and some evolved independently several times. Events of early embryology, which often can be consulted to help clarify such phylogenetic uncertainties, fail us in this instance because opinions differ about even basic comparative details of embryonic events in living forms.

Perhaps it is best to begin with a generalized view of the structure of the vertebral column and the companion terminology. Next we look to each vertebrate class for actual examples. The original function of the vertebrae was to protect the spinal cord and dorsal aorta. Later, vertebrae became important as attachment sites for body musculature. In tetrapods, their roles expanded to include suspension of the body and locomotion on land.

Basic Components

Vertebrae

The first components of the vertebra to appear were the dorsal and ventral arches that rested upon the notochord (figure 8.1). The dorsal arches, **neural** and **interneural** (intercalary) **arches,** protected the neural tube. The ventral arches, **hemal** and **interhemal arches,** enclosed blood vessels. The next stage in the evolution of the basic elements of a vertebra was the formation of two **centra:** an **intercentrum** (hypocentrum) and a **pleurocentrum.** The bases of the ventral arches expand to form these centra where they meet the notochord. The centra served to anchor and support these arches.

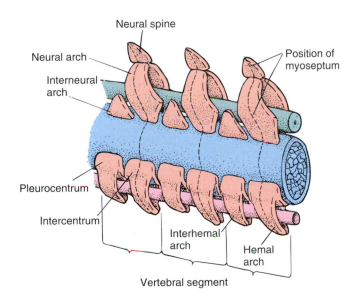

FIGURE 8.1 **The axial skeleton of a generalized primitive gnathostome.** Three vertebral segments from the tail region are illustrated. The notochord is prominent. Vertebral elements are represented by a pair of centra (intercentrum, pleurocentrum), by their associated ventral arches (hemal, interhemal), and by dorsal arches (neural, interneural) that often support a neural spine. Dashed lines indicate the location of myosepta, connective tissue sheets that bound each section of body musculature.

Regions of the Vertebral Column

Each **vertebral segment** consists of arches and centra: up to two dorsal arches (neural and interneural), as many as two ventral arches (hemal and interhemal arches), and often two centra (intercentrum and pleurocentrum). The evolution of these basic vertebral elements is characterized first of all by enlargement of some elements at the expense of others. Second, the vertebral components generally come to displace the notochord as the primary mechanical axis of the body. Third, the vertebral segments composing the axial column tend to become regionally differentiated within the vertebral column they collectively define (figure 8.2). In most fishes, the vertebral column is differentiated into two regions: an anterior **trunk region** and a posterior **caudal region.** In tetrapods, the trunk becomes further differentiated anteriorly into the neck, or **cervical region,** and posteriorly into the hip, or **sacral region.** In early tetrapods, a **postsacral region** of up to five or six vertebrae, bearing the fused, proximal bases of their respective ribs, typically continues behind the sacral region. In some tetrapods, there is further differentiation of the trunk into the chest, or the **thoracic region,** and into the area between the thorax and the hips, the **lumbar region.**

Centra

Among vertebrates there is great variation in the structure of the centra, in the relative importance of the pleurocentrum compared with the intercentrum, in the extent of ossification, and in the degree to which centra supplement or replace the notochord as mechanical elements of the axial column. Each centrum constitutes the **body** of the vertebra. In some vertebrates, centra may be absent **(aspondyly).** Others exhibit one **(monospondyly)** or two **(diplospondyly)** centra per segment. In many anamniotes, the caudal vertebrae may be diplospondylous, and the trunk vertebrae monospondylous. In some groups, the single centrum in trunk vertebrae is derived from the pleurocentrum; in others, it is derived from the intercentrum. In Holocephali and Dipnoi, the number of centra may secondarily increase to five or six per segment **(polyspondyly).** In amniotes, the pleurocentrum predominates and becomes the body of each vertebral segment. Remnants of the intercentrum contribute to cervical vertebrae and perhaps in a minor way to other vertebral elements, but for the most part, the intercentrum becomes the intervertebral cartilage (disk) of the amniote vertebral column.

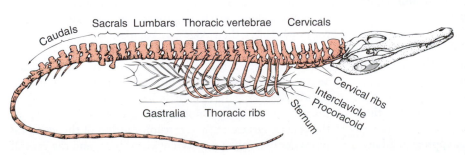

FIGURE 8.2 **Regions of the vertebral column.** The chain of vertebrae composing the axial column differentiates into two regions (trunk and caudal) in all fishes, but becomes differentiated into up to five regions (cervical, thoracic, lumbar, sacral, and caudal) in amniotes. The vertebral column of an alligator depicted in the figure illustrates these five regions.

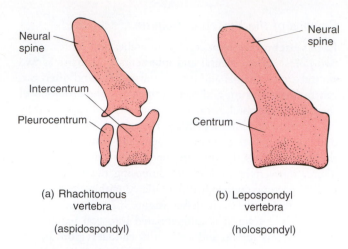

Neural spine

Intercentrum

Pleurocentrum

Centrum

Neural spine

(a) Rhachitomous vertebra

(aspidospondyl)

(b) Lepospondyl vertebra

(holospondyl)

FIGURE 8.3 **General vertebral types.** (a) An aspidospondyl vertebra is characterized by ossified elements that remain separate. The specific type illustrated is a rhachitomous vertebra that has three discrete parts: pleurocentrum, intercentrum, and neural spine. (b) A holospondyl vertebra is characterized by fused construction of all components. The specific type shown is a lepospondylous vertebra, a holospondyl vertebra with a husklike centrum.

In tetrapods, a descriptive terminology has grown up to characterize the two general anatomical relationships between centra and their neural arches (figure 8.3a, b). In one condition, termed **aspidospondyly,** all elements are separate. Specifically, the three arch elements (intercentrum, pleurocentrum, and neural arch), either paired or single, remain as separate ossified elements. The **rhachitomous vertebra,** found in some rhipidistians and some early tetrapods, is a specialized type of aspidospondylous vertebra. The term means "cut-up spine," a reference to the numerous separate parts that constitute each vertebral segment (figure 8.3a). Both large intercentrum and small pleurocentrum are usually paired, and a separate neural arch is associated with them. Several derived aspidospondylous vertebral types are recognized. In the **embolomerous vertebra,** separate but approximately equal-sized centra are present. In the **stereospondylous vertebra,** the vertebra consists of a single body derived entirely from the intercentrum.

In the other general vertebral condition of tetrapods, termed **holospondyly,** all vertebral elements in a segment are fused into a single piece. Typically, the fused centrum derives from a pleurocentrum. The intercentrum, if present, remains as an unossified contribution to the intervertebral cartilage between vertebrae. In the **lepospondylous vertebra,** a specialized type of holospondylous vertebra, the centrum of the solid vertebra is husk shaped and usually perforated by a notochordal canal (figure 8.3b).

At one time, the vertebral type was used as the major criterion to define tetrapod taxa, and each type was thought to characterize a separate phylogenetic trend. With this taxonomic emphasis came a proliferation of descriptive terminology to track supposed vertebral phylogeny; however, problems became evident with this approach. Many

early tetrapods evolved from an aquatic ancestry into new terrestrial habitats, and their vertebrae became modified to accommodate life on land, where walking predominated. But other derived tetrapods reinvaded or secondarily returned to aquatic habitats, where swimming received renewed emphasis. The vertebrae of these later, but secondarily aquatic tetrapods are similar to the vertebrae of the most primitive and predominantly aquatic tetrapods. Thus, the morphologically similar types of vertebrae represent multiple evolutionary trends, testimony to functional convergence but not evidence of close phylogenetic unity. Consequently, much of the elaborate terminology, based on the mistaken assumption of close phylogenetic affinity, has been largely abandoned, although a few terms linger on in taxonomic usage. For example, a temnospondylous vertebra formally designated a vertebra of several parts with a separate arch, but in descriptive usage, this meaning has now been assigned to other terms. Further, other surviving terms, although developed in reference to tetrapods, are now often applied to vertebrae of fishes as a descriptive convenience.

The centra are linked successively into a chain of vertebrae, the **axial column.** The shapes of surfaces at the articular ends of the centra affect the properties of the vertebral column and the way in which forces are distributed between vertebrae. A functional scheme for classifying centra might be desirable, but analysis of their complicated mechanical functions has proved difficult and remains incomplete. Thus, the more traditional anatomical criteria employing articular shape are used more often, yielding several types of centra.

Centra with flat ends are **acoelous (amphiplatyan)** and seem especially suited to receive and distribute compressive forces within the vertebral column (figure 8.4a). If each surface is concave, the centrum is **amphicoelous,** a design that seems to allow limited motion in most directions (figure 8.4b). Centra that are concave anteriorly and convex posteriorly are **procoelous** (figure 8.4c). The reverse shape, concave posteriorly and convex anteriorly, characterizes centra that are **opisthocoelous** (figure 8.4d). Centra that are **heterocoelous** bear saddle-shaped articular surfaces at both ends (figure 8.4e). In both procoelous and opisthocoelous centra, the convex articular surface of one centrum fits into the concave surface of the next to form a kind of ball-and-socket joint, permitting extensive motion in most directions without stretching the nerve cord that their neural arches protect. By comparison, if acoelous or amphicoelous vertebral series are flexed, adjacent centra hinge about their edges. If rotation is extensive, then like opening a door, the space will tend to widen between centra and stretch the central nerve cord running dorsally above them (figure 8.5a). However, in procoelous and opisthocoelous centra, with ball-and-socket articulation, the point of rotation is not at the edge but at the center of the convex surface of the centrum. Flexion of the vertebral series does not open a space between them, and the central

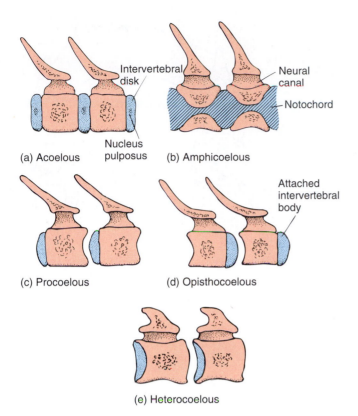

FIGURE 8.4 General centra shapes. The shapes of articulating centra ends, as viewed in sagittal section, define specific anatomical types: (a) acoelous, both ends are flat; (b) amphicoelous, both ends are concave; (c) procoelous, anterior end is concave; (d) opisthocoelous, posterior end is concave; (e) heterocoelous, saddlelike articulating ends. Anterior to the right.

After Kent.

This anatomical classification includes only the criterion of centrum shape, but soft tissues are often associated and are usually extremely important in affecting function. The notochord, or its adult derivatives, often runs through and fills the concavities at the articular ends of centra, which are capped by cartilaginous pads. The term *intervertebral disk* has been broadly used to designate any pad of tissue between articular surfaces of centra. However, strictly speaking, an intervertebral disk in the adult is a pad of fibrocartilage whose gel-like core, the **nucleus pulposus,** is derived from the embryonic notochord. By this strict definition, intervertebral disks are found only in mammals, in whom they reside between successive surfaces of adjacent centra. In other groups, the pad between centra is called an **intervertebral cartilage** or **body.** Joining the rims of adjacent centra is the **intervertebral ligament,** which is important in controlling the stiffness of the vertebral column when it flexes.

Apophyses, processes that will be described more fully as we meet them during the survey of the axial column later in the chapter, project from centra and from their arches. In general, apophyses include **diapophyses** and **parapophyses,** both of which articulate with ribs. The **basapophyses** are paired ventrolateral processes, remnants of the hemal arch bases that may receive the articulation with ventral ribs. Apophyses also form anti-twist, interlocking processes, or **zygapophyses,** between successive vertebrae. The term **transverse process** generally applies to any process extending from the centrum or the neural arch, but historically it has been used so loosely that it retains no exact morphological meaning.

Ribs

Ribs are struts that sometimes fuse with vertebrae or articulate with them. Ribs provide sites for secure muscle attachment, help suspend the body, form a protective case around viscera (rib cage), and sometimes serve as accessory breathing devices. Embryologically, ribs preform in cartilage within **myosepta** (myocommata), that is, within the dorsoventral

nerve cord is not unduly stretched (figure 8.5b). Heterocoelous centra allow great lateral and vertical flexion, but they prevent wringing or rotation of the vertebral column about its long axis (figure 8.5c). Heterocoelous centra are most common in turtles that retract their necks and in cervical vertebrae of birds.

FIGURE 8.5 Centra functions. (a) Amphicoelous or acoelous centra flex about a point on their rims, tending to stretch the centrally located dorsal nerve cord. (b) Opisthocoelous and procoelous centra eliminate this potentially damaging stretching tendency with ball-and-socket ends that establish a centrally located point of rotation instead of one at the rims. (c) In heterocoelous centra, opposite, saddle-shaped surfaces fit together, allowing extensive lateral and dorsoventral rotation. Ventral view of two vertebrae of the ostrich *Struthio.*

(a) Fish ribs

(b) Trunk vertebra

(c) Caudal vertebra

FIGURE 8.6 **Ribs.** (a) In fishes, dorsal ribs develop where myosepta intersect with the horizontal septum, and ventral ribs develop where myosepta meet the wall of the coelomic cavity. (b) Cross section of trunk vertebra of a fish. (c) Cross section of caudal vertebra of a fish. Ventral trunk ribs are serially homologous with caudal hemal arches.

sheets of connective tissue that partition successive blocks of segmental body musculature (figure 8.6a–c). In labyrinthodonts, the postsacral region bears short ribs. But in most tetrapods, the postsacral region is absent, and caudal vertebrae never develop ribs.

In many fishes, there are two sets of ribs with each vertebral segment, a dorsal and a ventral set. The **dorsal ribs** form at the intersection of each myoseptum with the **horizontal septum** (horizontal skeletogenous septum), a longitudinal sheet of connective tissue (figure 8.6a). The **ventral ribs** form at points where the myosepta meet the walls of the coelomic cavity. They are serially homologous with the hemal arches of the caudal vertebrae (figure 8.6c). In tetrapods, one of these sets of ribs is lost and the other, apparently the dorsal ribs, persists to become the trunk ribs of terrestrial vertebrates. Ribs of primitive tetrapods are **bicipital,** having two heads that articulate with the vertebrae. The ventral rib head, or **capitulum,** articulates with the **parapophysis,** a ventral process on the intercentrum. The dorsal head, or **tuberculum,** articulates with the **diapophysis,** a process on the neural arch (figure 8.7). If these vertebral processes fail to develop, the articular surface persists, forming a small concavity, the **facet,** to receive the rib. In amniotes, the intercentrum is lost or incorporated into other elements, so the capitulum must shift its articulation to the pleurocentrum (in most reptiles and birds) or between centra (in mammals).

Although ribs function in locomotion in tetrapods, they become an increasingly important part of the respiratory system to move air through the lungs. Classification of tetrapod ribs is based on the type of association they establish

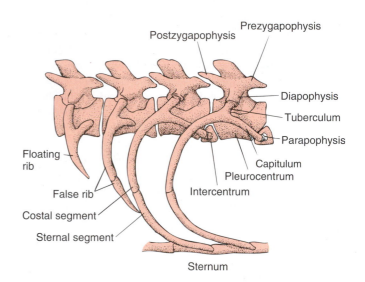

FIGURE 8.7 **Amniote ribs.** Ribs are named on the basis of their articulation with the sternum (true ribs), with each other (false ribs), or with nothing ventrally (floating ribs). Primitively, ribs are bicipital, having two heads, a capitulum and a tuberculum, that articulate respectively with the parapophysis on the intercentrum or the diapophysis on the neural arch. The body of the rib may differentiate into a dorsal part, the vertebral rib or segment, and a ventral part, the sternal rib or segment that articulates with the sternum.

with the sternum. Ribs that meet ventrally with the sternum are **true ribs.** Those that articulate with each other but not with the sternum are **false ribs.** Those false ribs articulating with nothing ventrally are **floating ribs.** True ribs consist of two jointed segments, the **vertebral (costal) rib,** a proximal segment articulated with the vertebrae, and the **sternal rib,** a distal segment that is usually cartilaginous and meets the sternum. The joint between vertebral and sternal segments accommodates changes in chest shape during respiratory expansion and compression.

In birds, cervical ribs are reduced and fused to the vertebrae. In the thoracic region, the first several ribs are floating ribs, followed by true ribs that articulate with the sternum. Some floating and most true ribs bear **uncinate processes,** projections that extend posteriorly from proximal rib segments. Uncinate processes, like the rib cage generally, offer sites of attachment for respiratory and shoulder muscles. In birds, these processes act primarily as lever arms for inhalatory muscles that flare the rib cage. Similar rib projections are also found in some living and fossil reptiles as well as in the early labyrinthodonts, *Acanthostega* and *Ichthyostega,* where they project posteriorly to overlap with the next adjacent rib. This overlap between successive ribs may introduce some overall firmness into the thoracic ribs, giving them the functional integrity to act as a unit during lung ventilation and locomotion.

In mammals, ribs are present on all thoracic vertebrae and define this region. Some are floating (posterior) and others are false. Most are true ribs, however, and they meet the sternum through cartilaginous sternal rib segments. Within cervical and lumbar regions, ribs exist only as remnants fused with transverse processes, forming what should properly be termed **pleurapophyses** (transverse process plus rib remnant).

Sternum

The sternum is a midventral skeletal structure that is endochondral in embryonic origin and arises within the ventral connective tissue septum and adjacent myosepta (figure 8.8a–f). The sternum offers a site of origin for chest muscles. As noted, it also secures the ventral tips of true ribs to complete the protective chondrified or ossified rib cage. The **rib cage** consists of ribs and sternal elements that embrace the viscera. Size and shape changes in the rib cage also act to compress or expand the lungs, promoting ventilation. The sternum may consist of a single bony plate or several elements in series.

Fishes lack a sternum. When it first appears in tetrapods, the sternum is apparently not a phylogenetic derivative of either the ribs or the pectoral girdle, although in many groups it has become secondarily associated with each. A sternum is absent in the first fossil tetrapods, but it is present in modern amphibians. In many urodeles, the sternum is a single midventral **sternal plate** grooved along its anterior borders to receive the ventral elements of the shoulder girdle, the **procoracoid plate** (figure 8.8a). In anurans, a single element, the **xiphisternum,** often tipped with the **xiphoid cartilage,** lies posterior to the pectoral girdle;

and in some, a second element, the **omosternum** capped by the **episternal cartilage,** lies anterior to the girdle (figure 8.8b). The sternum is absent in turtles, snakes, and many limbless lizards, but is common in other reptiles, where it consists of a single, midventral element associated with the shoulder girdle (figure 8.8c). During locomotion, the reptilian sternum confers stability on weight-bearing girdle elements. In flying birds, the massive flight muscles arise from a large sternum that bears a prominent ventral keel, the **carina.** The carina provides additional surface for muscle attachment (figure 8.8d). In most mammals, the sternum consists of a chain of ossified elements in series, the **sternebrae** (figure 8.8e, f). The first and last of these are often modified and are called the **manubrium** and **xiphisternum,** respectively.

Thus, a sternum occurs in some modern amphibians and most amniotes. However, its absence in the common ancestors to these groups means that it has arisen independently several times within the field of the midventral connective tissue.

Gastralia

Posterior to the sternum in some vertebrates is a separately derived set of skeletal elements, the **gastralia,** or abdominal ribs (figure 8.2). Unlike the sternum and unlike ribs, the gastralia are of dermal origin. Gastralia are restricted to the sides of the ventral body wall between sternum and pelvis and do not articulate with the vertebrae. They are common in some lizards, crocodiles, and *Sphenodon,* serving as an accessory skeletal system that provides sites for muscle attachment and support for the abdomen. Ventral dermal scales in the abdominal region of labyrinthodonts preceded the gastralia functionally and perhaps gave rise to them anatomically. In turn, these are probably related to the ventral scales of rhipidistian ancestors. As with gastralia, these abdominal scales helped support the viscera, as they were organized into tight, chevron-shaped rows in many labyrinthodonts. In *Acanthostega* they are very prominent, and the chevrons are interrupted by a system of transverse scale rows.

Within turtles, the **plastron** is a composite bony plate forming the floor of the shell (figure 8.9a–c). It consists of a fused group of ventral dermal elements, including contributions from the clavicles (epiplastrons) and interclavicle (entoplastron) as well as dermal elements from the abdominal region (possibly the gastralia). Such ventral dermal bones are usually absent in birds and mammals, but in many fishes, bones form within the dermis of the belly region. In fishes and other vertebrates, the dermis exhibits potential for producing independently skeletal derivatives like the gastralia in different phylogenetic lineages. Because of such multiple but independent derivations from the dermis, it is perhaps best to restrict the term *gastralia* to rib-shaped elements in the abdominal region rather than to apply this term to all abdominal dermal bones.

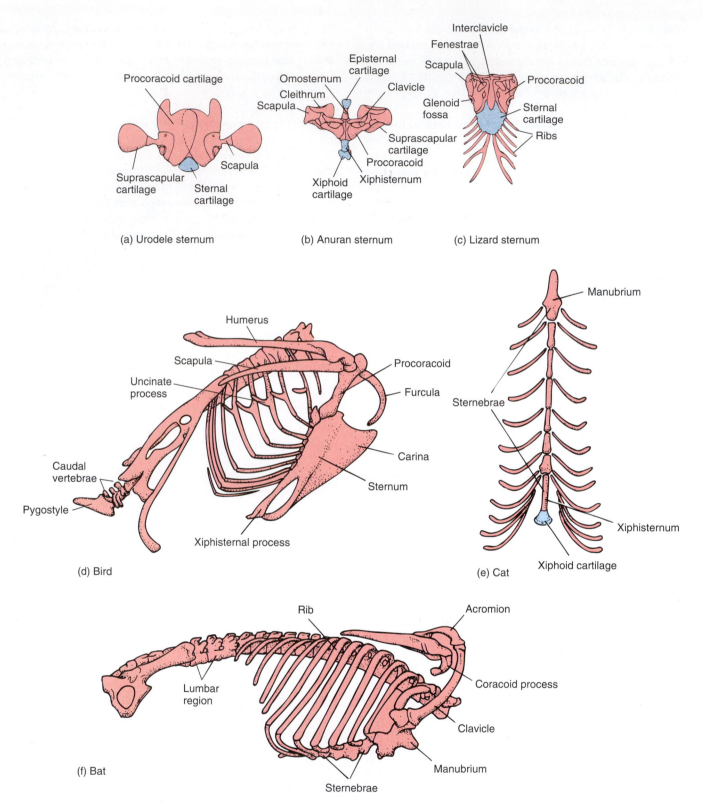

FIGURE 8.8 **Tetrapod sterna.** (a) Urodele, ventral view. (b) Anuran, ventral view. (c) Lizard sternum, ventral view. (d) Bird sternum, lateral view. In birds, the sternum is deeply keeled, forming a carina that offers increased attachment area for enlarged flight muscles. Within the axial column, the tail is short, ending in a specialized pygostyle that supports a fan of tail feathers; the pelvic bones and many of the vertebrae are fused; and the shoulder is braced by the large procoracoid. (e) Mammal (cat) sternum, ventral view. (f) Bat sternum, lateral view. In bats, the sternebrae are robust and fused. Within the axial column, the lumbar region and neural spines are short, the ribs are broad, and the coracoid process and clavicle are large, reflecting enhanced roles in flight.

(a–e) After Smith; (f) after Hildebrand.

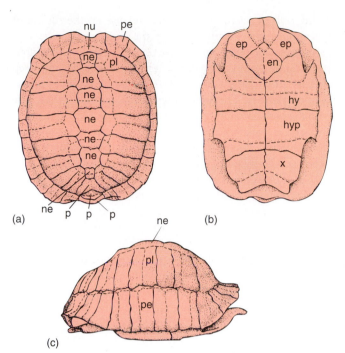

(a)

(b)

(c)

FIGURE 8.9 **Shell of the tortoise *Testudo*.** Solid lines indicate sutures between bony plates; dashed lines represent outlines of more superficial epidermal scales. (a) Dorsal view of convex carapace. (b) Ventral view of flat plastron. (c) Lateral view of whole shell. Carapace consists of numerous peripheral plates (pe) along the margin, eight paired pleural plates (pl), and a single nuchal plate (nu) followed by a series of neural plates (ne) down the dorsal midline, ending with three pygal plates (p). The three plates at the anterior margin of the plastron represent the epiplastron (ep), or paired clavicles, and the entoplastron (en), or the single interclavicle. The hypoplastrons (hy, hyp) and the xiphiplastron (x) are the remaining plastral plates to be incorporated into the shell.

After Romer and Parsons.

Embryonic Development

For the most part, vertebrae arise embryologically from mesenchyme. Early in embryonic development, the paraxial mesoderm becomes segmentally arranged into somites. As they differentiate, somites form vertebrae (and ribs), skeletal muscle, and the dermis of the dorsal skin. Commitment of cells within the somite to these respective fates occurs relatively late, after the somite forms. Soon after formation, lateral cells nearest the ectoderm differentiate into a dermatotome (dermis) and myotome (skeletal muscles). Medial cells depart and migrate toward the nearby notochord, forming streams of mesenchymal cells that settle segmentally along the notochord in cell groups called sclerotomes. These sclerotomes in turn contribute to the vertebrae and ribs.

Somite differentiation (p. 169)

In some groups, the steps in development have been abbreviated, amended, or deleted, and this has complicated the interpretation of developmental events. Homologous structures and parallel developmental events cannot always be easily determined. Consequently, interpretation and accompanying terminology applied to developmental steps and adult derivatives vary greatly. Without becoming embroiled in the esoteric parts of these controversies, let us see what we can state with some confidence.

Fishes

Among chondrichthyans and many primitive bony fishes, the inward streams of cells arriving from the sclerotomes first congregate in discrete clusters and differentiate into paired cartilages rather than into ossified vertebrae directly. Up to four pairs of cartilages are formed per segment. The embryologist Hans Gadow called these paired cartilages **arcualia.** The developmental fate of each arcualium could be charted from embryo to adult, and its specific contribution to the adult vertebra identified (figure 8.10). Although such developmental steps commonly occur in elasmobranchs and many primitive bony fishes, discrete cartilages do *not* always appear in later groups and certainly are always absent in tetrapods. Although arcualia are

Basidorsal → Neural arch
Interdorsal → Interneural
Interventral → Pleurocentrum
Basiventral → Intercentrum

Embryonic arcualia

Adult derivatives

FIGURE 8.10 **Arcualia.** During embryonic development in some primitive fishes, mesenchymal cells that gather around the notochord first form discrete blocks of cartilage, up to four pairs per segment. These blocks are called arcualia. In such fishes, each arcualium can be followed through its subsequent embryology to the part of the vertebra it forms in the adult.

(a) Intervertebral ligament

Chordal center

Dorsal arch center

Perichordal center

Chordal center

Ventral arch center

(b)

(c)

FIGURE 8.11 **Embryonic formation of teleost vertebrae.** Successive stages in embryonic development are shown in lateral views (left of each figure) and cross-sectional views (right of each figure) through the middle of a forming centrum. (a) A chordal center (chordacentrum) forms within the notochordal sheath. (b) Paired cartilaginous rudiments, or anlagen, of dorsal and ventral arches form at myosepta from condensations of mesenchyme. Within the perichordal tube, itself formed from sclerotomal cells, perichordal centers (autocentra) of ossification appear. (c) The chordal center becomes incorporated within the ossified perichordal center, forming the centrum. Arches often, but not always, fuse with the centrum to ossify along with it. The notochord may persist as intervertebral cartilaginous pads, as intervertebral ligaments, and as a constricted notochord running through the centers of the centra.

absent in these derived groups, Gadow nevertheless proposed that they were the underlying pattern in all later groups. He attributed their absence in amniotes to developmental shortcuts, reductions, and elimination of intervening developmental steps. However, this seems to force an interpretation on vertebral formation that does not fit tetrapods and is even inaccurate in teleosts. Let us look at the events of vertebral formation in teleosts and tetrapods to see the departures from Gadow's theory.

In most teleosts, embryonic formation of vertebrae proceeds in three steps. First, the sheath of the notochord itself differentiates into a chain of cartilaginous elements, the **chordal centers** (or chordacentra; figure 8.11a). Between successive chordal centers, the undifferentiated notochordal sheath is destined to become the intervertebral ligament between vertebrae of the adult. Second, local mesenchyme condenses at the level of the myosepta. These condensations become cartilaginous anlagen called **arch centers** (arcualia by some accounts) that give rise to the dorsal and ventral arches.

Third, cells of sclerotomal origin condense on the surface of the notochordal sheath, forming the perichordal tube that becomes ossified without first passing through a cartilaginous stage (figure 8.11b). As vertebral formation proceeds, the deep chordal centers fuse with their respective perichordal centers on their surfaces. The arch centers often, but not always, fuse with the perichordal tube and ossify (figure 8.11c). Therefore, although arcualia precede and then contribute to vertebrae in elasmobranchs and some primitive bony fishes, this pattern is not strictly followed in derived fishes such as teleosts. In teleosts, cartilaginous anlagen are the source of the arch centers, but the perichordal tube and notochordal sheath, not the arcualia, are the sources of the centra.

In tetrapods, vertebrae do not develop from modified arcualia, not even in part. Tetrapod centra arise from a perichordal tube of mesenchymal origin, not from discrete blocks of cartilage (arcualia). Consequently, Gadow's comprehensive view of a common arcualial pattern underlying vertebral development in all vertebrates is not accepted today.

Tetrapods

In the trunk and tail of tetrapods, each of the segmental chains of somites becomes internally subdivided into discrete layers of cells. Laterally, somites form the dermatome; beneath it, the myotome; and medial to both, the sclerotome. In most tetrapods, streams of mesenchymal cells depart from this inner sclerotome (figure 8.12a). These cells migrate inward toward the midline and cluster along the sides of the notochord, but not yet up the sides of the nerve cord. These arriving streams of cells initially form enlarged clusters called **perichordal rings,** arranged serially along the notochord. Intervertebral bodies or disks will eventually arise within these rings. Arriving mesenchymal cells condense to connect these rings and form a more or less continuous coat or **perichordal tube** of varying thickness enclosing the notochord (figure 8.12b). Next the general outline of the future vertebrae appears as mesenchymal condensations extend up along the sides of the neural tube, the future neural spine (figure 8.12c), and into an extended dorsal process, the future neural arch. The intervertebral bodies or disks differentiate within the rings of the perichordal tube, thereby delineating the boundaries of each vertebral segment. By now, the full, general outline of each vertebra is recognizable, and the mesenchymal condensations become chondrified. Bone formation is endochondral, so that in most tetrapods, subsequent ossification replaces these cartilaginous precursors, producing the bony vertebrae of the adult.

Note that the bony vertebra is not formed directly or exclusively from its adjacent sclerotome, one-to-one. Instead, halves of adjacent sclerotomes fuse, forming resegmented sclerotomes that produce the basic vertebra (figure 8.12c). As cells stream out of the segmental **primary sclerotomes** on their way to form the perichordal tube, they first regroup, or regroup along the way (figure 8.12a). This cellular regrouping

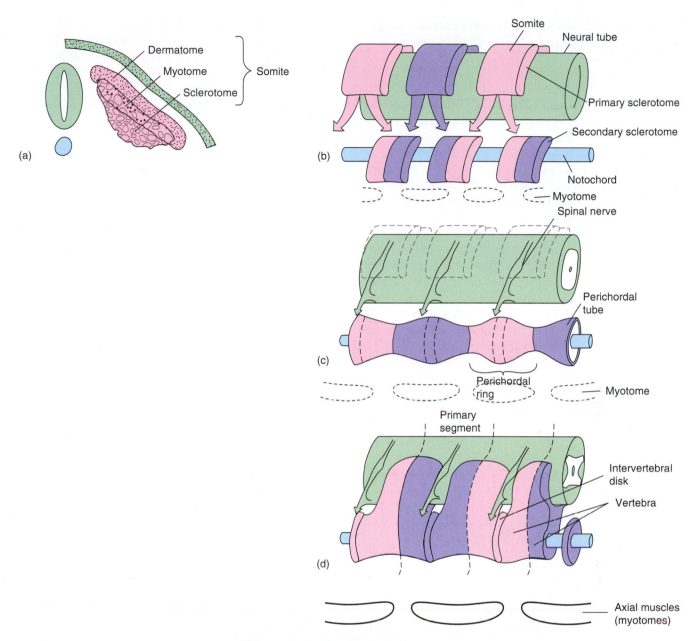

FIGURE 8.12 **Development of vertebrae in a generalized mammal.** (a) Section of one somite, next to neural tube and notochord, showing its beginning differentiation into the dermatome (skin), myotome (axial muscle), and primary sclerotome (vertebra). (b) The primary sclerotome forms from cells on the medial side of the somite that depart and stream downward toward the notochord (solid arrows). These arriving streams of cells form secondary sclerotomes, represented here as blocks, half derived from one somite, and the other half from the next adjacent somite. (c) The arriving sclerotomal cells soon amass as repeating perichordal rings that grow into contact, forming a more or less continuous perichordal tube. (d) The amassed cells of the perichordal tube grow upward, around the nerve cord, then above it, forming the outline of neural arches and spines. Chondrification, followed usually by ossification, produces the adult bony vertebrae. The intervertebral disks differentiate between vertebrae within the former perichordal rings. Note that the myotomes, which eventually give rise to axial musculature, appear first in register with the somites (b). But as resegmentation proceeds, the secondary sclerotomes come to lie between adjacent myotomes (c, d). This means that axial muscles that form from myotomes cross the intervertebral joint, instead of attaching to the same vertebra, thereby giving the muscles useful actions. Somites are colored alternately in pink and purple to aid in following their contributions to the shared secondary sclerotomes.

is accomplished by the caudal half of one sclerotome fusing with the cranial half of the next to form resegmented blocks of cells, the **secondary sclerotomes.** These regrouped cells continue to the notochord to form the perichordal tube (figure 8.12b). This provides a developmental mechanism by which

the sclerotome and its myotome, initially in register with one another, become offset or staggered before differentiating into respective vertebrae and muscles. Thus, the musculature forms across adjacent vertebrae, rather than within the same vertebra. In this way, muscles act on adjacent vertebrae in a suitable functional position. If a muscle were attached to only one vertebra, the muscle would, of course, have no functionally significant role. Spinal nerves sprout and grow out between successive vertebrae.

Some scientists, working with serial sections of developing vertebrae from mammals, dispute the claim that resegmentation occurs. They argue that cells migrate obliquely from the primary sclerotome directly to locations between myotomes without any prior regrouping. However, the results of experimental embryology support the view that the sclerotomes do, in fact, undergo resegmentation. Sclerotomal cells from a chick and a quail were used because each is microscopically distinctive and thus can be recognized. Every other sclerotome was surgically replaced in chick hosts with a sclerotome from a quail donor before resegmentation. Development was then allowed to proceed normally. The visible differences between chick and quail cells made it possible to determine the contribution of each to the resulting vertebrae. In these experiments, individual vertebrae contained both chick and quail cells, suggesting that initially alternate chick-quail sclerotomes did resegment prior to differentiation into vertebrae.

Phylogeny

Fishes

Agnathans

Among ostracoderms, the notochord is large and prominent, a major contributor to the functional axial skeleton. Vertebral elements embracing the notochord are harder to document, in large part because the internal skeleton is often poorly preserved. Or more probably, vertebrae were uncommon. In some heterostracans, osteostracans, and galaeaspids, trace impressions of vertebral elements have been observed in fossil specimens. These elements probably were small, unossified pieces of vertebrae resting upon a prominent notochord. Thus, among ostracoderms, a strong notochord provided the central mechanical axis for the body.

Among living hagfishes and lampreys, the situation is similar. Hagfishes possess a prominent notochord but adults lack any hint of vertebral elements. However, some larval hagfishes exhibit ventral vertebra-like elements (arcualia) in the post-anal tail. These may have some inductive influence but supply no adult structure themselves upon metamorphosis. Lampreys possess vertebral elements, but these are small, cartilaginous elements resting dorsally upon a very prominent notochord that gives primary axial support for the body (figure 8.13a, b).

Gnathostomes

Primitive Fishes In most placoderms and acanthodians, the axial column consisted of a prominent notochord. There is no evidence of vertebral centra, although dorsal and ventral arches were usually present. Some placoderms preserve evidence of a prominent notochord supporting ossified neural and hemal arches (figure 8.14a). Fossil impressions of most acanthodians also show clear evidence of an ossified series of neural and hemal arches (figure 8.14b). These arches rode upon a prominent notochord. In primitive chondrichthyans a prominent notochord provided axial support (figure 8.15a–c); a vertebral column was represented by only cartilaginous neural and hemal arches. However, in

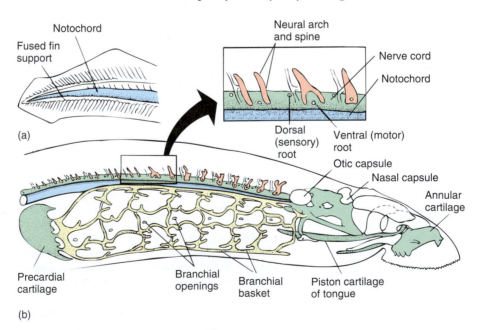

FIGURE 8.13 **Lamprey skeleton.** (a) Enlargement of caudal body section of the lamprey. (b) Anterior end of the lamprey with enlargement of the axial skeleton illustrating the prominent notochord. Note that only a few cartilaginous vertebral elements are present.

(a) After Remane; (b) after Jollie.

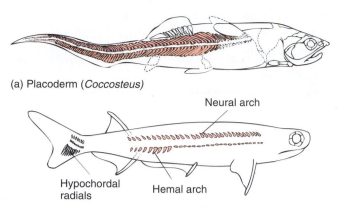

(a) Placoderm (*Coccosteus*)

Neural arch

Hypochordal radials

Hemal arch

(b) Acanthodian (*Acanthodes*)

FIGURE 8.14   Axial skeleton of primitive fishes. (a) Placoderm *Coccosteus,* with prominent notochord supporting dorsal and ventral vertebral elements. (b) Acanthodian *Acanthodes,* with neural and hemal arches that presumably rode upon a notochord.

After Moy-Thomas and Miles.

advanced sharks, these vertebral elements enlarge to become the predominant structural elements of the body axis, and the notochord persists as only a constricted element enclosed within the vertebral centra (figure 8.15d). Among palaeoniscoids, the notochord was unconstricted and reached from the skull nearly to the tip of the tail. A series of neural spines lay dorsally along the notochord, and ventral hemal arches accompanied it into the trunk and tail regions.

Among living primitive bony fishes, such as sturgeons and paddlefishes, the vertebral column is unossified, presumably a secondary condition, but several elements of the vertebrae are present in each segment (figure 8.16a). In more derived bony fishes, such as the bowfin (figure 8.17a–d) and teleosts (figure 8.16b, c), the vertebral column typically is ossified and its centra more prominent to replace the notochord as the major mechanical support for the body. Neural spines and ribs become more developed, as do accessory bony elements that help internally stabilize some of the unpaired fins.

Mechanically, the axial column of fishes represents an elastic beam. Lateral bending movements produced by the body musculature place the column in compression (figure 8.18a, b). Even during peak bursts of speed, the fish's notochord or ossified vertebrae experience stresses well within their capacity to withstand without breaking or collapsing (figure 8.18c). However, when laterally flexed, the vertebral column is in danger of buckling, and its separate vertebrae could become disarticulated if they were too loosely joined (figure 8.18d). The intervertebral ligaments resist this and return stiffness to the vertebral column. Thus, centra seem to function as compression members, and the stiffness that resists buckling is controlled by the degree of lateral flexure permitted by these ligaments between centra (figure 8.18e).

Although compression seems to be the most prevalent force, the axial column in some fishes must be able to resist torsion, the tendency to twist or "wring" the axial column. Torsional forces are especially acute in fishes with asymmetrical tails, where one lobe is quite long. In these fishes, oscillation of the asymmetrical tail produces desirable lift but also tends to twist the axial column, possibly even affecting the trunk vertebrae. In these fishes, and in later tetrapods in whom twisting, or **torsion,** places the integrity of the axial column at risk, several features of axial column design seem to address the mechanical demands of torsion. Consolidation of separate vertebral elements into a holospondylous vertebral column of solid vertebrae helps withstand torsional forces. Long neural spines extending over several segments functionally tie together adjacent regions to resist twisting. If the notochord remains prominent, its sheath is often quite thick and invested with bands of fibrous connective tissue oriented in such a way as to resist excessive torsion.

Caudal Skeleton and Fins   In most fishes, the axial skeleton continues into the tail, where it can take several forms. In many fishes, the tail is asymmetrical, with a long dorsal and a small ventral lobe separated by a notch. If the posterior end of the vertebral column turns upward and into this dorsal lobe, forming its central axis, a **heterocercal tail** forms (figure 8.19). In the **diphycercal tail,** the vertebral column extends straight back, with the fin itself developed symmetrically above and below it. Living lungfishes and bichirs are examples. The **homocercal tail,** characteristic of teleosts, has equal lobes and appears to be symmetrical, but the narrowed vertebral column that runs to its base slants upward to form the support for the dorsal edge of the fin. The hemal arches below expand into supportive struts, known as hypurals, to which the rest of the fin attaches (figure 8.19a–c). Among the earliest vertebrates, the tail was not commonly symmetrical. Rather, most ostracoderms show the heterocercal condition (figure 8.19a), or even a "reversed" heterocercal condition, termed a **hypocercal tail,** in which the vertebral axis enters the tail and turns down into an extended ventral lobe. The symmetrical diphycercal and homocercal tails (figure 8.19b, c) are usually derived from ancestors with asymmetrical heterocercal tails. They are common among fishes with lungs or air bladders that give their dense bodies neutral buoyancy. In sharks, which lack lungs or air bladders, lift to the posterior body is apparently provided by the extended dorsal lobe of the heterocercal tail.

When the heterocercal tails of sharks are removed and tested separately in experimental tanks, they have a tendency to push downward against the water, resulting in an upward reaction force on the tail, which produces lift (figure 8.20). In fact, removal of the dorsal or the ventral lobe alone reveals that within the tail, lift produced by the two lobes differs in magnitude and direction (figure 8.20b). In general, as the tail sweeps back and forth, the small ventral lobe deflects water upward, causing a small downward component of force,

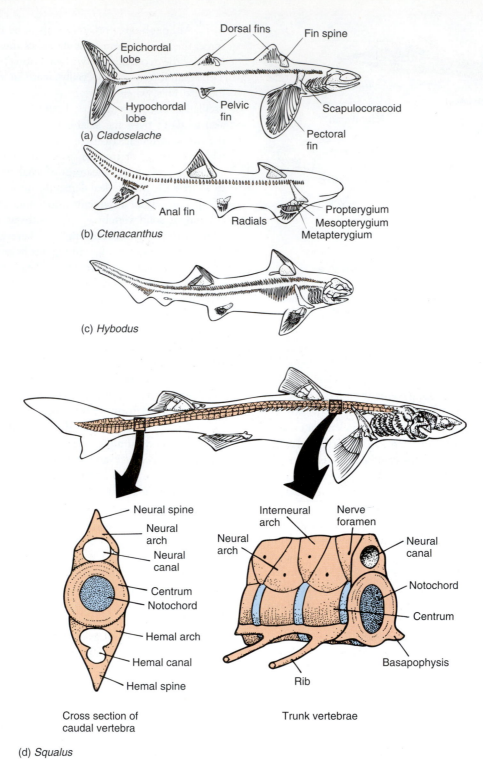

FIGURE 8.15 **The axial skeleton in sharks and their ancestors.** (a) Paleozoic shark *Cladoselache* with a chain of neural arches presumably riding upon a notochord that extended into the tail. (b) *Ctenacanthus* from the late Paleozoic. (c) *Hybodus* from the Mesozoic. (d) Modern shark *Squalus*. The vertebral elements tend to enlarge in elasmobranchs, surpassing the notochord as the major mechanical support for the body in modern forms.

(a–c) After Carroll.

(a) Sturgeon vertebrae
(lateral)

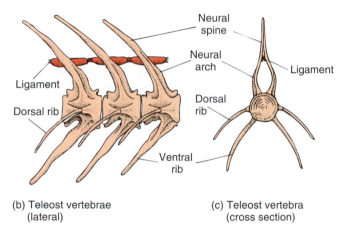

(b) Teleost vertebrae
(lateral)

(c) Teleost vertebra
(cross section)

FIGURE 8.16 **Actinopterygian trunk vertebrae.**
(a) Sturgeon vertebrae, lateral view. (b) Teleost vertebrae, lateral view. (c) Teleost vertebra, cross section.

After Jollie.

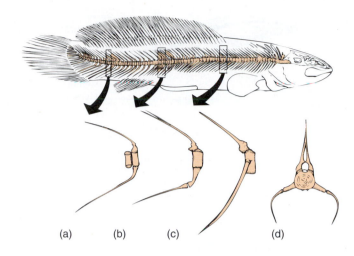

FIGURE 8.17 **Axial skeleton of the bowfin *Amia calva.*** (a–c) Representative lateral sections of the vertebral column. (d) Cross section of a trunk vertebra. Note the predominance of ossified vertebrae.

After Jarvik.

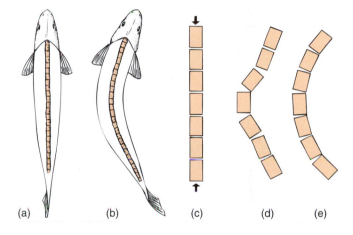

FIGURE 8.18 **Function of amphicoelous vertebrae in a teleost.** (a, b) Swimming involves development of lateral flexion of the vertebral column induced by contractions of the body musculature. (c) Chain of vertebrae shown under axial loads. Even during maximum bursts of speed, the ossified vertebrae are strong enough to withstand maximum compressive loads. (d) When flexed, the chain of vertebrae might buckle and fail. (e) Firm intervertebral ligaments that resist buckling return stiffness to the vertebral column.

Based on the research of J. Laerm, 1976.

whereas the large dorsal lobe deflects water downward, resulting in an opposite large upward force (figure 8.20b). The overall effect is for the tail to produce a resultant force directed forward and upward. Although at first it might seem strange that the ventral lobe produces forces contrary to the overall upward lift generated by the tail, this action of the ventral lobe might represent a method for fine-tuning the lift. In sharks that have just eaten a large meal or in gravid females, the center of body mass might shift unfavorably, tilting or angling the body out of its line of travel. The ventral lobe might help level the shark in a more direct body orientation. In nautical terminology applied to submarines, adjustment for vertical tilt is called "trimming." Small radial muscles reside in the ventral lobe of the shark's tail. Their contraction might alter stiffness, change the forces produced in the tail, help trim the body, and adjust the shark about its center of gravity. This lift imparted to the posterior body would tend to rotate the shark tail-up and nose-down. This is counteracted by lift generated by the entire streamlined head of the shark together with its pectoral fins. In a swimming shark, this cranial lift, together with caudal-produced lift, compensates overall for the sinking density of the shark.

If this interpretation of the function of a heterocercal tail is correct, then the reversed heterocercal tail, the hypocercal tail

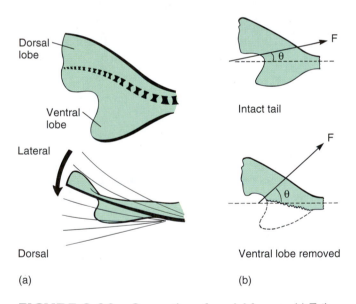

FIGURE 8.19 **Caudal fins of fish.** (a) Sturgeon. (b) Bichir *Polypterus*. (c) Bowfin *Amia*. Note the positions of the vertebral column and the conditions of the remaining notochord. Sequence leading to the homocercal tail is shown to the right of each figure.

After Kent.

FIGURE 8.20 **Generation of caudal forces.** (a) Tail of a Port Jackson shark as seen from the side and from above, moving in the direction of the arrow. The vertebral column extends into the dorsal lobe. The thick black lines indicate the stiff edges that lead the more flexible parts of the lobes that lag behind. Because of this bend, the dorsal lobe produces a large upward force, and its ventral lobe produces a small downward component. (b) Without the ventral lobe, the thrust is inclined at a greater angle (θ) with the body axis. Resultant force of intact tail (top), and tail with the ventral lobe removed (bottom).

Modified from J. R. Simons, 1970.

of ostracoderms, would have pitched the nose up counteracting the weight of the anterior body, thereby lifting the mouth from the substrate as the fish moved to a new location, where it might again find foods buried in soft sediments. This would aid the animal in feeding on foods buried in soft sediments.

Sarcopterygians The notochord continues to serve as the major supportive element within the axial skeleton of sarcopterygians, including the rhipidistian ancestors of early tetrapods. In living sarcopterygians, the vertebral column can be rudimentary and cartilaginous (figure 8.21a, b). However, in many early species, such as the rhipidistians, vertebral elements were usually ossified and exhibited a rhachitomous type of aspidospondyly in which each vertebra consisted of three separate vertebral elements: a neural arch, a hoop- or crescent-shaped intercentrum, and paired pleurocentra (figure 8.22a). In the tail, the intercentrum expanded into the continuous hemal arch and spine. Although differing in some details, an aspidospondylous condition occurred in many early rhipidistians, including *Eusthenopteron* (from late in the Devonian) and *Osteolepis* (from the mid-Devonian). In the tail of these early rhipidistians, each aspidospondylous vertebra included the small, paired pleurocentrum and dorsal (neural) and ventral (hemal) arches. In the trunk region, the hemal arch became reduced, and its base expanded into the prominent intercentrum (figure 8.22b, c). The segmental myosepta mark the borders of the earlier embryonic segments and attach to the neural arch and intercentrum medially in the adult. In later and usually larger rhipidistians,

Neural
spine

Notochord

(a) *Latimeria*

Neural spine

Notochord

(b) *Neoceratodus*

FIGURE 8.21 **Axial skeletons of living sarcopterygians.** (a) Enlarged lateral view of posterior axial skeleton of the coelacanth *Latimeria*. (b) Enlarged lateral view of trunk vertebrae and notochord of the lungfish *Neoceratodus*.

After Andrews, Miles, and Walker.

fusion of central elements produced a derived aspidospondylous condition in which each segment consisted of a single centrum of ringlike bone to which the neural arch may or may not have fused.

Tetrapods

Early Tetrapods

The vertebrate transition to land brought considerable changes in the selection pressures acting on design. As animals evolved from water to air, their bodies went from a buoyant support design to a design in which bodies were suspended between limbs. All systems, including respiration, excretion, and body support, were affected. Changes in the axial skeleton are especially indicative of these new mechanical demands.

Lepospondyls are named for their distinctive type of holospondylous vertebra, termed a lepospondylous vertebra, in which the vertebral elements are fused. They presumably arose from ancestors with aspidospondylous vertebrae. Thus, the single, solid vertebra typical of lepospondyls represents the fusion of vertebral elements that were separate originally (figure 8.23).

Many lepospondyls had long, deep tails, suggesting that they, like modern salamanders, were swimmers. Modern amphibians also have a vertebral column composed of single, solid vertebrae at each segment, suggesting that they might have evolved from these early lepospondyls. However, a silent gap in the fossil record extends from the last lepospondyls (in the Permian) to the first frogs (in the Early Jurassic) or first salamanders (Late Jurassic), almost 40 million years without fossils, to connect modern amphibians and late lepospondyls. Their similar vertebrae may reflect convergence of

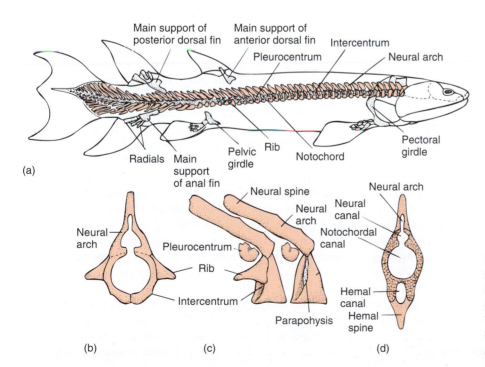

(a)

FIGURE 8.22 **Axial skeleton of the fossil rhipidistian *Eusthenopteron*.** (a) Restored axial skeleton. Cross section (b) and lateral view (c) of trunk vertebrae. (d) Cross section of a caudal vertebra.

(a) After Moy-Thomas and Miles; (b–d) after Jarvik.

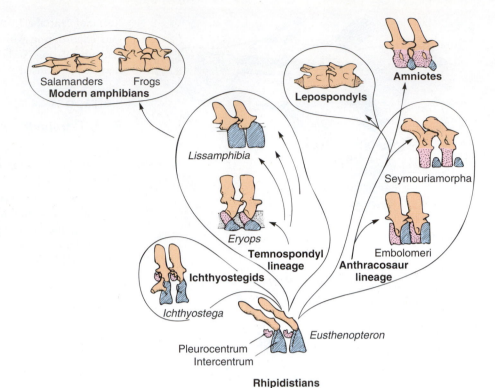

FIGURE 8.23 **Traditional view of the evolution of tetrapod vertebrae.** Lepospondyls are named for this solid vertebral type. The rhachitomous vertebra, inherited from rhipidistian fishes, evolved along two major lines: temnospondyl and anthracosaur. In the temnospondyl line, the intercentrum (blue) enlarged at the expense of the pleurocentrum (pink). In the anthracosaur line, however, the pleurocentrum came to predominate. (Neural arch and spine orange.)

morphological design to parallel functional roles in swimming. Consequently, vertebrae of solid construction may have been derived independently in one group or in all groups of modern amphibians.

Labyrinthodonts evolved directly from rhipidistians, taking over their aspidospondylous type of vertebra as well. The characteristic mode of fish progression in which locomotion depends on lateral waves of undulation in the vertebral column has been retained in modern salamanders and was probably present in early lissamphibia as well (figure 8.24a–c). Swimming in most fishes depends on the production of lateral body bends that sweep posteriorly as traveling waves that push the sides of the fish against the surrounding water (figure 8.24a, b). These traveling waves produce lateral undulations of the fish body and are also the basis for terrestrial locomotion in salamanders and even most reptiles. Synchronized with these lateral body swings are limb movements that lift and plant the foot to establish points of rotation about which the tetrapod vertebral column undulates (figure 8.24c).

Along with the vertebrae, such lateral body undulations were carried forward from fish ancestors into early tetrapods as well, constituting the basic mode of locomotion of early tetrapods walking on land. What was mechanically new in this early mode of terrestrial locomotion was a tendency to twist the vertebral column, placing it in torque. Without surrounding water to support the body and with planted feet establishing pivot points, walking on land placed new torsional stresses on the vertebrae. Several features of the design of early tetrapod vertebrae can be interpreted as functional modifications addressing these new stresses.

As in *Acanthostega* and *Ichthyostega,* most early labyrinthodont vertebrae were aspidospondylous. Although this tended to give way to derived conditions in later species, the vertebrae of these early species consisted of separate components applied to a still prominent notochord. Such a loose confederation of bony elements might at first seem ill suited to address the torsional forces introduced in the axial column when the first tetrapods ventured onto land. However, a functional hypothesis incorporating the fibrous nature of the notochord and solid structure of the vertebral elements suggests otherwise. If fibrous bands within the outer sheet of the notochord were wound in opposite spirals crossing at about 45°, they would create a kind of geodesic, or warp-and-weft, framework resistant to torsional forces (figure 8.25a). Rigid, bony pieces of the vertebra might have been so placed as to occupy the spaces between these fibrous bands. Lateral flexions of the body during terrestrial locomotion would bring the edges of these bony vertebral pieces into contact, checking further torque. Up to that point, however, the elastic sheath of the notochord would allow the flexibility required to produce these lateral body bends during locomotion. This functional model depicts the axial column of early labyrinthodonts as consisting of two mechanical components, the notochordal sheath introducing limited flexibility and the hard vertebral elements preventing excessive torque (figure 8.25b, c).

Various groups of labyrinthodonts were characterized by differences in the relative prominence of each vertebral centrum. In the **temnospondyl lineage,** the intercentrum became predominant. In the **anthracosaur lineage,** the

FIGURE 8.24 Lateral undulatory locomotion from fishes to tetrapods. Lateral swimming motions of fishes are incorporated into the basic pattern of terrestrial locomotion of primitive tetrapods. (a) Lateral undulations of a salamander do not press the body against its terrestrial surroundings, but these undulations serve to advance each foot forward, plant it, and then rotate the body about this point of pivot for locomotion on land. (b) Similar lateral undulations of a shark's body push against the water and drive the fish forward. (c) Side-to-side sweeps of the body of an eel exert a force against the surrounding water as the fish travels forward.

After Gray.

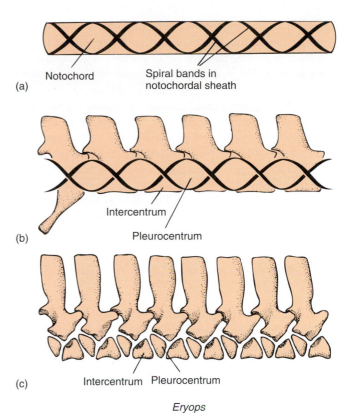

FIGURE 8.25 Geodesic model of the axial skeleton of an early tetrapod. (a) Bands of connective tissue wound in opposite directions, each at 45° to the long axis of the notochord, form a geodesic framework that resists both flexion and torque. (b) A rhachitomous vertebral column has much the same pattern; here it is overlaid by the geodetic framework. (c) Vertebral column of *Eryops* wherein the spaces of the geodetic framework are occupied by rigid bone around the notochord, permitting controlled flexion that resists torque.

After Parrington.

pleurocentrum became predominant. In both, the notochord became reduced as respective centra enlarged to assume the central role in axial support.

In the early temnospondyls, the vertebrae were rhachitomous, a type of specialized aspidospondylous vertebra. This type consisted of a neural arch, a crescent-shaped intercentrum below the notochord, and a pair of bony pleurocentra above the notochord. However, in later temnospondyls, the intercentra became preeminent, large, and completely ossified cylinders upon which the neural arch rested. Conversely, the pleurocentrum became much reduced or was lost entirely. The term *stereospondyl* has in the past been applied to these later temnospondyls. This was based on the view that all shared a common ancestry because all shared a common vertebral design (prominent intercentrum). However, it now seems more likely that these late

temnospondyls evolved independently from separate earlier groups and that the similar vertebral designs are instead a consequence of convergent evolution.

Within the anthracosaur lineage, the opposite centrum, the pleurocentrum, enlarged. Initially, in aquatic anthracosaurs, the pleurocentrum was about the same size as the intercentrum; subsequently, in terrestrial anthracosaurs, it came to predominate.

Vertebral evolution within these labyrinthodont lineages poses several questions. Why, for instance, does the intercentrum predominate in one lineage (temnospondyls) and the pleurocentrum in the other (anthracosaurs)? Or we might ask, why in *both* lineages does one centrum become predominant at all? Unfortunately, the relationship between vertebral structure and function remains poorly understood, so let us start with what is known.

With continued commitment by early labyrinthodonts to life on land, consolidation of the separate vertebral elements occurred in both lineages, leading to vertebrae composed of a single predominant centrum. In the two lines of labyrinthodont evolution, different elements are reduced, but the functional advantages are equivalent, namely increased strength. Locomotion on land imposed significantly greater weight-bearing stresses on the axial column. Thus, terrestrial locomotion required a vertebral column characterized by firmness and strength to suspend and support the body. Enlargement of the ossified centra at the expense of the notochord brought firm support to the body. Enlargement of one centrum at the expense of the other has the overall effect of reducing the number of centra per segment from two to one. This reduces flexibility, firms the axial column, and therefore increases its ability to support the weight of the body on land. Conversely, the more centra per segment, the greater the flexibility of the vertebral column, an advantageous design for an aquatic organism employing lateral flexions of its vertebral column during swimming.

Radiations in both temnospondyls and anthracosaurs included reinvasion of aquatic habitats as well as entry into semiaquatic and terrestrial habitats. The vertebral column, central to locomotion, was consequently as varied as the emerging lifestyles. If swimming was favored, the trunk and tail regions were usually elongated and the number of vertebrae increased. This was especially true in the later embolomeres, a group of anthracosaurs that apparently returned secondarily to swimming as the primary mode of locomotion, and in the early lepospondyls, a group that seems to have been swimming specialists from their first appearance. However, in early temnospondyls, such as *Eryops*, and later anthracosaurs, such as *Seymouria*, emphasis was on terrestrial locomotion. This was accompanied by a reduction in the number of vertebrae, extensive vertebral ossification, centra enlargement, reduction of the notochord, and greater overall firmness of the vertebral column.

It is not known why the intercentrum (temnospondyls) or the pleurocentrum (anthracosaurs) came to predominate. Simple chance events in the independent pathways of evolution might have tipped the advantage different ways on two occasions. However, it is more likely that the differences reflect functional differences in the two lines of labyrinthodont evolution. Increasing prominence of the intercentrum in temnospondyls may have been favored by their emphasis upon aquatic locomotion. Of the two centra, the intercentrum was more closely associated with axial muscles and ribs that served swimming. Enlargement of the intercentrum might then have accompanied the increased functional demands of aquatic locomotion. On the other hand, increasing prominence of the pleurocentra in anthracosaurs and later amniotes might have been favored by an opposite trend toward terrestrial locomotion. Pleurocentra supported neural arches, successively interlocked through their zygapophyses, that became more important with increased load-bearing function. Thus, enlarged pleurocentra might then have accompanied enlargement of their associated neural spines and zygapophyses as these came to play more prominent mechanical roles during terrestrial locomotion.

Certainly one of the vertebral innovations of tetrapods was these zygapophyses seen first in labyrinthodonts. Terrestrial vertebrates faced a new mechanical problem, a tendency for excessive twisting of the vertebral column. In fishes, the axial skeleton receives more or less continuous and even support along its entire length, whereas in tetrapods only two pairs of points, the fore- and hindlimbs, provide support. As opposite feet plant themselves on a surface to establish points of support during locomotion, the intervening vertebral column is wrung or twisted, placing shearing stress on the fibrous connections between successive vertebrae. The bony zygapophyses reach across these vertebral joints to interlock gliding articulations. They are oriented to allow bending in a horizontal or vertical plane, but they resist twisting.

The other new feature of the axial skeleton that also appears first in labyrinthodonts is delineation of a sacral region, the site of attachment of the pelvic girdle to the vertebral column. The earliest labyrinthodonts show such a region. Presence of a sacral region joining pelvic girdle and vertebral column is taken as evidence that direct transfer of propulsive forces in the hindlimbs to the axial skeleton became an important component of the terrestrial locomotor system very early in tetrapod evolution.

Other changes in the axial skeleton, related to extended exploitation of land, are evident for the first time in labyrinthodonts as well. Connection between the pectoral girdle and back of the skull was lost. This occurred in both *Acanthostega* and *Ichthyostega*, for example. Accompanying this loss was the redesign of the first vertebra that became a cervical vertebra, allowing greater freedom of head rotation upon it. For early tetrapods, life on land meant that the lower jaw rested on the ground. Opening the jaws required lifting of the head because the lower jaw could not be dropped. Uncoupled from the pectoral girdle, the head could be lifted without restraint or interference from the shoulder. This uncoupling, along with the appearance of a cervical vertebra, allowed the tetrapod to turn its head to one side without reorienting the rest of its entire body. Further, when the head was uncoupled from the pectoral girdle, it experienced less jarring as the feet struck the ground during terrestrial locomotion. This was advantageous because the head carried most of the sensory organs.

Amniotes

Amniotes phylogenetically receive their vertebrae from the anthracosaur line, so their major centrum is a pleurocentrum, and the small centrum is an intercentrum. The intercentrum contributes to the intervertebral cartilages. But, in many reptiles and birds and in all mammals, the intercentrum is usually lost to the vertebral column as a bony contribution, being remembered only by the rib's capitulum that still articulates between vertebrae where the intercentrum would occur. In some amniotes, the intercentrum contributes to parts of the cervical vertebrae. Most specialized textbooks, such as those on human anatomy, trouble themselves only to call the surviving pleurocentrum just the "centrum" or sometimes just the "body" of the vertebra, a reference to its fused unity with the neural spine. After this long, thought-provoking, and intriguing evolutionary history, such a bland name is unequal to the pleurocentrum's phylogenetic service.

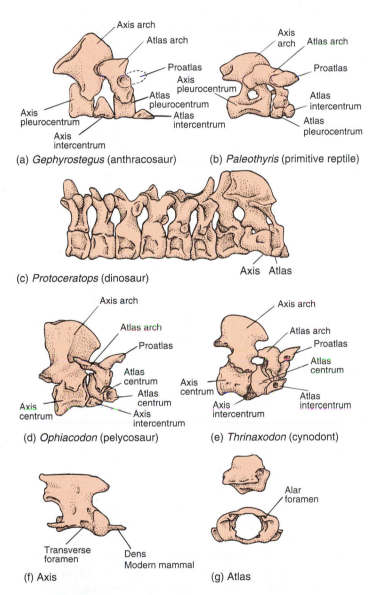

FIGURE 8.26 **Cervical vertebrae.** Fusions and reductions in the first few vertebrae produce the distinctive cervical vertebrae.
(a) Anthracosaur *Gephyrostegus*. (b) Primitive reptile *Paleothyris*. (c) Ornithischian *Protoceratops*. (d) Synapsid pelycosaur *Ophiacodon*. (e) Therapsid cynodont *Thrinaxodon*. (f) Axis of a modern mammal. (g) Atlas of a modern mammal.

(c) After Romer; others after Carroll.

In amniotes, the head rotates primarily on two anterior cervical vertebrae specialized to the function, an apparent answer to the problem of maintaining bony strength while retaining cranial mobility (figure 8.26a–g). The first cervical vertebra is the **atlas,** the second, the **axis** (figure 8.26f, g). Vertical (nodding) and horizontal (tilting) movements of the head are largely limited to the skull-atlas joint, whereas twisting movements occur largely within the atlantoaxial joint. This divides the labor between two joints yet maintains bony strength in the neck.

In turtles, the shell into which the limbs and head retreat is a composite unit made of expanded ribs, vertebrae, and dermal bones of the integument that fuse into a protective bony box that harbors the soft viscera (figure 8.27a–c). Turtles are unique in that the appendicular skeleton lies *within* the rib cage rather than on the outside as in all other vertebrates (figure 8.28a, b).

This curious morphological feature of slow and plodding turtles is easy to understand functionally. It brings the limbs and girdles, vulnerable to predators, inside a tough, protective bony case. But how to smoothly evolve it from outside to inside? The fossil record provides little help. The most primitive turtle known is from the Late Triassic, already housing limbs and girdles inside the shell, just like modern turtles. In fact, it is hard to imagine what gradual intermediate stages from outside to inside might look like. The answer comes from modern molecular genetics. Apparently this evolutionary step did not occur in gradual steps, but more suddenly. Turtles possess a suite of *Hox* genes, present in other tetrapods, but in turtles they preside over turtle-specific features of axial column patterning. This implies that a modification of a few *Hox* genes has produced a series of rapid changes in the developmental patterning programs of turtles that found adaptive favor in the survival value bestowed. This produced the radical morphological innovations inaugurating the successful turtle-specific body plan in the Triassic, but based on only a few simple, quick genetic changes.

The vertebral column of amniotes is often specialized. In snakes, where twisting forces might be even greater because they are legless, additional sets of zygapophyses, anterior **zygosphene** and posterior **zygantrum,** provide additional checks on torsion but do not significantly restrict normal lateral bending of the vertebral column (figure 8.29a, b). In birds, numerous cervical vertebrae have highly mobile heterocoelous articulations between them, giving the skull, which rides on this flexible chain of vertebrae, great freedom of movement and reach (figure 8.30). At the other end of the vertebral column, the posterior thoracic, lumbar, sacral, and, occasionally, caudal vertebrae fuse into a unit, the **synsacrum.** Similarly, adjacent bones of the pelvic girdle fuse into the **innominate** bone, which in turn fuses with the synsacrum (figure 8.31). The overall result is the union of pelvic and vertebral bones into a sturdy but light structure supporting the body during flight.

In mammals, the vertebral column is differentiated into distinct regions. Typically, mammals have seven cervical

Genes, especially *Hox* genes, exert global control over patterning of the vertebrate body. Specific gene effects on morphology can be revealed by "knockout" experiments wherein targeted genes of interest are eliminated or their expression suppressed. This creates essentially a laboratory produced gene "mutant." For example, in a normal mouse (box figure 1a) its axial skeleton includes a ribcage of 13 thoracic (ribbed) vertebrae, followed by a lumbar region with 6 vertebrae devoid of ribs, 6 sacral vertebrae with short ribs fused to transverse processes that meet the pelvic girdle, and a caudal region of various vertebral number. If, however, *Hox10* gene (and its duplicate copies) is knocked out (-*Hox*10), thereby eliminating its effect on axial design, no lumbar vertebrae are formed; instead ribs project from all posterior vertebrae (box figure 1b). If *Hox11* (and its duplicate copies) alone is knocked out (-*Hox*11), lumbar vertebrae appear normal but sacral vertebrae are not formed; instead these vertebrae assume a lumbar-like identity (box figure 1c). These knockout experiments together tell us how some of the axial column patterning may be effected by gene control. *Hox10*

BOX FIGURE 1   Axial Patterning by *Hox* **Genes** (a) Normal mouse axial pattering showing ribcage, lumbar region, and sacral region. (b) Knockout of -*Hox10* alone removes its regulatory influence and ribs form on vertebrae in the expected lumbar region. (c) In the absence of -*Hox11,* sacral vertebrae fail to form. Asterisks identify the lumbar region; arrow denotes position of 13th thoracic rib. Girdles and limbs not included. Figure kindly supplied by Mario Capecchi based on his research in *Science* (2003).

acts to repress rib formation, but *Hox11* partially suppresses *Hox10* locally checking its expression and thereby locally permitting formation of sacral vertebrae. From this we might postulate the genetic basis for the lumbar region in mammals and further how changes in *Hox* gene expression account for variation in axial number in the lumbosacral region by shifts in boundaries of gene expression.

**Hox genes and their kingdoms
(p. 204);**
Hox genes (p. 729)

vertebrae, beginning with an atlas and axis, which permit the head great freedom of movement. Even the long-necked giraffe and "neckless" whale have seven cervical vertebrae, although exceptions occur in sloths (with six to nine) and sirenians (with six). In armadillos and many jumping mammals such as kangaroo rats, the seven cervical vertebrae may fuse. The number of vertebrae within the thorax and lumbar regions ranges from about 15 to 20, and there are usually two or three sacral vertebrae, although humans have five. The caudal vertebrae are quite variable in number. The mammalian tail is much less massive than the reptilian tail. Arches, zygapophyses, and transverse processes diminish toward the posterior tip of the tail so that most caudal vertebrae near the end of the series consist only of centra.

Form and Function

Most phylogenetic changes in the form of the vertebral column address new functions. Transition from water to land was one significant change in vertebrate lifestyle, and it was accompanied by considerable change in the mechanical demands experienced by the axial skeleton. To understand these mechanical forces and their impact on design, we should first compare the general problems faced by aquatic and terrestrial vertebrates.

Fluid Environment

In an aqueous medium, such as fresh water or marine seas, an organism does not depend primarily on the endoskeletal framework for support. Instead, the body takes advantage of its buoyancy in the surrounding water (figure 8.32a). For an active aquatic organism, two problems are uppermost. The first is drag on the body as it slips through a relatively dense medium, water. The answer is streamlining, contouring of the body to reduce drag forces. It is no accident that the general body shapes of fast-swimming fishes and supersonic aircraft are both streamlined. This shape improves the performance of both fishes and aircraft as they meet common physical demands while traveling through a medium that resists their passage.

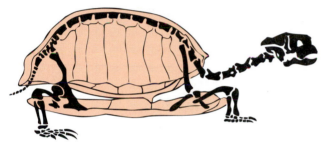

FIGURE 8.27   Turtle skeleton. (a) The skeleton of this fossil turtle shows how expanded vertebrae, ribs, and peripheral dermal plates fuse to form the shell. (b) Silhouette of the cranial, appendicular, and axial skeleton within the shell. (c) Head of the softshell turtle *Trionyx* retracted within its shell. Flexible articulations between cervical vertebrae permit this extensive movement.

(a) After Bellairs; (b) after Radinsky; (c) after Dalrymple.

FIGURE 8.28   Cross section of turtle body. (a) Showing the unusual position of the appendicular skeleton inside the rib cage (dark) compared with the skeleton of other vertebrates, which lies outside the rib cage, illustrated by this cross section of an alligator (b).

Streamlining (p. 145)

The second problem for an active aquatic organism is orientation in three-dimensional space. Any streamlined body has a tendency to tip and deviate from its line of travel, rotating about its center of mass. In fishes, these perturbations are countered by stabilizing fins appropriately positioned along the body.

Three-dimensional stability (p. 327)

Terrestrial Environment

Land generally presents a two-dimensional surface across which to maneuver. Because tetrapods live on land without the buoyancy of a dense medium such as water, gravity presents a problem. When remaining in place, the tetrapod's body either rests on the ground between sprawled legs, or it is suspended between the pairs of legs, as in most mammals and quadrupedal dinosaurs. The pairs of legs function as abutments that support the body between them. The vertebral column serves as a bridge between the support posts, the legs, and suspends the body from it (figure 8.32b). A convenient mechanical analogy has been drawn between this posture and engineered structures such as bridges.

What bridge engineers call the Forth Bridge is a two-armed bridge in which both extensions are balanced against each other, or cantilevered, and carry the weight of the railbed to the pier (figure 8.33a). Compressive forces are borne by solid structural members, tensile forces by cables. The bridge carries the extended railbed between them. The weight of each section of the railroad is transferred to the nearest pier. The point between piers where weight transfer changes is the **nodal** (figure 8.33b).

Mechanics of loading (p. 147)

The mammalian vertebral column, if viewed in engineering terms, might be represented by two suspension bridges, with the body suspended from them. The spines and centra represent the compression members; the ligaments and muscles, the tension members; the two pairs

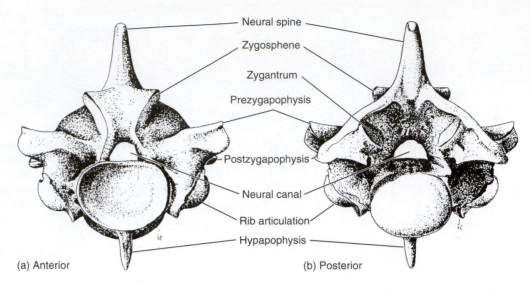

FIGURE 8.29 Trunk vertebrae from a snake in anterior (a) and posterior (b) views. In addition to interlocking pre- and postzygapophyses, snakes have an additional set of processes, the zygosphene and zygantrum, that engage to further prevent wringing of the long serpentine vertebral column.

Neural spine
Zygosphene
Zygantrum
Prezygapophysis
Postzygapophysis
Neural canal
Rib articulation
Hypapophysis

(a) Anterior

(b) Posterior

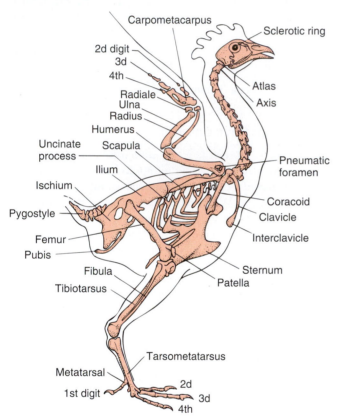

FIGURE 8.30 Skeleton of a chicken.

Carpometacarpus
Sclerotic ring
2d digit
3d
4th
Radiale
Atlas
Ulna
Axis
Radius
Humerus
Uncinate process
Scapula
Ilium
Pneumatic foramen
Ischium
Pygostyle
Coracoid
Femur
Clavicle
Pubis
Interclavicle
Fibula
Tibiotarsus
Sternum
Patella
Tarsometatarsus
Metatarsal
2d
1st digit
3d
4th

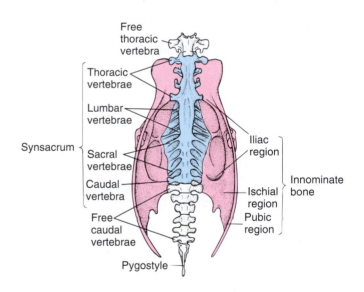

FIGURE 8.31 Synsacrum and innominate of a pigeon, ventral view. Notice how the synsacrum (blue) is fused to joined elements of the pelvis, the innominate bone (pink).

Free thoracic vertebra
Thoracic vertebrae
Lumbar vertebrae
Synsacrum
Iliac region
Sacral vertebrae
Innominate bone
Caudal vertebra
Ischial region
Free caudal vertebrae
Pubic region
Pygostyle

of legs, the piers. The point of the nodal depends on the relative weight distribution between the two piers, the two pairs of legs (figure 8.33c). Where the nodal occurs, force distribution within the vertebral column changes, and structural members that receive these forces become modified as well. Such an engineering analogy helps explain the reverse orientation of the neural spines midway along the length of the vertebral column between the two pairs of legs. The point at which the neural spines reverse might correspond to the biological nodal, and thus structurally reflects the underlying mechanical forces that the vertebral column must address.

If the body is heavy, one region is often cantilevered against another. In the iguanodon, a bipedal dinosaur, the heavy tail helps balance the weight of the thorax and anterior body across the hindlimbs (figure 8.34).

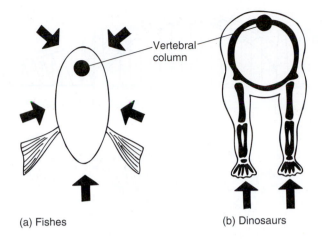

Vertebral
column

(a) Fishes

(b) Dinosaurs

FIGURE 8.32 **Body support.** (a) In fishes, the surrounding water (arrows) supports and buoys the weight of the body. (b) In tetrapods, the limbs support and the vertebral column suspends the weight of the body.

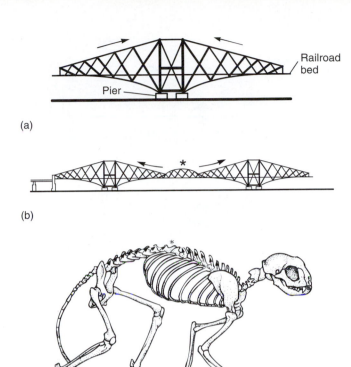

Railroad bed

Pier

(a)

(b)

(c)

FIGURE 8.33 **Engineering analogies and design of the vertebral column.** (a) The Forth Bridge works by resisting compression in its rigid members and tension in its flexible members. Each section of the bridge rests on piers. (b) If sections and piers are combined, the weight of the railroad can span the distance between the nearest piers. The "nodal" (*) marks the point of trade-off in weight distribution between two piers. (c) By analogy, the vertebral column might be viewed as serving roughly the same function, spanning the distance between fore- and hindlimbs. The bones resist compression; the muscles and ligaments resist tensile forces. Change in orientation of the neural spines marks the point of the nodal.

After Dubrul.

Other engineering analogies help clarify the ways in which some features of biological form might represent solutions to problems of mechanical stress. For example, to carry weight, any arch must maintain its bowed shape and avoid flattening out. Arch suspension bridges suspend the weight of the roadbed (figure 8.35a). The same mechanical principle seems to be incorporated into the design of mammals. Between the pairs of limbs, the abdominal muscles and sternum keep the arched vertebral column from sagging and so effectively maintain its structural and, thus, functional integrity. The neck forms a reversed arch, with ligaments and muscles holding the head (figure 8.35b–d).

Design of Vertebrae

Not all vertebrae are morphologically alike even within the same vertebral column. Differences in design reflect different mechanical demands within parts of the column as well.

Direction of the Neural Spine

The angle that the neural spine makes with its centrum often varies from vertebra to vertebra. This angle may represent a structural way to orient the spine so that it receives the suite of mechanical forces in the least stressful direction. Local mechanical forces on the spine arise largely from contraction of the axial musculature. The complex axial musculature originates at distant sites along the vertebral column and reaches to the ends of the neural spines, applying forces on these spines. Rostral muscles that are inserted on the neural spine pull it forward; more caudal muscles pull it backward. If these groups contract together, then the spine experiences the single resultant force of both acting together, not one or the other force acting separately. Recall that bones, like most

structures, are weakest in tension and sheer, but strongest when loaded in compression. If this resultant force bends the spine, it would place parts of the spine in tension or shear, which is worse, and so expose it to forces it is least able to withstand. Thus, the neural spine appears to be oriented in such a way that its long axis is in parallel with the resultant forces imposed collectively by all axial muscles inserted on it. This orientation means that the spine experiences these forces as a compressive force, the direction of stress loading in which it is strongest (figure 8.36a).

Height of the Neural Spine

The height of a neural spine is apparently proportional to the mechanical leverage the muscles must exert to move or stabilize the vertebral column. In a sense, the neural spines

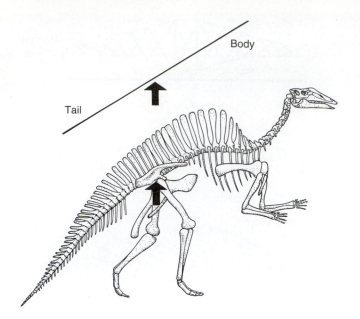

FIGURE 8.34 **Bipedal balance.** In bipedal animals, such as in this iguanodontid dinosaur *Ouranosaurus,* the weights of the heavy tail and the upper body are balanced like a simple seesaw on the fulcrum of the hips. Firmness of the vertebral column was apparently maintained by networks of strong ligaments that tied together the tall neural spines.

After Carroll.

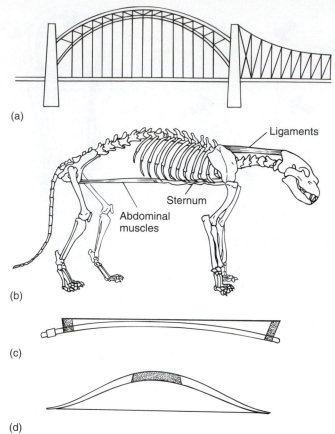

FIGURE 8.35 **Holding the arch.** (a) The roadbed and piers of an arch suspension bridge hold the span between them from an arch. As long as arch integrity is maintained, it supports the weight of the bridge. (b) Similarly, muscles and ligaments hold the vertebral column in arches. (c) The arch of the cervical vertebrae is like the reverse arch of a violin bow. (d) The other arch, formed by the trunk vertebrae, resembles an archer's bow.

After Dubrul.

are levers that transmit the force of muscle contraction to centra (figure 8.36b). This force is proportional to the physiological cross section of the muscle and to its lever arm, its perpendicular distance to the centrum. To increase this force, the muscle could be enlarged or the neural spine could be lengthened. Increasing spine length increases the lever arm from centrum to line of muscle action, and thus effectively increases the mechanical advantage of the muscle.

Vertebral designs incorporate modifications to meet these mechanical problems. For instance, in many reptiles, the spines on trunk vertebrae are about equal in height and similar in orientation (figure 8.37a). Compared with mammals, the axial musculature of reptiles is less specialized for rapid locomotion. In many mammals, height and direction of the spines vary within the same vertebral column, indicating the specialized functions performed by different sections of the vertebral column (figure 8.37b).

Regionalization of the Vertebral Column

We can now step back and take an overall look at the vertebral column with a view to summarizing how the transition from water to land has changed the mechanical demands placed on vertebrae and how the changes in design have followed.

In fishes, the vertebral column is differentiated into two regions, the caudal and trunk regions (figure 8.38a). Zygapophyses and similar interlocking projections are generally absent. The centra are unspecialized except if they receive ribs or hemal or neural arches. The relatively undifferentiated vertebral column of fishes reflects the fact that it is not used to support the body. Support comes generally from the buoyancy of the surrounding water. The vertebral column mainly offers sites of attachment for the swimming musculature. It serves as a mechanical replacement of the notochord, resisting the tendency of the body to telescope, yet it allows lateral flexibility for swimming.

In tetrapods, however, the vertebral column supports the body against gravity and receives and transmits the propulsive forces that limbs generate during locomotion. Diverse functional demands are placed on the vertebral column, so we might expect to find delineation of specialized regions.

In early tetrapods, caudal, sacral, trunk, and modest cervical regions of the vertebral column are delineated (figure 8.38b). Most early tetrapods are not strictly terrestrial and return frequently to water. Much of the musculature and axial skeleton still retain similarities to their fish

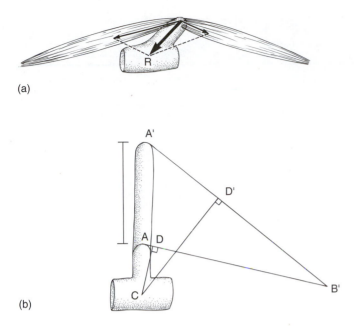

(a)

(b)

FIGURE 8.36 **Orientation and height of the neural spine reflect mechanical forces acting on or through it.** (a) Axial muscles develop forces, the resultant of which is coincident with the direction of the neural spine. If the resultant produced a shearing or bending of the spine, it would introduce forces that the neural spine, like most supportive structures, is least able to withstand. (b) Muscles that are inserted in the neural spine act on it like a lever to bring a force to the centrum. One way to increase the mechanical leverage of this force is to increase the length of the neural spine. Increasing the length from A to A′ changes the length of the lever arm, the perpendicular distance from the line of action to the centrum. In this example, the lever arm increases from CD to CD′ and so it increases the effective force on the centrum.

(a) Pelycosaur

(b) Bison

FIGURE 8.37 **Variation in the height of neural spines can be seen in vertebrates in which relatively heavy weights must be supported by the vertebral column.** (a) Skeleton of a pelycosaur, with most of the neural spines of similar height and orientation. (b) Skeleton of a bison, illustrating the tall neural spines in the shoulder. Through ligaments to the skull and cervical vertebrae, these neural spines help support the weight of the heavy head.

(a) After Carroll; (b) after Romer.

ancestors. For instance, the long tail often supports a broad fin, and the trunk region is relatively undifferentiated, as in fishes. However, locomotion on land is important, especially among adults. Through the pelvic girdle, the hindlimbs are directly attached to the adjacent region of the vertebral column to define the sacral region. The cervical region also is differentiated, allowing some freedom for the skull to turn independently of the body.

In primitive amniotes, cervical, thoracolumbar, sacral, and caudal regions are present (figure 8.38c). The sacral region is stronger than that in early tetrapods, designed to support more habitual existence on land. Most retain a trunk (**dorsal, thoracolumbar** region). Ribs on the vertebrae immediately in front of the hindlimbs may be shortened, and in some fossil and modern reptiles, the trunk may differentiate into two regions, the thorax with ribs and the lumbar region without ribs. The appearance of a lumbar region within the posterior thorax deserves some notice because it reflects an increase in locomotor performance. As hindlimbs swing forward to take long strides during rapid locomotion, the vertebral column usually flexes laterally on itself. This can cause ribs on adjacent vertebrae to

crowd one another. Loss of ribs in the area of greatest flexion addresses this crowding, producing a presacral section of the vertebral column without ribs, the lumbar region. Consequently, the appearance of a lumbar region marks a point at which tetrapods begin to experiment with more rapid forms of locomotion. This occurs in many archosaurs, some modern reptiles, and synapsids. Locomotor behavior cannot be easily determined from fossils directly. But behavior can be inferred from morphology. In chapter 9, we see that the morphology of the appendicular skeleton confirms this interpretation. Presumably more active lifestyles accompanied these more rapid forms of locomotion in later tetrapods.

Five distinct regions are differentiated within the vertebral column of mammals: cervical, thorax, lumbar, sacral, and caudal (figure 8.38d). The musculature is attached to the vertebral column in complex ways, corresponding to the demands that active locomotion places upon the individual vertebrae. In aquatic mammals and in other tetrapods that secondarily resume an aquatic lifestyle, the axial column returns, at least in part, to the compression girder of fishes. Hindlimbs are often reduced, and forelimbs form paddles (figure 8.39). In

The human vertebrate is one of the most interesting pieces of personal body engineering. Our posture is upright when we walk. In other words, we are bipedal. We depend upon two legs rather than four like our quadrupedal ancestors. Our unusual posture has required some reengineering to restabilize our upright carriage. Few other mammals are built to stand and walk comfortably on two legs. Prairie dogs sit up on their hindlegs, some deer lift up their forelegs, other primates become bipedal for short distances, but humans are built to be comfortable bipeds.

Upright instability comes from two features of our bipedal posture. First, we use half the number of support posts, two limbs, compared with the four of quadrupeds. Second, upright posture places the thorax and much of the rest of our body well above our center of gravity. For these reasons, unusual adaptations to upright posture have been incorporated into our design.

We might imagine that our upright posture is built up of three changes, each elevating the torso in increments of about 30° (box figure 1a–e). First, the front of the body is lifted about 30° (box figure 1a). This change is seen in some primates and can be accomplished without much redesign of the leg muscles. Second, the upper part of the pelvis is tilted back, rotating the vertebral column an additional 30° (box figure 1b). Third, the vertebral column in the lumbar region curves the last 30° to bring the upper body fully upright (box figure 1c).

To accommodate birth of an infant with a relatively large head, the birth canal through the pelvis is expanded by shifting the sacral

BOX FIGURE I **Human bipedal posture.** Compared with quadrupedal primates (a), human upright posture is engineered by changes in the slant of the pelvic girdle (b), increased curvature in the lower back (c), and broadening (d) and then shortening of the hips (e).

Based on the research of G. Krantz.

region farther back (box figure 1d). In reaching this upright posture, the lines of muscle action from hips to femur are altered. A broad pelvis is required to accomplish the same spread of muscle orientations and favorable mechanical angles on the femur. Shortening and broadening of the pelvis restore a more advantageous muscle leverage of the gluteal muscles during a striding gait (box figure 1e).

Birth canal (p. 354); human striding gait (p. 388)

The column of vertebrae projecting above the hips, like the mast of a ship projecting above the hull, is stabilized by a system of ligaments and muscles that act like rigging to support the column (box figure 2a). Flaring and shortening the pelvis also broaden the base of support (box figure 2b). If the pelvis had remained tall, the upper body would have been placed well above its balanced support on the femur (box figure 2c). By shortening the distance between the hips and the head of the femur, the weight of the upper body is brought close to and directly above the femur upon which it is balanced and situated in a less precarious position.

---

the porpoise, for example, anti-twist zygapophyses are absent from the centra, although interlocking notches of successive neural spines resist torsion (figure 8.40a, b).

Birds are an interesting example, exhibiting a close match of form and function within the vertebral column. Cervical vertebrae are flexibly articulated to give the head great freedom of movement and reach when a bird preens its feathers or probes for food. On the other hand, most of the vertebrae in the middle and posterior part of the column are fused to each other and to the pelvic girdle (figure 8.41). This brings rigidity to the vertebral column and establishes a firm and stable axis for control while a bird is in flight. Indirectly, this fusion of elements decreases the weight of the body because less muscle is required to control individual vertebrae. Muscles otherwise required to bring stability can be reduced, economy of

(b)

Chimpanzee
Pan

Human
Homo

(a)

(c)   Cat   Human

BOX FIGURE 2   **Bodies and biomechanics.** Upright posture of humans brings with it some instability as the weight of the upper body is brought above the hips. Several features of the human skeleton are remodeled to return some stability. (a) Muscles and ligaments, like the rigging of a sailing ship, span the distance between vertebrae and up from the hips to the lower vertebrae and ribs to stabilize the rising vertebral column. (b) The blade of the ilium is widened to broaden the base of support at the hips. (c) The distance between the sacrum and the head of the femur is shortened in humans, thus bringing the base of the vertebral column closer to its eventual support by the acetabulum (socket of the hip), and offering the same spread of hip muscle attachments to maintain a favorable line of action on the femur.

After Dubrul.

design is realized, and weight of the bird is lightened. The vertebral columns of most birds show much uniformity, a likely indication of the overriding significance of flight and its demands on the biological design of these vertebrates that have mastered life in the air.

Overview

The axial skeleton includes the notochord and vertebral column. The notochord is the oldest, predating vertebrates, having appeared within early chordate evolution. It is a slender rod that develops from mesoderm, lying dorsal to the coelom. Typically, it is composed of a core of fluid-filled cells wrapped in a fibrous sheath. Mechanically, it laterally flexes but does not axially compress, thereby turning contraction of axial muscles into lateral undulations during locomotion. It is prominent well into derived fishes, serving as the major means of axial support. Even when replaced by the vertebral column, it still appears as an embryonic structure, inducing the neural tube to develop above it and serving as a scaffold for the growing embryonic body. Chains of articulated

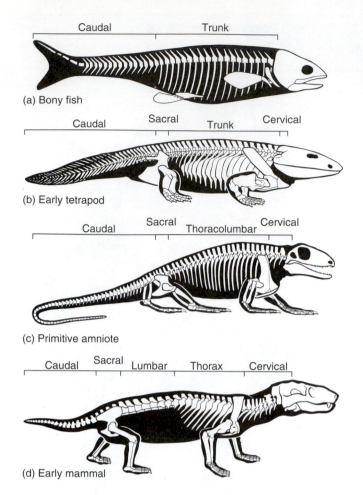

FIGURE 8.38 Regionalization of the vertebral
column. (a) Bony fish. (b) Labyrinthodont. (c) Primitive amniote.
(d) Early mammal.

FIGURE 8.39 Axial skeletons of whales in lateral
silhouettes. Shown are two fossil species (*Basilosaurus,
Zygorthiza*), two odontocetes (*Hyperoodon, Physeter*), and two
mysticetes (*Balaenoptera, Balaena*). Notice the reduction in size of
the limbs and girdles and the proportionate increase in size of the
vertebral column compared with the size of limbs, girdles, and
vertebral columns of quadrupeds.

After Kent.

vertebrae, cartilaginous or bony, constitute the vertebral column. Each vertebra is composed of a centrum, supports a
neural arch and spine, and often is associated with processes,
including ribs. Intervertebral bodies or disks occur between
successive vertebrae. The shapes of articular surfaces determine functional roles. Ribs, which often meet the sternum
midventrally, protect viscera and contribute to respiratory
movements in tetrapods. In turtles, the hip and shoulder
girdles are actually brought inside the rib cage beneath
the protective shell. Embryonic development of the vertebral column may involve prior formation first in arcualia
(chondrichthyan fishes, some other fishes), appearance of
specialized anlaga (teleosts), or perichordal tube (most tetrapods), before ossification occurs.

Vertebrae are rare in agnathan and primitive gnathostome fishes, in which axial support more commonly depends
upon a prominent notochord (figure 8.42). The caudal fin
delivers propulsive forces, driving the fish forward, but also
can produce lift forces. Especially in bony fishes and tetrapods, the ossified vertebral column largely replaces the notochord as the major source of axial support and locomotion.

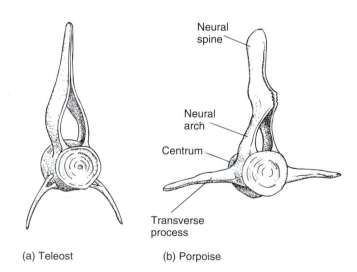

FIGURE 8.40 Vertebrae of aquatic vertebrates.
(a) Vertebra of a teleost fish. (b) Vertebra of a porpoise. Notice
the reduction in zygapophyses in the vertebra of the aquatic
porpoise.

FIGURE 8.41 **Bird vertebral column.** Regions of extensive vertebral fusion are indicated posteriorly within the synsacrum. The numerous heterocoelous cervical vertebrae allow for great mobility of the head.

Intervertebral cartilages or intervertebral disks are composed of fibrous connective tissue and fluid, and lie between successive vertebrae. The various types of collagen that make up the fibrous connective tissue are arranged in such a way as to resist tension and shear forces. Especially within the core of intervertebral disks are proteoglycans, proteins attached to special chains of carbohydrates. Their special chemical structure gives them the property of binding water, which therefore brings it into the functional service of resisting compressive forces.

In the buoyancy of water, the axial column serves primarily as a compression girder, resisting telescoping of the body during locomotion and translating axial muscle forces into lateral swimming undulations. These same lateral undulations of fishes are carried into early tetrapods on land as the early basis of terrestrial locomotion.

In a terrestrial environment, the axial column assumes the additional function of suspending the weight of the body, without the aid of aquatic buoyancy, as it is carried over the surface of the land. The tetrapod vertebral column often incorporates elements of design that are analogous to humanly engineered structures such as bridges, whereby weight is cantilevered or suspended about or between supportive columns (limbs). Torque becomes a feature of quadrupedal locomotion, favoring the appearance in tetrapods of anti-twist features of the vertebrae, such as the zygapophyses. The height and direction of neural spines reflect their role as levers, delivering forces to the vertebral centra and thereby moving or stabilizing the vertebral column. The vertebral column is regionalized, reflecting functional demands. In fishes, the vertebral column is relatively undifferentiated trunk and caudal regions. It lacks zygapophyses. It is not used in support of the weight of the body, but instead is used basically for muscular support and attachment, a compression girder. In tetrapods, the vertebral column is used to support the body proper, limbs provide the propulsive force for locomotion, and these forces are transmitted to the body through the vertebral column. A cervical region differentiates in early tetrapods, for cranial mobility, as does a sacral region, for direct attachment of the hips to the axial column. These regions are even more distinct in reptiles, where posteriorly the trunk may exhibit rib reduction, yielding a lumbar or lumbarlike region. In mammals, locomotion based on flexions of the vertebral column vertically (cf. laterally) is accompanied by the appearance of a distinct lumbar region. This provides mammals with five distinct regions—cervical, thorax, lumbar, sacral, and caudal. In birds, the overriding dynamic demands of aerial locomotion are accompanied by accentuated fusions and flexions of the vertebral column. Fusion of the synsacrum (sacrum plus adjacent vertebrae) with the innominate (ilium, ischium, pelvis) produces a stable, firm platform while in flight; multiple, heterocoelous cervical vertebrae return flexibility when deploying the head.

Overall, the axial skeleton contributes, with its musculature, to bending of the body, storing elastic energy, and transmitting useful forces for locomotion generated by appendages (fins or limbs) or by the tail (aquatic). Associated ribs help ventilate the lungs. The axial skeleton also holds the position of the body steady against gravity.

The form and function of the vertebral column are related directly to the static and dynamic demands placed upon it. In turn, these are related to the general environments in which it serves—aquatic or terrestrial—and the type of locomotion in which the vertebral column is involved. As we shall see in chapter 9, the design of the appendicular skeleton is similarly affected by such functional demands.

FIGURE 8.42 **Phylogeny of the Axial Column.** This diagrammatic summary of axial column occurrence illustrates the early and continuing importance of the notochord in chordate evolution. Neural and hemal arches and their spines may be present early in the phylogeny, but do not become prominent until placoderms. Vertebral centra are found late in shark evolution, late in bonyfish evolution, and in tetrapods where centra replace most of the notochord, which contributes only to the core of the intervertebral disks. Notochord, shaded; vertebral elements, white. The general presence of neural spines in ostracoderms is uncertain, hence indicated with?.

WEB LINK

Visit the text website at *www.mhhe.com/kardong7e* for additional study aids including selected references, functional laboratories, and web links to chapter topics.

CHAPTER 9

Skeletal System: The Appendicular Skeleton

INTRODUCTION

BASIC COMPONENTS
Fins
Limbs

ORIGIN OF PAIRED FINS
Gill-Arch Theory
Fin-Fold Theory
Embryonic Development of Tetrapod Limbs

PHYLOGENY
Fishes
Agnathans
Placoderms
Chondrichthyans
Acanthodians
Bony Fishes
Tetrapods
Pectoral Girdle
Pelvic Girdle
Manus and Pes

EVOLUTION OF THE APPENDICULAR SYSTEM
Dual Origin of the Pectoral Girdle
Adaptive Advantage of Lobe Fins
Onto the Land

FORM AND FUNCTION
Swimming
Terrestrial Locomotion
Early Gaits
Early Modes of Locomotion
Cursorial Locomotion
Aerial Locomotion
Origin of Bird Flight
Fossorial Locomotion

OVERVIEW

Introduction

From components of the appendicular skeleton, evolution has fashioned some of the most elegant and specialized locomotor devices, from the fins of fishes to the limbs of tetrapods. Like the rest of the skeletal system, the appendicular system is represented well in the fossil record. This brings us into direct contact with the structural details of extinct animals and helps us to track the general course of phylogenetic modifications of these skeletal elements. Within the appendicular skeleton, the relationship between structure and biological role is direct, at least in a general way. We need not be aerodynamic engineers to understand that the wings of birds give them access to the air and the special lifestyles that follow, that the limbs of tetrapods serve them on land, and that the fins of fishes are suited to water. Transitions from water to land and from land to air have had an impact upon the design and redesign of the appendicular system.

What we see here in chapter 9, however, is that form and function can be closely matched. Not all birds use the air in equal ways. Some, in fact, such as penguins and ostriches, do not fly at all. Tetrapods use the land differently. Some lumber, some dash, some dig, some climb trees. For some fishes, fins provide lift for cruising through the water; for others, fins become specialized for maneuvering in tight places. Form and function are slightly different for each, and biological design reflects these differences.

Basic Components

The appendicular skeleton includes the **paired fins** or **limbs** and the **girdles**, the braces within the body that support them. The anterior girdle is the **shoulder** or **pectoral girdle**, to which dermal and endochondral skeletal elements contribute, and that supports a pectoral fin or limb. The posterior girdle is the **hip** or **pelvic girdle**, consisting of endochondral skeletal elements that support the pelvic fin or limb.

Fins

Particularly in primitive fishes, the body is apt to carry projecting spines, lobes, or processes. Unlike these projections, fins are membranous or webbed processes internally strengthened by radiating and thin **fin rays.** They form initially at the interface between dermis and epidermis, like scales, but then sink into the dermis, and so are sometimes called *dermal* fin rays. In elasmobranchs, these dermal fin rays, termed **ceratotrichia**, are slender keratinized rods (figure 9.1a). Fin rays, or **lepidotrichia**, in bony fishes are usually an ossified or chondrified series of tiny elements that strengthen this web (figure 9.1b). In some bony fishes, the tip of the fin may be further stiffened by additional keratinized rods, the **actinotrichia.** The proximal part of the fin close to the body is supported by **pterygiophores** of two general types: the enlarged **basals** within the proximal part of the fin, and the slender **radials** that extend support from the basals into the middle region of the fin (figure 9.2a).

Fins occur singly, except for a pair near the head and a second pair posterior to this, the pectoral and pelvic fins, respectively. The basal pterygiophores of these projecting paired fins articulate with and are braced by girdles inside the body wall. These paired fins will receive our greatest attention because they are the phylogenetic source of the tetrapod limbs.

Limbs

Formally, a limb is termed a **chiridium,** a muscular appendage with well-defined joints bearing digits (fingers and toes), not a fin, at its end. The chiridia of tetrapods, forelimbs and hindlimbs are all built on the same pattern, composed of three recognized regions. The **autopodium,** the distal end of the limb, consists of numerous elements composing the wrist and ankle, which in turn support their respective digits (figure 9.2b). The special term *hand* implies a structure modified for grasping, and *foot* suggests a role in standing. But these two terms do not apply logically across tetrapods. For instance, the terminal part of a horse forelimb is not a hand, nor is the terminal part of the hindlimb of a dolphin a foot. Therefore, the terms **manus** and **pes,** respectively, are preferred for the autopodium of fore- and hindlimbs. The middle limb region is the **zeugopodium,** with two internal supportive elements: ulna and radius of the forearm, tibia and fibula of the shank. The limb region closest to the body is the **stylopodium,** with a single element: humerus of the upper arm, femur of the thigh.

(a) Fin

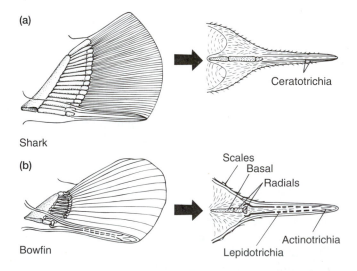

(a)

Ceratotrichia

Shark

(b)

Scales
Basal
Radials

Actinotrichia
Lepidotrichia

Bowfin

FIGURE 9.1   Fin rays. (a) Ceratotrichia are keratinized rods that radiate out like vanes in a fan to internally support the fins of chondrichthyan fishes. (b) Lepidotrichia are cartilaginous or ossified supports within fins of bony fishes.

Girdle

Morphological term	Forelimb	Hindlimb
Stylopodium	Upper arm	Thigh
Zeugopodium	Forearm	Shank (crus)
Autopodium	Manus (wrist-palm-fingers)	Pes (ankle-sole-toes)

(b) Limb

FIGURE 9.2   Basic components of the fin and the limb. (a) The fin is composed of pterygiophores, basals and radials, and dermal fin rays. Fin rays are called *lepidotrichia* in bony fishes and *ceratotrichia* in elasmobranchs. (b) The limb, either fore- or hindlimb, includes three regions: stylopodium (upper arm/thigh), zeugopodium (forearm/shank), and autopodium (manus/pes).

A depression within the pectoral girdle, the **glenoid fossa,** articulates with the humerus. A deep socket in the pelvis, the **acetabulum,** receives the femur.

Origin of Paired Fins

Like any object traveling in three-dimensional space, the body of a fish is susceptible to deflections from its line of travel about its center of mass. It may swing from side to side **(yaw),** rock about its long axis **(roll),** or buck forward and back **(pitch;** figure 9.3a). Wind tunnel tests using shark models with selected fins removed have helped clarify how fins bring stability to a streamlined body. It appears that dorsal and lateral fins control the body by resisting perturbations of the body about its center of mass. Separate tests of pectoral fins show that these fins do not produce significant lift as in a fixed-winged aircraft. Instead, pectoral fins are used for close maneuvering within enclosed vegetation (some teleosts) and during steady, horizontal swimming (sharks) for initiating rising or sinking turns.

Dynamics of heterocercal tails (pp. 307–308)

As early fishes became more active, they would have experienced instability while in motion. Presumably, just such conditions favored any body projection that resisted pitch, roll, or yaw, and led to the evolution of the first paired fins. The associated girdles stabilized the fins, served as sites for muscle attachment, and transmitted propulsive forces to the body.

In gnathostome fishes, two fundamental types of fins developed from two different arrangements of the **metapterygial stem** or **axis,** a chain of endoskeletal basals. One fin type is the **archipterygial fin,** in which the metapterygial stem runs down the middle of the fin (figure 9.3b). From this central stem, endoskeletal radials project outward to support the **preaxial** (anterior) and **postaxial** (posterior) sides of the fin evenly. Slender dermal fin rays extend to the edges of the fin to complete this support. Externally, the archipterygial fin looks leaf-shaped and narrowed at its base. The second basic fin type is the **metapterygial fin,** in which the metapterygial stem of basals is located posteriorly. Most radials project from this posterior axis into the preaxial side of the fin, and dermal fin rays extend from the ends of the radials to the edges of the fin (figure 9.3b). These two fin types have influenced theoretical work on the origin of paired fins. To track the phylogenetic source of early fish fins, the gill-arch and the fin-fold theories have been put forth.

Gill-Arch Theory

During the second half of the nineteenth century, morphologist C. Gegenbaur proposed that paired fins and their girdles arose from gill arches (figure 9.4). Specifically, the endoskeletal girdle arose from the gill arch, and the primitive archipterygial fin arose from the gill rays. Initially, Gegenbaur based his theory on fin anatomy in sharks. However, the discovery (1872) of the Australian lungfish, *Neoceratodus,* convinced him that the primitive fins were archipterygial

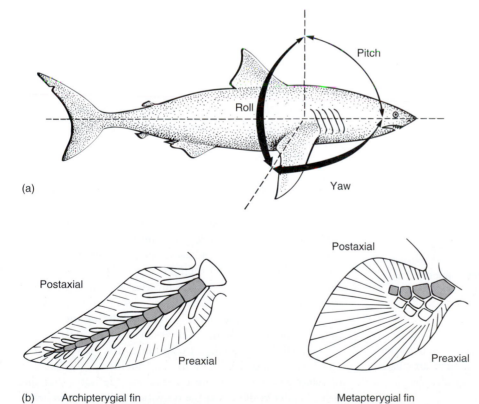

(a)

(b) Archipterygial fin   Metapterygial fin

FIGURE 9.3 Fins as stabilizers. (a) The body of a fish can deviate from its intended line of travel in three ways. A roll rocks the fish about its long axis, a yaw swings it from side to side, and a pitch bucks it up and down about its center of mass. (b) There are two types of fin structure, the archipterygial fin with a symmetrical central axis (left) and the metapterygial fin with the asymmetrical axis displaced toward the postaxial side (right). The metapterygial axis or stem present in both is shaded.

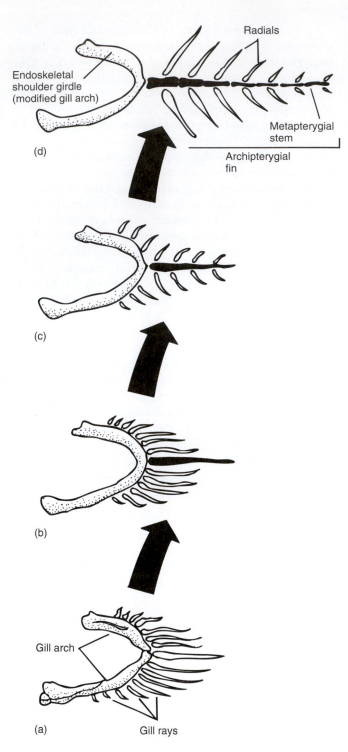

FIGURE 9.4 Gill-arch theory proposed by Gegenbaur for the origin of paired fins. Gill rays expand (a, b) and proliferate (c), forming a long central support for an external fin, not unlike the archipterygial condition (d) found in some present-day lungfishes.

fins similar to the paired fins in *Neoceratodus*—a central stem supporting a series of radials. This central stem articulated with the lungfish's endoskeletal shoulder girdle, the future scapulocoracoid.

Nonetheless, the gill-arch theory left much unexplained. Although it accounted for the evolution of the pectoral girdle, it does not explain (1) the appearance of a posterior pelvic girdle distantly placed from the gill arches, nor (2) the presence of dermal bone in the pectoral girdle, nor (3) the different embryologies of pectoral girdle and gill arches.

Fin-Fold Theory

At about the same time, the second half of the nineteenth century, morphologists F. M. Balfour and J. K. Thacher independently put forth the fin-fold theory, an alternative view expanded by later scientists. With this view, paired fins arose within a paired but continuous set of ventrolateral folds in the body wall that were stiffened by a transverse series of endoskeletal pterygiophores, proximally the basals and distally the radials (figure 9.5a–c). Additional stability came from the inward extension of basals and their eventual fusion across the midline to produce the supportive girdles (figure 9.10c). Dermal bone, a contribution of the overlying bony armor, was later added to the pectoral girdle to strengthen the paired fins further.

In support of the fin-fold theory, several indirect pieces of evidence are usually cited. The pregnathostomes with the first fin-folds were *Myllokunmingia* and *Haikouichthys* (figure 3.6). Both of these early Cambrian agnathans were equipped with fin-folds. Further, many surviving fossils of early fishes carry hints or presumed remnants of these earlier fin-folds. For instance, some primitive ostracoderms possessed lateral continuous folds along the ventral body wall. Acanthodians possessed a paired row of spines to mark where a paired fin-fold presumably resided in their ancestors (figure 9.6a–c). Furthermore, if pectoral and pelvic fins arose from a fin-fold, then they likely arose at the same time (figure 9.6b). In this context, it is significant that the paired fins of embryonic sharks develop together from a continuous thickening of ectoderm along the lateral body wall (figure 9.6e). This has been interpreted as an embryonic recapitulation of the phylogenetic transition from fin-folds to paired fins.

More recently, the fin-fold theory has been given greater detail. For example, Erik Jarvik has emphasized the segmental contributions to the archipterygial fins in some living fishes. If fins initially were steering keels, contributions from the adjacent segmental myotomes muscularize the fin-folds, making them movable. Jarvik further suggested that the endoskeletal basals and radials developed from the mesenchyme within the core of the fold, supporting the projecting fin and offering attachments to muscles. Within the fin, the supportive dermal fin rays developed from modified rows of scales, an event that seems to be repeated during the embryonic development of dermal fin rays in many living fishes.

Jarvik has also taken exception to the view that dermal bone is added initially to the shoulder girdle because of selection forces favoring fin stability. He notes that the fish shoulder girdle lies at the transition from trunk to head.

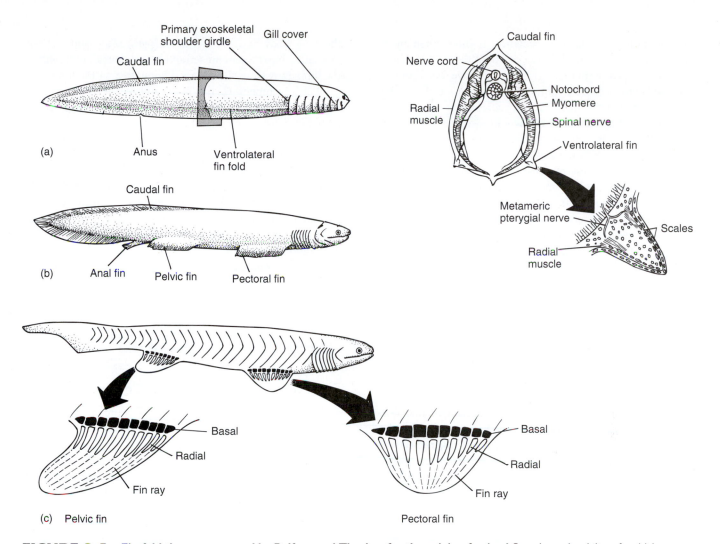

FIGURE 9.5 **Fin-fold theory proposed by Balfour and Thacher for the origin of paired fins.** Lateral stabilizer fins (a) become divided into specialized pectoral and pelvic fins (b). Basals and radials enlarge to support the fins (c). The metapterygial stem is solid black.

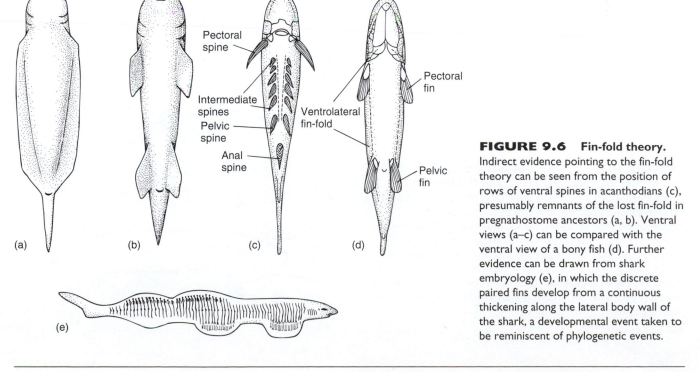

FIGURE 9.6 **Fin-fold theory.** Indirect evidence pointing to the fin-fold theory can be seen from the position of rows of ventral spines in acanthodians (c), presumably remnants of the lost fin-fold in pregnathostome ancestors (a, b). Ventral views (a–c) can be compared with the ventral view of a bony fish (d). Further evidence can be drawn from shark embryology (e), in which the discrete paired fins develop from a continuous thickening along the lateral body wall of the shark, a developmental event taken to be reminiscent of phylogenetic events.

At this point, the axial musculature is interrupted by the pharyngeal slits. Consolidation of small dermal skin bones into a composite dermal girdle may have been initially advantageous because it offered an anterior site for attachment of the interrupted axial musculature at this point of transition. This dermal girdle would also form the posterior wall of the buccal cavity, protect the heart, and be the attachment site of some sets of jaw and gill-arch muscles. For one or all of these reasons, an anteriorly placed dermal girdle may have arisen and only secondarily joined with endoskeletal elements in support of the fin. Of course, no similar selection forces would be acting posteriorly, where the axial musculature runs uninterrupted from the trunk to the tip of the tail. This would help account for the presence of dermal contributions to the pectoral girdle and their absence from the pelvic girdles.

Experimental analysis of limb development in chick, mouse, and shark embryos provides an additional perspective on the underlying genetic basis for the evolution of fin-folds to fins. In tetrapods, the embryonic body is compartmentalized with respect to the dorsoventral axis. The *Engrailed-1* gene expression is restricted to the ventral compartment of the embryonic body, helping to establish a dorsoventral patterning. Within this ventral expression zone *T-box* genes, *Tbx5* and *Tbx4*, determine the identity of each paired appendage, anterior and posterior appendages, respectively. Sharks also have these two *T-box* genes specifying anterior and posterior fins attached along their length parallel with

the long body axis. The *sonic hedgehog (Shh)* gene, absent in sharks, but present in tetrapods, promotes outgrowth of the limb away from the tetrapod body, freeing the limb from parallel attachment to the body and establishing a proximal-to-distal-limb axis. On the other hand, the prevertebrate amphioxus has only one such *T-box* gene, *AmphiTbx4/5*. These discoveries suggest the following scenario for evolution of the lateral fin-folds into paired appendages.

The hypothetical ancestor (figure 9.7a) possessed lateral fin-folds with *T-box* genes *(Tbx)* expressed within the *Engrailed-1* domain along the ventral compartment of the body. The presence of pectoral, but not pelvic, fins as in some ostracoderms could be the result of this single *Tbx* set of genes. Next, and before chondrichthyan fishes arose, duplication of the *Tbx* cluster produced two sets of genes, one anterior *(Tbx5)* and one posterior *(Tbx4)*, expressing pectoral and pelvic fins, respectively (figure 9.7b). The subsequent acquisition of *Shh* expression established a proximal-distal axis and promoted outgrowth of the fins from the body wall (figure 9.7c), as for example occurs in sarcopterygian fishes and in later tetrapods.

Embryonic Development of Tetrapod Limbs

Although Gegenbaur considered archipterygial fins the most ancient type of fin, this now seems unlikely. The archipterygial fins of lungfishes are more likely modified from metapterygial fins, which are common in gnathostome fishes. Neither type shows derivation from embryonic gill

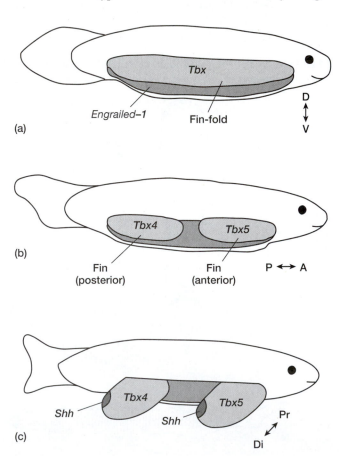

FIGURE 9.7 Evolution of fin-folds to paired fins. (a) Hypothetical ancestor with lateral fin-folds. *T-box* gene, *Tbx*, is expressed within the domain of the patterning gene *Engrailed-1*. (b) Duplication of *Tbx* results in two gene complexes, one influencing anterior fin development *(Tbx5)* and the other presiding over posterior fin development *(Tbx4)* in the shark. Between the two, the flank is undifferentiated. (c) The acquisition of *Shh* promotes the outward growth of the fins from the body wall, a feature carried into tetrapods. Abbreviations: *T-box* genes *Tbx5* and *Tbx4*, dorsal (D), ventral (V), anterior (A), ventral (V), proximal (Pr), distal (Di), *sonic hedgehog (Shh)*.

After Tanaka et al.

arches as Gegenbaur envisioned. The most constant and perhaps most ancient part of the fin is the posteriorly placed metapterygial stem, which we recognize in the paired fins of gnathostome fishes as well as in the tetrapod limb.

In fact, recent embryonic studies detect what seems to be a common developmental pattern underlying most tetrapod fore- and hindlimbs. This pattern is characterized by the predominance of elements arising posteriorly on the postaxial side of the limb. First, a stylopodium appears, which consists of a single proximal element (humerus or femur). The stylopodium subsequently branches to yield an adjacent set of preaxial (radius or tibia) and postaxial (ulna or fibula) elements that represent the intermediate section of the limb, the zeugopodium. Through subdivision and budding of embryonic primordia, the postaxial element alone extends the emerging embryonic pattern farther distally, forming most of the primordia of the autopodium (manus or pes), which includes all the digits. The preaxial element contributes only a few wrist or ankle elements but none of the digits to the emerging embryonic autopodium (figure 9.8a).

This embryonic pattern of limb formation is asymmetrical. The postaxial series of skeletal elements branch to produce most of the limb and all its digits. The preaxial series never branch. Limb asymmetry in tetrapods seems to result from retention of the postaxial metapterygial stem in fishes (figure 9.8b). In chondrichthyans, primitive actinopterygians, and fossil rhipidistians, the branching metapterygial stem forms the series of postaxial elements in their fins. Elements of the metapterygial stem become incorporated into the tetrapod limb along with the basic asymmetry; therefore, the metapterygial axis of the tetrapod limb is asymmetrical, running through the postaxial elements and digits. Variation occurs from group to group in adult limbs, but this variation can be traced back to developmental modifications within the underlying embryonic pattern. Modifications of the basic pattern include fusion or loss of its fundamental elements, expansion of existing elements, and occasional appearance of new skeletal components, or **neomorphs.**

The basic configuration of supportive elements is generally the same in pectoral and pelvic appendages; however, some differences arise from changes in or to the pelvic girdle. For example, in chondrichthyans and in many placoderms, the pelvic fins of males are often endowed with **claspers,** modifications of the pterygiophores used during mating to engage the female. In many teleost fishes, the pelvic girdle and associated fins move forward to reside with the pectoral fins within the shoulder region.

The fin-to-limb phylogenetic transition, informed from embryological studies of tetrapod limb development, has drawn recent support from analysis of *Hox* genes involved in the growth and patterning of the limb. Embryological studies reveal that the major straight, metapterygial axis of most fish species is in the tetrapod limb bent around anteriorly to form the foundation for the distal limb such that digits arise from the posterior margin of this arched axis (figure 9.8c). *Hox* gene expression domains, in fish and mice, confirm that the major axis is arched anteriorly and that digits arise from its posterior margin. Other *Hox* genes that control the formation of digits on tetrapod limbs are also known to control the formation of the fin skeleton in primitive actinopterygian fishes. Thus, digits are not a tetrapod neomorph, but instead represent the co-opting of an old, underlying genetic mechanism present in basal actinopterygians brought forward into early tetrapods to contribute to the appearance and differentiation of digits.

Phylogeny

Fishes

Agnathans

Two early vertebrates from the Cambrian, *Myllokunmingia* and *Haikouichthys,* possessed ventrolateral fin-folds, but lacked discrete paired fins. Paired fins are also absent in living agnathans, hagfishes, and lampreys.

Ostracoderms had unpaired medial fins: a caudal fin on the tail, and often unpaired anal and dorsal fins. Anaspids usually possessed a paired sharp spine in the shoulder region, and some genera had long and stabilizing lateral fin-folds running the length of their bodies. Heterostraci and Galeaspida fossils lacked all trace of paired fins. Only among some osteostracans were paired fins present and only in the pectoral region. The corners of the head shield bore indented fossa into which lobe-shaped pectoral fins fitted, and the margins of the fossa offered attachment sites for associated fin musculature. Details of the fins themselves are incomplete, but there is evidence of an endoskeleton and associated muscles. Thus, all ostracoderms lacked pelvic fins, and most lacked even rudimentary pectoral fins.

Like acanthodians, chondrichthyans, and placoderms, ostracoderms lacked lungs or air bladders. Their bony armor surface gave them a density greater than the surrounding water, so they tended to sink to the bottom when they stopped swimming. Pectoral fins or spines provided anterior lift, as did the flattened head shield. Both gave ostracoderms some devices to generate modest lift when they swam. However, absence or slight development of pectoral fins, small body musculature, and absence of jaws suggest that these fishes were bottom dwellers and feeders, only occasionally becoming active swimmers in open waters.

Placoderms

Placoderm fishes first appear in the early Silurian and enjoyed quite an extensive radiation, apparently taking advantage of powerful jaws and active lifestyles. Both pectoral and pelvic girdles were present. The pelvic girdle seems to have been a single endoskeletal element. The more complex pectoral girdle consisted of various fused dermal elements that contributed to the walls of the thoracic bony armor and braced the endoskeletal scapulocoracoid. Indented within the scapulocoracoid was an

FIGURE 9.8 Hypothesized developmental pattern underlying most limb development. (a) The stylopodial element appears and then divides into a preaxial and a postaxial element in the zeugopodium. The preaxial element (radius/tibia) does not branch but gives rise to more distal elements contributing to the autopodium. The postaxial element (ulna/fibula) branches to form the carpals or tarsals and the digital arch, which yields the fingers and toes. This postaxial side of the limb is thought to be derived from the metapterygial stem of fishes. (b) Position of the metapterygial stem in representative fishes. (c) Fin-to-limb transition. Both embryological work (a) above, and recent *Hox* gene studies support the view that tetrapod limbs are built upon the fish-fin major axis curved around to produce the digits along its posterior margin. Abbreviations: anterior (A), posterior (P), humerus (H), radius (R), ulna (U), radiale (r), prepollux (pp), intermedium (i), ulnare (u), centrales (c1–c5), carpals (1–5), digits (I–V).

(a, b) Based on Shubin and Alberch; (c) after Jarvik.

articular fossa that received the basal pterygiophores of the fin. In some placoderms, such as the antiarchs, the pectoral "fin" was quite specialized, forming a tapered appendage of endochondral elements encased in dermal bone (figure 9.9a, b).

Chondrichthyans

Primitive chondrichthyans, such as early sharks, possessed pectoral and pelvic fins that were primarily stabilizers. They consisted of basal elements and tightly packed radials supporting the fin; the girdle was a single, enlarged basal element (figure 9.10a).

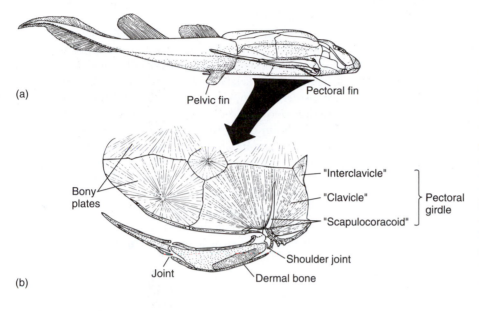

(a)

Pelvic fin   Pectoral fin

Bony plates

"Interclavicle"
"Clavicle"
"Scapulocoracoid"

} Pectoral girdle

Shoulder joint

Joint   Dermal bone

(b)

FIGURE 9.9   The antiarch _Bothriolepis,_ a placoderm from the Late Devonian. (a) Lateral view. (b) Ventral view detailing the pectoral fin. The pelvic fin was only slightly developed, and the pectoral fin, although distinct, was more of a specialized spine articulated with the girdle. Endochondral elements were encased in dermal bone. Pectoral fin is sectioned to show endochondral elements within the dermal exoskeleton. Homologies of bones contributing to the pectoral girdle are uncertain, so they are placed in quotes.

After Stensiö, 1969.

Pelvic fin

Pectoral fin

(a) Primitive shark

Pelvic fin

Pectoral fin   Radials

(b) Modern shark

FIGURE 9.10   Primitive (a) and modern (b) sharks. One trend in the evolution of the shark appendicular skeleton was the fusion of separate basal girdle elements across the midline (c). These fusions of pterygiophores produced the pubioischiac and scapulocoracoid bars.

(a, b) After Carroll; (c) after Hyman.

Radials   Basals

(c) Origin of girdles

In later sharks, the paired basal components of the pectoral and pelvic girdles became extended across the midline of the body to fuse into U-shaped **scapulocoracoid** and **pubioischiac bars,** respectively (figure 9.10b, c). Even the earliest chondrichthyans show no evidence of dermal contributions to the shoulder girdle. Modern sharks possess three enlarged pterygiophores at the base of the pectoral fin. The most posterior of these three is the **metapterygium,** arising within the series of basals of the metapterygial stem, followed by the **mesopterygium** and **propterygium,** enlarged derivatives of the radials. The metapterygial stem within the pelvic fin consists of a postaxial series, usually with one long element supporting a stand of radials (figure 9.11).

Acanthodians

In acanthodians, large spines formed the leading edge of dorsal, anal, and paired fins. Often, additional spines ran in rows between the pectoral and pelvic fins. In life, a web of skin usually stretched across these spines and was covered by rows of delicate scales (figure 9.12a–c). If present at all, basal and radial elements tended to be quite small. In some acanthodians, the pectoral spine articulated with a scapulocoracoid, but the pelvic spine is not known to articulate with an endoskeletal girdle.

Bony Fishes

Actinopterygians The pectoral girdle of actinopterygians is partly endochondral but mostly dermal. An air bladder is common throughout the group, so most members are neutrally buoyant. The fins function mainly as small oars for close maneuvering or slight adjustments of body position, or for braking.

The dermal shoulder girdle, well established even in primitive bony fishes, forms a U-shaped collar of bone around the posterior border of the gill chamber and braces the small endoskeletal scapulocoracoid. The largest element of the dermal shoulder girdle is the **cleithrum,** upon which the scapulocoracoid usually resides (figure 9.13). Ventrally, the cleithrum meets the **clavicle,** which bends medially to meet the opposite clavicle at the midline beneath the gill chamber. Where they meet they form a **symphysis.** Dorsally, the cleithrum supports a **supracleithrum** and, in turn, a **posttemporal** through which the dermal girdle is attached to the back of the skull. In some actinopterygians, additional dermal bones may join this girdle (e.g., the postcleithrum = anocleithrum), whereas in others, dermal bones may be lost (e.g., the clavicle is often lost in teleosts). However, this basic set of dermal bones is common in the shoulder girdle of actinopterygians.

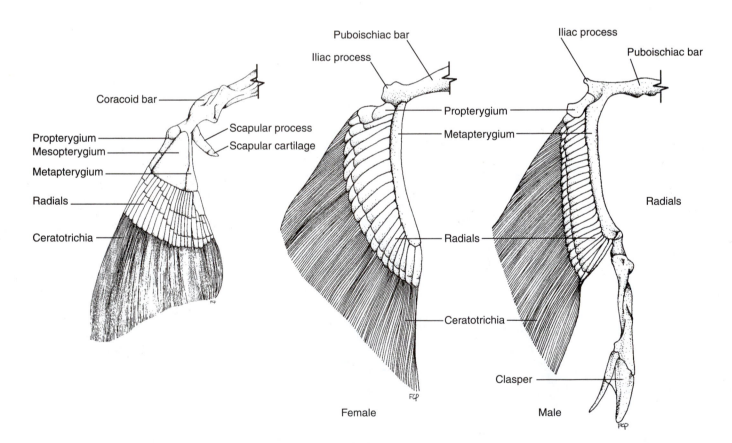

FIGURE 9.11 Appendicular elements in the pectoral and pelvic fins and girdles of the modern shark *Squalus.*

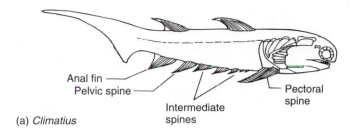

(a) *Climatius*

Anal fin
Pelvic spine
Intermediate spines
Pectoral spine

(b) *Acanthodes*

Anal fin
Pelvic fin
Pectoral fin

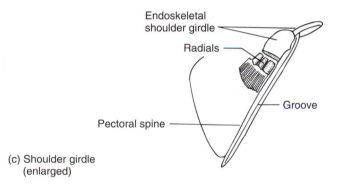

Endoskeletal shoulder girdle
Radials
Groove
Pectoral spine

(c) Shoulder girdle (enlarged)

FIGURE 9.12 **Acanthodian fishes.** (a) *Climatius*, showing the row of spines between pectoral and pelvic elements. (b) *Acanthodes*. (c) Enlargement of restored endoskeletal shoulder of *Acanthodes*. Note the pectoral spine with the small radials and fused elements at its base.

(a, b) After Carroll; (c) after Jarvik.

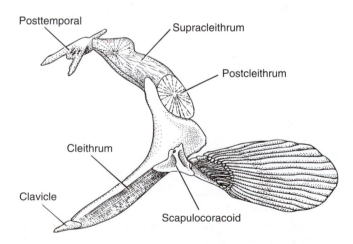

Posttemporal
Supracleithrum
Postcleithrum
Cleithrum
Clavicle
Scapulocoracoid

FIGURE 9.13 **Pectoral girdle of Amia, a primitive actinopterygian.**

We use the name "postcleithrum" for the dermal element found in the shoulder girdle of most actinopterygians through sarcopterygians, and into very early tetrapods where it is eventually lost. However, this breaks with some naming conventions, wherein the name changes from postcleithrum to anocleithrum in sarcopterygians and early tetrapods. The student should be prepared for this name shift in some scientific references.

Sarcopterygians The sarcopterygians are sometimes referred to as the lobe-finned fishes, a reference to the muscles and internal supportive elements that project from the body to form the fleshy base of the dermal fin. Among the sarcopterygians, the rhipidistians are of particular interest because these primitive fishes possess certain fin features that approach limb features in early tetrapods.

Surviving sarcopterygians include three living genera of lungfishes and a single genus of coelacanth, *Latimeria*. Lungfishes dating from the Early Devonian show some skeletal specializations that characterize their subsequent evolution. Among the living genera, the fins are considerably reduced; only the skeletal elements of the Australian lungfish *(Neoceratodus)* exhibit features that can be homologized to those of other fishes. The dermal shoulder girdle includes a cleithrum and clavicle together with a dorsal postcleithrum (= anocleithrum) (figure 9.14a). The endoskeletal girdle includes a scapulocoracoid that supports the projecting series of fin elements in an archipterygial pattern. Paired pelvic fins are also archipterygial and rest on a single cartilaginous girdle element.

Despite their endoskeletal structure, these fins do not carry lungfishes on trips across land. The thin, threadlike fins of the South American and African lungfishes are unsuited for terrestrial transport of the body. The Australian lungfish has slightly stronger fins, but it is, in fact, the most aquatic of the three genera. Fins serve these lungfish to maneuver in shallow water or crawl about aquatic vegetation and obstacles on the bottom.

Coelacanths first appeared during the Middle Devonian, although the surviving *Latimeria* is a more recent taxon within the coelacanths. Skeletal elements within the fins of *Latimeria* form a long, unbranched axis. The dermal shoulder girdle lacks an interclavicle but includes a crescent of four bones, a ventral clavicle, a cleithrum and probable postcleithrum that support the scapulocoracoid, and a dorsal bone believed to be the supracleithrum (figure 9.14b). The pelvic girdle consists of a single element bearing several processes. Direct observation of living *Latimeria* in their natural habitats at depths of about 150 m reveals that the paired fins are used to stabilize and control their position in underwater currents in which they drift slowly or remain suspended.

Some fossil rhipidistians have left a remarkably detailed record of the structure of their lobed fins. One of the best studied of these is *Eusthenopteron*, a Late Devonian

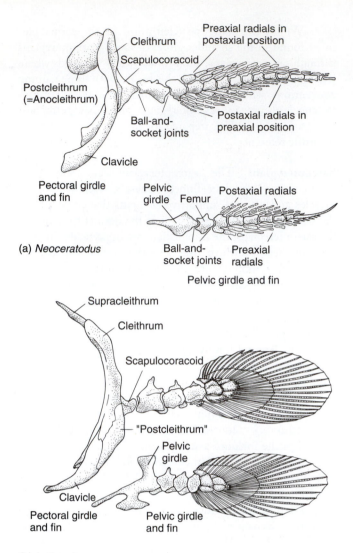

FIGURE 9.14 Appendicular skeletons of living sarcopterygians. (a) The Australian dipnoan *Neoceratodus*. (b) The coelacanth *Latimeria*.

(a) After Jarvik; (b) after Millot and Anthony.

rhipidistian. Pectoral and pelvic appendages support dermal fins but internally possess bones above the wrist/ankle homologous to those of early tetrapod limbs (figure 9.15a, b). The pectoral fin articulates with a scapulocoracoid and a series of supporting paired dermal elements of the girdle: clavicle, cleithrum, postcleithrum (or anocleithrum), supracleithrum, posttemporal. In addition to these, a single unpaired dermal element is present midventrally, which overlaps both lower tips of the two halves of the girdle (figure 9.15b). This oval bone, essentially an enlarged oval scale, is the **interclavicle.** The interclavicle is a new member of the fish girdle, joining it first here in rhipidistians. It is retained in the dermal girdle of later tetrapods as well.

The pelvic fin articulates with a single endoskeletal girdle bone. Left and right members of this paired girdle do not meet at the midline, nor do they articulate with the axial column. Instead, they are embedded within the body wall, offering a bony base from which the fleshy fin projects outward from the sides of the fish (figure 9.15a).

Tetrapods

The first tetrapods retained or quickly displayed changes in the appendicular skeleton correlated with locomotion on land and exploitation of the terrestrial environment. Appearing in the sarcopterygian fish *Tiktaalik* and retained in tetrapods was the loss of the pectoral girdle attachment to the skull, a feature allowing increased cranial mobility and perhaps reduced jarring of the head. Girdles and limbs became stronger, more robust, and more completely ossified (figure 9.16a, b). One of the earliest was *Ichthyostega.* Its pelvic girdle was composed of a single bone with three parts: **pubis, ischium,** and **ilium.** Through the ilium, the pelvic girdle became attached to the vertebral column and so defined the sacral region (figure 9.17). The pectoral girdle lost its attachment to the skull. The fins of fish ancestors were replaced by digits.

Eogyrinus, a secondarily aquatic tetrapod of the Carboniferous, reached over 2 m in length. The limbs and girdles, although strengthened, were relatively small for its size and mostly cartilaginous, so they were unlikely to be used to romp across the land. Instead, like modern salamanders and some fish, *Eogyrinus* must have used its limbs as pivots around which it moved in water (figure 9.18a). Limbs and girdles of early tetrapods became progressively larger and strengthened, reflecting an increased terrestrial habit and exploitation of the land. *Eryops,* a temnospondyl of the Permian, also reached lengths of almost 2 m. But the limbs and girdles of *Eryops* were robust, extensively ossified, and strong supportive structures of a tetrapod committed to life on land (figure 9.18b, c). Skulls of young *Eryops,* unlike those of adults, show evidence of a lateral line system, an aquatic sensory system. This suggests that adults were predominantly terrestrial, whereas young were predominantly aquatic, a life cycle not unlike that of many modern amphibians.

Lateral line system (p. 694)

Pectoral Girdle

Tetrapods carry over from rhipidistians a shoulder girdle consisting of dermal and endoskeletal elements; however, unlike their fish ancestors, tetrapods have a shoulder girdle that is structurally and functionally detached from the skull. Because the pectoral girdle is no longer connected to the back of the skull, the dorsal series of dermal bones, which previously were involved in establishing this connection in fish, are lost as

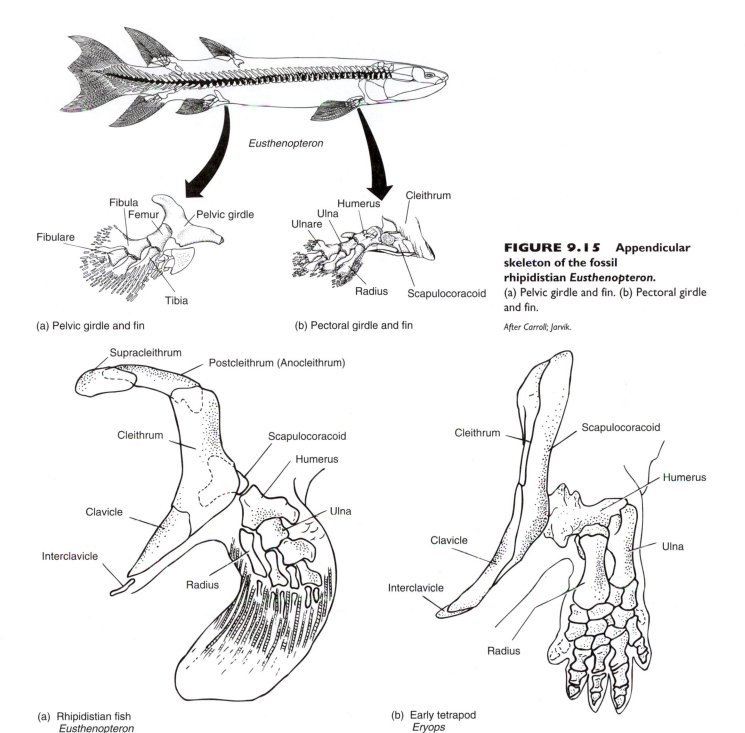

Eusthenopteron

Fibula
Femur
Pelvic girdle
Fibulare
Tibia

(a) Pelvic girdle and fin

Humerus   Cleithrum
Ulna
Ulnare
Radius   Scapulocoracoid

(b) Pectoral girdle and fin

FIGURE 9.15 **Appendicular skeleton of the fossil rhipidistian *Eusthenopteron*.** (a) Pelvic girdle and fin. (b) Pectoral girdle and fin.

After Carroll; Jarvik.

Supracleithrum
Postcleithrum (Anocleithrum)
Cleithrum   Scapulocoracoid
Humerus
Clavicle   Ulna
Interclavicle
Radius

(a) Rhipidistian fish
Eusthenopteron

Cleithrum   Scapulocoracoid
Humerus
Clavicle   Ulna
Interclavicle
Radius

(b) Early tetrapod
Eryops

FIGURE 9.16 **Appendicular skeleton of a rhipidistian, showing left girdle and appendage.** (a) *Eusthenopteron*, rhipidistian fish from the Late Devonian. (b) *Eryops*, a temnospondyl from the Carboniferous.

After Romer, Jarvik, and other sources.

well. Thus, in early tetrapods, the connecting skull bone, the posttemporal, and adjoining shoulder bones, supracleithrum and postcleithrum (= anocleithrum), are absent, leaving a dermal shoulder girdle composed of the remaining ventral elements: the paired cleithrum and clavicle, and an unpaired midventral interclavicle that joins both halves of the girdle across the midline (figure 9.19). In modern amphibians, the dermal bones are usually lost entirely, as in salamanders, or reduced in prominence, as in frogs. The endoskeletal scapulo-procoracoid becomes the predominant girdle element, yet retains its fidelity to the cleithrum (figure 9.19). In fishes, the scapulocoracoid, as its composite name suggests, tends to be a

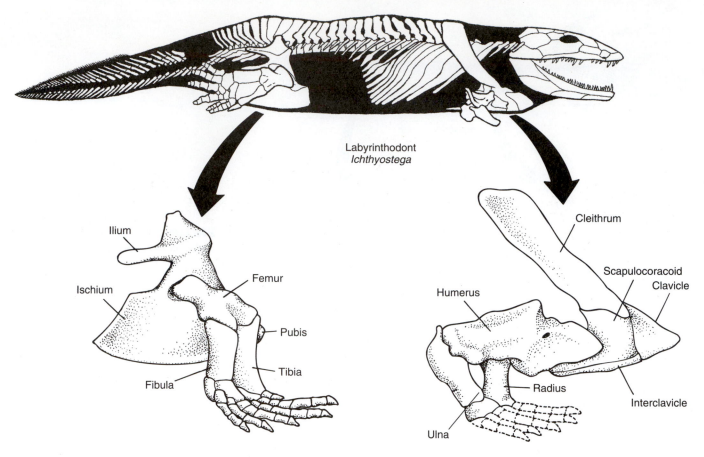

FIGURE 9.17 **Primitive tetrapod *Ichthyostega*, showing components of the appendicular skeleton.** The number of digits carried by the forelimb is not known. Seven digits were present on the hindlimb.

Top, after Ahlberg, Clack, Coats, Blom; bottom, after Jarvik.

single element. However, in early tetrapods, it actually arises from two distinct embryonic centers of endochondral ossification and produces two distinct bones, the **scapula** and the **procoracoid** (figure 9.20).

In primitive amniotes, the clavicle and interclavicle persist, but the cleithrum is usually absent. The clavicle is lost in some modern reptiles, but it is retained in many. In turtles it is incorporated into the plastron as the **entoplastron.** In birds, the paired clavicle usually fuses with the unpaired interclavicle, producing the composite wishbone, or **furcula.** Both scapula and procoracoid of the endochondral girdle persist. In fact, they now become a more prominent part of the shoulder girdle in birds and modern reptiles.

A single, unified scapulocoracoid accompanies the first vertebrates onto land. But, soon we begin to see in the shoulder of some early tetrapods two articulated but distinct endochondral elements. These develop from two centers of ossification, giving rise to a separate scapula (dorsally) and a separate "coracoid" (ventrally). Still later in basal amniotes, the shoulder girdle undergoes another modification to now exhibit three primary endochondral elements—dorsally the scapula but ventrally two elements, the old "coracoid" retained from earlier tetrapods, and a new

"coracoid." To avoid confusion and to track the fate of each, separate names are given to these two coracoids. The older element homologous to the coracoid of early tetrapods is the **procoracoid** (= anterior coracoid). The new coracoid is called just **coracoid** (= posterior coracoid, = metacoracoid) (figure 9.19). In synapsids, this three-part shoulder girdle is retained in pelycosaurs, therapsids, and monotreme mammals. In therian mammals, the procoracoid becomes vestigial (marsupials) or incorporated into the manubrium of the sternum (eutherians). The coracoid becomes reduced and joins the scapula as the coracoid process. Although anatomically united, genetic patterning of the therian coracoid process and of the scapula is under the control of different *Hox* genes, lending further support to the view that each is a separate phylogenetic derivative.

In sauropsids, the endochondral three-part shoulder girdle is also retained but its evolution becomes complicated in derived groups, partly because of the current uncertain phylogenetic placement of turtles and partly because of apparent ontogenetic fusions within the girdle, where the procoracoid is often present in subadults but fuses with the scapula in adults. Most, but not all, basal sauropsids exhibit a three-part shoulder girdle,

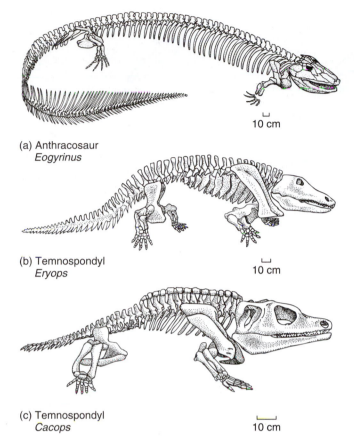

(a) Anthracosaur
Eogyrinus

10 cm

(b) Temnospondyl
Eryops

10 cm

(c) Temnospondyl
Cacops

10 cm

FIGURE 9.18 **Primitive tetrapods.** (a) *Eogyrinus,* from the Carboniferous, was over 2 m in length. Although limbs and girdles were present, they were built slightly for an animal of this size, perhaps serving as points of pivot rather than actual weight-bearing structures. (b, c) By comparison, the more robust limbs and girdles of *Eryops* and *Cacops,* both from the Permian, reflect increased use of the limbs for terrestrial locomotion.

After Carroll.

even in adults—scapula, coracoid, procoracoid. In turtles, a distinct scapula and distinct coracoid are present. The acromial process represents the fused procoracoid (although some anatomists still hold that the process is an outgrowth of the scapula and the procoracoid is absent). In most lizards, all three elements are present—scapula, coracoid, procoracoid—although the procoracoid is reduced to more of a process joined to the other bones. Among other reptiles, a scapula and a coracoid are present in the ossified skeleton, but the procoracoid has various fates. For example, in crocodilians the procoracoid does not detach during development from its early embryonic anlagen to form a separate bone in the adult, but instead becomes lost during development or becomes incorporated into the ventral portion of the scapula, losing identifying sutures in the process. In birds, the rudimentary procoracoid may contribute to the portion of the scapula around the glenoid fossa and to one end of the coracoid as the procoracoid process.

Several dermal elements of the shoulder persist in early synapsids. The clavicle and interclavicle are present in therapsids and monotremes, but in marsupials and eutherians, the interclavicle is absent, the clavicle often is reduced in size, and the scapula becomes the predominant shoulder element. On the other hand, the coracoid (posterior coracoid) is reduced and fused to the scapula as the **coracoid process** (figures 9.19, 9.20).

Pelvic Girdle

The pelvic girdle is never joined by contributions of dermal bone. From its first appearance in placoderms, the pelvic girdle is exclusively endoskeletal. It arose from pterygiophores, perhaps several times, in support of the fin. In most fishes and very early tetrapods, it is formed of a single element, but in tetrapods, three endochondral bones contribute: ilium, ischium, and pubis (figures 9.21 and 9.22). Through the ilium, the pelvic girdle is attached to the vertebral column first in the earliest tetrapods, establishing and therefore defining the sacral region. Throughout later amniotes, these three bones of the pelvic girdle persist, although their general pattern varies. For example, two distinctive patterns, the saurischian and ornithischian pelvic girdles, define two respective groups of dinosaurs. In birds, all three bones appear embryologically as distinct centers of ossification, but then they fuse to form the composite **innominate bone,** usually with no trace of sutures between them. Further fusion between the composite innominate and composite synsacrum introduces considerable firmness in the posterior avian skeleton.

Manus and Pes

The autopodium at the end of each tetrapod limb has undergone a complex evolution of its own. Tracking this evolution has been difficult, largely because of the numerous elements that participate. There are several **digits,** each beginning proximally with a **metapodial element** (**metacarpals** on the forelimb, **metatarsals** on the hindlimb) followed by a chain of **phalanges** (sing., *phalanx*). The digits rest upon several separate bones, collectively known as the **carpals** in the wrist and the **tarsals** in the ankle. In some marine vertebrates (ichthyosaurs, plesiosaurs, cetaceans, sirenians, and marine carnivores, for example), the major trend has been toward **polyphalangy,** a proliferation in the number of phalanges. It is very unusual to find species with more than five digits, a condition known as **polydactyly.** But in many groups, such as ungulates and some terrestrial carnivores, the opposite trend has occurred, namely, toward reduction in the number of phalanges and loss or fusion of associated carpals and tarsals.

Traditionally, the basic tetrapod limb was thought to consist of five digits that were named and numbered (Roman

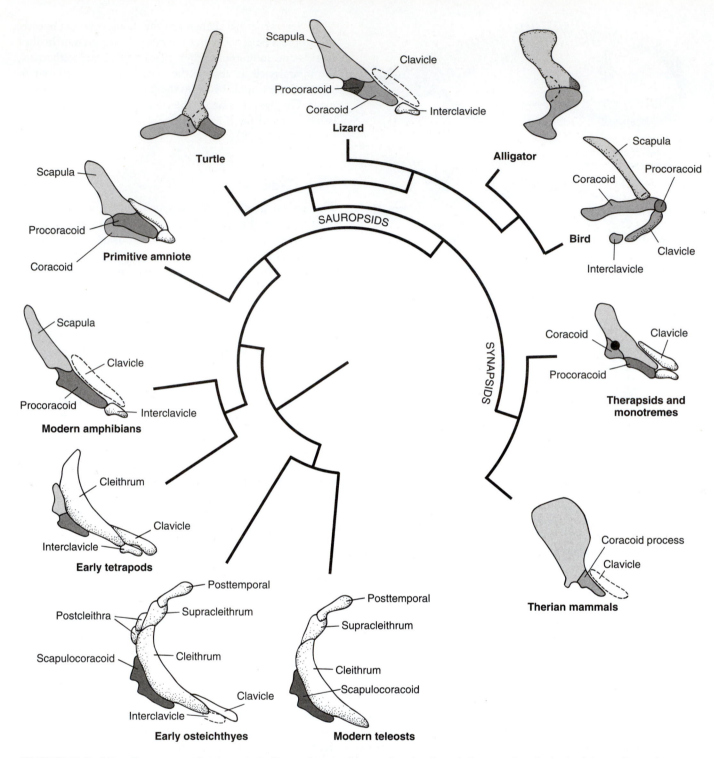

FIGURE 9.19 Summary of pectoral girdle evolution. Notice that the dermal elements (no shading) of the girdle tend to be lost, and endochondral elements (shaded elements) tend to assume a greater prominence. A single endochondral element, the scapulocoracoid, is present in fishes, but in early tetrapods, two distinct bones, the scapula and the procoracoid, are present. In primitive amniotes, a third endochondral bone appears, the coracoid, to join with the phylogenetically older scapula and procoracoid. The three persist into primitive mammals, although only two remain in therian mammals (scapula and coracoid as a process). In modern reptiles and birds, the scapula and coracoid persist; the procoracoid is reduced or lost from the adult shoulder girdle.

(a) *Dimetrodon* (b) *Morganucodon* (c) *Ornithorhynchus* (d) *Tachyglossus* (e) *Didelphis*

(f) *Seymouria* (g) *Sphenodon* (h) *Alligator* (i) *Chelonia* (j) *Sturnus*

(k) *Ichthyostega* (l) *Eryops* (m) *Ascaphus* (n) *Salamandra*

(o) *Cornuboniscus* (p) *Acipenser* (q) *Lepisosteus* (r) *Amia* (s) *Salmo*

FIGURE 9.20 **Pectoral girdles of selected vertebrates.** (a) Pelycosaur. (b) Primitive extinct mammal. (c) Platypus, a monotreme. (d) Echidna, a monotreme. (e) Opossum, a marsupial. (f) Labyrinthodont. (g) *Sphenodon*, a reptile. (h) Alligator, a reptile. (i) Turtle, a reptile. (j) Bird. (k) Ichthyostegid, early tetrapod. (l) Temnospondyl. (m) Frog. (n) Salamander. (o–s) Actinopterygians. Abbreviations: procoracoid (Ac), clavicle (Ca), cleithrum (Cl), coracoid (Co), glenoid (G), interclavicle (Ic), scapula (Sc), supracleithrum (Scl). Cartilage elements are shaded.

After Romer and Parsons; Jollie.

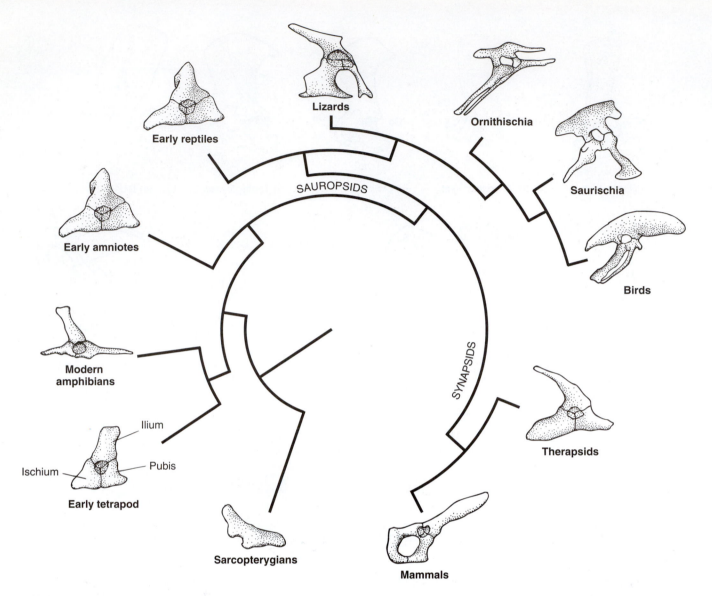

FIGURE 9.21 **Summary of pelvic girdle evolution.** Three endochondral elements—ilium, ischium, pubis—characterize the pelvic girdle in early tetrapods. This basic pattern persists into later tetrapods.

numerals) by their **pentadactylous** pattern (figure 9.23). Some were lost in specialized lines, but this underlying five-digit pattern is a reasonable hypothesis because the limbs of most primitive amniotes commonly carry five digits. Unfortunately, fossil specimens of the very earliest tetrapods, such as *Ichthyostega*, did not preserve enough of the digits to permit a confident count, at least not until recently. New discoveries of early tetrapod specimens test this five-digit hypothesis. The hindlimb of *Ichthyostega* had seven digits (its manus remains unknown still), the manus of *Acanthostega* included eight digits (the pes number is uncertain), and the fore- and hindlimbs of *Tulerpeton* had six digits (figure 9.24). These Late Devonian fossils are the earliest tetrapod remains available. Collectively, they indicate that the primitive tetrapod pattern was polydactylous and the five-digit pattern is a later stabilization.

Recent speculation also suggests that two independent tetrapod limb designs may have emerged from the polydactyl condition in primitive tetrapods. One was the amniote lineage in which digit number stabilized at five on each limb. The other lineage leads to modern amphibians, with five toes on the hindlimbs but only four on the forelimbs. The older view is that this four-digit pattern is derived from a five-digit ancestor. If we trace modern amphibian ancestry back to these earliest tetrapods, as some now suggest, then amphibians independently derived their reduced digit number directly from polydactylous ancestors.

Such speculations are tantalizing and refreshing but still quite tentative. The earliest tetrapods were clearly polydactylous. But the significance of this to later tetrapod evolution awaits further study. For our purposes, the

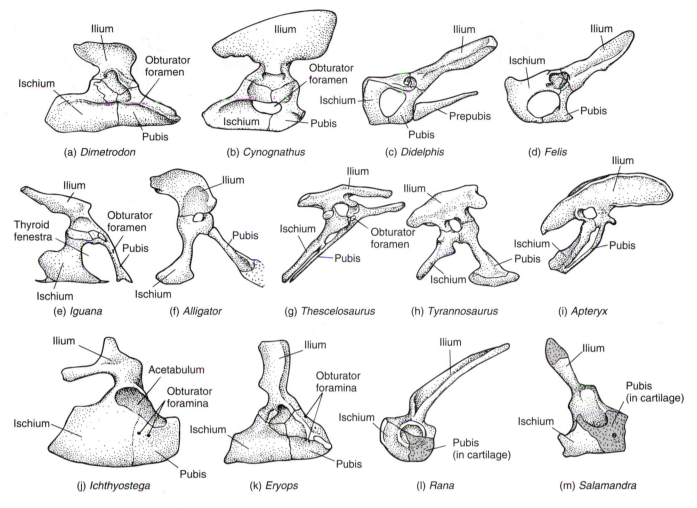

FIGURE 9.22 **Pelvic girdles of selected vertebrates.** Stippling represents cartilaginous areas. (a) Pelycosaur. (b) Therapsid reptile. (c) Opossum, a marsupial. (d) Cat, a placental. (e) Lizard. (f) Alligator, a reptile. (g) Ornithischian, a dinosaur. (h) Saurischian, a dinosaur. (i) Bird. (j) Early tetrapod. (k) Temnospondyl, tetrapod. (l) Frog. (m) Salamander. Cartilage elements are shaded.

After Romer and Parsons.

five-digit pattern represents a useful basis for discussing limb evolution and changes in functional design.

In the manus, a digit consists of several phalanges with a metacarpal at its base. In turn, each of the five metacarpals articulates with a carpal. The wrist bones that articulate with the radius and ulna are, respectively, the **radiale** and the **ulnare.** The **intermedium** lies between these two wrist bones. Within the middle of the wrist are one to three **centrales.** In the pes, the primitive number of digits is also taken as five, each with a metatarsal at its base. In turn, each digit articulates proximally with the following sequence of bones: tarsal, centrale, tibiale, intermedium, and fibulare, the last three meeting the tibia and fibula of the shank (figure 9.23).

Although this stately pattern of expected manus and pes elements gives us a starting point when we look at distal limb anatomy, the actual morphology is often considerably modified by fusions, elongations, eliminations, and additions of apparently new elements to this pattern (figures 9.25a–h and 9.26a–g). For example, the **pisiform** is a sesamoid bone

that can lie to the outside of the carpus, especially in reptiles and mammals. In birds, fusion of the forelimb elements produces an autopodium with three digits (II, III, IV). The ulnare regresses during ontogeny, and in its space a neomorph arises from a new embryonic condensation (figure 9.25e). This new avian wrist bone has not been named. Some still call it an ulnare; others say "ulnare" in quotation marks. Some number it; others call it a neomorph. All will be found in new references in the avian literature for a while.

In the hindlimb, lateral digits may be lost in cursorial mammals, and the medial metatarsals (III and IV) fuse into a composite ankle bone, or are reduced to a single prominent bone (metatarsal III), commonly called the **cannon bone** (figure 9.26e). In birds, fusion of elements in the hindlimb produces a composite bone, the **tarsometatarsus,** named for the elements contributing (figure 9.26c). In mammals, the fibulare is the specific tarsal that articulates with the fibula, but it is more commonly called the **calcaneum.** The tibiale, commonly called the **astragalus,** actually fuses with

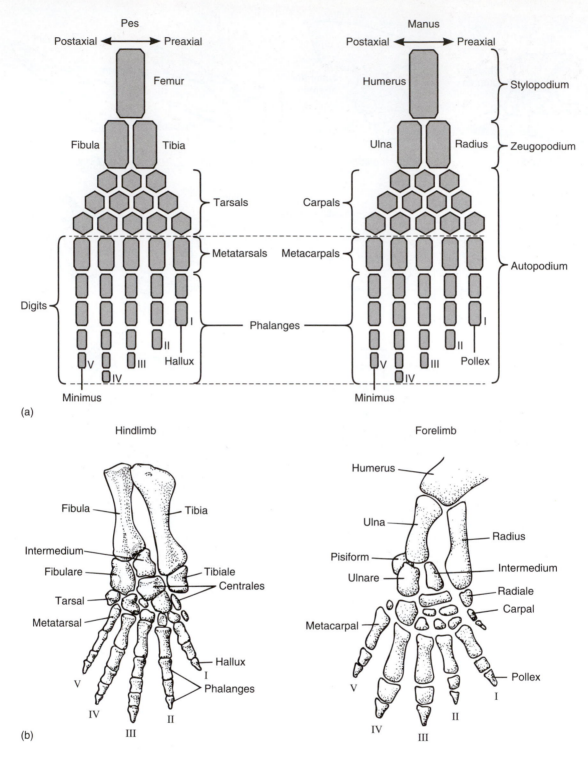

FIGURE 9.23   Basic organization of the fore- and hindlimbs. (a) Manus and pes have five digits; each digit includes its metacarpal or metatarsal and chain of phalanges. These digits in turn articulate with various wrist and ankle bones. (b) Fore- and hindlimbs of primitive tetrapod.

(a) After Smith; (b) after Jarvik.

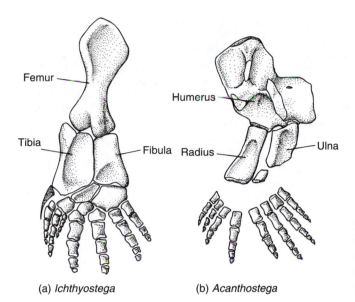

(a) *Ichthyostega*

(b) *Acanthostega*

FIGURE 9.24 **Polydactylous limbs of the earliest tetrapods (dorsal views).** (a) Hindlimb of *Ichthyostega* with seven digits. (b) Forelimb of *Acanthostega* (an ichthyostegalian) with eight digits.

Source: After Coates and Clark, 1990.

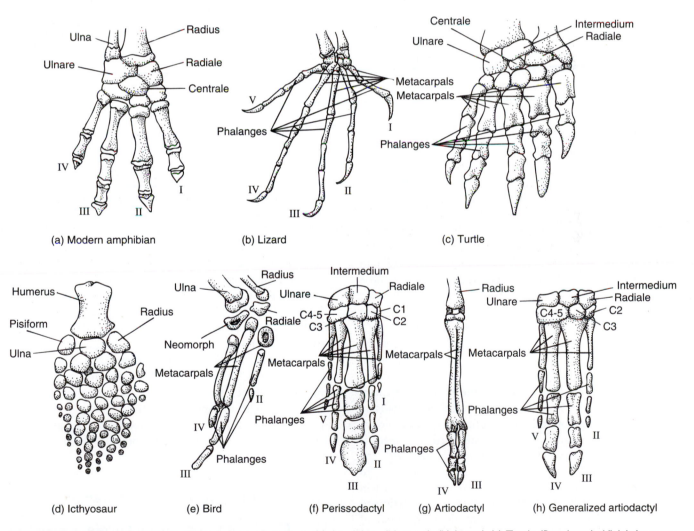

(a) Modern amphibian

(b) Lizard

(c) Turtle

(d) Icthyosaur

(e) Bird

(f) Perissodactyl

(g) Artiodactyl

(h) Generalized artiodactyl

FIGURE 9.25 **Variations of the tetrapod manus.** (a) Amphibian (*Necturus*). (b) Lizard. (c) Turtle (*Pseudemys*). (d) Ichthyosaur. (e) Bird. (f) Hypothetical primitive perissodactyl. (g) Artiodactyl. (h) Generalized artiodactyl.

After Smith.

FIGURE 9.26 **Variations of the tetrapod pes.** (a) Amphibian (*Necturus*). (b) Lizard. (c) Bird (chicken). (d) Fossil horse. (e) Modern horse. (f) Cat. (g) Human.

the intermedium and together they articulate with the tibia (figure 9.26f). Although originating in mammals, the terms *calcaneum* (= fibulare) and *astragalus* (= tibiale) have been carried over to refer to the ankles of reptiles and birds; however, differences in embryonic contributions to these two bones in reptiles, birds, and mammals perhaps betray differences in homology and, strictly speaking, call for different names. But some uncertainties regarding embryonic contributions remain, and the convenience of the terms *calcaneum* and *astragalus* for tarsal bones of similar function may see these terms used for all groups, at least for the present.

Within amniotes, two types of ankle joints may be formed between the shank (crus) and between or within proximal tarsals (astragalus and calcaneum). A **mesotarsal joint** is a simple hinge joint passing between proximal (astragalus and calcaneum) and distal tarsal elements (figure 9.27a). The mesotarsal joint applies to most amniotes. The ankle in pterosaurs, dinosaurs, and birds is a specialization of this joint where proximal tarsals fuse with the tibia (forming the tibiotarsus) and distal tarsals fuse to the metatarsals (forming the tarsometatarsus). At least in dinosaurs and birds, it likely represents an adaptation to their bipedal locomotion. A tight interlocking union of astragalus and calcaneum ensures that the mesotarsal joint will restrict hingelike flexion to the same plane of motion as the overall limb. Further, the mesotarsal joint brings the

calcaneus back into tight contact with the fibula, where it becomes part of the shank. In this instance, the calcaneus assumes a greater weight-bearing function within the hindlimbs that are themselves now brought under the body to help support the animal.

A **crurotarsal joint,** a second joint type, forms such that the line of ankle flexion passes between the calcaneum and astragalus (figure 9.27b). In crocodilians and most advanced thecodonts, the main joint bends between the distal end of the astragalus and fibula (figure 9.27b); in the therian crurotarsal ankle, the main joint bends between the distal end of the tibia and distal end of the calcaneum.

Changes in the primitive pattern of manus and pes correlate with changes in functional demand arising from the biological roles in which the limbs participate. We look at these changes in function shortly, but first let us reflect on the overall significance of the morphological patterns we have encountered so far in the appendicular system.

Evolution of the Appendicular System

Dual Origin of the Pectoral Girdle

Throughout its evolution, the pelvic girdle is endochondral. In tetrapods, it comprises three distinct bones, the ilium,

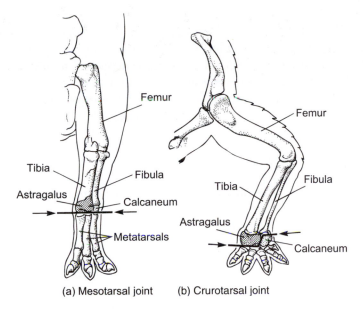

(a) Mesotarsal joint   (b) Crurotarsal joint

FIGURE 9.27   **Ankle types among archosaurs.**
(a) Mesotarsal joint. The joint occurs directly between proximal tarsals (astragalus and calcaneum) and distal metatarsals (hence *mesotarsal*). Birds, dinosaurs, and several other groups of Mesozoic reptiles had mesotarsal joints. (b) Crurotarsal joint. The line of flexion, indicated by heavy black line and arrows, occurs between the astragalus and the calcaneum. Crocodiles, shown here, some therapsids, and most therians have a crurotarsal joint. The solid line identified with arrows indicates the flexion axis of each joint hinge.

ischium, and pubis, instead of the single element characteristic of fishes.

However, the pectoral girdle of osteichthyans is clearly of dual origin, composed of dermal as well as endochondral bones. The endochondral component, the scapulocoracoid, evolved by fusion or enlargement of several basal fin elements. It functions as the articular surface of the fin and later of the limb. The appendicular musculature of the forelimb is attached securely to it. The dermal component of the shoulder girdle evolved from dermal bones of the body's surface. In ostracoderm fishes, these bones composed the outer protective armor. As the early fish girdle evolved, some dermal armor may have sunk inward to join the existing endochondral components of the fish girdle. Alternatively, before becoming associated with the pectoral fins, these dermal bones may have developed the additional role of securing the anterior point of transition between axial musculature and the branchial chamber. Whichever their evolutionary route, dermal bones became an important brace for the endoskeletal pectoral girdle. Like endochondral bones, these dermal bones were passed along to tetrapods, where a new dermal element, the interclavicle, debuted in some rhipidistians. The interclavicle was incorporated at the midventral symphysis between both halves of the shoulder girdle. In general, the dermal girdle offered additional area for muscle attachment and protected the heart, but it functioned primarily as a firm brace for the endochondral elements of the shoulder.

Adaptive Advantage of Lobe Fins

The transition from water to land led to significant changes in the appendicular system. Fortuitously, the rhipidistian predecessors to the tetrapods possessed lobe fins preadapted to become tetrapod limbs. But, lobe fins were present in rhipidistians not in anticipation of future roles on land, but to serve these fishes in the aquatic environments in which they lived. What, then, might have been the immediate biological role of lobe fins in rhipidistian fishes?

We know from the types of geological deposits in which their bones are found that many early rhipidistians lived in fresh water, as their dipnoan relatives do today. Living lungfishes use their rudimentary lobe fins to "walk" along the bottom of slow-moving streams or backwaters; that is, they use their fins as points of pivot about which their buoyant body moves. If the freshwater pools in which they live dry up during the hot, rainless seasons of the year, lungfishes apparently do not venture a short distance over mud to remaining pools of water. Instead they burrow into the mud, form around them a protective cocoon, and estivate. Their metabolic rate drops, their breathing slows, and they slumber until rains return to replenish the dried pools.

Lungfish estivation (p. 101)

It is not clear when fin rays were lost and digits arose. Perhaps there was a transition in which both were present. Some fossil rhipidistian fishes used station-holding, wherein they fixed themselves in stationary ambush for prey. Their appendages included robust endoskeletal elements with a webbed fin at their ends. Such a design has been suggested to be a precursor to weight-bearing limbs of later tetrapods. In the earliest tetrapods, an array of digits arose before wrists or ankles. Such a fan of digits could have spread the weight-bearing load on land before the more stable wrist and ankle joints evolved.

It would be helpful to consult living rhipidistians directly to see how lobe fins are actually used. But, of course, no freshwater rhipidistian survives to the present. Possibly early rhipidistians, like their lungfish cousins today, used their maneuverable lobe fins as points of pivot to move them through aquatic vegetation near the edges of ponds and streams. In shallow waters, aquatic plants and fallen debris from surrounding forests would produce a complex underwater environment offering retreat when a rhipidistian was threatened or prey when it was foraging. Lobe fins offer one solution to maneuvering within such a freshwater "jungle." With this view, lobe fins are aquatic adaptations that serve a fish in shallow waters. Terrestrial environments present different challenges. How well lobe fins might serve there is another issue.

Onto the Land

The musculature associated with the fins of early rhipidistians was probably too weak to have supplied propulsive thrust directly for transport on land or to have borne the

(a)
Water

(b)
Land

FIGURE 9.28 Transition from water to land. The same swimming motions used in water could have served rhipidistians venturing onto land. (a) In water, typical lateral undulations of the fish body provide propulsion for swimming. Horizontally held fins may have functioned like hydrofoils to produce lift. Fins rotated vertically and, drawn backward, would have served as oars, adding to forward propulsion. (b) On land, these fishes could have used the same lateral body undulations to place fins as points of pivot (x) about which the body "swam." Limbs would not have needed the strength of fully developed tetrapod limbs because they were not used to carry weight or produce locomotor force. They were needed only as pegs about which the strong body musculature could pivot.

weight of the body out of water. However, the slight musculature was sufficient to fix the fins on the body like pegs. This allowed the well-developed axial musculature to produce lateral undulation and the peglike fins to act as pivots around which the body could rotate. Thus, the same trunk undulations used in swimming could be used with little modification for short journeys across land (figure 9.28a, b). Existing morphology and swimming behavior provided the basis for gradual transition to land. The Devonian tetrapod *Ichthyostega*, even though it had limbs, still seems to have had aquatic habits. Its tail bore a caudal fin; its lateral line system was present, even in adults; internal gills were present; and its vertebrae had not yet replaced the notochord as the predominant basis for axial support. Similarly, most lepospondyls and early temnospondyls appear to have been predominantly aquatic. Not until the Permian, 50 million years after vertebrates first invaded land, did communities of more fully terrestrial tetrapods appear.

But why leave the water at all? What advantages might there have been for rhipidistians that left their aquatic world and ventured onto land? Several ideas have been proposed. One hypothesizes that overland travel developed, ironically, to keep these fishes in water. The Devonian was a time of occasional droughts and floods, which suggests that rhipidistians might have used strengthened fins/limbs to move from a small drying pool to larger, permanent ones. However, this presupposes that the limb was already strong enough to be used during sojourns across land before such droughts occurred. Further, modern lungfishes respond to drought by estivating rather than by leaving their aquatic environment.

An alternative hypothesis stresses that movement to the land was favored by predation in the water. To escape predation from other species or from cannibalistic adults, young fishes may have frequented shallow waters where predators could not follow. Because the young had to maneuver in shallow water among thick shoreline vegetation, their lobe fins may have evolved into more supportive appendages. Thereafter, movement onto land would have been an option. In this view, the first movement to land did not involve long treks to alternative pools but a short step up onto the nearby beach. Lobe fins would have needed only to be strong enough (to serve as pegs?) to participate mechanically in such a first tentative exploitation of the land.

Others have suggested that limbs developed to enable fishes to get out of the water and breathe. But because of their lungs, these fishes could breathe, if that was required, by simply coming to the surface and gulping fresh air above. Food has been proposed as another enticement for movement onto land. Of course, no other vertebrates were on land, and labyrinthodont teeth were not suited for feeding on plants. But arthropods were abundant, having radiated into terrestrial environments much earlier (Silurian). They may have offered an alternative food source for early tetrapods that scampered up onto beaches or shores to seek them out. A further proposal is that early tetrapods came up on land to bask, like modern day crocodilians, then once warmed slipped back into the water.

Because no one was there to record events, we cannot be sure of the selective pressures that favored the transition to land. But, the fossil record clearly demonstrates that just such a transition in vertebrate evolution occurred during the Devonian. As a matter of interest, the transition from water to land has happened several times, although with a less lasting phylogenetic impact. In some teleost fishes today, for instance, there are species, such as the mudskipper, that use a strengthened skeleton to venture temporarily onto land to search for food and perhaps gain respite from predators left behind in the water.

Form and Function

Changes in the skeletal system, as in many other systems, have been extensive in the transition from aquatic environments to life on land. On land, the main contributors to locomotion are the limbs, not the tail. Consequently, the limbs undergo extensive and significant morphological alterations. In addition, the shoulder and hip generally establish different structural associations with the axial column as a result of the transition to land. In tetrapods, the axial column is slung from the shoulder girdle by muscles; however, the hip is attached directly to the column (figure 9.29a, b). The shoulder moves on the thorax via these muscles so that the forelimb's impact with the ground is softened and these sudden forces are not transmitted to the skull. The hip is firmly associated with the sacrum via a bony connection (figure 9.29b). The powerful hindlimbs transmit their propulsive force directly to the bony axial column.

Different modes of locomotion on land place different mechanical demands on the appendicular skeleton. Tetrapods specialized for running are **cursorial.** Those that dig or burrow are **fossorial,** and those that hop are **saltatorial** (ricochetal). The fastest modes of locomotion occur among flying, or **aerial** (volant), specialists. **Arboreal** locomotion refers generally to animals that live in trees. One form is **scansorial** locomotion, which applies to climbing in trees with the use of claws as well as being adept at terrestrial locomotion. Squirrels are an example. Gibbons and spider monkeys, for example, travel under branches using long arm swings and alternating hand grasps to navigate from branch to branch, a method of arboreal movement referred to as **brachiation.**

Each of these terrestrial specialties is accompanied by morphological modifications of the basic limb and girdle structure. To understand the different functional demands these specialized modes of terrestrial locomotion place on the appendicular skeleton, we need first to step back and view at what point the basic terrestrial structure and behavior began, namely, with fishes swimming in water.

Swimming

As we have already seen, the body of an active fish coursing through a viscous medium such as water experiences drag, which slows the fish's forward progress. Streamlining prevents flow separation, reduces drag, and improves performance. Lateral undulations passing along the body move the fish through the aqueous medium, producing backthrust against the water and providing forward force. The basic primitive tetrapod locomotion evolved from this characteristic lateral undulation that fishes use to swim.

This same mode of progress still serves most modern amphibians and reptiles quite well, giving them access to a great variety of habitats. In tetrapods that secondarily become aquatic, as in cetaceans, for example, the limbs again may become secondary to the tail and lose their prominence in aquatic locomotion (figure 9.30a). However, not all secondarily aquatic vertebrates have reduced

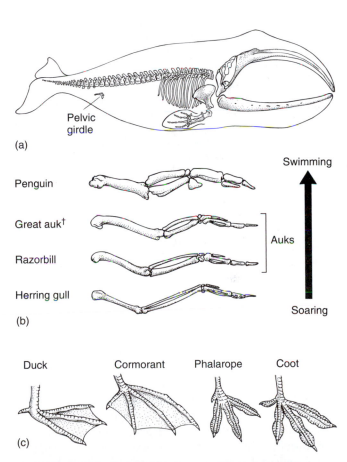

(a)

(b)

Swimming

Penguin

Great auk[†]

Razorbill

Herring gull

Auks

Soaring

(c)

Duck   Cormorant   Phalarope   Coot

FIGURE 9.30 **Adaptations of the appendicular skeleton in secondarily aquatic tetrapods.** (a) Skeleton of a right whale, showing reduction of the appendicular skeleton, especially the pelvic girdle and fin, in this aquatic mammal. (b) The gracile bones of the seagull contrast with the robust forelimb of the penguin. Forelimbs of several species of auks are shown between. Some are extinct (†). These changes reflect an increasing role in underwater swimming. (c) Swimming birds typically have webbed feet.

(c) *After Peterson.*

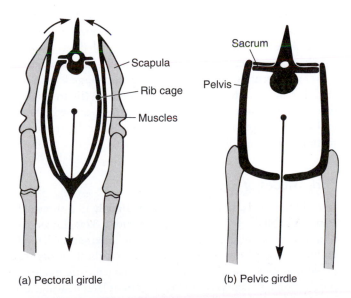

Scapula

Rib cage

Muscles

Sacrum

Pelvis

(a) Pectoral girdle

(b) Pelvic girdle

FIGURE 9.29 **Appendicular girdles of tetrapods.** (a) Muscles of the pectoral girdle support the anterior part of the tetrapod body in a muscular sling. (b) The pelvic girdle is attached directly to the vertebral column via the sacrum.

limbs. For example, pinnipeds clasp their hindlimbs together, forming a kind of "tail" that aids in swimming. The wings of aquatic birds often take on greater roles in swimming. Their forelimb bones become stouter and more robust, reflecting the increased strength required to provide the bird with flippers to propel them while swimming after food in the water. In penguins, the wings are flightless and used exclusively like flippers to enable the animal to swim underwater (figure 9.30b). The hindlimbs of swimming birds may become partially or completely webbed feet to increase pressure against the water when these birds paddle (figure 9.30c).

Terrestrial Locomotion

Early Gaits

The pattern of foot contacts, or **footfalls,** with the substrate during locomotion constitutes an animal's **gait.** One **cycle** is one complete use of all limbs before repeating a footfall pattern. The **duty factor** measures the percent of the total cycle a foot is in contact with the substrate. By convention, a duty factor of 50% or greater is a walk; less than 50% is a run. At faster speeds, the gait may include a **suspension phase** wherein all feet are momentarily out of contact with the substrate. Gaits differ in footfall patterns and in speed. As observed by the biologist Milton Hildebrand, the most important factors in gait selection are stability and economy of effort. These were likely important factors from the first tentative vertebrate steps on land to the very specialized methods of locomotion to evolve later. As noted earlier, lateral undulations of the body of fishes were carried forward onto land with the first tetrapods that incorporated lateral flexions of the body with footfall patterns to establish points of pivot. However, the early tetrapod gaits probably were also present in aquatic ancestors maneuvering in shallow waters.

Early tetrapod locomotion (p. 309)

One basic gait is a **diagonal sequence** where diagonally opposite feet strike the ground more or less in unison. A **trot** is based on the diagonal sequence and occurs in tetrapods, but also in some bottom-walking fishes, where lateral waves of undulation of the fish's body place diagonally opposite fins into contact with the substrate, helping to propel them along the bottom. The line between diagonal points of support passes under the center of mass, improving support. In quadrupeds, a third foot may be added to the footfalls, producing three points of contact based on a diagonal sequence. However, if the center of mass falls outside the triangle of support the feet establish, then this gait is unstable. Further, lateral undulations of the body cannot simultaneously contribute to both fore- and hindlimb excursions (figure 9.31a). One compensation bringing greater stability is to add this third point of support in such a pattern that the center of mass lies within the broad boundaries of the support triangle. In principle, this is much the same way a tripod establishes greater stability than a bipod. Bottom-walking fish or amphibians with long tails may incorporate the tail into a third point of support with two of the fins/limbs, bringing the center of mass within the triangle of support (figure 9.31a).

Another basic gait is the **lateral-sequence gait** where feet of the same side, hence lateral, move together and strike the ground more or less in unison. Adding a third limb of contact to this sequence establishes a triangle of support. During cycles of locomotion, the center of mass remains within this supportive configuration, never at its edge (figure 9.31b). Such a stable, lateral-sequence gait occurs in salamanders when walking on land and in reptiles.

Unlike the gait produced by the fins of bottom-walking fishes, these terrestrial gaits include significant longitudinal rotation of the stylopodium. This makes the limb more than just a way of establishing a point of contact with the ground about which the body pivots. The limb additionally contributes to locomotion by generating a tractive (pulling) and propulsive (pushing) force against the ground (figure 9.31c).

In rhipidistian fishes, lateral undulations passing along the body might have been the basis for positioning diagonally opposite fins as points of pivot, yielding a gait close to a walking-trot. Retained in early tetrapods, this gait would have also served them when submerged in shallow waters. But if they walked with their bodies above the water's surface, then the instabilities of a walking-trot would have become quite significant. The evolution of a lateral-sequence gait would have been a way to return stability to body carriage in these early tetrapods. Once this more stable mode of progression evolved, extended bouts of terrestrial locomotion would be biomechanically more efficient and therefore more likely. Salamanders retain both gaits, walking-trot and three-point lateral-sequence gaits.

This increase in gait types may have occurred rapidly or over a long period of time. It may have occurred in ancestors still in the water, but who frequented shallower water and increasingly emerged from the water. Limb retraction may have occurred early too, even in submerged walking rhipidistians. But, rotation of the stylopodium (figure 9.31c) would contribute to stride only after a right-angled elbow and knee were present, a point during or after the transformation of the rhipidistian fin into a limb.

Early Modes of Locomotion

In early tetrapods, limbs were placed laterally in a sprawled stance, establishing points of pivot (figure 9.32a). Locomotion was accomplished, as in fishes, by alternating lateral undulations of the vertebral column about these pivots. In modern amphibians and reptiles, the characteristic mode of progression still depends on this pattern of lateral swings by which the vertebral column moves about points of rotation established by the feet. However, in some terrestrial birds, in many species of dinosaurs, and in many groups of mammals, the trend has been toward cursorial locomotion.

Diagonal sequence
(three point)

Trot

Tripodal

Lateral sequence
(three point)

(a) Early gaits

(c) Propulsion

(b) Lateral sequence cycle (three point)

FIGURE 9.31   Primitive gaits. (a) Early gaits. The diagonal sequence establishes support among the three planted feet but the center of mass (open circle) falls outside the triangle of support, making it an unstable gait. During the trot, diagonally opposite feet (here right-front and left-hind feet) meet the ground together. The center of mass lies on or near the line connecting these two points of support. The same walking stance might be stabilized by adding a third point of support. A long tail pressed to the ground, together with two feet, produce a triangle of support (tripodal) within which is the center of mass. The lateral sequence similarly brings walking stability by establishing a triangle of support wherein the center of mass is supported within. (b) Lateral-sequence cycle. The center of mass (open circle) never leaves the triangle of support established at any one moment during a gait cycle by three of the four limbs. (c) Propulsion by limb rotation. Limb retractor muscles produce a force directed posteriorly (arrow) that rotates the long bone, and thereby retracts the foot to propel the animal forward.

FIGURE 9.32 Change in limb posture. (a) The sprawled posture exhibited by this salamander was typical of fossil amphibians as well as of most reptiles. (b) Eutherian mammal. This posture began to change in synapsids so that in late therapsids, the limbs were thought to be carried more under the body, a reflection of increased efficiency in locomotion.

(a) Salamander

(b) Eutherian mammal

Pelvic girdle

Pectoral girdle

Ilium

Femur

Fibula

Tibia

Tarsal bones

Ulna

Scapula

Humerus

Radius

Carpal bones

FIGURE 9.33 Digit orientation. Toes tended to point laterally in early tetrapods (dashed lines). However, accompanying more efficient terrestrial locomotion, the direction of digits changed along with limb position. Torsion of the humerus and femur brought the toes forward and more in line with the direction of travel. Note in particular how opposite ends of the humerus are rotated to bring the toes forward.

Parasagittal plane

(a) Salamander

(b) Dinosaur

FIGURE 9.34 Terrestrial locomotion. (a) Terrestrial but noncursorial salamanders achieve limb recovery by an overhand swing of the arm outside the parasagittal plane. (b) Cursorial dinosaurs achieved limb recovery by a pendulum-like swing in a parasagittal plane, which keeps the limbs directly below the body so that they support the body weight. The pendulum-like swing improves the ease and efficiency of limb recovery.

Fish swimming to early tetrapod walking (p. 309)

From the characteristic sprawled posture of early tetrapods, many later tetrapods have developed limbs drawn under their body, a change in posture that increases the ease and efficiency of limb swing during rapid loco-motion (figure 9.32b). Crocodiles and alligators use sprawled postures when resting on shore, but they can change their limb position when they move. If making a quick dash to water, they can draw their limbs under the body more directly beneath their weight. This allows the limbs to swing more easily beneath the raised body. In several lines of therapsids, in most eutherian mammals, and in many dinosaurs, this change in limb posture results from a structural change in limb design. The femur and especially the humerus of the limbs show torsion of their distal ends, which rotates the digits carried at the ends of the limbs forward and more in line with the direction of travel (figure 9.33).

Accompanying this change in posture was a tendency to restrict limb movement to one plane, the sagittal plane. Early tetrapods with a sprawled posture must use an overarm swing after each propulsive stroke to establish a new forward point of pivot (figure 9.34a). However, with legs positioned beneath the body, limb recovery after propulsive strokes can be accomplished efficiently by the animal swinging its limbs forward beneath its body in an easy pendulum fashion (figure 9.34b). In tetrapods with a sprawled posture, the adductor muscles, which run from girdle to limb, are massive in order to lift and hold the body in a push-up position. As the limbs move more directly under the body, the adductor musculature is reduced.

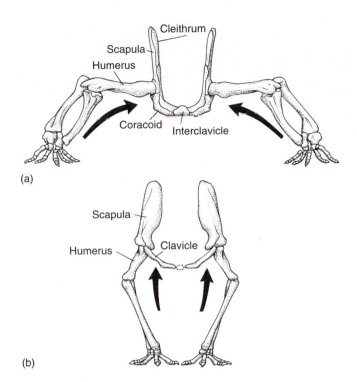

FIGURE 9.35 **Change in the role of the shoulder girdle with change in limb posture.** (a) Sprawled posture brings a medially directed force toward the shoulder girdle, conferring on medial elements a major role in resisting these forces. (b) As limbs are brought under the body, these forces are directed less toward the midline and more in a vertical direction. This position of the limbs might account for loss of some pectoral elements in phylogenetic lines in which limb posture shifted.

(a) Primitive reptile   (b) Therian mammal

FIGURE 9.36 **Changes in the pelvic girdle.**
(a) When limbs are sprawled, propulsive forces are transferred more vertically through the sacrum. (b) In mammals, in whom rapid locomotion becomes common, orientation of the pelvic girdle changes so that the forward thrust of the hindlimbs is brought more into alignment with the line of travel and transferred to the vertebral column.

In therapsids, the acetabulum and glenoid fossa shifted ventrally to follow the inward shift in limb posture. Most noticeably in the shoulder girdle, placement of the forelimbs directly beneath the scapula shifted mechanical forces away from the midline to the scapula (figure 9.35a, b). This gave the scapula a greater role in locomotion and weight bearing. Conversely, the medial elements—clavicle, interclavicle, coracoid, and procoracoid—with reduced supportive roles, became reduced in prominence. The hindlimbs were also drawn under the body, accompanied by a reduction in the adductor muscles. In turn, the pubis and ischium, sites of origin of these adductor muscles, were reduced as well. A shift in orientation of the pelvic girdle allowed for a forward thrust more aligned with the forward direction of travel (figure 9.36a, b).

A remarkable change in the functional way in which the vertebral column participates in locomotion appears first in early mammals. This change is characterized by a shift from lateral to vertical flexions. In animals with sprawled postures, lateral flexion of the vertebral column contributes to the sweeping overhand recovery of the limbs. With limbs carried under the body, lateral undulations contribute little to limb oscillations. Consequently, the structural changes

first seen in early mammals were accompanied by a shift from lateral to vertical flexion of the vertebral column, coordinating it with the limbs that are swung in the same plane. Loss of ribs from the posterior trunk, which produces a more distinct lumbar region, represents a structural specialization that allowed greater flexibility of the axial column in a vertical plane.

Generally, then, as locomotion was used for more sustained, efficient, and rapid transport on land, various structural modifications were incorporated into the appendicular skeleton. Torsion brought the digits forward and more in line with the direction of travel. Sprawled limbs were brought under the body. Vertical flexion of the vertebral column added its motions to the limb displacements. Together, these changes increase the ease and efficiency of limb oscillation and contribute to active lifestyles.

Similar changes in the appendicular skeleton appear in archosaurs as well. Limbs were positioned under the body to carry the body weight more efficiently as these animals moved or migrated in search of resources. However, locomotion was usually based on a bipedal posture with trunk and tail balanced across the hindlimbs.

Cursorial Locomotion

Beyond an increase in the ease and efficiency of limb oscillation, many later tetrapods became specialists at rapid locomotion, accompanied by further modifications that serve such a specialized mode of transport. Rapid locomotion has evolved in both predators and their prey, two sides to the evolutionary coin. It also provides an animal with the means

Human Engineering: Arms and Hands, Legs and Feet

Although a long 5 to 10 million years have passed since distant human ancestors swung through trees, we still retain evidence of this brachiating mode of locomotion.

In the forelimb, for example, brachiators are characterized by long arms with grasping hands. Although our arms are shorter than those of primates that still depend on locomotion through trees, nevertheless our arms are relatively long compared to those of other vertebrates. If we stand comfortably upright with arms to our sides, our fingers reach below our hips. By contrast, the forelimbs of a nonbrachiating animal such as a dog or cat, if pressed back, do not reach so far. In the manus of a brachiator, digits II through V form a hook, with which they grasp overhead branches. Without thinking it anything special, we use this same comfortable design to grip the handle of a suitcase carried at our side. The arm position changes from the overhead to the side, but the grip used is the same. In cursorial vertebrates such as cats, the clavicle is reduced. But, in brachiators like monkeys, the clavicle is a prominent structural element of the shoulder serving to transfer the weight of the body to the arm (box figure 1). *Homo sapiens* retains this prominent clavicle.

The design of our hindlimbs and pelvic girdle accommodates compromises to our upright bipedal posture (box figure 2a, b). The birth canal, the opening enclosed by left and right pelvic girdles through which the infant passes during birth, is wide, especially in human females. This canal accommodates the large size of the infant's cranium (box figure 2b). But widening the hips to accommodate an adequate birth canal places the heads of the femurs far apart and outside the center line of body weight. A bend in the femur just above the knee allows the limbs to swing directly beneath the body.

Our bipedal posture and pendulum-like leg motions result in changes in foot design. Apes retain a grasping hindfoot with a projecting large toe. In humans, the toe is aligned with the other digits of the foot so that as the limbs are swung beneath the body, they can be placed close to the line of travel without catching the projecting toe on the opposite leg. The human foot forms an arch, a way of broadening the base of support upon which the upper body stands. The arch also changes the geometry of the foot: As the foot pushes off during walking, it extends the ankle farther than if the arch were absent.

Clavicle

Scapula

BOX FIGURE 1 Brachiation, locomotion by swinging through trees suspended from the forelimbs, is accompanied by structural modifications. The clavicle is large and firmly attached to the sternum, thus bringing stability to the shoulder. The hands are modified for grasping. The toes of the hindlimb are opposable and also used for grasping.

Ilium

Arcuate line (birth canal)

Pubic symphysis

(a) Male

(b) Female

BOX FIGURE 2 **Birth canal.** Pelvises of human male (a) and female (b). In humans, the birth of a baby with a relatively large head requires a relatively large birth canal. The hips of females are correspondingly wider than those of males to accommodate the infant's head.

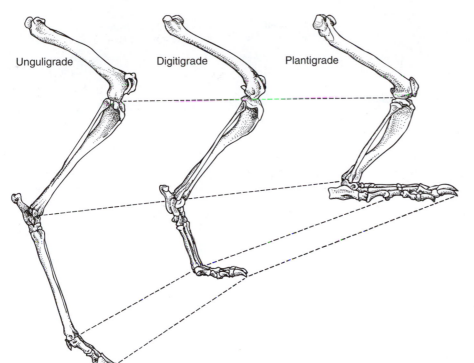

Unguligrade   Digitigrade   Plantigrade

FIGURE 9.37 Foot **postures.** Unguligrade, digitigrade, and plantigrade designs for feet. Note how changes in foot posture produce relatively longer limbs.

After Hildebrand.

to cruise from areas of locally depleted resources to new pastures and to locate dispersed resources in sparse lands.

The speed or velocity attained by a vertebrate is a product of its *stride length* and *stride rate*. Other things being equal, vertebrates with longer strides can cover more ground than those with short legs, so they attain greater speeds. The faster the rate of limb oscillation, the faster the animal travels. Let us consider adaptations that serve either stride length or stride rate, and hence contribute to cursorial locomotion.

Stride Length   One way to increase stride length is to lengthen the limb. Highly cursorial vertebrates exhibit a marked lengthening of their distal limb elements. A related modification is a change in foot posture. Humans walk with the entire sole of their foot in contact with the ground, exhibiting a **plantigrade** posture. Cats walk with a **digitigrade** posture in which only the digits bear the weight. Deer use an unguligrade posture, traveling on the tips of the toes (figure 9.37). Plantigrady is the primitive tetrapod condition from which all other foot postures derive. The change from plantigrade to digitigrade to unguligrade postures effectively lengthens the limb and increases the length of stride.

Another way to increase stride length is to increase the distance through which the limbs move while they are off the ground. For example, the cheetah, when increasing its speed from 50 to 100 km/hr (about 30 to 60 mph), does not appreciably change its rate of limb oscillation but instead increases its length of stride. With greater spring in each forward leap and with extreme flexion and extension in

the vertebral column, the cheetah limbs extend their reach during each stride to increase speed (see figure 9.42a).

Stride Rate   Velocity of travel also depends on the rate at which the limbs are moved. Larger, more mechanically efficient muscles increase the rate of limb movement. Shortening the limb would certainly make limb oscillation easier and increase stride rate, but it would also shorten the length of the stride and compromise speed. However, flexion of the limb during recovery effectively shortens it, increasing the rate of forward oscillation.

Another way to promote stride rate is to lighten the distal end of the limb in order to reduce the mass and thus the inertia that must be overcome due to mass. If the bulk of the powerful limb muscles is located close to the body and carries their force distally through light tendons to the point of force application at the end of the limb, inertia is reduced at the limb's distal end. By so lightening the limb, it can be more easily and efficiently moved with less energy. The bunched limb muscles in the shoulders and hips of deer, horses, and other fast animals are examples (figure 9.38). Another adaptation that increases stride rate is reduction in the number of digits. In highly cursorial mammals, one or two of the central digits are strengthened to receive the forces of impact with the ground. But the more peripheral digits tend to be reduced or lost (figure 9.39a, b). Overall, the result is to lighten the end of the limb and allow it to oscillate more rapidly. Birds specialized for rapid terrestrial locomotion, such as ostriches, show similar cursorial adaptations such as lengthening of hindlimbs and loss of digits (figure 9.39c).

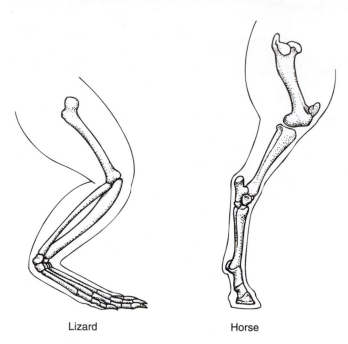

FIGURE 9.38 **Location of limb muscles in a lizard (left) and a horse (right).** In cursorial animals, such as horses, muscles acting along the leg tend to be bunched close to the body and exert their forces along the limb through long, light tendons. This design reduces the mass of the lower limb and thus reduces the inertia that must be overcome during rapid limb oscillation.

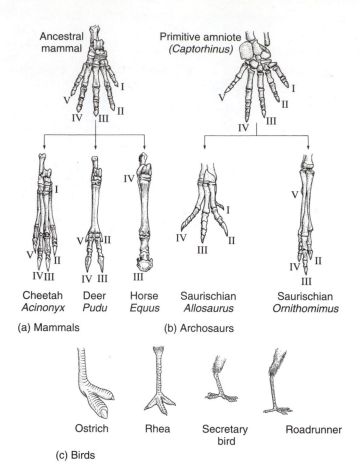

FIGURE 9.39 **Reduction of digits in cursorial animals.** Central digits at the ends of limbs tend to be strengthened while more peripheral digits are lost. The overall result is to lighten the distal portion of the limb. (a) The hindfoot of a cheetah, deer, and horse are shown. Note the varying degrees of digit reduction in comparison with a more general ancestral mammal. (b) A similar trend, apparently related to cursorial locomotion, occurs in archosaurs, although no archosaur foot is reduced so extremely as the horse foot, in which only a single digit remains. (c) Cursorial birds have slender hindlimbs and sometimes show a loss of digits, as in the ostrich.

(a) After Hildebrand; (b) after Romer; (c) after Peterson.

Gait The gait an animal selects depends on the rate of travel, obstructions in the terrain, maneuverability sought, and body size of the animal (figure 9.40). Subtle differences in footfall patterns and rhythms have led connoisseurs of animal gaits to recognize many types and subtypes, especially in mammals and particularly in horses. We will look at just a few general types.

The **amble** is an intermediate-speed gait with a distinct four-beat sequence with at least one foot in contact with the ground and without a suspension phase. It may be based on a speeded-up lateral sequence (elephants, horses) or diagonal sequence (primates). Ambling ensures continuous contact with the ground with at least one foot while reducing vertical displacements, thus accounting for why it is the only fast gait used by large elephants, and why it is used by arboreal primates dashing about on unstable branches.

In the **pace,** lateral pairs of limbs on the same side are moved together. Long-legged animals pace to avoid tangling their limbs, which might otherwise occur, especially at higher speed where there is also a suspension phase between oscillation of opposite sides. Some harness horses exhibiting a natural pace ability use this type of gait in racing.

The trot, another type of gait, is advantageous because the connecting line of support between diagonally opposite limbs runs directly under the center of mass. This makes it more stable than the pace, and the trot is the favored walking gait in salamanders, reptiles, and animals with broad bodies, such as hippopotamuses.

In the **pronk,** common in some artiodactyls, all four feet strike the ground in unison. Although this gait abruptly jars and decelerates the animal, it gives great four-footed stability each time the feet contact the ground. Conversely, all four feet are off the ground during the elevated suspension phase of each gait cycle, perhaps an advantage to an animal that must clear low brush. Alternatively, some propose that the pronk is not so much a gait as it is a social display or defensive behavior, wherein the pronking artiodactyl vaults to a height, giving it elevated surveillance to spot crouching predators hiding in the grass, or signals to conspecifics of its alert behavior.

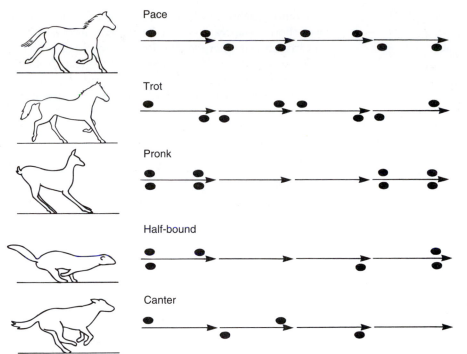

Pace

Trot

Pronk

Half-bound

Canter

FIGURE 9.40 **Gait or footfall patterns of various mammals.** The particular gait selected depends on speed of travel, size of the animal, and structure of the terrain. The footfall patterns produced by each gait when the feet strike the ground are indicated.

After Hildebrand.

In the **bound,** both hindfeet together launch the animal, the spinal column flexes to extend the reach during suspension, and landing is on both forefeet in unison. The half bound and gallop are also used at high rates of speed. When a pair of feet approaches the ground, the **leading foot** strikes in front of the **trailing foot.** In the **half bound,** the hindfeet make contact more or less in unison, but the forefeet make contact with a distinct leading and trailing pattern. In the **gallop,** both fore- and hindfeet display distinct lead and trailing pattern. At slow speeds with a slightly different footfall pattern, the gallop is called a **canter.** Galloping and half bounding are said to be **asymmetrical gaits** because footfalls of a pair, forefeet or hindfeet, are unevenly spaced during a cycle. These gaits may be less stable than **symmetrical gaits,** such as the trot or pace. The suspension phase, added to the faster gaits, inserts greater reach of the limbs and results in increased stride length.

Humans, like other running animals, observe economy of effort as their speed increases by changing gait pattern. As we gradually up our speed, we break from a walk into a run when we reach about 5 mph (2.4 m s^{-1}). Below that speed a walking gait requires less energy; above that speed a running gait requires less energy. By changing gaits, we keep our energetic costs at a minimum for the given speed at which we are traveling. Horses, too, voluntarily change gaits as their speed increases, switching from a walk, to a trot, to a gallop. The metabolic energy required is related to oxygen (O_2) use. To standardize energy (oxygen) use over different speeds so we can compare them

across gaits, we convert oxygen use to the milliliters of oxygen consumed over a 1-meter distance. If we measure and standardize oxygen use through different speeds and gaits, we discover that within each gait there is a minimum speed that consumes the least amount of energy. If the horse goes faster within the particular gait, more energy is consumed; if it goes slower, energy use similarly increases. There is a speed within each gait where economy of effort is best. This produces a relation between speed and energy use that is not linear (straight), but curvilinear (U-shaped) (figure 9.41). Thus, gait selection is certainly related to stability, but also to minimizing energy use through different speeds.

Uses of Cursorial Locomotion Cursorial locomotion is widespread among vertebrates, especially among mammals, but it is deployed in many ways. The lion and cheetah can use their speeds to accelerate quickly for short bursts, whereas the horse and pronghorn more often use their cursorial abilities to cruise open plains in search of dispersed resources or to put distance between themselves and potential predators. But it is even more complicated than this. Body size is a factor as well (table 9.1).

For example, both the horse and cheetah are adapted for cursorial modes of locomotion (figure 9.42a, b). The cheetah, in addition to using its speed differently, is also smaller than the horse. The horse can maintain a speed of 30 km/hr for over 30 km. But if it were built like a fox or cheetah, it could not maintain even moderate speeds for more than a few kilometers. The cheetah is not an

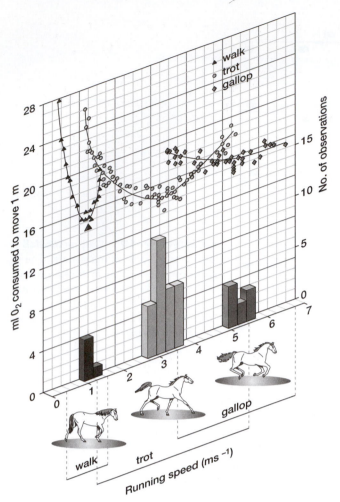

FIGURE 9.41 **Gait selection and energetics.** For the horse, three gaits at increasing speeds are experimentally examined for the metabolic energy cost incurred in each. Expressed as the milliliters of oxygen to move the horse 1 meter, the three gaits produce three curves of energy use. Note that within each gait there is a speed that is most economical. If the horse within a gait runs faster or slower, the metabolic costs generally go up.

Reprinted with permission from Macmillan Publishers, Ltd. Nature. D. F. Hoyt and C. Richard Taylor. 1981. Gait and the energetics of locomotion in horses. Nature 292: 239–240. Copyright 1981.

TABLE 9.1 Maximum Speeds and Sizes of Cursorial Animals

Animal	Maximum Speed km/hr	(mph)	Weight (kg)
Horse	67	(42)	540
Lion	80	(50)	180
Pronghorn	95	(59)	90
Human	35	(22)	85
Cheetah	102	(63)	35
Coyote	65	(40)	10
Fox	60	(37)	4.5

endurance machine; rather, it is designed for quick bursts over short distances. Were it larger with increased mass to carry, then its morphological design would also have to scale up accordingly. In the cheetah, the great flexion of the vertebral column extends and then gathers the limbs during suspension phases. The result is to increase the effective stride length. It is estimated that this extensive bending of the vertebral column alone, by increasing stride length, adds almost 10 km/hr (6 mph) to the speed of the animal. But such flexions also mean that much of the body mass is displaced vertically rather than in the direction of travel. The cheetah must expend significant energy to lift this mass during each set of strides. For a heavy animal like a horse, this design consumes too much energy to sustain long-distance travel. Consequently, the vertebral column of the horse bends very little even during full gallop. Vertical displacement of the vertebral column in the hips may be less than 10 cm and in the shoulder less than 5 cm. Proportionately less energy must be used to lift this mass and more of the weight of the animal is carried along the linear path of forward travel.

The horse and the cheetah represent cursorial animals that use their speed differently in different biological roles—endurance versus quick bursts, respectively. But their designs also represent different compromises with body size—large versus small size, respectively. If the larger horse were built like the smaller cheetah, it could not contend so well with its greater mass and be able to sustain its locomotor endurance on which it greatly depends. The dynamic demands of locomotion placed on the skeletal system depend both on the biological roles served by locomotion and on the demands imposed by body size.

Aerial Locomotion

Gliding and Parachuting Gliding and parachuting are fundamentally different ways of slowing a descent. Parachuting proper involves maximizing drag. It occurs commonly in small plant seeds, under 100 mg. Falling mice and cats may "parachute" to reduce impact speed but they have few other options as they have no significant surface area to produce lift. In contrast, gliding involves minimizing drag and using lift to produce a more favorable lift-to-drag ratio. It occurs in heavier seeds, as well as various aerial vertebrates. Gliding through the air occurs in at least a few species in all classes of vertebrates. "Flying" fish spread especially wide pectoral fins during short glides in the air above water (figure 9.43a). A tropical snake fleeing a threat flattens it body into a long narrow parachute to slow its falling descent from trees. A species of tropical frog spreads its long, webbed toes to slow its airborne fall (figure 9.43b). Lizards with special flaps of skin and squirrels with loose skin between fore- and hindlimbs spread these membranes to slow their drop through air or extend the distance of their horizontal travel (figure 9.43c–f). These tentative fliers are not really fliers at all, however. Instead they are gliders. True

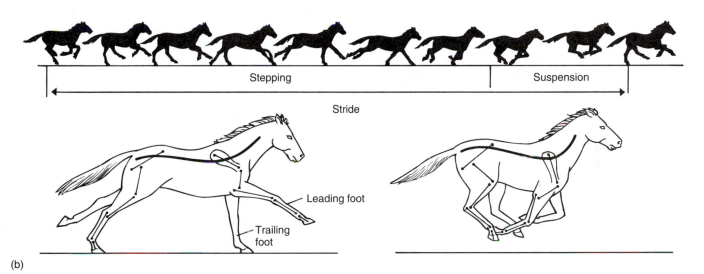

FIGURE 9.42 **Comparison of two cursorial mammals, a horse and a cheetah.** (a) The cheetah depends on quick bursts of speed to overtake prey. Notice the extensive flexion of the vertebral column that increases stride length and adds about 10 km/h (6 mph) to its overall speed. (b) The horse uses its speed for sustained locomotion; therefore, the vertebral column flexes much less to avoid the exhausting vertical rise and fall of the body mass characteristic of the cheetah. A less flexible vertebral column keeps the mass of the horse more linear along its line of travel. The lead foot and trailing foot change during sustained bouts of rapid running.

After Hildebrand.

powered flight occurs in just three groups: bats, pterosaurs, and most birds (figure 9.44). In each group, the forelimbs are modified into wings that both generate the force driving them forward through the air and provide lift against gravity.

Flight Most functional analyses of powered flight have centered on birds, taking advantage of the sophisticated aerodynamic equations engineers use to design aircraft. But borrowing directly from engineers has been especially difficult because the wings of birds have all the characteristics a designer of aircraft seeks to eliminate. Bird wings flap (whereas airplane wings are fixed), are porous (rather than solid), and yield to air pressure (rather than resisting it like airplanes). Although simplifying assumptions must usually be made, such analyses have yielded an understanding of several adaptations for flight in birds.

Feathers Contour feathers give the body of a bird its streamlined shape to help it cut efficiently through air. By filling the body out into a streamlined, aerodynamic silhouette, contour feathers help maintain a laminar airflow across the body and reduce friction drag. It has even been suggested that the body shape, which is similar to that of an airplane wing, also produces lift. However, most of the lift is produced by the wing. The primary feathers, attached to the manus, are responsible for providing forward thrust. Secondary feathers, attached to the forearm, provide lift (figure 9.45). Thus, the functions of flight are divided between these two types of flight feathers. Primaries provide forward thrust, and secondaries act like airplane wings, providing lift.

Aerodynamics (p. 145); feathers (p. 222)

(a) Flying fish

(b) Flying frog

(c) Draco

(d) Flying gecko

(e) Giant flying squirrel

(f) Colugo

FIGURE 9.43   Gliding and parachuting. In all classes of vertebrates, at least a few species can be found that occasionally take to the air. (a) Fish. (b) Amphibian. (c, d) Reptiles. (e, f) Mammals.

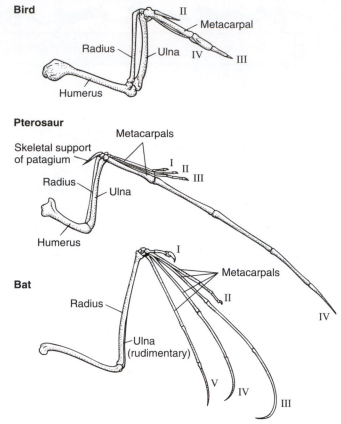

Bird

II
Metacarpal
Radius
Ulna   IV
III
Humerus

Pterosaur

Metacarpals
Skeletal support
of patagium
I   II
Radius   III
Ulna
Humerus

Bat

I
Metacarpals
II
Radius
Ulna
(rudimentary)   IV
V   IV
III

FIGURE 9.44   Modifications of the forelimb in bird, pterosaur, and bat to support the aerodynamic surface. Generally, the participating digits are lengthened and bones lightened.

After Hildebrand.

Skeleton   High-speed and radiographic films of birds in flight give a detailed view of wing motions and the role played by the pectoral girdle and rib cage. One wing-beat cycle can be divided into four phases: (1) upstroke-downstroke transition, (2) downstroke, (3) downstroke-upstroke transition, and (4) upstroke. During upstroke-downstroke transition, the leading edge of the wing is elevated above the body and lies nearly within a sagittal plane. The elbow and wrist joints are fully extended. From this position, the wing is forcefully brought downward (depressed) and forward (protracted) during the downstroke, producing thrust and lift. The wrist and elbow remain extended during downstroke. The wing continues downward and forward until its tip extends in front of the body. During the downstroke-upstroke transition, wing motion reverses, leading into the upstroke. The upstroke is complicated and apparently generates little lift but repositions the wing for the next downstroke. During the upstroke, the wing is folded and lifted upward (elevated) and backward (retracted), while the elbow and wrist are fully flexed (figure 9.46a–c).

Synchronous changes in the rib cage and shoulder girdle occur as well when these wings beat. During the downstroke, the U-shaped and flexible furcula, accompanied by the procoracoids, bends laterally. The sternum shifts upward and backward. During the upstroke, these

motions are reversed. The furcula springs back and the sternum drops downward and forward (figure 9.46d). These configurational changes within the rib cage alter the size of the thoracic cavity. In addition to whatever contributions such changes make to flight, it is hypothesized that they also are part of the respiratory mechanism ventilating the lungs. This coupling of locomotor and respiratory systems takes advantage of the muscle forces produced during flight to ventilate the lungs and air sacs at the same time.

Airflow through avian lungs (p. 442)

In most birds, the furcula probably functions like a spring when it bends and then recoils during flight. Energy is stored as elastic energy in bent bones during one part of the stroke and then recovered during recoil later in the cycle. But in birds such as parrots and toucans, the clavicles are not fused, and in flightless birds, they are typically vestigial or absent entirely. In some soaring birds, the furculae are quite rigid and probably resist bending. Although the functional significance of this structural diversity is not understood, the bird furcula might be expected to play additional roles in

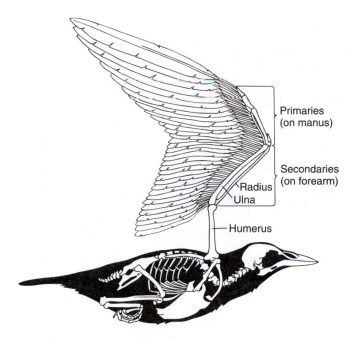

Primaries
(on manus)

Secondaries
(on forearm)

Radius
Ulna

Humerus

FIGURE 9.45 **Feathers along the wing divide the functions of flight among them.** Those at the tip, the primaries, attach to the manus and are primarily responsible for producing thrust; those more proximal, the secondaries, attach to the forearm and are primarily involved in producing lift.

flight other than just as a mechanism to store and return energy.

We have already mentioned that the innominate bone and the synsacrum fuse, which stabilizes the body in flight. The flexibility of the cervical vertebrae allows a bird to reach all parts of its body. These two design features, fusion and flexibility, are nearly uniform throughout birds, testimony to the influence of flying on biological design. The skeleton also exhibits other modifications for flight. The bones of birds, including *Archaeopteryx*, and pterosaurs, but not bats, are hollow rather than filled with blood-forming or fatty tissues like the bones of other vertebrates (figure 9.47). Absence of these tissues from bird bones results in overall lightening of the skeleton and reduces the weight that must be launched into the air. In pterosaurs, bats, and especially in birds, an expanded sternum serves as the origin for the powerful **pectoralis** flight muscles.

Sternum (p. 299)

Types of Flight Although flight is the common denominator in most birds, not all flight is the same. For hovering birds and strong fliers, emphasis is on maximum propulsive force and thus on the primary feathers attached to the manus. In such birds, the manus is proportionately the largest section of the forelimb (figure 9.48a). For soaring birds, emphasis is on lift and thus upon the secondary

feathers attached to the forearm. The forearm is proportionally the longest section of the wing in soaring birds (figure 9.48b, c).

Hummingbirds, swifts, and swallows depend on strong, frequent wing beats. Soaring birds take advantage of air in motion to gain altitude and stay aloft (figure 9.49a–d). Those birds that soar over open oceans take advantage of strong prevailing winds and have long, narrow wings like those of glider aircraft (figure 9.50a). These soaring birds may spend as much time during the day in the air as on the ground. In some soaring specialists, locking mechanisms of bones and ligaments of the wrist and shoulder, and not muscles, secure the wing in an extended position, reducing the active muscle energy required to soar. But the aerodynamic character of moving air can be different, so the character of soaring flight is different as well. Birds that soar over open country ride thermals, fountains of warm air rising upward. As the sun warms the Earth, the nearby air is warmed and begins to rise. Vultures, eagles, and large hawks find these rising thermals, circle to stay within them, and ride them to gain easy altitude. These birds have slotted wings (figure 9.50b). For flight in enclosed habitats such as woodlands and shrubby forests, elliptical wings give birds such as pheasants a quick, explosive takeoff and maneuverability within tight spaces (figure 9.50c). Birds of prey, migratory waterfowl, swallows, and others that depend on fast flight possess sweptback wings (figure 9.50d). To understand these general wing designs, we need to examine the aerodynamic basis of flight itself and the problems different wing designs address.

Aerodynamics During horizontal flapping flight, four forces act on a bird at equilibrium. The upward lift (L) is opposed by the weight (mg), tending to pull the bird down. Drag (D) acts in the direction opposite to the direction of travel and wings generate thrust (T), a forward force component (figure 9.51a). The angle at which the wing meets the airstream is its **angle of attack.** Increasing this angle increases lift, but only up to a point. As the angle of attack increases, drag increases as well because of the change in the wing profile meeting the airflow (from edge-on to broadside) and because of the increased flow separation across the wing. Thus, at some extreme angle of attack that depends on air speed and particular wing shape, airflow in the boundary layer separates from the top of the wing and lift drops dramatically. When this happens, the wings **stall** (figure 9.51b). Stalling can be delayed if the layers of air in the flow are prevented from separating. In birds, the small alula controls the airstream passing over the wing and prevents its early separation as the angle of attack initially increases. Thus, greater angles of attack can be reached before stalling, so greater lift can be produced, albeit with greater drag as well.

Aerodynamics (p. 145); feathers (p. 222)

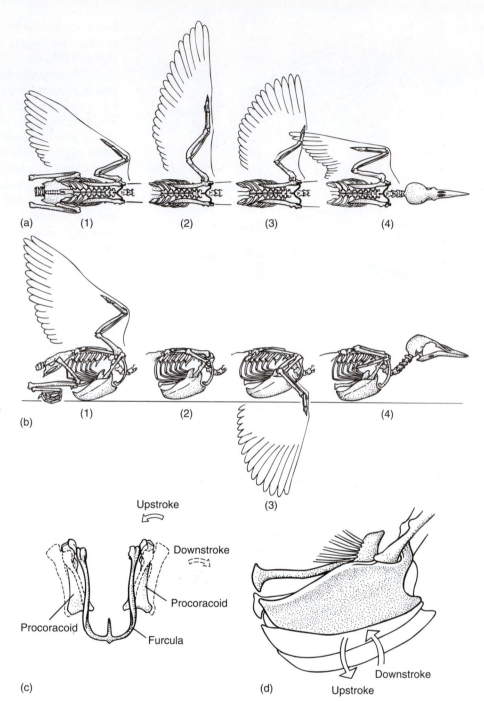

FIGURE 9.46 **Wing-beat cycle of a European starling.** Positions of shoulder girdle and wing are illustrated in dorsal (a) and lateral (b) views. (c) Anterior view of furcula and procoracoids that bend laterally on the downstroke (dashed lines) and recoil medially on the upstroke (solid lines). (d) Lateral view of the excursion of the sternum, which moves in a posterodorsal direction during the downstroke and reverses this motion to an anteroventral direction during the upstroke. In (b2) and (b4) only the humerus of the wing is shown so as not to block the view of the ribcage.

After Jenkins, Dial, and Goslow.

From F. A. Jenkins, Jr. et al., "A Cineradiographic Analysis of Bird Flight," Science, 16 Sept. 1988, 241:1495–98. Reprinted with permission from AAAS.

An **airfoil** is an object that, when placed in a moving stream of air, produces a useful reaction. Whether the airfoil is the wing of a bird, pterosaur, bat, or airplane, it generates lift as a consequence of its angle of attack, the details of its shape, and air speed. Airfoils act to speed up the air passing across one surface relative to the other, adjusting the angle of attack to produce the greatest lift relative to the cost in drag, thus the greatest lift-do-drag ratio. Since the late nineteenth century, we've recognized that **cambered wings**—wings curved upward on their upward surfaces—give superior results to simple, tilted, flat plates (such as kites). At zero angle of attack (zero-lift angle), the divided airstream travels at equal velocities across upper and lower surfaces, meeting again on the trailing edge (figure 9.51c) and producing no lift. However, as the angle of attack increases, the airfoil speeds up the airstream passing across the top of the wing, reaching the trailing edge before its divided lower half arrives (figure 9.51e) and producing lift. This lift results mainly from the reduced pressure, essentially "suction," of the fast-moving airstream passing across the upper surface of the wing. This lift is augmented by a smaller contribution from the increased pressure of the slower-moving airstream below the wing (figure 9.51e, f).

FIGURE 9.47 Long bones of birds. Many tissues that contribute to weight are reduced in birds. No tissue fills the marrow cavities of the long bones, and the bone walls are thinned. Thin struts stiffen the bone and prevent it from buckling. In lift, the spaces are filled with extensions of the air sacs.

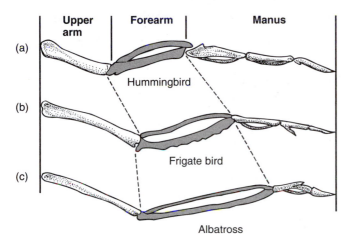

FIGURE 9.48 Differences in flight are reflected in differences in wing design. In a hovering bird, such as the hummingbird (a), emphasis is on the primary feathers and the distal part of the forelimb, offering attachment for the primaries, and consequently the manus is relatively lengthened. In soaring birds, such as the frigate (b) and especially the albatross (c), emphasis is on the secondary feathers, and the part of the forearm that supports these feathers is relatively lengthened.

Taking to the Air Formally, the wing creates a local pressure field that produces lift. Lift is produced in two major ways. First, the lower part of the wing deflects downward the air it meets, and this air mass in turn imparts upward momentum to the wing. Second, the upward-tilted wing creates a pocket of negative pressure, relative to ambient, on its upper surface. As the angle of attack changes, the pressure changes (figure 9.51d, f). At moderate angles of attack, the pressure profile of the wing exhibits, relative to ambient, positive pressures along its lower surface and negative pressures along its upper surface. The net result of both effects is lift: (a) the wing pushes air down and air pushes the wing up and (b) the negative pressure across the top of the wing pulls it up.

The actual pressure profile of a wing depends upon air speed, air density, and details of the wing itself, as well as angle of attack. In general, the front of the wing generates most of the lift. Notice how the negative pressure on top of

the wing is greater than the positive pressure acting on the underside of the wing (figure 9.51f). Some lift is due to diversion of the onrushing air by the bottom of the wing. But most of the lift is generated by events occurring on the top of the airfoil. Aerodynamic properties of this profile usually include **stagnation lines** where the relative air velocity falls to zero. A tiny bug walking on the wing of an airplane in flight could walk the stagnation line without experiencing any wind.

Airplanes are usually designed with cambered wings. With increasing angles of attack above zero (no-lift), lift increases, but only up to a point. At high angles of attack, the airflow across the wing starts to separate with a consequent loss of lift. The curved, cambered surface helps hold the streamlines and thereby prevents this flow separation at higher angles of attack. In animals, the wings may be thin and highly cambered, with the lower surface actually concave. As in airplanes, the upper camber helps prevent stall. The lower surface apparently is concave to more evenly distribute the imparted momentum over its surface. In 1903, the Wright brothers' airfoil was actually similarly designed: cambered on top, concave on the bottom. But this is seldom done today in aircraft because of manufacturing difficulties, and because large camber is beneficial only near stall. In aircraft, stall is an issue during takeoff and landing that is addressed by extending the flaps on the trailing edge of the wing. This effectively creates additional camber to reduce the chance of stall.

To calculate overall lift, **aerodynamic circulation,** a mathematical construct, must be included. To understand circulation, first consider the airflow patterns passing around an object such as a baseball. Without rotation, flow is symmetrical over both sides (figure 9.52a). However, when the ball spins, it pulls air around itself in the direction of spin because of viscosity and the tendency of air to stick to the surface, a kind of bound vortex circulating with the ball (figure 9.52b). Taken together, this circulation adds to airflow speed on one side and subtracts on the other, leading to an asymmetrical force (figure 9.52c), which causes the baseball to curve in one direction. (Spin is also what causes a golf ball to hook or slice, depending upon the direction of the spin.)

An airfoil does not spin to create an actual physical vortex circulating around the wing. Instead, because of its shape and effect on airflow, an airfoil affects the airstream *as if* it were spinning. This effect can be calculated or at least visualized with the separate effects summarized and then brought together. Without circulation, the airstream leaving the trailing edge of an airfoil does not come off smoothly as it attempts to turn the sharp edge (figure 9.52d). Alone, circulation (figure 9.52e) is the calculated mathematical flow needed to account for the smooth, departing airstream. When the two are summed (figure 9.52f), the air departs the trailing edge smoothly. Simply put, the effect of circulation is that the wing pushes air down, contributing to lift.

Airfoils have another curious but important consequence resulting from circulation. An airfoil that produces lift is encircled by a kind of vortex—one might say, a virtual

FIGURE 9.49 **Soaring and hovering flight.** (a) Soaring birds take advantage of updrafts from wind currents. Cold air moving off water slides under the lighter air warmed over land. Rising air creates updrafts that birds use to gain altitude. (b) Ridge soaring. Uplift is created when wind is forced upward by a low mountain. The wind rebounds again behind the mountain, creating repeated opportunities for ridge soaring. (c) Thermals. Local areas of the ground warmed by sunlight heat the adjacent air, which begins to rise. This bubble of rising warm air is a thermal with internal circulation as shown by the arrows. As the thermal rises, soaring birds enter and circle within the core to rise to a higher altitude. (d) Hovering birds, such as the hummingbird, must depend entirely upon the strength of their wing muscles to generate lift. Wings are rapidly swept along the dashed path, producing lift (arrows) on the backstroke as well as during the forward downstroke.

After Peterson.

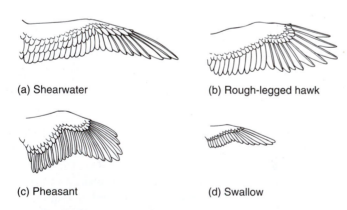

(a) Shearwater

(b) Rough-legged hawk

(c) Pheasant

(d) Swallow

FIGURE 9.50 **Wing shape differs with the type of flight.** (a) Soaring birds have long, narrow wings like glider planes. (b) Birds that soar over land, as hawks do, have slotted wings, with primaries slightly spaced at their tips. (c) For birds that must maneuver in close quarters, such as pheasants, the wings are elliptical in shape to allow quick bursts of flight in enclosed habitats such as forests. (d) Fast-flying birds, such as falcons and swallows, have swept-back wings.

After Pough, Heiser, and McFarland.

vortex. That is, if one subtracted the rearward velocity vector equal to the speed of the aircraft from the velocities at each point (figure 9.53a), one would be left with a vortex, flowing rearward above the wing and forward below it. It is virtual in that the air particles do not physically zip around the wing traveling in actual orbits, but as we saw in figure 9.52, the physical consequences of the airfoil on airflow act as if air particles do circumnavigate the wing.

Vortices don't normally have ends—they tend to turn back on themselves like donuts (toroids). So, these odd virtual vortices pass beyond the tips of wings, turning backward as a pair of trailing tip vortices (figure 9.53b). Such vortex shedding represents an energy loss; it is minimized by having long, skinny wings or wings with splayed outer primary feathers at their tips (figure 9.50). Sometimes, a bit of energy is recaptured. Thus, a bird taking part in a V-formation flight gets some lift from one of the tip vortices of the bird ahead of it (figure 9.53c).

Total drag is the overall force that resists movement of an animal through a fluid. Two general categories of drag contribute: parasitic drag and induced drag. **Parasitic drag** is an animal's resistance to passage through a fluid. In

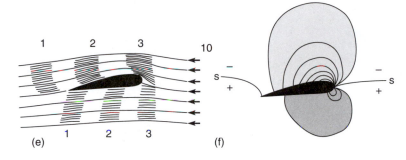

FIGURE 9.51 **Aerodynamics of lift.**
(a) During stable, flapping flight, four basic forces act on the animal. Active wing beats produce a forward thrust (T), and the wing produces lift (L). In opposition are the downward force due to gravity (mg) and the drag force (D) acting in the opposite direction to the line of travel. (b) Both lift and drag are affected by the angle of attack the wing makes with the oncoming airflow. At some critical angle of attack, flow separation occurs and the animal's wing stalls. (c) Laminar airflow is shown passing right-to-left across a wing at 0° angle of attack, where zero-lift is produced. Three successive cohorts of air in this laminar flow—1,2,3—are shown to illustrate how the divided airstreams that pass over and beneath the wing rejoin equally on the trailing edge. (d) The pressure profile of this wing indicates little pressure generation. (e) However, as angle of attack increases, here 10° angle of attack is shown, the upper airstream has speeded up so that the upper cohort reaches the trailing edge before its lower half (1), and the two halves thereafter do not rejoin. (f) The pressure profile indicates, with contour lines, the reduced (−) and increased (+) pressures now acting above and below the wing, respectively. Stagnation line (s).

(c–f) Modified from Denker.

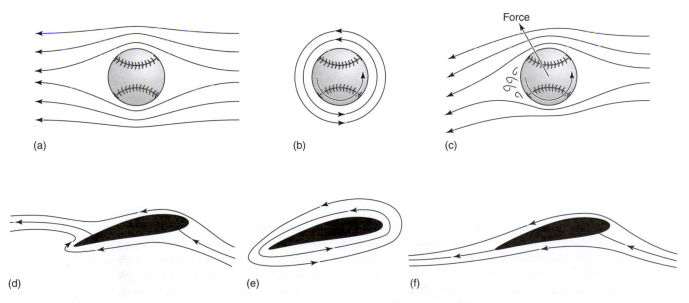

FIGURE 9.52 **Aerodynamic circulation.** Curve ball. A baseball in flight may experience two airflow patterns as a consequence of its motion. (a) Without rotation, the air flows symmetrically across both sides. (b) With rotation, the ball pulls the air near its surface, producing a bound, thin, circulating layer. (c) Acting together or in opposition, the symmetrical flow and the circulating thin bound layer produce flow speeded up on one side, slowed on the other, resulting in deflection of the overall airstream and an unbalanced force on the ball that causes it to curve. Lift of an airfoil. An airfoil does not spin, but its effect on airflow can be similar. (d) Airflow without circulation; (e) circulation only; (f) total, actual flow pattern around an airfoil.

(a–c) After Anderson and Eberhardt; (d–f) after NASA.

FIGURE 9.53 **Aerodynamics vortices and drag.** (a) As lift is produced, circulation is induced, producing a virtual vortex. Subtracting the speed of circulation (V′) below and adding it above the wing to the relative wing reveals the action of vortices around the wing. (b) Vortex system associated with a lift-producing wing. (c) V-formation flying birds take up a position and a wing-beat rhythm on the tips of others to benefit from energy in the shed vortices. (d) Induced drag. With increasing angle of attack, increased lift (L) is produced perpendicular to the surface. The vertical component of this lift (L$_e$) acts opposite to gravity (mg). The horizontal component is induced drag (D$_L$) acting opposite to the direction of travel.

(b, c) from Vogel, 2003.

turn, several types of resistance contribute to parasitic drag: Profile drag is the portion of this resistance caused by the shape of the animal moving through fluid; friction drag is caused by shear stress at the boundary layer; pressure drag is caused by adverse backflow in the wake. **Induced drag** is associated with the production of lift (figure 9.53d). A wing meeting the onrushing relative wind produces lift at right angles to the wing's surface. The useful part of this lift acts vertically, directly opposite to gravity. The vector difference between lift and its effective vertical component represents the induced drag. Therefore, induced drag is the vector component of the lift force acting opposite to the direction of travel. Paradoxically, in producing lift, the wing generates a retarding force component that slows the bird by increasing overall drag.

Friction and pressure drag (p. 145)

Animals that glide depend on the same aerodynamics as those that fly, except thrust is not produced. Performance differences have to do with relative differences between lift and drag forces. A gliding animal, such as the flying squirrel, spreads its body when airborne to present a broad surface to the air. If its glide is steady, several forces act upon it. Resistance of the outstretched body against the airstream produces lift (L). Drag (D) in the direction opposite to travel is also present, and weight (mg) acts as well. The descending glide path makes an angle (θ) with the ground. Relative to this angle, the component of force producing lift is mg cos θ, and drag is mg sin θ. Their L/D ratio is greater than one during a steady glide (figure 9.54a). This "flying" frog

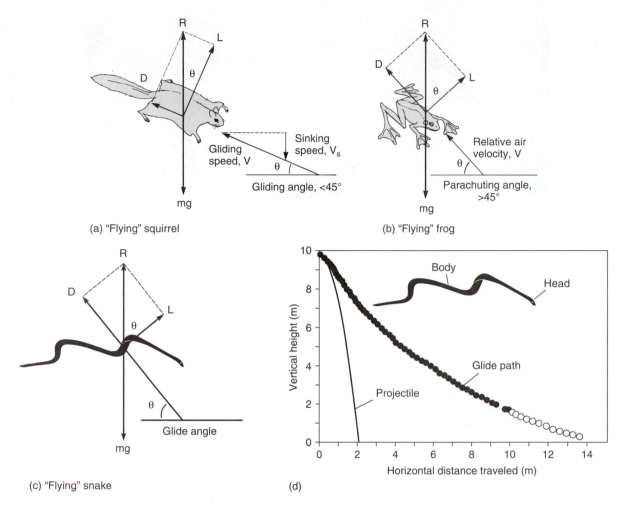

FIGURE 9.54 **Aerodynamics of nonflapping aerial locomotion.** (a, b) These two gliding mammals have different glide paths primarily because they differ in the area that supports their weight. The flying squirrel has special flaps of skin between fore- and hindlimbs that can be extended to produce some lift. The frog has increased webbing between its toes, which collectively produces lift. However, the frog's overall lift to drag ratio (L/D) is lower, resulting in a steeper gliding path and in a steeper angle with the ground (θ). (c) The paradise snake (*Chrysopelea*). (d) Glide path of snake (dots) compared to the falling path of a nongliding projectile (solid line) launched at the same initial velocity as the snake. Open dots at the end are extrapolated to the ground as the snake exited the field of view (solid dots) during filming of its glide. The force due to gravity (mg) is acting in a direction opposite to the resultant (R) of drag (D) and lift (L).

(a, b) After Norberg; (c, d) based on the research of J. J. Socha.

also glides (figure 9.54b), but it produces a lower lift-to-drag ratio. The force relationships are the same except that the area supporting weight is small, lift is small, and the L/D ratio is lower; thus, the glide path is steeper (figure 9.54b).

Even limbless snakes have invented a method to glide (figure 9.54c, d). The paradise snakes (*Chrysopelea*), forest inhabitants of Southeast Asia and Sri Lanka, launch themselves from high in the forest canopy and glide across open spaces to branches below or to the ground. The ecology of these snakes is not known, so the biological role of gliding is speculative. Many gliding animals live in forests. Moving to an adjacent tree for a nongliding animal would demand an arduous journey if the animal had to climb all the way to the ground, across the forest floor, then scamper up the tree of interest. However, a successful glide to a nearby tree but a few meters away would save the long trek to the ground and

then up again. The paradise snakes seem to fit this lifestyle, gliding from tree to tree directly. The snake produces lift by flattening its body and sweeping its body from side to side in undulations while in the air. It can also change direction along its glide path by pointing its head in the direction of turn just at the start of a new body undulation.

Origin of Bird Flight

Several contending theories currently attempt to account for the stages through which bird flight arose out of a reptilian ancestry. One, the **arboreal theory,** envisions the initial preflight step among bipedal, tree-dwelling ancestors (figure 9.55a); leaping from branch to branch is economical, saving a long journey to the ground below and back up an adjacent tree. Leaping establishes the behavior of taking to the air,

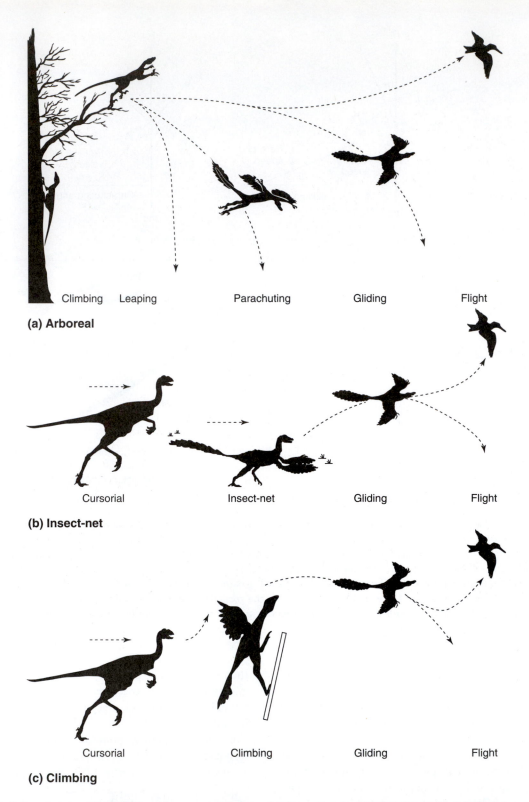

(a) Arboreal

(b) Insect-net

(c) Climbing

FIGURE 9.55 **Origin of bird flight.** Three major theories for the evolution of flight in birds. (a) Arboreal theory. Starting with life in trees, stages include leaping, parachuting, gliding, and flight. Initial factors favoring evolution of feather surfaces occur as a consequence of aerial descent. (b) Insect-net theory. Starting with a cursorial ancestor, the use of forearms to capture or chase insects favored feathered surfaces. (c) Also starting with a cursorial ancestor, protowings were favored as aids in dashing up inclines, then later, rudimentary gliding and flight.

With thanks to K. Dial for entertaining theories.

perhaps leaving pursuing predators behind. In parachuting, spread limbs favored nascent feather surfaces that slowed descent and softened impact upon landing. Gliding took advantage of larger feathered surfaces that deflected the line of fall even farther and extended the effective horizontal distance traveled in the air. Active flailing and, eventually, powered flight extended time in the air and produced a lifestyle exploited by birds. Such stages offer incremental, adaptive plateaus, eventually achieving aerial flight in birds.

Unlike the arboreal theory, which starts with an arboreal ancestor, the other two theories start with a ground-based ancestor. The **insect-net theory** (figure 9.55b) envisions the origin of feathers among quick, bipedal ancestors that dashed along the ground using their forearms to swat at and catch available insects. Here, feathers evolve to increase the swatting effectiveness or to sweep backwards, driving the preflight sprinter faster forward after prey. The habit of leaping into the air established the behavior of becoming temporarily airborne and, from such a stage, active flapping flight might next evolve. Despite the difficulty of imagining a flailing and swatting ancestor dashing after insects across a Mesozoic landscape, this theory is rather abrupt in its proposed stages. Further, as soon as the fly-swatter-equipped ancestor hopped into the air it would decelerate, reducing airborne momentum.

The **cursorial theory,** as first proposed, similarly envisioned the preflight ancestor to birds as a speedy, ground-dwelling, and bipedal reptile, dashing after food or fleeing predators. But for the same reason, deceleration upon liftoff, this theory had difficulty getting into the air, and there seemed no selective advantage to incipient feathers or wings. However, recent research suggests an answer. Some hatchling birds, before they can fly, can flap their still-developing wings to help them dash up inclines, a behavior termed **wing-assisted inclined running (WAIR).** Even small flapping wings press the young bird into the incline, preventing the chick from falling away, keeping its feet engaged in the inclined substrate, and permitting the strong hindlimbs to take advantage of the traction to drive it up its slope. Evolutionarily, in prebirds such an incipient feathered forelimb would provide similar locomotor advantages, perhaps representing an adaptive, intermediate stage in the origin of flight (figure 9.55c). Since such wing-assisted running inclines would require a dorsoventral forelimb stroke, this stage would also debut the transition to a dorsoventral flight stroke of later birds.

Central to most views of flight origin is that a gliding stage preceded the flight stage. This has been tested by Kevin Padian and Ken Dial by examining the comparative phylogenetic positions of vertebrate gliders and flyers. It turns out that the three clades of flyers—birds, bats, pterosaurs—are far removed from the 15 known living and extinct clades of gliders. Bats in particular are far removed from all eight clades of mammalian gliders. Further, no immediate outgroups or fossils of bats include bipedal members. Bats apparently evolved powered flight by modification of standard quadrupedal locomotion. Unlike the prevailing view, these contrary results in general suggest that powered flight is not usually preceded phylogenetically by a recent gliding ancestor.

Fossorial Locomotion

Animals that spend part or all of their lives underground are said to be **subterranean.** With such a lifestyle, an animal takes advantage of existing tunnels or holes into which it retreats. Snakes, lizards, turtles, and many amphibians escape down burrows or deep natural recesses in the earth to find relief from the harsh, cold winter or the excess heat of midsummer. Many fishes find relief from predators in tunnels, whereas predators often use tunnels to conceal themselves until they can pounce out on unsuspecting prey. Sleek predators such as snakes or weasels follow their prey beneath ground into subterranean chambers. Some subterranean animals store food in underground caches.

However, many subterranean vertebrates excavate their own tunnels by active digging. Such active earth movers are fossorial. The fossorial habit has evolved in every vertebrate class. Prairie dogs and rabbits excavate extensive, interconnecting tunnels; the subterranean living quarters of rabbits are **warrens.** These can include a maze of passageways with escape routes and snug nesting chambers in which young can be raised. Thus, digging may produce underground microhabitats with safer conditions, moderate climate, and more abundant food than offered on the surface.

Ways of Digging   A lungfish, seeking temporary retreat as a pool of water dries, digs into the soft mud using its body and fins. Flattened flounders wave pectoral fins to stir loose sand into suspension. As it settles, the sand covers the fish to conceal its body. Frogs back into a shallow burrow scooped out by their hindlimbs. Among reptiles, amphisbaenians use their pointed heads to penetrate soft soil. Body pressure against the walls of the tunnel compacts the substrate so the walls will hold and not immediately collapse on the animal. Some snakes push their way through loose sand to descend several inches beneath the surface away from the desert heat above (figure 9.56). Many rodents gnaw into soil to loosen it with their powerful incisor teeth before excavating it with their limbs.

Fossorial Adaptations   The Namib Golden mole lives in shifting sands of the Namid Desert of South Africa. Usually by noon in this toasty desert, the mole buries itself up to 50 cm deep to find comfortable temperatures and escape from the searing heat of the day above. After dusk, it emerges to cooler temperatures and forages on the surface. Burrowing is achieved in two phases. During the buttressing phase, the mole lifts its head upward against the sand, then the chest is pressed downward (figure 9.57a). The result is to compact the sand and open a pocket in front of the mole. This is quickly followed by repeated cycles of the digging-propulsion phase, where strokes of the hindlimbs and especially power strokes of the strong forelimbs sweep them backward and drive the body forward (figure 9.57b).

FIGURE 9.56 While buried, fossorial animals face special problems, not the least of which is obtaining sufficient air. The sand snake uses its head to push a hole slightly larger than its body in loose sand, thereby creating a sand-free space to facilitate breathing.

After Gans.

The appendicular skeleton, especially the forelimbs, can apply great force to move earth. Several structural modifications usually are involved. First, the limb bones of fossorial animals are especially stout and robust, and the muscles that are attached to them are relatively large (figure 9.58a, b). This produces a short and forceful bone-muscle system, unlike the long or delicate limbs of cursorial or aerial specialists. Second, the limb as a lever system is adapted for high-force output. The forearm and hand of fossorial vertebrates, which deliver the out-force, are relatively short; the elbow, which delivers the in-force, is lengthened to increase the lever input of muscle contraction. Third, the hand is usually broad and wide, like a shovel, and extended with stout claws that scoop soil with each stroke.

Lever mechanics (p. 142)

Overview

The appendicular skeleton includes paired fins or limbs and the girdles within the body that support them. The hip or pelvic girdle is exclusively endochondral; the shoulder or pectoral girdle is composed of dermal and endochondral contributions. Such a dual design suggests a dual evolutionary origin, endochondral elements arising from basal fin supports, and dermal elements arising from the encasing bones in the integument. Paired fins appeared early, bringing active fishes maneuverability and stability in a three-dimensional aquatic environment. They arose perhaps from parts of the gill arches, or more likely out of ventrolateral fin-folds of early agnathan vertebrates. Paired fins, only in the pectoral region, occurred in some ostracoderms. Early placoderms, acanthodians, and chondrichthyans were all equipped with paired fins and girdles, pectoral and pelvic. Forelimbs and hindlimbs are built on a common plan, proximal to distal—stylopodium, zeugopodium, autopodium. They arose from the fleshy fins of rhipidistian ancestors, which likely used their fins as points of pivot in water. The limbs of the earliest tetrapods carried

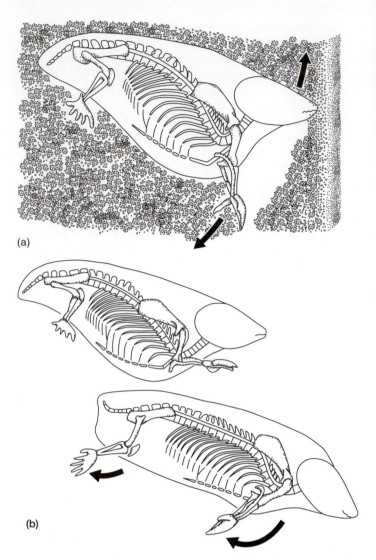

(a)

(b)

FIGURE 9.57 Burrowing by the Namib Golden mole. (a) Buttressing phase. Initially, the mole opens a space in front of it by lifting its head and depressing its chest to compact the sand. (b) Digging-propulsion phase. Hindlimbs and especially strong forelimbs stroke backward (arrow) to propel the body of the mole forward into the space produced. Several repeated cycles occur before another buttressing phase of the mole (actually not a true mole but a member of the Afrotheria).

Based on the research of J. P. Gasc, F. K. Jouffroy, S. Renous, and F. von Blottnitz.

From Gasc, J. P., Jouffroy, F. K., Renous, S. and Blottnitz, F. V. (1986). Morphofunctional study of the digging system of the Namib Desert Golden Mole (Eremitalpa granti namibensis): cineflurographical and anatomical analysis. J. Zool., Lond. 208, 9–35. Reprinted with permission from John Wiley & Sons.

multiple digits, up to seven or eight per chiridium, until reducing to the pentadactylous pattern in later tetrapods. The lateral undulations of the swimming fish were carried into early terrestrial locomotion, where limbs established points of support around which the body undulated across land. Loss of shoulder girdle attachment to the skull allowed increased cranial mobility, and was accompanied in early tetrapods by loss of the connecting dermal bones—posttemporal (skull) and dorsal girdle dermal bones.

(a) Pangolin forelimb

(b) Mole forelimb

FIGURE 9.58 **Skeletal adaptations for digging.** (a) The forelimb of a digging mammal, the pangolin, is short and robust, giving it a power advantage to move earth. (b) The mole is similarly designed, with powerful forelimbs and a broad shovel-like manus.

After Hildebrand.

Throughout its evolution, the pelvic girdle is composed of three processes (early) or actually three separate contributing bones (later)—ilium, ischium, and pubis. Evolution of the pectoral girdle is more complex; dermal elements tend to be lost, especially in derived teleosts and in tetrapods. The endochondral components of fishes, scapulocoracoid, come to predominate in tetrapods, where they form from two centers of ossification, the scapula and the procoracoid. In early amniotes, a new endochondral element joins the shoulder girdle, the coracoid (= posterior coracoid). Both "coracoids" persist in subsequent tetrapods, although the *procoracoid* is prominent as a ventral shoulder element in amphibians, reptiles, and birds; and, of the two, only the *coracoid* survives in therian mammals.

Early tetrapod locomotion on land was accomplished within a sprawled posture, where the body laterally undulated about alternating points of pivot established by the feet. As terrestrial locomotion became more important and specialized, morphological changes resulted in altered posture, with the limbs more under the body, thereby increasing the ease and efficiency of limb oscillations. Specialized cursorial locomotion was especially important in archosaurs and later in an independent trend in synapsids, starting with therapsids. Cursorial locomotion is served through adaptations that increase stride length (distal limb lengthening, change in foot posture) and in stride rate (limb lightening through changes in muscle mass, and loss of some digits). In early mammals, the vertebral column changes from lateral flexions to dorsoventral flexions, thereby adding to stride length. Differences in morphological design accompany the particular way in which cursorial locomotion is used, for quick acceleration or for long-distance cruising, and with the problems of scaling to size.

Aerial locomotion in birds illustrates another modification of morphological features to a specialized biological role—flight. Forelimbs typically are specialized, as airfoils producing lift to resist gravity and as thrust surfaces developing forward velocity. Just as there are different types of cursorial locomotion, there are different types of flight. Some birds hover (e.g., hummingbirds), with emphasis upon propulsive forces and primary feathers supported on the autopodium (manus). Others have long, narrow wings, with emphasis upon secondary feathers supported on the zeugopodium (forearm). These provide the lift used in soaring that takes advantage of rising thermals or natural updrafts of wind currents. Elliptical wings (e.g., pheasants) provide maneuverability in enclosed habitats. In penguins, the forelimbs have lost their role in flight altogether, and are robust in design to meet the demands of locomotion in water.

The appendicular skeleton is linked closely to the locomotor demands placed upon this system, especially in tetrapods. Consequently, it is a good system to illustrate the matching of form and function to changing environmental and, consequently, changing biomechanical demands. The transition from water to land was accompanied by modifications in girdles and limbs, which became weight-bearing and important in developing the propulsive forces moving the tetrapod over land. Generally within terrestrial locomotion, the appendicular system was brought into the service of specialized biological roles—cursorial, aerial, fossorial, and other modes of locomotion. These specialized modes of locomotion are reflected in the specific adaptations of form and function of the appendicular system.

WEB LINK

Visit the text website at *www.mhhe.com/kardong7e* for additional study aids including selected references, functional laboratories, and web links to chapter topics.

CHAPTER 10

The Muscular System

INTRODUCTION

ORGANIZATION OF MUSCLES
Classification of Muscles
Skeletal Muscle
Cardiac Muscle
Smooth Muscle
Structure of Skeletal Muscles
Tendons
Basis of Muscle Contraction
Resting and Active Muscle
Molecular Mechanisms of Contraction

MUSCLE FUNCTION
Muscle Fibers
Tension-Length Curves for a Single Muscle Fiber
Properties of Muscle Fibers
Muscle Organs and Fibers
Whole Muscle Force Generation
Tension-Length Curves for a Whole Muscle
Graded Force
Cross-Sectional Area
Fiber Orientation
Velocity of Shortening
Distance of Shortening

Bone–Muscle Lever Systems
Sequencing of Muscle Actions
Overview of Muscle Mechanics
Muscle Actions
Muscle Homologies

EMBRYONIC ORIGIN OF MUSCLES
Postcranial Musculature
Appendicular Musculature
Axial Musculature
Cranial Musculature
Jaw and Pharyngeal Musculature
Extrinsic Eye Muscles

COMPARATIVE ANATOMY
Postcranial Musculature
Axial Musculature
Appendicular Musculature
Cranial Musculature
Branchiomeric Musculature
Hypobranchial Musculature

OVERVIEW

Introduction

Muscles make things happen. They supply force for movement and together with the skeletal system are the movers and levers that make an animal act. Just as important, muscles restrain motion. When we stand comfortably or sit reflectively, muscles hold our body in position to keep it from toppling over. Muscles also act upon the viscera—blood vessels, respiratory channels, glands, organs—to affect their activity. For instance, muscles wrapping the tubular digestive tract contract in peristaltic waves that mix and move the food within. Muscles form sphincters, gatekeepers that control the passage of materials out of tubular ducts. Sheets of muscle within the walls of the respiratory tract affect the flow of air to and from the lungs. Muscles lining the walls of blood vessels affect circulation.

Secondarily, muscles play a role in heat production. As any exercising athlete knows, heat is a by-product of muscle contraction. Ordinarily, the human body generates enough heat, but if its core temperature should drop in cold weather, large muscles throughout the body strongly contract to produce shivering. Shivering muscles do no extra work but they give off extra heat, and the core temperature comes up to normal. In some species of fishes, the extrinsic eye muscles that rotate the eyeball take on the additional specialized function of producing heat. These enlarged muscles include biochemical pathways that generate heat. From these muscles, this heat is carried by blood vessels directly to the brain to warm it.

Two by-products of muscle contraction that usually go unnoticed are noise and very low voltage of electricity. However, many sharks and some other predaceous fishes have sensory receptors that detect such stray noise and electric signals at close range. Even when prey are hidden or buried, their muscles contract to pump water across their gills during regular breathing. The electric noise from these contracting muscles can betray their position to predators. In some species of fishes, these stray electric by-products have become a major function of specialized muscles. Blocks of these specialized muscles produce high levels of voltage, not force. Such blocks of muscle are **electric organs** and occur to varying degrees in over 500 species of fishes (figure 10.1). Electric organs have arisen in several species of

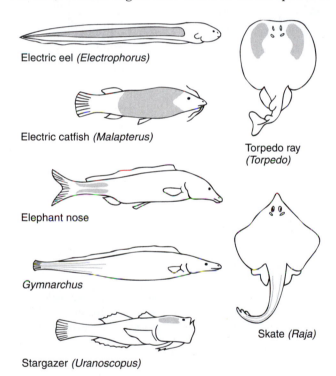

Electric eel (Electrophorus)

Electric catfish (Malapterus)

Torpedo ray (Torpedo)

Elephant nose

Gymnarchus

Skate (Raja)

Stargazer (Uranoscopus)

FIGURE 10.1   Electric organs. The electric organs are specialized blocks of muscle derived, for example, from branchial muscles in the torpedo ray and from axial muscles in the skate. In the torpedo, skate, electric eel, and electric catfish, these electric organs can produce a jolt of voltage sufficient to stun prey or discourage a predator's attack. In other fishes, the electric organs produce a weak electric field around the fish's body, allowing it to detect any object that disrupts the field. In this way, fishes with electric fields can navigate and find food in dark or silty waters. Each fish illustrated in the figure belongs to a different family. The torpedo and skate are elasmobranchs, the others are bony fishes. Thus, electric organs have arisen independently several times within different groups.

After Novick.

chondrichthyans as well as in teleost fishes belonging to different families. Such an independent appearance of electric organs represents an example of convergent evolution.

Electroreceptors (p. 709)

Electric organs generate bursts of energy to paralyze prey. Other fishes, such as the torpedo fish, employ jolts of voltage to protect themselves from predators. Still others use the electric organs to generate an electric field around their bodies. As they move through murky or dark waters, objects in the environment that come close disturb this surrounding electrical field and alert the fish to objects in its path. Thus, the specialized muscles of electric organs play a biological role in food capture, defense, and navigation.

However, in most vertebrates, muscles produce forces that control motion. Motion can drive the organism through its environment or control the actions of its internal body processes.

Organization of Muscles

Classification of Muscles

Because muscles have many functions and many scientists from diverse fields study them, it is not surprising to find different criteria used to classify muscles. The criterion picked depends on which property of muscles is of personal interest. The following criteria are most commonly used as the basis for distinguishing muscles:

1. Muscles are classified according to their color. There are **red** and **white** muscles. This classification has fallen out of favor because such a color distinction alone underestimates the complexity of muscles.

2. Muscles are classified according to their location. **Somatic muscles** move bones (or cartilages), and **visceral muscles** control the activity of organs, vessels, and ducts.

3. Muscles are classified by the way in which they are controlled by the nervous system. **Voluntary muscles** are under immediate conscious control, but **involuntary muscles** are not.

4. Muscles are classified according to their embryonic origin. This is discussed more fully later in this chapter.

5. Muscles are classified by their general microscopic appearance. There are **skeletal, cardiac,** and **smooth muscles.** Let us look at this general microscopic appearance next.

All muscle cells come with standard cellular equipment, namely nuclei, mitochondria, and the like, but specialized terms have grown up for familiar cell organelles. The **sarcolemma** is the term given to the cell membrane;

FIGURE 10.2 **Skeletal muscle.** Each muscle cell is internally composed of myofibrils, each myofibril is a chain of sarcomeres, and each sarcomere is composed at the molecular level of myofilaments: specifically, overlapping myosin (thick) and actin (thin) myofilaments. The underlying molecular arrangement of these filaments produces the pattern of striations within a myofibril. Because bundles of myofibrils are aligned in register within the muscle cell, they produce a visible striated pattern superficially on the muscle cell. Individual muscle cells are wrapped in connective tissue (endomysium), groups of cells are bundled by more wrappings (perimysium), and the entire muscle is covered by an outer sheet of fibrous connective tissue (epimysium). These layers of connective tissue extend beyond the ends of the muscle cells to form tendons that connect muscles to bones.

After Krstić.

sarcoplasmic reticulum to the elaborated smooth endoplasmic reticulum. Other specialized terminology will be defined as we meet it within the chapter.

Skeletal Muscle

Viewed with a microscope, skeletal muscle appears to have cross bands or striations that result from its underlying structure. Skeletal muscle is also under voluntary control and is usually associated with the bones and cartilages. Each skeletal muscle cell is multinucleate, with many nuclei distributed throughout its cytoplasm. Individual cells are usually

less than 5 cm long, but they can be attached end to end to form longer composite fibers. Internally, each skeletal muscle cell is packed with long units called **myofibrils.** Each myofibril is a chain of repeating units, or **sarcomeres.** In turn, each sarcomere itself is composed of underlying **myofilaments** of two kinds: **thick** and **thin filaments.** Under the electron microscope, thick and thin filaments appear as a highly ordered and repeated design within each sarcomere, giving each sarcomere a distinct banding pattern (figure 10.2). This ordered molecular arrangement is too small to be seen directly with the light microscope. However, because myofibrils within a muscle cell tend to be aligned in register with

FIGURE 10.3 **Cardiac muscle.** Found only in the heart, cardiac muscle cells are short and joined one to another by intercalated disks, specialized attachment sites. Cardiac muscle cells form sheets that constitute the thick pumping walls of the heart.

After Krstić.

FIGURE 10.4 **Smooth muscle.** On the right, a single smooth muscle cell is enlarged and isolated from the block of smooth muscle. Although smooth muscle lacks striations, the underlying contractile mechanism is thought to be based on sliding filaments of actin and myosin.

After Krstić.

one another, the overall effect produces a "striated" pattern on the muscle fiber that is visible even under a light microscope.

Each muscle cell is individually innervated by one branch of a single nerve cell. The terminus of the innervating nerve expands into a **motor end plate,** its point of contact with the sarcolemma of the muscle cell. The wave of electrical excitation arriving from the nerve cell spreads in turn along the sarcolemma, and is carried into the cell's interior by inward invaginations of the sarcolemma, the **transverse tubules.**

Cardiac Muscle

Cardiac muscle occurs only in the heart. Like skeletal muscle, it is characterized by a banding pattern. Unlike skeletal muscle, however, individual cardiac muscle cells are short, mononucleate, often branched, and joined to each other by distinct **intercalated disks** into sheets (figure 10.3). Cardiac muscle cells are involuntary. Waves of contraction that conduct electric impulses spread through cells and across the intercalated disks. These waves of contraction can be initiated by nerves or contraction arising intrinsically within the muscle tissue itself. Cardiac muscle tissue kept healthy and active outside the body can contract spontaneously and rhythmically without stimulation from the outside.

Smooth Muscle

Viewed through a light microscope, smooth muscle lacks striations, so it has been labeled as "smooth." Smooth muscle is almost entirely concerned with visceral functions—digestive tract, blood vessels, lungs—so it is a type of visceral muscle as well. Activity of smooth muscle is outside voluntary control. Typically, contractions are slow and sustained compared with the rapid contractions of skeletal muscle. Consequently, smooth muscle is suited for sphincters, where fatigue might mean untimely loss of control.

Each smooth muscle cell is mononucleate, short, and fusiform in shape (figure 10.4). Smooth muscle cells are joined to each other at junctions to form sheets. These sheets are wrapped around the organs on which they exert mechanical control. Smooth muscle cells in sheets are electrically coupled with each other. Nervous innervation to the surface usually spreads throughout the sheet. Hormones may also directly excite or inhibit contraction. The molecular mechanism of contraction is not as well understood as it is in striated muscle, but it is generally assumed to be based on a sliding filament mechanism.

In our discussion of muscles in this chapter, skeletal muscles receive our center of attention. Skeletal muscles supply the force that moves the skeleton. Cardiac and smooth muscles are discussed with the viscera in later chapters.

Heart (p. 473); digestive system (p. 520)

Structure of Skeletal Muscles

The term *muscle* has at least two meanings. Sometimes muscle refers to muscle tissue (muscle cells and their endomysium); other times muscle refers to the whole organ (muscle cells plus associated connective tissue, nerves, blood supply). The meaning intended must sometimes be decided from the context in which the term *muscle* is used. We use the specific term **muscle cell** to denote the active contractile component of a **muscle organ.**

Often, in place of the term *muscle cell, muscle fiber* is used. To the naked eye or under low microscopic magnification, a muscle organ teased apart looks like frayed rope (figure 10.2). This inspired the term *muscle fiber* for these tiny, frayed strands, which in fact are the long individual muscle cells. Because these are actually the striated muscle cells, the choice of the term *fiber* is unfortunate. Logically, the term *muscle cell* should be used for them, but this is not the customary practice. Because this usage is firmly established, we follow the convention of anatomists and physiologists and use the term *muscle fiber* to refer to a whole muscle cell, skeletal or smooth, but not a cardiac muscle cell.

The fleshy part of a muscle organ is its **belly (gaster),** and its ends that join the skeleton or adjacent organs form **attachments.** The biceps brachii muscle of your arm is composed, like all skeletal muscles, of packages of muscle cells. Each muscle cell is lightly wrapped first by a layer of connective tissue, the **endomysium.** Groups of muscle cells are wrapped in a **perimysium.** The entire muscle organ is surrounded by an outer coat of connective tissue, the **epimysium.** A **fascicle** refers to a bundle of muscle cells defined by its particular perimysium.

Tendons

The muscle organ is not actually attached to bones by the contractile muscle fibers that compose it. Instead, the various wrappings of connective tissue extend beyond the ends of the muscle fibers to connect with the periosteum of the bone. These connective components of the muscle organ that establish a cordlike attachment to bone are called **tendons.** Tendons drawn out into thin, flat sheets of tough connective tissue are **aponeuroses** (sing., *aponeurosis*). Sheets of fibrous connective tissues that wrap and bind parts of the body together are considered **fascia.**

Tendons serve various functions. Muscle mass may be located in one convenient location, yet muscle force can be transmitted to a distant point via tendons. For example, the limb muscles of cursorial animals are usually bunched close to the body, but through long tendons, their force is applied at the ends of the legs (figure 10.5). Tendons also permit delicacy of control by distributing forces to digits for precise movements. The long tendons that stretch from the forearm muscles to the tips of the fingers in raccoon or primate hands are examples.

Tendons are metabolically economical and vascular supply is modest. They require little maintenance and consume little energy compared with muscle fibers. Tendons allow the metabolically expensive muscle fibers to be just long enough to produce the required amount of shortening or force. The rest of the length of the muscle between its two sites of attachment is provided by tendons.

FIGURE 10.5 **Limb tendons of a giraffe.** Tendons distribute the forces of muscle contractions to sites distant from the muscle itself. The limb muscles of a giraffe are located close to the body, but tendons of these muscles extend outward along the leg bones and deliver their forces at the giraffe's hooves.

Basis of Muscle Contraction

Resting and Active Muscle

A muscle receiving no nervous stimulation is relaxed, or in a **resting state.** A muscle organ in a resting state is soft, and collagenous fibers surrounding the muscle maintain its shape during this phase. The muscle generates no force, and if tensile force is applied to it, it stretches. Resistance to the applied tensile force arises from the collagenous fibers. When nerves stimulate a muscle to its threshold level, contraction results and generates **tensile force,** constituting the **active state** of a muscle. The bone to which the muscle is attached and the mass that must be moved represent an external resistance called the **load.** Whether or not a muscle actually shortens upon contraction depends on the relative balance between the tensile force of contraction and the load to be moved.

Molecular Mechanisms of Contraction

Although muscle contractions are active and produce force, they cannot lengthen to push apart their sites of attachment. The underlying chemistry of muscle contraction is built on sliding filaments of muscle proteins that slip past one another to shorten the muscle. In skeletal muscle, where the contractile mechanism is best understood, contraction involves chemical cross-bridges that form and reform between thick and thin filaments to ratchet or slide these filaments past one another. We do not need to understand the biochemistry involved, but we must realize that the effect of their sliding is to shorten the sarcomere of which they are part. The terminus of the innervating neuron, the motor end plate, delivers the arriving electrical wave of depolarization to the sarcolemma (cell membrane). The sarcolemma spreads the propagating stimulus to all parts of the muscle fiber, and inward at regular intervals via the transverse tubules that in turn come into association with the sarcoplasmic reticulum. Within the muscle fiber, this electric wave of depolarization stimulates local chemical events, resulting in the sliding of molecular filaments. Because contraction occurs simultaneously throughout all sarcomeres within a muscle cell, the overall result is for the chains of sarcomeres to shorten together, which shortens the muscle fiber that in turn generates a tensile force.

Muscle Function

Muscle Fibers

Some of the major contractile characteristics of a muscle fiber include how rapidly it reaches maximum tension, and how long it can sustain this tension. A variety of properties interact to determine the amount of tension generated by the muscle fiber. One of these is the underlying molecular mechanism of shortening itself, namely, the sliding of thick and thin filaments. The consequences of this underlying contractile mechanism emerge in what are called tension-length curves.

Tension-Length Curves for a Single Muscle Fiber

Tension produced by a particular muscle fiber is not constant, but depends on the muscle's fixed length when it is stimulated. We can hold the two ends of a muscle fiber at a fixed length and then stimulate it and record the tension produced. If this is done for the same muscle fiber fixed at different lengths, different tensions are produced at different lengths. Tensions and lengths can be plotted in a tension-length curve that peaks at intermediate lengths but drops at both ends (figure 10.6a–c). This curve arises from the limitations of cross-bridging between the underlying muscle myofilaments.

When the muscle fiber is fixed in a lengthened position, filaments overlap very little, few cross-bridges form, and tension is low (figure 10.6b). When the muscle is fixed in the shortest positions, filaments overlap, interfering with cross-bridge formation, and tension is again low (figure 10.6a). Only at intermediate lengths is the number of cross-bridges maximized and tension at its peak (figure 10.6c).

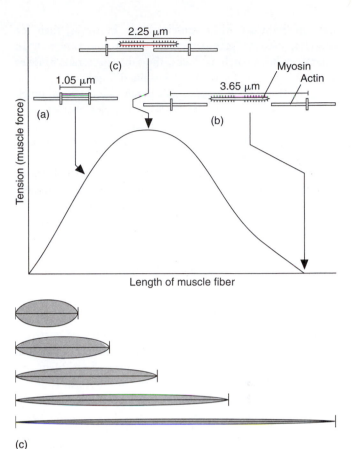

FIGURE 10.6 **Tension-length curves of a muscle fiber.** If a muscle fiber is fixed at set lengths and then stimulated, the force it produces will vary with the length. Somewhere between extreme lengths, its force peaks. (a) When muscle length is short, overlap of thick and thin filaments reduces total force. (b) When muscle length is stretched, filaments establish fewer cross-bridges and, thus, less force is generated. (c) The optimum occurs at intermediate lengths because the maximum number of cross-bridges is formed to achieve the maximum force. Below the graph, a muscle fiber is shown at five different fixed lengths corresponding to the tension shown above on the curve.

Properties of Muscle Fibers

Color Even omnivores like ourselves might notice occasionally during a festive feeding frenzy that meat from the same animal is colored differently. Turkey, for example, contains light and dark meat. In fishes, you may notice that most of the muscle is white, but occasionally in some species a small lateral strip of red muscle is present. The two types of muscle, which early experimenters conveniently termed red and white muscle, also have different physiological properties.

Muscles made up of red fibers tend to be highly vascularized and rich in **myoglobin,** a dark macromolecule that stores oxygen and looks red. Red muscle is resistant to fatigue. Muscles made up of white fibers are less vascularized and low in myoglobin, but they contract rapidly. Game birds, such as turkeys, fly in quick, rapid bursts. They do not migrate long, sustained distances, and their pectoralis or "breast" muscles used in flight are white muscle fibers. However, their leg muscles, used to scamper along the ground, are red. In migratory birds, such as ducks, the same pectoralis muscles are dark, capable of supporting sustained flight.

In fishes such as pike and perch, which make quick darts to catch prey, the lateral body muscles characteristically are white. In migratory fishes and those that swim in swift streams against a sustained current, the same lateral body muscles tend to be red.

If not pushed too far, this match of muscle color with contraction speed, endurance, and physiology helps us understand the basis of animal performance and the muscle types that serve it. However, muscle color alone does not always reveal underlying subtle differences in fiber physiology. For example, another important distinguishing feature of a muscle fiber is its capacity to establish sustained generated force.

Tonic and Twitch Fibers On the basis of a fiber's ability to establish and sustain generated force, muscles can be categorized as tonic or twitch fibers. **Tonic fibers** are relatively slow contracting and produce low force, but they can sustain contraction for prolonged periods of time. Such fibers are involved in postural support, so they compose much of the axial and appendicular musculature. Tonic fibers are common in amphibians and reptiles, less so in fishes and birds, and rare in mammals, where they are present in the extrinsic eye and middle ear muscles. **Twitch (phasic) fibers**, by contrast, generally produce fast contraction, so they often

make up muscles used for rapid movements. Twitch fibers are found in somatic muscles of all classes of vertebrates.

Twitch fibers have been most extensively studied, and, generally, are of two kinds: **slow twitch** and **fast twitch** fibers. As their names suggest, slow twitch fibers take longer to reach maximum force than fast twitch fibers, more than twice as long in some cases. But *fast* and *slow* are relative terms and are species specific. For example, fast-contracting muscles in hummingbirds have a twitch time of 8 ms. In guinea pigs, twitch time can be around 21 ms. Within mammals, fast twitch and slow twitch contractions in rats average around 13 and 38 ms, respectively. In cats, twitch times are 40 and 90 ms, depending on the muscles involved. Differences in contraction speeds seem related to differences in myosin types in the fibers, to differences in the breakdown of ATP (adenosine triphosphate), and to differences in nervous innervation. It has been shown, at least in fishes, that these two fiber types, slow and fast twitch, also differ embryologically arising from different populations of cells in the somites and in different cellular environments.

The staining reaction of a fiber can divulge its underlying biochemical character, which suggests its possible physiological function. These techniques reveal that some types of muscle fibers carry large stores of glycogen; other types contain enzymes that support either short bursts of activity or sustained activity. Some fiber types seem intermediate (figure 10.7). Muscles that appear red or white to the naked eye may have one or more of these fiber types present, depending on the species of vertebrate. This mixing of fiber types is

(a)

(b)

FIGURE 10.7 **Histochemical profiles of muscle fibers in cross section.** Fresh muscle is removed, serially sectioned, and stained. (a, b) Adjacent sections of the same muscle treated with different stains. (a) Slow twitch (S) fibers are light; fast twitch fibers are dark. (b) Slow twitch fibers are known, leaving only fast twitch to be identified. Fast twitch glycolytic (FG) are light, and fast twitch oxidative glycolytic (FOG) are dark. (a) Stained for myosin ATPase, pH 10.4 preincubation. (b) Stained for NADH-D (nicotinamide adenine dinucleotide dehydrogenase).

Modified from and thanks to Young, Magnon, and Goslow, 1990.

TABLE 10.1	Muscle Fiber Types and Their Physiological Properties			
Fiber Type	Contraction Time	Force Output	Resistance to Fatigue	Histochemistry
Slow twitch				
S	Slow	Very small	Very high	Slow oxidative (SO)
Fast twitch				
FR	Fast	Small	High	Fast oxidative glycolytic (FOG)
FI	Fast	Medium	Medium	Fast intermediate (FI)*
FF	Fast	Large	Low	Fast glycolytic (FG)

Note: Abbreviations of fibers are: slow (S); fast twitch, resistant (FR); fast twitch, intermediate (FI); fast twitch, fatigable (FF).
*May be an artificial category.

why color alone is usually not sufficient to characterize the underlying physiology of the muscle. Many scientists now turn to histochemical characteristics or to molecular means to help identify fiber types.

In addition to contraction speed, fast and slow twitch fibers differ in their resistance to fatigue during sustained exercise. Slow twitch fibers (S) tend to be resistant to fatigue. For example, some slow twitch fibers studied in cats maintained constant tension over 60 minutes of sustained activity. Fast twitch fibers are more diverse in their contractile properties. In cat limb muscles, up to three fiber types are recognized (table 10.1). At one extreme are fast twitch fibers that produce a large force but fatigue quickly, their tension falling to zero after less than one minute of continuous stimulation. Such fibers are fast twitch, fatigable (FF). Other fast twitch fibers are fatigue-resistant (FR) and typically produce smaller force but can sustain prolonged contractions. Between these two in contractile properties are fast twitch, intermediate (FI) fibers, although this type has proved difficult to characterize confidently, and may not be a useful category.

There are many different types, or isoforms, of myosin, and their contraction characteristics are not always well predicted by such histochemical profiles. Consequently, other methods, in particular molecular techniques, are now often used to examine the physiology and contractile machinery of muscles.

The muscle organ is often a mixture of fiber types with different resistances to fatigue. It has been hypothesized that within a muscle the motor neurons establish a priority in recruiting the fiber types during prolonged exercise. As activity begins, slow twitch fibers are called on first to generate tension. As activity continues, fast twitch fibers resistant to fatigue are recruited. If activity persists, then finally fast twitch fibers that can produce large output but fatigue relatively quickly are recruited. Just when each fiber type is employed apparently depends on the strenuousness of the activity, the species of animal, and the prior conditioning of the muscle.

To summarize, contractile characteristics of muscle fibers depend on molecular properties of thick and thin filaments, fiber types, proportions of fibers within a muscle, and the pattern of fiber recruitment during muscle activity.

No single characteristic alone is diagnostic of the fiber's contractile speed, level of tension produced, or sustained force. Not all fiber types occur in any one vertebrate class. Muscle has the intrinsic ability to respond to different types of mechanical activity, but within limits. For example, strength training typically results in an increase in muscle mass and strength, but endurance training results in increases in efficient oxygen uptake and fatigue resistance, at the expense of muscle mass and strength. Muscle overload—chronic exercise—leads to hypertrophy of muscle; muscle immobilization leads to atrophy. Differences in different loads and endurance can switch on different genes that preside over the manufacture of different myosin isoforms appropriate to the physical demands. At least in humans, increased aerobic-endurance training leads to increase of FF (fast glycolytic) fibers; strength training leads to increase in both S (slow twitch) and FR (fast oxidative glycolytic) fibers. However, no schedule of training seems able to convert slow twitch (S) to fast twitch (F) fibers of any type, or vice versa, in significant numbers.

As we see next, the muscle organ itself might perform different functions in different species. Furthermore, differences in performance can result in part from changes in a muscle organ's overall design and internal architecture.

Muscle Organs and Fibers

Some muscle organs produce strong forces, and others move their loads quickly. Some move loads long distances; others displace loads only short distances. Some muscles produce graded movements when moving heavy loads, but when given a light load, they perform smoothly as well. These differences in muscle performance do not entirely result from modulation of molecular cross-bridging. How, then, is this variety of properties produced out of a common molecular mechanism?

Whole Muscle Force Generation

The total force output of a muscle organ is ultimately based on two functional components. The sliding of molecular filaments is responsible for the **active component** that

Origin   Contractile

Insertion

Elastic

Load

Elastic

Muscle organ

FIGURE 10.8 **Diagrammatic representation of functional components of a muscle organ.** Sliding filaments represent the active component of a muscle. Springs represent the elastic component and may lie next to (parallel) or follow (series) the contractile component. Total output force of the muscle arises from both components.

contributes to total force. The elastic component can contribute to overall muscle force as well (figure 10.8). As opposing muscle groups or gravity stretch a muscle, some of this energy is stored within the muscle organ. The elastic components act like a stretched rubber band. When the muscle shortens, this elastically stored energy adds to the active component of contraction to help shorten the muscle. The elastic energy is stored in the connective tissue in muscles and in muscle tendons. This elastic feature of muscles seems prevalent especially in muscles involved in repetitive events such as limb oscillation during running or trunk flexure during sustained swimming. For example, the long tendons in the lower legs of camels act like elastic "springs," storing energy when the limb touches the ground to bear the animal's weight as it walks or runs. When the limb pushes off, over 90% of this stored energy is recovered and contributes to forward momentum. It is suspected, but not yet proved, that the extensive sheets of tendons and aponeuroses in the dolphin tail act similarly to store energy when muscles are stretched during swimming. When the tail stroke is reversed, this energy is returned.

The muscle organ includes the active contractile machinery composed of the sliding of molecular units and the elastic component residing in connective tissue. Active contraction consumes chemical energy delivered by ATP, the cell's source of chemical energy. The elastic component depends on mechanical energy resulting from gravity or motion of body parts that load the muscle like a spring, storing energy until it is released. Thus, the total force output generated by a muscle organ arises from the combined action of active contraction and elastic recoil.

In some situations, as when running over level ground at constant speed, leg muscles need to shorten only very little to keep the animal moving. Muscles and tendons absorb and release mechanical energy to maintain cyclical movements of the limbs and support the body mass. Just as a pogo stick stores gravitational energy when contacting the ground and then releases it to aid in liftoff, a muscle organ similarly

may act as a spring to store and release gravitational energy. During the stance phase of running, leg muscles shorten very little and store energy, then elastically recoil to return this stored energy when the foot presses off the ground. Mechanically, it is similar to bouncing a basketball—little new force is needed to keep it going. However, when running up an incline, or slowing or speeding up, leg muscles must work harder. Under these conditions, more muscle is recruited and muscle shortening may be great to produce needed force.

Tension-Length Curves for a Whole Muscle

The tension-length curve for a whole muscle (figure 10.9) has different properties from the tension-length curve for one isolated muscle fiber (figure 10.6). This is because the muscle organ includes packed sheets of connective tissue that add an elastic component to the active component of force. Because a muscle's active and elastic components contribute to the force generated, its tension-length curve is a combination of both. The **passive tension** represents the force required to stretch the relaxed muscle to greater lengths and results from the elastic constituents of the muscle, mainly from its collagenous fibers. The curve for **total tension** is measured at different lengths when the muscle is contracting. The sum of active and elastic components constitutes total tension. The **active tension,** the contribution made by the active component alone, is derived from the difference between total tension and passive tension.

The shape of this tension-length curve becomes important when we think about the design of bone and muscle systems. Muscle fibers, being the component that produces active tension, have one length at which tension in them is greatest. Thus, any given muscle has a length at which it produces its maximum active tension (figure 10.9a), implying that a muscle that must be shortened over a relatively long distance cannot produce maximum force throughout its entire range of motion. Therefore, if a skeletal part is moved a long distance, several muscles are occasionally found acting to move the bone in the same direction (figure 10.9b). Each muscle reaches its peak tension at a slightly different point during motion, passing the responsibility of generating the maximum force from one muscle to the next as the bone rotates. This may explain why several muscles with duplicate actions might be part of a design where one large muscle might seem sufficient.

On the other hand, some joint positions allow more joint force to be produced at a particular instant, likely because the tension-length curves of different muscles are roughly lined up. This seems the case with human jaw closing, in which maximum bite force is produced at a half-open position of the jaws, and the major jaw-closing muscles concurrently reach peak force at that point.

Graded Force

The wave of electrical depolarization arriving from the nerve and spreading to the muscle fiber must reach or exceed a threshold level, otherwise no muscle twitch occurs.

FIGURE 10.9 **Tension-length curve of a muscle organ.** (a) The curve of passive tension represents the force required to stretch a relaxed muscle. The total tension curve is the force measured from an active muscle at various lengths. The difference between the resting and total tension curves is the active tension curve, representing the force only of the active contractile components of the muscle. Beneath the horizontal axis, note that five lengths of the muscle are depicted. (b) When this hawk strikes its prey, its limbs reach out to make contact. As they do, the distance between origin and insertion of leg muscles changes; thus, the tension produced by these muscles changes as well. Because each of the two muscles attains its peak force at different lengths, one of the two muscles is at or close to its peak tension through the range of leg extension.

Based on the research of G. E. Goslow.

Consequently, nervous excitation is **all-or-none.** Either it reaches this threshold or it does not; essentially the muscle fiber contracts maximally or it does not contract at all. However, the tension generated by the muscle fiber is *not* graded. It does not increase proportionally with the magnitude of the nervous stimulus. A moment's reflection on this characteristic of muscle fiber contraction will yield a puzzle. If activation of a muscle fiber is all-or-none, then how is graded motion produced?

For example, the same muscle organ can produce a graded motion, delivering a large force to a bone when the animal moves a heavy load or a small force when it moves a small load. For example, when you lift a heavy but not impossible weight in your hand, the biceps brachii is equal to the chore and produces the necessary large force. But when you lift a light pencil with your forearm, the same biceps brachii produces a lesser force to match. One way in which graded force is generated is by **rate modulation.** Up to a point, force increases as the **rate,** not the intensity, of arriving nerve impulses increases. Increasing force with increasing impulse rate constitutes rate modulation. Eventually, this force peaks

and does not increase further even if the impulse rate continues to increase, a condition known as tetanus. Within the range of graded response to rate modulation, motor nerves to a muscle fiber can develop a graded force output.

A second way this graded force can be matched to loads is accomplished by selective contraction of a few, many, or all muscle fibers within a muscle organ. How is this done? One motor neuron exclusively supplies a group of the muscle fibers. Another neuron does the same but innervates a different set of muscle fibers within the muscle organ. A single motor neuron together with the unique set of muscle fibers it innervates is called a **motor unit.** By recruiting additional motor neurons, the central nervous system can selectively increase the total output force a muscle generates until it matches the load (figure 10.10). Not surprisingly, if delicate movements are required, there are fewer muscle fibers for each innervating neuron. Motor units of laryngeal muscles that control vocalizations or extrinsic eye muscles that move the eyes can contain as few as ten muscle fibers, whereas a motor unit in the large gastrocnemius muscle of the leg can have several thousand muscle cells per motor neuron.

When muscles are exercised, they get bigger, or at least they get bigger if the exercise involves increased load and is continued on a regular basis over a period of time. The late humorist Robert Benchley had a couch he named "the track." As he described it, when friends urged him to exercise, he obliged by telling them as he graciously departed the room that he was on his way to spend some time at the track. They were always surprised to see how rested he seemed upon his return.

Exercising means more than this to most of us. But, strictly speaking, when a muscle contracts, even during slow walking, it is being "exercised." Thus, exercise physiologists prefer the term *chronic overload* to describe elevated levels of sustained muscle activity, or just "training" to recognize the elevated muscle demands and their consequences.

Muscles enlarge in response to training. This enlargement results from several changes within the muscle. Capillaries proliferate, and fibrous connective tissue increases to add to muscle bulk. However, increased muscle bulk results primarily from the enlargement of existing cells. Each cell adds more myofilaments. This results in an increase in the cross-sectional area of fibers, up to a 50% increase following some schedules of exercise training. Until recently, evidence of an accompanying increase in fiber number was less clear. Individual fibers got bigger, but they did not seem to increase in number. If a weight is applied to a quail wing, the stretched latissimus dorsi muscle shows a dramatic gain in mass through increases in both fiber number and size. But in this avian example, the overload is continuous, not intermittent as in most exercise training. In cats, a small increase in fiber (9% increase) can be induced by exercise. These additional fibers apparently do not arise by the splitting of existing fibers. Instead, new fibers are added from undifferentiated cells within the muscle. Therefore, present studies indicate that large increases in muscle mass induced by training may involve some change in the number of fibers, but predominantly result from an increase in individual fiber size.

When humans, or rats, or cats have been exercised to test the physiological responses of their muscles to training, the muscles responding to chronic overload are those actually involved in the increased activity. For example, if a human is trained on an exercise bicycle, muscles of both legs get larger over the extended period of training. If only one leg is strapped to the exercise bicycle, then only muscles of that leg get larger. The physiological response is localized.

How fiber types adapt to overload is complicated. Certainly muscle fibers change during growth. Nervous innervation seems to play a part in determining fiber type. Experimentally, if nerves stimulating slow and fast twitch fibers are switched, to some extent the muscle fiber takes on the contractile properties of its new nerve (i.e., slow twitch muscles become fast twitch and vice versa).

However, how muscle fibers respond to training is less clear. The response of muscle to training depends somewhat on the nature of the training, namely, on the loads moved and on the length of training. Pumping a bicycle against easy resistance has different effects from pumping a bicycle against high resistance.

Strength and endurance training produce quite different physiological adaptations in muscles. In general, strength training increases muscle mass and strength; endurance training increases oxygen uptake and metabolic changes, enhancing endurance. But oddly, simultaneous strength and endurance training sometimes produce interference of one with the other. Apparently, at the molecular level, different training regimes promote antagonistic genetic signaling mechanisms, which in turn interfere with the muscle's response to strength training, in particular.

World-class athletes have been examined in order to study the response of muscles to training. Small samples of muscle can be biopsied, removed, and examined. When this is done in the quadriceps muscles of elite marathon runners, the muscle shows a large proportion of slow twitch fibers. When the same muscle is biopsied from elite sprinters, the quadriceps show a large proportion of fast twitch fibers. This suggests that the physiological characteristics of the leg muscles are matched to the athletic event. Endurance runners have more slow twitch but fatigue-resistant fibers. Sprinters have more fast twitch fibers.

Are these muscle fiber proportions the result of training or inheritance? To answer this, humans were trained on an endurance program consisting of intense bicycling twice a day. The proportion of slow twitch fibers in the vastus lateralis was biopsied before and after this six-month endurance training. The oxidative potential of the muscle, its ability to utilize oxygen in synthesizing ATP, nearly doubled. This increased oxidative capacity occurred not just in slow twitch or fast twitch fibers but in both kinds. Further, there was an increase in the concentration of capillaries that served the muscle. Finally, the muscles got bigger with training, as might be expected, but the proportions of slow to fast twitch fibers did not change significantly. Training improved performance but did not change the basic fiber types composing the muscle. Training alone does not seem to be enough to make a world-class marathon runner out of a world-class sprinter.

This has raised the prospect of future athletes, be they animal athletes such as racehorses or greyhounds or human Olympic athletes, being discovered early, by biopsy, and set on a course to greatness. It is not, however, so simple, even if it were desirable. Peak performance of an athlete is more than just muscle. The extent of muscular vascularization, the ability of the respiratory system to deliver oxygen, the rate of conversion of stored energy to available energy, and so forth also affect the performance of an athlete, to say nothing of "motivation," a major factor as well. Muscle physiology and fiber type certainly set boundaries to performance, but in complex ways. We are not yet able to forecast with certainty the future athletic performance attainable by a human or a racehorse.

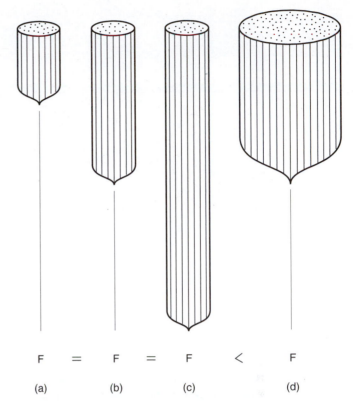

FIGURE 10.10 **Two motor units.** Many motor neurons supply a muscle, but each neuron innervates only a few muscle fibers. Selective recruitment of additional motor units can increase muscle force. Thus, the number of contracting muscle fibers can be increased in graded amounts to equal the force required to accomplish the job.

F = F = F < F

(a)   (b)   (c)   (d)

FIGURE 10.11 **Muscle force is proportional to cross-sectional area.** (a–c) Muscles that differ in lengths but have equal cross-sectional areas therefore produce the same force. (d) A muscle with greater cross-sectional area, and hence with more muscle fibers, produces a greater tension than the other muscles, other things being equal.

Cross-Sectional Area

The maximum force produced by a muscle is proportional to the total cross-sectional area of all its myofibrils. Two terms express this relation between tension and muscle fibers. The cross-sectional area of a *muscle* perpendicular to its longitudinal axis at its thickest part constitutes its **morphological cross section.** A muscle's **physiological cross section** represents the cross-sectional area of all muscle *fibers* perpendicular to their longitudinal axes. In muscles in which all fibers run parallel to one another and to the long axis of the muscle, morphological and physiological cross sections are equal. If fibers are oblique to the long axis of the muscle, the muscle's physiological cross section represents a more accurate index of its ability to generate tension. The physiological cross section is related to the number of muscle fibers present. Reasonably, the more fibers present, the greater the tension and the maximum force produced. Thus, contrary to what you might at first expect, a long muscle and a short muscle, of equal physiological cross sections, generate the *same* forces, not different forces (figure 10.11a–d).

This is somewhat analogous to the properties of a chain. A chain is no stronger than its weakest link. Increasing a chain's length does not make it stronger. To increase strength, more parallel sections of chain are added adjacent to one another. Similarly, a muscle cell might be thought of as a package containing chains of sarcomeres, with its tension limited by the weakest sarcomere. Therefore, increasing length does not increase tension. To increase tension, the number of adjacent chains of sarcomeres must be increased by increasing the number of parallel muscle fibers (figure 10.11d).

Fiber Orientation

Other factors being equal, tension generated by a muscle varies with the orientation of fibers within it. Muscle fibers may be arranged in one of two general ways, each conferring different mechanical properties. A **parallel muscle,** in which all fibers lie along the line of tension generated, has, as the name implies, fibers parallel to one another. A **pinnate muscle** has fibers that lie oblique to the line of force generated, and it is inserted on a common tendon that receives the inclined muscle fibers (figure 10.12a). Each type of muscle has mechanical advantages and disadvantages.

Parallel muscles are best at moving a light load through a long distance. The sternomastoid that turns the head or the long sartorius that adducts the hindleg are examples. Pinnate muscles are best suited for moving a heavy load through a short distance (figure 10.12b). The strong gastrocnemius muscle of the calf is an example. It is inserted on the calcaneus and exerts considerable force to extend the foot and lift the weight of the body, but it can be shortened only through a small distance.

For both parallel and pinnate muscles, generation of tension is based on the contracting mechanism of the sliding filaments. Their mechanical properties arise from differences in fiber arrangement. Pinnate muscle permits the packing of more muscle fibers in the same space. Consider two muscles of equal size and shape but with different fiber arrangements. Within a given volume, pinnate muscles have shorter fibers, but more of them, than parallel muscles have. Being shorter and inclined at an angle to the line of muscle action, pinnate muscles shorten less, so their insertion tendon

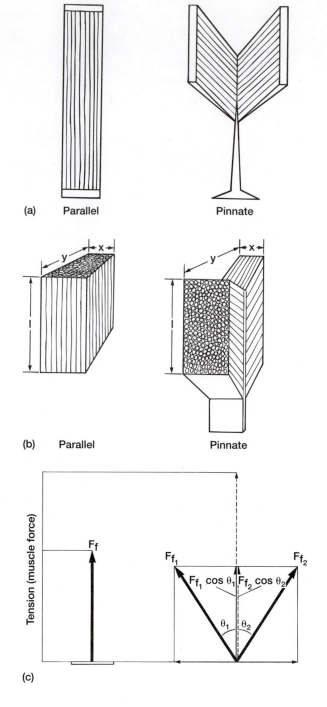

(a) Parallel   Pinnate

(b) Parallel   Pinnate

(c)

FIGURE 10.12 **Parallel and pinnate muscles.**
(a) Muscles with fibers aligned along the line of action are parallel. Those with fibers oblique to the line of action are pinnate.
(b) The pinnate orientation permits the packing of more fibers within the same volume than the parallel arrangement does. The oblique orientation of fibers in pinnate muscles reduces the effective distance through which the insertion can be moved and slightly reduces the force that each fiber can direct along the line of action. However, the greater overall number of fibers compensates and makes pinnate muscles especially suited for moving heavy loads short distances. Both blocks of muscles are of equal three-dimensional size, xyl. The force the parallel muscle can produce is proportional to its cross section, xy. The force of the pinnate muscle is greater, being proportional to its physiological cross section, xl of one side plus xl of the other.
(c) The force a fiber produces along the line of action in a parallel muscle is equal to the force of that fiber. In a pinnate muscle, the useful force of a fiber lies along the line of action of the muscle organ. This useful force is the trigonometric component of the force of the fiber (F_f) times the cosine of the angle the fiber makes with the line of muscle action (θ). Because more fibers can be packed within the same volume of a pinnate muscle, the useful force from two fibers is additive, giving a total force greater than in a similar-sized parallel muscle.

moves a shorter distance. However, because more fibers are packed in the same space, the usable force produced along the line of action is greater (figure 10.12b).

More formally, the physiological cross-sectional area of a pinnate muscle is greater than that of a comparable parallel muscle. In a pinnate muscle, the force produced by individual fibers can be separated into its vector components, one a useful component aligned with the tendon, the other at right angles contributing no useful force (figure 10.12c). The useful force component calculated from the vector trigonometry equals the fiber force (F_f) times the cosine of the angle of pinnation (θ), or F_f cos. So long as this angle of pinnation does

not become too large, most of the force of fiber contraction will yield a large useful component vector along the line of tendon action. This useful force component is a bit less than the fiber's tension, but this is compensated for by having more fibers than a comparable-sized parallel muscle.

In practice, most muscles are compromises between the two specialized extremes, parallel and pinnate. What should be emphasized, however, is that the different properties of parallel and pinnate muscles do not arise from differences in their mechanism of contraction at the molecular level. It is at the tissue level of organization that different properties of overall performance arise.

Velocity of Shortening

Other factors being equal (such as fiber physiology and angle of pinnation), the absolute velocity of shortening is greater in a long muscle than in a short muscle (figure 10.13a, b). Assume that we have two muscles identical in all properties and dimensions except that one is long and the other short. The relative time it takes for each to contract to half its resting length is equal, but the absolute velocity traveled by the point of insertion is greater in the longer muscle. The longer muscle has more sarcomeres in series, and their velocities are additive. Thus, the longer a muscle, the greater is the velocity of its insertion.

Distance of Shortening

Other factors being equal, the absolute distance through which a muscle contracts is greater for a long muscle than a short muscle. This property, like velocity, is a consequence of the additive effect of chains of sarcomeres. As individual sarcomeres shorten, their distance of travel is added to that of adjoining sarcomeres in series. Because there are more sarcomeres in each chain of a long muscle, the additive effect is greater in a long muscle than in a short muscle. As a result, the insertion of a long muscle is displaced a greater distance than the insertion of a short muscle.

Bone–Muscle Lever Systems

The action of a muscle is more than just a property of its underlying physiology or fiber arrangement. Theoretical work indicates that the performance of a muscle may depend on how it is attached to bones of the lever system. For example, a muscle crossing a single joint can be attached close to (proximal) or away from (distal) the point of rotation at the joint. Each site of insertion, proximal or distal, results in different mechanical properties. Inserted distally, the muscle is best suited for strong movements; inserted proximally, it is best suited for fast movements. If you wish to swing a heavy gate or one that is stuck at the latch, you apply force farthest from the hinges to get more leverage. If you wish to open a light gate rapidly, applying force close to the hinges yields best results. These differences arise from differences in simple mechanical advantage, but one is at the expense of the other. Rapid motion (proximal insertion) comes at the expense of strong motion (distal insertion), and vice versa.

The site of insertion also affects the distance through which a moving part swings. For example, the biceps brachii muscle generates a long sweep of the end of the forearm if it is inserted close to the elbow joint. If the biceps is inserted distally and shortened the same length, it provides a much shorter sweep of the forearm (figure 10.14a, b).

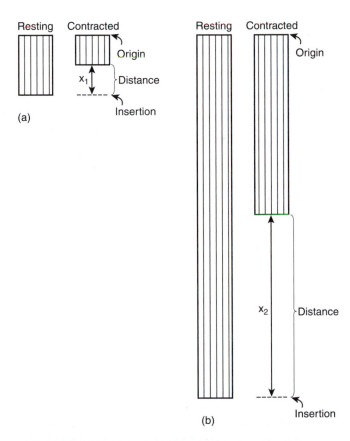

FIGURE 10.13 **The point of insertion in a long muscle has greater velocity and moves a greater distance than in a short muscle.** (a) A short muscle that contracts to half its resting length shortens by the distance x_1. (b) A long muscle contracting to half its length will do so in the same amount of time if it is made up of the same type of sarcomeres. But the contraction speed of chains of sarcomeres is additive, producing a faster speed over a longer distance (x_2). When fast displacement is required, long muscles are often involved.

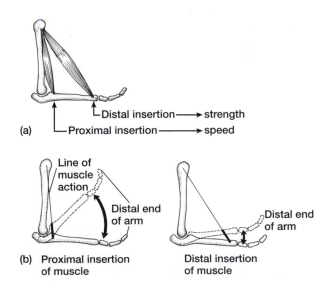

FIGURE 10.14 **Strength versus speed.** A muscle inserted on different points in a lever system produces different mechanical advantages. (a) If inserted near (proximal) the point of rotation, the muscle favors speed. If inserted distal to its point of rotation, it favors strength. (b) Proximal insertion also favors greater excursion of the distal end of the part rotated. The thick, solid bar represents the distance of muscle shortening, which is equal in both.

FIGURE 10.15 **Strength versus speed in the design of forelimbs.** (a) In the armadillo, the teres major is inserted distally on the humerus. (b) In the cat, the teres major is inserted closer to the point of rotation. The resulting change in the lever arm in (L_i) and lever arm out (L_o) changes the mechanical advantage from one favorable to strength (the armadillo) to speed (the cat). Both forelimbs are viewed laterally and drawn to the same overall length. The teres major inserts on the medial side of the humerus, just out of view here. The solid line down its long axis indicates its line of action, and its lever arm is perpendicular to the point of limb rotation (solid black dot within the glenoid joint). In the plantigrade armadillo, the lever arm out acts at the sole of the foot; in the digitigrade cat, the lever arm out acts at the tips of the metacarpals.

After the research of Hildebrand.

(a) Armadillo

(b) Cat

A hypothetical analysis of limbs indicates that trade-offs must inevitably be struck between designs that favor strength and those that favor speed of limb displacement. For example, the teres major, which runs from the scapula to the humerus, is inserted more distally in the forelimb of an armadillo (strong digger) than in the forelimb of a cat (rapid runner; figure 10.15a, b). This proximal and distal relationship can be expressed as the **lever advantage,** the ratio of lever arm *in* to lever arm *out*. The perpendicular distance from the point of bone rotation to the line of muscle action is the lever arm in (L_i). From the point of rotation to the point at which motion is applied is the lever arm out (L_o). For the armadillo, this ratio is about 1:5, and for the cat it is about 1:9. The armadillo has a greater strength advantage in its forearm system than the cat does. In the armadillo, the point of muscle insertion is more distally placed and the leg is relatively shorter to give the teres major a strength advantage. With respect to velocity, however, the ratios tell us that the cat's forearm muscles have the speed advantage. The 1:9 ratio indicates that the toes move nine times faster than the point of muscle attachment. A higher rate of limb oscillation is achieved but at the expense of strength. These hypothesized changes in performance with changes in ratios still await experimental verification in a variety of animals. Such analyses provide provisional insight into the mechanical advantages and disadvantages of alternative limb designs.

Lever arms (p. 142)

Even within the same limb of the same individual, different muscles can be inserted in bone in such a way as to enjoy different lever advantages and make different contributions to strength or speed during limb oscillations. Many cursorial animals have **low** and **high gear muscles.** Low gear muscles, such as the hamstrings, enjoy a mechanical strength advantage to help overcome inertia when accelerating or moving the mass of the limb. High gear muscles enjoy a speed advantage and produce rapid limb oscillation.

In addition to effects on strength or speed, the proximal insertion of a muscle has further consequences on performance. If a muscle is inserted near the point of rotation, it can produce a long excursion of the distal end of a bone while shortening very little itself around the peak force of its tension-length curve. If inserted distally, the muscle would have to be shortened much more to produce the same distal displacement of the bone. Generally, the more a muscle is actually shortened, the more energy it consumes even if the force remains the same. Therefore, proximally inserted muscles use less energy and provide a more economical design for rotating limb segments during locomotion.

Some precautions should be highlighted. First, most of these principles of performance and design are based on theoretical arguments taken from the presumed mechanical consequences of muscular attachment sites and lever arm advantages or disadvantages. It has proved difficult to confirm many of these by direct experimental tests. Second, most of the arguments assume that changes in the lever arm will not become too short for the job, or else the muscle would not be up to the task. Obviously, even a big muscle working with a very small lever arm could not move a part effectively. Third, we have assumed that the muscle and its mechanical advantage are comfortably matched to its external load. For example, we noted that the longer the muscle fibers, the faster is the velocity and the greater is the distance of shortening. However, if a *heavy* load is to be moved, then the reverse may be true. A short but strong muscle with many fibers provides more force to address the large load than a long but weaker muscle with fewer fibers. We are simply stating the obvious.

Human track athletes with below-the-knee amputations face an engineering problem when being fitted with prosthetic replacements. In a way, running is a bouncing gait, like dribbling a basketball. A firmly inflated basketball bounced against the floor stores mechanical energy elastically in the compressed air and stretched skin of the ball as it strikes the floor. This energy is quickly returned so that the rebounding ball reaches its former level, and you need only supply a little extra push with the dribbling hand, compensating for loss of energy to friction. In a running human, when the foot strikes the ground, mechanical energy is absorbed in the elastic components, primarily in the muscles and tendons of the leg; as the foot next pushes off, this elastic energy is returned to the leg and, along with additional forces from contracting muscles, pushes the runner forward (box figure 1a). When the leg is amputated below the knee, mechanically this removes the supportive bones, of course, but this also removes the muscle and tendon system participating in elastic storage and in generating active forces. The prosthetic approach has been to design a flexible, carbon blade that attaches to the lower leg and is specialized for fast forward running. Upon contact, the blade absorbs the mechanical energy and returns it at push-off (box figure 1b). In part because the lower leg muscles are missing, the blade, to compensate, is longer than the normal leg—hence, it generates longer stride length. This has sparked discussion in Olympic circles of whether the bionic prosthetic makes the runner better. As replacements for damaged or amputated parts get better, this controversy is likely to widen.

(a)

(b)

BOX FIGURE 1

Running blades. (a) During a normal running gait, the lower leg stores mechanical energy on footfall and elastically returns it at foot push-off. Active muscle contraction adds to this elastically returned force. (b) The curved blade acts mechanically in a similar way, storing mechanical energy on contact and elastically returning it at push-off.

Both the speed of shortening and the distance of shortening depend not only on the length of muscle fibers but also on the relationship of the force generated and the size of the external load. From cautious generalizations about force velocity and distance of shortening, we can recognize the trade-offs between internal muscle design and size of external load.

Sequencing of Muscle Actions

Muscles, of course, do not act in isolation. Any movement involves various muscles, and each reaches its peak force at different times during movement. For example, the striding gait of humans is composed of two movement phases:

FIGURE 10.16 Sequential action of muscles.
During activity, muscles reach peak activity at slightly different times to distribute and spread the forces producing motion. The human striding gait includes a stance phase (a–c, right leg), during which the foot is in contact with the ground, and a swing phase (d, e, right leg), during which the limb freely swings forward. Groups of leg muscles reach their peak forces at different times during each phase. Most are inactive during the swing phase because gravity pulls the limb forward in pendulum fashion. Greatest activity of hamstring, quadriceps, and pretibial muscles occurs early in the stance phase. The calf group reaches peak activity and presumably peak force just before the end of the stance phase.

Modified from J. V. Basmajian.

the **stance phase,** when the heel strikes the ground until the toe is lifted from it, and the **swing phase,** when the toe is lifted from the ground until the heel strikes it again (figure 10.16a–e). Muscles that move the limb become active at different points during these phases. As the heel strikes the ground, the hamstrings and pretibial muscles reach peak activity; thereafter, the quadriceps increase in activity as the torso is carried forward over the limb. With the heel off the ground, the calf group (gastrocnemius, soleus) increases in activity. During the swing phase, most of these muscles are electrically silent or show low activity as gravity swings the bent leg forward like a pendulum beneath the body. Selective use of muscles greatly reduces muscle contractions and overall energy expenditure.

During normal locomotion, the central nervous system coordinates this elaborate pattern of selective muscle deployment. Part of the reason it is difficult to design artificial human limbs to replace amputations is because of this complexity of muscle sequence and action. Not only must the prosthetic limb produce the necessary forces, but these forces must be generated in the proper order if they are to simulate normal locomotion faithfully.

Recruitment of muscle types is also selective. For example, in elasmobranchs and some primitive teleosts, red fibers are fatigue resistant, whereas white fibers fatigue

more easily but contract faster. Electromyographic studies of sharks show that the block of red muscles along the body produces swimming undulations at slow speeds. As speed increases, the white blocks of axial musculature are recruited (figure 10.17a). Carp and a few other teleosts have "pink" muscle, a third type of fiber that is intermediate in physiological character between red and white muscle. As swimming speed increases in carp and some teleosts, there is an orderly recruitment of blocks of axial musculature, first red, and then pink, and finally white muscles at high rates of speed (figure 10.17b).

Overview of Muscle Mechanics

All muscle contraction is fundamentally based on the mechanism of sliding filaments, the formation of cross-bridges between the molecules of primarily actin and myosin. As the cross-bridges form, they slide past one another to shorten the sarcomere, which collectively shortens the muscle fiber. Despite this universal mechanism, muscles perform a variety of tasks and are involved in quite a range of functions. Changes in performance and function are brought about by alterations at higher levels of organization, not by changes in the basic sliding filament mechanism. For instance, the properties of a muscle are influenced by the length or orientation of a muscle fiber, by physiology, by the way muscles attach to the lever system they move, and by the sequence of muscle actions relative to one another.

By reducing the question to underlying actin and myosin molecules, we can only partially understand muscle design and function. The particulars of organization at the cell, tissue, and organ level are also required to explain the basis of muscle performance.

Muscle Actions

A **motor pattern** generally means any repetitive movement activated by the nervous system. Muscles can act independently, concurrently, or in sequence to produce complex motor patterns that control the skeletal system. Even seemingly simple motor patterns may involve many muscles. During a strong cough to clear the throat, 253 named muscles contract. Muscles that act together to produce motion in the same general direction are **synergists.** The biceps brachii and brachialis muscles of the upper human arm are synergists in the action of flexing the forearm. Muscles that produce opposing motions are **antagonists.** The biceps brachii on one side of the upper arm and the triceps brachii on the other are antagonists. They contract in opposite directions during rapid rotation of the forearm, not to frustrate each other but to balance each other and control and coordinate rapid or powerful movements.

Under different conditions, the same muscle can have several actions. The primary action a muscle produces is its **prime motion.** Its two points of attachment to the skeletal

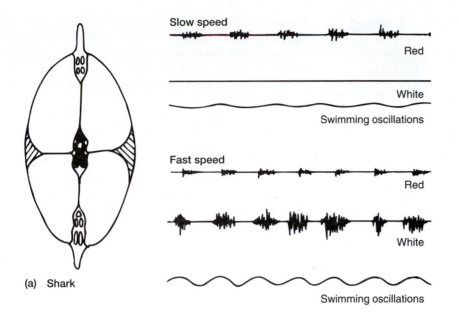

Slow speed

Red

White

Swimming oscillations

Fast speed

Red

White

Swimming oscillations

(a) Shark

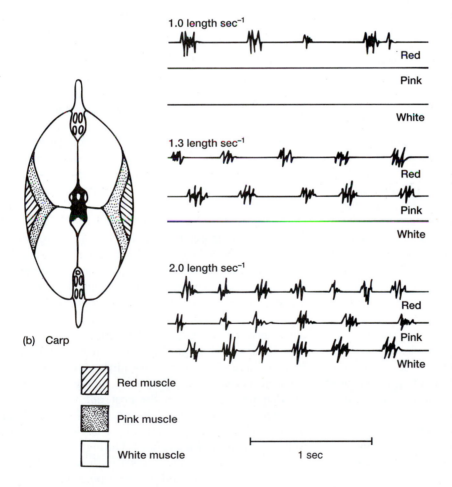

1.0 length sec^{-1}

Red

Pink

White

1.3 length sec^{-1}

Red

Pink

White

2.0 length sec^{-1}

Red

Pink

White

(b) Carp

Red muscle

Pink muscle

White muscle

1 sec

FIGURE 10.17 Sequential recruitment of axial musculature in fishes. Cross sections of a shark (a) and carp (b) show the positions of muscle types in the tails. The axial musculature of fishes is characterized by up to three types of muscle fibers. These types, on the basis of appearance, are termed red, pink, and white muscle. They represent a physiological sequence from slow twitch (endurance) to fast twitch (fatigable) fibers. Because these fibers are organized into discrete regions within a body, electrodes can be inserted to record the swimming speed at which each set of fibers is recruited and begins contributing to swimming undulations. (a) In sharks, only red and white muscles are present. At slow speed, red fibers contract, and at higher speeds, white fibers join in. The wavy tracings under each set of electromyograms represent the shark's swimming oscillations. (b) In some teleost fishes, like carp, all three types of fibers are present. Red, and then pink, and then white muscle fibers are recruited sequentially as swimming speed increases. Each myogram represents electrical activity and hence reveals muscle contraction at that point. Body lengths per second are used to express swimming speed.

Modified from Johnson et al., 1977.

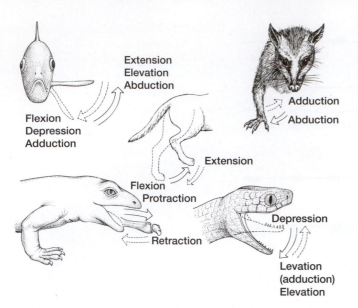

FIGURE 10.18 Muscle actions. Muscle adduction draws an appendage toward the ventral midline, and muscle abduction moves it away. Although these terms apply to tetrapod limbs and fish fins, the terms *depression* and *flexion* are sometimes used synonymously with *adduction* in fishes; *extension* and *elevation* are synonymous with *abduction*. In tetrapods, flexion means bending a part, extension means straightening it. Protraction sends a part forward from its base, retraction pulls it back. Opening the jaws is depression or abduction, and closing them is elevation or levation or adduction.

system are defined accordingly. A muscle's **origin** is its relatively fixed point of attachment, and its **insertion** is its relatively movable point of attachment. Each site of origin of a muscle is a **head,** and each site of insertion is a **slip.** Occasionally a muscle may act synergistically and have a secondary motion as well as its prime motion. For example, the geniohyoid of mammals runs between the hyoid and chin. Its prime motion is to move the hyoid, which in turn moves the larynx forward. Secondarily, when the larynx is held fixed in position, the geniohyoid acts in support of the digastric muscle to lower the mandible and help open the jaws. Muscles may also act as **fixators** to stabilize a joint or lever system. If you make a gentle fist, only your forearm muscles contract. Your biceps and triceps remain relaxed. However, if you clench your fist vigorously, your biceps and triceps of the upper arm involuntarily contract as well, not to help close your fingers directly but to stabilize the elbow joint while the nearby forearm muscles tighten your fist.

Other terms describe muscle actions as well (figure 10.18). Flexion and extension apply chiefly to limbs. **Flexors** bend one part relative to another about a joint. **Extensors** straighten a part (e.g., straightening the knee). Adduction and abduction are most often used to describe motion of the limb relative to the body. **Adductors** draw the limb toward and **abductors** move the limb away from the midline of the body. Applied to jaw action,

levators (a special kind of adductor) close and **depressors** (a special kind of abductor) open the jaws. Contraction of **protractors** results in projection of a part, such as the tongue of a frog, away from its base, whereas **retractors** bring it back. A limb can be turned by **rotators,** specifically **supinators** if they rotate the palm or sole up, or **pronators** if they rotate it down. Rotation is sometimes used in a general way to mean overall limb oscillation or swing as well. The **constrictor** or **sphincter** muscles surround tubes or openings (e.g., gill constrictors around the pharynx, intestinal sphincters around the anus) and tend to close them; **dilators** act antagonistically to open the orifice. As we meet them later in this chapter, we will define other muscle actions.

Muscle Homologies

During their evolution, some muscles have fused with one another, others have split into distinct new muscles, some have become reduced in prominence, and others have changed their points of attachment and hence their function. Unlike the evolutionary history of bones, muscles leave no direct trace in the fossil record. Their positions must be inferred from attachment scars on fossilized bones. Tracking these changes in the fossil record to establish homologies is difficult. Consequently, alternative criteria are often used. One such criterion is attachment similarity. Similar attachments in different muscles are assumed to attest to their homology. However, sites of attachment for the same muscle can vary in different groups (figure 10.19a). In mammals, the gastrocnemius on the posterior side of the shank is inserted in the calcaneus of the heel. In frogs, the tendon of the gastrocnemius instead stretches across the bottom of the foot as the plantar aponeurosis.

Another criterion is functional similarity. Similar function of two muscles is assumed to represent retention of a common ancestral pattern. This criterion can be misleading as well. For example, the depressor mandibulae opening the lower jaw in reptiles has a single belly. In mammals, the digastric serves roughly the same function, but it has two bellies (hence *di-* and *-gastric*), and its parts arise from separate embryonic sources (figure 10.19b).

Another criterion often used is nervous innervation, because there seems to be some phylogenetic permanence between a muscle and its nerve supply. In mammals, the diaphragm muscle of the posterior thorax might be expected to be innervated by nearby posterior thoracic nerves. Instead, it is innervated by a cervical nerve arising well anterior to it near the head. During embryonic development, the muscular predecessor to the diaphragm arises in the cervical region and migrates posteriorly. The cervical nerve that comes to innervate it, the phrenic nerve, grows from the cervical region as well and accompanies the diaphragm muscle back to its eventual posterior site of residence.

Embryonic origin is often used to establish muscle homologies. As with other structures, similar embryonic development is suggestive of similar phylogenetic ancestry. However, even this criterion can present difficulties. For example, vertebrae of primitive fishes and tetrapods arise by a different sequence of embryonic events, perichordal tube or arcualia. Apparently, embryonic events have been modified in tetrapods. This makes embryonic criteria less useful as a consistent standard for assigning homologies between fish and tetrapod vertebrae.

Vertebra development (p. 303)

Embryonic Origin of Muscles

In general, muscles arise from three embryonic sources. One source is the **mesenchyme,** a loose confederation of cells that become dispersed throughout the embryonic body. Smooth muscles within the walls of blood vessels and some viscera arise from mesenchyme. The second source of muscles is the paired **hypomere.** As the hypomere becomes distinct from the rest of the body mesoderm, its medial walls (splanchnic) embrace the gut and differentiate into the smooth muscle layers of the

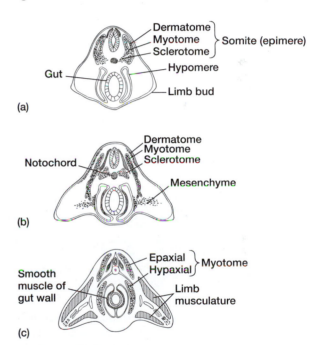

FIGURE 10.19 Criteria for muscle homology. Muscle homologies can be based on several criteria, although each may have its uncertainties. (a) Similar function between muscles suggests homology. The posterior shank muscles of the frog and reptile extend the foot. Although this muscle is thought to be the gastrocnemius, the insertions are slightly different in each animal. (b) The depressor mandibulae and the digastric muscles depress the lower jaws of reptiles (alligator) and mammals (opossum), but the nerve supply to these muscles is different, suggesting that the two are not homologous. (c) A common embryonic pattern of development is often used to establish muscle homologies. For example, during the embryonic development of fin muscles in a shark, the ventral tips of myotomes grow downward along the body, forming a low ridge and eventually entering the fin. In tetrapods, limb muscles arise in a similar way directly from myotomes.

(c) After Goodrich.

FIGURE 10.20 Embryonic origin of postcranial muscles in a generalized tetrapod. (a) Cross section illustrating each of the three regions of the mesoderm—epimere, mesomere, and hypomere—as they differentiate during embryonic development. The epimere becomes segmented to form the somite, which in turn forms the dermatome, myotome, and sclerotome. (b) Cells of the dermatome move beneath the skin and there differentiate into the dermis layer of the skin. Mesenchyme cells derived from the somatic layer of lateral plate mesoderm (skeletal precursors) and from the somitic myotome (limb muscle precursors) move into the limb bud. Cells of the sclerotome grow medially to form around the notochord and differentiate into the vertebrae. (c) Interactions between surface ectoderm and migrating mesenchymal cells promote the overall development of the limb. Longitudinal division of the myotome produces epaxial and hypaxial muscles of the body (shown in cross section).

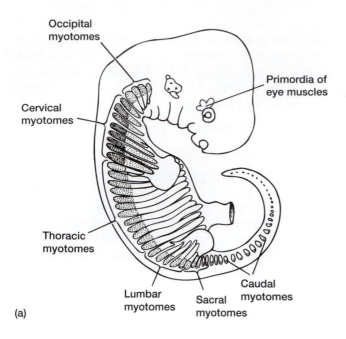

Occipital myotomes

Cervical myotomes

Primordia of eye muscles

Thoracic myotomes

Lumbar myotomes

Sacral myotomes

Caudal myotomes

(a)

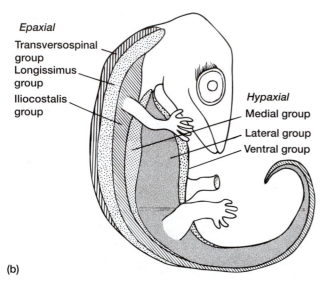

Epaxial

Transversospinal group
Longissimus group
Iliocostalis group

Hypaxial

Medial group
Lateral group
Ventral group

(b)

FIGURE 10.21 Muscles derived from embryonic myotomes in a reptile (lizard). (a) During embryonic development, the myotomes expand into respective areas of the body. (b) Differentiation of muscle groups of the superficial trunk and tail is shown.

alimentary tract and its derivatives (figure 10.20a). Cells of the hypomere also form the cardiac muscle of the tubular heart. The third embryonic source is the **paraxial mesoderm** from which most skeletal muscles develop (figure 10.21). During or shortly after neurulation, the paraxial mesoderm, as its name suggests, forms next to the neural tube along the axis of the embryonic body (figure 10.22a). In the trunk, the paraxial mesoderm becomes segmentally arranged into anatomically separate **somites.** In the head, the paraxial mesoderm does not differentiate into discrete somites, but rather forms clusters of mesoderm, called **somitomeres,** in series with the separate somites that follow behind them (figure 10.22a).

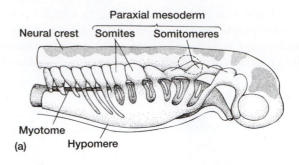

Paraxial mesoderm

Neural crest   Somites   Somitomeres

Myotome

Hypomere

(a)

Otic capsule   Optic capsule

Preotic myotomes

Nasal capsule

Hypobranchial musculature

(b)

FIGURE 10.22 Embryonic origin of cranial muscles in a shark embryo. (a) Paraxial mesoderm, the dorsal mesoderm next to the embryonic notochord, divides into discrete somites in the trunk but forms somitomeres—localized swellings that remain connected—in the head. Somitomeres contribute to much of the head musculature, including that which is associated with the branchial arches. The trunk somites differentiate into axial musculature and contribute to the fin musculature. (b) At a slightly later stage in embryonic development, cervical myotomes derived from the somites grow ventrally into the throat to give rise to the hypobranchial muscles beneath the gill arches.

After Goodrich.

In amniotes, there are usually seven pairs of somitomeres in the head, but sometimes fewer. The somites of the body divide into populations of cells that contribute to the skin (dermatome), vertebral column (sclerotome), and body musculature (myotome; figure 10.20b). The somitomeres of the head form head and pharyngeal muscles.

Differentiation of hypomere (p.168);
Differentiation of mesoderm (p. 178)

Postcranial Musculature

Appendicular Musculature

In many fishes, the ventral tips of adjacent myotomes grow downward into the emerging fin bud and differentiate directly into the fin musculature. In amniotes, limb muscles derive from mesenchyme cells shed from the ventral tips of adjacent somites (myotomes) rather than from direct contribution of embryonic myotomes to the muscles. These mesenchyme cells migrate into the limb bud to differentiate subsequently into appendicular musculature. The adjacent lateral plate mesoderm also sheds mesenchymal cells that enter the limb bud but, generally, these give rise to the bone or cartilage of the limb, along with the tendons, ligaments, and vasculature (figure 10.20). In a pattern analogous to that of amniotes, the fin musculature of teleosts also arises from a small number of mesenchyme cells migrating out of the adjacent somites and into the fin to form its musculature. Therefore, chondrichthyan and perhaps other fishes use a primitive embryonic mechanism to derive fin musculature—direct myotomal contributions (figure 10.19c). Teleosts (and amniotes) use a more derived embryonic mechanism—migratory mesenchymal cells, which give rise to appendicular musculature. At the genetic level, expression of the embryonic patterns in teleosts and amniotes is based on similar *Hox* genes, but different genetic cues direct formation of fin muscles in chondrichthyans.

Axial Musculature

The axial musculature arises from myotomes that differentiate from somites. These myotomes grow and expand along the sides of the body, forming the musculature associated with the vertebral column (or notochord), ribs, and lateral body wall (figure 10.21a, b). In gnathostomes, a longitudinal sheet of continuous connective tissue, the **horizontal septum,** divides the myotomes into dorsal and ventral regions, each destined to become, respectively, the **epaxial** and **hypaxial musculature** (figures 10.20c and 10.21b).

Cranial Musculature

Jaw and Pharyngeal Musculature

The jaw musculature arises from two distinct embryonic sources, each with a different nerve supply. These are functionally integrated to work the jaws cooperatively. One set is the **hypobranchial musculature** (figure 10.22b). It arises from myotomes of trunk somites whose ventral tips grow downward and forward into the throat along the ventral side of the branchial arches, hence *hypo-* (beneath) and *-branchial* (arches). Although they grow forward into the throat, these myotomes are accompanied by nerves emanating from the cervical region of the spinal column adjacent to the original trunk somites. Consequently, the hypobranchial musculature is supplied by spinal nerves. The hypobranchial musculature runs between the ventral elements of the gill arches and between the gill arches and the pectoral girdle. It also contributes to the tongue.

The other set of jaw and pharyngeal muscles, the **branchiomeric musculature,** is derived from somitomeres in the head and supplied by cranial nerves. Notice that the branchiomeric muscles originate from somitomeres, negating an older view. Because the gill arches and their branchial muscles are located within the wall of the pharynx, they were once thought to be serially homologous with the smooth muscles that are also located within the wall of the digestive tract. It seemed reasonable to conclude that head structures arose embryologically from the visceral or splanchnic part of the hypomere, like the smooth muscle differentiating in the gut wall. Cells of the hypomere, to the extent they could be followed in microscopic sections of the embryo, seemed to confirm this view. The term *visceral skeleton,* based on that view, survives in use today.

This now abandoned view—that much of the head is derived from the hypomere—fueled debate over the organization of the head itself. One hypothesis suggested that all cranial components of the head were serially homologous with the segmental plan of the trunk. The opposite hypothesis stressed the degree to which cranial tissues were distinct and fundamentally different from the trunk. With the advantage of recent techniques for marking individual cells and groups of cells, the embryonic sources of the head can be identified with greater confidence. Not all groups of vertebrates have been sampled and analyzed with these new techniques. However, we now have enough evidence to abandon the idea that skeletal jaw muscles arise from the hypomere. Instead, we realize that the two sets of jaw muscles arise from serial parts of the paraxial mesoderm: The hypobranchial muscles develop from somites, the branchiomeric muscles from somitomeres. Neither, of course, is part of the hypomere.

Extrinsic Eye Muscles

Tiny muscles that move or shape the lens to focus light on the retina are intrinsic, within the eyeball, and are discussed in chapter 17 with sensory organs. The extrinsic muscles on the outside of the eyeball rotate it within the ocular orbit to direct the eye's gaze at objects of interest (figure 10.23). The six extrinsic eye muscles originate from the walls of the orbit and are inserted on the outer surface of the eyeball. Their attachments allow rotation of the eye to desired positions. These six muscles arise from three (or perhaps four) different somitomeres. The most anterior somitomere of the three gives rise to the superior, inferior, and medial rectus and to the inferior oblique muscles, all supplied by the third (III) cranial nerve. The next somitomere gives rise to the superior oblique, supplied by the fourth (IV) cranial nerve. A third somitomere gives rise to the lateral rectus, supplied by the sixth (VI) cranial nerve. The contribution, if any, of a fourth somitomere to the extrinsic eye musculature is still under study.

The argument over the number of somitomeres that contribute to the extrinsic eye musculature has always been contentious, perhaps because it is now clear that the number arising within the head varies between groups of vertebrates. The more general term **preotic myotomes** is sometimes used to recognize, but without commitment to number, the somitomeres contributing to the extrinsic eye musculature.

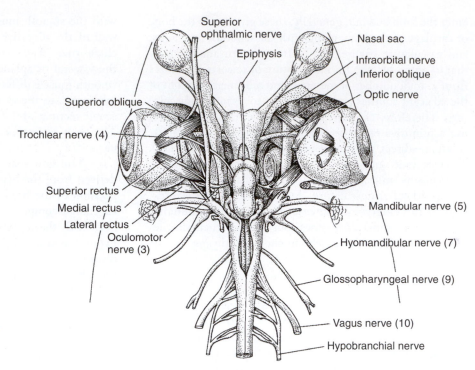

Superior ophthalmic nerve
Epiphysis
Nasal sac
Infraorbital nerve
Inferior oblique
Optic nerve
Superior oblique
Trochlear nerve (4)
Superior rectus
Medial rectus
Lateral rectus
Oculomotor nerve (3)
Mandibular nerve (5)
Hyomandibular nerve (7)
Glossopharyngeal nerve (9)
Vagus nerve (10)
Hypobranchial nerve

FIGURE 10.23 **Extrinsic eye musculature of a shark (dorsal view).** The extrinsic eye muscles are derived from somitomeres and rotate the eyeball within the orbit in order to direct the gaze. The roof of the chondrocranium over the eyeball has been removed to expose several extrinsic muscles (left). The superior oblique and superior rectus muscles have been cut to expose the deeper extrinsic muscles (right).

Comparative Anatomy

Generally, cranial and postcranial somatic musculatures arise from paraxial mesoderm. Thus, they are serially homologous with each other. This section looks at this system within fishes and then examines the complex remodeling of the muscular system in tetrapods as they meet the quite different set of functional demands on land.

Postcranial Musculature

Axial Musculature

Fishes In fishes, the axial musculature arises directly from the embryonic and segmental myotomes. Once fully differentiated into the adult musculature, the blocks of axial musculature retain their segmentation, but they are termed **myomeres** to distinguish them from the formative embryonic myotomes from which they arose. Successive myomeres are separated from each other by connective tissue sheets, the **myosepta** (myocommata). Myosepta extend inward, become attached to the axial column (vertebral column or notochord), and join successive myomeres into muscle masses. The **horizontal septum** of the skeleton is absent in cyclostomes (figure 10.24a) but present in all gnathostome fishes, where it divides the myomeres into epaxial and hypaxial muscle masses (figure 10.24b). Each spinal nerve that supplies a myomere bifurcates. The first branch, the **dorsal ramus,** supplies the epaxial division, and the second branch, the **ventral ramus,** supplies the hypaxial division. Dorsal ribs, when present, develop at the intersection of the horizontal septum with successive myosepta.

The axial musculature of fishes supplies the major propulsive forces for locomotion and, not surprisingly, constitutes the bulk of the body's musculature. Viewed from the lateral surface, the myomeres are folded into zigzag blocks that often look V- or W-shaped (figure 10.24b). Muscle fibers that compose the myomeres are short, but this folded shape of each myomere extends over several axial segments, giving it and its short fibers control over an extended length of the body. A contraction spreading within the axial musculature alternates from side to side, developing characteristic waves of lateral undulation. These powerful bends produced by the axial musculature are responsible for developing the body's lateral thrusts against the water and driving the fish forward. The axial column, be it a jointed vertebral column or a flexible notochord, receives the attachments of these muscles and acts as a compression girder, resisting telescoping of the body that might otherwise result.

The propulsive force of lateral undulation is perpendicular to the surface of the section of fish generating the force. Because the undulating body increasingly bends toward the tail, the direction of the force relative to the line of travel becomes inclined more posteriorly. Acceleration of the tail is also greater than that of the body sections near the head; hence, the force is greater in the tail. This increase in force and its more posteriorly directed inclination help to explain why the tail is important in generating swimming forces.

More formally, this can be explained by propulsive and reaction forces. The reaction or normal force that is returned by the water is equal and opposite to the propulsive force of the fish's body against the water (figure 10.25a). The normal

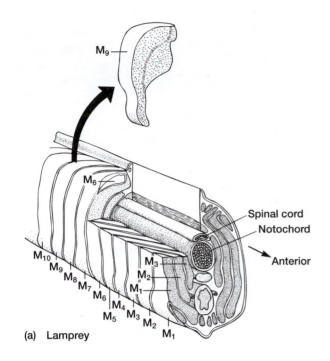

(a) Lamprey

M$_9$

M$_6$

Spinal cord
Notochord

Anterior

M$_{10}$
M$_9$
M$_8$
M$_7$
M$_6$
M$_5$
M$_4$ M$_3$ M$_2$
M$_1$

M$_3$
M$_2$
M$_1$

(b) Teleost

FIGURE 10.24 **Axial musculature of fishes.**
(a) Lamprey trunk in cutaway view showing the arrangement of segmental and numbered myomeres. (b) Lateral view of teleost showing the arrangement of myomeres that form the extensive trunk musculature. Sections of the trunk musculature have been removed to reveal the arrangement of the folded myomeres. A block of segmental muscle is enlarged and shown in isolation.

(a) Modified from Hardisty, 1979; (b) from Peters and Mackay.

force is resolved into two vector components, one laterally directed and the other at right angles forwardly directed. The lateral force vector adds nothing to forward progression, but the forward force vector drives the fish (or at least that section of the fish) forward. The size of this forward force vector represents the size of the forward thrust generated at that point on the fish. Because the normal force is larger in the tail and its forward-directed component vector is larger than that in the trunk, the tail is the most important part of the body generating useful swimming forces (figure 10.25b, c).

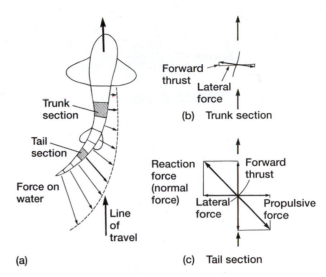

Trunk section

Tail section

Force on water

Line of travel

(a)

Forward thrust
Lateral force

(b) Trunk section

Reaction force (normal force)

Forward thrust

Lateral force

Propulsive force

(c) Tail section

FIGURE 10.25 **Swimming forces in a shark.**
(a) Traveling waves of undulation pass backward along the body, throwing it into curves. These press against the water to produce a propulsive force. The water returns a reaction, or normal, force. The body is arbitrarily divided into sections. The level and direction of the propulsive force generated by each section are indicated by the vectors (arrows). Notice the change in angle of the propulsive force along the body relative to its line of travel. (b) Diagram of trunk forces. The propulsive and normal forces from an anterior trunk section are shown. Notice that the forward component of the normal force is much reduced compared with the normal force that the tail experiences. (c) Diagram of tail forces. The lateral and forward component vector forces of the normal force are shown in this tail section. The forward vector along the line of travel drives the fish forward.

Posteriorly, the axial musculature continues from the trunk into the tail. Anteriorly, the axial musculature is attached to the skull and pectoral girdle. Some fishes use these attachments during feeding to lift the neurocranium or stabilize the pectoral girdle from which jaw opening muscles originate.

Tetrapods In tetrapods, the appendicular muscles generally take on more responsibility for locomotion and accordingly account for more muscle bulk. Although the axial musculature tends to be reduced, that which remains differentiates into specialized muscles, a reflection of the more complicated control exerted over flexion of the vertebral column and movement of the rib cage (figure 10.26a–c).

In salamanders, epaxial muscles are still essentially one muscle mass, the **dorsalis trunci** (figure 10.26b). The hypaxial musculature has differentiated into a few muscles, but compared with other tetrapods, the axial musculature is still quite simple and constitutes a large proportion of the overall body musculature. Continued prominence of the axial musculature in salamanders is thought to reflect the continued central role of the axial column in locomotion and the limbs' modest contribution to propulsion. In frogs, in whom the hindlegs serve the specialized saltatorial

Epaxial

Hypaxial

(a)

Epaxial
Dorsalis trunci

Hypaxial
Transversus
Internal oblique
Oblique externus profundus

Rectus abdominis

(b)

Epaxial
Transversospinalis group
Longissimus group
Iliocostalis group
Hypaxial
Dorsomedial
Subvertebral group
Medial
Transversus
Obliquus internus
Ventral
Rectus abdominis

Lateral
Intercostalis externus

External obliques
Superficialis
Profundus

(c)

FIGURE 10.26   Organization of axial musculature (cross sections). (a) Teleost fishes have regions of relatively undifferentiated epaxial and hypaxial muscle masses. (b) Salamanders have an epaxial musculature that is a relatively undifferentiated muscle mass, the dorsalis trunci. Hypaxial muscles differentiate into several discrete muscles. (c) Lizards have both epaxial and hypaxial muscle masses that have differentiated into several specialized groups of muscles. The horizontal septum is not easily recognized, but distribution of the spinal nerve branches permits identification of derivatives of epaxial (dorsal branch of spinal nerve) and hypaxial (ventral branch) musculature.

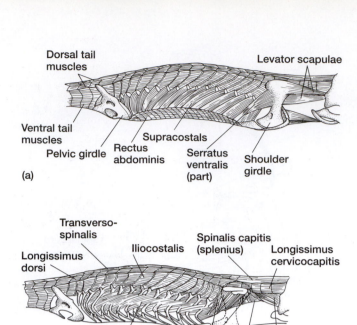

Dorsal tail muscles

Levator scapulae

Ventral tail muscles
Pelvic girdle
Rectus abdominis
Supracostals
Serratus ventralis (part)
Shoulder girdle

(a)

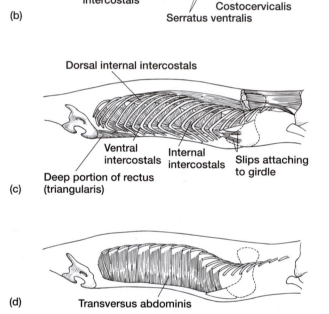

Transversospinalis
Longissimus dorsi
Iliocostalis
Spinalis capitis (splenius)
Longissimus cervicapitis

External intercostals
Serratus ventralis
Costocervicalis
Rectus capitis

(b)

Dorsal internal intercostals

Ventral intercostals
Internal intercostals
Slips attaching to girdle
Deep portion of rectus (triangularis)

(c)

(d)   Transversus abdominis

FIGURE 10.27   Lateral views of the axial musculature of the reptile *Sphenodon*. (a–d) Muscle layers are successively removed to reveal deeper muscles beneath.

locomotion, the appendicular musculature of the hindlimbs is large and the axial musculature is reduced in prominence.

In reptiles, the horizontal septum is lost or indistinct, although supply by the dorsal and ventral rami of the spinal nerve still betrays which muscles are of epaxial and hypaxial origin (figure 10.26c). Even though lateral undulations of the

vertebral column contribute to locomotion, the limbs become much more important in producing the propulsive forces central to locomotion. The epaxial musculature associated with the vertebral column is reduced. The hypaxial musculature forms much of the body wall and is associated with breathing because these muscles are attached to the rib cage. Because the rib cage in turtles is rigid, the hypaxial muscles are reduced or lost. In snakes, which are of course without limbs, the axial column figures significantly in lateral undulation, and the axial musculature is prominently developed.

The axial muscles in reptiles tend to split into several layers, forming many differentiated muscles that span several segments (figure 10.27a–d). There are three general divisions of the epaxial musculature: the **transversospinalis,** the **longissimus,** and the **iliocostalis** muscle groups. Muscles of these three groups usually attach to vertebrae, and in some species, they further split into additional muscle groups.

The hypaxial musculature attached to the rib cage controls breathing and also aids in moving the trunk. As mentioned, this can be a prominent role in snakes. Most descriptive work on reptile hypaxial musculature recognizes three embryonic precursors giving rise to three direct and a fourth composite group of muscles. One group is the **dorsomedial** musculature that runs beneath the vertebral column as the subvertebralis and extends anteriorly as the longus collis that aids in moving the neck. The second group is the **medial** musculature, which is distributed along the inside of the rib cage, and includes the transversus abdominis and internal obliques. The third group is the **lateral** musculature spanning the outside of the rib cage; it includes the external obliques and external intercostals. Apparently, derivatives of medial and lateral musculature contribute to the **ventral** musculature along the belly, which includes the rectus abdominis that extends from the sternum and ribs to the pelvis. It is divided midventrally along its length by the linea alba and is crossed at regular intervals by connective tissue inscriptions, a pattern suggesting a basic segmental nature.

In birds, the same divisions of the axial musculature are represented, but they are reduced, especially in regions where associated vertebrae are fused.

In mammals, the three reptilian divisions of the epaxial and four divisions of the hypaxial musculature are present, although they tend to form numerous splits that yield additional muscles. Absence of ribs in the abdominal region changes the segmental intercostals into a continuous sheet of obliques.

Appendicular Musculature

Fishes In fishes, two opposing muscle masses extend over the dorsal and ventral surfaces of the fins from girdle to pterygiophores. Embryologically, these arise from myotomes that grow out into the fins and differentiate into these dorsal and ventral muscle masses. Dorsal muscles elevate the fin; ventral muscles depress or adduct the fin. Occasionally, these muscles produce distinct muscle slips that aid in fin rotation. Compared with the massive axial musculature, the fin musculature of fishes is relatively slight.

Tetrapods In tetrapods, these dorsal and ventral appendicular muscles tend to be more prominent as the limbs assume more of the task of producing locomotor forces and the axial musculature becomes less involved. In addition to becoming more prominent, these muscle masses also tend to split and divide, forming many distinct muscles that increase considerably the complexity of the adult limb musculature. This story is further complicated by the fact that the tetrapod limb musculature receives phylogenetic contributions from other regions. The axial musculature along the

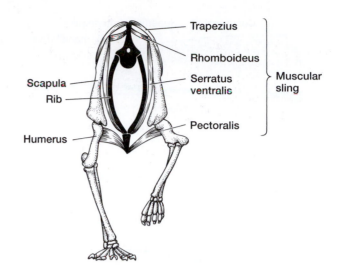

FIGURE 10.28 **The "muscular sling" of mammals**
Appendicular muscles of the forelimbs suspend the anterior body from the shoulders. Some of these muscles arise from axial muscles (rhomboideus, serratus ventralis), some from branchial muscles (trapezius), and some from the forelimb musculature itself (pectoralis).

body and the branchiomeric musculature of the gill arches also contribute to the tetrapod limb muscles, especially to the shoulder muscles. Further, the hip and shoulder muscles transmit locomotor forces to the vertebral column differently. The pelvic girdle is attached directly to the sacral region of the vertebral column, but the pectoral girdle of terrestrial vertebrates is hung in place by a **muscular "sling"** (figure 10.28). This is a set of muscles that run from thorax to shoulder to suspend the anterior part of the body through muscular ties from the blades of the pectoral girdles. Finally, many specialized tetrapods depart from the general trend. Frogs, for example, are specialized to leap and have complicated hindlimb musculature; birds are specialized for flight, and their limb musculature serves the special demands of aerial locomotion.

Pectoral Girdle and Forelimb Contributions to tetrapod shoulder and forelimb muscles come from four sources: branchiomeric, axial, dorsal limb, and ventral limb muscles (figure 10.29a–c).

1. **Branchiomeric muscles.** The branchiomeric muscles contribute the **trapezius** and **mastoid** groups derived from the cucullaris of primitive fishes such as chondrichthyans. In mammals, the trapezius group includes the clavotrapezius, acromiotrapezius, and spinotrapezius; the mastoid group includes the cleidomastoid and sternomastoid muscles (table 10.2).

2. **Axial musculature.** The axial musculature contributes the **levator scapulae, rhomboideus complex,** and **serratus** muscles. These three derivatives of the axial musculature, together with the trapezius of branchiomeric

(a) Shark

(b) Salamander (*Necturus*)

(c) Cat

FIGURE 10.29 Cranial and shoulder musculature.
Lateral views of shark (a), salamander *Necturus* (b), and cat (c).

origin, form the muscular sling that suspends the body between the two scapular blades. The pectoral girdle of turtles is an exception because it is directly attached to the shell. In some tetrapods, such as pterosaurs, birds, and bats, the pectoral girdle rests on the sternum. In bony fishes, the pectoral girdle is usually attached to the back of the skull, but in most tetrapods it is not. Freeing of the shoulder from

the skull is established in early tetrapods during the transition to land and in part seems related to increased cranial mobility. As the shoulder girdle became freed from the skull, the nearby branchiomeric and axial muscles were pressed into serving as part of the muscular sling through which the forelimbs are attached to the body. Most of the remaining pectoral and forelimb muscles of tetrapods arise from the dorsal and ventral muscle masses (table 10.2).

3. **Dorsal muscles.** The dorsal muscles of the shoulder are inserted on the humerus and function to oscillate it during movement or fix it in position while an animal stands. Of these muscles, only the **latissimus dorsi** originates outside the limb, from the body wall. In mammals, a tiny slip of the latissimus within the scapula separates as the **teres major.** The other dorsal muscles that act on the humerus are the **teres minor, subscapularis,** and **deltoideus,** which may form two distinct muscles. The prominent **triceps,** often showing several heads, is also a derivative of the dorsal musculature, but it acts to extend the forearm. Dorsal muscles of the forearm form most of the extensor musculature, which extends or straightens the digits via tendons (table 10.2).

4. **Ventral muscles.** The **pectoralis** is a very prominent ventral muscle of the chest. From a long origin along the sternum, its fibers converge on the humerus (figure 10.30a–c). This muscle tends to split into four more or less distinct derivatives in mammals: the **pectoantebrachialis,** the deeper **pectoralis major,** the deeper still **pectoralis minor,** and the deepest of all, the **xiphihumeralis.** The ventrally positioned **supracoracoideus** of reptiles extends from its origin on the coracoid laterally to its insertion on the humerus. However, in therian mammals, the supracoracoideus originates dorsally on the lateral face of the scapula. The bony scapular spine divides this muscle into the **supraspinatus** and **infraspinatus** muscles, which are inserted on the humerus as well. The **coracobrachialis** from the coracoid runs along the underside of the humerus. In mammals, the **biceps brachii** has two heads, representing the apparent fusion of two muscles that have their insertions on the forearm and flex it in lower vertebrates. Forearm flexors from ventral muscles act through tendons on the digits (table 10.2).

Pelvic Girdle and Hindlimb Unlike the shoulder, the tetrapod hip has no muscular sling in which it "floats." Instead, the pelvic girdle is fused to the vertebral column. Consequently, few extrinsic muscles control the hindlimbs. The **psoas minor** from the axial musculature is an

TABLE 10.2 Homologies of Axial and Appendicular Musculature

Muscle Groups	Salamander	Reptiles	Mammals
PECTORAL GIRDLE AND FORELIMBS			
Branchiomeric	Cucullaris	Trapezius	Clavotrapezius / Acromiotrapezius / Spinotrapezius
	Levatores arcuum	Sternomastoid	Cleidomastoid / Sternomastoid
Axial	Levator scapulae / Thoraciscapularis	Levator scapulae / Serratus ventralis	Levator scapulae / Rhomboideus / Serratus ventralis
Dorsal	Latissimus dorsi	Latissimus dorsi	Latissimus dorsi / Teres major
	Subcoracoscapularis	Subcoracoscapularis	Subscapularis
	Dorsalis scapulae / Procoracohumeralis longus	Dorsalis scapulae / Deltoideus clavicutans	Deltoideus (Acromiodeltoid and Scapulodeltoid)
	Triceps	Triceps	Triceps
	Forearm extensors	Forearm extensors	Forearm extensors
Ventral	Pectoralis	Pectoralis group	Pectoralis (4)
	Supracoracoideus	Supracoracoideus	Supraspinatus / Infraspinatus
	Coracoradialis	Biceps brachii	Biceps brachii (part)
	Humeroantebrachialis	Brachialis inferior	Biceps brachii (part)
	Coracobrachialis	Coracobrachialis	Coracobrachialis
	Forearm flexors	Forearm flexors	Forearm flexors
PELVIC GIRDLE AND HINDLIMBS			
Axial	Subvertebralis	Subvertebralis	Psoas minor
Dorsal	Puboischiofemoralis internus	Puboischiofemoralis internus	Psoas / Iliacus / Pectineus
	Ilioextensorius / Puboischiofemoralis externus	Iliotibialis / Femorotibialis	Rectus femoris / Vasti
	Iliotibialis	Ambiens	Sartorius
	Iliofemoralis	Iliofemoralis	Tensor fascia latae / Gluteus minimus / Gluteus medius / Pyriformis
	Tibialis anterior	Tibialis anterior	Tibialis anterior
	Extensor digitorum communis	Extensor digitorum communis	Extensor digitorum longus / Extensor hallucis longus / Peroneus tertius
	Peroneus longus	Peroneus longus	Peroneus longus
	Peroneus longus and brevis	Peroneus brevis	Peroneus brevis
	Extensor digitorum brevis	Extensor digitorum brevis	Extensor digitorum brevis
Ventral	Puboischiofemoralis externus	Puboischiofemoralis externus	Obturator externus / Quadratus femoris
	Adductor femoris / Pubotibialis	Adductor femoris / Pubotibialis	Adductor femoris brevis / Adductor femoris longus
	Caudofemoralis / Ischioflexorius	Caudofemoralis / Flexor tibialis externus / Flexor tibialis internus II / Flexor tibialis internus I	Caudofemoralis / Dorsal semitendinosus / Ventral semitendinosus / Biceps femoris / Semimembranosus
	Puboischiotibialis	Puboischiotibialis	Gracilis
	Flexor digitorum sublimis and longus	Gastrocnemius internus	Gastrocnemius medialis / Flexor hallucis longus
	Popliteus / Fibulotarsalis	Gastrocnemius externus	Gastrocnemius lateralis / Soleus / Plantaris

Careful Turns, Fast Starts, and Cruising in Open Waters

To best generate swimming thrust, the body of a fish should be deep (tall from dorsal to ventral surface) to present a broad, oarlike surface to the water. The sculpin is an example (box figure 1a). Not only is the body deep, but the lateral silhouette is enlarged further by expansive fins along the dorsal and ventral surfaces. Thus, the total lateral area of the fish pressed against the water during swimming is extensive and helps generate thrust. But not all fish are designed in this way because swimming is employed significantly differently among fishes.

For fishes that must perform careful maneuvers, the body is often disk shaped (box figure 1b). Angelfish, popular in tropical fish shops, slip carefully through crowded vegetation and position their bodies to find food along blades of aquatic plants. Butterfly fish, inhabitants of coral reefs, maneuver at slow speed among the coral shelves, poking their mouths into tiny crevices for food. Their disk-shaped bodies keep the anteroposterior body axis short so it can be turned in tight spaces. The edge of the body is often rimmed by a fringe of fins that control precise and specific adjustments.

Other fishes are designed for quick bursts in which they accelerate to surprise prey or dart from sudden approach of danger. The pike is an example (box figure 1b). Deploying swimming for quick dashes requires that the fish overcome inertia. Consequently, almost 60% of a pike's mass is axial muscle. This gives it a powerhouse of contractile units to generate sudden large forces. Further, its body is relatively flexible, at least through the tail, so it can bend into large-amplitude curvatures to produce a normal force more in line with its intended line of travel.

Although less specialized than the pike, the trout also has a relatively large mass of axial musculature and a flexible body. When threatened, it can quickly form a C-shaped bend and use its large mass of axial musculature to "push off" suddenly, accelerating in a quick burst in some other direction to make its escape (box figure 1c).

Still other fishes, such as tuna, are designed for cruising (box figure 1b). The trailing edge of the tail is expanded to deliver the tail thrust to the surrounding water. But the **peduncle** that connects the tail to the body is very narrow.

Consequently, the overall mass of the tail and its peduncle is quite low, and thus the inertia that must be oscillated during swimming is low as well. On the other hand, most of the axial musculature is bunched more anteriorly in the trunk, increasing its inertia. There is less tendency for the tail to impart its oscillations on the more massive body and cause wasteful lateral body swings. Overall, these changes in location of muscle mass together with the streamlined shape make the swimming design of the tuna especially efficient for sustained cruising, an advantage in a fish that covers great stretches of open ocean in search of schools of small fishes, its principal prey.

Of course, most fishes are not such specialists and must compromise between these specialized extremes. Further, the design of a fish requires more than just attention to the mass of the axial musculature. For example, the sculpin has a relatively large head in order to dislodge and pick up benthic animals with its powerful suction feeding. Consequently, the head is a relatively large part of the sculpin; thus, its overall optimum design is a compromise between the requirements of feeding and swimming.

---

exception. However, most hindlimb musculature derives from dorsal and ventral muscles that differentiate into the complex assortment of hip, thigh, and shank muscles (table 10.2).

1. **Dorsal muscles.** The **puboischiofemoralis internus** of lower tetrapods is a dorsal muscle that runs from the lumbar region and girdle to the femur, making it an important limb rotator. Three muscles differentiate from it in mammals. All three are inserted on the femur but originate from the lumbar region: **(psoas),** the ilium **(iliacus),** and the pubis **(pectineus).** The **iliofemoralis** of lower tetrapods extends from the ilium to the femur and functions to extend the limb. In mammals, it divides into the **tensor fascia latae, pyriformis,** and **gluteus complex.** The *quadriceps* is a collective term for the **rectus femoris** and the three heads of the **vastus** (lateralis, medialis, intermedius). These muscles lie along the anterior margin of the femur and are usually quite prominent. Through their common insertion on the patella, they are very powerful shank extensors. The long **sartorius** originates on the ilium but crosses two joints, the hip and the knee, before being inserted on the tibia. The **ambiens** of reptiles and the **iliotibialis** of amphibians are likely homologues of the sartorius. The **tibialis anterior** and various other dorsal muscles of the shank constitute the shank extensors that dorsiflex the ankle via long tendons (table 10.2).

2. **Ventral muscles.** In lower tetrapods, the **puboischiofemoralis externus** is a ventral muscle that extends from the pubis and ischium to the femur (figure 10.31a–c). In mammals, the **obturator externus** and **quadratus femoris** are derivatives. In lower vertebrates, the **caudofemoralis,** extending from the base of the tail to the femur, is a powerful muscle that retracts the hindlimb. When the hindlimb is fixed,

(a) Sculpin

Butterfly fish

Pike

(b) Tuna

(c) Trout

BOX FIGURE I
Swimming specializations.
(a) Broad lateral profile of a sculpin, which undulates against the water to produce thrust. The tail and prominent dorsal and ventral fins enlarge the thrusting surface. (b) Three body shapes illustrating three specialized roles for swimming—top, maneuvering (butterfly fish); middle, quick lunges (pike); bottom, sustained cruising (tuna). (c) Trout seen from above. In response to a threat, it sharply flexes its body and then straightens the tail in order to accelerate quickly and change direction.
After Webb.

the caudofemoralis has the opposite action of swinging the tail. In mammals, it is reduced in prominence. Similarly, the **obturator internus** and **gemelli muscles** in mammals are relatively reduced compared to their homologue, the **ischiotrochantericus** in reptiles. The **adductor femoris,** which is large in most tetrapods, extends the thigh. The name *hamstrings* is a collective term for three muscles: the **semimembranosus, semitendinosus,** and **biceps femoris.** All arise from the pelvis, run along the posterior margin of the femur, and have insertions on the shank or nearby on the distal end of the femur. Together these prominent muscles flex the shank. The **puboischiotibialis** of lower tetrapods covers much of the ventral surface of the thigh and retracts it. Its mammalian homologue is the **gracilis.**

The most prominent ventral muscle of the shank is the **gastrocnemius,** the "calf" muscle. In mammals, it has two heads, resulting from the fusion of two different phylogenetic predecessors. The mammalian **gastrocnemius medialis** and the **flexor hallucis longus** arise from the reptilian **gastrocnemius internus.** The **gastrocnemius lateralis** along with the **soleus** and **plantaris** arise from the reptilian **gastrocnemius externus** (table 10.2).

Specializations among Tetrapods   Locomotion among tetrapods is based on alternating limb displacements at moderate rates. Departure from this generalized mode of progression usually depends on modifications of the musculature that powers it. In anurans, for example, locomotion is saltatorial, and both limbs are activated simultaneously by contraction of powerful hindlimb extensors. At the end of a

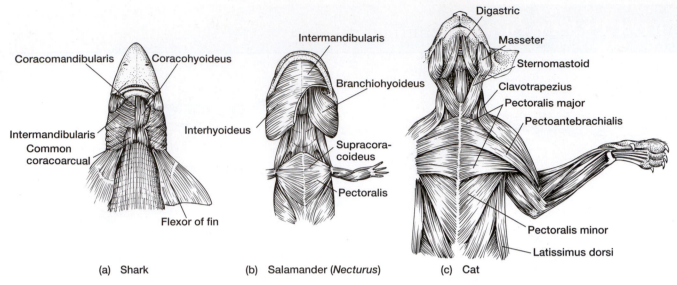

FIGURE 10.30 **Cranial, hypobranchial, and shoulder muscles.** Ventral views of shark (a), salamander *Necturus* (b), and cat (c).

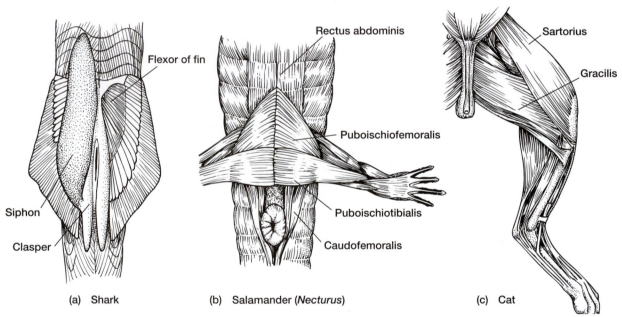

FIGURE 10.31 **Pelvic musculature.** Ventral views of shark (a), salamander *Necturus* (b), and cat (c).

leap, the pectoral girdle and forelimbs of anurans absorb the impact of landing. The muscular sling of the pectoral girdle suspending the body in other tetrapods probably functions to absorb the jolts and jars during locomotion. But this role in anurans is clearly accentuated, and the mode of progression is a marked departure from the alternating limb swings of other tetrapods. The muscles of the anuran forelimb are stout to help during landing, and the extensor muscles of the hindlimb are prominent to launch the animal. This specialized mode of locomotion might account

for the relatively complex and differentiated musculature of anurans compared with salamanders. Figure 10.32a, b illustrates the superficial musculature of the frog.

In tetrapods specialized for cursorial locomotion, the appendicular muscles tend to be bunched proximally near the trunk and their forces distributed distally through long tendons to the ends of limbs. This design reduces the mass carried by the limb itself, which in turn reduces the inertia that needs to be overcome during reciprocating limb oscillation. Among mammals, the perissodactyls (figure 10.33)

FIGURE 10.32 **Superficial musculature of a frog.** Dorsal (a) and ventral (b) views.

Modified from J. Z. Young, The Life of Vertebrates, Clarendon Press, Oxford.

and artiodactyls exhibit the most well-developed proximal repositioning of limb muscle mass, but similar trends can be found in most other mammalian groups that depend on rapid locomotion. Parallel trends apparently arose in reptiles of the Mesozoic, especially within fleet archosaurs. The appendicular muscles of the hindlimbs of these bipedal reptiles show evidence of proximal bunching, presumably with long tendons that extended to the ends of the limbs.

Cursorial trends (p. 353)

In birds, the general trends of muscle evolution characteristic of tetrapods are evident (figure 10.34). The axial musculature tends to decline in prominence, whereas the appendicular musculature increases. Muscle masses become differentiated into a more complex set of discrete muscles. Birds also show changes in musculature design related

to the specialized demands of powered flight and secure landing. Fusion of the posterior vertebral column with elements of the pelvic girdle into a rigid bony support reduces the requirement for large masses of axial muscles to firm up the vertebral column. Consequently, there is a reduction in the posterior axial musculature. At the anterior end of the axial column, the chain of cervical vertebrae controlled by a complex set of cervical muscles provides flexibility and very precise control of head movement. Musculature of the pelvic girdle and hindlimb is differentiated into a prominent muscle mass. When a bird lands, the hindlimb muscles catch and balance the mass of the body on impact as it comes to rest on the ground or a branch. Most muscles are bunched proximally and extend to the toes through long tendons. The proximal bunching of muscles keeps the mass close to the midline of the body, a feature important during flight, but the long tendons reaching to the toes give more

FIGURE 10.33   **Superficial musculature of a horse.**

After Goody.

precision to toe placement, a feature especially significant in perching birds and raptors.

Avian skeletal adaptations (p. 313)

Muscles of the pectoral girdle and forelimb (wing) are particularly well developed and specialized for powered flight. Most wing muscles are bunched proximally, especially the massive pectoralis that resides near the midline of the sternum from which it originates. The pectoralis is inserted on the humerus and provides a powerful downstroke during flight. Deep to the pectoralis is the supracoracoideus. Recall that in reptiles this muscle runs from its origin on the pectoral girdle to the humerus (figure 10.35), a course that makes the supracoracoideus a limb adductor (depressor). However, in birds, the strong tendon of the supracoracoideus runs over the pulleylike end of the coracoid and is inserted on the dorsal surface of the humerus (figure 10.36a–c). This reorientation of the point of insertion allows the supracoracoideus to lift the wing, thus

turning this muscle into a wing elevator. Consequently, the depressor (pectoralis) and the elevator (supracoracoideus) that are responsible for producing opposite motions during wing downstroke and upstroke both lie on and originate from the sternum (figure 10.36c). Because of the phylogenetic shifts in points of insertion, their actions in birds are quite different.

When a bird is flying, especially gliding, its forearm is extended to draw the overlying skin taut. The anterior region of skin between the shoulder and wrist is the **patagium,** and within its leading edge is the **patagialis** muscle (figure 10.34). The patagialis, which may form several small slips, arises from the clavicle and extends via a long tendon to the metacarpals of the wrist. Like a clothesline, the leading edge of the patagium is hung from this cordlike muscle, which forms the anterior edge of the flight surface of the wing. If the patagialis or its long tendon is cut, the patagium loses its aerodynamic shape and the wing becomes almost useless for flight.

Flexor digitorum profundus

Flexor digitorum superficialis

Flexor retinaculum

Cucullaris (dermotemporalis)

Flexor carpi ulnaris

External metacarpi radialis

Pronator superficialis

Latissimus dorsi

Serratus metapatagialis

Iliotibialis cranialis

Femorotibialis externus

Iliofibularis

Levator caudae

Flexor crurus lateralis

Flexor crurus medialis

Gastrocnemius

Cucullaris (dermotensor patagii)

Tensor patagialis

Biceps brachii

Triceps brachii

Pectoralis

Patellar ligament

Tibialis cranialis

FIGURE 10.34 **Superficial musculature of a parakeet, with wing elevated.**

After Evans.

The forearm muscles of a bird are small but numerous. They probably act to refine the positions of wing feathers, forming and controlling the aerodynamic surface that the wing presents to the airstream.

Flight (p. 353); Aerodynamics (p. 364)

Cranial Musculature

Until recently, the skeletal muscles associated with the jaws, pharynx, and branchial arches were thought to arise embryologically from two quite different sources: the branchiomeric musculature and the hypobranchial musculature. However, as discussed in the section on postcranial muscles (figure 10.22a, b), the current view envisions both branchiomeric and hypobranchial jaw musculature arising from a common source, namely, from the paraxial mesoderm. The branchiomeric musculature arises from the cranial paraxial mesoderm (somitomeres) and the hypobranchial musculature from the trunk paraxial mesoderm (somites). Having emphasized the unity of the jaw musculature rather than its embryonic distinctions, we nonetheless follow this historical division for convenience in describing comparative aspects of cranial musculature (table 10.3).

Branchiomeric Musculature

Cranial nerves supply the branchiomeric musculature associated with the sides of the branchial arches (figure 10.37a–d). In fishes, the branchial arches together

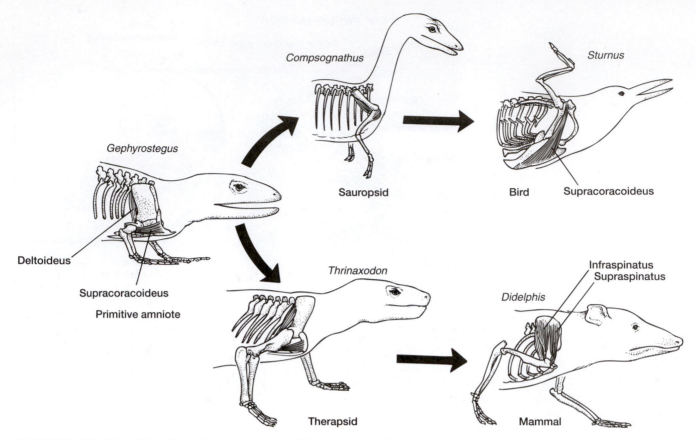

FIGURE 10.35 **Evolution of the supracoracoideus muscles.** In primitive reptiles such as *Gephyrostegus*, this muscle likely originated on the coracoid and was inserted on the proximal head of the humerus to adduct the limb. This muscle probably ran in a similar course in therapsids (*Thrinaxodon*). In mammals, such as the opossum (*Didelphis*), the supracoracoideus originates on the scapula and becomes divided by the scapular spine into the supraspinatus and infraspinatus. In the theropod *Compsognathus*, an archosaur, the supracoracoideus probably followed a course similar to that of primitive reptiles, having an origin on the coracoid and an insertion on the proximal end of the humerus. In modern birds, the supracoracoideus originates at the sternum, runs over the coracoid, and is inserted on the dorsal, proximal end of the humerus. The deltoideus and its derivatives are also shown.

Redrawn from G. E. Goslow, Jr., et al., "The avian shoulder: An experimental approach," American Zoologist, 29: 287–301. Reprinted by permission of the Society for Integrative and Comparative Biology (Oxford University Press).

with their branchiomeric muscles function as a pumping device to move water across the gills, replacing the ciliary system of protochordates. As the anterior branchial arch evolved into the jaws, the associated musculature accompanied the bony or cartilaginous elements to become part of the jaw-closing and -opening system of gnathostome fishes.

In general, each branchial arch is endowed with its own set of branchiomeric musculature, enlarged or reduced as the function of the particular arch changes. The presumed ancestral condition is seen in fish (figure 10.37d). Here, a sheetlike **constrictor** muscle extends laterally from each branchial arch within the core of the gill, and may continue onto the body surface under the skin. The most medial part of this muscle sheet is separated from the rest of the arch, and is called the **adductor.** The constrictors squeeze water through the pharynx; adductors bend the arch. Small, deep muscles attach to the dorsal and ventral tips of the arches: the **dorsal** and **ventral branchial muscles.** They are involved in a variety of functions related to moving elements within the branchial arch.

Phylogenetically there is great fidelity between a cranial nerve and its branchiomeric muscles, and in turn fidelity between a set of branchiomeric muscles and its respective arch. Consequently, by tracking gill arches, we can track muscle homologies as well discover what each muscle, or its derivatives, becomes in different groups.

Mandibular Arch In sharks, both the mandibular constrictor and adductor lie at the body surface. The **adductor mandibulae,** the largest of the jaw muscles, is located at the angle of the jaws where it provides powerful closing forces. Joining the adductor in sharks is an oral muscle, the **pre-orbitalis,** which arises near the orbit and tapers as it passes posteriorly to its insertion on the adductor mandibulae or lower jaw. In bony fishes, the adductor is composed of several derivative muscles that act on selected parts of the highly kinetic skull. In tetrapods, the adductor mandibulae persists as a strong jaw adductor. It often has several prominent and distinctive heads that converge as a pinnate muscle on a common tendon. In mammals, the **masseter**

(a)

Procoracoid

(b)

Archaeopteryx ──→ Hypothetical ──→ Modern birds
                     intermediates

Elevator
(supracoracoideus)

Depressor (pectoralis)

(c)

FIGURE 10.36   Supracoracoideus in modern birds. (a) The supracoracoideus is shown as it originates at the sternum beneath the pectoralis, which has been removed. Its tendon passes over the coracoid and is inserted at the dorsal surface of the humerus; therefore, this muscle elevates the wing. (b) Proposed changes in the supracoracoideus from *Archaeopteryx* through hypothetical intermediates to modern birds. (c) Separate sets of pectoralis muscles originating at the sternum produce both the powerful downstroke of the wing and the return recovery. Recovery muscles originate at the sternum; pass up through an opening defined by scapula, furcula, and coracoid; and are inserted on the dorsal side of the humerus. As the pectoralis muscles contract, they lift the humerus and elevate the wing. Deltoideus muscles arising from the scapula may also aid in wing elevation. Powerful depressor muscles inserting on the ventral side of the humerus pull the wing downward.

After Goslow, Dial, and Jenkins.

and **temporalis** are jaw-closing muscles with different lines of action, and both are derived from the adductor mandibulae, as are the **pterygoideus** muscles.

The dorsal and ventral parts of the mandibular constrictor muscle are separated by the jaw itself. The ventral part is the **intermandibularis,** a transverse sheet of muscle, extending between the ventral edges of the paired mandibles. In tetrapods, the intermandibularis persists as a transverse muscle sheet, usually located between the lower jaws. In mammals, it is called the **mylohyoid;** the anterior part of the **digastric** muscle, involved in opening the jaws, is also derived from the intermandibularis.

In sharks, the dorsal derivative of the mandibular constrictor is the **levator palatoquadrati,** running from the chondrocranium to the palatoquadrate cartilage. In some fishes such as the chimaera, and in tetrapods, the palatoquadrate becomes fused to the braincase, becoming part of it, and the levator palatoquadrati muscle is absent.

Hyoid Arch   The hyoid arch begins as a separate gill arch in primitive fishes (a condition now seen only in chimaeras), but elements of the hyoid arch become secondarily involved in suspension of the mandibles in some vertebrates (other jawed fishes) and separate as the hyoid apparatus in others (tetrapods). Associated muscles shift their positions as well. The hyoidean constrictor muscles are prominent in fishes, where they form the main muscles of the water-breathing pump; but they are reduced or lost in tetrapods. In sharks, the largest of the hyoidean constrictors is the **levator hyomandibulae,** which reaches from the chondrocranium to the hyomandibular cartilage. The second of these, often closely fused with the first, is the **epihyoidean,** which is inserted on connective tissue behind the angle of the jaw. In bony fishes, the equivalent of the epihyoidean is the **levator operculi,** with its insertion on the operculum. The **depressor mandibulae** of tetrapods, which opens the jaws, is the homologue of the levator operculi and epihyoidean. In mammals, the depressor mandibulae evolves into the **stapedius,** but the **digastric** functions in opening the jaw, not the stapedius, which protects the inner ear from loud sounds. The posterior section of the digastric is derived from the ventral hyoid musculature, the interhyoideus.

In fishes, the ventral part of the hyoidean constrictor is the **interhyoideus.** This muscle runs transversely between the lower tips of the paired hyoid bars. In tetrapods, it forms additional thin sheets of muscles, the **constrictor colli,** which become rather extensive layers of facial muscles in mammals. One of these, the **platysma,** is an unspecialized muscle derived from the hyoid arch. Generally, it is a thin, subcutaneous muscle layer spanning the throat, fastening skin in the neck. Other muscles derived from the hyoid arch have more specialized functions, including control of facial expression and lips during feeding. Suckling in mammals may have led to initial differentiation of platysmal muscles in mammals.

Arch	Cranial Nerve Supply	Shark	*Necturus*	Alligator, Lizard	Cat, Mink
BRANCHIOMERIC MUSCULATURE					
I	V	Levator palatoquadrati Spiracularis Adductor mandibulae Preorbitalis	Adductor mandibulae (levator mandibulae)	Adductor mandibulae Pterygoideus (4 present)	Masseter Temporalis Pterygoids Tensor veli palati Tensor tympani
		Intermandibularis	Intermandibularis	Intermandibularis	Mylohyoid Anterior digastric
2	VII	Levator hyomandibulae	Depressor mandibulae Branchiohyoideus	Depressor mandibulae Branchiohyoideus	Stapedius Platysma and facial muscles (part)
		Interhyoideus	Interhyoideus Constrictor colli	Interhyoideus Constrictor (sphincter) colli (gularis) in part	Platysma and facial muscles (part) Posterior digastric Stylohyoid
3	IX, X *XI	Cucullaris	Cucullaris Levatores arcuum	Trapezius Sternomastoid	Trapezius complex Sternocleidomastoid complex
		Interarcuals	—	—	
		Superficial constrictors and interbranchials	Dilatator laryngis Subarcuals Transversi ventrales Depressores arcuum	Some intrinsic muscles of the larynx and pharynx	Some intrinsic muscles of the larynx and pharynx
HYPOBRANCHIAL MUSCULATURE					
	*XII	Coracoarcuals Coracohyoid	Rectus cervicis	Rectus cervicis Sternohyoid Omohyoid	Sternohyoid Omohyoid Thyohyoid
		Coracomandibularis	Genioglossus Geniohyoid	Genioglossus Geniohyoid	Geniohyoid, others of the tongue and larynx
		Coracobrachialis	—	—	

*In tetrapods.

Thereafter, facial control during feeding became important in herbivorous mammals, because the lips are used to help grasp and break off parts of plants. In the rhinoceros (figure 10.38), the **levator labii superioris** and **levator nasolabialis** move the upper lip, and the **depressor labii mandibularis** moves the lower lips. The **zygomaticus** controls the corner of the mouth. The **orbicularis oris** closes the lips. Contraction of the **caninus** flares the nostril. The **buccinator** flattens the cheeks, thus pressing food between the tooth rows. Our own faces use similar muscles and are highly expressive.

Branchial Arches As mentioned, the constrictor muscle of a typical gill of fishes consists of a deep part in the core of the gill, and in elasmobranchs, a superficial constrictor that forms a simple gill cover on the body surface. Furthermore, short dorsal and ventral branchial muscles lie on the tops and bottoms of the arches and act with the adductors to control local movements of the arches during gill ventilation (figure 10.39). Related to the dorsal branchials is the cucullaris, formed by the fusion of slips from several successive

branchial muscles. It runs from the dorsal body surface down to the last branchial arch and to the scapula. In tetrapods, it extends from the axial musculature to the scapula, usually forming a pair of muscle complexes, the trapezius and mastoid groups.

The branchial arches are important structural components of the pumping and feeding apparatus in fishes. They become reduced in tetrapods and contribute only to the larynx and other parts of the throat. Associated constrictor muscles similarly become reduced to laryngeal muscles; however, some levators take on enlarged roles, contributing as the trapezius and mastoid muscles to the muscular sling supporting the shoulder girdle.

Hypobranchial Musculature

Hypobranchial muscles arise from cervical somites, whose ventral ends migrate to the floor of the pharynx (figure 10.37d). The hypobranchial musculature is supplied with spinal nerves and runs below the lower ends of the

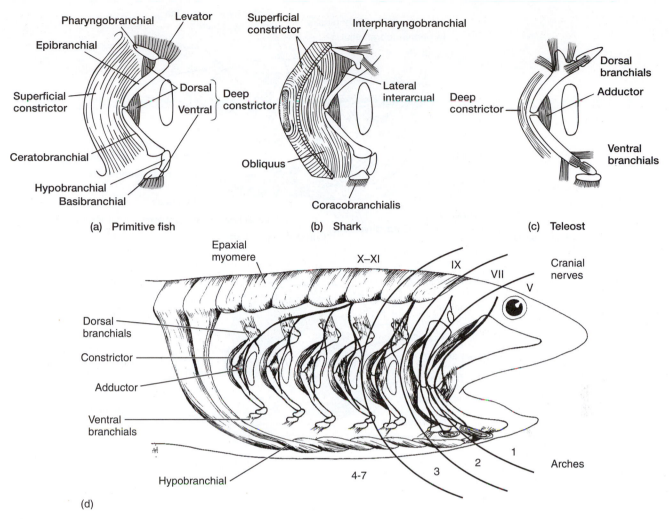

FIGURE 10.37 Branchiomeric musculature. (a) Lateral view of a primitive branchial arch illustrating the basic sets of levator and constrictor muscles together with their skeletal structures. (b) Shark branchial arch. (c) Teleost branchial arch. (d) Jaw muscles and their cranial nerve supply tend to stay with their respective branchial arch during the course of subsequent evolution. Each arch has levator and constrictor muscles that, respectively, elevate and close the articulated elements. Cranial nerves V, VII, IX, and X–XI supply muscles of arches 1, 2, 3, and 4–7, respectively. Fidelity of muscles, nerves, and arches generally was maintained as the branchial arches evolved and subsequently became modified into components of the jaws.

After Jollie; Mallatt.

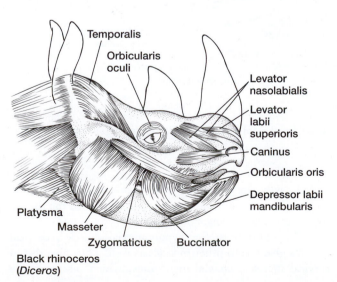

Black rhinoceros
(*Diceros*)

FIGURE 10.38 Head of black rhinoceros illustrates facial muscles that act on the borders of the mouth and nostril.

(a) Shark (*Squalus*)

(b) Lungfish (*Neoceratodus*)

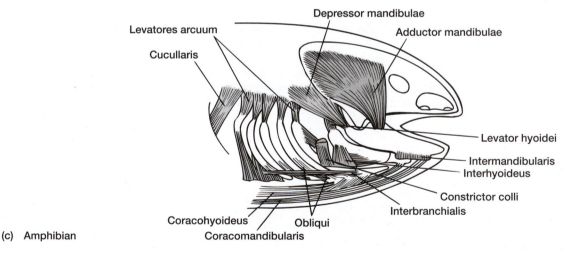

(c) Amphibian

FIGURE 10.39 **Lateral view of head musculature.** (a) The shark *Squalus*. (b) The lungfish *Neoceratodus*. (c) Amphibian (composite of adult muscles of an anuran and urodele).

After Jollie.

branchial arches in an anteroposterior course. In fishes, these muscles originate from the coracoid region of the shoulder girdle. They consist of the coracomandibularis and a sternohyoideus; although in sharks, the sternohyoideus is divided into an anterior coracohyoideus and a posterior coracoarcualis. They are prominent jaw openers and expanders of the buccal cavity. In tetrapods, they accompany the branchial arches with contributing muscles associated with the throat, hyoid apparatus, larynx, and tongue.

In addition to the hypobranchial muscles, cervical somites give rise to other muscles as well. They give rise to cervical muscles, **epaxial myotomes,** which insert on the

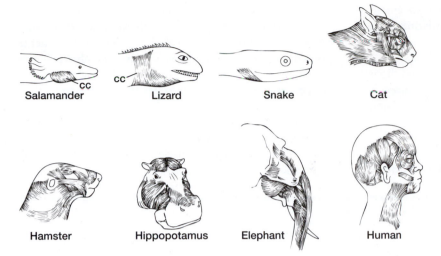

Salamander   Lizard   Snake   Cat

Hamster   Hippopotamus   Elephant   Human

FIGURE 10.40 Evolution of facial muscles. In tetrapods, the hyoid musculature expands over and partially encircles the neck as a thin sheet, the constrictor colli (cc). This muscle also tends to adhere to the dermis of the skin. In mammals, the musculature derived from the hyoid arch expands radically over the head and differentiates into a suite of facial muscles. They are best differentiated around the eyes, lips, and ears, where they serve to accent facial expression.

posterior wall of the neurocranium and may lift it during jaw opening. In sharks alone, some of the cervical somites also contribute to the **interpharyngobranchials,** which join the successive branchial arches in the pharynx.

Overview

Skeletal muscles move bones and cartilages; cardiac muscles move blood; smooth muscles control viscera. All are based on the underlying molecular machinery of sliding filaments, actin and myosin. Skeletal muscles attach to bones through tendons. Despite this common basis for contraction, skeletal muscles achieve different levels of performance in various ways. Tonic and twitch fibers produce relatively slow and fast contraction, respectively. In the muscle organ, active (sliding filaments) and elastic components contribute to overall force (tension). Graded force is generated by rate modulation and by selective recruitment of motor units. Features of performance, such as force and velocity of the insertion, are also affected by cross-sectional area, fiber orientation (parallel versus pinnate), muscle length, and insertion sites within a lever system (proximal versus distal). Sequencing of muscle contraction brings individual muscles into action at favorable moments during complex motion and permits recruitment as speed and endurance change.

In response to chronic exercise, muscles hypertrophy due to increased vascularization, additions of connective tissue, and enlargement of muscle cells due to addition of myofilaments. However, the number of individual muscle cells does not increase significantly. Weight training may stimulate conversion of one type of fast twitch to another type of fast twitch. However, whatever the training schedule, no significant conversion occurs between *slow* twitch and *fast* twitch fibers. With aging, muscle organs lose mass due to loss of muscle fibers. Both fiber types, fast and slow, are lost, up to 10% by age 50. However, aging is harder on fast twitch fibers, which are lost at a higher rate than are slow twitch fibers.

Within fishes, the axial musculature predominates and is represented by segmentally arranged blocks of muscle. In tetrapods, the axial musculature is relatively reduced, and the appendicular muscles increase in prominence. Tetrapods also exhibit greater complexity within the muscle masses, as illustrated by the several hundred different muscles that differentiate from these masses.

Phylogenetic differentiation of muscles occurs in many ways. The **migration** of muscle primordia, termed **anlagen,** to other regions of the body is one way. Muscularization of the diaphragm in mammals begins with the appearance of muscle anlagen in the neck that migrate posteriorly during embryonic development into the body septum anterior to the liver. There they differentiate into the skeletal muscles of the diaphragm. The hyoid musculature spreads throughout the neck and head in some tetrapods to form the platysma and other muscles of the face (figure 10.40). Another way muscles differentiate phylogenetically is by **fusion.** The rectus abdominis extending along the belly of tetrapods is formed by fusion of the ventral regions of several myotomes that grow into the abdominal region during early embryonic development. In this example, a single muscle arises from fusion of parts of successive myotomal segments. Muscles can also differentiate by **splitting.** The pectoralis muscle in the chest is a large, fan-shaped muscle in lower tetrapods, but in mammals it divides into as many as four discrete muscles that act on the forelimb.

Diaphragm (p. 197)

The greater differentiation of the tetrapod musculature compared with the fish musculature reflects the changing demands of terrestrial support and locomotion. Running and flying and other tetrapod activities involve more than just the mechanics of swinging limbs or flapping wings. These are complex motor patterns that require precise control. A cheetah dashing across an irregular landscape not only must oscillate its limbs rapidly, but its foot placement must quickly accommodate slight surface irregularities each time the foot touches the ground. As the changing airstream and wind gusts suddenly alter the aerodynamic flow across a bird's wing, the bird must quickly adjust its wings and flight feathers. The more differentiated musculature of tetrapods is an indirect indication of this greater variety and precision of movements they can and must perform.

Muscle phylogeny also illustrates the remodeling character of evolution. Muscles arising initially in the jaws (e.g., trapezius, mastoid) and in the axial musculature (e.g., serratus) become incorporated into the muscular system of the shoulder and forelimb. On the other hand, we also see within muscle evolution a remarkable fidelity between muscles and their nerve supply. The phrenic nerve to the diaphragm arises anteriorly, like the muscle anlagen, in the cervical region and accompanies this muscle to its final destination, posteriorly within the body. Sets of muscles associated with the branchial arches are, in general, supplied faithfully throughout vertebrates by the same cranial nerve, despite the fact that these muscles of the branchial arch are often remodeled to serve new roles in tetrapods.

WEB LINK

Visit the text website at *www.mhhe.com/kardong7e* for additional study aids including selected references, functional laboratories, and web links to chapter topics.

CHAPTER **11**

The Respiratory System

INTRODUCTION

RESPIRATORY ORGANS
Gills
Gas Bladders
Lungs
Swim Bladders
Cutaneous Respiratory Organs
Accessory Air-breathing Organs
Breathing and Embryos

VENTILATORY MECHANISMS
Cilia
Muscular Mechanisms
Water Ventilation: Dual Pump
Air Ventilation: Buccal Pump
Air Ventilation: Aspiration Pump

PHYLOGENY
Agnathans
Elasmobranchs
Bony Fishes
Overview of Fish Respiration
Gills
Lungs and Swim Bladders
Amphibians
Amphibian Larvae
Amphibian Adults

Reptiles
Mammals
Ventilation
Gas Exchange
Birds

FORM AND FUNCTION
Patterns of Gas Transfer
Rates of Gas Transfer
Breathing in Water
Breathing in Air

EVOLUTION OF RESPIRATORY ORGANS
Acid-Base Regulation
Ventilation
Ciliary Pumps
Muscular Pumps
Water-to-Land Transition
Air-breathing Organs
Advantages of Movement to Land
Air-breathing Mechanisms
Bird Lungs and Air Sacs

OVERVIEW

Introduction

To metabolize effectively and survive, cells within the body of a vertebrate must replenish the oxygen used and rid themselves of the by-products accumulated during metabolism. These chores fall primarily to two transport systems, the circulatory and respiratory systems. The circulatory system basically connects cells deep within the body with the environment and is discussed in chapter 12. The respiratory system, our focus in this chapter, involves gas exchange between the surface of an organism and its environment. At their simplest, these two systems aid in the process of **passive diffusion,** the random movement of molecules from an area of high to an area of low partial pressure (figure 11.1a). Oxygen is usually (but not always) at high partial pressure in the environment and tends to diffuse into the organism. Carbon dioxide collects in tissues and tends to diffuse out.

Yet, if unaided, passive diffusion alone is insufficient to meet the needs of large multicellular organisms. For example, a hypothetical, spherical aquatic organism could

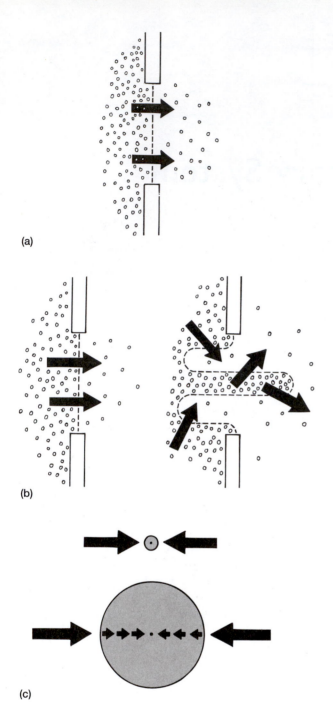

FIGURE 11.1 **Passive diffusion.** (a) Molecules of gas move from an area of high partial pressure to an area of low partial pressure. Eventually an equilibrium is reached when the concentration of molecules becomes equal on both sides of the surface across which diffusion occurs. (b) The rate of movement of diffusing molecules depends upon available surface area. Increasing the surface area increases the rate at which diffusion occurs, although the final equilibrium concentration will eventually be the same regardless of the surface area available. (c) The time it takes for molecules to reach deep tissues depends upon the distance they must travel. Molecules moving into cells at the center of the small circle reach the core much sooner than molecules that must traverse thicker tissues to reach those cells within the center.

have a radius no larger than 0.5 mm for its core tissues to receive adequate oxygen by diffusion alone, even if the surrounding water were saturated with oxygen. For oxygen to diffuse passively from your lungs to your extremities, the journey would require several years! Obviously this would not do. In large multicellular organisms, transport systems aid passive diffusion. The circulatory and respiratory systems speed the process.

Major modifications in the design of respiratory organs have occurred during animal evolution to optimize the diffusion of important gases. The rate of passive diffusion between an organism and its environment depends on several factors. One is surface area. The greater the available surface area, the greater is the opportunity for molecules to move across an epithelial surface (figure 11.1b). For instance, the gas exchange organs of vertebrates are highly subdivided to increase the surface available to transfer gases between air and blood. Another factor is distance. The greater the distance, the longer it will take for molecules to reach their destinations (figure 11.1c). Thick tissues slow diffusion, and thin barriers aid the process. The thin walls of the respiratory organs reduce the distance between the environment and the blood. A third factor is the resistance to diffusion by the tissue barrier itself. The moist skin of living amphibians facilitates gas transfer. In contrast to this situation, the skin of most mammals is cornified and thick, a feature that slows gas diffusion with the environment.

One of the most important factors affecting diffusion rate is the difference in partial pressures across the exchange surface. The gills of most fishes experience a high partial pressure of oxygen relative to the blood; therefore, oxygen diffuses across the gills into the blood. Occasionally a fish living in warm, stagnant water may encounter a partial pressure of oxygen in the water below that of its blood. Under these unusual conditions, oxygen may actually diffuse in the reverse direction, and the fish is in danger of losing oxygen to the water!

Both respiratory and circulatory systems have "pumps" that move fluids, such as air or water (respiration) or blood (circulatory). The heart is a pump that circulates blood. In fishes, the predominant respiratory pump is the branchial apparatus that drives water across the gills (figure 11.2a). In tetrapods, one familiar pump is the rib cage, sometimes assisted by a diaphragm, that moves air through the lungs (figure 11.2b). As we will see, many types of supplementary pumping devices in vertebrates are also part of the respiratory mechanism. By moving fluids that contain gases, these pumps function to maintain high partial pressure gradients across exchange surfaces.

The respiratory and circulatory systems, although anatomically distinct, are functionally coupled in the process of **respiration,**[1] the delivery of oxygen to tissues and the removal

---

[1]Biochemists have usurped the term and use it to refer to something quite different, namely, *chemical* respiration, the aerobic degradation of substrates in biochemical pathways.

Respiratory pump

Systemic tissues

Circulatory pump

(a) Fish

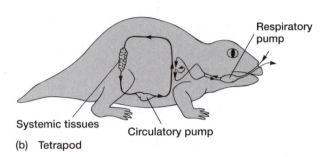

Respiratory pump

Systemic tissues

Circulatory pump

(b) Tetrapod

FIGURE 11.2 **Respiratory and circulatory systems cooperate to deliver oxygen to deep tissues and carry away carbon dioxide.** Both systems are diagrammed in the figure. During external respiration, air or water is inhaled and transported to the exchange capillaries of the blood. Thereafter, blood circulates oxygen to all systemic (body) tissues, represented here by a small patch of tissue, where internal respiration occurs. Oxygen is delivered to these tissues and carbon dioxide is carried away. (a) In fishes, the respiratory pump usually includes the branchial arches and their musculature. External respiration occurs in the gill capillaries. The heart, being the primary circulatory pump, drives blood through the gills and then to the systemic tissues. (b) In tetrapods, this respiratory pump can include the buccal cavity, which forces air into elastic lungs against resistance, and a rib cage around the lungs. External respiration occurs in the lungs. The circulatory pump, or heart, drives blood through vessels. Internal respiratory exchange occurs between blood and systemic tissues.

of waste products, principally carbon dioxide. **External respiration** refers to gas exchange between the environment and blood via the respiratory surface. **Internal respiration** refers to gas exchange between the blood and the deep body tissues.

During external respiration, gases diffuse between the environment and the organism—oxygen enters, carbon dioxide departs. **Ventilation,** or breathing, is the active process of moving the respiratory medium, water or air, across the exchange surface. Ceasing the movement of the respiratory medium is **apnea,** or breath holding. Pumping of blood through an organ via capillaries is known as **perfusion.** The **respiratory organs** specialize in ventilation to deliver oxygen and remove carbon dioxide accumulated during perfusion. The demands on the respiratory organ vary, depending on whether the medium is water or air. This is partly due to differences in density. Water, being denser than air, requires more energy to set it in motion. Other things being equal,

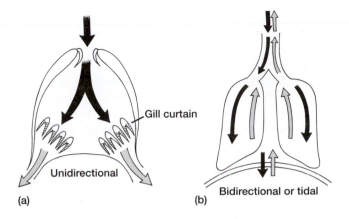

Gill curtain

Unidirectional

(a)

Bidirectional or tidal

(b)

FIGURE 11.3 **Unidirectional and bidirectional flow.** (a) In fishes and many aquatic amphibians, water movement is unidirectional because water flows through the mouth, across the gill curtain, and out the lateral gill chamber. (b) In many air-breathing vertebrates, air flows into the respiratory organ and then reverses its direction to exit along the same route, creating a bidirectional or tidal flow.

ventilation that involves moving water is energetically more expensive than ventilation that involves moving thin air. In addition, because water is more dense, structures are more buoyant in water than in air. Gills supported by water tend to collapse in air and therefore fail as respiratory organs on land. Lungs are structurally reinforced to work better in air.

But it is not just differences in the physical properties of air and water that affect ventilation and the devices that serve ventilation. The solubility of gas in air differs from that in water. This means that the availability of gases to respiratory organs differs in air and water. Atmospheric air is composed of oxygen (about 21%), nitrogen (about 78%), and carbon dioxide (less than 0.03%). Trace elements make up the rest. In microenvironments, such as animal burrows, the composition may change slightly. But, in general, the partial pressure of the physiologically important gases at sea level is extremely constant worldwide. Although partial pressure varies with altitude, the composition of gases in air is relatively unchanged up to over 100 km, thanks to mixing by winds and air currents. However, in water, the situation is quite different. When brought into contact with water, these gases go into solution. The amount of gas that dissolves in water depends on the chemistry of the gas itself, its partial pressure in air, the temperature of water, and the presence of other dissolved substances. As a result, the amount of oxygen in water can be quite variable; furthermore, it is never as concentrated as it is in air.

In most fish gills, ventilation is **unidirectional.** Water enters the buccal cavity through the mouth, passes across the row of gills known as the **gill curtain,** and exits flowing in one direction only (figure 11.3a). In active fishes, ventilation is almost continuous to keep a more or less steady stream of new water washing across the exchange surfaces of the gills. Lung ventilation, however, is usually **bidirectional (tidal),** with air entering and exiting through the same channels (figure 11.3b). A fresh breath of air that is **inhaled** into the lungs mixes with spent air and is **exhaled.** The exchange capillaries of the lung are replenished intermittently, not continuously, with air.

Vertebrates that live in aqueous environments most often encounter too little oxygen, termed **hypoxia,** partly because water is already low in dissolved oxygen. For this reason, most organs that supplement respiration are found among aquatic rather than strictly terrestrial animals.

One of the major transitions in vertebrate evolution was the change from water breathing to air breathing. This major evolutionary event, together with the physiology of the respiratory system, has made respiration the focus of much research. Let us begin by looking at the various organs that have arisen to facilitate respiration. They have something to tell us about the evolutionary forces at work in designing the respiratory system in water, in air, and in between.

Respiratory Organs

Gills

Vertebrate gills are designed for water breathing. Specifically, they are dense capillary beds in the branchial region that serve external respiration. They are supported by skeletal elements, the branchial arches. The mechanism of gill ventilation depends on whether the gills are located internally or externally. **Internal gills** are associated with **pharyngeal slits** and **pouches.** Often they are covered and protected laterally by soft skinfolds, such as the **interbranchial septum** in chondrichthyan fishes, or by a firm **operculum,** as in many osteichthyan fishes (figure 11.4a–c). Ventilation usually involves the muscular pump of the buccal cavity actively driving water across the internal gills. **External gills** arise in the branchial region as filamentous capillary beds that protrude into the surrounding water (figure 11.4d). They are found in the larvae of many vertebrates, including lungfishes, some actinopterygians, and amphibians. Water currents flow across their projecting surfaces, or in still water specialized muscles sweep external gills back and forth to ventilate them.

Gas Bladders

Many actinopterygian fishes possess a **gas bladder,** an elongated sac filled with gas, usually air that enters via a **pneumatic duct** connected to the digestive tract or with gas secreted directly into the bladder from the blood. Gas bladders participate in buoyancy control (swim bladders) and sometimes in respiration (lungs).

Lungs

Vertebrate lungs are designed for air breathing. Lungs are elastic bags that lie within the body. Their volume expands when air is inhaled and decreases when air is exhaled. Embryologically, lungs arise as endodermal outpocketings from the pharynx. In primitive fishes and most tetrapods,

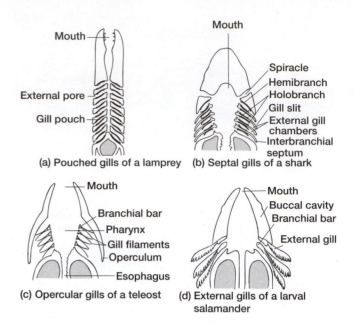

(a) Pouched gills of a lamprey
(b) Septal gills of a shark
(c) Opercular gills of a teleost
(d) External gills of a larval salamander

FIGURE 11.4 **Gill coverings.** (a) Branchial pouch in lampreys. No cover protects the lateral opening of the gill chamber. (b) Septal gills in sharks. Individual flap valves formed from individual gill septa guard each gill chamber. (c) In most teleosts and some other species, a common operculum covers their several gills. (d) In larval salamanders, the branchial arches support vascular external gills that project into the surrounding water.

the lungs of adults are usually paired. They lie ventral to the digestive tract and, in amniotes, are connected to the outside environment through the **trachea.** Entrance into the trachea is gained through the **glottis,** which is guarded by tiny sets of muscles that open and close it. Usually the trachea branches into two **bronchi,** one to each lung. In some species, each bronchus branches into successively smaller **bronchioles** that eventually supply air to the respiratory surfaces within the lung. In tetrapods with slender bodies, one lung may be reduced in size; and in some amphisbaenids and most advanced snakes, only a single lung is present.

The trachea, bronchi, and bronchioles can hold a significant volume of air. Although exhalation forces most of the spent air from the lungs, some remains in these passageways. Upon inhalation, this "spent" air is drawn back into the lungs before fresh air from outside reaches the lungs to mix with the used air. This volume of used air within the respiratory passageways is called the **dead space.** The total volume inhaled in a single breath is referred to as the **tidal volume.** In a chicken, the dead space may represent up to 34% of the total tidal volume. Normal tidal volume of a human at rest is about 500 ml. Because the dead space is about 150 ml (30%), 350 ml (500 ml–150 ml) of fresh air actually reaches the lungs.

Swim Bladders

If the gas bladder is used to control the buoyancy of the fish in the vertical water column, it is referred to as a **swim bladder.** Occasionally gas bladders may be heavily vascularized as well to participate in supplementary respiration and are called **respiratory gas bladder** or just **lungs.** The internal vascular walls of lungs are subdivided into many partitions that increase the surface area available for external respiratory exchange.

Swim bladders differ from lungs in three ways. First, swim bladders are usually situated dorsal to the digestive tract, whereas lungs are ventral. Second, swim bladders are single, whereas lungs are usually paired. *Neoceratodus,* the Australian lungfish, is an exception, because as an adult it has a single lung dorsal to the digestive tract; however, its trachea originates ventrally from the digestive tract. Its embryonic lung arises initially as a paired primordium, suggesting that the single lung of *Neoceratodus* is a derived condition. Third, in swim bladders, returning blood drains to the general systemic circulation (cardinal veins) before entering the heart. In lungs, venous return enters the heart separately from the general systemic circulation.

Despite their differences, swim bladders and lungs share many basic similarities of development and anatomy. Both are outpocketings from the gut or pharynx and have roughly equivalent nerve and muscle supplies. Some morphologists take these similarities as evidence that lungs and swim bladders are homologous. Even if they are homologous, it is not clear which function came first, gas transfer or buoyancy control. The two functions are not mutually exclusive. A filled swim bladder that aids gas transfer also makes the fish more buoyant, and a swim bladder used for buoyancy can also be tapped as a temporary source of oxygen. Among fishes, evolutionary reversals between respiratory and buoyancy functions have occurred repeatedly. Lungs have evolved into nonrespiratory swim bladders that in subsequent evolutionary processes have reverted to lungs.

Phylogenetically, neither lungs nor swim bladders are present in agnathans, elasmobranchs, or placoderms (figure 11.5). Lungs, ventral in position, likely arose in the immediate common ancestor to actinopterygians and sarcopterygians. Swim bladders of actinopterygians, if homologous, would be later derivatives of lungs (figure 11.5). The behavioral habit of gulping air may have been in place even before lungs. Tipping up to gulp surface air occurs in fishes without specialized air-breathing organs, even in some sharks. Thus, the habit of gulping air bubbles pressed into gill chambers may have preceded specialized anatomical devices designed to efficiently extract oxygen from air and exploit such behaviors.

Although I follow the view that double ventral lungs are the primitive condition in bony fishes (figure 11.5), we should allow that lungfishes may argue differently, with a single dorsal lung being the primitive condition. The Australian lungfish has such a single dorsal lung and may be plesiomorphic compared to the more derived African and South American lungfishes. The Australian lungfish has large, fleshy fins, a large operculum, a hefty fish body, well-developed gills, and until recent overfishing, ventured into salt water as well as fresh water like early bony fishes. Other lungfishes are more derived with diminutive fins, small opercula, elongated eel-like bodies, obligate air-breathers, and strictly freshwater in habitat. By this view, the double, ventral lung of the African and South American lungfishes would represent a derived state.

Cutaneous Respiratory Organs

Although lungs and gills are the primary respiratory organs, the skin can supplement breathing. Respiration through the skin, referred to as **cutaneous respiration,** can take place in air, in water, or in both. In the European eel and plaice, oxygen uptake through the skin may account for up to 30% of total gas exchange (figure 11.6). Amphibians rely heavily on cutaneous respiration, often developing accessory skin structures to increase the surface area available for gas exchange. In fact, in salamanders of the family Plethodontidae, adults lack lungs and gills and depend entirely on cutaneous respiration to meet their metabolic needs. Like most mammals, humans respire very little cutaneously, although our skin is permeable to some chemicals applied topically (spread on the surface). In fact, many medicinal ointments are absorbed through the skin. Bats take advantage of cutaneous respiration across their well-vascularized wing membranes to eliminate as much as 12% of their total carbon dioxide waste, but they take up only 1% or 2% of their total oxygen requirement through this cutaneous route (figure 11.6). Feathers and poorly vascularized skin of birds preclude cutaneous respiration. Similarly, in reptiles, the surface covering of scales limits cutaneous respiration. However, in areas between scales (at the hinges of scales) and in areas with reduced scales (e.g., around the cloaca), the skin is heavily vascularized to allow some cutaneous respiration. Sea snakes can supplement up to 30% of their oxygen intake via cutaneous respiration across the skin on their sides and back. Many turtles pass the cold winter in hibernation safely at the unfrozen bottom of ponds where the limited respiration around their cloaca is sufficient to meet their reduced metabolic needs.

The newly hatched larva of the teleost fish *Monopterus albus,* an inhabitant of southeast Asia, uses predominantly cutaneous respiration during its early life. At hatching, the large and heavily vascularized pectoral fins beat in such a fashion as to drive a stream of water backward across the surface of the larva and its yolk sac. Blood in superficial skin vessels courses forward. This establishes a countercurrent exchange between water and blood to increase the efficiency of cutaneous respiration in this larva (figure 11.7a). Such a respiratory organ allows the larva to inhabit the thin layer of surface water into which nearby oxygen from the air has dissolved. Similarly, in many amphibians, increased surface area allows for increased cutaneous gas exchange (figure 11.7b, c).

Countercurrent exchange (p. 156)

Accessory Air-breathing Organs

Lungs and skin are not the only organs that tap sources of oxygen in the air. Many fishes have specialized regions that take up oxygen from the air. *Hoplosternum,* a tropical catfish

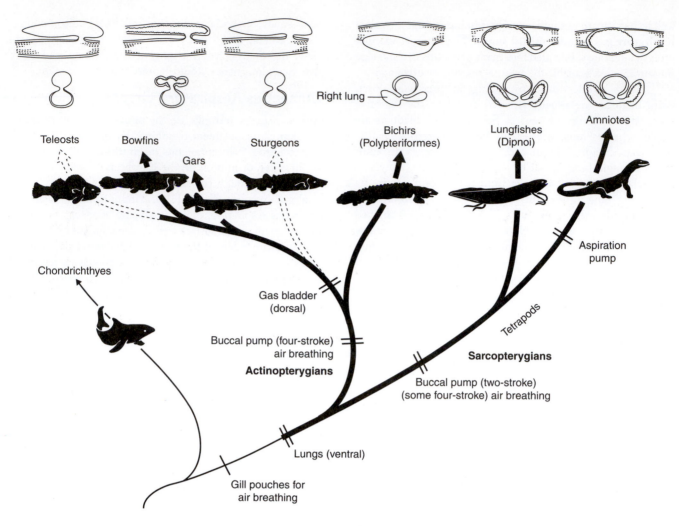

FIGURE 11.5 **Evolution of gas bladders.** Lungs, ventral in position, evolved in the common ancestor to actinopterygians and sarcopterygians. Swim bladders in actinopterygians may have evolved independently, or they may have been modified from earlier lungs. Some gas bladders are respiratory in function. Above the dendrogram outlining the evolutionary rise of each group, there are sagittal (top) and cross-sectional (bottom) views of the lung and its connection to the digestive tract. In Polypteriformes (*Polypterus*), paired lungs open through a common muscular glottis into the right floor of the pharynx. The left lung is reduced, the right one long, but the epithelial lining of both is smooth. Swim bladders of sturgeons originate from the stomach and those of primitive teleosts from the esophagus, suggesting that these nonrespiratory gas bladders may be of independent origin in these two groups. Dashed arrows indicate points where the respiratory function of the gas bladder is lost.

Based on Liem; Perry et al.

found in fresh waters in South America, gulps air and swallows it into its digestive tract (figure 11.8a). Oxygen in the gulped air diffuses across the wall of the digestive tract into the bloodstream. The digestive tract is richly supplied with blood vessels that supplement gill respiration. The electric eel *Electrophorus* gulps and holds air in its mouth to expose capillary networks of the mouth to oxygen (figure 11.8d).

Gills ordinarily are unsuitable organs for air breathing. The moist, leaflike exchange surfaces stick together in air and collapse without the buoyant support of water. However, in some fishes, gills are used in air breathing (figure 11.8b). The rockskipper *Mnierpes*, an inhabitant of wave-swept rocky shores of the tropical Pacific coast

of Central and South America, occasionally makes brief sojourns onto land to scrounge for food, to evade aquatic predators, and to avoid periods of intense wave action. During these sojourns, it holds gulped air against its gills to extract oxygen. Its gills are reinforced to prevent their collapse during these bouts of air breathing.

Breathing and Embryos

Among anamniotes, respiration generally takes place directly between the surrounding environment and the embryo across the skin. In birds and most reptiles, the embryo is wrapped in extraembryonic membranes and

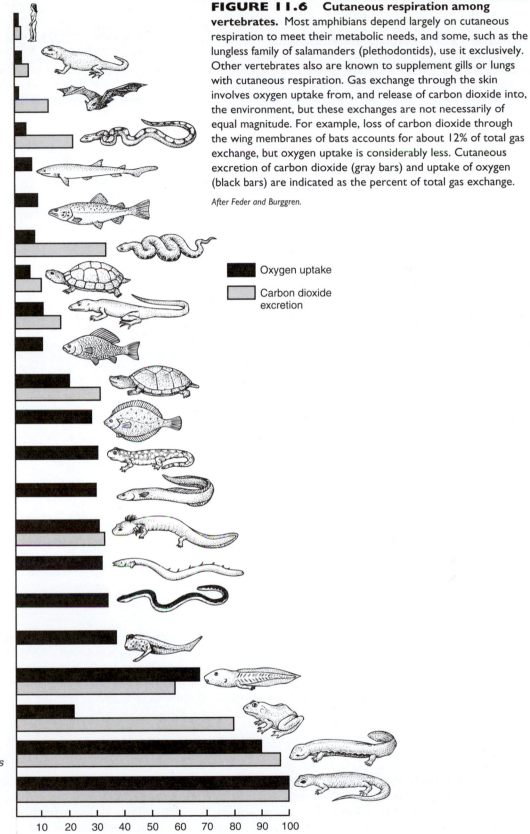

FIGURE 11.6 Cutaneous respiration among vertebrates. Most amphibians depend largely on cutaneous respiration to meet their metabolic needs, and some, such as the lungless family of salamanders (plethodontids), use it exclusively. Other vertebrates also are known to supplement gills or lungs with cutaneous respiration. Gas exchange through the skin involves oxygen uptake from, and release of carbon dioxide into, the environment, but these exchanges are not necessarily of equal magnitude. For example, loss of carbon dioxide through the wing membranes of bats accounts for about 12% of total gas exchange, but oxygen uptake is considerably less. Cutaneous excretion of carbon dioxide (gray bars) and uptake of oxygen (black bars) are indicated as the percent of total gas exchange.

After Feder and Burggren.

Oxygen uptake

Carbon dioxide excretion

Human
Homo sapiens

Chuckwalla
Sauromalus obesus

Big brown bat
Eptesicus fucus

Boa constrictor
Boa constrictor

Cat shark
Scyliorhinus canicula

Brown trout
Salmo trutta

Elephant trunk snake
Acrochordus javanicus

Red-eared turtle
Pseudemys scripta

Green lizard
Lacerta viridis

Goldfish
Carassius carassius

Southern musk turtle
Sternotherus minor

Plaice
Pleuronectes platessa

Tiger salamander
Ambystoma tigrinum

European eel
Anguilla anguilla

Mudpuppy
Necturus maculosus

Reedfish
Calamoichthys calabaricus

Pelagic sea snake
Pelamis platurus

Mudskipper
Periophthalmus cantonensis

Bullfrog (larva)
Rana catesbeiana

Bullfrog (adult)
Rana catesbeiana

Hellbender
Cryptobranchus alleganiensis

Lungless salamander
Ensatina eschscholtzii

10 20 30 40 50 60 70 80 90 100

Percent of total gas exchange

FIGURE 11.7 Adaptations for cutaneous respiration.

Many vertebrates exhibit complex or elaborate specializations that enhance the efficiency of gas exchange through the skin. (a) While still small, this fish larva, *Monopterus albus*, occupies the thin layer of water adjacent to the surface where oxygen levels are relatively high. Its pectoral fins beat, forcing water to flow across its body surface. Blood circulating through the skin flows in the opposite direction from the water, establishing a countercurrent exchange between blood and water. (b) In the Lake Titicaca frog, *Telmatobius culeus*, prominent loose skinfolds on its back and limbs provide extensive surface area for cutaneous respiration. (c) In the male hairy frog, *Astylosternus robustus*, numerous papillae appear during the breeding season, forming a ruffled supplementary respiratory organ on its sides and hindlimbs.

(a) After Liem; (b, c) after Feder and Burggren.

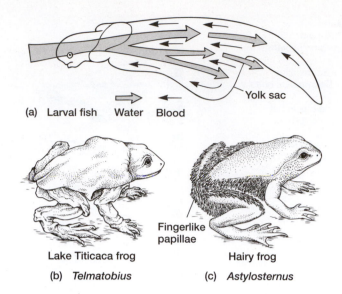

(a) Larval fish   Water   Blood   Yolk sac

(b) *Telmatobius* — Lake Titicaca frog

(c) *Astylosternus* — Hairy frog — Fingerlike papillae

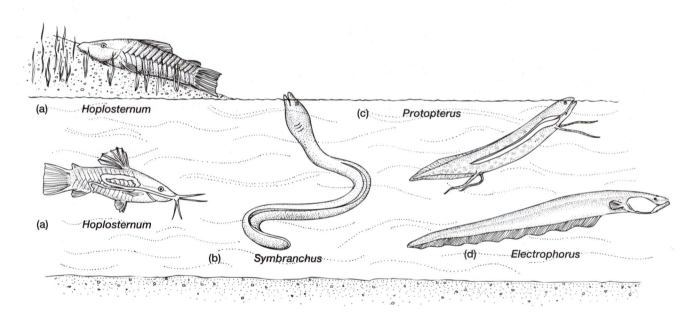

(a) *Hoplosternum*

(a) *Hoplosternum*

(b) *Symbranchus*

(c) *Protopterus*

(d) *Electrophorus*

FIGURE 11.8 Air-breathing fishes.

Fishes that temporarily breathe air usually live in waters where oxygen depletion occurs seasonally or frequently. Gulping air supplements depressed oxygen uptake through gills and helps a fish through short periods of hypoxia. (a) *Hoplosternum*, a carplike fish, swallows air into its intestine, where extra capillary beds take up this supplemental oxygen. (b) *Symbranchus* holds a gulped air bubble against its reinforced gills to take up extra oxygen. (c) *Protopterus*, a lungfish, has well-developed lungs for breathing air. (d) *Electrophorus*, an electric eel, gulps air into its mouth and takes up oxygen through the wall of its mouth.

After Johansen.

enclosed in a shell. One of these membranes, the chorioallantois, lies directly beneath the shell and acts as the respiratory organ. The porous shell allows oxygen to be picked up by the blood circulating within the chorioallantois and carbon dioxide to be eliminated from it. The chorioallantois sustains the respiratory needs of the chicken embryo for most of its time in the egg (figure 11.9a, b). About six hours before hatching, the chick pokes through the inner shell membrane to push its beak into a small air space within the egg. This allows its lungs to fill for the first time and begin to participate, along with the chorioallantois, in air breathing. When the chick breaks through the outer shell several hours later, its lungs breathe atmospheric air directly and the chorioallantois quickly shuts down (figure 11.9c).

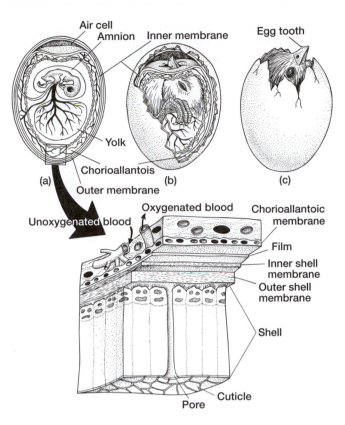

FIGURE 11.9 **Respiration in the chicken embryo.** (a) While the chick embryo is enclosed in its shell, it respires through this porous shell. The chorioallantois carries blood to the inner surface of the shell to exchange gases at this interface. The shell proper is made up of calcite crystals pierced by tiny pores. Inner and outer shell membranes separate the shell from the vascularized chorioallantois. The chick embryo meets all its respiratory needs, up to day 19 of incubation, as air passes through the porous shell and exchanges gases with blood in the chorioallantois. (b) On day 19, the embryo pokes its beak through the inner shell membrane into the air space between both membranes. Its lungs inflate, and the chick breathes air in addition to continued respiration via the chorioallantois. (c) Six hours later, the chick pecks through the shell proper, a process termed *pipping*, to breathe atmospheric air directly. Thereafter, chorioallantoic respiration declines and the chick further cracks the shell and soon steps out.

After Rahn and Paganelli.

Ventilatory Mechanisms

Whatever the exchange organ—lungs, gills, skin, or accessory structures—water or air moves actively across the respiratory surfaces to increase the rate of diffusion. A few ventilatory mechanisms are based on cilia, but most involve the action of muscles.

Cilia

If an animal is small and its metabolic demands modest, microscopic cilia are sufficient to move water across the respiratory surfaces and support the exchange of gases between tissues and the environment. Cilia line the routes along which the water current flows. Their coordinated sweeps drive water, a relatively viscous medium, through the pharynx and across the gills. Cilia, like oars, are ineffective against a relatively thin medium such as air. Furthermore, cilia are surface structures, so they are limited by available surface area. As an animal's size increases, mass increases faster than surface area, and cilia become less suitable as a mechanism for moving the ventilatory current that delivers oxygen to the organism. Thus, cilia, as part of the ventilation system, are found in small aquatic organisms with low metabolic demands, such as the protochordates.

In large vertebrates, the respiratory channels often retain cilia, but they are involved in clearing surface debris that can foul the breathing device. Although "inside" the body, lungs are continuously exposed to fresh air from the outside environment. Ciliated and mucous cells are specialized to remove impurities from this air. They are interspersed throughout the lining of the lungs and secrete mucus over the lining to trap dust and particulate material. Cilia beat in coordinated patterns to move this mucous blanket laden with foreign material up the airways and into the pharynx where it is swallowed unnoticed.

Another secretion that lines the lungs (and gas bladders) is **surfactant.** Surfactant reduces surface tension at the water-air interface. This becomes an increasingly important function where there is partitioning of the interior respiratory surface. Surface tension can collapse the resulting microscopic compartments in which gas exchange takes place. Surfactant lowers this surface tension, helps stabilize these compartments, and maintains their structural integrity as elaborated surfaces for respiratory exchange.

Muscular Mechanisms

Ventilation in vertebrates usually depends on muscle action. Water moving across gills ventilates them. In amphibians with external gills, muscles within or associated with the bases of the projecting gills contract to wave the gills back and forth through the water. Some swimming fishes take advantage of their forward progress through water. They open their mouths slightly, allowing water to enter and irrigate the gills. This technique by which a fish's own forward locomotion contributes to gill ventilation is known as **ram ventilation.** It is characteristic of many large, active pelagic fishes such as tuna and some sharks. More commonly, muscular pumps actively drive water or air through the respiratory organ. There are three principal types of pumps, one common in water-breathing and two found among air-breathing vertebrates.

Water Ventilation: Dual Pump

In water-breathing fishes, the most common pump is a **dual pump** (figure 11.10). This gnathostome system, as the name suggests, is two pumps in tandem, buccal and opercular, that work in a synchronous pattern to drive water in a nearly continuous, unidirectional flow across the gill curtain between

Dual pump

Buccal cavity

Gill curtain

Opercular cavity

Oral valve

Opercular valve

Suction phase

Force phase

FIGURE 11.10 **Water-breathing fishes: dual pump.**
In most fishes, the buccal and opercular cavities form dual pumps on opposite sides of the gill curtain. Muscle action expands both cavities, represented by the falling pistons (downward dark arrows, left) in the suction phase. During the force phase, muscles contract to compress the cavities, represented by the rising pistons (upward dark arrows, right). As pressure within each cavity falls and rises, more water (suction phase) is drawn in and expelled (force phase). Because of the slight difference in pressure between buccal and opercular cavities, water is almost continuously moving from buccal to opercular cavity. The valves of the mouth and operculum prevent reverse flow of water. Thus, a one-way and more or less continuous flow of water across the gills is established.

them. This mechanism of gill irrigation can be viewed as a two-stroke pump. The first stroke, or *suction phase,* begins with compressed buccal and opercular cavities and closed oral and opercular valves. As the buccal cavity expands, creating a low intraoral pressure, the oral valves open and outside water rushes in following the pressure gradient. The simultaneous expansion of the more posterior opercular cavity with its closed valve also creates a pressure that is even lower than in the adjoining buccal cavity. Consequently, water that first enters the buccal cavity is encouraged by the pressure differential to continue on across the gill curtain and into the opercular cavity.

During the second stroke, or *force phase,* the oral valves close and the opercular valves open. Simultaneous muscle compression of the buccal and opercular cavities raises pressure in both, but because of the open opercular valve, pressure in the opercular cavity is slightly lower. Consequently, water flows from the buccal cavity across the gill curtain and exits via the open opercular valve. The timing of the suction and force phases, together with the pressure differentials between them, results in a unidirectional, nearly continuous flow of new water across the gills.

Air Ventilation: Buccal Pump

Air-breathing fishes and amphibians utilize a buccal pump to ventilate their lungs. The **buccal pump** (pulse pump) employs the mouth cavity, which expands to fill with fresh

air and then compresses to pump this air into the lungs. Spent gases from the lung exit in synchrony under buccal forces. In the **two-stroke buccal pump** (figure 11.11a), initial expansion of the buccal cavity brings fresh air and expired gas from the lungs into the mouth where they mix during the first stroke. In the second stroke, buccal compression forces these mixed buccal gases into the lungs, with the excess gas expelled via the nares or mouth. Exhalation and inhalation of gases may also be based on a four-stroke mechanism. The **four-stroke buccal pump** (figure 11.11b) begins with buccal expansion that brings gas from the lungs into the mouth, so upon the second stroke, buccal compression, this gas is forced out the nares. In the third stroke, buccal expansion draws fresh air into the mouth, via the nares, so that in the fourth stroke, buccal compression forces this air into the lungs. Tracheal sphincters and narial valves are synchronized with buccal displacements to help control gas movement.

Air-breathing Fishes Fishes that occasionally gulp atmospheric air, such as lungfishes, are no different from other fishes when they are actively water breathing. They use the same dual pump mechanism to irrigate their gills. However, when the lungfish breathes air, the dual pump is modified into a buccal pump to move air in and out of the lungs. The four-stroke buccal pump can be summarized as an exhalation phase and an inhalation phase. The **exhalation phase** begins with the *transfer* (expansion 1) of spent air from the lungs into the buccal cavity. In some fishes, relaxation of a sphincter around the glottis permits this transfer from the lungs to the buccal cavity. Exhalation concludes with *expulsion* (compression 2) of air from the buccal cavity to the outside either through the mouth or under the operculum. As the fish rises and its snout breaks the surface, its mouth opens to *intake* (expansion 3) atmospheric air, the first step in the **inhalation phase.** Inhalation concludes with *compression* (4), which forces a bubble of fresh air from the buccal cavity into the lungs (figure 11.12).

Theoretically, this bidirectional or tidal exchange of air to and from the lungs of air-breathing fishes could be aided by the hydrostatic pressure of the water column surrounding the fish. Because surrounding hydrostatic pressure increases with depth, a fish rising to the surface with its head tipped upward experiences slightly greater pressure on its deeper body than on the buccal cavity near the surface. During exhalation, this could help force air from the lungs into the buccal cavity and out the mouth. In reverse, after the fish has gulped atmospheric air and turned downward, air in the deeper buccal cavity would be under slightly greater pressure than air in the slightly shallower lung. This could help move the bubble of freshly gulped air into the lung.

In practice, some fishes do take advantage of the hydrostatic differential in water pressure on their bodies when transferring or expelling air during exhalation. Usually this is augmented by muscle contractions within the buccal

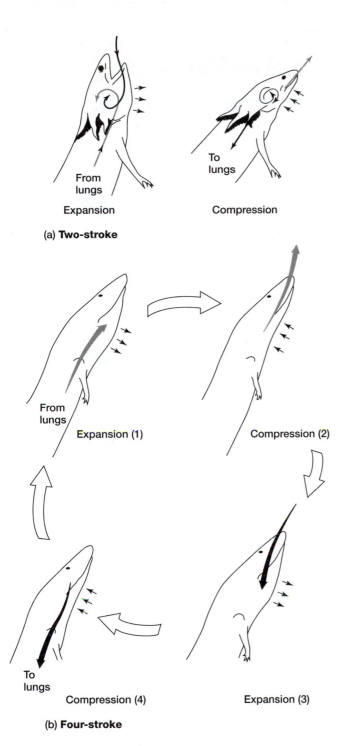

(a) **Two-stroke**

(b) **Four-stroke**

FIGURE 11.11 **Buccal pump: Two-stroke and four-stroke.** Air-breathing fishes and amphibians use an expansion and compression of the buccal cavity to move gases to and from the lungs. (a) Two-stroke buccal pump (based on *Ambystoma tigrinum*). Expansion mixes gas from the lungs and fresh air in the mouth; upon compression, this mixture is forced into the lungs, and the excess expelled through the nares. (b) Four-stroke buccal pump (based on *Amphiuma tridactylum*). Initial buccal expansion (1) draws air into the mouth, to be expelled next during buccal compression (2); now buccal expansion (3) draws in fresh air, to be forced into the lungs upon buccal compression (4).

Modified from Simons, Bennett, and Brainerd.

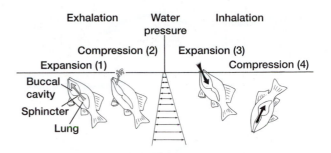

FIGURE 11.12 **Air-breathing fishes: buccal pump.** The four-stroke pump occurs in two phases—exhalation and inhalation. During exhalation, surrounding water pressure on the body walls forces air from the lungs and out the open mouth. During inhalation, the fish's head breaks the surface and its buccal cavity expands, drawing in air. Muscle compression of the buccal cavity forces the valve in the mouth closed, and positive pressure moves air into the lungs. A sphincter between the mouth and lungs closes to prevent escape of air from the lungs.

cavity and with striated muscles around the lung. However, inhalation seems to be based primarily on active contractions of the branchial musculature.

Amphibians Like air-breathing fishes, amphibians use a pulse pump to ventilate their lungs. Airflow is bidirectional. The two-stroke buccal pump method of ventilation is primitive for amphibians and is found in most. It has been reported in some lungfishes (e.g., African lungfish). A few aquatic amphibians use the four-stroke buccal pump. When an amphibian is in water, the hydrostatic pressure against the sides of its partially submerged body presses on the lung to produce a pressure that is higher than atmospheric pressure. When the amphibian exhales air, this hydrostatic pressure aids in venting the lungs. In addition to the passive effects of water pressure, hypaxial muscles (especially the transverses abdominis) contract to actively aid exhalation. This helps clear the lungs and produces a larger tidal volume during infrequent breathing. However, active muscle contraction apparently does not contribute to inhalation. During inhalation, the buccal cavity must work against this water pressure to refill the lungs. A strong, muscular buccal cavity addresses this problem of taking breaths of air while the animal is immersed in water. On the other hand, the buccal cavity on which this buccal pump is centered is also involved in feeding. As we will see later in the chapter, the dual roles of the buccal cavity in feeding and ventilation can lead to conflicting demands and compromises in design.

Air Ventilation: Aspiration Pump

The **aspiration pump** is a third type, after dual and buccal pumps, that does not push air into the lung against a resisting force. Rather, air is sucked in, or aspirated, by the low pressure created around the lungs (figure 11.13). The lungs are located *within* the pump so that the force required to ventilate them is applied directly. The "pump" includes

FIGURE 11.13 **Air-breathing amniotes: Aspiration pump.** In most amniotes, the buccal cavity has little to do with forcing air in or out of the lungs. Instead, a rib cage expands and compresses and/or a diaphragm moves forward and back within the body cavity to create a positive pressure that expels air or negative pressure that draws air into the lungs.

the rib cage and often a muscular diaphragm. A movable diaphragm in the thorax causes pressure changes rather than the action of the buccal cavity. The diaphragm, like a plunger, alters the pressure on the lungs to favor entry or exit of air.

The aspiration pump is bidirectional and moves air tidally. It is found in amniotes—reptiles, mammals, and birds. In birds especially, the aspiration pump is highly modified. The buccal cavity is no longer part of the amniote pumping mechanism. Unlike the buccal pump, feeding and ventilation are decoupled in vertebrates using an aspiration pump. This functional decoupling increases the opportunities for independent diversification of the feeding and ventilation mechanisms.

Phylogeny

Agnathans

As in cephalochordates, the ammocoete larva of the lamprey depends on cilia-lined channels to gather food collected. However, unlike cephalochordates, the feeding-ventilation current of water is produced by pumps composed of muscular **velar folds,** or **velum,** and by compression and expansion of the branchial apparatus (figure 11.14a). Closure of the velum and muscle compression of the branchial apparatus drive water across the gills and out the pharyngeal slits. Relaxation of these same muscles allows the elastic branchial apparatus to spring back into its expanded shape, thus drawing in exterior water past the open velum. The pharyngeal openings are small and round, not long slits like those in amphioxus. There are usually seven pairs of slits. Flaps of skin cover these openings, which act as valves. Although water can exit through them, inward water movement forces them closed; thus, reverse flow is prevented (figure 11.14b).

Unlike the lateral gills of gnathostomes, the gills of ammocoetes lie medial to the branchial arches. Each gill includes a central partition, the interbranchial septum,

FIGURE 11.14 **Ventilation in the ammocoete larva.** (a) The muscular velum draws water into the mouth and forces this water through the pharyngeal slits and across the gills before exiting to the outside. (b) Frontal section through three branchial arches showing position of gills and direction of water flow.

After Mallatt.

that supports a set of **primary lamellae (gill filaments)** on its anterior and posterior sides. Each filament is extensively subdivided into numerous, tiny, platelike **secondary lamellae** that contain the respiratory capillary beds. The current of water is directed across the sides of these secondary lamellae. Blood flowing within the capillary beds of the lamellae courses in the opposite direction. Thus, water and blood passing in opposite directions establish a countercurrent system between them to improve gas diffusion.

In many species, the adult lamprey is a short-lived reproductive stage that does not feed and dies soon after

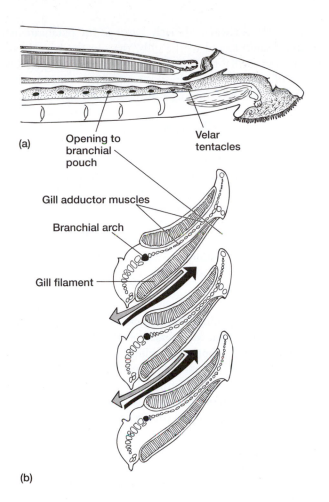

(a)

Opening to branchial pouch

Velar tentacles

Gill adductor muscles

Branchial arch

Gill filament

(b)

FIGURE 11.15 Ventilation in the adult lamprey.
(a) Longitudinal section. Because the adult lamprey's mouth often is attached to prey, water must alternatively enter as well as exit via pharyngeal slits. Thus, gill ventilation in the lamprey, unlike most fishes, is tidal. (b) Frontal section of three gill arches. Double arrows indicate tidal flow of water: black, inflow; gray, outflow.

After Mallatt.

breeding. In species with a prolonged adult stage, the adult feeds by attaching its circular mouth to the sides of living prey. The tongue is used to scrape flesh. In such species, the mouth grips the prey, prohibiting the entry of water to ventilate the gills. Instead, water exits *and* enters through the pharyngeal slits (figure 11.15a, b). Muscle compression and relaxation of the branchial apparatus drive this water, which moves tidally in and out of the branchial pouch via the associated slits, unlike in most fishes. A partition that divides the pharynx into a dorsal esophagus connected to the digestive tract and a ventral water channel that furnishes the branchial pouches prevents the mixing of food and water of respiration.

In hagfishes, no major expansions and contractions of the branchial apparatus occur. Instead, scrolling and unscrolling of the velum, one on each side, together with synchronized contractions and relaxations of the branchial pouches, produces a current of water that enters via the nostril and nasopharyngeal ducts, flows in one direction

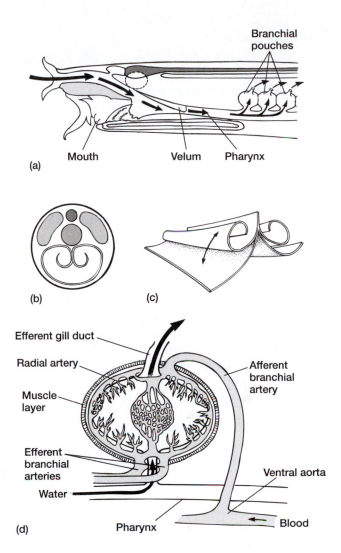

Branchial pouches

(a)

Mouth   Velum   Pharynx

(b)   (c)

Efferent gill duct

Radial artery

Muscle layer

Efferent branchial arteries

Water

Afferent branchial artery

Ventral aorta

(d)   Pharynx   Blood

FIGURE 11.16 Ventilation in the hagfish.
(a) Longitudinal section. Water (indicated by arrows) enters via the nostril, not the mouth, to reach the pharynx. The scroll-shaped velum rolls up and down as the branchial pouches contract to drive this current across the gills and out the gill pores. (b) Cross section of the scroll-shaped velum. (c) Lateral view of velum scrolling and unscrolling to move water through the pharynx. (d) An individual branchial pouch showing the sites of entry and exit of water and the position of the capillary beds within. The muscular walls of these pouches are compressed by contraction but expanded by elastic recoil.

After Liem.

across the gills, and exits (figure 11.16a). In cross section, the velum is shaped like an inverted T (figure 11.16b, c). Its sides furl and unfurl to produce the current of water that enters through the nostril and courses posteriorly through the branchial pouches. The branchial pouches are defined by an outer muscular wall that encloses the gill lamellae. Afferent blood vessels supply the lamellae, and efferent branchial vessels drain them (figure 11.16d). The current of water propelled by the velum and by the pumping action of these branchial pouches flows across the gill lamellae and out the common branchial duct.

Elasmobranchs

As in all gnathostome gills, elasmobranch gills lie lateral to the branchial arch. Each gill consists of a central partition, the interbranchial septum, covered on each face by primary lamellae (gill filaments). The primary lamellae are composed of standing rows of secondary lamellae. Across their sides flows water that irrigates the gills. Like ribs of a fan, **gill rays** within the septum give it support. The term **holobranch** refers to a branchial arch and the lamellae on both anterior and posterior faces of its septum. A gill arch with lamellae on only one face is a **hemibranch**. Facing plates of lamellae on adjacent gills constitute a **respiratory unit** (figure 11.17a, b).

Among chondrichthyans, the respiratory mechanisms of sharks has been studied most. Ventilation is based on a dual pump mechanism that creates alternating negative (suction) and positive pressures to draw water in and then drive it across the gill curtain. Pressures recorded on either side of the gill curtain within the buccal and **parabranchial** compartments reveal the efficiency of this dual pump. Although pressures rise and fall in each cavity, the pressure is always relatively lower in the parabranchial cavity, located lateral to the gills, than in the buccal cavity, located medial to the gills. In addition to bringing new pulses of water into the mouth, the dual pump mechanism of the shark also maintains a nearly constant pressure difference between buccal and parabranchial compartments. As a result, the pressure oscillations of the dual pump are converted into a smoother, almost continuous unidirectional irrigation of the gills (figure 11.18a–d). Blood coursing within the capillaries of the secondary lamellae sets up a countercurrent or perhaps a crosscurrent pattern, promoting efficient gas exchange.

Countercurrent and crosscurrent exchange (p. 156)

In sharks swimming in open waters, ram ventilation can add to gill irrigation and nearly replace the dual pump during such times.

What is embryologically the first gill slit is reduced to a small oval opening, the **spiracle,** which carries a much reduced hemibranch, sometimes referred to as a **spiracular pseudobranch.** In bottom-dwelling skates and rays, the ventral mouth may be partially buried, leaving the dorsally placed spiracle in an unobstructed position where it allows water for gill irrigation to enter. The spiracle may play a role in chemical sampling of the passing stream of water as well.

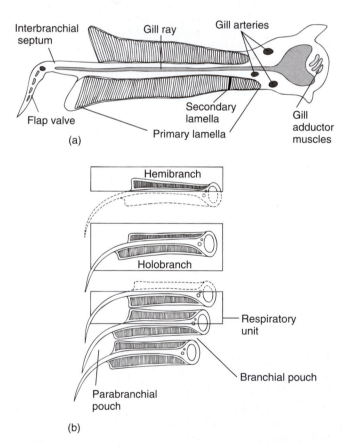

FIGURE 11.17 Shark gill. (a) The interbranchial septum has banks of lamellae supported by gill rays and a medial branchial arch. (b) Structural units include a hemibranch and a holobranch as well as a functional respiratory unit.

(a) After Mallatt.

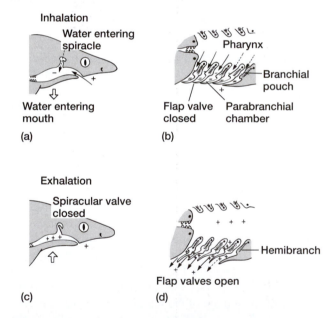

FIGURE 11.18 Gill ventilation in a shark. Lateral (a, c) and frontal (b, d) views. Relative positive and negative pressures are indicated by + and −, respectively. The ventilation mechanism consists of a buccal pump that draws water in and forces it across the gill curtain and out. Notice that the flap valves close during inhalation and that relative pressures are always lower in the parabranchial chamber than in the pharynx. Thus, water moves unidirectionally across the gills in a pulsing but continuous flow.

After Hughes and Ballintijn.

FIGURE 11.19 **Gill ventilation in teleost.**
(a) One gill bar is removed, showing the stack of gill lamellae. (b) Water flow is directed across the secondary lamellae opposite to that of blood flowing within each secondary lamella, establishing a countercurrent exchange between them.

For most other elasmobranchs, the function of the spiracular pseudobranch is unknown. In sharks, it probably does not have a respiratory function, because blood supplying the pseudobranch comes from an adjacent fully functional gill and is already oxygenated.

Holocephalians (ratfishes) lack spiracles altogether. They also differ from other elasmobranchs in having a single extensive flap of skin, or operculum, covering all the branchial arches, rather than individual flap valves over each pharyngeal slit.

Bony Fishes

The operculum of osteichthyans is bony or cartilaginous. It provides a protective cover over the branchial arches and gills they support. In addition, the operculum is part of the dual pump used to ventilate the gills.

In cross section, each gill is V-shaped and composed of primary lamellae (gill filaments) that are subdivided into secondary lamellae and supported on a branchial arch. Tiny adductor muscles cross between filaments to control the arrangement of adjacent gills that govern the flow of water across the secondary lamellae (figure 11.19a). As in most other fish gills, the blood in the secondary lamellae flows one direction and water flows in the opposite direction to establish a countercurrent exchange (figure 11.19b).

Fishes that ventilate a gas bladder do so by gulping and forcing fresh air through the pneumatic duct. Usually, a fish expels spent air as it approaches the water's surface, captures and swallows a new gulp of fresh air, and descends again. In the jeju, a freshwater fish of the Amazonian region, the anterior muscular compartment of the gas bladder is connected to a posterior compartment through a sphincter. As the jeju breaks the surface, the fresh air it gulped into the buccal cavity is forced along the pneumatic duct and preferentially enters the anterior chamber of the gas bladder (figure 11.20a, b). The sphincter closes, and spent air in the posterior chamber exits. Finally the sphincter opens, and the muscular walls of the anterior chamber contract, forcing the new air into the vascularized posterior chamber (figure 11.20c, d).

Overview of Fish Respiration

Gills

In water-breathing fishes, different devices have arisen to serve a common function—driving a stream of water across vascularized gills. Compression and expansion of the branchial apparatus irrigates lamprey gills tidally. Scrolling of a velum moves water across the gills of hagfishes. Ram ventilation occurs in active fishes swimming forward through the water. They open their mouth, permitting oncoming water to enter and pass across the gills. In gnathostomes, the most common device serving gill irrigation is the dual pump. The branchial arches and their associated muscles are the central components of this pump. Because they are also involved in feeding, the design of the branchial apparatus represents a compromise between the demands of feeding and ventilation.

(a)

(b)

(c)

(d)

FIGURE 11.20 **Air-breathing fishes.** Most air-breathing fishes use a buccal pump to fill their air bladders or lungs, which are able to separate spent and incoming air during ventilation. The mouth breaks the surface (a) so that air drawn in along the pneumatic tube preferentially enters the anterior air chamber (b). Spent air in the posterior chamber is forced out through the pneumatic tube and exits under the operculum (c). The sphincter between the anterior and posterior chambers opens, allowing air to replenish the posterior chamber as well (d).

After Randall, Burggren, Farrell, and Haswell.

Lungs and Swim Bladders

Air-filled sacs arise early in bony-fish evolution and serve respiratory and hydrostatic functions. In lungfishes and tetrapods, the respiratory function predominates. In the Australian lungfish, the trachea arises from the floor of the esophagus, but bends around the right side of the esophagus to join a single lung in a dorsal position within the body cavity, a location that is also favorable to buoyancy control. In the African lungfish, *Protopterus*, the trachea also arises from the floor of the esophagus, but serves about equally sized, paired lungs (figure 11.21a). The lungs are subdivided into faveoli (figure 11.21a, b). Air forced into these lungs exchanges with capillary blood circulating in the walls of the faveoli.

In actinopterygian fishes, the hydrostatic function became more pronounced as these fishes entered new adaptive zones of the marine environment and encountered a new array of selective forces. To understand this, we need to examine why a fish might require a hydrostatic organ.

Most fishes are denser than the water in which they live, so they tend to sink. If their skeletons are highly ossified, as in bony fishes, the high density of bone makes this

FIGURE 11.21 **Lungs of the lungfish *Protopterus*.**
(a) View of the lungs from the right side and in cross section. (b) Enlargement of the internal wall of the lung. The lung is subdivided internally, forming small compartments, or faveoli. Faveoli are most numerous in the anterior part of the lung. Approximate location of the lungs is indicated by the darkened area (top) in the lateral view of the fish's body.

sinking tendency more pronounced. It is not surprising that almost all osteichthyans possess some form of gas bladder (or lung). Air-filled gas bladders give buoyancy to the fish body and help resist its tendency to sink. Swim bladders are usually absent among bottom-dwelling bony fishes and fishes of open water, such as tuna and mackerel, that swim continuously.

In primitive teleosts, the swim bladder is **physostomous**, retaining its connection to the digestive tract via the pneumatic duct through which air is released and taken in (figure 11.22a). In most advanced teleost fishes, this connection is lost and the swim bladder is a closed bag of gases called a **physoclistous** swim bladder (figure 11.22b). Both types adjust the buoyancy of the fish to varying water depth.

The volume occupied by the swim bladder determines its buoyancy and its ability to compensate for the greater density of the fish body. Because water pressure increases with depth, the thin-walled swim bladder tends to be compressed when a fish descends and expanded when it rises. Thus, if the swim bladder is to maintain a constant volume, gas must be added when a fish dives and removed when it surfaces. Fishes with physostomous swim bladders can do this by gulping extra air or releasing spent air via the pneumatic duct. More commonly, gas secretion occurs directly across the walls of the bladder. Some swim bladders have special **gas glands** from which gas from the blood is released into the bladder. In the gas gland, blood vessels form a countercurrent capillary

In most bony fishes, gill irrigation is based on a dual pump that draws water into the mouth, across the gill curtain, and out under the operculum. However, some fishes with specialized feeding habits exhibit a modified ventilation mechanism, similar to that of the parasitic lamprey. An example is the sturgeon *Acipenser,* whose mouth is used as a protrusible suction tube for probing and feeding in muddy bottom sediments. When the sturgeon is not feeding, gill ventilation occurs as it does for a bony fish—water enters the mouth, moves across the gill curtain, and exits out the opercular opening (box figure 1a). However, when feeding, the sturgeon's mouth is buried in bottom sediments, so it cannot breathe. Under these circumstances, water enters the buccal cavity not through the feeding mouth, but through a permanent aperture at the upper margin of the operculum. The water then turns and passes across the gill curtain in the normal direction to exit out the customary opercular opening (box figure 1b). Curiously, although it is present and theoretically available, the spiracle accounts for very little of the water that enters during these alternative ventilatory movements during feeding.

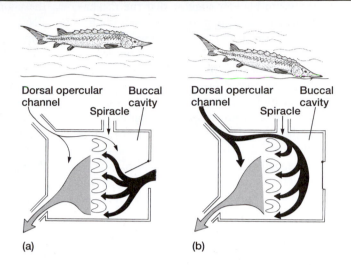

BOX FIGURE 1   **Gill ventilation in the sturgeon.** (a) In sturgeons, as in most fishes during normal breathing, water (indicated by solid, branched arrows) moves into the mouth, across the gill curtain, and out under the operculum. (b) When the sturgeon feeds on detritus, however, its mouth cannot serve as an entrance portal for water. During these times, water instead enters along a dorsal opercular channel to sweep across the gill curtain (column of open, U-shaped images) and then out the normal ventral channel under the operculum.

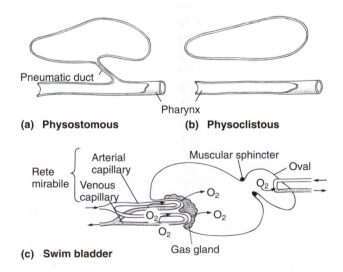

(a) **Physostomous**   (b) **Physoclistous**

(c) **Swim bladder**

FIGURE 11.22   **Swim bladders.** (a) Physostomous swim bladders retain their connection to the pharynx via the pneumatic duct. Air volume can be controlled if a fish gulps in more air or releases extra through the pneumatic duct. (b) In the physoclistous swim bladder, the connecting pneumatic duct has been lost. Air volume, and hence buoyancy, is controlled if more gas is released into the bladder at the rete mirabile or if some is removed at the oval. (c) The rete mirabile is a knot of capillaries. As blood leaves the gas gland of the swim bladder via the venous capillaries of the rete, lactic acid is added. This reduces hemoglobin's affinity for oxygen. Oxygen, therefore, tends to diffuse out and enter adjacent arterial capillaries passing blood to the rete. Consequently, the oxygen concentration builds in the arterial blood as it approaches the gas gland so that the partial pressure of oxygen in the arterial capillaries of the rete is high when it reaches the gas gland. This encourages oxygen release into the swim bladder.

arrangement, the **rete mirabile** (figure 11.22c). Incoming arterial and outgoing venous capillaries within this rete lie next to one another in the gas gland. Experiments on gas secretion into the swim bladder suggest that the mechanism involves lactic acid. During passage through the gas gland, lactic acid is added to the blood leaving the gland, increasing this blood's acidity. Increased acidity reduces the solubility of gases and the affinity of hemoglobin for oxygen. As a result, the partial pressure of oxygen in the venous capillaries is higher than the partial pressure of oxygen arriving in the adjacent arterial capillaries. Oxygen diffuses into the arterial capillaries, raising its partial pressure before the arterial blood flows into the gas gland. As the process is repeated, the partial pressure of oxygen in the arterial capillaries of the rete builds until it exceeds the partial pressure of oxygen in the swim bladder; therefore, oxygen is released into the bladder (figure 11.22c).

Resorption of gas often involves specialized regions. In many advanced teleosts, there is an **oval,** a pocket at one end in which gas is absorbed back into the blood. During resorption, blood vessels to the oval dilate and the smooth muscle sphincter dividing it from the rest of the bladder opens. Gas at high partial pressure in the bladder can now come into contact with the vascular walls of the oval and be taken up by the blood removed from the bladder.

Generally, the gases in swim bladders (78% nitrogen, 21% oxygen) are similar in composition to the gases in air, at least when the bladder is filled initially with gulped air. Among fishes with physoclistous swim bladders, which do not gulp air, the gas composition varies. In fishes living at great depths, the gas in the swim bladder is mainly oxygen. In trout and other salmonids, nitrogen is at very high proportions in the swim bladders, regardless of the depth at which they live.

Swim bladders also have secondary functions. In some fishes, the bladder is connected to the hearing apparatus and aids in sound detection. Some fishes produce sounds within the swim bladder or use it as a resonator. Releasing air by belching is one source of sound. Grinding of the teeth is another. Sounds can cause the swim bladder to vibrate, or the bladder may amplify or resonate them. Other fishes have specialized muscles that strum the bladder itself to produce a sound. Because males have specialized muscles that females do not, it is thought that the resulting sounds are part of territorial or courtship displays.

Sound detection by the swim bladder (p. 701)

No chondrichthyan has a swim bladder. The tendency of these fishes to sink is addressed in a different fashion. A cartilaginous skeleton avoids the added density of extensive ossification. In addition, two other sources counteract the tendency to sink. One is the fins. Elasmobranchs have broad pectoral fins and can change their angle to the flow of water in order to steer their body up or down. The heterocercal tail, as it sweeps back and forth during swimming, produces lift and compensates for the fish's density along

with the pectoral fins. A second source of lift is generated by an oil **(squalene)** consisting of lipids and hydrocarbons. Oils are lighter than water, so they reduce the density of the chondrichthyan. As any student who has dissected a shark knows, the copious oil permeates the large liver. In some sharks, liver oil alone can constitute 16 to 24% of the body weight. Squalene, by reducing the density, reduces the energy needed for swimming because the body and heterocercal tail need not devote as much of their efforts to compensating lift.

Heterocercal tails (p. 307)

Amphibians

In modern amphibians, the skin is a major respiratory organ, and in some species, it is the exclusive respiratory organ. The skin is moist and the layer of keratin relatively thin, allowing easy diffusion of gases between the environment and the rich supply of capillaries within the integument.

The significance of cutaneous respiration in modern amphibians is almost certainly greater than it was in early tetrapods. Many early tetrapods had scales that would have obstructed gas exchange through the skin. Ancient tetrapods likely depended on lungs for respiration. Many, including *Ichthyostega,* had prominent ribs encircling the thorax, although these ribs were not mobile but more likely used only for support. However, in modern amphibians, ventilation depends not on ribs but on pumping movements of the throat to irrigate gills or fill lungs.

In aquatic amphibians, pharyngeal slits often persist with internal gills. Feathery external gills are often present as well, especially among larval amphibians. Most, but not all, amphibians have lungs for breathing air. The respiratory surface within the lungs is usually developed best anteriorly, and it decreases posteriorly along the inner walls. This surface is **septal,** meaning that partitions form and subdivide to increase the surface area exposed to incoming air. The interconnecting septa divide the internal wall into compartments, **faveoli,** that open into the central chamber within each lung. Faveoli differ from the alveoli of mammalian lungs in that they are not found at the end of a highly branched tracheal system. Faveoli are internal subdivisions of the lung wall that open into a common central chamber. Inspired air travels along the trachea into the central lumen of the lung and from here diffuses into the surrounding faveoli. Capillaries located within the thin septal walls of the faveoli take up oxygen and give up carbon dioxide.

Amphibian Larvae

Salamander larvae typically have both internal and external gills. Pumping action of the throat irrigates the internal gills with a unidirectional stream of water across their surfaces. Feathery external gills are held out in the passing current, allowing water to flow across them. If there is no current

It [spermaceti] had cooled and crystallized to such a degree, that when, with several others, I sat down before a large Constantine's bath of it, I found it strangely concreted into lumps, here and there rolling about in the liquid part. It was our business to squeeze these lumps back into fluid. A sweet and unctuous duty! No wonder that in old times this sperm was such a favorite cosmetic. Such a clearer! such a sweetener! such a softener! such a delicious mollifier! After having my hands in it for only a few minutes, my fingers felt like eels, and began, as it were, to serpentine and spiralize.

　　　　　　Herman Melville, *Moby Dick*

"Thar she blows" was the call of whalers seeking their quarry, sperm whales spouting at the surface. Besides its blubber, the colossal spermaceti organ in its nose was a special prize because of the large quantity of high-quality, milky oil harvested from it, up to 4 tons in some large males.

When at the surface, a whale replenishes respiratory air by venting and then refilling its lungs through long nostrils that open at the blowhole near its bow end. When the whale exhales, the exiting warm air condenses, giving the illusion of a spouting stream of water. When it dives, the sperm whale may reach a depth of a mile or more in search of a favored food, giant squid. Deep diving gives the whale access to resources unavailable to most other large predators.

The spermaceti organ holds unique lipids and collagen fibers. It represents the greatly hypertrophied soft tissues on the right side of the face, producing an extremely asymmetrical nose. Beneath the spermaceti organ is the so-called junk of the sperm whale, fatty tissue with lens-like, lipid compartments homologous to the "melon" of dolphins. In dolphins, this tissue acts like a lens to form and shape sonar beams used in dolphin navigation along with targeting and tracking of prey. In sperm whales, the spermaceti organ and the junk represent an enormous energy investment, but an investment that cannot be drawn upon to meet metabolic needs because the chemical constituents are toxic to the whale.

The two nasal passages are different. From the blowhole, the left nasal passageway is specialized for breathing.

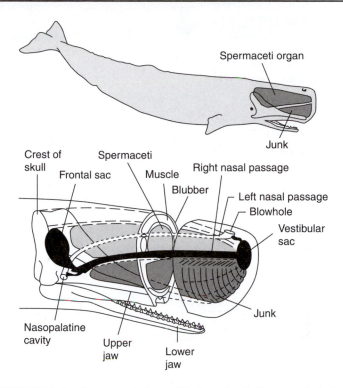

BOX FIGURE 1   Sperm whales.   The large head of sperm whales holds the spermaceti organ, which is impregnated with oil. The two nasal passages are asymmetrical. The left nasal passage runs eventually to the internal nares; the right runs beneath the spermaceti organ. The spermaceti organ most likely acts like a lens to focus produced sonic booms to stun and then capture prey. Its larger size in males suggests a further role in acoustical sexual selection.

After Clarke; based on the research of and thanks to Kenneth Norris and Ted Cranford.

It runs along the left side of the spermaceti organ to the superior internal nares. The unique larynx plugs into the inferior internal nares to establish complete continuity of the airway to fill and vent the lungs, separating it from the food channel of the esophagus. However, the right nasal passage is specialized for sound production. It runs forward from the blowhole into a vestibular sac. From this sac, the nasal passageway broadens into a wide tube as it courses posteriorly between the spermaceti organ and the junk, expanding dramatically into a second large sac (frontal sac) that covers the entire face of the amphitheater-shaped skull before entering the right internal naris (box figure 1). The functional significance of this huge, asymmetrical nose is debated.

One view holds that the spermaceti organ is a buoyancy device. By warming or cooling it, the oil melts or solidifies to become less or more dense, and thereby assists the sperm whale in ascent or descent, respectively. However, this is an unlikely function as the spermaceti organ lacks the vascularization necessary to heat or cool the oil. Further, this store of oil seems a huge investment for so small a return when the whale's wonderfully efficient swimming system could do the job. More plausibly, the sperm whale nose is a bioacoustical machine, generating and focusing a sonic beam on susceptible prey, to stun it and then scoop it up. The sperm whale nose is especially large in males, leading to the view that it might also act as a battering ram between combative males or is used in acoustical sexual selection.

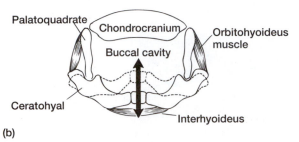

FIGURE 11.23 **Ventilation of tadpole gills.** (a) The chondrocranium and major components of the visceral cranium are illustrated. (b) The floor of the buccal cavity is raised and lowered (double-headed arrow) to produce the movement of water. Two sets of muscles are primarily responsible. The orbitohyoideus depresses the floor, and the interhyoideus elevates it.

After Wassersug and Hoff.

or if water is stagnant, the larvae can wave their gills back and forth through the water to irrigate the capillary beds they carry.

Larvae of anurans employ buccal and pharyngeal force pumps to produce a unidirectional flow of water across the gills and generate a food-bearing current. The "piston" for the buccal part of this pump includes enlarged elements of the splanchnocranium (ceratohyal, copula, hypobranchial plate). These elements articulate with the palatoquadrate, which acts as a fulcrum about which they rotate to expand and compress the buccal cavity (figure 11.23a, b). The action of muscles on the pharyngeal pump is not yet understood but seems to involve compression and expansion of this cavity.

The basic mechanism of amphibian gill ventilation includes a buccal cavity and a pharyngeal cavity separated from each other by a valve, the velum. The buccal cavity is separated from the mouth by the **oral valve** and from the nares by an **internal narial valve.** Inhalation depresses the

floor of the buccal cavity, which lowers the pressure within it. The velum closes temporarily to prevent entry of water into the pharyngeal cavity, but water fills the buccal cavity through the mouth and nares. Near the end of the inhalation stage, pharyngeal constriction causes a rise in pressure within the pharyngeal cavity relative to the buccal cavity. This keeps the velum closed and pushes water across the gill curtain. The exhalation stage begins with elevation of the floor of the buccal cavity, raising the pressure within it and forcing the oral and narial valves closed. Nearly simultaneous expansion of the pharyngeal cavity drops the pressure within it relative to the buccal cavity. Consequently, water in the buccal cavity pushes open the velum and refills the pharyngeal cavity, displacing the water within it. As with water-breathing fishes, the gills of frog tadpoles see an almost continuous unidirectional stream of water across their surfaces.

In some tadpoles, such as those of the tailed frog *Ascaphus truei*, the prominent oral sucker around the mouth is used to grip the surface of rocks in the fast-flowing streams in which these tadpoles live. A sucker that is firmly attached prevents entry of water through the mouth. However, action of the floor of the buccal cavity draws water in via the nares and then forces it across the gills before exiting (figure 11.24a). This same action of the buccal cavity, together with valves guarding the mouth, removes water from the area of the oral sucker to produce the low pressure that helps hold the tadpole to the rock (figure 11.24b).

Amphibian Adults

When the amphibian larva undergoes metamorphosis into an adult, gills are lost. Cutaneous respiration continues to play an important role in meeting respiratory demands after metamorphosis, and lungs, if present, are ventilated by a buccal pump.

The four stages of lung ventilation are best understood in frogs. In the first stage, the buccal cavity expands to draw fresh air in through the open nares (figure 11.25a). In the second stage, the glottis opens rapidly, releasing spent air from the elastic lungs. This air streams across the buccal cavity with little mixing and is vented through the open nares (figure 11.25b). In the third stage, the nares close, and the floor of the buccal cavity rises, forcing the fresh air held in this cavity into the lung through the open glottis (figure 11.25c). In the fourth stage, the glottis closes, retaining the air that has just filled the lungs, and the nares open again. Between cycles, the buccal cavity may oscillate repeatedly (figure 11.25d). This rapid oscillation was once thought to turn the lining of the mouth temporarily into an accessory breathing organ. However, experimental evidence refutes this. The capillaries lining the mouth do not serve in gas exchange. Instead, such buccal oscillations between lung fillings serve mainly to flush the buccal cavity of any stray residue of expired air in the mouth following each ventilatory cycle.

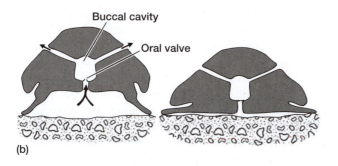

FIGURE 11.24 Gill ventilation in the tailed frog larva. The tadpole uses the extensive oral sucker around its mouth to establish a secure attachment to the undersurface of a rock in a fast-moving stream (solid arrows). (a) When the oral sucker is attached, water (solid arrows) to irrigate the gills enters through the nares, passes through the buccal cavity across the gill curtain, and then exits. (b) Water removed from the area to which the oral suction was attached creates a vacuum that helps the sucker hold the rock. The oral valve prevents a break in this seal.

After Gradwell.

FIGURE 11.25 Lung ventilation in the frog. (a) The frog's throat drops to replenish the air in the buccal cavity. (b) As the glottis opens, the thorax is compressed, forcing spent air from the lungs past that held in the buccal cavity and expelling it (open arrows). (c) Elevation of the throat and closure of the nares forces fresh air from the buccal cavity into the lungs. (d) Repeated pumping of the throat (multiple arrows) flushes the buccal cavity.

Modified from Gans, De Jongh, and Faber.

The buccal pump, and hence the buccal cavity, in frogs is also deployed in producing vocalizations that play a key role in the social organization and breeding success of frogs. Evolutionary modifications of the buccal cavity consequently affect three significantly different functions.

Opinions differ about how close in function the buccal pump of frogs is to the buccal pump of lungfishes. Certainly they differ in subtleties. For example, exchange of spent air in the lungs and fresh air held in the mouth seems to be more efficient in frogs. However, the similarities are striking. In both frogs and lungfishes, movement of the hyoid apparatus aids in filling of the buccal cavity, and spent air expelled from the lungs crosses this same chamber. In both groups, fresh air is pushed into the lungs against pressure. To some extent then, frogs have retained the basic pattern of lung filling deployed by lungfishes. However, all of this changes in reptiles, birds, and mammals. The mechanism of ventilation in these groups is the aspiration pump, a departure from that of amphibians and earlier air-breathing fishes.

Reptiles

Pharyngeal furrows and occasionally pharyngeal slits appear during the early embryonic development of reptiles, but they never become functional after birth. In some groups, supplemental cutaneous respiration is significant, but for the most part, paired lungs meet their respiratory needs.

The lungs of snakes and most lizards typically include a single central air chamber into which faveoli open (figure 11.26a, b). Like purse strings, cords of smooth muscle define and encircle the opening into each faveolus. The thin walls of each carry capillary beds and may be subdivided by even smaller internal septa. Sometimes the faveoli are reduced in the posterior part of the lung, leaving it as a nonexchange region. In monitor lizards, turtles, and crocodiles, the single central air chamber itself is subdivided into numerous internal chambers that receive air from the trachea. These internal chambers are ventilated by respiratory movements, whereas the exchange of gas between the faveoli and these chambers appears to occur by diffusion.

Filling of the lungs in all reptiles is based on an aspiration pump mechanism, although the anatomical parts that actually participate may differ. The aspiration pump acts on the walls of the lung to change its shape and induce airflow in or out. Ribs alter the shape of the body walls around the lungs, and intercostal muscles running between these ribs move them. In lizards, for instance, sets of intercostal muscles actively move the ribs forward and outward during inhalation.

FIGURE 11.26 Lung ventilation in a lizard.
(a) The lungs are located in the thorax, surrounded by ribs and connected to the trachea. Compression and expansion of the rib cage force air in or out of the lungs. (b) Cutaway view of the internal lining of the lungs showing numerous faveoli that collectively give the lining a honeycomb appearance. The internal faveoli of the lungs increase their respiratory surface area and function in gas exchange with capillaries lining their walls.

After Duncker.

The result is to enlarge the cavity around the lungs, decrease pressure within them, and draw air into the lungs. During active exhalation, different sets of intercostal muscles contract to fold the ribs back and inward, thus compressing the lungs within their cavity and expelling air. Occasionally, exhalation is passive. In this instance, muscle contraction is minimal, and gravity (and some elastic recoil) acts on the ribs, causing them to compress the lung cavity. Between breaths, the glottis is closed to prevent premature escape of air.

In snakes, the long, narrow lungs extend through most of the length of the body. In primitive snakes, as in other reptiles, the lungs are paired, but in many advanced snakes, the left lung is reduced and often lost entirely. In most snakes, faveoli are prominent anteriorly, but they decrease gradually and become absent posteriorly, producing two regions of the lung, an anterior respiratory portion (faveoli) and a posterior saccular portion (avascular) (figure 11.27a–c). Ribs and associated muscles run the entire length of the thorax so that regional compression and expansion of the body wall expand or deflate the lung. Opening and closing of the glottis are synchronized with these movements. Gas exchange occurs in the respiratory portion of the lung. The saccular portion of the lung acts as a bellows when the anterior body is occupied with different functions and unavailable to compress or expand the lung. For instance, when a snake swallows prey, the body becomes distended as food passes slowly through the esophagus, yet ventilation of the lungs must continue. Although the trachea, reinforced with semicircular rings of cartilage, stays open, the anterior body cannot act as the aspiration pump. Instead, the posterior body behind the prey expands and contracts, causing the saccular lung to fill and empty the lungs.

In caimans and other crocodiles, the liver assists the aspiration pump by acting like a "piston" to ventilate the lungs. During inhalation, the ribs rotate forward and outward, expanding the cavity around the lungs. In addition, the liver, located immediately behind the lungs, is pulled posteriorly by the action of **diaphragmatic muscles.** These muscles are derived from internal abdominal musculature. They extend forward from the pelvis and gastralia to the **posthepatic septum,** a thin sheet connected to the posterior side of the liver. Contraction of the diaphragmatic muscles draws the liver back, increasing the volume of the lung cavity and dropping the pressure within the lungs. This draws in atmospheric air. Exhalation reverses these movements. The ribs fold back into position, and the liver moves forward against the lung as a result of the contraction of **abdominal muscles.** Because pressure on the walls of the lung increases, air is expelled (figure 11.28). Overall, addition of the action of the diaphragmatic muscles to respiration increases the volume of air held in lungs and, hence, helps extend the dive time.

Ventilation in turtles represents a special problem in design. The shell around the lungs prevents changes in shape and precludes aspiration pumping using the ribs. In soft-shelled turtles, movements of the hyoid apparatus draw water in and out of the pharynx. Oxygen is absorbed in the pharynx to sustain the turtle while it is submerged. In snapping turtles, the plastron is reduced, permitting deformations of the body wall that contribute to lung ventilation. More commonly, in-and-out movements of the limbs alter pressure on the lungs, and special sheets of muscles within the shell change pulmonary pressure (figure 11.29a). Turtle lungs and other viscera reside in a single fixed cavity, so any change in volume alters pressure on the lungs. A limb extended from or pulled into the shell affects pressure in this cavity and aids the aspiration pump (figure 11.29b). In addition, the posterior visceral cavity is closed by a **limiting membrane,** connective tissue to which the **transversus abdominis** and **obliquus abdominis** muscles are attached. Contraction or relaxation of these muscles alters the volume of the cavity within the shell and contributes to the inhalation or exhalation of air (figure 11.29c). The diaphragmaticus muscle, although absent in tortoises, is present in most other turtles. The diaphragmaticus together with the transversus abdominis compress the visceral cavity to act as exhalation muscles. The glottis opens and the obliquus abdominis expands the visceral cavity to act as an inhalation muscle.

As in other tetrapods, ventilation of the lungs and locomotion are coupled. Locomotion imposes configurational changes on the rib cage and, hence, on the lungs housed within. In the 7-foot bipedal dinosaur *Deinonychus,* the caudotruncus muscle originates on the base of the tail, slips around the pulley-like end of the pubis, and inserts on the gastralia. Its contraction acts on the rib cage but is synchronized with rhythmic, cyclical forces generated during locomotion. As the hindlimb of the running dinosaur makes contact with the ground, the inertia of the neck and

In addition to the buccal cavity and lungs, frog vocalizations involve a third compartment, the vocal sac, a chamber opening off the floor of the buccal cavity. Access to it is gained through a muscle-controlled slit. Contractions of the body wall force air from the lungs, through the larynx, into the buccal cavity, and through the open slit into the vocal sac, inflating it. Next, contractions of muscles on the floor of the buccal cavity reverse the path of air so it flows back from the vocal sac to the buccal cavity, through the larynx, and to the lungs to reinflate them (box figure 1a–c).

In the toad *Bufo valliceps*, the larynx consists of a pair of arytenoid cartilages embraced by the circular cricoid cartilage. The arytenoid cartilages form a unit between the cornua of the hyoid. The laryngeal constrictor muscle originates from the hyoid cornua and is inserted on the arytenoid cartilage near the glottal opening. Upon contraction, it spreads the arytenoids to widen the opening. Anterior and posterior laryngeal muscles form a strap across the front and back of the arytenoids. When they both contract, they slip across the arytenoids toward their middle, having the greatest mechanical advantage at this spot to close these cartilages. Cooperative action of this dilator and these constrictor muscles affects airflow and modulates sound production.

As air is forcefully shifted back and forth between lungs and vocal sac, the nares are closed to prevent temporary escape of air.

BOX FIGURE I   **Frog songs.** (a) Musculature in the body wall forces air out of the lungs, through the larynx, and into the buccal cavity. From the buccal cavity, air enters the vocal sac via an aperture. Compression of the throat forces this air back along the reverse route into the lungs. (b) Larynx opened. (c) Larynx closed.

Based on research of C. Gans.

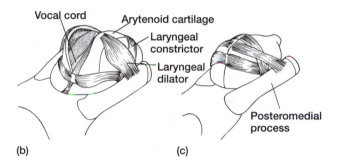

If the vocal sac is large, as in some species, then several pulses of filling are often used to inflate it fully. The paired vocal cords are two thin strips of tissue within the larynx, each held by an arytenoid cartilage and stretched across the airflow. As air rushes out of the lungs across the vocal cords, the cords and often nearby margins of the larynx are vibrated. The inflated vocal sac serves as a resonating chamber to modulate the sound produced. In a few species, sound is produced as the lungs fill, but in most species, sound is produced when air exits from the lungs.

---

tail draws them downward to compress the rib cage, contributing to exhalation (figure 11.30). As the limb pushes off, neck and tail rebound upward, increasing thoracic volume to promote inhalation (figure 11.30).

Mammals

An aspiration pump ventilates the lungs of mammals. Changes in the shape of the rib cage and pistonlike action of a muscularized **diaphragm** contribute to this pumping mechanism. The diaphragm consists of **crural, costal,** and **sternal** parts, all of which converge on a **central tendon.** Unlike the diaphragmatic muscles of crocodiles, which are located posterior to the liver, the diaphragm of mammals lies anterior to the liver, and acts directly on the **pleural cavities** in which the lungs reside (figure 11.31a, b). Intercostal muscles run between the ribs. The transversus abdominis,

serratus, and rectus abdominis that are inserted on the ribs and originate outside the rib cage (figure 11.31c, d) all aid in mammalian lung ventilation.

Ventilation

Mammalian **ventilation** is bidirectional and involves the rib cage and diaphragm. Upon inhalation, the external intercostal muscles contract to rotate the adjacent ribs and medial sternum forward. Because the ribs are bowed in shape, this rotation includes an outward as well as a forward swing of each arched rib. The result is to expand the space that the rib cage encloses around the lungs. Contraction of the dome-shaped diaphragm causes it to flatten, further enlarging the thoracic cavity. The elastic lungs expand to fill the enlarged thoracic cavity, and air is drawn in (figure 11.32a, b).

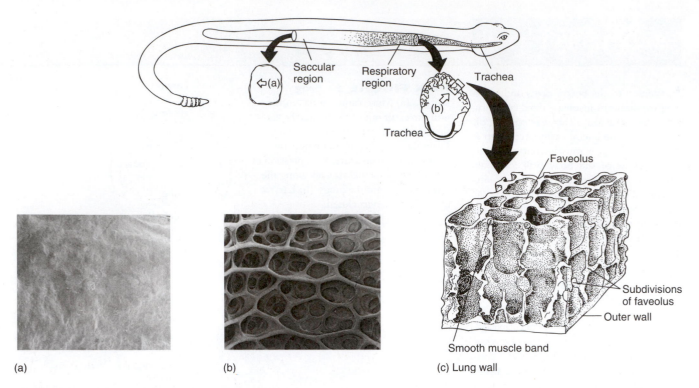

FIGURE 11.27 **Snake lung, rattlesnake.** Like the snake body, the rattlesnake's one lung is long and attenuated. Air travels down the long trachea to the lung. Most snakes have two lungs of unequal length, but in many venomous snakes, the left lung is lost. The trachea of the rattlesnake lung becomes an open trough where it meets the lung. The anterior lung is heavily vascularized and functions in respiratory exchange. The posterior part of the lung basically is a saccular, avascular region. Ribs along the sides of the body compress and expand to empty or fill the lungs. As the snake swallows prey, the tip of the trachea is pushed in front of the prey, so breathing continues. As the prey moves along the esophagus, which parallels the trachea, the anterior ribs expand to allow passage. At this time, they cannot compress and expand the anterior lung. Therefore, the posterior ribs act upon the saccular region of the lung, working like a bellows to move air across the respiratory surfaces. Representative cross sections of saccular and respiratory regions are illustrated at the top of the figures. Views shown in photos (a) and (b) are indicated in the cross section of the snake lung at the top. (a) Luminal view of the surface of the saccular region. (b) Luminal view of the respiratory region showing the faveoli. The entrance to each faveolus is defined by a honeycombed network of smooth muscles. (c) A section of wall from the respiratory region showing further subdivisions within the faveoli.

After Luchtel and Kardong.

During active exhalation, internal intercostal muscles slant in the opposite direction of the relaxed external intercostals and pull the ribs back. Relaxation of the diaphragm causes it to recoil and resume its arched, dome shape. Rib retraction and diaphragm relaxation decrease chest volume, forcing air from the lungs. Elastic energy stored in the lung and gravity acting to fold or collapse the rib cage may aid exhalation (figure 11.32c).

Although scientists agree on the muscles that control mammalian breathing, their precise functions have proved elusive, partly because of the surprisingly complex pattern of rib movement and partly because the rib cage and diaphragm are not equally involved in ventilation at all times. For example, during quiet breathing, only inhalatory muscles may show activity. At such times, exhalation muscles may not contract, and compression of the rib cage results from elastic and gravitational forces. As you can confirm for yourself, it is even possible to ventilate your lungs moving only the diaphragm and not the rib cage. When supporting vigorous ventilation during exercise, the rib cage, diaphragm, and most muscles are involved. To complicate the matter further, there appears to be a coupling of breathing cycles with locomotor cycles so that both are synchronized.

The mammalian diaphragm lies immediately posterior to the lungs and separates the thoracic cavity containing the lungs from the abdominal cavity containing other major viscera. When an animal is at rest, the muscularized diaphragm is the principal component in mammalian lung ventilation. However, during locomotion in quadrupedal mammals, the rib cage may receive ground reaction forces through the forelimbs that slightly change its shape. Further, the abdominal viscera, somewhat free to move within the body cavity, slide forward and backward in synchrony with the rhythm imposed on the body by the pattern of limb oscillation. The abdominal viscera act as a kind of "piston," first pressing anteriorly on the thoracic cavity and then sliding posteriorly, releasing pressure on the lungs. A running mammal takes advantage of this rhythmic movement of the viscera, expelling air when

FIGURE 11.28 **Ventilation in the crocodile.** In addition to a rib cage, the aspiration pump in the crocodile uses back-and-forth movements of the liver like a piston to act on the lungs. During inhalation, the rib cage expands and the liver is pulled back while the crocodile aspirates fresh air into its lungs. During exhalation, the rib cage and forward-moving liver compress the lungs and the crocodile expels spent air.

After Pooley and Gans.

the viscera press against the thorax and inhaling when they move away. Thus, in cursorial mammals, breathing patterns and locomotor gait are often coupled (figure 11.33a–c).

Gas Exchange

As we have seen in reptiles, faveoli along the interior walls of the lungs form the respiratory exchange surface. Air is drawn into the center core of the lung and diffuses outward into the faveoli. However, in mammals, the sites of respiratory exchange are reached via a different route. The respiratory passageway (including trachea, bronchi, bronchioles) repeatedly divides, producing smaller and smaller branches until they finally terminate in blind-ended compartments, the **alveoli,** which characterize the respiratory bronchioles and air sacs (figure 11.34a–c). The trachea, bronchi, and terminal bronchioles that transport gas to and from the alveoli are called the **respiratory tree** in recognition of their branching pattern. No gas exchange occurs along the conducting passageway of the respiratory tree until air reaches the respiratory bronchioles and alveoli. In mammals, the total alveolar area is extensive, perhaps over ten times that of amphibians of similar mass. Such a large exchange area is essential in mammals to sustain the high rate of oxygen uptake required by an active endotherm. The nasal passages not only form part of this conducting system but serve to warm and moisten the entering air.

Birds

Cutaneous respiration is insignificant in birds. The almost exclusive respiratory organ is the lungs. Like mammals, birds have two lungs connected to a trachea and ventilated

(a)

(b)

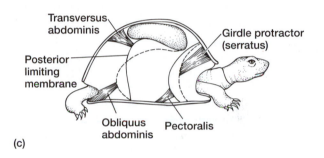

(c)

FIGURE 11.29 **Ventilation in the turtle.**
(a) Location of the lung inside the turtle shell. (b) Cutaway view of the lung showing its internal structure. Turtle lungs lie within a protective, rigid shell. Consequently, the fixed rib cage cannot act in ventilating the lungs. Instead, turtles have sheets of muscles within the shell that contract and relax to force air in and out of the lungs. Turtles also have the ability to alter air pressure within the lungs by moving their limbs in and out of the shell. (c) In the specialized tortoise, a diaphragmatic muscle is absent but other respiratory muscles take its place. Within the rigid shell, the viscera are enclosed by limiting membranes that under muscle action alter their position during exhalation (solid line) and inhalation (dashed line). During active exhalation, contraction of the transversus abdominis pulls the posterior limiting membrane up against the lung and contraction of the pectoralis draws the shoulder girdle back into the shell, further compressing the viscera. During active inhalation, exhalation muscles relax and contraction of the obliquus abdominis and girdle protractor expand the visceral cavity by pulling the posterior limiting membrane outward and the shoulder girdle forward, respectively.

(a, b) After Duncker; (c) after Gans and Hughes.

by an aspiration pump. Beyond this, however, the structural similarities are few. For example, there are no blind-ended alveoli in and out of which air moves. Instead, the conducting passages branch repeatedly and eventually form numerous tiny, one-way passageways, the **parabronchi,** that permit air to flow through the lungs. Small **air capillaries** open off the walls of each parabronchus, and gas

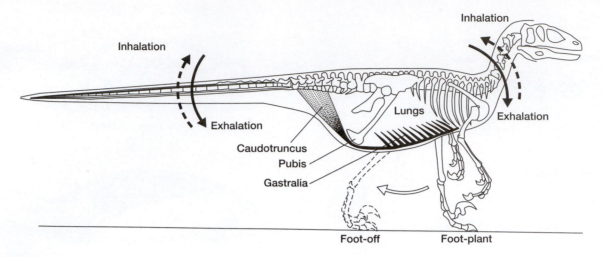

Inhalation

Inhalation

Exhalation

Exhalation

Lungs

Caudotruncus

Pubis

Gastralia

Foot-off   Foot-plant

FIGURE 11.30   Coupling of ventilation and locomotion. As the foot strikes the ground, neck and tail continue downward, compressing the rib cage (exhalation) (solid arrows); limb excursion swings the leg (dashed) backward (open arrow), leading to foot-off, which causes neck and tail to rotate upward, expanding the rib cage (inhalation) (dashed arrows). These locomotor effects were presumably synchronized with contractions of the caudotruncus to ventilate the lungs.

Based on the research of D. Carrier and C. Farmer.

exchange with the blood actually occurs in the air capillaries. Further, nine avascular **air sacs** are connected to the lungs, although they are tucked in among the viscera and extend into the cores of most large bones (figure 11.35a, b). Thus, the bones of birds contain air, not marrow. From 6 (house sparrow) to 12 (shorebirds) air sacs may be present. Generally, the anterior air sacs include the single **interclavicular sac** and the paired **cervical** and **anterior thoracic air sacs.** The posterior air sacs include the paired **posterior thoracic** and paired **abdominal air sacs** (figure 11.35a).

The trachea is divided into two **primary bronchi,** (= mesobronchi) that do not enter the lung but extend posteriorly to reach the posterior air sacs. Along the way, the primary bronchi give rise to numerous branches, the most prominent of which include **latero-, ventro-,** and **dorsobronchi** as well as **secondary bronchi.** These lead to the parabronchi (figure 11.36a–c). During passage through the parabronchus, gases diffuse between the lumen of the parabronchus and the connecting, blind-ended air capillaries. Oxygen diffuses in turn from the air capillaries into the adjacent blood capillaries that give up carbon dioxide to the air capillaries. Thus, the walls of air and blood capillaries constitute the sites of gas exchange.

Within this vast system of connecting passageways, there are no valves to suggest what the pattern of airflow might be. This has led to much speculation about the roles played by the different parts of the respiratory system. Without giving it much thought, some have proposed that air sacs function to lighten the bird like helium balloons to help lift the bird into the air. But because air in the sacs has the same density as air outside the bird, the air sacs provide no lift. Adding air sacs does not make the bird lighter. Others propose that air sacs serve to cool hot testes. This may be a secondary and later

function, but as female birds also have similar sacs this seems not be have been an original selective advantage. Certainly, air sacs are not a prerequisite for flight, because bats, who have typical mammalian lungs, are good fliers and can even, on occasion, migrate long distances.

Current evidence suggests that some terrestrial dinosaurs had air sacs. Some modern earthbound lizards and snakes possess air sacs, where they act as bellows in coordination with vascular lungs. This may be their derived role in birds where they act as a more sophisticated system of bellows. Details of this mechanism are still debated, but some aspects are understood. If we follow a single breath, its passage through the sacs and lungs includes two complete cycles of inhalation and exhalation (figure 11.37a, b). During the first inhalation, air enters the trachea, passes along the primary bronchi, and then is divided: Some air passes directly to the lungs, and the rest fills the posterior air sacs (posterior thoracic and abdominal air sacs). Upon the first exhalation, air from these posterior air sacs now flows through the lungs, displacing the spent air that exits via the trachea. As the second inhalation begins, the entering air again divides, some refilling the posterior air sacs and the rest flowing through the lungs, pushing the remainder of the spent air of the previous cycle out and temporarily into the anterior air sacs (anterior thoracic and interclavicular air sacs). With the second exhalation, air in these anterior sacs now exits along with air from the lungs, replaced by air from the posterior air sacs, which now flows through the lungs. Thus, this pattern of ventilation produces a nearly continuous, unidirectional flow of fresh air across the lungs. Speculating further, such a unidirectional flow may also establish a crosscurrent exchange within the lung, with air flowing from posterior to anterior air sacs as circulating blood flows next to it in the opposite direction (figure 11.38).

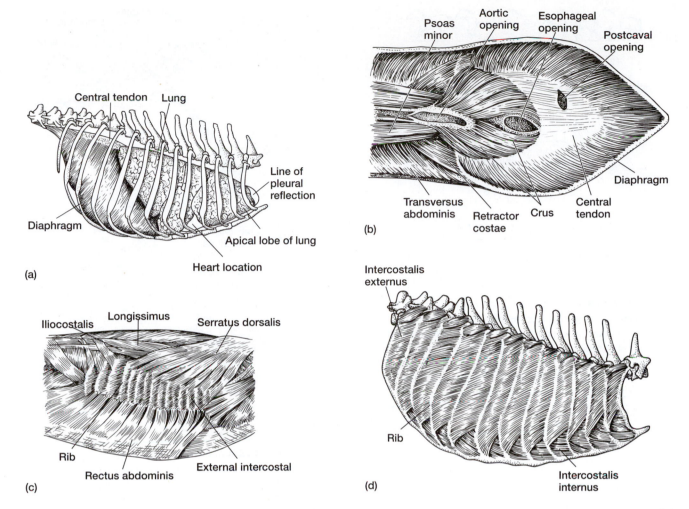

Central tendon Lung

Line of pleural reflection

Diaphragm

Apical lobe of lung

Heart location

(a)

Psoas minor

Aortic opening

Esophageal opening

Postcaval opening

Transversus abdominis

Retractor costae

Crus

Central tendon

Diaphragm

(b)

Iliocostalis Longissimus Serratus dorsalis

Rib

Rectus abdominis

External intercostal

(c)

Intercostalis externus

Rib

Intercostalis internus

(d)

FIGURE 11.31   Ventilation in the dog. Generally, ventilation of mammalian lungs involves expansions and contractions of the rib cage along with depression and elevation of the diaphragm. The details are remarkably complex. (a) Location of lungs and diaphragm within the rib cage of the dog (right lateral view). (b) Ventral view of the diaphragm (cranial to the right), which lies behind the lungs and has a dome shape. Notice the openings that allow anterior-posterior passage of the aorta, esophagus, and postcava. Superficial (c) and deep (d) muscles of the rib cage (right lateral views).

After Miller, Christensen, and Evans.

Subcostal

Sternum

Internal intercostal

External intercostal

(a)

Vertebral column

External intercostal muscle

A
a'
a
b'
b
B

(b)

C
c'
c
d'
d
D

(c)

Internal intercostal muscle

Sternum

FIGURE 11.32   Rib cage movements in humans.
(a) Various muscles run between adjacent ribs at slanted angles. (b) During inhalation, external intercostals contract, causing adjacent ribs to be drawn forward, expanding the pleural cavities around the lungs, and aspirating air into them. (c) Exhalation is often passive. Gravity pulls the ribs down (posteriorly), compressing the lungs and expelling air. During vigorous respiration, exhalation may be active. When this occurs, internal intercostals, slanted in an opposite direction, contract to compress the rib cage.

FIGURE 11.33 Locomotor and ventilatory cycles in mammals. During rapid locomotion, cycles of inhalation and exhalation are often synchronized with phases of the locomotor cycle. (a) Body positions of a horse at five successive points in a canter, indicated by circled numbers below. (b) Bursts of sound recorded at the nostrils reveal points of inhalation (circled I) and exhalation (circled E). (c) The footfall pattern indicates times of foot contact: left forelimb (LF), right forelimb (RF), left hindlimb (LH), right hindlimb (RH).

After Bramble and Carrier.

Form and Function

Patterns of Gas Transfer

In a general sense, a respiratory organ couples blood flow with ventilation. One function of the respiratory organ is to orient blood flow in relation to ventilation. The orientation is important because it affects the efficiency of gas exchange. One common pattern is countercurrent flow, illustrated in the gills of some fishes, in which water flows across the secondary lamellae in one direction and blood flows through capillaries in the opposite direction (figure 11.39a). This arrangement maintains high partial pressure gradients of gases while water and bloodstream pass each other. As noted, crosscurrent flow is thought to occur between air and blood capillaries in avian lungs. Airflow and blood flow cross each other obliquely rather than lying in parallel. Blood capillaries are in series with each other as they cross a gas gradient of air capillaries. Oxygen is efficiently loaded into the blood before it departs this exchange system. The gills of some fishes may operate on a crosscurrent pattern as well (figure 11.39b). Mammalian lungs illustrate gas exchange involving a **uniform pool.** Lung ventilation tends to keep the partial pressures of gases within the alveolar spaces uniform thanks to frequent breathing, mixing of gases, and absence of significant barriers to diffusion. The circulating blood in the alveolar capillaries encounters more or less uniform partial pressures (figure 11.39c).

FIGURE 11.34 Mammalian lung. The lungs of mammals are blind-ended, terminating in small alveoli. (a) The trachea leads to the pleural cavities and branches into bronchi to supply left and right lungs. Repeated bronchial branchings produce smaller and smaller bronchioles that eventually lead to alveolar sacs. (b) Enlarged alveolar sac. Arteries and veins supply the alveoli to accommodate gas exchange within them. (c) Internal subdivisions of the alveolar sacs are shown. Each small compartment is an alveolus where actual respiratory exchange between blood and air occurs. Note the smooth muscle bands at the openings to the alveolar sacs.

The respiratory area within vertebrate lungs has often been described as *alveolar,* a term inspired by the structure of mammalian lungs. However, the term is inappropriate for other groups. The respiratory compartments of most nonmammalian vertebrates do not form at the terminus of a bronchial tree. Instead, most compartments are subdivided by secondary and tertiary septa and should be called faveoli. This pattern should be called *faviform* to distinguish it from the mammalian alveolar pattern. In birds, this type of subdivision yields a third structural pattern, *parabronchial lungs,* in which blind sacs surround and open to a central parabronchus. Faviform (or faveolar) lungs have less surface area and less elasticity than alveolar lungs. But they are simple and economical, sufficient to meet the generally lower metabolic demands of most reptiles. Alveolar lungs have more surface area to support the greater metabolic requirements of mammals. Alveolar lungs are also more elastic and are ventilated differently, allowing mammals to ventilate their

(a)

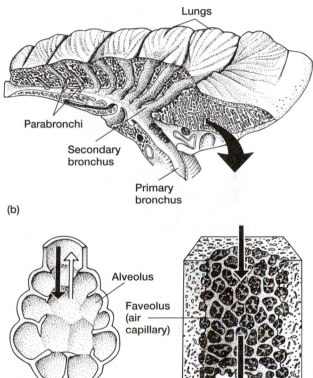

(c)

FIGURE 11.35 Avian respiratory system. (a) The respiratory system of birds consists of paired lungs located in the dorsal wall of the thoracic cavity. Air sacs that lie among the viscera and extend into the cores of adjacent bones are attached to the lungs. Apparently the lungs themselves do not change shape with rib cage motion. Rather, compression and expansion of the rib cage act on the air sacs, drawing air through them and then into the lungs. (b) Ventilation of the avian lung. Ribs are hinged to each other and to the sternum in such a way that lowering of the sternum results in expansion of the rib cage and inhalation. Elevation of the sternum compresses the air sacs and air is expelled (see figure 11.37).

FIGURE 11.36 Avian lung. (a) Right, laterial view. Lungs and air sacs are located within the body cavity between the sternum and the axial column. The lung is cut away to show the primary bronchus and parabronchial network inside. Inflated air sacs are indicated. (b) Right, laterial view. The isolated lung is sectioned. The small pores in the exposed lung are parabronchi. The trachea branches into two primary bronchi (mesobronchi) that extend to the posterior air sacs. Along the way, they open into secondary bronchi. These lead to parabronchi that open into the highly subdivided respiratory tissue, the air capillaries. In the bird lung, flow through the parabronchi is one way, unlike the mammalian airflow, which ends in blind aveoli. (c) Comparison of mammalian and avian respiratory surfaces. In the mammalian lung, the alveoli are blind-ended. For gas exchange to take place, the air must move tidally (open and solid arrows). In the avian lung, air passes one way (solid arrows) through the parabronchi, replenishing the air capillaries that surround and open into the parabronchi.

After Duncker.

(a)

Lung
(parabronchial
network)

Abdominal
sac

Posterior
thoracic
sac

Anterior
thoracic
sac

Interclavicular
sac

Primary
bronchus
(mesobronchus)

Trachea

Inhalation 1

Exhalation 1

Cycle 1

Cycle 2

Exhalation 2

Inhalation 2

(b)

FIGURE 11.37 **Schematic representation of hypothesized lung ventilation in birds.** (a) The avian respiratory system includes anterior (interclavicular and anterior thoracic) and posterior (posterior thoracic and abdominal) air sacs that connect to the parabronchial network and hence to the respiratory tissue. (b) Airflow patterns during breathing. Movement of one draft of air in, through, and out requires two cycles of inhalation/exhalation. During the first inhalation (1), the draft of air (solid arrow) enters and is divided, with some filling the posterior air sacs and the rest passing to the parabronchial network. Upon exhalation (1), fresh air from the posterior air sacs passes into the lungs, pushing spent air out. The second breath (dashed arrow) of inhalation (2) enters and is divided, with some refilling posterior air sacs and the rest passing to the parabronchial network, pushing the remainder of the air from the first breath into the anterior air sacs. During exhalation (2), spent air is pushed out of the lungs and exits with air from the anterior air sacs.

Based on Scheid and Piiper, 1989.

FIGURE 11.38 Crosscurrent gas exchange in the avian lung. Diffusion of gases between the air capillaries and the parabronchus (open arrows) replenishes the gases available for exchange between the lungs and the blood capillaries (solid arrows). It is hypothesized that oxygen is progressively loaded into the blood (and carbon dioxide given up) based on an efficient crosscurrent system.

FIGURE 11.39 Patterns of gas transfer. Orientation of ventilation (solid arrows) to blood flow (open arrows) is established by the respiratory organ. (a) Countercurrent. (b) Crosscurrent. (c) Uniform pool.

lungs constantly with low metabolic cost. The crosscurrent gas exchange system and the use of air sacs in the bird parabronchial lung gain birds support of their higher metabolic scope, and allow extraction of oxygen while at high altitudes.

Rates of Gas Transfer

The respiratory organs must also be designed to match the *rate* at which air or water passes over the respiratory surface (ventilation) with the *rate* at which blood moves through the respiratory organ (perfusion). When the lungs are functioning efficiently, the rates of ventilation and perfusion are balanced so that the amount of oxygen available to diffuse across the respiratory surface from one side is exactly matched by the ability of the blood perfusing the opposite side to carry this oxygen away (figure 11.40a). For carbon dioxide, the reverse is true. The amount of carbon dioxide carried by the blood must be matched to the ability of the respiratory medium to carry it away. If perfusion is too rapid relative to ventilation rate, blood moves through the organ too quickly and departs before it is fully saturated with oxygen (figure 11.40c). On the other hand, if perfusion is too slow, blood lingers too long in the organ after it has become saturated and can no longer take up additional oxygen (figure 11.40b). In either case, the metabolic cost of oxygen extraction will be higher than optimum, and respiration will be inefficient.

The ratio of ventilation to perfusion depends on the species. Within a species the ratio changes with activity levels and with availability of oxygen in the environment. In mammals, the ratio can be 1:1; in some reptiles, it is as much as 5:1. Some fishes have shown a ratio of 35:1. As a relative measure of the interaction of the respiratory and circulatory systems within a species, gas transfer ratios give us insight into the problems a species faces as well as its physiological response.

For example, water, even when it is saturated with dissolved air, still contains considerably less dissolved oxygen than an equal volume of air. Further, water is 1,000 times denser and more viscous than air, so gases diffuse much more slowly. Consequently, relatively large volumes of water must be moved across gill surfaces to match the high oxygen affinity of the perfusing blood; therefore, ventilation-perfusion ratios are generally large in fishes. Water flow may be up to 35 times blood flow. Water flows almost continuously and in a countercurrent fashion. In contrast to this, inactive reptiles with low metabolic demands may take a breath only every minute or so. In mammals, with high metabolic demands and tidal ventilation, breathing is more or less continuous, so the blood flowing from the alveoli is saturated. In exercising humans, metabolic demands of active tissues increase further. Both ventilation and perfusion increase in step at such times. You breathe faster (ventilation), and your heart rate accelerates (perfusion).

Respiratory
medium
(air or water)

Blood

Respiratory organ

(a) Normal ventilation

(b) Fast ventilation

(c) Slow ventilation

FIGURE 11.40 Ventilation: perfusion ratios. The air- or water-breathing organ balances blood flow (perfusion) with movement of the respiratory medium (ventilation). (a) If perfusion and ventilation are matched properly, blood departs from the respiratory organ just as it becomes saturated with oxygen. (b) If ventilation is too fast, blood lingers longer than necessary in the respiratory organ and becomes saturated early but takes up no more oxygen. (c) If ventilation is too slow, blood is only partially oxygenated when it departs from the exchange organ. Breathing that is too fast or too slow is inefficient. Widths of arrows are proportional to flow rates. Shading on arrows passing through the respiratory organ indicates degree of oxygen saturation.

Many subtle adjustments help to optimize gas exchange. For example, fishes respond to a drop in available oxygen in the water in several ways. As might be expected, gill ventilation increases, as does output of blood by the heart. Other adjustments occur as well. Gill filaments are repositioned to allow more secondary lamellae to participate in respiration (figure 11.41a–c). The transit time of water moving through the gills increases, and the distance of diffusion across the lamellae probably decreases. Acting in concert, these collective changes maintain oxygen uptake by maintaining, as well as possible, favorable exchange ratios during times of low oxygen availability.

As we have seen, the respiratory organs involved in breathing water and air are necessarily different in design because of the different problems faced with gas exchange in the two media. We explore these differences in the next two sections, Breathing in Water and Breathing in Air.

Breathing in Water

At 15°C, water holds about 1/30 as much oxygen as air. Furthermore, water is considerably denser than air. Nevertheless, water-breathing fishes can usually maintain sufficient delivery of oxygen to their tissues. In part, this is possible because of their high ventilation rate. Water flow is usually over ten times that of blood flow. But it is also due to the dual pump that keeps a nearly continuous stream of new water washing across the gills and to the efficient countercurrent pattern of flow. As a consequence of this design, water leaving the gills may have given up 80% to 90% of its oxygen. This is a rather high oxygen extraction. In mammals, for example, only about 25% of the oxygen in the lung is taken up before it is exhaled. Although fish ventilation

extracts more oxygen, the metabolic cost of moving the denser water is higher, so the high level of oxygen extraction is won only at a cost.

Breathing in Air

Water is dense, so its tidal movement would be a relatively expensive method of respiration as the water mass was accelerated in first one direction and then in the other. On the other hand, air is light, so its tidal movement requires relatively less energy. However, exchange surfaces are exposed to evaporation in vertebrates that breathe in air. For this reason, air-breathing organs, such as lungs, are usually recessed in cavities, precluding unidirectional flow or countercurrent exchange and requiring a tidal method of ventilation. The exception, of course, is the unusual arrangement that evolved in birds in which air-breathing organs are ventilated by unidirectional flow and gas exchange involves a crosscurrent pattern.

Evolution of Respiratory Organs

Acid-Base Regulation

The evolution of respiratory organs is in large part related to problems involving the *extraction* of oxygen from water or air to fuel metabolism, but not entirely. Often the design of a respiratory organ depends on its opposite role, namely, *elimination* of metabolic waste products. Mechanisms that regulate the acid-base balance of the blood illustrate how the body handles by-products of metabolism.

During respiration, oxygen is transported to active tissues of the body. Simultaneously, by-products of metabolism

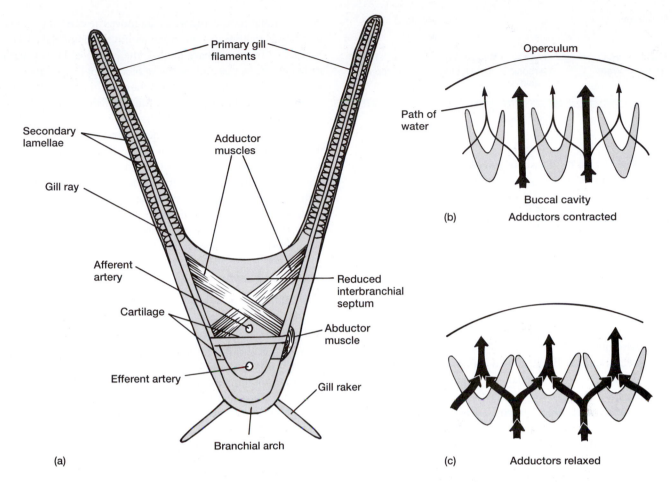

FIGURE 11.41 **Fish gills.** (a) Positions of gill filaments are controlled by crossing adductor muscles. (b) When resting, primary gill filaments may part to allow excess water to bypass the respiratory surfaces. (c) When active, the primary filaments are moved more directly into the water flow to increase irrigation of secondary gill filaments and increase the opportunity for gas exchange.

are removed. Ammonia (NH₃), a toxic by-product, is excreted via the gills or the kidneys in the form of less harmful urea and uric acid. But carbon dioxide excretion is another matter. Surprisingly, carbon dioxide itself is not very toxic, although the hydrogen ions (H⁺) it generates can be a problem. Elimination of carbon dioxide from the body is related to its effects on acid-base levels of the blood, or **pH balance.**

When carbon dioxide enters the blood, it combines with water to dissociate reversibly into carbonic acid, which in turn forms a hydrogen ion and a bicarbonate ion (figure 11.42). An increase in hydrogen ions in the blood causes a decrease in blood pH. As more hydrogen ions accumulate, the blood becomes more acidic. As hydrogen ions are removed, blood becomes less acidic (more basic). This is critical. The affinity of hemoglobin for oxygen decreases with decreasing pH. More fundamentally, protein enzymes that control essential cell metabolism operate within a narrow pH range. If blood pH is too high or too low, they become nonfunctional. Control of pH centers on control of hydrogen ions, and this in turn is affected by carbon dioxide levels in the blood. Elimination of carbon dioxide shifts the equation shown in figure 11.42 to the left. A hydrogen ion recombines with a bicarbonate ion, so the hydrogen ion concentration in the blood is reduced,

and acid levels fall. The buildup of carbon dioxide in the blood has the reverse effect—blood becomes more acidic.

When a vertebrate exercises, lactic acid accumulates in the blood as a by-product of protein metabolism. Eventually it is broken down by chemical degradation, but not immediately. Thus, the buildup of lactic acid threatens to change unfavorably the blood pH. A compensatory increase in carbon dioxide elimination counters this lactic-acid-induced change in blood pH. Hydrogen diffuses out of the blood and pH returns to normal levels. Carbon dioxide is part of a complex buffering system that prevents drastic swings in pH levels. In mammals, breathing rate increases with exercise; therefore, more oxygen is taken into the lungs to support aerobic metabolism, and more carbon dioxide is eliminated to buffer blood pH.

In fishes, some carbon dioxide is eliminated via the skin, but most exits across the gills. In adult amphibians, oxygen is taken in through the lungs, but carbon dioxide is almost exclusively eliminated through the skin. In addition, the amphibian kidney participates in regulating acid balance by secreting hydrogen ions, but it does so by employing a secretory mechanism that depends on an immediate supply of water. Because amphibians usually frequent sources of fresh water, this is simple and easy. Thus, acid balance in amphibians is maintained indirectly by carbon dioxide

$$\underset{\substack{\uparrow \\ \text{Build} \\ \text{up}}}{\underset{\substack{\text{Carbon} \\ \text{dioxide}}}{CO_2}} + \underset{\text{Water}}{H_2O} \rightleftharpoons \underset{\substack{\text{Carbonic} \\ \text{acid}}}{H_2CO_3} \rightleftharpoons \underset{\substack{\text{Hydrogen} \\ \text{ion}}}{H^+} + \underset{\text{Bicarbonate}}{HCO_3^-}$$

FIGURE 11.42 Dissociation equations for carbon dioxide and its effects on pH balance. Buildup or elimination of carbon dioxide (CO_2) affects blood pH. When CO_2 blood levels are low, hydrogen ions (H^+) combine with bicarbonate ions (HCO_3^-) to form carbonic acid (H_2CO_3), which dissociates into water (H_2O) and CO_2. The equation shifts to the left. When CO_2 accumulates in the blood, the equation shifts to the right, resulting in the buildup of H^+ in the blood and a more acidic pH.

elimination via the skin (which affects the dissociation equation and hence the concentration of hydrogen ions) or directly by hydrogen ion secretion via the kidneys.

Amphibians, however, pay a price for this simple system of elimination. Because acid balance via the kidneys is based on a secretory mechanism requiring large influxes of water, they must have immediate access to a supply of water. Water for elimination, and not the demands of oxygen uptake, is one of the major reasons why amphibians are so closely linked to aquatic environments.

The loss of fish gills in the early tetrapods meant the loss of one major route used in regulating pH balance. As we have just seen, the kidneys and skin of modern amphibians take over this function. It is not known how *Ichthyostega*, which had some thick skin scales, and other fossil lissamphia coped with this problem. In amniotes, the lungs take on the function of regulating pH balance by eliminating carbon dioxide. In the mammalian kidney, elimination is based on a different mechanism from that of amphibians, one that conserves water. In chapter 15, we discuss the kidney's role in water balance and in acid-base regulation. This short journey into blood chemistry reminds us that the evolution of life on land required more than the appearance of limbs. New physiological problems had to be solved as well.

Ventilation

The evolution of respiratory organs is also a story of mechanical devices that move water or air. Some respiratory pumps depend on cilia; however, most are based on muscle contraction.

Ciliary Pumps

Cutaneous gas exchange may have played an important part in early vertebrate respiration, and in some groups, such as modern amphibians, still does. Direct exchange of gases between tissues and the environment through the skin is a simple and direct way to meet modest metabolic needs of small organisms. The small larvae of some fishes still depend on cutaneous respiration. In some cases, such as the Australian lungfish larva, surface cilia are used to develop respiratory currents across the organism's surface.

Among protochordates, cilia move the current of water that brings food to mucous traps within the branchial basket. Ventilated by this "ciliary pump," the branchial basket, with its large surface area and extensive blood supply, also assumes a large portion of the respiratory chores of the skin. If such a condition existed in ancestors to the first fishes, it would have had consequences for their subsequent evolution.

First, such a branchial basket, based on an active ciliary pump specialized for ventilation and feeding, would allow the evolution of larger and more active species than would be possible with cutaneous respiration alone. Second, by reducing a vertebrate's dependence on cutaneous respiration, ciliary pumps made possible the evolution of the thick bony armor that prevents cutaneous respiration. Presence of dermal armor in ostracoderm fishes might reflect this evolutionary opportunity.

Muscular Pumps

If an animal is large or active, the ventilatory abilities of cilia fall behind the increase in metabolic requirements. Muscular pumps that replace cilia as the mechanism to move currents of water are an answer to this problem. For example, the ammocoete larva of lampreys employs a muscular velum to pump water across its gills. In the adult, the branchial basket participates in the muscular movement of water across exchange surfaces.

The appearance of muscular pumps in early vertebrates was probably a prerequisite to the attainment of large size and active lifestyles. Without such respiratory mechanisms, the kinds of vertebrates that evolved would have been considerably restricted.

Water-to-Land Transition

No single change in lifestyle had more important effects on vertebrate design than the transition from life in water to life on land. In regard to the respiratory system, this transition included a change from water- to air-breathing organs and eventually a change in the type of ventilatory pump. Air-breathing lungs arose before this transition was under way, and the aspiration pump that efficiently fills the lungs arose much later, after terrestrial vertebrates had become established.

Air-breathing Organs

One prerequisite to life on land is the presence of a respiratory organ that can serve in gas exchange with air. The evolution of air-breathing organs occurred several times within different lines of bony fishes. These organs include vascularized swim bladders, parts of the digestive tract, specialized compartments to the gill chamber, and in dipnoans, lungs. One feature common to many fishes with air-breathing organs is that they live in fresh water susceptible to seasonal hypoxia. As a result of high temperatures, drought, decay of organic material, or stagnant water, the oxygen levels in water occasionally plummet. Hypoxia can be a time of intense stress for fishes, yet, ironically, oxygen is within easy reach in the atmosphere above them. Presumably it was just such conditions of seasonal hypoxia that favored the evolution of accessory organs capable of extracting oxygen from gulps of atmospheric air.

The lure to explore or exploit underwater worlds directly has tempted humans for centuries. The easiest route has been to hold your breath and dive in. An obvious limitation is that you can stay under only until your breath runs out. To extend the downtime, various devices have been used to pump air to divers wearing hoods or hard hats. The limitation to these, of course, is that the dive is restricted by the air line to the surface.

Scuba gear solved this. Scuba, or more correctly, S.C.U.B.A., means "self-contained underwater breathing apparatus." A large volume of air, compressed into a small cylinder, and a device to deliver it on demand, the regulator, allow the diver great freedom of movement while underwater (box figure 1a).

A diving suit with compressed air was invented in 1825 and a demand regulator in 1866. But no one seemed to make the connection between the two, at least for purposes of underwater exploration, except for Jules Verne, who brought these together in the world of fiction, *20,000 Leagues Under the Sea*. Credit for bringing these devices together in the real world goes to Jacques Cousteau of France and Emile Gagnan of Canada. They fitted a demand regulator to a compressed air tank and tested it during the summer of 1943. It worked. Diving was revolutionized, but with some added risks.

Most risks come from the effects of increased partial pressures of gases. At the surface of Earth, the column of air above weighs down on a person at sea level, producing 1 atmosphere of pressure, or 101,000 Pa (14.7 psi). In seawater, every 10 m of descent increases pressure on a diver by about 1 more atmosphere. Thus, at 20 m, pressure is 3 atmospheres, or 303,000 Pa (1 atmosphere from the column of air at sea level plus 2 more for the additional pressure of seawater at 20 m). And this is what creates problems. For the compressed air to fill the lungs at 20 m, the regulator must match the pressure on the lungs. Thus, air enters the lungs at a higher pressure than it would on dry land at sea level. The high pressure of air in the lungs forces high levels of gases into the blood. When the blood is saturated, it holds more gas than it does at lower pressures. As a result of these elevated saturation levels, problems can result if a diver goes too deep or comes up too fast.

On the one hand, if the diver continues to descend, nitrogen (about 78% of air) reaches unusually high levels in the blood. At depths of over 30 m, elevated nitrogen levels can cause dizziness, loss of judgment, and impaired simple motor functions (box figure 1b). This condition is known as **nitrogen narcosis.** Surprisingly, oxygen (about 21% of air) can become toxic if high partial pressures force it to reach high levels in the blood. Excess oxygen can cause lung injury and permanent damage to the central nervous system.

On the other hand, if the diver ascends too quickly, problems from nitrogen can arise as well. When dives are deep and prolonged, the elevated partial pressures push high levels of nitrogen into the blood. When ascending, the diver must take adequate time to allow the excess nitrogen to escape through the lungs as air is expelled. When you quickly pop a soft drink lid, gas under pressure is suddenly returned to atmospheric pressure, and it comes out of solution to fizz and form bubbles. The same thing happens in blood when a diver comes to the surface too quickly. Nitrogen comes out of solution too quickly and forms bubbles in the blood. These bubbles can lodge anywhere—lungs, joints, muscles, stomach, brain—and cause pain or death. The condition is known as **caisson disease,** or the "bends." Treatment calls for quickly placing the afflicted diver in a recompression chamber to once again elevate the pressure on his or her body and force the nitrogen bubbles back into the blood. Then, slowly, the individual is brought back to normal atmospheric pressures, allowing time for the excess nitrogen to escape by diffusion through the lungs.

Deep-diving marine mammals, such as dolphins and seals, are not immune to the bends, but they seem to depend on devices

Lung evolution (p. 420)

It is instructive to compare this situation faced by many bony fishes with that of elasmobranchs, which never developed an air-breathing capability. Chondrichthyans approach neutral buoyancy because their skeletons have been reduced to cartilage and their buoyant oils reduce their overall density. They do not have gas bladders. They frequent well-aerated marine waters, and some sharks cruise deep open water that is distant from the water-air interface.

The evolutionary transition from water to land occurred between sarcopterygians and ancient tetrapods. Today, all coelacanths, except for *Latimeria,* are extinct. Unfortunately, *Latimeria* occurs in deep marine waters and apparently is specialized. Its lung, inundated with fat, is a nonrespiratory organ. Among the lungfishes, the Australian (*Neoceratodus*) and the South American (*Lepidosiren*) lungfishes live in shallow freshwater streams, and the African lungfish (*Protopterus*) primarily occupies lakes. When

their aqueous environment becomes hypoxic or dries entirely, they use their lungs to tap atmospheric oxygen.

The lives of these lungfishes suggest that lungs evolved not in anticipation of life on land but for the immediate adaptive advantage they conferred, namely, as supplements to gill respiration when dissolved oxygen in water became inadequate. Lungs were preadapted. Their biological role was supplementary, allowing fishes to tap into an alternative source of oxygen in the atmospheric air above their aquatic world. When the first tetrapods began exploiting the terrestrial world, lungs were ready to serve in their new role as the primary respiratory organs. Terrestrial life came after the appearance of lungs, not before.

Advantages of Movement to Land

What conditions might have favored the movement to land? One suggestion has been that the seasonal drying of freshwater pools favored movement of stranded fishes across land

(a)

(b)

BOX FIGURE I **S.C.U.B.A.** (a) Early forms of underwater breathing apparatuses. Not for home use. (b) Symptoms of nitrogen narcosis at various depths.

that minimize the problem. Most obviously, their lungs are not filled with air under pressure. They take a breath at the surface and then dive. No more air is added during the deep dive. In fact, excess air retained in the lungs is usually expelled. Thus, as an animal descends and pressure on its rib cage and lungs increases, this pressure does not force residual air into the blood at a high pressure. High nitrogen levels in the blood do not develop, so there is less risk of nitrogen coming out of the blood upon return to the surface. Further, the

bronchial tree is supported by cartilaginous rings only up to the level of the respiratory bronchioles. Absence of supporting rings beyond this point in the bronchial tree permits the alveoli to collapse under high pressure. Consequently, not much air is trapped in the lungs, where it might make prolonged contact with capillary exchange surfaces during a prolonged dive at great pressure. Finally, although the mechanism is not yet well understood, it seems that the tissues of marine mammals are resistant to the bends. Their fat, in particular,

seems to be able to absorb excess nitrogen safely. Ichthyosaurs, dolphin-like reptiles of the Mesozoic, show evidence of the bends. When bubbles of nitrogen form inside bones they block local blood supply. This causes bone cells to die, thereby weakening the bone which sometimes results in visual damage. Such bone damage can be seen in ichthyosaur fossil bones, perhaps occurring when they surface too quickly to elude predator attacks as from large sharks or as were then prevalent, large marine crocodiles.

---

to pools that persisted. Perhaps. But modern lungfishes faced with similar conditions do not normally migrate across land to new water. Instead, they estivate by burying themselves in mud, where their metabolic rate drops. Encased in mud cocoons, they can survive several years until rains return to replenish their pools.

Another suggestion has been that low oxygen levels in water prompted fishes onto land in search of alternatives. But, as we have seen, hypoxia stimulates air breathing but not necessarily migration.

Fishes today that venture onto land, such as the teleost mudskipper, apparently do so to search for food and to leave behind water-bound predators. Similarly, these advantages may have favored the movements of the first rhipidistians onto land, thereby beginning the terrestrial phase of vertebrate evolution.

Skeletal modifications for land (p. 348)

Air-breathing Mechanisms

Although air breathing itself evolved before vertebrates moved to land, the air-breathing mechanisms carried onto land by early tetrapods were modifications of the dual pump, a water-breathing mechanism of fishes (figure 11.43). We have seen the evolutionary stages involved. The dual pump is modified into the buccal pump of air-breathing fishes to force air into their lungs or gas bladders. This same buccal pump mechanism is the basis on which living adult amphibians fill their lungs, but with modifications. Because amphibians do not have internal gills, the tandem opercular component of the buccal pump becomes redundant in adult amphibians, and in adult frogs and salamanders it is lost. The job of ventilating the lungs now falls mainly to the other component of the buccal pump, the buccal cavity, which becomes enlarged and broadened. This means that the buccal cavity of adult

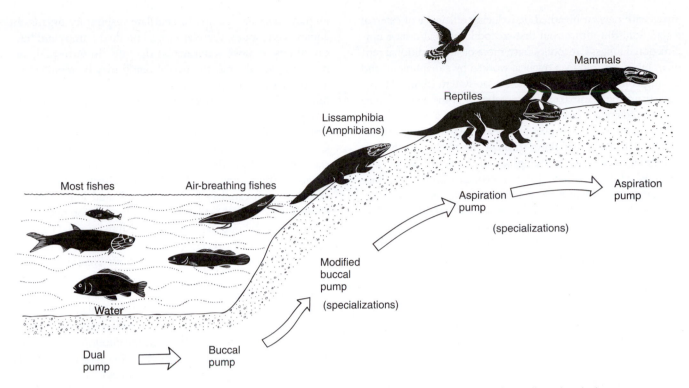

FIGURE 11.43 **Evolution of ventilatory mechanisms.** Water-breathing fishes irrigate their gills with a dual pump mechanism in which buccal and opercular cavities operate in tandem. Air-breathing fishes use a buccal pump mechanism, a modification of the dual pump in which the buccal cavity is the major mechanical component. In adult amphibians, lung ventilation is based on a modified buccal pump in which the opercular pump is lost entirely. However, the amphibian skull design is compromised because the buccal cavity must function in feeding and in lung ventilation. One solution is found in plethodontid salamanders. Gas exchange is taken over entirely by cutaneous respiration, and lungs are lost; therefore, the buccal cavity serves only feeding. In amniotes, the aspiration pump completely separates feeding from lung ventilation, uncoupling demands on the jaws for service in both activities. The basic dual and buccal pumps are specialized in many fishes. Bird respiration represents a specialization of the aspiration pump.

amphibians must serve two major functions, feeding and breathing, often with contradictory demands on design.

On the one hand, the amphibian buccal cavity moves a large volume of air to ventilate the lungs. To move a large tidal volume, the jaws should be light to reduce mass, and the skull broad. On the other hand, we would expect the jaws to be short and robust to serve feeding. It has been hypothesized that the plethodontid salamanders addressed these competing demands on the buccal cavity by losing their lungs. In this salamander family, respiration occurs entirely through the skin; therefore, the buccal cavity exclusively serves feeding and is accordingly narrow and robust.

The aspiration pump separates the respiratory apparatus (rib cage and diaphragm) from the feeding apparatus (jaws), offering another way to address opposing demands on the buccal cavity. The evolutionary consequences of this separation of functions are most evident in reptiles. Reptiles possess small heads and strong jaws. Feeding styles are varied and specialized. Unlike modern amphibians, they eliminate carbon dioxide through their lungs, not through their armored or thickened skin that resists water loss. These combined changes allowed reptiles to travel farther from sources of water and become more committed to terrestrial lifestyles.

Bird Lungs and Air Sacs

Occasionally, some birds reach remarkably high altitudes. During night migrations, most birds reach 1,800 m (6,000 feet). Airplanes and radar report bird sightings at 6,100 to 6,400 m (20,000–21,000 feet). Climbers on Mt. Everest mention observing birds at altitudes of 7,940 m (26,000 feet). The continuous one-way flow across avian respiratory tissue, perhaps in a crosscurrent pattern, gives birds an efficient respiratory system to support their high metabolic requirements. Mammals have a ventilatory system that supports a high metabolic rate as well. But the especially efficient bird lung, with its ability to take up oxygen even in thin air, gives birds a special advantage when they fly at high altitudes.

Why such a high-altitude ability evolved in birds is not known. Certainly, at high altitude, migrating birds could journey above cloud cover and secure access to navigational information from sun and star positions. They could also catch a ride on a high altitude jet stream and almost triple their speed. But it may be much more than this.

Countercurrent and crosscurrent exchange (p. 156)

The unidirectional airflow in bird lungs keeps inhaled oxygenated air separate from deoxygenated air departing the lungs. But in mammals, the spent air breathed out partially mixes with new air breathed in, reducing efficiency of external respiration. But it turns out that crocodiles may also have unidirectional lungs. Crocodiles have more or less traditional reptilian lungs serviced by a trachea, worked by a diaphragm, and lacking completely any avianlike air sac system (figure 11.28). Yet, surprisingly, initial experimental work using artificial ventilation and computer images seems to confirm a unidirectional flow of air through crocodile lungs. The hypothesis to explain this is that inhaled air flows through the trachea to the two primary bronchi, each servicing one of the paired lungs. The air initially bypasses the first-tier of secondary bronchi, because they are turned unfavorably at right angles to the entering air; the speeding air instead reaches a second-tier of secondary bronchi, which feed the parabronchi where gas exchange now occurs. The still moving air exits the parabronchi and enters the first-tier of secondary bronchi initially bypassed, before being exhaled. This effectively establishes a one-way flow of air through the air exchange components of the lung. Why crocodiles would benefit from such an efficient one-way breathing system is, as with birds, conjecture. We do note that crocodile ancestors, early archosaurs, lived on an Earth where atmospheric oxygen levels were half those of today—about what climbers on Mount Everest experience. Under such conditions, an efficient unidirectional pattern would have enjoyed adaptive favor and evolved before modern crocodiles and birds. More than birds and crocodiles, archosaurs in general may have, under such selective pressures of low atmospheric oxygen in the Early Mesozoic, evolved especially efficient lungs. This would have given them some advantages over their taxonomic contemporaries, the synapsids, who lost their early dominance to the archosaurs (see figures 3.42, 3.47). Finally, the bird air sacs, thought to be required for a unidirectional air flow, as I have described it a few pages earlier, in fact may be instead critical in redistributing weight to control pitch and roll during flight!

Overview

The respiratory system aids passive diffusion of gases between the organism and its environment (external respiration). Ventilation is the active process of moving the respiratory medium, water or air, across exchange surfaces. These respiratory surfaces may include the integument (cutaneous), gas bladders, accessory air-breathing organs (e.g., mouth, gut), or embryonic structures. However, the primary respiratory organs include gills and lungs. Water is denser and more viscous than air, and holds less oxygen. Because they differ in these physical properties, the vertebrate ventilatory pumps differ in water and air. Cilia work at a small scale in water, but are ineffective at a large scale or in thin air. Consequently, large vertebrates employ muscular pumps to ventilate respiratory organs. In aquatic vertebrates, these are based on dual pumps and buccal pumps to move water across the gills. In terrestrial, air-breathing vertebrates, the buccal pump may be modified, or a derived aspiration pump may replace it to move air through lungs. Gills are designed for water breathing, and usually move water in a unidirectional flow past the vascular exchange surfaces where oxygen and carbon dioxide exchange occurs. Gills are also important sites of osmoregulation and nitrogen excretion, usually in the form of ammonia. Lungs are designed for air breathing, and usually move air bidirectionally for tidal exchange with capillary beds. Both gills and lungs function efficiently to match ventilation rate with perfusion rate.

The vertebrate transition from fish to tetrapod included a transition from a water-breathing to an air-breathing respiratory system. However, much of the air-breathing organs, specifically the lungs, apparently evolved in fish ancestors, before ever setting fin or foot on land. Warm, stagnant waters would lose oxygen, become hypoxic, and place fish living there under selective pressure to develop and/or enhance air-breathing capabilities. However, the development of lungs alone was not sufficient to meet changing respiratory needs. Such stagnant, vegetation-choked waters also commonly experience buildup of carbon dioxide, hypercarbia. If this occurred, then the gills would become less effective in eliminating, by gradient diffusion, carbon dioxide to an aquatic environment. In fact, high levels of carbon dioxide and low levels of oxygen in the water might produce the reverse—an unfavorable flow of both gases. Under such stressed conditions, scaled or armored integument would provide a diffusion barrier to carbon dioxide influx from and oxygen loss to the water. Such a thickened integument would also be preadaptive for inhibiting water loss during terrestrial sojourns. The fish dual pump, coupling opercular and buccal mechanisms, underwent remodeling, losing the opercular and enlarging the buccal mechanism. The problem of the double, and sometimes conflicting, functions of the buccal region—ventilation and feeding—is addressed by the derived aspiration pump, dedicated to supporting ventilatory needs, leaving the buccal cavity to prey capture and swallowing.

Faviform (or faveolar) lungs of amphibians and reptiles have small surface area, but they are sufficient to meet the generally lower metabolic demands. Alveolar lungs have great surface area and support the great metabolic requirements of mammals. Alveolar lungs are elastic and ventilated by an aspiration pump, permitting ventilation to occur at low metabolic cost. In birds, the air sac system and the crosscurrent gas exchange within the parabronchial lungs support a high metabolic scope and uptake of oxygen.

WEB LINK

Visit the text website at *www.mhhe.com/kardong7e* for additional study aids including selected references, functional laboratories, and web links to chapter topics.

CHAPTER 12

The Circulatory System

INTRODUCTION

CARDIOVASCULAR SYSTEM
  Blood
  Arteries, Veins, and Capillaries
    Arteries
    Hemodynamics of Circulation
    Veins
    Microcirculation
  Single and Double Circulation
  Embryonic Development of the Cardiovascular System
  Phylogeny of the Cardiovascular System
    Arterial Vessels
    Venous Vessels
  Hearts
    Basic Vertebrate Heart
    Fishes
    Amphibians

    Reptiles
    Birds and Mammals
  Cardiovascular System: Matching Design to Environmental Demands
    Accessory Air-breathing Organs
    Diving Birds and Mammals
    Heart Flow
    Ontogeny of Cardiovascular Function
    Fetal Circulation in Placental Mammals
    Changes at Birth
    Heat Transfer

LYMPHATIC SYSTEM
  Lymphatic Vessels
  Lymphatic Tissue
  Form and Function

OVERVIEW

Introduction

Twelve miles per hour is faster than many people can run and a speed few can sustain. Yet, in the next Olympic marathon, the top finishers will average this speed for over 26 miles, lasting just over two hours! Whales can dive from the surface to depths of over 2,000 m and feed there for up to an hour. During that time, they experience immense pressures on their bodies, over 16 million Pa (about 2,300 psi) per square meter of body surface. That is about equal to a column of lead 150 m high pressing on each square meter of the body. Animals such as the oryx, an African antelope, can be exposed to searing ambient temperatures during the day, and their body temperatures can exceed 45°C (113°F). In large measure, human and animal athletes can adjust to changes in activity and physical stress because of adjustments orchestrated by the circulatory system.

In cooperation with the respiratory system, the circulatory system transports gases between the sites of external and internal respiration. But the circulatory system also has many other important functions. It adjusts to changes in pressure on or within the body. Blood transports excess heat produced within the body to the skin to dissipate it. Conversely, a cool reptile basking in the sun gathers surface heat to warm its blood, which is then circulated to the rest of the body. Glucose and other end products of digestion are carried to active organs for metabolic use or to other organs for temporary storage. The circulatory system transports hormones to target organs and waste products to the kidneys. Blood also carries cells and chemicals of the immune system to defend the body from invasion by foreign organisms.

The circulatory system of vertebrates is basically a set of connecting tubes and pumps that move fluid. The ability of the organism to adjust to immediate physiological changes in physical and metabolic activity depends on the rapid response of this system. The circulatory system includes the blood and lymph vascular systems. Lymphatic vessels and **lymph,** the fluid they circulate, collectively constitute the **lymphatic**

system, discussed later in this chapter. The vascular system includes the blood vessels that carry blood pumped by the heart. Together blood, vessels, and heart constitute the **cardiovascular system,** which we discuss first.

Cardiovascular System

Blood

Cells produced by hemopoietic tissues usually enter the circulation to become the **peripheral** or **circulating blood.** Circulating blood is comprised of plasma and formed elements. The **plasma** is the fluid component and can be thought of as the ground substance of blood, a special connective tissue. The **formed elements** are the cellular components of blood. **Red blood cells,** or **erythrocytes,** are one cell type of the formed elements. All erythrocytes have nuclei, except those in mammals. Mature red blood cells in mammals lack nuclei. **Hemoglobin,** the major oxygen transport molecule, is excreted by the kidneys if it is left free in the plasma. Thus, red blood cells function as containers for hemoglobin, preventing its elimination. Red blood cells vary in individual size, from 8 μm in humans, to 9 μm in elephants, to 80 μm in some salamanders. Most live three to four months in the circulating blood before being broken down and replaced.

White blood cells, or **leucocytes,** are a second major cellular constituent of the formed elements. Leucocytes defend the body from infection and disease. The **platelets** are a third formed element in the blood. They release factors that produce a cascade of chemical events leading to the formation of a **clot,** or **thrombus,** at sites of tissue damage.

Plasma and formed elements give blood a wide variety of roles in body processes. In addition to functioning in respiration and disease protection, blood also plays a part in nutrition (carries carbohydrates, fats, proteins), excretion (carries spent metabolites), regulation of body temperature (carries and distributes heat), maintenance of water balance, and transport of hormones.

Arteries, Veins, and Capillaries

Although they vary in size, there are three principal types of blood vessels: arteries, veins, and capillaries. **Arteries** carry blood away from the heart, **veins** carry blood toward the heart, and **capillaries** are the tiny vessels that lie between them. Most arteries carry blood high in oxygen and veins carry blood low in oxygen, but this is not always true. For example, the pulmonary artery carries blood low in oxygen from the heart to the lung to be replenished, and the pulmonary vein usually returns blood high in oxygen. Thus, the direction of blood flow with reference to the heart defines the type of vessel, not the oxygen content of blood it carries.

Arteries and veins have tubular walls organized into three layers that enclose a central lumen (figure 12.1). The innermost layer, the **tunica intima,** includes the lining of endothelial cells that face the lumen. On the outside is the **tunica adventitia,** composed mostly of fibrous connective tissue. Between these two layers is the **tunica media,** which differs the most in arteries and veins. Some smooth muscle contributes to the tunica media of large arteries, but elastic fibers predominate. In large veins, this middle layer contains mostly smooth muscle with almost no elastic fibers. Veins

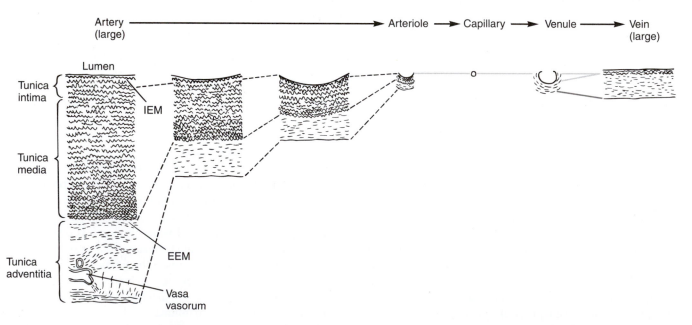

FIGURE 12.1 **Blood vessels.** The three layers of blood vessel walls change in relative thickness and size from large arteries to small arterioles, capillaries, venules, and veins. In large arteries, the tunica media is especially well impregnated with elastic fibers, including belts termed *internal* (IEM) and *external* (EEM) *elastic membranes.* Their walls allow arteries to stretch and receive the pulse of blood delivered suddenly from the heart. Large blood vessels receive their own blood supply from small vessels in their walls, the vasa vasorum.

usually have one-way valves within their walls, and arteries lack such valves. Very small arteries and veins are called **arterioles** and **venules,** respectively. In these small vessels, the tunica adventitia is thin, and the tunica media is composed mostly of smooth muscle; thus, arterioles and venules are quite similar in structure (figure 12.1).

Smooth muscles form sheets that surround the walls of arteries or veins. Smooth muscle cells respond to nervous and hormonal stimuli. When they contract, the caliber of a vessel narrows, a response termed **vasoconstriction.** When circular muscle contraction ceases, resident blood pressure forces the vessel to open and restore or enlarge the size of the lumen, a response called **vasodilation.** Obliquely oriented smooth muscle may also assist vasodilation.

Gases, nutrients, water, ions, and heat move across the walls of blood capillaries. To facilitate efficient exchange, capillaries are very small and have extremely thin walls. Capillaries lack a tunica media and a tunica adventitia. Only the endothelial wall of the tunica intima remains. Sets of capillaries serving one area of tissue constitute a **capillary bed.** Each tissue is invested with multiple sets of overlapping capillary beds. As activity of tissues increases or decreases, more or less of these overlapping capillary beds open or close to regulate blood supply to those tissues.

Arteries

The structure of arteries varies with their size. Large arteries have considerable amounts of elastic fibers in their walls; small arteries have almost none. Structural differences occur because of functional differences between large and small arteries. Arteries function primarily as a supply system that carries blood away from the heart and out to body tissues. They also absorb and distribute the sudden surge of blood through them when the heart contracts. Rhythmic contractions of the heart send spurts of blood into the large arteries. With their elastic walls, these large arteries expand to receive the sudden injection of blood, which can be felt in your wrist or neck arteries as a "pulse." Between contractions, the stretched arterial walls elastically recoil, driving this volume of blood smoothly along through the smaller arteries and into the arterioles. Arterioles direct this blood to local tissues. Humans are one of the few species prone to arterial disease, characterized by hardening of the arterial walls and loss of elastic recoil. As a consequence, afflicted arteries do not expand to blunt the sudden pulse of blood, nor do they move the column of blood along between heartbeats. The heart must work harder, and the smaller arteries and arterioles experience higher surges of blood pressures. Not designed for such pressures, these small vessels may rupture. If this occurs in a critical organ such as the brain, death may follow.

Hemodynamics of Circulation

The pressures and flow patterns of the blood circulating through vessels constitute the **hemodynamics** of circulation. Because of their different hemodynamics, the blood pressures associated with the arterial and venous sides of the

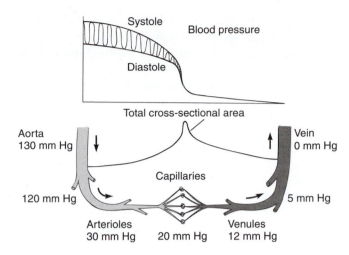

FIGURE 12.2 **Hemodynamics of blood flow.** The systemic flow of blood is depicted graphically at the bottom of the figure. Above the respective vessels is their total cross-sectional area. Note the blood pressure in the different vessels. As blood flows from large arteries, such as the aorta, to capillaries and veins, the initial pressure imparted by the force of heart contraction falls. This is due to the frictional resistance from the walls of the vessels and from the increasing total cross-sectional area. Capillaries have an especially large cross-sectional area. The difference between systolic and diastolic pressures declines as blood approaches the capillaries and is usually minimal thereafter in the venous flow. Normal blood pressure for an adult human is indicated in mm Hg.

circulation are considerably different. When the ventricles of the heart contract, the peak force produced is the **systolic pressure.** The **diastolic pressure** constitutes the lowest pressure within the blood vessels, which is reached between heartbeats. Diastolic pressure results from the force sustained by the elastic recoil of arteries. Blood pressure usually is expressed in a shorthand fashion, with the systolic pressure read and recorded first. For example, in most young adult humans, 120/80 are normal systolic and diastolic values, respectively, obtained from the vessels of the arm (figure 12.2). If the arteries start to show signs of disease, blood pressure will rise, a telltale indication that the major arteries are beginning to fail in absorbing cardiac pumping forces as a result of structural changes in their walls. Some of the highest blood pressures recorded for any vertebrate are in the giraffe, averaging resting pressures of 260/160 at the level of the heart. Such pressures are necessary to supply the brain with blood at sufficient pressure when the animal stands erect. Pressure of blood reaching the brain has dropped, due largely to gravity effects, to about 120/70, comparable to that of humans.

In most vertebrates, pressure characteristically declines as blood flows away from the heart (figure 12.2). This drop in pressure results from two factors: friction as the blood encounters resistance from the luminal walls of vessels, and increase in the total cross-sectional area of blood vessels. The flow of any fluid in tubes is resisted by the friction of the liquid against the walls. For blood to circulate, a force must be used to overcome this frictional resistance to blood flow; nevertheless, as a result of this resistance, blood pressure falls as circulation proceeds. Additionally, as blood flows from large to small arteries, arterioles, and capillaries, the total cross-sectional area of vessels increases, especially in capillaries. Like a fast stream entering a large lake, pressure declines as the larger volume is filled. As a result, blood reaching the venous side of the circulatory system retains very little pressure. In fact, in some of the large veins, forces moving the blood may fall to zero or even become negative. When this happens, blood wants to flow in a reverse, or retrograde, direction. Dealing with these unfavorable pressures is a task that falls to the veins.

Veins

Because veins return blood to the heart, they are collecting tubes. At any one moment, up to 70% of the circulating blood within the body may reside in veins. During times of stress, slight vasoconstriction of strategic veins effectively decreases "reserve" volume and moves some blood from this reservoir to the arterial side of the circulatory system.

Veins are also designed to address low blood pressures. One-way valves that prevent retrograde blood flow are common within their walls. If veins pass between active muscles or through parts of the body subjected to pressure changes (e.g., the pleural cavities containing the lungs), external forces impinge upon and squeeze their walls. These supplemental forces contribute to venous flow, and because of the one-way valves, blood moves only one way back to the heart (figure 12.3). Understandably, in veins that pass through body organs and tissues that offer no induced forces, such as those within the bones or the brain, one-way valves are absent, and return of the blood to the heart depends on any remaining intrinsic pressure and gravity.

Microcirculation

The specific component of the cardiovascular system that regulates and supports cell metabolism intimately is the **microcirculation.** Capillary beds plus the arterioles that supply them and the venules that drain them form the microcirculation. Blood flow to the capillary beds is controlled by smooth muscles. The **precapillary sphincters** are little rings of smooth muscle restricting the entrance to the capillary beds. The walls of both arterioles and venules include thin sheets of smooth muscles. Global nervous and hormonal control of these smooth muscles regulates the flow of blood to the capillaries, as do local events in the supplied tissues themselves. Whether by general events of the body (nervous, hormonal) or local activity (autoregulation), capillary beds adjust blood flow to match cell activity. Blood can be diverted through **shunts** that bypass some regions entirely (figure 12.4).

FIGURE 12.3 One-way valves in veins. The one-way valves within the lumina of veins prevent retrograde movement of blood and ensure its return flow toward the heart (vertical arrow). The pressure that drives blood flow (horizontal solid arrows) comes from surrounding organs, usually muscles, that impinge upon and squeeze veins. Dissected view of hindleg of a lion is shown.

As an animal lowers its head to drink from a stream, blood pressure within its tissues changes quickly (figure 12.5). Quick adjustments in the microcirculation serve to equalize and distribute these temporary pressure fluctuations to

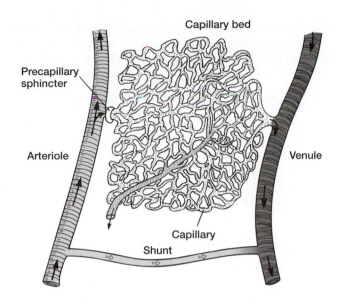

FIGURE 12.4 Microcirculation. The microcirculation includes the capillary beds as well as the arterioles supplying and venules draining them. The usual flow of blood to, through, and from a capillary bed is diagrammed (solid arrows). Smooth muscles of the walls of the arterioles form small bands, the precapillary sphincters, that control blood flow to the capillary bed. A direct shunt running from the arterial to the venous side of circulation allows for major diversions of blood (open arrows).

prevent undue stress on especially sensitive organs such as the brain and spinal cord. Heat distribution within the body is also influenced by the microcirculation. When an animal is active, excess heat transported by the blood reaches the body surfaces. Capillary beds of the skin open to increase blood flow, bringing more heat to the body's surface, where it can be dissipated. Humans with fair skin redden during exercise, indicating this increased peripheral blood flow. In cold weather, the opposite occurs. As body temperature drops, peripheral blood supply decreases, reducing heat loss and helping to maintain the core body temperature. Emotional states can also increase peripheral blood flow, opening more capillary beds. When this happens in the face, we blush.

The microcirculation is also involved in allocating blood to active organs. Capillaries are so small that it would take an hour for just a few drops of blood to pass through a single one. Yet, collectively, capillary beds represent an extensive volume, their linear extent being somewhere around 96,500 km (about 60,000 miles) of microtubing in all. No animal has enough ready blood volume to fill all its capillaries at once. If all the capillaries of a body opened simultaneously, all major blood vessels would be quickly emptied, and the circulatory system would fail. This does not happen because blood is selectively directed to open capillary beds only in active organs.

Ordinarily, not all body tissues are active simultaneously, so the supply of blood to those active tissues is sufficient. By selective deployment, the volume of blood required at any moment can be kept low. However, under some

FIGURE 12.5 Blood pressure changes. As posture changes, the head and extremities are raised or lowered relative to the heart. A giraffe lowering its head to drink quickly experiences increased pressure within its brain and cranial tissues. Adjustments in the microcirculation are one way to prevent such pressures from creating a problem.

circumstances, the microcirculation fails to deliver sufficient blood to meet tissue needs. For example, if there are more active organs than available blood, then the microcirculation gives preference to some and not to others. If strenuous exercise is undertaken soon after a large meal, then the digestive system and skeletal muscles compete for blood to support their activities. Preference is given to the skeletal muscles as more capillary beds open in them and the stomach receives less blood. Humans who exercise might complain of "side cramps." This results from **ischemia,** a localized lack of sufficient blood to the stomach to meet metabolic expectations. Following severe injury or trauma, the microcirculation may fail to regulate blood distribution. When this happens, a condition called shock, properly **hypotensive shock,** results from a cascade of events. Too many vessels open, not enough blood is available, pressure drops, and circulation fails. If shock is not quickly reversed, death can soon follow. The chemical arsenal in some snake venoms takes advantage of this general physiological feature of the cardiovascular system. When injected into prey, the venom induces shock, helping the snake to dispatch its meal quickly.

Single and Double Circulation

Blood travels in one of two general patterns. Most fishes have a **single circulation** pattern in which blood passes only once through the heart during each complete circuit. With this design, blood moves from the heart to the gills to the systemic tissues and back to the heart (figure 12.6a). Amniotes have a **double circulation** pattern in which blood passes through the heart twice during each circuit,

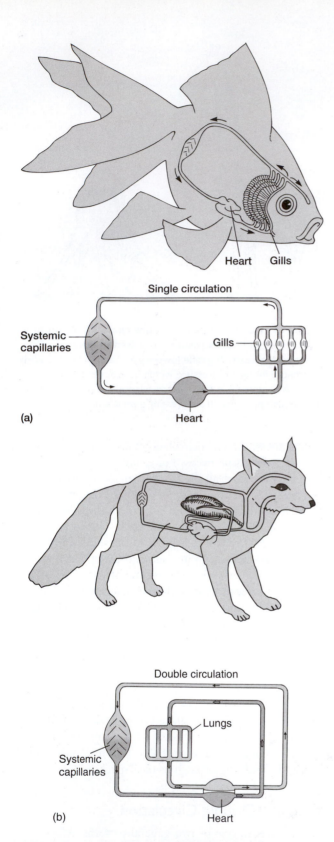

FIGURE 12.6 **Single and double circulation.**
(a) The single circulation of fishes includes heart, gills, and systemic capillaries in series with one another (arrows indicate path of the blood flow). (b) The double circulation of most amniotes includes heart, lungs, and systemic capillaries. Blood passes twice through the heart before completing one route. This places lungs and systemic tissues in separate circuits that parallel each other.

traveling from the heart to the lungs, back to the heart, out to the systemic tissues, and back to the heart a second time (figure 12.6b). The rise of this double circulation involving the addition of a pulmonary circuit was a major evolutionary event. Between those vertebrates with a single and those with a double circulation stand functional intermediates with characteristics of both conditions. The intermediates include lungfishes, amphibians, and reptiles. They suggest the adaptive advantages of transitional forms venturing onto land and highlight the evolution of the circulatory system design. Let us begin by examining the basic embryonic derivation of the cardiovascular system.

Embryonic Development of the Cardiovascular System

Most blood vessels arise within embryonic mesoderm (or from mesenchyme) almost as soon as this germ layer becomes established. Small clusters of mesodermal cells called **blood islands** mark the embryonic debut of the cardiovascular system (figure 12.7a–d). Embryonic blood islands yield both blood vessels and blood cells, so they are involved in **angiogenesis** (blood vessel formation) and **hemopoiesis** (blood cell formation). Blood islands merge, forming a connected vascular network that eventually links parts of the embryo to each other and connects it to its nutrient supply and respiratory organs. The embryonic heart is tubular. Early on, it has autonomous, rhythmic beats that drive blood through the developing vascular network. In the early embryo, such heartbeats help primarily to promote new blood vessel formation. As in the adult, the cardiovascular system of the mid- to late-term embryo also assumes an active and essential role in respiration, metabolism, excretion, and growth.

When first formed, the embryonic vertebrate heart is already contractile and includes four major adjoining chambers. The **sinus venosus** is the first chamber to receive returning blood. Blood flows next into the **atrium,** then into the **ventricle,** and finally into the fourth chamber, the **bulbus cordis.** From the bulbus cordis, blood leaves the heart to enter arteries departing for the body of the embryo. In most tetrapods, splanchnic mesoderm forms the basic four-chambered, tubular heart. Development of the heart begins when cells leave the splanchnic mesoderm to form a medial pair of **endocardial tubes** (figure 12.8a,b). Cells remaining in the splanchnic mesoderm proliferate, producing a thickened lateral region, the paired **epimyocardium.** Cells of the endocardial tube and epimyocardium grow toward the midline and fuse into the single, centrally located, tubular heart. Specifically, the fused endocardial tubes form the endothelial lining of the heart, called the **endocardium,** and the epimyocardium gives rise to the extensive cardiac muscle of the heart wall, the **myocardium,** together with the thin visceral peritoneum covering the heart's surface. With these fusions, the basic four-chambered embryonic heart is established (figure 12.8c).

Subsequent folding of the tubular heart twists the heart into different configurations, but the internal sequence

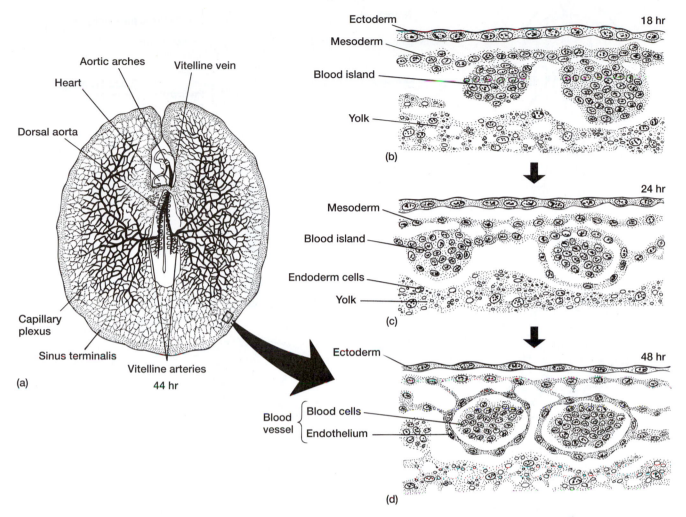

FIGURE 12.7 **Embryonic blood cell formation.** (a) Chick embryo after about 44 hrs of incubation (ventral view). The circulatory system is already well established. Peripheral blood islands have coalesced into major vitelline vessels. (b–d) Detailed developmental sequence of blood vessel formation. (b) Local clusters of mesoderm are organized into blood islands after 18 hrs of incubation. (c) Development at 24 hrs of incubation. (d) At about 44 to 48 hrs of incubation, distinct blood vessels and blood cells that are part of the vitelline vascular network have been formed.

After Patten and Carlson.

of blood flow remains the same (figure 12.9). In most fishes, adults retain this basic four-chambered embryonic heart. However, in lungfishes and tetrapods, varying degrees of internal subdivisions cordon off additional compartments within the heart, and some of the original chambers may become reduced or appropriated by other parts of the adult vascular system. We examine these anatomical modifications and their functional significance as we meet them in this chapter. First, let us review the basic layout of the major arteries and veins, the blood distribution system the heart serves (figure 12.10).

Phylogeny of the Cardiovascular System

The vessels of the cardiovascular system are as varied as the diverse organs they supply. However, these variations are based on modifications of a fundamental plan of organization common to vertebrates. Because it is usually highly modified in advanced forms, this fundamental organization of the cardiovascular system is most evident in primitive vertebrates. Blood leaving the heart first enters an unpaired **ventral aorta** and courses forward below the pharynx. Anteriorly, the ventral aorta divides into the **external carotids,** which carry blood into the ventral region of the head. Before producing these external carotids, however, the ventral aorta gives off a series of **aortic arches,** which pass dorsally within the branchial arches between pharyngeal slits. Above the pharynx, these aortic arches meet a paired **dorsal aorta.** Sprouting from the anterior end of the dorsal aorta are the **internal carotids,** which carry blood forward into the head and usually penetrate the braincase to supply the brain. The dorsal aorta itself, however, carries blood posteriorly (figures 12.10 and 12.11).

At about the level of the liver, the paired vessels of the dorsal aorta unite to form the unpaired **aorta,** which distributes blood to the posterior part of the body and eventually

FIGURE 12.8 Embryonic heart formation. Chick embryo in successive stages of incubation (25, 27, 29 hrs, respectively). Ventral (left) and corresponding cross-sectional (right) views of heart formation are illustrated. (a) Angiogenic cells emerge from the epimyocardium, a thickened splanchnic mesoderm. (b) Angiogenic cells differentiate into a pair of primordial endocardial tubes. (c) This pair of endocardial tubes fuses medially into the single endocardial tube, the future lining of the heart. The thickened epimyocardium forms the thin peritoneum on the surface of the heart, and the extensive myocardium forms the muscular wall of the heart.

After Patten and Carlson.

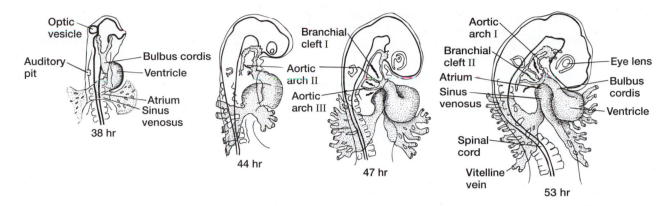

FIGURE 12.9 **Growth of the chick heart.** The four-chambered heart consists of sinus venosus, atrium, ventricle, and bulbus cordis. Once it forms, subsequent folding and enlargement shift the relative positions of these chambers. This process does not alter the route of blood flow through the functioning embryonic heart.

After Patten and Carlson.

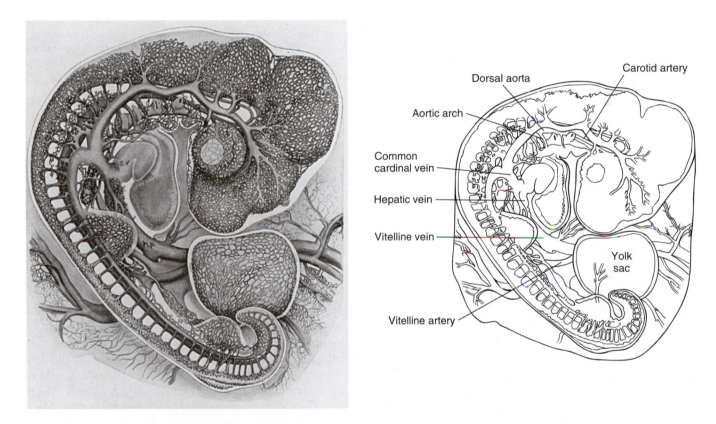

FIGURE 12.10 **Cardiovascular system of a four-day-old chick.** Venous circulation shows anterior and posterior cardinal veins draining into the common cardinal vein that enters the sinus venosus. Vitelline veins return through the forming postcava and travel through the liver sinusoids to enter the heart via the hepatic vein. Arterial circulation is also well established. Aortic arches pass around the pharynx to join above in the dorsal aorta, which supplies blood to the head via the carotid arteries. The dorsal aorta continues posteriorly, eventually forming the vitelline arteries to the yolk.

After Patten.

extends into the tail as the **caudal artery.** Along the way, the dorsal aorta gives off numerous small **parietal arteries** to the local body wall, as well as several major arteries, usually paired, to somatic tissues. Paired **subclavian arteries** supply the anterior appendages (fins or limbs) and usually branch from the dorsal aorta, as do the caudal **iliac arteries,** which supply the posterior appendages. The gonads receive blood from paired **genital arteries (ovarian** or **spermatic).** Paired **renal arteries** to the kidneys are large, major branches from the dorsal aorta. This ensures that the kidneys receive blood early in the arterial circuit while blood pressure is still relatively high, a feature of the hemodynamics that aids

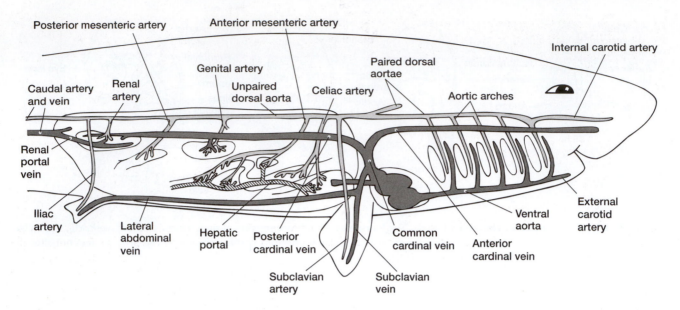

Posterior mesenteric artery
Anterior mesenteric artery
Internal carotid artery
Genital artery
Paired dorsal aortae
Caudal artery and vein
Renal artery
Unpaired dorsal aorta
Celiac artery
Aortic arches
Renal portal vein
External carotid artery
Iliac artery
Lateral abdominal vein
Hepatic portal
Posterior cardinal vein
Common cardinal vein
Anterior cardinal vein
Ventral aorta
Subclavian artery
Subclavian vein

FIGURE 12.11 **Basic vertebrate circulatory pattern illustrated in a shark.** The heart pumps blood to the ventral aorta, from which it is distributed to the paired aortic arches and then to the single dorsal aorta. From the dorsal aorta, blood flows forward to the head and posteriorly to the body, where major branches carry it to visceral and somatic tissues.

After Goodrich.

renal filtration. Typically in vertebrates, three unpaired arteries depart from the dorsal aorta to supply the viscera. These are the **celiac,** supplying the liver, spleen, stomach, and part of the intestines; the **anterior mesenteric,** supplying most of the small intestine; and the **posterior mesenteric,** supplying the large intestine.

Renal circulation (p. 546)

In primitive vertebrates, blood return to the heart includes several prominent veins. The **common cardinal vein,** or **sinus,** is the major vein that receives blood returning from the **anterior cardinal vein (precardinal)** and the **posterior cardinal vein (postcardinal).** These drain the anterior and posterior regions of the body, respectively. Tributaries from the anterior appendage usually empty into the common cardinal vein via the **subclavian vein.** Veins from the lateral body wall and posterior appendage also empty into the common cardinal vein via the **lateral abdominal vein.**

A **portal system** is a vascular pathway that begins in one set of capillaries and runs to another without coursing through the heart in between. There are two major portal systems of the venous circulation in vertebrates. The hepatic portal system begins in capillaries within the wall of the digestive tract and runs as the **hepatic portal vein** to the liver, where it empties into the capillaries and blood sinuses of the liver. This hepatic portal vein transports absorbed nutrients directly from the digestive tract to the liver for storage or processing of many end products of digestion. The renal portal system transports blood returning from capillary beds

within the tail or hindlimbs through the paired **renal portal veins,** which empty into capillaries within the kidneys.

The function of the renal portal system is not well understood. Because it carries caudal blood to the kidneys, some have suggested that it provides a direct route for delivering metabolic by-products to the kidneys that result from active locomotion involving the caudal musculature. Others suggest that it may represent a way of improving kidney filtration. Arterial blood entering the renal arteries directly from the dorsal aorta has high pressure; venous blood of the renal portal system has low pressure. Kidney filtration depends in part on an initial high pressure to move fluid out of the blood and into the kidney tubules, but low pressure in the renal portal veins aids in the recovery of water and other usable solutes, returning these fluids to the general circulation. The renal portal system is present in all classes of vertebrates except mammals. Even though mammals lack a renal portal system, the mammalian kidney nevertheless has a low-pressure vascular network that may be its counterpart and functions similarly to recover fluids from the forming urine.

Phylogenetic modifications within this basic pattern of arteries and veins are largely correlated with functional changes. In the transition from water to land, gills gave way to lungs, accompanied by the establishment of a pulmonary circulation. In some fishes and certainly in tetrapods, the cardinal veins become less involved in blood return. Instead, the composite, prominent **postcava (posterior vena cava)** arose to drain the posterior part of the body and the **precava (anterior vena cava)** developed to drain the anterior part of

FIGURE 12.12 **Aortic arches, gills, and anterior arteries of a lamprey.**

After Hardistry.

the body. Beginning first with arterial vessels, let us now follow the major phylogenetic modifications within the cardiovascular system in detail.

Arterial Vessels

Aortic Arches The number of primitive aortic arches and branchial arches through which they run are still debated. Some ostracoderms had as many as ten pairs of branchial arches and presumably ten pairs of aortic arches. The number of pairs of aortic arches varies in living forms. Lampreys have 8 (figure 12.12), hagfishes 15. Some species of sharks possess 10 or 12 pairs. Nevertheless, only up to six pairs customarily appear during embryonic development in most gnathostome fishes and in all tetrapods. Accordingly, six is the number of aortic arches usually taken as the basic embryonic pattern, and they are designated by Roman numerals (I–VI; figure 12.13). The phylogenetic variation within the aortic arches can be intricate (figure 12.14a–e; see also figure 12.17a–c). Thus, reference to a basic six-arch pattern brings a simplifying theme to a complex anatomy. However, you should be prepared for personal preferences of other authors who may use different numbers for aortic arches. Some insist on using numbers up to ten in recognition of the presumed primitive number. Some abandon the effort to assign homologies, ignore those arches lost phylogenetically, and simply number the arches as they find them in the adult—1, 2, and 3, for example. In this book, Roman numerals are used to track the presumed phylogenetic fate of the arches, taking six pairs to represent the basic embryonic pattern.

Fishes Soon after branching from the ventral aorta, the aortic arches divide into capillary beds within the gills. The section of the aortic arch delivering blood to the gills is the **afferent artery,** and the dorsal section carrying it away is the **efferent artery.** The capillary beds between them partially or completely encircle the gills and empty first into the **collecting loop,** which joins the efferent artery.

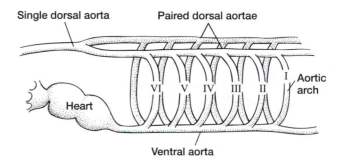

FIGURE 12.13 **Primitive pattern of aortic arches.** Diagram of the basic six-arch pattern.

In chondrichthyans, the first pharyngeal slit becomes reduced but not lost, forming the small **spiracle.** During embryonic development, the ventral section of the first aortic arch, which is ordinarily expected to supply the first pharyngeal slit, does not appear. Instead, a vascular sprout from the adjacent collector loop grows to the spiracle, feeding a small capillary bed in its wall. This vessel constitutes the **afferent spiracular artery.** The dorsal section of the first arch forms the **efferent spiracular artery,** which drains this small capillary bed. Because of its small size and the fact that it receives oxygenated blood via the afferent spiracular artery, the spiracle is thought to play little role in respiratory exchange. Instead it may develop as part of a secretory or sensory organ.

The remaining aortic arches (II–VI) form small sprouts halfway along their lengths. These merge and cross connect as the collecting loops serving the vascular capillary beds within the gills that form adjacent to the enlarged pharyngeal slits. The anterior and posterior halves of each collecting loop are its **pretrematic** and **posttrematic branches,** respectively (figure 12.15a). Although the external carotid artery arises embryologically from the anterior end of the ventral aorta, it becomes associated with the collecting loop, an understandable change if it is to carry oxygenated blood to the lower jaw. The internal carotid artery supplying the

(a) Teleost fish

(b) Lungfish

(c) Salamander (larval)

(d) Salamander (adult)

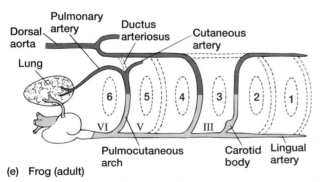

(e) Frog (adult)

FIGURE 12.14 **Aortic arches of anamniotes and some derivatives.** Diagrams of the basic six-arch pattern. (a) Teleost fish. (b) Lungfish (*Protopterus*). (c) Neotenic and larval salamander. (d) Adult salamander. (e) Adult frog. Dashed lines and ovals represent ancestral features lost in the group.
After Goodrich.

brain receives oxygenated blood from the first fully functional collecting loop (pharyngeal slit II) via the efferent branchial artery (II) (figure 12.15b).

In most actinopterygian fishes, four pairs of aortic arches (III–VI) arise from the ventral aorta. They service the gills associated with five pharyngeal slits. In sturgeons and a few other species, the first pharyngeal slit persists as a small spiracle, but in most fishes, even this modest slit is absent in the adult. The first aortic arch is lost along with this first slit (figure 12.14a).

In most fishes with supplementary air-breathing organs, oxygenated blood departing from these organs enters the general venous circulation, boosting the overall oxygen level of the blood returning to the heart. However, in lungfishes, blood leaving the highly vascularized lungs directly reenters the heart via a separate pulmonary vein (figure 12.14b). Phylogenetically, separate left and right atria appear first in lungfishes, thereby establishing a separate pulmonary circuit to and from the lung.

In lungfishes, as in other bony fishes, the first pharyngeal slit is reduced to a spiracle that has no respiratory function. Its associated aortic arch (I) is reduced as well. In the Australian lungfish, *Neoceratodus*, the remaining five pharyngeal slits open to fully functional gills supplied by four aortic arches (III–VI). In the African lungfish, *Protopterus*, the functional gills are reduced further. The third and fourth gills are absent entirely, but their aortic arches (III–IV) persist (figure 12.14b). In all lungfishes, the efferent vessel of the most posterior aortic arch (VI) gives rise to the **pulmonary artery** but maintains its connection to the dorsal aorta via the short **ductus arteriosus.**

If you think carefully about the blood flow implied by this anatomical pattern, you will understand the misconceptions it invites. For example, notice that if blood low in oxygen in the ventral aorta flows along its presumed route through arches II, V, and VI, it passes through the capillary beds of the gills, is replenished with oxygen, and enters the dorsal aorta as oxygenated blood. But notice that in the African lungfish, blood low in oxygen in the ventral aorta seems to have an alternative route through arches III and IV, which lack gills. Theoretically, blood could reach the dorsal aorta unaltered, still lacking oxygen. If that happened, as the anatomy alone might suggest, then both oxygenated and deoxygenated blood would mix in the dorsal aorta, reduce the overall oxygen tension in blood flowing to the systemic tissues, and apparently defeat most advantages that air-breathing lungs might bring.

Because of this, the aortic arch pattern seemed inefficient to early anatomists, but they excused it, and perhaps even expected such a pattern, because they thought that lungfishes were neither complete water breathers nor air breathers. Lungfishes held a middle ground capable of a bit of both, but doing neither particularly well. This mistaken view—that lungfishes were imperfectly designed compared to more advanced vertebrates—was tested when careful studies of their circulatory physiology were completed.

(a)

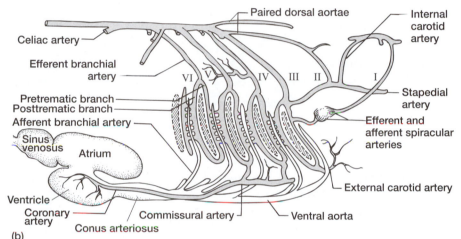

(b)

FIGURE 12.15 **Aortic arches of a shark.** (a) Embryonic modifications of aortic arches. New additions (white) to the arches establish the pretrematic and posttrematic parts of the collecting loops that receive afferent and supply efferent branchial arteries, derivatives of ventral and dorsal sections of the aortic arches, respectively. (b) Adult shark aortic arch derivatives. Roman numerals indicate aortic arches.

After Kent.

In fact, very little mixing of oxygenated and deoxygenated blood actually occurs, thanks largely, as we see later in this chapter, to the role played by the partially divided heart.

Amphibians The same mistaken view that the cardiovascular system was imperfectly designed was also held about amphibians and for much the same reasons. The anatomical arrangement of their aortic arches suggests that some mixing occurs between oxygenated blood from the gills and deoxygenated blood returning from the body.

In amphibians, the first two aortic arches (I, II) disappear early in development. The pattern of the remaining arches differs between larvae and metamorphosed adults. In most larval salamanders, the next three aortic arches (III–V) carry external gills, and the last aortic arch (VI) sprouts the pulmonary artery to the developing lung. A notable exception is the neotenic salamander *Necturus,* in which part of the sixth arch disappears and only its dorsal section persists, forming the base of the pulmonary artery (figures 12.16 and 12.19). In most species of salamanders, the external gills are lost following the larva's transformation into the adult, but the aortic arches are retained as major systemic vessels.

The short section of dorsal aorta between aortic arches III and IV, termed the **carotid duct,** usually closes at metamorphosis. This forces the carotids to fill with blood from a derivative of the ventral aorta. The section of ventral aorta between arches III and IV becomes the **common carotid artery,** which feeds the external carotid (from the anterior ventral aorta) and the internal carotid (the anterior section of the dorsal aorta together with the third aortic arch). The **carotid body** is a small cluster of sensory cells associated with capillaries, usually located near the point at which the common carotids branch. Its functions are not completely known. Certainly the carotid body plays a role in sensing the gas content or pressure of the blood as well as having some endocrine functions.

The next two arches (IV, V) constitute major systemic vessels that join the dorsal aorta. The final aortic arch (VI) also joins the dorsal aorta, its last short section forming the ductus arteriosus. Shortly before joining the dorsal aorta, the sixth aortic arch gives off the pulmonary artery, which itself divides into small branches to the floor of the mouth, pharynx, and esophagus before actually entering the lungs. In lungless salamanders, the pulmonary artery, if it persists, supplies the skin of the neck and back.

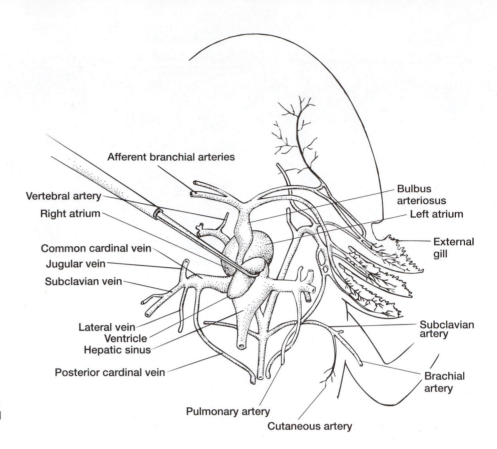

FIGURE 12.16 Aortic arches
of the salamander *Necturus* (ventral
view).

In frogs, the larva usually has internal gills that reside on the last four aortic arches (III–VI), and the embryonic pulmonary artery buds from arch VI. At metamorphosis, these gills are lost together with the carotid duct and all of arch V. The aortic arches that persist (III, IV, and VI) expand to supply blood to the head, body, and pulmonary circuits, respectively. The third arch and associated section of anterior dorsal aorta become the internal carotid. The anterior extension of the ventral aorta is the external carotid. Internal and external carotids both branch from the common carotid, the section of ventral aorta between arches III and IV. A carotid body can usually be found at the root of the internal carotid. The next enlarged aortic arch (IV) joins with the dorsal aorta, the major systemic artery supplying the body. The last arch (VI) loses its connection to the dorsal aorta because the ductus arteriosus closes and becomes the **pulmocutaneous artery.** One branch of the pulmocutaneous artery is the now well-developed pulmonary artery that enters the lung. The other branch is the **cutaneous artery,** which delivers blood to the skin along the dorsal and lateral body wall.

Some early morphologists concluded that inefficiencies in amphibian blood flow would result from these anatomical patterns. For example, oxygenated blood returning to the heart was thought to mix with deoxygenated blood returning from systemic tissues. Suffice it to say, this does not happen. In fact, little mixing occurs. But this mistaken view gained popularity and, to some extent, still prevails in some scientists' attitudes about the physiology of these lower vertebrates.

Reptiles Beginning in reptiles, but carried into birds and mammals, the symmetrical aortic arches of the embryo tend to become asymmetrical in the adult. Aortic arches III, IV, and VI persist in reptiles, but most of the changes center on enhancements and modification of the fourth arch. Perhaps the most significant anatomical modification of the arterial system in reptiles is the subdivision of the ventral aorta. During embryonic development, the ventral aorta splits to form the bases of three separate arteries leaving the heart: the left aortic arch, the right aortic arch, and the pulmonary trunk (figure 12.17a).

The pulmonary trunk incorporates the bases of the paired sixth arch and their branches as part of the **pulmonary arch** to the lungs. The base of the left aortic arch, the left aortic arch (IV) itself, and the curved section of the left dorsal aorta into which it continues constitute the **left systemic arch.** The **right systemic arch** includes the same components on the right side of the body: the base of the right aortic arch, the right aortic arch itself, and the arched section of the right dorsal aorta. The two systemic arches unite behind the heart to form the common dorsal aorta. The right systemic arch tends to be the most prominent of the two, primarily because of the additional vessels that it supplies. For example, the carotid arteries, originating from the ventral aorta in more primitive vertebrates, arise in reptiles from the right systemic arch. Blood passing through the right systemic arch might flow to the body or enter the carotid arteries to supply the head. In most reptiles, the subclavian arteries branch from the dorsal aorta, but in some reptiles, they branch from the systemic arches.

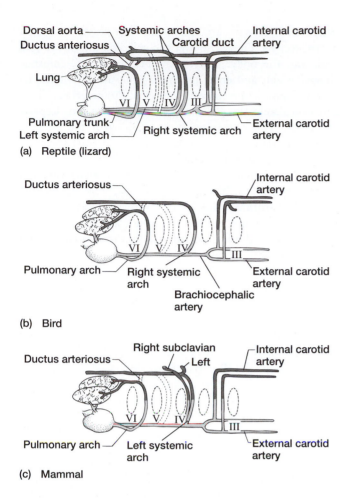

FIGURE 12.17 **Aortic arches of amniotes.** Diagrams of derivatives of the basic six-arch pattern. (a) Reptile. (b) Bird. (c) Mammal.

After Goodrich.

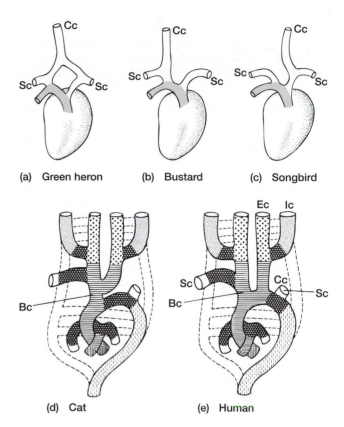

FIGURE 12.18 **Ventral views of aortic arches.** In birds, various alternative configurations of the departing arches can be found. In the green heron (a), two carotid arches unite. A single carotid persists on the right side of the bustard *Eupodotis* (b) and on the left side of the songbird *Passeres* (c). In mammals, formation of major anterior arteries can vary between species, such as the cat (d) and the human (e). Abbreviations: brachiocephalic (Bc), common carotid (Cc), external carotid (Ec), internal carotid (Ic), subclavian (Sc).

These modifications of the aortic arches in reptiles produce one pulmonary circuit and two systemic circuits, each of which arises independently from the heart.

Birds In birds, the right systemic arch becomes predominant (figure 12.17b). The bases of the aortic arch, the right aortic arch (IV), and the adjoining section of the right dorsal aorta form the right systemic arch during embryonic development. Its opposite member, the left systemic arch, never fully develops. The carotids arise generally from the same components of the aortic arches as in reptiles (aortic arch III and parts of the ventral and dorsal aortae), and they branch from the right systemic arch. However, the paired subclavians to the wings arise from the internal carotids and not from the dorsal aorta. The common carotids and subclavians supply the head and forelimbs, respectively. The common carotids can branch from the right systemic arch separately, or both can join to form a single carotid (figure 12.18a–c). A short but major vessel, the **brachiocephalic artery,** is present in a few reptiles, especially turtles, but serves as the major anterior vessel in many

birds. It too branches from the right systemic arch. Beyond this junction of the brachiocephalic artery, the systemic arch curves posteriorly to supply the rest of the body. In birds as in reptiles, the pulmonary arch forms from the bases of the paired sixth arch and their branches to supply both lungs.

Mammals Up to six aortic arches arise in the mammalian embryo, but only three persist in the adult as the major anterior arteries: the carotid arteries, the pulmonary arch, and the systemic arch (figure 12.17c). The carotid arteries and pulmonary arch are assembled from the same arch components as those of reptiles. The mammalian carotids arise from the paired aortic arches (III) and parts of the ventral and dorsal aortae. The pulmonary arch forms from the bases of the paired sixth arch and its branches. The systemic arch arises embryonically from the left aortic arch (IV) and left member of the paired dorsal aorta, and therefore is a *left* systemic arch in mammals. The common carotids may share a brachiocephalic origin or branch independently from different points on the aortic arch (figure 12.18d,e). The other notable difference in mammals is in the formation of the subclavian arteries. The left subclavian departs from the left systemic arch in mammals. The right subclavian, however, includes the right aortic arch (IV), part of the adjoining right dorsal aorta, and the arteries that grow from these into the right limb (figure 12.18d,e).

Overview of Aortic Arch Evolution   In most fishes, the aortic arches deliver deoxygenated blood to the respiratory surfaces of the gills and then distribute oxygenated blood to tissues of the head (via the carotids) and remainder of the body (via the dorsal aortae). In lungfishes and tetrapods, the aortic arches contribute to the pulmonary arch, the arterial circuit to the lungs, and the systemic arches, the arterial circuits to the rest of the body (figure 12.19). The carotid arteries still bear the primary responsibility for supplying blood to the head in tetrapods, but now they usually branch from one of the major systemic arches. The double systemic arches (left and right) present in amphibians and reptiles (figure 12.20a,b) become reduced to a single systemic arch—the right in birds, the left in mammals (figure 12.20c,d). Although birds and mammals share many similarities, including endothermy, active lives, and diverse radiation, they arose out of different reptilian ancestries. Any similarities in their cardiovascular anatomies represent independent evolutionary innovations.

The basic six-arch pattern of aortic arches is a useful concept that allows us to track aortic arch derivatives and organize the diversity of anatomical modifications we encounter. Furthermore, the appearance of six aortic arches during the embryonic development of living gnathostomes suggests that this is the ancestral pattern. However, as we have seen, the actual adult anatomy can be quite varied among different species.

Venous Vessels

The major veins that return blood to the heart are complicated and highly variable. Within each vertebrate group, the veins compose a few main functional systems that arise embryologically from what seems to be a common developmental pattern. Before examining the anatomy of veins in each group, we turn first to these basic systems of venous circulation. In vertebrates with an established double circulation, there are two general functional systems of venous circulation: the systemic system draining the general body tissues, and the pulmonary system draining the lungs. Within the systemic system, hepatic portal veins serve the liver, renal portal veins serve the kidneys, and general body veins drain the remaining systemic tissues.

Systemic System   Early in development, three major sets of paired veins are present: the **vitelline veins** from the yolk sac, the **cardinal veins** from the body of the embryo itself, and the **lateral abdominal veins** from the pelvic region. The paired vitelline veins are among the first vessels to appear in the embryo. They arise over the yolk and follow the yolk stalk into the body. They then turn anteriorly, continue along the gut, and enter the sinus venosus. The liver primordium grows into the vitelline veins. Proliferation of liver cords breaks up the associated vitelline veins into **hepatic sinusoids.** The remaining short sections of the vitelline veins that drain these hepatic sinusoids and enter the sinus venosus are the **hepatic veins.**

The cardinal veins include the **anterior cardinal veins,** which drain blood from the head region, and the **posterior cardinal veins,** which return blood from the embryo's body. Both pairs of anterior and posterior cardinals unite at the level of the heart into short **common cardinal veins** that open into the sinus venosus. The anterior cardinals consist of several parts that develop as vessels receiving tributaries from the brain, cranium, and neck. The posterior cardinals develop primarily as vessels of the embryonic kidneys.

The lateral abdominal veins are present in fishes, but they are usually merged or absent in tetrapods. In fishes, each vein joins the **iliac vein** from the pelvic fin and travels forward in the lateral body wall. At the level of the shoulder, the iliac vein joins the **brachial vein** and thereafter becomes the **subclavian vein,** which turns medially to enter the common cardinal vein. However, in tetrapods, the subclavian returns separately to the heart, and the lateral abdominal veins enter the liver. In amphibians, left and right lateral abdominal veins may unite into a single median vein, the **ventral abdominal vein,** which runs along the floor of the body coelom. In alligators, birds, and mammals, the abdominal vein is absent.

Subsequent venous development involves changes in these early paired vessels accompanied by anastomoses between them, loss of parts by atrophy, and appearance of additional embryonic vessels. The alterations are usually more extensive than those seen among arteries and produce major, often asymmetrical, adult venous routes of blood return to the heart.

Hepatic Portal Vein   The hepatic portal vein runs from the digestive tract to the liver and forms a direct route to transport absorbed end products of digestion immediately to the liver. It is common to all vertebrates and develops mostly from the embryonic **subintestinal vein,** an unpaired vessel originating in the caudal vein (figure 12.21a). The subintestinal vein loops around the anus and extends forward, running along the ventral wall of the intestine from which it collects blood. It passes through the liver from the intestine and finally joins the left vitelline vein. Within the proliferating liver cords, the vitelline veins become broken into a network of small hepatic sinusoids. The anterior end of the subintestinal vein empties blood into these hepatic sinusoids, and its posterior end regresses to lose contact with the caudal vein. This modified subintestinal vein is now properly called the **hepatic portal vein** (figure 12.21b). It

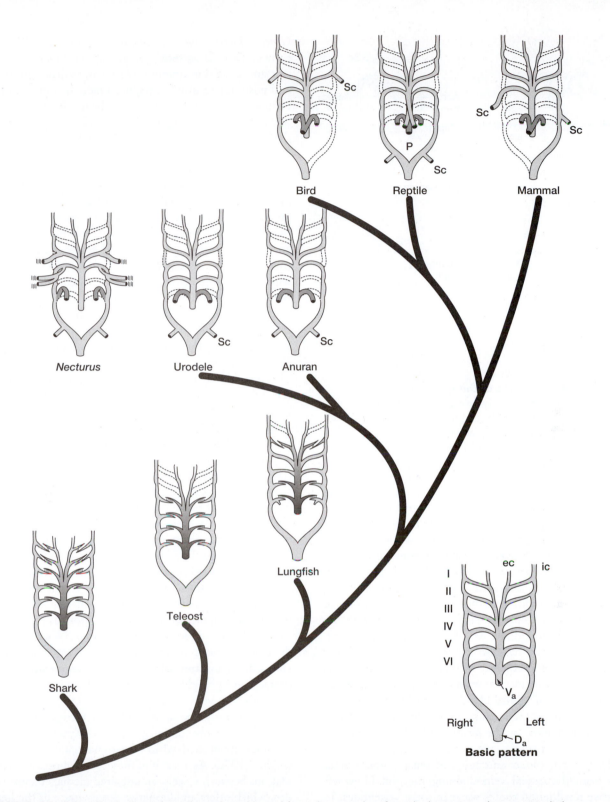

FIGURE 12.19 **Evolution of aortic arches.** Viewed from the ventral surface, the basic six-arch pattern includes a ventral aorta, paired aortic arches, and paired dorsal aortae. Selective loss or modification of this underlying pattern produces the derived aortic pattern of adult vertebrates. Vertical lines on the aortic arches represent gills. Dashed vessels are lost from the basic pattern in the adult. Abbreviations: dorsal aorta (D_a), external carotid (ec), internal carotid (ic), paired dorsal aorta (Pa), subclavian (Sc), ventral aorta (V_a).

collects blood not just from the intestines but also from the stomach, pancreas, and spleen and delivers it to the vascular sinusoids within the liver.

Renal Portal System Early in development, blood returning in the caudal vein from the tail flows through the subintestinal vein or through the posterior cardinals, the posterior

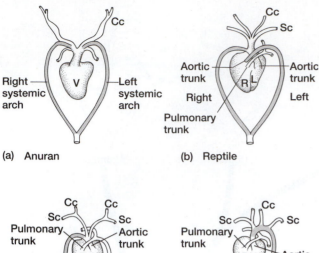

(a) Anuran

(b) Reptile

(c) Bird

(d) Mammal

FIGURE 12.20   Fate of the systemic arches in tetrapods (ventral views). Systemic arches of both sides persist in the adult in anurans (a) and reptiles (b). The right systemic arch persists in birds (c), and the left in mammals (d). Abbreviations: common carotid (Cc), left ventricle (L), right ventricle (R), subclavian (Sc), ventricle (V).

cardinals being the more usual route (figure 12.21a). The posterior cardinals travel dorsal to the kidneys, drain blood from them, and then continue forward to empty into veins entering the heart. Subsequently, a set of **subcardinal veins** arise ventral to the kidneys, drain them, and run forward to empty into the posterior cardinals. Once this route via the subcardinals becomes established, the short section of posterior cardinal atrophies between its junction with the subcardinal and the kidney. At this point in development, the subintestinal vein has also lost its connection with the caudal vein. As a consequence of these vascular alterations, blood from the tail must now pass through the kidneys. With the routing of caudal blood through the kidneys, the caudal vein becomes the renal portal system. From the kidneys, blood is drained by the newly established subcardinal veins (figure 12.21b).

In general, blood entering the renal portal system arrives from the caudal vein draining the tail. However, alternative renal portal routes occur in some vertebrates. In cyclostomes and some teleosts, blood of the renal portal system enters the kidneys via segmental veins from the body wall. In some lungfishes, additional blood from the pelvic fins and the posterior abdominal region contributes to the renal portal flow entering the kidneys. The caudal vein of these lungfishes does not supply but instead drains the kidneys and then continues forward to join the posterior cardinals or postcava.

General Body Veins   In primitive vertebrates, the basic early embryonic pattern is retained, and blood from anterior and posterior systemic tissues is returned in anterior and posterior cardinal veins, both pairs of veins uniting in common cardinal veins near the heart. In derived vertebrates, the cardinals appear but usually persist only in the embryo, being functionally replaced by alternative adult vessels, the precava and postcava (anterior and posterior venae cavae).

The embryonic derivation of the precava and postcava from precursor veins reveals the extensive modification on which the adult venous system is based. Formation of the precava is preceded by the early embryonic appearance of the anterior, posterior, and common cardinal veins (figure 12.22a). Formation of the precava itself begins with the enlargement of small intersegmental veins into the subclavian veins that empty into the anterior cardinals (figure 12.22b). Next, an **intercardinal anastomosis** develops between the anterior cardinals (figure 12.22c). With growth of the embryo, these newly established channels become used increasingly, especially on the right side, to return blood from the head. The common cardinal of the right side enlarges to receive this returning blood and becomes the precava of the adult (figure 12.22d). The common cardinal of the left side regresses, persisting only as a small vein from the atrium of the adult heart.

Formation of the postcava is even more elaborate. Initially, the paired posterior cardinal veins return blood from the embryonic body behind the heart. However, subsequent consolidation of parts of three embryonic vessels—hepatics, subcardinals, and supracardinals—and extensive anastomoses between them result in a progressive shift of returning blood away from the posterior cardinals to an emerging single medial channel made up of parts of several veins. Contributing vessels to this return channel eventually merge into a single vein, the postcava. Its developmental history begins with the appearance of the posterior cardinals that drain the early embryonic kidneys (mesonephros). Next, the subcardinal veins arise and connect with each other through the **subcardinal anastomosis.** Finally, the **supracardinal veins** develop and provide supplementary drainage of the posterior body.

Anteriorly, the right vitelline vein (right hepatic vein) joins the right subcardinal, and posteriorly a new connection becomes established with subcardinals and supracardinals (figure 12.22c). As a result of these anastomoses and consolidations between vessels, an unpaired medial channel develops, which offers an alternative return route to the heart as the earlier return route via the posterior cardinals regresses. This channel is modest at first, but it enlarges as more blood seeks this path of return to the heart, and it eventually becomes the adult postcava. Thus, the precava and postcava are mosaics of preceding vessels, parts of which are pirated during embryonic development to produce the definitive adult vessels that drain the anterior and posterior parts of the body, respectively.

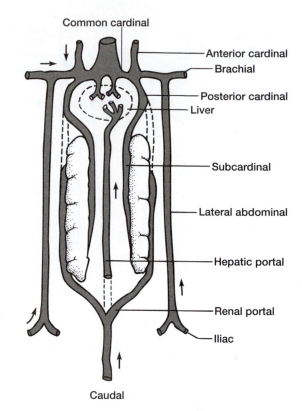

(a) Basic embryo pattern

(b) Adult fish

FIGURE 12.21 **Major veins.** The basic embryonic (a) and modified adult (b) pattern of major veins in the shark. The hepatic portal vein forms from the embryonic subintestinal vein. Anterior parts of the embryonic vitelline veins give rise to the short hepatic veins draining the liver. Addition of a subcardinal vein drains the kidney, and the renal portal becomes established from posterior derivatives of the posterior cardinal veins. The lateral abdominal vein drains the pelvic appendage, receives blood from the body wall as it runs anteriorly, and joins with the subclavian vein from the pectoral appendage and the anterior cardinal vein from the head to enter into the heart.

Pulmonary System Many fishes have supplementary air-breathing organs, but only fishes with lungs possess a pulmonary system. Among living fishes, only dipnoans have true lungs. If the ancient placoderms had lungs, a possibility mentioned earlier, then the pulmonary system would have evolved early in vertebrate evolution.

Lungs and gas bladders (p. 416)

Pulmonary Veins The pulmonary veins return blood from the paired lungs to the heart. Before entering the heart, they usually unite into a single vein. Embryologically, the pulmonary vein does not arise by conversion of existing vascular channels. Instead, numerous small vessels originate separately within and drain the embryonic lung buds. They then converge into several common vessels that become the pulmonary veins entering the left atrium.

Lung evolution (p. 418)

Fishes The head is drained by the paired anterior cardinal veins and small **inferior jugular veins,** which join the common cardinal veins just before they empty into the sinus venosus of the heart. The subclavian and iliac veins

drain the appendages via the lateral abdominal vein. Both also join the common cardinal. In most fishes, modification of the posterior cardinal diverts all returning blood from the tail so that it flows through the kidneys before emptying into the remaining sections of the posterior cardinal. The hepatic portal vein transports blood from the digestive tract to the capillaries in the liver. From the liver, blood flows to the heart via the short hepatic veins (figure 12.23a,b).

In actinopterygians, the lateral abdominal veins are usually lost and the pelvic fins are drained by the posterior cardinal. Blood from gas bladders empties into the hepatic or common cardinal veins.

In lungfishes, venous return to the heart is similar to that of other fishes except that the right posterior cardinal vein enlarges to assume most of the responsibility for draining blood from the posterior part of the body and, accordingly, is usually called the postcaval vein (figure 12.23c). The paired lateral abdominals fuse to form the unpaired ventral abdominal vein that drains the pelvic fins and empties into the sinus venosus. Blood returning from the lungs enters the atrium of the heart directly.

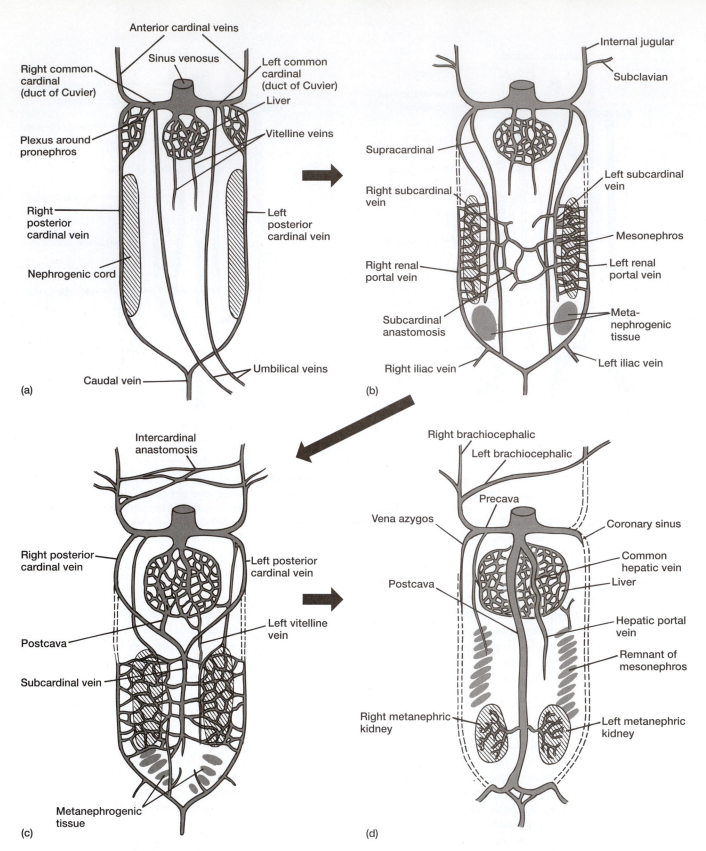

FIGURE 12.22   Embryonic development of mammalian veins. (a) Early in development, the anterior, posterior, and common cardinals become established. (b) Intersegmental veins close to the pectoral limbs come to empty into the anterior cardinals. The subcardinals arise between the kidneys and pass forward to enter the posterior cardinals. (c) Intercardinal anastomosis becomes established between the anterior cardinals. Returning blood from the posterior body now includes a route through the liver because part of the right vitelline vein has been incorporated within the right subcardinal. (d) The precava receives blood from left and right brachiocephalic veins (intercardinal anastomosis and right anterior cardinal, respectively). The postcava is the major channel returning blood from the posterior region of the body.

After Ballinsky.

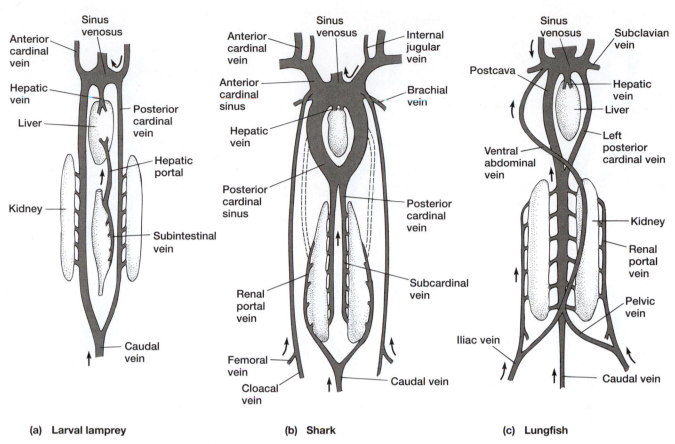

(a) Larval lamprey

(b) Shark

(c) Lungfish

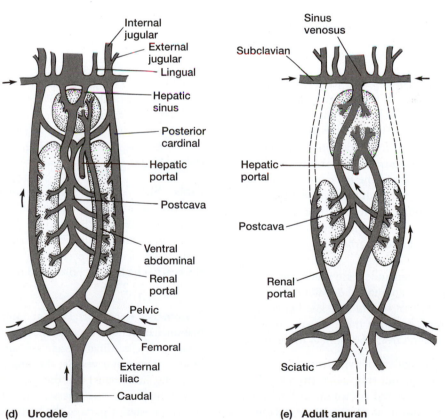

(d) Urodele

(e) Adult anuran

FIGURE 12.23 **Major venous channels of vertebrates.** (a) Larval lamprey. (b) Shark. (c) Lungfish (*Protopterus*). (d) Urodele. (e) Adult anuran.

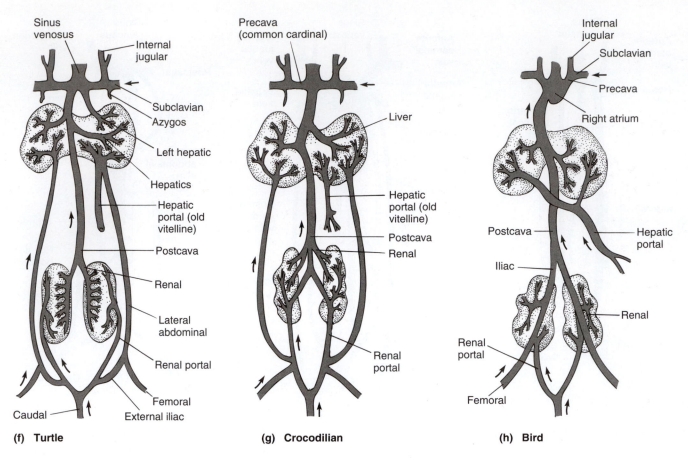

FIGURE 12.23 (f) Turtle. (g) Crocodilian. (h) Bird.

Amphibians In larval salamanders, such as *Necturus,* the **internal jugular** (derived from the anterior cardinal) and **external jugular veins,** together with the small **lingual vein** from the tongue, return blood from the head. The large postcava offers one route for blood returning from the kidneys. The **ventral abdominal vein** transports blood primarily from the hindlimbs to the liver sinusoids. Blood from the tail is offered several alternative return routes: through the ventral abdominal, the posterior cardinal, or the postcava via the kidneys. The hepatic portal persists. Numerous hepatic veins that enter the postcava drain the liver (figure 12.23d,e).

The veins of adult salamanders are much the same as those of the larvae. In adult anurans, the major difference is in the posterior cardinal, part of which is lost between the kidney and the common cardinal; therefore, the posterior cardinal cannot return blood from the kidneys to the heart.

Reptiles The internal jugular (derived from the anterior cardinal), external jugulars, and subclavian from the forearm are tributaries of the paired common cardinals. The enlarged and modified common cardinals are customarily called the precava in reptiles. The posterior cardinal is considerably reduced to small **azygos veins** draining the inner wall of the thorax. Paired lateral abdominal veins are

present, as is a single postcava. The hepatic portal joins the capillaries of the digestive tract with the sinusoids of the liver. Blood from the liver sinusoids returns by short hepatic veins that join the postcava. The precava and postcava enter the much-reduced sinus venosus of the heart (figure 12.23f,g).

Birds Short external jugulars join long internal jugulars (anterior cardinals) to return blood to the common cardinals, which are modified into the paired precava. The femoral, caudal, and renal veins are tributaries of the extensive postcava, which also receives hepatic veins before entering the heart. The hepatic portal and renal portals are also present (figure 12.23h).

Mammals Renal portal and abdominal veins are absent in mammalian venous circulation, but a hepatic portal vein is present. The cardinal vessels are substantially modified to produce two major vessels: the single precava (the superior vena cava in humans) and the single postcava (inferior vena cava in humans). These vessels collect blood from the anterior and posterior parts of the body, respectively, and return it to the right atrium of the heart. The posterior vena cava is divided into several sections, including the hepatic, renal, and subcardinal veins (figure 12.23i).

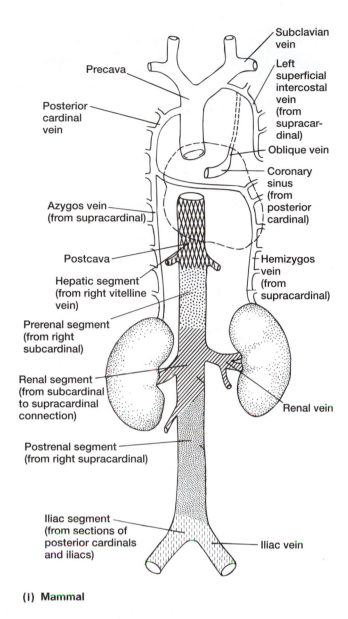

Precava

Subclavian vein

Left superficial intercostal vein (from supracardinal)

Posterior cardinal vein

Oblique vein

Coronary sinus (from posterior cardinal)

Azygos vein (from supracardinal)

Postcava

Hemizygos vein (from supracardinal)

Hepatic segment (from right vitelline vein)

Prerenal segment (from right subcardinal)

Renal segment (from subcardinal to supracardinal connection)

Renal vein

Postrenal segment (from right supracardinal)

Iliac segment (from sections of posterior cardinals and iliacs)

Iliac vein

(i) Mammal

FIGURE 12.23 (j) Mammal.

Hearts

The heart is a pump that moves blood through vessels both by pushing blood through the circulatory system and also by aspiration—creation of negative pressure that sucks blood into the heart. In a slow-swimming dogfish shark, the heart can move 7.5 liters of blood per hour; in a resting hen, 24 liters per hour; in a human, 280 liters (about 75 gallons) per hour. In a giraffe, almost 1,200 liters of blood can circulate throughout the body per hour. If heart rate increases, a response known as **tachycardia,** these values can increase fivefold. If heart rate decreases, **bradycardia** results and these values can fall precipitously. For example, when a turtle dives, its cardiac output can drop to less than one-fiftieth (1/50) of its prior output. In addition to functioning as a pump, the heart also serves to channel deoxygenated and oxygenated

blood to appropriate parts of the circulation, thereby preventing their mixing. Before discussing the special functions of the heart, we first look to its structure in vertebrates.

Basic Vertebrate Heart

Phylogenetically, the heart probably began as a contractile vessel, much like those found within the circulatory system of amphioxus. In most fishes, the heart is part of a single circulation. Vessels serving gas exchange in the gills and systemic capillary beds are in series with each other. The embryonic fish heart consists of four chambers, which are also in series, so that blood flows in sequence from the **sinus venosus,** to the **atrium,** to the **ventricle,** and finally to the fourth and most anterior heart chamber, the **bulbus cordis,** before entering the ventral aorta. Differences in structure, doubts about homology, and loose use of terms have led to confusion about nomenclature for this fourth chamber. We use the term *bulbus cordis* for this chamber in embryos. In adults, the ventricle empties into the ventral aorta, or to this intervening fourth chamber, the bulbis cordis, termed in the adult the **conus arteriosus** if its contractile walls possess cardiac muscle, **bulbus arteriosus** if its elastic walls lack cardiac muscle. Internally, each may contain various numbers of **conal valves.**

A conus arteriosus is generally present in chondrichthyans, holosteans, and dipnoans. Although absent as a distinct chamber in adult tetrapods, during development its embryonic forerunner, the bulbus cordis, divides into the bases of the major arteries leaving the heart. In some fishes, most notably in teleosts, the bulbus arteriosus is thin walled with smooth muscle and elastic fibers, but it lacks both cardiac muscle and conal valves. The adult bulbus arteriosus, like the conus arteriosus, arises generally from the embryonic bulbus cordis, but in some fishes the adult bulbus arteriosus may incorporate part of the adjoining ventral aorta as well. Another term often used ambiguously in the older literature is **truncus arteriosus,** which should apply only to the ventral aorta or its immediate derivatives but not to any part of the heart proper. In tetrapods, the ventral aorta often becomes reduced, sometimes persisting only as a small section of vessel at the base of major departing aortic arches. In these cases, the term *truncus arteriosus* is most apt.

Like any active muscle, the heart requires gas exchange (oxygen, carbon dioxide) to support its metabolism. In many fishes and primitive tetrapods, this demand is met by direct gas exchange between the myocardium and blood passing through its lumen. The inner wall of the myocardium, especially of the ventricle, often forms projecting cones of muscle termed **trabeculae** that are set off by deep recesses. The resulting texture, when viewed from the lumen, looks spongy and is referred to as **trabeculate. The coronary vessels** perfuse the heart wall, usually only the outer part of the myocardium. They are especially well developed in elasmobranchs, crocodiles, birds, and mammals, in which they supply most of the myocardium. In fishes, the coronary arteries are derived from the efferent arches or collecting loops of the gills, which carry oxygenated blood. The coronary veins enter the sinus venosus.

In addition to the conal valves, the endocardium develops sets of valves between its chambers: The **sinoatrial (SA) valves** form between the sinus venosus and the atrium, and the **atrioventricular (AV) valves** form between the atrium and ventricle. During normal flow, the valves are pushed open, although blood reversal immediately forces them closed, thereby preventing retrograde blood flow. The heart lies within the **pericardial cavity** lined by a thin epithelial membrane, the **pericardium.** In many fishes, the pericardial cavity lies within bone or cartilage, forming a semirigid compartment that holds the heart (figure 12.24a). Sequential contraction of the heart chambers helps move blood from one chamber to the next and finally drives it from the heart into the ventral aorta. Normal muscular movements impinging on nearby veins raise internal pressure and help drive venous blood back to the heart. But refilling of the sinus venosus and atrium by returning blood is often aided by the low pressure produced within the confines of the semirigid compartment holding the heart. This is termed the **aspiration effect.** As the large muscular ventricle contracts, blood exits through the conus into the ventral aorta to empty the ventricle. This temporarily reduces the volume the ventricle occupies within the pericardial cavity, which lowers the pressure throughout the pericardial cavity surrounding the thin-walled atrium and sinus venosus. Negative pressure surrounding the relaxed sinus venosus and atrium causes them to expand; in turn, they develop a negative pressure that aspirates or sucks in venous blood. Once refilled, the atrium and sinus venosus contract to fill the ventricle (figure 12.24b).

Contraction, as mentioned earlier, is an intrinsic property of cardiac muscle. Individual cells even show rhythmic contractions if they are isolated outside the body in a suitable culture medium. Cardiac cells tend to beat in synchrony. Contraction of the entire heart usually begins within a restricted region in the sinus venosus called the **pacemaker,** or **sinoatrial (SA) node,** and then spreads through a conducting system of fibers into the ventricle and other contracting regions of the heart. In mammals, the conducting system includes, in addition to the SA node, a second node, the **atrioventricular (AV) node** in the wall of the heart. The AV node consists of **Purkinje fibers,** neuronlike fibers that are modified cardiac muscle cells. The Purkinje fibers depart from the AV node, divide into left and right bundles traveling within the interventricular septum to the apex of the heart, then turn and sweep around the respective sides of the ventricles. The rate at which heartbeats are initiated is under the influence of the nervous and endocrine systems. Heart rate also responds to the rate of venous filling. During exercise, venous return to the heart increases in part due to the increased pressure veins experience from active muscles surrounding them. As returning venous blood fills the heart chambers, they are stretched, resulting in a stronger subsequent contraction. This response is called the **Frank-Starling reflex,** named after the physiologists who first documented it. This

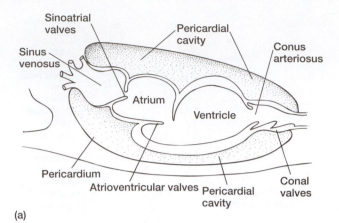

(a)

(b)

FIGURE 12.24   Basic heart structure and aspiration filling. (a) The four chambers of the fish heart are enclosed within the pericardial cavity. One-way valves between each chamber prevent reverse flow of blood as successive chambers contract. (b) When the ventricle contracts, the volume occupied by the ventricle within the pericardial cavity is momentarily reduced (this is exaggerated in the diagram). Reduced ventricular volume creates a negative pressure around the other chambers. Because the walls of the sinus venosus and atrium are thin, this low surrounding pressure causes them to expand, creating within their lumina a negative pressure that sucks in or aspirates blood from the returning veins.

reflex self-adjusts the strength of heart contraction, thereby adjusting stroke volume, increasing it as venous return increases and decreasing it as venous return slows.

Birds and mammals have four-chambered hearts, but of the original four fish chambers, only two persist as major receiving compartments, the atrium and the ventricle, both of which are divided into left and right compartments to produce four anatomically separate chambers. Although the hearts of birds and mammals are both derived from early tetrapods, they arose independently from different tetrapod ancestors. Phylogenetically, living amphibians and reptiles stand between these derived tetrapods and fishes, a position that seduced many into believing that these intermediate vertebrates possessed hearts that should be evaluated in light of how well they anticipated the hearts of birds or mammals. Certainly the evolutionary route to birds and mammals led through primitive tetrapods. But living amphibians

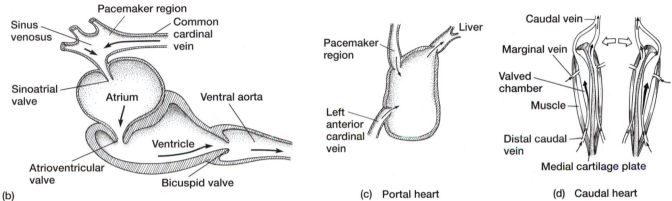

FIGURE 12.25 **Hagfish circulation.** (a) Diagram of the cardiovascular system. (b) Branchial heart illustrating the three chambers. Accessory hearts. The walls of the cardinal (not shown) and portal hearts (c) pulsate to help drive blood. The caudal hearts (d) have paired striated muscles and a central flexible support. They lie almost at the end of the hagfish's tail. Contraction of the left muscle bends the medial cartilage plate, expelling blood from the compressed right chamber and allowing the left to fill. Contraction of the right muscle has the opposite effect. Alternating contractions of the caudal heart musculature enlarge and compress the veins, causing them to fill and then empty.

(a,b) After Jensen.

and reptiles are themselves millions of years removed from these earliest ancestors. Their hearts, like their cardiovascular systems generally, should be examined for the special functional roles they serve in amphibians and reptiles today. To do this, we next examine heart structure and its relationship to the functions it serves in all vertebrates.

Fishes

Hagfishes are primitive fishes, descendants of early agnathans, but today live the life of slow, prowling scavengers. Their cardiovascular system holds some surprises. Like all vertebrate hearts, the hagfish heart lies within the anterior trunk region, is composed of cardiac muscle, and receives blood returning from the general systemic circulation (figure 12.25a). It includes three chambers in series: the sinus venosus, the atrium, and the ventricle (figure 12.25b). Some biologists take a slight microscopic thickening of the base of the ventral aorta as evidence for a fourth compartment, the bulbus arteriosus. But it is not an obvious chamber morphologically, so this

thickening is more simply interpreted as part of the ventral aorta. Blood returning from the two common cardinals and liver first enters the sinus venosus. It flows through the atrium and then the ventricle; finally it is pumped directly into the ventral aorta and hence to the gills. One-way valves between heart chambers prevent reverse blood flow. No major nerves innervate the hagfish heart to stimulate contraction. Instead, filling of the sinus venosus by returning blood elicits the Frank-Starling reflex, which stimulates stronger contractions that originate in the pacemaker and then spread sequentially to the other chambers.

Occasionally, the hagfish heart is called a **branchial heart** to distinguish it from unique accessory blood pumps elsewhere in its circulation (figure 12.25b). These supplementary circulatory pumps are sometimes called **accessory "hearts,"** in quotation marks because they contract but usually lack the cardiac muscle of true branchial hearts (figure 12.25c,d). In most fishes, systemic tissues drain via discrete venules and veins. However, in the hagfish, venous drainage of some regions, such as the head and

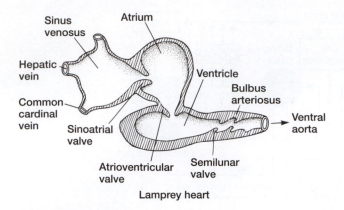

FIGURE 12.26 Lamprey heart. The four chambers characteristic of most fishes are present in the lamprey.

(a) Shark heart

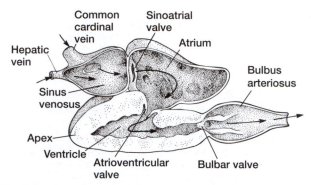

(b) Teleost heart

FIGURE 12.27 Fish hearts. (a) Shark. (b) Teleost. Blood leaves the shark heart through the muscular conus arteriosus, a chamber that is absent in many teleost fishes. Instead, in the teleost heart, the base of the ventral aorta is swollen, creating the elastic bulbus arteriosus.

After Lawson.

the subcutaneous caudal regions, is provided by large open sinuses. A probable consequence is that venous blood pressure is especially low. Accessory hearts on the venous side of the circulation are an apparent answer to the problem of returning low-pressure venous blood.

The **cardinal hearts** lying within the anterior cardinal veins are like sacs whose pumping action is initiated by skeletal muscles around their outer walls. The paired **caudal hearts,** which are located in the tail, represent a unique blood-pumping mechanism among vertebrates. They are composed of a central cartilaginous rod, skeletal muscles on the side, and veins between. Alternative contraction of these muscles bends the rod back and forth, which presses on the walls of the vessels and pumps blood into the caudal vein (figure 12.25d).

The **portal heart** is a single, expanded vascular sac that receives venous blood from one anterior and one posterior cardinal vein, and then it contracts to drive the blood through the liver (figure 12.25c). Only hagfishes have such an accessory heart in the course of the hepatic portal vein, which elevates blood pressure prior to the blood's entry into the liver sinusoids. Furthermore, the portal heart is the only accessory heart to have walls of cardiac muscle like the cardiac muscle of true branchial hearts.

The lamprey heart (branchial heart) also includes three compartments through which blood flows sequentially—sinus venosus, atrium, and ventricle (figure 12.26)—but in contrast to the hagfish heart, it is innervated and, further, the ventricle empties into the bulbus arteriosus, whose walls lack cardiac muscle but contain smooth muscle cells arranged longitudinally and circumferentially. One-way valves separate compartments. The sinoatrial and atrioventricular valves prevent retrograde blood flow. The luminal walls of the bulbus arteriosus are thrown into leaflets, collectively forming the **semilunar valves,** which prevent reverse blood flow and possibly aid in distributing blood to the aortic arches. From the arches, blood flows to the delicate gill capillaries next in line in the circulation.

The hearts of chondrichthyans and bony fishes consist of four basic chambers—sinus venosus, atrium, ventricle, and conus arteriosus (or bulbus arteriosus)—with one-way valves

stationed between compartments (figure 12.27a,b). Like the other chambers, the muscular conus arteriosus contracts, acting as an auxiliary pump to help maintain blood flow into the ventral aorta after the onset of ventricular relaxation. Its contraction also brings together the conal valves located on its opposing walls. When these valves meet, they prevent the backflow of departing blood. In teleosts, the conus arteriosus may regress, leaving only remnants of a myocardial conus, or be replaced entirely by an elastic, noncontractile bulbus arteriosus, lacking cardiac muscle but invested with smooth muscle, collagen, and elastic fibers. A single pair of **bulbar valves** at the juncture of the bulbus arteriosus and the ventricle prevents retrograde flow. When receiving blood following ventricular contraction, the bulbus arteriosus stretches and then gently undergoes elastic recoil to maintain blood flow into the ventral aorta. The result is a **depulsation,** or dampening of the large oscillations in blood flow and pressure introduced by ventricular contractions. This has been proposed as a means of protecting the delicate gill capillaries up next in the circulation from exposure to sudden spurts of blood at high pressure that would otherwise occur.

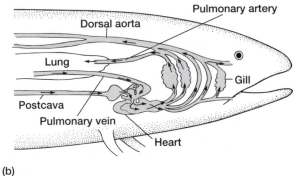

FIGURE 12.28 **Contraction cycle of the teleost heart.** (a) Relaxation of the sinus venosus and atrium draws in blood via the hepatic and common cardinal veins. (b) Contraction of the atrium closes the sinoatrial valve and forces blood into the ventricle. (c) As the atrial walls relax again, blood enters the atrium. (d) Contraction of the ventricle forces blood through the bulbus arteriosus and distributes it to the aortic arches. The aspiration effect (negative signs) contributes to and completes the refilling of the sinus venosus and atrium to start the cycle again.

The S-shaped arrangement of chambers in the fish heart places the thin-walled sinus venosus and atrium dorsal to the ventricle, so that atrial contraction assists ventricular filling. Blood flows from posterior chambers to anterior chambers in the following sequence. First, venous blood fills the sinus venosus and pushes open the sinoatrial valve to fill the atrium. The aspiration effect drives this movement of venous blood and encourages the initial filling of the sinus venosus and atrium (figure 12.28a). Second, the atrium contracts, boosting pressure within its lumen. Atrial contraction forces the sinoatrial valve closed and opens the atrioventricular valve, thus allowing blood to flow into and fill the ventricle (figure 12.28b). Third, the atrium relaxes, dropping the pressure in it and in the sinus venosus. Consequently, blood is drawn in by the aspiration effect and begins to fill both chambers again (figure 12.28c). Fourth, the ventricle contracts to drive the blood it holds forward into and through the conus arteriosus, which now begins its contraction (figure 12.28d).

Lungfishes The lungfish heart is modified from that of other bony fishes. The first chamber to receive returning blood is still the sinus venosus. In all three lungfish genera, the single atrium is partially divided internally by an **interatrial septum** (pulmonalis fold) that defines a larger **right** and smaller **left atrial chamber** (figure 12.29a). Pulmonary veins conveying blood from the lungs empty into the

FIGURE 12.29 **Heart of the African lungfish** *Protopterus.* (a) Internal structure of the heart. (b) Path of blood. When the lungfish breathes air, venous blood returning from systemic tissues flows through the heart and tends to be directed to the last aortic arch. The pulmonary artery carries most of the deoxygenated blood to the lung. Blood high in oxygen returning from the lung passes through the heart and then tends to enter the aortic arches without gills. In this manner, blood is shunted directly to the general circulation. Thus, when lungfishes breathe air, they display the beginnings of a double circulation system. The five aortic arches phylogenetically represent the second through the sixth (Roman numerals). The first (II) and last two (V, VI) of these carry gills.

sinus venosus (Australian lungfish, *Neoceratodus*) or directly into the left atrial chamber (South American lungfish, *Lepidosiren*, and African lungfish, *Protopterus*). The sinus venosus conveying systemic venous blood opens into the right atrial chamber (figure 12.29a). In place of the atrioventricular valves is the **atrioventricular plug,** a raised cushion in the wall of the ventricle. It moves into and out of the opening from the atrium, like the AV valves, to prevent retrograde flow of blood into the atrium. The ventricle is also divided internally, but only partially, by an **interventricular septum.** Within dipnoans, the South American lungfish shows the greatest degree of both ventricular and atrial internal subdivision. The Australian lungfish shows the least. Alignment of the interventricular septum, atrioventricular plug, and interatrial septum establishes internal channels within and through the heart. When the lungfish breathes air, the left channel tends to receive oxygenated blood returning from the lungs. The right channel tends to carry deoxygenated systemic blood (figure 12.29b). Thus, despite the anatomically incomplete internal septation of the lungfish heart, blood entering from the sinus venosus does not tend to mix with blood returning from the lungs.

The **spiral valve** within the conus arteriosus aids in separating oxygenated and deoxygenated blood. Apparently derived from conal valves, the spiral valve consists of two endocardial folds whose opposing free edges touch but do not fuse. The conus makes a couple of sharp bends and rotates about 270°, thus turning these folds into a spiral within its lumen. Although unfused, these twisting folds internally divide the conus into two spiraling channels. Because the conus is attached directly to the ventricle, oxygenated blood entering the left channel and deoxygenated blood entering the right channel tend to flow through different spiraling channels within the conus and remain separate. As the oxygenated and deoxygenated streams of blood exit from the conus arteriosus, they enter different sets of aortic arches.

When a lungfish surfaces to gulp fresh air into its lungs, pulmonary blood flow to the lungs increases. When this oxygenated blood returns from the lung, it is shunted through arches III and IV, which lack gills, and flows to systemic tissues directly. Venous blood returning from systemic tissues is shunted through the posterior arches, V and VI, and then diverted to the lung. The blood supply to these posterior arches is derived from the spiral channel that itself received deoxygenated blood from the right side of the heart. Oxygenated blood traveling through the left side of the heart is channeled along the opposite spiral of the conus to enter the anterior set of aortic arches.

These cardiovascular adjustments of the lungfish to breathing air are nicely matched to environmental demands. Under most conditions, oxygen tension is high in the rivers and ponds in which lungfishes live. Deoxygenated blood flowing through the arches with gill capillaries picks up sufficient oxygen from the water to meet metabolic demands. However, as a result of seasonal drought, high temperatures, or stagnant waters, oxygen levels in the water can significantly decline, leaving little to diffuse across gills into the blood. During such times, the lungfish comes to the surface to gulp fresh air into its lungs. Under these deteriorating conditions, physiological changes take full advantage of this added source of oxygen. In *Protopterus*, deoxygenated blood returning from systemic tissues tends to be diverted to the lungs (not to the gills), and about 95% of the oxygenated blood from the lungs tends to be directed via anterior aortic arches to the systemic tissues (not through the gills). The fraction of blood that passes from the lungs to the anterior arches steadily declines to about 65% just before the next breath, following which the fraction returns again to 95%.

This air-breathing system has several physiological advantages. First, streams of oxygenated blood (from the lungs) and deoxygenated blood (from the systemic tissues) tend to be kept separate. Thus, the stream of oxygenated blood on its way to active systemic tissues is not diluted by blood depleted of oxygen, and blood passing through the exchange surfaces of the lung is low in oxygen, promoting rapid uptake of oxygen.

Second, the adjusted blood flow in an air-breathing fish prevents loss of oxygen to the water. Paradoxically, if oxygenated blood from the lungs passed through the gills, it could actually lose oxygen by diffusion to the oxygen-poor water. However, oxygenated blood is preferentially directed along aortic arches without gill capillaries to flow directly to systemic tissues. In addition to the preferential shunting of oxygenated blood into the anterior arches, a secondary mechanism, involving a shunt at the base of the gill capillaries, prevents oxygenated blood from being exposed to water low in oxygen in the gills. Some lungfishes have thick, muscular arteries that connect afferent and efferent aortic arches. When these shunts are opened, blood entering arches with gills can bypass the gill capillaries entirely and avoid exposure to oxygen-poor water irrigating the respiratory beds.

Not all modern lungfishes exhibit the same ability to adjust to air ventilation. The gills of the Australian lungfish are well developed and its lung less so. This lungfish does quite well in oxygenated water, but if kept out of water, it cannot maintain sufficiently high oxygen levels through its lung to sustain itself for long. However, the African lungfish has gills not as well developed, but its lung is more so. If kept out of water, its circulatory system and lung can sustain the lungfish for extended periods of time. Thus, the degree of physiological response to breathing water or air depends on the species of lungfish.

Amphibians

Amphibians rely on cutaneous gas exchange (plethodontid salamanders lack lungs entirely), on gills (many larval forms), on lungs (most toads and frogs), or on all three modes (most amphibians). Because the sources of oxygenated and deoxygenated blood vary, heart structure varies as well. Generally, in amphibians with functional lungs, the heart includes a sinus venosus, right and left atria divided by an anatomically complete interatrial septum, a ventricle lacking any internal subdivision, and a conus arteriosus with a spiral valve (figure 12.30a). Except for salamanders of the genus *Siren*, which have a partial interventricular septum, amphibians are unique among air-breathing vertebrates in lacking any internal division within the ventricle.

The cardiovascular system is perhaps best studied in frogs. The conus arteriosus of the frog heart arises from a single trabeculate ventricle (figure 12.30b). Semilunar valves lie at the base of the conus and prevent retrograde flow of blood back into the ventricle. Internally, a spiral valve twisting through nearly a complete rotation establishes two channels within the conus, each of which guides blood to specific sets of systemic and pulmocutaneous arches. The systemic and pulmocutaneous arches both arise from the truncus arteriosus, a remnant of the ventral aorta, but the two sets of arches receive blood from different sides of the spiral valve.

In lungless salamanders or those with reduced lung function, the interatrial septum and spiral valve may be much reduced or absent entirely. Unlike frogs, in which the pulmocutaneous artery branches give rise to the cutaneous

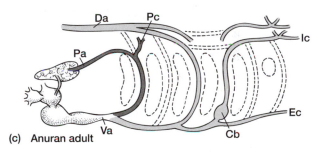

FIGURE 12.31 **Blood flow to aortic arches in amphibians.** (a) Larval salamander. (b) Adult salamander. (c) Anuran. Notice the pulmocutaneous (Pc) branch to the skin. In frogs, a sphincter prevents blood flow to the lung during diving, thus diverting blood flow to the skin to increase cutaneous respiration. Abbreviations: carotid body (Cb), dorsal aorta (Da), external carotid (Ec), internal carotid (Ic), pulmonary artery (Pa), pulmocutaneous artery (Pc), ventral aorta (Va).

After Goodrich.

FIGURE 12.30 **Amphibian hearts.** (a) Diagram of typical amphibian heart. Notice that the atrium is divided into left and right chambers but the ventricle lacks an internal septum. (b) Bullfrog (*Rana catesbeiana*) heart. Although lacking internal septa, the wall of the ventricle folds into numerous trabeculae. The small compartments between these trabeculae are thought to aid in separating bloodstreams that pass through the heart.

(b) After Lawson.

artery, salamanders lack a cutaneous artery. Instead, branches from vessels supplying the systemic circulation carry blood to the salamander skin. The pulmonary artery and the systemic arches in salamanders arise from the truncus arteriosus (figure 12.31a,b).

The two different streams of blood returning from the systemic and pulmonary circuits of amphibians are kept mostly separate as they pass through the heart (figure 12.31c). As in lungfishes, deoxygenated blood is selectively directed to the lungs via the pulmonary artery, and oxygenated blood is directed to systemic tissues via the aortic arches. In frogs breathing air, oxygenated and deoxygenated blood are mostly separated and distributed

by the heart. What is somewhat surprising about this ability is that the ventricle of the frog heart, like that of other amphibians, lacks even a partial internal septum. The trabeculate topography produces deep recesses in the walls of the ventricle that are thought to separate streams of blood that differ in oxygen tension. It is hypothesized that as one stream enters the ventricle, it preferentially fills the recesses between trabeculae, whereas the second stream occupies the center of the ventricle. Because of their different positions, the oxygenated and deoxygenated streams depart by different exits to reach appropriate sets of arteries. Thus, the trabeculae apparently are the structures in the frog ventricle that separate pulmonary and systemic venous streams of blood flowing through the heart (figure 12.30b).

When a frog dives, a sphincter at the base of the pulmonary artery constricts, resulting in reduced blood flow to the lung and increased flow to the skin. Thus, while a frog is submerged, loss of pulmonary respiration is somewhat offset by increased cutaneous respiration (figure 12.31c).

In adult salamanders, pulmonary and systemic circuits are similarly separated in the heart. In specialized species, heart design is modified. For example, in the lungless plethodontids, in which 90% of the respiratory needs are met through the skin and 10% through the buccal cavity, the heart lacks a left atrium, the compartment that would receive blood returning from the lungs. Where gills predominate

over lungs as respiratory organs (e.g., *Necturus*), the interatrial septum is reduced or perforated.

Reptiles

Reptiles have entered more fully terrestrial environments and adopted more active lifestyles than have amphibians. The cardiovascular system of reptiles supports accompanying higher metabolic rates and elevated levels of oxygen and carbon dioxide transport. It is capable of generating elevated blood pressures, higher cardiac output, and efficient separation of oxygenated and deoxygenated bloodstreams. The diversity of hearts and heart function in reptiles is

FIGURE 12.32 **Lizard heart, ventral view.** (a) Part of the ventral wall of the heart has been removed, as has the apex of the ventricle, to show its three interconnected compartments—cavum venosum separated by a muscular ridge from the cavum pulmonale, and deeper cavum arteriosum. The solid arrow indicates the route of blood flow from the cavum arteriosum via the interventricular canal into the cavum venosum entering at the bases of the aortic arches. (b) The wall of the cavum pulmonale has been cut away to better reveal the association of the deeper cavum arteriosum. Trimming of the atria and left aortic arch permits better viewing of the sinus venosus and pulmonary artery.

becoming better understood. It is clear that no single reptile heart can represent all others. Further, looking at reptile hearts as evolutionarily incomplete and imperfect bird or mammal hearts does little justice to the specialized, elegant, and rather effective cardiovascular designs of reptiles that support their specialized and distinctive lifestyles. In general, two basic reptilian heart patterns are recognized. One is found in chelonians and squamates, the other in crocodilians. We shall take them in that order.

Chelonian/Squamate Hearts In these reptiles, the sinus venosus is reduced in comparison to amphibians but it retains the same functions. It is still the first chamber to receive venous blood and contains the pacemaker. The atrium is completely divided into right and left atria. Prominent atrioventricular valves guard the entrance to the ventricles. The conus arteriosus (or bulbus cordis) appears during early embryonic development but becomes divided in the adult to form the bases (trunks) of three large arteries leaving the ventricle: the **pulmonary trunk** and the **right** and **left systemic trunks.** In snakes, a valved interaortic foramen connects the bases of adjacent aortae. But the shunting of blood made possible by this foramen has not been explored. Usually, the brachiocephalic artery, delivering blood to the subclavians and carotids, emanates directly from the right aortic arch, but in some turtles, it arises directly from the ventricle, crowded in with the trunks of the three aortic arches (figure 12.32a,b). The conus also gives rise to a band of contractile muscle tissue at the base of the pulmonary trunk to control the resistance blood meets as it flows to the lungs. Strictly speaking, the ventricle is a single chamber functioning as a single fluid pump to drive blood into the major arteries leaving the heart. Internally, however, it has three interconnected compartments: the **cavum venosum** and the **cavum pulmonale** separated from each other by a **muscular ridge,** and the **cavum arteriosum** connected to the cavum venosum via an **interventricular canal.** The cavum arteriosum fills with blood from the left atrium but has no direct arterial output. During systole, the blood it receives flows through the interventricular canal to the aortic arches. The cavum pulmonale does not receive blood directly from the atria. Instead, blood from the cavum venosum moving across the muscular ridge fills the cavum pulmonale. In turn, much of the blood filling the cavum venosum is deoxygenated blood from the right atrium. Thus, the heart has five chambers, composed of two atria and three compartments of the ventricle, or six chambers if you count the sinus venosus.

The pattern of blood flow through the hearts of Chelonia and squamates differs depending on whether they breathe air or hold their breath. For example, in an air-breathing turtle on land, most deoxygenated blood returning from systemic tissues is directed to the lungs, and most oxygenated blood from the lungs is directed to the systemic tissues via the aortic trunks.

Specifically, from the sinus venosus, the right atrium receives deoxygenated blood returning from the body. The left atrium receives oxygenated blood returning from the lungs. When the atria contract, deoxygenated blood in the right atrium flows into the cavum venosum and then across the muscular ridge to the cavum pulmonale. Additionally, when the right AV valves open, they lie across the opening to the interventricular canal and temporarily close it. Oxygenated blood in the left atrium enters the cavum arteriosum and temporarily remains there while the AV valves occlude the interventricular canal. When the ventricle contracts, the muscular ridge is compressed against the opposite wall to separate the cavum venosum from the cavum pulmonale. The AV valves close to prevent retrograde backflow into the atria, but in so doing, the right AV valve opens the interventricular canal and allows blood to flow through it. Thus, blood leaves the ventricle via the most accessible routes: Deoxygenated blood in the cavum pulmonale exits primarily through the pulmonary artery to the lung, although some also squirts across the muscular ridge to enter the left aortic arch; oxygenated blood in the cavum arteriosum moves through the interventricular canal to reach the bases of the aortic trunks, through which it then exits (figures 12.33a and 12.34a,b).

There is also a slight asynchrony in the timing of contractions in the walls of the ventricle. As a result, deoxygenated blood is driven into the pulmonary artery before oxygenated blood is set in motion. When the adjacent ventricular walls contract, oxygenated blood finds high resistance in the mostly filled pulmonary artery. Consequently, oxygenated blood exits through the systemic arches because they offer the least resistance.

Measurements of oxygen content in the major arteries confirm that the distribution of systemic and pulmonary blood is highly directional—deoxygenated blood flows to the lungs, oxygenated blood to the systemic tissues. In turtles breathing air, 70 to 90% of all blood reaching the left systemic arch is oxygenated blood coming from the pulmonary circuit; 60 to 90% of deoxygenated blood reaching the lungs comes from systemic tissues. Isolation of oxygenated and deoxygenated streams occurs in spite of the fact that the compartments of the ventricle are not anatomically separate. Notice that this functional separation extends to the aortic trunks. The left aortic arch fills mainly with oxygenated but also with some deoxygenated blood; but when the turtle breathes air, the right aortic arch carries only oxygenated blood to ensure a flow of highly oxygenated blood to the brain through the carotids from the brachiocephalic artery (figure 12.35).

When the turtle dives beneath the water, the physiological problems with which the circulatory system must cope change significantly. The heart responds with a **right-to-left** or **cardiac shunt.** Ordinarily, systemic blood returns to the *right* side of the heart (right atrium/right ventricle), then is pumped to the lungs; next this pulmonary blood returns to the *left* side of the heart (left atrium/left ventricle) before being pumped back to systemic tissues. When using a cardiac shunt, returning blood to the right side of the heart instead goes directly to the

(a) Air breathing

(b) Diving

FIGURE 12.33 **Blood flow through the squamate (and turtle) heart.** (a) When squamates breathe air on land, venous blood from the right atrium enters the cavum venosum of the ventricle and crosses a muscular ridge to fill the cavum pulmonale momentarily. Upon ventricular contraction, most of this blood exits via the pulmonary artery. Simultaneously, blood from the left atrium enters the deep cavum arteriosum. Contraction of the ventricle squirts this blood through the interventricular canal, and then the blood departs via the left and right systemic arches. (b) When squamates dive, resistance to pulmonary blood flow encourages blood that would normally exit to the lungs to move instead across the muscular ridge and depart primarily via the left aortic arch.

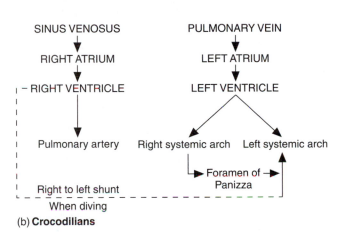

(a) **Chelonia and squamates**

(b) **Crocodilians**

FIGURE 12.34 Reptilian hearts. This flowchart compares the path of blood in the hearts of Chelonia and squamates (a) and crocodilians (b). Dashed lines indicate the cardiac diving shunts that divert blood from the pulmonary circuit directly to the systemic circuit.

Squamate aortic circulation

FIGURE 12.35 Squamate systemic circulation. Blood flows to major arteries when squamates breathe air. Oxygenated blood (light gray arrows) is directed to the systemic arches. Most, but apparently not all, of the deoxygenated blood (black arrows) enters the pulmonary artery. The small amount of deoxygenated blood flowing to the systemic circulation enters the left systemic arch. Thus, oxygen tension of blood in this arch is slightly lower than it is in the right systemic arch. It may be significant that carotid vessels supplying the head and brain branch from the right systemic arch.

left side and departs for systemic tissues, thereby bypassing the lungs. In the diving turtle, blood entering the cavum venosum is directed to the opposite side and out the aortae rather than out the pulmonary circuit (figure 12.33b). Differences in the resistances of systemic and pulmonary circuits are believed to control this shunting. A sphincter at the base of the pulmonary artery contracts to increase the pulmonary resistance to blood flow after a turtle dives. Because blood tends to follow the path of least resistance, it flows into the systemic circulation. When a turtle dives, blood that would pass to the lungs during air breathing is shunted instead through the aortic arches to the systemic circuit (figure 12.35).

A diving turtle makes the best of a difficult situation. The air held in its lungs is soon depleted of oxygen, so there is little advantage in circulating large quantities of blood to it. The energy used to pump blood to the lungs

would return no physiological benefit. Instead, by diverting blood to the systemic circuit, this increases the blood volume that can remove metabolites or gather oxygen stored in the tissues.

Crocodilian Hearts In many respects, the hearts of alligators and crocodiles are structurally similar to those of other reptiles. The conus arteriosus (bulbus cordis) produces the bases of the trunks of the three departing arteries—pulmonary and left and right aortic trunks. One-way **lunar valves** at the bases of each trunk permit blood to enter the conus but halt reverse backflow into the ventricle. The sinus venosus is reduced but still functions as the receiving chamber for returning systemic blood. The atrium is completely subdivided into two distinct left and right chambers, and the sinus venosus empties into the right atrium. The pulmonary vein enters the left atrium in adults, but it does not open into the left atrium during embryonic development. What are initially separate pulmonary veins, one from each lung, unite as a single stem, the pulmonary vein, that enters the

Lizard hearts invite misunderstandings. Two anatomically separate atria are present, but the ventricle is a single chamber with interconnecting compartments. Three major arteries depart directly from the ventricle. Such a design suggested to early anatomists that bloodstreams mixed as oxygenated and deoxygenated blood entered the common ventricle. The subtle assumptions behind this interpretation were as much an obstacle to understanding heart function as was the complex anatomy itself. Lizards were seen as primitive, and anatomists were looking ahead to the cardiovascular systems of advanced endotherms. As stated by one such scientist, the "perfect solution" to the separation of bloodstreams "was not attained until the avian and mammalian stages were reached." Recent experimental research on the blood flow through the lizard heart has shown just how wrong this earlier physiology and philosophy were.

Several techniques have been used to clarify blood flow within the cardiovascular system of living lizards. One technique uses radiology and takes advantage of contrast fluids that are generally nontoxic and compatible with blood. These contrast media are **radiopaque**—that is, visible when viewed by X ray. By the introduction of a radiopaque medium into selected veins, the subsequent route taken by the blood can be followed, usually through a sequential series of photographs or on a video monitor. Because the radiopaque medium is within the blood vessels of a live animal, its course appears to represent the normal circulation of blood. One such experiment was performed by Kjell Johansen on the lizard *Varanus niloticus,* a large member of the family Varanidae (Johansen, 1977). The radiopaque contrast medium was injected into the right jugular vein and postcava.

The right atrium, ventricle, and pulmonary arteries could be seen to fill with contrast medium in successive stages.

Although the lizard ventricle shows no complete internal division, none of the contrast medium entered the systemic arches (box figure 1a–c). This is experimental confirmation that the heart of this varanid lizard keeps returning deoxygenated blood separate from oxygenated blood and targets deoxygenated blood to the pulmonary circuit for oxygenation within the lung.

Another technique uses narrow-diameter tubes called cannulae that are inserted into vessels to directly measure pressure, draw samples of blood, or inject tracer-dyes. Cannulae have been used in experiments on the savannah monitor lizard, *Varanus exanthematicus,* to clarify the hemodynamics of blood pressure and the flow of oxygenated blood (Burggren and Johansen, 1982). Using anesthesia and surgical procedures, Burggren and Johansen inserted small cannulae connected to pressure transducers and recorders into the lumina of selected arteries of the monitor lizard. Additional cannulae in other arteries permitted the researchers to withdraw tiny samples of blood and measure the oxygen tension. As in other squamates and Chelonia, the varanid lizard heart was able to separate oxygenated and deoxygenated bloodstreams, targeting them to systemic and to pulmonary circuits, respectively. However, the researchers discovered that in the varanid, unlike in other squamates and Chelonia, blood pressure in the systemic circuit reached levels over twice that in the pulmonary circuit during systole (box figure 2a–c). In most other lizards, the systolic pressures in both circuits are quite similar. Thus, the heart of the varanid lizard not only diverts separate streams of oxygenated and deoxygenated blood to systemic and pulmonary circuits, but it also generates separate pressures within each circuit as well.

The cavum venosum within the varanid ventricle is considerably reduced, but otherwise the varanid heart is anatomically similar to that of other squamates and Chelonia. However, generation of high systemic and low pulmonary blood pressures makes it hemodynamically similar to the hearts of crocodilians, birds, and mammals. Why this should be so in varanids is not clear. Once they reach their preferred body temperatures, varanids have a higher metabolic rate than most other lizards. It has been suggested that the high systemic pressure might permit perfusion of a larger number of capillary beds than the systemic pressure in other lizards, without a consequent drop in capillary pressure. High systemic pressure would allow delivery of high levels of oxygenated blood to support the high oxygen requirements of active varanid muscles. However, if lung capillaries experienced such high pressures, they might leak excess fluid that would collect in the lung tissues and interfere with gas exchange. Because the pulmonary capillaries are part of the low-pressure pulmonary circuit, they are protected in varanid lizards. This suggests that by being separated into two pressure pumps, the varanid ventricle protects the pulmonary capillaries from excess pressure while meeting the high-pressure demands of active muscles.

Lizard hearts did not evolve in anticipation of the "perfect" hearts of birds and mammals. The lizard heart proved to be functionally complex and finely adapted to the special demands of a squamate lifestyle. The experimental research has settled much of the physiological function, but it should also invite a reassessment of the philosophy behind interpretation of systems in primitive vertebrates. We should abandon the view that lower vertebrates, because they arose early in vertebrate evolution, are imperfectly adapted. Modern research in functional morphology demonstrates just the opposite. The cardiovascular system, like other morphological systems, is surprisingly sophisticated both in primitive vertebrates and in their derived descendants.

BOX FIGURE I   **Tracing the course of blood through the heart of the lizard *Varanus niloticus*.** (a) Radiopaque contrast medium was injected into the right jugular vein and forced back slightly into the postcaval vein. The contrast medium has already entered the right atrium, which has contracted to fill part of the ventricle (the cavum venosum and cavum pulmonale). (b) The ventricle is undergoing contraction, and the pulmonary artery is filled. (c) Near completion of ventricular contraction, the contrast medium has been almost entirely expelled, and the branches of the pulmonary artery to each lung are clearly filled. Note in the last two stages that closure of the atrioventricular valves prevents retrograde blood flow into the right atrium. Notice also that this deoxygenated medium does not enter the systemic arches; therefore, these arches do not show up on the radiographs.

After Johansen.

(a)

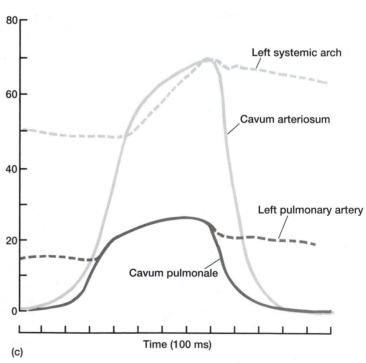

(b)

(c)

BOX FIGURE 2   **Hemodynamics of blood flow through the heart of the lizard *Varanus exanthematicus.***   (a) Cannulae are placed in the lizard heart to allow continuous monitoring of blood pressure. In this example, pressure cannulae are inserted into the left aortic arch (LAA) and the cavum arteriosum (CA), as well as into the left pulmonary artery (LP) and the cavum pulmonale (CP). (b) Tracings of pressures recorded at these locations are shown from one individual. (c) Tracings of pressures from different vessels are superimposed. During ventricular contraction, blood pressure in the left aortic arch (LAA) rises along with pressure in the cavum arteriosum (CA), from which it receives blood, until they reach maximum and equal pressures. Pressures in the left pulmonary artery (LP) and the cavum pulmonale (CP) also rise to similar peaks. However, the peak pressure in the aortic arch is over twice that in the pulmonary arch. This provides experimental evidence that the ventricle operates as a dual pressure pump, simultaneously producing high pressures in the systemic circuit and low pressures in the pulmonary circuit. Abbreviations: right pulmonary artery (RP); right aortic arch (RAA).

After Burggren and Johansen.

sinus venosus. However, as embryonic development proceeds, this part of the sinus venosus together with the associated pulmonary vein become incorporated into the developing left atrium (figures 12.34b and 12.36).

In other respects, the crocodilian heart is quite different from what we have seen so far. The ventricle is divided by an anatomically complete interventricular septum into distinct left and right chambers. The pulmonary trunk and *left* aortic arch open off the thick-walled *right* ventricle. The *right* aortic arch opens off the *left* ventricle. A narrow channel called the **foramen of Panizza** connects the left and right aortic arches shortly after they depart from the ventricle (figure 12.36a).

In a crocodile breathing air, right and left atria fill with deoxygenated systemic and oxygenated pulmonary blood, respectively. Contraction of the atria delivers blood to the respective ventricles. When the ventricles contract, blood flows through the nearest portals of least resistance. At the moment of systole, pressure is greatest in the left ventricle. The oxygenated blood it holds enters the base of the right aortic arch, but because of its high pressure, it also enters the left aortic arch via the foramen of Panizza. High pressure in the left aortic arch keeps the lunar valves at its base closed, leaving only the pulmonary route of exit for blood in the right ventricle. As a result, both aortic arches carry oxygenated blood to systemic tissues, and the pulmonary artery carries deoxygenated blood to the lungs (figure 12.36a).

When a crocodile dives, this pattern of cardiac blood flow changes because of a cardiac shunt. Resistance to pulmonary flow increases due to vasoconstriction of the vascular supply to the lungs and partial constriction of a sphincter at the base of the pulmonary artery. Finally, a pair of coglike, connective tissue valves in the pulmonary outflow close. As a result, systolic pressure within the right but not the left ventricle rises substantially, matching and somewhat exceeding that within the left aortic arch. Blood in the right ventricle now tends to exit through the left aortic arch rather than through the pulmonary circuit, which presents high resistance to blood flow. Diversion of blood in the right ventricle to the systemic circulation represents a right-to-left cardiac shunt. Blood in the right ventricle, which would flow to the lungs in an air-breathing crocodile, instead travels through the left aortic arch, joining the systemic circulation and bypassing the lungs (figures 12.34b and 12.36b). This lung bypass confers the same physiological advantages we have seen in turtles, namely, an increase in efficiency of blood flow while fresh air is unavailable.

Apnea, breath holding, occurs not only during diving. Most reptiles at rest on land can go for long intervals without taking a breath. As apnea continues, oxygen from the lungs becomes depleted, and pulmonary perfusion declines until just before another breath. Thus, for reptiles breathing air but ventilating intermittently, the cardiac shunt allows pulmonary perfusion to match air ventilation. In reptiles in temperate (or desert) regions, the cardiac shunt probably diverts blood during times of hibernation (or estivation)

when metabolic needs are reduced, ventilation rate declines, and high levels of pulmonary perfusion would bring few physiological benefits.

But there is more—much more. Some very recent and creative research by Colleen Farmer points to other functions of these cardiac shunts. She suggests they are also or primarily the basis for a very sophisticated support for ectothermic digestion and possibly skeletal growth. To date, cardiac shunts have been studied experimentally in only a few of the thousands of sauropsid species, making it difficult to decipher their general functional significance. Further, cardiac shunts have historically been viewed as diving adaptations, which they certainly might be. But, now we see that they might be much more. For example, in chelonians, crocodilians, and some large lizards, left and right systemic arches are asymmetrical in their sizes. Before joining the right systemic arch to form the single dorsal aorta, the left systemic arch supplies gastrointestinal vessels to the digestive viscera, including stomach, liver, pancreas, spleen, and small intestine (figure 12.37). These viscera are important in the biosynthesis of gastric acid, fatty acids, amino acids, and in precursors (e.g., purines) to RNA and DNA. It may be surprising to realize, but CO_2 is not just a waste product carried in the blood on its way to elimination from the body—it also contributes to the biosynthesis of these products. In water, CO_2 dissociates to form bicarbonate (HCO_3^-) and acid (H^+) (see figure 11.42). In the stomach, H^+ is pumped into the lumen to acidify the contents; HCO_3^- is delivered to the pancreas, liver, and small intestine, where it helps neutralize the acidic chime, provides a blood-borne source of CO_2 to supplement acid/base balancing, and contributes to the synthesis of some molecular participants in cellular anaerobic respiration.

Based on this vascular arrangement and cellular physiology, it is hypothesized that these sauropsids can adjust blood acid/base levels, allowing relatively alkaline blood to flow to the somatic and appendicular regions (right systemic arch) and relatively acidic blood flow to the digestive viscera (left systemic arch). Alkaline blood, high in oxygen content, favors increased osteoblast activity and, hence, bone deposition; acid blood, high in CO_2 content, flowing to digestive viscera favors their participation in biosynthesis (figure 12.37). When deployed, the cardiac shunt may be important in shifting CO_2 to or from these systemic circuits. This boost may be especially important for ectotherms, where low temperatures can otherwise reduce effective growth or gastric function.

Except for some large lizards (e.g., varanids), most lizards differ in vascular anatomy from chelonians and crocodilians and therefore may use a cardiac shunt differently. Left and right systemic arches are of about the same size at their confluence, forming the single dorsal aorta. Gastric and celiac arteries depart from the dorsal aorta downstream of this confluence, not from one arch or the other. Further, oxygen levels remain high (and CO_2 levels low) during digestion, implying that the cardiac shunt, if then present, is rather small. The functional significance of the cardiac shunt may eventually prove to include many roles in the lifestyles of sauropsids.

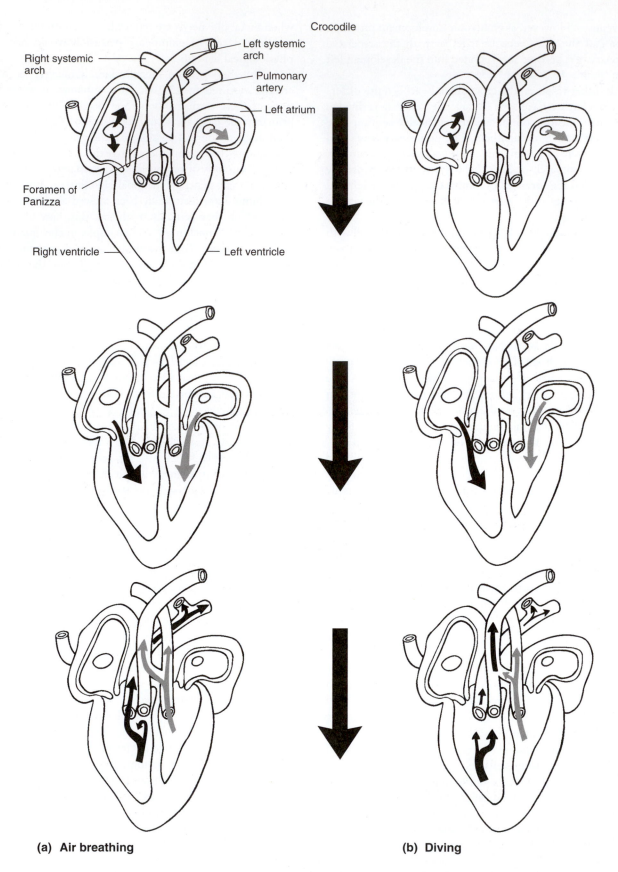

Crocodile

Right systemic arch

Left systemic arch

Pulmonary artery

Left atrium

Foramen of Panizza

Right ventricle

Left ventricle

(a) Air breathing

(b) Diving

FIGURE 12.36 **Blood flow through the crocodile heart.** (a) Systemic and pulmonary blood flow when the crocodile breathes air. (b) Internal changes that result in decreased pulmonary flow when the crocodile dives.

FIGURE 12.37 **Hypothesized digestive significance of the cardiac shunt, based on crocodilians.** (a) No cardiac shunt. Blood returning from systemic tissues is directed to the lungs, unloading CO_2 and replenishing with oxygen, then back to the heart, and then out to systemic tissues primarily via the right systemic arch, carrying blood high in oxygen content in support of metabolic activity and promoting bone deposition. (b) Cardiac shunt. Following a meal, some returning blood low in oxygen content, and high in CO_2 levels, may bypass the lungs via a cardiac shunt, and be directed to the left systemic arch, delivering acidified blood to digestive viscera, thereby supporting digestive functions and biosynthesis. (Not shown is the other stream of blood high in oxygen returning from the lungs and sent out the right systemic arch.)

Based on the research of Colleen Farmer.

Birds and Mammals

As noted, the hearts of birds and mammals have four chambers that arise from the two chambers (atrium and ventricle) of the fish heart. In birds, the sinus venosus is reduced to a small but still anatomically discrete area. The conus arteriosus (bulbus cordis) is only a transient embryonic chamber that gives rise to the pulmonary trunk and a single aortic trunk in the adult (figure 12.38). In mammals, the sinus venosus is reduced to a patch of Purkinje fibers, or sinoatrial node, in the wall of the right atrium. The sinoatrial node functions as a pacemaker, initiating the wave of contraction that spreads across the heart as in all other vertebrates. As in birds, the conus arteriosus splits during embryonic development in mammals to produce the pulmonary trunk and single aortic trunk of the adult (figure 12.39).

Although structurally similar, bird and mammal hearts arose independently from different groups of tetrapod ancestors. This difference is reflected in their embryonic development. Appearance of the interventricular and interatrial septa that form the paired chambers occurs quite differently in the two groups. Bird and mammal hearts function similarly as well. Both consist of parallel pumps with double circulation circuits. The right side of the heart gathers deoxygenated blood from systemic tissues and pumps it into the pulmonary circuit. The left side of the heart pumps oxygenated blood from the lungs through the systemic circuit. The hearts of birds and mammals are anatomically divided into left and right compartments; thus, there is no cardiac shunting with changing ventilation rates. Therefore, unlike in amphibians and reptiles, a cardiac shunt cannot be used in birds and mammals to decouple perfusion of the lung and systemic tissues. Although the reasons are not well understood, some propose that endothermic animals (birds and mammals) may require complete anatomical separation of the cardiac chambers to

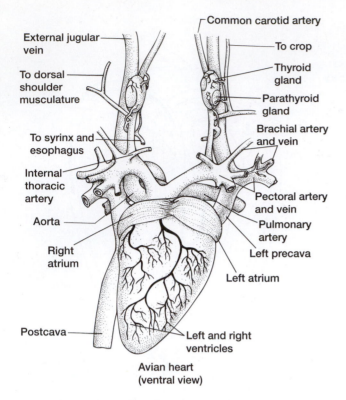

Avian heart
(ventral view)

FIGURE 12.38   Avian heart, ventral view.

After Evans.

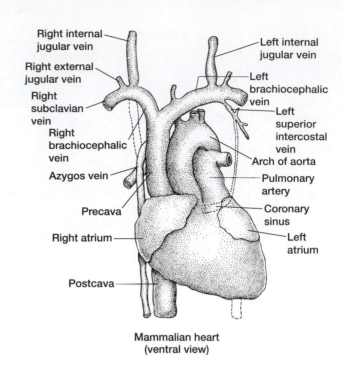

Mammalian heart
(ventral view)

FIGURE 12.39   Mammalian heart, ventral view.

After Lawson.

prevent blood being sent to the lungs at the same high pressure as blood sent to systemic tissues (see Box Essay 12.1).

Cardiovascular System: Matching Design to Environmental Demands

It is tempting to measure heart performance of lower vertebrates by how well their cardiovascular systems might serve mammalian needs. Partial internal heart septa have been termed "incomplete" in comparison with the "complete" anatomical divisions in mammalian hearts. Hearts and aortic arches of lower vertebrates have been interpreted as "imperfect" designs because the mammalian design was considered optimal. As we have seen, the lungfish cardiovascular system was mistakenly thought to significantly mix oxygenated and deoxygenated blood.

If we begin with the view that lower vertebrates are designed imperfectly, then such a naive conclusion is bound to follow. In lungfishes, if oxygenated blood from the respiratory gills (II, V, and VI) met with deoxygenated blood from arches without gills (III, IV) in the dorsal aorta, then the two would mix. And if mixing occurred, blood perfusing active systemic tissues would have a lower oxygen tension. Certainly this would be an inefficient design. Experimental work coupled with knowledge of basic vertebrate anatomy now proves this interpretation incorrect.

The "incomplete" internal heart divisions and associated aortic arch arrangements enable lungfishes to adjust their physiological patterns of circulation to changes

in the availability of oxygen in their environment. The cardiovascular systems of lower vertebrates are extraordinarily flexible, permitting adjustments to air and water ventilation patterns. Their cardiovascular systems are no less adaptive for the environments in which they serve than the systems of more "advanced" birds and mammals. Evolution of the cardiovascular system represents not a progressive improvement of design but rather an equally adaptive alternative way of addressing the demands that different lifestyles place on the circulatory system.

The cardiac shunt, taking advantage of features of heart design such as incomplete septation, also permits the heart to produce different pressures within the systemic and pulmonary circuits, high and low, respectively. Low pressures within the lungs help prevent edema, the collection of fluid leaking out of capillaries. High pressure in the systemic circulation keeps blood pressure high in the renal arteries to the kidneys and thereby facilitates excretory filtration.

The precision of this cardiac shunt and the morphological structures facilitating it can be very specialized. In the crocodile, a special "cog-tooth" valve, derived from connective tissue, lies at the base of the pulmonary artery. The valve can close to restrict blood flow to the lungs and thus contribute to the cardiac shunt, which may divert a large proportion of the cardiac output away from the lungs. During prolonged dives, blood flow through the foramen of Panizza may actually reverse, with blood passing from left to right aortic arches, thereby ensuring filling of the right aortic arch, which supplies the coronary (heart) and cephalic (brain) circulation.

Significant cardiac shunting may occur while the reptile is on land, resting. This also may be a way to keep renal filtration high without unnecessarily keeping the pulmonary circulation also high when metabolic scope is low. It might also serve digestion. In crocodiles, the left aortic arch supplies much of the stomach and intestine. If carbon dioxide buildup raises the acidity of the blood, the shunting of acidic blood to the gut may be advantageous in the secretion of HCl into the stomach after a meal.

Accessory Air-breathing Organs

Many fishes do not experience the extremes of hypoxia lungfishes face; nevertheless, they occasionally endure temporary stress from low oxygen availability. Vascularized gas bladders seem to be one answer. Blood is diverted from the general circulation to the gas bladder, where oxygen is taken up and circulated back to the general circulation (figure 12.40a). In *Hoplosternum*, a carplike fish, branches from the dorsal aorta carry blood to areas of the digestive tract enriched with capillary beds. Blood in these capillary beds is exposed to a bubble of air swallowed into the digestive tract. Oxygen from the air is taken up and added directly to the systemic circulation to boost overall oxygen tension (figure 12.40b).

In fishes with accessory air-breathing organs, oxygenated blood joins the general circulation before entering the heart. The result is to raise oxygen levels in the blood enough to compensate for low levels in the water and get the fish through temporary periods of hypoxia. This design is sufficient and adaptive for the limited stresses introduced by occasional low levels of oxygen. Lungfishes are unique among living fishes in possessing a discrete pulmonary vein that returns blood directly to the heart. Under more severe conditions of frequent, prolonged hypoxia and during droughts, this cardiovascular design allows dipnoans to survive. Thus, this group can occupy habitats and tolerate conditions for which other fishes are much less suited.

Diving Birds and Mammals

The hearts of diving birds and mammals do not offer the physiological options that amphibians and reptiles employ to adjust to the demands of diving. In a sense, birds and mammals are locked into a design unsuited to the intermittent breathing that accompanies aquatic life. When they dive, oxygen in the lungs is quickly depleted. It soon becomes disadvantageous for the heart to pump the customary large volume of blood to the nonfunctioning lungs. Yet, because of the complete internal division of the heart, diversion of blood away from the lung cannot occur within the heart. Adjustments must occur by other means.

When a tetrapod dives, three major physiological adjustments take place within the circulatory system. First, bradycardia occurs. The decreased heart rate reduces the energy spent on pumping blood to lungs depleted of oxygen. Second, anaerobic metabolism in skeletal muscles increases. Third, the microcirculation alters the blood flow to major

(a) *Hoplerythrinus*

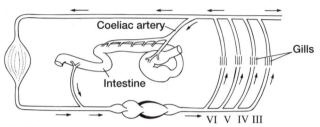

(b) *Hoplosternum*

FIGURE 12.40 **Blood supply to accessory air-breathing organs of fishes.** (a) The jeju, *Hoplerythrinus*, has evolved a vascularized swim bladder supplied by a branch of the coeliac artery. When this fish breathes air, blood flow to the swim bladder almost doubles, but there is no anatomical structure to separate the returning oxygenated blood from the venous circulation before both enter the heart. (b) *Hoplosternum* gulps air into its intestine, from which oxygen enters the systemic circulation. As in the jeju, oxygenated blood mixes with deoxygenated blood on its way back to the heart.

organs and tissues. For example, blood flow to the brain and adrenal glands is maintained, but blood flow to the lungs, digestive tract, and appendicular muscles (which function under anaerobic conditions) is decreased.

Overall, a diving bird or mammal makes the best of a stressful condition. When it is submerged, there is little oxygen available. The lungs are depleted and often collapse under the water pressure. And although a large amount of oxygen is stored in myoglobin, it is soon exhausted. Consequently, energy-consuming activities switch to metabolic pathways that require no immediate oxygen (muscles change to anaerobic metabolism), energy is conserved (bradycardia occurs), and available blood is shunted to priority organs (microcirculation shifts). None of these physiological responses is unique to birds and mammals. All tetrapods show similar responses when they dive in water. But because birds and mammals lack a heart with a cardiac shunt, these are the only major cardiovascular adjustments available.

In contrast to birds and mammals, the hearts of reptiles and amphibians function as two independent pumps. During dives, pulmonary resistance increases, and more blood can be diverted to the systemic circulation. But because the pumps are independent, independent pressures can be produced. It may, for example, be important to keep systemic pressure high so renal filtration of the blood does not decline; this can be accomplished within the systemic circuits without requiring simultaneous elevation of lung pressure as well.

Heart Flow

The heart not only produces the initial pressure that moves blood, but it also separates oxygenated and deoxygenated bloodstreams and directs blood into appropriate aortic or pulmonary trunks. Separation of oxygenated and deoxygenated blood depends on many features of heart structure and function, including septation, position of entry and exit portals, dynamics of blood flow, and texture of the heart lining. Internal septa, whether complete or incomplete, aid in separating oxygenated from deoxygenated bloodstreams flowing through the heart. Locations of entrance and exit portals of the heart also assist in maintaining separate arterial and venous blood flow. For instance, in the crocodile heart, if the left aortic arch arose from the left rather than the right ventricle, there would be no cardiac shunt. In the lizard heart, delivery of blood from the left atrium into the cavum arteriosum places oxygenated blood at favorable locations within the heart so that it is strategically positioned to take the proper exit. Further, the balance of resistance between pulmonary and systemic circuits also influences the direction of blood flow from the heart. In the amphibian heart, the dynamics of blood flow through the ventricle partly explain why so little mixing of oxygenated and deoxygenated blood occurs in this common chamber. The recesses of the trabeculate myocardium may provide temporary sites in which blood entering from one stream is momentarily sequestered from another. The heart and its lining, in ways not yet understood, probably produce laminar rather than turbulent flow, further reducing the churning that might induce mixing of oxygenated and deoxygenated blood. For the heart to function properly, many subtleties of design must interact, even though we sometimes do not fully understand their indispensable contributions.

Ontogeny of Cardiovascular Function

Embryo and adult often live in quite different environments. It is not surprising, then, that the circulatory system is different in these two stages of an individual's life history. The embryonic heart begins to beat in its first few days. In the chick, it consists of a pump with a single, undivided ventricle that has the same hemodynamic demands as an adult fish heart. In both the adult fish and the chick embryo, gas exchange tissues and systemic tissues are in series. They are served by a single cardiac pump that must generate sufficient pressure to drive blood through both. Similar structural designs serve common functional demands. The beating embryonic heart, no less than the adult heart, serves the needs of the embryo, although it is a transitory stage. For most vertebrates, critical changes in the circulatory system must quickly accommodate sudden changes in physiological demands that occur at birth or hatching. These changes are most extensive and perhaps best known in placental mammals.

Fetal Circulation in Placental Mammals

In eutherian mammals, the fetus depends exclusively on the placenta for its oxygen supply (figure 12.41). A single **umbilical vein** carrying oxygenated blood away from the placenta flows to the liver, where approximately half the blood enters the sinusoids of the liver and the other half bypasses the liver via the **ductus venosus** and enters the hepatic vein. Blood in the hepatic vein joins the large volume of blood returning via the precava and postcava to the right atrium. Pulmonary circulation to the nonfunctional lungs is reduced. About 90% of the blood that reaches the pulmonary artery bypasses the lung via the **ductus arteriosus** and is diverted instead to the dorsal aorta. Within the heart, the interatrial septum is incomplete. The **foramen ovale,** an opening between right and left atria, allows most blood entering the right atrium to flow directly to the left atrium without first passing through the lungs. Thus, the foramen ovale, like the ductus arteriosus, diverts most blood away from the nonfunctional lungs and into the systemic circulation. Blood returns to the placenta via the paired **umbilical arteries,** which branch from the iliac artery (figure 12.42a).

Placentae (p. 192)

Near the end of gestation, the mammalian fetus has a specialized and complex circulatory system. Blood entering the right atrium is a mixture of deoxygenated blood (from the liver, precava, postcava, and coronary sinus) and oxygenated blood from the placenta (via the umbilical vein and ductus venosus). However, even with this mixing in the right atrium, oxygenated blood from the placenta tends to be shunted through the foramen ovale to the left atrium. From the left atrium it flows in turn to the left ventricle, the dorsal aorta, the carotids, and to the head. Therefore, the fetal brain preferentially receives blood that is higher in oxygen partial pressure compared with blood sent to organs elsewhere in the body.

Because pulmonary resistance is high, pressures are higher on the right side of the heart than on the left side. This pressure differential and the one-way action of the foramen ovale ensure that blood flows only from the right to the left atrium.

Changes at Birth

When a human is born, several changes in the circulatory system occur nearly simultaneously. As maternal and fetal tissues separate in the birth process, placental circulation ceases. The neonatal lungs expand with the first vigorous breaths and become functional for the first time (figure 12.42b). When breathing begins, the sudden rise in blood oxygen partial pressure stimulates contraction of smooth muscle in the walls of the ductus arteriosus, immediately closing it. Over a period of several weeks, fibrous tissue invades the lumen and obliterates the ductus arteriosus, which becomes a cord of tissue, the *ligamentum arteriosum* (ligament of Botallus). Because more blood enters the functional lungs after birth, more blood returns to the heart,

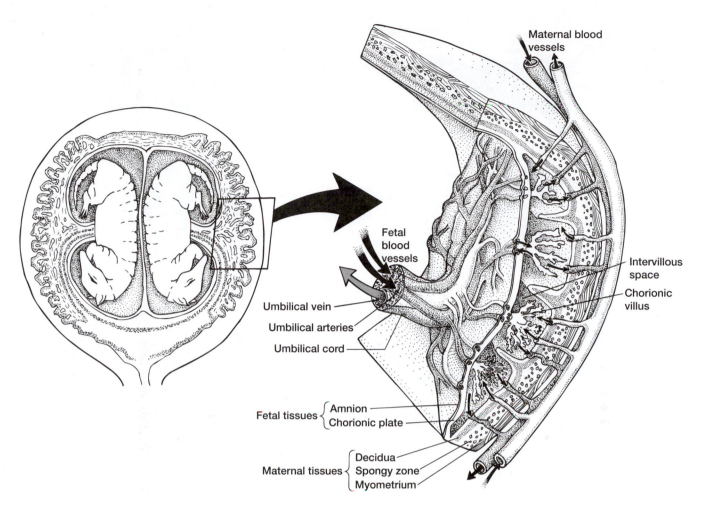

FIGURE 12.41 **Mammalian placenta.** Extraembryonic membranes of the fetus produce the chorionic plate associated with the maternal tissues of the placenta. At parturition, the placenta separates from the uterus at the spongy zone. Fetal blood low in oxygen tension flows through two umbilical arteries into a dense, branching network of capillaries in the chorionic villi. Fetal blood takes up oxygen from maternal blood in the chorionic villi. Oxygenated blood flows from these capillaries through the umbilical vein to enter the fetal circulation. Maternal blood flows through the placenta via branches of the uterine artery. It saturates the intervillous spaces and bathes the walls of the chorionic villi, giving up oxygen to the fetal capillaries. Maternal blood flows from these spaces through tributaries to the uterine vein.

After Mossman, 1937.

increasing the pressure in the left atrium and forcing the septa of the foramen ovale closed. In most individuals, the septa gradually fuse so that an anatomically complete wall forms between the atria when a human is about a year old. However, in about one-third of human adults, this anatomical fusion fails. Instead, the septa are kept closed by pressure differences between the atria, usually leading to no clinical symptoms.

Smooth muscles within the walls of the umbilical vessels contract and gradually become invaded by fibrous connective tissue. This continues through the first two or three months of postnatal life. Occluded sections of the umbilical arteries become the lateral umbilical ligaments. Other sections of the umbilical arteries contribute to the common and internal iliac arteries. The umbilical vein persists only as a cord of connective tissue, the ligamentum teres. Over a two-month period, the ductus venosus atrophies into a fibrous mass, the ligamentum venosum (figure 12.42b).

As a result of these changes at birth, a double circulation pattern is quickly established and becomes anatomically consolidated over the first few months of neonate life. Failure of one or several of these changes to occur can result in inadequate oxygenation or distribution of blood. As poorly oxygenated blood reaches the peripheral circulation, the infant's skin darkens, taking on the bluish tint of deoxygenated blood, a condition known as **cyanosis** (blue-baby syndrome). The severity of the condition and the appropriate medical response depend on which and how many of these changes fail to occur.

Marsupials are born at an early stage of development after a brief period of gestation, as few as 13 days (Virginia opossum) or as many as 37 days (eastern gray kangaroo). As with eutherians, marsupial young during intrauterine development possess cardiovascular shunts that allow the bypass of the lungs. A ductus arteriosus is present. An interatrial communication exists between right and left atria, but this

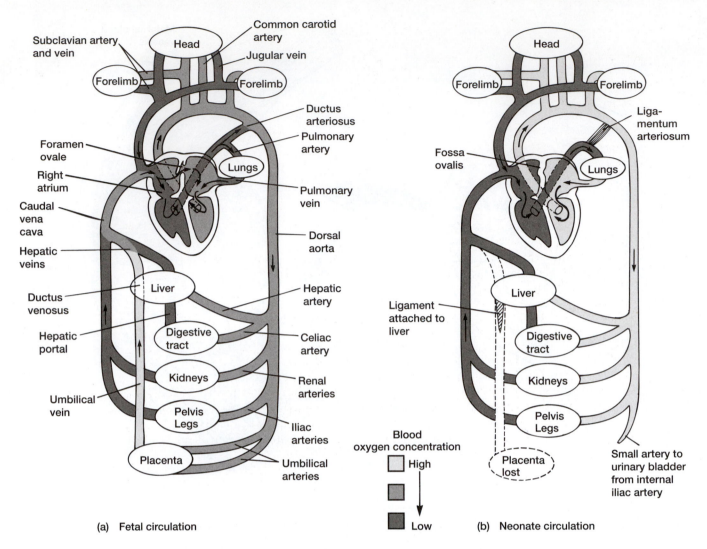

FIGURE 12.42 Mammalian (eutherian) circulatory changes at birth. (a) Fetal circulation. Because the lungs are nonfunctional, uptake of oxygen and nutrients occurs through the placenta. The ductus venosus is a liver bypass. The foramen ovale and ductus arteriosus are lung bypasses. (b) Neonatal circulation. Following birth, the lungs become functional, the placenta departs, and the ductus venosus, foramen ovale, and ductus arteriosus close.

After Walker.

is through a fenestrated septum between the two fetal heart chambers. This fenestrated septum may correspond to part of the foramen ovale of eutherians. Unlike in eutherians, this marsupial septum does not develop a one-way flap valve, but allows blood flow in either direction between atria. In some marsupials, an interventricular foramen appears during very early development.

At the end of gestation, the young marsupial must journey from the birth canal into the pouch, and meet the physical challenges of extrauterine existence. At birth, the ductus arteriosus constricts to close rapidly within a few hours, and closure of the interventricular foramen, if not already complete, also occurs within a few hours of birth. Proliferation of tissue invades the interatrial, fenestrated septum, but full closure can take several days. The timing and somewhat gradual closure of these shunts at birth may provide a way of fine-tuning the pressure differences between the systemic and newly functional pulmonary circulations.

Heat Transfer

In addition to delivering gases and metabolites, the circulatory system also functions in heat transfer. For example, reptiles basking in the sun absorb heat in their peripheral blood vessels. As this heated blood circulates throughout the body, it warms deeper tissues. Conversely, blood transports heat produced as a by-product of exercising muscles to the surface of the body (figure 12.43a,b). It can more easily be dissipated through the integument to prevent overheating. Whether blood transports warmth to deep tissues or carries away excess heat to the surface, the appropriate changes in circulation are largely mediated by changes in the microcirculation. During cooling, more capillary beds open in the peripheral circulation of the skin to increase blood flow and transfer heat to the environment. When heat is conserved, peripheral blood flow is reduced.

Compared to air, water has a higher thermal capacitance (ability to hold heat). Consequently, animals in water often face

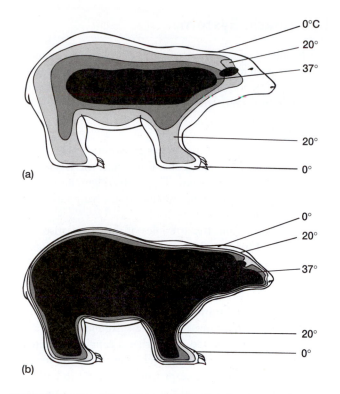

(a)

(b)

FIGURE 12.43   Heat distribution in mammals.
(a) Mammals, such as polar bears, preferentially maintain a relatively high core temperature. As blood nears the body's surface, it reaches cooler areas of the animal. (b) When an animal overheats, as during exercise or on hot days, excess heat is circulated to the skin surface, where it can be dissipated more easily.

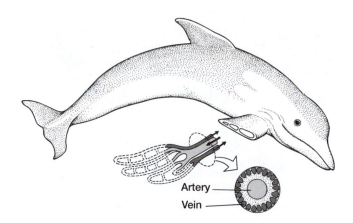

FIGURE 12.44   Heat blocks. Birds and mammals that live in cold waters are in danger of losing heat to their surroundings continuously. This is especially true in the extremities that have large surface area in relation to their mass. Warm blood flowing to the extremities loses heat to the environment. Potentially this could be a severe heat drain on the organism. In dolphins, warm arterial blood circulating to the flipper passes returning venous blood, which is cold. Because returning veins of the flipper surround its central artery, a countercurrent system of flow is established. Arterial blood flowing into the flipper gives up its warmth to the returning venous blood, reducing the dolphin's heat loss to its environment. This countercurrent arrangement in the upper flipper serves as a heat block, preventing excess loss of heat to the environment.

After Schmidt-Nielsen.

special problems with respect to controlling heat loss or gain. The flippers or webbed feet of whales, seals, or wading birds are bathed in cold water. Blood circulating to these extremities is warm, but the water can be icy; therefore, much heat would be lost to the environment were it not for specialized features of the circulatory system. In the upper regions of flippers or legs, an elaborate, intertwining network forms between outgoing arteries and returning veins. These adjacent networks of arteries and veins are termed *retes*. Blood in a rete establishes a countercurrent pattern between outgoing arteries and returning veins. Before reaching the flipper or foot, warm blood passes through the rete. Heat carried in the arteries is transferred almost completely to the returning blood in the veins. By the time blood reaches the extremity, little heat remains to be dissipated to the environment. Such retes function as **heat blocks** to prevent body heat from being lost through the extremities.

Countercurrent exchange (p. 156)

In dolphins and whales, an additional mechanism is used to control heat loss. Deep within the core of the fin, the single central artery is surrounded by numerous veins. The outgoing central artery and numerous returning veins form a countercurrent exchange system that functions as a heat block (figure 12.44). However, when the animal is active and excess heat must be dissipated, the same circulatory mechanism participates. Additional blood flows to the fin via the central artery. As the blood dilates this artery, it puts pressure on the surrounding veins, causing them to collapse. Because these veins are closed, the warm blood seeks alternative return routes in veins near the surface of the fin. The overall result is to close the deep heat block temporarily and simultaneously divert blood to the surface, where excess heat can be transferred to the water.

Pursuing prey or evading enemies generates excess metabolic heat; exposure to solar radiation increases body temperature. The brain is especially susceptible to such temperature extremes. If it overheats even slightly, the result can be lethal. In many animals, a special **carotid rete** at the base of the brain addresses this problem (figure 12.45a). For example, in the nose of a dog, highly folded *turbinate bones* support an extensive area of moist nasal membranes cooled by evaporation. Cooled venous blood returning from this nasal membrane enters the carotid rete to absorb heat from the blood in the carotid artery before it enters the brain (figure 12.45b). Of course, not all heat is blocked to the brain because the brain must be kept warm. But the carotid rete serves as another heat block. In this instance, the heat block provides a mechanism for absorbing excess heat and preventing harmful thermal extremes in the brain (figure 12.45c).

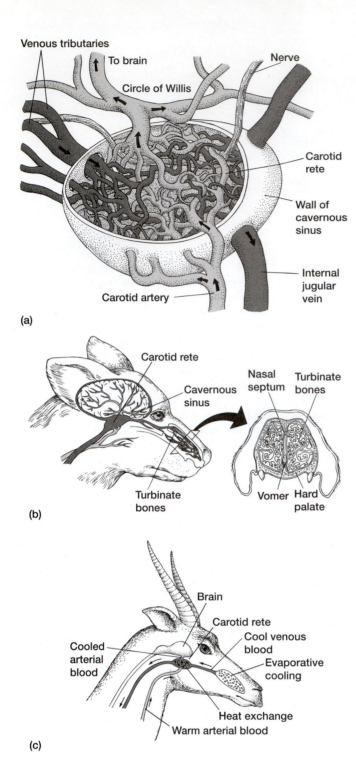

(a)

(b)

(c)

FIGURE 12.45 **Cooling.** (a) In many mammals, a carotid rete is found at the base of the brain. This rete brings arteries and veins close together, allowing heat exchange between them. (b) The dog's nose includes a highly convoluted set of turbinate bones supplied with blood. Air moving through the nose cools the venous blood before it flows through the carotid rete. In the rete, arterial blood on its way to the brain gives up its heat to this cool venous blood, a process that protects the brain from overheating. (c) In the eland, a desert mammal, notice the location of the carotid rete in relation to the nasal passages and the brain.

After Baker.

Lymphatic System

The lymphatic system is partner with the circulatory system. It aids fluid return to the circulatory system and is engaged in several special functions. Structurally, there are two components of the lymphatic system: lymphatic vessels and lymphatic tissue.

Lymphatic Vessels

Collectively, the lymphatic vessels constitute a blind-ended, tubular system that recirculates fluid from the tissues back to the cardiovascular system. The walls of lymphatic vessels are similar to those of veins and, like veins, lymphatic vessels contain one-way valves.

Pressure within the arterioles of the blood arises from two sources. **Hydrostatic pressure** represents the remaining force generated initially by ventricular contraction. It tends to favor the flow of fluid from the blood into the surrounding tissue. **Osmotic pressure** results from unequal concentrations of proteins within the arteriole and outside in the surrounding tissue fluid, so fluid moves from the surrounding tissue into the blood. As an arteriole approaches a capillary bed, residual hydrostatic pressure is usually higher than osmotic pressure. Consequently, fluid seeps from the blood to bathe the surrounding cells. This fluid that has escaped from the blood capillaries is called **tissue fluid.** At the venule side of the capillary bed, most of the hydrostatic pressure has dissipated, leaving the osmotic pressure to predominate. The net inward pressure results in recovery of almost 90% of the original fluid that leaked from the arterial blood. The remaining 10%, if not recovered, would build up in connective tissues, causing them to swell with excess fluid, a condition termed **edema.** Edema does not usually occur because excess tissue fluid is picked up by the lymphatic tubules and eventually returned to the general blood circulation (figure 12.46).

The fluid carried by the lymphatic vessels is **lymph.** It consists mostly of water and a few dissolved substances such as electrolytes and proteins, but it contains no red blood cells. The main vessels of the lymphatic system collect the lymph resorbed by the tiny, blind lymphatic capillaries and return it to the venous circulation near the precaval and postcaval veins (figure 12.47). Lymphatic vessels form a network of anastomosing channels. The major vessels that generally compose the lymphatic network and the parts of the body they drain are the **jugular lymphatics** (head and neck), **subclavian lymphatics** (anterior appendage), **lumbar lymphatics** (posterior appendage), and **thoracic lymphatics** (trunk, viscera of body cavity, tail; figures 12.47a and 12.48a,b).

The low pressure within the lymphatic vessels aids them in taking up tissue fluid but presents a problem in moving lymph along. In some vertebrates, such as teleost fishes and amphibians, **lymph "hearts"** occur along the route of return. These are not true hearts because they lack cardiac muscle, but striated muscles in their walls slowly develop pulses of pressure to drive the lymph. Spinal nerves supply lymph hearts, although the hearts can also pulse rhythmically on their own if innervation is severed. In teleost fishes,

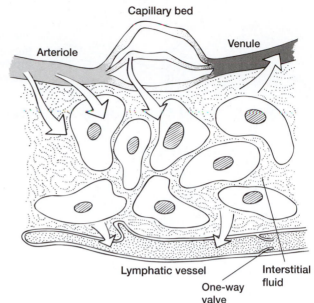

Capillary bed

Arteriole

Venule

Lymphatic vessel

One-way valve

Interstitial fluid

FIGURE 12.46   Formation of lymph. The relatively high pressure in capillaries results in fluid from the blood leaking into surrounding tissues. Some of this interstitial fluid returns to the blood on the low-pressure venous side of the circulation. Blind-ended channels called lymphatic vessels collect the excess fluid (lymph) and return it to the general circulation, usually through one of the large thoracic veins (arrows indicate movement of fluid).

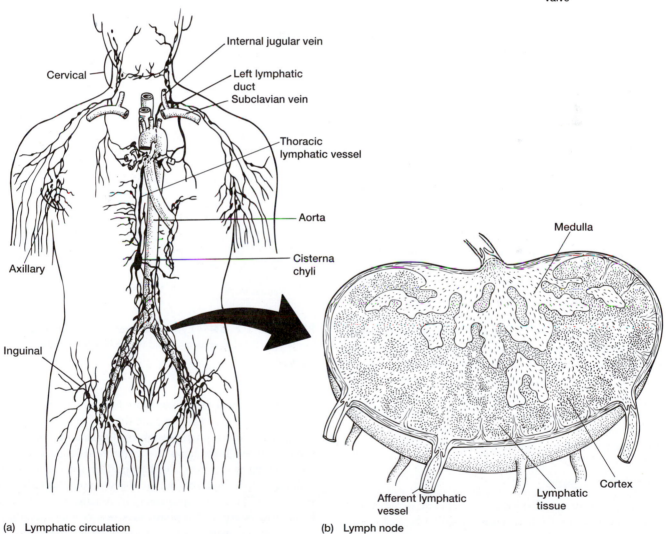

Internal jugular vein

Cervical

Left lymphatic duct

Subclavian vein

Thoracic lymphatic vessel

Aorta

Cisterna chyli

Axillary

Inguinal

Medulla

Cortex

Afferent lymphatic vessel

Lymphatic tissue

(a)   Lymphatic circulation

(b)   Lymph node

FIGURE 12.47   Lymphatic circulation and lymph nodes. (a) Lymphatic vessels returning from all parts of the body join to form major lymphatic vessels, the largest being the thoracic duct, which empties lymph into the postcaval or subclavian veins. (b) Cross section of a lymph node. In mammals and in a few other species, small swellings or nodes occur along lymphatic vessels. These lymph nodes house lymphatic tissue, which functions to remove foreign materials from the lymph circulating through them. Lymph nodes have a cortex and medulla bounded by a fibrous connective tissue capsule. Notice the entering and departing lymphatic vessels.

From J. V. Basmajian, Primary Anatomy, *7th ed. Copyright © 1976 Williams and Wilkins, Baltimore, MD. Reprinted by permission.*

Auricular sinus
Submandibular lymphatic vessel
Jugular trunk

Tracheal trunk
Lymphatic ring
Subclavian trunk
Axillary sinus
Jugular sinus
Lateral trunk
Lung
Thoracic trunk
Subvesical sinus
Lumbar trunk
Iliac trunk
Pelvic and sacral trunks

Collectors (posterior limb)

(a)   (b)

FIGURE 12.48   Lymphatic vessels in the turtle *Pseudemys scripta* (ventral view). (a) The plastron and most of the viscera have been removed to show the lymphatic vessels. (b) Diagrammatic view of major lymphatic trunks in turtles.

After Ottaviani and Tazzi.

lymph hearts are found in the tail and empty into the caudal vein. They also occur in some amphibians (figure 12.49), reptiles, and embryonic birds. Often they are found where lymphatic vessels enter veins. One-way valves in the lymph hearts help ensure the return of lymph to the cardiovascular system.

The mechanism of lymph return also takes advantage of general body movements, such as inhalation and exhalation pressure differences in the thorax and contractions of nearby muscles that impinge on the walls of the lymphatic vessels to force the flow of lymph. In many vertebrates, the lymphatic vessels form sheaths around major pulsing arteries. Pulse waves traveling within the arterial walls impart their energy to the surrounding lymph (figure 12.50a). The one-way valves within lymphatic vessels ensure that these forces move the lymph back to the blood circulation (figure 12.50b).

Lymphatic Tissue

The lymphatic system also includes lymphatic tissue, a collection of connective tissue and free cells. The free cells comprise mostly leucocytes, plasma cells, and macrophages, all of which play a part in the body's immune system. Lymphatic tissue can be found almost anywhere in the body: in diffusely distributed tissue, in patches (e.g., lymph nodules), or encapsulated in lymph nodes. A **lymph node** (figure 12.47b) is a collection of lymphatic tissue wrapped in a capsule of fibrous

Dorsal trunk   Head trunk
Lymph hearts

Ventral trunk   Sinus   Lateral   Sinus
trunk

FIGURE 12.49   Lymphatic vessels of a salamander. Lymph hearts aid in returning fluid to the blood circulation.

After Smith.

connective tissue. Lymph nodes are located within channels of the lymphatic vessels along the route of returning lymph. This position in lymphatic vessels ensures that the lymph will percolate through the lymphatic tissue held in the node and be presented to the free cells. Lymph nodes occur in mammals and some waterbirds but are absent in other vertebrates. In reptiles, dilation or expansion of lymphatic vessels, termed **lymphatic cisterns** or **lymphatic sacs,** occurs at locations usually occupied by true lymph nodes in birds and mammals.

Form and Function

The lymphatic vessels function as an accessory venous system, absorbing and returning escaped fluid to the general circulation. They also absorb lipids from the digestive tract.

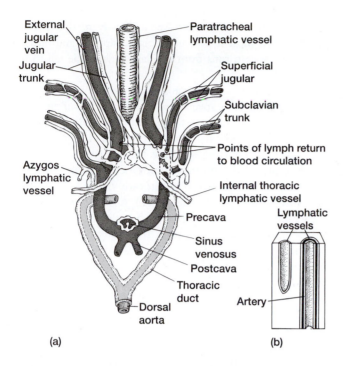

(a)   (b)

FIGURE 12.50 **Lymphatic system in reptiles.**
(a) Anterior lymphatic vessels in the crocodilian *Caiman crocodilus*. The heart has been removed to reveal the major blood vessels and associated lymphatic vessels more clearly. The pressure necessary to move lymph through lymphatic vessels is derived from the action of surrounding organs. Many lymphatic vessels lie next to active muscles as well. The walls of the lymphatic vessels passing through the thoracic cavity are compressed by the rhythmic respiratory movements. One-way valves in these vessels ensure that this pressure drives lymph back into the general blood circulation. (b) In addition, lymphatic vessels often surround major arteries, deriving force from the pulse waves of these arteries. Most lymphatic vessels form an extensive network of interconnected blind channels that gather and return lymph to the systemic blood circulation.

After Ottaviani and Tazzi.

Numerous lymphatic vessels in the digestive tract, termed **lacteals,** pick up large-chain fatty acids and return them to the blood circulation (figure 12.51).

High pressure on the arterial side of the circulation results in fluid loss to tissues in the capillaries. Return of this fluid depends upon the low pressure of the venous system, together with lymphatics and a favorable osmotic pressure of blood proteins. But the lymphatics, returning this collected fluid, must enter the circulation at a point where the low pressure encourages return of the fluid. In most fishes, the venous side of the circulation provides this opportunity, especially if low pressure occurs just as the veins enter the heart, and the effects of heart aspiration contribute to lymph and blood return into the sinus venosus. However, with the evolution of lungs, venous blood returning from the lungs was under high pressure. The division of the atrium into right (systemic) and left (pulmonary) receiving chambers, and hence division of returning venous pressures, permitted maintenance of the low pressures of the venous

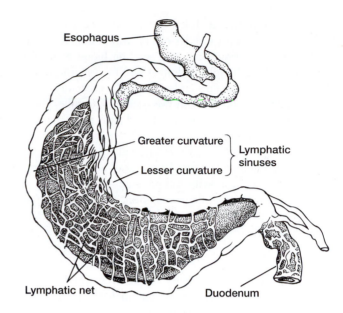

FIGURE 12.51 **Lymphatic vessels associated with the stomach of the turtle *Pseudemys scripta*.** Microscopic lymphatic vessels within the walls of the stomach are lacteals that primarily take up large-chain fatty acids absorbed through the wall of the stomach. Lacteals empty into extensive lymphatic nets that drain into large lymphatic sinuses along the lesser and greater curvatures of the stomach. Lymphatic sinuses empty into lymphatic trunks that travel through mesenteries, often receiving tributaries from other visceral lymphatic vessels and entering the thoracic lymphatic duct.

Based on Ottaviani and Tazzi.

systemic system. Hence, the initial septation of the heart occurred in the atrium to establish a low-pressure system, which made possible the return of collected fluid. Lymphatic tissue is involved in the removal and destruction of harmful foreign material, such as bacteria and dust particles. Plasma cells produce some antibodies that circulate in blood. Macrophages cling to leucocytes as they function to destroy bacteria. Lymphatic tissue also intercepts cancer cells migrating through the lymph nodes, although free cells cannot destroy cancer cells. Eventually, lymph nodes are successively overwhelmed by rapidly dividing cancer cells. If a cancer is detected early, surgical intervention usually can cure a patient. Follow-up tests should be performed to detect the extent to which cancer cells have spread through the lymphatic vessels, and then all affected nodes should be removed.

Overview

The cardiovascular system aids passive diffusion of gases between internal tissues and blood (internal respiration). Therefore, it is the complement to the respiratory system (external respiration). The cardiovascular system also carries heat and hormones, components of the immune system, end products of digestion, and molecules contributing to or derived from active metabolism. It is a system of connecting tubes and pumps that are filled with blood. The major pump is the branchial heart, a series of one-way chambers receiving blood, generating force, and sending blood to respiratory organs (usually gills or lungs) and to systemic tissues. General body movements and occasionally accessory pumps also help drive blood through the system. The tubes (blood vessels) include arteries that carry blood away from the branchial heart, veins that return it, and the microcirculation (capillary beds) between, where internal respiration occurs. Control over the microcirculation also helps harmonize blood flow to tissue activity and pressure.

Along with the microcirculation, the lymphatic system collects and returns excess tissue fluid to the general circulation, aided by body movements and in some species by lymph "hearts." A specialized set of lymphatic vessels, the lacteals, gather principally long-chain fatty acids from the alimentary canal and carry these to the liver. The lymphatic system is without erythrocytes, but includes lymphocytes and other components of the immune system.

A major evolutionary transition was from a single (gills in series with systemic tissues) to a double (lungs in parallel with systemic tissues) circulation, resulting in major changes in the cardiovascular system. Looking back from our vantage point as mammals, we might be tempted to interpret the cardiovascular systems of prior groups as anticipating the fully double circulation we possess. But, of course, that would be a serious misreading of vertebrate phylogeny. We should divest ourselves of the view that earlier vertebrate groups were "on their way" to becoming mammals, as clearly they were not such visionaries. Neither were their systems "imperfect" as earlier anatomists thought. Instead, their circulatory systems served them well to address the ecological demands arising from their lifestyles. Lungfishes and later amphibians breathed water and air, and thanks to blood shunting, could make the best of both. Similarly, reptiles adjust physiologically to the demands of diving, or of just breath-holding, as part of an economical ectothermic metabolism. Notice also (figure 12.52a) that the anatomically complete double circulation arose independently, twice—once in birds and a second time in mammals. To remind us of the independent evolution of some cardiovascular systems, we should remember fishes such as the specialized teleost, *Haplosternum* (figure 12.52b). Although its invention of intestinal respiration to supplement its cardiovascular system did not lead to later groups based on such a design, it suits it nicely in the immediate environment, where occasional low oxygen levels in the water can be supplemented by gulping air.

The aortic arches are a major set of blood vessels characterized by a fundamental pattern of six aortic arches, connecting ventral to dorsal aortae. The adult system is built from this by deletions and expansions of the available six-arch pattern, producing the basic fish or tetrapod system. The arterial supply to a region or organ is usually matched by a venous drainage, returning blood toward the heart. A major portal system, the hepatic portal system, principally carries end products of digestion from the alimentary canal directly to the liver for processing. In most fishes, amphibians, reptiles, and birds, a renal portal system runs from capillary beds in the tail to the kidneys.

Lungs or lunglike organs were present in some early fishes, evolving into specialized gas bladders supplementing gill respiration in bony fishes and later replacing gills in tetrapods as the primary respiratory organ. Lungs brought advantages in supplying systemic tissues with oxygen and, in fishes, provided a means of buoyancy control. However, lungs may have initially evolved in early fishes to supply the heart with oxygen (see Farmer, 1997). A well-developed coronary supply to the heart muscle is absent in most bony fishes and primitive tetrapods. Instead, the spongy heart myocardium receives oxygen directly from blood passing through its lumen. The need for cardiac oxygen is greatest in active fishes. In just such fishes (e.g., gars, bowfins, and the *Tarpon*, a primitive pelagic teleost), oxygen-rich blood returning from the respiratory gas bladder enters the venous circulation upon approach to the heart, thereby ensuring it will not be spent first on systemic tissues, but will instead be available to the myocardium as it flows through the cardiac lumen.

In the transition to land, gills became lost and the air-breathing lung of fish ancestors expanded its physiological role. In particular, getting oxygen to systemic capillary beds became a larger role for the lung and heart. Heart septation, along with modifications of the vascular system, helped meet these physiological needs by selective channeling of the now two streams of blood, systemic and pulmonary. Returning blood high in oxygen could be used effectively if sent directly, and undiminished, to systemic tissues. Division of the fish atrium and ventricle produces effectively a two-sided heart. The right side receives returning systemic deoxygenated blood and sends it to the lungs; the left side receives oxygenated blood from the lungs and sends it to systemic tissues. Lungfishes and many tetrapods take full physiological advantage of the extraordinary cardiac shunt to efficiently adjust systemic and pulmonary blood flows to changing metabolic and environmental demands. Cutaneous respiration (e.g., modern amphibians) adds oxygen to blood entering the right side of the heart; the intracardial shunt allows oxygen-rich blood from the left to wash into the right side. By such means, the myocardium on the separated right side of the heart is exposed to luminal blood high in oxygen. In birds and mammals, which lack significant cutaneous respiration and whose hearts are completely divided anatomically, well-developed coronary arteries are present, largely replacing luminal supply to the myocardium.

Gas bladders (p. 416)

The fully anatomically divided hearts of birds and mammals have several advantages. One is that such hearts prevent mixing of oxygen-rich blood (from the lungs) with oxygen-poor blood (returning from systemic tissues). Consequently, undiluted oxygen-rich blood is delivered to

FIGURE 12.52 **Evolution of the circulatory system.** (a) The single circulation of most fishes gives rise independently to the double circulation in birds and mammals. (b) As a reminder, many specialized circulations have evolved, such as this intestine "breathing" fish, but without further phylogenetic diversification on such a theme.

active tissues and oxygen-poor blood is sent to the lungs for replenishment. Another advantage of a divided circulatory system is that blood pressures can be separated. High systemic pressures can be generated without exposing delicate pulmonary tissues to these same high and possibly damaging pressures. Tetrapods with incompletely divided cardiac systems are more of a puzzle. Blood, via different mechanisms, may employ the open anatomical connections to flow from the right side of the heart to the left side without passing through the lungs, the right-to-left shunt. Such an anatomically incomplete division of the cardiovascular system does not prevent physiological separation of oxygen-rich from oxygen-poor blood, nor does it preclude generation of high systemic blood pressure uncoupled from pulmonary pressures. It may give the animal some advantages when diving, as I have argued earlier. This shunt may also aid digestion.

Following a meal, gastric acid secretion is increased in the gastrointestinal tract, stomach, and intestines, to facilitate digestion. The left systemic arch primarily supplies first these organs of digestion—stomach, pancreas, liver, and other organs of digestion. Gastric secretion is energy rich, so oxygen delivered by the left systemic arch supports this important digestive event. But through the right-to-left shunt, CO_2-rich (and oxygen-poor) blood is added to the left systemic arch. Carbon dioxide is the substrate for the formation of acid in the stomach in particular, and this shunt is a way of adding CO_2 to the blood, supporting gastric digestion. This is also a way to bring carbon to the synthesis of lipids, hemoglobin, and indirectly to proteins in the liver, small intestines, and spleen.

The process of gastric acid secretion is temperature sensitive, being less efficient at low temperatures. This might account for the presence of such a right-to-left shunt in ectotherms (reptiles) but its absence in endotherms (birds, mammals).

One of the most dramatic changes in the cardiovascular system occurs in amniotes at birth. To meet its respiratory needs, the amniote embryo relies on vascularized derivatives of the extraembryonic membranes (e.g., placenta, chorion). However, at birth, the neonate switches to air breathing via lungs, a conversion that must be completed quickly. Vascular bypasses of the nonfunctional embryonic lungs quickly close under changing pressure differences (foramen ovale), smooth muscle contractions (ductus arteriosus), or atrophy (ductus venosus); flow to the fetal respiratory membranes ceases. In birds and mammals, the result is to quickly convert the embryonic cardiovascular system into a double circulation.

The lymphatic system supplements the venous system, returning escaped fluids from the blood into tissues back into the general circulation (as lymph). It also is a partner in the immune system and contributes to the absorption (lacteals) of end products of digestion otherwise resistant to uptake from the intestine.

WEB LINK

Visit the text website at *www.mhhe.com/Kardong7e* for additional study aids including selected references, functional laboraotories, and web links to chapter topics.

CHAPTER 13

The Digestive System

INTRODUCTION
Preview

COMPONENTS OF THE DIGESTIVE SYSTEM
Buccal Cavity
Boundaries
Palate
Teeth
Tongue
Pharynx
Alimentary Canal
Esophagus
Stomach
Intestines
Cloaca
Specializations of the Alimentary Canal
Vascularization of the Gastrointestinal Tract
Fishes
Tetrapods

Associated Glands of Digestion
Oral Glands
Liver
Pancreas

FUNCTION AND EVOLUTION OF THE
DIGESTIVE SYSTEM
Absorption
Feces
Mechanical Breakdown of Food
Mastication
Gizzards
Chemical Breakdown of Food
Foregut Fermentation
Hindgut Fermentation
Foregut versus Hindgut Fermenters
Size and Fermentation
Digesting Toxins
Feeding and Fasting

OVERVIEW

Introduction

In the nineteenth century, Alfred Lord Tennyson served up the grim description of "Nature, red in tooth and claw," a poetic reminder that animals must procure food to survive, a sometimes harsh but practical necessity. For predators, food means another animal; for herbivores, food means plants. A quick chase and overpowering kill might characterize prey capture by a carnivore; prolonged browsing or migration to fresh sources of succulent plants might characterize feeding by a herbivore. But such a hard-won meal is initially unusable. The process of turning a meal into usable fuel for the body is the business of the digestive system. The digestive system breaks up the large molecules contained in a succulent meal so they can be absorbed into the bloodstream and made available for use throughout the body.

A lump of food in the mouth is called a **bolus.** Both mechanical and chemical processes go to work to digest this bolus. Initially, mechanical chewing with teeth and churning of the digestive tract break up the bolus, reducing it to many smaller pieces and thereby increasing the surface area available for chemical digestion by enzymes. Muscles encircling the walls of the digestive tract produce kneading and traveling waves of contraction, termed **peristalsis,** that constrict food in the lumen, mixing and forcing it from one section of the tract to the next. As mechanical and chemical action work on the bolus, it soon becomes a pulpy mass of fluid more commonly called **chyme,** or **digesta.**

Preview

The adult digestive system includes the digestive tract and accessory digestive glands. The **digestive tract** is a tubular passageway that extends through the body from the lips of

Urinary bladder
Cloaca
Cloacal opening
Large intestine
Small intestine
Pancreas
Gallbladder
Stomach
Liver
Esophagus
Oral glands
Alimentary canal
Pharynx
Buccal cavity
Digestive tract

FIGURE 13.1 **Vertebrate digestive system.** The digestive system consists of the digestive tract plus associated glands of digestion. The digestive tract includes the buccal cavity, pharynx, and alimentary canal. The alimentary canal is divided into esophagus, stomach, intestines, and cloaca.

the mouth to the anus or cloacal opening. Glands embedded in the walls lining the tract release secretions directly into the lumen. On the basis of histological differences among these intrinsic **luminal glands** and differences in size, shape, and embryonic derivation, three regions of the digestive tract are recognized: the **buccal cavity,** or mouth; the **pharynx;** and the **alimentary canal** (figure 13.1). From histological differences in the luminal wall of the alimentary canal, up to four regions are identified: **esophagus, stomach, small intestine,** and **large intestine** (figure 13.1).

In most vertebrates, the alimentary canal ends in a **cloaca,** a terminal chamber receiving both fecal materials from the intestines and products of the urogenital tract. The exit portal of the cloaca is the **cloacal opening,** or **vent.** However, in some fishes and most mammals, a cloaca is absent and the intestines and urogenital tract have separate exit portals. A coiled large intestine often straightens into a **rectum** with an anal opening **(anus)** to the outside. The **accessory digestive glands** are extrinsic glands located outside the walls of the digestive tract, but they secrete chemical enzymes of digestion and emulsifying salts into the lumen via long ducts. Principal glands of the digestive tract are the **salivary glands, liver,** and **pancreas.**

The digestive system, together with its associated glands, is built around a simple tube with regionalized functions from buccal cavity to cloaca. During embryonic development, the endodermal lining of the gut is surrounded by mesenchymal cells shed from the mesoderm of the embryonic lateral plate. This regionalization is established early in embryonic development via localized *Hox* gene expression (see figure 13.45) in both the endodermal and mesodermal layers, which pattern the gut, giving it its differentiated regions.

In the embryo, the **gut** is a simple tube of endoderm that retains, via the yolk stalk, an anatomical connection to the yolk (figure 13.2). However, yolk does not directly enter the gut through this stalk, but instead yolk is absorbed into and transported by the vitelline vessels developing as part of the extraembryonic membranes of the yolk sac. The simple embryonic gut gives rise to the pharynx and alimentary canal along with their associated digestive glands. During embryonic

development, invaginations of the surface ectoderm come into contact with the endodermal gut at opposite ends of the body. The anterior invagination, or **stomodeum,** meets the anterior gut, or **foregut.** Between stomodeum and foregut, a temporary **buccopharyngeal membrane** forms and eventually ruptures to join the lumina of both. The stomodeum gives rise to the buccal cavity. The posterior invagination of ectoderm, the **proctodeum,** meets the posterior gut, or **hindgut.** Between hindgut and proctodeum, the **cloacal membrane** forms and then ruptures, creating the outlet for the hindgut. The proctodeum becomes the adult cloaca (figure 13.2a–d).

Extraembryonic membranes (p. 190)

Components of the Digestive System

Terms that describe parts of the digestive tract are often used casually. Some take the term *alimentary canal* to mean the entire digestive system, mouth to anus; others apply it in a restricted sense. The term *gastrointestinal tract,* or *GI tract,* literally means stomach and intestines, but most apply it to the whole digestive tract from buccal cavity to anus. Clinical physicians speak of small and large bowels, slang terms for small and large intestines, respectively. There is nothing sinister or sloppy about this, just different professionals seeking terminology to serve their needs. In this text, we use the terminology defined in this section.

Buccal Cavity

The buccal cavity contains the teeth, tongue, and palate. Oral glands empty into it. Salivary glands help moisten food and secrete enzymes to begin chemical digestion. In some species, mastication begins the mechanical breakdown of food.

Boundaries

The **oral opening,** the margins of which are the lips, creates the buccal cavity entrance. Generally, upper and lower lips follow the line of the tooth rows to meet posteriorly at the angle of the jaws. In mammals, upper and lower lips meet

FIGURE 13.2 Embryonic formation of the digestive system.
(a) Early amniote embryo in sagittal section showing initial position of the gut. Notice the embryo's connections to the yolk via the yolk stalk and to the allantois. (b) Generalized amniote embryo in sagittal section. Note regions of the gut and invaginations destined to form associated glands of the digestive tract. (c) Ventral view of the isolated gut together with the embryonic kidneys. Notice the extensive pouches produced by the pharynx, each of which contributes to specific adult structures. (d) Lateral view of differentiating gut.

well forward of the jaw angle near the front of the mouth, thus creating a skin-covered **cheek** region. Cheeks prevent loss of food out the sides of the mouth during chewing. In some rodents and Old World monkeys, they expand into **cheek pouches,** small compartments in which gathered food can be temporarily held until it is chewed or carried to caches. Lips are usually pliable, although birds, turtles, some dinosaurs, and a few mammals have rigid beaks with firm margins. In most mammals, lips are fleshy, a feature helping an infant form a seal around the nipple during nursing. Human lips help form the vocalizations of speech.

The lips define the anterior border of the mouth, and the palatoglossal arch is a fold that marks the posterior border of the mouth and lies between the mouth and pharynx (figure 13.3a). However, if this or other anatomical markers are absent, the mouth and pharynx form a collective chamber called the **oropharyngeal cavity** (figure 13.3b).

The stomodeum not only forms the buccal cavity, but it also contributes to surface features of the head in some vertebrates. Two good embryonic landmarks present within the stomodeum, the **hypophyseal pouch** (adenohypophyseal placode) and the **nasal placode,** make this evident (figure 13.4a–c). In cyclostomes, only the posterior part of the stomodeum turns inward, contributing to the mouth. The anterior part turns outward, contributing to the outer

surface of the head. In sharks and actinopterygians, the hypophyseal pouch lies within the mouth, the nasal placodes differentiate on the outside of the head, and the margins of the mouth form between these anatomical landmarks. In rhipidistians and tetrapods, both landmarks lie within the adult buccal cavity (figure 13.4d).

Palate

The roof of the buccal cavity is the palate, formed from the fusion of ventral skull bones overlying the mouth. In osteichthyans and tetrapods, the **primary palate** includes a medial series of bones (vomers, pterygoids, parasphenoid) and a lateral series (palatines, ectopterygoids). In most fishes, the primary palate is a low vault with no openings. In rhipidistians and tetrapods, the nasal passages reach the mouth through paired openings in the primary palate, the **internal nares,** or **choanae** (figure 13.5a,b). The **palatal folds** are inward growths of lateral bones that meet at the midline and form a second horizontal roof that separates the nasal passages from the mouth. This new roof, present in mammals and crocodiles, is called the **secondary palate.** The anterior part of the secondary palate is the **hard palate,** comprising paired bony contributions of the premaxilla and maxilla. In some species, the palatine and pterygoids contribute as well. In mammals,

(a) *Rhesus*

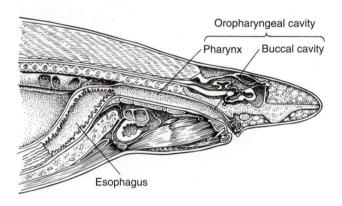

Oropharyngeal cavity

Pharynx   Buccal cavity

Esophagus

(b) Shark

FIGURE 13.3 **Sagittal view of the buccal cavity, pharynx, and developing esophagus.** (a) Head and neck of *Rhesus*. (b) Shark.

(a) After Geist; (b) after Wischnitzer.

the posterior margin of the secondary palate is the fleshy **soft palate,** which extends the position of the internal nares even farther to the back of the buccal cavity (figure 13.5c).

Secondary palate (p. 281)

Teeth

Teeth are unique among vertebrate animals. They are usually capped with enamel or enameloid, a mineralized coat found only in vertebrates. Inductive interaction between embryonic epidermis and neural-crest-derived mesenchyme is required to form teeth. In general, cells derived from epidermis make the tooth enamel whereas the mesenchyme makes the dentine. The "outside-in" theory holds that teeth arose phylogenetically from bony armor of primitive fishes, probably from surface denticles that rode inward as the invaginated stomodeum (ectoderm) moved inward to the buccal cavity with the newly developed jaws.

The presence of teeth, but not dermal armor, in conodonts complicates this picture of tooth evolution. Conodont teeth were mineralized, but it is not certain if the

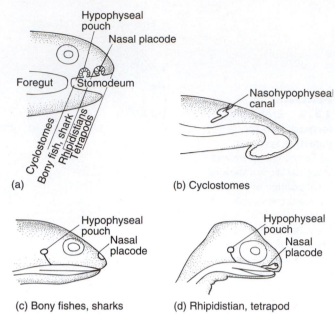

(a) Cyclostomes
Bony fish, shark
Rhipidistians
Tetrapods

Foregut   Stomodeum

(b) Cyclostomes

Nasohypophyseal canal

(c) Bony fishes, sharks

Hypophyseal pouch
Nasal placode

(d) Rhipidistian, tetrapod

Hypophyseal pouch
Nasal placode

FIGURE 13.4 **Boundaries of the buccal cavity.** The extent to which the embryonic stomodeum contributes to the mouth can be followed by two markers: the nasal placode and hypophyseal pouch. (a) Comparative positions of the anterior margin of the mouth in various groups of vertebrates. (b–d) Diagrammatic view of each group. These two markers remain outside the mouth in cyclostomes. In sharks and bony fishes, the hypophyseal pouch is pinched off from the mouth. In rhipidistians and tetrapods, both nasal placode and hypophyseal pouch open to or are derived from the mouth.

mineral was enamel. Further, some teeth are present within pharyngeal regions of many jawless fishes. This prompted the alternative view of an "inside-out" theory whereby teeth arose in the pharynx (endoderm) and subsequently progressed forward into the buccal cavity. The key to settling this may reside in discovery of the yet-unknown molecular programs that regulate tooth development.

Conodonts (p. 89)

Teeth help catch and hold prey. They also offer strong opposing surfaces that jaws work to crush hard shells of prey. In mammals and a few other vertebrates, mechanical digestion begins in the mouth. After each bite, the tongue and cheeks collect food and place it between the upper and lower tooth rows, and the teeth break down the bolus mechanically, reducing it to smaller chunks to make swallowing easier. By breaking the large bolus into many smaller pieces, chewing also increases the surface exposed to chemical digestion. Even in vertebrates that do not chew their food, sharp teeth puncture the surface of the prey, creating sites through which digestive enzymes penetrate when food reaches the alimentary canal. For vertebrates that feed on insects and other arthropods, punctures through the chitinous exoskeleton are especially important in giving proteolytic enzymes access to the digestible tissues within.

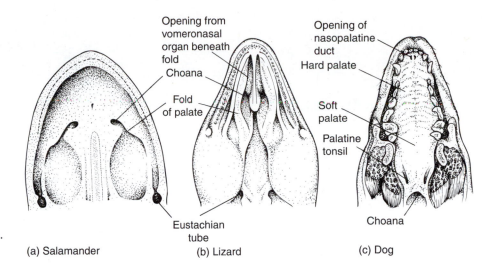

FIGURE 13.5 Roof or palate of the mouth in tetrapods.
(a) Salamander. (b) Lizard. (c) Mammal (dog). Notice the point of entry of the choana, or internal naris, in each animal.

After Romer and Parsons.

(a) Salamander   (b) Lizard   (c) Dog

Tooth Anatomy The part of the tooth projecting above the gum line, or **gingiva,** is the **crown;** the region below is the **base.** If the base fits into a hole, or **socket** (alveolus) within the jaw bone, the base is referred to as a **root.** Within the crown, the **pulp cavity** narrows when it enters the root, forming the **root canal,** and opens at the tip of the root as the **apical foramen.** Mucous connective tissue, or **pulp,** fills the pulp cavity and root canal to support blood vessels and nerves that enter the tooth via the apical foramen. The **occlusal surface** of the crown makes contact with opposing teeth. The **cusps** are tiny, raised peaks or ridges on the occlusal surface (figure 13.6a,b).

Three hard tissues compose the tooth: enamel, dentin, and cementum. **Enamel** is the hardest substance in the body and forms the surface of the tooth crown. Concentric rings seen under microscopic examination are believed to result from pulses of calcium salt deposits before tooth eruption, and no further enamel is deposited on the crown after the tooth erupts.

Dentin resembles bone in chemical composition but it is harder. It lies beneath the surface enamel and cementum and forms the walls of the pulp cavity. Even after tooth eruption, new dentin is laid down slowly throughout the life of an individual. Growth occurs by daily apposition along the walls of the pulp cavity, so that in very old animals, dentin may almost fill the entire cavity. The daily layers of dentinal growth are called the **incremental lines of von Ebner.**

Cementum, like bone, has cellular and acellular regions. Cementum rests upon the dentin and grows in layers on the surface of the roots. In many herbivores, cementum can extend up along the crown to between the enamel folds and actually contribute to the occlusal surface of high-crowned teeth. Cells within the cementum, termed **cementocytes,** elaborate the matrix but in seasonally related pulses, so that cementum increases irregularly with age. The result is the production of **cemental annuli,** concentric rings that characterize the cementum layer. The appearance of these annuli changes predictably with mechanical properties of food (hard), nutritional

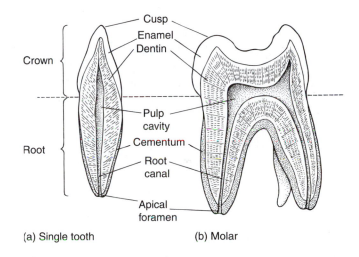

(a) Single tooth   (b) Molar

FIGURE 13.6 Tooth structure. (a) Tooth with single root. (b) Molar tooth with three roots.

state (lean times), and season (winter). In fossils, cementum appearance can thereby answer questions about diet and even the season of the animal's death.

The **periodontal membrane** (periodontal ligament) consists of thick bundles of collagenous fibers that connect the cementum-covered root to the bone of the socket.

In lower vertebrates, teeth are usually **homodont,** similar in general appearance throughout the mouth. Modern turtles and birds lack teeth altogether, but some tetrapods, especially mammals, have **heterodont** teeth that differ in general appearance throughout the mouth. Most lower vertebrates have **polyphyodont** dentition; that is, their teeth are continuously replaced. A polyphyodont pattern of replacement ensures rejuvenation of teeth if wear or breakage diminish their function. However, most mammals are **diphyodont,** with just two sets of teeth. The first set, the **deciduous dentition,** or "milk teeth," appears during early life. It consists of incisors, canines, and premolars, but no molars (figure 13.7a). As a mammal matures, these are shed and replaced by the **permanent dentition,** consisting

(a) Puppy

(b) Adult dog

FIGURE 13.7 **Deciduous (a) and permanent teeth (b) in a dog.** The carnassials (shaded teeth) are specialized carnivore teeth derived from the last premolar (upper) and first molar (lower).

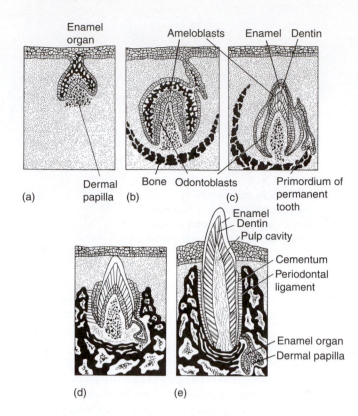

FIGURE 13.8 **Mammalian tooth development.** (a) Enamel organ (from the epidermis) and dermal papilla (from the dermis) appear. (b) Ameloblasts are the source of tooth enamel and form from the enamel organ. Odontoblasts are the source of dentin and form from the dermal papilla. Bone appears and begins to delineate the socket in which the tooth will reside. (c) The primordium of the permanent tooth appears. (d) Tooth growth continues. (e) The deciduous tooth erupts and is anchored in the socket by a well-established periodontal ligament. The enamel organ and dermal papilla of the permanent tooth primordium will not begin to form the tooth until shortly before the deciduous tooth is lost.

After Kardong.

of a second set of incisors, canines, and premolars and now molars, which have no deciduous predecessors (figure 13.7b).

Tooth Development Teeth are embryonic derivatives of the epidermis and dermis and develop beneath the surface of the skin initially. When mature, fully formed teeth **erupt** through the skin and extend into the buccal cavity. The epidermis produces the **enamel organ,** and mesenchyme cells of neural crest origin collect nearby within the dermis to produce the **dermal papilla** (figure 13.8a). Cells within the enamel organ form a specialized layer of **ameloblasts,** which secrete enamel. Cells within the dermal papilla form the **odontoblasts,** which secrete dentin (figure 13.8b). Thus, neural crest cells directly contribute to producing dentin through odontoblasts and indirectly induce the overlying ameloblast cells to deposit enamel. The crown of the tooth forms first, and then shortly before eruption, the root begins to develop (figure 13.8c,d). The cementum and periodontal ligament develop last.

Neural crest (p. 190)

In mammals, teeth erupt in a sequence, which is why our "wisdom" teeth come in later in life. Growth of the permanent tooth begins from a separate primordium of the enamel organ and dermal papilla that is usually adjacent to or deeper than the newly erupted deciduous tooth (figure 13.8e). Through similar steps, expansive growth of the permanent tooth against the roots of the deciduous tooth gradually cuts off its nutrition, causing resorption of the root and eventual loss of the deciduous tooth, which is replaced by the emerging permanent tooth. Finally, appearance of the cementum and periodontal ligament ensures firm attachment of the replacement dentition.

In rodents and lagomorphs, incisors and cheek teeth continue to grow from their roots as the tooth crowns are worn down. Enamel coats the convex side of the tooth, dentine, the concave side (figure 13.9a). Enamel, which is harder than dentine, wears away more slowly, leaving a self-sharpening edge. In elephants, molars also erupt sequentially

Jaws of gnathostomes evolved from gill arches of agnathan ancestors. In particular, jaws evolved from the internal branchial arches, providing a device that moved predatory jawed fishes to the top of the food chain and set the stage for an explosive radiation of early placoderms, chondrichthyans, acanthodians, and bony fishes. However, Jon Mallatt argues that the first steps to a jawed condition were driven not by predation but by the respiratory demands of increasingly active lifestyles.

In his view, the earliest vertebrates possessed unjointed internal and external branchial arches, and musculature encircling the pharynx (box figure 1a,b). When breathing, active contraction of the encircling muscles squeezed water across the gills (exhalation phase); then passive elastic recoil of the branchial arches enlarged the pharynx to draw water in through the mouth (inhalation phase). Subsequently, increasingly active lifestyles favored stronger ventilatory movement of water in and across the gills. In pregnathostomes, these demands were addressed by a strengthening of the encircling exhalatory muscles and the evolution of new inhalatory muscles (hypobranchials from some myotomes) that produced active expansion of the pharynx. The internal branchial arches, on which these more forceful muscles acted, also became more robust and now jointed as a lever system delivering active inhalatory and exhalatory movements of the pharynx. The first (mandibular) arch, which anchored the series of branchial muscles, and its musculature enlarged the most, forming a "ventilatory jaw." These ventilatory jaws opened the mouth wide during inhalation and closed it forcefully during exhalation (box figure 1c).

Although evolved for ventilation of the gills, this set the stage for further evolution, and the ventilatory jaws quickly took on a feeding function. Rapid inhalatory expansion of the pharynx produced a suction

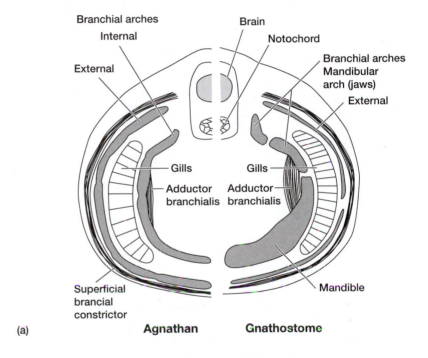

(a)   **Agnathan**   **Gnathostome**

BOX FIGURE I   **Evolution of gnathostome jaws and mouth.**
(a) Diagrammatic cross section of the pharynx. In agnathans (left), the gills lie between internal and external branchial arches; in gnathostomes (right), the jaws derive from enlarged internal branchial arches and their adductor muscles from adductor branchialis muscles.
(b) Ancestor. The agnathan ancestor is envisioned to have encircling muscles within the cheeks and pharynx. Unjointed external branchial arches ran between gill openings; unjointed internal branchial arches were present deep (not shown).
(c) Pregnathostome ("ventilatory jaws"). Increased participation in ventilation produces enlargement of internal branchial arches and of the associated adductor muscles. (d) Early gnathostome ("feeding jaws"). The enlarged mandibular arch, the first branchial arch, tilts forward and defines the new mouth.

(b) Ancestor

(c) Pregnathostome

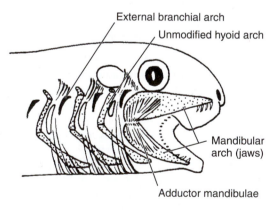

(d) Early gnathostome

BOX FIGURE I   *continued*

that could draw in prey animals. Rapid exhalation closed the jaws to grasp and bite the prey that had been inhaled. If these fish had previously fed on small, slow, benthic invertebrates, then the advent of suction feeding and a strong bite now allowed these newly jawed fishes to capture large, evasive, pelagic prey. Enlargement and strengthening of these structures followed as the "feeding jaws" became more prominent in capturing and grasping prey (box figure 1a,d).

In all fossil and living agnathans, a well-developed mouth (buccal cavity) lies anterior to the mandibular arch. This premandibular mouth has lips that define the oral opening and prominent, cheeklike walls. As they evolved into feeding devices, the jaws not only enlarged, but also tilted forward to a more favorable position where they could effectively grasp entering prey (box figure 1d). As this occurred, the cheeks of the premandibular mouth became considerably reduced, so as not to interfere with the mandibles during prey capture. As a result, in gnathostomes the jaws define the oral opening, and the buccal cavity lies immediately behind the jaws, in a postmandibular position.

The evolution of the new mouth (postmandibular) replacing the old mouth (premandibular) has left its traces in modern groups. Most modern jawed fishes suction feed by producing rapid expansion of the pharynx to inhale the prey. This is a remnant of the early ventilatory stage in the evolution of jaws, essentially an exaggerated ventilatory act now serving feeding. In the dogfish shark, lateral flaps of skin can be found tucked behind the corner of the mouth. These are parts of the side-cheeks of the premandibular mouth, now in front of the tooth-bearing mandibular arch. Mallatt's hypothesis of a "new" mouth in gnathostomes has sparked a fresh look at "old" issues.

Human speech is much more than loud grunts, at least when done well. Words are built up from carefully formed sounds called **phonemes.** By themselves, sounds have no meaning. Animal communication with sounds is mainly an emotional response to immediate circumstances. But to humans, phonemes in combinations carry ideas and thoughts about past events or future actions. We assign meaning to combinations of sounds rather than to individual sounds themselves. So functionally distinctive is our speech from the vocalizations of other vertebrates that some anthropologists mark the transition to *Homo sapiens* at the point in our ancestry where speech enters.

The relationships of sounds, not the sounds themselves, build words. And words placed in ordered sentences build an idea. But our speech apparatus can produce sounds quickly and shape them carefully only because it is redesigned. Anatomical changes serving speech centered on lengthening of the pharynx, which was accomplished by the separation of soft palate and epiglottis. By such lengthening, air can be effortlessly channeled on a sustained basis through the mouth, where it is shaped into sounds. Apes, with a short pharynx, must "bark" out sounds through short

BOX FIGURE I   **Human speech.** (a) In chimpanzees, the larynx is located high in the neck near the point at which it receives air via the internal naris. (b) In humans, the larynx is lower, serving to lengthen the pharynx used to produce speech sounds. However, this separates the larynx from easy connection with the air passageway, and air and food routes cross.

bursts of released air. Wolves can sustain a howl by lifting their heads, stretching their throats, and thereby temporarily lengthening their pharynx. Through control of its muscular walls, the redesigned pharynx of humans became the major vowel-producing chamber.

In nonhuman primates and in most other mammals, the larynx sits high in the neck and fits into the nasopharynx at the back of the nasal passage (box figure 1a). This establishes a direct route of air from nose to lungs without interfering with the food route from mouth to esophagus. In humans, the larynx has dropped down, lengthening the pharynx and forcing food and air routes to cross (box figure 1b). As in other mammals, you, as a human, cannot swallow and breathe at the same time. When a mix-up occurs, food headed for the esophagus becomes instead caught in the epiglottis, and you choke. To right matters quickly, residual air in the lungs can be forcibly expelled to shoot out the obstructing food. This is the basis of the Heimlich maneuver done on people who are choking.

FIGURE 13.9   **Specialized tooth growth in mammals.** In most mammals, molars do not grow nor are they replaced after they erupt. (a) One exception is found in rodents, whose incisor teeth continue to grow at their roots as their chisel-like crowns are worn away. The superficial bone is cut away to show the roots of upper and lower incisors. (b) In elephants, molar teeth erupt sequentially over a protracted period of time. A new molar that erupts at the back of the tooth row pushes older and worn molars to the front. (c) The superficial bone has been removed over the elephant's teeth to show their eruption and positional changes with age. Corresponding views of molar crowns (M_{1-6}) are shown at the bottom.

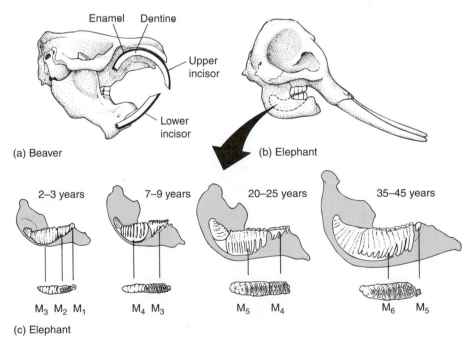

(a) Beaver

(b) Elephant

(c) Elephant

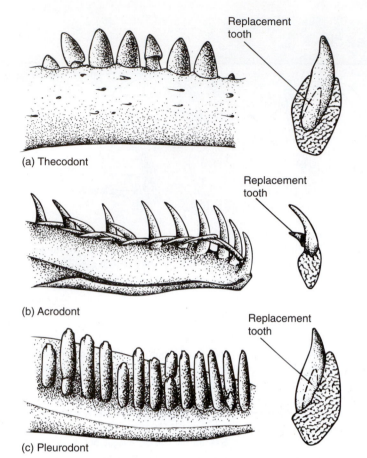

FIGURE 13.10   **Types of tooth attachment.**
(a) Thecodont teeth are set in sockets (alligator). (b) Acrodont teeth attach more or less on the occlusal surface of the bone (snake). (c) Pleurodont teeth attach to the side (lizard).

After Smith.

(a) Port Jackson shark

(b) Fossil teiid lizard

(c) *Sphenodon*

(d) Colubrid snake

(e) Alligator

FIGURE 13.11   **Heterodont dentition.** Heterodont dentition is most pronounced among mammals in whom distinct incisors, canines, premolars, and molars are discernible. However, among many ectotherms, tooth differentiation is also evident. (a) Lower jaw of Port Jackson shark. (b) Fossil teiid lizard. (c) *Sphenodon*. (d) Colubrid snake, a boomslang, exhibiting an enlarged set of grooved fangs on the posterior end of the maxilla. (e) Alligator with some large teeth.

After Smith; Kardong.

over a prolonged period of time. What is unique about elephants is that each hypsodont tooth is enlarged to the size of the entire erupted tooth row, and rotated in succession into the jaw as the one before it wears down. The newest molars erupt at the back of the jaws, and as they slowly emerge, they push older and worn molars to the front of the tooth row (figure 13.9b). Molars moving forward are sequentially worn down completely but are replaced from behind until very late in an elephant's life when the limited number of molars is spent (figure 13.9c). However, for most mammals, once the permanent teeth are in place, they are not replaced nor do they grow in length.

Specialized Teeth in Lower Vertebrates   Teeth are attached to supporting bones in three general ways. Archosaurian reptiles and mammals have **thecodont** teeth sunken into sockets within the bone (figure 13.10a). Other vertebrates exhibit an **acrodont** condition, with shallow sockets and teeth attached to the crest of the bone, or a **pleurodont** condition, with teeth attached to the medial side of the bone (figure 13.10b,c).

Among some herbivores and predators, teeth are often broadly flattened into anvil-like surfaces for crushing fibrous plant material or hard mollusc shells. The teeth of many teleost fishes form abrasive surfaces used to scrape encrusted algae from rocks, place it into suspension, and ingest it (figure 13.11a–e).

The oral cavity and its teeth also serve as a prey trap, an apparatus designed to snare unwary prey. Among most carnivores, teeth are simple sharp cones. They puncture the skin of the prey to give the jaws a firm grip on the captured and often still struggling animal. Skin is what engineers call a compliant material. Because of its great flexibility, or compliancy, it easily deforms or yields to attempts at puncture. To address this mechanical problem, the teeth of predators have pointed cusps to pierce or cut this compliant material. In addition, the teeth of some predators, such as sharks, have sharp, knifelike cutting edges along the sides of their teeth to help pierce skin. For slicing chunks from flesh, these edges are further serrated, like those on a bread knife, to cut the soft, compliant skin (figure 13.12a). The teeth of some rhipidistians and some early tetrapods have sharp, single cusps, and their enamel sides are complexly convoluted, inspiring the name *labyrinthodont* for such teeth. This infolded enamel produces surface ridges that may improve tooth penetration and strengthen the tooth internally (figure 13.12b).

In larval salamanders, most teeth are pointed cones, but the teeth of metamorphosed adults often show specializations. The crowns in some species are **bicuspid,** having two cusps,

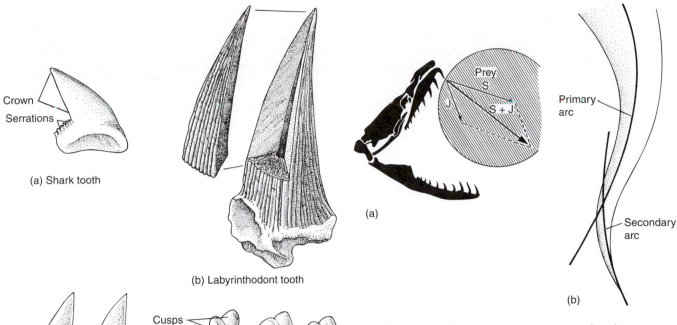

(a) Shark tooth

Crown
Serrations

(b) Labyrinthodont tooth

Cusps
Collagen fibers
Pedicel

Larva

Transformed adult

(c) Salamander teeth

FIGURE 13.12   Specializations of teeth.
(a) Shark tooth (blacknose shark, *Carcharhinus acronotus*). The pointed crown is nearly smooth edged for piercing prey; the base is serrated for cutting flesh. (b) Lateral view of a labyrinthodont tooth from a fossil amphibian. A wedge of tooth has been removed to show infolded enamel. (c) Teeth before metamorphosis (larval) and after metamorphosis (adult) in northwestern salamander (*Ambystoma gracile*). Larval teeth are pointed. Those of the transformed adult have divided cusps that sit on a basal pedicel. The cusps are thought to inflect with the struggling prey, thus resisting its escape from the mouth.

(b) After Owen.

and the crown itself sits upon a basal **pedicel** to which it is attached by collagenous fibers. When a tooth is replaced, the crown is lost and the pedicel quickly resorbed, leading some to argue that such pedicelated teeth represent a way for rapid tooth replacement. However, the main advantage of this design is that it aids in grasping prey. The "joint" formed between crown and pedicel allows the tooth tip to bend inward but not outward. Thus, when a struggling prey is advanced farther into the mouth of a salamander, the tooth tips relax and bend in the same direction, encouraging movement toward the throat. Because these tooth tips are slanted into the buccal cavity, they resist prey outward escape (figure 13.12c).

Like the teeth of most carnivorous reptiles, snake teeth generally taper to a sharply pointed cusp that

Prey
S
J
S + J

(a)

Primary arc

Secondary arc

(b)

FIGURE 13.13   Reverse curvature of snake teeth. (a) When a snake launches its head at and closes its jaws on prey, two forces are transmitted through the tip of the anterior teeth. These component forces are represented here by vectors. One vector represents the force arising from the forward momentum of the skull (S). The other represents the force of jaw closure (J). The resultant force at impact is S + J. (b) The forward inclination of the tooth's tip (secondary arc) relative to its base (primary arc) may bring the tip into closer coincidence with the line of this resultant force upon impact. The primary arc of the tooth's base orients the tooth posteriorly. When the snake swallows prey, this backward slant resists escape of the prey out of the mouth. During the strike, the reverse secondary arc helps the tooth penetrate the surface of the prey.

Based on the research of T. H. Frazzetta.

penetrates skin and gives it a firm hold on the prey. Some snake teeth are specialized and bear a bladelike edge or low ridges along their sides thought to aid in tooth penetration. When a snake strikes, it brings its mouth over the prey quickly. The series of needlelike teeth form a prickly surface that easily snags the surface of the prey. Teeth at the front of the snake's mouth are often **reverse,** with the tip inclined forward from the rest of the tooth (figure 13.13a,b). This gives the tooth a major posterior bend at its base and a forward slant at its tip. Forward inclination of the cusp means that during the strike, the sharp tip is brought more in line with the snake's line of approach to the prey. Alignment of teeth tips with the prey facilitates skin puncture upon impact. The posterior-directed bend at the base of the tooth works to hold the prey and facilitate its swallowing. If the prey should pull back in an attempt to escape, the teeth sink more deeply and securely into the skin because of their backward slant. Reverse-curved teeth are found in other vertebrates, such as sharks, and presumably function in a similar fashion. The cusp penetrates on impact and the base holds the struggling prey.

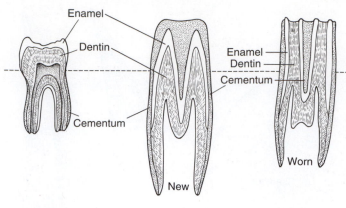

Incisor

Canine

Enamel

Dentin

Cementum

Enamel
Dentin
Cementum

Worn

New

(a) Brachydont tooth   (b) Hypsodont tooth

FIGURE 13.14   Skull of the mule deer (*Odocoileus hemionus*) with lower jaw lowered. The dental formula for the upper row is 0-0-3-3, and the formula for the lower is 3-1-3-3. The absence of upper incisors and canine is normal, indicated in the dental formula by zeros. The lower incisors and canine are present, and the canine is adjacent to the incisors at the front of the mandible.

Enamel

Enamel   Dentin

Enamel
Dentin

(c) Bunodont tooth   (d) Lophodont tooth   (e) Selenodont tooth

FIGURE 13.15   Crown height and occlusal surfaces. Tooth height: (a) Brachydont tooth. (b) Hypsodont tooth. When the occlusal surface of a newly erupted hypsodont tooth (left) becomes worn, alternating layers of dentin and enamel are exposed (right). The alternating layers of varying hardness ensure that ridges and depressions will form, producing a rough surface that does not become smooth even after prolonged use. Note that teeth continue to erupt (move above the gum line) as they get worn down from the top. In mammalian teeth, various occlusal surfaces occur: (c) Bunodont tooth (e.g., pigs, primates). (d) Lophodont tooth (e.g., horses, rhinoceros). (e) Selenodont tooth (e.g., camels, deer).

(a,b) After C. Janis. (d) Halasey

Maxillary teeth of some snakes have open grooves down which oral secretions flow during feeding. In venomous snakes, the edges of these grooves fuse, forming a hollow channel down the core of the tooth through which venom passes from the venom duct into the prey. The term **fang** is appropriate for such a hollow tooth modified for venom delivery. Being a modified tooth, the snake fang, like other teeth, is part of a polyphyodont system and is replaced on a regular basis. Thus, artificial removal of the fang will not render a venomous snake permanently "harmless," because within a day or less, a replacement tooth takes its place.

Specialized Teeth in Mammals   In mammals, the teeth not only capture or clip food but are specialized to chew it, producing a complex and distinctive dentition. In fact, the dentition in different groups is so distinctive that it is often the basis for identifying living animals and fossil species. Not surprisingly, an elaborate terminology has grown up to describe the particular features of mammalian teeth.

The heterodont dentition of mammals includes four types of teeth within the mouth: **incisors** at the front, **canines** next to them, **premolars** along the sides of the mouth, and **molars** at the back. The number of each type differs among groups of mammals. The **dental formula** is a shorthand expression of the number of each kind of tooth on one side of the head for a taxonomic group. For example, the dental formula of the coyote *(Canis latrans)* is:

I 3/3, C 1/1, PM 4/4, M 2/3

This means that there are three upper and three lower incisors (I), one upper and one lower canine (C), four upper and four lower premolars (PM), and two upper and three lower molars (M), 21 per side or 42 total. Sometimes the dental formula is written as 3-1-4-2/3-1-4-3, the first four

numbers indicating the upper teeth and the second four the lower teeth of the coyote. The dental formula for the mule deer (*Odocoileus hemionus*) is 0-0-3-3/3-1-3-3. Notice that the missing upper incisors and canines are indicated by zeros (figure 13.14).

Generally, incisors at the front of the mouth are used for cutting or clipping; canines, for puncturing or holding; premolars and molars, for crushing or grinding food. The collective term embracing both premolars and molars is **cheek teeth,** or **molariform teeth.** In humans and pigs, crowns are low, or **brachydont** (figure 13.15a). In horses, crowns are high, or **hypsodont** (figure 13.15b). If the cusps form rounded peaks, as in omnivores, the teeth are **bunodont** (figure 13.15c). Cusps drawn out into predominantly straight ridges, as in perissodactyls and rodents, produce **lophodont** teeth (figure 13.15d). Crescent-shaped cusps, as in artiodactyls, characterize **selenodont** teeth (figure 13.15e). Hypsodont teeth are typically found in herbivores that grind plant material to break tough cell walls, especially animals such as grazers where the silica particles in the grasses make their diet more abrasive. Their

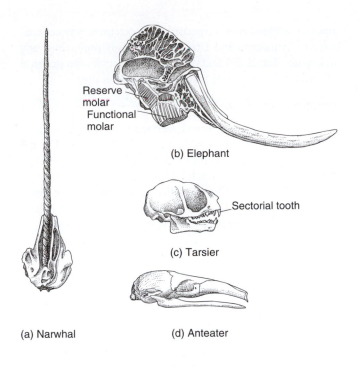

Reserve molar
Functional molar

(b) Elephant

Sectorial tooth

(c) Tarsier

(a) Narwhal

(d) Anteater

(e) Walrus

(f) Porpoise

(g) Right whale

FIGURE 13.16 **Specialized mammalian teeth.**
Lateral and frontal views of tusks. Tusks arise from the upper left incisor in the narwhal (a), from both upper incisors in the elephant (b), and from canines in walruses (e). Sectorial teeth in the primate tarsier (c) and peglike teeth of a porpoise (f) are shown. Teeth are absent in adult anteaters (d) and baleen whales (g).

(a,b,d) After Smith.

occlusal surface is worn unevenly because the minerals that form the surface—enamel, dentin, and cementum—differ in hardness. Occlusal surfaces are functionally important because they ensure that ridges and depressions persist throughout life, thereby maintaining a rough grinding surface that does not become smooth with continued use (figure 13.15b).

Mammals possess a variety of specialized teeth. In some, **sectorial teeth** are modified so that ridges on opposing teeth slice by one another to cut tissue. In some primates, cutting edges form on the upper canine and lower first premolar, the sectorial teeth. These teeth are deployed in fights between individuals or in defense. In the order Carnivora, the upper fourth premolar and lower first molar form **carnassials**, specialized sectorial teeth that slice against each other like scissors to cut sinew and muscle. **Tusks** arise from different teeth in different species. The single 3-m spiral tusk of the narwhal is the left upper incisor (figure 13.16a). Speculation of the tusk's function runs from spearing fish, to punching holes in polar ice, to stirring the ocean bottom to scare up buried food. All males, but only a few females, bear the tusk, suggesting rather that it is a secondary sexual characteristic used in male-male and/or courtship display. In elephants, the paired tusks are elongate incisors (figure 13.16b), and in walruses, the paired tusks are upper canines that protrude downward (figure 13.16e). In carnivorous mammals, canine teeth together with powerful jaws are used to kill prey. Sometimes these teeth puncture major blood vessels in the neck, causing the victim to bleed profusely and weaken. A practiced carnivore, such as an adult lion, is more likely to bite into the neck and collapse the trachea to suffocate its prey. Some mammals, such as anteaters and baleen whales, lack teeth altogether (figure 13.16d,g).

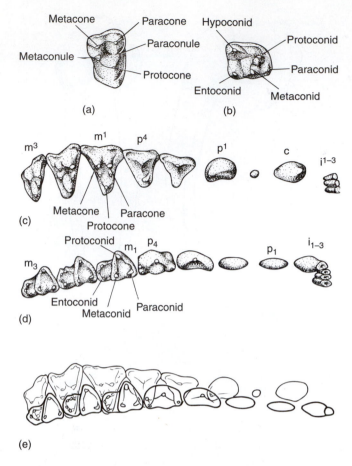

FIGURE 13.17 Molar patterns of placental mammals. (a) Upper right molar. (b) Lower left molar. (c) Occlusal view of right upper teeth. (d) Occlusal view of right lower teeth, same species as (c). (e) Upper and lower tooth rows placed in occlusion, with outlines of lower teeth (heavy outlines) superimposed on those of upper. Teeth include canine (c), incisor (i), molar (m), premolar (p) accompanied by their number within the upper (superscript) or lower (subscript) tooth row.

After Romer and Parsons.

The cusp patterns of mammalian cheek teeth are so distinctive that they are used in species identification. Cusps are termed **cones.** Major cones are identified by adding the prefixes *proto-, para-, meta-, hypo-,* or *ento-;* minor cusps are indicated by the suffix *-ul(e).* For connoisseurs, the terminology goes on: **Cingulum** is used for accessory enamel ridges on the margins of the crown, *-loph* denotes ridges across the crown connecting cusps, *-cone* signifies cusps on upper teeth, and *-conid* signifies cones on lower teeth. For example, meta*cone* and meta*conid* denote the same cusps, but on upper and lower teeth, respectively (figure 13.17a–e).

Much of this terminology was inspired by late-nineteenth-century paleontologists, who proposed that the three cusps on each molar of ancestral cynodonts spread across the expanded crown of mammalian descendants to become the paracone (paraconid), protocone (protoconid), and metacone (metaconid) of upper (and lower) molars. Other

cones were added later. Expecting such exactness between tiny cusps of cynodonts and later mammals was probably overly optimistic. But, it did and continues to provide one practical technique to characterize mammals taxonomically.

Tongue

Cyclostomes possess a tongue derived not from hypobranchial musculature, but from the floor of the pharynx. During feeding, cyclostomes protrude this soft tongue, bearing ridges of rasping keratinized "teeth." However, most gnathostome fishes lack a tongue. Occasionally, teeth borne on the lower ends of gill bars can be worked against those of the palate, but a fleshy, muscularized tongue is not usually present. A mobile tongue develops first in tetrapods from the hypobranchial musculature attached to and resting on the underlying **hyoid apparatus,** a skeletal derivative of modified lower ends of the hyoid arch and adjacent branchial arches.

The tongue of many tetrapods holds the **taste buds,** sensory organs responsive to chemicals entering the mouth. The **vomeronasal organ** (Jacobson's organ) is found within many tetrapods, where it is implicated primarily in the detection of pheromone signals (social chemicals of communication). Delivery of chemicals is usually from the mouth to the vomeronasal organ. This may be augmented by tongue action, as for example tongue flicking in lizards and snakes, which project their foretongue out of the mouth to retrieve air and/or substrate chemicals for evaluation by the vomeronasal organ. In many mammalian carnivores, the surface of the tongue is roughened, like a file, with numerous keratinous spiny projections, or **filiform papillae,** that help rasp flesh from bones.

Vomeronasal organ (p. 677)

Many tetrapods use their tongue in **lingual feeding.** They protract their tongue out of the mouth at prey. Its sticky surface holds the catch until jaws advance or tongue retraction brings it back into the mouth. Many terrestrial salamanders and lizards use this technique. In fact, some argue that development of just such a mobile and projectile tongue represents a major feeding innovation in the transition of early tetrapods to life on land. Woodpeckers use their long specialized tongues like a probe to obtain insects between cracks in tree bark or in holes they manufacture (figure 13.18a–c).

The tetrapod tongue can also transport captured prey, that is, move prey through the buccal cavity to the back of the pharynx where it is swallowed (figure 13.19a). This process is known as **intraoral transport** and proceeds in several steps. First, the jaws part slowly, and the tongue is advanced forward and beneath the food to fit partially around it (figure 13.19b). Second, the jaws open more rapidly, and the tongue draws the adhering food backward in the mouth (figure 13.19c). Third, the jaws close slowly and grip the more posteriorly positioned food, and the tongue comes away from its adhesive hold on the food. The cycle is repeated, with the tongue working in synchrony with the jaws to move the food in increments to the back of the buccal cavity and into the pharynx (figure 13.19d).

FIGURE 13.18 **Tongue protrusion in a woodpecker.** (a) The flexible and thin hyoid apparatus supports the fleshy tongue. (b) When the woodpecker protrudes its tongue, the hyoid apparatus slips forward and the tongue extends from it. (c) Ventral view of the hyoid apparatus from the woodpecker *Picus*.

(a,b) After Smith; (c) after Owen.

The tongue's grip on food during transport depends partially on its surface irregularities that interlock or physically engage the prey. The ability of the fleshy tongue to shape itself to the food may help it gain a physical grip on the prey. This tongue adhesion also depends on a "wet" adhesion, the sticky effects created by surface tension in air and by capillary action. Where tetrapods have returned to aquatic feeding, these physical phenomena are less effective. This may account for the much-simplified tongue devoid of intrinsic musculature found, for example, in crocodilians. Many aquatic tetrapods such as aquatic feeding turtles resort to modified suction feeding in which the tongue has little role to play. In fact, if the tongue of aquatic tetrapods were a large, fleshy structure occupying the floor of the buccal cavity, it could interfere with the sudden expansion of the buccal cavity necessary for suction feeding.

Pharynx

In adults, the pharynx is little more than a corridor for passage of food and air. But, phylogenetically it is the source of many organs, and developmentally its history is complex (table 13.1). During embryonic development, the stomodeum (ectoderm) opens into the pharynx, but the pharynx itself forms from the anterior foregut (endoderm) and is relatively prominent compared with the rest of the still-forming digestive tract. A series of bays, or **pharyngeal pouches,** form on its lateral walls and grow out to meet inpocketings of the skin ectoderm,

FIGURE 13.19 **Intraoral transport of food in a generalized tetrapod.** (a) At rest. (b) Slow opening. The jaws begin to open slowly, and the tongue moves forward to make contact with the food, after which it becomes fitted to the food. (c) Fast opening. The jaws part, and the retracting tongue transports the food to the back of the mouth. (d) Closing. The jaws close to establish a purchase on the food, after which the tongue comes free of the food. If an animal repeats this sequence, food is moved successively to the back of the mouth and into the pharynx, where it is swallowed into the esophagus.

Based on the research of K. Schwenk, D. Bramble, and D. Wake.

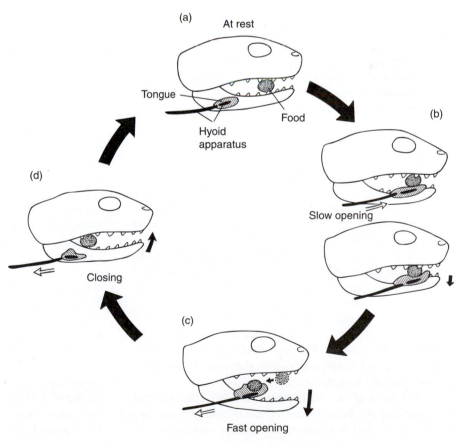

TABLE 13.1   Pharyngeal Pouch Derivatives in Vertebrates

Pharyngeal Pouch	Position	Lamprey	Elasmobranch	Urodele	Anuran	Reptile	Bird	Mammal
1	Dorsal	Thymus	Spiracle	Tubotympanic recess[a]	Tubotympanic recess[a]	Tubotympanic recess[a]	Tubotympanic recess[a]	Tubotympanic recess[a]
	Ventral	Branchial pouch	—	—				
2	Dorsal	Thymus	Thymus	—	Thymus	Thymus[1]	—	Tonsil (Palatine)
	Ventral	Branchial pouch	—	—	—	Parathyroid	—	—
3	Dorsal	Thymus	Thymus	Thymus	—	Thymus[1,2]	Thymus	Parathyroid
	Ventral	Branchial pouch	—	Parathyroid	Parathyroid	Parathyroid	Parathyroid	Thymus
4	Dorsal	Thymus	Thymus	Thymus	—	Thymus[2,3]	Thymus	Parathyroid
	Ventral	Branchial pouch	—	Parathyroid	Parathyroid	Parathyroid	Parathyroid	Thymus
5	Dorsal	Thymus	Thymus	Thymus	—	Thymus[3]	—	—
	Ventral	Branchial pouch	Ultimobranchial body	Ultimobranchial body	Ultimobranchial body	Ultimobranchial body	Ultimobranchial body	Ultimobranchial body

[a]Auditory cavity of middle ear and eustachian tube
[1]Lizard
[2]Turtle
[3]Snake

termed **branchial grooves.** At their point of contact, these pouches and grooves establish a partition, or **closing plate,** between them. In fishes and larval amphibians, closing plates are perforated to form the functional gill slits. In other vertebrates, these closing plates fail to rupture, or if they do, they are soon sealed over so that functional gill slits do not develop.

The subsequent contribution of the embryonic pharynx to adult structures is staggering. In mammals, the first pharyngeal pouch expands into an elongated **tubotympanic recess** that envelops the middle ear bones, giving rise to the narrow eustachian tube and part of the tympanic cavity. The second pharyngeal pouch gives rise to the palatine tonsil. The third and fourth pouches contribute to the parathyroid. The fifth pharyngeal pouch gives rise to what are termed the ultimobranchial bodies. The ultimobranchial bodies are separate glands in fishes, amphibians, reptiles, and birds, but in mammals, they become part of the thyroid gland and apparently form its internal population of C cells. C cells are involved in controlling blood calcium levels. All pouches contribute to the thymus in fishes, variable numbers contribute in amphibians, and pharyngeal pouches III and IV contribute in mammals. The roof of the pharynx gives rise to the pharyngeal tonsil; the floor, to the thyroid, part of the tongue, lingual tonsil, and lung primordium.

Swallowing, or **deglutition,** involves the forceful movement of the bolus from the mouth and pharynx into the esophagus and then the stomach. Most vertebrates bolt

FIGURE 13.20   A snake swallowing prey. A snake steps its jaws alternately left and right along its prey, "walking" them along its surface until the prey reaches the back of the snake's throat. Smooth muscles in the wall of the esophagus aided by striated muscles in the lateral body wall move the prey toward the stomach. While the prey is in the snake's mouth, the trachea slips beneath it and forward to maintain an open route through which the snake breathes even while swallowing.

After Kardong.

their food, swallow it whole, and the esophagus expands to accommodate food size. Seabirds catch fishes in their bills and toss them to the back of the throat. The esophagus becomes distended as food enters it. Contractions of muscles within its walls squeeze the fish along and into the stomach. Snakes work their flexibly articulated jaw bones over the prey, drawing the sides of their mouth over the prey to engulf it. When food enters the esophagus, waves of contraction in its walls and general neck movements force the bolus along to the stomach (figure 13.20).

Saber-tooth "Cats"

In some mammals, the upper canines evolved into curved, saberlike teeth. This occurred independently on four occasions: three times within placentals—once in primitive carnivores, the creodonts, and twice in Carnivora, the fossil nimravids and the felids (cats) (box figure 1)—and one time within a Pliocene marsupial family (Thylacosmilidae). All saber-tooth mammals are extinct, but fossil canines from saber-tooth carnivores are well known. Their canine teeth were long and curved, and the posterior edge of the blade carried a faint serration. One saber-tooth nimravid bore evidence of a stab wound inflicted by another saber tooth. A fossil dire wolf was found with part of the saber tooth from the felid *Smilodon* embedded in its skull. Modern carnivores feed mostly on herbivores and only rarely dine on each other. Thus, fossil evidence of saber-tooth attacks on other carnivores likely represents a cat defending its kill from scavenging or marauding carnivore competitors seeking to steal the dispatched carcass.

How these canine sabers were deployed during feeding is not well understood. Saber-tooth cats could open their jaws wide, but these jaws were not strong enough to rip out a large mouthful of prey. It seems more likely that the sabers made slashing wounds that bled profusely, but they were not used to rend large sections of flesh from the bodies of the prey.

Why, then, don't cats today possess similar saber teeth to serve them in dispatching prey? Tigers, lions, cougars, and smaller cats make rapid dashes from ambush and use claws to grasp and control prey while they bite into its neck. Biting inflicts puncture wounds and clamps

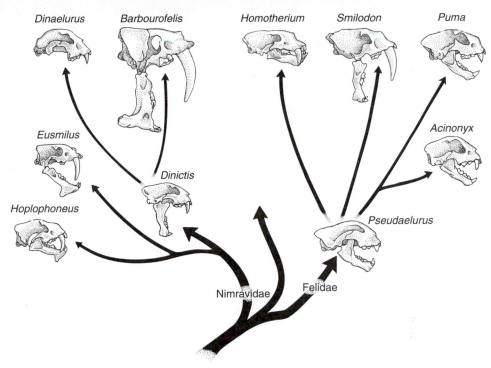

BOX FIGURE 1 **Possible phylogeny of catlike Carnivora.** One major eutherian branch produced the Nimravidae, the second, the Felidae. Saber tooths also evolved independently in creodonts, an extinct group of eutherian Carnivoria, and in marsupials, for a total of four origins.

After Martin.

off the trachea, suffocating prey. It is unknown how saber teeth would have been advantageous if hunting strategies were similar. Alternatively, some suggest that teeth were different because prey were different and required a different hunting strategy. If saber-tooth mammals fed on large ground sloths or other large, slow herbivores, then this type of prey might have presented problems different from the swift herbivores that most large cats prey on today. Without a living saber-tooth cat as reference, the special function of these teeth is difficult to determine. As of yet, there is no consensus. But, if saber teeth represent a specialization for specialized prey, then the absence of large, slow herbivores today might also account for the absence of saber-tooth predators as well.

Three temporary seals form as most mammals chew and swallow their food. The anterior oral seal is formed by the lips. The middle oral seal develops between the soft palate and back of the tongue. The third posterior oral seal occurs between the soft palate and the **epiglottis.** As an animal chews, food tends to gather temporarily in the **vallecula,** the space in front of the epiglottis, and the **pyriform recess,** the passageways around either side of the **larynx.** The epiglottis sits above the larynx, and the trachea sits below it. When the animal swallows, the back and sides of the tongue expand against the soft palate

(you can notice this yourself when you are eating), forcing food out of the vallecula, through the pyriform recess, and into the esophagus. The **glottis** is a muscular slit that closes momentarily across the larynx to prevent inadvertent aspiration of food into the trachea and lungs. In most mammals, the posterior seal (soft palate–epiglottis) is in place during swallowing to ensure passage of food into the esophagus without blocking the air passage (figure 13.21a). In human infants, the posterior seal directs milk to the esophagus, but in adults, this seal is lost because the pharynx descends in order to accommodate the onset of speech.

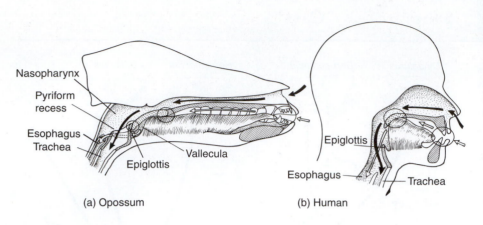

FIGURE 13.21 **Food and air passages.** (a) Sagittal view of an opossum head showing three oral seals: anterior (lips), middle (soft palate and tongue), and posterior (back of tongue and epiglottis). Air (solid arrows) flows directly from the nasal passages into the trachea. Food (open arrows) passes around the sides of the larynx to reach the esophagus. (b) Sagittal view of a human head. The posterior oral seal is absent because the pharynx drops lower in the neck to accommodate sound production for speech. Thus, food and air potentially cross in the lengthened human pharynx.

After Hiiemae and Crompton.

Labels for (a) Opossum: Nasopharynx, Pyriform recess, Esophagus, Trachea, Epiglottis, Vallecula

Labels for (b) Human: Epiglottis, Esophagus, Trachea

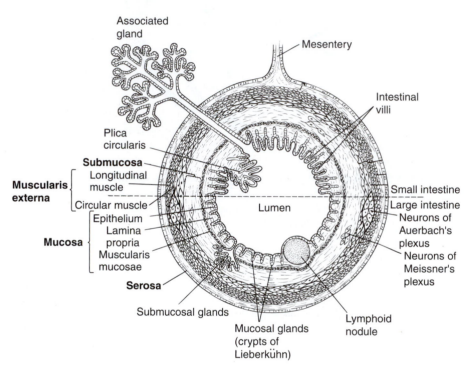

FIGURE 13.22 **General organization of the alimentary canal.** The concentric layers of mucosa, submucosa, muscularis externa, and serosa (or adventitia) are common to all regions of the alimentary canal. Within the mucosa are folds or intestinal villi. Lymphoid tissue occurs throughout, although it may form discrete nodules within the mucosa. Glands may occur within the mucosa, submucosa, or even outside the digestive tube.

After Bloom and Fawcett.

Labels for Figure 13.22: Associated gland, Mesentery, Intestinal villi, Plica circularis, Submucosa, Muscularis externa, Longitudinal muscle, Circular muscle, Mucosa, Epithelium, Lamina propria, Muscularis mucosae, Serosa, Lumen, Small intestine, Large intestine, Neurons of Auerbach's plexus, Neurons of Meissner's plexus, Submucosal glands, Mucosal glands (crypts of Lieberkühn), Lymphoid nodule

Adult humans rely on the middle oral seal (soft palate–back of tongue) to keep food and air passages separate during swallowing (figure 13.21b).

Alimentary Canal

In some vertebrates, **digestion** begins in the buccal cavity. But we get down to the serious business of food processing in the alimentary canal, which encompasses further breakdown of the bolus, absorption of its available constituents, and elimination of indigestible remnants. The design of the alimentary canal suits the diet of the organism. Because diet can differ even between related groups, alimentary canals can differ significantly between phylogenetically related vertebrates. Most vertebrates have an alimentary canal made up of an esophagus, stomach, intestines, and cloaca. As distinctive as these might appear, all share an underlying unity of design (figure 13.22).

Each region is built on a common plan of organization, namely, a hollow tube with walls composed of four layers. The innermost layer is the **mucosa,** which includes the epithelium that lines the lumen, the thin smooth muscle fibers of the muscularis mucosae, and the region of loose connective tissue, the lamina propria between the epithelial lining and the muscularis mucosae. The **submucosa,** consisting of loose connective tissue and nerve plexes of the autonomic nervous system, forms the second layer of the digestive tract. Outside this layer lies the **muscularis externa,** composed of circular and longitudinal sheets of smooth muscle. The surface layer is the **adventitia,** consisting of fibrous connective tissue.

If a mesentery envelops the alimentary canal, this outer layer of connective tissue plus mesentery is called the **serosa.**

During embryonic development, the endoderm gives rise to the lining of the gut, and the surrounding mesoderm forms smooth muscles, connective tissue, and blood vessels. In most tetrapods, a series of positional changes transform the relatively straight gut of the embryo into the coiled digestive tube of the fetus (figure 13.23a–c). First, the primary gut loop forms (figure 13.23d,e), a large bend in what was up to this point a straight gut. Next, rapid elongation of the gut twists this loop into the first major coil (figure 13.23f). Thereafter, continued elongation and coiling produce the compact digestive tube, and distinctive regions within it become delineated (figure 13.23g). The end result of embryonic twisting, growth, and looping of the gut is to solve a packing problem—stuffing the long, active gut into a confined space, the body cavity. Remarkably, this long tube is packed into a small space in such a way as to avoid kinking the coiled gut that, like kinking a garden hose, would interrupt flow through the tube. When food arrives, the section of the gut becomes active mechanically (churning, peristalsis) and chemically (digestive secretions) as it is suspended by and controlled in its activity by the mesenteries.

Esophagus

The esophagus connects the pharynx with the stomach. It is a slender tube that easily becomes distended to accommodate even a large bolus of food. Mucus, to aid in the passage of food, is often secreted, but the esophagus seldom produces enzymes that contribute to chemical digestion. In some vertebrates, the esophageal mucosa is lined with ciliated cells that control the flow of lubricating mucus around the food. The ciliated epithelium may also help gather small crumbs from the meal and move these along to the stomach. In others, the mucosa is stratified epithelium; this may even be keratinized in animals ingesting rough or abrasive foods. In vertebrates that swallow large quantities of food at once, the esophagus serves as a site of temporary storage until the rest of the alimentary canal begins digestion. Anteriorly, muscle coats tend to be composed of striated muscle that becomes replaced posteriorly by smooth muscle.

Stomach

The esophagus delivers the bolus of food to the stomach, an expanded region of the alimentary canal. Stomachs are absent in cyclostomes and in protochordates, except for some urochordates in which a stomach is present to receive mucus that is laden with collected food particles from the branchial basket. Animals that take in large quantities of food on an irregular basis, such as many carnivores, have stomachs that serve as storage compartments until the processes of mechanical and chemical digestion catch up. Such food storage may have been an initial function of the stomach when early vertebrates evolved from suspension feeding to feeding on larger chunks of food. Hydrochloric acid produced by the stomach may have functioned to retard food putrefaction by bacteria, thus preserving it until digestion was under way. In most vertebrates, the stomach performs an expanded role. Some absorption of water, salts, and vitamins occurs in the stomach, but predominantly it serves to churn and mix food mechanically and add digestive chemicals collectively called **gastric juice.** Gastric juice includes some enzymes and mucus but is primarily composed of hydrochloric acid released from the mucosal wall of the stomach.

The stomach's expanded size sets it apart from the narrow esophagus that enters it and the small intestine into which it empties. When not distended with food, the internal stomach wall relaxes into folds known as **rugae,** which also help delineate its boundaries (figure 13.24). However, gross external morphology does not always reliably mark internal differences in structure of the mucosal wall. Consequently, the histological character of the mucosal wall is often used to distinguish important functional regions within the stomach.

On the basis of mucosal histology, two regions of the stomach can be distinguished. The stomach's **glandular epithelium** is characterized by the presence of **gastric glands.** These are branched, tubular glands, several of which empty into the bases of surface indentations, or **gastric pits.** There are three divisions of the stomach—cardia, fundus, and pylorus—based on the relative position and type of gastric gland. The **cardia** is a very narrow region found only in mammals, and it marks the transition between the esophagus and the stomach. Its gastric glands, termed **cardiac glands,** are composed predominantly of mucus-secreting cells. The **fundus** is usually the largest region of the stomach and contains its most important gastric glands, the **fundic glands.** Mucous cells are present in fundic glands, but these glands in mammals are distinguished by their abundance of **parietal cells,** the source of hydrochloric acid, and **chief cells,** the presumed source of several proteolytic enzymes. Other vertebrates possess instead oxyntopeptic cells that produce both HCl and pepsinogen. Upon release into the stomach's lumen, pepsinogen is cleaved by HCl to produce pepsin, an active proteolytic enzyme. Before emptying into the intestine, the stomach usually narrows into a **pylorus,** whose mucosal walls hold distinct gastric glands called **pyloric glands.** The pyloric glands are predominantly composed of mucous cells whose secretions help to neutralize the acidic chyme as it moves next into the intestine. Thus, most of the chemical and mechanical processes of gastric digestion occur in the fundus. The cardia (when present) and pylorus add mucus. Smooth muscle bands in their walls act as sphincters to prevent the retrograde transfer of food (figure 13.24).

In addition to a region of glandular epithelium, the stomach of some vertebrates also has a second region characterized by **nonglandular epithelium,** devoid of gastric glands. As in some herbivores, the nonglandular region may develop from the base of the esophagus. In other species, such as rodents, loss of gastric glands in the mucosa leaves a nonglandular epithelial stomach in which smooth muscle contractions knead and mix digesta. This nonglandular

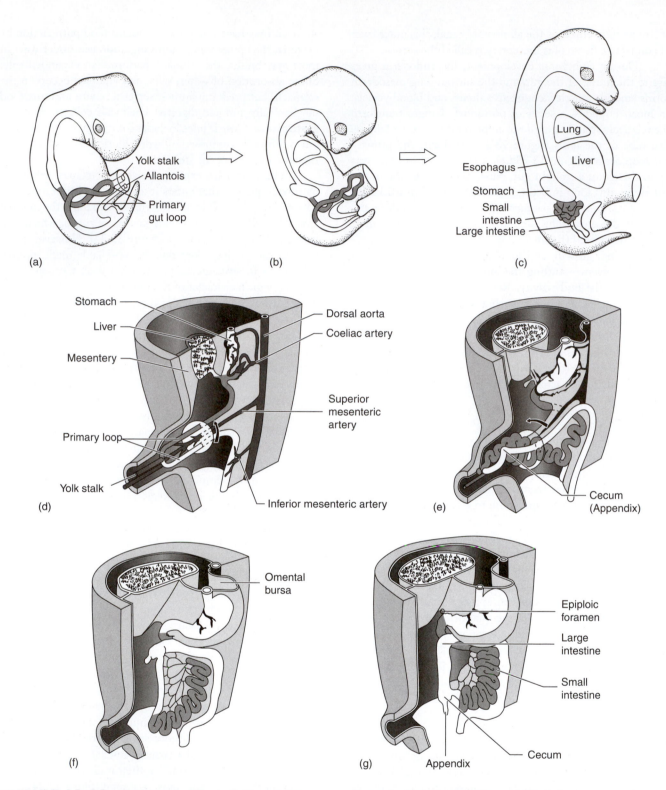

FIGURE 13.23 **Embryonic differentiation of the mammalian gut.** (a–c) The intestines coil (shaded), elongate, and become tucked behind the stomach within the body cavity. Left side view, cutaway (d–g). (d) Differentiation begins with formation of the primary loop. Note the three primary arteries branching from the dorsal aorta to supply different portions. (e) The allantoic stalk, which marks the point between large and small intestines, regresses. Growth, elongation, and coiling of the digestive track are under way, and continue (f). Note the secondary looping (arrows). (g) The result is a long but tightly coiled intestine and differentiated stomach.

After Moor.

Fundus

Right gastric
artery and vein

Rugae

Left gastric
artery and vein

Smooth muscle
layers

Antrum

Sphincter Canal

Pylorus

Gastric pit

Gastric area

Gastric pits

Necks of gastric pits

Fundic gland

Muscularis
mucosae

Parietal cell

Submucosa

Chief cell

Smooth
muscle layer

Muscularis
mucosae

FIGURE 13.24 **Anatomy of the stomach.** Up to three regions are usually discernible in the stomach. The largest of these is the fundic region. Gastric pits open into fundic glands having parietal and chief cells at their bases. The two other regions of the stomach are the cardia and the pylorus, which contain cardiac and pyloric glands, respectively, at the bases of their gastric pits. Various types of mucous cells predominate in these glands.

epithelium in rodents also can be keratinized, perhaps as a result of mechanical abrasion from rough foods such as seeds, grasses, and insect chitinous exoskeletons. Chemical insult from digestive enzymes added in the mouth may also cause a keratinized nonglandular epithelium.

Intestines

The mucosa of the intestines is distinctive. First, it contains an epithelium whose free surface facing the lumen has many **microvilli,** perhaps up to several thousand per cell. These tiny, fingerlike projections of the apical surface substantially increase the overall absorptive surface area of the alimentary canal. Their surfaces also seem to harbor a microenvironment away from the large central lumen, where digestive enzymes can more favorably act upon food. Most digestive enzymes are bound to these microvilli and include disaccharidases, peptidases, and likely some lipases. Second, the intestinal mucosa also includes **intestinal glands** (crypts of Lieberkühn). These form a reservoir of immature cells that divide and migrate up, taking their places as mature absorptive epithelial cells lining the lumen of the intestines.

Generally, there are two major regions of the intestines, the small and large intestines. The small intestine may be quite lengthy, but it is usually smaller in diameter than the large intestine. It possesses **villi,** small surface projections that increase the surface area of the mucosa (not to be confused with the much smaller *micro*villi, tiny projections of individual cells; figure 13.25). There can be up to three successive parts of the mammalian small intestine: **duodenum, jejunum,** and **ileum.** The duodenum receives chyme from the stomach and exocrine secretions primarily from the liver and pancreas. The jejunum and ileum are best delineated in mammals on the basis of histological features of the mucosal wall (figure 13.26). Such distinct regions are absent or less well defined in other vertebrates. The **ileocolic valve** (ileocecal valve) is a sphincter between the ileum of the small intestine and the large intestine. This valve regulates the movement of food into the large intestine.

The large intestine, so named for its large diameter, is usually a straight tube passing to the cloaca or anus. Its mucosa lacks villi. It may be pushed to one side of the body cavity or, as in many mammals, form a large, gentle loop called the **colon.** The large intestine often straightens near its end, forming a distinctive terminal portion before exiting. If this terminal section also receives products of the urinary and/or reproductive systems, it is properly a **cloaca,** exiting via a cloacal opening or vent. If it only receives products from the alimentary canal, it is a **rectum,** exiting via an anus. The

FIGURE 13.25 **Anatomy of the small intestine.** Villi project above the level of the mucosal wall; intestinal glands are sunken within the mucosal wall. In addition to having a blood supply, each villus houses within its core a system of lacteals, specialized lymphatic vessels.

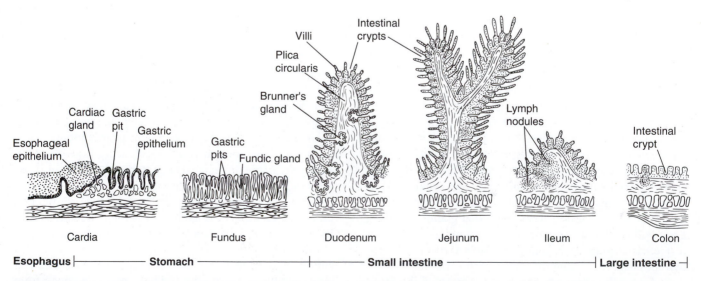

FIGURE 13.26 **Comparative histology of the mucosa along the alimentary canal of a mammal.** Note that the large intestine, like the small intestine, contains intestinal glands, but villi are absent.

At Fort Mackinac, in June 1822, in what was then Michigan Territory, an incident occurred that changed the life of the victim and the course of biology. It started with an accidental gunshot that sent powder and duckshot into Alexis St. Martin, a French-Canadian trader who was standing but 3 feet away. What was St. Martin's misfortune was the good fortune of the army surgeon who attended him, William Beaumont. St. Martin, not expected to survive, in fact lived to the next day, and then a week, and eventually recovered his health.

However, the large hole caused by the gunshot did not heal properly. Instead, the edges of his torn stomach and the hole in the rib cage formed an open fistula, an abnormal passage leading from the stomach through the side of the body to the outside. After many months of convalescence, St. Martin was declared a pauper and refused further treatment. William Beaumont took

the patient into his own home, dressed his wounds, and continued to nurse him. Beaumont also began what he called his "experiments," taking advantage of the fistula that gave a direct view and access into the stomach. Beaumont drew samples of the gastric juice, tossed in various foods on a string and withdrew them later to see what had happened, and observed the mechanical action of the stomach during digestion.

Physiologists of the day thought of the stomach as a vat or stewpot or as an organ that worked by allowing putrefaction of food. Because Beaumont was able to draw samples of gastric juices and observe the process of digestion, he correctly reported the chemical nature of gastric digestion based on the release of hydrochloric acid and the churning action of the stomach. He also massaged bile from the duodenum backward through the pylorus into the stomach. Although rare, this was not

the first gastric fistula opening the process of digestion to direct viewing. But, to his credit, Beaumont was the first to observe digestion carefully and place the physiology of digestion on a sound basis.

As for St. Martin, he lived to the ripe old age of 83, actually outliving Beaumont. But his special stomach had by then become a valuable commodity within scientific circles. On several occasions, he escaped back to the life of the trader he knew best, only to be found and returned to Beaumont. He was pursued (hounded would better describe it) by many physiologists seeking their fame from his special gastric talents. When St. Martin died, his family, by now quite fed up (no pun intended) with his peculiar stomach, refused permission for an autopsy. To make sure he remained unmolested in death, they allowed his body to rot to uselessness for 4 days and then buried him 8 feet deep.

---

rectum narrows into the anal canal in which a transition occurs from simple columnar epithelium to stratified epithelium within its mucosal wall. A smooth muscle sphincter within the muscularis of the anal canal controls the release of waste products from the digestive tract.

Generally, the intestines serve several functions. First, peristalsis within the intestinal walls moves food along the digestive tract. Second, the intestines add secretions to food being digested. Mucous secretions protect the epithelial lining from digestive enzymes and lubricate it to facilitate passage of food. The **intestinal juice** produced by the intestinal glands includes enzymes for digestion of proteins, carbohydrates, and lipids. Accessory glands add some secretions as well. For example, the **duodenal glands (Brunner's glands),** located in the submucosa, empty their secretions into the duodenum to help neutralize the acidity of the chyme entering from the stomach. The pancreas releases its proteolytic enzymes into the duodenum also. Third, the intestines selectively absorb the final products of digestion—amino acids, carbohydrates, and fatty acids. Water is also absorbed, especially in the large intestine.

Cloaca

As already mentioned, the proctodeum at the end of the embryonic gut gives rise to the cloaca, a common chamber that receives products from the intestines and urogenital tracts. In some fishes and most mammals, the cloaca is absent. Instead, the intestine opens to the outside through the anus, which is a separate opening from that of the urogenital system.

Cloacae (p. 582)

Specializations of the Alimentary Canal

Structural modifications accommodating specialized diets occur in several ways within the digestive tract. First, the path that food travels can be lengthened according to the time required for digestion. A **spiral valve** within the lumen of the alimentary canal is one way to increase the length of the route through the digestive tract. This valve creates a helical partition that forces passing food to wind through a spiral channel, thereby increasing the amount of time the bolus spends in the intestines and prolonging digestion (figures 13.27 and 13.29). Another way to increase the path traveled by food is by increasing the length of the alimentary canal. Carnivores have relatively short intestines, but herbivores that must extract nutrients from resistant plant cells usually have long intestines (figure 13.28).

In larval lampreys, a prominent longitudinal fold, the **typhlosole,** projects from one wall of the intestine into the lumen. The typhlosole is a site of hemopoiesis, blood formation, in the larva, but it also increases the surface area available for absorption of digestive end products. In the adult lamprey, the typhlosole is generally lost and is replaced by numerous longitudinal folds in the mucosal wall, which dramatically increase the intestinal absorptive area.

Second, expansions or extensions of the alimentary canal also may develop to accommodate specialized diets. A **crop** is a baglike expansion of the esophagus, found only in birds, often used to store food temporarily during processing. One of the most common extensions is a **cecum,** a blind-ended outpocketing from the intestines positioned at

FIGURE 13.27 Variations within the stomach and intestines of lower vertebrates and birds. Amphibians, snakes, caimans, forest chameleons, and red-tailed hawks are carnivores with relatively short, unspecialized intestines. To prolong passage of digesta, various specializations occur, such as the spiral valve of the bowfin, ceca of some herbivores, or the double ceca of the grouse and rhea.

After Hume; Stevens; Degabriele.

FIGURE 13.28 Variations within the stomach and intestines of mammals. Relatively long small intestines are found in the ant-eating echidna and pangolin, and also in dolphins. Terrestrial mammals that are strict carnivores, such as the mink and dog, have relatively short, unspecialized intestines. Kangaroos, koalas, sloths, rhinoceroses, rabbits, and zebras are herbivores with intestinal specialization that promotes fermentation. Note the relatively simple stomach and intestines of the platypus, whose diet is not well known but is thought to consist of aquatic insects and worms.

After Stevens; Harrop and Hume.

the junction of the small and large intestines through which food circulates as part of the digestive process (figure 13.28).

Third, differentiation of the alimentary canal may occur through regionalization as well. Occasionally, new regions form within the gut. What is a single intestinal tube in some species becomes differentiated into small and large intestines in other species. As we will see, regions are sometimes divided or lost. For example, the cloaca receives the contents of the intestines and urogenital tracks, but as these two systems develop their own separate outlets, the cloaca is lost.

Vascularization of the Gastrointestinal Tract

Blood supply to the gut is as simple anatomically as it is functional. The dorsal aorta successively sends off several major branches, such as the celiac and mesenteric arteries, which supply sections of the gut in series. Such a sequential blood supply supports sequentially active segments of the gut, while flow to inactive sections may be reduced by peripheral vasoconstriction within the gut wall. Arrangements of blood and lymph vessels to the walls of the stomach and intestines are basically similar. The arteries penetrate the gut wall to reach the submucosa, where they break up into a plexus of smaller vessels. From here they distribute outward to the muscularis layer and inward to the mucosa where extensive capillary beds are formed. Within the mucosa, these capillary beds embrace active glands and fill the cores of the villi within the small intestine. As they do for other active tissues, blood vessels support the metabolic needs of gut tissues. But they are also important in transporting away from the gut the end products of digestion—carbohydrates, proteins, small-chain fatty acids—for use and processing elsewhere within the body. Accompanying the blood vessels are the lymphatics, which form an extensive system of lymphatic capillaries reaching the mucosa. The lymphatic vessels are thought to be important in absorption of long-chain fatty acids, especially in the small intestine. Here, in the small intestine, the lymphatic capillaries are termed **lacteals** in recognition of this specialized function.

Fishes

Little is known directly about the digestive system of ostracoderm fishes. Coprolites, fossilized fecal castings, from some of these fishes suggest a diet of detritus selected from the substrate. The coiled shape of these castings implies the presence of a scroll intestine, one characterized by an ingrowth of an elongate mucosal flap of the intestinal wall that was rolled into a coiled **scroll valve** within the lumen.

In cyclostomes, the alimentary canal is a straight tube leading from mouth to anus without coils, folds, or major bends. The ciliated esophagus runs directly from the pharynx to the intestine. No distinct stomach is present (figure 13.29a). Diet includes small particulate matter, blood and tissue rasped from prey, and detritus. Storage in an expanded stomach before

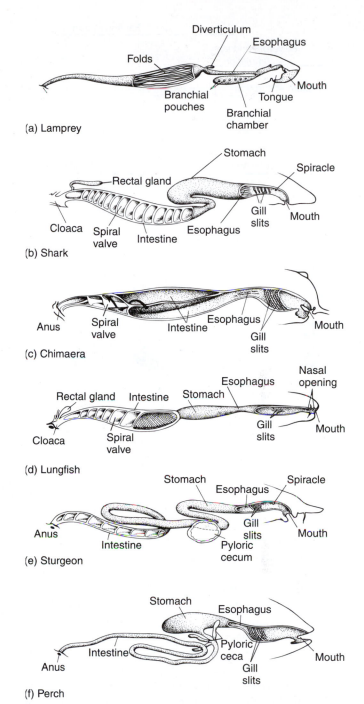

FIGURE 13.29 **Digestive tracts of selected fishes.**
(a) Lamprey. (b) Shark. (c) Chimaera. (d) Lungfish. (e) Sturgeon.
(f) Perch. When a spiral valve is absent, the intestine is often lengthened, as in the perch.

(a) After Youson; (b–f) after Dean.

entering the intestine would be of little value, so food passes directly from the esophagus into the intestine. In lampreys (figure 13.29a), metamorphosis from ammocoetes to adults is usually accompanied by the appearance of a "new" esophagus. A cord of cells evaginates from the dorsal surface of the pharynx, acquires a lumen, and offers a new esophageal route for passage of food from the buccal cavity to the intestine. This

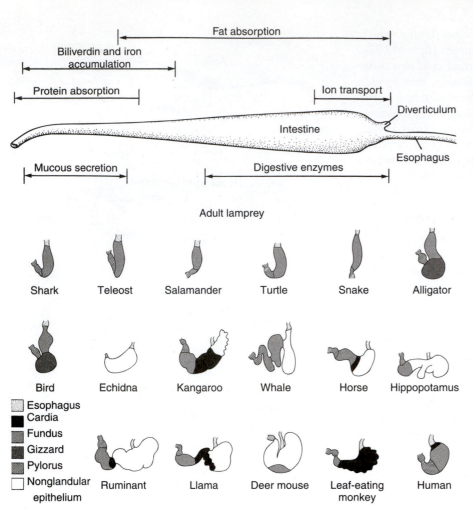

FIGURE 13.30 Alimentary canal of adult lamprey. Absorption, elimination, and transport from the various regions of the alimentary canal are at the top of the diagram. Regions holding digestive enzymes and releasing mucus into the alimentary canal are shown at the bottom.

After Youson.

FIGURE 13.31 Stomachs of various vertebrates. Two regions of the stomach may be recognized, glandular and nonglandular regions. The glandular region of the stomach includes gastric glands and often exhibits three divisions: cardia, fundus, pylorus. The nonglandular region of the stomach is lined with an epithelium devoid of gastric glands that in some species also may be keratinized. The walls of the stomach are composed of layers of smooth muscle, but in some species, these muscular coats are enlarged into a specialized gizzard.

After Pernkopf.

metamorphic change accommodates changes in adult feeding habits and use of the oral disk for attachment. The pharynx attains an independent role in tidal ventilation. The new esophagus, often with numerous longitudinal folds, maintains digestive continuity from the mouth to the alimentary canal. Part of the larval esophagus regresses and part becomes incorporated into the anterior intestine of the adult. The cranial end of the intestine bears one or two (depending on species) diverticula near the point of entry of the esophagus. Products from the liver enter the cranial end via a bile duct. The adult intestine is lined with epithelium that contains numerous gland cells dispersed along numerous longitudinal folds. Digestive enzymes are released into the anterior intestine, and mucus is secreted into the posterior intestine. In parasitic lampreys, the anterior intestine is especially important in absorption of fats. In addition, this region of the intestine of marine forms holds swallowed salt water and is important in osmoregulation. The posterior section of the intestine is important in protein absorption and elimination of biliverdin, a pigment of bile (figure 13.30).

Within gnathostome fishes, there is considerable variation in the design of the alimentary canal, perhaps as much as among all the rest of the terrestrial vertebrates (figure 13.29b–f). Generally, an esophagus, stomach, and intestine are present, although a stomach is usually not differentiated in chimaeras, lungfishes, and some teleosts. When present, the stomach is commonly J-shaped, consisting of a wide fundic and a narrow pyloric region (figure 13.31). In sharks, the muscle layers within the wall of the fundus are composed of striated muscle anteriorly and replaced by smooth muscle posteriorly. A spiral valve is found in the intestines of elasmobranchs and many primitive bony fishes, but it is absent in teleosts (figure 13.29). In teleosts, it is more common to find elongated intestines folded back on themselves in coils. The terminal section of the intestines is usually slightly widened into a cloaca or, more commonly, into a rectum. In elasmobranchs, and coelacanths, a **rectal gland** opens into the cloaca. Although the rectal gland is not directly involved in digestion, it eliminates excess salt ingested during feeding.

In most bony fishes, **pyloric ceca** that open into the duodenum form at the junction between the stomach and the intestine. These number from several to nearly 200 in some teleosts. They are primary areas for digestion and absorption of food, not fermentation chambers.

Tetrapods

In amphibians, the esophagus is short and its transition to the stomach is gradual, but both regions are discernible. The esophageal epithelium is a single or double layer of mucous

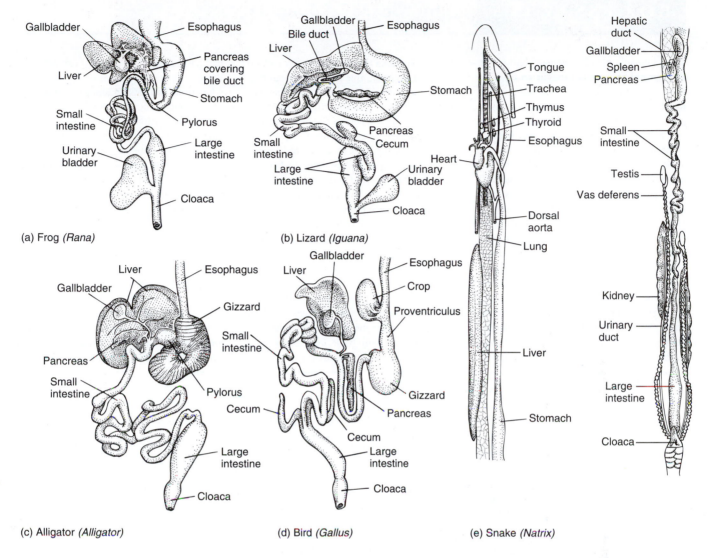

FIGURE 13.32 **Ventral views of the alimentary canals in tetrapods.** (a) Frog. (b) Lizard. (c) Alligator. (d) Bird (chicken). (e) Snake.

(a–d) After Romer and Parsons; (e) after Bellairs.

(goblet cells) and ciliated cells. The stomach mucosa contains characteristic gastric glands, including fundic glands throughout most of the stomach and pyloric glands at its narrowed approach to the intestine. The intestines are differentiated into a coiled small intestine, the first part of which is the duodenum, and a short, straight large intestine that empties into a cloaca (figure 13.32a).

In reptiles, the alimentary canal is similar to that of amphibians, except that in some reptile species it is larger and more elaborate. In many lizards, the stomach is heavy walled and muscular (figure 13.32b). Crocodiles and alligators possess a **gizzard,** a region of the stomach endowed with an especially thick musculature that grinds food against ingested hard objects, usually small stones deliberately swallowed into the stomach (figure 13.32c). The thin-walled glandular region of the crocodilian stomach lies in front of the gizzard, where gastric juices are added.

A distinct large intestine is usually present in reptiles. In some herbivorous lizards, a cecum is present between the small and large intestines (figure 13.32e). The cloaca is partially differentiated into the **coprodeum,** a chamber into which the large intestine empties, and the **urodeum,** a chamber into which the urogenital system empties.

In birds, the esophagus produces an inflated crop, in which food is held temporarily before proceeding along the digestive tract or regurgitated as a meal for nestlings. In pigeons, the crop secretes a nutritional fluid called "milk," which is fed to the young for several days after hatching. The esophagus joins the thin-walled glandular section of the stomach, the **proventriculus,** which is connected to the posterior gizzard (figures 13.32d and 13.33a,b). The proventriculus secretes gastric juice to help digest the bolus, and the gizzard, together with selected pieces of hard grit and pebbles, grinds large food into smaller pieces. The long,

The human appendix receives bad press. Like an old sock, it is thought to have no further function, so it is expendable. Its full name is vermiform appendix. Certainly, a person can get along without an appendix. For reasons not particularly clear, it occasionally becomes infected and inflamed. When this happens, it may rupture, spilling out intestinal contents (digestive juices, bacteria, partially digested food, and pus from the inflammation) into the surrounding viscera and there create a life-threatening condition. So if the appendix becomes infected, a reasonable surgeon will quickly relieve you of it. But what of its function?

The human appendix, at the junction of small and large intestines, is a much reduced cecum. Its small size reflects its small, in fact negligible, role in cellulose fermentation. No longer does our cecum harbor large-scale microbial fermentation.

But just because the appendix performs no digestive function, this does not mean that it lacks a function. The walls of the appendix are richly endowed with lymphoid tissue, much like the rest of the intestine. Just like lymphoid tissue elsewhere, lymphoid tissue of the appendix monitors the passing food, detecting and responding to harmful foreign materials and potential pathogenic bacteria. In short, the human appendix is part of the immune system.

Yet a further function has been proposed. In human societies, before modern medicines, the natural and necessary symbiotic bacterial community of the gut occasionally may have been devastated by disease or foul toxic foods. Under such conditions the appendix may have served as a safe harbor for these symbiotic microbes broadcasting them back into the gut to repopulate it once the disease or trauma

had passed. The fact that you can get along without an appendix does not mean it lacks a function. You can get along without some of your fingers, but that does not mean they lack a function. Many senior and near-senior citizens may recall when it was the fashion to surgically remove a child's chronically inflamed tonsils, because it was thought this would improve an infant's health. Tonsils are stationed around your throat and are the first members of the lymphatic system to detect the arrival of foreign pathogens entering with food. Children certainly got along without their tonsils, probably because the lymphatic system underwent a compensatory enlargement elsewhere, but tonsils do perform a function. There was no malice in the surgeon's urge to remove a child's tonsils. It was well intended but, with the advantage of hindsight, perhaps one of the misplaced fashions in medicine.

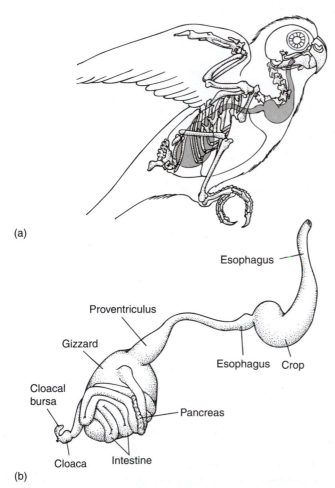

(a)

(b)

FIGURE 13.33 **Alimentary canal of a parakeet.**
(a) Approximate position of the alimentary canal within the bird.
(b) Alimentary canal enlarged.

After Evans.

coiled small intestine consists of a duodenum and ileum. A short, straight large intestine empties into the cloaca. In many species, one or several ceca can sprout from the intestine, usually near the junction of large and small intestines.

In mammals, the esophagus usually lacks a crop, and the stomach shows no tendency to form a gizzard. In some cetaceans, the stomach or esophagus may expand into a pouch that apparently serves, like an avian crop, to store food temporarily, although some gastric digestion may begin as well in this pouch. The mammalian small intestine is long and coiled and usually can be differentiated histologically into duodenum, jejunum, and ileum. The large intestine is often long, although not as long as the small intestine. In herbivores, a cecum is usually present at the junction between large and small intestines. In humans, this much reduced cecum is called the appendix, or more specifically the **vermiform appendix.** In monotremes and a few marsupials, the large intestine terminates in the cloaca. In eutherian mammals, it opens directly to the outside through the anal sphincter.

In ruminants, the stomach is highly specialized. It has four chambers, although the first three—**rumen, reticulum,** and **omasum**—arise from the esophagus and only the fourth—**abomasum**—is an actual derivative of the stomach (figure 13.34a). The large rumen, which gives its name to these mammals, receives the food after it is clipped by teeth and swallowed. The reticulum is a small accessory chamber with a honeycombed texture. Like the first two chambers, the omasum is lined with esophageal epithelium, although it is folded into overlapping leaves. The three types of mucosa distinctive of the mammalian stomach (cardia, fundus, pylorus) are found only in the abomasum, the presumably "true" stomach. Taxonomically, camels are not ruminants but they do practice rumination

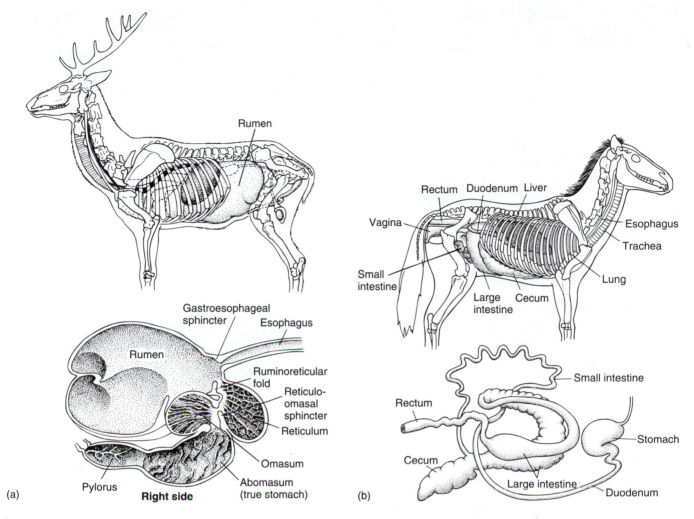

FIGURE 13.34 **Alimentary canal of foregut and hindgut fermenters.** (a) Ruminants ferment food in their foregut. Note the position of the alimentary canal (top) in this deer. The sagittal section of a sheep's "stomach" (bottom) is illustrated below the deer. Note the series of four chambers. Rumen, reticulum, and omasum are derivatives of the esophagus. The fourth compartment, the abomasum, is the actual or true stomach. (b) Hindgut fermenters. Position of the alimentary canal (top) in a horse. Isolated view (bottom) of the large cecum near the juncture of small and large intestines. There is no four-chambered stomach. In hindgut fermenters, the cecum and/or large intestine are the major sites of fermentation.

even though they lack a true omasum, leaving them with a three-chambered "stomach"—rumen, reticulum, abomasums.

In many herbivores, digestion of plant cellulose is enhanced by a cecum situated between small and large intestines (figure 13.34b). The cecum contains additional microorganisms effective in cellulose digestion and provides an expanded region, prolonging the time available for digestion.

Functional ruminant stomach (p. 537)

Associated Glands of Digestion

Oral Glands

The epithelium lining the buccal cavity contains a rich source of cells that secrete mucus and serous fluid. When these secretory cells are gathered together and emptied by a common duct, they constitute an **oral gland.** Such discrete glands are rare in fishes.

In tetrapods, oral glands are more prevalent, perhaps reflecting the absence of a watery medium to moisten food. The most common are the **salivary glands,** a term loosely applied to most major oral glands in tetrapods. These glands are not all homologous, however. In salamanders, mucous glands are found on the tongue, and a large intermaxillary gland is located within the palate. Reptiles also possess oral glands. Usually, strips of glandular tissue, called **supralabial** and **infralabial glands,** are present along the upper and lower lips. In addition, glands may occur within the tongue **(lingual glands)** or below it **(sublingual glands),** in association with the snout **(premaxillary** and **nasal glands),** and along the roof of the mouth **(palatine gland).** These glands release mucus to lubricate the prey during intraoral and esophageal transport. The **lacrimal** and **Harderian glands** release secretions that bathe the eye and vomeronasal organ.

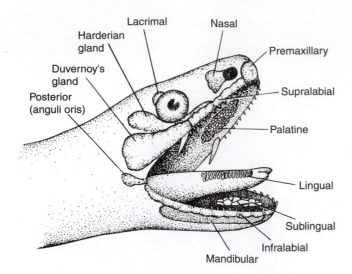

FIGURE 13.35 **Oral glands of reptiles.** Not all oral glands are present in all reptilian species. The venom gland of advanced snakes is a phylogenetic derivative of Duvernoy's gland and is located within the temporal region behind the eye, a position similar to that of Duvernoy's.

After Kochva.

Duvernoy's gland, situated along the posterior upper lip, is found in many nonvenomous snakes and releases its serous secretion via a duct adjacent to the posterior maxillary teeth (figure 13.35).

Secretions from one or most of these glands, in addition to lubrication of food, may also help maintain healthy oral membranes, neutralize toxins carried by prey, and perhaps initiate the chemical stages of digestion. In venomous snakes, the **venom gland,** a homologue of Duvernoy's gland, secretes a cocktail of different chemicals with various functions—some toxic, some digestive (figure 13.36). Thus, the injected secretion of the poison gland not only functions to dispatch prey quickly, but in some snakes, it also contains a suite of enzymes introduced along with the toxins during the strike. These enzymes are injected deep into the prey to process the prey's tissues from within.

When a venomous snake strikes defensively, these enzymes and toxins are delivered into the victim. Medical treatment of a snakebite in domestic animals, pets, and humans must include not only neutralizing the toxic components of the venom but also inactivating the proteolytic enzymes. Otherwise, even if a patient recovers, extensive scarring may persist from local tissue damage inflicted by enzymes at the site of the bite.

Most birds, particularly those feeding in water, lack oral glands, but there are exceptions. Some passerine birds use mucus from oral secretions to help bind together materials composing their nests.

The most common oral glands in mammals are the salivary glands. There are usually three primary pairs of

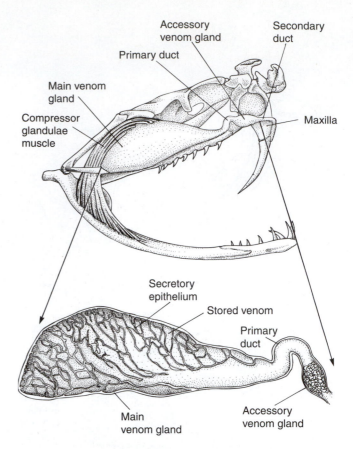

FIGURE 13.36 **Internal structure of a venom gland in a viperid snake.** The secretory epithelium releases venom into the lumen of the gland, where large quantities accumulate, ready for a strike. During the strike, contraction of the compressor glandulae muscle applies pressure on the gland, forcing a charge of venom through the ducts and into the prey. During normal hunting of small rodents, the snake usually does not expend all of its venom reserves within the lumen in a single strike. If a snake is artificially forced to expend all its reserves of venom, full replacement of the venom takes about two days.

After Kardong; after Mackessy.

salivary glands, named for their approximate positions: **mandibular** (submandibular or submaxillary), **sublingual,** and **parotid.** They form the **saliva,** which is added to food in the mouth. These three pairs of glands lie at the angle of the jaws, usually at about the juncture between the head and the neck, but they are positioned superficial to the neck musculature. Ducts from the mandibular and sublingual glands run anteriorly and release secretions into the floor of the buccal cavity. The duct from the parotid gland opens into the roof of the buccal cavity. In some species, additional salivary glands may be present. In dogs, cats, and some other carnivorans, a **zygomatic** (orbital) **gland** is present, usually beneath the zygomatic arch (figure 13.37). Like most digestive secretions, saliva contains mucus, salts, proteins, and a few enzymes, most notably **amylase,** which initiates starch digestion. Saliva also aids swallowing by lubricating food.

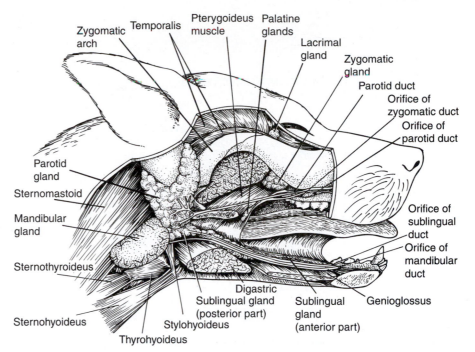

FIGURE 13.37 **Salivary glands of a mammal, dog.** Note the locations of the main salivary glands (sublingual, mandibular, and parotid) along with their ducts leading to the buccal cavity. All mammals possess these three salivary glands. In dogs and cats, a zygomatic gland is also present.

After Miller, Christensen, and Evans.

Liver

The liver is the second largest organ in humans, exceeded in size only by the skin. It functions in a wide variety of roles. Early in fetal life, the liver is directly involved in the production of red blood cells, and later it is involved in the destruction of old blood cells. Throughout life, it detoxifies and removes toxic substances from the blood. Bile is manufactured in the liver and released into the intestine to **emulsify** fats, or break them up into smaller droplets. Carbohydrates, proteins, and fats are stored and metabolized in the liver.

The liver is one of the most heavily vascularized organs of the body, being supplied with arterial blood via the hepatic artery. However, unlike most organs, it is also supplied with venous blood via the hepatic portal vein that runs directly from the intestines and spleen to the liver, delivering absorbed products of digestion.

During embryonic development, the liver appears as a ventral evagination, or **hepatic diverticulum,** of the gut floor that grows forward into the surrounding mesenchyme (figure 13.38a). The mesenchyme does not directly contribute to the liver, but it induces the endoderm of the hepatic diverticulum to proliferate, branch, and differentiate into **hepatocytes,** glandular cells of the liver. As the hepatic diverticulum continues to grow, it makes contact with embryonic blood vessels, the vitelline veins. These veins form the **hepatic sinusoids,** blood vessels within the spaces between the sheets of hepatocytes (figure 13.38b).

All vertebrates possess a liver. Among the protochordates, a cecum from the gut in amphioxus can be found in the approximate position in which the embryonic liver forms in vertebrate embryos. This cecum has a venous portal system similar to a hepatic portal system. Consequently, the cecum in amphioxus is sometimes called the hepatic cecum. However, it is a site of enzyme production and food absorption, quite unlike the vertebrate liver, so it is unlikely a literal antecedent of the liver of vertebrates.

At the gross level, the vertebrate liver is bulky and situated within the rib cage, conforming to the available shape of the body cavity. In snakes, it is long and narrow within the tubular body cavity. Although details differ, the microscopic structure of the liver is basically similar throughout vertebrates. It is composed of sheets of hepatocytes separated by blood sinuses, through which courses venous blood returning from the intestines and arterial blood from the hepatic artery (figure 13.39).

The exocrine product of the liver is bile, which is delivered to the intestine, where it serves principally to emulsify fats. In most vertebrates, bile is stored in the **gallbladder** and released in sufficient quantities when digesta enter the intestine. The gallbladder is absent in cyclostomes, most birds, and a few mammals, but otherwise is present throughout vertebrates, including elasmobranchs and bony fishes, amphibians, reptiles, a few birds, and most mammals. Why it might be absent in some vertebrates and present in most others is not clear. For example, among ungulates, the gallbladder is absent in cervids but present in bovids (except for duikers, which lack one).

Pancreas

Embryonic development of the pancreas is closely associated with liver development. The pancreas arises from two unpaired diverticula: the **dorsal pancreatic diverticulum,**

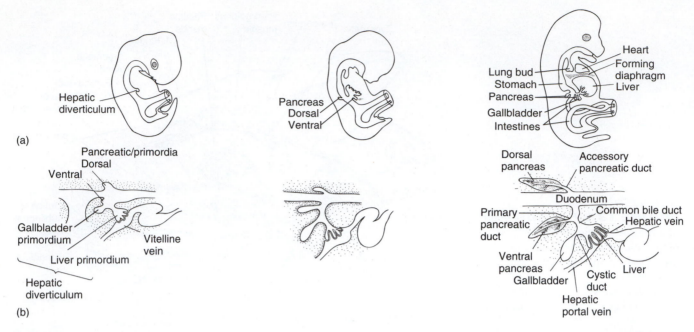

(a)

Hepatic
diverticulum

Pancreas
Dorsal
Ventral

Heart
Forming
diaphragm
Liver

Lung bud
Stomach
Pancreas
Gallbladder
Intestines

Pancreatic/primordia
Dorsal
Ventral

Dorsal
pancreas

Accessory
pancreatic duct

Duodenum

Gallbladder
primordium

Vitelline
vein

Primary
pancreatic
duct

Common bile duct
Hepatic vein

Liver primordium

Ventral
pancreas
Gallbladder

Cystic
duct

Liver

Hepatic
diverticulum

Hepatic
portal vein

(b)

FIGURE 13.38 **Embryonic formation of the liver.** (a) Growth of the liver in a mammalian embryo. (b) Dorsal and ventral pancreatic primordia appear about the same time as the liver primordium. As the liver bud grows, it comes in contact with the vitelline vein, out of which arises the lining to the liver sinusoids.

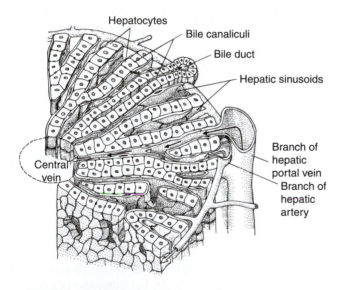

Hepatocytes
Bile canaliculi
Bile duct
Hepatic sinusoids
Central
vein
Branch of
hepatic
portal vein
Branch of
hepatic
artery

FIGURE 13.39 **Blood and bile flow in the liver.** About three-quarters of the blood that reaches the periphery of each liver lobule comes from the hepatic portal vein. The other quarter comes from the hepatic artery. Blood empties into the sinusoids between cords or stacks of hepatocytes (liver cells) and eventually reaches the central vein. From the central vein, it enters the postcaval vein. Solid and open arrows indicate the flow of blood through the liver. Bile is manufactured by hepatocytes, collected in the bile ducts, stored in the gallbladder, and emptied into the duodenum via the common bile duct as it is needed to emulsify fats.

After Bloom and Fawcett.

a bud directly from the gut; and the **ventral pancreatic diverticulum,** a posterior bud of the hepatic diverticulum. These dorsal and ventral pancreatic rudiments may have independent ducts to the intestine, as in some fishes and amphibians, or they may merge, as in amniotes, to form a common pancreatic gland. Even if they merge, each rudiment may retain separate ducts to the intestine, as in horses and dogs, or share a single duct, as in humans, pigs, and cows.

Whether one or two, the ducts empty into the duodenal portion of the intestine and release an alkaline exocrine product, **pancreatic juice,** composed primarily of the proteolytic enzyme trypsinogen, converted within the intestine to the active protease, **trypsin.** Amylases for carbohydrate digestion and lipases for fat digestion are also secreted. Embedded in the pancreas are small **pancreatic islets** (islets of Langerhans) that produce the hormones **insulin** and **glucagon,** both of which regulate the level of glucose in the blood. The pancreas is thus both an exocrine gland, producing pancreatic juice, and an endocrine gland, producing insulin and glucagon (figure 13.40). Both the exocrine and endocrine epithelia of the pancreas arise embryologically from endoderm induced by surrounding mesenchyme.

The pancreas is present throughout vertebrates, both as an exocrine (pancreatic cells) and endocrine (pancreatic islets) gland, although it is not always organized into a discrete organ. In cyclostomes, the exocrine pancreas is dispersed throughout the submucosa of the intestine and on the liver

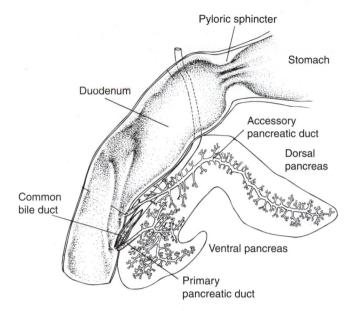

FIGURE 13.40 **Pancreatic ducts of the giant panda.** Both dorsal and ventral pancreas join as a common organ but retain their separate ducts that enter the duodenum. The accessory pancreatic duct drains the dorsal pancreas. The primary pancreatic duct drains the ventral pancreas and enters the duodenum with the common bile duct.

After D. D. Davis.

also. In larval cyclostomes, the endocrine pancreas (islets) apparently has ductless follicles located in the submucosa at the anterior part of the intestine. In adult hagfishes, the endocrine follicles develop as discrete encapsulated clumps near the opening of the bile duct into the intestine. They receive their own rich vascular supply. In adult lampreys, the endocrine pancreas is a distinct patch of tissue close to the bile duct and separate from the dispersed exocrine pancreas along the intestine itself. In elasmobranchs, the pancreas is sometimes dispersed along the course of blood vessels within the liver or, as in sharks, it forms a discrete gland with associated exocrine and endocrine components. In bony fishes, a distinct exocrine and endocrine pancreas are present, and the pancreatic islets are clearly delineated. In tetrapods, an exocrine and endocrine pancreas is always present as a discrete organ located near the duodenum.

Endocrine pancreas (p. 601)

Function and Evolution of the Digestive System

Absorption

Absorption of food begins in the stomach. Water, salts, and simple sugars often cross the mucosa and are absorbed in blood capillaries. However, in most vertebrates, the end products of digestion are usually formed and absorbed in the intestine.

Absorption of food depends on the area available and time spent in the alimentary canal. Both microscopic and gross anatomical features can increase surface area. The numerous villi on the epithelial lining of the intestine increase ten- to twentyfold the area available for absorption. The microvilli, in turn lining the apical surface of luminal cells, may further contribute to a 100-fold overall increase in surface area. At the gross level, the spiral valve found in the intestines of many fishes serves to force food through the winding channel and increases the time it is exposed to digestion. In herbivorous vertebrates, the intestines may be quite long, and their intestinal ceca extended. These modifications prolong the time food takes to traverse the intestines and allow microbial fermentation to digest cellulose more completely.

Long intestines present a packing problem. In leaf-eating monkeys and other herbivores, the abdominal space expands after a large meal, resulting in an enlarged belly. Occasionally, structural reorganization is required. In ornithischian dinosaurs, for example, the pubic bone is rotated backward, enlarging the abdominal area, perhaps as an accommodation to the extensive intestines of a herbivorous dietary style.

Appearance of a long, distinct large intestine in terrestrial vertebrates correlates with greater requirements to conserve water. The mucosa of the large intestine contains mainly mucous glands so that here digestion is brought about by action of resident microorganisms. The large intestine retains digesta so that the electrolytes and water secreted in the upper digestive tract can be resorbed by the body. In lower vertebrates, the large intestine resorbs electrolytes and water secreted by the kidneys. The kidneys of amphibians, reptiles, and birds are limited in their ability to concentrate urine. Much of the urinary sodium and water are resorbed in the cloaca, into which ducts from the kidneys empty. In addition, retrograde peristaltic waves can reflux material from the cloaca back into the large intestine and cecum, providing a further opportunity to resorb these by-products.

Retrograde peristalsis essentially prolongs the time digesta spend in the digestive tract. In some warblers, retrograde peristalsis forces intestinal contents back into the gizzard. This seems especially characteristic of birds feeding on fruits with waxy coatings of saturated fats. When the waxy digesta reach the duodenum, high levels of bile salts and pancreatic lipases are added. This mix is refluxed back into the efficient emulsification mill, the gizzard, for further processing. Saturated fatty acids in the wax can be more efficiently broken down and assimilated.

Feces

For some animals, feces are a resource. Tinamous, a family of neotropical birds, and rabbits, hares, many rodents, and even gorillas eat their feces, a behavior termed **coprophagy,** or refection. But the feces eaten usually come only from the cecum, not from the main intestine. The cecum is emptied in the early morning, and only these droppings are consumed. Within the alimentary canal, there is also a selection process at the cecal-intestinal junction. Liquids and fine particles are diverted into the cecum for extended fermentation and coarser fibers are excluded. Thus, the coarse fiber that bypasses the cecum is not reingested, and only a small percentage of the digesta from the cecum is consumed a second time. Coprophagy allows reingestion, an additional opportunity for the full length of the alimentary canal to capture the products of fermentation, namely, vitamins (K and all B vitamins), amino acids, and volatile fatty acids. If normal coprophagy is prevented, the animal may require vitamin supplements to remain healthy. Coprophagy has been reported in captive apes, cervids, and some other animals, but it is not known if such behavior is important in the wild among these groups.

For the marsupial koala of Australia, feces are eaten by the growing neonate as a transitional food between milk and leaves. The koala mother feeds her own droppings to the six-month-old neonate to begin the process of weaning it from milk to feces to eucalyptus leaves.

The odor from feces can alert a predator to the presence of vulnerable young. Among many herbivores that hide from predators, the young animal does not pass feces until it is licked by its mother. Licking stimulates elimination of feces, which the mother eats so that feces do not collect at sites where young hide and leave a telltale odor. Many young birds bundle their feces. As feces move into the cloaca, its walls secrete a mucous bag that holds the digesta. Parents carry off these bundles of dirty feces, both contributing to good housekeeping (the nest is not fouled) and removing any smelly feces that might attract the attention of a predator.

Mechanical Breakdown of Food

The purpose of mechanical manipulation of food is to improve the access of digestive enzymes. Biting teeth can puncture an otherwise impermeable exoskeleton (arthropods) or protective armor (bony armor) of prey and allow digestive enzymes to invade the tissue. Some fishes and aquatic salamanders often spit out captured prey only to snatch it with the jaws again. As they repeat this process, their tiny teeth tear down the tough outer layer of the prey.

Mastication

Mastication, or chewing, occurs in some fishes and lizards but is characteristic of mammals. The mastication process reduces a large bolus to smaller particles so that digestive enzymes can work on more surface area.

The physical properties of food govern the mastication process. Soft but sinewy foods, such as muscle and skin, are best cut up by the blades of specialized carnassial teeth—of carnivores, for example. As upper and lower sets of carnassials close, they tightly slide by each other, like scissors, slicing the food into smaller pieces (figure 13.41a). Fibrous foods, such as grasses and other plant material, are best broken down by grinding. The molar teeth of ungulates, subungulates, and rodents are corrugated on their working surfaces. As the jaws move from side to side, these tooth surfaces slide past one another to tear plant fibers. Chewing mechanically shreds tough plant fibers and breaks down cell walls, thereby exposing the cytoplasm within to digestive enzymes (figure 13.41b). Hard, brittle foods, such as nuts and seeds, yield best to compression, like that of a mortar and pestle. Molar teeth that roll over each other pulverize this type of food into smaller pieces (figure 13.41c).

Gizzards

Reduction of food by mechanical action is not restricted to teeth. The churning action of the stomach and intestines also contributes, and the gizzard represents a specialized region dedicated to this function. Hard stones are selected, swallowed, and held in the gizzard, where repeated use over time smooths them down. The muscularized gizzard works these swallowed stones against the bolus and grinds it into smaller pieces. Eventually, more gritty stones are swallowed to replace those ground up along with the food. The gizzard is especially important in animals that process plant materials with tough cellulose walls, although the gizzard is an archosaurian feature, including some dinosaurs. "Gizzard stones" have been found within the abdominal regions of some fossil herbivorous dinosaurs. Smooth and polished, these stones provide indirect evidence that some large dinosaurs had specialized gizzards. Crocodiles and alligators also possess gizzards. The gizzard is well developed in birds, especially in those that feed on seeds.

Chemical Breakdown of Food

Intestinal enzymes embedded in the microvilli act on food as it passes through the digestive tract. In some species, chemical digestion begins in the mouth and usually involves amylase digestion of carbohydrates.

The end products of digestion are amino acids, sugars, and fatty acids, as well as vitamins and trace minerals indispensable to fuel the organism for growth and maintenance. Most of these end products come from the breakdown of three classes of macromolecules: proteins, carbohydrates, and fats. Digestive enzymes are themselves proteins that are usually sensitive to pH and temperature. Most are inactivated at temperatures above 45°C. Many enzymes are named for the substrate on which they act, to which the suffix -ase is added.

Proteases digest proteins by splitting their peptide bonds. Fat digestion begins with emulsification of large globules into many smaller ones. (Household detergents act in this way to break up grease into small droplets.) Bile, produced by the liver, is one of the body's major emulsifying agents. Emulsification is a physical process, not a chemical one, because it does not break chemical bonds within fats. This is done by **lipases** that chemically break down long-chain fat molecules into smaller-chain fatty acids. Emulsification increases the surface area of fats exposed to these lipases.

(a) Carnassial teeth

(c) Compression teeth

(b) Grinding teeth

FIGURE 13.41   Mastication in mammals.
(a) Carnivore skull showing position of carnassials (shaded). Carnassials function like scissors to slice through soft but sinewy foods. (b) Artiodactyl skull showing position of grinding teeth (shaded). Corrugated occlusal surfaces of these teeth grind fibrous foods. Teeth rows of either side may grind food but the lower set is moved past the upper set. (c) Primate skull showing position of compression teeth (shaded) that pulverize hard foods.

After Hiiemae and Crompton.

Digestion of carbohydrates produces simple sugars. One of the most important carbohydrates is cellulose, a structural component of all plants. Cellulose is insoluble and extremely resistant to chemical attack. Many herbivores depend on it as a major energy source, yet surprisingly, no vertebrate is able to manufacture **cellulases,** the enzymes that can digest cellulose. Symbiotic microorganisms, bacteria and protozoans, that live in the digestive tract of the host vertebrate produce cellulases to break down cellulose from ingested plants. The microbial process of breaking down cellulose is known as **fermentation,** which yields organic acids that are absorbed and utilized in oxidative metabolism. Carbon dioxide and methane (CH_4) are unusable by-products released by belching.

In many vertebrate herbivores, parts of the digestive tract are specialized as fermentation chambers in which symbiotic microorganisms eventually digest cellulose. Because microbial fermentation is relatively slow and cellulose relatively resistant, these chambers are often quite extensive and lengthy. Microbial fermentation may occur in specialized stomachs or pouches that open off the intestine, and is known, respectively, as foregut and hindgut fermentation.

Foregut Fermentation

When the digestion of cellulose is centered in or near a specialized stomach, it is called **foregut fermentation (= gastric fermentation).** Herein microbial fermentation arises within the esophagus as well as the stomach (figure 13.42a,b).

The old adage, "Don't look a gift horse in the mouth," means that if you are kindly given a free horse, don't insult the donor by looking for defects in the gift. It is based on the fact that as horses grind away at fibrous food, their teeth wear down. Eventually, even hypsodont teeth are reduced to small stubs. Because wear increases progressively with age, tooth height is proportional to age. Looking at a horse's mouth, then, is a not too subtle way of assessing the age and, thus, the value of the gift horse.

Estimating the age of animals is of interest to others besides the owners of horses. Generally, larger animals live longer than smaller animals, but not always. Some bats may live as long as bears. Small vipers may live almost as long as pythons. The ages of members of a population are important

to wildlife biologists. A scarcity of young individuals may imply a decline in potential new breeding members of the population. On the other hand, if there are low numbers of older individuals, there may be too few young produced to breed and sustain the population. Management decisions are based on such information, but how do biologists age members of a population?

One way, although only a rough way, is to examine tooth height. The lower the crown, the older is the individual. Another way is to examine the width of the pulp canal. Inside the tooth, odontoblasts persist and continue to add layers of dentin slowly to the inner walls of the pulp cavity throughout the life of the individual. Therefore, the pulp cavity progressively narrows with advancing age. On the outer

surface of the root, additional cementum is added usually on a seasonal basis, to produce annuli, or rings, within the cementum (box figure 1a,b). These three features of teeth—wear, narrowing of the pulp cavity, and deposition of cemental annuli—are roughly proportional to age. Techniques for estimating the ages of individuals are most reliable when age estimates are compared against similar characteristics from a sample containing individuals of known age. In lower vertebrates, where new teeth replace worn teeth, these techniques do not work, of course. But in mammals, who have only one set of permanent teeth to last a lifetime, these three dental features are rough indicators of age.

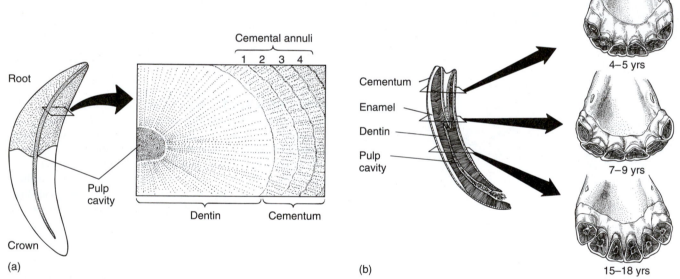

BOX FIGURE 1   **Techniques used to determine age of mammalian teeth.** (a) Upper canine tooth from a carnivore. A cross section through the root reveals cemental annuli in the cementum superficial to the dentin. The rate of rhythmic formation of these annuli can be calibrated if biologists first study canines from known age groups in the same species. Then, by counting annuli in teeth from free-ranging individuals, they can determine the age of individuals in the population and estimate the age structure of the population. (b) Horse teeth. Progressive use with age wears horse teeth down through successive levels, exposing different layers within the teeth. By examining these teeth and comparing them with teeth of known age, biologists can identify the distinctive pattern on the crowns and determine an approximate age of the horse.

(b) After DeLahunta and Habel.

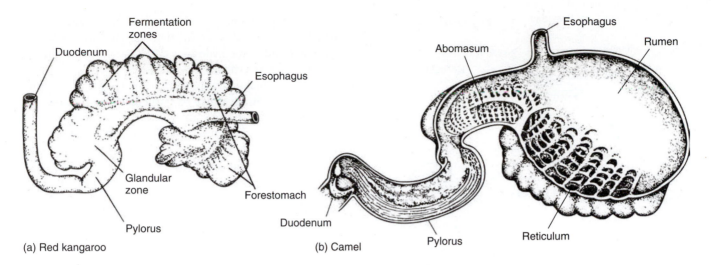

Fermentation zones

Duodenum

Esophagus

Glandular zone

Forestomach

Pylorus

(a) Red kangaroo

Esophagus

Abomasum

Rumen

Duodenum

Pylorus

Reticulum

(b) Camel

FIGURE 13.42 **Fermentation in the stomach.** (a) Kangaroo stomach. (b) One-hump camel stomach. Bacteria in the stomach release cellulases that break down cellulose, making it available for absorption. Both the marsupial kangaroo and the placental camel independently evolved stomachs that harbor bacteria to ferment the fibrous plants that constitute a large part of their diets.

(a) After Dawson; (b) after Pernkopf.

In birds, foregut fermentation is known only in the hoatzin, where it occurs in an enlarged crop and wide lower esophagus. Among mammals, many employ such fermentation but ruminants are specialists and take their taxonomic name from the process. When a ruminant feeds, food initially collects in the saclike rumen, the first of four chambers. The rumen is thin walled and lined with numerous projecting papillae that increase its absorptive surface area. It serves as a large holding and fermentation vat. Later, food in the rumen is regurgitated back into the mouth, remasticated, and swallowed again. This process is repeated until there has been thorough mechanical breakdown of plant material (mastication) and chemical attack on cellulose (fermentation).

Rumination involves complicated waves of contraction that sweep through the rumen and that are synchronized with remastication and with passage of food along the digestive tract. Initially, ruminant animals clip plant material, mix it with saliva, roll it into boli, and swallow it into the rumen (figure 13.43a). Cycles of contraction pass through the rumen and reticulum to circulate and mix the ingested food with microorganisms. In grazers, this mixing also results in physical separation of coarse and fine food particles. Small particles sink into the fluid that accumulates ventrally within the rumen. Large, undigested plant fibers float on top of this fluid. Methane gas that forms during fermentation collects above this fluid and plant fiber (figure 13.43b). In browsers, the reticulum collects much finer particles than does the rumen. However, no such physical separation of coarse and fine food particles occurs within the rumen. Instead, muscle contractions in the rumen are very strong, less fluid is present, food particles are generally small, and methane gas produced by fermentation is quickly eliminated by belching. Consequently, the rumen is not large and gas inflated, but constitutes a relatively small compartment. The methane gas belched up constitutes a substantial by-product of foregut

digestion. Worldwide, ruminants contribute up to 60 tons of methane per year, about 15% of the total atmospheric methane, making ruminants the second major source of atmospheric methane, after natural plant fermentation.

Incompletely masticated food is regurgitated for remastication in the mouth (figure 13.43c). Were such food finely ground in the mouth first, it would, when swallowed, pass quickly through the complex sieves of the stomach without being fermented. Instead, the large size of the initial digesta keeps it in the rumen long enough to be fermented, followed by remastication, thereby controlling particle size and processing.

Three steps are involved in regurgitation. First, the ruminant contracts its diaphragm as if taking in a breath but keeps the glottis (entrance into the trachea) closed. This produces a negative pressure in the thorax around the esophagus. Second, the gastroesophageal sphincter is relaxed and digesta are aspirated from the rumen into the esophagus. Third, peristaltic contractions sweep the digesta up the esophagus into the mouth, so the animal can rechew the undigested plant material. The process of regurgitation and remastication, termed **ruminating,** occurs repeatedly until most of the material is broken down mechanically. The amount of time an animal spends ruminating depends proportionately on the fiber content of the food. In grazing cattle, this may occupy up to eight hours per day and involve rumination of each bolus 40 to 50 times.

The reticulum contracts to slosh digesta between itself and the rumen. Possibly it also separates coarse from fine plant material, making the fine plant material available for further transit. The omasum operates like a two-phase pump to transfer digesta from the reticulum to the abomasum (figure 13.43d). First, relaxation of the muscular walls of the omasum aspirates fluid and fine particles from the reticulum into the lumen of the omasum.

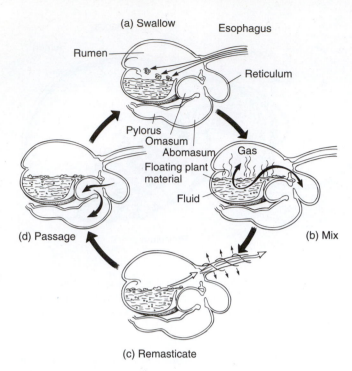

(a) Swallow
Esophagus
Rumen
Reticulum
Pylorus
Omasum
Abomasum
Floating plant
material
Gas
Fluid
(d) Passage
(b) Mix
(c) Remasticate

FIGURE 13.43 **Foregut fermentation in the bovine rumen (based on domestic cow).** (a) In ruminants, food is clipped, rolled into a bolus, mixed with saliva, and swallowed. (b) Contractions spread through rumen and reticulum in cycles that circulate and mix the digesta. Contents separate into fluid and particulate material. Floating, fibrous plant material and a pocket of gas form during fermentation. (c) Poorly masticated boli of plant material are regurgitated and rechewed later to break down fibrous cell walls mechanically and expose further plant tissue to cellulases. Respiratory inhalation, without opening the trachea, produces negative pressure around the esophagus to draw some of this poorly masticated material into the esophagus through the gastroesophageal sphincter. Peristaltic waves moving forward in the wall of the esophagus carry the bolus into the mouth for rechewing. (d) The omasum transports reduced digesta from the reticulum to the abomasum in two phases. First, relaxation of omasal walls produces negative pressure that draws fine particulate material from the reticulum into its own lumen. Next, contraction of the omasum forces these particulates into the abomasum, the stomach region rich in gastric glands. Thus, the abomasum is the first true part of the stomach.

Second, the omasum contracts to force this digesta into the abomasum. The abomasum is the fundic part of the stomach in which further digestion occurs before digesta pass to the intestines.

In animals that feed upon fibrous plants, the combination of remastication and the fermentation process is very efficient. In cattle, organic acids produced in the rumen alone make up 70% of their total energy requirements. Eventually, rechewed food travels through the reticulum into the omasum. The omasum absorbs volatile fatty acids and ammonia and water and at the same time separates the fermenting contents of the rumen and reticulum from the highly acidic

contents of the abomasum. The omasum moves smaller food particles into the abomasum, the true stomach, in which enzymatic and acidic hydrolysis takes place. Finally, digesta enter the intestine (figure 13.43d).

In the suckling ruminant neonate, the abomasum and intestine digest milk, so fermentation in the rumen is unnecessary. Milk bypasses the rumen of a neonate through a **reticular groove** that reflexively closes when the neonate swallows milk. Thus, milk passes directly from the esophagus to the abomasum via the reticular groove.

Foregut fermentation has arisen independently in groups other than ruminants. Only slightly less elaborate is the stomach of some nonruminants, including leaf-eating sloths, langur monkeys, elephants, hippopotamuses, and many rodents. Among marsupials, wallabies, wombats, and kangaroos depend on microbial fermentation in a specialized region of the stomach to digest plants on which they feed (figure 13.42a). However, only ruminants and camels regurgitate stomach contents and remasticate food in a regular cycle. Such regurgutation and remastication is called **rumination** in ruminants and **merycism** in nonruminants.

Hindgut Fermentation

Microbial digestion of cellulose centered in the intestine is **hindgut fermentation (= intestinal fermentation).** Extensive elongation of the intestine and large ceca extend the volume available for fermentation. Rabbits, pigs, horses, and koalas are examples of hindgut fermenters (figure 13.28), as are many amphibians, reptiles, and birds. Herbivorous and omnivorous birds tend to have large ceca; fish- and grain-eating birds have small ceca.

Foregut versus Hindgut Fermenters

In both foregut and hindgut fermenters, microorganisms of the digestive tract release enzymes that digest the plant cellulose (figure 13.44a,b). However, the physiological advantages of such fermentation differs between foregut and hindgut fermenters. At first glance, it might seem that foregut fermenters, such as ruminants, camels, kangaroos, and wallabies, enjoy all the advantages when it comes to efficient digestion. First, fermentation takes place in the anterior part of the alimentary canal, yielding end products of digestion early in the digestive process so they are ready for uptake next in the intestine (figure 13.44a). Second, the ruminant system allows rechewing and more complete mechanical breakdown of the cell walls. By shuttling food between mouth and rumen via the esophagus, the ruminant can keep grinding away at plant fibers. The distant ceca of intestinal fermenters make such shuttling impossible. Third, the ruminant system turns much of the nitrogen, which in most vertebrates is a waste product, into a resource. This is particularly useful in ruminants that consume low-protein

(a) Foregut fermentation

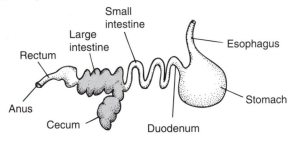

(b) Hindgut fermentation

FIGURE 13.44 Foregut versus hindgut fermentation. (a) Ruminants and other foregut fermenters depend upon microbial activity early in the digestive process as the ruminant "stomachs" chemically attack cellulose in plant cell walls. (b) In hindgut fermenters, fermentation takes place in the long intestines and extended cecum.

diets. Initially, the ruminant system converts nitrogen into waste products, such as ammonia and urea. Nitrogen in the form of ammonia is a by-product of fermentation of protein in the rumen (and of deamination of amino acids in the liver). Microorganisms take up ammonia, combine it with organic carbon compounds, and use it to make their own cell proteins as they proliferate. Periodically, the rumen contracts, flushing these microorganisms into the abomasum and intestine, where, like any food, the microorganisms themselves are digested and their high-quality proteins are absorbed.

Foregut fermenters are also able to turn urea, another waste product, into a resource. For example, a camel fed nearly protein-free foods excretes almost no urea in its urine. Urea certainly forms during metabolism, but it reenters the rumen, partly by direct transfer across the rumen wall and partly in the camel's saliva. In the rumen, urea is broken down into carbon dioxide and ammonia. Finally, in addition to utilizing ammonia, microorganisms also aid in the breakdown of cellulose, a carbohydrate in the cell walls of plants. Cellulose fermentation produces carbon dioxide, water, and volatile fatty acids.

Thus, foregut fermentation is especially efficient at extracting the most nutrients, even from food of poor quality. Ruminant and ruminant-like animals have been especially successful in habitats where the forage is scarce, fibrous, and poor for at least part of the year, such as in alpine regions (goats), deserts (camels), and harsh winter areas (bison). Further, action of fermentation in foregut animals early in the passage of digesta destroys or neutralizes potential plant toxins.

However, hindgut fermentation has some advantages. For the hindgut fermenter, the bolus passes through the major absorptive regions of the alimentary canal *before* reaching the principal sites of fermentation, usually the ceca (figure 13.44b). Soluble nutrients such as carbohydrates, glucose, and proteins can be safely absorbed before fermentation begins. By contrast, among foregut fermenters, fermentation occurs early, and many of these necessary nutrients are disposed of before they can be absorbed. To compensate for premature digestion, foregut fermenters must rely on occasional flushing of microorganisms into the intestine, where they are digested to replace nutrients lost in fermentation.

Further, although foregut fermentation is thorough, it is also slow. To ferment fibrous plants, food must occupy the rumen for extended cellulose processing. Where forage is abundant, the hindgut fermenter can move large quantities of food through the digestive tract, process the most easily digestible component of the forage, excrete the low-quality component, and replace that which has been excreted with fresh forage. Horses are hindgut fermenters. They depend on a high food intake and rapid transit to meet their nutritional needs, as anyone who has walked behind horses in a parade knows. In addition, coprophagy allows rabbits and some rodents to reingest feces containing fermentation products produced on the first pass through the digestive tract. This gives these hindgut fermenters a second chance to extract some of the undigested material.

Size and Fermentation

The body size of a herbivore affects the relative advantages of foregut versus hindgut fermentation. For example, small herbivorous mammals have higher metabolic rates than large mammalian herbivores relative to body size. Consequently, a small herbivore must digest food rapidly to meet the demands of its high metabolic rate. Thus, most small herbivores are hindgut fermenters with digestion based on the rapid passage of abundant and relatively high-quality foliage. By contrast, large herbivores have low metabolic rates and proportionately more volume in which to process food. For large ruminants, therefore, the slower transit of food does not cause digestion to lag behind metabolic demand because metabolic rate is also lower and the rumen volume devoted to digestion is relatively greater. But similarly, large nonruminants can also enjoy some advantages. In fact, if body size is large enough, hindgut fermenters can attain relatively complete digestion, approaching that of ruminants.

Thus, large herbivores, be they ruminants or nonruminants, can extract more energy from plant material than can small herbivores. Ruminants of intermediate body sizes seem to enjoy an advantage over nonruminants of intermediate body sizes, but only if forage quality is high.

The earliest ruminants for which we have a fossil record were small. But as we have just seen, slow digestion in the rumen of small herbivores is a disadvantage compared with rapid food passage in nonruminants. This has led to the suggestion that the rumen probably evolved for other functions initially, such as detoxification or synthesis of proteins. Later, when grasslands expanded, the foregut was preadapted for medium-sized herbivores requiring more efficient processing of this fibrous forage.

Digesting Toxins

The digestive tract has evolved to do much more than reduce food to its end products, making them available to the organism. Many animals have digestive tracts involved in the detoxification of potentially poisonous chemicals in food. The koala exists on a diet exclusively of eucalyptus leaves. There are over 500 species of eucalyptus in Australia, and the koala favors perhaps a dozen of these as food. Because plants cannot run from herbivorous animals, many produce chemical defenses to make their tissues unpalatable or toxic to herbivores. Examples include tannins in grapes, caffeine in coffee plants, and cannabis in marijuana leaves. These chemicals are bitter or distasteful, causing animals to avoid them. Some produce altered states of alertness, leading perhaps to a herbivore that is less attentive to its own safety and thus an easy mark for a predator not on drugs. Such unpalatable or toxic antiherbivore compounds are called **secondary plant compounds** because they are not part of the plant's primary metabolic activity. Similarly, many prey animals have evolved toxins of their own, analogous to secondary plant toxins, that discourage predators. The toxic skin glands of most amphibians are examples.

Eucalyptus produces volatile oils that are toxic to most animals if eaten. Thanks to these secondary compounds, few herbivores can safely eat eucalyptus. The koala is an exception. Its digestive system can detoxify the harmful oils and exploit a resource that is, for the most part, unavailable to other herbivore competitors.

For many ruminants that browse on foods high in tannins, the ruminant saliva plays a central role in neutralizing the detrimental effects. In addition to being toxic, tannins bind with proteins to reduce their absorption, thus reducing the digestibility of plants. However, ruminants that consume plants rich in tannins produce salivary proteins that bind firmly with the tannins when food first enters the mouth to reduce the toxicity of tannins immediately. But salivary proteins further neutralize such secondary compounds to reduce their later detrimental effects upon digestibility. Not surprisingly, the salivary glands of such ruminants are usually enlarged. In particular, their parotid gland is three times the size of that of ruminants not browsing on tannin-laden plants.

Feeding and Fasting

The alimentary canal is a dynamic organ that responds dramatically to immediate changes in quantity and quality of food. With the arrival of digesta, the receiving section of the gut becomes mechanically (churning, peristalsis) and chemically (secretion) active. Digesta can be shunted to or between specialized compartments for futher processing (e.g., fermentation). In birds and mammals that feed daily, cell proliferation within the intestinal crypts supplies new cells constantly that make their way to the surface and are shed eventually into the lumen, replaced by new cells beneath. Over the period of a week, the entire intestinal mucosa may be so replaced on a continuous basis. For vertebrates that fast, as during long-distance migration (birds) or hibernation (mammals), the intestinal mucosa may respond to renewed daily feeding primarily by enlargement of individual lining cells, thereby increasing the mass of the digestive tract.

In species that frequently fast between large meals, such as many snakes, the response of the alimentary canal can be quite dramatic. In response to renewed feeding after fasting, the snake intestinal mucosa answers through increased size of lining cells and secondarily by cell proliferation, outpacing loss of surface cells to the lumen. Further, microvillar length may increase fivefold so that overall the mucosa may double or even triple in area to meet immediate physiological demands of renewed digestion. This overall response is referred to as **up-regulation.** This includes the apical surfaces of the intestinal cells, filling of the lining cells with lipid droplets, and increased prominence of the blood and lymphatic capillaries within the mucosal walls. The liver, pancreas, and even the kidneys, too, can up-regulate, doubling in mass and activity. Such up-regulation can be in place within 24 to 48 hours of ingesting a new meal.

These dynamic responses of the alimentary canal are fully reversible in species such as snakes, once food has cleared the stomach and small intestine. After the meal is digested, the intestinal mucosa returns to a resting state—a decrease in mucosal surface area via decreased intestinal cell proliferation or epithelial folding, decreased blood and lymphatic capillaries, and reduction in liver size—an overall response termed **down-regulation.**

For species with long periods of fasting, down-regulation of the gut and related digestive organs may represent a way to conserve maintenance energy between meals. Nevertheless, this can be expensive. In the Burmese python, up-regulation may cost as much as a third of a meal's energy. But, for vertebrates that feed frequently, the cost of up- and down-regulation may outweigh the energetic savings gained during very short periods of fasting. For most vertebrates, such regulation with feeding and fasting is common only in vertebrates with predictable long episodes of fasting.

Overview

The endodermal lining of the embryonic gut is surrounded by mesenchyme shed from mesoderm of the lateral plate. This regionalization is established early in embryonic development via localized *Hox* gene expression in both the endodermal and mesodermal layers, which pattern the gut, giving it its differentiated regions (figure 13.45).

The alimentary canal is a convoluted highway carrying foods yielding necessary chemical end products to support the energy needs of the organism—proteins, carbohydrates, and fatty acids. This breakdown from meal to molecules is the primary function of the digestive system (figure 13.46). At its simplest, the digestive system is a tube that progresses the bolus of food along by peristalsis, absorbs what can be taken up directly, and moves along the remaining digesta for further processing. The digestive system is made up of specialized regions. Tongue and teeth may grasp or catch the food and bring it into the mouth. In the mouth, food is prepared and compacted in the pharynx, then swallowed down the esophagus to reach the stomach. The esophagus in amphibians and reptiles may temporarily store food, and in birds include a crop. The churning stomach establishes an acidic environment to neutralize pathogens and denature proteins, beginning the process of their breakdown. A muscular wall, the gizzard, present in some vertebrates, mechanically grinds up the meal. Upon entry into the intestines, the acidic digesta is neutralized,

as digestive enzymes and emulsifying chemicals are added. More digestive chemicals are added, further breakdown of the digesta continues, and the undigested material eventually voids at the end of the digestive tract.

The breakdown of food includes both chemical and mechanical means. Digestive chemicals moisten the food, emulsify it, and act directly to reduce the food into its end products. These are added by associated exocrine glands along the route of food passage that empty their products through ducts into the lumen and by microscopic mucosal glands lining the lumen, especially of the stomach and intestines. Mechanical digestion includes ways of mixing the digesta and physically breaking it down into smaller units. This may include grinding teeth, gizzards, and the peristalsis and churning of the gut wall.

These end products of digestion, beginning in the stomach and increasing in the intestine, are absorbed across the gut wall into associated capillary beds and into the lacteal (lymphatic) system. The lining of the alimentary canal includes specializations that increase the absorptive area, from folds in the luminal wall, to villi in the mucosa, and microvilli on cell surfaces.

Foods that resist digestion or are of low quality meet various digestive adaptations. One strategy is to lengthen the time such food is in the gut, thereby increasing its exposure to the digestive process. Spiral valves wind food around the lumen; long intestines lengthen the route; ceca offer blind-ended outpocketings for special processing of

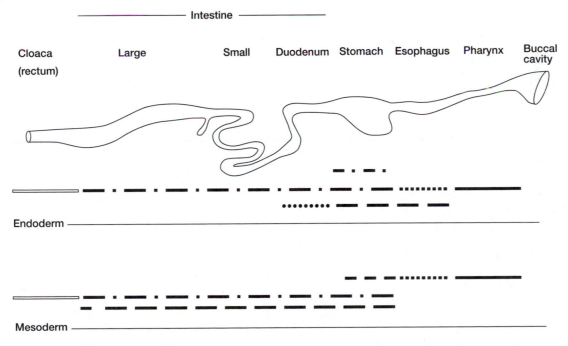

FIGURE 13.45 *Hox* gene expression regions in the embryonic gut. *Hox* genes are expressed in the gut endoderm and mesoderm during early embryonic development, patterning it, and establishing its distinctive regions. Some *Hox* genes continue to be expressed into the adult. Perhaps over 100 *Hox* genes are directly involved, of which only a representative few are shown here. The different (or same) line styles represent regions where different (or the same) *Hox* genes are expressed and thereby preside over gut differentiation.

FIGURE 13.46 Regions of a generalized tetrapod gut. Areas of foregut and hindgut fermentation, when present, are shown (stippling). Digestive glands (liver, pancreas, gallbladder) release their products at the juncture between stomach and small intestine. Birds possess a crop where a large food bolus may be temporarily stored.

food; coprophagy sends food a second time through the alimentary canal. All are ways of increasing the exposure of food to the digestive process. Fermentation represents another strategy. Foregut and hindgut regions hold symbiotic microorganisms that are especially effective at breaking down cellulose in plant material and in turning nitrogen waste products into usable form.

Unfortunately, riding along on this journey through the body are also pathogenic bacteria and viruses, parasites, and toxins. Consequently, located in the walls of the alimentary canals are components of the lymphatic system, bringing the immune system into close proximity to these health risks. The immune function of the gut is a first line of defense against pathogens, parasites, and poisons. Lymphocytes traffic into and out of the mucosal epithelium, picking up antigens, while other cells help to neutralize potential toxins. The heightened vigilance and detoxification ability of the gut's immune system in vultures accounts for why they can eat putrefying foods that would kill us. Without the gut immune system immediately intercepting these challenges, vertebrates would be easily colonized by all kinds of unsavory microorganisms.

Secretions released by salivary and other digestive glands provide herbivores with the ability to deal with pathogens and secondary plant toxins. The acidic environment within the stomach, even during fasting, may intercept and guard against the establishment of pathogenic microorganisms. On the other hand, microorganisms may play a positive role in the gut, as we have seen with fermentation. But even nonherbivores maintain a complex community of microorganisms within their intestines and ceca, whose role in the dynamics of the gut is yet to be understood.

The dynamic responses of the alimentary canal are evident during feeding and fasting. The arrival of a meal in the gut is met with increased mechanical and chemical activity. In animals that regularly fast between meals, such as many snakes, up-regulation and down-regulation are common, accompanied by dramatic increases or decreases, respectively, in the mass and activity of digestive organs. Migrating birds may fast while in flight with down-regulation of organ size. Up-regulation returns during stopover and refueling. Lactating mammals increase food intake and guts enlarge; when food intake decreases, the gut undergoes reduction in mass and metabolic activity.

WEB LINK

Visit the text website at *www.mhhe.com/kardong7e* for additional study aids including selected references, functional laboratories, and web links to chapter topics.

CHAPTER 14

The Urogenital System

INTRODUCTION

URINARY SYSTEM
  Embryonic Development
    Nephrotome to Nephric Tubules
    Tripartite Concept of Kidney Organization
  Kidney Phylogeny
    Fishes
    Tetrapods
  Kidney Function and Structure
    Excretion: Removing the Products of Nitrogen
      Metabolism
    Osmoregulation: Regulating Water and Salt Balance
  Evolution
    Preadaptation
    Origin of Vertebrates

REPRODUCTIVE SYSTEM
  Structure of the Mammalian Reproductive System
  Embryonic Development
    Gonads and Gametes
    Reproductive Tracts

  Overview
Female Reproductive System
    Ovary
    Genital Ducts
    Oviduct
    Uterus
Male Reproductive System
    Testis
    Genital Ducts
    Copulatory Organs
Cloaca
Urinary Bladder
Function and Evolution
    Potency and Fertility
    External and Internal Fertilization
    Delays in Gestation

OVERVIEW

Introduction

Evolutionary survival depends on doing many things successfully: escaping from predators, procuring food, adjusting to the environment, and so on. All of this comes down to reproducing successfully, which is the primary biological role of the genital system. On the other hand, the urinary system is devoted to quite different functions: namely, to the elimination of waste products, primarily ammonia, and to the regulation of water and electrolyte balance. Although urinary and reproductive functions are quite different, we treat both systems together as the urogenital system because both share many of the same ducts.

Anatomically, the urinary system includes the kidneys and the ducts that carry away their product, **urine.** The genital system includes the gonads and their ducts that carry away the products they form, **sperm** or **eggs.** Embryologically, urinary and reproductive organs arise from the same or adjacent tissues and maintain close anatomical association throughout the organism's life.

Urinary System

The vertebrate kidneys are a pair of compact masses of tubules situated dorsal to the abdominal cavity. Urine produced by the tubules is ultimately released into the **cloaca** or its derivative, the **urogenital sinus.** We discuss the urinary ducts in some detail later in this chapter when we consider the reproductive system. In this section, we examine the kidney.

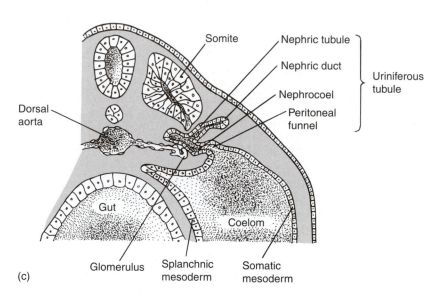

FIGURE 14.1 **Embryonic appearance of nephric tubules.** (a) Nephric tubules develop within the nephric ridge. (b) Preceding this, segmental nephrotomes appear in the posterior part of the intermediate mesoderm. (c) The medial end of nephrotomes differentiates into the first part of the nephric tubule, the renal capsule, into which the glomerulus grows. Arteriol sprouts from the dorsal aorta ramify to form the glomerulus. The lateral ends of nephrotomes grow outward and fuse with each other into the nephric duct. Sometimes the nephrotome remains connected to the coelom via the ciliated peritoneal funnel.

Embryonic Development

The embryology of the vertebrate kidney and its ducts is basically similar across groups, namely development within the intermediate mesoderm. However, modulation of this basic, underlying embryonic pattern produces noteworthy differences in kidney derivation, and differences in the fates of its ducts in males and females. This has produced a vast terminology with many synonyms, coming out of different scientific backgrounds, and differences between males and females. [Sorry, but in my defense just remember this vast and duplicate terminology is not my fault—AU.] Let's sort this out by beginning with the embryology.

Intermediate mesoderm (p. 169)

Nephrotome to Nephric Tubules

The kidneys form within the intermediate mesoderm located in the dorsal and posterior body wall of the embryo. At the onset of its differentiation, this posterior region of

the intermediate mesoderm expands, forming a **nephric ridge** that protrudes slightly from the dorsal wall of the body cavity (figure 14.1a). The next structure to appear usually is the paired **nephrotome** (figure 14.1b). The nephrotome is often segmental and contains the **nephrocoel,** a coelomic chamber that may open via a ciliated **peritoneal funnel** to the coelom. Next, the medial end of the nephrotome widens into a thin-walled renal capsule into which grows the **glomerulus,** a tuft of arterial capillaries. The lateral end of the nephrotome grows outward. This outgrowth fuses with similar outgrowths from successive nephrotomes to form the common **nephric duct** (figure 14.1c) passing in the intermediated mesoderm posteriorly to reach the cloaca. From this point in embryonic development on, the modified nephrotome is more properly called a uriniferous tubule to now include the nephric duct, plus its connection by the nephric tubule to the nephrocoel, which may or may not retain a connection with the coelom via a persistent peritoneal funnel (figure 14.1c).

Thus, the fundamental plan underlying the excretory system is envisioned to consist of paired and segmented uriniferous tubules that open on one end to the coelom and on the other end to the nephric duct, with a glomerulus in between. The ciliated peritoneal funnel seems to drive fluid from the coelom into the tubule, the associated glomerulus adds fluids from the blood, and the tubule itself modifies this collected fluid before it flows into the nephric duct. Although this structure represents the primitive or fundamental plan of excretory tubule organization, in fact, tubules opening to the coelom via a peritoneal funnel are rarely found in the kidneys of adult vertebrates.

Tripartite Concept of Kidney Organization

Developmental and structural differences in the nephric tubules that arise within the nephric ridge inspired a view of kidney formation known as the **tripartite concept.** This concept envisions formation of nephric tubules in one of three locations within the nephric ridge. Subsequent loss, merger, or replacement of these tubules constitutes the developmental basis for the definitive adult kidneys. Specifically, nephric tubules may arise within the anterior, middle, or posterior region of the nephric ridge, giving rise to a *pronephros, mesonephros,* or *metanephros,* respectively (figure 14.2a–c). In addition to positional differences, the three regions vary with respect to connections to the coelom. In the pronephros, tubules usually retain their connections to the coelom through the peritoneal funnel; however, tubules arising within the middle or posterior regions lose this connection in adult vertebrates via this peritoneal funnel to the coelom, ending this direct route of communication with coelomic fluids.

Pronephros The anterior pronephros is usually only a transient embryonic developmental stage in all vertebrates. Tubules that appear within the anterior part of the nephric ridge are called **pronephric tubules.** These tubules join to form a common pronephric duct. This duct grows posteriorly in the nephric ridge, eventually reaching and opening into the cloaca (figure 14.2a). Glomeruli may protrude into the roof of the body coelom, and fluid filters across epithelial membranes and from them into the body cavity. Pronephric tubules then take up this coelomic fluid through ciliated peritoneal funnels, act on it, and eventually excrete the fluid as urine. However, in most pronephric kidneys, glomeruli make direct contact with pronephric tubules.

Pronephric tubules become associated with glomeruli to form functional kidneys in larval cyclostomes, some adult fishes, and embryos of most lower vertebrates. Fluid filtered from the blood enters the tubules directly, and the peritoneal funnels may or may not remain open, depending on the species. In a few amniotes, usually only several pronephric tubules appear during embryonic development. They are not connected to the coelom and do not become functional. In most vertebrates, the embryonic pronephros regresses, and

as it does, it is replaced by a second type of embryonic kidney, the mesonephros.

Mesonephros Tubules of the mesonephric kidney arise in the middle of the nephric ridge. These **mesonephric tubules** do not produce a new duct but instead tap into the preexisting **pronephric duct.** To be consistent, the pronephric duct is now properly renamed the **mesonephric duct** (figure 14.2b).

The mesonephros becomes functional in the embryo, but by the adult, it is modified through incorporation of additional tubules arising within the posterior nephric ridge. This extended mesonephric kidney with additional posterior tubules is termed the **opisthonephros** (figures 14.2d and 14.3). The opisthonephros is found in most adult fishes and amphibians. Although usually lost in the adult, the embryonic peritoneal funnel of the mesonephros may be retained in a few adult species (e.g., *Amia,* sturgeons, some toads). In amniotes, the mesonephros is replaced in later development by a third type of embryonic kidney, the metanephros.

Metanephros The first embryonic hint of a metanephros is the formation of the metanephric duct that appears as a **ureteric diverticulum** arising at the base of the preexisting mesonephric duct. The ureteric diverticulum grows dorsally into the posterior region of the nephric ridge. Here it enlarges and stimulates the growth of **metanephric tubules** that come to make up the metanephric kidney. The metanephros becomes the adult kidney of amniotes, and the metanephric duct is called the ureter (figures 14.2c and 14.4).

Overview The nephric ridge is a **nephrogenic** region, meaning that it is the embryonic source of the kidneys and their ducts. Anterior, middle, or posterior parts of the nephric ridge may contribute to kidneys and ducts. Transient stages often yield to later urinary structures. The tripartite concept that we have used as the conceptual framework in which to discuss these events envisions development of the adult vertebrate kidney as stemming from one of the three regions of the nephric ridge. These three regions are treated as anatomically discrete, and the kidneys they yield as distinct types—pronephros, mesonephros, or metanephros. Additionally, the ontogenetic appearance of such kidneys seems to retrace their phylogenetic origins.

However, anatomical demarcations between these three regions of the nephric ridge are not always apparent, and the entire nephric ridge may be more a unit than composed of three parts. Consequently, many morphologists prefer to use an alternative conceptual framework to interpret kidney development and evolution. This alternative view stresses the unity of the entire nephric ridge and is termed the **holonephric concept.** Morphologists who take such a view emphasize that the three types of kidneys arise as parts of one organ, the **holonephros,** which produces

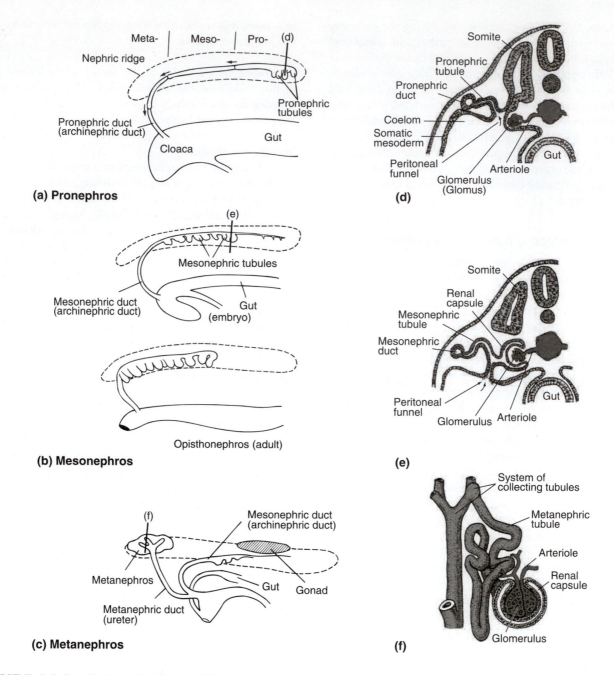

FIGURE 14.2 Embryonic kidneys. Tubules forming the kidney arise in one of three regions of the nephric ridge: anterior (pro-), middle (meso-), or posterior (meta-). (a) Pronephros. Tubules arise in the anterior part of the nephric ridge. They produce a pronephric duct that grows posteriorly in the nephric ridge and empties into the cloaca. Of the three types of kidneys, the pronephros is the first to arise during embryonic development. It becomes the adult kidney in a few fishes but is usually replaced during embryonic development by the mesonephros. (b) Mesonephros. Tubules arise in the middle of the nephric ridge and tap into the existing pronephric duct, now appropriately renamed the mesonephric duct. The mesonephros is usually embryonic and transient. Tubules of the opishonephros arise from the middle and posterior nephric ridge to form an extended kidney that may persist into the adult kidney of fishes and amphibians. (c) Metanephros. Sprouting from the mesonephric duct, the ureteric diverticulum (later the ureter) grows into the posterior section of the nephric ridge, where it stimulates differentiation of tubules that form the metanephros. In males, the mesonephric duct usually takes over the duties of sperm transport and is called the vas deferens. In females, the mesonephric duct degenerates. (d–f) Respective embryonic cross sections, locations indicated by vertical lines at left, of three kidney types showing internal structure.

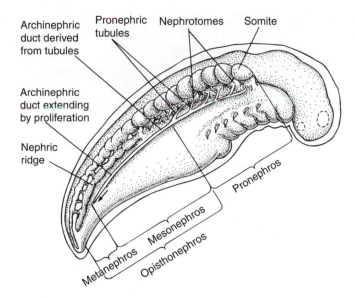

FIGURE 14.3 **Three-part kidney.** Within the nephric ridge, which is derived from intermediate mesoderm, up to three sets of tubules may arise. An extended adult kidney, drained by the mesonephric duct and composed of mesonephric and metanephric posterior tubules, is called the opisthonephros.

After Pough, Heiser, and McFarland.

tubules in anterior to posterior succession during development. There is no anatomical discontinuity marking separate kidney types. Thus, the holonephros is that part of the nephric ridge that produces the kidney.

Experimental embryology is provocative. For example, transplantation of mesonephros-forming or metanephric-forming mesoderm to the "pronephros" region of mesoderm results in differentiation of these transplanted tissues into pronephric tubules and not into what they would have become if left in place. This indicates that tissues within the nephric ridge are flexible and not committed to one type of kidney or another. Differentiation of the nephric ridge into pronephric, mesonephric, or metanephric tubules is induced by tissue location or by interactions with adjacent tissues and not by intrinsic regionalization within the intermediate mesoderm itself. Because the nephric ridge is nonspecific and developmentally pliable, it has the capacity to form different types of nephrons; therefore, some morphologists argue that the term *holonephros* should be used to describe the unity of the nephric ridge. A holonephric kidney seems to characterize the early development of some hagfishes, elasmobranchs, and caecilians.

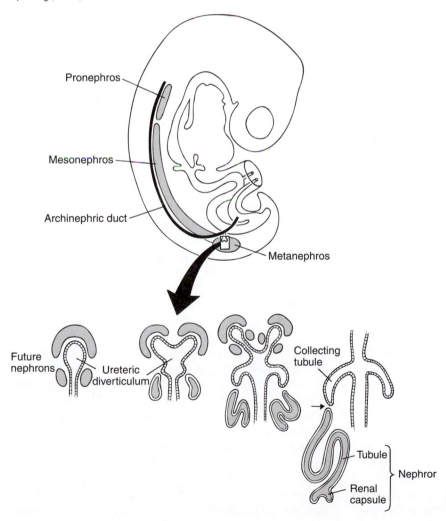

FIGURE 14.4 **Stages in formation of the amniote kidney.** Cross section of the forming metanephros shows that the ureteric diverticulum stimulates surrounding tissue in the nephric ridge to differentiate into nephrons. The ends of the ureteric diverticulum form the collecting tubules.

(a) Hagfish

(b) Lamprey

FIGURE 14.5 **Cyclostome kidneys.** (a) Hagfish. The adult kidney includes aglomerular anterior tubules and a few posterior glomerular tubules. (b) Lamprey. The adult kidney includes a posterior opisthonephros. In some species, a few anterior pronephric tubules with peritoneal funnels may persist. Several pronephric tubules share a glomus, and each can be composed of proximal, intermediate, and distal sections.

(b) After Goodrich; Youson and McMillan.

However, no adult vertebrate retains a holonephros. Absence of examples from adults seems contrary to what we would predict from the holonephric concept and leads other morphologists to retain the tripartite concept.

As a practical matter, in this book we use the descriptive richness of the tripartite concept to examine the evolution of vertebrate kidneys. To characterize the kidney, I use terms that indicate which sections of the nephric ridge contribute to its formation. If the kidney forms just from the anterior region, it is a pronephros; if from the middle, it is a mesonephros; and if from the posterior region, it is a metanephros. The opisthonephros forms from the middle plus posterior regions of the nephric ridge.

Kidney Phylogeny

Fishes

The most primitive vertebrate kidneys are found among cyclostomes. In the hagfish *Bdellostoma*, pronephric tubules arise in the anterior (cranial) part of the nephric ridge during embryonic development. These tubules unite successively with one another, forming the archinephric or pronephric duct (figure 14.5a). Anterior tubules lack glomeruli but open to the coelom via peritoneal funnels, whereas posterior tubules are associated with glomeruli but lack connection to the coelom. In the adult, anterior aglomerular tubules together with several persisting posterior glomerular tubules become the compact pronephros. Although the adult pronephros may contribute to formation of coelomic fluid, the mesonephros is considered to be the functional adult kidney of hagfishes. Each paired mesonephros consists of 30 to 35 large glomerular tubules arranged segmentally along the excretory duct (pronephric duct) and connected to it by short tubules.

In lampreys, the early larval (ammocoete) kidneys are pronephric, consisting of three to eight coiled tubules served by a single compacted bundle of capillaries called a **glomus.** A glomus differs from a glomerulus in that each vascular glomus services several tubules. Each pronephric tubule opens to the coelom through a peritoneal funnel and empties into a pronephric duct. The pronephros is the sole excretory organ of the young larva. Later in larval life, it is joined by additional mesonephric tubules posteriorly. Upon metamorphosis, additional tubules are recruited from the hindmost part of the nephric ridge, yielding an opisthonephros

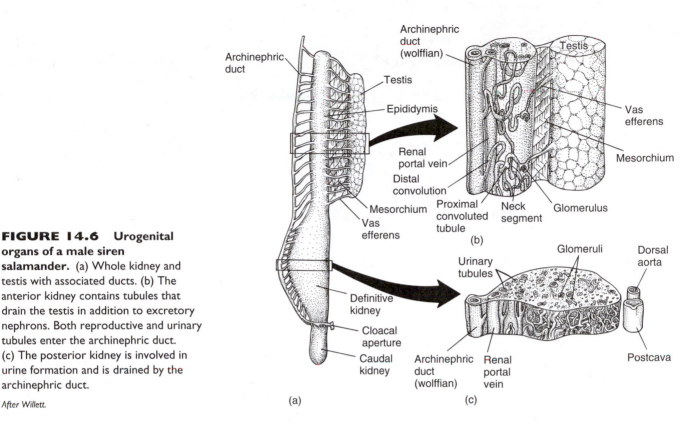

FIGURE 14.6 **Urogenital organs of a male siren salamander.** (a) Whole kidney and testis with associated ducts. (b) The anterior kidney contains tubules that drain the testis in addition to excretory nephrons. Both reproductive and urinary tubules enter the archinephric duct. (c) The posterior kidney is involved in urine formation and is drained by the archinephric duct.

After Willett.

(a)

Labels in figure:

(a): Archinephric duct; Testis; Epididymis; Renal portal vein; Distal convolution; Proximal convoluted tubule; Mesorchium; Vas efferens; Definitive kidney; Cloacal aperture; Caudal kidney

(b): Archinephric duct (wolffian); Testis; Vas efferens; Mesorchium; Glomerulus; Neck segment

(c): Urinary tubules; Glomeruli; Dorsal aorta; Archinephric duct (wolffian); Renal portal vein; Postcava

that becomes the functional adult kidney. The pronephros degenerates, although a few tubules appear to persist into the adult in some lamprey species (figure 14.5b).

In larval fishes, the pronephros often develops initially and for a time may become functional, but it is usually supplemented by a mesonephros. In most fishes, the pronephros degenerates as more tubules are added caudal to the mesonephros to form a functional ophisthonephric kidney in the adult. In a few teleost species, the pronephros persists as the functional adult kidney.

Tetrapods

Among amphibians having active, free-living larvae, a pronephros may develop and become functional for a time. One or two pronephric tubules may contribute to the adult kidney as well. In caecilians, as many as a dozen pronephric tubules have been reported in the adult kidney. However, the early embryonic pronephros is usually succeeded by the larval mesonephros, which upon metamorphosis is replaced by an opisthonephros in most amphibians. Nephrons within the opisthonephros tend to differentiate into proximal and distal regions before joining the urinary ducts. In amphibians, as in many sharks and teleosts with opisthonephric kidneys, the anterior kidney tubules transport sperm, illustrating again the dual use of ducts that serve both genital and urinary systems (figure 14.6a–c).

In amniotes, the anterior end of the nephric ridge rarely produces pronephric tubules. When present, these are few in number and without excretory function. The predominant embryonic kidney is a mesonephros, but in all amniotes, it is supplemented in late development and then completely replaced in the adult by the metanephros drained by a new urinary duct, the ureter. Metanephric tubules tend to be long with well-differentiated proximal, intermediate, and distal regions.

The overall mammalian kidney has a distinctive structure. A cutaway view of the mammalian kidney reveals the two regions: an outer **cortex** surrounding a deeper **medulla** (figure 14.7a). Urine produced by the kidney enters the **minor** and then the **major calyx,** which joins the **renal pelvis,** a common chamber leading to the **urinary bladder** via the **ureter.** Elimination of urine from the body occurs through the **urethra.** Within the kidney, the functional unit that forms urine is the microscopic **uriniferous tubule** (figures 14.7b and 14.8). The uriniferous tubule consists of two parts: the **nephron (nephric tubule)** and the **collecting tubule** into which the nephron empties. The number of uriniferous tubules varies from only a few hundred in the kidneys of cyclostomes to over a million per kidney in mammals, in whom the tubules of both kidneys combined constitute over 120 km of tubing. The nephron forms urine. The collecting tubule affects the concentration of urine and conveys it to the minor calyx, the beginning of the excretory duct.

The renal artery, one of the major branches from the dorsal aorta, delivers blood to the kidneys. Through a series of subsequent branches, it eventually forms tiny capillary beds known as **glomeruli,** each being associated with a **renal capsule (Bowman's capsule)** constituting the first part of the nephron. Collectively, the glomerulus and renal capsule form the **renal corpuscle.** An ultrafiltrate without

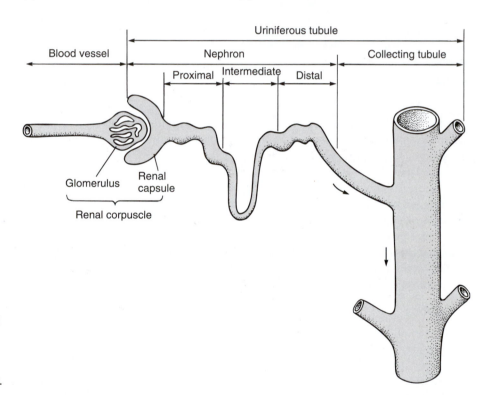

FIGURE 14.7 **Structure of the mammalian kidney.** (a) Section of kidney showing cortex, medulla, and departure of ureter. (b) The uriniferous tubule begins in the cortex, loops through the medulla, and then returns to the cortex, where it joins with the collecting tubule.

FIGURE 14.8 **Uriniferous tubule.** The nephron (nephric tubule) and collecting tubule make up the uriniferous tubule. In turn, the nephron comprises the renal (Bowman's) capsule and the proximal, intermediate, and distal tubules. The glomerulus is the capillary bed associated with the renal capsule. The excretory duct carries away waste products from several uriniferous tubules.

blood cells and proteins is forced through the capillary walls and collects in the renal capsule before it passes through the **proximal** convoluted tubule, **intermediate** tubule, and **distal** convoluted tubule of the nephron, eventually entering the collecting tubules. During transit, the composition of the fluid is altered and water is removed. After circulating through the glomerulus, blood flows through an extended capillary network entwined about the rest of the uriniferous tubule (figure 14.7b). Thereafter, blood is collected in progressively larger veins that join the common renal vein leaving the kidneys.

In mammals, in particular, the intermediate section of the tubules is especially elongated, constituting the major part of the **loop of Henle.** This term refers to both a positional and a structural feature of the nephron. Positionally, the loop includes the part of the nephron that departs from the cortex and dips into the medulla (the descending limb), makes a sharp turn, and returns to the cortex (the ascending limb). Structurally, three regions contribute: the straight portion of the proximal tubule, the thin-walled intermediate region, and the straight portion of the distal tubule (figure 14.7b). Notice that the terms

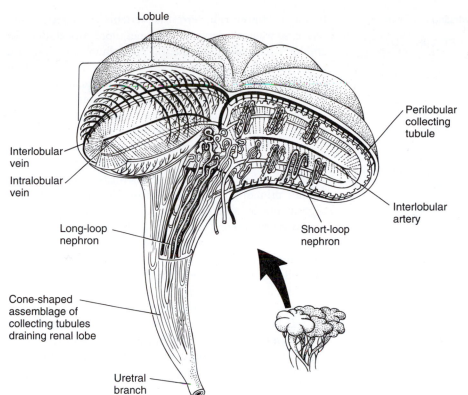

Lobule

Perilobular
collecting
tubule

Interlobular
vein

Intralobular
vein

Interlobular
artery

Long-loop
nephron

Short-loop
nephron

Cone-shaped
assemblage of
collecting tubules
draining renal lobe

Uretral
branch

FIGURE 14.9 **Avian kidney.**
A section of the kidney is enlarged and
cut away to reveal the arrangement of
nephrons within and the blood supply to
the nephrons.

After Braun and Dantzler.

descending and *ascending limbs* refer to the parts of the loop that are departing or entering the cortex, respectively. The terms *thick* and *thin* refer to the height of the epithelial cells forming the loop. Cuboidal cells are thick, and squamous cells are thin.

Loops of the nephron occur only in groups capable of producing concentrated urine. Among vertebrates, only the kidneys of mammals and some birds can produce urine in which solutes in the urine are more concentrated than in the blood, and only these two groups possess nephrons with loops. All mammalian nephrons have loops, specifically loops of Henle. Mammalian kidneys produce urine 2 to 25 times more concentrated than blood. Further, the ability to concentrate urine is correlated with loop length, and loop length is correlated with availability of water. The beaver has short loops and excretes urine only about twice the osmotic concentration of its blood plasma, but some desert rodents have long loops and can produce urine that is about 25 times as concentrated as their blood.

In a few species of birds, the kidneys contain some nephrons with short, distinct loop segments (figure 14.9). Although analogous to the loops of Henle in mammals, these short avian loops evolved independently. These avian kidneys exhibit a modest ability to produce concentrated urine. Their product is about two to four times more concentrated than their blood. However, the nephrons of most birds do not have loops. In the absence of a loop, the avian nephron is similar to the nephron of reptiles.

Kidney Function and Structure

Nephron structure can be quite different from one taxonomic group to the next and may appear at first to have no obvious correlation with the phylogenetic position of the taxon. In hagfishes, the nephron is quite simple. A short tubule connects the renal capsule to the excretory duct (figure 14.10a). In lampreys and freshwater bony fishes, the nephron is more differentiated. It includes a renal capsule, proximal and distal tubules usually joined by an intermediate segment, and a collecting tubule (figure 14.10c). However, the nephron of saltwater teleosts is usually reduced because the distal tubule is lost, and in some, the renal capsule is lost (figure 14.10a). In amniotes, the nephron is again quite differentiated, and the intermediate segment that contributes to the loop of Henle in mammals is often elaborated (figure 14.10b).

To understand kidney design, the adaptive basis of its excretory and regulatory functions, and the evolution of the nephron, we must look at the demands placed on the kidneys. In general, the vertebrate kidney contributes to the maintenance of a constant, or nearly constant, internal environment, termed **homeostasis,** so that active cells (e.g., striated muscle, cardiac muscle, neurons) are not stressed by radical departure from optimum operating conditions. To accomplish this, the kidney performs two fundamental physiological functions, **excretion** and **osmoregulation.** Both are related to maintaining a constant internal environment in the face of accumulating metabolic by-products and perturbations in salt and water concentrations.

Excretion: Removing the Products of Nitrogen Metabolism

Most excreted components in the urine are metabolic by-products that collect within the organism and must be voided so they will not interfere with the organism's physiological balance.

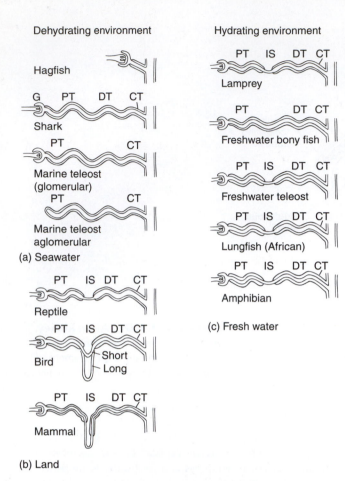

Dehydrating environment

Hagfish

G   PT   DT   CT

Shark

PT   CT

Marine teleost
(glomerular)

PT   CT

Marine teleost
aglomerular

(a) Seawater

PT   IS   DT   CT

Reptile

PT   IS   DT   CT

Bird
          — Short
          — Long

PT   IS   DT   CT

Mammal

(b) Land

Hydrating environment

PT   IS   DT   CT

Lamprey

PT   DT   CT

Freshwater bony fish

PT   IS   DT   CT

Freshwater teleost

PT   IS   DT   CT

Lungfish (African)

PT   IS   DT   CT

Amphibian

(c) Fresh water

FIGURE 14.10 Nephrons from major groups of vertebrates. The segments contributing to the vertebrate nephron depend in large part on whether the animal lives in a dehydrating environment, such as seawater or land (a,b), or in a hydrating environment such as fresh water (c). Nephrons are diagrammed and not rendered to scale among groups. Abbreviations: collecting tubule (CT), distal tubule (DT), glomerulus (G), intermediate segment (IS), proximal tubule (PT).

Energy to support growth and activity comes from the metabolism of food. Carbon dioxide and water are end products of carbohydrate and fat metabolism, and both are easily eliminated. But metabolism of proteins and nucleic acids produces nitrogen, usually in the reduced form of ammonia (NH_3). Because ammonia is highly toxic, it must be removed from the body quickly, sequestered, or converted into a nontoxic form to prevent accumulation in tissues. Three routes of eliminating ammonia, sometimes in combination, exist in vertebrates. Direct excretion of ammonia is **ammonotelism.** Excretion of nitrogen in the form of uric acid is called **uricotelism.** The third route is **ureotelism,** excretion of nitrogen in the form of urea (figure 14.11). Ammonotelism is common in animals living in water. Ammonia is soluble in water, and a great deal of water is required to flush it from body tissues. For vertebrates living in an aqueous medium, water is plentiful. Thus, ammonia is eliminated through the gill epithelium, skin, or other permeable membranes bathed

by water. However, in terrestrial vertebrates, water is often scarce, so water conservation becomes more critical. Because amniotes have lost gills, the gill epithelium is no longer a major route for ammonia excretion. Given these terrestrial constraints, ammonia is converted into urea or uric acid, both being nontoxic forms that address the immediate problem of ammonia toxicity. Furthermore, less water is required to excrete urea or uric acid, so water is conserved as well.

In advanced tetrapods, two evolutionary routes have been followed in addressing the related problems of water economy and nitrogen elimination. Birds and most living reptiles primarily depend on uricotelism. Uric acid, only slightly soluble in water, is formed in the kidneys and transported via the ureters to the cloaca. In the cloaca, uric acid joins with ions and forms a precipitate of sodium, potassium, and ammonium salts. The water not used diffuses through the walls of the cloaca back into the blood. A concentrated, nearly solid uric acid "sludge" forms, allowing nitrogen elimination with little accompanying loss of water.

It is hypothesized that the synthesis of uric acid arose first as an embryonic adaptation, but because of its advantages in water conservation, it was carried over into the adult physiology. The cleidoic egg that first evolved in reptiles is usually laid in dry sites, making water conservation a factor in the embryo's survival. Embryonic adaptations that conserve water include (1) the eggshell, which retards water loss, (2) internal production of water through metabolism of stored yolk, and (3) uricotelism. Because uric acid precipitates out of solution, it does not exert osmotic pressure within the embryo; therefore, it is safely sequestered within the egg without requiring large volumes of water to remove it.

Mammals have followed a different evolutionary route in dealing with nitrogen elimination. They depend largely on ureotelism, the conversion of ammonia into urea. Mammalian kidneys accumulate urea and excrete it as a concentrated urine, thus also detoxifying ammonia and conserving water.

Within an individual, routes of nitrogen excretion can vary in relation to the availability of water. For example, the African lungfish excretes ammonia when it swims in rivers and ponds. But, during droughts, when ponds dry and the lungfish estivates, ammonia is transformed into urea, which can accumulate safely in the body during times of scarce water. With the return of rain, the lungfish rapidly takes up water and excretes the accumulated urea. Similarly, many amphibians eliminate ammonia in water and then excrete urea when they emerge onto land after metamorphosis. In alligators, both ammonia and uric acid are excreted. Turtles excrete primarily ammonia in aquatic habitats but eliminate urea or uric acid when on land (figure 14.11).

Osmoregulation: Regulating Water and Salt Balance

The second major physiological function of the kidneys is osmoregulation. Osmoregulation involves the maintenance of water and salt levels. The external world may vary considerably

Dehydration threatens all vertebrates that venture onto land, but it is especially severe for animals living in hot, dry deserts. To manage dehydration, the kangaroo rat (*Dipodomys spectabilis*) has developed several physiological adaptations that allow it to inhabit desert habitats. Even rainwater is scarce for the kangaroo rat, so it doesn't depend on drinking water to replace water that is evaporated during the day. During lush springs, the vegetation it eats contains some water, but late in summer when the diet consists largely of dry seeds, food is not an important source of water. Instead, kangaroo rats depend on water produced as a by-product of carbohydrate and fat metabolism. When metabolized, food yields carbon dioxide and water. In fact, up to 90% of the kangaroo rat's water budget may come from the oxidation of food. On the other hand, less water is excreted in the urine than in most other mammals. The loops of Henle are elaborated in the kidneys of kangaroo rats. The long loops allow the production of a concentrated urine, up to four times as concentrated as that of humans. Thus, the kangaroo rat recovers some water from metabolism of its food and loses little in its urine. These adaptations allow it to maintain water balance even under desert conditions.

For amphibians moving from land to seawater, dehydration is also a problem. They are hyposmotic to the salty medium; therefore, water is drawn from their bodies. If water loss is not regulated, they will dehydrate and die. Most amphibians live in fresh water or on land. One of the few exceptions is the Southeast Asian frog, *Rana cancrivora*. At low tide, it ventures into saltwater pools to feed on crabs and crustaceans, a habit that has led to its common name, the crab-eating frog. It tolerates these salty conditions through an increase in blood concentrations of sodium and chloride ions and, especially, of urea, as in sharks. At least for short periods in tide pools, it is able to keep its blood levels hyperosmotic to seawater, thus preventing severe water loss and dehydration.

FIGURE 14.11 **Mechanisms of eliminating nitrogenous wastes.** Among many fishes, amphibians, and some reptiles, nitrogen is excreted in the form of ammonia (ammonotelism). Excretion of nitrogen as uric acid (uricotelism) occurs in some reptiles and all birds. In mammals and some amphibians and fishes, nitrogen is eliminated as urea (ureotelism).

After Schmidt-Nielsen.

for an active vertebrate, but cells within see a relatively constant environment. A steady-state intracellular environment is maintained largely by exchange of solutes between the body fluids and the blood and lymph. In turn, the kidneys in large part regulate the constant volume and composition of blood and lymph in terrestrial vertebrates. In aquatic vertebrates, the gill epithelium and digestive tract are as important as the kidneys in addressing problems of salt balance.

(a) Terrestrial environment

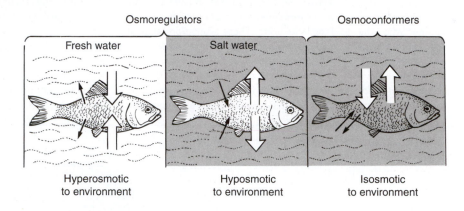

(b) Aquatic environment

FIGURE 14.12 **Water balance.** (a) In terrestrial vertebrates, the relatively dry surrounding environment tends to draw water from the body, posing the problem of dehydration. (b) In aquatic vertebrates, the tendency to gain, lose, or be in balance with the surrounding water depends on the relative concentration of solutes in the animal compared with those of the surrounding water. Osmoregulators control the concentrations of salt and water in their bodies. In fresh water, an animal is usually hyperosmotic to the medium, and the osmotic gradient leads to an influx of excess water. In salt water, most vertebrates are hyposmotic; therefore, water tends to flow from their bodies into the surrounding environment. As with terrestrial vertebrates, dehydration is the result. In both aquatic situations, the vertebrate must make physiological adjustments to eliminate or take up water in order to maintain homeostasis. In a third aquatic situation in which the level of solutes in body tissues rises to meet that of the surrounding salt water, no significant osmotic gradient develops. Such vertebrates are called osmoconformers because they are isosmotic to seawater and no net flux of water occurs. Open arrows represent net direction of water fluxes; solid arrows represent net direction of solute movement. Shading indicates relatively high concentration of solutes in water.

Water Balance Most vertebrates require physiological vigilance to maintain internal balance because the external world constantly intrudes. This is particularly true for water, which may be drawn from an organism and dehydrate it or seep inward across permeable surfaces and dilute body fluids. For example, a terrestrial vertebrate is usually in danger of losing water from its body. To counter dehydration, drinking can help replace lost water (figure 14.12a). Some groups, such as reptiles, control water loss with a thick integument that reduces the permeability of their skin to water. In addition, the kidneys, the cloaca, and even the urinary bladder are **water conservers,** meaning that they recover water before nitrogen is eliminated from the body.

On the other hand, an aquatic life presents other challenges in the management of **water fluxes.** Water may move in or out of the body. In freshwater fishes, the osmotic problem results from a net tendency for an *inward* flux of water. Relative to fresh water, the body of the fish is **hyperosmotic,** meaning that its body fluids are osmotically more concentrated (hence *hyper-*) than the surrounding water. Because fresh water is relatively dilute and the body is relatively salty, water flows into the body (figure 14.12b). If allowed to continue, the net influx of water would substantially dilute body fluids and thus create an imbalance in the extracellular environment. For freshwater fishes, the major homeostatic problem is ridding the body of this excess water. To address this problem, the kidneys are designed to excrete large quantities of dilute urine, about ten times the volume excreted by their marine counterparts.

For most saltwater fishes, the osmotic problem is just the reverse. There is a tendency for a net *outward* flux of water from the body tissues, dehydrating them. Relative to salt water, the bodies of most marine fishes are **hyposmotic,** meaning that the body is osmotically less concentrated (hence *hypo-*) than seawater. Water tends to be drawn from the body, and dehydration of the body will result if this condition is not controlled physiologically. In this respect, a fish in salt water faces a physiological problem much like that of a tetrapod on land, loss of body water to the environment (figure 14.12a). For marine fishes, osmoregulation is complex. They can drink to recover water, but if they do, they must excrete the excess salt ingested along with the seawater. To aid in water conservation, the kidneys are designed to excrete very little water, thus reducing water loss. To address the problem of excess salt, the gills and sometimes special glands become partners with the kidneys in the business of osmoregulation.

The body of some animals is **isosmotic,** meaning that the osmotic concentrations of the internal environment and surrounding seawater are approximately equal (hence *iso-*). Because of this balance, there is no net tendency for water to move in or out of the body, so the animal faces no special problems from excess water or dehydration (figure 14.12b). Dissolved molecules and ions, known as solutes, in the body increase in concentration until the osmotic concentration in the body equals that of the surrounding seawater. Such an animal is called an **osmoconformer.** Among vertebrates, hagfishes are osmoconformers. Concentrations of sodium and other ions are close to those of the surrounding seawater. Chondrichthyans and coelacanths (*Latimeria*) also have tissue fluids osmotically close to seawater, but this is due to elevated levels of the organic compound urea circulating in the blood. As a result, the osmotic concentration of the blood approaches that of seawater. Although this reduces the physiological problems of dealing with water fluxes, it requires that cells of hagfishes, elasmobranchs, and coelacanths operate efficiently in a fluid environment that is higher in osmotic concentration than that of other vertebrates. It is believed that such elevated concentrations may incur energetic costs in osmoregulation.

All vertebrates except for hagfishes, chondrichthyans, coelacanths, and some amphibians are **osmoregulators.** Despite fluctuations in external environmental osmotic levels, they maintain body fluids at constant osmotic levels through active physiological adjustments. Adjustments may involve conservation or elimination of body water to compensate for osmotically driven water loss or uptake relative to the external environment. Solutes also are regulated through excretion and uptake to maintain homeostasis of body fluids. Thus, osmoregulation involves adjustments of water *and* solutes. Next we meet structures designed to

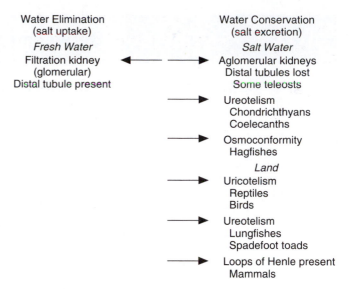

FIGURE 14.13 **Summary of kidney adaptations to two environmentally imposed homeostatic problems: water elimination and water conservation.** In fresh water, most vertebrates need to eliminate excess water. A filtration kidney with fully developed glomerular apparatus and distal tubules can produce copious amounts of dilute urine and rid the body of excess water. In salt water and in terrestrial environments, water-conserving kidneys are advantageous. In marine fishes, aglomerular kidneys that lack distal tubules, ureotelism resulting in elevated levels of blood solutes, and osmoconformity represent three different adaptive routes to water conservation. Vertebrates in terrestrial environments conserve water through structural changes in the nephron (loop of Henle) that promote recovery of water or through more economical means of ridding the body of nitrogen, such as uricotelism or ureotelism, that require less water than ammonotelism.

move both. Let us begin by looking at kidney adaptations that serve water balance. There are two problems that the environment imposes—water elimination and water conservation (figure 14.13).

Water Elimination Water elimination is a problem for hyperosmotic vertebrates living in fresh water. The vertebrate mechanism of urine formation seems especially well suited to address such a problem. The kidneys of most insects and some other invertebrate animals are **secretion kidneys.** Urine is formed by secretion of constituents into the tubules along their length. However, vertebrate kidneys, like the kidneys of most crustaceans, annelids, and molluscs, are **filtration kidneys.** Large quantities of fluid and solutes pass immediately from the glomerulus into the renal capsule to form a **glomerular filtrate.** As this filtrate moves along the tubule, selective secretion adds constituents, but most of the initially filtered water and solutes are absorbed back into the capillaries entwined about the tubules. In humans, for example, each day the kidneys form about 170 liters (45 gallons) of glomerular filtrate in their 2 million renal capsules. This is four to

Most fishes are stenohaline; they can tolerate only a narrow range of salinities. A few fishes are euryhaline; they tolerate wide swings in salinity and may in fact migrate between fresh and salt water. **Anadromous** fishes hatch in fresh water, migrate to salt water where they mature, and then return to fresh water to spawn. Salmon are an example. Depending on species, anadromous fishes spend one to several years at sea, feeding and growing,

then return to their natal stream, where they breed. **Catadromous** fishes migrate in the opposite direction, from salt to fresh water. European and American eels, *Anguilla*, are examples. They mature in streams and migrate to the ocean to breed.

Although euryhaline fishes pass part of their lives in fresh water and part in salt water, the transition from one to the other cannot be abrupt. A period of adjustment, usually involving several weeks in brackish

water, is often required to allow acclimation. When these fishes swim into fresh water, the major physiological challenge is coping with salt loss across the gills. Marine stenohaline fishes placed in fresh water cannot compensate for the high permeability of their gills to salt. Salt continuously leaks out, and the fishes die. Euryhaline fishes develop reduced physiological permeability to salt and survive.

---

five times the total volume of water in the body. If this volume were voided each day, there would be little time for anything else, to say nothing of the large volumes of water we would need to drink to replace water that was excreted. In fact, all but about 1 liter of filtrate is resorbed back into the blood along the uriniferous tubules.

In freshwater fishes and aquatic amphibians, the kidneys characteristically have large, well-developed glomeruli. Consequently, relatively large volumes of glomerular filtrate are produced. The prominent distal tubule absorbs solutes (salts, amino acids, etc.) from the filtrate to retain these in the body, but it absorbs only a third to a half of the filtered water. In this instance, a large proportion of the water is eliminated in the urine. Thus, the kidney is designed to produce large amounts of dilute urine and address the main osmotic problem of excess water in freshwater vertebrates.

Water Conservation As emphasized, water conservation is a problem not just for terrestrial vertebrates facing a hot, drying environment but also for vertebrates in salt water. A variety of structural and physiological adaptations have arisen to address the problems of desiccation in salt water and in terrestrial environments.

The filtration kidney is disadvantageous for hyposmotic fishes in seawater because it is designed to form large volumes of urine. These fishes must conserve body water, not eliminate it. Consequently, in many species of marine teleosts, parts of the nephron that contribute to water loss are absent, specifically the glomerulus and the distal tubule. Absence of the glomerulus and associated renal capsule reduces the quantity of tubular fluid that initially forms. These marine teleosts have **aglomerular kidneys** that, by not producing copious amounts of glomerular filtrate, never face the problem of resorbing it later. Essentially, aglomerular kidneys conserve water by eliminating the filtration process in the renal capsule.

Loss of the distal tubule also contributes to water conservation. The distal tubule absorbs salt from urine but allows water to be excreted. Loss of the distal tubule therefore favors

water retention by the fishes. Without glomeruli and distal tubules, these teleosts depend largely on selective secretion of solutes into the aglomerular tubules to form a concentrated urine.

Terrestrial vertebrates have alternative adaptations to conserve water. In mammals, and to a lesser extent in birds, water conservation is based on modification of the loop of Henle. The loop creates an environment around the tubules that encourages the absorption of water before it can be excreted from the body. Consequently, urine becomes concentrated, and kidney design serves water conservation.

In the mammalian kidney, the relationship between tubule design and water conservation is complex. The first step in urine formation is formation of glomerular filtrate. Circulating blood cells, fat droplets, and large plasma proteins do not flow into the nephron, but most water and solutes from the blood plasma pass from the capillaries of the glomerulus into the renal capsule. Second, most of the sodium ions, nutrients, and water are reabsorbed in the proximal tubule. Absorption is facilitated by the large surface area of proximal tubule cells and depends on active transport of sodium. Usable proteins that were part of the glomerular filtrate are also absorbed in the proximal tubule. Third, the filtrate enters the intermediate tubule of the loop of Henle. Contrary to earlier theories, the loop of Henle is not an additional site in which water is extracted from the filtrate. Instead, the loop actively pumps sodium ions from the filtrate out into the interstitial space to create hyperosmotic interstitial fluid around the collecting ducts. Fourth, as collecting ducts carry the modified filtrate to the renal pelvis, they pass through a region that, thanks to the loops of Henle, is hyperosmotic to the filtrate. The osmotic gradient between surrounding tissue fluid and dilute urine entering the collecting ducts provides the driving force that moves water out of the collecting ducts and into the surrounding fluid. When the body is dehydrated, the permeability of cells of the collecting duct changes under hormonal influence and water is drawn from the tubular fluid into the surrounding interstitial fluid. Here

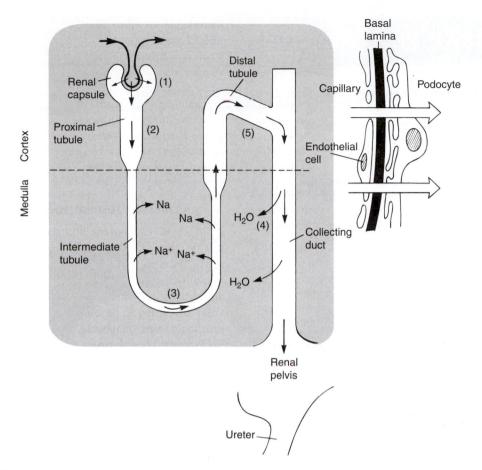

FIGURE 14.14 Mammalian kidney function. At the start of urine formation, high pressure in the glomerulus encourages fluid in the blood to flow from the capillaries into the renal capsule, forming a glomerular filtrate. (1) As the glomerular filtrate passes through the rest of the nephron, some constituents are added, but most of the water is absorbed back into the capillaries. In mammals (and birds), this absorption occurs primarily in the proximal tubule (2) and collecting ducts (4). The intermediate tubule of the loop of Henle produces a salty environment (3) in the kidney medulla. As urine flows in the collecting duct through the medulla (4), it is carried through this hyperosmotic region, and water follows the osmotic gradient out of the tubule into the surrounding tissue. Blood vessels of the vasa recta (not shown) take up this water and return it to the systemic circulation. This produces a concentrated urine in the collecting ducts that is excreted from the kidney via the ureter (5). The insert (upper right) is an enlarged view of the renal corpuscle showing the endothelial wall of the glomerulus, the specialized endothelial cell of the renal capsule (podocyte), and the thick basal lamina between these endothelial layers. Arrows indicate the direction of flow of fluid from the blood into the renal capsule to form glomerular filtrate.

blood capillaries, collectively termed the **vasa recta,** take up water, together with some solutes, and return them to the circulation. The urine that remains in the collecting ducts thus becomes concentrated before it flows into the renal pelvis and ureter (figure 14.14).

Blood flow to the uriniferous tubules is necessary for filtration and resorption to occur. The glomeruli sprout from the renal arteries, which branch directly from the dorsal aorta. Blood pressure is still high in the renal arteries; therefore, blood pressure in the glomeruli is high and promotes the flow of fluid into the renal capsules. On the other hand, pressure in the vasa recta is low, as these vessels arise from the arterioles beyond the glomeruli and pressure drops as blood flows through the glomeruli. The lower pressure in the vasa recta encourages uptake of the water that collects around the loops of Henle.

Notice that unlike the water-conserving kidney of aglomerular teleosts, the distal tubule is retained in the water-conserving kidney of mammals. In mammals, part of the distal tubule is incorporated into the loop of Henle, where its ability to absorb salts contributes to the production of a hyperosmotic interstitial environment around the collecting ducts. Thus, in aglomerular teleosts, water conservation is accomplished by elimination of parts of the uriniferous tubule that allow water loss, whereas in mammals, those homologous parts of the uriniferous tubule are retained but become incorporated into a totally different mechanism of concentrating urine.

Osmoconformers In a sense, one way to address the problem of water fluxes is to avoid the problem in the first place. This is the strategy of osmoconformers, whose body fluids have the same osmotic concentration as that of the surrounding

The kidney seems to bring out the philosopher in all of us. Homer W. Smith, who spent a lifetime in the study of kidney physiology, produced the reflective book, *Man and His Gods,* examining the effects of religious and secular myths on human thought and human destiny. No less than Albert Einstein found it an intensely interesting book and wrote the foreword. Isak Dinesen, perhaps today best known for her book turned into a movie, *Out of Africa,* similarly reflected upon the kidney. In a 1934 collection of writing, *Seven Gothic Tales,* her character, an Arab sailor on deck of his ship cruising off the African coast, philosophizes as follows,

What is man when you come to think upon him, but a minutely set, ingenious machine for turning, with infinite artfulness, the red wine of Shiraz into urine?

Urine comes from the Latin word, *urina,* entering into English usage about the fourteenth century. Before then, the French word "pissier" gave the English "piss," used comfortably by Geoffrey Chaucer (fourteenth century) and even by proper Elizabethan ladies and gentlemen. Not until Oliver Cromwell and puritanism (seventeenth century) did the term fall out of favor. Only recently has it enjoyed rediscovery and use once again in public.

Urine has been put to a variety of household uses: as a hairdressing, as a fermentor of bread, to flavor cheese, and to macerate tobacco leaves. Wealthy French ladies of the seventeenth century could often be found in urine-enriched baths to beautify the skin. In various cultures it has been tried as a mouthwash and gargle. For centuries, it was considered proper and humane to wash battlefield injuries by urinating on the wounds of comrades (no more sterile nor antiseptic elixir was available). In the early nineteenth century, uroscopy, or "water casting," was in great vogue in the medical profession throughout North America and Europe. This involved inspection of the "piss pot," as Elizabethans called it, or urinal, and these medical devices were often elaborately decorated with flowers in middle-class households and by gold and silver in finer families. So prominent was nineteenth-century uroscopy that the urinal became an emblem of the medical profession.

Students too have come to appreciate their kidneys. When loading up with coffee and great thoughts the night before a final exam, or when celebrating with drink and great excuses after the exam, we are reminded of our kidneys and the volumes of urine they can produce when required. Although evolved for their water-conserving abilities, our kidneys possess the physiological flexibility to rid our bodies of excess water when we overindulge. The collecting ducts become impermeable to the egress of water (ADH, a pituitary hormone, changes its permeability), less moves from collecting ducts into the interstitial space, less is available to be absorbed by the vasa recta, and more fluid is left behind to be excreted in copious amounts. Drinking establishments around the world serve different beers, wines, coffees, and soft drinks, but all have restrooms.

Such inspiration between physiology and philosophy has made public watering holes the site where we celebrate our kidneys. Perhaps it was just such a homage to the human kidney that prompted Samuel Johnson, himself a legendary raconteur and heavy user of his kidneys, to observe,

There is nothing which has yet been contrived by man by which so much happiness is produced as by a good tavern or inn.
Life of Dr. Johnson, James Boswell

---

medium. Isosmotically balanced with their environments, osmoconformers do not have to cope with the problems of water entry or loss. Osmoconformer vertebrates are all marine. In hagfishes, unlike in the hyposmotic body fluids of most marine fishes, concentrations of Na^+ and Cl^- in blood and extracellular fluid are elevated, so they are close to those of seawater. Hagfish tissues tolerate these relatively high levels of solutes. Because the hagfish is in osmotic equilibrium with its environment, the nephron does not need to excrete large volumes of urine. Consequently, the nephron is reduced to little more than a renal capsule connected to the archinephric duct by a short, thin-walled duct (figure 14.10a). Surprisingly, the renal corpuscles are quite large. Because water elimination is not a problem for the hagfish, the well-developed renal corpuscle probably functions in regulating divalent ions such as Ca^{++} and SO_4^{--}.

Elasmobranchs and the coelacanth *Latimeria* are also approximate osmoconformers, but this is achieved via ureotelism. Urea accumulates at high concentrations in the blood and elevates the blood osmolarity up to that of seawater. For these fishes, large fluxes of water do not occur, and maintenance of water balance presents no special problem. Excess salts that enter the body fluids from seawater are eliminated through special glands, such as the *rectal gland* of sharks, or through the gills.

Tolerance of Fluctuations Changes in kidney structure and osmoconformity are not the only ways of dealing with osmotic stress. Some aquatic vertebrates can tolerate wide variations in salinity. Those that are osmotically tolerant are **euryhaline** (*eury-,* wide; *haline,* salt) animals. Some euryhaline animals can move from marine to brackish water and even to fresh water. Other vertebrates that can withstand only a narrow range of environmental salinities are **stenohaline** (*steno-,* narrow; *haline,* salt) animals.

Salt Balance Although we have focused on renal mechanisms that eliminate or conserve water, striking an osmotic balance involves moving salts as well as water. Various structures are devoted to the task of regulating

(a) Elasmobranch rectal gland

Rectal gland

Cloaca

Stomach

Intestine

Mitochondria

Lumen

Microvillus

Basal lamina

Rectal artery

Arteriole

Duct

Cloaca

Vein

Intestine

Capsule

Secretory tubule

Stratified epithelium

Venous sinus

Gland lumen

Na⁺

Outside of body

Inside of body

(b) Amphibian integument

S. corneum   S. granulosum   S. spinosum   S. basale

FIGURE 14.15   Regulating salt levels. (a) Rectal glands of sharks and other elasmobranchs. These glands have evolved to eliminate salts efficiently from the body without the expenditure of large volumes of water. The outer capsule of the elasmobranch rectal gland consists of connective tissue and smooth muscle. Blood enters via the renal artery, circulates around the secretory tubules, enters a venous sinus, and then flows into the renal vein. Salt collected by the secretory tubules passes into the lumen of the rectal gland and is then forced into the intestine to be eliminated with the feces. (b) Diagrammatic cross section of amphibian integument. Salt tends to diffuse from amphibians into fresh water. They have evolved the ability to take up replacement salts, especially sodium ions, through the skin by means of active transport. Sodium is taken up across the stratum granulosum and moved by active transport into the spaces between cells. Eventually it makes its way into capillaries within the dermis.

After Berridge and Oschman.

salt balance. As mentioned, the distal tubule within the kidney recovers salts from the urine. Gills address ion imbalances by pumping salts out of (marine bony fishes) or into (freshwater fishes) the body. The rectal gland of elasmobranchs also collects, concentrates, and rids the body of salts (figure 14.15a).

Marine reptiles and birds that eat salty foods or drink seawater to replace lost fluids also ingest high levels of salt. Because their kidneys cannot handle this excess salt, it is excreted by special **salt glands.** In response to a salt load, salt glands intermittently produce a highly concentrated secretion containing Na⁺ and Cl⁻ primarily. In reptiles, these salt glands can be specialized nasal glands (in some marine lizards), orbital glands (in some marine turtles), sublingual glands (in sea snakes), or glands on the tongue's surface (in Asiatic saltwater crocodiles and North American crocodiles).

In marine birds, paired nasal salt glands are present. These large, specialized glands are usually located within shallow depressions on the dorsal surface of the skull and release their concentrated secretion into the nasal cavity. Marine mammals lack specialized salt glands. Their kidneys produce urine that is much more concentrated than seawater, so most salt is eliminated through the kidneys. Many terrestrial mammals have sweat glands in the integument primarily serving thermoregulation, but they also eliminate some salt.

In fresh water, the problem is totally different. Salt tends to be lost to the environment. Freshwater fishes absorb salts through their gills. In aquatic amphibians, the skin aids in the regulation of salt balance (figure 14.15b).

Balancing Competing Demands The cloaca, urinary bladder, and large intestine also aid in the regulation of both salt and water balances. Managing salt and water balance must be compromised with other demands. We have already seen that the demands of nitrogen excretion must sometimes be balanced with the need for water conservation. Furthermore, amniotes often incur a heat load if they live in hot climates or lead active lives. Birds pant and mammals sweat to help dissipate heat through the process of evaporative cooling. Water also is lost in this process. Although reptiles lack sweat glands, they possess a thick, water-resistant skin and exhibit only a modest panting mechanism, so they cannot regulate their body temperature through evaporative cooling. Instead, they move out of the sun (into the shade or into burrows) or become active at night. Behavioral thermoregulation and lower metabolic rates reduce evaporative water loss and contribute to water conservation in reptiles.

Evolution

Vertebrate kidneys illustrate preadaptation, a theme we have seen within other systems. But the preadaptation of the urinary system raises an issue we have not addressed—the freshwater origins of vertebrates.

Preadaptation

Excretion of urea or uric acid conserves water and is adaptive for life on land; however, conversion of ammonia into urea or uric acid probably arose well before vertebrates actually ventured onto land. In chondrichthyans and coelacanths, the formation of urea answers the problem of water balance by turning these fishes into osmoconformers. Detoxification of ammonia, by converting it to urea, allows lungfishes to address the immediate problem of surviving droughts. The amniote embryo, confined to a cleidoic egg, converts ammonia into uric acid so it can safely sequester nitrogenous wastes without requiring large quantities of water to flush them away. One or more of these conditions may have preceded life on land and been preadaptive. When vertebrates eventually ventured onto land, they entered an environment in which water was scarce, making water conservation especially important. But as such a transition occurred, the metabolic means of conserving water may already have been in place.

Origin of Vertebrates

Homer Smith, a physiologist, was the first person to notice that the vertebrate kidneys appeared to be better suited to life in fresh water than in salt water. In fact, he argued that the kidneys were so well designed for fresh water that vertebrates must have evolved in fresh water and only later entered salt water. His reasoning went like this: The kidneys of vertebrates are filtration kidneys that can produce large volumes of glomerular filtrate. Such a design would be a liability in marine environments in which water must be conserved, but it would be an asset in freshwater environments in which fishes must rid their bodies of influxes of excess water.

Marine invertebrates are osmoconformers. The levels of salts in their blood are close to those of seawater, making them isosmotic. They are in no danger of dehydration; however, this is not true for marine vertebrates. Compared with marine invertebrates, the levels of salt in the blood of marine vertebrates is almost two-thirds lower. Consequently, vertebrates are hyposmotic to seawater and can become dehydrated. To make matters worse, vertebrates have a filtration kidney capable of producing large volumes of water, not conserving it.

Such disadvantageous features of marine vertebrates can be explained, Smith reasoned, if vertebrates originated in fresh water. If vertebrate ancestors lived in fresh water, evolution of filtration kidneys and low solute levels would be adaptive to cope with water influxes an animal experienced in such an environment. However, when these vertebrates later radiated from fresh to salt water, their filtration kidney was disadvantageous and modifications were required. In chondrichthyans and coelacanths, solute levels rose in the blood to address this problem. Other fishes developed adaptations, such as drinking seawater, that recovered water, and salt glands and gills that eliminated excess salt along with loss of glomeruli and distal tubules. Smith felt that the fossil record available in 1931 also supported a freshwater origin for the earliest vertebrates.

Others have taken issue with Smith's hypothesis and favor instead a marine origin for vertebrates. First, the filtration kidneys of vertebrates are characteristically high-pressure kidneys that produce large volumes of glomerular filtrate. Large volumes of fluid moving from blood to kidney tubules give the kidneys a greater chance to act on the constituents within the circulating fluids of the body. A high-pressure system produces a high volume of filtrate, which aids in processing nitrogenous wastes. Thus, the filtration kidney could represent an efficient system for eliminating nitrogenous and other wastes by moving large volumes of filtrate through the kidney. Second, the filtration kidney is not unique to vertebrates. Crustaceans and many other invertebrates possess filtration kidneys, yet clearly they evolved from marine ancestors. Moreover, many are marine osmoconformers today. Finally, reexamination of early vertebrate fossil deposits suggests that they came from marine seas and not from freshwater habitats as Smith supposed. Contrary to Smith's views, the filtration kidney of the marine vertebrates was preadapted to fresh water, but it did not arise there.

In this debate, the hagfish poses a problem for everyone. Hagfishes are osmoconformers like most marine *invertebrates*, but unlike most vertebrates. They are members

Water, Water, Everywhere, Nor Any Drop to Drink*

Adrift on the ocean, seamen who have survived the loss of their ships face a paradox. Exposed to heat, they dehydrate. Yet, they are surrounded by water, but to drink it would only make matters worse. The reason is that seawater is hyperosmotic to body fluids. If a person drinks seawater, the salt is absorbed, and blood osmotic levels rise. Now to flush the excess salt from the body, the kidney must spend as much or more water than was originally gulped in by the thirsty castaway. The net result is to make the body even more dehydrated. Furthermore, there is another problem. Seawater also contains magnesium sulfate, an ingredient used in laxatives. It stimulates diarrhea, and hence even more fluid is lost via the digestive tract.

Many marine animals address this problem differently. They drink seawater but excrete the excess salt by active transport in special salt glands rather than by flushing it through the kidneys with water. This allows them to use the seawater but not fall behind like humans in their water balance.

*The Rime of the Ancient Mariner, Samuel Taylor Coleridge.

---

of the oldest surviving vertebrate group, the cyclostomes, and possess a filtration kidney, yet they live in salt water. If Smith were correct, then these primitive vertebrates would live in fresh water. They do not, of course. If the marine origin of vertebrates is to hold, then hagfishes should be osmoregulators like other vertebrates. Of course, they are not. Perhaps it is best to recognize that the hagfish, although a representative of the earliest group of vertebrates, is very ancient and may have diverged significantly in its physiology from an ancestral condition.

Reproductive System

The reproductive system includes the gonads, their products, hormones and gametes, and the ducts that transport gametes. Reproductive hormones facilitate sexual behavior and parental care, prepare the reproductive ducts to receive the gametes, support the zygote, and perform other functions that we turn our attention to in chapter 15 on the endocrine system. Now we look at the gametes and the ducts that provide a home for and convey gametes during reproduction. The eutherian mammal is used again to introduce the terminology applied to the reproductive system.

Structure of the Mammalian Reproductive System

In mammals, each ovary consists of an outer connective tissue capsule, the **tunica albuginea,** that encloses a thick **cortex** and deeper **medulla.** The **ova,** or eggs, occupy the cortex and are wrapped in layers of **follicle cells** derived from connective tissue. An ovum plus its associated follicle cells is termed a **follicle.** Some follicles remain rudimentary, never change, and never release their ova. Others pass through a series of growth, or **maturation,** stages, at the end of which the ovum and some of its clinging follicle cells are cast out of the ovary in the process of **ovulation,** and become ready for fertilization. If fertilization occurs, the ovum continues down the **oviduct** and becomes **implanted** in the wall of the prepared **uterus,** where subsequent growth of the embryo occurs. If fertilization does not occur, the undeveloped ovum continues down the oviduct and is flushed out of the uterus during the next menstruation (figure 14.16).

Embryonic implantation (p. 193)

Each mammalian testis also consists of an outer tunica albuginea, which encloses the **seminiferous tubules** that produce sperm. Within the walls of the seminiferous tubules, stem cells multiply and grow into sperm that eventually are released into the lumen. The coiled seminiferous tubules straighten, forming **tubuli recti** just before they join the **rete testis.** Via the **efferent ductules,** the rete testis joins the **epididymis,** where sperm are temporarily stored. Upon ejaculation, sperm travel along the **vas deferens (ductus deferens)** into the urethra. Along the way, three accessory sex glands, the **seminal vesicle, prostate,** and **bulbourethral** (Cowper's) **gland,** respectively, add their secretions as sperm move from the testes to the urethra. This fluid and the sperm it contains constitute **seminal fluid,** or **semen** (figure 14.17).

Embryonic Development

Gonads and Gametes

The paired gonads arise from the **genital ridge,** initially a thickening in the splanchnic mesoderm to which adjacent mesenchyme cells contribute (figure 14.18). The early gonad is little more than a swelling on the dorsal wall of the coelom with a thick outer cortex around a deeper medulla (figure 14.19a,b). Because the gonad shows neither unique male nor female characters at this early stage, it is termed an **indifferent gonad.** The gonads of both sexes initially contain **germ cells,** the future sperm or eggs. Surprisingly, germ cells themselves do not arise in the genital ridge nor even in the adjacent mesoderm. In fact, they do not arise in the embryo at all. They first debut in remote sites outside the embryo in the extraembryonic endoderm. From the extraembryonic endoderm, they undergo a journey that takes them eventually to the indifferent gonad, where they

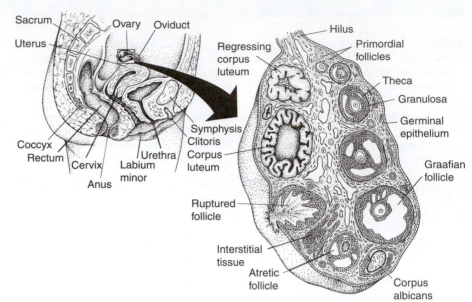

Female reproductive system
(human)

FIGURE 14.16 **Female reproductive system (human).** This sagittal section of the female pelvis shows the reproductive organs and their relationships to the urinary and digestive systems. The ovary is enlarged and sectioned at the right. The successive stages in follicle maturation, beginning with the primordial follicles and then clockwise to the Graafian follicle and corpus luteum, are summarized within the representative ovary. Atretic follicles and other regressing stages are included.

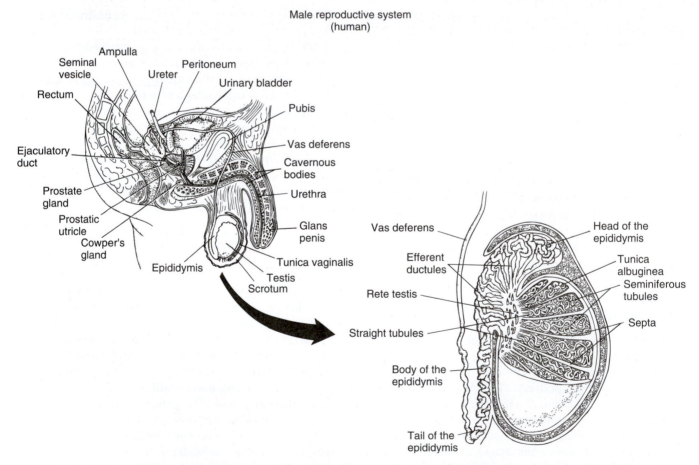

Male reproductive system
(human)

FIGURE 14.17 **Male reproductive system (human).** This sagittal section of the male pelvis shows the reproductive organs and their relationships to the urinary and digestive systems. The enlarged and cutaway view of the testis and its duct system is shown at the bottom. Spermatozoa produced in the seminiferous tubules eventually pass through the straight tubules into the rete testis and enter the epididymis. Fluid is added as spermatozoa are moved through the vas deferens by contractions of sheets of smooth muscle in its walls.

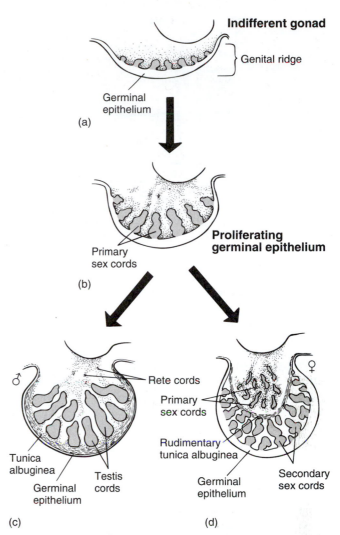

FIGURE 14.18 Urogenital ridge. In the posterior part of the developing embryo, paired urogenital ridges arise within the roof of the coelom. The medial ridges are the genital ridges and give rise to the gonads. The lateral nephric ridges give rise to the kidney and its ducts. Primordial germ cells that develop into eggs or sperm arise outside the gonads, migrate to them, and colonize the early gonad rudiments.

take up a permanent residence. In females, germ cells establish residence in the cortex. In males, arriving germ cells establish residence in the medulla, which develops into the seminiferous tubule (figure 14.19c,d).

Reproductive Tracts

Parts of the embryonic urinary system are salvaged by or shared with the genital system. In female mammals, the mesonephric duct **(wolffian duct)** drains the embryonic mesonephros, but it regresses later in development when the metanephros and its ureter become the kidney of the adult. However, a second parallel **Müllerian duct** arises next to the embryonic mesonephric duct before it regresses. The Müllerian duct, rather than the wolffian, forms the oviduct, uterus, and vagina (figure 14.20). A few mesonephric tubules may persist as the **Paroöphoron** and **Epoöphoron.** In male mammals, the mesonephric duct becomes the vas deferens. Mesonephric tubules and some of the associated ducts contribute to the epididymis. A rudimentary Müllerian duct occasionally arises in embryonic males but never assumes a significant role in the adult male (figure 14.20).

Overview

The urogenital system of vertebrates certainly does not heed the Shakespearean warning, "Neither a borrower nor lender be." Parts evolved first to serve the kidneys (e.g., pronephric duct) later ended up serving the testis in males (e.g., vas

FIGURE 14.19 Embryonic formation of the gonad. (a,b) Thickening of the genital ridge and inward movement of adjacent mesenchymal cells give rise to a swelling, the genital ridge, from the roof of the coelom. Because this early developmental stage is similar in both sexes, it is referred to as the indifferent gonad, which includes cortex and medulla. Primordial germ cells arriving from distant locations outside the embryo usually take up residence in the indifferent gonad. (c) In males, the medulla enlarges to become the testis cords that will form the seminiferous tubules. (d) In females, the cortex expands, forming secondary sex cords that house the follicles.

deferens). In some species, a given duct is shared between the urinary and reproductive systems. In others, the same duct functions in only one of these systems. Even within the same species, homologous parts perform different roles in opposite sexes. Keeping track of these anatomical differences is no simple matter. A prolific terminology that has developed to track these anatomical and functional differences can obscure the underlying unity of the system. In this book, we select a set of terms applicable throughout the vertebrate urogenital system (noting synonyms) and apply it consistently (figure 14.21). When we examine phylogeny, we use

FIGURE 14.20 Duct derivatives male and female (a) The Müllerian and Archinephric ducts have different fates in females and male.
(b) Embryonic formation of the male and female genital systems in mammals. Via an early indifferent stage to male and female anatomies.

(a) After Kalthoft; (b) After Tuchmann-Duplessis and Haegel.

Ducts of the Urogenital System	
General term	*Alternative term*
Archinephric duct	Pronephric duct/mesonephric duct Wolffian duct Opisthonephric duct Ductus deferens (vas deferens)
Müllerian duct	Oviduct
Metanephric duct	Ureter

FIGURE 14.21 Terminology of the urogenital system. Associations of ducts change during evolution and development. Sometimes the same duct performs different roles in males and females. The result has been a proliferation of synonyms, which are summarized in this figure. The duct serving the early pronephros is the pronephric duct, but when the mesonephros replaces the pronephric kidney, this duct now serves the new mesonephros and is called the mesonephric duct. With the advent of the metanephros, this duct degenerates in the female but becomes the vas deferens of the testis in the male. Some authorities prefer the term *archinephric duct* or *wolffian duct* for this structure. Although the term *metanephric duct* might parallel the terms *pronephric* and *mesonephric* ducts, more often the term *ureter* is used for the metanephric duct.

terminology that applies to the homology of reproductive parts throughout vertebrates and place the more common or functional term for a given species and sex in parentheses.

As we saw earlier in our discussion of the kidney, the pronephric duct usually persists and drains the mesonephros or extended opisthonephros. It is renamed the mesonephric duct or the opisthonephric duct, respectively. In some males, this duct transports sperm and is called the vas deferens. In females, it is known embryologically as the wolffian duct. Because this duct plays different roles in different groups, the more general term **archinephric duct** is preferred. The **metanephric duct** is commonly called the ureter. In some males, the kidney divides its services between reproductive and excretory roles. To recognize this, it is common to speak of the **reproductive kidney** and the **uriniferous kidney.**

In females, the archinephric (mesonephric) ducts tend to function only within the urinary system. The Müllerian duct arises embryologically next to the archinephric (wolffian) duct. In males, the Müllerian duct regresses if it appears at all, but in females, the Müllerian ducts become the oviducts of the reproductive system. Released ova enter the oviduct through the **ostium,** which typically flares into a **funnel** (infundibulum) in many vertebrates. The fringed margins of the ostium are the **fimbria** that embrace the ovary. The ovary and ostium are sometimes enclosed in a common peritoneal sac, but usually the oviducts are not connected to the ovaries directly. Instead, the ciliated fimbria

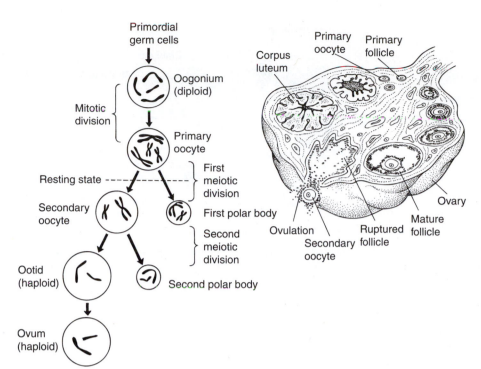

FIGURE 14.22 Oogenesis. Diploid primordial germ cells colonize the ovary of the embryonic female. When they arrive in the ovary, these germ cells are called primary oocytes. They gather around themselves a layer of connective tissue cells to form an ovarian follicle. Most oocytes begin meiosis but do not complete this process until ovulation or later, depending upon the species. Of the hundreds or thousands of oocytes residing in follicles within the ovary, only a few will ever mature, be released at ovulation (after which they are usually termed ova), and become fertilized.

and infundibulum gather released ova and move them into the oviduct. Fertilization, if it is internal, usually occurs soon after the ovum enters the oviduct. Shortly before their terminus, the oviducts may expand into the uterus, the organ in which an embryo is housed and nourished. If the fertilized egg is wrapped in a shell, **shell glands** or **shell-secreting regions** may be evident in the oviduct.

Female Reproductive System

Ovary

The ovary produces both hormones and mature ova (sing., *ovum*). **Oogenesis** is the process of egg maturation, which occurs from the time of its appearance in the ovary until it completes meiosis. Oogenesis is a complex process involving **mitotic** as well as **meiotic** cell division, growth in egg size, and changes in cytoplasmic composition (figure 14.22). Once germ cells take up residence in the ovary, they are called **oogonia.** The diploid oogonia undergo mitotic division, yielding diploid cells. At the end of this phase of development, they are **primary oocytes.** Primary oocytes then begin meiotic cell division for the first time. As a result of the first meiotic division, each ovum yields a first **polar body** and a **secondary oocyte.** Although the first polar body may divide again, its role in helping to reduce the chromosome number is complete; thereafter, it is of little importance. The secondary oocyte undergoes a second meiotic division, yielding a second polar body and a haploid ovum.

A capsule of supportive connective tissue cells termed *follicle cells* forms around the primary oocyte. Follicle cells and the oocyte they embrace form an **ovarian follicle.** The follicle cells contribute to nutritional support and help

build up yolk within the ovum. During the breeding season, selected follicles and the oocytes they contain resume maturation under hormonal stimulation. As meiosis is completed, a **secondary oocyte** is formed. Release of the oocyte from the ovary is termed ovulation.

Much variation occurs during the time before meiosis takes place. These events of oogenesis may occur largely before or after sexual maturity, depending on the species. At the birth of a mammalian female, the primordial germ cells have already migrated into the ovary and started to undergo meiosis, but further oogenesis is usually arrested until the onset of sexual maturity. In fact, not all the primary oocytes mature. For example, the human female is born with half a million primary oocytes in her ovaries, but perhaps only several hundred of these complete oogenesis. The rest eventually degenerate. In some mammalian species, meiosis occurs before ovulation. In other species, it does not occur until after fertilization.

The ovary is suspended from the dorsal wall of the coelom by a mesentery, the **mesovarium** (figure 14.23). Except for cyclostomes, in which eggs escape through secondary pores in the body wall, vertebrate eggs travel through genital ducts after they are released from the ovaries. In most vertebrates, the ovaries are paired; however, in cyclostomes, some reptiles, most birds, the platypus, and some bats, only a single ovary is functional (table 14.1).

Oviparity, viviparity (p. 164)

Genital Ducts

Fishes In cyclostomes, the single large ovary is suspended from the middorsal wall. In lampreys, as many as 24,000 to over 200,000 ovarian follicles can develop in

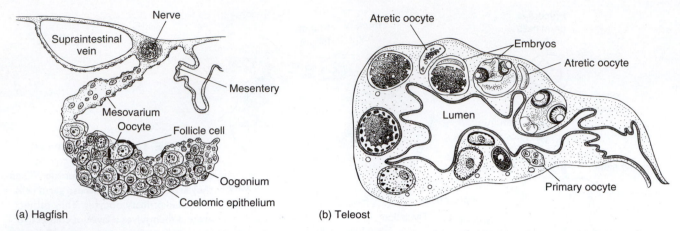

(a) Hagfish

(b) Teleost

FIGURE 14.23 **Fish ovaries.** (a) Hagfish. Oocytes and surrounding follicle cells are held within the ovary. (b) Teleost. Section of the ovary of the guppy *Poecilia reticulata*. Ova are fertilized while they are within the ovary, and they are retained well into embryonic development. There can be one to seven oocytes in progressive stages of development. Atretic oocytes that fail to develop and developing embryos are illustrated.

(a) After Hardisty; (b) after Lambert.

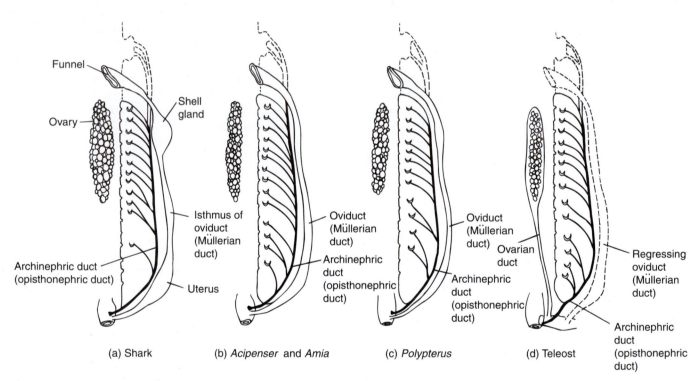

(a) Shark

(b) *Acipenser* and *Amia*

(c) *Polypterus*

(d) Teleost

FIGURE 14.24 **Oviducts of female fishes.** (a) Shark. (b) Sturgeon and bowfin. (c) Bichir. (d) Teleost. The oviduct (Müllerian duct) arises adjacent to and parallel with the archinephric duct in most fishes. In teleosts, the oviduct is usually replaced by an ovarian duct that is derived separately.

synchrony, and are ovulated during a single breeding season. Most lampreys spawn once and die shortly thereafter. Few follicles are present in hagfishes, and little is known about their reproductive behavior. Cyclostome ovaries lack ducts. Instead, eggs are shed into the coelom. From the coelom, they reach the cloaca (in lampreys) or anus (in hagfishes) through secondary pores. The archinephric ducts drain the kidneys exclusively.

In elasmobranchs, the ovaries are initially paired, but in some species, only one may develop. The Müllerian duct, or oviduct, differentiates into four regions: funnel, shell gland, **isthmus,** and uterus (figure 14.24a). The funnel collects the eggs shed from the ovary. Anterior ends of the paired oviduct may fuse into a single funnel, or asymmetric development may leave only one primary funnel. In some species, the shell gland (nidamental gland) stores sperm, but

TABLE 14.1 Vertebrates with One Functional Ovary

Species	Explanation for One-Ovary Condition
Agnatha	
Lampreys	Fusion of two gonads
Hagfishes	One gonad fails to develop
Osteichthyes	
Perches, *Perca*	Fusion of two gonads
Pike perch, *Lucia-Stizostedion* sp.	Fusion of two gonads
Stone loach, *Noemacheilus* sp.	Fusion of two gonads
European bitterling, *Rhodeus ararus*	Fusion of two gonads
Japanese ricefish, *Oryzias latipes*	One gonad fails to develop
Guppy, *Poecilia reticulata*	One gonad fails to develop
Chondrichthyes	
Sharks	
Scyliorhinus	Left ovary becomes atrophic
Pristiophorus	Left ovary becomes atrophic
Carcharhinus	Left ovary becomes atrophic
Galeus	Left ovary becomes atrophic
Mustelus	Left ovary becomes atrophic
Sphyrna	Left ovary becomes atrophic
Rays	
Urolophus	Left ovary functional
Dasyatis	Right ovary absent
Reptilia	
Blind worm snakes, *Typhlops*	Left ovary and oviduct absent
Aves	
Birds	Left ovary functional in most species; right ovary regresses in embryos
Mammalia	
Platypus *Ornithorhynchus anatinus*	Left ovary functional
Bats	
Miniopterus natalensis	Left ovary functional
Miniopterus schreibersi	Right ovary functional
Rhinolophus	Right ovary functional
Tadarida cyanocephala	Right ovary functional
Molossus ater	Right ovary functional
Mountain viscacha, *Lagidium peruanum*	Right ovary functional
Water buck, *Kobus defassa*	Left ovary functional

In female bony fishes, like most other anamniote females, the archinephric ducts serve the kidneys, and the paired oviducts (Müllerian ducts) serve the paired ovaries (figures 14.24b–d and 14.25a–c). In some teleosts, such as salmonids, eggs released from the ovaries fill the body cavity. Eventually they reach short, funnel-like remnants of the oviducts situated at the posterior part of the coelom. However, in many teleosts, the oviducts regress entirely, leaving egg transport to new **ovarian ducts** (figure 14.26a–c). These ovarian ducts are not homologous to the oviducts (Müllerian ducts) of other vertebrates. Instead, they are derived from peritoneal folds that embrace each ovary and have grown posteriorly to form new ducts.

Most teleost fishes lay eggs, but some bear live young. Among these viviparous teleosts, maternal tissues may nourish the embryo. One extreme case is found in the family of teleosts that includes the guppy. In this group, fertilization occurs while the ova are still in the ovarian follicles. The ovary continues to hold the embryos during subsequent development until they are released as tiny fry. Oocytes that fail to reach a point in maturation where they can be fertilized usually undergo involution and are called **atretic oocytes** (figure 14.23b). Recycling of atretic tissue provides nutrition for the surviving oocytes.

Tetrapods Amphibian ovaries are paired, hollow structures that usually show a prominent cortex covered by germinal epithelium. The genital ducts of female amphibians are usually simple and consistent. The archinephric ducts serve the opisthonephric kidneys; the oviducts (Müllerian ducts) serve the ovaries.

In amniotes, remnants of the mesonephros may persist in larval stages, but adults have metanephric kidneys drained exclusively by new paired ducts, the ureters (metanephric ducts). In females, the archinephric ducts are rudimentary. The oviducts (Müllerian ducts) persist in their roles of transporting ova from the ovaries and supporting the embryo while it is in transit. The tubular oviducts (Müllerian ducts) of amniotes often have prominent sheets of smooth muscle within their walls and a lumen lined by a secretory mucosa. In oviparous amniotes, a shell gland may be prominent; in viviparous amniotes the uterus may be distinct (figures 14.27a–c and 14.28a–d).

Oviduct

After ovulation, the fimbria move the ovum into the oviduct. If fertilization is internal, the ovum and sperm meet almost immediately in the upper reaches of the oviduct. If fertilization is external, the smooth muscle and cilia lining the oviduct drive the ovum to the outside, where it is fertilized.

In addition to transporting the ovum, the oviduct in some vertebrates may add layers of membrane or a shell. In many species, parts of the oviduct are specialized as distinct shell glands that add these coats. Because membranes and shells are impervious to sperm, they are added

in most elasmobranchs, it secretes albumen and mucus. In oviparous species, the shell gland produces the egg case as well. In viviparous species, in particular, the shell gland may be indistinguishable. The isthmus connects the shell gland to the uterus. The uterus nutritionally supports embryos if they are held in the oviduct for an extended period. Oviducts may join before they enter the cloaca, or they may enter separately. The genital ducts of chimaeras are similar to those of sharks, except the oviducts always share a common funnel and each oviduct opens separately into the cloaca. The archinephric duct drains the female opisthonephric kidney.

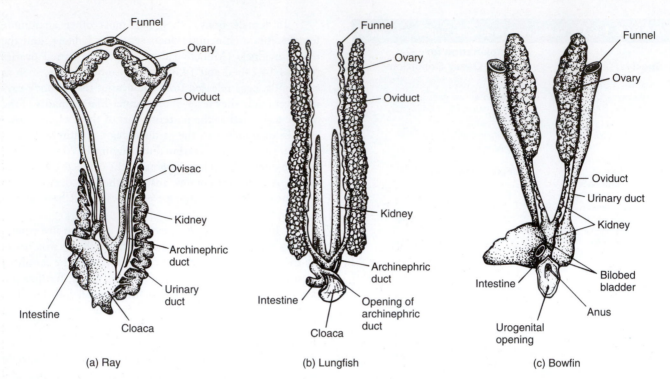

FIGURE 14.25 **Urogenital systems of female fishes** (a) Ray, *Torpedo*. (b) Lungfish, *Protopterus*. (c) Bowfin, *Amia*.

After Romer and Parsons.

(a) Ray (b) Lungfish (c) Bowfin

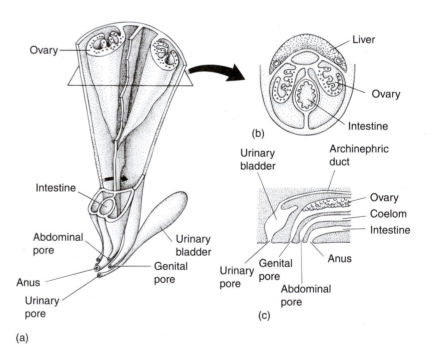

FIGURE 14.26 **Urogenital system of a teleost female.** (a) Ventral view, partially cut away, of the urogenital system in a generalized teleost fish. Ovaries are suspended from the dorsal wall, and release ova into the genital funnels formed from folds in the peritoneal wall. The coelom connects with the outside through abdominal pores. Feces are eliminated via the anus, and urine via the urinary pore of the bladder. (b) Cross section at the level of the ovaries. (c) Sagittal section.

after fertilization. In birds and egg-laying reptiles, a layer of albumen, and then a shell membrane, and finally a calcareous outer shell are added as the fertilized ovum slides along the oviduct (figure 14.29). The encapsulated egg is then held within the oviduct until a suitable environmental site in which to lay it is prepared.

Uterus

The uterus is the terminal portion of the oviduct. Shelled eggs waiting to be laid or embryos completing their development are held within the uterus. In eutherian mammals and a few other vertebrates, the walls of the uterus and extraembryonic membranes of the embryo establish a close vascular

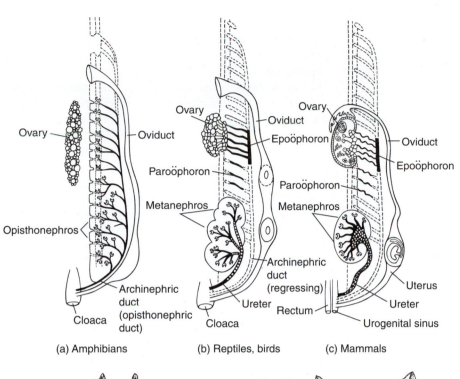

Ovary

Ovary — Oviduct

Opisthonephros

Archinephric duct (opisthonephric duct)

Cloaca

(a) Amphibians

Ovary — Oviduct

Epoöphoron

Paroöphoron

Metanephros

Archinephric duct (regressing)

Ureter

Cloaca

(b) Reptiles, birds

Ovary — Oviduct

Epoöphoron

Paroöphoron

Metanephros

Uterus

Ureter

Rectum

Urogenital sinus

(c) Mammals

FIGURE 14.27 Urogenital anatomy of tetrapod females.
(a) Amphibians. (b) Reptiles and birds. (c) Mammals.

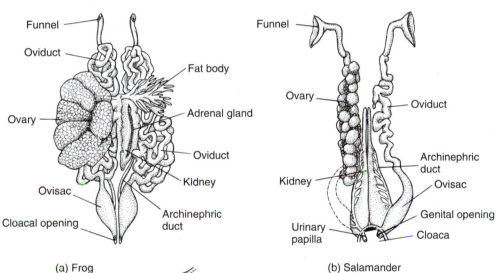

Funnel

Oviduct

Fat body

Ovary

Adrenal gland

Oviduct

Kidney

Ovisac

Archinephric duct

Cloacal opening

(a) Frog

Funnel

Ovary — Oviduct

Archinephric duct

Kidney

Ovisac

Urinary papilla

Genital opening

Cloaca

(b) Salamander

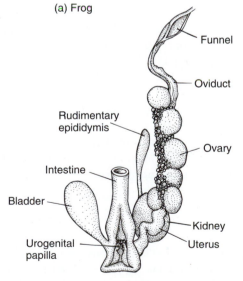

Funnel

Oviduct

Rudimentary epididymis

Ovary

Intestine

Bladder

Urogenital papilla

Kidney

Uterus

(c) Reptile

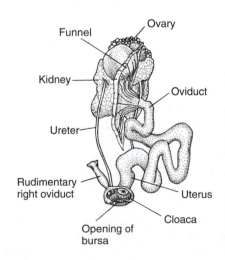

Funnel — Ovary

Kidney

Oviduct

Ureter

Rudimentary right oviduct

Uterus

Cloaca

Opening of bursa

(d) Bird

FIGURE 14.28 Urogenital systems of tetrapod females, ventral views. (a) Frog, *Rana.* The intestine, urinary bladder, and left ovary have been removed to reveal underlying structures. Urinary ducts of the right side are pulled away from the kidney to show their course. (b) Salamander, *Salamandra.* (c) Reptile, *Sphenodon.* (d) Bird, *Columba.*

After Romer and Parsons.

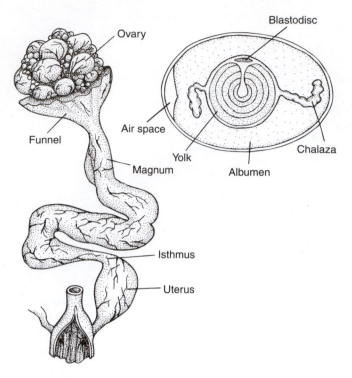

FIGURE 14.29 **Oviduct of a hen.** After ovulation, ova are gathered by the infundibulum of the oviduct. If egg membranes are to be added, fertilization occurs here in the upper reaches of the oviduct. The oviduct adds a coat of albumen, a shell membrane, and eventually a calcareous shell.

association through a **placenta.** Nutrients and oxygen are transported to the developing embryo, and carbon dioxide is given up to the maternal circulation via the placenta.

Placentae (p. 192)

In therian mammals, the terminal ends of the oviduct tend to merge into a single uterus and **vagina** that lie along the body's midline. The vagina receives the male penis or intromittent organ during copulation. The female homologue of the male penis is the **clitoris.** Unlike the penis, the clitoris participates neither in gamete transfer nor in urination.

Male Reproductive System

Testis

Except in cyclostomes and some teleosts, testes are paired, and each is suspended from the dorsal wall of the coelom by a mesentery, the **mesorchium.** The testes of vertebrates have two functions—sperm production and hormonal secretion. The hormones of the testes are steroids collectively called **androgens.** The principal androgen is **testosterone,** secreted primarily by the **interstitial cells** (Leydig cells) of the testes. Testosterone controls the development and maintenance of secondary sexual characteristics, elevates the sex impulse (or libido), and helps maintain the genital ducts and accessory sex organs. More will be said about the endocrine role of the testes in chapter 15.

During the breeding season, primordial germ cells in the testes begin the process termed **spermatogenesis,** whereby selected germ cells eventually become spermatozoa. Spermatogenesis (like oogenesis) involves both mitotic and meiotic divisions, as well as cytoplasmic reorganization (figure 14.30). In vertebrates, there are two general patterns of spermatogenesis, one in anamniotes and the other in amniotes.

Amniotes In reptiles, birds, and mammals, sperm form within the luminal wall of the seminiferous tubules, which lack subcompartments. Resident primordial germ cells, more commonly called **spermatogonia** at this stage, divide by mitosis. One member of the resulting pair of cells stays within the wall of the seminiferous tubule to propagate further spermatogonia, while the other grows in size. At the end of this growth, the diploid spermatogonium is called a **primary spermatocyte** and begins meiotic division. During meiosis, it briefly becomes a **secondary spermatocyte** and then a haploid **spermatid;** thereafter, it undergoes no further division. However, spermatids undergo cellular reorganization in which nuclear DNA condenses and excess cytoplasm and organelles are jettisoned to form sleek **spermatozoa,** or sperm.

For a time, **Sertoli cells** embrace and nutritionally support spermatids, perhaps promoting further maturation. Most sperm are stored in the lumina of the seminiferous tubules and in the connecting epididymis. At orgasm, sheets of smooth muscle in the walls of the ducts rhythmically contract, forcibly expelling sperm in the process of **ejaculation.** Sperm are transported in a thick, composite fluid secreted by accessory sex glands. In mammals, there are three such glands. The bulbourethral gland discharges mucus during erection and ejaculation. The prostate gland secretes an alkaline substance during ejaculation to protect the sperm from the acidity of any urine remaining in the male urethra. Finally, the seminal vesicle adds a thick secretion rich in the sugar fructose as a source of nutritional support for the sperm.

Anamniotes In fishes and amphibians, sperm are produced in clones, each located within a cyst or follicle, all of which are housed in separate tubular compartments within the testes (figure 14.31a,b). Generally, a spermatogonium is engulfed by one or several connective tissue cells, called (as in females) follicle cells, that become functional Sertoli cells as maturation proceeds. Proliferation of a spermatogonium within the follicle (Sertoli) cells produces a nested clone of many spermatogonia, sometimes called a spermatocyst. Cells within this **spermatocyst** undergo spermatogenesis in unison, eventually producing mature sperm.

Genital Ducts

Fishes In cyclostomes, the large unpaired testes are not served by any genital ducts. Sperm are shed into the coelom and exit via abdominal pores. The archinephric

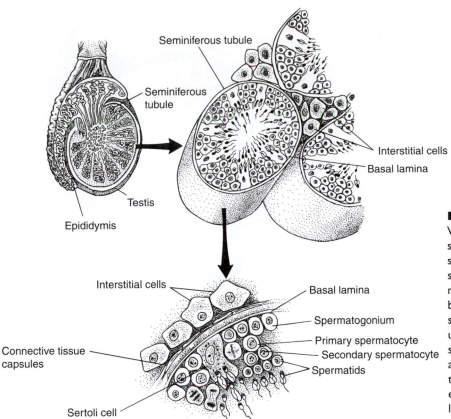

Seminiferous tubule

Seminiferous tubule

Interstitial cells

Basal lamina

Testis

Epididymis

Interstitial cells

Basal lamina

Spermatogonium

Primary spermatocyte

Secondary spermatocyte

Spermatids

Connective tissue capsules

Sertoli cell

FIGURE 14.30 Spermatogenesis.
Within the walls of the seminiferous tubules, spermatogonia divide, giving rise to cells that stay in place and preserve the population of spermatogonia as well as to cells that undergo meiotic and cytological reorganization. These become first primary and then secondary spermatocytes. Secondary spermatocytes undergo changes that transform them into spermatozoa. Sertoli cells hold spermatozoa and then release them into the lumen of the seminiferous tubules and connecting epididymis. Interstitial cells (cells of Leydig) lying between the seminiferous tubules secrete male hormones.

ducts drain the kidneys exclusively (figure 14.32a). In elasmobranchs, the prominent Müllerian ducts of the female are rudimentary in the adult male (figure 14.32b). The **accessory urinary ducts,** distinct from the archinephric ducts, are usually present to service the posterior uriniferous kidney (figure 14.33a). Each anterior reproductive kidney has short tubules that join the testis to the archinephric duct, which because of its role in sperm storage and transport may be termed a vas deferens (figure 14.32b). These tubules within the anterior part of the kidney function as an epididymis, connecting the rete testis to the vas deferens and perhaps storing sperm. Adjacent Leydig cells in this cranial region secrete seminal fluid into the genital ducts.

In bony fishes, the archinephric ducts drain the kidneys and may receive sperm from the testes. However, the testes tend to develop separate sperm ducts and routes of exit (figure 14.33b,c). In most teleosts, this separate duct system forms a **testicular duct,** which is not homologous to the archinephric duct and may even establish its own opening to the exterior (figure 14.33d). Some teleosts, such as salmonids, lack sperm ducts entirely. Sperm are released into the body cavity and exit the body through pores near the posterior part of the coelom.

Tetrapods In male amphibians, several genital duct configurations can occur (figure 14.34a,b). In *Necturus* and a few other species, the archinephric ducts transport both sperm from the testes and urine from the uriniferous kidneys. However, this is likely a specialized condition of the paedomorphic *Necturus*. In general, this condition occurs only in larval salamanders. In some salamander families, new accessory urinary ducts service the caudal kidneys, and sperm are transported from the testes through tiny ducts in the cranial kidneys to the archinephric ducts (vas deferens) to be stored. In all frogs and a few species of salamanders, tiny ducts that reach directly from the testes to the archinephric ducts bypass the anterior part of the kidneys. Elimination from the uriniferous kidneys occurs exclusively via the accessory urinary ducts. Thus, in some adult amphibians, the archinephric ducts may have both reproductive and excretory roles, whereas in other species, these ducts may be involved exclusively in sperm transport and new accessory urinary ducts may drain the opisthonephros (figures 14.35a,b and 14.36a–c).

In male amniotes, the archinephric duct (vas deferens) transports sperm exclusively (figures 14.35c,d and 14.36c,d). Several mesonephric tubules of the embryonic kidney may contribute to the epididymis that connects each testis to a vas deferens (figure 14.37). Each amniote kidney is drained by a new duct, the ureter (metanephric duct).

In most vertebrate males, the testes reside within the abdomen; however, the testes of most mammals descend into the **scrotum,** a coelomic pouch suspended outside the body but connected to the abdominal coelom via an **inguinal canal** (figure 14.35d). In other mammals, the testes either remain in the body cavity (e.g., monotremes, some primitive

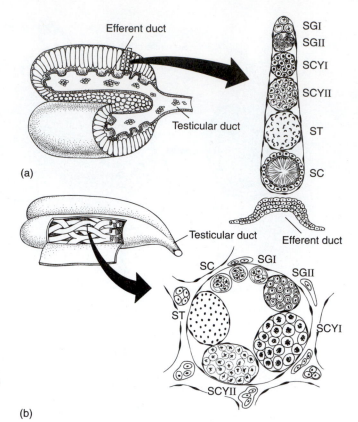

FIGURE 14.31 Sperm production in the teleost testis. Sperm may develop within compartments (a) or tubules (b). During copulation, mature sperm pass into the testicular duct. Primary spermatogonia (SGI) become, in succession, secondary spermatogonia (SGII), primary spermatocytes (SCYI), secondary spermatocytes (SCYII), and spermatids (ST). Sertoli cells (SC) form part of the epithelium lining the compartments or tubules.

After van Tienhover, based on van den Hurk, 1975.

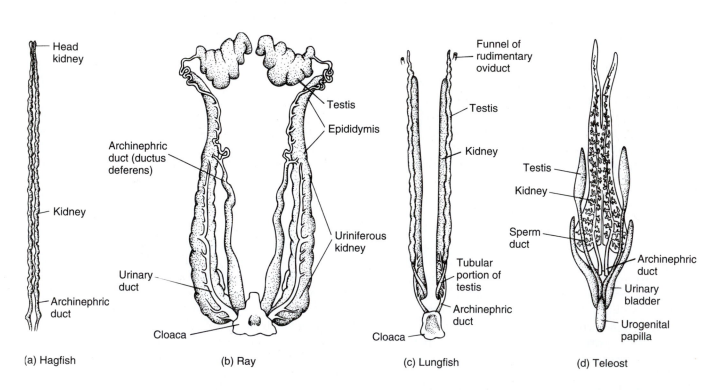

(a) Hagfish   (b) Ray   (c) Lungfish   (d) Teleost

FIGURE 14.32 Urogenital systems of male fishes. (a) Hagfish, *Bdellostoma.* The single testis of the hagfish hangs in the dorsal body wall between the kidneys. (b) Elasmobranch, *Torpedo.* (c) Lungfish, *Protopterus.* (d) Teleost, sea horse *Hippocampus.*

After Romer and Parsons.

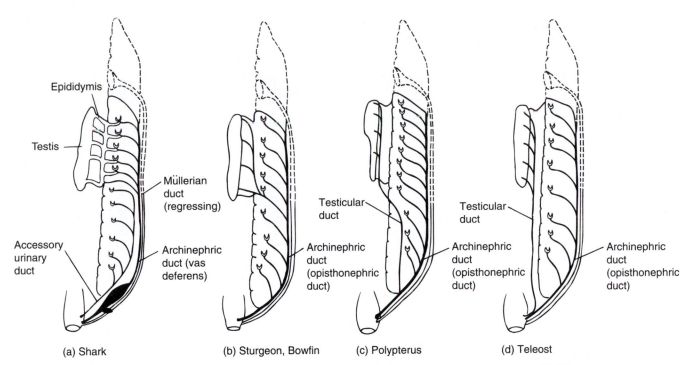

FIGURE 14.33 **Urogenital ducts of male fishes.** (a) Shark. (b) Sturgeon and bowfin. (c) Bichir. (d) Teleost. In sharks, an accessory urinary duct develops to drain the kidney, and the archinephric duct is concerned with sperm transport. In other groups, additional ducts that develop to drain the testis sometimes join the archinephric duct. In teleosts, these exit independently.

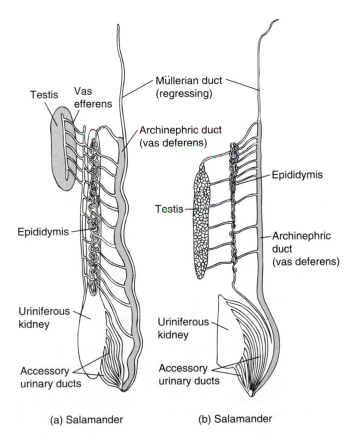

FIGURE 14.34 **Urogenital systems of male amphibians.** (a) Salamander, *Ambystoma.* (b) Salamander, *Gyrinophilus.*

(a) After Baker and Taylor; (b) after Strickland.

insectivores, sirenians, elephants, sloths, cetaceans, armadillos) or descend into a muscular pouch but not a true coelomic scrotum (e.g., moles, shrews, many rodents, lagomorphs, pinnipeds, hyenas). Some mammals have testes that descend temporarily into the scrotum during the breeding season (e.g., chipmunks and squirrels, some bats, some primates). Absence of a scrotum in monotremes, as in sauropsids, means that testes are held internally within the abdominal cavity. This likely represents a primitive condition in monotremes. The presence of a scrotum in later therian mammals is a derived condition. Many marsupials have a scrotum, but it is prepenile, in front of the base of the penis. When present in eutherian mammals, it is postpenile. The functional significance of this difference is not known, nor is it known why some eutherian mammals lack a scrotum. However, absence of a scrotum in streamlined cetaceans is likely an adaptation to hydrodynamics. Although they are within their warm body cavity, the cetacean testes are cooled by special vascularization dedicated to the function. Most other eutherian mammals have testes that descend permanently, which typically occurs during embryonic development.

The testes migrate from the body cavity, through the abdominal wall via the inguinal canal, and into the scrotum where the temperature is cooler, often up to 8°C cooler than in the abdomen. The external cremaster muscles lift the testes closer to the body under cool conditions and allow them to descend under warm conditions, thus warming or cooling the testes as required. Further, arteries and veins entering and leaving the testes intermingle in a **pampiniform plexus**, a countercurrent exchange mechanism that serves as a heat

Amphibians

(Most urodeles and anurans)

(Some urodeles and anurans)

Reptiles, birds

Mammals

Testis

Müllerian duct

Archinephric duct (vas deferens)

Cloaca

(a)

Testis

Cloaca

Accessory urinary duct

Archinephric duct (vas deferens)

(b)

Testis

Archinephric duct (vas deferens)

Cloaca

Ureter

(c)

Testis

Epididymis

Archinephric duct (vas deferens)

Ureter

Rectum   Urethra

(d)

FIGURE 14.35 **Urogenital ducts of tetrapod males.** (a) Most urodeles and most anurans (adults). (b) Some urodeles and some anurans (adults). (c) Reptiles and birds. (d) Mammals.

block to the testes. If the testes fail to descend (a condition called **cryptorchidism**) or are artificially warmed in the scrotum, sperm production falls or even ceases in these species. Thus, the testes in mammals with scrotums seem to have lost the capacity to function at body temperature. However, why some mammals have evolved a scrotum and others have not is still not understood.

Copulatory Organs

In most water-dwelling vertebrates, fertilization is external. Eggs and sperm are shed simultaneously from the body into the water, where fertilization occurs. However, if the female uterus houses the embryo or if a shell seals an egg, sperm must fertilize the egg before it descends from the oviduct. In these instances, fertilization is internal. Sperm deposited within the female genital tract journey to the upper reaches of the oviduct to fertilize the egg. In many vertebrates, **copulation (coitus)** involves the direct, momentary apposition of the male and female cloacae to transfer sperm. Often, however, the male possesses external **intromittent organs** specialized

to deliver sperm during coitus. In salamanders, sperm transfer is external and involves a **spermatophore,** but fertilization is internal.

In male sharks, rays, chimaeras, and some placoderms, the pelvic fins are specialized as **claspers** (figure 14.38a–c). During copulation, one clasper is inserted into the female cloaca and its terminal cartilages spread by muscle action to help hold the clasper in place. Sperm leave the male cloaca, enter a groove on the clasper, and are flushed by water squirted from siphon sacs within the body wall of the male into the female cloaca. In the killfish (*Fundulus*), a teleost, pelvic and anal fins interlock during spawning, holding male and female cloacae close together as gametes are released (figure 14.39a). In a few species of teleosts, the anal fin is fashioned into a grooved intromittent organ, termed a **gonopodium,** that deposits sperm into the female during copulation (figure 14.39b).

Fertilization in almost all frogs is external. The male grasps the female from above in a behavior called **amplexus** and releases sperm from his cloaca as eggs leave the female's cloaca. An exception among frogs is the tailed frog, *Ascaphus.*

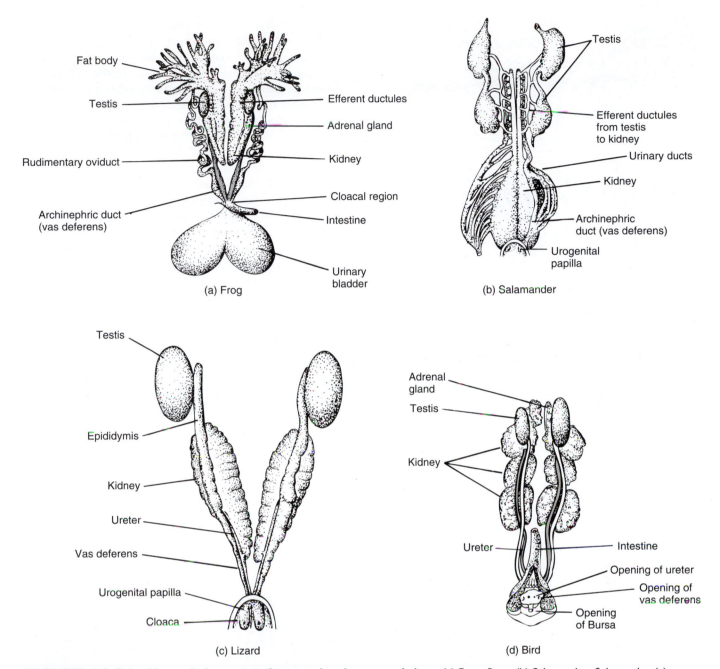

FIGURE 14.36 **Urogenital systems of tetrapod males, ventral view.** (a) Frog, *Rana.* (b) Salamander, *Salamandra.* (c) Lizard, *Varanus.* (d) Bird, *Columba.*

After Romer and Parsons.

The male possesses a short, grooved, tail-like extension of the cloaca used to transfer sperm directly into the cloaca of the female. The males of most salamander species produce a spermatophore, which consists of a cap of sperm on top of a gelatinous pedestal (figure 14.40a–c). The spermatophore is deposited in front of the female at the culmination of a stylized courtship. The female nips off the sperm cap with the lips of her cloaca to gather in the sperm (figure 14.41). Females of some species collect only a portion of each spermatophore sperm cap, but they sample from as many as 20 or 30 different spermatophores. Sperm are stored in a dorsal pocket of the cloaca, the **spermatheca,** until they are released to fertilize eggs internally as ova travel from the oviducts and out the cloaca. This method of reproduction decouples sperm transfer from fertilization. Thus, sperm transfer may occur at a time and place favorable to courtship but not to egg deposition. In caecilians, the male everts the posterior part of his cloaca through the vent and fits it into the female cloaca to aid sperm transfer.

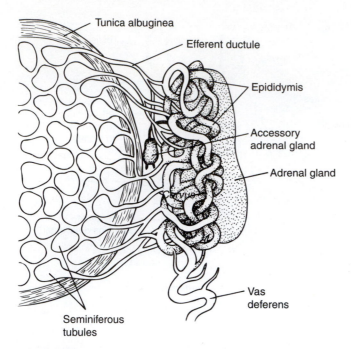

FIGURE 14.37 Avian testis and epididymis in the jackdaw *Corvus*.

After Lake.

Well-developed intromittent organs are absent in *Sphenodon*, although paired out-pouchings, supplied with thin strips of muscle, have been reported stationed at corners of the cloaca near the base of the tail. During courtship, the cloacae are pressed together and sperm are transferred directly. Some male birds and male turtles, crocodiles, and mammals have a single **penis,** an intromittent organ down the midline of the body (figure 14.42a–c). The evolutionary origin of the penis is unknown, but it seems to be a derivative of the cloaca. When not in use, the penis is usually flaccid and may be retracted into a sheath or returned to the cloacal chamber. It becomes engorged with blood or lymph that fills its specialized compartments and makes it erect. When the penis is erect, it penetrates the female and holds the channel open to ejaculate sperm. Erection achieved by blood infiltration is termed **hemotumescence.** In turtles, the midventral penis consists of two parallel bands of sinusoidal tissue, the **corpora cavernosa.** Between them lies a groove, the sulcus spermaticus (figure 14.42a,b). When engorged with blood, the corpora cavernosa enlarge, protrude the penis from the cloacal wall through the vent, and shape the sulcus spermaticus into a duct that receives and transfers sperm from each vas deferens. The females of some turtle species possess a homologue to the penis. Although this structure

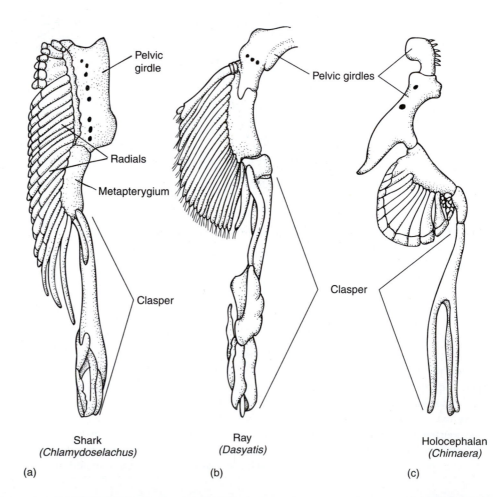

FIGURE 14.38
Intromittent organs of chondrichthyans.

(a) Shark, *Chlamydoselachus*. (b) Ray, *Dasyatis*. (c) Holocephalan, *Chimaera*.

After van Tienhoven.

(a) Shark (*Chlamydoselachus*)

(b) Ray (*Dasyatis*)

(c) Holocephalan (*Chimaera*)

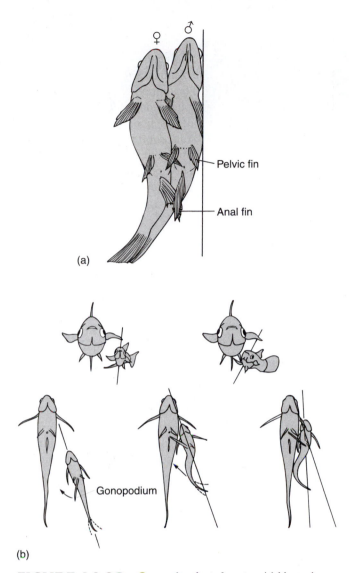

(a)

Pelvic fin

Anal fin

(b)

Gonopodium

FIGURE 14.39   Spawning in teleosts. (a) Ventral view of interlocking anal and pelvic fins of *Fundulus.*
(b) Gonopodium of male is inserted into anal region of female.

may be functionless, it possibly completes the other half of the male's sperm groove and therefore contributes to the sperm transfer channel.

The penis of male crocodilians is similar to that of turtles, except that it is relatively longer and the whole organ projects farther from the cloaca (figure 14.42c–e). Although the mechanism of erection is not clear, hemotumescence that defines a sulcus spermaticus seems to be involved. Female crocodilians also possess a rudimentary homologue of the male penis, but it remains within the cloaca and does not protrude.

In lizards and snakes, males possess a pair of intromittent organs, the **hemipenes.** Each hemipenis is usually grooved to allow for sperm transport. It is rough or spinous at its tip to ensure secure engagement when the male inserts it into the female's cloaca. A retractor muscle returns each hemipenis to the body by turning it outside in, a process called invagination. The retractor pulls it into a pocket located at the base of the tail, behind the vent. During erection, muscle action and hemotumescence force each hemipenis through the cloaca and balloon it out through the vent, turning it inside out—this is evagination (figure 14.43a,b). A sulcus spermaticus is defined in each hemipenis, which is sometimes Y shaped. During copulation, only one hemipenis is inserted in the cloaca of the female (figure 14.43c).

In birds, two types of intromittent organs are found. In the domestic turkey, little more than the edges of the cloaca swell during copulation (figure 14.44a). Male and female cloaca are pressed together at coitus. Semen flows between the lateral penile swellings of the male and is ejaculated into the female cloaca. Ostriches and some other groups have another intromittent organ. It is a true penis with an erectile shaft that the male inserts into the female cloaca. In the male ostrich, the erect penis is conical and widened at its base. It bears a sulcus spermaticus along its length (figure 14.44b). In ducks, the erect penis may be

---

BOX ESSAY 14.5   From da Vinci to Viagra

The Renaissance scientist, artist, and inventor Leonardo da Vinci (1452–1519) quietly dissected, at some risk of censure from religious voices, the bodies of the human deceased. This improved his art of human figures but also his understanding of biological function. Penises from men recently hung inspired his discovery of the basis of erection—the filling of blood sinuses produced erection. Perhaps from his own experiences, he observed that "the penis does not obey the order of its master, the penis erects

freely while its master is asleep." In fact, episodes of nightly arousal also occur in women, usually during dreaming. Actually these arousals are presided over by central nervous system activity, but in complex ways. Excitatory nerves to the penis release chemicals that cause constricted arteries to the penis to relax, thus increasing blood flow to penis sinuses, which in turn fill and expand within, and the penis erects. Viagra works by slowing or preventing the breakdown of these natural relaxing chemicals; hence prolonging

penile filling. As the penis fills, veins draining it are compressed, reducing emptying, and erection continues. An erection lasting over four hours is considered a medical emergency, because the turgid erection essentially deprives the penis of new blood carrying critical oxygen. Normally, after a man climaxes, other nerves limit blood flow, sinuses drain, and the penis becomes flaccid, and in cool weather experiences further "shrinkage," as the comedian Jerry Seinfeld made known to the world.

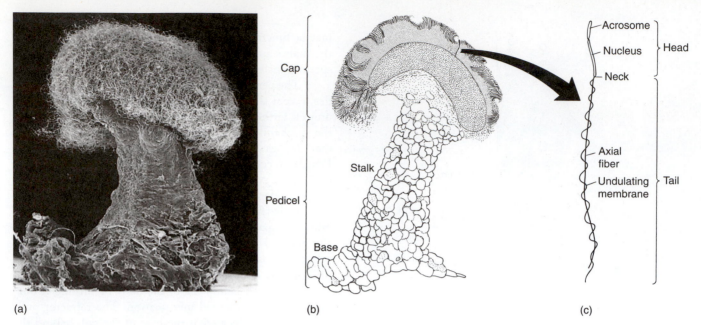

(a) (b) (c)

FIGURE 14.40 **Spermatophores of amphibians.** (a) Whole spermatophore deposited by male *Ambystoma macrodactylum*. (b) Longitudinal section of a spermatophore from *Ambystoma texanum*. Generally, sperm heads point outward, tails point inward. (c) An enlarged spermatozoon.

(a) Kindly supplied by E. Zalisko.

FIGURE 14.41 **Courtship in the blackbelly salamander *Desmognathus quadramaculatus*.** (a) The male approaches the female and rubs his head under hers. (b) The male slides under the female's chin and moves forward; she follows and straddles his tail, now laterally undulating (solid arrows). (c) The male deposits the spermatophore (open arrow) on the substrate and moves forward. (d) The female follows, and the cap of the spermatophore is taken into her cloaca. Sperm are stored in specialized tubules until they are used to fertilize ova internally, just before they are laid several weeks later. (About five minutes elapsed time.)

Based on the research of P. Verrell.

(a) (b) (c) (d)

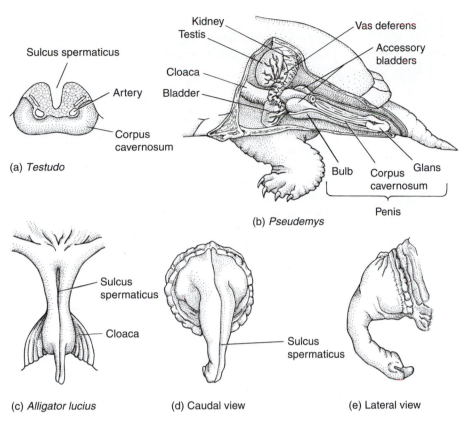

FIGURE 14.42 **Penises of reptiles.** (a) Turtle, *Testudo*: cross section of the penis within the cloaca. (b) Turtle, *Pseudemys*: sagittal section of the penis. (c) Alligator, *Alligator lucius* penis. Caudal (d) and lateral (e) views of the penis of the crocodile, *Crocodylus palustris*.

(a,c–e) After A. S. King; (b) after van Tienhoven.

(a) *Testudo*

Sulcus spermaticus
Artery
Corpus cavernosum

(b) *Pseudemys*

Kidney
Testis
Vas deferens
Accessory bladders
Cloaca
Bladder
Bulb
Corpus cavernosum
Glans
Penis

(c) *Alligator lucius*

Sulcus spermaticus
Cloaca

(d) Caudal view

Sulcus spermaticus

(e) Lateral view

Crocodylus palustris

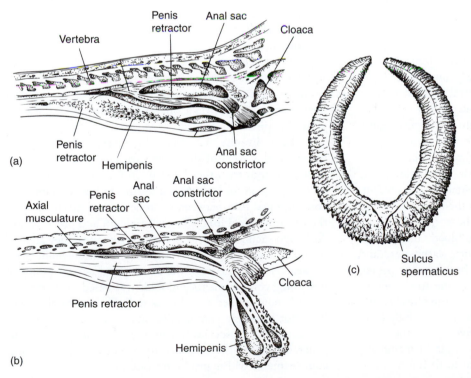

FIGURE 14.43 **Hemipenis of a snake.** Lizards and snakes have paired hemipenes, but usually only one is used during copulation. (a) The hemipenis is pulled back into the body by the retractor muscle (sagittal view). (b) When erect, the hemipenis's internal sinuses become engorged with blood, and it pops through the vent (sagittal view). During copulation, the male inserts its hemipenis into the cloaca of the female. Sperm travel down the sulcus spermaticus into the female. (c) One of the two hemipenes from the rattlesnake *Crotalus atrox* is shown everted. This single hemipenis is divided, which gives it a horseshoe shape. Note the divided sulcus spermaticus that runs along each arched branch of the hemipenis.

Modified from Dowling and Savage, 1960.

(a)
Vertebra
Penis retractor
Anal sac
Cloaca
Penis retractor
Hemipenis
Anal sac constrictor

(b)
Axial musculature
Penis retractor
Anal sac
Anal sac constrictor
Penis retractor
Hemipenis
Cloaca

(c)
Sulcus spermaticus

Semen

Penile swelling

(a) Turkey

Sulcus
spermaticus

(b) Ostrich

Lymphatic
chambers

Lymphatic
cavity

Ligament

Sulcus
spermaticus

Sulcus
spermaticus

(c) Duck

(d) Duck

FIGURE 14.44 **Intromittent organs of birds.**
(a) Domestic turkey with penile swellings. The margins of the
cloaca form the central gully down which sperm flows during
copulation. (b) Erect ostrich penis. (c) Erect duck penis with
bird in standing position. Cross section shows lymphatic cavities
thought to be responsible for eversion of the penis from the
cloaca. (d) Diagrammatic lateral view of lymphatic chambers
whose filling is thought to be responsible for penile erection.

After A. S. King.

quite elaborate, with the sulcus spermaticus spiraling along
the tapering shaft. When relaxed, the penis is coiled and
tucked within the cloaca along the ventral wall. Lymphatic
channels within the penis connect to expanded chambers.
The mechanism of erection is thought to involve filling
of these internal chambers. As a result, the penis projects
from the cloaca and bends forward (figure 14.44c,d).

All mammals copulate with a penis. In addition to the
paired corpora cavernosa, a third sinusoidal tissue is present,
the **corpus spongiosum** that surrounds the closed sulcus, or
cavernous urethra (figure 14.45a). These spongy sinuses
in the penis become engorged with blood and stiffen.
In addition, insectivores, bats, rodents, carnivores, and
most primates except humans have a **baculum** (os penis),
a permanent bone located within the connective tissue of
the penis to stiffen it. In these mammals, the already stiffened penis becomes engorged with blood into a fully erect
position (figure 14.45b,c). The sensitive tip of the penis
is the **glans penis.** The male penis is single in mammals,
although in marsupials the tip is forked to fit into the two

lateral vaginas of the female. As a result, ejaculated sperm
move into each lateral vagina and then into the vaginal
sinus, a chamber that receives both uteri (figure 14.51).

Cloaca

The cloaca has already been defined as a common chamber
receiving products from kidneys, intestines, and often gonads.
It opens to the outside through a cloacal opening or vent.
(It is customary to point out that in Latin *cloaca* means
"sewer.") The cloaca arises at some point during embryonic development in all vertebrates, but in many it becomes
subdivided, lost, or incorporated into other adult structures (figure 14.46a–f). A well-developed cloaca occurs in
adult sharks and lungfishes (figure 14.46b,d). But in teleosts, distinct urinary, anal, and genital openings are present,
replacing the cloaca (figure 14.46f). Among tetrapods, a cloaca is present in amphibians, reptiles, birds, and monotremes.
A shallow cloaca persists even in marsupials (figure 14.47a–k).

A cloaca is apparently a primitive vertebrate feature
because it occurs in most primitive gnathostomes and persists
in the embryos of almost all vertebrates. Its absence in chimaeras (Holocephali), ray-finned bony fishes (Actinopterygii),
Latimeria (coelacanth), and most eutherian mammals may
represent independent losses.

Embryologically, the cloaca arises from hindgut
endoderm and proctodeal ectoderm. Structurally, three
functions influence it: defecation, urination, and copulation. Each function tends to be associated with a compartment, and each compartment is controlled by muscles
that regulate the entry and departure of products from
the intestines, kidneys, and gonads. The most proximal
compartment is the **coprodeum** into which the intestine
empties. The **urodeum** receives products from urinary
and genital ducts. The most distal compartment is the
proctodeum, which functions in copulation and in many
amniotes develops a penis (figure 14.47a). Many urogenital ducts, upon approach to the cloaca, inflate slightly to
form an expanded **urogenital sinus.** These ducts often
open into the cloaca via a small projection called the **urogenital papilla.**

Late in the nineteenth century, Hans Gadow
suggested that each of the three cloacal compartments was
separated from the other by folds in the mucosal wall: the
rectocoprodeal fold between intestine and coprodeum, the
coprourodeal fold between coprodeum and urodeum, and
the **uroproctodeal fold** between urodeum and proctodeum.
Although such folds occur in many vertebrates, they are low
or absent in some, making it difficult to delineate boundaries
between compartments of the cloaca. Gadow's terminology
describing compartments and folds was based on tetrapods,
but it is now applied to fishes as well. Unfortunately, there
have not been any comparative studies of fishes in which
a large sample of species was examined, so it is difficult to
generalize about the presence or absence of these cloacal
compartments within groups of fishes.

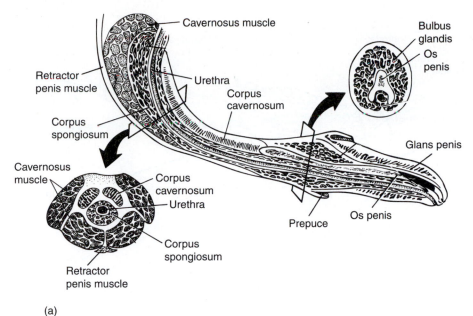

(a)

FIGURE 14.45 Penile erection in the dog. (a) Sagittal view and cross sections of the penis. (b) Flaccid penis. Arterial blood enters the internal pudendal artery, circulates through capillaries of the penis, and flows from the penis through the pudendal vein. (c) Erect penis. Stimulation of the nerves of erection causes increased blood flow to the penis (1). In addition, partial inhibition of venous drainage (solid arrows at (2) results in diversion of blood into the cavernous bodies (3) (corpus cavernosum and bulbus glandis), which fill, stiffen the penis, and result in erection. The os penis (baculum) also helps firm the penis.

After Miller, Christensen, and Evans.

(b) Flaccid

(c) Erect

(a) Lamprey

(b) Shark ♀

(c) Chimaera ♀

(d) Australian lungfish

(e) Sturgeon ♀

(f) Salmon ♀

FIGURE 14.46 Cloacal and anal regions of fishes. (a) Lamprey. (b) Female shark. (c) Female chimaera. (d) Australian lungfish. (e) Female sturgeon. (f) Female salmon. Structures of the urogenital system include the anus (A), abdominal pore (Ap), cloaca (Cl), genital pore (Gp), oviduct (Ovd), rectum (R), urinary ducts (U), urogenital opening (Ug), urogenital sinus (Us), and urinary papilla (Up).

After Romer and Parsons.

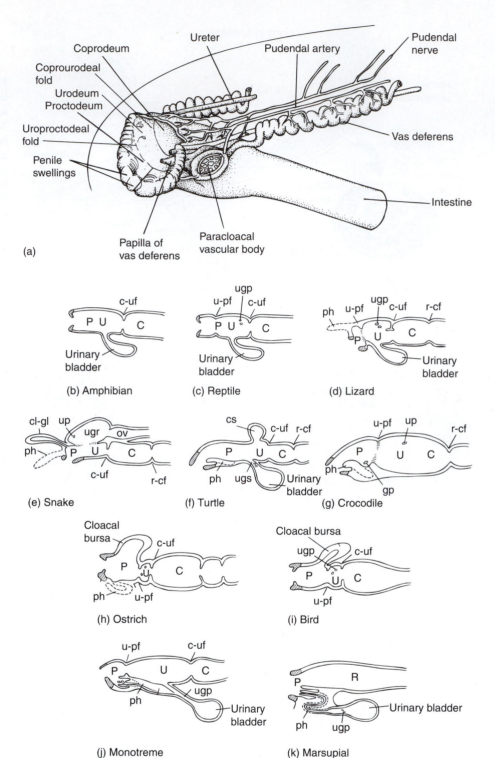

FIGURE 14.47 Diagrams of sagittal sections of tetrapod cloacae. (a) Bird cloaca with ducts and organs that open into each of the three chambers. (b) Amphibian. (c) Reptile, *Sphenodon*. (d) Lizard, *Lacerta*. (e) Snake, *Tropidonotus*. (f) Turtle, *Pseudemys*. (g) Crocodile. (h) Ostrich. (i) Bird. (j) Monotreme. (k) Marsupial. Parts of the cloacae include the coprodeum (C), proctodeum (P), rectum (R), and urodeum (U). Other abbreviations: coprourodeal fold (c-uf), cloacal gland (cl-gl), cloacal sac (cs), genital pore (gp), oviduct (ov), penis (ph), rectocoprodeal fold (r-cf), urinary pore (up), urogenital pores (ugp), uroproctodeal fold (u-pf), urogenital sinus (ugs), urogenital reservoir (ugr).

(a) After Lake; (d–k) after A. S. King.

The cloaca of most amphibians is simple. Folds usually delineate the coprodeum and urodeum, but in the absence of an intromittent organ or a uroproctodeal fold, the proctodeum is not anatomically demarcated from the rest of the cloaca (figure 14.47b). Among reptiles, the cloaca of *Sphenodon* is subdivided by folds into three compartments; the proctodeum is simplified and lacks a penis. The cloaca of snakes and lizards also has three compartments,

but the proctodeum is usually reduced (figures 14.47c–e and 14.48a,b). The internal subdivision of the cloaca is much less distinct in turtles (figure 14.47f), and in crocodilians the coprodeum and urodeum and, to a lesser extent, the proctodeum are more or less united into a single large chamber (figure 14.47g). In birds, the cloacal folds are quite variable. The ostrich cloaca has a rectocoprodeal fold (figures 14.47h and 14.49a,b), but this is apparently lacking in other groups

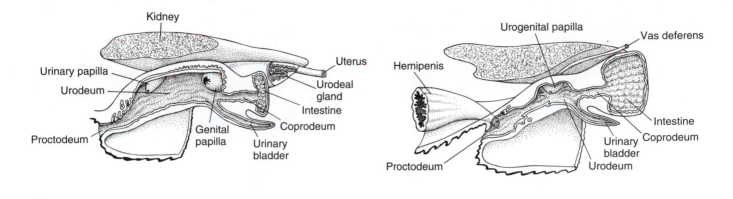

(a) Female lizard   (b) Male lizard

FIGURE 14.48   **Cloaca of the lizard *Coleonyx*.** (a) Female. (b) Male.

After Gabe and Saint-Girons.

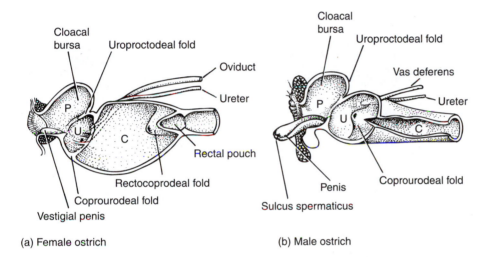

(a) Female ostrich   (b) Male ostrich

FIGURE 14.49   **Bird cloacae.** (a) Female ostrich cloaca, longitudinal view. (b) Male ostrich cloaca. Chambers include coprodeum (C), proctodeum (P), urodeum (U).

After A. S. King.

(figure 14.47i). In birds, the proctodeum is associated with a **cloacal bursa** (bursa of Fabricius) that has an immune function.

The cloaca persists in monotremes, where distinct coprourodeal and uroproctodeal folds demarcate the urodeum from other compartments (figure 14.47j). The ureter and vas deferens open to a urogenital sinus, but urine flows directly into the urodeum, and semen flows through a sperm duct within the penis. Marsupials possess a reduced cloaca that is represented primarily by the proctodeum (figure 14.47k). The ectodermal part of the cloaca persists in some rodents and insectivores; however, in all other eutherian mammals, the cloaca divides in the sexually indifferent stage and forms separate orifices from the coprodeum and urodeum (figure 14.50a–c). Generally, the coprodeum becomes the rectal region of the digestive tract with an anal opening. The urodeum yields separate structures, depending on the sex. In the male, the urogenital sinus becomes the urethra that transports sperm and urinary products (figure 14.50d).

In most eutherian mammals, the female urethra remains conjoined with the vagina to form a urogenital sinus. In others, the urogenital sinus divides again to produce a urethral opening for the urinary system and a vaginal opening for the reproductive system (figure 14.50c).

Two patterns are evident in the reproductive organs of marsupial females. In opossums, the oviducts enter a vaginal sinus that loops symmetrically around the viscera to form lateral vaginas (figure 14.51a). In kangaroos, the vaginal sinus, via an unpaired central vaginal canal, joins the lateral vaginal loops in the common urogenital sinus (figure 14.51b). In therian females, one end of each oviduct narrows into a slender fallopian tube, which receives the egg released from the ovary. At the other end, the oviducts expand into the uterus to support the young during their embryonic development. In some eutherian species, the oviducts join the vagina separately, forming a **duplex uterus.** In **bipartite** and **bicornuate uteri,** the uteri partially fuse. If the uteri fuse entirely, a **simplex uterus** is formed (figure 14.52).

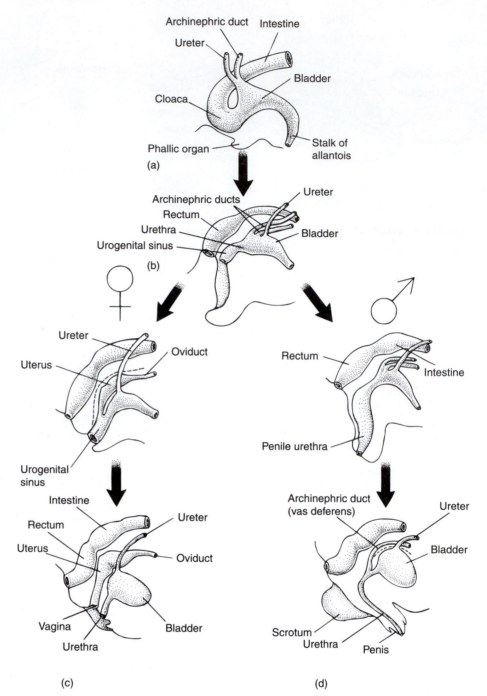

FIGURE 14.50 **Embryonic derivatives of the urogenital sinus in some eutherian mammals.** (a) In the indifferent stage, the cloaca is undivided. (b) The first step toward differentiation is separation of the urogenital sinus from the rectum. (c) In the female, the urogenital sinus divides to form the urethra and the vagina, both with separate external openings. (d) In the male, the urogenital sinus becomes the urethra of the penis and transports both sperm and urine.

Urinary Bladder

Before being excreted, urine is usually stored in specialized regions of the urogenital system. In this way, the vertebrate can void urine at opportune times rather than continuously as it is formed. If water conservation is important, the bladder sequesters the concentrated urine so that it does not create osmotic pressure that draws water out of the tissues of the animal.

In fishes, urine is usually stored within the ends of the urinary ducts where they join the cloaca or open to the outside. A urinary bladder of this type is mesodermal and noncloacal in origin. It is found among elasmobranchs, holocephalians, and most teleost fishes (figure 14.53a).

In tetrapods, the urinary bladder arises as an outpocketing of the cloaca. Urine flowing from the urinary ducts usually empties into the cloaca first and then fills

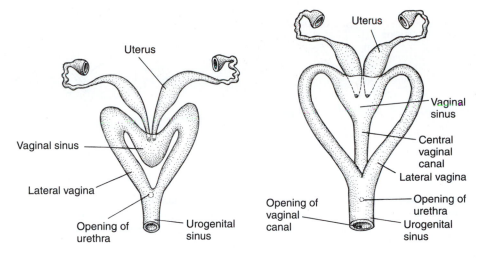

Uterus

Vaginal sinus

Lateral vagina

Opening of
urethra

Urogenital
sinus

(a) Opossum

Uterus

Vaginal
sinus

Central
vaginal
canal

Lateral vagina

Opening of
vaginal
canal

Opening of
urethra

Urogenital
sinus

(b) Kangaroo

FIGURE 14.51 Reproductive
organs of female marsupials.
(a) Opossum. (b) Kangaroo.

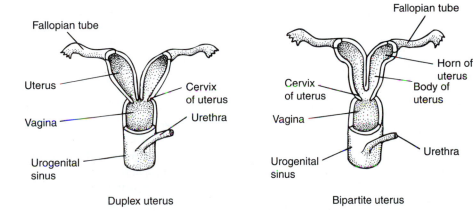

Fallopian tube

Uterus

Vagina

Urogenital
sinus

Cervix
of uterus

Urethra

Duplex uterus

Fallopian tube

Cervix
of uterus

Vagina

Urogenital
sinus

Horn of
uterus

Body of
uterus

Urethra

Bipartite uterus

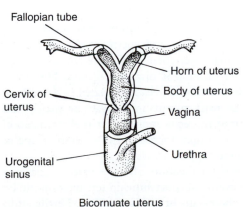

Fallopian tube

Cervix of
uterus

Urogenital
sinus

Horn of uterus

Body of uterus

Vagina

Urethra

Bicornuate uterus

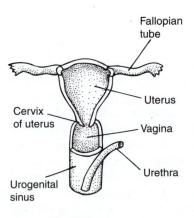

Fallopian
tube

Cervix
of uterus

Urogenital
sinus

Uterus

Vagina

Urethra

Simplex uterus

FIGURE 14.52 Reproductive
organs of female eutherian
mammals. The uterus is characterized
by the degree of fusion of the paired uteri.

FIGURE 14.53 **Evolution of the urinary bladder.** (a) In teleosts, the intestine and urinary ducts establish separate exits, anus and urinary pores, respectively. As a consequence, the embryonic cloaca is lost in the adult. The teleost urinary bladder, when present, is formed from the expanded ends of the urinary ducts. (b,c) In tetrapods, the urinary bladder is an outgrowth of the cloaca. It empties into the cloaca in amphibians and reptiles (b) but exits via the urethra in mammals (c).

After M. Wake.

the baglike urinary bladder (figure 14.53b). In therian mammals, the urinary ducts (ureters) empty directly into the urinary bladder (figure 14.53c). The tetrapod urinary bladder appears first among amphibians and is present in *Sphenodon*, turtles, most lizards, ostriches among birds, and all mammals. The urinary bladder has been lost in snakes, some lizards, crocodilians, and all birds except the ostrich.

Function and Evolution

In most vertebrates, reproduction is seasonal. Courtship and copulation are usually restricted to a brief annual breeding season. During the breeding season, hormone-readied genital ducts receive and transport released eggs and sperm. Onset of reproductive readiness is called **recrudescence.** Only among humans is breeding a year-round affair.

Potency and Fertility

Fertility refers to the ability of the female to produce fertilizable eggs or of the male to produce sperm in sufficient numbers to achieve fertilization. A male producing insufficient numbers of sperm is **infertile** or **sterile.** In a human male, ejaculated semen can contain 200 million sperm. And although it just takes one sperm to fertilize an egg, a drop in the sperm count to 50 million may result in sterility. Although millions of sperm may be ejaculated into the vagina, the number of sperm that eventually survive the journey to the upper reaches of the oviduct rarely exceeds a few hundred. Considering that the spermatozoon is small in comparison to the volume of the oviduct, it is not surprising that only a very modest number of sperm arrive at the site of fertilization. Finally, many sperm interact to break through follicle cells or surface mucus clinging to the egg so that one sperm can penetrate the egg cell membrane. Thus, fertilization is performed by the fusion

of a single spermatozoon with a single egg, but this comes after much attrition and cooperation among many sperm to promote penetration of the egg.

Potency refers to the ability of the male to engage in copulation. **Impotence** results from the failure to achieve an erection. Impotence is different from sterility. Castrated males are sterile because they lack testes and produce no sperm. However, if the testes are removed after puberty, there has often been enough time for androgens to masculinize the individual so that some secondary sexual characteristics, sex drive, and ability to engage in sexual intercourse (potency) are retained.

Spermatogenesis is under hormonal control. In a seasonal breeder, sperm are produced only certain times of the year. **Follicle-stimulating hormone (FSH),** a pituitary gonadotropic hormone, stimulates the multiplication of spermatogonia in the seminiferous tubules as the breeding season approaches. With advancing age, there may be a slow decline in the ability of the seminiferous tubules to produce mature sperm, but there is no abrupt cessation comparable to the female **menopause** that occurs in some mammals.

External and Internal Fertilization

External fertilization is common among invertebrates and primitive vertebrates. Eggs and sperm meet outside the body. However, many vertebrates live in environments in which external fertilization is disadvantageous. The tailed frog *Ascaphus*, for instance, lives and mates in fast-moving streams where swift currents might wash away eggs and sperm released into the environment. Internal fertilization via an intromittent organ increases the success of sperm transfer under these conditions.

But internal fertilization offers a further adaptive advantage. The events of courtship and fertilization can be separated from the events of egg deposition. Fertilization does not always occur in an environment that is also suitable for egg deposition. For example, some salamanders mate on land where courtship displays are visible, but dry land offers few favorable sites for the development of their water-dependent eggs. In most salamanders, a spermatophore is taken up by the female during courtship, but eggs are not released at that time. Instead, the sperm are held in the spermatheca until she has found a suitable location for deposition. The eggs are fertilized as they are laid (figure 14.54a–c).

Physiological constraints can restrict the evolution of viviparity in some groups. Among amniotes, calcium for ossification of the embryonic skeleton can be stored in the yolk (e.g., squamates) or in the eggshell (e.g., turtles, crocodiles, and birds). In viviparity, the calcareous eggshell is lost, allowing for efficient exchange between fetal and maternal tissues. However, the shell's calcium reservoirs are lost as well. This may help explain why viviparity is absent among turtles, crocodiles, and birds, groups in which the eggshell is used for calcium storage. Viviparity is common among lizards and snakes that do not use the shell as a calcium reservoir.

In both oviparity and viviparity, the young are carried internally, extending the time between courtship and birth or egg deposition and giving the female a chance to seek safe sites for young to be born or hatch. In vertebrates that regulate their temperature internally or behaviorally, the females retain their embryos, allowing them to develop at a stable temperature. If an ectothermic reptile deposits her eggs under a rock, the eggs will be subjected to environmental fluctuation in temperature. But if she retains them in her body, she can shuttle between sites and on cool days bask in whatever warmth is available to elevate the temperature of the developing embryos within her body.

Delays in Gestation

Gestation lasts from conception to hatching or birth. It includes fertilization, implantation (in some species), and development. In some species of mammals, the onset of each stage may be prolonged or delayed. For example, **delayed fertilization** occurs in some bats. Copulation occurs in autumn just before hibernation, but females do not ovulate at that time. Instead, sperm are stored either in the uterus or the upper vagina. When bats emerge from hibernation several months later, eggs are released, sperm become active, and fertilization finally occurs. Young are born in early summer, a season that is usually characterized by an abundance of insects for food.

In **delayed implantation,** known only in mammals, fertilization and early development occur, but the embryo fails to implant in the uterus. Development is arrested for an extended period, until implantation finally occurs and gestation resumes. Delayed implantation occurs in many members of the weasel family (Mustelidae), bears (Ursidae), and a few other groups (figure 14.55). In most cases,

delayed implantation is tied to the annual seasonal cycle. In some marsupials, such as kangaroos and wallabies, however, delayed implantation of the **blastocyst** is tied to the presence of a young kangaroo in the pouch, termed a joey. Suckling by an older joey in the pouch inhibits implantation of the next blastocyst, a type of delay referred to as **embryonic diapause.** In **delayed development,** known from several species of bats, fertilization and implantation occur on schedule, but subsequent growth of the embryo is slow.

Delay in fertilization, implantation, or development increases the time between mating and birth to ensure that young will not be born at an inopportune time (e.g., during migration) or when food is scarce (as in the middle of winter). Female caribou give birth immediately after their migration from winter forests to summer tundra. Many species of whales give birth after they migrate from polar seas and arrive in temperate or tropical oceans. Seals give birth when they reach their breeding beaches after an extended migration at sea.

Overview

The urinary and genital systems arise as neighbors from adjacent regions within the embryo and share some of the same ducts in the adult. Functionally, however, the two systems are quite distinct. The urinary system includes the kidneys and the ducts that carry away urine, a watery waste product. But urine is a by-product of the urinary system's primary function, the internal regulation of body fluid composition. Other organs may also participate—the skin, gills or lungs, alimentary canal, liver—but the kidneys are specialized for controlled maintenance of the internal fluid levels of water and solutes, osmoregulation, and for elimination of the waste products of metabolism, excretion. The functional unit of the kidney, the uriniferous tubule, collects in the renal capsule an ultrafiltrate of blood plasma diffusing in from the vascular glomerulus under favorable osmotic and hemodynamic pressures. As the ultrafiltrate moves along the tubule, regions of the tubule add and subtract fluid and solutes, eventually producing urine.

Organisms live in environments that may dehydrate them, critically lowering internal fluid levels, or in environments that result in an influx of fluid following an osmotic gradient, swelling tissues with excess fluid. By controlling the water composition of the ultrafiltrate, fluid may be recovered (producing a concentrated urine) or added (producing a dilute urine) to compensate for environmental stress upon internal fluid and solute balance. One end product of metabolism is nitrogen, usually in the form of ammonia, which can be toxic. Its elimination may be direct, usually via skin or gills (ammonotelism), or indirect, via the kidneys by first converting ammonia to uric acid and then eliminating (uricotelism) or to urea and then eliminating (ureotelism).

The genital system includes the gonads and the ducts that carry away their products, eggs and sperm. Therefore, the genital system is most directly involved in implementing

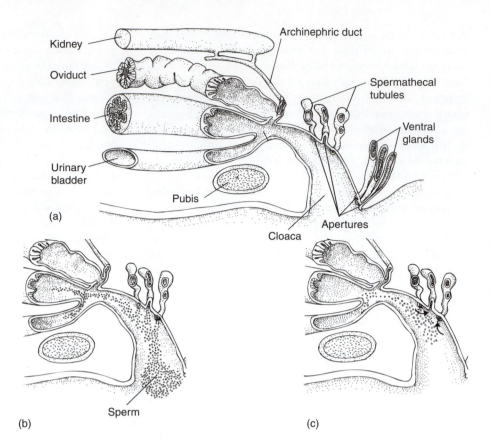

FIGURE 14.54 **Sperm storage within the spermatheca of the salamander *Notophthalmus*.**
(a) Diagram of the urogenital system. A few hours after sperm enter the cloaca (b), they move into the spermathecal tubules (c), where they are stored. In this species, the ova are not released for several months. When they are released, stored sperm are discharged into the cloaca to fertilize the passing eggs.

After Hardy and Dent.

Labels in figure: Kidney, Oviduct, Intestine, Urinary bladder, Archinephric duct, Spermathecal tubules, Ventral glands, Pubis, Apertures, Cloaca, Sperm

(a) (b) (c)

Family	Species	Common name	Months (J F M A M J J A S O N D J F M A M J J)
Mustelids	*Martes americana*	Marten	
	Lutra canadiensis	Otter	
	Mustela erminea	Ermine	
	Taxidea taxus	American badger	
	Meles meles	European badger	
Ursids	*Ursus americanus*	Black bear	
	Ursus maritimus	Polar bear	
Pinnipeds	*Phoca vitulina*	Harbor seal	
	Callorhinus ursinus	Fur seal	
	Mirounga leonina*	Elephant seal	
Artiodactyla	*Capreolus capreolus*	Roe deer	
Edentata	*Dasypus novemcinctus*	Armadillo	
Chiroptera	*Eidolon helvum*	Fruit bat	

* Southern hemisphere

■ Breeding season   ---- Preimplantation   —— Postimplantation   ▭ Birth season

FIGURE 14.55 **Delayed implantation.** Seasons of breeding, preimplantation, postimplantation, and birth for several species of mammals are shown. To ensure that young are born when resources are most likely available, many mammals have evolved methods to lengthen gestation beyond harsh seasons or times of migration so that birth occurs when conditions are favorable. Delayed implantation occurs following fertilization when the embryo does not immediately implant itself in the uterine wall. Instead, the embryo goes into a stage in which further development is slowed or arrested. Later, after implantation occurs, the pace of embryonic development picks up. Note that a few species give birth during the winter.

After Sadleir.

successful reproduction, its primary biological role. Besides producing eggs and sperm, the gonads are endocrine organs that preside over the paced development of gametes, embryo, and breeding. They also initiate development of secondary sexual characteristics, prepare for pregnancy, maintain physiological support for the embryo, and activate accompanying reproductive behavior. Within amniotes, fertilization is usually internal, with sperm transfer occasionally aided by a male intromittent organ.

The vertebrate transition from water to land eliminated one convenient route of waste elimination, the gills, thereby increasing the role of the kidneys. As vertebrates became more terrestrial, water economy became more important during nitrogen elimination. Formation of uric acid, including resorption of water in the cloaca, is one answer. The other is formation of urea, a means of converting ammonia to a nontoxic form, and requiring less water in the process. The transition to land also initially favored internal fertilization as an alternative to spreading gametes externally in the surrounding aquatic environment. Internal fertilization permits decoupling in time of several reproductive activities—courtship, fertilization, and egg deposition. Thereby, these activities may occur at a time and in an environment most suited to each.

WEB LINK

Visit the text website at *www.mhhe.com/Kardong7e* for additional study aids including selected references, functional laboraotories, and web links to chapter topics.

15

The Endocrine System

SURVEY OF ENDOCRINE GLANDS
Thyroid Gland
Structure and Phylogeny
Function
Ultimobranchial Body and Parathyroid Gland
Ultimobranchial Body
Parathyroid Gland
Form and Function
Adrenal Gland
Structure and Phylogeny
Function
Pancreatic Islets
Structure and Phylogeny
Function
Pituitary Gland
Structure
Phylogeny
Function

Gonads
Pineal Gland
Secondary Endocrine Glands
Gastrointestinal Tract
Kidneys

ENDOCRINE COORDINATION
Mammalian Reproduction
Male
Female
Metamorphosis in Frogs
Fundamentals of Hormonal Control
Functional and Structural Linkage
Target Tissue Responses
The Endocrine System and the Environment

EVOLUTION

OVERVIEW

Survey of Endocrine Glands

Two major control systems preside over activity levels within the body. One is the nervous system, which is covered next in chapter 16; the other is the endocrine system covered here. These control systems often working together, are responsible for coordinating activities between organs, increasing organ activity in response to increased physiological needs, and maintaining steady-state conditions.

The endocrine system includes the **endocrine glands,** the chemical messengers or **hormones** they produce, and the **target tissues** they affect. Endocrine glands are located throughout the body. Hormones are not transported in ducts; instead they are carried by the blood. Although hormones circulate throughout the body, each one usually affects selected target tissues, so its influence is localized.

Endocrine glands are as varied as the target tissues they control. They preside over reproduction, metabolism, osmoregulation, embryonic development, growth, metamorphosis,

and digestion. We begin by looking at the distribution of endocrine glands and the hormones they produce among vertebrate groups.

Thyroid Gland

Structure and Phylogeny

The thyroid gland produces, stores, and releases two separate **thyroid hormones** that regulate metabolic rate, metamorphosis, growth, and reproduction. The thyroid hormones are said to be **permissive,** meaning that they "permit" target tissues to be more responsive to stimulation by other hormones, by the nervous system, or possibly by environmental stimuli (such as light or temperature). The thyroid secretes hormones containing iodine. In 1915, **thyroxine,** the first thyroid hormone, was isolated and identified. Another name for this hormone is **tetraiodothyronine,** or T_4 for short (so named because

TABLE 15.1 Endocrine Glands and Primary Secretions in Mammals

Hormone	Source of Hormone
Adenohypophysis	
Growth hormone (GH)	Pars distalis
Prolactin (PRL)	Pars distalis
Thyrotropin (TSH)	Pars distalis
Follicle-stimulating hormone (FSH)	Pars distalis
Luteinizing hormone (LH)	Pars distalis
Corticotropin (ACTH)	Pars distalis
Melanotropin (MSH)	Pars intermedia
Neurohypophysis	
Vasopressin (VADH)	Neurons projecting to neurohypophysis from paraventricular and supraoptic
Oxytocin (OXY)	nuclei of hypothalamus
Parathyroid	
Parathyroid hormone (PTH)	Chief cells
Thyroid	
Thyroxine (tetraiodothyronine) (T_4)	Principal cells
Triiodothyronine (T_3)	Principal cells
Calcitonin	Parafollicular cells
Adrenal gland	
Cortex	
Aldosterone	Zona glomerulosa
Glucocorticoids	Zona fasciculata, zona reticularis
Androgens	Zona reticularis
Medulla	
Norepinephrine	Chromaffin cells
Epinephrine	Chromaffin cells
Pancreatic islets	
Insulin	B cells
Glucagon	A cells
Somatostatin	D cells
Pancreatic polypeptide	PP cells
Duodenum	
Cholecystokinin (CCK)	Intestinal mucosa
Secretin	Intestinal mucosa
Testis	
Testosterone	Interstitial cells
Ovary	
Estradiol	Theca interna, interstitial cells, granulosa cells (?)
Progesterone	Corpus luteum, theca interna
Other follicular estrogenic steroids in smaller amounts	
Placenta	
Chorionic gonadotropin	Syntrophoblast
Estradiol	Syntrophoblast
Estriol	Syntrophoblast
Adrenal corticoids	Syntrophoblast
Placental lactogen (PL)	Syntrophoblast
Prolactin	Syntrophoblast
ACTH-like substances	Syntrophoblast

each molecule contains four iodine atoms). A second thyroid hormone identified in 1952 is **triiodothyronine,** or T_3 (three iodine atoms; table 15.1). Initially isolated in mammals, both T_3 and T_4 are now known to be synthesized in all vertebrates. In cyclostomes, these hormones are stored *intra*cellularly. However, in gnathostomes, the thyroid stores large quantities of hormones *extra*cellularly within the lumina of hundreds of tiny irregular spheres, or **follicles.** This condition is unique compared with all other vertebrate endocrine glands. A single layer of epithelial cells called **principal cells** (follicle cells) forms the walls of these follicles (figure 15.1a–c). Principal cells produce a gelatinous **colloid** in which these hormones are stored within the follicles. Principal cells also release thyroid hormones on demand (figure 15.1b). In all vertebrates,

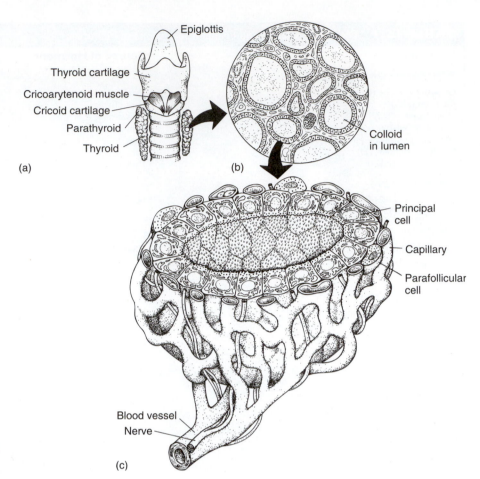

Epiglottis

Thyroid cartilage

Cricoarytenoid muscle

Cricoid cartilage

Parathyroid

Thyroid

(a)

Colloid
in lumen

(b)

Principal
cell

Capillary

Parafollicular
cell

Blood vessel

Nerve

(c)

FIGURE 15.1 Mammalian thyroid gland. The thyroid gland is composed of numerous spherical follicles. Principal cells in the wall of each follicle produce thyroid hormones and secrete them on demand into capillaries. (a) Ventral view of larynx and trachea of a dog showing paired thyroid and parathyroid glands. (b) Enlarged histological section of the thyroid illustrates follicles and colloid that fills the lumina. (c) Cutaway view of a single thyroid follicle showing the arrangement of principal cells and parafollicular cells (C cells) composing the follicular wall. Note the nerve supply and capillaries embracing the basal regions of these cells.

After Krstić.

the thyroid arises as an outgrowth from the floor of the pharynx. This outgrowth may initially be solid or hollow but soon breaks free of the pharynx (figure 15.2a–d). In teleosts, it fragments into dispersed masses of follicles. In most other vertebrates, it forms a single- or double-lobed gland in the throat enclosed in a connective tissue capsule (figure 15.3a–l).

A biosynthetic pathway for thyroid hormone production is reported in urochordates and cephalochordates. In amphioxus and lampreys, the endostyle is similar in function to a thyroid gland in that it secretes iodine-rich products, but it releases these directly into the digestive tract. During metamorphosis of the ammocoetes, the endostyle is converted into the follicle cells of the thyroid gland, which releases its hormones into the circulatory system. The endostyle of protochordates, like the thyroid gland of vertebrates, collects iodine, adding further to the view that the endostyle is the phylogenetic predecessor of the thyroid. But, the role that iodinated compounds play in the metabolism of protochordates is not yet clear.

Function

When being stockpiled, the protein thyroglobulin is secreted by the principal cells under the influence of the pituitary hormone **thyrotropin,** or **thyroid-stimulating hormone (TSH),** and stored in the colloid (figure 15.4a). Specifically, the amino acid tyrosine is incorporated into the protein and iodinated, used to form T_4 (two iodinated tyrosines are coupled), linked to the protein backbone by peptide bonds. When mobilized,

principal cells become taller and form apical extensions that envelop the stored protein, allowing these cells to phagocytize and then hydrolyze the colloid in lysosomes. Specifically, thyroglobulin is taken back up into the principal cells (again under the influence of TSH), where hydrolytic enzymes cleave the thyroglobulin to produce T_4. Some T_4 is partially deiodinated to T_3 prior to release from the gland so that some T_3, but mostly T_4, enters the circulation (figure 15.4b). Much of the circulation T_4 is either degraded or converted by liver enzymes into the more active form of the hormone, T_3.

Thyroid hormones are present in cyclostomes, where they inhibit metamorphosis, but their function in the adult is not known. The effects of thyroid hormones on target tissues are best known in mammals and birds.

Metabolism In endotherms, thyroid hormones elevate oxygen consumption and heat production by tissues. Injections of thyroid hormones can increase basal metabolic rate several fold. Thyroid hormones affect metabolic activity specifically through increase of plasma membranes, especially in mitochondria, and by increases in molecular activities of membrane proteins. Ectotherms have no "basal" metabolic rate, although of course they do have a body temperature dependent upon ambient conditions and activity levels. Thyroid hormones in ectotherms do affect metabolism and this effect is body-temperature sensitive. An effect the thyroid has on ectotherm metabolism can be

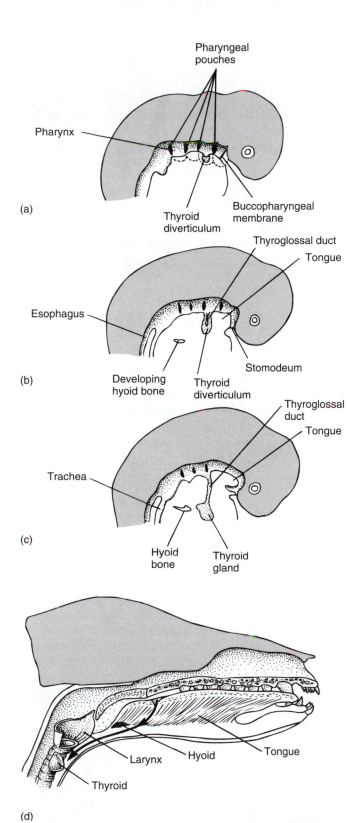

FIGURE 15.2 Embryonic development of the mammalian thyroid. (a) Sagittal section through the embryonic pharynx. (b,c) Successive stages in the appearance and growth of the thyroid diverticulum. (d) Location of the thyroid in an adult mammal, which migrates embryologically along the route indicated by the arrows.

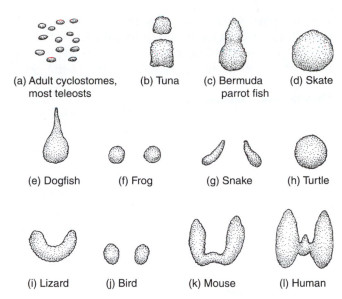

FIGURE 15.3 Vertebrate thyroid glands.
(a–e) Fishes. (f) Amphibians. (g–i) Reptiles. (j) Bird. (k,l) Mammals.

After Gorbman and Bern.

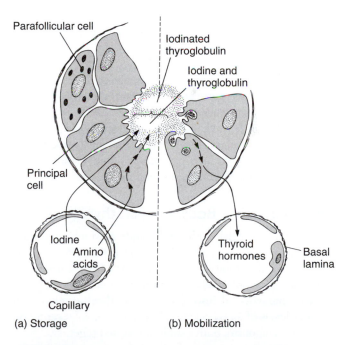

FIGURE 15.4 Thyroid secretion and mobilization.
(a) Thyrotropin (TSH) stimulates the principal cells to take up iodine and amino acids, combine them with thyroglobulin, and secrete the resulting colloid into the lumen. (b) Stimulated by TSH from the pituitary, principal cells mobilize hormones stored in the colloid and release them into adjacent capillaries. The capillary walls are usually fenestrated, but the basal lamina is complete.

seen in reptiles when their temperature becomes environmentally elevated. At low temperatures (20°C), lizard tissues are unresponsive to thyroid hormones; however, at preferred temperatures (30°C), tissues respond to thyroid hormones.

Growth and Metamorphosis In birds and mammals, normal growth depends on normal levels of thyroid hormones. **Hypothyroidism**, underproduction of these hormones, results in stunted growth and mental retardation in human infants, a syndrome known as **cretinism**. In adults, hypothyroidism results in lethargy and impaired cognitive ability. **Hyperthyroidism**, overproduction of thyroid hormones, results in heightened activity, nervousness, bulging eyes, and rapid weight loss, a medical condition called Graves' disease.

The growth of reptiles and fishes similarly depends on thyroid hormones. For example, enlargement of the thyroid gland occurs when a young salmon (termed a **parr**) is transformed into a **smolt**, its migratory stage in which it travels downstream to the sea. Amphibians differ from most vertebrates in that their thyroid hormones arrest growth of the larvae and promote metamorphosis.

Molt Thyroid hormones affect loss and subsequent replacement of hair or feathers when mammals and birds **molt.** Thyroxine promotes sloughing or shedding of the skin, which suggests a general effect of thyroid hormones on the vertebrate integument. If thyroid hormones are deficient in birds or mammals, hair or feather growth is impaired, pigment deposition is reduced, and the skin tends to thin. The skin of fishes, amphibians, and reptiles is also adversely affected by thyroid hormone deficiencies.

Reproduction In most vertebrates, elevated levels of thyroid hormones are correlated with gonad maturation and oogenesis or spermatogenesis. Again amphibians seem to be an exception because their thyroid hormones apparently arrest physiological processes that promote reproduction. Surgical removal of the amphibian thyroid glands is followed by accelerated gonadal development.

Ultimobranchial Body and Parathyroid Gland

The ultimobranchial body and parathyroid gland release hormones with opposite, or antagonistic, effects. The ultimobranchial body secretes **calcitonin** (thyrocalcitonin), which lowers blood levels of calcium. The parathyroid gland secretes **parathyroid hormone** (= **parathormone**), which elevates levels of blood calcium. Because their roles center on calcium metabolism, both glands are treated together.

Ultimobranchial Body

Embryonic primordia from the fifth pharyngeal pouches form the **ultimobranchial bodies** (figure 15.5). These bodies are separate, usually paired cell masses located in the throat region of fishes, amphibians, reptiles, and birds. Cyclostomes do not appear to have ultimobranchial bodies. In mammals,

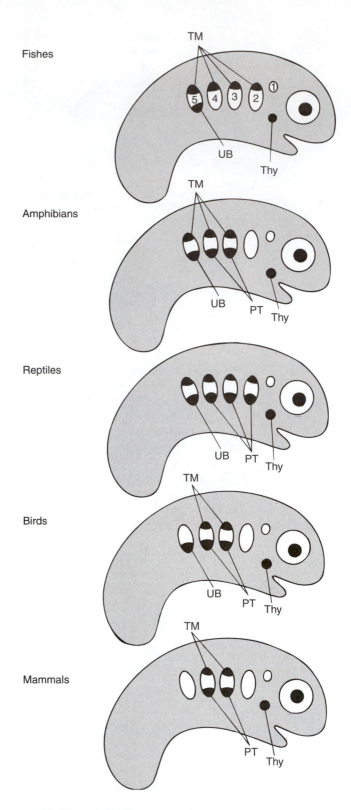

FIGURE 15.5 **Embryonic contributions of vertebrate pharyngeal pouches to thyroid (Thy), parathyroid (PT), thymus (TM), and ultimobranchial bodies (UB).** The reptilian thymus develops from pouches 2 and 3 in lizards, pouches 3 and 4 in turtles, and pouches 4 and 5 in snakes. The ultimobranchial bodies in mammals settle into the thyroid gland as the parafollicular cells (C cells). Pharyngeal pouches are numbered, the first usually being reduced in embryonic development.

their distribution is unique, the primordia being incorporated directly into the thyroid to form a small, dispersed population of **parafollicular cells** (C cells) scattered among the principal cells in the walls of the thyroid follicles (figure 15.1c).

The neural crest is the embryonic source of ultimobranchial cells. It is not yet clear whether neural crest cells enter the pharyngeal primordium before it migrates to its site of differentiation or whether neural crest cells colonize the primordium later during differentiation.

Parathyroid Gland

The ventral edges of the embryonic pharyngeal pouches are the source of the **parathyroid glands.** The pouches that contribute vary among species (figure 15.5). The term *parathyroid* describes the close association of this gland with the thyroid gland in mammals, which is embedded in (e.g., mouse, cat, human) or near (e.g., goats, rabbits) the thyroid gland. One or two pairs may be present. However, in amphibians, reptiles, and birds, the parathyroid may be located either on the thyroid or dispersed along the major veins in the neck (figure 15.6a–c). In fishes, the parathyroid is absent. Because it is absent in fishes and absent in at least some obligate neotenic salamanders (e.g., *Necturus*) where gills persist, it has been suggested that the role of the parathyroid is preceded phylogenetically by cells in the gills.

Within the parathyroid gland, cells fall into a cord-and-clump arrangement. **Chief cells,** the most abundant cell type, are probably the source of parathormone. In humans and a few other mammalian species, **oxyphil cells** of unknown function are present as well.

(a) Salamander   (b)

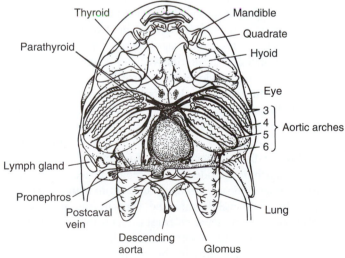

(c) Frog

FIGURE 15.6 **Locations of thyroid and parathyroid glands in amphibians.** (a) Exposed ventral view of throat of the salamander *Triturus viridescens.* The left intermandibularis, interossial quadrate, and quadratopectoralis muscles are removed to show deeper muscles and glands. (b) Enlarged area showing thyroid and parathyroid glands with surrounding arteries and veins. (c) Ventral view of larval bullfrog *(Rana catesbeiana)* throat. Note paired thyroid glands anterior to the heart and paired sets of parathyroids at the bases of the branchial (aortic) arches.

(a,b) After Stone and Steinitz; (c) after Witschi.

Form and Function

Immediate access to calcium is important in most vertebrates. When birds secrete calcified eggshells or deer grow a new rack of antlers, large amounts of calcium must be rapidly mobilized and transported from one site to another. Maintenance of normal bone strength depends on calcium levels. If levels of calcium in the blood fall too low, skeletal muscles can go into uncontrolled spasms. If blood levels rise too high, osteogenic cells cannot retain calcium in the bone matrix to maintain bone density and strength.

Parathormone secreted by the parathyroid acts to raise blood levels of calcium by promoting kidney retention of calcium, encouraging its absorption across the walls of the digestive tract, and affecting bone deposition. The competing processes of bone deposition and bone removal occur simultaneously and continuously, but they are usually dynamically balanced. Parathormone tips the balance toward net bone removal. As a result, more bone matrix is removed than is deposited; therefore, calcium is liberated from the matrix and taken up by the circulation, causing blood levels of calcium to rise. Calcitonin from the parafollicular cells has the opposite effect. It shifts the balance toward net bone deposition. Calcitonin causes calcium to be extracted from the blood and used to build new bone matrix, causing blood levels of calcium to fall.

Details of the mechanism controlling calcium levels in tetrapods are still debated, but generally three organs are involved: intestines, kidneys, and bones. The interaction of

FIGURE 15.7 **Homeostasis of calcium in tetrapods.** Arrows indicate major routes by which calcium is withdrawn or added to the blood plasma and extracellular fluid. Calcium from food is absorbed in the intestines. In the kidney it initially enters the ultrafiltrate that forms in the renal capsule of the kidney, but all calcium ions are recovered and returned to the blood. Calcium moves passively out of the supersaturated blood and crystallizes to form bone. Active bone resorption, under parathyroid hormone stimulation, returns some calcium to the blood.

these organs is depicted in the diagram in figure 15.7. Soft tissues, such as muscles, also require calcium, but their net effect on blood levels of calcium is usually minimal. Calcium in food is absorbed by the intestines. The kidneys can recover all calcium from glomerular filtrate and return it to the extracellular fluid. Control of calcium levels in bone is more complicated. Calcium is incorporated into bone in a crystalline form. The calcium saturation level in bone is lower than it is in the blood, so the net flux of calcium is from blood to bone. Formation of new crystals of bone is a passive process. Although elevated levels of calcitonin correlate with falling blood levels of calcium, the details of how this is accomplished are not yet clear. Parathormone promotes the opposite reaction—the efflux of calcium back into the blood—by promoting osteoclasts that resorb bone. Removal of calcium from bone is an active process. It is not known if the two hormones interact directly or indirectly to inhibit each other's actions.

Fishes regulate calcium by different mechanisms. For example, teleosts secrete somatolactin from the pars intermedia (pituitary), which influences calcium homeostasis. Teleosts generally have acellular bone, which is therefore not a good source from which to dynamically mobilize calcium, so instead they rely on scales (and endolymphatic sacs of the ear) as calcium reservoirs.

Adrenal Gland

Structure and Phylogeny

The **adrenal gland** is a composite gland derived from two separate phylogenetic sources. One is the **adrenocortical tissue** (= interrenal tissue, or interrenal bodies), which produce **corticosteroid hormones.** Corticosteroids belong to a class of organic compounds called **steroids.** There are three categories of steroids: those involved in (1) water reabsorption and sodium transport by the kidney (**mineralocorticoids),** (2) metabolism of carbohydrates (**glucocorticoids),** and (3) reproduction (**estrogens, androgens,** and **progestogens**). In adults, estrogens stimulate development and vascularization of the female reproductive tract; androgens are masculizing agents that promote development of male secondary sexual characteristics; and progestogens (= progestins) maintain pregnancy and the uterine wall during its secretory phase.

The other phylogenetic source of the adrenal gland is **chromaffin tissue,** or **chromaffin bodies,** which produce **catecholamines.** Catecholamines are **chromaffin hormones** such as **epinephrine** (adrenaline) and **norepinephrine.** The embryonic origins of these tissues, like their phylogenetic origins, are distinct (figure 15.8). Adrenocortical tissue arises from splanchnic mesoderm in the region adjacent to the urogenital ridge. Chromaffin tissue arises from neural crest cells.

In adult cyclostomes and teleosts, the adrenocortical tissue remains separate from the chromaffin bodies. In cyclostomes, the adrenocortical tissue is scattered along the posterior cardinal veins in the vicinity of the pronephros. Chromaffin cells reside in clusters near but not in contact with the adrenocortical tissue. In teleosts, the adrenocortical tissue occurs within the pronephros in scattered clusters or in a strip of tissue around the posterior cardinal veins. The adrenals of teleosts show considerable variation in the anatomy of chromaffin tissue. Usually associated with the anterior kidney, the chromaffin cells may be intermingled among the adrenocortical tissue and also may form clumps that are entirely separate, or both. In elasmobranchs, the adrenocortical tissue forms distinct glands along the edges of the kidneys, but the chromaffin tissue is still separate, consisting of arrays of cell clusters between and anterior to the kidneys (figure 15.9a,b). In amphibians, adrenocortical and chromaffin tissues mingle or reside adjacent to each other and form strands or rows of adrenal tissue, now lying on or near the kidneys (figure 15.9c). The two tissues also mingle in reptiles and birds, although the adrenal glands in

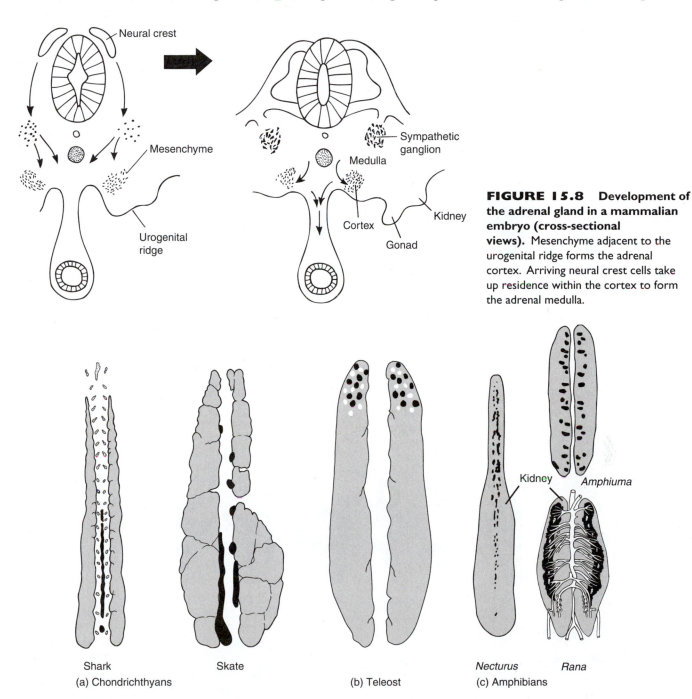

FIGURE 15.8 **Development of the adrenal gland in a mammalian embryo (cross-sectional views).** Mesenchyme adjacent to the urogenital ridge forms the adrenal cortex. Arriving neural crest cells take up residence within the cortex to form the adrenal medulla.

FIGURE 15.9 **Vertebrate adrenal tissues.** (a) Chondrichthyans. (b) Teleost. (c) Amphibians. Adrenals are on the ventral surface of the kidneys. (d) Reptiles. (e) Bird adrenals. (f) Bird adrenals, in cross section. (g) Mammals, showing position of adrenals (solid black) relative to kidneys. (h) Mammal adrenals, in cross section. Note adrenocortical tissue forms the cortex (black) and chromaffin cells reside in the core, forming the distinct medulla (white) of the adrenal. Adrenocortical tissue in black; chromaffin tissue in white; kidney in gray. Because chromaffin cells may be thinly scattered or embedded, they cannot always be indicated in gross anatomy of the adrenal system.

After Bentley.

Chelonia
(Pseudemys)

Crocodile
(Crocodylus)

Snake
(Thamnophis)

Vena
cava

Lizards
(Ctenosaura) *(Heloderma)*

Ovary

(d) Reptiles

Adrenal

Kidney

Vein

Artery

Herring gull *(Larus)*

(e) Bird

Woodpecker   Pelican   Warbler   Quail   Redstart   Hawk

(f) Bird adrenals, cross sections

Dorsal
aorta   Kidney

Opossum (marsupial)   Cat   Armadillo   Raccoon

(g) Mammals

Echidna
(monotreme)

Lemur

Shrew

Kangaroo
(marsupial)

Monkey   Orangutan   Capybara   Chimpanzee   Human

(h) Mammal adrenals, cross sections

FIGURE 15.9 *continued*

amniotes tend to be distinct structures located on or close to the kidneys (figure 15.9d–f). In reptiles, for the first time phylogenetically, the adrenocortical tissue receives its own arterial and venous blood supply, and does not rely on the kidney and renal portal system for distribution of its secretory products. In mammals, for the first time, adrenocortical and chromaffin tissues form a **cortex** (from adrenocortical tissue) and a **medulla** (from chromaffin tissue) to create the composite adrenal gland **(suprarenal)** (figure 15.9g,h).

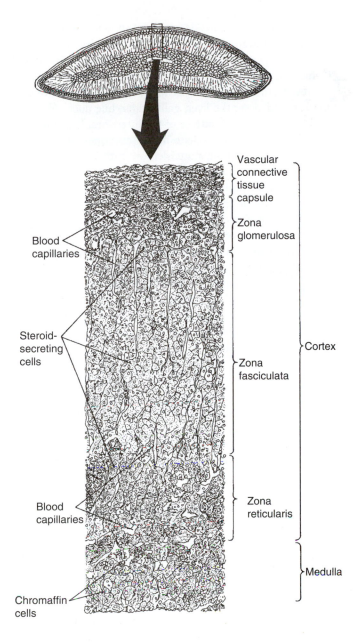

Vascular
connective
tissue
capsule

Zona
glomerulosa

Blood
capillaries

Steroid-
secreting
cells

Cortex

Zona
fasciculata

Blood
capillaries

Zona
reticularis

Medulla

Chromaffin
cells

Adult mammal

FIGURE 15.10 **Zones within the adrenal glands of adult mammals.** Three zones are recognized within the adrenal cortex: zona glomerulosa, zona fasciculata, and zona reticularis. The medulla is composed of nonregionalized chromaffin tissue of neural crest origin.

Function

In mammals, the adrenal cortex produces corticosteroids. Histological studies have shown three zones within the adult adrenal cortex (figure 15.10). Cells of the outermost **zona glomerulosa** region are small and compact. The kidney releases the hormone **renin,** which leads to a series of events that ultimately stimulate cells of the zona glomerulosa to release mineralocorticoids (e.g., **aldosterone**). Mineralocorticoids affect the reabsorption of sodium, increasing it in the kidney,

producing a concentration gradient that favors water retention, thereby reducing urine volume and helping to restore fluid volume in the blood and tissues. Cells of the middle **zona fasciculata** region of the adrenal cortex are arranged in rows or cords with blood sinuses between them. **Adrenocorticotropic hormone (ACTH)** released by the pituitary stimulates cells of the zona fasciculata to secrete glucocorticoids, which include cortisol and principally corticosterone. These occur in most amphibians, all reptiles and birds, and in some mammals. Cells of the third and innermost cortical region, the **zona reticularis,** are small and compact. They are controlled by the pituitary to secrete androgens and additional glucocorticoids.

In many mammals (e.g., primates), an extensive **fetal zone** occupies the periphery of the adrenal cortex prior to birth. This zone is responsible for producing circulating steroids that are chemical precursors of the estrogens synthesized in the placenta. Failure of the fetal zone to function terminates gestation and results in premature birth. Normally, the fetal zone of the adrenal gland ceases to function at birth and dramatically declines in size thereafter.

Thirty or more corticosteroids have been isolated from the mammalian cortex, but most of these are not secreted. Those not secreted seem to be intermediates in the synthesis of definitive hormones released into the blood. In nonmammalian vertebrates, zonation of the interrenal tissue is less conspicuous. Distinct histological regions have been found in anurans, reptiles, and birds, but these may be seasonal.

In vertebrates other than mammals, the cortical hormones primarily regulate sodium transport as well as metabolism. In addition to transport of sodium through the walls of kidney tubules, cortical hormones are thought to control sodium transport through the rectal glands in chondrichthyans, the gills and digestive tract in teleosts, the skin and urinary bladder in amphibians, and the salt glands in reptiles and birds.

One of the most important functions of the adrenal gland is coordinating the whole organism's response to stress. Environmental life-threatening stressors, such as the sudden appearance of a predator or territorial competitor, trigger immediate physiological upgrading to a "fight-or-flee" readiness via release of adrenal catecholamines, and action of the sympathetic nervous system. But, distinct from this, is a shifting to an **emergency life history state.** Unlike fight-or-flee, this long-term physiological response takes minutes to hours to develop and can be triggered by unexpected environmental stressors that do not necessarily pose immediate threats to survival, although overall lifetime fitness may be served. Many hormones may be involved, but primarily adrenal glucocorticoids (e.g., cortisol, corticosterone) preside over the establishment of long-term life history states. For example, environmental stressors not necessarily life-threatening (e.g., loss of young to food shortages, storms, or habitat destruction) may act via the brain-pituitary-adrenal axis (see figure 15.26) to promote emergency life history physiology/behavior wherein the affected individual abandons parental care, migrates to more favorable climates, or locates new suitable habitat. This allows the individual to physiologically adapt to immediate, nonsurvival challenges and be ready to return to normal life history pursuits when these stressors pass.

Fight-or-flee (p. 642)

The stress itself plays a natural role in the physiological adjustment to environmental threats and traumas. It mobilizes the endocrine and nervous systems to meet short- or long-term challenges to survival, after which the organism returns to a more normal and comfortable physiological state. However, in humans and some captive animals, the stress may be more intense or prolonged beyond its usefulness. This can lead to pathological conditions adversely affecting the immune system, the cardiovascular system, the gut, and general metabolism, producing stress diseases. In captive animals, relief from stress often involves improving husbandry; in humans, as students appreciate, relief from stress comes from graduating.

The medulla, composed of chromaffin tissue, forms the core of the mammalian adrenal gland (figure 15.10). Unlike the cortex, no distinct histological regions are recognized in the adrenal medulla. Catecholamines produced in this region prepare the organism to meet threats or short-term challenges.

Blood supply reaches the cortex through the connective tissue capsule. Blood percolates through the sinuses, bathing cords of cortical tissue, and enters the veins within the medulla. In addition, the medulla is supplied by blood vessels from the capsule that pass without branching through the cortex but break up into a rich capillary network around the cords and clumps of chromaffin cells in the medulla. Thus, the mammalian medulla receives a dual blood supply: one directly from the capsule and one from the cortical sinuses. This second vascular supply via the cortical sinuses places cells within the medulla downstream from the cortex; therefore, cortical hormones released into the blood sinuses are carried first to the medulla and act on it before they leave the adrenal gland. The advantages of this chemical support of chromaffin cell function are not entirely clear. In amniotes, adrenocortical tissue produces glucocorticoids that control protein, lipid, and carbohydrate metabolism. In this capacity, adrenocortical tissue, like chromaffin tissue, affects metabolic activity, although in different ways. Thus, the close association of adrenocortical and chromaffin tissues established by blood vessels that serve both might be of some advantage in synchronizing their activities.

Pancreatic Islets

Structure and Phylogeny

The **pancreas** is a composite gland consisting of exocrine and endocrine portions (figure 15.11a). The **exocrine** portion consists of **acini** that secrete digestive enzymes into ducts. The endocrine portion, the **pancreatic islets** (islets of Langerhans), consists of masses of endocrine cells embedded within the exocrine pancreas (figure 15.11b,c). In cyclostomes and most teleost fishes, exocrine and endocrine portions of the pancreas are adjacent to each other, although they are separate groups of glandular tissue (figure 15.12). In hagfishes, islets are found at the base of the common bile duct, and in

lampreys, they are embedded within the mucosal wall of the intestine and even within the liver. In chondrichthyans and coelacanths *(Latimeria)*, islets occur around the ducts of and within the exocrine pancreas. In most bony fishes, isolated masses of pancreatic islet tissue known as **Brockmann bodies** are scattered along the liver, gallbladder, bile ducts, abdominal blood vessels, and surface of the intestines. In a few bony fish species, islets accumulate in the exocrine pancreas, but in many teleosts, the separate islets are gathered into a single mass called the **principal islet.** In most tetrapods, the endocrine islets are typically distributed evenly in small clumps. In many birds and in the toad *Bufo*, they form lobes embedded within the exocrine portion of the pancreas.

Both exocrine acini and endocrine islets differentiate within the **pancreatic diverticulum,** which grows out from the embryonic gut and pushes its way through surrounding mesenchyme. Transplants of marked neural crest cells from quail into early chick embryos reveal that these transplanted neural crest cells give rise to parasympathetic ganglia in the chick pancreas, but apparently the neural crest cells make *no* contribution to the pancreatic islets.

Function

With **special stains,** up to four cell types can be distinguished within the pancreatic islets of most vertebrates (table 15.2). **Insulin** is produced by **B cells** of the islets. Insulin, in cooperation with other hormones, controls the overall metabolism of carbohydrates, fats, and proteins. It is especially important when these end products of digestion are plentiful, as it promotes, often indirectly, their conversion into storage forms. One action of insulin is to inhibit fat breakdown, promote fat synthesis, and, consequently, lower blood levels of fatty acids. Insulin augments intracellular metabolism of glucose and inhibits the breakdown of glycogen in the liver, but its most important function is to bind to cell membranes and promote the entry of glucose into cells, especially into skeletal and cardiac muscle cells. Thus, blood levels of glucose fall, a condition known as *hypo*glycemia, as intracellular levels of glucose rise. If insulin production is too low, glucose is unable to enter cells, builds up in the blood, and is excreted in the urine. This condition is termed **diabetes mellitus,** meaning "sweet urine" disease. In earlier days, physicians used their taste buds to diagnose this disorder.

Because glucose levels in the blood are high, recovery of glucose by the kidney falters, the osmotic balance is upset, and the kidney is less able to recover water. Consequently, urine is produced in great volume, leading to the descriptive title of the Middle Ages for the disease, the "pissing evil." A further result of inadequate insulin production is to allow increased conversion of stored fats (triglycerides) and proteins into glucose to replace its loss. As a consequence, ketone bodies (a family of by-products from lipid metabolism) and urea (protein metabolism) are produced, enter the blood, and during kidney filtration further osmotically encourage an increased volume of urine production. Overall, the patient

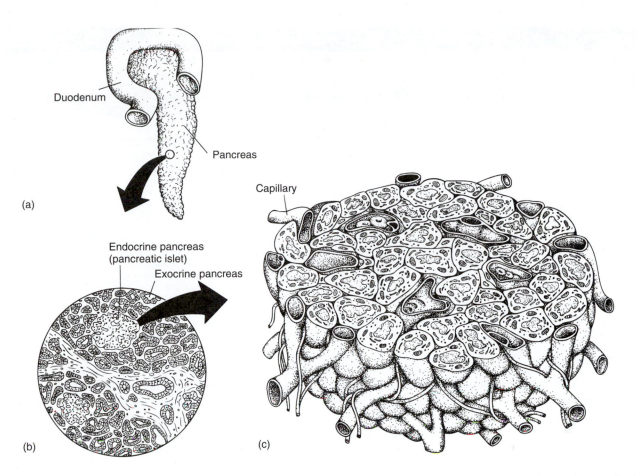

FIGURE 15.11 **Mammalian pancreatic islets.** (a) The pancreas is composed of an exocrine gland and an endocrine gland. (b) Patches of heavily vascularized tissue called pancreatic islets collectively constitute the endocrine pancreas. (c) Enlarged section of a pancreatic islet.

After Krstić.

FIGURE 15.12 **Distribution of pancreatic islets among vertebrates.** Shading indicates endocrine tissue (pancreatic islets), and dotted areas indicate exocrine tissue that secretes digestive enzymes. Two general arrangements of pancreatic tissue are known in actinopterygians, type i and type ii.

After Epple.

TABLE 15.2	Distribution of the Endocrine Pancreas Cell Types among Vertebrate Groups			
Class	A Cells	B Cells	D Cells	PP Cells
Agnatha				
Hagfishes	−	++++	+	?
Chondrichthyes				
Elasmobranchs	+++	++++	+	+
Osteichthyes				
Teleosts	+++	++++	+	+
Amphibia				
Anura and Urodela	+++	++++	+	+
Reptilia				
Lepidosauria	++++	+	+	+
Crocodilia	+++	+++	+	?
Aves	+++++	+++	+	+
Mammalia	++	+++++	+	+

Note: The number of plus signs (+) represents the relative abundance of each cell type in a group and should not be construed as a precise ratio. The minus sign (−) indicates no cells present.

stricken with such diabetic disease faces dehydration, coma, cardiac failure (from loss of blood volume), and death if untreated.

Glucagon is produced by **A cells** of the pancreatic islets. It mobilizes stored products into more readily usable chemicals. Consequently, its primary actions are opposite to those of insulin because glucagon results in elevated blood levels of glucose, termed *hyper*glycemia, through its stimulation in the liver of the conversion of stored glycogen to glucose. Glucagon has opposite effects on lipid metabolism, breaking down fats with the consequent rise in blood levels of fatty acids.

Glucagon is one of several hyperglycemic hormones, although it is especially critical for metabolic regulation in herbivores and in fasting carnivores. In birds and lizards, it is more important than insulin in regulating the fates of the end products of digestion.

Somatostatin is produced by **D cells** of the islets. It inhibits secretion of both insulin and glucagon, but the physiological significance of this is not known. **Pancreatic polypeptide** (PP) is secreted by **PP cells** in the islets and is usually released into the blood following a protein-rich or fat-rich meal. Apparently, this hormone aids in controlling such gastrointestinal activities as promoting the flow of gastric juice, especially hydrochloric acid, in the stomach.

Pituitary Gland

Structure

The **pituitary gland,** or **hypophysis,** is found in all vertebrates. The name *hypophysis* is a recent term inspired by its position beneath the brain (*hypo-* means under and *-physis* refers to growth). The name *pituitary* is centuries old and refers to the mistaken view that this gland produces slime or viscous mucus called *pituita* (phlegm). Although small, this gland

has pervasive effects over most of the body's activities. The pituitary has two embryonic sources. One source is the **infundibulum,** a ventral outgrowth from the **diencephalon** of the brain. The other is Rathke's pouch, a diverticulum from the **stomodeum,** which grows dorsally and becomes associated with the infundibulum (figure 15.13a,b). The infundibulum retains its connection to the brain and becomes the **neurohypophysis.** Rathke's pouch (adenohypophyseal placode) is pinched off from its connection to the stomodeum, and becomes the **adenohypophysis** (figure 15.13b–d).

The adenohypophysis and neurohypophysis in turn differentiate into regions that we recognize by their tissue arrangements (cords and clumps), staining properties (acidophils, basophils, and chromophobes), or anatomical position. Three distinct regions subdivide the adenohypophysis: the **pars distalis,** the **pars tuberalis,** and the **pars intermedia** (figure 15.13e). In all vertebrates, the pars distalis is the major portion of the adenohypophysis and the source of a variety of hormones. Often it is differentiated into lobes (cephalic and caudal) or subregions (proximal and rostral). The pars tuberalis is located anterior to the pars distalis. Its function is not well understood, but it is found only in tetrapods. In mammals at least, it responds to melatonin and releases a hormone that in turn is linked to the photoperiodic control of prolactin secretion on a circadian rhythm. The pars intermedia adjoins the neurohypophysis, often associated with a cleft, a remnant of the embryonic lumen of Rathke's pouch.

The neurohypophysis consists of up to two subdivisions: the **pars nervosa** and the more anterior **median eminence.** Each of these regions has its own vascular supply. A short portal system between them places the adenohypophysis downstream from the median eminence. The pars nervosa has an extensive blood supply from the general body circulation, which is separate from the supply to

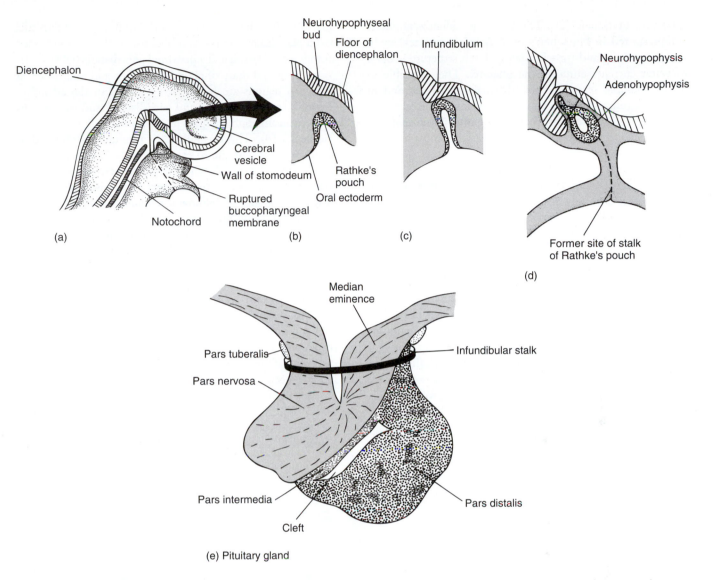

FIGURE 15.13 **Development of the vertebrate pituitary.** (a) Sagittal section of a young embryo showing formation of Rathke's pouch and rudimentary infundibulum. (b–d) The two diverticula make contact during embryonic development, and Rathke's pouch breaks free from its source in the stomodeum. (e) Anatomy of the adult pituitary gland. Note how the two embryonic sources are combined.

TABLE 15.3	Divisions of the Pituitary Gland		
Embryonic Source	**Embryonic Divisions**		**Anatomical Divisions**
Rathke's pouch	Adenohypophysis	Pars tuberalis Pars distalis Pars intermedia	Anterior lobe
Infundibulum	Neurohypophysis	Pars nervosa Infundibular stalk and median eminence	Posterior lobe

the adenohypophysis of the pituitary. The descriptive terms **anterior** and **posterior lobes** are avoided in this book because they are *not* synonymous with the embryonic divisions of the pituitary. Instead they refer to anatomical divisions. The term *posterior lobe* actually includes parts derived from both embryonic sources (table 15.3). The preferred terms *adenohypophysis* and *neurohypophysis* divide the

pituitary according to its embryonic origin from Rathke's pouch and the infundibulum, respectively.

Among protochordates, the vertebrate infundibulum is found in cephalochordates (but apparently not in urochordates) and is represented by a right, ventral lobe of the anterior neural tube that extends downward along the right side of the notochord, and ends near the dorsal surface of Hatschek's pit. The vertebrate adenohypophysis is represented in Hatschek's pit (cephalochordates) or in the subneural gland (urochordates). Each structure is open to water currents entering the pharynx. They are able to directly sample seasonal cues (thermal, chemical), and in turn synchronize reproductive activity through the release of hormones (gonadotropins) that affect gonad development.

Phylogeny

Fishes The size and organization of the pituitary are quite variable even among vertebrates of the same class. In hagfishes, the embryonic sources of the pituitary differ from those of other vertebrates. As in other vertebrates, the hagfish neurohypophysis is a hollow, elongated sac that extends from the diencephalon of the brain, but a median eminence is absent. The hagfish adenohypophysis appears to arise from endoderm rather than from stomodeal ectoderm. It consists of patches of cells embedded in a dense connective tissue layer but undifferentiated into regions. Thus, the hagfish adenohypophysis may not be homologous with other vertebrate pituitaries.

Although a median eminence is absent in hagfishes and lampreys, in most other respects their pituitary closely resembles that of other fishes (figure 15.14). The neurohypophysis of lampreys extends from the ventral part of the brain and contacts the adenohypophysis. The adenohypophysis arises as an ectodermal pocket but usually retains its connection with the

olfactory organ until metamorphosis. Both a pars intermedia and a pars distalis are present. The pars distalis is subdivided further into a **rostral** and a **proximal pars distalis** (table 15.4).

In the pituitary of chondrichthyans and lungfishes, at least two regions are recognized typically in the adenohypophysis (pars intermedia and pars distalis) and two regions in the neurohypophysis (pars nervosa and median eminence) (figure 15.14). The elasmobranch pituitary exhibits additional features. Unique to elasmobranchs is a forward projection from the pars distalis termed the **ventral lobe,** which some endocrinologists call the **pars ventralis.** The function of the ventral lobe is unknown, although because it secretes some of the same hormones, it is likely just an extension of the pars distalis. The **saccus vasculosus** of the elasmobranch pituitary is a structural specialization derived from the hypothalamus and located above the neurohypophysis, but its function is still unknown. A vascular portal system is present between the median eminence and the pars distalis. As in lampreys, the pars distalis of elasmobranchs is subdivided into a rostral pars distalis and a proximal pars distalis. In osteichthyans other than dipnoans, rostral and proximal subdivisions are recognized within the pars distalis, the saccus vasculosus is present, and the ventral lobe is absent (figure 15.14 and table 15.4). In teleosts, two primary regions are also recognized, the adenohypophysis and neurohypophysis, although unlike those of other bony fishes, they lack a median eminence. Neurons in the teleost

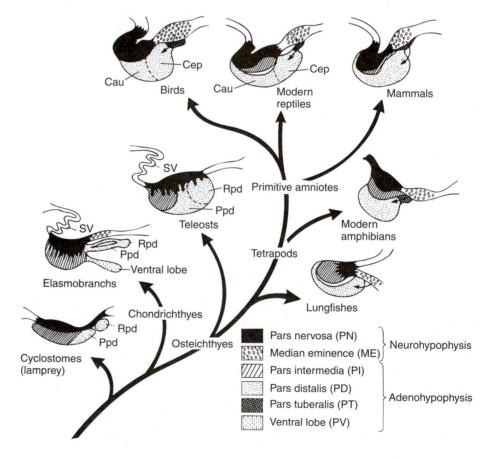

FIGURE 15.14 Phylogeny of the vertebrate pituitary. Thin, solid arrows within the pituitaries designate the vascular portal connection from the median eminence to the pars distalis. The pars distalis often exhibits anterior and posterior regions: the rostral pars distalis (Rpd) and the proximal pars distalis (Ppd) or the cephalic pars distalis (Cep) and caudal pars distalis (Cau). In mammals, the pars distalis is not subdivided. The ventral lobe, a projection of the adenohypophysis, is unique to elasmobranchs. In some lower vertebrates, the saccus vasculosus (SV) is present and is derived from the hypothalamus of the brain.

Cep
Cau
Birds
Cau
Cep
Modern reptiles
Mammals

SV
Rpd
Ppd
Teleosts

Primitive amniotes

Modern amphibians

SV
Rpd
Ppd
Ventral lobe
Elasmobranchs

Tetrapods

Lungfishes

Chondrichthyes

Rpd
Ppd
Cyclostomes (lamprey)

Osteichthyes

Pars nervosa (PN)
Median eminence (ME) } Neurohypophysis
Pars intermedia (PI)
Pars distalis (PD)
Pars tuberalis (PT)
Ventral lobe (PV) } Adenohypophysis

TABLE 15.4 Summary of Anatomical Features of the Vertebrate Pituitary

Group	PT	Pars distalis RPD (= caudal)	Pars distalis PPD (= cephalic)	PI	PV	ME	PN	SV
Agnatha								
Hagfishes		*	*				+	
Lampreys		+	+	+			+	
Chondrichthyes		+	+	+	+	+	+	+
Osteichthyes								
Polypterus		+	+	+		+	+	+
Teleosts		+	+	+			+	+
Latimeria		+	+	+		+	+	+
Dipnoans				+		+	+	
Amphibia	+	*	*	+		+	+	
Reptilia, most	+	×	×	+		+	+	
Crocodilia	+	×	×			+	+	
Birds	+	×	×			+	+	
Mammalia	+	×	×	+		+	+	

Note: Plus signs (+) indicate that the part is present. (*) Pars distalis present but no regionalization into RPD and PPD. (×) Region of pars distalis present and homologous. Abbreviations: pars tuberalis (PT), rostral pars distalis (RPD), proximal pars distalis (PPD), pars intermedia (PI), pars ventralis (PV), median eminence (ME), pars nervosa (PN), saccus vasculosus (SV).

hypothalamus directly reach into the adenohypophysis to activate its secretory cells.

Tetrapods The pars tuberalis appears in early tetrapods and persists in most later amniotes (figure 15.14). Thus, the tetrapod pituitary characteristically consists of an adenohypophysis with three subdivisions (pars intermedia, pars distalis, and pars tuberalis) and a neurohypophysis that retains two subdivisions (pars nervosa and median eminence). In amphibians, the adenohypophysis establishes a basic tetrapod pattern of pars tuberalis, pars distalis (no regionalization), and pars intermedia; the neurohypophysis is comprised of a median eminence and pars nervosa. Reptilian pituitaries conform generally to the tetrapod pattern but are remarkably varied in size and shape. The adenohypophysis of snakes is lobed and a cleft may be present in some reptiles. Within the reptilian pars distalis, **cephalic** and **caudal lobes** are recognized. The pars tuberalis is well developed in most reptiles but reduced in lizards and absent in snakes. Both the crocodile and the bird pituitaries are similar to those of other tetrapods, although the pars intermedia is absent in both. The pars distalis again consists of cephalic and caudal lobes. The well-developed median eminence is sometimes divided into anterior and posterior regions. A number of mammals also lack a pars intermedia, and in most monotremes and therians the basic pattern is evident—adenohypophysis with pars tuberalis, pars intermedia, and pars distalis and a neurohypophysis with pars nervosa and median eminence (figure 15.14 and table 15.4).

Function

Strictly speaking, cells within the neurohypophysis do not produce pituitary hormones. Instead, axons of **neurosecretory neurons** of the hypothalamus dorsal to it project into the neurohypophysis, where their secretions are released into blood vessels or temporarily stored. In addition to these axons, **pituicytes** within the neurohypophysis are thought to support neurosecretory neurons, but they do not synthesize or secrete hormones.

In contrast to the cells of the neurohypophysis, cells of the adenohypophysis synthesize pituitary hormones. In teleosts, neurosecretory neurons project directly into the adenohypophysis to directly preside over its activity. In all other vertebrates with a median eminance, the hypothalamus indirectly influences their activity. Neurosecretory neurons from the hypothalamus project into the region of the median eminence and there secrete their **neurohormones** into capillaries. Through a tiny vascular portal link, these neurohormones are transported a short distance through a capillary plexus and then diffuse into the adenohypophysis (figure 15.15). These neurohormones are **releasing hormones** or **release-inhibiting hormones,** depending on whether they stimulate or inhibit cells of the adenohypophysis.

From early staining methods, cell types were identified on the basis of their reactions with dyes. **Acidophils** and **basophils** have affinities for acidic and basic dyes, respectively. **Chromophobes** do not react with dyes. Although these terms are still useful for descriptive purposes, new stains

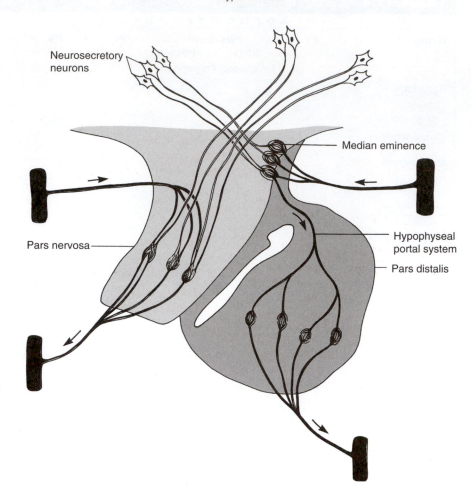

Neurosecretory neurons

Median eminence

Pars nervosa

Hypophyseal portal system

Pars distalis

FIGURE 15.15 Vascular supply and circulation within the pituitary gland. Note the short hypophyseal portal shunt between the median eminence and the pars distalis. A separate capillary supply to the pars nervosa arises from the general circulation. Neurosecretory neurons release neurohormones into both of these capillary networks. Neurohormones entering the median eminence are carried to cells within the pars distalis. Neurohormones released in the pars nervosa enter the general body circulation.

and better techniques for identifying hormones have shown that one cell type may produce several different hormones.

Neurohypophysis   Two hormones synthesized by neurosecretory cells of the hypothalamus have been identified in the mammalian pars nervosa. One hormone is **vasopressin,** which acts on smooth muscle in the walls of peripheral arterioles, causing them to constrict. Resistance to blood flow increases and brings about a rise in blood pressure. If an organism sustains a fair amount of blood loss, pressure sensors in the carotid artery detect declines in blood pressure and stimulate increased secretion of vasopressin through reflex control of the hypothalamus.

Vasopressin is also called **antidiuretic hormone (ADH)** because it promotes water conservation within the kidneys (figure 15.16a). If a mammal becomes dehydrated, neurosecretory neurons of the hypothalamus release ADH into the neurohypophysis, where it is picked up in the blood and carried to the kidneys. ADH acts on the walls of the renal collecting ducts, making them highly permeable to water; therefore, water flows from the tubules into the hyperosmotic interstitial fluid and creates a concentrated urine. In the absence of ADH, walls of the collecting ducts remain impermeable to water.

Less water is reabsorbed, and the urine is copious and dilute. Under pathological conditions in which disease or tumors prevent sufficient release of ADH, large volumes of dilute urine are passed, a medical condition known as **diabetes insipidus.** As a result, the individual experiences constant thirst and drinks large amounts of water to compensate.

The second hormone found in the pars nervosa is **oxytocin.** Its target tissues are the **myometrium,** the smooth muscle layer of the uterus, and the contractile **myoepithelial cells** of the mammary gland. Late in pregnancy, the level of oxytocin in the blood increases, which gives it a role in uterine contractions during parturition. A suckling neonate initiates a reflex through sensory nerves that eventually stimulates neurosecretory neurons of the hypothalamus to release oxytocin at their terminus in the pars nervosa. The bloodstream transports the hormone to the mammary gland, where it promotes contractions of the myoepithelial cells on the walls of the exocrine milk glands. About one minute after the onset of suckling, milk begins to flow from the nipple or teat (figure 15.16b).

Oxytocin's role in natural uterine contractions during birth led to its medical use in artificially inducing labor in women. Throughout vertebrates it promotes rhythmic contractions of the oviducts during ovoposition or birth, and in

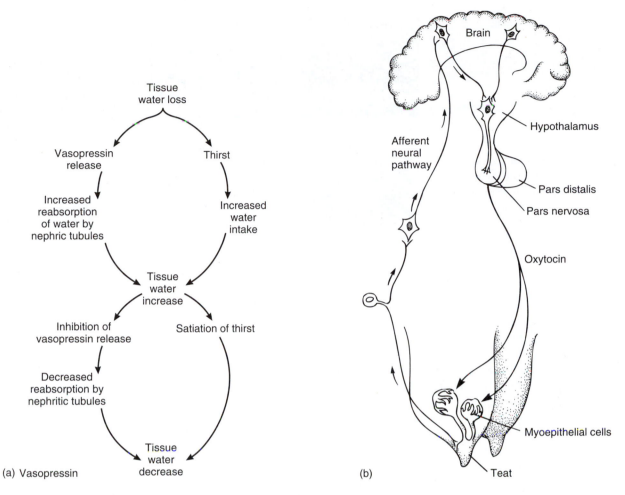

Tissue water loss → Vasopressin release → Increased reabsorption of water by nephric tubules → Tissue water increase

Tissue water loss → Thirst → Increased water intake → Tissue water increase

Tissue water increase → Inhibition of vasopressin release → Decreased reabsorption by nephritic tubules → Tissue water decrease

Tissue water increase → Satiation of thirst → Tissue water decrease

(a) Vasopressin

Brain
Afferent neural pathway
Hypothalamus
Pars distalis
Pars nervosa
Oxytocin
Myoepithelial cells
Teat

(b)

FIGURE 15.16 **Hormones found in the mammalian pars nervosa.** (a) Vasopressin restores water balance through a complex series of steps. (b) Oxytocin promotes the release of milk during suckling. Impulses from tactile stimulation of the nipple are transmitted by afferent nerves first to the brain and then to the hypothalamus. As the neurosecretory cells are activated by these afferent nerves, they manufacture and release oxytocin. Oxytocin released by the pars nervosa is transported by the blood to the mammary gland, where it stimulates contractile myoepithelial cells that cause the release of milk.

males it stimulates contraction of the sperm ducts. Oxytocin deserves our gratitude for its rhythmic contractions of smooth muscle of reproductive organs in men and women, which are responsible for the sensation of orgasm.

Adenohypophysis Six major hormones have been identified within the adenohypophysis. **Growth hormone (GH)** may target the liver, which responds by secreting an insulin-like growth factor that mediates some GH effects on growth and metabolism. GH also produces effects throughout the body, including increased protein synthesis, increased mobilization of fatty acids, and decreased glucose utilization. In young animals, deficient levels of growth hormone lead to **pituitary dwarfism,** and excess levels lead to **pituitary gigantism. Acromegaly** is a condition that occurs in adults in which a disproportionate proliferation of cartilage results from an excess of growth hormone released after puberty.

In mammals, **prolactin (PRL)** promotes development of the mammary glands and lactation during pregnancy. In birds, prolactin stimulates lipid synthesis during premigratory fattening and supports brooding behavior. In some species, prolactin stimulates the appearance of a **brood patch,** a defeathered, highly vascularized region of the breast skin placed against the incubating eggs to warm them. In pigeons and related birds, prolactin promotes the secretion of **crop milk,** a nutritional fluid produced in the crop and fed to fledglings. In lizards, prolactin affects tail regeneration, and in amphibians it affects growth. In teleosts, prolactin is important in osmoregulation, especially in migratory fishes moving from salt water to fresh water during spawning.

Thyrotropin, or **thyroid-stimulating hormone (TSH),** stimulates the thyroid gland to synthesize and release T_3 and T_4 into the blood.

The adenohypophysis releases **gonadotropins,** typically two hormones that affect the gonads and reproductive tracts. The principal gonadotropins produced by the adenohypophysis are follicle-stimulating hormone and luteinizing hormone. Rising levels of **follicle-stimulating hormone (FSH)** induce the development of selected ovarian follicles. In males, FSH initiates and helps maintain spermatogenesis,

FIGURE 15.17 **Melanophores of frog skin.** Melanophores, located in the integument, respond to melanophore-stimulating hormone by dispersing pigment granules to darken skin color (left) or in its absence by concentrating them to lighten skin color (right).

although the term may seem illogical for this situation. **Luteinizing hormone (LH)** acts in females to finalize maturation of the ovarian follicles. A rise in LH level promotes ovulation. Following ovulation, it promotes reorganization of the follicle cells into the **corpus luteum.** In males, luteinizing hormone, more aptly termed **interstitial cell-stimulating hormone (ICSH),** stimulates the interstitial cells of the testis to secrete testosterone. Overall, LH and FSH stimulate androgen and estrogen synthesis in both males and females. Occasionally, females produce higher androgen levels than do males of the same species; and males, such as stallions, produce estrogens. The evidence increases for a role, although not yet clearly defined, of androgens in females and estrogens in males.

Corticotropin, or **adrenocorticotropic hormone (ACTH),** stimulates the cortex of the adrenal gland to release glucocorticoids.

Melanophore-stimulating hormone (MSH) is located in the pars intermedia. Its targets are the **melanophores,** pigment cells of the skin. Within a few minutes, MSH affects melanin distribution within melanophores, changing the darkness of the skin in lower vertebrates. Stimulation causes the pigment **melanin** to disperse into fixed cytoplasmic **pseudopods** of the melanophores, which darkens the skin. In the absence of MSH (or through inhibition by melatoin produced in the pineal), pigment granules gather at the center of the cell. The overall effect is to lighten the skin (figure 15.17). In birds and mammals, skin pigmentation results from the release of melanin granules into skin, feathers, and hair. MSH may act to increase the production of pigment over the long term or on a seasonal basis.

Chromatophores (p. 236)

At one time, the term melano*cyte* was used for pigment cells in which MSH caused increased melanin synthesis but no pigment movement within the cell. Melano*phore* designated another cell type in which melanin moved about within the cell in response to MSH. However, discovery of melanocytes in which both synthesis and movement occur casts doubts on the usefulness of such a sharp distinction. I will use **melanophore** as encompassing both, but you should be prepared to meet different usage in other textbooks.

Gonads

In addition to producing gametes, the gonads produce hormones that support secondary sex characteristics. In humans, these include pubic hair, male facial hair, female mammary glands, preparation of the sexual ducts for reproduction, and maintenance of sex drive. In males, the **interstitial cells** (Leydig cells) that cluster between seminiferous tubules produce androgens. The principal androgen is **testosterone.** In females, the endocrine tissues of the ovary include the follicles, corpus luteum, and interstitial tissue. The principal hormones produced are estrogens (e.g., estradiol) and **progestogens** (e.g., progesterone). Endocrine coordination of reproduction is discussed in more detail later in the chapter.

Reproductive system (chapter 14)

Pineal Gland

The unpaired **pineal gland,** or **epiphysis,** is a dorsal evagination of the midbrain. It is part of a complex of evaginations from the roof of the midbrain that we meet in more detail in chapter 17 when we examine photoreceptive organs. In some vertebrates, the pineal gland affects perception of photoradiation. For example, in some fossil vertebrates, the pineal gland was inserted into an opening in the bony cranium, known as the **pineal foramen,** and covered only by a thin layer of integument. This may have allowed the pineal to respond to changes in photoperiod. In some living vertebrates, this gland is still located just under the skin, but more often it resides beneath the bony cranium. Nevertheless, the presence of light-sensitive cells within the pineal gland of lower vertebrates indicates that this gland may be involved in detecting seasonal or daily light schedules. The pineal has also been shown to regulate reproductive cycles in a variety of vertebrates.

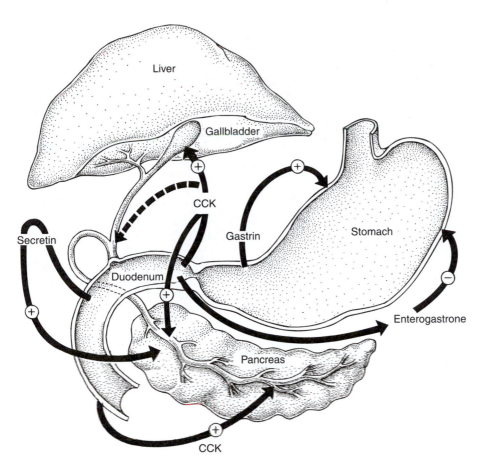

FIGURE 15.18 Some mammalian gastrointestinal hormones. Sites of release and effects on target tissues are indicated. Hormonal promotion or inhibition of secretory activity is indicated by plus (+) and minus (−) signs, respectively. Cholecystokinin (CCK).

Early Greek anatomists speculated that the pineal gland regulated the flow of thoughts. Absence of evidence did not deter later speculation about the pineal being the seat of the soul. The first experimental hint of an endocrine function came in 1927 when an extract prepared from ground pineal gland was placed in an aquarium with frog tadpoles. The tadpoles' skin blanched, suggesting that the extract affected melanophores. Later, the hormone responsible for this effect was isolated and called **melatonin.** However, subsequent research has proved frustrating. The pineal seems to modulate activities already in progress rather than to initiate activities. In lower vertebrates, it clearly affects melanophores in the skin, but in birds and mammals, this role is less important. As mentioned, considerable research suggests that the pineal gland regulates seasonal reproductive patterns. In reptiles and birds, the pineal may aid in the organization of daily, or **circadian,** rhythms. As ambient temperatures warm, the pineal mediates the initiation of courtship behavior in garter snakes. In mammals, experiments in which the pineal was removed or injections of pineal extracts were administered provide circumstantial evidence that the pineal may be involved in release of ACTH from the adenohypophysis, in heightened vasopressin secretion, in inhibition of thyroid activity, and even in stimulation of components of the immune system.

Secondary Endocrine Glands

Some glands that play a central role in activities other than endocrine regulation also release chemicals carried by the vascular system to responsive tissues. Such glands function secondarily in the endocrine system. Usually the hormones they release help these **secondary endocrine glands** regulate their own primary activities. Two examples are the digestive tract and the kidneys.

Gastrointestinal Tract

The alimentary canal functions primarily, of course, in digestion. The walls of the digestive tract produce chemicals that stimulate or inhibit target tissues in the gastrointestinal tract or related digestive organs (e.g., liver, pancreas). These chemicals are secreted directly, rather than being discharged through ducts. Thus, the digestive tract functions secondarily as an endocrine organ.

When food enters the stomach of amniotes, the gastric mucosa releases the hormone **gastrin** (figure 15.18). Gastrin enters the blood and is transported to the stomach, where it stimulates the secretion of gastric juice. When the stomach empties the churned and acidified food into the duodenum, the intestinal mucosa releases **secretin.** Secretin stimulates the pancreas to release highly alkaline pancreatic juice that buffers the acidic **chyme** arriving from the stomach. **Enterogastrone,**

FIGURE 15.19 **Endocrine control of digestion.** As food enters the alimentary canal (1), the release of gastrointestinal hormones is stimulated (2). These hormones enter the circulatory system, travel through the hepatic portal vein (3) to the liver, and then travel to the heart. From the heart, they are transported back to the alimentary canal via the celiac artery (4). When arriving in the pancreas, these gastrointestinal hormones stimulate the release of pancreatic enzymes. Foremost among these gastrointestinal hormones is cholecystokinin, which triggers the release of digestive enzymes from the exocrine pancreas and bile from the gallbladder. The gastrointestinal hormone secretin causes the pancreas to release bicarbonate to help neutralize the acidic chyme that enters the duodenum from the stomach. Another gastrointestinal hormone, glucose insulinotropic peptide (GIP), stimulates the endocrine pancreas to release insulin, indicated by solid arrowheads (5). One end product of digestion is glucose (6), which is absorbed across the intestinal wall (7) and transported to the liver via the hepatic portal vein (3).

After Elias and Pauly.

also released by the intestinal mucosa, inhibits further gastric secretion and mobility. Fats, proteins, and acids stimulate the secretion of **cholecystokinin (CCK),** or **cholecystokinin-pancreozymin (CCK-PZ),** from the intestinal mucosa. Originally, cholecystokinin was thought to be two hormones (hence its hyphenated alternative name) because it has two functions. It stimulates the relaxation of the sphincter at the base of the bile duct, the contraction of the gallbladder, and the ejection of bile that flows into the duodenum, where the bile acts on fats. Cholecystokinin also stimulates the pancreas to secrete pancreatic juice containing digestive enzymes (figure 15.19).

Since the discovery of these gastrointestinal hormones, others with more restrictive actions have been discovered. For example, **enterocrinin,** released by the intestinal mucosa, increases the production of intestinal juice. We examine the endocrine function of the digestive organs in more detail when we consider the evolution of endocrine regulation later in this chapter.

Kidneys

Primarily, the kidneys excrete nitrogenous wastes and function in osmoregulation, but they act as an endocrine organ as well (figure 15.20). When blood pressure drops, the **juxtaglomerular cells** wrapped around renal arterioles release the hormone renin. **Renin** sets in motion a cascade of changes that eventually result in elevating the blood pressure. It catalyzes the transformation of **angiotensinogen** in blood to

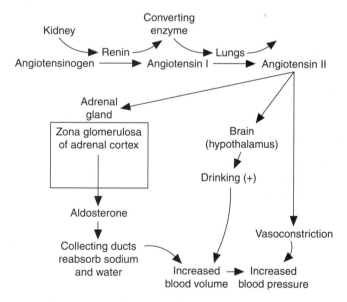

FIGURE 15.20 **The kidney as an endocrine organ.** A drop in blood pressure in blood vessels serving the kidney results in release of renin, the hormone that catalyzes the transformation of angiotensinogen to angiotensin I. In the lungs and elsewhere, angiotensin I is converted to angiotensin II, which indirectly causes the kidney to retain more water and, through its stimulation of the hypothalamus, promotes increased drinking behavior. Both increase blood volume and, hence, blood pressure. Angiotensin II also acts as a vasoconstrictor, further contributing to increased blood pressure.

angiotensin I, which is converted into **angiotensin II** in the lungs and in other organs also. Angiotensin II is a vasoconstrictor that also increases blood volume by stimulating the release of aldosterone from the adrenal gland. Aldosterone causes the distal tubules of the kidneys to reabsorb more sodium, which causes increased water reabsorption and a subsequent rise in blood volume. Together, vasoconstriction and increased blood volume elevate the blood pressure.

Reduced oxygen levels in blood passing through the kidneys stimulate cells there to produce erythropoietin (EOP). EOP is a hormone that stimulates red blood cell production by the hemopoietic tissues in mammals. Occasionally a human athlete, especially one in an endurance sporting event, will surreptitiously and illegally take artificial doses of EOP to boost the number of oxygen-carrying blood cells and thereby enhance performance. Such cheaters are caught by detection of abnormally high levels of EOP. But because EOP is a natural hormone, such tests can be controversial.

Endocrine Coordination

So far we have surveyed endocrine glands, their hormones, and their target tissues. Next we examine two ways in which endocrine glands interact to coordinate activities. Let us begin by considering reproduction in mammals.

Mammalian Reproduction

Male

In males, the adenohypophysis releases FSH and LH, which have immediate effects on the testes (figure 15.21). FSH plays a prominent role in controlling spermatogenesis. LH acts on the interstitial cells in the testis to promote the production of androgens, especially testosterone. First, testosterone regulates the development and maintenance of the secondary sexual characteristics (including antlers and brightly colored plumage), the sex impulse, and the accessory sex glands. Second, it promotes spermatogenesis. Third, testosterone has a negative feedback effect on the adenohypophysis to limit the production of LH, and hence prevents overproduction of this gonadotropic hormone (figure 15.21).

Female

In females, oocytes within the chordate ovary are coated with follicle cells derived from ovarian epithelium. In most tetrapods, each ovary houses hundreds or thousands of oocytes wrapped in follicle cells. However, only a few follicles actually undergo **maturation** to release their ova during **ovulation,** making **fertilization** possible. As maturation of an ovum progresses, the enveloping, single inner layer of follicle cells proliferates, becoming the thickened, multilayered **granulosa.** Later in maturation, spaces filled with fluid appear within the granulosa and coalesce into the **antrum,** a single fluid-filled space. In addition, connective tissue cells within the ovary form an outer coat, called the **theca,** around

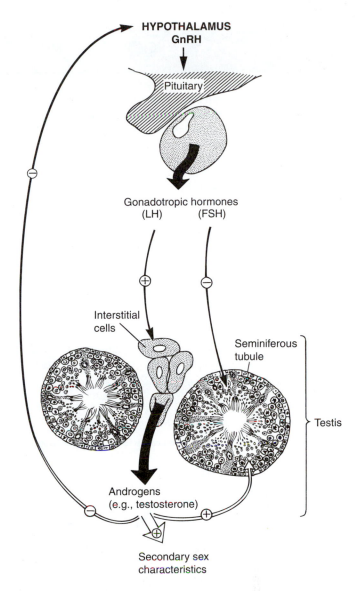

FIGURE 15.21 **Gonadotropic hormones in male mammals.** Follicle-stimulating hormone (FSH) promotes spermatogenesis. Luteinizing hormone (LH) stimulates interstitial cells of the testes to release testosterone. Testosterone in turn helps maintain secondary sex characteristics in males and inhibits secretion of gonadotropic hormones. Hormonal promotion and inhibition of activity are indicated by plus (+) and minus (−) signs, respectively.

the follicle. Following ovulation, the follicle becomes the **corpus luteum.** The outer layer remains a connective tissue capsule, but the inner layer becomes endocrine, a source of the androgens needed by cells of the granulosa to synthesize estrogens. Cells of the granulosa become **granulosa lutein cells,** which constitute most of the corpus luteum, and thecal cells persist as **theca lutein cells,** which form the outer capsule of the corpus luteum. Eventual regression of the corpus luteum yields the **corpus albicans** in progressive stages of degeneration. Regression of follicles before ovulation yields **atretic follicles** (figure 15.22) (see chapter 14).

In large part, medical successes in developing oral contraceptives for females but not for males can be attributed to intrinsic differences in the natural hormonal control of reproductive functions in the two sexes.

Contraception, mediated by sex hormones, is a *normal* monthly phenomenon in the human female. Progesterone, produced by the corpus luteum, suppresses further ovulation by inhibiting the release of new follicle-stimulating hormone (FSH). Thus, while progesterone is secreted, no FSH is produced, no more follicles mature, no ova are released, and no fertilization occurs. Oral contraceptives seek to mimic this natural series of events.

Oral contraceptives for women contain progesterone, which prevents ovulation by suppressing FSH secretion. But this means that women taking an oral contraceptive do not form corpora lutea; therefore, progesterone and smaller amounts of estrogen produced by the corpus luteum are absent. One modification of the original recipe for oral contraceptives was to add estrogen to the progesterone to compensate for the missing corpus luteum and its secretion of both. Another modification has been to modify the levels of progesterone administered. Usually they are lower in oral contraceptives to match the woman's individual needs.

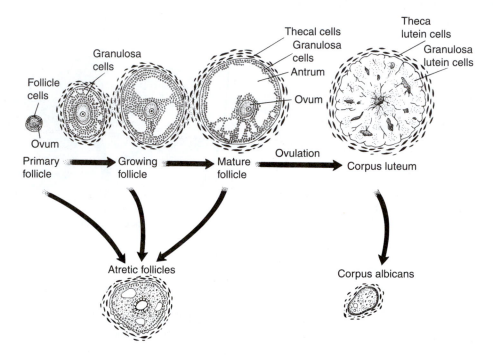

FIGURE 15.22   Maturation of an ovarian follicle in mammals. In a female entering her first reproductive season, the ovary is populated by oocytes wrapped in follicle cells that have remained quiescent since primordial germ cells first colonized the ovary during her embryonic development. At the onset of the breeding season, hormones stimulate maturation of some of these follicles. The follicle cells proliferate, mature, and enclose the antrum, a fluid-filled space in the follicle. Upon ovulation, the mature follicle ruptures, releasing the ovum and some clinging follicle cells. After the ovum is released, the walls of the follicle form the corpus luteum, which continues to play an endocrine role for a while. If pregnancy does not occur, the ovary quickly recycles to bring the animal back into a reproductive-ready mode. The corpus luteum then undergoes involution and remains as a remnant patch of connective tissue, the corpus albicans. If follicles regress before ovulation, the involuted follicles form atretic follicles, once considered to be without further function, but now recognized as perhaps retaining an endocrine role as they contribute to the ovarian interstitial gland involved in steroid synthesis.

The events of follicle maturation are best understood in mammals, especially in humans. Hormones promote follicle maturation and simultaneously prepare the uterus to receive a fertilized ovum (figure 15.23a). There are four major steps involved. First, falling levels of progesterone are accompanied by rising levels of FSH. As levels of FSH rise, selected follicles begin to mature. Why some and not other follicles in the ovary respond is not known. In those that respond, the thin layer of follicle cells divides to produce a thickened coat of cells. Fluid-filled spaces within, the forerunners of the antrum, appear as well.

Second, as follicles grow under continued FSH stimulation, the inner granulosa cells secrete increased amounts of estrogen. At this point, estrogen has two actions. It stimulates the endometrium of the uterus to proliferate, and indirectly promotes secretion of luteinizing hormone (LH)

Sperm production in the human male, however, does not follow monthly rhythms. Sperm are produced more or less continuously, so that cyclic, hormone-mediated contraception does not occur in the human male as it does in females. This precludes any mimicking in the male of a natural contraceptive process. Instead, strategies for producing a male oral contraceptive have sought to inhibit FSH secretion directly. Of course, this cannot be accomplished with progesterone. If it could, the side effects would produce a sterile but feminized male. Other hormonal disruptions of testicular function sufficient to stop sperm production have usually had similar unwanted side effects. But there is some recent hope.

Researchers testing an experimental cancer drug wanted to be sure it would not have negative side effects on healthy cells. So they tried it in mice and looked for any side effects. To their surprise, they discovered that as a side effect, it worked as a male birth control drug. When injected in mice, their testes stopped making sperm; when the drug was stopped, sperm production resumed. Follow-up experimental trials are underway. If eventually it works in people, and is safe, it would provide a hormone-free way to give men control of their procreative efforts.

(a)

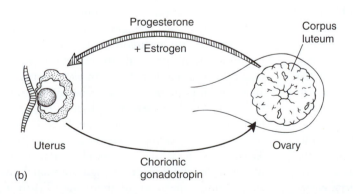

(b)

FIGURE 15.23   Ovarian and uterine cycles of human females. (a) Follicle maturation and accompanying thickening of the uterine lining. The endometrium of the uterus thickens and then enters a secretory phase in anticipation of receiving a fertilized ovum. The follicle is an endocrine gland that releases increasing levels of estrogens in response to rising levels of FSH and LH from the adenohypophysis. Rising levels of estrogen act through the hypothalamus, releasing a GnRH surge, which stimulates the LH surge, which causes ovulation. Following ovulation, the ruptured follicle persists as a corpus luteum. This modified endocrine tissue still produces some estrogens; however, it primarily produces progesterone, which inhibits FSH production and temporarily prevents the maturation of additional follicles. Progesterone also stimulates the uterus to maintain an environment hospitable to the implanted embryo. If implantation fails, the corpus luteum disintegrates at about day 28 of the cycle, progesterone levels fall, and FSH is released, allowing the cycle of follicular development to begin again. (b) If pregnancy occurs, chorionic gonadotropin released by the placenta initially supports the corpus luteum, which secretes progesterone to maintain the uterine wall during and following implantation. This mutual hormonal support lasts until about the third month of pregnancy, when the placenta begins to secrete progesterone. The corpus luteum undergoes slow involution at this time.

through its effects on the hypothalamus, which secretes **gonadotropin-releasing hormone (GnRH)** that actually stimulates LH release.

Third, LH release causes ovulation. A mature follicle ruptures and releases its ovum. Thereafter, LH promotes consolidation of the ruptured follicle into the corpus luteum.

Fourth, the corpus luteum takes over the function of secreting estrogen that was initiated by the follicles, although secretion is now at lower levels. In addition, the corpus luteum produces progesterone. Progesterone is an "optimistic" hormone, promoting the final stages of preparing the uterus for a fertilized ovum. Further, progesterone inhibits the secretion of FSH from the pituitary; therefore, no more follicles mature at this time.

In humans, if pregnancy does not occur, hormonal support for growth of the corpus luteum drops after 10 to 12 days and it deteriorates. When this happens, it involutes, becoming a patch of scar tissue, the corpus albicans. With the decline of the corpus luteum, estrogen and progesterone levels drop, FSH and LH secretion rises, and the cycle begins again.

If pregnancy occurs, **chorionic gonadotropin (CG)** hormone stimulates the growth of the corpus luteum. CG is produced by the rudimentary placenta established by the embryo implanted in the uterine wall (figure 15.23b). Eutherian CG functions to maintain the corpus luteum, which in turn produces progesterone to maintain the uterus housing the implanted embryo and its rudimentary placenta. In humans, the corpus luteum, the placenta, and the growing embryo are mutually maintained in this reciprocal fashion until about the second month of pregnancy. Thereafter, the corpus luteum undergoes a slow involution. At this point in the pregnancy, however, involution of the corpus luteum and the consequent drop in its output of progesterone (and estrogen) do not bring about menstruation and loss of the implanted embryo because by now, the placenta itself is producing progesterone (and estrogen) to maintain itself.

The reproductive cycle of the red kangaroo *(Megaleia rufa)* illustrates how the endocrine and nervous systems coordinate reproductive processes (figure 15.24a–c). Like most marsupials, the red kangaroo has a short gestation period. The female kangaroo may support up to three young at staggered stages of development. Her reproductive tract is designed to accommodate embryos at different stages of maturation. Ovulation alternates between the two ovaries. The blastocyst enters the central vaginal canal, where it develops during its brief gestation. Sperm from a subsequent mating travel along the lateral vaginal canals without encountering the embryo. After birth, the infant, now called a **joey,** migrates into the pouch and begins suckling from a teat. Via afferent nerves to the pituitary, suckling stimulates the release of prolactin and causes a decrease in gonadotropin. As a result of these hormonal changes, the ovarian corpus luteum is inhibited and its progesterone output declines. Without progesterone, the uterus no longer promotes development of the next blastocyst. Its development is temporarily arrested, and the blastocyst enters **embryonic diapause** (see chapter 14). When the growing joey

begins to make tentative forays away from the female pouch, the intensity of its suckling stimulus decreases, levels of gonadotropic hormone rise, stimulating the corpus luteum, and progesterone levels rise. The female comes into estrus and mates. The blastocyst in diapause resumes its development and completes gestation. The newborn moves to the pouch and attaches itself to an available teat. Again the suckling stimulus arrests the development of the new blastocyst, and it enters embryonic diapause. During suckling, the composition of the milk also changes. It increases in fat content as the joey grows.

Death or premature removal of a joey results in decreased prolactin and increased gonadotropic hormonal secretion by the pituitary. Consequently, the corpus luteum is reactivated, progesterone secretion is increased, and development of the blastocyst is resumed. The female enters estrus and usually mates. However, environmental events, such as short photoperiods, can have an effect similar to that of the suckling young. If a young joey is removed in the fall, the blastocyst may not resume development until the spring.

Metamorphosis in Frogs

Metamorphosis in frogs is an excellent example of the coordination of a complex physiological process involving nervous, secretory, and vascular responses mediated by the endocrine system. The frog tadpole undergoes three developmental stages (figure 15.25a–c). The first **premetamorphosis** stage is characterized by growth in body size. In the second **prometamorphosis** stage, the most conspicuous change is development of the hindlimbs, although some growth in body size still continues. The third stage is **metamorphic climax,** at which time the tadpole is transformed into the young froglet. The forelimbs emerge, the beak is lost, the mouth widens, and the tail is resorbed. Hormones, developmental events, and the nervous system are involved in each stage.

During premetamorphosis, the adenohypophysis produces high levels of prolactin, which stimulates growth but inhibits metamorphosis. The adenohypophysis also produces small amounts of thyroid-stimulating hormone (TSH) autonomously, without any prompting from the hypothalamus. TSH stimulates the thyroid to secrete thyroxine but not in sufficient levels to initiate metamorphosis. During this early stage of development, the median eminence of the pituitary does not respond to thyroxine and remains undeveloped. Thus, during premetamorphosis, the tadpole grows in size but few other changes occur (figure 15.25a).

During prometamorphosis, the median eminence becomes responsive to thyroxine and begins to develop, establishing a modest but complete portal system that allows neurohormones to be transported from the hypothalamus to the adenohypophysis. The neurohormone **corticotropin-releasing hormone (CRH)** stimulates the secretion of increasing amounts of TSH. Rising levels of TSH stimulate the thyroid to produce more thyroxine. Now, metamorphosis advances through additional changes. The accelerating conversion of thyroxine (T_4) to T_3, together with the appearance of receptors on target tissue responsive

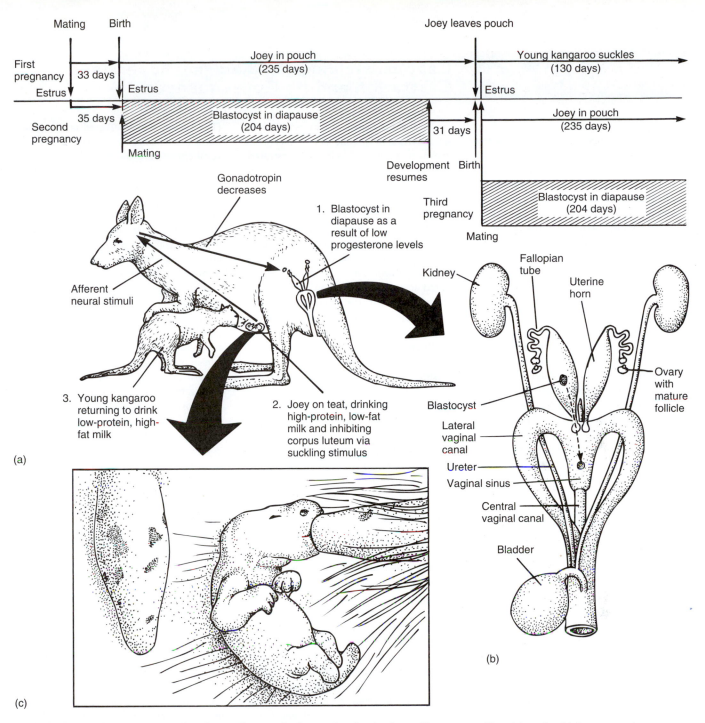

FIGURE 15.24 **Hormonal and neural control of reproduction in the red kangaroo, *Megaleia rufa.*** (a) At any one time, the female may support up to three young at various stages of development: a blastocyst, a joey, and a young kangaroo. This staggering of three pregnancies is diagrammed. At the end of the first 33-day gestation period, the first birth occurs. The female enters estrus again and mates, resulting in pregnancy. However, the first young, now termed a joey, enters the pouch and attaches its mouth to a teat. Its suckling stimulates (via afferent neural stimulation to the pituitary) increased levels of prolactin and decreased levels of gonadotropic hormones. Consequently, progesterone secretion by the corpus luteum decreases, so the uterus is no longer able to support development of the second embryo. Now the second embryo enters embryonic diapause. As the first joey becomes more independent and begins to make forays from the pouch, the suckling stimulus wanes, the levels of prolactin decline, and the levels of gonadotropic hormones rise. High gonadotropic levels stimulate the corpus luteum again. High progesterone levels return, and the second embryo is reactivated to complete gestation and take up residence in the pouch. At this time, the female may enter estrus, mate, and become pregnant again. Development of this third pregnancy is controlled by the suckling stimulus of the second joey in the pouch. If a joey is prematurely removed from the pouch or dies, the blastocyst resumes development and the female enters estrus again. If she becomes pregnant, the process is repeated. The first young is not fully weaned until about four months after it begins its forays from the pouch. (b) Reproductive tract of female red kangaroo. Following mating, sperm migrate up the lateral vaginal canal. The central vaginal canal usually holds the blastocyst. (c) Young joey attached to the teat. As the joey grows, the mammary gland enlarges to supply greater volumes of milk. The larger teat is still being suckled by a young kangaroo. Note the large forearms of the joey, which it uses to pull its way from the uterus to the pouch.

(a) Modified from Short, 1972; (b) modified from G. G. Sharman, 1967.

Capons, Geldings, and Singing Superstars: Castration and Its Consequences

Roosters are succulent, stallions sedate, and men possibly better singers if they are castrated when young. Castration means the removal of the gonads, ovaries, or testes, and therefore it applies to females as well as males. But in mammals at least, testes are hung externally, easily accessible. Radical removal extirpates the gametes, leaving the individual infertile, but this also extracts the intimately associated endocrine tissue, thereby depriving the individual of some of the hormones that ordinarily control physiology and implement behavior. The physiological and behavioral consequences of castration depend upon the age at which it is performed. In general, the earlier the castration, the greater the later consequences. Certainly the fundamental sexuality of individuals arises from their basic genetic character, but as they mature, the imprint of their gender spreads to somatic tissues. Secondary sexual characteristics appear together with the appropriate behaviors to deploy these pubescent anatomical characteristics.

In birds, access to the gonads is gained through the lateral side of the body. A thin slit is made in the skin, and the gonad scooped from its position on the dorsal wall of the body cavity. Young castrated roosters are capons. The operation, besides producing birds with more flavorful meat, also eliminates their aggressive, cocky behavior, and makes them less of a nuisance around the barnyard.

A fully endowed stallion can be a determined beast, especially around an estrous female. Breeders of horses find this an annoyance. Unwanted breeding can lead to unwanted traits in unwanted foals. Occasionally mares are put on progesterone to suppress estrous, reduce their coquettish flirtations, and diminish their seductive attraction to stallions. But more often the problem is solved by castrating the stallion when a young colt, producing a *gelding*. This also reduces the male's belligerence around stallions, a great aid for those raising and racing track horses. Geldings may have more manageable personalities, but of course they are sterile, and lose their economic value as studs after their racing days are over. Occasionally, a gelding will win a large money purse or prestigious race, and the owner, considering its lost value as a stud, can only think, "oops."

Throughout human history, accidents of war or work, or intentional punishments of thieves or traitors, have left men castrated. Men were sometimes castrated to produce harem guardians, making them into what were thought to be safe human geldings. But the deliberate castration of prepubescent boys became a new fashion in the late seventeenth century in Europe. In fact, by the mid-eighteenth century, the practice was so prevalent, over 4,000 boys were castrated annually in Italy alone. What drove this practice was the discovery that castrated boys sometimes developed unique singing voices when adults. The demand for these voices grew—first in choirs, then in opera. These singing eunuchs were the "castrati" (*castrato,* sing), with voices distinctly their own, neither tenor (male) nor soprano (female). Many may have been among the greatest vocal artists in human history. Those few who developed these unique voices sang before large, adoring public crowds and generous royalty. They reaped lavish financial rewards, and enjoyed a "rock star" fame across all of Europe.

Where did castrati get their voices? Vocal cords are composed of two parts: a firm cartilaginous portion and a more pliable membranous part. The shorter and thinner the membranous part, the higher and more flexible is the resulting singing voice. The vocal cords of young children of both sexes are about the same size and length. When children pass through puberty, many physiological and anatomical changes occur, driven by accompanying endocrine changes. For boys, elevated levels of androgens (interstitial cells) in the testes stimulate the development of secondary sexual characteristics, including lengthening and thickening of the vocal cords, which deepens the male voice. In girls, vocal cords also thicken and lengthen, but not so much as in boys, leaving women with a higher natural voice. However, the vocal cords of a castrated boy develop very differently from those of normal boys and girls. The membranous part of the vocal cords does not enlarge, remaining as short as those in a child. Castrati were often described as having other-worldly voices. They could sing higher and sweeter and with greater agility than even most sopranos. It is not known exactly how these anatomical features produced a unique singing voice.

(The society was interested in the sound, not the science.) Most likely, the castrati's vocal cords were shorter and thinner than even those of most adult women. This, along with intensive musical training in the conservatories of Italy, produced those high, "angelic" voices. The voices were sweet but also very powerful, due, in part, to the larger rib cages, lung capacity, and physical stamina of the castrati compared to most smaller female singers.

But there were also side effects of emasculation. Castrati tended to develop increased subcutaneous fat in areas more typical of the female form—hips, buttocks, and breasts. Pubic hair was distributed in a female pattern (patch) rather than the male pattern (dispersed). They did not develop beards. Most distinctively, the arms and legs were usually quite long. At puberty, the epiphyseal plates normally ossify, long bones no longer lengthen, and growth stops. But in the absence of androgens, the long bones of castrati continued to elongate, resulting in the disproportionately tall "eunuchoid appearance" often ridiculed in caricatures appearing in the seventeenth-century press (box figure 1).

It is not known why young boys were first castrated. One view holds that women were forbidden to sing in church choirs in the papal states, owing to Saint Paul's dictum "Let your women keep silent in the churches." But women did sing in some churches, especially in rural areas. The demand for castrati seems more likely related to the appeal of their voices. Around the middle of the fifteenth century, a greatly complex form of *a capella* singing developed that demanded extremely competent and proficient "soprano" level voices. Prepubertal boys were often trained to sing those parts, but by the time they had become expert singers, they went through puberty, and their voices broke. As this musical style developed further during the next century, castrati began to replace the child singers and came to be much sought after for their special, "spiritual" voices honoring God in the churches of Italy.

The castrati soon found much favor and better financial rewards in the secular world. Opera in the late 1600s was a new, expanding, and popular entertainment. As more castrati appeared in operas, music was written to feature their voices. A new form

BOX FIGURE I   An eighteenth-century caricature of two castrati performing in a Handel opera. The artist depicts them looming over their female co-star and thus ridicules one of the physical deformities that often occurred in the castrated male—abnormally long arms and legs.

of opera, *opera seria,* showcased the power, flexibility, and excitement of the castrato voice. Monteverdi, Scarlatti, and Handel all wrote operas for the castrati. Mozart composed parts for castrati in *Indomeneo* (1781) and *La Clemenza di Tito* (1791, the year of his death). Today there is a world shortage of castrati (thank goodness). But these and many Baroque operas are still produced for appreciative audiences. In current productions, the castrati parts are often sung by female sopranos and mezzo sopranos as "trouser roles." But, the music is not being played on the original instrument. Without castrati, audiences of today are not hearing the same operas the composers intended and Baroque audiences enjoyed.

Imagine what it must have been like to hear Farinelli (1705–1782, born Carlo Broschi), one of the most famous of the castrati. He enjoyed a quasi-mythical status. He took full advantage of his androgynous beauty and full height to gain a commanding stage presence. But it was his voice that marked him as possibly one of the greatest singers of all time, much praised for

its beauty, purity, agility, and power. His voice register reportedly spanned three and a half octaves, and it was said he could sing 250 notes on one breath. Contemporary accounts waxed eloquently about the marvels of his voice and may explain the cry "eviva il coltello" ("long live the knife"), which went up from enraptured audiences after a performance.

Despite the numbers and popularity of the castrati, intentional castration was officially banned. The Catholic Church punished, by excommunication, anyone deliberately castrating a young boy. But fame and fortune were too tempting. To satisfy the Church, each castrato had his own story of how he became parted from his gonads. A fall from a horse left one castrated; another told of an attack by a wild boar. As simple explanations were used up, cover stories became more fanciful. One castrato claimed to have been attacked by a gaggle of geese that nibbled off his testicles. Musical scholars estimate that at the height of the Baroque period, 70% of all male opera singers were castrati.

Because of the Church's ban, the materials and methods of castration are not well documented. Apparently, either the testicles were removed, or their blood supply was pinched off. Following administration of a drug, such as opium, the boy was placed in a hot bath and the operation performed on the nearly insensible child.

Not all emasculated boys made it big in the opera world. Most never developed the big voice that commanded big money. They ended up singing in local choirs or were hired as music coaches. But the successful, like Farinelli, enjoyed great wealth and international fame. And like modern sports and music heroes, they enjoyed the attention of and were surrounded by ever-present groupies, many of whom wore medallions with pictures of their favorite castrato. Adored by women for both their glorious voices and their romantic, androgynous appearance, each castrato had a reputation to uphold not only on stage but also in the bedroom. But, could castrated men really rise to the occasion—erection and orgasm? Probably, but it is hard to say for sure. Sexual prowess is affected by the mind and social expectations, as well as by the gonads. Castrati, who were considered highly socially desirable, must have been under considerable pressure to deliver a satisfactory performance. But they undoubtedly differed from each other in their sexual abilities as much as other, unmutilated men, and much depended on the age at which they were castrated—the closer to puberty, the more likely they could successfully engage in sexual activities. Certainly the castrati were sterile. No sperm, no fertilization. But that might have been precisely their appeal; women could carry on a dalliance without the fear of becoming pregnant.

Castrati continued to sing in Italian churches up to the end of the nineteenth century. In 1922, the last of the castrati, Alessandro Moreschi, director of the Sistine Choir, died. He was 64. We have no direct recording of his voice in its prime. He left only a scratchy recording of an aging voice. Perhaps synthesized voices on computers will re-create electronically what opera audiences heard eloquently and directly for over two centuries. In the meantime, young boys are advised to beware of marauding gaggles of geese.

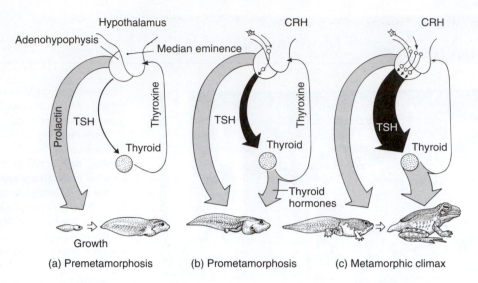

Hypothalamus
Adenohypophysis
Median eminence
CRH
CRH

Prolactin
TSH
Thyroxine
Thyroid

TSH
Thyroxine
Thyroid

TSH
Thyroid

Thyroid hormones

Growth

(a) Premetamorphosis   (b) Prometamorphosis   (c) Metamorphic climax

FIGURE 15.25 **Frog metamorphosis.** (a) Premetamorphosis is characterized by high levels of prolactin, which promote growth of the tadpole, and low levels of thyroid-stimulating hormone (TSH) and thyroid hormones. (b) Prometamorphosis includes elaboration of the median eminence and its portal system, which allows the hypothalamus to influence the adenohypophysis. As a result, levels of corticotropin-releasing hormone (CRH) rise, which promotes rising levels of thyroid hormones. Thyroid hormones stimulate hindlimb development. (c) In metamorphic climax, additional vascular routes via the enlarging median eminence stimulate increased secretion of TSH. The resulting elevated level of thyroxine promotes metamorphosis of the tadpole into a froglet. Not shown are the pathways stimulating the adrenals and, in turn, the accelerating effects of released adrenal corticoids on metamorphosis in older tadpoles. Also not shown are the rising levels of thyrotropin-releasing hormone (TRH) from the hypothalamus that progressively inhibit prolactin.

to thyroid hormone, brings about hindleg development (figure 15.25b).

These events generate a positive feedback system in which rising levels of thyroxine promote the more responsive median eminence to develop a more extensive portal connection so that more CRH is delivered to the adenohypophysis. The arrival of CRH stimulates the secretion of even higher levels of TSH and in turn more thyroxine. As these events snowball, thyroxine levels continue to increase, leading to metamorphic climax. Adrenal corticoids (steroids) inhibit early metamorphosis, but as levels of thyroid hormones rise and receptors on target tissues develop, adrenal corticoids also stimulate accelerated metamorphosis in older tadpoles. Early models of endocrine control of frog metamorphosis envisioned levels of prolactin falling as levels of thyroxine rise, but this seems not to be true. Levels of prolactin remain high through metamorphic climax, at least in frogs, but its inhibitory effects on metamorphosis are apparently overridden by the rising levels of thyroxine.

Fundamentals of Hormonal Control

Frog growth and metamorphosis highlight some basic features of hormonal control. First, hormones act not only by exerting a positive influence on target tissues; they also control events by inhibiting target tissues; Second, a target tissue, such as the median eminence, responds to hormones only after earlier stages of development have been completed. Third, endocrine control is exerted not just on the basis of the presence or absence of a hormone but also

on changes in its level. Fourth, the endocrine system is also responsive to environmental conditions and can, within limits, extend or shorten metamorphosis. If a tadpole is placed in an environment that is unusually cold or without sufficient nutrients, growth and metamorphosis are retarded.

Functional and Structural Linkage

The endocrine and nervous systems are functionally linked through the hypothalamus within the forebrain. This places the endocrine system under the influence of the central nervous system; thus, through the endocrine system, the nervous system indirectly extends its control to target tissues.

The physiological bridge between nervous and endocrine systems is mediated by neurosecretory neurons, so named because they exhibit properties of both nerve cells (they carry electrical impulses) and endocrine cells (they secrete chemicals into blood vessels). Under the influence of higher brain centers, neurosecretory cells in the hypothalamus secrete hormones into the short portal system that begins in the median eminence. When they arrive in the pars distalis, these neurosecretory hormones stimulate or suppress the secretion of other pituitary hormones. Hormones secreted by the pituitary can in turn directly affect target tissues, or they may stimulate another endocrine gland to produce a third hormone that is then carried to target tissues. For example, the neurohormone CRH stimulates the release of TSH, which stimulates the thyroid gland to release thyroxine, which affects target tissues (figure 15.26).

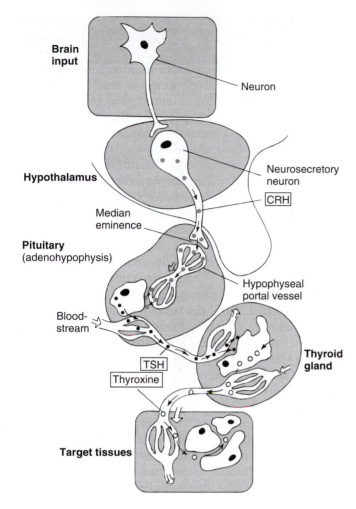

FIGURE 15.26 Schematic representation of the paths by which neurons from the brain influence target tissues through intermediate endocrine glands.

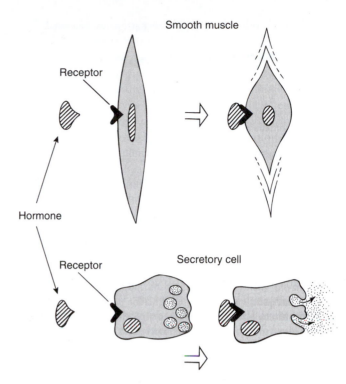

FIGURE 15.27 Hormonal receptor sites. Smooth muscle cells and secretory cells possess receptor sites. Receptors for some hormones are located on the plasma membrane. Receptors for others, such as steroids, are located in the cytoplasm and cell membranes. The appropriate hormone links with its receptor in the target cell, and through a chain of coupled metabolic pathways, they promote a cellular response. Evolution often involves changes in receptor sites rather than changes in hormones. Evolution of the appropriate receptor site makes a target cell responsive and brings it under the influence of the endocrine system. Loss of the receptor removes it from immediate endocrine control.

Target Tissue Responses

The actions of a hormone on tissues are usually selective, and the ability of a tissue to respond to a hormone depends on cell **receptors** that recognize it (figure 15.27). These receptors may be located in the cell membrane or in the cytoplasm. Unresponsive tissues lack cell receptors. To produce an effect, a hormone must be bound to receptor chemicals in or on cells that are selective for specific hormones. The hormone-receptor complex exerts an influence by promoting synthetic or catabolic reactions. For example, androgen levels rise at puberty in human males, but the selective response to these rising levels depends on the presence of receptors in target tissues that promote cellular differentiation of secondary sexual characteristics. Hair follicles in the axillary, pubic, facial, and chest regions respond with increased hair growth. In human females, mammary gland cells and ducts have receptors that allow them to respond to rising levels of the circulating hormone estradiol, but gland cells elsewhere in the body do not.

Ultimately, hormones influence target tissues by altering rates of cell division or by initiating or inhibiting synthesis of new products. Cell types differ in their responses to a given hormone. For example, a smooth muscle may respond to a hormone by contracting, whereas a gland may respond to the same hormone by releasing a secretory product (figure 15.27). Although the character of hormones is important in endocrine control of metabolism, so is the character of the target tissue itself. LH in males and females is chemically identical, yet it initiates different processes. LH stimulates ovulation in females, whereas the same hormone in males promotes interstitial cell growth within the testes. These functional differences result primarily from differences in target tissues, not from differences in the triggering hormone.

The Endocrine System and the Environment

The endocrine system regulates internal physiology, coordinates embryonic development, balances levels of minerals and nutrients to match demands, stimulates growth and metabolism, and synchronizes activities between distant parts of the organism. Yet the endocrine system itself is influenced by the external environment. Many physiological events such as reproduction, migration, and hibernation are seasonal. The environment acts through the nervous system to alter the endocrine system, which in turn brings about specific physiological and/or behavioral changes. Together with the nervous system, the endocrine system acts as an intermediary between the environment and the internal physiology of an organism to coordinate internal changes with external conditions.

For ectotherms, environmental temperature is central to activity. Cooling autumn temperatures may foster reductions in metabolic rate and send temperate reptiles into hibernation. Warming spring temperatures may bring them out of hibernation. Similarly, changing lengths of daylight hours affect the endocrine system, apparently via the eyes or the pineal gland. For many tetrapods, lengthening days can promote the onset of reproduction. Shortening days often result in internal physiological changes that lead to fat deposition and hibernation or migration to warmer climates. The social environment can also affect the endocrine system. For instance, female lizards show signs of accelerated ovarian activity, or **recrudescence,** if they are exposed to a courting male, but recrudescence is delayed if the female sees male-male territorial displays. The endocrine system thus links physiological changes, especially those based on a seasonal cycle, to changes in the surrounding environment. In this way, physiology and behavior respond optimally to environmental conditions.

Evolution

The evolution of the endocrine system includes phylogenetic changes in hormones, endocrine glands, and target tissues. As you know from the first part of this chapter, the structure of endocrine glands is quite varied. In anamniotes,

some endocrine glands tend to be distributed in patches and dispersed compared with a more compact arrangement in amniotes. For example, in anamniotes, the components of the adrenal gland appear as separate glands, each containing adrenocortical or chromaffin tissue. In amniotes, these components form the cortex and medulla of a composite adrenal gland, respectively. Incorporation of parafollicular cells into the thyroid is another example of an evolutionary merging of what in anamniotes are separate glands. Location of the pancreatic islets within the pancreas is yet another example of a composite organ in which endocrine and exocrine tissues are combined. Little is known about the functional significance of these mergers. Combining different glands gives them immediate influence over one another, and this would seem to make coordination of activities more convenient. However, it is not yet clear what adaptive advantages might have favored the phylogenetic emergence of separate glands.

Adaptive changes in the endocrine system often involve changes in the responsiveness of local tissues to existing hormones rather than changes in the hormones themselves. Consequently, similarities between hormones among different classes do not necessarily imply similar function. For example, prolactin has a wide variety of roles in different classes, including stimulation of milk production in mammals, inhibition of metamorphosis and promotion of growth in amphibians, development of dermal pigmentation in amphibians and reptiles, and modulation of parental care and water balance in fishes. Even within the same class, the role of a hormone may change. For example, in most birds, prolactin initiates behavior that leads to incubation. In addition, in some bird species, the integument of the lower part of the breast develops a brood patch in response to elevated

---

| BOX ESSAY 15.4 | The Rabbit Died: A Word on Pregnancy Tests |

It was and occasionally still is a fixture of many comedy routines to infer positive results for pregnancy if "the rabbit dies." The test is an old one no longer used. In fact, the rabbit never died from a positive test!

Most pregnancy tests are tests for chorionic gonadotropin (CG), a hormone produced by the trophoblast almost as soon as the embryo is implanted in the uterine wall when it is about six days old. Like all hormones, CG circulates in the blood, and some is excreted in the urine. The urine can be tested for the presence (implying pregnancy) or absence (implying no pregnancy) of CG. Early pregnancy tests used a rabbit, hence the source of the anecdote.

The test took advantage of the fact that ovulation in the rabbit does not occur until luteinizing hormone (LH) is secreted. As it turns out, CG mimics LH and produces ovulation in the rabbit. Thus, urine from a woman was injected into a rabbit's blood. If the woman was pregnant, then CG would be present in her urine and stimulate ovulation when injected into the rabbit. The rabbit's ovaries were inspected following injection of the woman's urine. If evidence of ovulation was found, the woman was pregnant. Technically, the rabbit did die when its ovaries were examined, but death was not the indicator of pregnancy.

Frogs were used in another early pregnancy test. CG mimics hormones that

cause female frogs to lay eggs. Contemporary tests are simpler. Antibodies produced against CG are mixed with urine from the woman. If agglutination occurs, CG is present, and the test is positive, indicating pregnancy. These tests were not very sensitive and could detect only high levels of CG reached at two or three months of pregnancy. Nowadays, the tests detect pregnancy in the first month. Test kits can be purchased in pharmacies or groceries, and a rabbit or frog need not be directly involved.

levels of prolactin, an adaptation that does not involve a new hormone but merely a change in the responsiveness of the breast integument to an existing hormone.

Another example of the evolution of tissue responsiveness to an existing hormone can be found in some hormones of the digestive tract. Cholecystokinin (CCK) stimulates the release of digestive enzymes at least as far back phylogenetically as the protochordates; however, later in phylogeny, the gallbladder also became responsive to this ancient hormone.

Interactions between digestive hormones and their target tissues have undergone a complicated sequence of evolutionary changes (figure 15.28a–e). In bony fishes, for example, two hormones are especially important in controlling the secretion of acid from the walls of the stomach. **Bombesin,** a blood-borne hormone, is secreted by endocrine cells residing in the stomach. When food arrives in the stomach, it prompts these cells to secrete bombesin, which promotes the release of gastric acid. As food passes from the stomach to the small intestine, CCK cells of the intestine are stimulated to release CCK. Transported by the blood to the stomach, CCK acts to inhibit the release of additional gastric acid into the empty stomach (figure 15.28a,b).

In amphibians, CCK cells are located in the stomach as well as in the intestine. Cells that produce bombesin are located in the stomach, but instead of entering the circulatory system as it does in fishes, bombesin directly stimulates adjacent CCK cells or even gastric cells in the wall of the stomach. Gastric cells respond to CCK or to direct bombesin stimulation by secreting acid into the stomach (figure 15.28c). In mammals, in birds, and probably in reptiles, CCK cells occur only in the intestine. In their place in the stomach are cells that produce the hormone gastrin (figure 15.28d,e). In most amniotes, gastrin rather than CCK stimulates the release of acid from the stomach walls, and stomach cells that secrete bombesin stimulate adjacent gastrin-secreting cells. In mammals, bombesin-secreting cells become neurosecretory in that they contain short axons that extend to the gastrin-secreting cells and stimulate them to produce gastrin.

Several phylogenetic changes are evident in this sequence. First, CCK inhibits gastric acid secretion in fishes, whereas in amphibians CCK promotes the release of gastric acid. In contrast to these two situations, gastrin in amniotes replaces CCK as the hormone that activates the release of stomach acid. Second, bombesin-secreting cells that release their bombesin into the blood in fishes become neurosecretory cells that activate local gastric target tissues in mammals. Bombesin has a wide repertoire of effects in addition to its action on the stomach, including effects on thermoregulation, pituitary activity, and digestive tract mobility. The advantage of this phylogenetic change in bombesin cells from an endocrine to a neurosecretory role is probably related to more localized and precise delivery of stimulation that does not interfere with the other endocrine effects of bombesin.

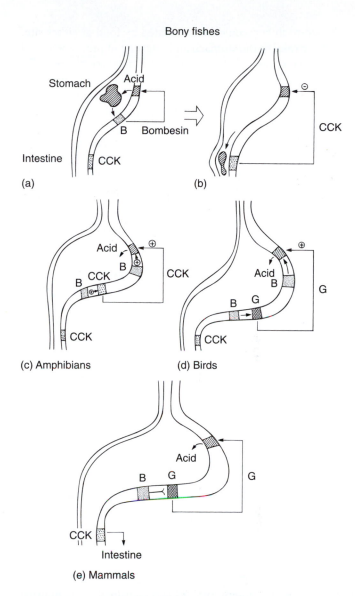

FIGURE 15.28 Evolution of gastrointestinal control by the endocrine system. (a) Bony fishes. Food in the stomach of a bony fish stimulates bombesin cells to release blood-borne bombesin (B), which stimulates the secretion of gastric acid. (b) As food moves into the teleost intestine, CCK cells are stimulated to release CCK, which inhibits gastric secretion. (c) Amphibians. Note that CCK cells reside both in the intestine, as in bony fishes, and in the stomach. Bombesin cells release a secretion that diffuses through adjacent epithelium, directly stimulating CCK cells to promote gastric acid secretion. (d) Birds (and probably reptiles). CCK cells are restricted to the intestine. In the stomach, they are replaced by cells that secrete gastrin (G). (e) Mammals. Bombesin cells, or their derivatives, stimulate gastrin cells through direct chemical contact.

Replacement of CCK by gastrin as the hormone controlling gastric secretion probably made digestion more efficient. CCK arose first as an intestinal hormone important in processing food. When a distinct stomach appeared, CCK cells were located in both the intestine and the stomach and could be stimulated by food at either site. But the roles of the stomach and intestine in processing food are different, especially in later vertebrates (see chapter 13). With the restriction of CCK cells to the intestine and the appearance of gastrin cells in the stomach, the gastric and intestinal phases of digestion could be controlled separately.

Mammalian CCK is chemically similar to that of fishes, so there has been little evolution of the hormone. However, there have been significant alterations in the endocrine control of gastric and intestinal digestion. Gastrin arose, and bombesin-secreting cells changed their routes of action from delivery via the blood to direct neural stimulation.

In some instances, evolution has involved important changes in hormonal structure, or old molecules have been co-opted for new hormonal roles. For example, epinephrine is modified from a single amino acid (tyrosine). Epinephrine is placed in service as a neurotransmitter that is released locally by axons into synaptic spaces. Extending this, the adrenal gland deploys epinephrine as hormone, releasing it into the blood for dispersed effects on distant target tissues. In the endocrine system, chemical messages coordinate internal activities by traveling long distances through the circulatory system. In the nervous system, chemical messages travel short distances across the spaces between neurons and responding cells. Thus, the nervous system, like the endocrine system, regulates activities of the body, and its functional basis is much the same—the release of chemical messages that affect responses. Therefore, we turn next to the nervous system (see chapter 16).

Overview

The endocrine system initiates and coordinates the internal activity of the organism. It is comprised of ductless glands, endocrine glands, that release hormones into the blood vessels that carry these chemical messengers to the target tissues they affect. In general, each major endocrine gland may regulate a diversity of body activities—at different times in the development of the individual, and in different ways in different phylogenetic groups. The thyroid gland presides over metabolic rate, metamorphosis, growth, and reproduction. The ultimobranchial body and parathyroid gland act antagonistically to build bone matrix or to remove bone matrix, respectively, by their affects on calcium deposition/resorption. The adrenal gland is composed of two tissue types: Adrenocortical tissue produces corticosteroid hormones that affect water retention, metabolism, and reproduction; chromaffin tissue produces catecholamines that prepare the organism for strenuous activity. Pancreatic islets primarily produce insulin, which is involved in glucose and fat metabolism. The pituitary gland develops from the merger of embryonic neural (infundibulum) and ectodermal (Rathke's pouch) contributions, producing in the adult, respectively, the neurohypophysis, which influences, via neurosecretory neurons, the adenohypophysis. Pituitary hormones affect smooth muscle contractions and water conservation (neurohypophysis), as well as growth, reproduction, and melanophores (adenohypophysis). The digestive activity of the alimentary canal is regulated by hormones released from its walls to coordinate food passage and release of digestive chemicals. Hormones released from the kidney participate in blood pressure regulation and red cell production.

Hormones are signaling compounds released by dedicated glands, carried in the blood, and targeting specific organs. Sometimes, the endocrine "glands" are spread more diffusely throughout the body. For example, as adipose cells (adipocytes) throughout the body fill, they release leptin, which travels in the bloodstream to hypothalamic receptors serving as a chemical signal to reduce food consumption.

Generally, the endocrine system regulates the activity of internal organs such as the digestive system. It helps to initiate and pace developmental events such as metamorphosis or the onset of secondary sexual characteristics. The endocrine system also adjusts the organism to its environment. Immediate challenges are met with internal adjustments in metabolic rate, water balance, and alertness. Seasonal changes are matched by preparations for migration or hibernation, or by preparation for reproductive activity.

Evolution of the endocrine system of vertebrates includes the appearance of new molecules that participate in hormonal roles. Molecules that are by-products of normal physiological activities are often co-opted for service as chemical messengers. Endocrine glands have evolved, mostly by anatomical changes in association with each other and with other organs. By evolving responsiveness to circulating hormonal signals, target tissues are the most active evolutionary part of the endocrine system.

WEB LINK

Visit the text website at *www.mhhe.com/Kardong7e* for additional study aids including selected references, functional laboraotories, and web links to chapter topics.

CHAPTER 16

The Nervous System

INTRODUCTION
  Types of Cells within the Nervous System
    Neuroglia
    Neurons
  Transmission of Information
  Neurosecretory Cells

PERIPHERAL NERVOUS SYSTEM
  Spinal Nerves
  Cranial Nerves
  Evolution
  Functions of the Peripheral Nervous System
    Spinal Reflexes
    The Autonomic Nervous System

CENTRAL NERVOUS SYSTEM
  Embryology
  Spinal Cord
    Spinal Reflexes
    Spinal Tracts
  Brain
    Phylogeny
    Form and Function
    *Functional Associations of Parts of the Central Nervous
      System*

OVERVIEW

Introduction

The nervous system is divided into the **central nervous system (CNS),** which includes the brain and spinal cord, and the **peripheral nervous system (PNS),** which consists of all nervous tissue outside the CNS. The nervous system *receives* stimuli from one or more **receptors** and *transmits* information to one or more **effectors** that respond to stimulation. Effectors include **mechanical effectors,** such as muscles, and chemical effectors, such as glands. Thus, responses of the nervous system involve muscle contractions and glandular secretions. The nervous system regulates an animal's performance by integrating immediate incoming sensory information with stored information, the results of past experience, and by then translating past and present information into action through the effectors.

The nervous system includes millions of nerve cells, each of which establishes thousands of contacts with other nerve cells, so the total number of interconnections is astronomical. That is why analysis of the function of the nervous system often includes as much philosophy as it does science. The task is formidable, but not hopeless. Let us begin by looking at the fundamental cellular components of the nervous system.

Types of Cells within the Nervous System

There are two types of cells within the nervous system: **neurons** and **neuroglial cells,** or **glia.**

Neuroglia

Neuroglial ("nerve" and "glue") cells do not transmit impulses. They support, nourish, and insulate neurons. All neuroglia bind together nervous tissue, and they can be specialized (figure 16.1). **Microglia** engulf foreign material and bacteria; **oligodendroglia** and **Schwann cells** insulate the axons of neurons; **ependymal cells** line the central canal of the brain and spinal cord; and **astrocytes** pass nutrients between blood capillaries and neurons. Additionally, they guide neuron development, regulate levels of chemical communication between cells (synapses), and control blood flow to active neurons.

Neurons

Neurons are specialized for long-distance transmission of electrical stimuli throughout the body. The neuron is the structural and functional unit of the nervous system. It consists

FIGURE 16.1 **Four types of neuroglia found within the central nervous system.** Astrocytes form cytoplasmic connections to transport nutrients between blood capillaries and neurons. Phagocytic microglia engulf stray or foreign materials. Ependymal cells line the central canal of the central nervous system. Oligodendroglia insulate axons within the central nervous system (not shown).

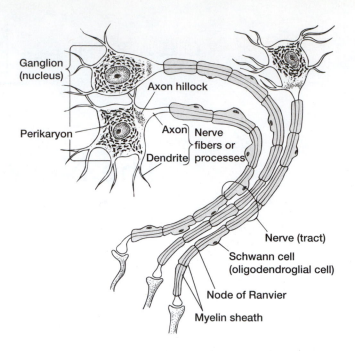

FIGURE 16.2 **Structure of a neuron.** The cell nucleus and surrounding cytoplasm form the cell body of a neuron (perikaryon). Nerve fibers or processes are cytoplasmic extensions from the perikaryon. Axons carry impulses away from the perikaryon and dendrites carry impulses toward it. The same structures are given different names in the peripheral and the central nervous systems. Central nervous system terms are given in parentheses.

of the nerve cell body, the **perikaryon,** the **body** or **soma** of the neuron, and thin cell processes called **nerve fibers** (or neurites) if they are long (figure 16.2). The processes are of two types. There usually is one **axon** per neuron and one or many **dendrites.** Dendrites transmit incoming electrical impulses toward the perikaryon. Axons carry impulses away from the perikaryon. Neurons are grouped by the number of their processes. **Unipolar neurons** have a single stem that divides into a dendrite and axon. **Bipolar neurons** have two processes, usually at opposite ends. **Multipolar neurons** have many processes associated with the cell body (figure 16.3a–h).

Neurons and their processes are often known by different terms, depending on whether they occur in the CNS or the PNS. For example, a collection of nerve fibers traveling together is a nerve **tract** in the CNS and a **nerve** in the PNS. A collection of nerve cell bodies is a **nucleus** in the CNS and a **ganglion** in the PNS. Neuroglial cells wrap some axons in a thick **myelin** sheath. Such fibers are called **myelinated nerves,** and those without sheaths are **unmyelinated nerves** (figure 16.4a,b). A neuroglial cell that produces the myelin sheath is an oligodendroglial cell in the CNS and a Schwann cell in the PNS. The **nodes of Ranvier** are indentations between adjacent neuroglial cells in the myelin sheath.

Some peripheral nerves, if not damaged too severely, can regrow, sprouting a new axon from the severed axonal stump or from the perikaryon that slowly grows down the Schwann cell tube to reestablish innervation of the effector organ. Once formed, nerves of the central nervous system were thought to lack the ability to replace themselves. More recent evidence now suggests otherwise. In fact, across vertebrates so far studied, neurons of the CNS are replaced regularly. Even in the adult mammalian brain, thousands of new neurons are added daily. Although such new neurons are a tiny proportion of the total population, over a lifetime of addition this can be considerable. Additional neurons in adults are especially evident in parts of the brain important for learning and memory. In birds, new neurons are added seasonally to areas concerned with courtship.

Transmission of Information

Information traveling through the nervous system is transmitted in the form of electrical and chemical signals. Electrical signals are **nerve impulses** that travel within the plasma membrane of the neuron and are of two kinds: graded potentials and action potentials. A **graded potential** is a wave of electrical excitation proportional to the magnitude of the stimulus that triggers it. The graded potential declines in magnitude as it travels along a nerve

Late in the 1990s, it was known that a process termed "neurogenesis" (neuron + producing) added new neurons to the brains of rats, cats, and singing canaries. But what of human brains? Could humans add back brain neurons lost to age or injury? To answer this would require invasive experimentation and, in humans, that of course raised special ethical limitations. About that time, cancer researchers were using special dyes that were picked up preferentially by rapidly dividing cells, such as cancer cells, thus aiding detection of the disease. Fred Gage, a neuroscientist, realized that if new brain cells were being added, these too would pick up the dye in chronically ill cancer patients. When patients eventually died, Gage examined their donated brain tissue and found, in fact, that the human brain was making new neurons, even in these patients who were sick and tended to be elderly.

Turning to animal studies, subsequent work has found that exercise, injury, stroke, estrogen, and high social status stimulate neurogenesis; but neurogenesis is inhibited with aging, stress, loss of sleep, boring environments, and falling asleep in lecture (just kidding). Most promising is that these new neurons come from dormant stem cells stashed away in different regions of the brain, ready to be activated. One region is near the hippocampus, which plays a key role in memory. New neurons seem to be especially adaptable in forging new connections, perhaps the anatomical basis for establishing new stores of information.

See also Vastag, B. 2007. Brain gain: Constant sprouting of neurons attracts scientists, drugmakers. Sci. News 171:376–77, 380.

FIGURE 16.3 **Types of neurons.** (a) Bipolar neuron. (b) Unipolar neuron. (c–h) Multipolar neurons.

From R. V. Krstić. General Histology of the Mammal. © 1984 Springer-Verlag. Reprinted by permission.

FIGURE 16.4 **Myelinated and unmyelinated nerve fibers.** (a) Despite their name, unmyelinated nerve fibers are associated with neuroglial cells. Usually there are several fibers per neuroglial cell, but these neuroglial cells are not wrapped repeatedly around the fibers as they are in myelinated nerves. (b) The myelin sheath is formed by a neuroglial cell that is repeatedly wrapped around a section of the nerve fiber. In the peripheral nervous system, the neuroglial cell is a Schwann cell. In the central nervous system, it is an oligodendroglial cell. Successive neuroglial cells collectively form the myelin sheath. The boundaries between them are termed nodes of Ranvier.

(a)

(b)

FIGURE 16.5 **Transmission of information within the nervous system.** (a) Neurons transmit and receive stimuli as electrical impulses along their fibers. Synapses are junctions between nerve cells. Axons release chemical messengers (neurotransmitters) that diffuse across the synapse. When they arrive at the dendrite in sufficient concentration, the neurotransmitter molecules initiate an electrical impulse in the next neuron. (b) Synapses assist in processing information. Electrical input can converge or diverge. Inputs of some neurons inhibit or reduce the sensitivity of other neurons.

fiber. An **action potential** is an all-or-none phenomenon. Once initiated, it propagates without decrement along a nerve fiber. Action potentials are often used for long-distance signaling in the nervous system. Within the dendrites and the perikaryon, nerve impulses are usually graded potentials, but they become action potentials as they travel out the axon.

Chemical signals are generated at **synapses,** gaps between the junctions of neurons (figure 16.5a). These gaps occur between the processes of one neuron and the next, and between axons and perikarya. Upon arrival at the terminus of an axon, an electrical impulse stimulates the release of stored **neurotransmitters** into the tiny space between processes. Neurotransmitters diffuse across this synaptic junction and settle on the associated cellular process of the next neuron. When collected in sufficient concentration, neurotransmitters initiate an electrical impulse in the next neuron. Excess and spent neurotransmitter molecules are rapidly inactivated to prevent prolonged effects. Neurotransmitters must quickly reach a threshold level to initiate an electrical impulse in the next neuron. Thus, passage of information through chains of connected neurons includes alternating events of electrical and chemical transmission involving nerve impulses and neurotransmitters, respectively (figure 16.5b).

Synapses introduce control into the processing of information transfer. If synapses were absent and neurons were in direct contact with each other, excitation in one neuron might spread inevitably throughout an entire network of interconnected neurons like ripples on a pond, without any local control. Synapses break up a network of neurons into information-processing units. Whether an impulse is transmitted to the next neuron in a sequence depends on whether or not there is a sufficient concentration of neurotransmitters at the synapse. Where neurons converge, transmission of a single impulse from one neuron might be insufficient. Several impulses might need to arrive simultaneously in order to release enough neurotransmitter molecules to trigger an electrical impulse in the next neuron. Convergence promotes **summation** of information. Conversely, if a neuron sends branches to several circuits, information diverges and is distributed to appropriate areas. Branches of a single axon are called **collateral branches.** Inhibition also affects the flow of information by decreasing the responsiveness of neurons to incoming information. Convergence, divergence, and inhibition are modes of information processing that take advantage of the character of the synapse (figure 16.5). Furthermore, the structure of neurons at a synapse ensures that transmission across the gap occurs in only one direction.

In humans, the number of synapses peak between 2 months and 2 years of age. Thereafter some sculpting of synapses occur wherein some are reduced, others strengthened. Strengthening is stimulated by electrical discharge at the synapse, which promotes secretion of special proteins that strengthen the synapse.

Neurosecretory Cells

Neurosecretory cells are specialized neurons. Most neurons release neurotransmitters at the ends of their axons. Neurosecretory cells also release secretions at the ends of their axons, but these secretions are delivered into a blood capillary and transported to a target tissue. Neurosecretory cells are thus endocrine in function.

Peripheral Nervous System

The terms used to describe the components of the peripheral nervous system refer to the anatomical or functional properties of the nerves (figure 16.6). Peripheral nerves serve either somatic or visceral tissues and carry sensory or motor information. **Somatic nerves** pass to or from somatic tissues—skeletal muscle, skin, and their derivatives. **Visceral nerves** pass to or from viscera—involuntary muscles and glands. Nerves carrying information from tissues *to* the central nervous system are **afferent,** or **sensory, neurons.** Nerves carrying information *away* from the CNS to effectors are **efferent,** or motor, neurons. Thus, a somatic sensory nerve might carry information about touch, pain, or temperature from the skin to the central nervous system. A somatic motor nerve carries impulses from the CNS to a striated muscle to stimulate its contraction. A visceral

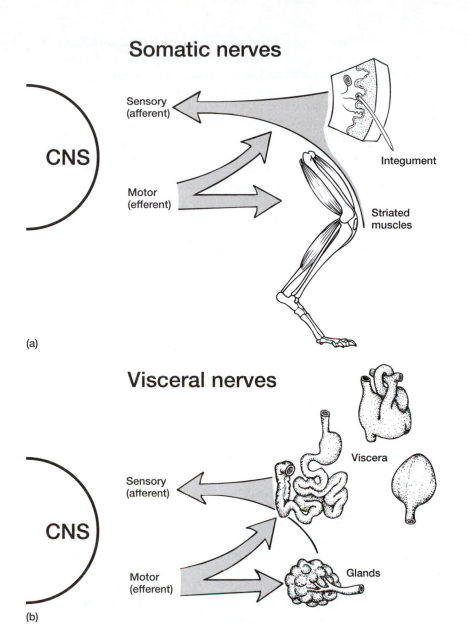

Somatic nerves

Sensory
(afferent)

CNS

Motor
(efferent)

Integument

Striated
muscles

(a)

Visceral nerves

Sensory
(afferent)

CNS

Motor
(efferent)

Viscera

Glands

(b)

FIGURE 16.6   Functional categories of neurons of the peripheral nervous system. Some neurons supply somatic tissues, others visceral tissues. They can be sensory and respond to stimuli from these tissues, or they can be motor and deliver stimuli to these tissues. CNS, central nervous system.

sensory nerve delivers information about the condition of internal viscera to the CNS. A visceral motor nerve innervates visceral effectors (cardiac muscle, smooth muscle, or glands). The components of the PNS that control visceral activity constitute the **autonomic nervous system (ANS).**

Nerves have two additional properties based on their distribution. Neurons are termed *general* if the innervated tissues are widely distributed, or *special* if the tissues are restricted in location. Thus, general somatic neurons innervate sense organs or supply effectors to the integument and most striated muscles. Special somatic neurons are associated with somatic sense organs (e.g., eyes, olfactory organs, inner ears) or effectors (e.g., branchiomeric muscles, ciliary eye muscles, extrinsic ocular muscles) that are limited in distribution. General visceral neurons innervate sensory organs or supply effectors in glands or smooth muscles of the digestive tract, heart, and other viscera. Special visceral neurons concerned with sensory input innervate the taste buds and olfactory epithelium.

From anatomical criteria, the peripheral nervous system can be divided into **spinal nerves** emanating from the spinal cord and **cranial nerves** emanating from the brain. We begin by looking at these anatomical divisions of the peripheral nervous system.

Spinal Nerves

Spinal nerves are sequentially arranged and numbered (C-1, T-1, L-1, S-1) according to their association with regions of the vertebral column (cervical, thoracic, lumbar, sacral). Early anatomists recognized **dorsal** and **ventral roots** of each spinal nerve. Afferent fibers enter the spinal cord via the dorsal root, and efferent fibers leave by way of the ventral root. The dorsal root ganglion, a swelling in the dorsal root, is a collection of neuron bodies whose axons contribute to the spinal nerve. Parallel to the spinal cord and attached to each spinal nerve through the **ramus communicans** is the **sympathetic chain** of ganglia (paravertebral ganglia), a paired series of

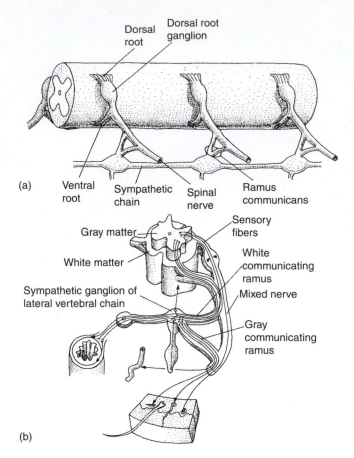

Dorsal root

Dorsal root ganglion

(a)

Ventral root

Sympathetic chain

Spinal nerve

Ramus communicans

Gray matter

White matter

Sympathetic ganglion of lateral vertebral chain

Sensory fibers

White communicating ramus

Mixed nerve

Gray communicating ramus

(b)

FIGURE 16.7 **Spinal nerve anatomy.** (a) Dorsal and ventral roots connect spinal nerves to the spinal cord. A dorsal root is enlarged into a dorsal root ganglion. Spinal nerves join with the sympathetic chain through communicating rami. (b) Configuration of sensory and motor neuronal routes in an adult mammal.

(b) After Tuchmann-Duplessis et al.

linked ganglia adjacent to the vertebral column or notochord (figure 16.7a,b). Other peripheral ganglia form the **collateral ganglia** (prevertebral ganglia). The paired **cervical, coeliac,** and **mesenteric ganglia** are examples of the collateral ganglia. The **visceral ganglia** occur within the walls of visceral effector organs (figure 16.7b). Thus, there are three types of ganglia: sympathetic, collateral, and visceral.

Peripheral nerves in the trunk arise during embryonic development from two sources (figure 16.8a–c). One source is the neurons that differentiate within the spinal cord. Axonal processes sprout from these neurons and grow outward to the ganglia or to the effectors they supply (figure 16.8b). The other source is the neural crest. Cells migrate from the neural crest to specific locations and sprout processes that grow back to the central nervous system and out to the tissues they innervate (figure 16.8a). The ventral roots arise from neurons in the spinal cord that send fibers out from the spinal cord. The dorsal root arises from cells of neural

crest origin that send fibers into the spinal cord. In gnathostomes, the two roots usually merge to form the composite spinal nerve and attached sympathetic chain.

The fibers of each spinal nerve innervate restricted structures at that level of the cord. This is especially pronounced with spinal nerve innervation of somatic tissues. Each growing spinal nerve tends to accompany its adjacent embryonic **myotome,** the source of somatic muscles, and its **dermatome,** the source of dermal connective tissue and muscle, as they spread and differentiate during development (figure 16.9a–c). Once it is differentiated, a spinal nerve supplies the skeletal muscles derived from its adjacent myotome and receives somatic sensory input from the restricted area of the body surface differentiated from its dermatome. Strictly speaking, a **dermatome** refers to an embryonic structure, but the term is often used to denote the region of the adult body derived from it. The fidelity between a dermatome and its spinal nerve permits mapping of the body surface in terms of the corresponding spinal nerves that supply each region. Loss of sensation in a dermatome can be diagnostic for the particular specific spinal nerve involved.

Cranial Nerves

Cranial nerves have roots enclosed in the braincase. Most are named and numbered by Roman numerals from anterior to posterior. The conventional system for numbering these nerves is sometimes inconsistent. For instance, most anamniotes are said to have ten numbered cranial nerves plus six pairs of unnumbered, lateral line cranial nerves. A few anamniotes and all amniotes are said to have 12. In fact, there is an additional terminal nerve at the beginning of this series. If counted at all, it is numbered 0 to avoid renumbering the conventionally numbered sequence. Further, the second cranial nerve (II) is not a nerve at all but an extension of the brain. Nevertheless, by convention it is called the optic "nerve." The eleventh cranial nerve (XI) represents the merger of a branch of the tenth cranial nerve (X) with elements of the first two spinal nerves (C-1 and C-2). Despite its composite structure, it is called the spinal accessory nerve and designated by Roman numeral XI. In addition to these numbered cranial nerves, up to six pairs of unnumbered, lateral line cranial nerves are present in jawed fishes and many amphibians.

Phylogenetically, the cranial nerves are thought to have evolved from dorsal and ventral nerves of a few anterior spinal nerves that became incorporated into the braincase. Dorsal and ventral nerves fuse in the trunk but not in the head, and they produce two series: dorsal cranial nerves (V, VII, IX, and X) and ventral cranial nerves (III, IV, VI, and XII). Like spinal nerves, the cranial nerves supply somatic and visceral tissues and carry general sensory and motor information. Some cranial nerves consist of only sensory or only motor fibers. Other nerves are **mixed,** containing both types. Cranial nerves concerned with localized senses (e.g., sight, hearing, lateral line, olfaction, taste) are called **special cranial nerves** to distinguish them from

FIGURE 16.8 **Embryonic development of afferent and efferent spinal nerves.** (a) The neural crest forms from ectoderm during neurulation and becomes organized as segmental populations of cells arranged dorsally along the neural tube. (b) From this dorsal location, some cells migrate (open arrows) to specific sites within the body, forming distinct populations of neural cells at these sites. Neurons differentiating within the primordial dorsal root sprout cell processes that grow back to the neural tube and out to both somatic and visceral tissues. Neuronal bodies that remain in position constitute the dorsal root ganglion. Neurons differentiating within other populations grow cell processes to effectors, and their bodies constitute ganglia. Motor neurons differentiate within the neural tube and grow cell processes to these peripheral ganglia or directly to effectors. (c) Diagrammatic representation of established afferent and efferent neurons within spinal nerves.

(a) After Krstić; (b,c) after Tuchmann-Duplessis et al.

those concerned with the sensory or motor innervation of the more widely distributed viscera, **general cranial nerves.**

Primitively, all cranial nerves serving the branchial pouches formed three branches per pouch: **pretrematic, posttrematic,** and **pharyngeal** (figure 16.10). In amniotes, these tend to be lost or their homologies become uncertain.

Most anamniotes possess 17 cranial nerves. The first few spinal nerves behind the braincase become housed in the skull of later derived groups. But in anamniotes, these anterior spinal nerves are still partially outside the skull. In cyclostomes, these anterior spinal nerves outside the skull are called **occipitospinal nerves.** In other fishes and amphibians, the anterior spinal nerves become partially incorporated into the braincase. They exit via foramina in the occipital region of the skull and are called **occipital nerves.** Occipital nerves unite with the next few cervical spinal nerves to form

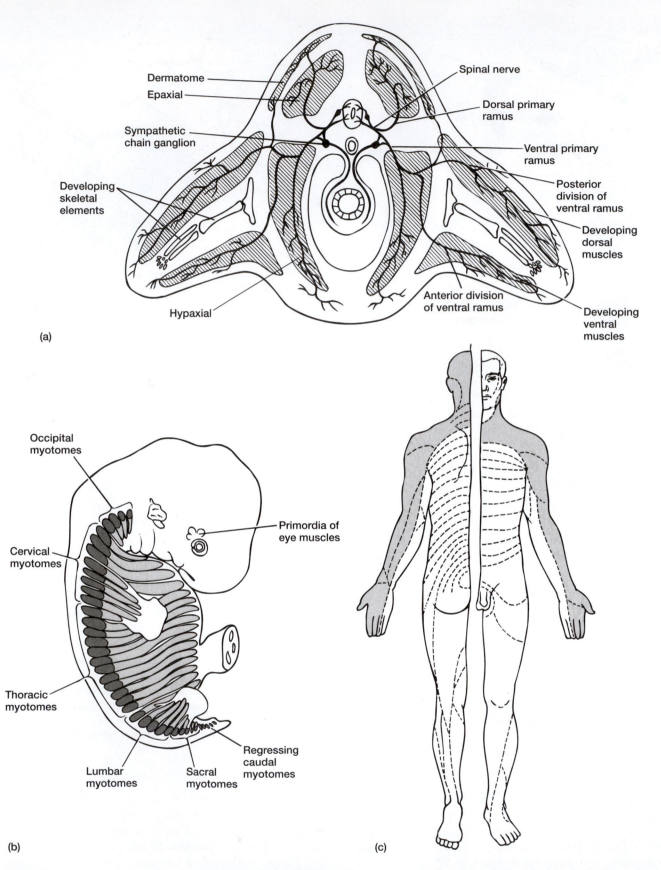

(a)

(b)

(c)

Dermatome

Epaxial

Sympathetic
chain ganglion

Developing
skeletal
elements

Hypaxial

Spinal nerve

Dorsal primary
ramus

Ventral primary
ramus

Posterior
division of
ventral ramus

Developing
dorsal
muscles

Anterior division
of ventral ramus

Developing
ventral
muscles

Occipital
myotomes

Cervical
myotomes

Thoracic
myotomes

Lumbar
myotomes

Sacral
myotomes

Regressing
caudal
myotomes

Primordia of
eye muscles

FIGURE 16.9   **Spinal nerves supply vertebrate limbs and body walls.**   (a) Cross section of a generalized vertebrate. Note
the distribution of spinal nerves to axial and appendicular muscles. (b) Human embryo illustrating the distribution of myotomes served by
segmental spinal nerves. (c) Split image showing dorsal and ventral distribution of dermatomes within the human body.

After Patten and Carlson.

Shingles is the common name for a disease caused by the virus *Herpes zoster*, the same virus that produces chicken pox. Shingles is characterized by a line of blisters that usually radiate out along one side of the body following one of the spinal or cranial nerves to a dermatome.

For most people, chicken pox is a childhood disease that lasts several weeks. It causes itchy blisters over the body and also provides a few days of respite from school. Eventually, the immune system forces the virus into remission. Subsequent events are not well known, but it is thought that the virus retreats to the perikarya of neurons and is held in check there by the immune system. For most people, that is the end of *Herpes zoster*. But in some, the immune system lets down a bit and the virus proliferates, except that during this second flare-up, its spread is more restricted. The virus migrates along a nerve to the dermatome it supplies (see figure 16.9b,c). The tissue along this pathway reacts by forming the characteristic but very painful blisters. When the immune system responds again, the virus is beaten back again, usually but not always, for the last time, and the symptoms of the disease abate. If nerves to the face are involved damage may be inflicted on the eyes, for example, lingering on permanently even after the virus goes into remission.

Our knowledge of nerve anatomy and corresponding dermatome associations aids in the diagnosis of shingles. In most cases, it is possible to determine from the pattern of blisters which spinal or cranial nerve the virus spreads along.

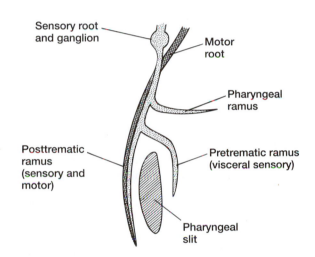

FIGURE 16.10   **Components of a cranial nerve in a fish.** The pharyngeal ramus, to the lining of the pharynx, and the small pretrematic ramus, to the front of the pharyngeal slit, both carry visceral sensory fibers. The dorsal ramus from the skin is composed of somatic sensory fibers. The posttrematic ramus running down the back of the pharyngeal slit includes both sensory and motor fibers. Rostral is to the right of the figure.

the composite **hypobranchial nerve** that supplies hypobranchial muscles in the throat (figure 16.11a,b).

Latimeria (coelacanth) and many amphibians have 17 cranial nerves. In amniotes, the lateral line nerves are lost and the occipitospinal nerves are incorporated into the skull and modified. Their roots shift from the spinal cord forward into the medulla. In this way, amniotes derive the eleventh and twelfth cranial nerves. The 12 numbered cranial nerves are illustrated in figures 16.12 through 16.15. They are described in more detail next, and their functions are summarized in table 16.1.

Nervus Terminalis (0) The terminal nerve may be testimony to an ancient anterior head segment that has been lost. The terminal nerve is a nerve, or perhaps a complex of nerves, that arises from olfactory placodes. It is present in all classes of gnathostomes except birds. It runs to blood vessels of the olfactory epithelium in the olfactory sac and carries visceral sensory and some motor fibers. A role in reproduction is suspected.

Olfactory Nerve (I)   The **olfactory nerve** is a sensory nerve concerned with the sense of smell. Olfactory cells lie in the mucous membrane of the olfactory sac. A short axon leads from each cell to the olfactory bulb. Each axon constitutes an olfactory fiber. Collectively, the olfactory fibers form the short olfactory nerve, which is the only cranial nerve composed of the axons of the receptor cells themselves.

Optic Nerve (II)   Strictly speaking, the **optic nerve** is not a nerve but a sensory tract. That is, it is not a collection of peripheral axons; it is a collection of fibers in the CNS. Embryologically, it develops as an outpocketing of the brain. However, once it is differentiated, it lies outside the brain. Its fibers synapse in the thalamus and midbrain.

Oculomotor Nerve (III)   The **oculomotor nerve** primarily supplies extrinsic eye muscles (superior rectus, medial rectus, inferior rectus, and inferior oblique muscles) derived from preotic myotomes. It is a motor nerve that also carries a few visceral motor fibers to the iris and ciliary body of the eye. Fibers arise in the oculomotor nucleus in the floor of the midbrain.

Trochlear Nerve (IV)   The **trochlear nerve** is a motor nerve that supplies the extrinsic, superior oblique eye muscle. Fibers arise in the trochlear nucleus of the midbrain.

Trigeminal Nerve or Trigeminus (V)   The **trigeminus** is so named because it is formed of three branches: **ophthalmic (V_1)**, **maxillary (V_2)**, and **mandibular (V_3)** in amniotes (figures 16.12c and 16.15). The ophthalmic nerve, sometimes called the deep ophthalmic (= profundus) nerve to distinguish it from a more superficial nerve, usually merges with the other two branches. However, in anamniotes, the ophthalmic nerve often emerges from the brain separately.

Cranial Nerve		Somatic Sensory		Visceral Sensory		Visceral Motor		Somatic Motor	
		General	Special	General	Special	General	Special	General	Special
0	Terminal	X		X					
I	Olfactory				X				
II	Optic		X						
III	Oculomotor					(X)		X	
IV	Trochlear							X	
V₁	Trigeminal	X							
V₂,₃	Trigeminal proper	X							X
VI	Abducens								
VII	Facial	(X)		X	X	X			X
VIII	Auditory		X						
IX	Glossopharyngeal	(X)		X	X	X			X
X	Vagus	X		X		X			X
XI	Spinal accessory								X
XII	Hypoglossal							X	
	Lateral line		X						

Note: Parentheses indicate variable or negligible function in the category indicated.

This independent emergence was once taken as evidence that it anciently supplied an anterior branchial arch that had since been lost. However, the ophthalmic nerve (V₁) arises embryologically from a placode, unlike the maxillary (V₂) and mandibular (V₃) that arise from neural crest, suggesting an independent origin for the ophthalmic altogether without the need to postulate an association with a lost branchial arch. The other two branches, the maxillary ramus (V₂) to the upper jaw and the mandibular ramus (V₃) to the lower jaw, presumably represent pretrematic and posttrematic rami of a typical branchial nerve to the mandibular arch.

The mixed trigeminus includes sensory fibers from the skin of the head and areas of the mouth and motor fibers to derivatives of the first branchial arch. Sensory fibers of the trigeminus return to the brain from the skin, teeth, and other areas through each of the three branches. The mandibular branch also contains somatic motor fibers to muscles of the mandibular arch.

Abducens Nerve (VI)   The **abducens** is the third of the three cranial nerves that innervate muscles controlling movements of the eyeball. It is a motor nerve that supplies the extrinsic, lateral rectus eye muscle. Fibers arise in abducens nucleus located in the medulla.

Facial Nerve (VII)   The mixed **facial nerve** includes sensory fibers from taste buds as well as motor fibers that service derivatives of the second (hyoid) arch. This nerve also carries a substantial number of somatic sensory fibers to the skin. In fishes, the skin of the entire opercular complex is innervated by the facial nerve.

Octaval (= Auditory) Nerve (VIII)   The sensory **octaval nerve** (acoustic, vestibulocochlear, statoacoustic) carries sensory fibers from the inner ear, which is concerned with balance and hearing. The nerve synapses in several regions of the medulla.

Glossopharyngeal Nerve (IX)   The mixed **glossopharyngeal nerve** supplies the third branchial arch. It contains sensory fibers from the taste buds, the first gill pouch, and the adjacent pharyngeal lining. Motor fibers innervate muscles of the third branchial arch.

Vagus Nerve (X)   The term **vagus** is Latin for wandering and aptly applies to this mixed nerve. The vagus meanders widely, serving areas of the mouth, pharynx, and most of the viscera. It is formed by the union of several roots across several head segments. Occasionally, additional nerves of the lateral line merge with the vagus.

Spinal Accessory Nerve (XI)   In anamniotes, the **spinal accessory nerve** is probably composed of a branch of the vagus nerve and several occipitospinal nerves. In amniotes, especially in birds and mammals, it is a small but distinct motor nerve that supplies derivatives of the cucullaris muscle (cleidomastoid, sternomastoid, trapezius). A few of its fibers accompany the vagus nerve to supply part of the pharynx and larynx and perhaps the heart. Fibers arise from several nuclei within the medulla.

Hypoglossal Nerve (XII)   The **hypoglossal nerve** is a motor nerve of amniotes that innervates hyoid and tongue muscles. Fibers originate in the hypoglossal nucleus within

(a) Colubrid snake

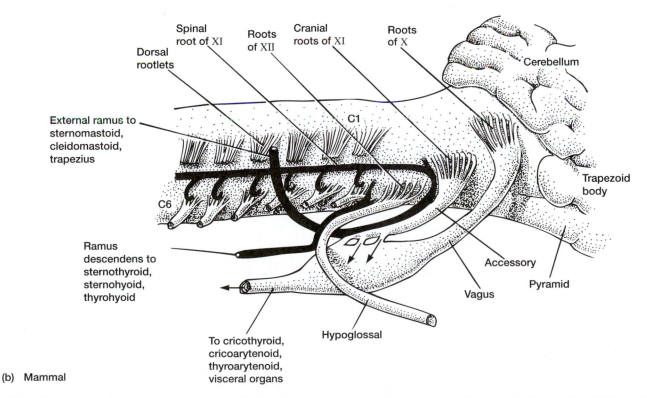

(b) Mammal

FIGURE 16.11 **Posterior cranial nerves.** (a) Colubrid snake. The glossopharyngeal (IX), vagus (X), hypoglossal (XII), and one of the spinal nerves join to form the craniocervical trunk. Unlike most other amniotes, snakes appear to lack a spinal accessory nerve (XI). (b) Mammal. The roots of the hypoglossal nerve are in series with the ventral roots of the preceding spinal nerves. Spinal nerve contributions to the accessory (XI) and hypoglossal (XII) nerves are shown in solid black. The vagus receives contributions from the accessory nerves (arrows).

(b) After Kent.

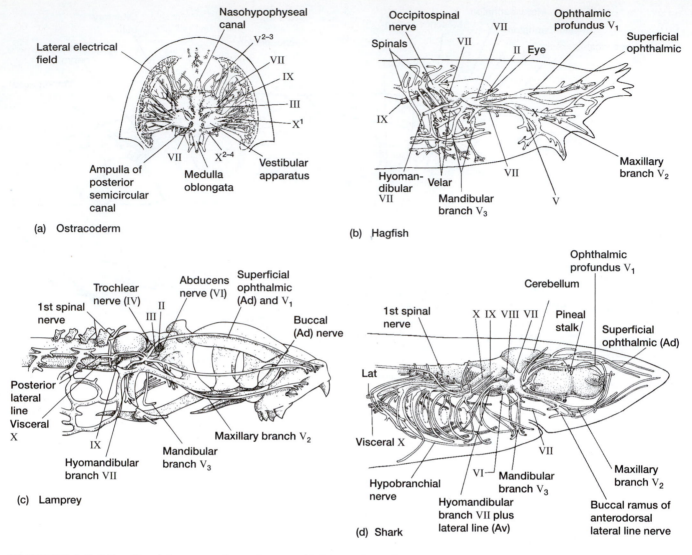

FIGURE 16.12 **Cranial nerves of vertebrates.** (a) Ostracoderm, *Kiaeraspis*. (b) Hagfish, *Myxine*. (c) Lamprey. (d) Lateral view of cranial nerves in the shark, *Squalus*. Abbreviations: anterodorsal lateral line nerve (Ad), anteroventral lateral line nerve (Av).

(a) After Stensiö; (b–d) after Jollie.

the medulla. In fishes and amphibians, the confluence of one or several occipital nerves (ventral roots of original spinal nerves) and often modified spinal nerves form the hypobranchial nerve. In amniotes it is incorporated into the skull and therefore more appropriately recognized as a cranial nerve, the hypoglossal nerve.

Lateral Line Nerves In addition to formally numbered cranial nerves, fishes possess preotic and postotic **lateral line cranial nerves** that are rooted in the medulla and supply the lateral line system. They were once thought to be components of the facial, glossopharyngeal, and vagal nerves, but they are now recognized as independent cranial nerves, derivatives of the dorsolateral placodes (octavolateralis placodes) (figure 16.13a). Unfortunately, this late recognition as distinct cranial nerves has left them without an identifying Roman numeral. In most jawed fishes and

some amphibians, up to six pairs of lateral line nerves are present. Three are preotic (rostral to the otic vesicle, future inner ear): the anterodorsal, anteroventral, and otic lateral line nerves; three are postotic: the middle, supratemporal, and posterior lateral line nerves (figure 16.13a,b). Each of the dorsolateral placodes initially gives rise to a distinct sensory ganglion whose distal fibers innervate the lateral line receptors (neuromasts and ampullary organs) that themselves also arise from the same placode. Only rarely are any of the ganglia of the lateral line nerves fused with cranial nerves V, VII, IX, and X, but all preotic and all postotic lateral line nerves converge, entering the brain rostral and caudal to the otic vesicle, respectively. Frequently each converged nerve divides into a dorsal and ventral root. If present, the dorsal root only carries fibers that innervate ampullary organs and the ventral root only carries fibers that innervate the neuromast organs. Because the lateral

Dorsolateral Placodes

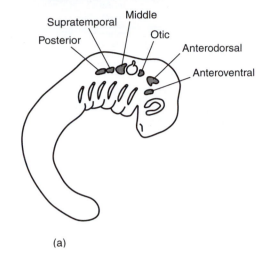

(a)

Lateral Line Nerves

(b)

(c) Fetal dog

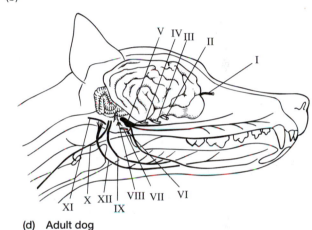

(d) Adult dog

FIGURE 16.13 **Embryonic development of cranial nerves.** (a) The lateral line cranial nerves arise from the dorsolateral placodes. (b) Generalized number and innervation pattern of lateral line nerves in jawed fishes. Ganglia are represented by expanded areas with small circles inside. Relative position of eye (dashed circle) and otic vesicle (dashed pear-shape) are indicated. Hypobranchial trunk includes the lateral line nerve and the facial cranial nerve VII. (c) Fetal dog. (d) Adult dog. The posterior lateral line nerve has three rami: dorsal (d), lateral (l), and ventral (v). The anterodorsal lateral line nerve has two major rami: the superficial ophthalmic (SO) and buccal (B). The anteroventral lateral line nerve produces two major rami: the anterior (Ad) and posterior (Pd) divisions.

(b) After Northcutt.

line and the octaval nerves share organizational similarities, they are often spoken of as the octavolateralis system (= acouticolateralis).

Neural crest and ectodermal placodes (p.189); mechanoreceptors, neuromasts (p. 693); electroreceptors, ampullary organs (p. 709); lateral line system (p. 694)

Evolution

In early vertebrates, each head segment may have been innervated by anatomically separate dorsal and ventral roots in much the same way that separate dorsal and ventral spinal nerves supply each trunk segment in lampreys. Each segment was perhaps innervated by a mixed dorsal nerve and a motor ventral nerve. It has been suggested that the

cranial nerves are derived from losses or mergers of these separate dorsal and ventral nerves. But complex fusions and losses make it difficult to determine the distribution of ancient nerves to their respective head segments. The mandibular arch incorporates the deep ophthalmic nerve into its own dorsal root branches (the maxillary and mandibular branches), forming the composite trigeminal nerve. Other persisting dorsal nerves include the facial, glossopharyngeal, vagus, and accessory nerves. Ventral nerve derivatives include the oculomotor, trochlear, abducens, and occipital nerves.

When associated with a branchial arch, each cranial nerve exhibits fidelity to that particular arch and its muscles. Consistently throughout vertebrates, the first arch, the mandibular, is innervated by the trigeminal nerve (V); the

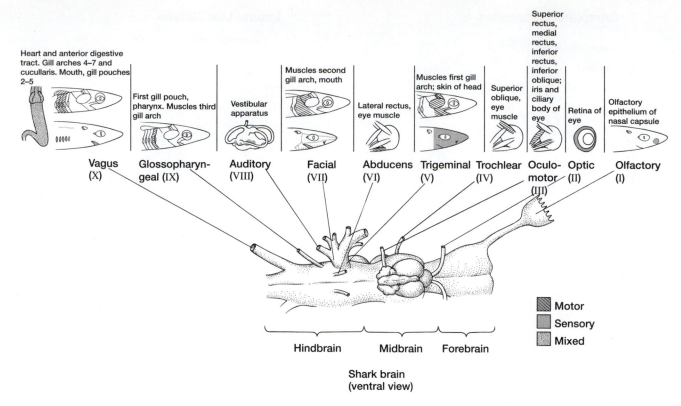

Heart and anterior digestive tract. Gill arches 4–7 and cucullaris. Mouth, gill pouches 2–5

First gill pouch, pharynx. Muscles third gill arch

Vestibular apparatus

Muscles second gill arch, mouth

Lateral rectus, eye muscle

Muscles first gill arch; skin of head

Superior oblique, eye muscle

Superior rectus, medial rectus, inferior rectus, inferior oblique; iris and ciliary body of eye

Retina of eye

Olfactory epithelium of nasal capsule

| Vagus (X) | Glossopharyn-geal (IX) | Auditory (VIII) | Facial (VII) | Abducens (VI) | Trigeminal (V) | Trochlear (IV) | Oculo-motor (III) | Optic (II) | Olfactory (I) |

Motor
Sensory
Mixed

Hindbrain   Midbrain   Forebrain

Shark brain (ventral view)

FIGURE 16.14   Distribution of cranial nerves in the shark *Squalus*. Enlarged views of the innervated structures of cranial nerves II, III, IV, VI, and X. Lateral views of the head with and without skin indicate the location of the first ten cranial nerves.

After Gilbert.

second, the hyoid, by the facial nerve (VII); the third, by the glossopharyngeal (IX); and the remaining arches by the vagus (X) and spinal accessory (XI) nerves (table 16.2 and figure 16.16a,b).

The olfactory (I), optic (II), and auditory (VIII) nerves and the lateral line cranial nerves are believed to be derived separately in conjunction with their respective special sense organs rather than in association with ancient head segments.

The shift from aquatic to terrestrial life is reflected in the cranial nerves. The lateral line system, devoted to detecting water currents, is completely lost in terrestrial vertebrates, as are the cranial nerves that served it. Pre- and posttrematic branches associated with the gills are modified as well. The spinal accessory and hypoglossal nerves enlarge or emerge as separate cranial nerves. The spinal accessory separates from the vagus nerve. It supplies the branchiomeric muscles that become more prominent in holding and rotating the head. The hypoglossal nerve to the tongue and hyoid apparatus becomes prominent as the role of these structures in terrestrial feeding and manipulating food in the mouth expands.

Functions of the Peripheral Nervous System

Spinal Reflexes

Spinal reflexes exhibit the simplest level of control within the nervous system. Although reflexes can disperse information to higher centers, all of their necessary and functional components reside or are rooted in the spinal cord. The **spinal reflex** is a circuit of neurons from a receptor to the spinal cord and out to an effector. Incoming sensory and departing motor information travels in circuits laid down by neurons in the spinal nerves. Within the spinal cord, **association neurons (interneurons, internuncial neurons)** connect these sensory and motor neurons to complete the circuit between them. There are two types of spinal reflex arcs, somatic and visceral (figure 16.17). The neuronal circuitry for each type of arc is distinctive, at least in mammals, where it has been most widely studied (table 16.3). The role of the central nervous system in modifying spinal reflexes is discussed later in this chapter.

Most **somatic reflex arcs** at the level of the spinal cord include three neurons: somatic sensory and somatic motor neurons, with an association neuron connecting them. The body of the somatic sensory neuron is located in the dorsal root. Its nerve fibers travel through the spinal nerve and synapse with an association neuron within the spinal cord. The association neuron may transmit impulses in several possible directions. It may synapse with a somatic motor neuron on the same side of the cord, on the opposite side of the cord, or travel up or down the cord to motor neurons at different levels. The motor neuron then transmits the impulse through the ventral root to a somatic effector. A somatic reflex arc may be even more simple. Spinal reflexes controlling posture involve only two neurons. The sensory neuron synapses

FIGURE 16.15 **Distribution of cranial nerves in a mammal,** *Homo sapiens.* Sensory (solid lines) and motor (dashed lines) nerve fibers are indicated. Enlarged views of innervated structures of cranial nerves are shown around the human brain in ventral view. Abbreviations: general somatic sensory (GSS), general visceral sensory (GVS), general somatic motor (GSM), general visceral motor (GVM), somatic motor (SM), special somatic sensory (SSS), special somatic motor (SSM), special visceral sensory (SVS).

After H. M. Smith.

TABLE 16.2	Cranial Nerves and Their Associated Branchial Arches		
Ancient Segment	**Current Arch**	**Dorsal Root Representative**	**Ventral Root Representative**
?		Terminal (0)	
0		Deep ophthalmic (V)	Oculomotor
—	Mandibular		
1		Superficial ophthalmic (skin; V) maxillary (pretrematic; V) mandibularis (posttrematic; V)	Trochlear
2	Hyoid	Facial (VII)	Abducens
3	Branchial 3	Glossopharyngeal (IX)	
4	Branchial 4	Vagus (X)	
5	Branchial 5	Vagus (X)	Hypoglossal
6	Branchial 6	Vagus (X)	Hypoglossal
7	Branchial 7	Spinal accessory (XI)	Hypoglossal

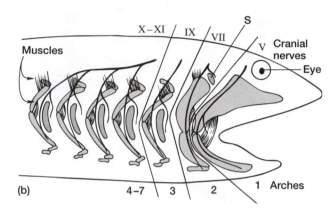

FIGURE 16.16 Phylogenetic derivation of cranial nerves. (a) Hypothesized primitive condition. Each pharyngeal slit was supplied by a nerve. The first, or terminal (T), nerve supplied an anterior arch that was lost early in vertebrate evolution. (b) Nerve supply to associated branchial arches. Cranial nerves V, VII, IX, and X–XI supply the following arches: mandibular (1), hyoid (2), third (3), and fourth–seventh (4–7), respectively. These associations between cranial nerves and their derivatives remain stable throughout teleosts and tetrapods. Abbreviations: gill slits lost in gnathostomes (0, 0'), gill slits usually present in gnathostomes (1–5), spiracular slit (S).

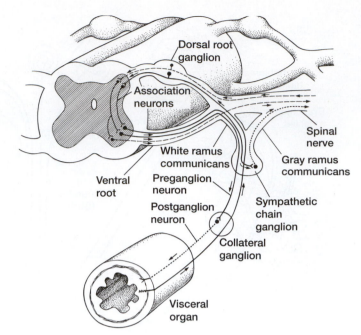

FIGURE 16.17 Somatic and visceral reflex arcs of mammals. Sensory input arrives in fibers of the dorsal root that synapse in the spinal cord. Motor output departs in fibers of the ventral root. An association neuron usually connects input and output within the spinal cord. Somatic sensory fibers reach the dorsal root via a spinal nerve. Visceral sensory fibers travel from the visceral organ through one or more ganglia and then through the ramus communicans and the dorsal root finally to synapse in the spinal cord. The somatic motor outflow includes a single neuron that sends its fiber out from the spinal nerve to the effector. The visceral motor outflow includes two neurons in series: a preganglionic neuron (solid line) and a postganglionic neuron (dotted line). The synapse between them can occur in a sympathetic chain ganglion, in a collateral ganglion, or in the wall of the innervated organ. If they synapse in the sympathetic chain, postganglionic fibers usually reach the effector via the spinal nerve.

TABLE 16.3	Reflexes in Mammals	
Components of a Reflex Circuit	**Somatic Arc**	**Visceral Arc**
Effector	Skeletal muscle	Cardiac and smooth muscle, glands
Number of neurons in circuitry	Three (or two) neurons: sensory (association) and motor	Four neurons: sensory (association), preganglionic motor, postganglionic motor
Neurotransmitters	Acetylcholine	Acetylcholine, norepinephrine

directly with the motor neuron. If an animal should start to deviate inadvertently from its normal posture, its muscles are stretched. These stretched muscles elicit a somatic reflex that causes the appropriate muscle to contract and restore the animal to its original posture (figure 16.18a,b).

The **visceral reflex arc** is structurally more complex. The body of a visceral sensory neuron also resides in the dorsal root, but its nerve fibers travel through one or more sympathetic chain ganglia and then through the ramus communicans. Its axons eventually synapse within the spinal cord with an association neuron (figure 16.17). Unlike with the somatic arc, the motor output of the visceral reflex arc includes two neurons in sequence. The first is the **preganglionic neuron,** which extends out the ventral root and synapses in the sympathetic ganglion, in a collateral ganglion, or in the wall of a visceral organ with a second neuron, the **postganglionic neuron.** The postganglionic neuron travels on to innervate the effector visceral organ. Thus, at its simplest, the visceral arc includes four neurons: one visceral sensory neuron, two

visceral motor neurons in series, and an interconnecting association neuron.

In summary, the somatic arc includes somatic afferent neurons that carry sensory impulses to the CNS from skin, voluntary muscles, and tendons. Somatic efferent neurons

(a)

(b)

(a) Lamprey

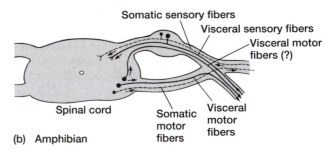

(b) Amphibian

FIGURE 16.19 **Somatic and visceral circuits in anamniotes.** (a) Lamprey. (b) Amphibian. It is unclear whether the amphibian dorsal root carries visceral motor output.

After Hildebrand.

FIGURE 16.18 **Somatic reflex arc.**
(a) Posture can be maintained through a spinal reflex involving a single sensory and single motor neuron connected directly within the spinal cord. When a tetrapod begins to deviate from its normal posture, sensory receptors within joints and muscles detect the deviation. (b) Sensory fibers that carry this impulse to the spinal cord synapse with appropriate motor neurons that stimulate skeletal muscle motor units to contract, straighten the limb, and restore normal posture.

deliver motor impulses to somatic effectors. The visceral arc includes visceral afferent neurons that carry sensory impulses to the CNS from the digestive tract and other internal structures. The visceral efferent neurons carry motor impulses to visceral organs; this part of the circuit includes two neurons: preganglionic and postganglionic.

In amniotes, the dorsal root carries predominantly sensory information, which can be somatic or visceral. The ventral root almost exclusively carries motor information, which can be somatic or visceral. In anamniotes, there is considerable variation in both the structure of the spinal nerve pathways and the information they carry. In lampreys, the dorsal and ventral roots do not join. The ventral root carries only somatic motor information transmitted to striated muscles at that level of the spinal cord. The dorsal root carries somatic and visceral sensory information as in amniotes, but it also carries visceral motor fibers (figure 16.19a). In fishes and amphibians, dorsal and ventral roots are joined, but visceral motor fibers depart via both the dorsal root, as in lampreys, and the ventral root, as in amniotes (figure 16.19b).

The Autonomic Nervous System

Early anatomists noticed that visceral activity did not appear to be under voluntary control. The peripheral nerves and ganglia associated with visceral activity seemed to be autonomous, or independent of the rest of the nervous system. Collectively, they were considered to constitute the autonomic nervous system, a functional division of the peripheral nervous system that presides over visceral activity. Both sensory and motor fibers are included. Autonomic sensory fibers monitor the internal environment of the organism—that is, blood pressure, oxygen and carbon dioxide tension, core and skin temperature, and activity of the viscera. Motor fibers are general visceral motor neurons that innervate cardiac muscle, smooth muscle, and glands. Thus, they control the digestive tract, blood vessels, respiratory tree, bladder, sex organs, and other general body viscera. Because the autonomic nervous system includes the general visceral motor circuit, pre- and postganglionic neurons in series characterize the motor innervation to each organ.

Conscious centers also can affect visceral activity controlled by the autonomic nervous system. For instance, through practiced meditation or through deliberate effort that brings a chilling thought to mind, it is possible to affect the heartbeat or the release of sweat. But, for the most part, the autonomic system operates subconsciously and is not under voluntary control. Reflexes control activities maintaining the internal environment. In its simplest form, the neuronal circuitry of the autonomic nervous system includes four neurons linked in a reflex loop: a sensory neuron that synapses with an association neuron that synapses with a preganglionic motor neuron in series with a postganglionic motor neuron.

Functional Divisions of the Autonomic Nervous System In mammals, the autonomic nervous system is divided into two contrasting, antagonistic systems of control over visceral activity: the sympathetic system and the parasympathetic system.

The **sympathetic nervous system** prepares the body for strenuous action by increasing activity of the viscera, although it slows digestive processes. Stimulation of the sympathetic system inhibits activity of the alimentary canal but promotes contraction of the spleen (causing it to release extra red blood cells into the general circulation), increases heart rate and blood pressure, dilates coronary blood vessels, and mobilizes glucose from glycogen storage in the liver. It is often said that the sympathetic nervous system prepares the individual to **fight or flee**, perhaps indicative of an organism's quotient of courage or wisdom (table 16.4).

The general visceral motor nerves that participate in sympathetic activity depart from the thoracic and lumbar regions of the mammalian spinal cord. This activity is referred to as the **thoracolumbar outflow.** The sympathetic preganglionic neuron is usually short and synapses in the sympathetic chain ganglion or in a ganglion located away from the vertebral column. The postganglionic fiber is usually long (figure 16.20).

The **parasympathetic nervous system** restores the body to a restful or vegetative state by lowering its activity level, although digestion is stimulated. The effects of the parasympathetic system are antagonistic to those of the sympathetic system. It enhances digestion, slows heart rate, drops blood pressure, constricts coronary vessels, and promotes glycogen formation.

Participating visceral motor neurons include cranial nerves VII, IX, and X together with spinal nerves departing from the sacral region. This is referred to as **craniosacral outflow.** Parasympathetic preganglionic fibers are long and reach to the wall of the organ they innervate and synapse with very short postganglionic fibers (figure 16.20).

Adrenergic and Cholinergic Control The sympathetic system is said to be **adrenergic** because the neurotransmitters released during stimulation are adrenaline or **noradrenaline** (also termed **epinephrine** and **norepinephrine**). The parasympathetic system is said to be cholinergic because the neurotransmitter released is **acetylcholine.** Acetylcholine is also released between pre- and postganglionic fibers in both systems (figure 16.20) and at junctions between nerves and skeletal muscles.

In mammals, almost every visceral organ has sympathetic and parasympathetic innervation (figure 16.21 and

TABLE 16.4 Functional Divisions of the Autonomic Nervous System		
Organ/ Activity	**Sympathetic Stimulation**	**Parasympathetic Stimulation**
Eye		
Ciliary muscle	Relaxation	Contraction
Pupil	Dilation	Constriction
Glands		
Salivary	Vasoconstriction	Vasodilation
	Slight secretion	Copious secretion
Gastric	Inhibition of secretion	Stimulation of secretion
Pancreas	Inhibition of secretion	Stimulation of secretion
Lacrimal	None	Secretion
Sweat	Sweating	None
Digestive tract		
Sphincters	Increase tone	Decrease tone
Walls	Decrease motility	Increase motility
Liver	Glucose release	None
Gallbladder	Relaxation	Contraction
Bladder		
Smooth muscle	Relaxation	Contraction
Sphincter	Contraction	Relaxation
Adrenal gland	Secretion[a]	None
Heart		
Muscle	Increase rate and force	Slowed rate
Coronary arteries	Dilatation	Constriction
Lungs (bronchi)	Dilatation	Constriction
Spleen	Contraction	Relaxation
Blood vessels		
Abdomen	Constriction	None
Skin	Constriction	None
Sex organs		
Penis	Ejaculation	Erection
Clitoris	?	Erection
Metabolism	Increased	None

[a]Preganglionic neuron innervation.

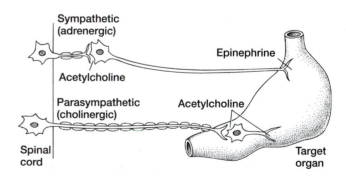

FIGURE 16.20 Neurotransmitters of the autonomic nervous system. Adrenergic and cholinergic neurotransmitters are released at the ends of the sympathetic and parasympathetic circuits, respectively. This is the basis for differential organ response.

FIGURE 16.21 **Autonomic nervous system in a mammal.** Subdivisions are symmetrical and paired but here for simplicity each is depicted singly, sympathetic (left) departing from the thoraco-lumber regions and parasympathetic (right) departing from the cranio-sacral regions of the central nervous system. Note the double innervation of most organs. Preganglionic and postganglionic fibers are indicated.

table 16.4). Exceptions to this double innervation include the adrenal gland, peripheral blood vessels, and sweat glands, all of which receive only sympathetic innervation. Cessation of sympathetic stimulation allows these organs to return to a resting state.

The adrenal gland is also exceptional in that it is innervated by the preganglionic fiber only; the postganglionic fiber is absent. Because epinephrine and norepinephrine serve both as adrenergic chemical signals of the sympathetic circuit and as hormones produced by the adrenal gland (see chapter 15), there is a possibility for chemical confusion. But the preganglionic neuron releases acetylcholine rather than adrenaline or similar chemicals, so direct innervation of the adrenal gland by preganglionic fibers removes the possibility of chemical ambiguity between parasympathetic innervation and hormonal stimulation by the gland.

Anatomical Divisions of the Autonomic Nervous System The division of the autonomic nervous system into sympathetic and parasympathetic functional components holds reasonably well for mammals; however, in other vertebrates, the comparative anatomy of the autonomic nervous system is poorly understood. Most viscera receive contrasting sympathetic and parasympathetic innervation, but these functional divisions do not always correspond to thoracolumbar and craniosacral outflow, respectively. Often, in nonmammalian vertebrates, the autonomic nerves departing from these regions have a mixed function. When examining the location of autonomic nerves of anamniotes, we cannot safely infer function from anatomical position. Therefore, we prefer anatomical distinctions alone, without implied functional significance, when describing the autonomic nervous system of nonmammalian vertebrates.

There are three anatomical divisions of the autonomic nervous system: cranial autonomic, spinal autonomic, and enteric autonomic systems (table 16.5). The **cranial autonomic system** includes the cranial nerves leaving the brain. The **spinal autonomic system** consists of all autonomic fibers departing from the central nervous system in the spinal segments, specifically all thoracic, lumbar, and sacral autonomic fibers.

The **enteric autonomic system** includes intrinsic sensory and motor neurons residing in the wall of the digestive tract. The enteric system consists of large numbers of neurons, probably as many neurons as there are in the central nervous system. Nerves formed from these neurons interconnect and mingle to form woven patches of nerve processes, termed **plexuses,** within the wall of the digestive tract. The **myenteric plexuses** (Auerbach's plexuses) are situated within the outer wall of smooth muscles, and the **submucosal plexuses** (Meissner's plexuses) are located deep within smooth muscles near their lumen. The enteric autonomic system is responsible for coordination of digestive tract activity. It is independent from but can be modified by the spinal and cranial autonomic systems. Food distending the smooth muscles of the digestive tract mechanically stimulates enteric neurons. These neurons, in turn, activate contractions of the circular and longitudinal smooth muscles in the wall of the digestive tract, resulting in synchronized **peristaltic waves** that propel food through the tract. The enteric autonomic system seems to be present in all classes of vertebrates, although it may be poorly developed in some.

Fishes In cyclostomes, the autonomic nervous system is fragmentary. Sympathetic chains are absent, but collateral ganglia, presumably part of the autonomic system, are scattered throughout the viscera. In hagfishes, cranial autonomic fibers apparently occur only in the vagus (X). However, in lampreys, in addition to the vagus (X), the facial (VII) and glossopharyngeal (IX) nerves include autonomic fibers that mediate events in the gills. In hagfishes, spinal autonomic fibers pass through the ventral roots of spinal nerves, but their subsequent distribution is poorly known. In lampreys, spinal autonomic fibers depart through dorsal roots of the spinal nerves to supply the kidneys, gonads, blood vessels, posterior digestive tract, cloaca, and other viscera.

In chondrichthyan and osteichthyan fishes, the autonomic nervous system is well represented. A sympathetic chain of ganglia is present within the spinal autonomic system. Nerves of the cranial autonomic system pass to the viscera. However, in elasmobranchs, collateral ganglia are absent and sympathetic chains apparently do not contribute fibers to the cranial nerves. The vagus is well developed with branches to the stomach and heart, but the heart apparently lacks a sympathetic counterpart to the inhibitory vagal innervation (figure 16.22a). Further, the sympathetic ganglia of elasmobranchs are associated with populations of chromaffin cells, neural crest derivatives that in teleosts

TABLE 16.5 Relationship between Functional and Anatomical Divisions of the Autonomic Nervous System		
Location	**Function (Mammals)**	**Anatomical Designation**
Cranial	Parasympathetic	Cranial autonomic
Thoracic	Sympathetic	Spinal autonomic
Lumbar	Sympathetic	Spinal autonomic
Sacral	Parasympathetic	Spinal autonomic
Intrinsic digestive tract	Enteric	Enteric autonomic

and most tetrapods (except urodeles) become separated from the ganglia. In most teleosts, collateral ganglia occur, and some spinal nerve fibers are shared with cranial nerves (figure 16.22b).

(a) Elasmobranch

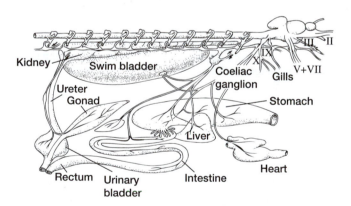

(b) Teleost

FIGURE 16.22 Autonomic nervous system of fishes. (a) Elasmobranch (shark). Notice that the vagus nerve (X), which supplies the pharynx, stomach, and heart, does not carry fibers from any of the spinal nerves. (b) Teleost. The vagus (X), which supplies most of the same viscera as in the shark as well as contributes to innervation of the swim bladder, is connected to the sympathetic chain.

Redrawn from S. Nilsson, 1983, "Autonomic nerve function in the vertebrates," in Zoophysiology. Ed. By D.S. Farmer, Springer-Verlag, NY, based on Young 1933, and Nilsson 1976. Reprinted by permission.

Tetrapods   The autonomic nervous system is well developed in tetrapods. The paired sympathetic chain is present, collateral ganglia are dispersed among the viscera, and cranial and spinal nerves are well delineated. Autonomic outflow in amphibian spinal nerves passes through the ventral roots, but it is still unclear whether motor fibers also occur in the dorsal root (figure 16.23a). In reptiles, birds, and mammals, the autonomic systems are quite similar in basic construction. Spinal autonomic motor fibers depart via the ventral roots of spinal nerves (figure 16.23b,c).

Overview   With the exception of cyclostomes, the anatomical organization of the autonomic nervous system is similar in

all vertebrate classes. The spinal autonomic outflow includes a paired sympathetic chain (except in cyclostomes and elasmobranchs), with some fibers contributing to cranial nerves. The cranial autonomic outflow includes the facial (VII), glossopharyngeal (IX), and vagus (X) cranial nerves, although the facial and glossopharyngeal can be reduced in fishes. In vertebrates with eyes, the oculomotor (III) nerve may send fibers to the iris and ciliary muscles in the eyes.

In mammals, especially in humans, the autonomic nervous system is better known, and the circuits for motor outflow can be traced with greater confidence. The human autonomic nervous system includes a sympathetic thoracolumbar outflow and a parasympathetic craniosacral

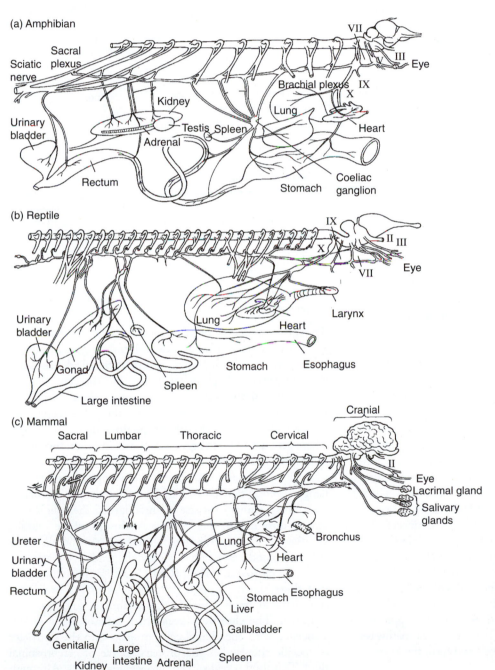

(a) Amphibian

(b) Reptile

(c) Mammal

FIGURE 16.23   Autonomic nervous system of various tetrapods. (a) Amphibian (anuran). (b) Reptile (lizard). (c) Mammal (eutherian).

Redrawn from S. Nilsson, 1983, "Autonomic nerve function in the vertebrates," in Zoophysiology. Ed. By D.S. Farmer, Springer-Verlag, NY, based on Young 1933, and Nilsson 1976. Reprinted by permission.

outflow. The existence of a sacral parasympathetic system in other vertebrates is still uncertain. The pelvic nerves of amphibians arising from the posterior end of the spinal cord and supplying the urinary bladder and rectum have traditionally been regarded as parasympathetic sacral outflow. However, even in mammals, sacral parasympathetic fibers mingle with thoracolumbar sympathetic fibers in the pelvic plexus, making it difficult to trace the posterior circuitry of these two systems on their way to visceral effectors. Until the comparative features of the autonomic nervous system are better known in vertebrates, the functional roles of autonomic nerves in nonmammalian classes must be inferred.

To summarize, somatic and visceral effectors receive motor information. Effectors and receptors are linked through the central nervous system. Control of much of the body's activity involves simple reflexes. The somatic reflex arc is primarily involved in controlling skeletal muscles. The visceral reflex arc is the basic component of the autonomic nervous system, which is responsible for monitoring internal visceral activity. We turn next to the central nervous system to examine its role in processing information.

Central Nervous System

The central nervous system primarily coordinates activities that enable an organism to survive and reproduce in its environment. In order to do this, the central nervous system must receive incoming information from several sources. Sensory receptors known as **interoceptors** gather information and respond to general sensations of organs within the internal environment. **Proprioceptors** are a type of interoceptor that inform the central nervous system about the position of the limbs and the degree to which joints are bent and muscles are stretched. This information-processing component of the nervous system is referred to as the **somatosensory system,** which includes proprioceptors and surface receptors within the skin. Sensations gathered by the somatosensory system are especially important in coordinating limb and body positions during locomotion. **Exteroceptors** gather information from the external environment. Sensations of touch, pressure, temperature, sight, hearing, smell, taste, and other stimuli from the external environment are transmitted via exteroceptors to the brain and spinal cord. A third source of information comes from memory, which allows an organism to adjust its activity on the basis of past experiences.

The central nervous system processes incoming information and returns instructions to the effectors (figure 16.24). These constitute the response of the organism. Entering information **diverges** to inform various areas of the brain and spinal cord about the state of affairs at that point. When a decision is made, instructions **converge** to the appropriate effectors. The spinal cord and brain carry the pathways through which this information travels and form the association areas where it is evaluated.

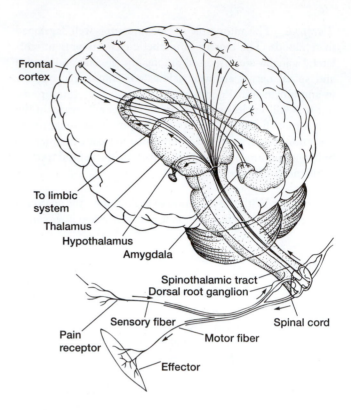

FIGURE 16.24 Sensory and motor circuits.
Sensory receptors in the skin respond to stimuli by generating an electrical impulse that travels to the spinal cord and synapses in the thalamus. This impulse is relayed by other neurons to other areas of the brain, which produces a response that travels down the spinal cord to a motor neuron and out to an effector.

Embryology

The vertebrate central nervous system is hollow, a result of the fusion of two raised neural folds within the ectoderm. In the brain, the central canal enlarges into fluid-filled **ventricles** that are connected spaces located within the center of the brain. Within the anterior neural tube, three embryonic regions of the brain differentiate into the **prosencephalon, mesencephalon,** and **rhombencephalon** (figure 16.25a–c). These give rise to three regions of the adult brain: **forebrain, midbrain,** and **hindbrain** (figure 16.25f).

The brain and spinal cord are wrapped in **meninges** (sing., *meninx*) derived in part from neural crest. In mammals, the meninges consist of three layers: the tough outermost **dura mater,** the weblike **arachnoid** in the middle, and the innermost **pia mater** (figure 16.26a). The pia mater contains blood vessels that supply the underlying nervous tissue. **Cerebrospinal fluid (CSF)** is a slightly viscous fluid that flows slowly through the ventricles of the brain, the subarachnoid space beneath the arachnoid, and the central canal. The **choroid plexus,** small tufts of blood vessels associated with ependymal cells, projects into the ventricles at specific sites and is the primary source of cerebrospinal fluid. This fluid is reabsorbed into venous sinuses. Although

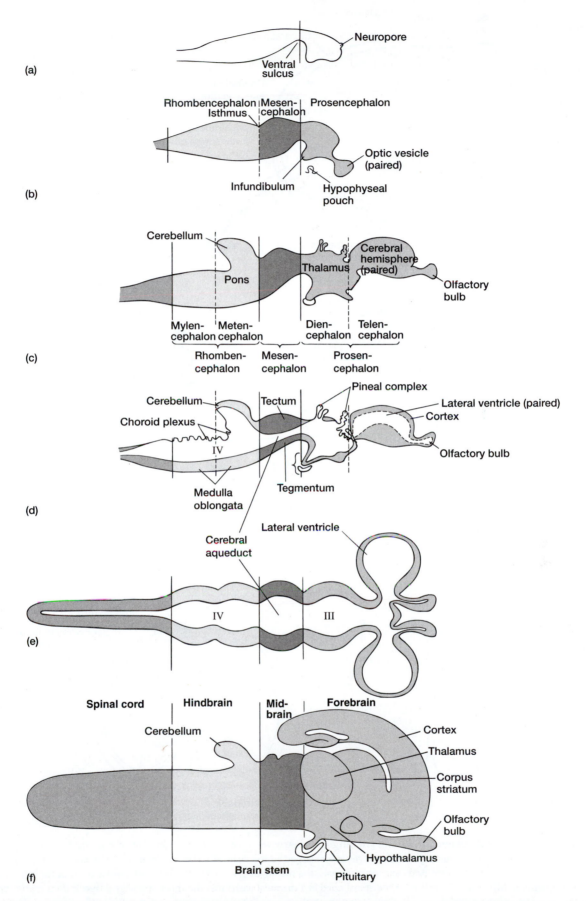

FIGURE 16.25 **Development of the central nervous system.** (a–d) Embryonic development. (e) Fluid-filled ventricles within the central nervous system. (f) Anatomical regions of the adult brain.

Arachnoidal
villi

Superior sagittal
sinus

Dura mater

Subarachnoid
space

Arachnoid

Choroid plexus of
lateral ventricle

Intraventricular
foramen (Monro)

Aqueduct of
Sylvius

Choroid plexus
of third ventricle

Foramen of
magendie

Choroid plexus of
fourth ventricle

Dura mater

Arachnoid

Subarachnoid
space

Pia mater

} Meninges

Gray matter

White matter

} Brain

Bone

(a)

Primitive
meninx

Dura mater

Secondary meninx

Gray matter

(b) Fishes

(c) Tetrapods
(nonmammalian)

FIGURE 16.26 Cerebrospinal fluid and meninges. (a) Arrows trace the circulation of cerebrospinal fluid through the brain and spinal cord of a mammal. The triple-layered meninges are enlarged to the right. (b) The meninges of fishes consist of a single thin layer, the primitive meninx. (c) In all tetrapods except mammals, the meninges are double layered and consist of an outer dura mater and inner secondary meninx. (d) Cutaway section of the spinal cord in a mammal illustrates the three meningeal layers: dura mater, arachnoid, and pia mater. Branches of the spinal nerve are shown along with their connections to the sympathetic chain. (*Continued on next page.*)

(a–c) After H. M. Smith; (d) after R. T. Woodburne.

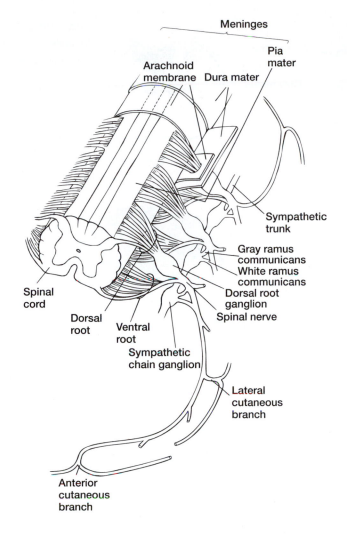

Meninges

Arachnoid
membrane Dura mater

Pia
mater

Sympathetic
trunk

Gray ramus
communicans

White ramus
communicans

Dorsal root
ganglion

Spinal nerve

Spinal
cord

Dorsal
root

Ventral
root

Sympathetic
chain ganglion

Lateral
cutaneous
branch

Anterior
cutaneous
branch

(d) Mammals

FIGURE 16.26 continued

cerebrospinal fluid is derived from the blood and returns to it, it is devoid of red blood cells or any other large formed elements. When a person is injured and trauma to the central nervous system is suspected, a procedure called a spinal tap is done to sample the cerebrospinal fluid. If it contains red blood cells, then the brain or spinal cord may be damaged. Cerebrospinal fluid forms a cushion of fluid around the brain and spinal cord to support the delicate nervous tissues and absorb shocks from concussions. The average human has about 150 ml of cerebrospinal fluid, less than a cup, that is replaced several times per day, flushing the central nervous system. Recent speculation suggests that it might carry chemical messages important in regulating the organism's daily circadian rhythms.

In fishes, the meninges consist of a single membrane, the **primitive meninx,** wrapped around the brain and spinal cord (figure 16.26b). With the adoption of terrestrial life, the meninges doubled. In amphibians, reptiles, and birds, the meninges include a thick outer dura mater

derived from mesoderm and a thin inner **secondary meninx** (figure 16.26c). With a double meningeal layer, cerebrospinal fluid may circulate more effectively and absorb shocks from the jolts sustained during terrestrial locomotion. In mammals, the dura mater persists, but division of the secondary meninx yields both the arachnoid and the pia mater from ectomesoderm (figure 16.26d).

Spinal Cord

The vertebrate spinal cord, like the brain, is organized into two regions and named because of their appearance in fresh preparations (figure 16.27a–f). The **gray matter** of the spinal cord includes nerve cell bodies that lie within the core of the spinal cord. Dorsal and ventral extensions of the gray matter are the **dorsal horns** and **ventral horns,** respectively. The dorsal horns contain the bodies of neurons receiving incoming sensory information. The ventral horns contain the bodies of motor neurons (figure 16.17). The **white matter** of the spinal cord surrounds the gray matter. It is predominantly composed of nerve fibers linking different levels of the spinal cord with each other and with the brain. Many of these fibers are myelinated, creating their white color.

The spinal cord functions in two capacities. It establishes simple reflexes and contains pathways of diverging and converging information.

Spinal Reflexes

As you know from the discussion of somatic and visceral reflex arcs, the spinal cord completes the reflex loop between sensory input and motor output. In doing this, the spinal cord selects the effectors to be activated or inhibited. Although the spinal cord operates at the reflex level, it also contains circuits that coordinate different parts of the cord.

Arriving sensory fibers synapse in the dorsal horn of the gray matter with association neurons (figure 16.28a). Association neurons carry the impulse to the ventral horn on the same side, to the opposite side, or to a different level of the spinal cord or brain. In the ventral horn, the association neuron synapses with a motor neuron whose axon travels out the ventral root to the effector. Dispersion of information within the spinal cord can produce complex responses to stimuli without involving higher centers. For example, if an animal should inadvertently place its foot on a sharp object, the reflex to withdraw it could involve as few as three neurons (figure 16.29). The first, the afferent sensory neuron, carries the painful stimulus to the spinal cord, where it synapses with an association neuron. The association neuron transmits the stimulus to the ventral horn, where it synapses with a motor neuron whose axon conveys the impulse to appropriate retractor muscles that contract and withdraw the foot. Association neurons that connect to appropriate levels on the opposite side of the cord reach motor neurons innervating extensor muscles in the opposite leg. These muscles contract, extend the leg, and prevent the animal's collapse when it lifts its other leg from the sharp object.

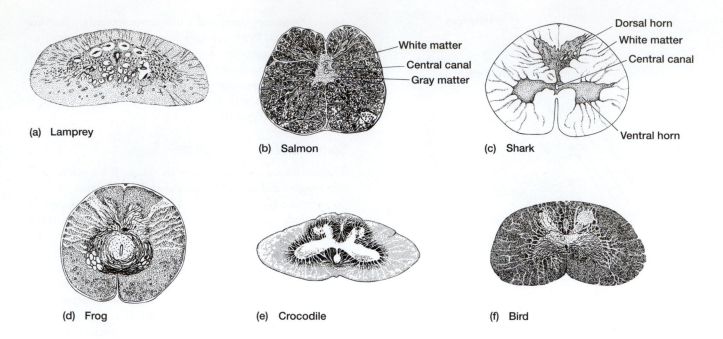

(a) Lamprey

(b) Salmon — White matter / Central canal / Gray matter

(c) Shark — Dorsal horn / White matter / Central canal / Ventral horn

(d) Frog

(e) Crocodile

(f) Bird

FIGURE 16.27 Cross sections of vertebrate spinal cords.

After bolk

The circuitry involved demands a connection between painful stimulus and appropriate effectors (retractor and extensor muscles). It need not involve higher brain centers. Usually, the association neurons also convey the painful stimulus to conscious centers of the brain, where it is perceived (figure 16.28b); however, by the time the higher centers become aware of the surprise trauma to the foot, the spinal reflex to retract it is already under way.

Spinal Tracts

Not all information is processed at the level of the spinal cord. Much, perhaps most, information is carried to higher levels of the nervous system for evaluation. The resulting decisions are carried down the spinal cord to appropriate effectors. Nerve fibers carrying similar information tend to travel together in nerve tracts, bundles of similar fibers that occupy a specific region of the spinal cord. Nerve tracts may be ascending or descending tracts, depending on whether they convey information up or down the cord, respectively (figure 16.30). They are usually named for their source and their destination. For example, the spinothalamic tract begins in the spinal cord and extends to the thalamus (table 16.6).

Wars, accidents, and diseases can lead to localized wounds of the spinal cord that sever the ascending or descending flow of information. In humans, such losses of function have been correlated with the specific region in which the wound occurred and used to map the positions of these nerve tracts. More precise information from animal studies has added to our understanding of spinal cord organization. For convenience, these mapped tracts are drawn in discrete locations. In practice, their precise positions may change slightly at different levels of the cord, and there is some overlap of tracts as well.

Ascending tracts carry sensory impulses from the spinal cord to the brain. Among the most prominent are the **fasciculus gracilis** and **fasciculus cuneatus,** located in the dorsal region of the spinal cord. Both carry proprioceptive stimuli and sensations associated with posture to the medulla. As each tract ascends, more axons are added. For instance, the gracilis is supplemented laterally to produce the cuneatus (figure 16.31). Thus, at higher levels of the cord, the more medial fasciculus gracilis carries sensations from the lower limb and the more lateral fasciculus cuneatus carries sensations from the upper limb.

The **spinocerebellar tracts** carry proprioceptive information concerning positions of the limbs and body to the cerebellum. This information is not consciously perceived, but it enables the cerebellum to coordinate movements of different parts of the body. The **lateroventral** (= anterolateral) **spinothalamic tract** transmits information to the thalamus concerned with sensations of pain and temperature.

Descending tracts transmit impulses from the brain to the spinal cord. One of the most important is the **corticospinal tract,** which runs directly from the cerebral cortex to motor neurons going to the limbs; thus, it places skeletal muscles under cerebral control. The **tectospinal tract** is associated with optic and auditory stimuli. It does not go through conscious centers, but extends a short distance directly down the spinal cord to cervical levels, where it ends

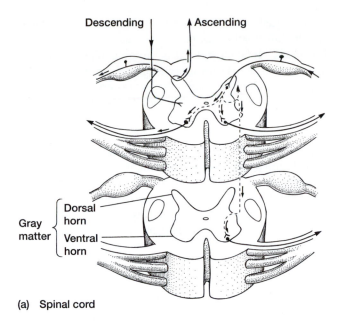

Descending **Ascending**

Gray matter { Dorsal horn / Ventral horn

(a) Spinal cord

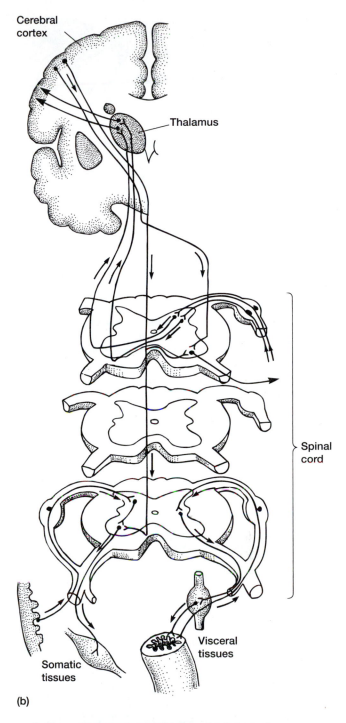

Cerebral cortex

Thalamus

Spinal cord

Somatic tissues

Visceral tissues

(b)

FIGURE 16.28 Spinal reflexes. (a) Association neurons (dashed lines) within the gray matter receive afferent signals and relay these across the cord, to the same side of the cord, or to a different level of the spinal cord. (b) Spinal reflexes place somatic and visceral effectors under the immediate control of sensory information. But motor neurons that travel to these tissues are also influenced by descending circuits from conscious centers of the brain.

on somatic motor neurons that innervate neck muscles. Its function is to turn the head quickly toward threatening or surprising visual or auditory stimuli. The **rubrospinal tract** conveys impulses from the midbrain to the spinal cord and is involved in initiating coordinated movements.

Brain

The brain forms embryologically from the neural tube anterior to the spinal cord. It includes three anatomical regions (figures 16.25a–f and 16.32). The most posterior region is the hindbrain, which includes the **medulla oblongata, pons,** and

cerebellum. Next is the midbrain, which includes a sensory **tectum** and a motor **tegmentum.** The **brain stem** includes all regions of the hindbrain and midbrain except for the cerebellum and colliculi. The most anterior region of the brain, the forebrain, includes the **telencephalon,** or **cerebrum,** and the **diencephalon,** which is the source of the **thalamus.**

Phylogeny

Independently, the forebrain tends to enlarge in various vertebrate groups, including hagfishes, some sharks, rays, teleost fishes, and tetrapods (figure 16.33). Some of this

---

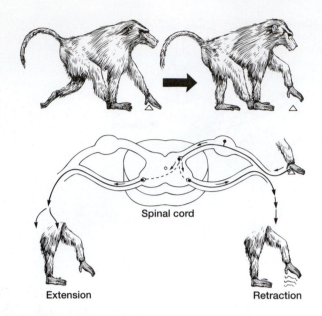

FIGURE 16.29 Spinal reflex. Association neurons (dashed lines) within the spinal cord deliver stimuli to motor neurons, causing the retractor muscles of this animal to lift its foot from a harmful object. These stimuli also spread to motor neurons in other areas of the cord that innervate extensor muscles of the opposite limb to contract and support the weight of the body.

Spinal cord

Extension

Retraction

is correlated with the increased importance of olfactory (smell) information, as occurs, for example, in hagfishes. Forebrain enlargement also accompanies increasingly complex behaviors and muscle control. In amniotes, limb posture and body carriage change as terrestrial modes of locomotion become predominant. The limbs move from a sprawled position to one in which they are carried more directly under the weight of the body, increasing the ease and efficiency of limb oscillation (see chapter 8). Coordination of limb oscillation and placement during rapid locomotion becomes especially complicated in bipedal archosaurs and birds. Increased input of somatosensory information and increased output of motor responses to skeletal muscles requires mediation. The enlargement of the amniote forebrain reflects its increasing role in this mediation within the locomotor system.

In advanced teleost fishes, the midbrain tends to enlarge rather than the forebrain. This seems to be correlated with the processing of visual information, as well as the increasing importance of sensory input from the lateral line system, and with greater movement of teleosts in the three-dimensional space of their aquatic environment.

Within these general patterns, the brain of each species reflects the demands of information processing

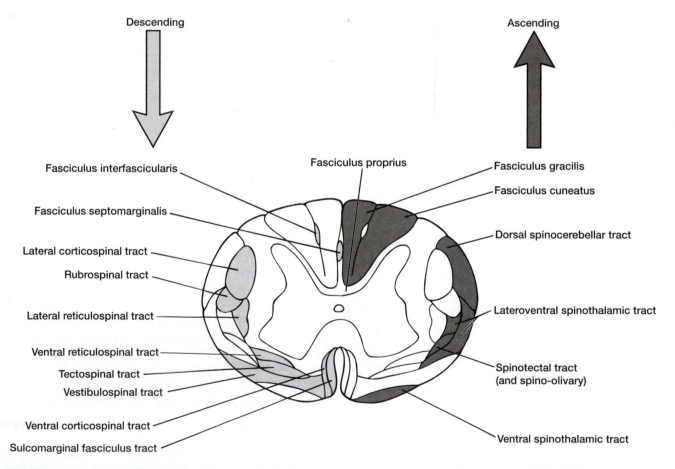

Descending

Ascending

Fasciculus interfascicularis

Fasciculus septomarginalis

Lateral corticospinal tract

Rubrospinal tract

Lateral reticulospinal tract

Ventral reticulospinal tract

Tectospinal tract

Vestibulospinal tract

Ventral corticospinal tract

Sulcomarginal fasciculus tract

Fasciculus proprius

Fasciculus gracilis

Fasciculus cuneatus

Dorsal spinocerebellar tract

Lateroventral spinothalamic tract

Spinotectal tract
(and spino-olivary)

Ventral spinothalamic tract

FIGURE 16.30 Cross section of the human spinal cord showing approximate locations of ascending (right) and descending (left) nerve tracts.

After Netter.

TABLE 16.6 Locations and Functions of Descending and Ascending Nerve Tracts of the Spinal Cord

Tract	Source	Destination	Function
Descending			
Lateral and ventral corticospinal tracts	Cerebral cortex	Spinal cord	Motor connections direct from cortex to primary motor neurons of arms and legs (places motor neurons under direct voluntary cortical control)
Rubrospinal tract	Midbrain (red nucleus of tegmentum)	Spinal cord	Motor connections in spinal cord
Lateral and ventral reticulospinal tracts	Medulla reticular formation	Spinal cord (dorsal horn)	Postural reflexes
Tectospinal tract	Midbrain (colliculus, roof)	Spinal cord	Visual and auditory stimuli to limbs and trunk
Vestibulospinal tract	Medulla (vestibular nucleus)	Spinal cord	Postural reflexes accomplished by axial and limb musculature
Ascending			
Fasciculus gracilis and fasciculus cuneatus	Spinal cord	Medulla	Fine tactile information, Golgi tendon organs, Pacinian corpuscles in joints
Dorsal and ventral spinocerebellar tracts[a]	Spinal cord	Cerebellum via peduncle	Proprioceptive information from muscles to cerebellum, intrafusal muscle fibers
Lateroventral spinothalamic tract	Spinal cord	Thalamus	Pain and temperature sensations to thalamus
Spinotectal tract	Spinal cord	Midbrain (tectum)	Proprioceptive information from neck and shoulders
Spinoreticular tract	Spinal cord	Medulla (reticular formation)	Pain and sensations from internal organs

[a]May be single tract.

FIGURE 16.31 Gray and white matter at various levels of the human spinal cord. Sections are identified by their region—cervical (C), thoracic (T), lumbar (L), sacral (S)—and by specific vertebrae numbered (Arabic number) within each of these regions from which they came. At the level of the arms (C-5) and legs (L-3), additional sensory and motor fibers enter and leave the spinal cord. This is reflected in the more extensive gray matter compared to other regions of the cord (e.g., T-2, T-8). Note the addition of the fasciculus cuneatus at the highest level on the cord. Primarily, this carries sensory information from the arms.

After Netter.

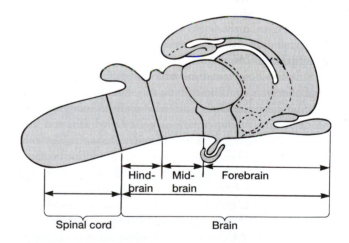

FIGURE 16.32 Regions of the vertebrate brain represented diagrammatically.

After Nauta and Feirtag.

required by its habitat and mode of life (figure 16.34). Cavefish, for instance, have reduced eyes and live in caves, a permanently dark subterranean environment. Correspondingly, the tectum of the midbrain, which normally receives visual input, is reduced as well. On the other hand,

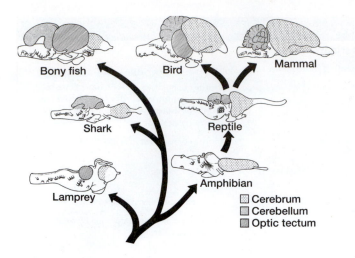

FIGURE 16.33 **Evolution of the vertebrate brain.** Note the phylogenetic enlargement of the cerebrum and cerebellum.

Bony fish · Bird · Mammal · Shark · Reptile · Lamprey · Amphibian

☒ Cerebrum
☐ Cerebellum
☐ Optic tectum

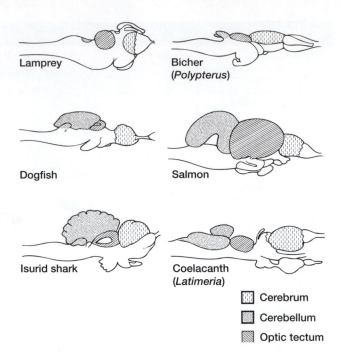

Lamprey · Bicher (*Polypterus*) · Dogfish · Salmon · Isurid shark · Coelacanth (*Latimeria*)

☒ Cerebrum
☐ Cerebellum
▨ Optic tectum

FIGURE 16.34 **Brains of fishes.** Note variations in the sizes of different regions of the brain. These reflect differences in the role that each region plays in processing information important to different species.

After Ebbesson and Northcutt; Roberts and Kremers.

when visual information constitutes a large part of the brain's sensory input, as in salmon, the tectum is enlarged. Thus, reduction or loss of sensory input from an exteroceptor or interoceptor results in a corresponding reduction or loss of brain nuclei that receive and process this information, whereas increased sensory input leads to increased prominence of the appropriate association.

Form and Function

Representative vertebrate brains are shown in figures 16.35 through 16.37.

Hindbrain The **medulla oblongata** operates primarily at the reflex level. It has three major functions. First, it houses the primary nuclei of cranial nerves (figure 16.36a–c). In sharks, the primary nuclei or roots of cranial nerves V through X are contained in the medulla, whereas in mammals, the primary nuclei of cranial nerves VII through XII reside in the medulla. Second, the medulla serves as a major route through which ascending and descending pathways run to and from higher centers of the brain. Third, the medulla contains centers for visceral, auditory, and proprioceptive reflexes, including reflex centers for respiration (figure 16.38), heartbeat, and intestinal motility. Damage to the medulla can be life-threatening because these centers control vital functions.

Medullary nuclei receive afferent signals from spinal and cranial sensory nerves as well as descending signals from higher centers such as the hypothalamus. All branchiomeric cranial nerves—trigeminal (V), facial (VII), glossopharyngeal (IX), vagus (X)—arise in the medulla. Within these medullary centers, arriving information is processed and efferent output is initiated to adjust visceral activity as well as rhythmic feeding and respiration patterns.

The floor of the hindbrain in amniotes becomes a crossroads of increasing importance for the flow of information. In mammals, it develops into a distinct enlargement, the **pons** (figure 16.37e). The pons is primarily formed by the pontine nuclei, which convey information to the cerebellar cortex from the cerebral cortex.

The **cerebellum** is present in gnathostomes, and is even quite large in some, but is apparently absent in cyclostomes and ostracoderms, although some uncertainties remain. In lampreys, a raised neural lip defines the anterior, dorsal wall of the medulla oblongata ("cerebellum"; figure 16.35a), once proposed to be a modest cerebellum. However, its cell types differ from cerebellar cells and it is more likely part of the medulla oblongata. In most gnathostomes, the cerebellum is a dome-shaped extension of the hindbrain. Its surface is often highly convoluted and folded. The cerebellum can be divided into a medial **corpus** (body) and a paired lateral **auricle.** The sides of the corpus expand into the cerebellar hemispheres in birds and mammals. The **flocculus,** or flocculo-nodular lobe, of tetrapods is homologous to the dorsal half of the auricle of fishes. The ventral auricle of fishes receives lateral line input.

The cerebellum modifies and monitors but does not initiate motor output. It operates at an involuntary level and has two primary functions. First, it is important in maintaining equilibrium (figure 16.39). Information pertaining to touch, vision, hearing, proprioception, and motor input from higher centers is processed in the cerebellum. Integration of these incoming sensations results in the maintenance of muscle tone and balance. For an organism to run, jump, fly, or swim in a three-dimensional world, it must be able to keep itself upright and orient itself in space relative to gravity. The cerebellum is involved in processing information that results in maintaining positional equilibrium of the organism.

The second primary function of the cerebellum is the refinement of motor action. The cerebellum compares incoming impulses and sends modified signals to motor centers. Direct electrical stimulation of the cerebellum does not produce muscle contractions. Following removal of the cerebellum, an organism can still move in space, but its movement is uncoordinated, exaggerated, or insufficient, and its motion is likely to be uneven. Thus, the role of the cerebellum is to monitor and modify rather than initiate action.

This said, we must note that much of what the cerebellum does is not well understood. It is now clear that the cerebellum is also heavily involved in forming memory related to motor events. It acquires detailed sensory

FIGURE 16.35 **Vertebrate brains.** Dorsal views are shown above, lateral views below. (a) Lamprey (*Lampetra*). (b) Shark (*Scymnus*). (c) Bowfin (*Amia*). (d) Frog (*Rana*). (e) Alligator (*Alligator*). (f) Insectivore (*Gymnura*). (g) Goose (*Anser*). (h) Horse (*Equus*).

(a,b) (d–g) After Romer and Parsons; (c) after Davis and Northcutt; (h) after Getty.

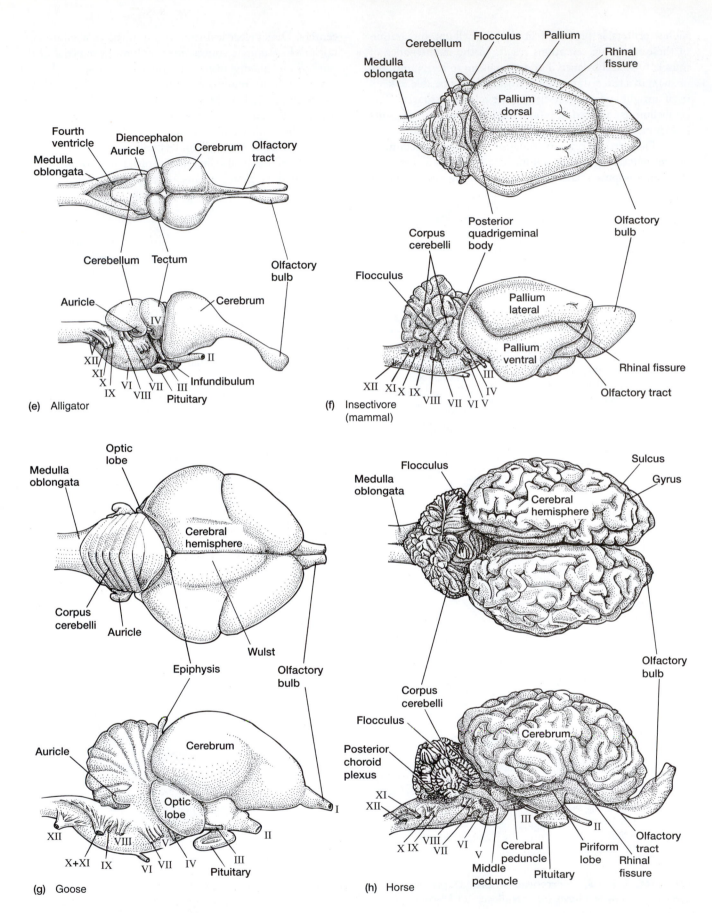

(e) Alligator

Fourth ventricle · Diencephalon · Auricle · Cerebrum · Olfactory tract
Medulla oblongata
Cerebellum · Tectum
Auricle · Cerebrum
IV
XII · XI · X · IX · VI · VIII · VII · III · Infundibulum · Pituitary
II
Olfactory bulb

(f) Insectivore (mammal)

Cerebellum · Flocculus · Pallium · Rhinal fissure
Medulla oblongata
Pallium dorsal
Corpus cerebelli · Posterior quadrigeminal body · Olfactory bulb
Flocculus
Pallium lateral
Pallium ventral
XII · XI · X · IX · VIII · VII · VI · V · III · IV
Rhinal fissure
Olfactory tract

(g) Goose

Optic lobe
Medulla oblongata
Cerebral hemisphere
Corpus cerebelli · Auricle · Wulst
Epiphysis · Olfactory bulb
Auricle · Cerebrum
Optic lobe
XII · X+XI · IX · VIII · VI · VII · V · IV · III · Pituitary · II · I

(h) Horse

Flocculus · Sulcus · Gyrus
Medulla oblongata
Cerebral hemisphere
Olfactory bulb
Corpus cerebelli
Flocculus
Posterior choroid plexus · Cerebrum
XI · XII
X · IX · VIII · VII · VI · V · III · II
Cerebral peduncle · Piriform lobe · Olfactory tract
Middle peduncle · Pituitary · Rhinal fissure

FIGURE 16.35 continued

Disease, War, Barroom Brawls—The Early Science of Neurology

Some of the first insights into the functioning of the nervous system came from the results of damage to it. Distribution of information in the nervous system is very orderly in mammals. Sensory information arrives via the dorsal root, motor responses depart via the ventral root, and association neurons intervene between them (see figure 16.29). Anatomical structure and functional activity are closely matched. Because form and function are closely matched in the nervous system, disruption of function can be used to identify the location of an anatomical injury. Because the spinal cord and brain are organized into discrete functional areas, damage to a part results in selective impairment of function. The earliest indication of this came from battle wounds that soldiers survived, but with persistent deficits in function (box figure 1a). Stab or bullet wounds causing restricted damage to the dorsal horn of the gray matter leave patients with more or less normal motor ability but impair their ability to feel sensations from the level of the body where a wound was inflicted (box figure 1b).

Other pathologies affect motor output rather than sensory input. In 1861, the German neurologist P. P. Broca performed a postmortem examination of the brain of a patient who suffered from a speech defect following injury to his head. While still alive, the patient's lips, tongue, and vocal cords were fully functional, but he could not speak intelligibly. His speech was slow, and many nouns and verbs were deleted. A lesion was found postmortem in a restricted area of the forebrain, a region still known as Broca's motor speech area (box figure 1c).

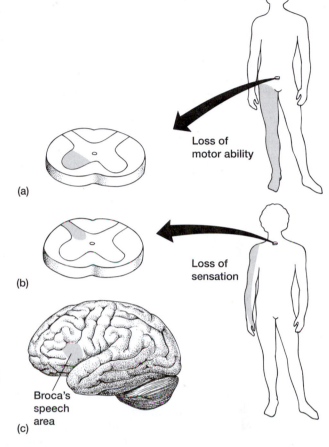

BOX FIGURE 1
Clinical evaluation of injuries to the nervous system.
(a) Loss of motor control to right leg muscles can imply selective injury to the ventral horn of the spinal cord at the level where motor neurons to the skeletal muscles of the leg reside. (b) Loss of sensation to the right arm can result from loss or injury to the dorsal horn of the cord at that level. (c) Injury to Broca's area of the brain leaves a person with an understanding of language but results in impaired speech.

Loss of motor ability

Loss of sensation

Broca's speech area

Poliomyelitis, once a common disease primarily afflicting children, struck motor neurons in the ventral horns of the spinal cord. If the disease settled low in the spinal cord, the lesion would likely cause paralysis in the leg on the same side.

In the twentieth century, car accidents were added to the list of events that inflict this type of damage. Experiments with animals have augmented our knowledge of the functional organization of the central nervous system.

---

information about external space and may take advantage of innate, prewired activity patterns, as well as the results of previous learned experience. Although involved in orientation, much of balance is also mediated by the vestibular and ocular nerves acting directly on motor nerves at lower levels of the spinal cord. In most animals, if the cerebellum is surgically removed, there are few lasting, severe effects on behavior. In humans, destruction of the midline region (vermis) results in ataxia, the loss of coordination of limbs, body, speech, or eye movements. Destruction of the lateral lobes of the cerebellum (hemispheres) results in a condition known as dysmetria, characterized by under- or overshooting a target that is reached for with either hands or feet.

As with other parts of the brain, the size of the cerebellum is proportional to its role. In fishes, the cerebellum is usually relatively large because of extensive input from the lateral line sensory system regarding water currents and electrical stimuli. Furthermore, active aquatic organisms must navigate and orient themselves in a three-dimensional medium. Equilibrium and balance are important; therefore, the cerebellum is well developed. As we would expect, in bottom-dwelling fishes (e.g., flounders) and in fishes that

FIGURE 16.36 **Vertebrate brains, ventral views.** (a) Shark (*Scymnus*). (b) Alligator (*Alligator*). (c) Horse (*Equus*).

After Romer and Parsons.

are not active swimmers (e.g., lampreys), the cerebellum has a reduced role and is relatively small (figure 16.35a). It should be noted that the region in fish termed the "cerebellum" is actually part of the octavolateralis nucleus, the primary target of electroreceptor fibers of the lateral line nerve.

With the advent of terrestrial life, the lateral line system is lost and sensory input to the cerebellum decreases. However, as robust limbs used in terrestrial locomotion develop, proprioceptive information and refinement of muscle action become important and place increased demands on the cerebellum. The cerebellum of terrestrial vertebrates thus remains large and prominent.

Midbrain The roof of the midbrain is the **tectum,** which receives sensory information. Specifically, the midbrain roof is divided into an optic tectum, receiving visual information, and a torus semicularis, receiving auditory and lateral line input. In mammals, the optic tectum is specialized into **superior** and **inferior colliculi.** The floor of the midbrain is the **tegmentum,** which initiates motor output usually via the trochlear (IV) and occulomotor (III) nerves, which arise in the midbrain.

In fishes and amphibians, the midbrain is often the most prominent region of the brain (figure 16.35a–e). The tectum receives direct input from the eyes. In addition,

information from the octavolateralis system, the cerebellum, and the cutaneous sensors is transmitted indirectly to the tectum. The tegmentum is also prominent in anamniotes. In some fishes, it seems to be an important learning center.

In reptiles, birds, and mammals, the tectum continues to receive visual and auditory input, which it relays to the telencephalon through the thalamus. Thus, visual information in all vertebrates reaches the telencephalon via the tectum. A second route by which visual information reaches the telencephalon is through the thalamus of the forebrain, without passing through the tectum (figure 16.40). This route is present in all vertebrates, if only modestly, but becomes enlarged and more important in tetrapods, especially mammals.

Forebrain The **diencephalon** includes four regions: epithalamus, hypothalamus, ventral thalamus, and **dorsal thalamus.** The roof of the diencephalon produces the epithalamus, which includes the **pineal gland** and the **habenular** nucleus at its base. The function of the habenular nucleus is uncertain. In anamniotes, the pineal gland affects skin pigmentation by acting on melanocytes, and it may be important in regulating photoperiod as well. In amniotes, the pineal plays a role in regulating biological rhythms (see chapter 15).

The floor of the diencephalon produces the hypothalamus. The **mammillary bodies,** which are very

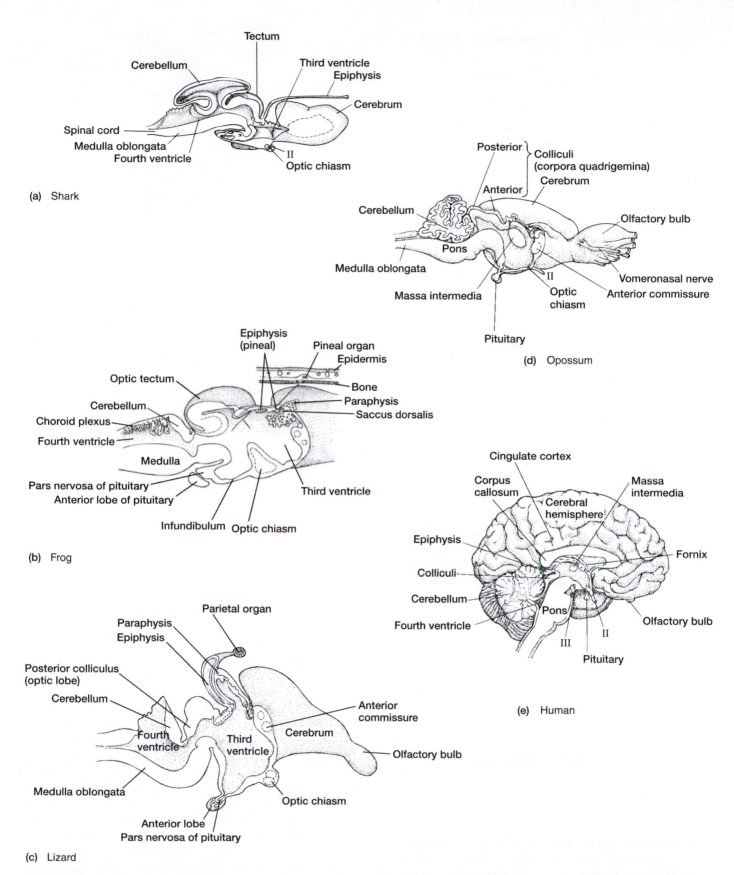

FIGURE 16.37 **Vertebrate brains, sagittal views.** (a) Shark (*Scyllium*). (b) Frog (*Rana*). (c) Lizard (*Lacerta*). (d) Opossum (*Didelphis*). (e) Human (*Homo*).

(a,d,e) After Romer and Parsons; (b,c) after Jollie.

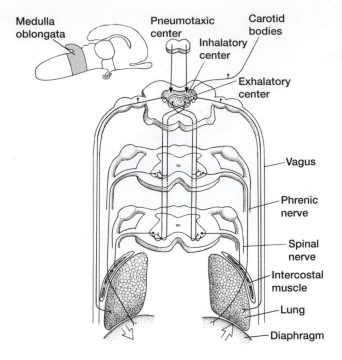

FIGURE 16.38 **Coordination of breathing by the mammalian medulla.** Reflex control of respiration is under the influence of three paired nuclei: the pneumotaxic center in the pons and the dorsal inhalatory center and ventral exhalatory center in the medulla. The inhalatory center receives information about gas composition and blood pH from the carotid bodies and about the degree of lung expansion from the vagus nerve. The inhalatory center excites descending neurons that terminate in motor neurons of the phrenic nerve to the diaphragm. It also stimulates a spinal nerve to the intercostal muscles. When these nerves are excited, inhalation and lung expansion result. The ventral expiratory center does not seem to function during quiet, normal breathing. This center is connected to motor neurons (not shown) serving antagonistic intercostal and accessory muscles of expiration.

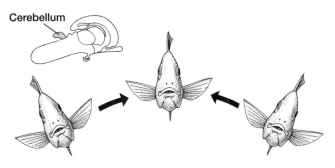

FIGURE 16.39 **Function of the cerebellum.** Balance and orientation are mediated through the cerebellum. As an animal changes its orientation in a gravitational field (left and right sketches), sensory organs that detect its altered position send impulses to the cerebellum. The cerebellum mediates responses that restore the animal's position (center sketch).

FIGURE 16.40 **Function of the amniote tectum.** The tectum receives visual information directly from the retina of the eye and relays this first to the thalamus and then to the cerebrum. In most vertebrates, visual information from the retina reaches the cerebrum via a second route without passing through the tectum. From the retina, visual information first reaches the thalamus and then is relayed to the cerebrum.

prominent in mammals, develop within the hypothalamus (figure 16.41). These bodies are part of the Papez circuit (figure 16.44), which is involved in reproductive behavior and short-term memory. The hypothalamus houses a collection of nuclei that regulate homeostasis to maintain the body's internal physiological balance. Homeostatic mechanisms adjusted by these nuclei pertain to temperature, water balance, appetite, metabolism, blood pressure, sexual behavior, alertness, and some aspects of emotional behavior. The hypothalamus stimulates the pituitary gland situated beneath it to regulate many homeostatic functions. The **limbic** and **reticular systems** influence the functions of the hypothalamus as well. These systems are discussed later in the chapter.

The ventral thalamus is a small area between the midbrain and the rest of the diencephalon. The largest part of the diencephalon is the dorsal thalamus, sometimes called just thalamus, an area comprising nuclei that receive sensory input. The thalamus is the major coordinating center of afferent sensory impulses from all parts of the body. Except for the olfactory tracts, which transmit stimuli directly to the cerebral cortex, all somatic and visceral sensory tracts, including those relaying sensations of touch, temperature, pain, and pressure, as well as all visual and auditory fibers, synapse in thalamic nuclei on their way to the cortex. Thus, the thalamus is a relay center for sensory information going to the cerebral cortex. The thalamus integrates sensory somatic impulses into a pattern of sensations that is projected to the somatic sensory area of the cerebral cortex.

The **telencephalon,** or cerebrum, includes a pair of expanded lobes known as **cerebral hemispheres,** plus the

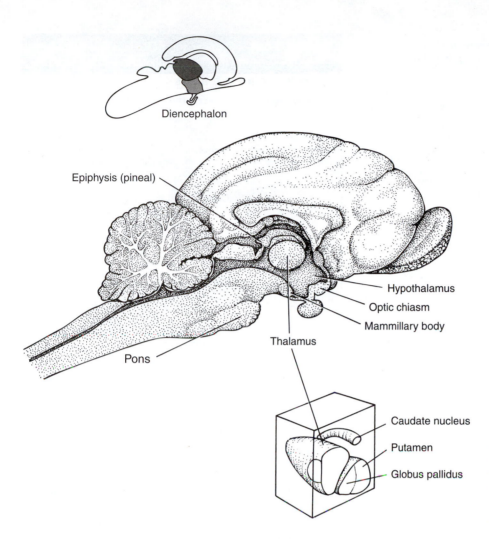

Diencephalon

Epiphysis (pineal)

Hypothalamus

Optic chiasm

Mammillary body

Thalamus

Pons

Caudate nucleus

Putamen

Globus pallidus

FIGURE 16.41 **Hypothalamus and its relationship to adjacent regions of the brain.** The region of the diencephalon is shaded in the small, top figure. The isolated dorsal thalamus in three-dimensional cutaway view is shown in the small block below the brain.

olfactory bulbs. The outer wall of these hemispheres forms the **cerebral cortex,** or **cortical region.** The **subcortical region** comprises the remaining cerebral tissue. The subcortical tissue immediately surrounding the corpus callosum is the **cingulated gyrus,** part of the limbic system (figure 16.44). The hemispheres appear embryologically at the most anterior end of the neural tube. In actinopterygian fishes, the embryonic telencephalon proliferates outward to form the everted adult cerebrum. In all other fishes and tetrapods, the embryonic telencephalon forms lateral swellings, which give rise to the cerebral hemispheres of adults (figure 16.42). The reason the hemispheres are everted in ray-finned fishes, but in no other vertebrates, is not known. But it has been suggested that it is an indirect consequence of embryonic packing of the hemispheres in the crowded space between the developing large nasal capsules and eyes.

Reception of olfactory information is a major function of the telencephalon. Even in basal vertebrates, however, ascending fibers arrive from the thalamus, suggesting that the telencephalon has assisted in regulating other sensory integrative functions as well since early in vertebrate evolution. In reptiles and especially in birds and mammals, the cerebral region enlarges 5- to 20-fold compared with most anamniotes of similar body size. This phylogenetic enlargement occurs, in part, because the cerebrum must process more sensory information from the thalamus. This is accompanied by an increased number of association centers within the cerebrum. Nevertheless, within any vertebrate class, the size of the telencephalon may vary considerably among species. For example, among chondrichthyan fishes, primitive sharks and rays possess cerebrums comparable in size with those of amphibians, but in advanced sharks and skates, the relative size of the cerebral hemispheres approaches that of birds and mammals.

In many mammals, the cerebral cortex is folded in a complicated fashion to accommodate its increased volume. The rounded folds are **gyri,** and the intervening grooves are called **sulci.** The term **fissure** is often used to note a deep sulcus that separates major surface regions of the cerebrum. Not all mammals show such folding. In the platypus, opossum, and many rodents, the cerebral cortex is smooth. In the echidna, kangaroos, and most primates, the degree of folding is variable. In all groups of mammals, the extent of folding seems to be more pronounced in large species. Left and right halves of the forebrain interconnect via **commissures,** bands of neurons transversely crossing the midline between respective brain regions. Most prominent of the commissures is the **corpus callosum,** found

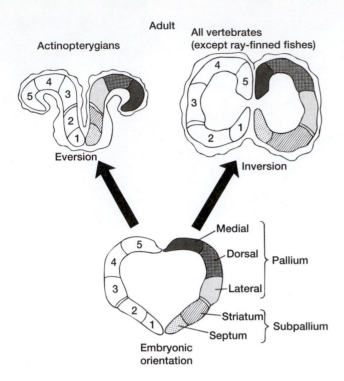

Adult

Actinopterygians

All vertebrates
(except ray-finned fishes)

Eversion

Inversion

Medial

Dorsal } Pallium

Lateral

Striatum

Septum } Subpallium

Embryonic
orientation

FIGURE 16.42 **Embryonic development of the telencephalon.** In actinopterygians, the telencephalon becomes everted during development, the pallium swings outward. In all other vertebrates, it becomes evaginated and inverted; walls of the hemispheres inflate outward (evagination) and the medial pallium and septum roll inward (inversion).

TABLE 16.7 Comparison of Recent and Former Terms Designating the Telencephalon

FORMER TERMS		Recent Terms
Morphological	Descriptive	
Roof of Telencephalon		
Pallium		PALLIUM
Archipallium	Hippocampus	**Medial** pallium
		Dorsal pallium
		Dorsomedial cortex (cingulate)
Neopallium	Cerebral cortex	
(Neocortex)		**Lateral** pallium
		Dorsal ventricular ridge
Paleopallium	Piriform cortex	**Lateral** cortex
Floor of Telencephalon		
Corpus striatum	Basal nuclei	SUBPALLIUM
Neostriatum	Caudate nucleus, Putamen	**Striatum**
Paleostriatum	Globus pallidus	**Pallidum**
Archistriatum	Amygdala (part)	
Septum	Septal area	**Septum**

only in eutherian mammals. It communicates between left and right cerebral hemispheres. In monotremes and marsupials, all commissural fibers between the halves of the cerebral hemispheres cross in the anterior commissure. In eutherian mammals, the anterior commissure contains fibers interconnecting the olfactory and piriform cortices. Other commissures connect paired regions and nuclei within the brain.

Early theories about the evolution of the cerebrum held that new regions progressively emerged out of preexisting regions. A recent "neostructure" was thought to arise from a former "archistructure" that evolved from an initial "paleostructure." The morphological terms coined attempted to express these presumed phylogenetic relationships. In addition, much early study of the brain centered on mammals, especially humans, in which descriptive terms were preferred. Instead of recognizing phylogenetic homologies, these terms expressed quaint or fanciful features. For example, *hippocampus* means horse tail, *amygdala* means almond, and *putamen* refers to a fruit pit. Some of these older terms, including *hippocampus* and *amygdala*, are still used today. However, within recent years, new experimental techniques have improved our understanding of comparative brain structure, leading to an ongoing reinterpretation of earlier ideas and introducing a new and broadly applicable terminology.

These differences in terminology are compared in table 16.7. Proliferation of terms has been especially noticeable in mammalian neuroanatomy and human medicine.

The revised view of cerebral evolution challenges not just the terminology but the assumptions on which the old terminology was based. The current view holds that the basic regions of the telencephalon did not emerge in a stepwise fashion. The pattern in which these regions are laid out is very ancient and was present in the common ancestor of all vertebrates. From this fundamental pattern, we see that the cerebrum has two regions: a dorsal **pallium** and a ventral **subpallium.** The pallium possesses **medial, dorsal,** and **lateral** divisions. The subpallium consists of a **striatum** and a **septum** (figure 16.42). All vertebrates have a cerebrum based on this basic plan. Major phylogenetic changes in the cerebrum center on loss, fusion, or enlargement of one or more of these regions.

Pallium The *medial pallium* receives a small primary olfactory input, but substantial auditory, lateral line, somatosensory, and visual inputs. The dorsal and lateral pallia receive ascending input, including visual information relayed from the thalamus. Agnathans possess a characteristic pallium (medial, dorsal, lateral) and subpallium (striatum, septum). However, in lampreys, the cerebral hemispheres incorporate only the lateral pallium and septum; the rest of the pallium (medial and dorsal) and subpallium (striatum) is located just posterior to this in the caudal telencephalon. The lamprey olfactory bulbs are large, about the same size as the cerebral hemispheres (figure 16.35a). Processing of olfactory information is an important role of the olfactory bulb, but it is unclear what additional sensory inputs reach the adjacent telencephalon from ascending tracts.

The elasmobranch pallium includes lateral, dorsal, and medial divisions, although these may in turn be subdivided. The lateral pallium receives the main olfactory input via the lateral olfactory tract. Parts of the dorsal pallium receive visual, lateral line, thalamic, and possibly auditory stimuli. Less is known about the medial pallia, but exchange of information between hemispheres is likely because they fuse across the midline. In ray-finned fishes, the characteristic pallial and subpallial regions can be recognized in a basal group, such as *Polypterus* (figure 16.43), even in the everted telencephalon (figure 16.42). However, in derived teleosts, the embryonic cells of the pallium disperse and intermingle, rather than differentiate into characteristic regions. Thus, the teleost pallium is generally homologous with that of *Polypterus* and other vertebrates, although many of its subdivisions may be unique.

The lungfish and amphibian pallia are similar to each other and to those of primitive sharks but less complex than those of reptiles. In both, the pallium consists of three regions—dorsal, lateral, and medial pallial divisions—which receive olfactory input as well as sensory input from the thalamus. In living amphibians, but not in lungfishes, the amygdala is another region of the pallium concerned with information from the vomeronasal organ.

The pallium of reptiles includes dorsal, lateral, and medial divisions as well as a hypertrophied region, the **dorsal ventricular ridge (DVR),** that dominates the central region of the cerebral hemisphere. Once thought to be part of the striatum, the DVR is now generally believed to be a derivative of the lateral pallium. In birds, the DVR expands further. It accounts for much of the relative increase in size of the cerebral hemispheres and crowds the lateral ventricle into a slit. The dorsal part of the DVR of birds hypertrophies into a region usually called the **Wulst** (figure 16.35g), containing highly organized visual information important in stereoscopic vision. The DVR receives visual, auditory, and somatosensory input from several major thalamic nuclei and projects this information to the striatum and to other parts of the pallium. Its size and central position in the flow of information suggest that the DVR may be a major higher association area in both reptiles and in birds. Both the lateral pallium (formerly termed the piriform lobe) and the medial pallium (the former hippocampus) persist as significant cortical areas in reptiles and birds, but the dorsal pallium is usually reduced in prominence, especially in birds.

Birds have a surprising number of sophisticated behaviors mediated through their enlarged forebrain. Some can memorize over 700 visual patterns (pigeons); make simple tools (crows); recall events that take place within a specific place and time (scrub-jays); have highly accurate sound localization for nocturnal hunting (owls); and exhibit vocal learning and even learn human words to communicate reciprocally with humans (parrots).

Mammals also show a dramatic increase in proportionate size of the cerebral hemispheres, but not because of an enlarged DVR as in reptiles and birds. Instead, the dorsal pallium is enlarged in mammals. In the course of this enlargement, the dorsal pallium thickens and differentiates into layers. The resulting mammalian cerebral cortex becomes proportionately an extensive area called the **cerebral cortex** or neocortex. In primates, approximately 70% of the neurons in the central nervous system are found in the cerebral cortex. It is devoted to deciphering auditory, visual, and somatosensory information as well as to controlling the function of the brain stem and spinal cord. All sensory areas are channeled or relayed to the cerebral cortex, bringing together sensory and recall information.

The mammalian medial pallium (hippocampus) receives sensory information and seems to initiate inquisitive or investigative behaviors. It is also concerned with memory of recent events. Olfactory information is shunted to the mammalian lateral pallium (piriform).

Subpallium As mentioned, the *subpallium* is divided into two regions: a medial septum and a more extensive lateroventral striatum. Both regions are distinct, even in the earliest fishes. In lampreys, the subpallium is split between the posterior telencephalon (striatum) and cerebral hemispheres (septum). In other fishes, including teleosts, the subpallium differentiates into homologous septal and striatal regions contained within the cerebral hemispheres. Lungfishes and all tetrapods retain this organization (figure 16.43).

The septum receives information from the medial pallium and is connected to the hypothalamus of the forebrain as well as to the tegmentum of the midbrain. It is an important part of the limbic system. The striatum has a more complicated phylogeny.

The **striatum,** together with a region termed the **pallidum,** are part of an assortment of nuclear groups at the base of the cerebral hemispheres known collectively as the **basal ganglia.** The basal ganglia have been best described in amniotes, especially mammals (table 16.7). Depending upon species, the striatum may form subdivisions, principally the **caudate nucleus** and the **putamen.** The pallidum may form several distinct subdivisions, principally the **globus pallidus.** Although homologies of some of these subdivisions have been difficult to confirm outside amniotes, basal ganglia were likely present in the brain of the ancestral jawed vertebrates. The basal ganglia receive sensory input to and from a nucleus called the **substantia nigra,** located in the midbrain tegmentum. The basal ganglia participate in the control of movement. They receive information on body position and motivational state, then integrate this into suitable motor activity or suppression of unwanted movement. Disruption of

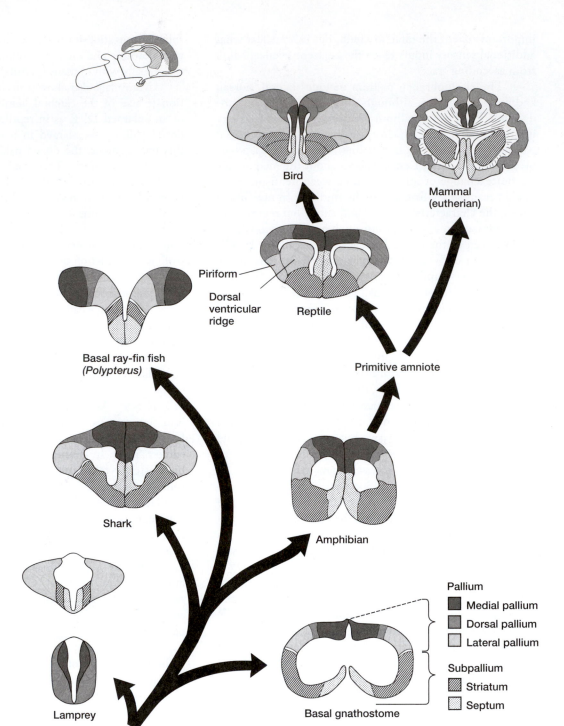

FIGURE 16.43
Evolution of vertebrate cerebral hemispheres. The cerebral hemispheres are shown in characteristic cross section. In the lamprey, the cerebral hemispheres incorporate (top) the components of the subpallium (septum, striatum), but only the lateral pallium; the rest of the pallium resides mostly in the unevaginated region of the telencephalon (bottom) posterior to the cerebral hemispheres. In basal gnathostomes, and thereafter, these five regions are within the cerebral hemispheres.

Bird

Mammal (eutherian)

Piriform

Dorsal ventricular ridge

Reptile

Primitive amniote

Basal ray-fin fish (*Polypterus*)

Shark

Amphibian

Lamprey

Basal gnathostome

Pallium
- ■ Medial pallium
- ■ Dorsal pallium
- □ Lateral pallium

Subpallium
- ▨ Striatum
- ▨ Septum

the basal ganglia leads to involuntary and purposeless motions known as dyskinesias. Parkinson's disease, characterized by an involuntary tremor that is often worse when the patient is at rest, is associated with degeneration of the basal nuclei. The **amygdala** is complexly derived and functionally integrated into major regions of the brain. One part arises ontogenetically from the pallium and receives vomeronasal input; the other part arises from other regions of the subpallium. The amygdala is linked with the limbic system.

In reptiles and birds, the striatum receives information from the DVR and transmits it first to the brain stem and then to the optic region of the tectum. Neurons within the avian striatum are often organized into layers or bands. Expansion of the DVR (reptiles and birds) and the dorsal pallium (mammals) is accompanied by a corresponding expansion of the striatum.

Vomeronasal (p. 677); olfaction (p. 675)

Phrenology to Neuroanatomy: Attempts to Decipher the Relationship between Brain Anatomy and Personality

Phrenology, popular in the nineteenth century, attempted to map moral qualities or personal appetites, which were thought to arise from specific areas of the brain. It was believed that by massaging the scalp, one could detect overdeveloped passions because the brain in that region would be enlarged and produce a swelling in the overlying skull (box figure 1a).

Careful stimulation of motor areas of the brain allows neurophysiologists to map the regions of the brain devoted to specific motor cortex functions (box figure 1b).

BOX FIGURE I **Function and phrenology.** (a) Map of the human skull used by nineteenth-century phrenologists. (b) In this example, areas of the motor cortex responsible for control of movements in different parts of the body are indicated.

After Netter.

Functional Associations of Parts of the Central Nervous System

Telencephalon The pallium receives direct sensory input, especially auditory, visual, and somatosensory information, from the thalamus, processes this information, and transmits responses to the striatum, hypothalamus, and brain stem. Thus, it indirectly controls locomotion. Major reorganization and expansion of the forebrain are correlated with changes in terrestrial locomotion and posture. In birds, upright posture and complex wing movements are served by the expanded DVR. In mammals, the dorsal pallium enlarges to assume an increasing role in the coordination of complex locomotion.

Especially important sensory signals may be duplicated several times within the telencephalon, giving multiple representations of the same information. For example, visual input, which is important in almost all vertebrates, has two parallel routes to the telencephalon. One is from the retina to the tectum and then to the telencephalon via a relay in the thalamus. The other is from the retina to the thalamus to the telencephalon. In some eutherian mammals in which vision is a major source of information, there may be a dozen areas in the telencephalon that decipher visual stimuli. Similarly, multiple visual areas are found in cats, squirrels, bats, and primates. Duplication of centers that process stimuli apparently improves comparison of sensory input within the nervous system and helps extract information it contains. The anatomical consequence is an increase in the size of the brain area to accommodate the reception and processing of multiple sets of similar sensory information.

(a)

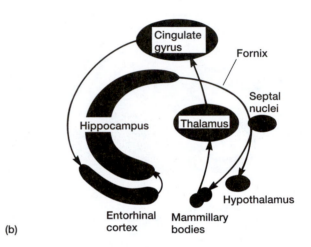

(b)

FIGURE 16.44 Limbic system. (a) Anatomical components of the limbic system. (b) Flow of information through the limbic system, the Papez circuit.

Limbic System

The limbic system was first described in the nineteenth century by Paul Broca, whose name is today mostly associated with speech centers of the brain (see box essay 16.3). The functional significance was unknown then, but Broca defined it anatomically. Based mostly upon human brains, Broca defined the limbic system as the cerebral cortex immediately encircling the corpus callosum and brain stem. This includes the deep parts of the cerebral cortex (cingulate gyrus) and medial surface of the temporal lobe of the cerebral cortex. It was not until the early twentieth century that James Papez realized the relationship between the limbic system and emotion. In particular, he and later scientists recognized a functional association of brain centers that include nuclei of the thalamus, hypothalamus, amygdala, hippocampus (medial pallium), **cingulate gyrus,** and septum. The **fornix** is a two-way fiber system that connects all nuclei of the limbic system (figure 16.44a,b and table 16.8). This circuit, the Papez circuit, receives stimuli from the cerebral cortex and returns responses to the cerebral cortex and to the autonomic nervous system. The hypothalamus contains nuclei that affect heart rate, respiration, and general visceral activity through the autonomic nervous system. Changes in these usually accompany strong emotion. The amygdala is active in the production of aggressive behavior and fear. The hippocampus (medial pallium) lies adjacent to the amygdala. Damage to it causes loss of recent memory. The cingulate gyrus and septum are other routes of input to this system. Damage to the cingulate gyrus results in disruption of the order of complex behaviors such as parental care. A female rat with such damage still attends to her pups but nurses, licks, repairs the nest, and so forth in no logical sequence and frequently switches randomly from one behavior to another.

The limbic system is involved in two functions. First, as mentioned, it regulates the expression of emotions. Experimental or accidental removal of parts of the

TABLE 16.8 Centers of the Brain Allied as the Limbic System	
Cortical Centers	**Subcortical Centers**
Telencephalon	Telencephalon
Pallium	Subpallium
Hippocampus	*Septum*
Dentate gyrus	*Amygdala (part)*
Parahippocampus	
	Diencephalon
Cingulate gyrus	*Habenular nucleus*
Amygdala (part)	*Thalamus*
Piriform cortex	*Hypothalamus*
Entorhinal cortex	*Mammillary bodies*

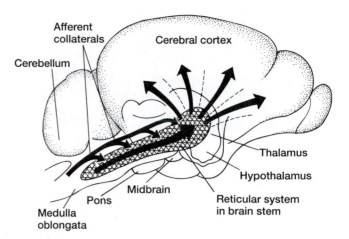

FIGURE 16.45 Reticular formation. The reticular formation lies in the medulla and midbrain and projects to the higher centers of the brain. When active, it seems to bring about general alertness. Sensory afferent pathways traveling to higher centers send branches, afferent collaterals, into the reticular system. Through the thalamus, the reticular system is then projected to the cerebral hemispheres, resulting in general arousal.

After T. E. Stize et al., 1951.

limbic system leads to emotional passiveness. This function is important to survival. To sustain itself, an animal must actively seek food, be alert to danger, and respond appropriately when threatened. Phylogenetically, the limbic system, or at least many of its centers, arises early in evolution even before there are many direct connections between the thalamus and the cerebral cortex. The limbic system has been called the "visceral brain" because of its substantial influence on visceral functions through the autonomic nervous system.

The second function of the limbic system involves spatial and short-term memory. Severe damage to the hippocampus (medial pallium) does not destroy the memory of events prior to the injury, but subsequent events are recalled only with great difficulty or not at all. Memory is probably resident in the cerebral cortex rather than in the limbic system, but the limbic system is involved in temporarily retaining the memory of a recent experience until the experience becomes established as long-term memory in the cerebral cortex. Destruction of the hippocampus, which is very large in food-hiding birds and mammals, severely impairs their ability to locate the hidden food items. In humans, persons with hippocampal damage retain good long-term memory (memory prior to damage), but poor short-term memory. If interrupted while speaking, they must be prompted about what they were saying. Persons they have met just a few minutes before must be reintroduced when returning to the room. One patient, who lived in his house for over 20 years, could not draw a diagram of the house because of impaired spatial and short-term memory loss due to hippocampal damage two decades earlier.

Reticular Formation The reticular formation resides in the medulla and midbrain (figure 16.45). This structure is defined in several ways, but it generally consists of enmeshed neurons and their fibers. The term *reticula* and the term *formation* referred to the microscopic appearance of this region using early methods. It seemed to lack delineated tracts or nuclei like a "center" or a "system." This diffuse arrangement of fibers resembled some parts of the nervous systems of anamniotes and thus inspired the idea that the

reticular formation was a phylogenetic retention of an earlier feature. More recent methods for tracing pathways and identifying groups of neurons have turned up almost 30 nuclei, all interconnected and in turn radiating to other areas.

The reticular formation has several functions. First, it is arousal in action via its awakening or stimulation of the cerebral cortex. An alert animal is more attentive to sensory input. Some anesthetics and tranquilizers act by suppressing transmissions through the reticular formation. Damage to the reticular formation can lead to a prolonged coma. Second, the reticular formation also acts as a filter. It selects information to be relayed to higher centers or down the spinal cord. It tends to pass along information that is novel or persistent.

Finally, neurons in the reticular formation act like association neurons in the spinal cord, except here the neurons connect sensory nuclei in the brain with motor neurons in the brain stem and spinal cord. Many complex innate functions such as feeding, vocalization, posture, respiration, and locomotion involve muscles that are innervated by several different cranial centers. The neurons of the reticular formation interconnect these centers and coordinate their separate control into cooperative motor output.

Spinocortical Associations So far we have looked at regions of the central nervous system that perform local functions: reflexes of the spinal cord, association centers of the brain, and systems of alliance. But the central nervous system shows a high degree of integration. Even reflexes completed at the level of the spinal cord are registered at higher centers, and events in higher centers influence lower levels in the

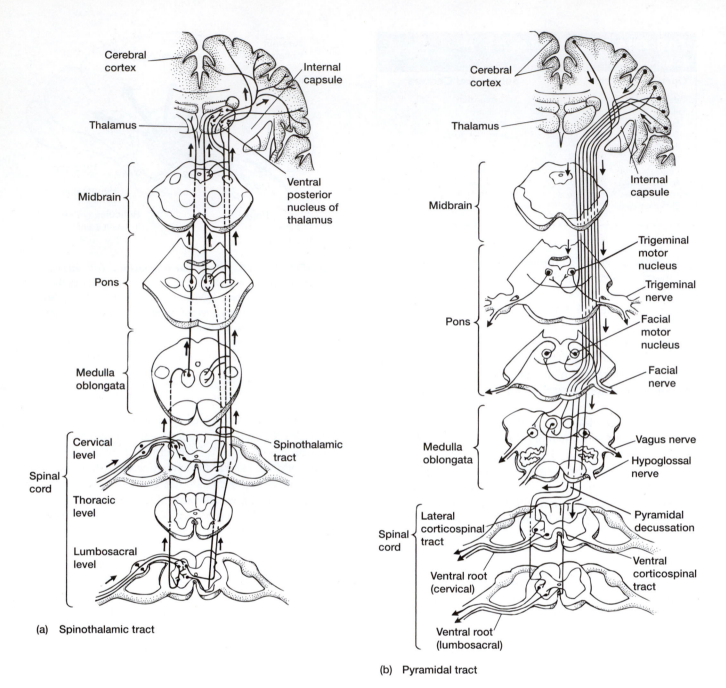

FIGURE 16.46 **Processing of sensory and motor information.** (a) The spinothalamic tract gathers sensory neurons that carry sensations of pain. The tract then courses to the thalamus, where sensations are relayed to higher brain centers. (b) Pyramidal tract. Decisions initiated in the cerebral cortex are conveyed along descending motor neurons, which form the pyramidal tract, to the appropriate level of the spinal cord. From the spinal cord, the response is passed along a motor neuron to the effector.

After Barr and Kiemm.

spinal cord. This flow of information tends to occur along distinctive tracts.

Sensory impulses travel up the spinal cord, but before they reach conscious centers in the cerebral cortex, they synapse in the gray matter, in the thalamus, and even in additional nuclei. Thus, information that reaches conscious centers has already been sifted and filtered (figure 16.46a). As discussed previously, all sensory fibers synapse in the thalamus on their way to the cerebral cortex except for olfactory tracts. In the thalamus, sensory impulses are coordinated into an integrated pattern of sensations that is then projected to specialized sensory areas in the cortex. In other words, the cortex receives information that has already been interpreted by subcortical centers. Information traveling down the spinal cord, even if it originated in the cerebral cortex, is modified by the cerebellum, by subcortical centers, and by reflexes at the level of the cord (figure 16.46b).

Overview

The study of nervous system function often includes as much philosophy as it does science. From its basic units, the nervous system builds into complex regions of bewildering and intricate information processing. The basic unit is the neuron, insulated and nourished by neuroglia. Chained together at synapses, neurons form circuits connecting one part of the organism to another—sensory input to motor output via effectors. Somatic and visceral organs pass sensory information to the CNS, and after processing, nerves here carry information, in the form of electrical signals, out to effectors. The peripheral nervous system consists of cranial and spinal nerves, carrying sensory (afferent) or motor (efferent) or both types of fibers. Roots of cranial nerves are usually enclosed in the braincase and supply a variety of organs. Simple reflex response to sensory input occurs at the level of the spinal cord, where arriving sensory information is distributed by association neurons to appropriate motor output. The autonomic nervous system exerts contrasting control over visceral activity to prepare an active state (sympathetic) or restore the organism to a vegetative state (parasympathetic).

The central nervous system includes the spinal cord and brain. The spinal cord is a corridor carrying similar sensory information up to higher levels and returning motor information down to motor neurons. Within the brain, nuclei, clusters of nerve cell bodies, receive ascending sensory information, pertinent to the function in which it is involved. Through association neurons (interneurons), nuclei pass this information to other nuclei involved in processing of similar information, until a response via motor output occurs. Thus, a similar, but more complex, passing of information from sensory input to motor output communicated via interconnecting neurons is established in the brain. Regions of the brain hold collections of nuclei. As functions enlarge phylogenetically, meeting adaptive demands in different environments, the nuclei responsible for processing associated information increase, and the respective brain region exhibits enlargement. The hindbrain includes the medulla oblongata, which houses the primary nuclei of cranial nerves, is a major corridor for ascending and descending tracts, and operates at a suprareflex level presiding over visceral activity. The cerebellum monitors and modifies rather than initiates motor output, thereby smoothing action of effectors. The midbrain roof includes the tectum, which receives sensory input, primarily visual, auditory, and lateral line. Its floor is the tegmentum, a site initiating motor output. The forebrain includes a variety of important regions, the thalamus being one of the most important. Except for olfactory tracts, all other ascending somatic and visceral sensory neurons synapse

first in the thalamus, which organizes this input into a pattern of sensations relayed to the cerebral cortex. The cerebrum is inflated into cerebral hemispheres made up of an outer wall, the cerebral cortex, and subcortical regions beneath. The cerebrum is based in all vertebrates upon a common plan that includes a dorsal region, the pallium, and a ventral region, the subpallium. These highly interconnected regions preside over integration of information arriving from lower centers. These regions are in turn organized into systems responsible for coordination of somatic and visceral activity, and even emotional states. In amniotes, the cerebrum proportionately enlarges phylogenetically but due to expansion of different regions of the cerebrum in mammals (dorsal pallium) and reptiles/birds (lateral pallium and its dorsal ventricular ridge).

One way to think about the function of the brain is to consider the vertebrate body as consisting of receptor sheets—two-dimensional layers holding sensory receptors. For example, the integumental surface of skin receives impinging points of environmental stimuli or the retinal layer of the eye responds to arriving light stimuli. Neurons from these and other receptor sheets project to corresponding regions of the brain, where this environmental information is mapped, landscaped into an organized geographic chart of the corresponding receptor sheet (e.g., box essay 16.4, figure 1b). In turn, brain maps connect with each other via extensive fiber highways. For example, the corpus callosum connects parts of your right and left brain across the midline, and contains about 200 million fibers. As modern neurobiology has discovered but not yet clarified, these parallel and reciprocal interconnections between mapped brain regions may be the basis for brain function above simple reflex responses leading to higher-order mental processes.

These maps are set up in two steps. The first is during embryonic development wherein genes direct formation of a general map of receptor sheets to brain regions. The second step occurs as the animal activates these pathways in its young life. As a consequence, the synapses between nerve cells change and shift to accommodate the neural pathways used by the active animal, with the result that these maps become modified and more refined. During this step, these connections are not precisely prespecified in the genes but arise as a consequence of the organism's activity and exposure to environmental information. This means that brain maps are not fixed, but instead their borders fluctuate over time, making each slightly different from individual to individual.

Short-term memories that initially form in the hippocampus are eventually transferred to long-term storage in other parts of the brain, residing in both areas for some time. Some of this new memory is thought to be established through addition or modification of new connecting synapses. Neurogenesis, the birth of new neurons, arising in the adult brain is also thought to participate. Within a couple of weeks, newborn neurons perish if not challenged to learn something new. If involved in learning something new with great effort, the new neurons persist. But these new neurons also weaken or boot old memories, a function thought to clear the hippocampus of old memories, making room for new memories.

The subtleties of the central nervous system are profound. We have seen that the nervous system gathers information about the body's internal status and the outside world and the results of previous experience and turns these into responses that might allow the organism to maintain itself in its environment. But there is more to this process than the mechanical processing of information. Emotions, goals, and conscious participation shape a response, at least in humans. To a large extent, our responses are shaped by our perceptions of physical stimuli, which are mediated by sensory receptors. In chapter 17, we take a closer look at these sensory receptors.

WEB LINK

Visit the text website at *www.mhhe.com/Kardong7e* for additional study aids including selected references, functional laboraotories, and web links to chapter topics.

Sensory Organs

INTRODUCTION

COMPONENTS OF A SENSORY ORGAN

GENERAL SENSORY ORGANS
    Free Sensory Receptors
    Encapsulated Sensory Receptors
    Associated Sensory Receptors
       Proprioception
    Mechanisms of Perceiving Stimuli from General
       Sensory Receptors

SPECIAL SENSORY ORGANS
    Chemoreceptors
       Nasal Passages
       Vomeronasal Area
       Mouth

Radiation Receptors
    Photoreceptors
    Infrared Receptors
Mechanoreceptors
    Lateral Line System
    Vestibular Apparatus
    Auditory System
    Functions of the Ear
Electroreceptors
    Structure and Phylogeny
    Form and Function
Additional Special Sensory Organs

OVERVIEW

Introduction

To survive, an organism must react to danger and take advantage of opportunity. Appropriate responses require information about the external environment, the body's internal physiology, and previous experience. The results of previous experience are recorded within the nervous system as memory, but **sensory receptors** monitor the external and internal environments (figure 17.1). Sensory receptors are specialized organs that respond to selected information. Sensory receptors code or translate environmental energies into nerve impulses that are transmitted to the central nervous system (CNS) via afferent fibers. These impulses may or may not be received at conscious levels of the brain.

Sensations that we as humans become conscious of are referred to as **perception.** Our view of the world is partially determined by the kinds of information our sensory receptors detect and by how that information is processed. Vertebrates differ in their ability to perceive stimuli. Bats and even dogs hear sounds at frequencies beyond our ears. Hawks can hunt high above the ground and detect tiny rodents scampering below. Rattlesnakes hunt in light too dim for humans

to see in. Yet humans see the world in color, while most other mammals see it mostly in black and white. Our perception of the world is limited or enhanced by the availability and sensitivity of our sensory receptors.

It should also be emphasized that conscious sensations are an organism's subjective interpretation of the environment. The environment contains chemicals and photons of light, but the senses of taste and color are interpretations of these phenomena. Similarly, pain does not exist in the environment. You cannot measure pain as you can measure temperature or force. Taste, color, and pain are perceptions arising out of events in the brain itself.

Nerve impulses carried by sensory nerves are electrical impulses. The optic nerve carries the same kind of electrical impulses as the auditory nerves, olfactory nerves, taste nerves, and so on. Different sensations result from different ways in which the nervous system interprets signals from different sensory receptors. Because the impulses are the same, the central nervous system can be fooled. Artificial stimulation of the auditory nerve is perceived as a sound. Artificial

FIGURE 17.1 **Sensory receptor.** A sensory receptor is usually composed of the dendrites of a neuron and can include tissue that amplifies the stimulus. The receptor is a transducer that transforms a stimulus into an electrical impulse or burst of electrical impulses that spread to the cell body and along its axon to other neurons, usually in the central nervous system.

stimulation of the optic nerve is perceived as light. Mechanical pressure on the eyeball stimulates the optic nerve to send electrical impulses to the brain. These impulses are interpreted as what they are expected to be, sensations of light, rather than as mechanical stimuli. This is why a mechanical blow to the eye can make one "see stars."

To be clear in discussing perception, we should distinguish the environmental stimulus from the way in which it is interpreted, but this is seldom done. In common usage and even in scientific research, convention rules. We speak of the senses of seeing, hearing, tasting, smelling, and so on as if the stimulus and the perception of the stimulus were the same. For example, the chemicals in our food carry no intrinsic taste. They are just chemicals. But we may sense these chemicals as sweet or sour or bitter. That perception of their taste is a product of our nervous system, a result of brain events. The treatment of sensory receptors in this chapter attempts to separate the kind of environmental energy or stimulus monitored from the nervous system's interpretation of that stimulus.

Components of a Sensory Organ

Sensory neurons are nerve cells specialized for detecting and transmitting information about the external or internal environment. Each sensory neuron sends out slender processes, or **nerve fibers.** Sensory receptors usually contain **dendrites,** processes that are responsive to stimuli and carry impulses toward the body of the nerve cell. A sensory neuron usually has an **axon** as well, a nerve process that transmits impulses away from the cell body to other neurons.

Neurons (p. 626)

The sensory receptor acts as a **transducer,** a device that translates energy from one form into another. The microphone of a public address system, which translates sound waves into electrical energy, is another example of a transducer. Most sensory receptors translate light, mechanical, or chemical stimuli into electrical impulses. Often the tip of the sensory nerve fiber is associated with accessory tissues that amplify the stimulus and thereby increase the

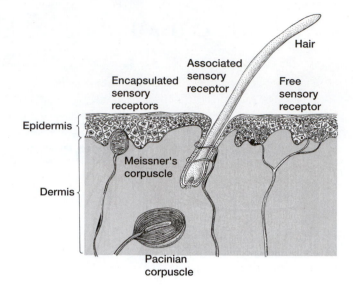

FIGURE 17.2 **General sensory receptors.** A free nerve ending, encapsulated sensory receptors (Meissner's and Pacinian corpuscles), and an associated sensory receptor of a hair follicle.

sensitivity of the receptor. A sensory nerve fiber with its associated tissues is termed a **sensory organ.**

Sensory organs can be classified according to several criteria. **Somatic sensory organs** refer to those of the skin, body surfaces, and skeletal muscles. **Visceral sensory organs** reside in the viscera. **Exteroceptors** receive sensations from the environment, and **interoceptors** respond to sensations from organs. The **proprioceptor,** a sensory organ located in striated muscles, joints, and tendons, is a special type of interoceptor. A third way to classify sensory organs, and the one used in this chapter, is on the basis of how extensively they are distributed. **General sensory organs** are widely distributed throughout the body and are concerned with sensations of touch, temperature, and proprioception. **Special sensory organs** are localized and often specialized.

General Sensory Organs

General sensory receptors may be placed in one of three anatomical categories: free, encapsulated, or with associated nerve endings. The structure of the nerve ending is designed to increase the effect of the stimulus (figure 17.2).

Free Sensory Receptors

When the terminus of a sensory process lacks any specialized association, it is termed a **free nerve ending,** or **free sensory receptor.** At its terminus, the free nerve ending may **arborize,** or branch, extensively to increase the area monitored. Free sensory receptors are primarily concerned with sensations interpreted as painful, but they can also be stimulated by extremes of heat or cold. Tissue damage can lead to swelling and direct stimulation. A toothache is an example. Tactile sensations—that is, sensations of pressure or touch—are often detected by free nerve endings as well. Free sensory receptors are abundant in areas in which sensitivity is highly developed, such as the skin, cornea, oral cavity, tooth pulp, and intestines. For example, the burn of hot chili peppers results from capsaicin molecules in the peppers, which target receptor channels on the sensory cells. When activated by the binding of capsaicin, ions flood the receptor cell, the cell fires, and its electrical impulse is interpreted by the brain as a burning sensation. Although used to spice up dull human foods, the capsaicin probably evolved in peppers to thwart herbivores with less of a discerning taste.

Encapsulated Sensory Receptors

When the terminus of a sensory process is enclosed in a specialized structure, it is called an **encapsulated nerve ending,** or **encapsulated sensory receptor.** For example, **Meissner's corpuscle** is a sensory ending wrapped in mesodermal cells located in the dermis of the skin just beneath the epidermis (figure 17.2). It responds to touch. The **corpuscle of Ruffini,** responsive to warmth, and the **end-bulb of Krause,** responsive to cold, are other encapsulated receptors located in the dermis. **Pacinian corpuscles** (corpuscles of Vater-Pacini) are located in the skin, joints, and deep tissues of the body. For instance, it is not uncommon to find them associated with the pancreas. They respond to pressure. In encapsulated receptors, the capsules enhance the deformation of the nerve endings, thus assisting in the initiation of the nerve impulse. In Pacinian corpuscles, for example, the nerve ending is enclosed in a series of concentric layers that form an "onion skin" capsule. This capsule acts as a tiny transducer that converts pressure into electrical depolarization of the nerve ending.

Associated Sensory Receptors

When the terminus of a sensory process is wrapped around another organ, it is called an **associated nerve ending,** or an **associated sensory receptor.** For example, nerve endings are associated with the base of a hair follicle (figure 17.2). When a hair is moved, the entwined nerve endings at the base of the hair are stimulated.

Proprioception

Proprioception is based largely on information gathered by associated sensory receptors located in muscles, tendons, and joints. These receptors monitor the state of limb flexion and the degree of muscle contraction. As a result, the central nervous system is kept informed about limb or body position. If a body part is moved, the muscles involved and the amount of contraction supplied will differ depending on the initial position of the part. Proprioceptive information is indispensable for determining the location of a part before and during its movement. If you are a sighted person, you probably do not take advantage of proprioceptive information relayed to your conscious centers. However, if you are blindfolded and a partner gently swings your extended arm to a new location, you will be aware of the new position to which your arm has been rotated. Nevertheless, most proprioceptive information is processed at subconscious levels of the nervous system to make automatic adjustments of posture or to synchronize body and limb movements.

Some proprioceptive fibers come from encapsulated Pacinian corpuscles located in joint capsules, but most come from two types of associated receptors: muscle spindles and Golgi tendon organs.

Muscle Spindles   Within skeletal muscles, the muscle fiber that produces the major force moving a part is the **extrafusal muscle cell** (extrafusal muscle fiber). Such fibers are innervated by **alpha motor neurons,** whose bodies are located in the gray matter of the spinal cord. Interspersed among the extrafusal muscle fibers are fusiform-shaped packages of **muscle spindles** that contain modified, striated **intrafusal muscle cells** (intrafusal muscle fibers). Unlike extrafusal fibers that work the lever system, intrafusal fibers are specialized sensory organs.

There are two types of intrafusal muscle fibers. The **nuclear bag intrafusal fiber** has nuclei clustered in a swollen region near the middle of the fiber and is associated with a **primary afferent sensory nerve** (annulospiral nerve). The **nuclear chain intrafusal fiber** has nuclei strung out along the fiber instead of clustered. It is associated with a **secondary afferent sensory nerve** (flower spray nerve). Both types of intrafusal fibers are innervated by **gamma motor neurons** (figure 17.3).

Muscle spindles function to maintain muscle tone. Normal muscle maintains a small amount of tension even when it is relaxed, a state in which the muscle has **tonus.** When a muscle relaxes more than normal, the muscle spindle sags. Primary and secondary afferent nerves wound around the intrafusal fibers sense this sag. Through reflex connections in the spinal cord, these afferent neurons synapse with alpha motor neurons to stimulate contraction of extrafusal fibers that stimulate muscle tension and restore muscle tone.

The stretching of a muscle lends to its reflex contraction. When postural muscles are stretched or load is added, muscle spindles lengthen, initiating the **stretch reflex** (figure 17.4a). Because gamma motor neurons cause the intrafusal fibers to contract, these neurons are thought to increase or decrease the sensitivity of this reflex (figure 17.4b,c).

The sensory function of the intrafusal muscle fibers is to inform the nervous system about the rate of change in the length of the extrafusal muscle fibers with which they are associated. This information can initiate a stretch reflex to adjust tonus. It is also relayed to the cerebellum, which modulates muscle activity.

Postural reflex (p. 653)

Golgi Tendon Organs   Golgi tendon organs are sensory receptors in the tendons that attach muscles to bone. Thus, they lie along the line of muscle action and function as tension recorders, supplying the central nervous system with information about the forces generated by muscles (figure 17.3).

Muscle spindle

Extrafusal
muscle cell

Nuclear chain
intrafusal
muscle cell

Nuclear bag
intrafusal
muscle cell

Intrafusal
muscle
cells

Sensory nerves
→ Primary afferent
→ Secondary afferent
Motor nerves
← Gamma efferent

Muscle spindle
Golgi
tendon
organ
Capsule
organ
Joint
capsule

FIGURE 17.3 **Muscle spindle.** Extrafusal muscle cells produce the contractile force of a muscle. Intrafusal muscle cells, which consist of nuclear chain and nuclear bag intrafusal fibers, are modified muscle cells. Intrafusal muscle cells are innervated by gamma efferent motor neurons and by primary and secondary afferent sensory neurons. The Golgi tendon organ is a sensory receptor associated with the tendon.

Mechanisms of Perceiving Stimuli from General Sensory Receptors

Two theories attempt to explain the relationship between the stimuli that general sensory receptors receive and the perception that the central nervous system produces from them. The **theory of specific nerve energies** proposes that the nerve endings of each sensory receptor are associated exclusively with a specific sense. For instance, stimulation of a Meissner's corpuscle sets up a volley of impulses to the nervous system that, because they come from this type of receptor, are interpreted as tactile stimuli. Impulses from the end-bulb of Krause are interpreted as cold, those from the corpuscle of Ruffini, as hot, and so on.

The alternative concept is the **pattern theory of sensation,** in which small complexes of nerve endings are associated with a particular location. Stimulation of the receptors within a specific location makes it possible for different combinations or patterns of sensations to be sent simultaneously to the nervous system, allowing for qualitative differences in interpretation. For example, we recognize that sensations of pain vary in quality and intensity. Some pains are "sharp," whereas others are "dull," or we may sense a "burning" pain. Both theories help to explain how the central nervous system interprets general sensory stimuli.

Special Sensory Organs

Special sensory organs are usually localized in their distribution, and their responses are restricted to specific stimuli. There are chemical, electromagnetic, mechanical, and electrical stimuli to which sensory organs respond.

Chemoreceptors

Sensory receptors sensitive to chemical stimuli are **chemoreceptors** and the chemicals to which they respond are odors. When a chemical contacts an appropriate receptor, it initiates an electrical impulse in the sensory neuron. Taste and smell are the most familiar chemoreceptive senses in humans, but this distinction is misleading. The sense of taste recognizes only five basic qualities: salty, sweet, sour, bitter, and savory. What we interpret as the rich "taste" of food is primarily due to the mechanical texture of the food and to the aroma that stimulates our sense of smell. If we are stricken with the flu, which clogs our noses and denies us these stimuli, food loses much of its "taste."

In aquatic vertebrates, the distinction between the sense of taste and smell is even less useful. Some fishes, for example, have chemoreceptors distributed across the outer surface of their bodies (figure 17.5). Should we say these are used to "smell" or to "taste" the water? Instead of making

FIGURE 17.4 **Stretch reflex.** (a) When posture changes or a load is placed on an animal's body, muscle spindles are stretched, stimulating associated nerves to generate continuous impulses that travel to the spinal cord (solid line). (b) These impulses travel via afferent fibers to the spinal cord and synapse with motor neurons that conduct these impulses to appropriate extrafusal muscles (dashed line), causing them to contract. (c) Muscle contraction tends to straighten the limb (small open arrow) and relieve the stretch on the muscle spindle. When the muscle spindle relaxes, the associated sensory nerves cease firing, and appropriate posture returns.

FIGURE 17.5 **Distribution of chemoreceptors in the catfish.** In many fishes, taste buds occur across the surface of the body and on the fins as well. Each dot represents approximately 100 taste buds.

Modified from Atema, 1971.

an arbitrary distinction between taste and smell, we classify chemoreceptors by their locations.

Chemical messages, termed **pheromones,** are a specific category of odors released into the environment by one individual and influence the behavior or physiology of another individual of its species. Produced in exocrine glands, many pheromones affect sexual activity or territoriality. Pheromones represent one avenue of communication between individuals and are often detected by specialized receptor organs.

For example, nepetalactone is a compound in the leaves and stems of the catnip plant that acts as an artificial cat sexual pheromone. When nepetalactone enters the cat's nasal passages, it binds to protein receptors on sensory nerves and activates nerve impulses passing to the central nervous system. Here, centers (amygdala) project to areas governing expression of sexual behaviors, including body or head rubbing on the catnip, rolling, and vocalizing often accompanied by salivation. Not all cats exhibit this behavior, but it has been seen in domestic cats to the largest lions.

Nasal Passages

The sense of smell, or *olfaction*, involves chemoreceptors usually located in the nasal passages. Anatomically, there are three components of the olfactory circuitry: the olfactory epithelium, the olfactory bulb, and the olfactory tract (figure 17.6).

The **olfactory epithelium** is a specialized patch of epithelium within the nasal cavity that collects pertinent chemicals in the airstream. It contains **basal cells,** which are probably replacement cells, and **sustentacular cells,** which secrete mucus and support the **olfactory sensory cells** (figure 17.7a). The actual chemoreceptor cells in the epithelium are olfactory sensory cells. Each olfactory cell sprouts a tuft of sensory cilia at its apical end. At its basal end, it sends an axon through the **cribriform plate** into the **olfactory bulb** (figure 17.7b). The term **olfactory nerve** is properly applied only to these short axons from the olfactory sensory cells.

Within the enlarged olfactory bulb reside several cell types, the most important of which is the **mitral cell.** Axons of the olfactory sensory cells synapse with mitral cells, which in turn send their long axons, collectively termed the **olfactory tract,** to the rest of the brain (figure 17.7b). Axons in the olfactory tract synapse primarily within the **piriform lobe** and **septum** in the cerebrum before being relayed to other regions of the brain (figure 17.7a). This gives the striatum and limbic systems direct olfactory input.

Olfactory centers (p. 658); limbic system (p. 666)

Embryology The olfactory system begins embryologically as a pair of **olfactory placodes,** thickenings of ectoderm that invaginate dorsally toward the overlying neural tube. The lateral walls of each placode form the respiratory epithelium lining the nasal passages. The central region of the placode forms the olfactory epithelium. Olfactory sensory cells differentiate within this epithelium and sprout axons that grow from the epithelium and through the mesenchyme to reach the forming telencephalon (figure 17.8a). These olfactory fibers induce the telencephalon to produce a swollen outgrowth, the olfactory bulb, that is connected to the rest of the telencephalon by the olfactory tract (figure 17.8b,c).

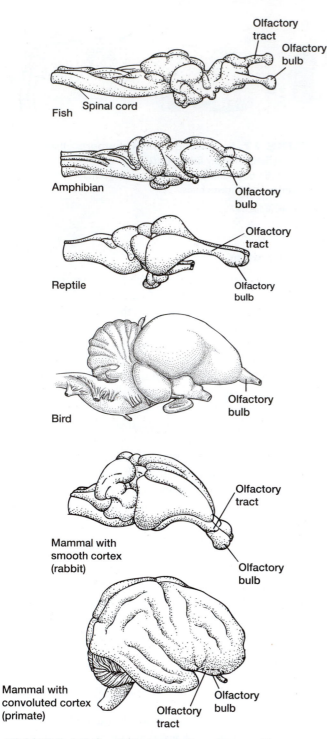

Fish
Spinal cord
Olfactory tract
Olfactory bulb

Amphibian
Olfactory bulb

Reptile
Olfactory tract
Olfactory bulb

Bird
Olfactory bulb

Mammal with smooth cortex (rabbit)
Olfactory tract
Olfactory bulb

Mammal with convoluted cortex (primate)
Olfactory tract
Olfactory bulb

FIGURE 17.6 Olfactory bulb and tract. The olfactory tract is an extension of the brain rather than a nerve. The tip of this tract is usually expanded into the olfactory bulb, which receives the short olfactory nerve (not shown) arriving from the olfactory epithelium.

After Tuchmann-Duplessis et al.

Although neural crest cells may migrate to the vicinity of the differentiating olfactory system, they apparently do not directly form the olfactory sensory cells.

Phylogeny In tetrapods, the nose is associated with breathing, but primitively it arose as an olfactory area. In fact, the projection of olfactory receptors to the medial pallium, one of the earliest regions of the cerebrum, suggests that the olfactory system is quite ancient.

In most fishes, olfactory sensory receptors are recessed in paired, blind-ended pits known as the **nasal sacs** (figures 17.9a and 17.10a). In living cyclostomes and many ostracoderms, these pits were secondarily fused, although the olfactory tracts remained paired. Water carrying chemicals flows in and out of these sacs as the fish swims. However, among some groups of fishes, a one-way flow of water through the nasal sacs is possible. A partial septum divides the sac into incurrent and excurrent apertures. In some lines, these openings become separated (figure 17.9b–d). The excurrent opening may become displaced to the margin of the mouth (e.g., in some sarcopterygians) or even open directly into the buccal cavity (e.g., in holocephalians). The excurrent opening into the mouth has happened several times independently in fish evolution and led to a reevaluation of homologies that is still under way.

Nasal sac, nasolacrimal duct (p. 263)

In tetrapods and their immediate fish ancestors, called choanate fishes, a small **external naris** provides access to each nasal passage. The back of the nasal passage opens into the mouth through the **internal naris,** or **choana** (figure 17.10b–e). In amphibians, the nasal sac enlarges between nares, from which a short recess, the **vomeronasal (Jacobson's) organ,** projects (figure 17.10c).

In plethodontid salamanders, a pair of sunken grooves connects the front of the mouth with each external naris. These **nasolabial grooves** transport aqueous material preferentially to the vomeronasal organ in the nasal cavities (figure 17.11). The grooves are unciliated, and fluid moves through them by capillary action. Chemicals carried in the fluid might be pheromones or come from food. By touching their nose to the substrate, a behavior termed "nose-tapping," salamanders gather and pass chemical cues important in identifying their home territory along the nasolabial grooves.

The limbless amphibians, caecilians, have small paired protrusible tentacles anterior to the eyes. These tentacles are chemosensory, apparently aiding in the transfer of chemicals to the vomeronasal organ in the roof of the mouth.

In reptiles, the nasal sac becomes differentiated into two regions: the anterior **vestibule** that first receives air entering via the external naris, and the posterior **nasal chamber** into which air next flows. In some reptiles, a lateral wall projects into the chamber to form **conchae,** or **turbinals,** which are folds that increase the surface area of the respiratory epithelium. Air departs the nasal chamber through the narrow **nasopharyngeal duct,** which leads to the internal naris (figure 17.10d). The nasal passages of birds are similar, and their conchae may develop into complicated scrolls.

(a)

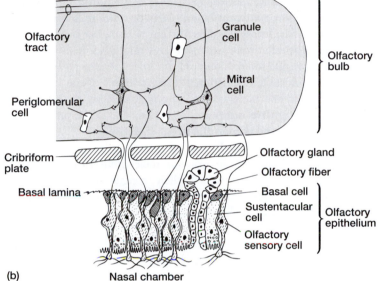

(b)

FIGURE 17.7 Olfactory epithelium.

(a) Nasal passages in mammals are lined with respiratory epithelium. The olfactory epithelium is a small region of this lining that contains specialized neuronal fibers that make contact with neurons of the olfactory tract. These processes relay impulses to the piriform lobe and septal area of the brain. (b) Histology of olfactory epithelium. The olfactory epithelium includes supportive sustentacular cells, basal cells, and olfactory sensory cells. The apical surface of each olfactory cell develops cilia that project into the air passage. Its basal end consists of a nerve fiber that travels through the cribriform plate into the olfactory bulb, where it synapses with periglomerular, mitral, and granule cells. Fibers of the mitral cells constitute the olfactory tract, which goes to the brain.

In mammals, the nasal chamber is large and usually includes extensive turbinals to ensure that the entering air will be warmed and moistened before it flows to the lungs. The olfactory epithelium occupies the posterior wall of the nasal chamber. The remaining lining consists of respiratory epithelium (figure 17.10e).

Form and Function The sense of smell is usually well developed among fishes, but it is one of the secondary senses in birds, bats, and higher primates, including ourselves. Olfactory information is important for fishes who hunt or follow a chemical gradient. In aquatic vertebrates, water-borne chemicals circulate across the olfactory epithelium lining the nasal sacs. Development of one-way flow improved olfaction, ensuring a fairly continuous flow of new water to wash away chemicals that have already been detected and to deliver new ones. Respiratory and olfactory currents are coupled in fishes in whom the excurrent opening enters their mouth. The motions that irrigate the gills also draw water across the olfactory epithelium. By coupling respiratory and olfactory functions, the nasal passages were preadapted for their later role in air breathing in tetrapods.

In tetrapods, air replaces water in chemical transport, although before reaching the sensory receptors, airborne particles are still absorbed into a mucous film covering the olfactory epithelium. Air entering the nostrils must flow past the olfactory epithelium on its way to the lungs; therefore, the olfactory epithelium can sample chemicals in the airflow. A terrestrial vertebrate may begin to sniff the air when it detects chemicals of special interest. Sniffing is independent of respiration and draws in quick pulses of air to replenish air in the nasal chamber. This increases the turnover of air in the nasal chamber and permits more frequent sampling of environmental odors (figure 17.12).

Vomeronasal Area

Named for the bones that usually house this chemosensory organ, the vomeronasal organ is known only in some tetrapods. It is absent in most turtles, crocodiles, birds, some bats, and aquatic mammals. In amphibians, it is in a recessed area off the main nasal cavity. In reptiles, the vomeronasal organ is a separate pair of pits to which the tongue and oral membranes deliver chemicals. In mammals possessing this organ, it is an isolated area of the olfactory membrane within the nasal cavity that is usually connected to the mouth via the **nasopalatine duct** (figure 17.13).

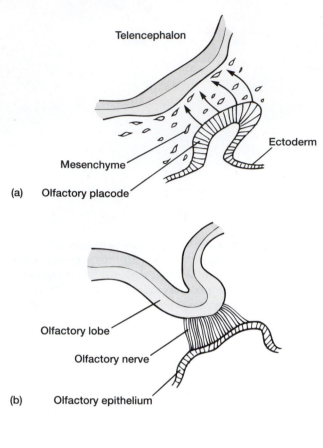

(a)

Telencephalon

Mesenchyme

Ectoderm

Olfactory placode

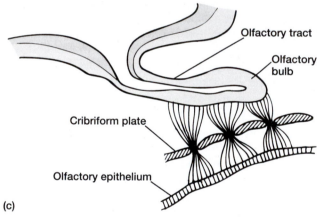

(b)

Olfactory lobe

Olfactory nerve

Olfactory epithelium

(c)

Olfactory tract

Olfactory bulb

Cribriform plate

Olfactory epithelium

FIGURE 17.8 **Embryonic development of the olfactory system.** (a) Thickening of ectoderm forms the olfactory placode. Cells within it sprout nerve fibers that grow into the nearby telencephalon. (b) These fibers collectively form the olfactory nerve. (c) The outgrowth of the telencephalon that receives the olfactory nerve is the olfactory bulb. The olfactory tract connects the olfactory bulb to the brain.

The vomeronasal organ is an accessory olfactory system. It includes basal, sustentacular, and bipolar sensory cells similar to those of the olfactory epithelium. Sensory receptor cells of the vomeronasal organ project into the lumen by means of microvilli, whereas olfactory sensory cells have cilia. The neural circuitry of the vomeronasal system runs parallel to but remains entirely separate from the main olfactory system. Like the main olfactory system, the vomeronasal system can be traced through the limbic system to the hypothalamus and thalamus.

Thalamus and hypothalamus (p. 658)

In many vertebrates with a vomeronasal organ, the respiratory airflow carries particles to the organ. But the vomeronasal organ may also establish an association with the mouth, leading some to suggest that it may sense the chemical composition of food in the buccal cavity. However, the vomeronasal also seems to be especially sensitive to chemicals important in social or reproductive behavior. Theoretically, individuals lacking this organ could increase their respiratory ventilation rate to draw more frequent samples of air across the olfactory epithelium, but this would be energetically more expensive and might result in problems with regulation of blood pH. In order for us to recognize the anatomical and functional distinctiveness of the nasal epithelium and the vomeronasal organ, we refer to the chemicals detected as **odors** and **vomodors** and to the processes as **olfaction** and **vomerolfaction,** respectively.

Snakes and lizards extend their tongue from the mouth, sweep air in front of their snout to collect vomodors, and then retract their tongue into the mouth to deliver these chemicals to the vomeronasal organ (figure 17.14a–d). In snakes and possibly in lizards, the returning tongue does not directly enter the lumen of the vomeronasal organ. Instead, the reptile wipes its tongue against the entry ducts to the organ and against small ridges in the lower part of the mouth. These ridges are lifted up to the ducts, apparently adding vomodors they have gathered to those the tongue delivered directly to the entry ducts. In snakes, removal or inactivation of the vomeronasal organ leads to deficits in courtship, pheromonal trailing, and prey detection.

Mouth

Taste, like olfaction, centers on the detection of chemical stimuli by chemoreceptors. But the chemoreceptors of taste are **taste buds** located in the mouth. In amphibians, reptiles, and birds, taste buds occur in the mouth and pharynx. Mammalian taste buds tend to be distributed throughout the tongue.

In mammals, three separate cranial nerves deliver sensory information from taste buds to the nervous system: the **facial, vagus,** and **glossopharyngeal nerves** (figure 17.15a). Each taste bud, which pokes through the epithelium via a **taste pore,** is a barrel-shaped collection of 20 or more cells of three types. Basal cells located at the base or periphery of taste buds are thought to be stem cells—that is, cells that replace the other cell types. The life span of taste bud cells is about a week; thus, replacement is a continuous process. Sustentacular cells (dark cells) are supportive and secretory in function. **Gustatory cells** (light cells) are thought to be the primary chemoreceptive cells of the taste bud. Taste bud cells do not have axons. Instead, sensory fibers of the three cranial nerves entwine around these three cell types and establish special, synapselike contacts with the gustatory cells (figure 17.15b,c).

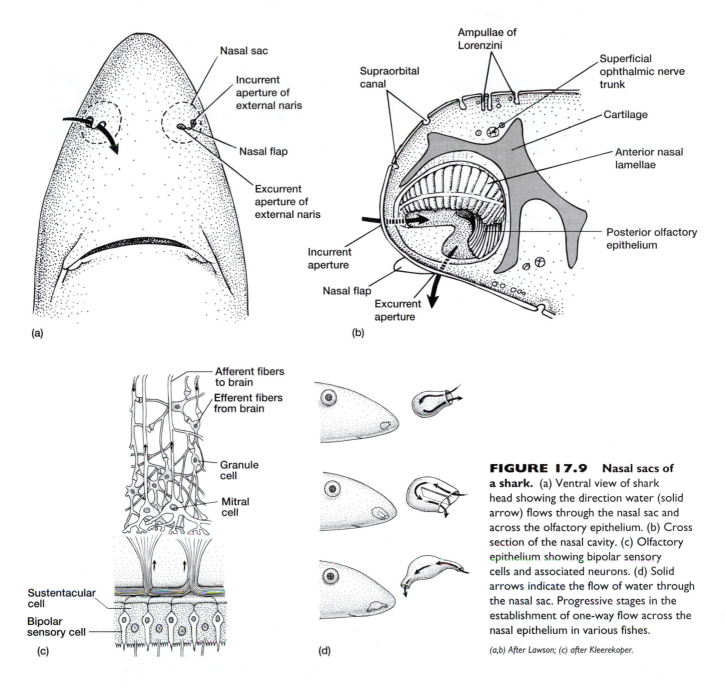

FIGURE 17.9 Nasal sacs of a shark. (a) Ventral view of shark head showing the direction water (solid arrow) flows through the nasal sac and across the olfactory epithelium. (b) Cross section of the nasal cavity. (c) Olfactory epithelium showing bipolar sensory cells and associated neurons. (d) Solid arrows indicate the flow of water through the nasal sac. Progressive stages in the establishment of one-way flow across the nasal epithelium in various fishes.

(a,b) After Lawson; (c) after Kleerekoper.

Collectively, the taste buds sense chemical stimuli and act as transducers to initiate electrical impulses that travel via the cranial nerves to the central nervous system. Sweet, sour, salty, bitter, and savory substances send different patterns of electrical volleys to the nervous system from distinct receptor cells. Once it was thought that these particular receptors were restricted to specific regions of the tongue, but it is now recognized that all taste sensations come from all regions of the tongue, although some tongue regions may be especially sensitive to certain flavors.

Radiation Receptors

Radiation travels in waves. Cosmic radiation has the shortest wavelength, and radio waves have the longest. Together with intermediate wavelengths, these constitute

the **electromagnetic spectrum.** From the standpoint of an organism, radiation carries information about its intensity, wavelength, and direction.

No organism taps the full range of information available throughout the electromagnetic spectrum. Organisms can perceive only a limited range of wavelengths. Insects, such as honeybees and butterflies, can see ultraviolet (UV) radiation, as do many fishes, reptiles, and birds. Most mammals have lost the ability to see UV light, except perhaps for some rodents. Their urine reflects UV, making its detection by social rodents a useful signal. On the other side of the visual range (figure 17.16), some vertebrates, such as vampire bats, pythons, and pit viper snakes, can detect infrared (IR) radiation using specialized surface receptors. However, most vertebrates can perceive only a narrow band of

(a) Nonchoanate fish

Nasal
sac

Nasal
sac

External
naris

Internal
naris

(b) Choanate fish

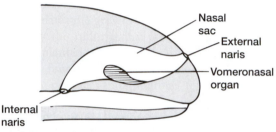

Nasal
sac

External
naris

Vomeronasal
organ

Internal
naris

(c) Amphibian

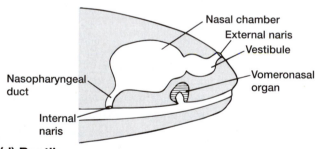

Nasal chamber

External naris

Vestibule

Vomeronasal
organ

Nasopharyngeal
duct

Internal
naris

(d) Reptile

Nasopharyngeal
duct

Nasal
chamber

External
naris

Internal
naris

(e) Mammal

FIGURE 17.10 **Phylogeny of olfactory organs.**
Note that the vomeronasal organ is absent in fishes but present
in most tetrapods. (a) Nonchoanate fish. (b) Choanate fish.
(c) Amphibian. (d) Reptile. (e) Mammal.

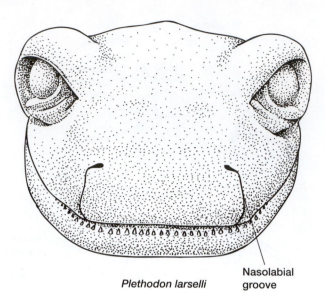

Plethodon larselli

Nasolabial
groove

FIGURE 17.11 **Head of Larch Mountain
salamander *Plethodon larselli*.** In plethodontid salamanders,
a nasolabial groove runs between the mouth and nostrils. It is
thought to convey chemicals from mouth to nose.

Based on photographs supplied by J. H. Larsen.

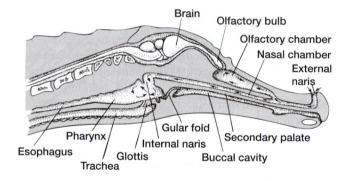

Brain   Olfactory bulb

Olfactory chamber

Nasal chamber

External
naris

Pharynx   Gular fold

Esophagus   Internal naris   Secondary palate

Glottis   Buccal cavity

Trachea

FIGURE 17.12 **Sniffing by the crocodile.** In
tetrapods, olfaction often depends on the arrival of new
chemicals with each respiratory exchange of air to the lungs.
Sniffing allows more frequent sampling of air without increasing
respiratory rate. The crocodile can close both the glottis and the
gular fold, momentarily isolating the nostrils and the mouth. By
depressing the floor of the pharynx, fresh air can be drawn just
into the olfactory chamber and new chemicals sampled without
respiratory ventilation.

After Pooley and Gans.

electromagnetic radiation between about 380 and 760 nm.
This restricted band is called "visible" light, meaning that
we can see it (figure 17.16). When we talk of "light," we
are being quite provincial because we really mean this very
narrow range of electromagnetic radiation out of a much
wider spectrum. Similarly, when we speak of the sense of
"vision," we are referring to the ability to perceive light
within this narrow range.

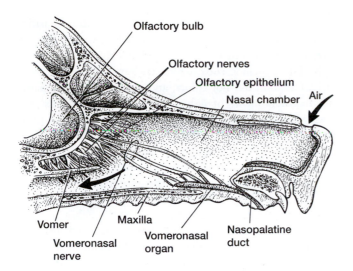

FIGURE 17.13 **Vomeronasal organ of a dog.** The nasopalatine duct passing through the incisive foramen joins the mouth with the nasal chamber. Olfactory nerves travel to a restricted area of the olfactory epithelium.

After Miller and Christensen.

Vertebrates have evolved a variety of sensory organs that gather electromagnetic radiation. Different regions of the spectrum represent different energies (figure 17.16) and present different levels of stimuli to sensory receptors.

Photoreceptors

Eye photoreceptors are sensitive to light, and include two specific categories of light-sensitive receptors, the rods (low light sensors) and cones (color sensors). The vertebrate eye can focus light on photosensitive cells to form an image of the environment. The capacity to focus light on objects at different distances is called **visual accommodation.**

The nervous system takes advantage of the physical properties of light to interpret the images presented to it. Differences in light intensity are interpreted as contrasts. Within the visible spectrum, different wavelengths are interpreted as different colors.

Structure of the Eye The mammalian eye has three layers: sclera, choroid, and retina (figure 17.17a).

Sclera The **sclera** is the outer layer of the eye. It forms the "white of the eye" and consists of a tough capsule of connective tissue to which the extrinsic ocular muscles attach. Contractions of these muscles rotate the eyeball in its orbit to direct the gaze toward an object of interest. The sclera helps define the shape of the eyeball. In birds, reptiles, and fishes, small plates of bone called **scleral ossicles** are often present to help hold the shape of the sclera. At the front of the eye, the sclera clears to become the transparent **cornea.**

Uvea The middle layer of the eye, the **uvea,** is composed of three parts. The **choroid** adjacent to the retina is

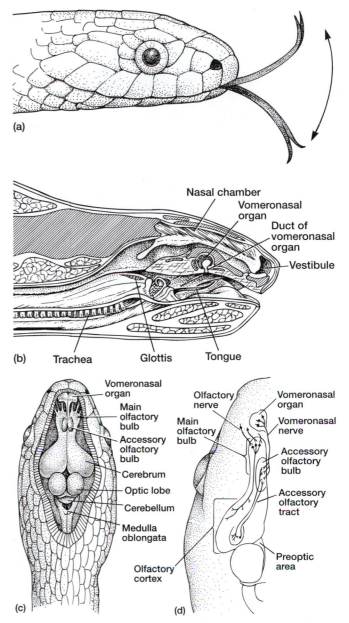

FIGURE 17.14 **Tongue flicking in snakes.** (a) Snakes, like lizards, extend their tongues to sweep air in front of them. The tongue collects and then transports airborne particles into the mouth. Probably along with other oral membranes, the tongue wipes these particles onto the vomeronasal organ on the roof of the mouth. (b) Sagittal section of the head of a boa constrictor. The vomeronasal organ is a blind pocket with a lumen that opens directly into the mouth via a duct. The tip of the retracted tongue projects from its sheath beneath the trachea. (c) Skull and overlying tissue have been cut away to reveal a dorsal view of the snake brain. (d) Neuroanatomy of a snake's olfactory organs. The main olfactory bulb receives input from the olfactory epithelium. The accessory olfactory bulb, via a separate tract, receives information from the vomeronasal organ. Vomeronasal and olfactory systems are separate chemoreceptive organs whose fibers travel separately within the olfactory tract. Thereafter, sensory information tends to be brought together in the olfactory cortex of the telencephalon.

After Kubie et al; Halpern and Kubie.

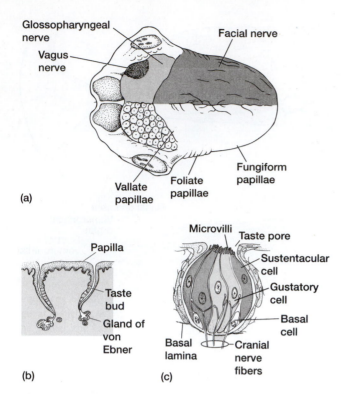

(a) labels: Glossopharyngeal nerve, Vagus nerve, Facial nerve, Fungiform papillae, Foliate papillae, Vallate papillae

(b) labels: Papilla, Taste bud, Gland of von Ebner

(c) labels: Microvilli, Taste pore, Sustentacular cell, Gustatory cell, Basal cell, Basal lamina, Cranial nerve fibers

FIGURE 17.15 **Chemoreceptors of the mouth.**
(a) Taste buds, distributed across the tongue, detect five basic qualities: savory (umani), sweet, sour, salty, and bitter. Areas supplied by facial, glossopharyngeal, and vagus cranial nerves are indicated. (b) Taste buds reside along the recesses of papillae within the surface epithelium. (c) Each taste bud is made up of several dozen cells, including supportive sustentacular cells, basal cells, and sensory gustatory cells. Microvilli on the apical surfaces of the cells project through the surface epithelium. Afferent nerve fibers are associated with all of these cells.

(a) After Bloom, Lazerson, and Hofstadler.

the largest part. Because it is highly vascular, the choroid provides nutritional support to the ocular tissues. The choroid is pigmented. In some nocturnal vertebrates, it includes a special reflective material, the **tapetum lucidum.** Under conditions of dim light, this structure reflects the limited light to restimulate the light-sensitive cells in the retina. The tapetum lucidum produces the "eye shine" of mammals seen at night in car headlights or in flashlight beams.

The second part of the uvea, the **ciliary body,** is a tiny circle of smooth muscle around the interior of the eyeball. The **ciliary muscle** controls visual accommodation. It is attached to the flexible **lens** through a circular **suspensory ligament.** Tension on the lens tends to deform it, whereas relaxation allows the lens to restore its shape elastically.

The third part of the uvea, the **iris,** is a thin continuation of the uvea across the front of the eyeball. The **pupil** is not a structure but an opening defined by the free edge of the iris. Tiny smooth muscles within the iris act like a diaphragm to reduce or enlarge the size of the pupil and regulate the amount of light that enters the eye.

Retina The innermost layer of the eye, the photosensitive **retina,** is itself composed of three cell layers. The deepest layer of cells within the retina contains the photoreceptor cells. These include **rods,** which are sensitive to low levels of illumination but not colors, and **cones,** which are sensitive to colors in bright light. The photoreceptor cells synapse with **horizontal** and **bipolar cells.** Proximal to these are layered **amacrine** and **ganglion cells.** This arrangement means that light entering the eye and falling on the retina passes sequentially through ganglion, amacrine, bipolar, and horizontal cells before it reaches the photoreceptive rods and cones (figure 17.17b). The functional significance, if any, of this arrangement is not known.

In a few vertebrates, the retina is indented, forming a **fovea** (figure 17.17a). The fovea is the point at the back of the eyeball where light converges. Composed entirely of cones, it forms the point of sharpest focus. Although rods are absent from the fovea, they increase peripherally.

Within the eye are three chambers. Two lie in front of the lens: the **anterior chamber** between the iris and the cornea, and the small **posterior chamber** between the iris and the lens. The third and largest **vitreal chamber** is located behind the lens. These chambers are filled with a transparent fluid that helps maintain the shape of the eyeball. The anterior and posterior chambers are filled with watery **aqueous humor.** The vitreal chamber contains a thick **vitreous humor,** sometimes called the **vitreous body** because it can be dissected from the eye as a single viscous plug.

Embryology Embryologically, the eye is a composite structure formed from surrounding mesenchyme and the optic placode, a thickened neuroectodermal tissue that gives rise to the optic vesicle. The retina is an extension of the brain, eventually containing 100 million light cells in humans. Development of the eye begins with the appearance of paired outgrowths, the **optic vesicles,** from the sides of the future telencephalon (figure 17.18a). As the optic vesicles approach the overlying ectoderm, it thickens to become the optic placode and invaginates to form the **lens primordium** (figure 17.18b). The optic placode pinches off to settle into an indentation, the **optic cup.** Mesenchyme surrounding the developing eyeball condenses to produce the outer coats of the eye (figure 17.18c,d).

Thus, the ectoderm gives rise to the eyelid, cornea, and lens. The mesenchyme forms the choroid and sclera, and the iris and retina develop from the optic cup. The optic vesicle retains its connection with the brain as the **optic stalk,** from which it initially arose. The optic stalk comes to carry the axons of ganglionic cells that project to optic areas in the diencephalon. Although this stalk is actually an extension of the brain and should therefore be termed a tract, in practice it is often termed the **optic "nerve"** and counted as the second cranial nerve.

Cranial nerves (p. 630)

Phylogeny During vertebrate eye development, specific molecules play crucial roles in eye development. The frontal eye of amphioxus expresses these same molecules, suggesting a

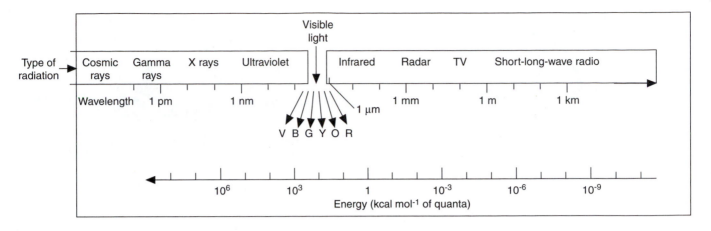

FIGURE 17.16 **The spectrum of electromagnetic radiation.** Between extremely short cosmic rays and long radio waves lies a narrow band of "visible light," to which human eyes are normally sensitive. Wavelength increases to the right. Energy within the electromagnetic radiation increases to the left. Abbreviations: violet (V), blue (B), green (G), yellow (Y), orange (O), red (R).

After Schmidt-Nielsen.

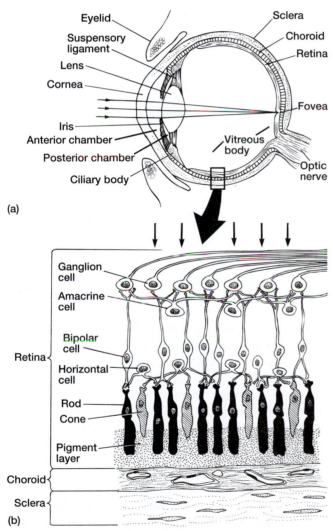

FIGURE 17.17 **Structure of a higher primate eye.** (a) Cross section. (b) Enlarged view of layers in the back wall of the eyeball. Neurons indirectly connect light-sensitive rods and cones to ganglion cells that form the optic nerve. Black arrows indicate the path of entering light.

general homology of its photoreceptor cells and pigment cells to the vertebrate eye. In lampreys, the striated corneal muscle of myotomal origin is attached to the spectacle, a clear area of skin over the cornea. Contraction of this muscle tends to draw the spectacle taut and flatten the cornea. This in turn pushes the lens closer to the retina. Accommodation is thus accomplished by deformation of the eyeball from outside. Upon relaxation of the corneal muscle, elasticity of the cornea and vitreous humor return the lens to a resting position (figure 17.19a).

In bony and cartilaginous fishes, the eyeball is often supported by scleral bones or cartilages. The lens is nearly round (figure 17.19b,c). It is held by the suspensory ligament and moved by the **retractor,** or **protractor lentis muscle,** which is inserted directly on the lens. To focus the image, the retractor lentis pulls the lens forward in elasmobranchs and backward in teleosts.

The amphibian eye is sometimes reinforced by a cup or ring of scleral cartilage. The lens is nearly round and held by a circular suspensory ligament (figure 17.19d). The retractor lentis muscle of amphibians is inserted on the base of the suspensory ligament instead of directly on the lens. The lens is normally focused on distant objects. In order for amphibians to view near objects, the retractor lentis pulls the lens forward.

Among amniotes, except snakes, the lens changes shape to accommodate the visual image. This usually involves contraction of the ciliary muscle, which may squeeze the lens to change its shape or act through the circular suspensory ligament to stretch the lens and make it flatter. Relaxation of the muscle allows the lens to return, thanks to its resilience, to a more rounded shape. In some reptiles and some birds, but not in mammals, scleral ossicles are present. They are particularly well developed in raptors that reach high rates of aerial speed. A **papillary cone** in reptiles or a **pecten** in birds is projected into the vitreous chamber from the back wall of the eye. These structures are thought to provide supplemental nutritional support for deep ocular tissues (figure 17.19e,f).

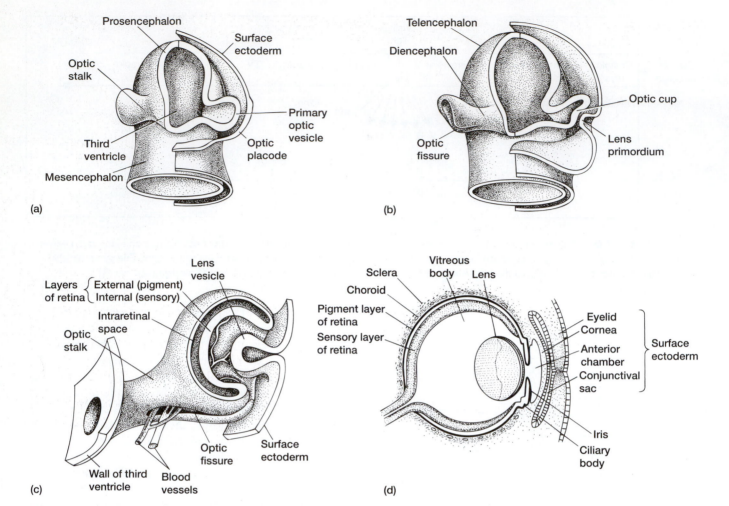

FIGURE 17.18 **Embryonic development of the vertebrate eye.** (a) Early embryo in whom outwardly growing optic vesicles eventually meet the thickened optic placode in the ectoderm. (b) Slightly later, interaction between optic vesicle and optic placode leads to initial differentiation of the lens and what becomes a double-layered secondary optic vesicle or optic cup. (c,d) Successive stages in separation of the lens within the optic vesicle. The eyelid and cornea form, and the wall of the eyeball differentiates into distinct layers. After the lens is pinched off, openings appear within the surface epithelium to delineate the cornea and anterior chamber.

After Tuchmann-Duplessis, Auroux, and Haegel.

Form and Function Water and air present fundamentally different physical challenges for vision. Water affects light in several ways. It carries dissolved and suspended materials that can block light. In marine waters, the limit of useful vision may be 30 m, but in turbid freshwater rivers and lakes, the limit often falls to only around 1 to 2 m. The deeper an animal dives, more sunlight from above is filtered out and the water becomes dimmer to eventually dark. Light intensity diminishes selectively with depth. The first wavelength to be absorbed is ultraviolet, and then infrared, red, orange, yellow, green, and finally blue. Virtually no sunlight penetrates below 1,100 m even in clear water. Nevertheless, most fishes at depths greater than 1,100 m possess large and complex eyes, but they do not detect sunlight, because light never penetrates to such great depths. Instead, they respond to flashes of bioluminescent light that the fishes themselves produce.

Ichthyosaurs, extinct marine reptiles of the Mesozoic, had larger eyes for their body size than found today in marine mammals and reptiles. Such relatively large eyes may have served them well in low-light conditions at ocean depth, perhaps over 500 m. One favored food was squid, also occupants of ocean depth and equipped as well with relatively large eyes. Harbor and spotted seals are sensitive to dim visual images equivalent to depths of 600 to 650 m, yet do not possess especially large eyes. Perhaps, then, ichthyosaurs with their large eyes fed at depths greater than 500 m, where large, acute eyes gathered what little light reached those depths.

Ichthyosaurs (p. 30; p. 112)

Visual Accommodation Water and air also differ in their effects on visual accommodation. To focus an image, light rays must be diverted, or "bent," from their normal parallel lines of travel to converge on the retina. For vertebrates on land, the difference in the **refractive indices** of the eye and air is pronounced, so that light striking the cornea is bent abruptly. As a result, the cornea of terrestrial vertebrates does most of the focusing (figure 17.20a). The lens merely refines the image to bring it

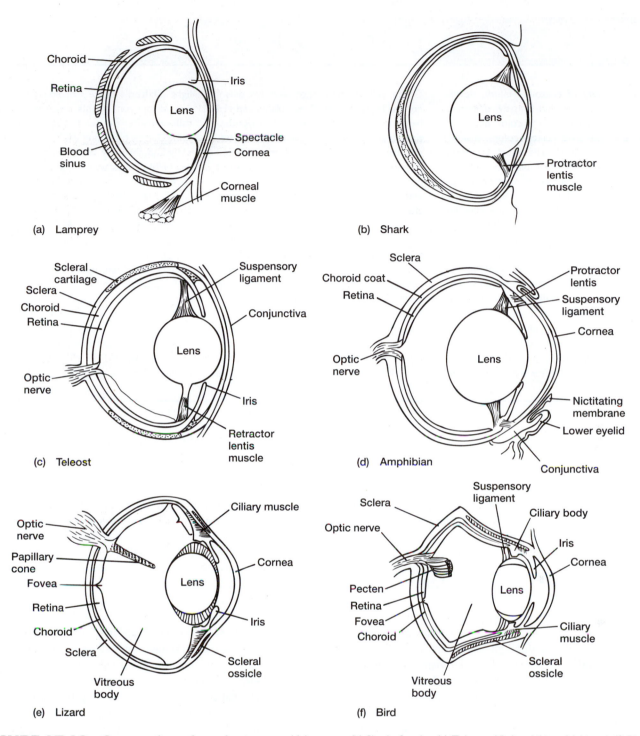

FIGURE 17.19 **Cross sections of vertebrate eyes.** (a) Lamprey. (b) Shark, *Squalus.* (c) Teleost. (d) Amphibian. (e) Lizard. (f) Bird.
After G. A. Walls.

into clear focus on the retina. However, in aquatic vertebrates, the cornea contributes very little to focusing. The refractive indices of water and the cornea are nearly equal. Therefore, the lens of fishes refracts most light. The refractive index of the lens is well above that of water thanks to its crystalline structure and considerable thickness (figure 17.20b).

Optics (p. 158)

Focusing in terrestrial vertebrates, except for snakes, depends on changes in lens *shape* (figure 17.21c–e). In contrast to this, focusing in fishes depends on changes in lens *position* within the eye (figure 17.21a,b). It is unclear whether the fish eye at rest is focused on close or distant objects. Different mechanisms of accommodation are employed by different groups of fishes. When electrically

For a variety of reasons—injury, developmental miscues, but mostly defective genes—some people are color deficient, unable to perceive certain colors, making them "color blind." The most common form is red-green color blindness due to lack of cones sensitive to medium wavelengths (green). The condition is sex-linked, such that males (about 5%) are affected more than females (less than 0.5%).

Such an afflicted individual was John Dalton (1766–1844), English chemist (he proposed the atomic theory of elements) and physicist. He probably suspected something earlier, but a defining moment came when he attended a formal black-robe meeting of Quakers, and an unsuspecting Dalton showed up wearing bright, immodest red. He (and his brother) could perceive blue and purple, but after that only yellow.

Green, yellow, and orange were but different shades of yellow. Green plants and red blood looked the same. He finally hypothesized that the gel core of his eyes, the vitreous body, instead of being clear was tinted blue, filtering out the shades he could not perceive. There was no easy way to test this without removing his eyes. So very soon after his death, his assistant followed Dalton's instructions, removed his eyes, poured the contents of one eye into a watch glass, shined a light on it, and recorded all was clear. No tinting. Then he cut a small window in the back of the other eye and peered through. Again, all clear. No tinting. Color blindness, sometimes called "Daltonism" in his honor, resulted from a defect elsewhere.

Today we know it is the inherited condition that leaves the retina without cones sensitive to middle wavelengths of visible light. Dalton's theory of color blindness was disproven by his own instructions followed after his death, but we can thank John Dalton for being the first to so honestly and carefully describe the condition in 1798 ("Extraordinary facts relating to the vision of colours, with observation," *Mem. Literary Phil. Soc. Manchester* 5: 28–45, 1798).

There is a footnote: After examination, Dalton's eyes were preserved, kept in a jar, and looked after by the John Dalton Society of Britain, up to the present day. In fact, Dalton, or at least his eyes, had one more contribution to make. In the mid 1990s, using methods of molecular biology, researchers extracted and examined DNA from his preserved eye cells. John Dalton was confirmed to be afflicted with just an ordinary form of inherited color blindness.

---

stimulated, the retractor lentis muscles move the lens posteriorly in some fishes, obliquely downward in others, and obliquely upward in still others.

In trout, the lens is round and the retina is ellipsoid, so the two are not concentric (figure 17.22a). This may mean that at rest the near field of view at the center and the far field at the periphery are in focus. During accommodation, the lens is pulled backward. This does not change the focus on peripheral objects much, but it focuses distant objects in the center of the field of view (figure 17.22b).

Photoreception Most vertebrates have two overlapping visual systems within the retina. One is **scotopic**—responsible for visual ability in dim light employing rod cells, and the other **photopic**—responsible for color vision in bright light employing cone cells. Rods provide night vision, sensitive to dim light, but lack color perception. Cones provide color vision. The light-absorptive pigments in rods and cones belong to a class of visual pigments termed **opsins,** which in turn are linked to the molecule retinal, related to vitamin A. When these pigments absorb photons of light, retinal molecules change shape and through a series of linked steps excite the cell to produce a nerve impulse that travels to higher centers.

The proportions of rods and cones vary considerably in different animals. In animals active under bright conditions, such as diurnal tetrapods and diurnal shallow-water fishes, both rods and cones are present. Most cones are concentrated near the fovea, and rods predominate at the periphery. Nocturnal animals or those that inhabit dimly lit waters are adapted to poor illumination. In these groups, cones are few or absent, and the retina is composed almost entirely of rods.

The retinas of all vertebrates possess rods and, hence, all vertebrates possess some degree of scotopic vision. However, many vertebrates such as chondrichthyans lack retinal cones, suggesting lack of color vision. Vertebrates with color vision may have up to four types of cone cells, recognized by the presence of distinctive opsin pigments (or their associated DNA), and named for the spectral region in which they are most sensitive. In fact, the earliest vertebrates had all four cone types, which are, with approximate maximum color sensitivity: violet (370 nm), blue (445 nm), green (508 nm), and orange (560 nm). Such tetrachromic (four-color) color vision is found in many fishes, turtles, lizards, and birds. In addition to opsin pigments, cones in these groups may also have oil droplets that act to filter short wavelengths, narrowing the spectrum and reducing overlap of spectral sensitivity. Cones report intensity of photon stimulation, but because there is still some overlap, a cone cannot tell which exact wavelength excites it. That is done in the brain, which makes comparisons of cone information and assigns colors.

Many amphibians lack color vision, but even in those with it, color vision is reduced to trichromatic (three-color)—they can distinguish blue from green, and green from red. Perhaps most intriguing is color vision in mammals—almost all are dichromatic (two-color), losing two cone cell types, retaining only violet and orange cone cells, and losing oil droplets as well. This is thought to be related to mammal history, which included mostly a nocturnal

(a) Tetrapod—air

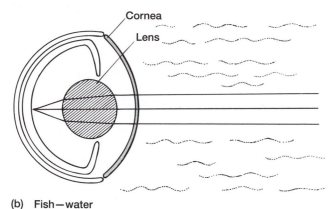

(b) Fish—water

FIGURE 17.20   Vision in air and water. (a) Light that passes through air is strongly refracted when it passes through the cornea; therefore, the cornea is primarily responsible for focusing light rays. The lens fine-tunes the focused image. (b) Because the cornea has refractive properties similar to those of water, incoming light is affected very little when it first enters the eye; therefore, the large lens bears primary responsibility for bringing light rays into focus on the retina.

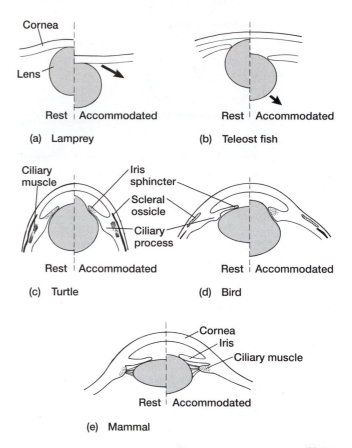

(a) Lamprey   (b) Teleost fish

(c) Turtle   (d) Bird

(e) Mammal

FIGURE 17.21   Accommodation in vertebrates. (a) Lamprey. Contraction of the corneal muscle pulls the cornea against the lens, forcing a change in its position and ability to focus incoming light. (b) Teleost. Accommodation in the teleost depends on a change in lens position. (c) Turtle and (d) bird. Accommodation in both involves the iris sphincter muscle squeezing the lens. (e) Mammal. Accommodation is brought about by relaxation of the suspensory ligaments of the eye.

After G. A. Walls.

lifestyle, requiring little color vision. This leaves most mammals today unable to contrast colors in the green, yellow, and red regions of the spectrum. Humans and other higher apes are exceptions.

In humans and apes, there are three color pigments that produce a full but secondarily derived trichromatic system. Genetically in apes, a gene duplication (mutation) in a pigment sensitive to long wavelengths added a third cone pigment to the two of other mammals, thereby forming color vision based on a trichromatic system of cones, consisting of approximately violet/blue (425 nm), green (530 nm), and orange (560 mn) cones. This gives humans a remarkable perceptibility of an estimated 2 million colors. Why such a color acuity and range would evolve in apes is debated. Some suggest it helped spot ripe fruits (change in color) in dense forests or an estrous female (enlarged pubic swellings) amongst the group.

Whatever the evolutionary value, trichromatic color vision arose in ancestral apes, which is why we as humans evolving within this group have such relatively good color vision (birds are still better). Hunters who wear orange or red hats and clothing make it easier for other humans to spot and identify them, but the dichromatic animals hunted are unable to see contrasting colors in the green, yellow, and red regions of the spectrum, which makes the red/orange-bedecked hunter noticeable to human eyes but difficult to see by the hunted mammal.

In diurnal species, single cones tend to synapse with single bipolar cells that synapse with single ganglion cells that project to the central nervous system. This direct one-to-one transfer of impulses is thought to increase the resolution of the retina and hence its acuity (figure 17.23a,b). In nocturnal species, large numbers of photoreceptor cells converge on a small number of interneurons; thus, there is a pooling of information. Acuity decreases, but sensitivity increases. The horizontal cells and perhaps the amacrine cells spread information laterally to help accentuate contrasts.

One last point about photoreception. Some vertebrates have receptors in their retina that are responsive to ultraviolet (UV) radiation. Insects have been known for over a century to have such UV sensitivity. But now it is documented in many fishes, turtles, lizards, and birds (but not mammals, except rodents). The functional significance of

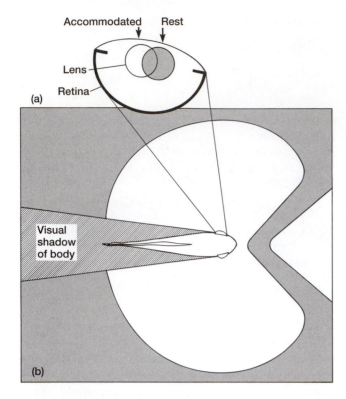

FIGURE 17.22 Accommodation in trout. (a) In order for the trout eye to accommodate light, the lens must be pulled posteriorly. Because the curvatures of the lens and the retina differ, it has been proposed that these differing curvatures provide dual focusing. (b) When the eye is at rest, the stippled area, consisting of the V-shaped region immediately in front of the fish and the distant lateral regions, is in focus; the white areas are out of focus. When the retractor muscle pulls the lens posteriorly, the fish can focus on the distant objects in front of it as well as on the adjacent lateral field.

After Pumphrey; Somiya and Tamura.

UV reception is not known, but it may be related to foraging behavior. Voles lay urine- and feces-scented trails that reflect UV wavelengths. Kestrels (small falcons) can apparently use these UV trails to track these rodents.

Depth Perception The position of the eyes on the head represents a trade-off between breadth of the visual field and depth perception. If the eyes are positioned laterally, then each eye scans separate portions of the surrounding world, and the total field of view at any moment is extensive. Vision in which the visual fields do not overlap is termed **monocular vision.** Because it allows an individual to see a large portion of its surroundings and to detect potential threats from most directions, it is common in prey animals. Strictly monocular vision in which the visual fields of the two eyes are wholly separate is relatively rare. It occurs in cyclostomes, some sharks, salamanders, penguins, and whales.

The visual field overlaps in animals with **binocular vision** (figure 17.24). Extensive overlap of visual fields characterizes humans. We have as much as 90° of

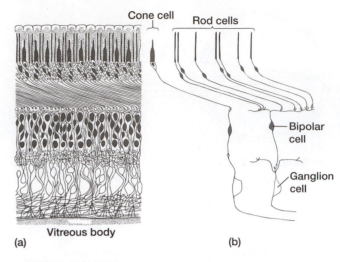

FIGURE 17.23 Retinal connections. (a) Diagram of retinal rods, cones, bipolar cells, and ganglion cells. (b) In diurnal species, the one-to-one circuit, in which one cone connects to a single bipolar cell that connects to one ganglion cell, is thought to increase visual acuity, whereas one bipolar cell that receives many rods blends different stimuli but increases sensitivity.

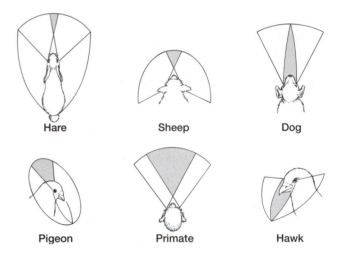

FIGURE 17.24 Monocular and binocular vision in birds and mammals. The degree of overlap of visual fields (indicated by shading) varies considerably. Large panoramic fields of view characterize animals susceptible to attack.

binocular vision, leaving 60° of monocular vision on each side. Binocular vision is important in most classes of vertebrates. Some birds have up to 70° of overlap, reptiles up to 45°, and some fishes as much as 40°. Within the area of overlap, the two visual fields merge into a single image, producing **stereoscopic vision.** The advantage of stereoscopic vision is that it gives a sense of depth perception. If a person with two functioning eyes closes one, she or he will lose much sense of depth.

Depth perception comes from the method of processing visual information. In binocular vision, the visual field seen by each eye is divided. Half the input goes to

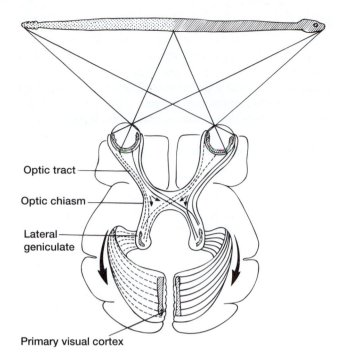

Optic tract

Optic chiasm

Lateral geniculate

Primary visual cortex

FIGURE 17.25 **Depth perception.** Overlapping fields of view in the eyes collect information about the same object (snake), but some of this information travels to the same side of the brain via the optic chiasm, whereas other information travels to the visual areas on the opposite side of the brain. This permits similar information from the slightly differently placed eyes to be brought together and compared. Such comparisons are thought to be the basis for depth perception. Notice the specific organization of visual information. Information from the left visual field (tail of snake) is received by the right side of each eye. The right visual field (snake head) is received by the left side of each eye. Nerve fibers from the medial retina cross in the optic chiasm. Lateral fibers do not cross. Overall, information from the left visual field (tail) is processed in the primary visual cortex of the right cerebral hemisphere. Information from the right visual field (head) is processed in the left cerebral hemisphere.

the same side, and the other half crosses via the **optic chiasm** to the opposite side of the brain. The result is to bring information gathered by both eyes to the same side of the brain. The brain compares the parallax of the two images. **Parallax** is the slightly different view one gets of a distant object from two different points of view. Look at a distant lamppost from one position, and then step a few feet laterally and look at it again. Slightly more of one side can be seen and less of the opposite side. The position of the post relative to background reference points changes as well. This is parallax. The nervous system takes advantage of parallax that results from differences in the positions of the two eyes. Each eye registers a slightly different image due to the distance between the eyes. Although this difference is slight, it is enough for the nervous system to produce a sense of depth from differences in parallax (figure 17.25).

Visual accommodation also contributes to depth perception. Even with a single eye, the degree of accommodation required to bring the object into focus can be used to interpret its distance.

Integration of Visual Information   In lower vertebrates, the optic nerve tends to run directly into the midbrain. In amniotes, the axons in the optic tract travel to one of three regions: the **lateral geniculate nuclei** of the thalamus, the tectum of the midbrain, and the pretectal area in the tegmentum of the midbrain.

Most fibers of the optic tract travel to the paired lateral geniculate nuclei of the thalamus. From here, fibers relay information to the primary visual cortex of the cerebrum (figure 17.26). Thalamic cells and photoreceptor cells and some cells of the visual cortex respond to light intensities, but other cells in the visual cortex are more specialized. Some of these respond to visual images in the shape of slits, bars, or edges. Others respond only to moving or to stationary visual edges. Such specialized cells add more information about the size, shape, and movement of the visual image. In other words, the retina responds to light intensity and wavelength, but in the visual cortex, the image gains contour, orientation, and motion. Image formation might be likened to movies composed of dots of discrete streaming videos. As edges, shades, and shapes are sorted, the image progressively emerges (figure 17.27a–c). Furthermore, visual information is thought to reach a conscious level in the visual cortex. When it reaches the cortex, the visual world is consciously perceived.

Before entering the thalamus, some ganglion cells of the optic tract send branches into the midbrain to contact the tectum. In addition to this visual input, neurons of the **superior colliculus** receive information about sound, head position, and feedback from the visual cortex. The tectum, in turn, produces motor output to muscles that rotate the eyes, head, and even the trunk toward the visual stimuli. Humans have two eyes and two visual fields. Mentally, these merge into one field, in part because of the intricately synchronized movements of both eyes as we scan our surroundings. The tectum also sends visual impulses through the **pulvinar nucleus** of the thalamus to the visual cortex. The function of this route of visual input to the cerebral cortex is not known. If the primary visual cortex is damaged or denied direct input from the thalamus, this alternate input from the midbrain may preserve a rudimentary response to visual stimuli.

A few fibers of the optic tract also send input to the **pretectal** area in the tegmentum of the midbrain. A reflex relay from the tegmentum to motor nerves controlling the iris muscles permits pupil size to be adjusted immediately to light intensity.

Thus, visual perception depends on the stimulation of photoreceptor cells in the retina and culminates in the central integrating regions of the brain. It should be emphasized, however, that the processing of visual information

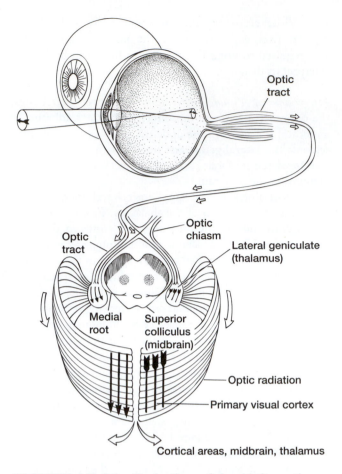

Optic tract

Optic chiasm

Optic tract

Lateral geniculate (thalamus)

Medial root

Superior colliculus (midbrain)

Optic radiation

Primary visual cortex

Cortical areas, midbrain, thalamus

FIGURE 17.26 **Projections of visual information to the primary visual cortex in primates.** The object of focus is a solid arrow. Output from the retina is conveyed via the optic tract to the lateral geniculate nuclei in the thalamus. Selective crossing of some fibers in the optic chiasm results in each lateral geniculate receiving the image (the arrow) from the opposite half of the visual field. The visual image is represented or mapped out in the lateral geniculate as a set of adjacent and highly ordered fields of stimulated neurons. From the lateral geniculate, axons project via an optic radiation to the cerebral hemispheres, where the corresponding parts of the image are now represented as a pile of superimposed fields in the primary visual cortex. This ordering of multiple representations of an image both in the thalamus (lateral geniculate) and in the cerebral hemispheres (primary visual cortex), although not entirely understood, is thought to contribute to neuroanalysis of the binocular image. Notice that the visual pathway does not end in the primary visual cortex but continues to the midbrain and even back to the thalamus. A small branch of the optic tract, the medial root, does not enter the thalamus but instead leads to the superior colliculus (optic tectum). The medial root synapses with motor neurons that control the visual orientation reflex involving eyeball movements, head turning, and rotation of the trunk.

begins in the retina itself, which you recall from earlier in this chapter is embryologically an extension of the brain. It has been suggested that as much as 90% of visual information is processed in the retina before it is transmitted to the thalamus.

Pineal Complex In most vertebrates, the roof of the diencephalon, termed the **epithalamus,** produces a single median photoreceptor, the **parietal organ.** However, there is great variation in this organ from group to group, and it is often joined by additional adjacent specializations of the epithalamus. To further complicate matters, a phylogenetic change occurs in the function of the parietal organ. It participates in photoreception among anamniotes but tends to become an endocrine organ in amniotes.

Pineal gland (p. 610)

Not surprisingly, a great deal of terminology has grown up around the parietal and adjacent organs of the epithalamus. Let us begin by sorting out this terminology.

Structure Depending on the species, the epithalamus may evaginate to produce up to four structures, each one a discrete organ (figure 17.28a). The most anterior is the **paraphysis,** followed by the **dorsal sac, parietal organ,** and **epiphysis cerebri (epiphysis).** If two or more of these are present together, the collective term **pineal complex** applies.

The functions of the paraphysis and dorsal sac are not well understood, but their structure suggests that they are glandular organs. The epiphysis is sometimes called the **pineal organ** or, if it is largely endocrine in function, the **pineal gland.** Because it is adjacent to the pineal organ, the parietal organ is sometimes called a **parapineal organ,** or a **parietal eye** if it forms a photoreceptive sensory organ. The parietal eye may include a modest cornea, lens, and area of photoreceptive cells that synapse with adjacent ganglion cells to form a nerve that travels to the nervous system (figure 17.28b,c).

Phylogeny A single dorsal parietal foramen through the skull of many ostracoderms testifies to the presence of a parietal organ. Living lampreys possess both an epiphysis and a parietal organ. Both of these organs exhibit some capacity for photoreception (figure 17.29a). In elasmobranchs and bony fishes, the epiphysis is prominent, but the parietal organ, if present at all, is only rudimentary (figure 17.29b,c).

Amphibians usually possess both an epiphysis and a parietal organ. In fossil lissamphibia, the parietal organ forms a distinct parietal eye. Living frogs retain this, but it is absent in salamanders (figure 17.29d,e).

In many reptiles, the parietal organ is present. Lizards and *Sphenodon* possess a parietal eye so distinctive that it is often referred to as a **third eye.** The epiphysis is present as well (figure 17.29f).

In birds and mammals, the parietal organ is absent. The epiphysis is present in both classes, but it is exclusively an endocrine organ and is usually referred to as the pineal gland (figure 17.29g,h).

Even within a given class of vertebrates, there is considerable variation in the pineal complex. For instance, the pineal gland is small in owls, shearwaters, opossums, shrews, whales, and bats but large in penguins, emus, sea lions, and seals. In hagfishes, crocodilians, armadillos, dugongs, sloths, and anteaters,

(a)   (b)   (c)

FIGURE 17.27 **Emergence of visual perception.** (a) This photo is an enlargement of a portion of the image to its immediate right. (b) This photo is an enlargement of a portion of the image on the right. (c) The full differentiated image emerges. The visual image in the nervous system begins with processing in the retina, is continued in the thalamus, and is synthesized in the visual cortex. Just as dots merge to form a legible image of a dog on the right, successive processing of visual information in the brain brings the individual stimuli recorded by individual cells together into a synthesized, composite image.

From Bloom, Lazerson, and Hofstadter, Brain, Mind, and Behavior, © *1985 by W. H. Freeman and Company. Reprinted by permission.*

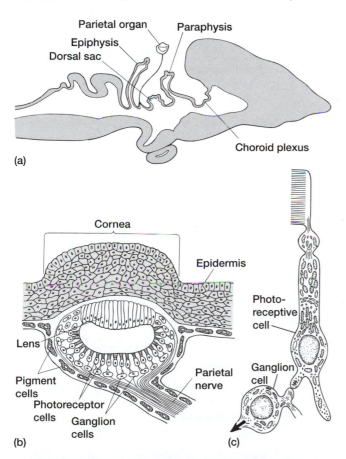

FIGURE 17.28 **Pineal complex.** (a) Sagittal section through the central nervous system of a generalized vertebrate. Up to four evaginations of the roof of the diencephalon may form. (b) Generalized parietal eye. (c) Photoreceptor cell from a parietal organ.

(b) After Northcutt.

the epiphysis is absent entirely. The parietal organ may develop independently of the epiphysis, the two can be closely associated, or, as in some fishes and anurans, they may fuse.

Form and Function The primitive pineal complex was a photosensitive organ. Among most cyclostomes, fishes, amphibians, and reptiles, photosensitivity has been experimentally demonstrated in the pineal organ complex. This lends support to the view that this complex is an ancient structure among vertebrates that arose as an accessory sensory system sensitive to photoradiation. Photoreceptive cells from the pineal complex are tall and columnar in shape, with specialized apical extension (figure 17.28c). The basal regions synapse with adjacent ganglion cells. These form the pineal nerve, which relays impulses to the **habenular** and other regions of the brain.

The change from photoreception to endocrine secretion occurs in birds and mammals. In some birds, pineal tract fibers have been reported, which suggests a lingering role in photoreception. But, for the most part, the bird epiphysis is glandular and thought to be involved in endocrine secretion. In mammals, pineal fiber projections to the brain are unknown. The mammalian epiphysis is exclusively endocrine in function and composed of secretory cells called **pinealocytes,** which may be modified photoreceptor cells.

Infrared Receptors

Infrared radiation lies just to the right of the visible band of light on the electromagnetic spectrum (see figure 17.16). Some vertebrates have special sense organs that respond to infrared radiation. This is especially useful at night when visible light is usually unavailable.

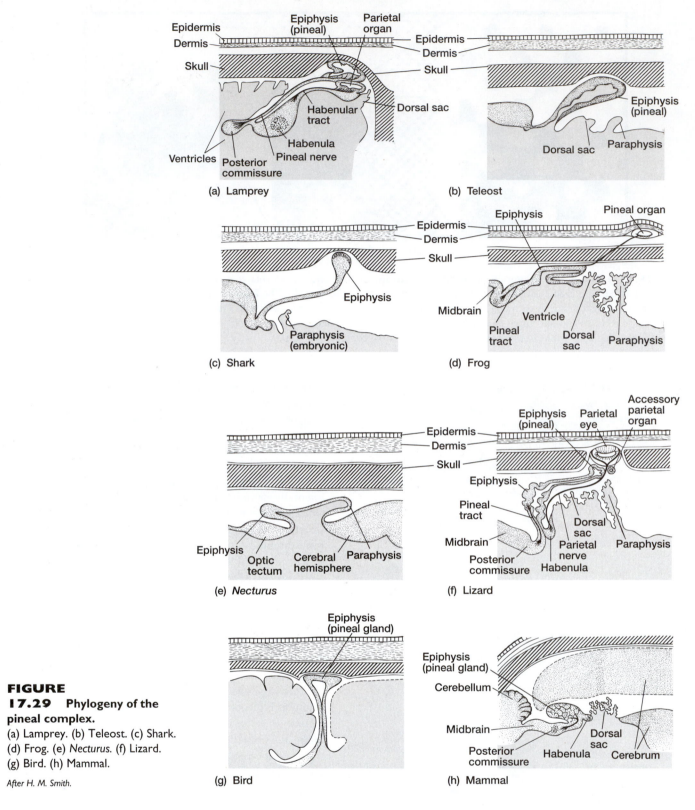

**FIGURE
17.29** **Phylogeny of the
pineal complex.**
(a) Lamprey. (b) Teleost. (c) Shark.
(d) Frog. (e) *Necturus*. (f) Lizard.
(g) Bird. (h) Mammal.

After H. M. Smith.

For us to see an object, visible light must strike it and be reflected from it. The natural source of visible light is sunlight. However, infrared light emanates directly from the surface of any object with a temperature above absolute zero, that is, any object warmer than −273°C. Obviously, the sun is well above this temperature, so infrared radiation is included in its spectrum. But all other natural objects are also above this extremely low temperature and give off infrared radiation from their surfaces, whether it is day or night. Infrared radiation can be detected by some species of snakes and be used to guide their search for prey in the dark.

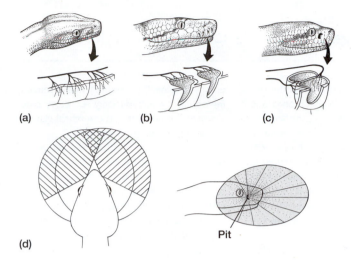

FIGURE 17.30 **Infrared receptors.** Sensors receptive to infrared radiation are located within the integument in some boas (a), at the bottom of recessed pits in pythons (b), and on a thin membrane halfway between the opening and bottom of the sensory organ in venomous pit vipers (c). In pit vipers, pit organs scan the shaded cone-shaped area (d) in front and to the side of the infrared receptor.

After deCock Buning.

Because infrared radiation is emitted from warm objects, and objects it shines on are warmed as they absorb it, this type of radiation is sometimes incorrectly termed "heat radiation." Strictly speaking, infrared radiation is a narrow range of wavelengths of electromagnetic radiation, not heat. Nevertheless, this warming effect of infrared radiation on bodies that absorb it is how infrared receptors are stimulated.

Special sense organs containing **infrared receptors (thermoreceptors)** occur in several groups of vertebrates. They are present on the faces of vampire bats that feed on ungulates. Infrared receptors apparently help these bats detect warm blood vessels beneath the thick skin of their prey. The most discrete infrared receptors are found in two groups of snakes: the primitive boas and the advanced pit vipers. In both cases, the sensory receptor is a free nerve ending located in the skin. As the skin absorbs infrared radiation, it is warmed; this excites the associated free nerve endings, which transmit this information to the optic tectum of the midbrain.

In boa constrictors, the free nerve endings lie within epidermal scales along the lips (figure 17.30a). In pythons, the nerve endings lie at the bottom of a series of several recessed **labial pits** along the lips (figure 17.30b). The venomous pit vipers take their name from the presence of a pair of infrared receptors called **facial pits** (loreal pits). Facial pits also are sunken, but they differ from the labial pits of pythons. Sensory nerve endings are suspended in a thin **pit membrane** halfway between the bottom and the top of the pit, instead of lying at the bottom of the pit as in pythons (figure 17.30c,d).

When free nerve endings are embedded in an epidermal scale, as in boa constrictors, or associated with tissue at the bottom of a pit, as in pythons, the surrounding tissue may dissipate heat and thus slow the local warming that stimulates the free nerve endings. However, in pit vipers, the nerve endings are warmed rapidly because the free nerve endings are suspended in the thin membrane away from the walls of the pit. This increases their sensitivity to infrared radiation. A pit membrane that is warmed by as little as 0.003°C reaches a threshold sufficient to excite its sensory receptors. For a pit viper, this translates into infrared sensitivity to a mouse about 30 cm away. Pythons and boas can detect mice at distances of about 15 cm and 7 cm, respectively.

Mechanoreceptors

Detecting water currents, maintaining balance, and hearing sounds may seem to be different sensory functions. Yet, all are based on **mechanoreceptors,** sensory cells responsive to small changes in mechanical force.

One basic mechanoreceptor is the **hair cell,** an unfortunate term because these cells have nothing to do with hair. The name comes from the microscopic "hairlike" processes at their apical surfaces. These tiny processes include a tight strand of **microvilli** of unequal lengths and a single long **cilium,** which is sometimes called a **kinocilium.** The microvilli are often constricted at their bases and rest on a dense **terminal web** or **cuticular plate.** Each microvillus includes a core of fine microfilaments with molecular cross-bridges so that it behaves like a rigid rod. Because they are nonmobile, microvilli are more appropriately called **stereocilia.** A tuft of stereocilia with a kinocilium is a **hair bundle** (figure 17.31a).

Hair cells are transducers that transform mechanical stimuli into electrical signals. Mechanical stimulus of the hair bundle triggers ionic changes in the hair cell. Hair cells are epithelial cells that originate embryologically from surface ectoderm. They lack axons of their own. Instead, each hair cell is embraced by the sensory fibers of neurons sensitive to ionic changes in the hair cell. Through synapses or synapselike contact points between them, electrical excitation is transmitted from the hair cells to their embracing neurons and then to the central nervous system. Hair cells also receive efferent nerves from the central nervous system. Efferent nerves can change the sensitivity of the hair cell or help focus its sensitivity to a restricted range of mechanical frequencies.

Hair cells respond selectively to mechanical stimuli. For example, stimuli applied from one direction will trigger a hair cell, whereas stimuli applied from the opposite direction will not (figure 17.31c). Selectivity is thought to result from the asymmetry of the hair bundle itself, due to the different lengths of its stereocilia.

A **neuromast organ** is a small collection of hair cells, supporting cells, and sensory nerve fibers, composing the most common arrangement of a mechanoreceptor. Neuromasts arise from migrating primordial cells that originate from embryonic placodes and follow defined pathways to

Lizards are the closest living relatives of snakes. Snakes may have evolved from lizards, or both groups may share a common ancestor. What is especially intriguing about the origin of snakes is the fact that their derivation from lizards or lizard-like ancestors may have included a fossorial phase in which photoreceptors were reduced in prominence. Following this extended fossorial phase, more modern snakes would have radiated back to aboveground habitats. The photoreceptors that regressed during the fossorial phase were in a sense rebuilt to meet the adaptive requirements of the species moving into diurnally active lifestyles.

The view that snakes went through a fossorial period was championed by Gordon Walls in the early 1940s. It was inspired by his monumental study of the vertebrate eye. Walls noted that the snake eye represented a marked departure from the eyes of all other reptiles, including lizards (box figure 1a). For example, accommodation of the lizard eye occurs through a change in lens shape. In snakes, it occurs by moving the lens forward or back. Typically, lizards have three movable eyelids: upper, lower, and nictitating membrane. Snakes have none. Instead, the snake cornea is covered by the **spectacle,** a transparent derivative of the eyelids that is fixed in position. Scleral ossicles occur in some lizards,

but they are absent in all snakes. In the retinas of diurnal lizards, distinct cones and rods are present, but in snakes, the "cones" seem to be modified rods that serve in color perception. Walls noted further differences in circulation, internal structure, and chemical composition that attest to the distinctiveness of the snake eye, not just among reptiles but among all other vertebrates as well. To Walls, the snake eye seemed to be unique.

Walls proposed that these anatomical peculiarities of the snake eye could not be easily explained if snakes evolved from surface-living lizards. Instead, he suggested that snakes evolved from forms in which the eyes were reduced in association with low light conditions. He proposed that snakes went through a burrowing phase. In his view, restructuring of the eye occurred as snakes radiated back to surface-living and diurnal conditions.

Other snake photoreceptors show similar evidence of reconstruction from a regressed reptilian pattern. For example, the parietal organ of many lizards is highly developed into a photoreceptive organ that includes a lens and a photoreceptive layer. The lizard parietal occupies an opening in the skull and resides beneath the skin, where it has direct access to natural light. However, in snakes, the parietal is lost entirely, and only the basal portion of the

epiphysis (pineal) is retained. In the remaining ophidian epiphysis, only secretory pinealocytes are present and the pineal gland is beneath the skull (box figure 1b).

The issue of whether the origin of snakes includes a fossorial phase or just a nocturnal phase during which snake ancestors lived under conditions of low light is still being debated. In any case, the special photoreceptors of snakes are certainly a radical departure from those of other reptiles.

BOX FIGURE 1   Comparison of photoreceptors of lizards and snakes. (a) Lizard and snake eyes in cross section. (b) Pineal complexes of a lizard and a snake.

(a) After G. A. Walls.

---

species-specific locations on the head and body. The projecting hair bundles are usually embedded in a gelatinous cap called the **cupula** (figure 17.31b). The cupula most likely accentuates the mechanical stimulation of the hair cells, thereby increasing their sensitivity. The neuromast organ, or a modification of it, is the fundamental component of all three types of mechanoreceptive systems: the lateral line system, which detects water currents; the vestibular apparatus, which senses changes in equilibrium; and the auditory system, which responds to sound.

Lateral Line System

The **lateral line system** is present within the skins of most cyclostomes, other fishes, and aquatic amphibians, but it is unknown in terrestrial vertebrates, including aquatic birds and mammals. It consists of long, recessed grooves, or **lateral line canals,** concentrated on the head and extending along

the sides of the body and tail (figure 17.32a,b). Neuromast organs are the sensory receptors of the lateral line system. Neuromast organs, and the sensory lateral line nerves that supply them, arise embryologically from epidermal placodes.

Neuromasts can occur separately on the surface of the skin, but they are usually found at the bottoms of the lateral line canals. The canal can be recessed in a valley or sunken and covered by surface skin that has pores through which currents of water flow over the neuromast organs.

The neuromasts respond directly to water currents. Hair cells are oriented with their most sensitive axis parallel to the canal. About half are oriented in one direction, the rest in the opposite direction. In the absence of mechanical stimulation, each neuromast generates a continuous series of electrical pulses. Water flowing in one direction stimulates an increase in this discharge rate. If flow is in the opposite direction, the discharge rate falls below its resting rate.

Lizard

Sclera
Choroid
Retina
Fovea
Lens (colorless)
Ciliary muscle
Cornea
Lids
Nictitating membrane
Lacrimal duct
Retina

(a)

Snake

Sclera
Choroid
Retina
Lens (yellow)
Cornea
Spectacle
Harderian gland
Skin

Epidermis
Dermis
Skull
Epiphysis
Parietal eye
Dorsal sac
Paraphysis

Epiphysis
Dorsal sac
Paraphysis

(b)

If water passes at right angles to these neuromasts, the resting electrical discharge rate is unaffected. This provides information about the animal's direction of movement and about disturbances in the water. Even cave fishes that are blind can navigate around obstacles in their environment because of their lateral line systems.

Some fishes use their lateral line canals as a type of "distance touch"; that is, it allows them to detect compression in the water in front of them as they approach a stationary object ahead. In some surface-feeding fishes, the lateral line system detects frequency oscillations produced by insects struggling on the water's surface. If the lateral line nerve is cut or if its grooves are covered, an affected fish loses its ability to navigate or perform schooling behavior.

There is some evidence that the lateral line canals can detect low-frequency sounds, at least from nearby objects. The uncertainty about its role in hearing is partly the result of the similarity between stimuli. Rapid oscillations of water and low-frequency vibrations resulting from sounds traveling through water are mechanically similar. It has also been difficult to isolate the lateral line system experimentally from the inner ear, which is also sensitive to sound. It is clear that navigation is the primary role of the lateral line system. Perhaps it detects water disturbances that prey produce, and we should allow for a possible secondary role in hearing as well.

Vestibular Apparatus

The **vestibular apparatus** (membranous labyrinth) is a balancing organ that arises phylogenetically from part of the lateral line system. It is suspended in the vestibular apparatus by loose connective tissue. The vestibular apparatus is filled with **endolymph** and surrounded by **perilymph.** Both fluids

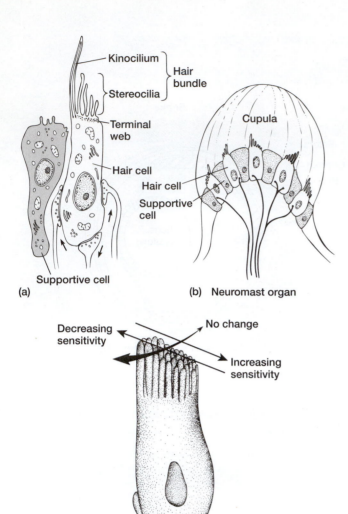

Kinocilium
Hair bundle
Stereocilia
Terminal web
Hair cell
Supportive cell
Supportive cell

(a)

Cupula
Hair cell
Supportive cell

(b) Neuromast organ

Decreasing sensitivity
No change
Increasing sensitivity

(c)

FIGURE 17.31   **Hair cells.** (a) A hair bundle, composed of stereocilia of unequal length and a single kinocilium, projects from the apical surface of a hair cell. Afferent and efferent nerve fibers are associated with each hair cell. A supportive cell, which is thought not to react to direct mechanical stimulation, is often adjacent to a hair cell. (b) Clumps of hair cells and supportive cells form a functional unit, the neuromast organ. The cupula is a cap of gelatinous material that fits over the projecting hair bundles. (c) A hair cell responds to direct mechanical stimulation, but its response is selective. Each cell is most responsive to forces from one direction. Mechanical forces in the opposite direction reduce sensitivity. Forces at right angles to the hair cell cause no change. By such selective responses, hair cells indicate the directions of the mechanical forces that impinge upon them.

have a consistency similar to that of lymph. Embryologically, the vestibular apparatus forms from the **otic placode,** which sinks inward from the surface to produce hair cells, neurons of the otic ganglia, and the vestibular apparatus. In elasmobranchs, the vestibular apparatus maintains continuity with the environment through the endolymphatic duct. In other vertebrates, it is pinched off to form a closed, fluid-filled system of channels.

The vestibular apparatus contains semicircular canals and at least two connecting compartments: the **sacculus** and **utriculus** (figure 17.33a). These are lined by **vestibular epithelium** within which arise neuromast organs that participate in sensing equilibrium and sounds. Gnathostomes have three **semicircular canals** oriented roughly in the three planes of space. Sensory receptors within the semicircular canals are called **cristae** (figure 17.33b). Each crista is an expanded neuromast organ composed of hair cells and cupula. The cristae lie within expanded **ampullae** at the base of each semicircular canal.

The semicircular canals respond to rotation, technically to angular acceleration produced when the head is rotated or turned. When the canals are accelerated, fluid inertia causes the endolymph to lag behind movement of the canal itself. The fluid deflects the cupula, stimulates hair cells, and alters their rhythmic discharge of electrical impulses to the nervous system.

The sensory receptor within the sacculus and the utriculus is the **macula,** or the **otolith receptor.** It too is a modified neuromast organ with hair cells and gelatinous cupula, but in addition, tiny calcium carbonate mineral concretions known as **otoconia** are embedded within the surface of the cupula (figure 17.33c).

Maculae respond to changes in orientation within a gravitational field. Moving the head or changing position of the body tilts the maculae and changes their orientation with respect to gravity. Acceleration of the body also causes the maculae to respond. The otoconia are inertial masses that accentuate shearing displacements of the hair bundles in response to linear acceleration or changes in head orientation.

The vestibular apparatus keeps the central nervous system informed as to whether an animal is at rest or in motion and conveys information about its orientation. The maculae sense gravity and linear acceleration, and the cristae respond to angular acceleration. In fishes, the vestibular apparatus is also responsive to sound, although this function does not seem to be well developed. In fishes, reptiles, birds, and especially mammals, the vestibular apparatus produces a specialized region of sound reception, the **lagena** (figure 17.34).

Auditory System

The lagena is involved in hearing. It develops as an enlargement of the sacculus, so it is part of the vestibular apparatus. In terrestrial vertebrates, it tends to lengthen, and in most mammals it becomes coiled into the **cochlea.** Within the lagena, or cochlea in mammals, lies the sensory receptor of sound, the **organ of Corti,** a specialized strip of neuromasts connected to the brain via the auditory nerve. The lagena, like the rest of the vestibular apparatus, lies within bone or cartilage of the skull. The "ear" includes up to three adjoining compartments: the external, middle, and inner ears.

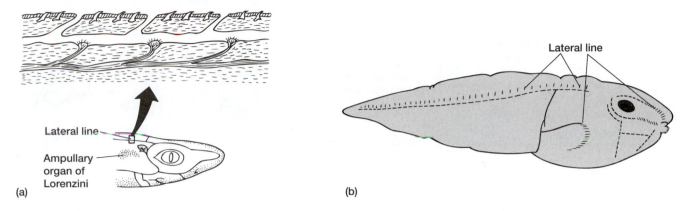

(a)

Lateral line

Ampullary organ of Lorenzini

(b)

Lateral line

FIGURE 17.32 **The lateral line system.** (a) Section through the skin of a shark showing the sunken lateral line canal opening to the surface via small pores. (b) Distribution of neuromasts on the side of a frog tadpole. The long axis of each neuromast is represented by the orientation of the short bars. Notice how this orientation changes.

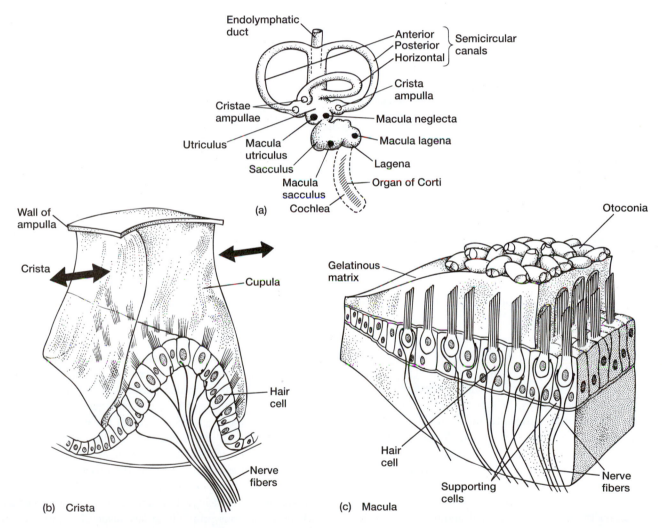

(a)

Endolymphatic duct

Anterior
Posterior
Horizontal
} Semicircular canals

Crista ampulla

Cristae ampullae

Macula neglecta

Utriculus

Macula utriculus

Macula lagena

Sacculus

Lagena

Macula sacculus

Organ of Corti

Cochlea

(b) Crista

Wall of ampulla

Crista

Cupula

Hair cell

Nerve fibers

(c) Macula

Otoconia

Gelatinous matrix

Hair cell

Supporting cells

Nerve fibers

FIGURE 17.33 **Vestibular apparatus.** (a) Generalized vestibular apparatus showing the three semicircular canals and major compartments: utriculus, sacculus, and lagena. (b) The crista is an expanded neuromast organ. One crista resides at the base of each semicircular canal in an enlarged region, the ampulla. The gelatinous cupula extends across the ampulla and is attached to the opposite wall. Acceleration of the head (arrows) produces a shearing force of endolymphatic fluid against the cupula, which bends and deforms hair cells embedded within it. (c) The maculae form a neuromast platform containing otoconia. These maculae reside in the three compartments of the vestibular apparatus. They derive their names from these compartments. In some species a fourth macula is present, the macula neglecta.

(b,c) After Parker.

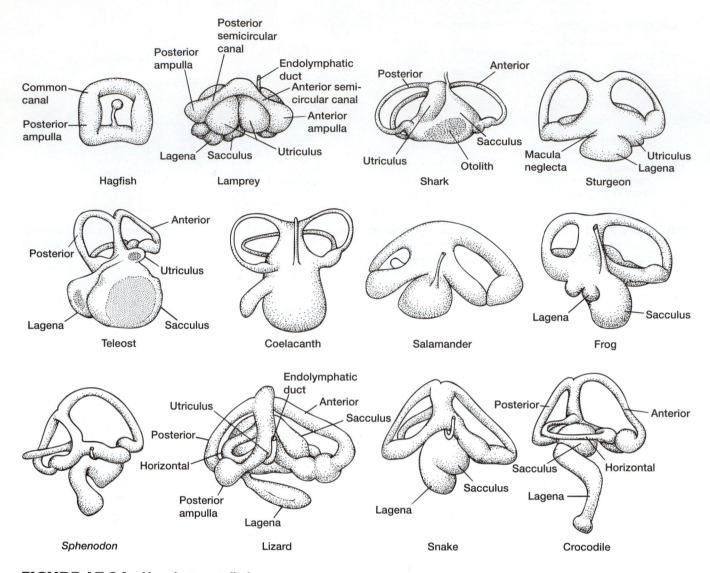

FIGURE 17.34 Vertebrate vestibular apparatuses.

Anatomy of the Ear The **external ear** is absent in fishes and amphibians, but present in reptiles, such as lizards and crocodilians. It consists of a short, indented tube, the **external auditory meatus,** which opens to the surface through the **external orifice.** In birds and mammals, the external auditory meatus is elongated. What most people call the "ear" is correctly termed the **pinna.** This external cartilaginous flap surrounding the external orifice is present in most mammals. The irregular shape of the pinna helps to differentiate sounds approaching from different directions and channel them into the external auditory meatus. Just as two eyes with overlapping visual fields give stereoscopic vision, paired ears provide stereophonic hearing.

The **middle ear** consists of up to three parts: a tympanum (= tympanic membrane), a middle ear cavity or meatus, and one to three tiny bones, or **middle ear ossicles.** The **tympanum,** or eardrum, is a thin, taut membrane at the body surface in frogs and a few reptiles, but it is often recessed at

the bottom of the external auditory meatus in living reptiles as well as in most birds and mammals. As we shall see, the tympanum evolved multiple times in tetrapods.

The first pharyngeal pouch enlarges as the tubotympanic recess. Its expanded end forms the **middle ear cavity** (figure 17.35a–e). The remainder of the recess stays open to form the **eustachian tube,** which maintains continuity between the middle ear cavity and the pharynx.

The first tetrapod middle ear ossicle to differentiate was the **stapes,** or columella, a derivative of the fish **hyomandibula,** where it functions primarily in jaw suspension in fishes. At first it was a stout bone, lodged firmly between braincase and palatoquadrate, continuing in its primary role as a mechanical buttress. Its contact with the braincase placed it strategically so that it might secondarily transfer sound vibrations to the nearby inner ear. It next became a specialization around the spiracle, helping to control the inhalation of respiratory water. In subsequent tetrapods, the stapes abandoned this role and

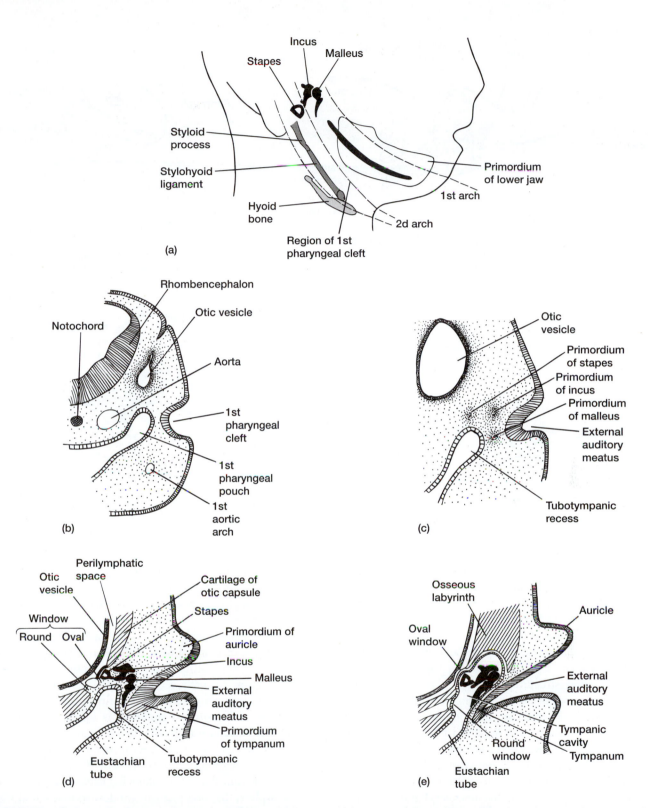

(a)

Incus
Stapes
Malleus

Styloid
process

Stylohyoid
ligament

Hyoid
bone

Region of 1st
pharyngeal cleft

Primordium
of lower jaw

1st arch

2d arch

(b)

Rhombencephalon

Notochord

Otic vesicle

Aorta

1st
pharyngeal
cleft

1st
pharyngeal
pouch

1st
aortic
arch

(c)

Otic
vesicle

Primordium
of stapes

Primordium
of incus

Primordium
of malleus

External
auditory
meatus

Tubotympanic
recess

(d)

Otic
vesicle

Perilymphatic
space

Window
Round   Oval

Cartilage of
otic capsule

Stapes

Primordium of
auricle

Incus

Malleus

External
auditory
meatus

Primordium
of tympanum

Eustachian
tube

Tubotympanic
recess

(e)

Osseous
labyrinth

Oval
window

Auricle

External
auditory
meatus

Tympanic
cavity

Round
window

Tympanum

Eustachian
tube

FIGURE 17.35 **Embryonic formation of the middle ear.** (a) Location of the middle ear ossicles relative to derivatives of the splanchnocranium. (b) The surface of the ectoderm thickens, forming an otic placode that sinks beneath the skin and gives rise to the otic vesicle. The otic vesicle moves into the vicinity of the first pharyngeal cleft and pharyngeal pouch. (c) Mesenchyme (indicated by heavy stippling) begins to condense and differentiate into the ear ossicles: the incus, the malleus, and the stapes (d,e).

Redrawn from H. Tuchmann-Duplessis et al., 1974. Illustrated Human Embryology, Vol. III, Nervous System and Endocrine Glands. © 1974 Springer-Verlag, NY. Reprinted by permission.

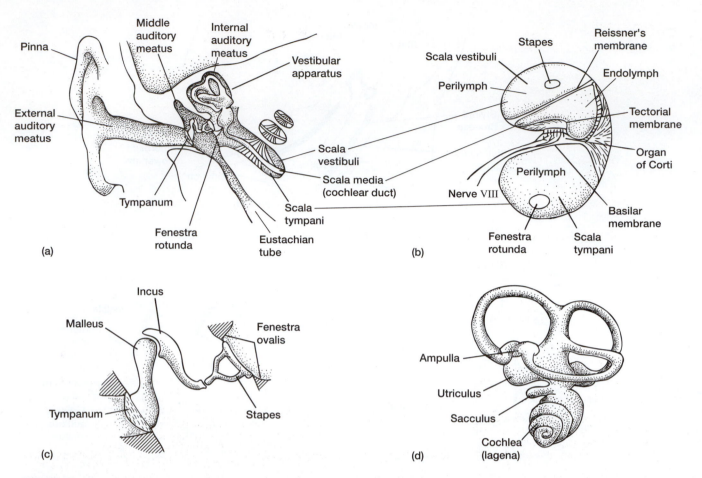

FIGURE 17.36 **Anatomy of the therian ear.** (a) External, middle, and inner ears. (b) Cross section of the cochlea. (c) Three middle ear ossicles of mammals. (d) Mammalian vestibular apparatus. Note that the lagena is lengthened and coiled to form the cochlea.

became dedicated to sound transmission. The stapes lightened, making it more responsive to vibrations. Independently, in various tetrapod lineages, it became associated with a tympanum, and enclosed within its own middle ear chamber, where it transfers sound from the tympanum to the inner ear. In some amphibians and in reptiles and birds, the stapes is tipped with a cartilaginous extension, the **extracolumella.** This cartilaginous structure, derived from the hyoid arch, rests on the undersurface of the tympanum.

In mammals, there are three middle ear bones. The **stapes** is much reduced and lightened from the early amniote condition. The **incus** and **malleus** are derivatives of the **quadrate** and **articular** bones, respectively. The malleus, incus, and stapes form an articulated chain that spans the middle ear cavity from the tympanum to the inner ear (figure 17.36c).

Evolution of middle ear bones (p. 276)

The **inner ear** includes the vestibular apparatus and the surrounding perilymphatic spaces. As mentioned, in birds and mammals, the auditory part of the vestibular apparatus is the tubular lagena. In mammals, the lagena forms a coiled cochlea (figure 17.36b,d). The organ of Corti runs along a central channel suspended within the lagena. Two

parallel perilymphatic channels run on either side. Thus, the cochlea consists of three coiled fluid-filled channels. The two perilymphatic channels are the **scala vestibuli** and the **scala tympani,** and the canal between them is usually called the **scala media** (cochlear duct). The **basilar membrane** separates the scala tympani from the scala media, and the organ of Corti vibrates with the basilar membrane in response to sound waves. In many vertebrates, the hair bundles of the organ of Corti are embedded in a firm plate, the **tectorial membrane. Reissner's membrane** is located between the scala vestibuli and scala media (figure 17.36a,b).

Sound enters the inner ear through the **fenestra ovalis** (= fenestra vestibuli), or oval window (figure 17.36c). One end of the thin stapes (or columella) expands into a pedicle, or **footplate,** that occupies this window so that sound waves pass from this ear ossicle to the fluid filling the chambers of the inner ear. The **fenestra rotunda,** or round window, at the end of the perilymphatic channels is sealed by a flexible membrane (figure 17.36a,b). The auditory apparatus, containing sensitive hair cells, floats in fluid between the fenestrae ovalis and rotunda. In all vertebrates, sounds conveyed to this fluid vibrate and mechanically stimulate the auditory receptors.

Functions of the Ear

Fishes The inner ear of fishes is very similar to that of tetrapods. There are three semicircular canals (except in hagfishes, which have one, and lampreys, which have two) and three otolithic regions: the sacculus, the utriculus, and the lagena. In addition, a few fishes and a few tetrapods have a **macula neglecta,** a supplementary sensory area near the utriculus.

In fishes, hearing generally involves the sacculus and the lagena, although sometimes one region predominates. The maculae within the sacculus and lagena are the sound receptors. When hair cells of the maculae are set in motion by sound vibrations, they move against the relatively dense and stationary otoliths. Differences in size and shape of the otoconia are thought to lead to slight differences in hair cell stimulation and thus allow different sound frequencies to be detected. Further, the hair cells within the sacculus and the lagena are oriented along perpendicular axes (figure 17.37). Therefore, motion in one direction stimulates one set of hair cells maximally and the other set minimally. Signals from different directions apparently produce different stimuli that are used by the nervous system to pinpoint the source of a sound.

Fish otoconia usually consist of secreted calcium grains, but in some bottom-dwelling elasmobranchs, sand grains enter through the open endolymphatic duct and settle on the maculae. As mentioned, the maculae of fishes are sensors of gravity and motion as well as of sound.

Sound reaches and stimulates the inner ear via several routes. The tissues of a fish and its environment are primarily composed of water and have similar frequency amplitudes. Thus, sound waves pass directly from water, through fish tissues, to the inner ear and set hair cells in motion. The dense otoconia oscillate at a different frequency from the associated hair cells, causing the hair bundles to bend and stimulate these sensory cells.

Sounds are transmitted to the inner ear of fishes through additional routes. In some species, extensions of the swim bladder come in direct contact with the inner ear. Vibrations picked up by the swim bladder are conveyed directly to the sound-detection apparatus (figure 17.38a). Furthermore, because the air-filled swim bladder and fish's body tissues are very different densities, the light swim bladder may be more responsive to sound and act as a resonator to enhance sound detection. In other species, the swim bladder is connected to the inner ear through a series of tiny bones called **Weberian ossicles** that carry vibrations to sound detectors within the sacculus and lagena (figure 17.38b).

Tetrapods Ears adapted for hearing in an aquatic environment were brought to an air environment during the vertebrate transition from water to land. Most anatomical changes in the tetrapod ear came about to deliver sound energy traveling in air to the small fluid spaces of the inner ear. At issue is a physical problem of **impedance matching** that arises from the different physical responses of water and air to vibrations. Fluid is thick

FIGURE 17.37 **Fish inner ear.** Orientations of hair cells within the sacculus and lagena are shown. Arrows indicate the side of the hair bundle on which the kinocilium lies. Dashed lines indicate the divisions between bundles with different orientations. These differences in the placement of hair bundles cause different responses to mechanical vibrations traveling in the surrounding fluid.

After Fay and Popper.

and air is thin, so the two media differ in the amount of sound energy necessary to vibrate molecules. For a fish in water, sound passes easily from the fluid aquatic environment into the fluid inner ear environment. But for an animal living in an air environment, the responsiveness of inner ear fluid to sound waves differs from the air's responsiveness to sound waves, producing a water/air boundary. Because of its greater viscosity, fluid resists being set in motion by arriving airborne sounds. Consequently, most airborne sounds are reflected away from the ear and go undetected. The structures of the middle ear are largely involved in impedance matching, which means that they gather, concentrate, and deliver airborne sounds to the inner ear at a sufficient level to impart these vibrations to the fluid-filled spaces of the inner ear. This allows the strategically positioned sensory hair cells within the inner ear to respond to oscillations traveling through the surrounding fluid. Without

FIGURE 17.38
Route of sound transfer to the inner ears of fishes. (a) In some fishes, the swim bladder includes anterior extensions that contact the inner ear. (b) In other fishes, the Weberian ossicles, a tiny series of bones, connect the swim bladder to the inner ear.

After Bone and Marshall; Popper and Coombs.

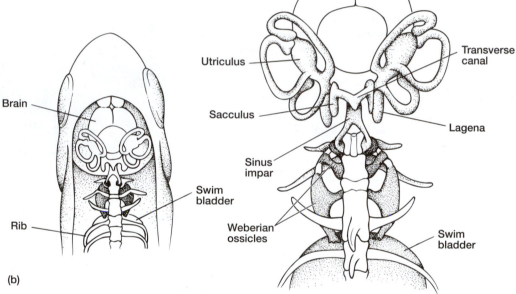

(b)

impedance matching by these middle ear ossicles, only about 0.1% of the sound energy in air would make it to the fluid of the inner ear. The rest would be reflected away.

Arriving sound waves set the tympanum in motion. It in turn affects the middle ear ossicles. These tiny bones function in three critical ways: (1) They act as a lever system to transmit vibrations to the fenestra ovalis; (2) they transform sound waves in air into sound waves in fluid; and (3) they amplify the sound. The tympanum may be over ten times the area of the fenestra ovalis to which the ossicles transmit sound. By collecting sound over a large area and focusing it on a much smaller area, the tympanum and fenestra ovalis effectively amplify sounds.

The stapes apparently does not act like a piston. Instead it rocks within the fenestra ovalis to disturb the perilymphatic fluid filling the inner ear. This fluid is incompressible and would resist such compression if it were not for the membrane-covered fenestra rotunda at the end of the peri-lymphatic channels. This is a specialized pressure relief window that allows some flexibility in the movements of the fluid as sound is transmitted to the inner ear. It has been suggested that the fenestra rotunda also helps dampen sounds once they have made a first pass through the cochlea and prevents these waves from ricocheting around the inner ear.

Amphibians There are two auditory receptors in amphibians: the **papilla amphibiorum** (amphibian papilla), unique to amphibians, and the **papilla basilaris** (basilar papilla), a possible forerunner of the organ of Corti in amniotes. Both are specialized neuromast organs (figure 17.39b).

Early physiologists concentrated on living amphibians. The frog, with its conspicuous tympanum and sensitivity to

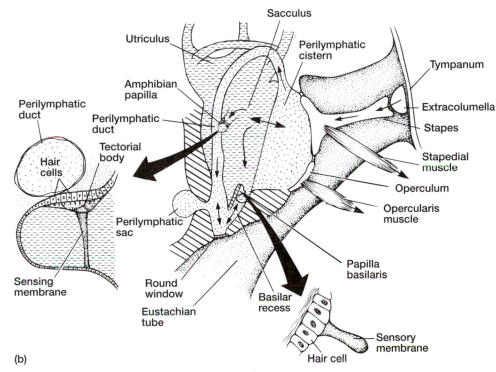

(a)

(b)

(a) After Romer and Parsons; (b) after Wever.

FIGURE 17.39 Hearing in frogs. (a) Cross section through the head of a frog. Because the eustachian tubes connect the two ears through the pharynx, a sound that sets one tympanum in motion also affects the ear on the opposite side by producing vibrations in the connecting air passageway. This is thought to allow frogs to localize the source of sounds. (b) Sound arrives at a frog's inner ear via two routes: One involves the tympanum-columella, and the other involves the opercularis muscle-operculum. Vibrations arriving by either pathway cause the fluid in the inner ear to vibrate. This vibration stimulates the auditory receptors. The way in which these receptors discriminate sounds is not well understood, but because they seem to be modified neuromast organs, it is thought that they respond selectively to shearing oscillations that arriving vibrations impart to the fluid of the inner ear.

airborne sounds, has received much attention. The lagena and utriculus are likely the vestibular receptors, and the sacculus, containing the papillae amphibiorum and basilaris, seems to be the primary site of sound detection. Typically the tympanum in frogs is flush with the surface of the skin. The extracolumella and stapes deliver these vibrations in series to the inner ear. The footplate of the stapes shares the fenestra ovalis with a small movable bone, the **operculum.** The **opercularis** is a tiny muscle that joins the operculum to the suprascapula of the pectoral girdle. The opercularis is a derivative of the levator scapulae muscle. The tiny **stapedial muscle** is also thought to be a derivative of the levator scapulae and runs from the suprascapula to the stapes (figure 17.39a). Both muscles connect the inner ear to the pectoral girdle and, indirectly via the limb, to the ground. In this way, the muscles introduce routes through which **seismic** sound waves may travel from the ground to the inner ear.

Thus, sound reaches the inner ear via the opercular and stapes pathways, and sound waves vibrate the fluid in the inner ear, stimulating the auditory receptors. One method frogs use to discriminate sound frequencies apparently takes advantage of these two routes. The papilla amphibiorum responds best to low-frequency sounds arriving via the operculum. The papilla basilaris responds to higher-frequency sounds arriving via the stapes. The middle ear muscle enhances this discrimination pattern. Contraction of the opercularis muscle (and relaxation of the stapedial muscle) leaves the stapes free to vibrate in the oval window. The reverse situation, contraction of the stapedial muscle and relaxation of the opercularis, immobilizes the stapes. In addition, these tiny muscles protect the auditory receptors from violent stimulation by selective contraction. In this way, they might also enhance the inner ear's ability to discriminate different sound frequencies or select sounds arriving along different acoustic pathways.

To localize sounds, frogs might take advantage of another feature of middle ear design. Eustachian tubes join the left and right middle ears through the buccal cavity. Pressure generated by the tympanum responding on one side of the head is transmitted to the buccal cavity, which may act as a resonator before it transmits the sound to the opposite tympanum. Such coupling of tympanum vibrations by a resonator between them means that sound reaching the left and right ears has different acoustic qualities. This difference allows frogs to localize sound sources (figure 17.39b).

In modern salamanders, vibrations transmitted through the stapes stimulate the inner ear; however, the tympanum is absent. Instead, the tip of the stapes is attached to the squamosal bone through a short **squamosal-stapedial ligament.** As in anurans, the footplate of the stapes shares the fenestra ovalis with the operculum. Only the opercularis muscle is present in salamanders. In most cases it is derived from the levator scapulae muscle, but in one family, it arises from the cucullaris muscle. Two routes of sound transmission to the salamander inner ear are thus available: one from the squamosal bone to the stapes, and another from the pectoral girdle to the operculum (figure 17.40a).

Although the routes that sound travels to the inner ear are known, the mechanism of reception and the method of processing sound waves in salamanders are not clearly understood. When a salamander is in water, impedance is low and sound reaches the inner ear with little reflected loss. However, lacking a tympanum, the salamander ear would seem to be poorly suited to detection of airborne sounds. Further, the fenestra rotunda is also absent, possibly making the fluids in the inner ear less responsive to pressure vibrations. These observations led to the suggestion that salamanders are deaf in air; however, experimental work refutes this. In fact, salamanders on land respond to seismic and airborne sounds, although their sensitivity to such sounds is less than that of anurans.

Sounds that reach the fenestra ovalis set the fluid of the inner ear in motion. In salamanders, these vibrations ripple across the skull via a channel filled with cerebrospinal fluid in order to reach the inner ear of the opposite side (figure 17.40b). Along this internal route lie the papillae amphibiora that these passing vibrations stimulate.

In summary, the inner ear of living amphibians includes two major auditory receptors. The primary auditory receptor is the papilla amphibiorum, a sensory receptor found only in living amphibians. The other is the papilla basilaris, which each order of living amphibians seems to have evolved independently. In anurans, the papilla basilaris resides in the sacculus. In salamanders, it is located in the lagena, and in caecilians, in the utriculus. The auditory receptor cells of amphibians are perched on an immobile base attached to the cranium. The hair bundles project into the fluid that sets them in motion. As we shall see next, the opposite is the case in reptiles. Sensory cells move on a flexible basilar membrane, and hair bundles are usually restrained by the tectorial membrane.

Reptiles   In most reptiles, airborne sounds vibrate the tympanum. This imparts motion to the extracolumella and stapes, which transmit these vibrations to the inner ear (figure 17.41a,b). The primary sound-sensitive area within the inner ear is the slightly expanded lagena. The primary receptor is the **auditory papilla,** which is stimulated by

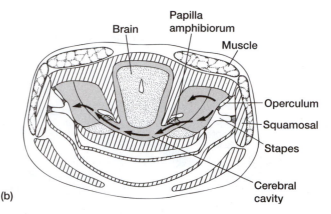

FIGURE 17.40   Hearing in salamanders.
(a) In many salamanders, sounds reach the inner ear via a squamosal-columella route and via the opercular muscle from the scapula. (b) The two inner ears on opposite sides of the head are connected via a fluid-filled channel that passes through the cerebral cavity. This channel may allow sonic vibrations to spread from one ear to the other (solid arrows).

sound vibrations transmitted to the fluid of the inner ear (figure 17.41c). In addition, various supplemental auditory receptors are present in some species.

Snake ears, unlike those of most reptiles, lack a tympanum. Through a short ligament, the stapes is attached at one end to the quadrate bone of the upper jaw, and the opposite end fits into the fenestra ovalis. Despite the popular opinion that snakes are deaf, experimental work disproves this. Recordings of electrical activity from the areas of the brain to which auditory nerves travel confirm that the inner ear of snakes is responsive to seismic and airborne sounds, although the range of sensitivity is somewhat restricted.

Birds   Because the lagena of the avian inner ear is lengthened in comparison to the reptile's, the stand of hair cells is drawn out into a long strip (figure 17.42a,b). Hair bundles are usually embedded in the tectorial membrane, a continuous sheet that acts to increase shear on these bundles as hair cells are agitated by sound vibrations.

In owls, the tightly packed, raised rim of facial feathers that gives owls their "cute" bespectacled look is the functional

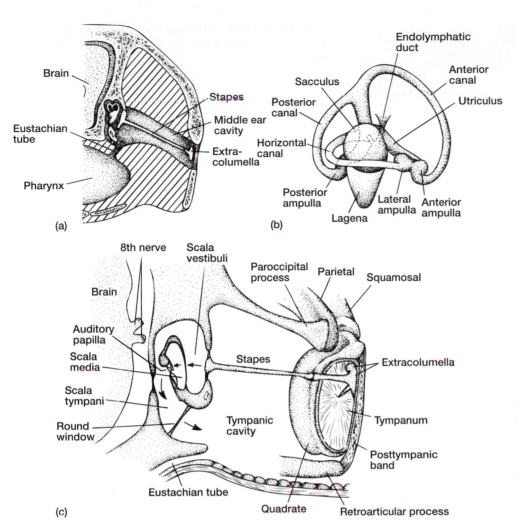

FIGURE 17.41 **Hearing in reptiles.** (a) Cross section through a reptilian head. (b) Vestibular apparatus of a lizard. (c) Section through the ear of an iguana showing the relationship of tympanum, extracolumella, stapes, and inner ear.

(a) After Romer and Parsons; (b,c) after Wever.

equivalent of the pinna. This facial ruff of feathers channels sounds into the auditory meatus (figure 17.43a). In an individual owl, the left and right external orifices and their associated auditory meatuses are different in size and shape (figure 17.43b,c), resulting in two ears that produce different acoustic qualities from many directions. The nervous system takes advantage of this to increase the precision with which sources of sound can be pinpointed in the habitat.

Mammals The pinna is likely a therian invention. It is absent in monotremes, but present in therian mammals, except where it has been secondarily lost, as in moles and whales. The pinna directs sounds into the external auditory meatus, where these sounds vibrate the tympanum. All mammals have three ear ossicles that amplify these vibrations and carry them to the fenestra ovalis (figure 17.44a,b). From the fenestra ovalis, vibrations ripple through the fluid in the extensive cochlea. As mentioned, the cochlea is composed of three parallel channels. The middle channel (cochlear duct) includes the organ of Corti, which consists of an outer and inner row of hair cells. There may be 20,000 to 25,000 of these cells. Hair bundles are embedded in the tectorial membrane (figure 17.44c).

Discrimination of Different Frequencies Sound waves stimulate the neuromast organs (e.g., auditory papilla, organ of Corti), the sites of sound sensation. In some species, the hair bundles protrude directly into the surrounding fluid. When this fluid vibrates in response to sound waves, the bundles are bent by the moving fluid and stimulated. In most higher vertebrates, hair bundles are embedded in the tectorial membrane. Sound waves impart motion to the tectorial membrane, which then deflects the hair bundles. This action stimulates hair cells.

In addition to being able to detect sound waves, auditory receptors can discriminate between different frequencies (tones) of sound. Thus, the inner ear is also a resonance analyzer. Hair cells are tuned into only a narrow range of frequencies. In mammals, differences in the orientation of hair cells in the inner and outer rows of the organ of Corti produce differences in sensitivity in different regions. Sequential grading of the tuned hair cells along the organ of Corti produces tone discrimination over a range of frequencies.

However, there may be more to tone detection than the tuning of hair cells. In mammals, the basilar membrane on which the organ of Corti resides changes gradually in width as the cochlea twists. The basilar membrane may be

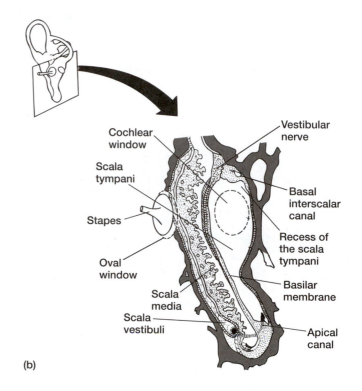

FIGURE 17.42 **Hearing in birds.** (a) Vestibular apparatus of a chicken. (b) Section through the lagena of a chicken.

After H. Evans.

Labels for figure (a) vestibular apparatus:

Inner ear

Middle ear cavity

Opening of pharyngotympanic tube

Anterior canal

Common crus

Ampulla

Posterior canal

Oval window

Round window

Stapes

Horizontal canal

Lagena

Extracolumella

Nerve

Tympanum

Labels for figure (b):

Cochlear window

Vestibular nerve

Scala tympani

Stapes

Basal interscalar canal

Oval window

Recess of the scala tympani

Scala media

Basilar membrane

Scala vestibuli

Apical canal

like a harp in that each section of the membrane may resonate only to a distinctive range of frequencies. If this is the case, tones that enter the cochlea impart the greatest motion to the section of the basilar membrane that corresponds to their frequency; thus, specific tones stimulate specific sections of hair cells. It has been suggested that such differential stimulation of hair cells could contribute to tone discrimination (figure 17.45).

Evolution of Hearing   Any organism moving through its environment produces sounds that betray its approach. It is not hard to imagine the adaptive advantages of a sound detection system that alerts an organism to another's presence. Although the importance of the sense of hearing is easy to understand for audiophiles like ourselves, who have a well-developed ability to discriminate a wide range of sounds in our environment, there are still many unanswered questions about the evolution of hearing.

One of the major evolutionary changes in the auditory system occurred during the transition from water to land. This transition involved the appearance or enhancement of the middle and external ears, structures that collect airborne sounds and match them to impedance properties of the inner ear fluid. In addition, the inner ears of early tetrapods were able to receive seismic sounds as well. As important as this water-to-land transition was, it has been difficult to study directly because there are no living forms close to those of the basal tetrapods.

However, based upon new fossil evidence, the earliest tetrapods lacked an auditory system specialized for detection of airborne sounds. This is not too surprising because most were still predominantly aquatic animals. The early, *primitive tetrapod ear* possessed an inner ear, carried forward from sarcopterygian ancestors, and a stapes, modified from the fish hyomandibula. In some derived sarcopterygian fishes, the stapes (modified hyomandibula) manipulated the spiracular opening for air intake. In primitive tetrapods, the stapes was massive, assuming a new role as primarily a mechanical brace to the back wall of the braincase. However, other otic specializations were not yet present, namely, missing were an enclosed middle ear cavity and tympanum.

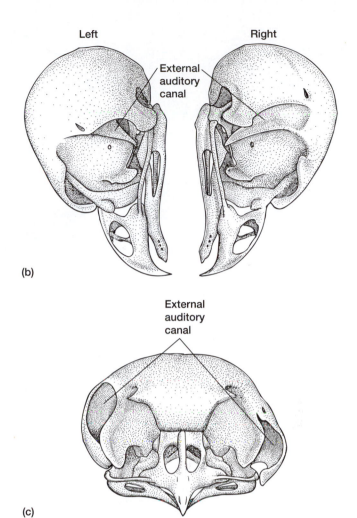

FIGURE 17.43 Auditory acuity in owls. (a) Facial disk feathers in normal position (inset) have been removed to reveal sets of auditory feathers. In this barn owl, tightly packed parabolic rims of feathers encircle the face and external orifice of the ear. This facial ruff of feathers, as it is called, collects and directs sounds to the external orifice of the ear. Note the asymmetrical positioning of the preaural flaps of feathers (dashed lines). (b) Left and right sides of the skull of a Tengmalm's owl. There are slight differences in the size of the external auditory canal. (c) Anterior view showing asymmetry of otic areas in Tengmalm's owl.

(a) After Knudson; (b,c) after Norberg. From "The Hearing of the Barn Owl" edited by E. I. Knudson, drawn by Tom Prentiss in Scientific American, December 1981. Copyright © 1981 by Scientific American, Inc.

As the early tetrapod braincase experienced fundamental remodeling and adjustments to an amphibious life, the massive stapes was relocated, bringing it into closer proximity with the inner ear. As the hyomandibula, it fitted into a facet on the wall of the fish braincase. In early tetrapods, this relocation settled it into an ossified gap, a hole in the braincase that was to become the fenestra ovalis. Although still a bracing bar at the back of the early tetrapod braincase, this shift brought the stapes into a position where any vibrations it carried could affect the inner ear.

From this primitive tetrapod ear, a *tympanic ear* (enclosed middle ear plus tympanum) subsequently evolved independently at least five times, perhaps more, in later tetrapods. Lissamphibia, and the related temnospondyls, were the earliest group to evolve a tympanic ear. The tympanum and the enclosed middle ear were inherited by frogs but lost in urodeles and caecilians. Within amniotes, the tympanic ear appears independently in turtles, sqamates, and archosaurs, as well as in mammals. In turtles, a cone-shaped excavation in the quadrate holds the tympanum. Within living lepidosaurs, only lizards exhibit the fully developed condition, as *Sphenodon* and snakes have secondarily lost both the tympanum and middle ear cavity. Archosaurs, including living crocodiles and birds, also separately evolved a tympanic ear, which correlates with increasing hearing acuity related to the advantages of isolating sound reception from other functions.

In mammals, the evolution of a tympanic ear was more complex. We begin with pelycosaurs, basal synapsids. The pelycosaur ear was simple, essentially a primitive tetrapod ear. There was no evidence of a specialized otic region for the reception of airborne sounds. The stapes was massive, still a bracing bar at the back of the braincase. There was no enclosed middle ear cavity, and a tympanum was absent. By the time mammals appear within the later synapsid radiation, a tympanic ear is present. The single stapes is joined in the middle ear by the incus and malleus, derivatives of the quadrate and articular, respectively. These three middle ear bones reach from tympanum to inner ear. A good series of fossils reveals the anatomical changes that occurred in the transition from pelycosaur to mammal within the synapsids. But the adaptive advantages favoring such changes are still being debated.

One suggestion correlates these changes with changing metabolic demands as later synapsids became endothermic. Recall that the lower jaw of pelycosaurs was formed of the main dentary and a collection of postdentary bones, articulated through the quadrate-articular jaw joint. The stout stapes contacted and braced the quadrate, which in turn articulated with the articular, and in turn contacted postdentary

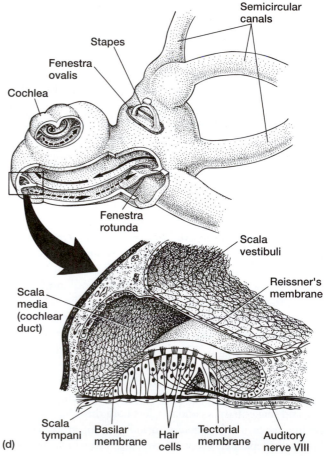

FIGURE 17.44   Mammalian ear. (a) Bony endocast in lateral view of left vestibular apparatus, capybara (South American rodent). (b) Cross section through a mammalian skull. (c) Internal structure of the cochlea. (d) Section through the organ of Corti showing inner and outer rows of hair cells and the tectorial membrane in which hair bundles are embedded. Sound waves travel first in the scala vestibuli (solid arrows) before passing at the apex of the cochlea into the scala tympani (dashed arrows).

(a) Image kindly supplied by Dr. Irina Ruf, Universität Bonn; (b) after H. M. Smith; (c) after Romer and Parsons; (d) after vanBeneden and vanBambeke.

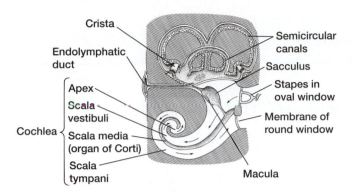

FIGURE 17.45 **Distribution of sound vibrations through the cochlea.** The stapes delivers vibrations at the fenestra ovalis. These vibrations spread through the perilymph within the chamber of the scala vestibuli and around the tip of the cochlea into the connecting chamber of the scala tympani. The cochlear duct, containing the organ of Corti, lies between these two chambers. Passing vibrations are thought to stimulate appropriate sections within this organ. The flexible membrane across the fenestra rotunda serves to dampen sound waves and prevent their rebound back through the cochlea.

After vanBeneden and vanBambeke.

bones. But this also meant that these bones were secondarily part of an auditory chain. Any sound vibrations, to which the postdentary bones were responsive, would pass via the articular-quadrate-stapes chain to the inner ear. In cynodont synapsids, the masseter muscle first appears, perhaps representing a general increase in jaw muscles, accommodating greater use of jaws in feeding and mastication. This brought greater stresses to the jaw joint, favoring the strengthening of the postdentary bones. But consolidation and firming of the postdentary bones would be disadvantageous to their role as vibrating auditory ossicles. A conflict of functions arose. The solution was to transfer jaw muscle insertions to the dentary along with its enlargement, at the expense of postdentary bones, to receive these muscles. The subsequent development of a separate dentary-squamosal jaw joint provided the opportunity for final loss of the postdentary bones, and movement of the quadrate and articular to the middle ear, where they joined the stapes, dedicated to just sound transmission.

Evolution of middle ear bones (p. 276)

Electroreceptors

Structure and Phylogeny

Most fishes and living monotremes, but not other tetrapods, possess **electroreceptors,** sensory receptors that are responsive to weak electrical fields. Electroreceptors are modified neuromast organs located in pits within the skin that are predominantly concentrated on the fish's head. There are two types of electroreceptors. An **ampullary receptor** contains supportive cells that lie at the bottom of a narrow channel filled with a gelatinous mucopolysaccharide (figure 17.46a). Afferent neurons embrace the receptor cells.

The **tuberous receptor** lies buried under the skin in an invagination beneath a loose layer of epithelial cells (figure 17.46b). This loose layer of epithelial cells may differentiate into **covering cells** over the sensory cells and a superficial set of **plug cells.** This type of electroreceptor is responsive to higher frequencies than the ampullary receptor and is generally adapted to receive electric discharges of the fish's own electric organ. Hence, tuberous receptors are known only in electric fishes so far.

Electroreceptors appeared very early in fish evolution, judging from the presence of pits in the dermal bone of acanthodians and some groups of ostracoderm fishes. Among living fishes, electroreceptors are found in all elasmobranchs, catfishes, sturgeons, some teleosts, and the lungfish *Protopterus* (see illustration of electroreceptors of a skate in figure 17.46c).

Form and Function

In electric fishes, specialized blocks of muscle called **electroplaques** form the electric organ. In some, the electric organ can generate a sudden jolt of voltage to stun prey or thwart a predator.

Electric organs (p. 373)

In most electric fishes, however, the electric organ produces a mild electrical field around the fish. Electrically conductive and nonconductive objects that enter this field have different effects on the flow of current produced. Living animals, such as other fishes, are relatively salty, making them conductive and causing the lines of the electric field to converge. Rocks are usually nonconductive and cause the current lines to diverge (figure 17.47a). Electroreceptors are sensitive to these distortions in the surrounding electric field. This type of electroreceptor is common in freshwater fishes that live in murky waters or hunt at night. They use this information to navigate and detect prey. The electric fish *Gymnarchus* holds its body rigid to align generating and receiving receptors throughout the body (figure 17.47b).

Patterns of electric organ discharge are also used in intraspecific communication. These patterns change with social circumstances. For example, fishes use electrical communication to recognize the sex and species of other fishes. From these electric signals, they detect threat, submission, and courtship.

Many fishes that do not actively generate their own electrical field nevertheless possess electroreceptors. Elasmobranchs, some teleosts, sturgeons, catfishes, and others are endowed with abundant electroreceptors across their heads, especially concentrated around the mouth. These organs are sensitive to stray electrical fields produced by muscle contractions of prey animals. In sharks, these receptors are termed **ampullae of Lorenzini.** These electroreceptors can localize the weak electrical fields produced by the breathing muscles of prey animals buried out of sight in sand or loose sediment (figure 17.47c). Further, the gel that packs these shark electroreceptors produces voltage changes in proportion to

Many fishes carry about their own light source in specialized structures termed light organs or photophores. The fishes themselves do not produce light; rather, they carry symbiotic bioluminescent bacteria in specially designed skin pouches. These pouches of bacteria glow continuously, so to turn off the light, the fish flips them into a black-lined pocket, lifts a shutterlike cover over them, or pulls the pouches into its body.

Freshwater fishes lack light organs, but such organs are present in many marine species. At ocean depths not reached by natural light, many fishes have light organs. However, light organs are not restricted to deep-water species. Many shallow-water marine fishes that are presumably active at night also possess light organs.

Light organs are deployed in a variety of roles. By blinking the shutter over their light organs, some species produce characteristic flashes used in signaling members of the same species as part of sexual communication or schooling behavior. During daylight hours, fishes in the upper 1,000 m of the ocean are easily silhouetted from beneath against the light sky above. To camouflage their shapes to predators or prey below, they produce a glow that matches their body color to the light from above. This is accomplished by protruding light organs from gut diverticula. The soft light that is emitted illuminates the fish's ventral surface (box figure 1a). Light organs are also used extensively in feeding. They form lures on the tips of barbels around or in the mouth. Flashlight fish use light organs on their heads to illuminate their prey. Besides lures attracting prey, the photophores may spew clouds of bioluminescent bacteria into the surrounding water to confuse predators. Some fishes have carried this a step further. At great depths, the visual pigments of most fishes are sensitive only to blues and greens. However, *Pachystomias* has a red-emitting light organ with red-sensitive retinal pigments that allow it, unlike its prey, to see light. Consequently, it can red-illuminate its unsuspecting prey without alerting it (box figure 1b).

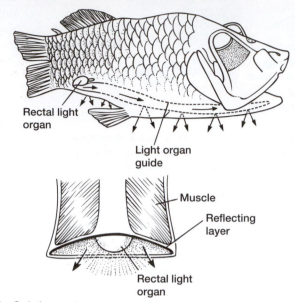

Rectal light organ

Light organ guide

Muscle

Reflecting layer

Rectal light organ

(a) *Opisthoproctus*

Red-emitting light organ

(b) *Pachystomias*

BOX FIGURE 1 **Bioluminescent light organs of fishes.** (a) The pelagic *Opisthoproctus* has a single bacterial rectal light organ. Radiated light is spread evenly by a long tubular light guide. A reflective layer through the transparent skin of the belly directs the light across the whole ventral surface of the fish. By illuminating the ventral surface, the fish is less silhouetted by downwelling light. Ventral illumination is thought to make this fish more difficult to detect by deeper predators attempting to locate its silhouette above them. A cross-sectional diagram of the tubular light guide projects a view of the rectal light organ at the far end. (b) The deep-ocean fish *Pachystomias* has a light organ near its eye that emits red light to illuminate its prey.

(a) After Herring; (b) after Guenther.

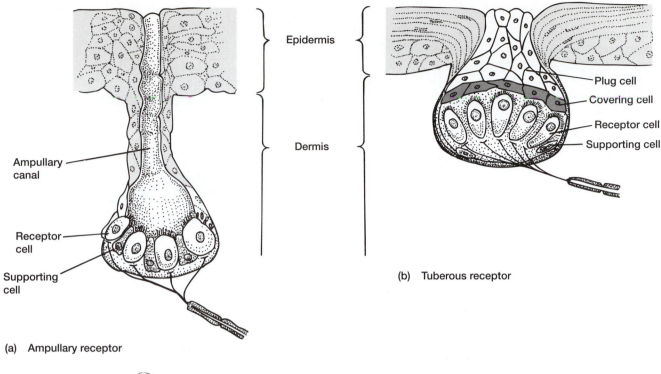

Epidermis

Dermis

Ampullary
canal

Receptor
cell

Supporting
cell

(a) Ampullary receptor

Plug cell

Covering cell

Receptor cell

Supporting cell

(b) Tuberous receptor

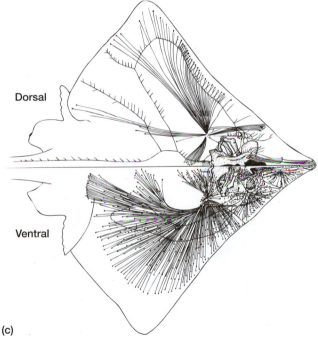

Dorsal

Ventral

(c)

FIGURE 17.46   Electroreceptors. (a) Ampullary
receptor. In ampullary receptors, the electroreceptor
(or receptor) cells lie at the bottom of a deep ampullary canal filled
with a mucopolysaccharide. Ampullary receptors are common in
fishes that are sensitive to electric energy in the environment.
(b) Tuberous receptor. In tuberous receptors, electroreceptor
cells form bunches near the body surface within a slight skin
depression. They are covered by a loose layer of covering cells
and plug cells, both cells being specializations of the epidermis.
Tuberous receptors are found only in electric fishes—that is,
those capable of producing electric signals. (c) Distribution of
electroreceptors (black dots) in the skate *Raja laevis.* Note that
distributions across the dorsal and ventral surfaces differ.

(c) After Raschi.

temperature differences and may be sensitive to changes less
than 0.001°C. This suggests a role in thermoreception as
well. It is guessed that tracking marine temperature gradients
may help sharks detect subtle thermal boundaries where prey
congregate.

Electroreceptors send spontaneous impulses to the
central nervous system at a regular rate. This rate increases
or decreases with distortions in the electrical field or with
stimuli from stray electrical discharges that prey generate.
The mechanisms by which such electrical fields stimulate
receptor cells is not clear. Impulses are thought to be transmitted from electroreceptors across synapses to afferent

neurons and then to the central nervous system. Most information that electroreceptors gather is transmitted directly to
the cerebellum. In fishes that depend on such information,
the cerebellum is often enlarged (figure 17.48a,b).

Some electroreceptors respond very fast to stimuli.
They are designed to relay information rapidly to the central
nervous system. Rapid relay about the electrical discharge
patterns of other electric fishes suggests that electroreceptors
play a role in communication. Other electroreceptors are
sensitive to changes in electrical amplitude. These receptors
seem suited to respond to changes in a fish's own electrical
field and thus play a role in navigation.

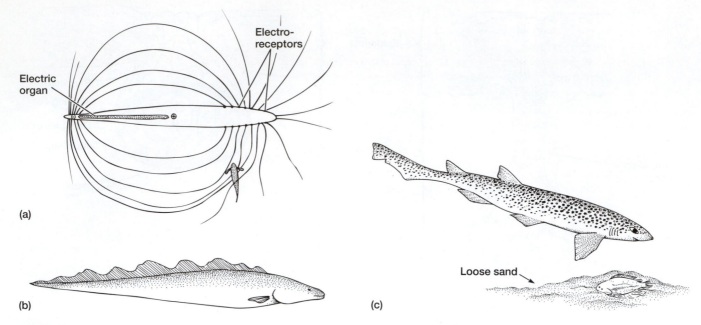

(a)

(b)

(c)

FIGURE 17.47 **Functions of electroreceptors.** (a) Navigation. This electric fish generates its own low electric field. Environmental objects entering the field distort it. Electroreceptors concentrated on the head detect this distortion, and the nervous system interprets it to aid the fish in navigating around such objects. (b) The electric fish *Gymnarchus* holds its body rigid to align generating and receiving receptors throughout the body. Undulating fins produce movement. (c) Prey detection. Electroreceptors (ampullae of Lorenzini) around the mouth and head of this shark can detect the low levels of electric discharge given off by the active respiratory muscles of its prey. When the prey is buried beneath shallow sediment, these low levels of electric charge can betray its presence.

After Lissmann.

FIGURE 17.48 **Cerebellum of electric fish.**
(a) Dorsal view of the brain of a mormyrid fish. Note the large cerebellum. (b) Dorsal view of the same brain with the cerebellar hemispheres removed.

After Northcutt and Davis.

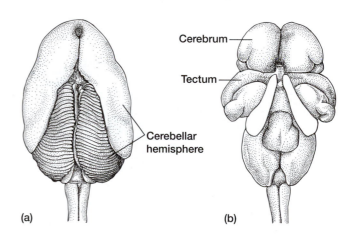

(a)

(b)

Additional Special Sensory Organs

Electromagnetic radiation and mechanical and electrical stimuli may not be the only types of information to which vertebrates are sensitive. Sea turtles and birds, for example, seem to use, among other stimuli, the orientation of the Earth's magnetic field to navigate. The Earth's magnetic field runs north/south, and near the poles the lines of the field are increasingly inclined toward the surface of the Earth. Thus, the magnetic field provides information about both direction and latitude. Sea turtles may use this information in an otherwise featureless ocean to reach preferred feeding areas. Years later, they may use magnetic fields again, to guide them back to their breeding grounds. Although experiments have provided evidence of sea turtles using the Earth's magnetic field for navigation, we have not discovered sensory receptors that gather such information.

Overview

Sensory organs gather information about the internal and external environment, and code this information as electrical signals reaching the central nervous system. Our conscious perceptions are based on this information. But the "taste" we experience, the "sounds" we hear, the "colors" we see, and the "pain" we feel, are brain events, interpretations of this coded information. Strictly speaking, sensory organs do not register "taste," or "sounds," or "colors," or "pain." These do not exist as such, independently in the environment, and so cannot be measured directly. The environment provides only chemicals, airborne vibrations, electromagnetic waves, and mechanical stimuli, and it is these that are gathered and coded by the sensory organs, and a perceptual interpretation applied by the brain.

General sensory organs are widely distributed throughout the body. They are concerned with external and internal sensations of touch, temperature, pressure, and proprioception. Often the nerve endings are encapsulated in tissue or associated with other organs that enhance deformation and thereby increase the sensitivity of the receptor. Free nerve endings lack such associations and respond directly to stimuli. Proprioceptors detect limb position, joint angle, and the state of muscle contraction. This information, usually processed at a subconscious level, helps in coordination of movement.

Special sensory organs are usually localized in their distribution and specialized for response to specific stimuli, principally chemical, radiation, mechanical, and electrical environmental energies. Chemoreceptors are responsive to chemical stimuli, which include prey and predator odors as well as pheromones, chemical messages between individuals. In tetrapods, taste buds are restricted to the mouth, but in fishes may be distributed externally over the body. Olfaction includes chemoreceptors within the nasal passage of fishes and tetrapods that detect passing odors. Vomerolfaction includes chemoreceptors in the separate vomeronasal organ, known only in some tetrapods, that represent an accessory chemosensory system that detects vomodors.

Various radiation receptors are specialized for responsiveness to different energy levels in the electromagnetic spectrum. The vertebrate eye is the best studied of the radiation receptors. It is responsive to wavelengths of light within the "visible" spectrum. Various methods of accommodation, through adjustments in lens and/or cornea, focus light on the photoreceptors within the retina, rods and cones. Paired eyes bring monocular or binocular vision, allowing interpretation of depth within the central nervous system. The unpaired pineal complex participates in photoreception in lower vertebrates, but tends to become endocrine in function in derived groups. Infrared radiation lies outside the visible band of light. Its detection, by specialized infrared receptors (thermoreceptors), occurs in some groups of snakes, via a specialized recessed facial pit, and in vampire bats.

Mechanoreceptors, based upon a neuromast organ, find their way into a variety of special sensory organs. The lateral line system, traversing the surface of fishes and aquatic amphibians, is a system of recessed grooves holding neuromast organs that are responsive to water currents. The vestibular apparatus is a balancing organ composed of fluid-filled channels affecting modified neuromast organs at their base, the cristae. The macula (otolith receptor) is responsive to accelerations or changes in body posture. The ear is a specialized mechanoreceptor that is responsive to sounds, which occur in the inner ear within the lagena (cochlea). In tetrapods, a tympanic ear arose several times wherein ear ossicle(s), enclosed within a middle ear chamber, connect a tympanum to the inner ear. Tetrapod ears are especially important in impedance matching, addressing the physical resistance of the fluid in the inner ear to respond to the airborne vibrations.

Electroreceptors occur mostly in fishes, where they are part of a navigational system and occasionally may be used in social signaling.

Every sensory organ projects to its own space in the brain, and is organized into a hierarchical system such that perception is a progressive process. Visual receptors, especially important in humans and the most studied of sensory organs, perhaps best reveal to us how this works. The eye forms an image but it is not a camera; it takes incomplete information from the world, not a faithful exact copy. The accumulative process of producing a perceived image is done progressively, starting in the retina and continuing in the thalamus and then the visual cortex (figure 17.27). Specific neurons in the visual cortex are specific to particular features of the image—line, movement, position, orientation, light, edges. From limited and incomplete information the brain reconstructs, and sometimes discards, information to produce the eventual perceived image. In the end, the brain also makes guesses which show up to us as optical illusions (figure 17.49).

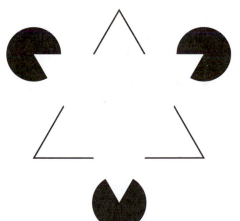

FIGURE 17.49 Optical Illusion, a Kanizsa triangle. Within this image we perceive a white triangle, which of course does not exist out in the environment, but is a brain event, produced in the brain itself during progressive processing of visual sensory signals.

WEB LINK

Visit the text website at *www.mhhe.com/Kardong7e* for additional study aids including selected references, functional laboraotories, and web links to chapter topics.

CHAPTER 18

Conclusions

INTRODUCTION

STRUCTURAL ANALYSIS

FUNCTIONAL ANALYSIS
    How Does It Work?
    Functional Coupling, Functional Compromise
    Multiple Functions
    Performance

ECOLOGICAL ANALYSIS

EVOLUTIONARY ANALYSIS
    Historical Constraints
    Primitive and Advanced

Diversity of Type/Unity of Pattern
Mosaic Evolution
Morphology and Modules

MODE AND TEMPO OF EVOLUTION
    Remodeling
    Embryonic Changes
    Hox Genes
    Evolutionary Significance

THE PROMISE OF VERTEBRATE MORPHOLOGY

Introduction

Morphology holds a central place in the intellectual development of modern biology. However, most persons are unaware of this and think of morphology for its practical applications. For example, in medicine, a knowledge of normal anatomy and function is necessary for physicians to recognize pathological conditions and restore a diseased or injured individual to a healthy condition. Heart valves, kidney machines, artificial limbs, repair of torn ligaments, setting of broken bones, realignment of occlusal surfaces of a tooth row, and so forth are all intended to replace or restore deficient parts to their normal form and function.

As valuable as this may be, morphology is more than just a practical sidekick to medicine. Morphology is also the analysis of animal architecture (figure 18.1a). In addition to naming parts, morphology delves into the study of why animals are designed the way they are. Certainly architecture, like morphology, has its practical side. Architecture too could be reduced to a mundane and everyday discipline if we saw it only as a means to put a roof over our heads. But for an inquiring mind, it offers much more. The great Gothic cathedrals are more significant and offer much more than just a place to put a congregation (figure 18.1b). To analyze cathedral architecture, we might begin by learning the

building's parts—apse, nave, clerestory, entablature, jamb, plinth, spandrel, triforium—but soon we realize that more goes into the architectural design of a cathedral than just the rocks and mortar of its anatomy. The design expresses many things about the people who produced it, the history out of which it came, and the variety of functions it performs. In a similar way, the architectural design of an animal expresses something about the processes that produced it, the history out of which it came, and the functions its parts perform.

When we ask why a particular part of an organism is designed the way it is, we formally approach this question from three integrated, analytical perspectives: structure, function, and ecology (figure 18.2). We begin with a description of the architecture of the part, its structure, extend this to include its function, and then look to the ecological setting in which form and function serve. If this is extended through time, we add the evolutionary dimension to our analysis. These three perspectives—form, function, environment—are parallel lines of design analysis. If we seek to understand the evolutionary changes in design, then each of these three becomes an internal critique of the series of morphological changes of interest. Function and

(a) Bird bone

(b) Gloucester Cathedral

FIGURE 18.1   Animal architecture. (a) Cross section of a bird bone showing the internal supportive struts. (b) Vault of Gloucester Cathedral (circa 1355) showing the supportive ribs that carry the weight of the ceiling vault to the sidewall piers.

environment are not just another set of characters to add to a pot of anatomical characters, but they represent parallel lines of distinct analysis.

An example of the use of anatomical, functional, and ecological series might be in the study of the evolution of birds (figure 18.3). Feathers arose through a series of anatomical modifications starting with scales in ancestral reptiles. Parallel to these anatomical transformations, a series of functional changes that feathers served can be proposed: protection, thermoregulation, leaping, parachuting, gliding, and flight. The ecological settings in which form and function serve might involve transitions from terrestrial to arboreal to aerial life. Attempts to understand the origin of birds center on hypotheses about these parallel changes in form, function, and accompanying ecology.

One way to test the vitality of such hypotheses is to examine how compatible the three series are with each other. For example, we might question the literal transformation of a reptilian scale into a bird feather, noting differences in structure (see chapter 6, Overview). In reply, we might instead attribute the two—scale and feather—not to a literal homology but to an underlying common homology of developmental interaction between dermis and epidermis. We might question the initial proposed functional role—thermoregulation—noting the absence of nasal turbinates in the earliest birds and their immediate dinosaurian ancestors. The absence of turbinates suggests the absence of endotherm, and hence the absence of a thermal insulating role for early feathers.

Evolution does not anticipate the future. Structures evolve not for roles they will serve millions of years in the future, but to address current survival roles in current environments. Preadaptation describes the happy accident of structures in ancestors finding adaptive favor in later descendants. But the initial success of a structure is determined by its survival value in current environments, no matter what the future may promise. Hypotheses regarding stages in the evolution of birds and the attainment of flight must identify the forms, functions, and ecological relationships that went together at each stage. Each stage represents an adaptive condition brought about by natural selection. Form, function, and ecological setting must be congruent at each stage. As one envisions a series of evolutionary stages, parallel lines of analysis give some idea of how transformations might have occurred in the evolution of birds.

As a practical matter, such an extensive analysis of vertebrate design is usually too much work for one person to do in one lifetime. This has led in recent years to more interdisciplinary efforts. It is now common to find paleontologists, with a knowledge of fossil anatomy, teamed up with physiologists and ecologists to approach the analysis of design in a fossil series.

Analysis of animal design has taken many forms. Scientists tend to emphasize in their research those techniques that immediately address the problem of analysis of design. Descriptive morphologists, interested initially in structural characteristics, give greatest attention to the anatomy itself. Functional morphologists are interested in how parts work and then tend to give their greatest attention to the physiology or biomechanics of the structure. Ecological morphologists seek to examine the interrelationships of an organism in its natural environment to discover how the structures are actually deployed. Evolutionary morphologists usually take advantage of all this information to form hypotheses about the historical development of structures through time.

Scientists from a variety of disciplines have brought a wide range of techniques and philosophical concepts to the study of animal design. At present, new and diverse analytical tools are being incorporated into the study of animal design and expanding it rapidly. Let us look at each of the four perspectives that morphology offers—structure, function, ecology, and evolution—to see what each contributes to the analysis of vertebrate design and what special insights each returns.

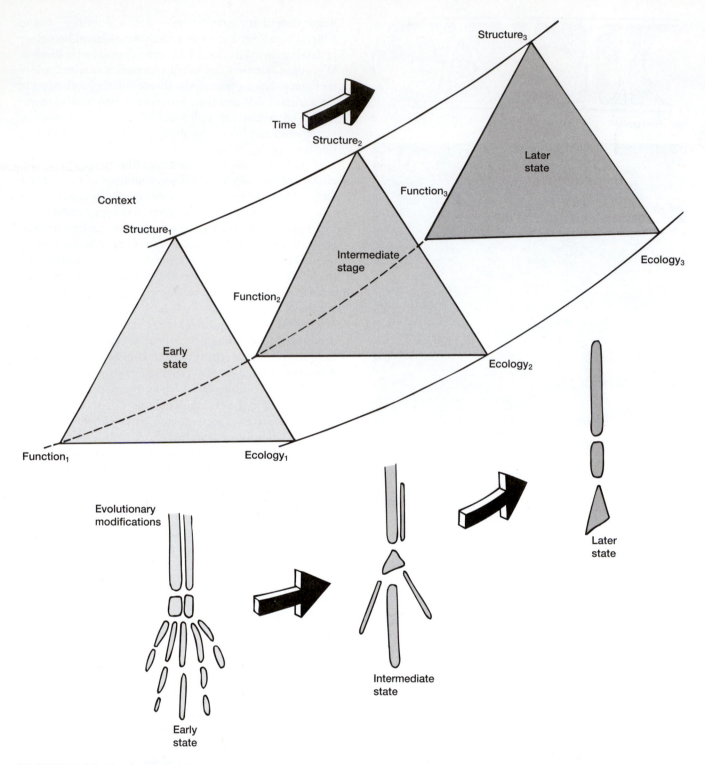

FIGURE 18.2 Analysis of design. At any point in time, the particular structure of a part serves a function or functions in a particular ecological setting. Through time, the structure may change along with its function and ecological setting; therefore, analysis of design includes structure, function, and ecology. If followed through time, these three change and help explain the evolutionary modifications of overall design.

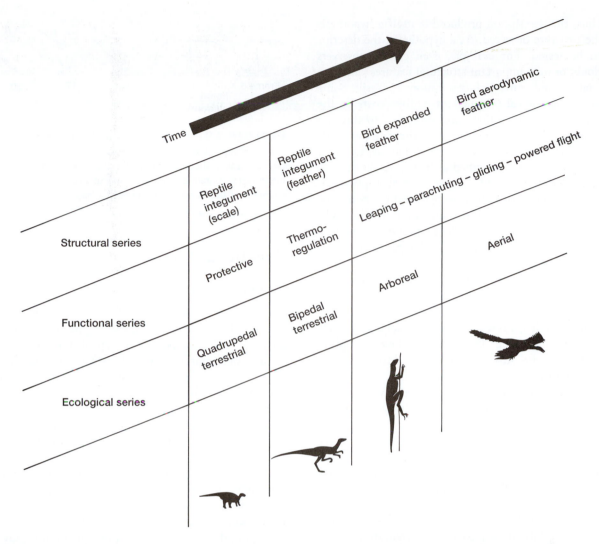

FIGURE 18.3 **Hypothetical evolution of flight feathers in birds.** Parallel structural, functional, and ecological series are indicated. The characteristic reptile scale affords protection from abrasion and desiccation to the ground-dwelling reptile. The active bipedal reptile takes advantage of a modified scale to thermoregulate its body temperature. Climbing up trees and occasionally leaping onto adjacent branches would make expanded scales advantageous in slowing the descent and in extending the aerial distance traveled. Gliding and powered flight would characterize a derived aerial organism, conditions favoring feathers as aerodynamic structures.

Structural Analysis

Analysis of animal architecture usually begins with a description of the organism or the part of interest. If our interest centers on feeding, we might expect to begin with a careful description of the jaws, teeth, skull, and articulations of cranial structures. Usually we do this in a straightforward manner by dissecting the organism carefully or by examining small parts under a microscope. From these observations, we produce a description of the anatomy.

As a practical matter, the description of structure begins by making a choice. We choose what level of organization to describe—cell, tissue, organ—and we choose what specifically to highlight. The choice of what level to study is often a matter of personal interest. The choice of what specifically to describe is best determined by considering which structural attributes are significant for the functional properties of the part being considered. Our description must serve the purpose of our study. If we are interested in the lines of muscle action, then points of origin and insertion should be included in our description. If force is of interest, then description of muscle fiber type and perhaps the angle that muscle fibers make with the line of action should be added. Descriptive analysis is a distinct endeavor, but it cannot be done in isolation. Knowledge of form and function proceed together. An understanding of function helps us make choices about what to describe. In turn, the resulting description provides detailed information about the elements that we hypothesize to be central to function.

Thus, the description produced is itself a hypothesis about the structure observed. As a hypothesis, the description can be tested. For example, even careful observers might doubt or miss important structural features. This can be especially troublesome with descriptions of fossils. Some claim to see anatomical evidence for hairlike coats on the surface of pterodactyls. Others take exception to this and suggest that these hairlike impressions in the rock surrounding the fossils have another explanation. Similarly, for many years, *Archaeopteryx* was described as having feathers; thus, it was considered to be a bird. Recently this description was challenged, and the fossil feathers were thought to be a hoax. This prompted an anatomical reexamination of the fossils, testing the hypothesis. The challenge proved groundless, and the descriptive hypothesis that *Archaeopteryx* possessed feathers was corroborated.

Even among living forms, new descriptions replace old ones. The primitive bony fish *Amia* has been a fixture in comparative anatomy courses for decades and studied by generations of students. It was described as lacking a clavicle within the pectoral girdle. That long-standing description has proved inaccurate. Although the bone is difficult to find, recent work indicates that it is, in fact, present.

Another approach to forming descriptive hypotheses involves the construction of a hypothetical sequence of events. Glenn Northcutt (1985) took this approach with a study of early vertebrate brains. He could have examined the brains of living hagfishes and lampreys, the most primitive living vertebrates, but he worried that they might have been radically altered during their long evolutionary history. If this were the case, they would not fairly represent a truly primitive condition. Or, he could have directly examined fossils of the earliest vertebrates, but the internal details of ostracoderm brains were not preserved. Instead Northcutt looked for features of the brain shared by hagfishes, lampreys, and gnathostomes. He reasoned that those characteristics common to all these groups must have been present in their common ancestor, and he used this suite of shared characteristics to construct a hypothetical description of the primitive vertebrate brain. This was used as his point of reference when he analyzed subsequent changes in the soft internal anatomy of the vertebrate brain.

Although descriptive morphology is not a glamorous part of science, the importance of description in the analysis of animal architecture cannot be overstated. A description sets forth a hypothesis that is to be analyzed further. If the morphological description is in error, then subsequent analyses of function and ecological role can be sidetracked. Descriptive morphology carries its own implications. Whether or not the fossil *Archaeopteryx* had feathers is not a trivial anatomical issue. A description of *Archaeopteryx* with feathers implies quite a different sort of animal than a description without feathers implies. Careful descriptions are the centerpieces of the analysis of animal design.

FIGURE 18.4 **Mechanics of dog jumping.** When a force is applied against the force platform, a reactive force (arrow) against the hindlimb of the dog is produced. This can be measured to obtain an indication of the stresses on the structural elements of the skeleton during such a leap.

Based on the research of R. McNeil Alexander.

Functional Analysis

How Does It Work?

An analysis of function addresses the question "How does it work?" For some, such a question gives life to descriptive anatomy, and these persons are apt to think of themselves as functional morphologists. To study function, functional morphologists borrow the analytical tools from engineering and physics to look at the biomechanics of animal function. Physiology also figures prominently in the functional analysis of many vertebrate systems. With such tools of analysis, the mechanical and physiological relationships between structural elements can be deciphered.

Forces acting on structures when they perform can sometimes be measured directly. For example, Lanyon (1974) glued strain gauges to bones to record fluctuations in forces during loading. Alexander (1974) measured forces that a jumping dog produced on a force platform (figure 18.4). Because reactive forces within the dog must equal those of the force platform, Alexander was able to evaluate the stresses experienced by the dog's hindlimbs.

Swimming in water and flying in air raise problems of fluid mechanics. One approach to the study of the fluid mechanics of fishes feeding was done by van Leeuwen (1984). He put tiny polystyrene spheres into water in which fishes were feeding and then filmed the ensuing events. During suction feeding, the nearby spheres accelerated along with the captured food. From their pattern of movements, van Leeuwen was able to calculate the velocity of water entering the mouth and map the surrounding area from which it came (figure 18.5a). To study the slow forward flight of birds, Spedding, Rayner, and Pennycuick (1984) floated neutrally buoyant, helium-filled soap bubbles in the path of a flying pigeon. As the bird flew through this cloud of bubbles, photographs recorded the effects of wing beats on the bubbles in the bird's wake (figure 18.5b). The swirling pattern of these bubbles confirmed that the wake is composed of small vortex rings, but oddly enough the forces implied by the observed patterns could not account for the lift produced by the wings. It appeared to be half what would be required to carry the weight of the bird's body, yet the bird obviously

FIGURE 18.5 **Fluid mechanics.** (a) Trout feeding. Small polystyrene spheres floating around the food respond to the suction developed when the trout feeds. Analysis of movements of these spheres reveals the volume of water drawn into the mouth and the shape of the water pulse carrying the food. (b) Slow pigeon flight. Helium-filled bubbles swirl in a vortex as the pigeon flies through them (left). This pattern can be used to calculate the components of motion that wing beats impart and the momentum they deliver (right). A_1 and A_2 are the cross sections of vortices produced by wing beats. Movements of one selected bubble (B) are shown as v and u components of an axis, X'Y', between vortex centers. Vortex structures in the wake are inclined at an angle, Ψ, with the horizontal.

Modified from van Leeuwen, 1984.

FIGURE 18.6 **Form-function models.** (a) The actual strike of a poisonous snake indicates the sequence of events that result in raising of the fang, injection of the venom, and release of the prey. (b) The kinematic model of the jaws simulates the structural elements and functions of the jaw elements. The accuracy of the form-function model is tested against the actual sequence of events.

After Kardong.

could and did fly. This emphasizes the point that biological design is often very subtle, difficult to analyze, and full of surprises to ruffle preconceptions.

This type of analysis yields a form-function model that represents the primarily structural and functional elements of the part studied. This form-function model is then tested against observations of the actual part in action. For example, a kinematic linkage model of the jaw of a cottonmouth snake was compared to high-speed films of the actual strike to see if the model accurately simulated fang rotation during envenomation (figure 18.6a,b). Movement and control of jaw elements in the model matched those of the snake. Lombard and Wake (1976) proposed a form-function model of salamander tongue protrusion and predicted that the rectus cervicis profundus muscle was responsible for retraction of the tongue. When the muscle was cut, the salamander could not retract its tongue, thus corroborating their prediction. Zweers (1982) proposed a form-function model of how pigeons drink. He predicted that the bird's esophagus drew in water by suction and tested this by inserting a fistula into the esophagus to prevent it from developing the negative pressure necessary to produce a suction. However, the pigeon could still drink, falsifying Zweers's form-function model. (The pigeon pumped in water rather than sucking it in; Zweers, 1992.)

Relationships between form and function can be diagrammed to illustrate the mutual influences between units of an organism. Parts that are closely linked form a *functional unit*. The jaws of a shark are structurally united for feeding; the elements of the limbs of a horse, for support and locomotion; the wing bones of a bird, for flight; the penguin's flippers, for swimming. Yet each of these functional units is connected and integrated with other parts of the body as well to bring unity to the overall functional performance of the organism. This grid of relationships within and between functional units of an organism places internal constraints on structures. To maintain the functional integrity of a part and to ensure its proper performance, these constraints are necessary. But these constraints also limit the changes possible and thus limit or at least restrict the course of subsequent evolutionary change as well. It is in such a way that the union of form and function is thought to affect likely evolutionary events.

Functional Coupling, Functional Compromise

The salamander mouth serves both in respiration (lung ventilation) and in feeding (prey capture). Plethodontid salamanders lack lungs altogether and depend entirely on cutaneous respiration. Because this group is without lungs, the mouth no longer participates in lung ventilation. Respiration and feeding are uncoupled in this group. Plethodontid jaws serve feeding almost exclusively and are quite different

FIGURE 18.7 **Structural and functional coupling between units.** The coupling of lung ventilation and locomotion may either constrain or enhance opportunities for subsequent evolution of alternative designs. (a) Lizard. Lateral bends of the body during locomotion alternatively impinge upon the lungs, interfering with ventilation. (b) Mammal. Dorsoventral bends of the body alternately compress and expand both lungs, serving to aid exhalation and inhalation. Plus and minus symbols indicate positive and negative pressures on the lungs (dashed lines), respectively.

Based on the research of D. Carrier, 1987.

in design from the jaws of other salamanders where feeding and breathing are coupled. The plethodontid hyoid elements, which were previously required for lung ventilation, have been lost.

Carrier (1987) has argued that a functional coupling exists between locomotion and respiration in some tetrapods. When a lizard runs, for example, it bends laterally, placing alternate compression on first the left and then the right lung. Air may be pumped back and forth between the lungs, but little new air enters to replenish the spent air (figure 18.7a). By contrast, a galloping mammal first compresses and then expands both its lungs simultaneously. This results in efficient inhalation and exhalation synchronized with flexions of the body (figure 18.7b). Carrier argued that in ancestral tetrapods, as in modern lizards, prolonged rapid locomotion interfered with breathing; this functional coupling constrained subsequent evolution. Ectothermic descendants of ancestral tetrapods became specialized for bursts of activity based on anaerobic metabolism, but such constraints precluded more active modes of locomotion. On the other hand, birds and mammals arose from ancestors that developed morphological changes circumventing these constraints and allowing endothermic tetrapods to adopt prolonged locomotion based in part on aerobic metabolism.

Recognition of this constraint inspired reevaluation of older work and new research on running lizards. In some species, ventilation did decline with speed. But in the monitor lizard, oxygen levels in blood leaving the lungs remained high, suggesting that ventilation remains adequate to support sustained running. In monitor lizards, lung ventilation is supplemented during rapid locomotion by a gular pump. The floor of the buccal cavity (gular region) actively rises and falls, producing positive pressures to draw air into the mouth and then

drive it into the lungs. Disabling the gular pump reveals that the constraint Carrier predicted is in fact present, but is usually masked by the successful supplementary lung ventilation by the gular pump. At least here, in monitor lizards, a way around the constraint has been found. The mechanical limitations of locomotion imposed on respiration have been circumvented.

Or consider feather color in a functional conflict between its role in courtship and in thermoregulation of body temperature. Some male birds present elaborate body displays of brightly ornamented, colored feathers to females during courtship to win favorable responses and successful reproductive outcomes. But outside a role in courtship, feathers also absorb valuable solar radiation on cool days to warm the body or reflect away excess radiation on hot days to prevent overheating. The two biological roles—courtship and thermoregulation—may conflict. Brightly colored feathers that impress a female may also reflect useful solar radiation advantageous in warming the chilled body. Sometimes there is no way around competing and conflicting biological roles. Biological design is a balance of compromises between competing biological roles.

Multiple Functions

Any structure is likely to have multiple functions, and its design is a compromise among them. The feathers of birds serve flight, but they also insulate, and they can carry colors displayed during courtship. The tail feathers of the peacock function during courtship and may actually hinder flight. Snake jaws snatch vigorously struggling prey, but bone articulations must be supple to permit subsequent swallowing of the dispatched animal. Male bighorn sheep use their horns during combat when they butt heads, but the curl of the horns is also used as a visual display to other males. The limbs of a cheetah carry it swiftly in pursuit of prey but are also the instruments used to snag prey when they are within reach.

All functions of a structure should be examined because each function influences design. The vertebrate digestive tract plays a central role in digestion, but it also houses lymphoid tissue and is therefore part of the immune system as well. The walls of the cecum, like most of the rest of the digestive tract, carry lymphoid tissue. The vermiform appendix of humans is homologous to this cecum, but it is much reduced. This has led some to suggest that it is without a function, a vestige only. Certainly, the human appendix has lost one major function, cellulose digestion, but it has not lost all function because it still houses lymphoid tissue.

Parts may also have a role to play in what Melvin Moss (1962) has termed a *functional matrix*. The mammalian lung, for example, functions in gas exchange. But, it also supports the rib cage and surrounding tissues. If the lung is removed on one side, the ribs on that side tend to be deformed because they lose the internal "scaffold" the lung provided. The snake lung is another example. In snakes, the long tubular lung runs down the center of the body. Its main function is in respiration. But it also functions to hold

(a)

(b)

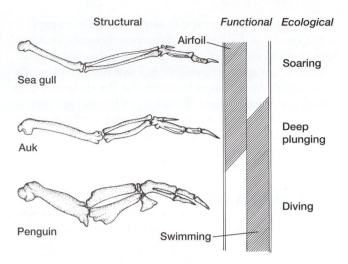

FIGURE 18.8 **Functional matrices.** In addition to serving their primary function, most internal structures support surrounding tissues. (a) Sagittal (top) and cross-sectional (bottom) views of a snake. Small muscles close the glottal opening into the lungs between breaths. The weight of the snake's body rests upon the inflated, elongate lung, which bears the body mass (solid arrow). (b) Frontal (left) and cross-sectional views (right) of developing human skull and brain. The growing brain carries the bony elements outward (arrows). These bones undergo compensatory growth at their margins (shading) to maintain contact and produce the structurally complete braincase.

Based on the research of M. Moss, 1962.

FIGURE 18.9 **Functional compromise.** The wings of the gull serve flight; those of the penguin serve swimming. As a result, the demands on each are quite different, which is reflected in design. The auk, which flies but also uses its wings when diving below the water, is a morphological intermediate. The design of its wing is a compromise between these different mechanical demands.

and give shape to the snake's body. Snakes lack a sternum to complete the rib cage and, of course, they have no limbs to support the body. Thus, most of the weight of the body rests on the inflated lung (figure 18.8a). The mammalian brain is another example. As the brain grows during development, the outer case of surrounding skull bones conforms to its expanding shape (figure 18.8b). If congenital defects occur during growth and the brain overexpands or underexpands, then the molding of the surrounding bony braincase is affected as well. Thus, lungs, brains, and other parts form matrices around which other elements take shape.

No structure can serve all of its functions equally because functional demands often are contradictory. Thus, compromises in design might be expected. Many marine birds, such as auks, use their wings for flight but also for maneuvering underwater in pursuit of fish. In comparison to shorebirds such as gulls, who use their wings very little for swimming, the wings of auks are short and robust. However, in comparison to penguins, who do not use their wings for flight, the auk wing is more slightly built. Between these two extremes, the auk wing is structurally intermediate in design, neither specialized for flight alone nor for swimming alone. Auk wings represent a compromise between the two (figure 18.9).

Compromise in design can occur within one sex when the biological role of that sex poses specific design requirements. The human pelvis is an example. In females, the birth canal is necessarily large to accommodate the passage of the

relatively large-headed infant during parturition. However, this widening of the hips to enlarge the birth canal results in a spreading of the limbs well to the side of the midline of the trunk above. Consequently, the limbs are not as well placed beneath the weight of the upper body that they must carry. This results in a tendency for increased asymmetrical loading of the limbs, which increases the chances of overstressing the bones. As a consequence, the quadriceps inserts on the knee at a more oblique angle, which makes wide-hipped human females more prone to knee dislocation. Hip design in human females seems to represent a compromise between the demands of locomotion and the demands of reproduction.

Performance

If the study of function tells us how a part works, then the study of performance answers the question "How *well* does it work?" One way to rate performance is to compare the structure to an engineering simulation serving the same function. If the engineering model represents the best design that is theoretically possible, it can be considered an *optimum design.* The difference in performance between the actual structure and the ideal model would represent the difference in efficiency.

There are several limitations to such an approach. First, most structures represent a biological compromise among several functions rather than an optimum design for any one function. Thus, a structure might fail to meet expectations of an optimal engineering design but still function quite well. Second, a structure need not be perfect for it to be preserved by natural selection. An organism requires structures that allow it to survive at a frequency equal to or slightly better than that of its competitors. The parts of an organism need not be perfect, only adequate to meet this minimum requirement for survival. Thus, the differences between an actual structure and an optimal one might have no biological significance. To measure performance in a way that is biologically meaningful, the part must be assessed in the environment in which the organism lives. Eventually, what matters most is not how close to a theoretical optimum a structure comes but how well it performs under the ecological conditions in which the part actually serves.

Ecological Analysis

Living animals are more or less suited to the current environments in which they reside. To complete an analysis of animal architecture, animals must be studied in their current environments to see what biological roles the structures play. Seeing how a part is used helps us understand why it might be designed in the way it is. The long legs of a giraffe increase stride length and, consequently, they would be expected to increase running speed. But an ecological analysis would show instead that the primary survival value of a giraffe's long legs is to lift the body high above the ground so that, along with the extended neck, the giraffe can reach browse that is inaccessible to short-legged competitors. Field studies often begin with basic natural history information on migrations, diets, reproductive patterns, and so forth, but they can expand into experimental studies that test ideas about the roles of particular structures in the life of an organism.

Evolutionary Analysis

Historical Constraints

Some morphologists study the structure, function, and ecology of an organism and consider these a sufficient basis on which to analyze animal design. However, what this deletes is the history out of which the design comes, namely, its evolutionary source. The history behind a structure must be included in the analysis of design; otherwise we cannot address the question of why this particular design has come to characterize this particular organism. Suppose that we analyze the form and function of the tails of a porpoise and an ichthyosaur. We could further relate these to the environment of the porpoise and at least to the likely environment of the ichthyosaur. But, our analysis would not explain why the flukes of the porpoise are *horizontal* and the flukes of the ichthyosaur were *vertical* (figure 18.10a,b). These differences likely result from the different evolutionary backgrounds of these organisms. Porpoises descended from terrestrial mammals in which locomotion included vertical flexions of the vertebral column. Ichthyosaurs descended from terrestrial reptiles that used horizontal undulations of the vertebral column. When porpoises and ichthyosaurs became adapted to aquatic environments, they developed

(a)

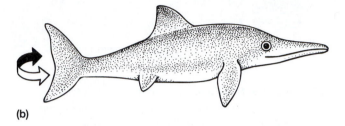

(b)

FIGURE 18.10 **Convergent design.** Both porpoises (a) and ichthyosaurs (b) are designed for swimming in similar aquatic habitats. However, orientation of the tail flukes is horizontal in porpoises and vertical in ichthyosaurs. These differences are probably explained by differences in the evolutionary histories out of which each comes rather than from functional or ecological factors alone.

flukes with horizontal and vertical orientations, respectively. These orientations took advantage of the preexisting patterns of vertebral column bending. Past history constrains future directions of structural and functional change. If we do not include the historical dimension in our analysis of structure, we limit our ability to explain such differences in the designs of organisms.

Primitive and Advanced

We have already noted that the respective terms *primitive* and *advanced* are used to distinguish species arising early from those that emerged later in a phylogenetic lineage. But, it should be said again that these terms are unfortunate choices because they feed the view that advanced also means better. To replace the term *advanced* with the term *derived* is helpful. Primitive species are those that retain the early features present in the first members of the lineage. Derived species are those with modified characteristics, representing a departure from the primitive condition (figure 18.11).

However, replacing terms will not entirely eliminate the mistaken view that evolving species become progressively better, because the misunderstanding lies not with the terms but with the bias most of us bring to the subject of evolution. Many students expect biological evolution to be driven by the same purposes as technological change. Human inventions are usually progressive; they attempt to make life better.

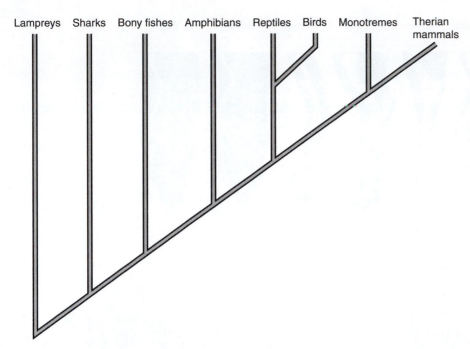

Lampreys Sharks Bony fishes Amphibians Reptiles Birds Monotremes Therian mammals

FIGURE 18.11 **Primitive and derived groups.** Lampreys are primitive, and mammals are derived or advanced, but this means only that lampreys arose before mammals. The lamprey carries forward much of the early anatomy of ancestral groups; however, it has been around a very long time, and during that interval, it has probably undergone considerable changes in many of its systems. In its own way, the lamprey is highly adapted to its environment and lifestyle, so it would be a mistake to think of it as less well adapted or imperfect in design compared with the mammal.

Antibiotics improve health. Trains speed travel over the horse and airplanes over the train. Computers replace the slide rule, which replaced counting on fingers and toes. Certainly there is a price to pay in pollution and expenditure of resources, but most people view these technological changes as progressive improvements. Life gets better.

But this is not the way to look at biological innovation. Mammals are not improvements over reptiles, nor are reptiles improvements over amphibians, nor are amphibians improvements over fishes. Each taxon represents a different, but not necessarily a better, way of meeting the demands of survival and successful reproduction. Each group is equally well adapted to the tasks and lifestyles required for survival. Advanced groups are not better adapted than primitive groups.

This idea of progress is a deep and difficult bias to set aside, even for scientists. As pointed out regarding the evolution of aortic arches in chapter 12, many scientists succumbed to the mistaken view that aortic arches, incomplete internal septation of the heart, and lungs in lungfishes represented imperfect structures. This presumed mediocrity was excusable in "primitive" animals. Oxygenated and deoxygenated bloodstreams were thought to mix, and this was seen as an unsolved "problem" until advanced groups such as birds and mammals evolved. As one morphologist rejoiced, "the perfect solution" was reached in the advanced avian and mammalian stages, but he conceded that lungfishes and amphibians had made some progress in separation of the two bloodstreams.

Recent research shows that there is in fact little mixing of these bloodstreams, but the point here is more subtle. The mistake is to view lungfishes and amphibians as ineptly designed animals in comparison with birds and mammals. Certainly lungfishes would make poor mammals, but on the other hand, mammals would make poor lungfishes. The design of each must be measured against the environment in which it serves, not against what it might become. Evolution does not look to the future. It is not progressive in the same way that technological changes make life better.

Often students will ask why a primitive fish such as a lamprey continues to survive alongside advanced fishes such as trout and tuna. Behind such a question is the assumption that advanced means better, which leads to the mistaken view that the "superior" (i.e., advanced) should have replaced the "imperfect" (i.e., primitive). The notion of "better" does not apply to biological changes. Primitive and advanced species represent *different* ways of surviving, not *better* ways of surviving.

Diversity of Type/Unity of Pattern

The evolution of animal architecture usually proceeds by remodeling, not by new construction. Certainly, novel mutations arise to provide fresh variety, but their effect is usually to modify an existing structure rather than to replace it with something entirely new. As a result, the anatomical specializations that characterize each group are modifications of a common underlying pattern. It is unity of plan and diversity of execution, as T. H. Huxley said in 1858. The bird wing is a modified tetrapod forelimb, which is a modified fish fin (figure 18.12a). Similarly, we have seen the diversity of aortic arches derived from a basic six-arch pattern (figure 18.12b). The basic five pharyngeal pouches are the common source for a variety of glands, and the basic tubular digestive tract became diversified into specialized regions in various groups. This remodeling feature of evolution accounts for the structural similarities from one group to the next.

FIGURE 18.12 Diversity of type/unity of pattern. (a) The forelimbs of seven tetrapods show great diversity, but all are modifications of a common underlying pattern. (b) The aortic arches of various groups are quite different; however, they seem to be derived from a common six-arch pattern.

(a) After Mayer; (b) after Goodrich.

Mosaic Evolution

Natural selection acts on individual organisms. But different parts of the same organism are under selection pressures of different intensities, so parts change at different rates. For example, in the evolution of horses, the limbs changed considerably between the four- or five-toed ancestors and the single-toed horses of today (figure 18.13a). Teeth and skull changed also but perhaps less radically than the limbs, and relative brain size changed very little. In any evolving lineage, some parts change rapidly, some slowly, some almost not at all. Gavin de Beer (1951) termed such a pattern *mosaic evolution* because of the uneven rates at which parts of an organism undergo modification within a phylogenetic lineage.

When we study evolution of species, we must keep such a mosaic feature of evolution in mind because the selection pressures acting on one part of an organism might be quite different in intensity from those acting on another part. For example, the advanced group of snakes, the Caenophidia, includes a family of mostly nonvenomous species, the Colubridae, and generally two families of highly venomous snakes derived independently from them: the Viperidae (vipers and pit vipers) and the Elapidae (cobras, sea snakes, and their allies). In the evolution of venomous species from nonvenomous ancestors, the jaw apparatus has undergone rather extensive modification, becoming the instrument for delivery of the poisonous toxins produced in a specialized venom gland (figure 18.13b). However, other parts have changed much less dramatically. The vertebrae of venomous snakes are modified slightly from those of nonvenomous ancestors. The basic structure of the scale has hardly changed at all. Because of these different evolutionary rates, our view of the tempo of evolution within a group is likely to be slanted by the particular system we examine.

Failure to recognize the mosaic nature of evolution has led to what might be termed the This is the mistaken expectation that evolutionarily intermediate species should in all respects be halfway between the ancestral and the descendant group. For example, *Archaeopteryx* from the mid-Jurassic certainly stands near the transition between reptiles on the one hand and modern birds on the other. Yet, it is not intermediate between these two groups in all respects. It had feathered wings and a furcula like its avian descendants, but it still possessed teeth and a hindlimb like the reptilian ancestors from which it came. A living example of mosaic evolution is the platypus of Australia (figure 18.14). In some ways, it is specialized. It has webbed feet, bears a spike on its hindfeet, and has a broad snout. In some ways evolution has lagged. The pectoral girdle retains elements (e.g., distinct coracoid, interclavicle, and procoracoid) from tetrapod ancestors, and reproduction, as in most reptiles, includes a shelled egg. In other ways, evolution has been rapid. The tetrapod scale yielded hair, and the young are nursed at mammary glands.

The prevalent expectation that transitional forms should be halfway between their ancestors and descendants in all respects does not conform to the mosaic character of evolution. The missing-link fallacy has even hindered study of our own evolution. A "missing link" that was intermediate between modern apes and *Homo sapiens* was envisioned, but none was found. However, because our own evolutionary history, like that of most species, proceeds in a mosaic pattern, we should not expect to find such an intermediate. In fact, the anatomical features that characterize humans evolved at different rates, and modern traits appeared at different times (figure 18.15). Bipedal locomotion evolved early, perhaps occurring as early as 6 to 7 million years ago in the hominid *Sahelanthropus tchadensis*. Grasping hands, opposable thumbs, and a firm clavicle emerged even before this when earlier antecedents to hominids swung through trees; however, large brains, relatively hairless bodies, and speech came much later in our evolution in *Homo habilis, H. erectus,* and primitive *H. sapiens.* No single ancestral species possessed all characteristics halfway in transition between ourselves and apes.

Morphology and Modules

Embryos and adults are composed of modules, subunits of the overall organism that act semi-independently either developmentally or functionally. The components within a module are highly integrated within themselves, yet are part of the larger unit, the organism. For example, the skeletal system consists of at least three major modules—axial skeleton, skull, and appendicular elements (figure 18.16a–c). Within the skeletal system, limbs form under the influences of internal signals and controls residing in the limb bud itself. Within the musculoskeletal system there are highly integrated locomotor modules. In primitive tetrapods, the body axis and all four limbs act as an integrated unit during locomotion (figure 18.16d). However, many basal dinosaurs were comfortable or obligate bipeds. Herein, the tail and hindlimbs comprised a single locomotor module (figure 18.16e). But birds, which arise out of this ancestry, possess three locomotor modules. The bird forelimb acquired locomotor function as a wing, and the tail decoupled from the hindlimb to specialize in the control of the tail feathers, forming a new alliance of pectoral and caudal modules (figure 18.16f).

During evolution, major changes occur by processes that affect developmental modules and, in turn, adult design—dissociation, duplication and divergence, and co-option. For example, heterochrony dissociates the development of one portion of the body from another by accelerating or retarding one part's development. Duplication and divergence occur with repeated parts of the adult that produce structures such as the dolphin fin with numerous additional phalanges supporting the fin; or *Hox* genes may duplicate and subsequent divergences produce major new groups. Co-option occurs when components of a

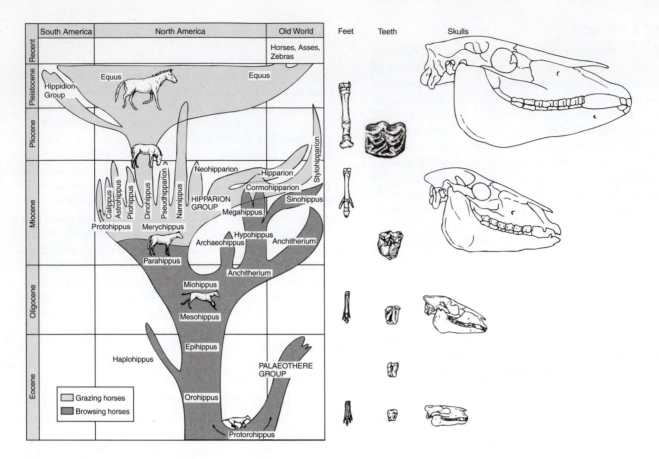

(a)

FIGURE 18.13 **Mosaic evolution.** (a) Evolution of the horse has been characterized by relatively rapid changes in foot and tooth structure but little substantial changes in other systems, such as the integument. Note also that the evolution in horses is "bushy," especially in the Miocene, not "ladder"-like. (b) Evolution of highly venomous snakes has been characterized by relatively rapid and extensive modifications of the maxilla and fang, but the vertebral column and especially the integument changed much less extensively. The maxilla is removed from its position in the skull and enlarged. Transformation grids are used to illustrate its changes within these families of advanced snakes.

(a) After MacFadden; (b) after Kardong.

(b)

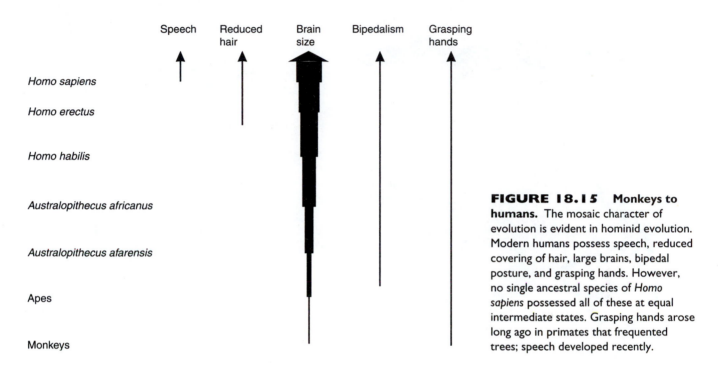

FIGURE 18.14 Missing-link fallacy. The platypus carries some advanced characteristics such as hair and mammary glands, but evolution of its other systems lags. Transitional forms are not necessarily intermediate between ancestors and descendants in all characteristics; instead, they display a mosaic of characters at different stages of evolutionary modification.

Speech   Reduced hair   Brain size   Bipedalism   Grasping hands

Homo sapiens

Homo erectus

Homo habilis

Australopithecus africanus

Australopithecus afarensis

Apes

Monkeys

FIGURE 18.15 Monkeys to humans. The mosaic character of evolution is evident in hominid evolution. Modern humans possess speech, reduced covering of hair, large brains, bipedal posture, and grasping hands. However, no single ancestral species of *Homo sapiens* possessed all of these at equal intermediate states. Grasping hands arose long ago in primates that frequented trees; speech developed recently.

module, destined for one function, are diverted to alternative functions. For example, the bones offering jaw articulation (quadrate, articular) in reptiles become commandeered into service as middle ear bones. Modularity allows evolution to build new structures and functions and associations, to thereby produce new designs that meet the culling process of natural selection.

Heterochrony (p. 198); *Hox* genes (p. 204); jaw evolution (p. 245)

Mode and Tempo of Evolution

If a new species were built from the foundation up, trait by trait, then the process could be immeasurably long, even if each new trait derived from a new gene mutation were immediately favored by natural selection. Vertebrates are complex and traits are integrated into a coherent unit, so evolving a new species may at first seem improbable—each new part would await the simultaneous arrival of thousands of new and favorable mutations, functionally integrated, producing each new body part. Each body part includes thousands of associated tissues—various nerves, muscles, bone, connective tissue, and so forth—that must be redesigned

to building new species one piece at a time. For example, moving from water to land—rhipidistian fishes to first tetrapods—would require modifications in respiration, locomotion, feeding, integument, and behavior, to mention just a few affected systems. Hundreds, if not thousands, of favorable mutations must occur in rhipidistian fishes to prepare them for their pioneering trek to land. And these new traits must appear in the same individual or at least in the same local population so as to enjoy the favorable benefits of the traits acting in concert. Even accepting the extreme and unlikely assumption that this transition required only six traits highly favored by selection to move these fishes closer to life on land, such a combination might be expected to occur about once in every billion years (Frazzetta, 1975). Yet, the fossil record tells us differently. This transition occurred in the space of about one-hundredth of that time, much faster than the simple prediction that envisions evolution being the additive total of all new gene mutations that sum up to a new species. Clearly something is wrong—a mismatch between the actual rate of evolution and the human expectation that evolution should be a slow process awaiting the timely arrival of all new genes, one at a time, to build an all-new species. How is it that evolutionary modification often occurs faster than at first we might expect?

FIGURE 18.16 **Modules and morphology.** (a–c) Skeletal modules are shown wherein the semi-independence of the axial skeleton, skull, and appendicular skeleton is tracked. (d–f) Within the locomotor system of basal tetrapod to bird, the basic musculoskeletal system, including body axis and all four limbs, acts as an integrated module. (d) In bipedal dinosaurs, the comparable module includes hindlimb and tail (e). In birds, the tail musculoskeletal system is decoupled from the hindlimb and acts in alliance with the forelimb system (f).

(d–f) After Gatesy and Dial.

Remodeling

Evolution is remodeling—descent with modification. Consequently, each new species is built on the old, not from scratch. The refashioned ancestor with a few new, adaptive traits is the foundation of a new species. This means that often old traits serve new functions. We met this earlier when preadaptation was discussed. Feathers in birds, or in their immediate ancestors, likely evolved initially as insulation to conserve body heat. Like hair in mammals, feathers were an indispensable, energy-conserving feature. Flight came later in birds. Immediate ancestors to birds were ground- or tree-dwelling, reptilelike animals. As flight became more important, feathers already present for insulation were co-opted into aerodynamic surfaces—wings—to serve flight. With birds, feathers arose from reptile scales, to serve as insulation. But once present, and playing a favorable role in heat conservation, they were, in a sense, available for subsequent chores. The idea of preadaptation does not imply anticipation. Feathers did not evolve at one time for service millions of years later in flight. They evolved initially for their advantages of the moment—insulation—not for their role in the distant future—flight. Looking back on this, we see that subsequent changes in lifestyles led to appropriation of feathers (insulation) into aerodynamic surfaces (flight).

Vertebrate jaws evolved from the gill arches that preceded them; legs evolved from fins; penguin flippers evolved from wings of ancestors; dolphin fins evolved from legs; and so it goes. Examples of preadaptation abound, and the term captures an essential feature of evolution—namely, remodeling. If new features were built from the ground up, it would take immeasurably longer stretches of time for evolution to build new species. Each new species would await the arrival, simultaneously, of thousands to millions of new mutations producing each new body part, building the new species one piece at a time—integrating it, testing it, trying it. By comparison, descent with modification is faster. If new traits fare well in old bodies, then a new species arises from the ancestor.

Preadaptation (p. 18); evolution of flight (p. 367)

Embryonic Changes

Another way to produce rapid changes is through major adjustments during embryonic development, based on genetic mutations that affect embryology. Some lizard species are legless. Of course, this alone does not turn the legless lizards into snakes. Legless lizards still have lizard features of underlying bony anatomy, and they have eyelids and external ear openings that snakes lack. Arms and legs are so much a part of our anatomy, it is hard to imagine the benefits of living without them. But ground-dwelling lizards negotiate tight, crowded spaces, slipping between loose rock and dense brush. Absence of obstructing limbs, under such conditions, allows the sleeked-down body to slip efficiently through such tight habitats. This major anatomical change in lizards from limbed ancestors to limbless descendants seems based on a major change in the underlying embryology.

In lizards with limbs, an early embryonic gathering of cells into a *somite* grows downward along the sides of the embryo at sites where fore- and hindlimbs are to form (figure 5.43). Here, the somite's lower growing tip meets special cells—mesenchymal cells—and together they initiate a "limb bud." Subsequently, as the lizard embryo matures, these limb buds grow outward to sprout the limbs, which are ready by the time the young lizard is born. In legless lizards, early embryonic events unfold similarly except the lower tip of the somite fails to grow downward into the area of the prospective limb. This single change at this critical embryonic moment denies the limb bud the stimulation it requires to grow. These embryonic limb buds subsequently regress, and the lizard is born limbless.

Here, a major adaptive change occurs from limbed to limbless, but the foundation of this change is basically a single, critical change in the underlying embryology. Within limbed lizard ancestors, a mutation occurred that interrupted the downward growth of the somites during embryology. Young were born without limbs, a new variation within the population. In the environment, the limbless young realized some competitive advantages (sleekness) over others with limbs (obstructions), and survived. Other changes were to follow, including changes in movement. And, of course, not all lizard environments would be favorable to limbless individuals, and they would perish. However, where limbless features are advantageous, the shift from limbed to limbless occurred rapidly thanks to this fortuitous but critical mutation affecting the embryology.

Epigenomics (p. 205)

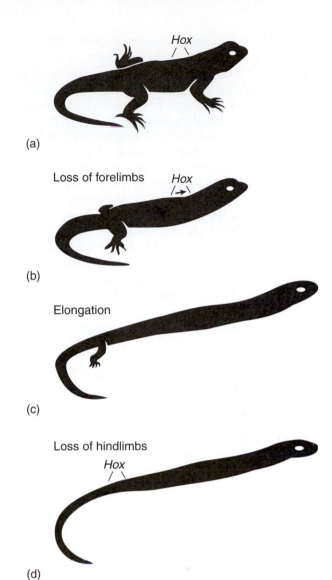

(a)

Loss of forelimbs

(b)

Elongation

(c)

Loss of hindlimbs

(d)

FIGURE 18.17 *Hox genes and rapid evolution: lizards to snakes.* *Hox* genes associated with the chest region in lizards (a), expand their influence, leading to loss of forelimbs (b). By other changes in embryology, more vertebrae are added to the vertebral column, producing an elongated body (c). Either by a shift in influence of other *Hox* genes and/or by changes in limb bud growth (for example, see figure 5.43), hindlimbs are lost and an essentially modern snake body is produced (d). These steps may have occurred in a different order. Certainly, other changes accompanied these three basic steps to consolidate and integrate them. But apparently, the major steps from lizard to snake are built upon only a few gene or embryonic modifications. Different *Hox* genes (*Hox*) are indicated at locations wherein mutations in them are hypothesized to produce a change in body design.

Hox Genes

Another way to produce rapid, major changes in morphology is more directly by gene action, especially by the master control *Hox* genes (see chapter 5). In animals, these genes, like military generals, control large armies of other genes.

Through their action on other suites of genes, *Hox* genes regulate the appearance of major body parts such as body regions, legs, arms, and other modular parts. A simple change in one of these master-control *Hox* genes can produce a major change in body design. For example, limbless snakes arose from limbed lizard ancestors. Pythons are primitive snakes, legless of course, but with rudimentary hindlimbs. The *Hox* genes that regulate forelimb development have, in snakes, deactivated normal forelimb development, leading to absence of forelimbs. Specifically, it is proposed that the *Hox* genes controlling the expression of the chest region in lizard ancestors expanded their domain or sphere of influence. In a sense, the body of a snake is an expanded chest (figure 18.17). As the *Hox* gene for chest zipped its domain through the body, limb development was simultaneously suppressed, producing the characteristic limbless condition of snakes today. As with limbless lizards, other traits were to follow, thereby consolidating the limbless condition of snakes. But the basic snake body was produced with just a few, but major, gene changes.

Hox genes and their kingdoms (p. 204)

Evolutionary Significance

Such large-scale changes in design—limbed lizards to limbless snakes—need not be built slowly, one small, single gene mutation at a time. Evolutionary modifications need not wait for a hundred gene mutations, each eliminating one finger, one joint, one muscle, one nerve, one connective tissue, one part of the forelimb at a time; then another hundred to do the same for the hindlimb, and so on. Instead, big changes in morphology can be initiated by the relatively few but important master-control genes, upon which rapid and big evolutionary changes are based.

The Promise of Vertebrate Morphology

The flourishing field of vertebrate morphology is proving to be one of the few disciplines that takes a holistic and comprehensive approach to the study of the individual. Molecules alone are not enough. Human beings are more than the molecules of which they are made. The individual is too complex and the effects of molecules that make up the genes too distant from the finished product to account entirely or even mostly for the extraordinary features of design that characterize an individual organism. Because the designs of individuals reflect functional demands, environmental pressures, and constraints of history out of which they evolved, these designs are the result of natural events accessible to discovery and understanding. As the Nobel Prize–winning geneticist Hermann Muller put it in 1946, "To say that a man is made up of certain chemical elements is a satisfactory description only for those who intend to use him as a fertilizer."

Animal architecture holds a mystery that any lively mind must notice. In a congested world in which day-to-day survival predominates, much personal delight is to be found in an intellectual pursuit of questions deeper than the ordinary and everyday. As has been said before, the discipline of vertebrate morphology with the individual at its center holds the promise of helping us to see who we are and what we might become.

WEB LINK

Visit the text website at *www.mhhe.com/Kardong7e* for additional study aids including selected references, functional laboraotories, and web links to chapter topics.

Vector Algebra

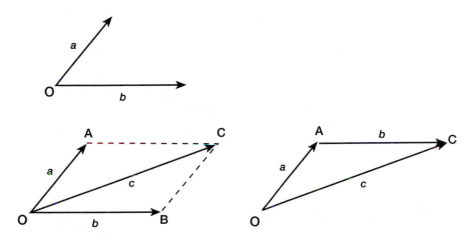

FIGURE A.I **Graphic addition of vectors.** Two vectors, *a* and *b*, can be added together to give a *resultant*, the single vector *c*, which summarizes the effect of the two separate vectors. One way to determine the resultant is by graphic construction. Graphic addition of vectors can be done by completing a parallelogram (left) or by triangle construction (right).

By using the two vectors as initial sides of a parallelogram, we can determine the opposite sides (dashed lines) to complete the parallelogram. The diagonal, OC, then gives the resultant, *c*.

Construction of a triangle involves adding vectors heads to tails. In this example, vector *b* is moved graphically with its tail positioned at the tip of vector *a*. Although transferred to a new position, its direction and length are, of course, preserved. The connecting distance drawn from the tail of *a* to the head of *b* gives the resultant, *c*.

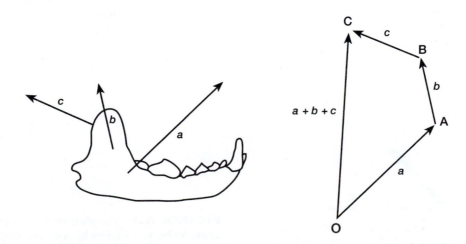

FIGURE A.2 **Addition of multiple vectors.** Several vectors, *a, b, c,* acting simultaneously on the carnivore lower jaw can be added graphically (right) to determine the single resultant, *a + b + c,* of their collective action (left).

FIGURE A.3 Mathematical addition of vectors. Trigonometric addition of the same vectors takes advantage of the law of cosines and sines. If the magnitude of the component vectors a and b and the angle between them, θ, are known, then the resultant, c, is calculated from the law of cosines:

$$c = \sqrt{a^2 + b^2 + 2^{ab} \cos \theta}$$

The angle (ø) of the resultant with respect to b is given by the law of sines:

$$ø = \sin^{-1}(a \sin \theta / c)$$

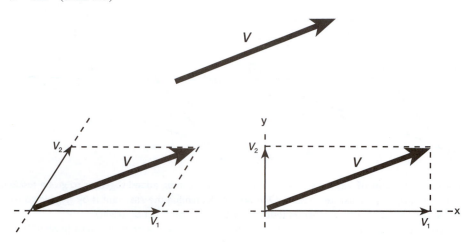

FIGURE A.4 Resolution of a vector. By definition, a vector has magnitude and direction. Thus, a car traveling at 55 miles per hour in a northeasterly direction would be a vector quantity. Graphically, a vector is represented by an arrow (V) whose length is proportional to its magnitude, and its orientation indicates direction. As discussed in the two preceding figures, adding several vectors gives a single resultant. The reverse process takes a single vector and resolves it into several *vector components*. This process takes advantage of the parallelogram construction. The vector, V (left), is projected to each side of the parallelogram to produce its component vectors, V_1 and V_2. When initially constructing the parallelogram, we are free to choose a slope to the sides that is convenient for our purposes. So long as opposite sides are parallel, projection of the resultant to each side faithfully gives one possible set of component vectors. Usually, orthogonal components are preferred in which all sides of the parallelogram are at right angles (right). This makes it easier to overlay the components on a rectangular Cartesian reference system.

FIGURE A.5 Resolution of a vector to multiple axes. So far, vectors have been discussed only in two-dimensional space, but a vector (V) can be resolved into three components on the same principle if a third axis, z, is used. This then represents the vector (V) by three components: V_x, V_y, V_z.

International Units (SI)

The SI system is based on six primary units:

Quantity	Name of Unit	Symbol
length	meter	m
mass	kilogram	kg
time	second	s
electric current	ampere	A
thermodynamic temperature	kelvin	K
luminous intensity	candela	cd

All other SI units are derived from these six basic units. For example, force is $kg\ m\ s^{-2}$. Some units are named for the person historically associated with them, for example, the newton (N) or watt (W). Some of these derived SI units are:

Quantity	SI Unit	Symbol	SI Base Units
area	square meter	m^2	
volume	cubic meter	m^3	
velocity	meter per second	$m\ s^{-1}$	
density	kilogram per cubic meter	$kg\ m^{-3}$	
acceleration	meter per second squared	$m\ s^{-2}$	
frequency	hertz	Hz	$1\ Hz = 1\ s^{-1}$
force	newton	N	$1\ N = 1\ kg\ m\ s^{-2}$
pressure, stress	pascal	Pa	$1\ Pa = N\ m^{-2}$
energy, work, heat	joule	J	$1\ J = 1\ N\ m$ $= 1\ kg\ m^2\ s^{-2}$
power	watt	W	$1\ W = 1\ J\ s^{-1}$ $= 1\ kg\ m^2\ s^{-3}$
electric charge	coulomb	C	$1\ C = 1\ A\ s$
electric potential	volt	V	$1\ V = 1\ W\ A^{-1}$
electric resistance	ohm	Ω	$1\Omega = 1\ V\ A^{-1}$
viscosity	pascal-seconds	Pa s	$Nm^{-2} = kg\ m^{-1}\ s^{-1}$

Notice that the units themselves are written in lowercase, but the symbol may be capitalized, for example, newton (N), joule (J), and watt (W). Index notation is recommended. Thus, 8 m s^{-1} is preferred rather than 8 m/s.

To indicate multiples, the appropriate prefix is attached to the basic unit. Some examples are:

Multiple		Prefix	Symbol	Example
1 000 000 000	$= 10^9$	giga	G	gigawatt (GW)
1 000 000	$= 10^6$	mega	M	megawatt (MW)
1 000	$= 10^3$	kilo	k	kilogram (kg)
100	$= 10^2$	hecto	h	hectare (ha)
0.1	$= 10^{-1}$	deci	d	decimeter (dm)
0.01	$= 10^{-2}$	centi	c	centimeter (cm)
0.001	$= 10^{-3}$	milli	m	millimeter (mm)
0.000 001	$= 10^{-6}$	micro	μ	microsecond (μs)
0.000 000 001	$= 10^{-9}$	nano	n	nanometer (nm)
0.000 000 000 001	$= 10^{-12}$	pico	p	picosecond (ps)

Constants	English	SI
g–acceleration due to gravity at surface of the Earth	32.17405 ft s^{-2}	9.80665 ms^{-2}

Conversion of Units

Dimensions

1 centimeter = 0.3937 inch

1 meter = 39.37 inches

1 meter = 3.281 feet

1 meter = 1.0936 yards

1 kilometer = 0.62137 mile

1 square kilometer = 0.386 square mile

1 inch = 2.54 centimeters

1 foot = 30.48 centimeters

1 foot = 0.305 meters

1 yard = 0.9144 meter

1 mile = 1.6094 kilometers

1 square mile = 2.59 square kilometers

1 square mile = 640 acres

1 acre = 4,840 square yards = 4,047 square meters

1 hectare = 10,000 square meters = 2.47 acres

football field (U.S.) = 360 × 160 feet = 120 × 53.33 yards = 6,399.6 square yards

Weights

1 gram = 0.03527 ounce

1 kilogram = 2.2046 pounds

1 ton (English) = 2,000 pounds

1 ton (English) = 1.016 metric ton

1 ounce = 28.35 grams

1 pound = 0.4536 kilogram

1 ton metric (= tonne) = 0.98421 ton (English)

Volume

1 cubic centimeter = 0.61 cubic inch

1 liter = 0.2642 gallons

1 gallon = 231 cubic inches = 0.1337 cubic foot

1 cubic inch = 16.39 cubic centimeters

1 gallon = 3.785 liters

Velocity

1 foot per second (ft s^{-1}) = 0.3048 meters per second (m s^{-1})

1 mile per hour (mile h^{-1}) = 1.609344 kilometers per hour (km h^{-1})

1 UK knot = 1.00064 international knots (kn)

= 1.15152 miles per hour (mile h^{-1})

Force

1 dyne = 1×10^{-5} N

1 pound-force (lbf) = 4.448 22 newtons (N)

1 pascal = 0.00014504 psi

1 watt = 1 newton-meter per second (Nm s^{-1})

1 newton = 100,000 dynes

1 newton = 0.224809 pound-force (lbf)

1 psi = 6894.65 pascals

1 horsepower = 746 watts

Angle

90 degrees (°) = $\pi/2$ radians (rad)

= 1.57080 radians (rad)

1 degree (°) = 0.0174533 radian (rad)

SI Conversions

1 newton = 10^5 dynes

13.55 mm H$_2$O = 1 mm Hg (at 4°C)

1 dyne = 0.00001 newton

Common Greek and Latin Combining Forms

a-(an)	L., without, not	cervix	L., neck
ab-	L., from	chlor	Gr., green
acinus	L., grape	choan	Gr., funnel
acro-	Gr., extremity, point	chrom-	Gr., color
ad-	L., to	chyl-	Gr., juice
adeno-	Gr., gland	cilium	L., eyelid
adipo	L., fat	clast-	Gr., to break
ala-	L., wing	cleistos	Gr., see kleistos
alb-	L., white	coel-	Gr., hollow
-algia	Gr., pain	collum-	L., neck
alveolus	L., hollow, cavity	conch-	Gr., a shell
ambi-	L., both	corn	L., horn
amphi-	Gr., both	cortico-	L., bark
amyl-	L., starch	crine (krino)	Gr., secrete, separate
an-	Gr., without, not	cross-	Fr., see crusi
ana-	Gr., up	crusi-	L., cross, ridge
ankylo-	Gr., bent	crypt-	Gr., hidden
ante-	L., before	cten-	Gr., comb
anti-	Gr., against	cumulus	L., a heap
apo-	Gr., from	cyan-	Gr., dark blue
aqua-	L., water	cyn-	Gr., dog
archi-	Gr., first	cyt-	Gr., cell, hollow
areolar	L., small, open space	de-	L., down, from
arthro-	Gr., joint	dent-	L., tooth
-ase	L., enzyme	di-	Gr., double
aspid-	Gr., shield	dia-	Gr., through
auto-	Gr., self	diplo	Gr., double
bi-	L., two	dis	see de
bio-	Gr., life	dramein	Gr., to run
blast-	Gr., germ, bud	duct-	L., convey
bothri-	Gr., pit	duo-	L., two
brachi-	Gr., arm	dura-	L., hard
brady-	Gr., slow	dys-	Gr., bad, ill
brevis-	L., short	ecto-	Gr., outside
caec-	L., blind	ella, -us, -um	L., diminutive
-campus	see kampos	emia-	Gr., blood
capit-	L., head	endo-	Gr., within
cata-	Gr., down	entero-	Gr., intestine
-cele	Gr., swelling or tumor	ento-	Gr., within
cephalo-	Gr., head	epi-	Gr., on, above
cer-	Gr., horn	ergaster-	Gr., workman

erythro-	Gr., red	-lysis	Gr., divide, destroy
eury-	Gr., wide	macro-	Gr., large
ex-	L., out	macula-	L., spot
extra-	L., outside, beyond	mal-	L., bad, ill
fenestra	L., window	mast-	Gr., breast
fer-	L., to carry, bear	medi-	L., middle
flay-	L., yellow	mega-	Gr., large
follicle	L., little bellows, small bag	-mere	Gr., a part
fossa	L., pit, cavity	mes-	Gr., middle
fug (e)	L., flee	meta-	Gr., change, after
gamo-	Gr., marriage	-meter	L., measure
gastro	Gr., stomach	micro-	Gr., small
-gen	Gr., to produce	mono-	Gr., single
-glia	L., glue	morph-	Gr., form
glossi-	Gr., tongue	morti-	L., death
glyco-	Gr., sweet, sugar	myelo-	Gr., narrow
gnath	Gr., jaw	myi-	Gr., fly
-gogue	Gr., leading	myo-	Gr., muscle
gon-	Gr., angle, seed	myxo-	Gr., mucus, slime
graph-	Gr., to write	necro-	Gr., dead
haemo-	Gr., blood	nemo-	Gr., thread
halos-	Gr., salt	neo-	Gr., new
hemi-	Gr., half	nephro	Gr., kidney
hepato-	Gr., liver	neuro-	Gr., nerve
hetero-	Gr., different	nid-	L., nest
hex-	Gr., six	nigr-	L., black
hippo-	Gr., horse	noct-	L., night
histo-	Gr., tissue	noto-	Gr., back
homeo-	Gr., same	nuc-	L., nut
homo-	L., human	nud-	L., naked
horm-	Gr., to excite	nyssus-	Gr., to prick
hyalo-	Gr., glassy, clear	ocell-	L., small eye
hydro-	Gr., water	oct-	L., sight
hymen-	Gr., a membrane	odont-	Gr., tooth
hypo-	Gr., under	-oid	Gr., like
in-	L., not, without	olig-	Gr., few
in-, en-	L., into	-oma	Gr., tumor
inter-	L., between	omma-	Gr., eye
interstitium	L., space between	omphalo	Gr., navel
intra	L., within	oneh-	Gr., barb
iso-	Gr., equal	oo-	Gr., egg
-itis	L., inflammation	opercul-	L., a cover
kampos	Gr., sea monster	ophthalm-	Gr., eye
kino-	Gr., movable, flexible	opisth-	Gr., behind
kleistos	Gr., closed	or-, os-	L., mouth
lact	L., milk	orchi-	Gr., testicle
lacun-	L., pit, lake	ortho-	Gr., straight
lamin-	L., layer, plate	-ose	L., sugar
leio-	Gr., smooth	ossi-	L., bone
lemma-	Gr., skin	osteo-	Gr., bone
leuco-	Gr., white	osti-	L., door
lingua-	L., tongue	ot (o)	Gr., ear
lipo-	Gr., fat	-otic	Gr., condition
lith-	Gr., stone	ov (i)	L., egg
-logy	Gr., discourse	oxy-	Gr., sharp
luteus	L., golden yellow	pataeo-	Gr., ancient

pan-	Gr., all	ram (i)	L., branch
par-	L., to beget	re-	L., again
para-	Gr., beside	rect-	L., straight
pariet-	L., wall	ren- (i)	L., kidney
path-	Gr., disease	ret- (e,i)	L., net
pathy-	Gr., thick	rhabdo-	Gr., a rod
ped-	L., foot	rheo-	Gr., to flow
penicillus	L., paintbrush	rhino-	Gr., nose
penta-	Gr., five	rhizo-	Gr., root
peri-	Gr., around	rhyncho-	Gr., snout
pertro-	Gr., stone	rostri-	L., beak
phago-	Gr., to eat	rumin-	L., throat
phil-	Gr., loving	sagitta-	L., arrow
phlebo-	Gr., vein	salpi-	Gr., trumpet
phon-	Gr., voice, sound	sapro-	Gr., putrid
phoro-	Gr., bearer of	sarco-	Gr., flesh
phot-	Gr., light	scalar-	L., ladder
phyll-	Gr., leaf	schizo-	Gr., cleft, split
-physis	Gr., growth	sclera-	Gr., hard
phyto	Gr., plant	sclerosis	L., hardening
pia-	L., tender	scoli-	Gr., bent
pituita	Gr., phlegm, slime	-scope	Gr., see
plasm-	Gr., formed	scut-	L., a shield
platy-	Gr., flat	seb-	L., tallow
pleuro-	Gr., side, lung	sect-	L., to cut
plexus	L., twine, braid	sella-	L., saddle
pneumo	Gr., air	semi-	L., half
pnoi	Gr., breath	sept-	L., wall
pod (i-o)	Gr., foot	septic-	L., putrid
poly-	Gr., many	serra-	L., a, was
pons	L., bridge	seti-	L., bristle
porta-	L., gate	sialo-	Gr., saliva
post-	L., after	siphon-	Gr., tube
pre-	L., before	siphuncal-	L., small tube
pro-	Gr., before	soma-	Gr., body
proct-	Gr., anus	somn-	L., sleep
protero-	Gr., former, earlier	sperm-	Gr., seed
proto-	Gr., first, before, beginning	spheno-	Gr., wedge
psalter-	Gr., book	splanchno-	Gr., viscera
pseudo-	Gr., false	squam-	L., scale
psor-	Gr., itch	stat-	L., standing
psych-	Gr., breath, soul	stella-	L., star
psyll-	Gr., flea	steno-	Gr., narrow
ptero-	Gr., wing	sterco-	Gr., dung
ptyl-	Gr., saliva	stero-	Gr., solid
pubic-	L., flea	stetho-	Gr., breast
pulmo-	L., lung	stigmo-	Gr., point
pupa-	L., baby	stoma	Gr., mouth
pygo-	Gr., rump	strati-	Gr., layered
pyknosis	Gr., dense mass	strepto-	Gr., turned
pyl-	Gr., gate	stria-	L., furrowed
pyo-	Gr., pus	strongylo-	Gr., round
pyri-	L., pear	stylo-	L., column, pillar
pytro	Gr., fire	sub-	L., under
quadr-	L., four	super-	L., over
rachi-	Gr., spine	supra-	L., above

sym-	Gr., together	troch-	Gr., pulley
syn-	Gr., together	trop-	Gr., turning
syringo-	Gr., pipe	trophy-	Gr., nutrition
tachy-	Gr., rapid	trypano-	Gr., an auger
tact-	L., touch	tumor-	L., swelling
taen-	Gr., ribbon	tunic-	L., a garment
tapet	L., carpet	tympano-	Gr., a drum
tele-	Gr., far, end	tyro-	Gr., cheese
tenui-	L., thin	ultra-	L., beyond
terato-	Gr., wonder	unc-	L., hook
tetra-	Gr., four	-uncula	L., little
thalam-	Gr., chamber	unguli	L., hoof
theco-	Gr., case, covering	uni-	L., one
theli-	Gr., nipple	uro-	Gr., tall, urine
therio-	Gr., breast	vaso-	L., vessel
thermo-	Gr., heat	ven-	L., vein
thromb-	Gr., clot	ventra-	L., belly
thrombid	Gr., timid	vermi-	L., worm
thyro-	Gr., door, shield	vesicul-	L., blister
thysan-	Gr., fringe	via-	L., way
tok-	Gr., birth	villi-	L., shaggy
tomo-	Gr., to cut	vita	L., life
-tonos	Gr., tone, tension	vitr-	L., glassy
toxo-	Gr., a bow	vivi-	L., alive
trabecul-	L., a small beam	vora-	L., to devour
trachel-	Gr., neck	xantho-	Gr., yellow
trachy-	Gr., rough	xero-	Gr., dry
trema	Gr., hole	xylon-	Gr., wood
tremat-	Gr., hole	zoo-	Gr., life, animal
tri-	Gr., three	zygo-	Gr., yoke
tricho-	Gr., hair	zym-	Gr., ferment

Classification of Chordates
Linnaean

A traditional, or Linnaean, scheme of classification for selected chordates follows. Here groups are placed in a hierarchical system of categories.

Phylum Chordata
  Subphylum Cephalochordata
  Subphylum Urochordata
  Subphylum Vertebrata (Craniata)
    Superclass Agnatha
      Class Myxini
      Class Petromyzontiformes
      Class Conodonta
      Class Pteraspidomorphi
        Order Heterostraci
        Order Arandaspida
      Class Cephalaspidomorpha
        Order Osteostraci
        Order Galeaspida
        Order Anaspida
    Superclass Gnathostomata
      Class Placodermi
        Order Stensioellida
        Order Pseudopetalichthyda
        Order Rhenanida
        Order Ptyctodontida
        Order Phyllolepidida
        Order Petalichthyida
        Order Acanthothoraci
        Order Arthrodira
        Order Antiarchi
      Class Chondrichthyes
        Subclass Elasmobranchii
          Order Cladoselachimorpha
          Order Xenacanthimorpha
          Order Selachimorpha—sharks
          Order Batoidea—rays (and skates)
        Subclass Holocephali
          Order Chimaeriformes
      Class Acanthodii
      Class Osteichthyes
        Subclass Actinopterygii
          Superorder Palaeonisciformes
          Order Palaeoniscoids
          Order Acipenseriformes—sturgeons, paddlefish
          Order Polypteriformes (Cladistia)

          Superorder Neopterygii
            Division Ginglymodi
            Order Lepisosteiformes—gars
            Division Halecostomi
            Subdivision Halecomorphi
            Order Amiiformes—bowfin
            Subdivision Teleostei
        Subclass Sarcopterygii
          Superorder Actinistia (Coelacanths)
          Superorder Dipnomorpha
            Order Dipnoi
            Order Porolepiformes
          Superorder Rhipidistia
            Order Osteolepiformes
            Order Rhizodontomorpha
      Class Labyrinthodontia
        Order Ichthyostegalia
        Order Temnospondyli
        Order Anthracosauria
          Suborder Embolomeri
          Suborder Seymouriamorpha
      Class Lissamphibia
        Subclass Amphibia
          Order Gymnophiona
           (Apoda)
          Order Urodela (Caudata)
          Order Anura (Salientia)
      Class Lepospondyli
          Suborder Aïstopoda
          Suborder Nectridea
          Suborder Microsauria
          Suborder Lysorophia
      Class Reptilia
        Subclass Mesosauria
        Subclass Parareptilia
          Order Pareiasauria
          Order Testudinata–turtles
            Suborder Chelonia
        Subclass Eureptilia
          Order Araeoscelida
          Order Captorhinomorpha
      Primitive Diapsids
      Infraclass Ichthyopterygia
        (Ichthyosauria)

Infraclass Lepidosauromorpha
  Superorder Lepidosauria
    Order Eosuchia
    Order Sphenodonta
    Order Squamata—lizards, snakes
  Superorder Sauropterygia
    Order Nothosauria
    Order Plesiosauria
Infraclass Archosauromorpha
  Superorder Archosauria
    Order Thecodontia
    Order Crocodylia
    Order Pterosauria
      Suborder Rhamphorhynchoidea
      Suborder Pterodactyloidea
    Order Saurischia
      Suborder Theropoda
      Suborder Sauropodomorpha
    Order Ornithischia
      Suborder Ornithopoda
      Suborder Pachycephalosauria
      Suborder Stegosauria
      Suborder Ankylosauria
      Suborder Ceratopsia

Class Aves
  Subclass Sauriurae
    Infraclass Archaeornithes *(Archaeopteryx)*
    Infraclass Enantiornithes
  Subclass Ornithurae
    Infraclass Odontornithes *(Hesperornis, Ichthyornis)*
Infraclass Neornithes
  Superorder Palaeognathae
    Order Apterygiformes—kiwis
    Order Tinamiformes—tinamous
    Order Struthioniformes—ostriches
    Order Rheiformes—rheas
    Order Casuariformes—emus, cassowaries
  Superorder Neognathae
    Order Opisthocomiformes—hoatzin
    Order Cuculiformes—cuckoos
    Order Falconiformes—falcons, hawks
    Order Galliformes—grouse, pheasants
    Order Columbiformes—pigeons
    Order Psittaciformes—parrots
    Order Podicipediformes—grebes
    Order Sphenisciformes—penguins
    Order Procellariiformes—albatrosses, petrels
    Order Pelecaniformes—pelicans, cormorants
    Order Anseriformes—ducks, geese
    Order Phoenicopteriformes—flamingos
    Order Ciconiiformes—herons
    Order Gruiformes—rails, cranes
    Order Charadriiformes—gulls, plovers
    Order Gaviiformes—loons
    Order Strigiformes—owls
    Order Caprimulgiformes—oilbirds
    Order Apodiformes—hummingbirds, swifts
    Order Trogoniformes—trogons
    Order Coliiformes—mousebirds
    Order Coraciiformes—kingfishers
    Order Piciformes—woodpeckers, toucans
    Order Passeriformes—songbirds
Class Synapsida
  Order Pelycosauria
  Order Therapsida
    Suborder Biarmosuchia
    Suborder Dinocephalia
    Suborder Gorgonopsia
    Suborder Cynodontia

Class Mammalia
  Subclass Prototheria
    Order Triconodonta—morganucodonts
    Order Docodonta
    Order Multituberculata
    Order Monotremata
  Subclass Theria
    Infraclass Symmetrodonta
    Infraclass Metatheria
      Superorder Marsupialia—marsupials
        Order Didelphimorpha—opossums
        Order Dasyuromorphia—marsupial "mice"
        Order Peramelemorphia—bandicoots
        Order Diprotodontia—possums, koalas
        Order Paucituberculata—shrew opossum
        Order Microbiotheria—Monito del Monte
        Order Notoryctemorpha—marsupial "mole"
    Infraclass Eutheria—placentals
      Superorder Afrotheria
        Order Afrosoricida—golden moles
        Order Macroscelidea—elephant shrews
        Order Tubulidentata—aardvarks
        Order Hyracoidea—hyraxes
        Order Proboscidea—elephants, mastodons
        Order Sirenia—sea cows
      Superorder Xenarthra
        Order Pilosa—anteaters, sloths
        Order Cingulate—armadillos
      Superorder Euarchontoglires
        Order Scandentia—tree shrews
        Order Dermoptera—colugos
        Order Primates
          Suborder Strepsirrhini—lemurs, lorises
          Suborder Haplorrhini—higher primates
            Infraorder Platyrrhini—Old World monkeys, marmosets
            Infraorder Catarrhini
            New World monkeys
            Apes—gibbons, orangutans, chimpanzees, gorillas, hominids
        Order Rodentia
          Suborder Sciurognathi—beavers, mountain beavers, etc.
            Infraorder Sciuromorpha—squirrels, gophers, chipmunks
            Infraorder Myomorpha—voles, mice, rats

Suborder Hystricognathi—
    porcupines
Suborder Caviomorpha—cavies
Order Lagomorpha—pikas, rabbits
Superorder Laurasiatheria
Order Erinaceomorpha—hedgehogs
Order Soricimorpha—shrews, moles
Order Pholidota—pangolins
Order Chiroptera—bats
Order Carnivora
    "Fissipeds"
        Family Viverridae—civets
        Family Herpestidae—mongooses
        Family Hyaenidae—hyaenas
        Family Felidae—cats
        Family Canidae—dogs, wolves,
            foxes
        Family Ursidae—bears
        Family Procyonidae—racoons
        Family Mustelidae—skunks,
            weasels, otters, badgers

"Pinnipeds"
    Family Odobenidae—walrus
    Family Orariidae—eared seals,
        sea lions
    Family Phocidae—earless seals
Order Perissodactyla—horses, tapirs,
    rhinos
"Ungulates"
Order Artiodactyla
    Infraorder Suiformes—pigs, hippos
    Infraorder Tylopoda—camels,
        llamas
    Infraorder Ruminantia
        Family Cervidae—deer, elk
        Family Bovidae—bison, sheep
        Family Giraffidae—giraffes
        Family Antilocapridae—pronghorns
Order Cetacea
    Suborder Odontoceti—toothed
        whales, dolphins
    Suborder Mysticeri—baleen whales

Classification of Chordates Cladistic

One possible scheme of classification within a cladistic context follows in an abbreviated form. Each indented group belongs to the one above it. Therefore, each set of successively indented taxa represents a clade. (The quotation marks remind us that the taxon is likely paraphyletic but is included for its familiarity from traditional classifications, as an interim name until currently unresolved relationships are established, or as a simple convenience.)

Chordata
  Urochordata
  Cephalochordata
  Vertebrata (Craniata)
    Agnatha
      Myxinoidea
      Petromyzontiformes
      Conodonta
      Pteraspidomorphi
      "Other ostracoderms"
    Gnathostomata
      Placodermi
      Chondrichthyes
        Elasmobranchii
        Holocephali
      Teleostomi
        Acanthodii
        Actinopterygii
          Palaeonisciformes
          Neopterygii
        Sarcopterygii
          Actinistia (coelacanths)
          Dipnoi
          "Rhipidistia"
          Tetrapoda
            "Labyrinthodontia"
            Temnospondyli
            Lissamphibia
            Cotylosauria
            Amniota
          Lepospondyli

Amniota (continued)
  Mesosauria
  Sauropsida
    Reptilia
      Parareptilia
      Eureptilia
      Diapsida
        Archosauromorpha
          Pterosauria
          Thecodontia
          Dinosauria
            Ornithischia
            Saurischia
            Aves
        Ichthyosauria
        Lepidosauromorpha
          Sauropterygia
          Lepidosauria
  Synapsida
    "Pelycosauria"
    Therapsida
      Biarmosuchia
      Dinocephalia
      Cynodontia
      Mammalia
        Stem Groups
        Monotremata
        Theria
          Metatheria
          (marsupials)
          Eutheria
          (placentals)

Glossary

abomasum The last of four chambers in the complex ruminant stomach; homologous to the stomach of other vertebrates. *See* omasum, reticulum, and rumen.

acceleration Rate of change of velocity, or how fast velocity is changing.

accommodation The eye's ability to bring an object into focus.

acoustic Pertaining to hearing or perception of sound.

action potential An all-or-none membrane depolarization propagated along a nerve fiber without loss of amplitude.

activation Changes in an egg initiated by fertilization that begin cell division.

acuity The sharpness or keenness of sensory perception, as in sharp vision and keen hearing.

adaptation A phenotypic feature of an individual that contributes to that individual's survival; a feature's form or function and associated biological role with respect to a particular environment.

adrenergic Pertaining to nerve fibers that release adrenaline or adrenaline-like neurotransmitter.

advanced Referring to an organism or species that is derived from others within its phylogenetic lineage. *See* derived and compare with primitive.

aerial locomotion Active flapping in flight; volant.

aerobic Using or requiring oxygen.

afferent Refers to the process of bringing to; for example, sensory afferent fibers convey impulses to the central nervous system. Compare with efferent.

agnathan A vertebrate lacking jaws.

air bladder A gas bladder for respiratory exchange or buoyancy control.

air capillary A small compartment within a bird lung in which gas exchange occurs. Compare with alveolus and faveolus.

airfoil Any object that produces lift when placed in a moving stream of air (as a bird wing).

akinetic skull A skull lacking cranial kinesis, that is, movable joints between skull bones.

allantois An extraembryonic extension of the hindgut of amniote embryos that functions in excretion and sometimes in respiration.

allometry The study of a change in size or shape of one part correlated with a change in size or shape of another part; this relationship can be followed during ontogeny or phylogeny.

alpha motor neuron A nerve cell that innervates extrafusal muscle cells. Compare with gamma motor neuron.

alveolus The smallest subdivision of respiratory tissue in mammalian lungs located at the ends of the branching respiratory tree. Compare with air capillary and faveolus.

amble A slow gait characterized by the two feet on the same side coming in contact with the ground simultaneously; a slow pace. Compare with pace.

ammonotelism Excretion of ammonia directly through the kidneys.

amnion A saclike membrane that holds the developing embryo in a compartment of water.

amniote A vertebrate whose embryo is wrapped in an amnion.

amphystyly Jaw suspension via two major attachments: the hyomandibula and the palatoquadrate.

amplexus A mating embrace of male frogs.

anadromous Characterizing fishes that hatch in fresh water, mature in salt water, and return to fresh water to breed; for example, salmon. Compare with catadromous.

anaerobic Not requiring oxygen.

analogy Features of two or more organisms that perform a similar function; common function.

anamniote A vertebrate whose embryo lacks an amnion.

anastomoses A network of connections between blood vessels.

angiogenesis Blood vessel formation.

angle of attack The orientation of the edge of a wing as it meets the oncoming airstream.

angular acceleration Rate of change of velocity around a point of rotation; rotational acceleration.

anlage (pl., anlagen) A primordium or formative embryonic precursor to a later development structure.

antagonist A muscle with an action opposite to that of other muscles. Compare with fixator and synergist.

antler A branched, bare bone that grows outward from skull bones on some artiodactyl species; usually grows annually in mature males and is shed during the nonreproductive season. Compare with horn.

apnea Temporary cessation of breathing.

aponeurosis A broad, flat tendon.

arboreal locomotion Movement through trees. Compare with brachiation and scansorial locomotion.

archetype The fundamental type or basic underlying blueprint or model on which a definitive animal or animal part is thought to be based.

archinephric duct A general term for the urogenital duct; alternative names (wolffian duct) are given it at different embryonic stages (pronephric duct, mesonephric duct, opisthonephric duct) or in different functional roles (vas deferens).

archipterygial fin A basic fin type in which the axis (metapterygial stem) runs down the middle of the fin. Compare with metapterygial fin.

arcualium An embryonic, cartilaginous anlage to parts of the adult vertebra.

artery Blood vessel carrying blood away from the heart; blood carried may be high or low in oxygen tension. Compare with vein.

artificial taxon A group of organisms not corresponding to an actual unit of evolution. Compare with natural taxon.

aspidospondyly The condition in which the centra and spines of vertebrae are anatomically separate. Compare with holospondyly.

aspiration Drawing in by suction.

aspondyly The condition in which centra are absent from vertebrae.

atavism Evolutionary throwback; reappearance of a lost ancestral trait. Compare with vestigial.

atmosphere The weight that a column of air exerts on an object at sea level; 1 atmosphere = 101,000 Pa = 14.7 lb/sq in.

atrophy A decrease in size or density. Compare with hypertrophy.

auditory Pertaining to the perception of sound.

auricularian hypothesis The view that the basic chordate body plan originated through modification of an echinoderm larva.

autostyly Jaw suspension in which the jaws articulate directly with the braincase.

axon A nerve fiber of a neuron carrying an impulse away from the cell body.

background extinctions *See* uniform extinctions.

baculum A bone within the penis.

baleen Keratinized straining plates that arise from the integument in the mouth of some species of whales.

benthic Bottom dwelling. Compare with pelagic and planktonic.

bilateral symmetry A body in which left and right halves are mirror images of each other.

biogenetic law Ernst Haeckel's claim that ontogeny recapitulates (repeats) phylogeny; now discredited.

biological role How the form and function of a part perform in an environmental context in order to contribute to the organism's survival. Compare with function.

biomechanics The study of how physical forces affect and are incorporated into animal designs.

bipedal Walking or running by means of only two hindlegs. Compare with quadrupedal.

blastocyst The mammalian blastula.

blastopore The opening into the primitive gut formed at gastrulation.

blastula The early embryonic stage that follows cleavage and consists of a hollow, fluid-filled ball of cells.

bolus Soft mass of food in the mouth or stomach. Compare with chyme.

bound A gait in which all four feet strike the ground in unison; a pronk in artiodactyls. Compare with half bound.

boundary layer The fluid layer closest to and flowing over the surface of a body.

brachiation Arboreal locomotion by means of arm swings and grasping hands, with the body suspended below the underside of branches. Compare with scansorial locomotion.

brachyodont Pertaining to teeth with low crowns. Compare with hypsodont.

bradycardia Abnormally slow heart rate. Compare with tachycardia.

braincase That part of the skull containing the cranial cavities and housing the brain.

brain stem The posterior part of the brain comprising the midbrain, pons, and medulla.

branchial basket The expanded chordate pharynx that functions in suspension feeding.

breaking strength The maximum force a structure reaches just before it fails or breaks.

budding A form of asexual reproduction wherein parts separate from the body and then differentiate into a new individual.

bunodont Pertaining to teeth with peaked cusps. Compare with lophodont and selenodont.

bursa A pouch or sac.

calcification A specific type of mineralization involving calcium carbonates (invertebrates) or calcium phosphates (vertebrates) in the matrix of special connective tissue.

cannon bone Hindlimb bone resulting from fusion of metatarsals III and IV (as in horses).

canter A slow gallop.

capillary The smallest blood vessel, which is lined only by endothelium.

carapace The dorsal, dome-shaped bony part of a turtle shell. Compare with plastron.

carnassials Sectorial teeth of carnivores, including upper premolars and lower molars.

catadromous Characterizing fishes that hatch in salt water, mature in fresh water, and return to salt water to breed; for example, some eels. Compare with anadromous.

catastrophic extinctions *See* mass extinctions.

catecholamines Epinephrine and norepinephrine hormones produced by chromaffin and other tissues.

caudal Toward the tail or back end of the body; posterior.

cecal fermentation Process by which microorganisms digest food in the ceca of the intestines. *See* intestinal fermentation.

cecum A blind-ended outpocketing from the intestines.

cell signaling Cell-to-cell communication mediated via direct contact or transported molecules between cells.

cementum Cellular and acellular layers that usually form on the roots of teeth, but in some herbivores, may contribute to the occlusal surface. *See* dentin and enamel.

central nervous system Nervous tissue comprising the brain and spinal cord.

centrum The body or base of a vertebra.

cephalization The gathering of sensory organs in the anterior body usually associated with swellings of the neural tube, the brain.

ceratotrichia Fan-shaped array of keratinized rods internally supporting the elasmobranch fin. Compare with lepidotrichia.

cheek The lateral fleshy wall of the mouth, especially in mammals.

chemoreceptor A sense organ that responds to chemical molecules. Compare with radiation mechanoreceptor and receptor.

chiasma Crossing of fibers.

choana The internal naris; the openings of the nasal passage into the mouth.

cholinergic Nerve fibers that release the neurotransmitter acetylcholine.

chondrocranium That part of the skull formed by endochondral bone or cartilage that underlies and supports the brain; also includes the fused or associated nasal capsules.

chromaffin tissue Endocrine tissue and source of catecholamines (e.g., epinephrine); becomes the medulla of the adrenal gland.

chromatophore General term for a pigment cell.

chyme The liquified bolus of partially digested food after it leaves the stomach and enters the intestine; digesta. Compare with bolus.

ciliary body The tiny ring of muscle in the eye that focuses the lens.

clade A natural evolutionary lineage including an ancestor plus all and only its descendants.

cladogram A branching dendrogram representing the organization and relationships of clades.

claw A sharp, curved, laterally compressed nail at the end of a digit; talon.

cleavage A rapid series of cell divisions that follows fertilization and produces a multicellular blastula.

cleidoic egg The shelled container in which the fetus is laid, as in reptiles, birds, and primitive mammals. Compare with egg.

coelom The fluid-filled body cavity formed within the mesoderm.

collagen Protein fibers secreted by connective tissue cells.

collateral Accompanying, ancillary, or subordinate.

colloid A gelatinous or mucoid substance.

composite theory The hypothesis that jaws evolved from several fused anterior branchial arches.

compressive force The direction of an applied force that tends to press or squeeze an object together.

concurrent Flow of adjacent currents in the same direction.

contralateral Occurring on the opposite side of the body. Compare with ipsilateral.

coprophagy The eating of feces, a behavior performed usually to process undigested material again; refection.

copulation Coitus involving an intromittent organ.

coracoid Posterior coracoid; an endochondral bone of the shoulder that first evolved in early synapsids or their immediate ancestors. Compare with procoracoid.

cornified Having a layer of keratin; keratinized.

cortex The outer portion or rim of an organ.

corticosteroids Steroid hormones.

cosmine An older term designating a derivative of dentin that covers some fish scales; cosmoid scale.

countercurrent Flow of adjacent currents in opposite directions.

cranial Toward the head or front end of the body; anterior or rostral.

cranial kinesis Movement between the upper jaw and braincase about joints between them; in restricted sense, skulls with a movable joint across the roofing bones. Compare with akinetic skull, mesokinesis, metakinesis, prokinesis.

cranial nerve Any nerve entering or leaving the brain. Compare with spinal nerve.

crista A mechanoreceptor within the semicircular canals of the vestibular apparatus of the ear; specialized neuromast organ detecting angular acceleration. Compare with macula.

crop A baglike expansion of the esophagus.

crown group The smallest clade encompassing the living members of a group and the extinct taxa nested within. Compare with stem group, total group.

cursorial locomotion Rapid running.

cutaneous respiration Gas exchange directly between the blood and the environment via the skin.

dead space The volume of used air not expelled upon exhalation. Compare with tidal volume.

decussation A crossing tract of nerve fibers passing from one side to the other side of the central nervous system.

deglutition The act of swallowing.

delamination Splitting of sheets of embryonic tissues into parallel layers.

dendrite A nerve fiber of a neuron carrying impulses toward the cell body.

dendrogram A branching diagram that represents the relationships or the history of a group of organisms.

dental formula Shorthand expression of the characteristic number of each type (incisor, canine, premolar, molar) of upper and lower teeth in a mammalian species.

dentin A material that forms the bulk of the tooth and is similar in structure to bone but harder; yellowish in color and composed of inorganic hydroxyapatite crystals and collagen; secreted by odontoblasts of neural crest origin. *See* cementum and enamel.

dentition A set of teeth.

derived Denoting an organism or species that evolved late within its phylogenetic lineage; advanced; opposite of primitive.

dermal papilla The part of the tooth-forming primordium that is derived from neural crest cells, becomes associated with the enamel organ, and differentiates into odontoblasts that secrete dentin. *See* enamel organ.

dermatocranium That part of the skull formed from dermal bones.

dermatome An embryonic skin segment.

dermis The skin layer that lies beneath the epidermis and is derived from mesoderm.

design The structural and functional organization of a part related to its biological role.

deuterostome An animal whose anus forms from or near the embryonic blastopore; the mouth forms at the opposite end of the embryo.

diarthrosis A joint permitting considerable rotation of articulated skeletal elements, and characterized by a joint capsule, synovial membrane, and articular cartilages on joined elements; synovial joint.

differentiation During embryonic development, the process of cellular diversification.

diffusion The movement of molecules from an area of high concentration to an area of low concentration; if the movement is random and unaided, it is known as passive diffusion.

digestion The mechanical and chemical breakdown of foods into their basic end products—usually simple carbohydrates, proteins, fatty acids—that are absorbed by the bloodstream.

digitigrade A foot posture in which the balls of the feet (middle of the digits) support the weight, as in cats and dogs. Compare with plantigrade and unguligrade.

dikinetic skull A kinetic skull with two joints passing transversely through the braincase. Compare with monokinetic skull.

dioecious Pertaining to female and male gonads in separate individuals. Compare with monoecious.

diphyodont A pattern of tooth replacement involving only two sets of teeth, usually milk teeth and permanent teeth.

dipleurula A hypothetical invertebrate larva proposed as the common ancestor of echinoderms and hemichordates.

diplospondyly The condition in which a vertebral segment is composed of two centra. Compare with monospondyly.

discoidal cleavage Early mitotic divisions restricted to the animal pole; extreme case of meroblastic cleavage. Compare with holoblastic cleavage.

dissection The careful exposure of anatomical parts, allowing students to discover and master the extraordinary morphological organization of an animal in order to understand the processes these parts perform and the remarkable evolutionary history out of which they come. Pronounced *dis*-section, as opposed to *di*section, which means chopping into two halves.

distal Toward the free end of an attached part, such as the limb. Compare with proximal.

diurnal Active during daylight. Compare with nocturnal.

dorsal Toward the back or upper surface of the body; opposite of ventral.

down-regulation Return of the gut, and other organs, to a resting state following digestion. Compare with up-regulation.

drag The force that resists the movement of an object through a fluid; total drag includes parasitic drag and induced drag.

ecomorphology The study of the relationship between the form and function of a part and how it is actually used in a natural environmental setting; the basis for determining biological role.

ectomesenchyme Loose association of cells derived from neural crest.

ectotherm An animal that depends on environmental sources of heat to reach its preferred body temperature. Compare with endotherm.

edema Swelling due to collection of fluid in body tissues.

effector An organ, such as a muscle or a gland, that responds to nervous stimulation.

efferent Refers to the process of carrying away; for example, motor neurons carry impulses away from the central nervous system.

egg The haploid cell produced by the female; ovum.

elastic In physical terms, the measure of a structure's ability to return to its original size following deformation.

electric organ A specialized block of muscles producing electrical fields and often high jolts of voltage.

electromagnetic radiation Energy waves through a spectrum that includes radio waves, infrared light, visible light, ultraviolet light, X rays, and gamma rays.

electromyogram The electrical record of a muscle contraction.

electromyography The study of the pattern of muscle contraction based on detecting its electrical activity.

electroreceptor A sensory organ that responds to electrical signals or fields.

emargination Large notches in the bony braincase. Compare with fenestra.

embolomerous vertebra A dispondylous vertebra in which both centra are separate (aspidospondylous) and of about equal size. Compare with stereospondylous vertebra.

emergency life history stage Long-term physiological adaptation triggered in minutes to hours to unusual environmetal stressors, via glucocorticoids. *See* fight-or-flee.

emulsify To break up fats into smaller droplets. Compare with digestion.

enamel Forms the occlusal cap on most teeth; hardest substance in vertebrate body, consisting almost entirely of calcium salts as apatite crystals; secreted by ameloblasts of epidermal origin. *See* cementum and dentin.

enamel organ The part of the tooth-forming primordium that is derived from epidermis, becomes associated with the dermal papilla, and differentiates into the ameloblasts that secrete enamel. *See* dermal papilla.

encapsulated sensory receptor The terminus of a sensory nerve fiber that is wrapped in accessory tissue. Compare with free sensory receptor.

endochondral bone formation Embryonic formation of bone preceded by a cartilage precursor that is subsequently ossified; cartilage or replacement bone. Compare with intramembranous bone formation.

endocrine Denoting a gland that releases its product directly into blood vessels. Compare with exocrine.

endocytosis A phagocytic process in which materials, such as food particles and foreign bacteria, are engulfed by a cell.

endoskeleton The supportive or protective framework within the body that lies beneath the integument. Compare with exoskeleton.

endothelium The single-celled inner lining of vascular channels. Compare with mesothelium.

endotherm An animal capable of maintaining an elevated body temperature with heat produced metabolically from within. Compare with ectotherm.

enteric nervous system The network of nerves intrinsic to the digestive system.

enterocoelom The body cavity formed within outpocketings of mesoderm. Compare with schizocoelom.

ependyma The layer of cells lining the central canal of the chordate spinal cord.

epiboly The spreading of surface cells during embryonic gastrulation.

epidermis The skin layer over the dermis that is derived from ectoderm.

epigenetics The study of developmental events above the level of the genes; embryonic processes not directly arising from the genes that contribute to the developing phenotype.

epiphysis 1. The secondary center of ossification on the end of a bone; also refers to the end of a bone. 2. The pineal gland.

estivation A prolonged resting state or hibernation during times of heat or drought that is characterized by lowered metabolic levels and breathing rates.

euryhaline Having a wide tolerance to salinity differences.

evolutionary morphology The study of the relationship between the change in anatomical design through time and the processes responsible for this change.

excretion Removal of wastes and excess substances from the body.

exocrine Denoting a gland that releases secretions into ducts. Compare with endocrine.

exocytosis A process by which the cell releases products.

exoskeleton A supportive or protective framework lying on the outside of the body. Compare with endoskeleton.

extant Living.

exteroceptor A sensory receptor that responds to environmental stimuli. Compare with interoceptor and proprioceptor.

extinct Dead.

extraembryonic Pertaining to a structure formed by or around the embryo but not retained by or directly contributing to the adult body.

extrafusal muscle cell The fiber of striated muscles that actually contributes to the force moving a part. Compare with alpha motor neuron and intrafusal muscle cells.

extrinsic Originating outside the part on which it acts. Compare with intrinsic.

failure In mechanics, loss of functional integrity and ability to perform; a material may fail but not break. Compare with fracture.

fascicle A bundle of muscle fibers defined by a connective tissue coat within a muscle organ.

fatigue fracture Reduced breaking strength of an object after prolonged use.

faveolus A tiny respiratory compartment within the lung that opens to a central air chamber and results from the subdivisions of the lung lining. Compare with alveolus.

fenestra An opening within the bony braincase.

fermentation A process in which microorganisms anaerobically extract energy from food in vertebrates by releasing cellulase enzymes that break down plant material.

fertility The ability to produce viable eggs or sufficient sperm to propagate offspring; infertility results from nonviable eggs or insufficient sperm. Compare with potency.

fetus The embryo at a later stage in development.

fight-or-flee Short-term, immediate physiological response to a survival threat, via endocrine and sympathetic systems.

fin An external plate or membrane that projects from the body of an aquatic animal (as in fishes).

fixator A muscle that functions to stabilize a joint. Compare with antagonist and synergist.

flight Aerial locomotion accomplished by active flapping of wings. Compare with gliding and parachuting.

follicle A small bag that holds cells containing hormones (e.g., thyroid follicle) or one that holds an ovum (e.g., ovarian follicle).

footfall Foot contact with the ground during locomotion.

foramen A perforation or hole through a tissue wall.

foramen of Panizza A connecting vessel between the bases of the left and right aortic arches in crocodilians.

foramen ovale The one-way connection between the right and left atria of an embryonic mammal; closes at birth.

foregut Anterior embryonic gut that gives rise to the pharynx, esophagus, stomach, and anterior intestine. Compare with hindgut.

foregut fermentation See gastric fermentation.

formed elements The cellular components of blood, excluding the plasma. Compare with plasma.

fossorial locomotion Active removal of soil to produce a burrow; digging.

fracture In mechanics, a break or loss of structural integrity; actual separation of material under load. Compare with failure.

free sensory receptor The terminus of a sensory nerve fiber that lacks any associated structures. Compare with encapsulated sensory receptor.

friction drag A consequence of friction between the fluid and the surfaces over which it flows.

frontal plane A plane passing from one side of an organism to the other so as to divide the body into dorsal and ventral parts. Compare with transverse plane.

fulcrum The point of pivot or the axis of rotation.

function How a part performs within an organism. Compare with biological role.

functional morphology The study of the relationship between the anatomical design of a structure and the function or functions it performs.

fusiform Refers to a narrow shape tapering toward each end.

gait The pattern or sequence of foot movements during locomotion.

gallop A gait characterized by a high rate of speed and an uneven footfall pattern.

gametogenesis The production of mature male and female reproductive cells.

gamma motor neuron A nerve cell that innervates intrafusal muscle cells. Compare with alpha motor neuron.

ganglion A gathering of nerve cell bodies within the peripheral nervous system.

ganoin An older term for a derivative of the enamel that covers some fish scales; ganoid scale.

gas bladder A gas-filled bag in fishes derived from the gut. Because the composition of the gases can vary, the term *air bladder* is less appropriate. Compare with respiratory bladder and swim bladder.

gastralia Rib-shaped dermal bones located in the abdominal region.

gastric fermentation Process in which microorganisms digest food in a specialized stomach; also known as foregut fermentation. *See* intestinal fermentation.

gastrocoel The cavity within the early embryonic gut of the gastrula.

gastrula Early embryonic stage during which the basic gut is formed.

general cranial nerve A bundle of fibers that detect sensations from the widely distributed viscera. Compare with special cranial nerve.

general to specific (embryology) Usually attributed to von Baer, the observation (law) that early embryos possess general features of the group that give way in later development to specialized features of the species.

genotype The genetic makeup of an individual. Compare with phenotype.

gestation The time that elapses from conception to hatching or birth.

gill Aquatic respiratory organ.

gill slit Pharyngeal slit associated with a gill.

gizzard An especially well-muscularized region of the stomach used to grind hard foods.

gliding A gradual airborne descent that can be extended by the action of the body and limbs on the relative wind but is not self-powered. Compare with flight and parachuting.

glomerulus 1. A small bed of capillaries associated with the uriniferous tubule. 2. A small cluster of capillaries on the stomochord of hemichordates.

gnathostome A vertebrate with jaws.

grade A level or stage of evolutionary attainment; a paraphyletic group.

graded potential A nerve impulse proportional to the intensity of the stimulus that produces it and declining thereafter. Compare with action potential.

gyrus A swollen ridge on the surface of the brain. Compare with sulcus.

hair cell A mechanoreceptive cell with a projecting hair bundle composed of a kinocilium and several stereocilia.

half bound A gait in which the hindfeet contact the ground simultaneously, but the forefeet do not. Compare with bound.

hemocoel Blood-filled channels within connective tissue that lack a continuous endothelial lining.

hemodynamics The forces and flow patterns of blood circulating within vessels.

hemopoietic tissue Blood-forming tissue. Compare with lymphoid tissue and myeloid tissue.

hepatocyte A liver cell.

heterochrony Within an evolutionary lineage, the change in time at which a characteristic appears in the embryo relative to its appearance in a phylogenetic ancestor; usually concerned with the time of onset of sexual maturity relative to somatic development. Compare with paedomorphosis.

heterodont Dentition in which the teeth are different in general appearance throughout the mouth.

hindgut Posterior embryonic gut that gives rise to the posterior intestines. Compare with foregut.

hindgut fermentation *See* intestinal fermentation.

holoblastic cleavage Early mitotic planes pass entirely through the cleaving embryo. Compare with discoidal cleavage and meroblastic.

holonephros A single kidney arising from several regions of the nephric ridge rather than the three types of kidneys (pronephros, mesonephros, metanephros) arising from the nephric ridge.

holospondyly The condition in which the centra and spines of vertebrae are anatomically fused into a single bone. Compare with aspidospondyly.

homeostasis The constancy of an organism's internal environment.

homeothermy The condition of maintaining constant body temperature, without regard to the method.

homodont Dentition in which the teeth are similar in general appearance throughout the mouth.

homology Features in two or more organisms derived from common ancestors; common ancestry. Compare with serial homology.

homoplasy Features in two or more organisms that look alike; similar in appearance.

hoof An enlarged cornified plate on the end of an ungulate digit.

hormone A chemical messenger that is secreted into the blood by an endocrine organ and affects target tissues.

horn An unbranched keratinized sheath with a bony core located on the head; usually occurs in both males and females and is retained year-round. Compare with antler.

hydrofoil An object that produces lift when placed in a moving stream of water, as, for example, the pectoral fin of a shark.

hydrostatic organ A structure whose mechanical integrity depends on a fluid-filled core enclosed by walls of connective tissue.

hydrostatic pressure Fluid force, as in blood, resulting from heart contraction. Compare with osmotic pressure.

hyostyly Jaw suspension primarily through attachment with the hyomandibula.

hyperosmotic Refers to a solution whose osmotic pressure is greater than the solution that surrounds it; for example, the pressure of the tissue fluid within some fishes is greater than the pressure from the fresh water surrounding them. Compare with hyposmotic and isosmotic.

hyperplasia An increase in the number of cells as a result of cell proliferation; usually occurs in response to stress or increased activity. Compare with hypertrophy and metaplasia.

hypertrophy An increase in the size or density of an organ or part, which does not result from cell proliferation. Compare with hyperplasia and atrophy.

hyposmotic Refers to a solution lower in osmotic pressure than the solution that surrounds it. Compare with hyperosmotic and isosmotic.

hypoxia Inadequate levels of oxygen to support metabolic demands.

hypsodont Teeth with high crowns. Compare with brachyodont.

impedance matching Adjustments of the sound conduction system to address the physical resistance that sound waves meet as they travel from the air to the fluid of the inner ear.

implantation The process by which the embryo establishes a viable residence within the wall of the uterus.

incus The middle ear bone of mammals derived phylogenetically from the quadrate.

index fossil A fossil animal widely distributed geographically but restricted to one rock layer or time horizon; defining species indicator of a stratum.

induced drag The resistance to forward travel resulting from the lift produced by an airfoil.

ingression A process by which individual surface cells migrate inward to the interior of the embryo.

ingroup The group of organisms actually studied. Compare with outgroup.

insertion The relatively movable site of attachment of a muscle. Compare with origin.

integument The skin covering the body.

internode In taxonomy, the line connecting nodes on a cladogram representing at least one ancestral species from a speciation event. Compare with node.

interoceptor A sensory receptor that responds to internal stimuli. Compare with exteroceptor and proprioceptor.

interrenal tissues Endocrine tissue that produces corticosteroids and becomes the cortex of the adrenal gland. Compare with chromaffin tissue.

interstitial Pertaining to the fluid-filled space between cells.

intervertebral body A pad of cartilage or fibrous connective tissue between articular ends of successive vertebral centra.

intervertebral disk A pad of fibrocartilage in the adult mammal that has a gel-like core derived from the notochord and is located between articular ends of successive vertebral centra. Compare with intervertebral body.

intestinal fermentation Process in which microorganisms digest food in the intestines; also known as hindgut fermentation and cecal fermentation. Compare with gastric fermentation.

intrafusal muscle cell The fiber of a striated muscle that is specialized as a sensory organ of proprioception; housed in a muscle spindle. Compare with extrafusal muscle cell, gamma motor neurons, and muscle spindle.

intramembranous bone formation Embryonic formation of bone directly from mesenchyme without a cartilage precursor; dermal bone. Compare with endochondral bone formation.

intratarsal joint An archosaur ankle in which the line of flexion passes *between* the calcaneus and the astragalus. Compare with mesotarsal joint.

intrinsic Belonging entirely to a part; that is, inherent to the part. Compare with extrinsic.

intromittent organ The male reproductive organ that delivers sperm into the female reproductive tract; penis or phallus.

invagination An indentation or infolding of the surface.

involution The turning of embryonic surface cells inward to spread across the interior of the embryo.

ipsilateral Occurring on the same side of the body. Compare with contralateral.

ischemia Insufficient blood flow to meet the metabolic demands of a tissue.

isocortex The cerebral cortex or outer layer of the mammalian cerebrum.

isolecithal Pertaining to an egg in which the yolk is evenly distributed throughout the cytoplasm. Compare with telolecithal.

isometry Geometric similarity in which proportions remain constant with changes in size.

isosmotic Refers to two solutions with equivalent solute levels. Compare with hyperosmotic and hyposmotic.

jaws Skeletal elements of bone or cartilage that reinforce the lower borders of the mouth.

keratin Fibrous protein.

keratinization The process by which the skin forms proteins, especially keratin.

kinesis Movement; usually refers to the relative movement of skull bones. *See* cranial kinesis.

kinocilium A modified, rigid cilium of the ear. Compare with microvillus.

lactation The release of milk from mammary glands to suckling young.

lacuna A small space.

lamina A thin sheet, layer, or plate; for example, gill lamella.

laminar flow The movement of fluid particles along smooth paths through layers that glide over one another. Compare with turbulent flow.

larva An immature (nonreproductive) stage that is morphologically different from the adult.

lateral Toward or on the side of the body.

lecithotrophic Pertaining to the nutrition that the embryo receives from the yolk of the ovum. Compare with matrotrophic.

lepidotrichia A fan-shaped array of ossified or chondrified dermal rods that internally supports the fin of bony fishes. Compare with ceratotrichia.

lepospondyly A holospondylous vertebra with a husk-shaped centrum usually pierced by a notochordal canal.

lever arm The perpendicular distance from the point at which the force is applied to the point about which a body rotates (moment arm). *See* moment.

lift The force produced by an airfoil perpendicular to its surface.

lines of arrested growth (LAGs) A period when annuli or rings are deposited in bones as a result of cessation or even resorption of bone.

lingual feeding The capture of prey with the tongue.

load In mechanics, the forces to which a structure is subjected.

loop of Henle A region of the mammalian nephron that includes parts of the proximal and distal tubules (thick limbs) and all of the intermediate tubule (thin limb).

lophodont Teeth having broad, ridged cusps useful in grinding plant material. Compare with bunodont and selenodont.

lumen The space within the core of an organ, especially a tubular organ.

lymph Clear fluid carried in lymphatic vessels.

lymphoid tissue Blood-forming tissue outside of bone cavities; found, for example, in the spleen and lymph nodes.

macrolecithal Pertaining to eggs with large quantities of stored yolk.

macula A mechanoreceptor within the vestibular apparatus of the ear; specialized neuromast organ detecting changes in body posture and acceleration. Compare with crista.

malleus One of the three middle ear bones in mammals, phylogenetically derived from the articular bone.

mass extinctions Loss of many species from different groups over relatively short periods of geological time. Syn., catastrophic extinction.

mastication The mechanical breakdown of a large bolus of food into smaller pieces, usually with the teeth; chewing of food.

matrotrophic Pertaining to the nutrition the embryo receives through the placenta or from uterine secretions. Compare with lecithotrophic.

meatus A canal or opening.

mechanoreceptor A sense organ that responds to small changes in mechanical force. Compare with chemoreceptor and photoreceptor.

medulla The inner portion or core of an organ.

meroblastic cleavage Early mitotic planes that do not complete their passage through the embryo before subsequent division planes form. Compare with discoidal and holoblastic cleavage.

merycism Remastication together with microbial fermentation of food in nonruminants. Compare with rumination.

mesenchyme Loosely associated cells of mesodermal origin.

mesokinesis Skull movement via a transverse joint passing through the dermatocranium posterior to the ocular orbit. Compare with metakinesis and prokinesis.

mesolecithal Pertaining to eggs with moderate amounts of stored yolk. Compare with macrolecithal and microlecithal.

mesonephros A kidney formed of nephric tubules arising in the middle of the nephric ridge; usually a transient embryonic stage that replaces the pronephros, but is itself replaced by the adult metanephros or opisthonephros. Compare with metanephros, opisthonephros, and pronephros.

mesotarsal joint An archosaur ankle in which the calcaneous and astragalus fuse and the line of flexion passes between them and the distal tarsals. Compare with intratarsal joint.

mesothelium A single-celled lining of body cavities.

metakinesis Skull movement via a transverse hinge that lies posterior between the deep neurocranium and outer dermatocranium. Compare with mesokinesis and prokinesis.

metamorphosis An abrupt transformation from one anatomically distinct stage (juvenile) to another (adult).

metanephric duct Ureter; distinct from the pronephric and mesonephric ducts.

metanephros A kidney formed of nephric tubules arising in the posterior region of the nephric ridge and drained by a ureter; usually replaces the embryonic pronephros and mesonephros. Compare with mesonephros, opisthonephros, and pronephros.

metaplasia Change of a tissue from one type to another type. Compare with hypertrophy.

metapterygial fin Basic fin type in which the axis (metapterygial stem) is located posteriorly in the fin. Compare with archipterygial fin.

metapterygial stem The chain of endoskeletal elements within the fish fin that define the major internal supportive axis.

microcirculation The capillary beds and arterioles that supply them and venules that drain them.

microlecithal Pertaining to eggs that contain small quantities of stored yolk. Compare with macrolecithal and mesolecithal.

microvillus A small cytoplasmic projection from a single cell. Compare with villus.

midsagittal plane Median parallel plane passing dorsoventrally through the long central axis of the body.

mineralization A general process wherein various metallic ions are deposited, under the control of living cells of the tissue, in the organic matrix of the connective tissue.

molariform A general term describing premolar and molar teeth that appear similar; cheek teeth.

molt The shedding of parts or all of the cornified layer of the epidermis; loss of feathers or hair that usually occurs annually; ecdysis.

moment The measure of the tendency of a force to rotate a body; the product of force times the perpendicular distance from the point at which the force is applied to the point of rotation (lever arm).

moment arm A lever arm.

monoecious Refers to female and male gonads within the same individual; hermaphrodite.

monokinetic skull Skull movement via a single transverse joint passing through the braincase.

monophyletic group A clade, all organisms in a lineage plus the ancestor they have in common, therefore a natural group. Compare with paraphyletic and polyphyletic groups.

monospondyly The condition in which a vertebral segment is composed of one centrum. Compare with diplospondyly.

morph A term referring to the general form or design of an animal; for example, juvenile morph (tadpole) and adult morph (sexually mature stage) of a frog.

morphogenesis During embryonic development, the processes that result in the reorganization of tissues into organs and the basic body configuration of the embryo.

morphological cross section A plane or cut through the area of a muscle perpendicular to its longitudinal axis at its thickest part. Compare with physiological cross section.

morphology The study of anatomy and its significance.

motor end plate The neuromuscular junction; specialized ending through which the axon of a motor neuron makes contact with the muscle it innervates.

motor neuron A nerve cell carrying impulses to an effector organ. Compare with sensory neuron.

motor pattern A defined local pattern of activity produced by muscles that shows little variation when repeated.

motor unit One motor neuron and the subset of muscle fibers that it supplies; important in producing graded muscle force.

mucous gland An organ secreting a protein-rich mucin that is usually a thick fluid. Compare with serous gland.

muscle fiber A muscle cell, that is, the contractile part of a muscle organ.

muscle organ Muscle cells together with the noncontractile tissues that support them (connective tissue, blood vessels, nerves).

muscle spindle A fusiform bundle in striated muscles that houses specialized sensory receptors known as intrafusal muscle cells.

myeloid tissue Blood-forming tissue housed inside bones.

myoepithelial cell A cell lining (hence, *epithelial*) a channel or gland and possessing contractile abilities (hence, *myo-*).

myofibril A contractile unit of a muscle cell; a chain of repeating sarcomeres composed of myofilaments.

myofilaments Thick and thin filaments in the fine structure of muscles composed predominantly of myosin and actin, respectively.

myomeres Differentiated segments of a muscle in an adult.

myotomes Undifferentiated embryonic blocks of presumptive muscles.

naris A nostril.

natural selection The process by which organisms with poorly suited features, on average, fare less well in a particular environment and tend to perish, thereby leaving (preserving) those individuals with more favorable adaptations; survival of the fittest.

natural taxon A group of organisms that represent an actual result of evolutionary events. Compare with artificial taxon.

neomorph A new morphological structure in a derived species that has no equivalent evolutionary antecedent.

neoteny Paedomorphosis produced by delayed onset of somatic development that is overtaken by normal sexual maturity. Compare with progenesis.

nephric ridge The posterior region of the intermediate mesoderm.

nephridium A tubular excretory organ.

nephron That portion of the uriniferous tubule in which urine is formed; composed of proximal, intermediate, and distal regions; nephric tubule.

nephrotome Segmental forerunner of the nephron in the urinary structure of the early embryo.

nerve A collection of nerve fibers coursing together in the peripheral nervous system.

nerve fiber The cytoplasmic process of a neuron; an axon or a dendrite.

network Any structure reticulated or decussated at equal distances with interstices between the intersections.

neural crest A paired strip of tissue that separates from the dorsal edges of the neural groove as it forms the neural tube.

neurocranium That part of the braincase that contains cavities for the brain and associated sensory capsules (nasal, optic, otic).

neuroglia Nonnervous supportive cells of the nervous system.

neurohormone A chemical secreted directly into a blood capillary by a neurosecretory neuron at the terminus of its axon.

neuromast organ A mechanoreceptive organ composed of several hair cells, as in the lateral line of the inner ear.

neuron A nerve cell.

neurotransmitter A chemical released at the synapse of a nerve fiber, usually an axon.

nocturnal Active at night. Compare with diurnal.

node In taxonomy, the branching point on a cladogram representing a speciation event. Compare with internode.

notochord A long axial rod composed of a fibrous connective tissue wall around cells and/or a fluid-filled space.

nucleus 1. A membrane-bound organelle within the body of a cell. 2. A group of nerve cell bodies within the central nervous system.

occlusion The meeting or closure of the upper and lower tooth rows.

odor A chemical detected by sensory cells in the nasal epithelium through the olfactory process. Compare with vomodor.

olfaction The act of smelling.

omasum The third of four chambers in the complex ruminant stomach; a specialization of the esophagus. *See* abomasum, reticulum, and rumen.

ontogeny Changes in an organism from zygote to death, although often focused on events from zygote to maturity.

operculum A lid or cover, as over the gills of fishes.

opisthonephros The adult kidney formed from the mesonephros and additional tubules from the posterior region of the nephric ridge. Compare with mesonephros, metanephros, and pronephros.

origin The relatively fixed site of attachment of a muscle. Compare with insertion.

osmoregulation The active maintenance of water and solute balance.

osmotic pressure The tendency for fluid solutes to move across a membrane in order to equalize the concentrations of solutes on both sides. Compare with hydrostatic pressure.

ossification A specific type of mineralization, unique to vertebrates, wherein hydroxyapatite (calcium phosphate) is deposited on the collagenous matrix, leading to bone formation.

osteoderm A dermal bone located under and supporting an epidermal scale.

osteon A highly ordered arrangement of bone cells into concentric rings, with bone matrix surrounding a central canal through which blood vessels and nerves run; the Haversian system.

otoconia Small calcareous crystals on the maculae of the inner ear; small otoliths.

otolith A single calcareous mass in the cupula of hair cells.

outgroup Any group used for taxonomic comparison but not part of the taxon under study. Compare with ingroup.

oviduct A urogenital duct transporting ova and often involved in protection and nourishment of the embryo; Müllerian duct.

oviparity The reproductive pattern of egg laying.

oviposition The act of laying eggs.

ovulation The release of the ovum from the ovary.

pace A high-speed gait characterized by the two feet on the same side coming in contact with the ground simultaneously; fast amble. Compare with amble.

paedomorphosis The retention of general juvenile features of ancestors in the late developmental stages of descendants. Larval stages of ancestors become the reproductive "adult" stages of descendants. *See* neoteny and progenesis.

parachuting An airborne fall slowed by the use of canopy-like membranes or body shape that increase drag.

paradaptation The concept that some aspects of a feature may not be adaptive or owe their properties to natural selection.

parallax The difference in the appearance of an object when it is viewed from two different points.

parallel muscle A muscle organ in which all its muscle fibers lie in the same direction and are aligned with its long axis. Compare with pinnate muscle.

paraphyletic group An incomplete clade resulting from removal of one or more component lineages. Compare with monophyletic and polyphyletic groups.

parasagittal plane A sagittal plane parallel with the midsagittal plane.

parasitic drag Resistance to the passage of a body through a fluid as a result of the body's surface friction and adverse backflow in the wake.

paraxial mesoderm Paired strips or mesodermal populations forming along the neural tube; in the head, it remains as strips of mesoderm called somitomeres, but in the trunk it becomes segmentally arranged as somites.

parition The general term for parturition and oviposition.

partial pressure The pressure one gas contributes to the total pressure in a mixture of gases.

parturition The act of giving birth via viviparity. Compare with oviposition.

patagium A stretched fold of skin that forms an airfoil or flight control surface.

patterning The process of establishing the main topographical regions and body axes in an embryo; dorsoventral, anteroposterior, for instance.

pelagic Living in open water. Compare with benthic and planktonic.

pentidactyl Having five digits per limb; thought to be the basic pattern characteristic of tetrapods but modified by functional demands.

perfusion The driving of blood through capillary beds of an organ. Compare with ventilation.

perichondrium The sheet of fibrous connective tissue around cartilage.

perikaryon The nucleus of a cell and its adjacent cytoplasm, especially applied to nerve cells.

periosteum The sheet of fibrous connective tissue around bone.

peripheral nervous system The cranial and spinal nerves and their associated ganglia composing that part of the nervous system outside the central nervous system.

peristalsis Progressive waves of muscle contractions within the walls of a tubular structure, as within the walls of the digestive tract.

permissive Pertaining to hormones that relax insensitive target tissues, allowing them to respond to hormonal, neuronal, or environmental stimuli; permitting target tissues to respond.

pharyngeal slit An elongated opening in the lateral wall of the pharynx.

phasic Recurring cycles of muscle contractions.

phenotype The physical and behavioral characteristics of an organism; somatic features. Compare with genotype.

pheromone A chemical released into the environment by one individual that influences the behavior or physiology of another individual of the same species.

photopic vision Color vision in bright light. Compare with scotopic vision.

photoreceptor A radiation sensory receptor responsive to visible light stimuli.

phylogeny The course of evolutionary change within a related group of organisms.

physiological cross section A plane or cut through the area of all muscle fibers perpendicular to their long axes. Compare with morphological cross section.

piezoelectricity Low-level electrical charges arising on the surface of stressed crystals; load stress on bones may produce surface electrical charges.

pinnate muscle A muscle organ in which all muscle fibers are aligned obliquely to its line of action. Compare with parallel muscle.

pituicyte A nonendocrine cell of the neurohypophysis.

placenta A composite organ formed of maternal and fetal tissues through which the embryo is nourished.

placode A distinct thickened plate of embryonic ectoderm.

planktonic Pertaining to a free-floating microscopic plant or animal that is passively carried about by currents and tides. Compare with benthic and pelagic.

plantigrade A foot posture in which the entire sole comes in contact with the ground. Compare with digitigrade and unguligrade.

plasma The fluid component of blood without any formed elements.

plastron The ventral bony part of a turtle shell. Compare with carapace.

platysma An unspecialized muscle derived from hyoid arch musculature that spreads as a thin subcutaneous sheet into the neck and over the face.

plesiomorphy An ancestral trait present at the base of a clade . Compare with synaphomorphy.

plexus A network of intermingling blood vessels or nerves.

podocytes Specialized excretory cells associated with blood capillaries of the kidney.

polydactyly An increase in the number of digits over the basic pentidactylous number. Compare with polyphalangy.

polyphalangy An increase in the number of phalanges in each digit. Compare with polydactyly.

polyphyletic group An artificial group characterized by features that are not homologous. Compare with monophyletic and paraphyletic groups.

polyphyodont A pattern of continuous tooth replacement. Compare with diphyodont.

polyspondyly The condition in which a vertebral segment is composed of two or more centra.

portal system A set of venous vessels beginning and ending in capillary beds or sinuses of the liver.

potency The ability of a male to engage in copulation; impotency is an inability to copulate. Compare with fertility.

power The amount of work that can be done per unit of time.

preadaptation The concept that features possess the necessary form and function to meet the demands of a particular environment before the organism experiences that particular environment. Compare with paradaptation.

prehension The rapid grasping and capturing of the prey, usually with the jaws or claws.

preservationism (embryology) The observation that embryonic development is conservative in that some features (e.g., gill slits) of ancestral embryos are retained in embryos of descendants.

pressure drag A consequence of laminar flow separation, setting up the eddying motion of the disturbed fluid.

primary neurulation Formation of the embryonic neural tube via folds in the edges of the neural plate that uplift and subsequently fuse, thereby defining the neuralcoel. Compare with secondary neurulation.

primitive Denoting an organism or species that appeared early within its phylogenetic lineage; opposite of derived.

procoracoid Anterior coracoid (or precoracoid); endochondral bone of the shoulder that first evolved in fishes. Compare with coracoid.

proctodeum The embryonic invagination of surface ectoderm that contributes to the hindgut, usually giving rise to the cloaca.

progenesis Paedomorphosis produced by precocious onset of sexual maturity in an individual still in the morphologically juvenile stage. Compare with neoteny.

project In the nervous system, to transmit neural impulses to.

prokinesis Refers to skull movement via a transverse joint that passes through the dermatocranium anterior to the ocular orbit. Compare with mesokinesis and metakinesis.

pronephros A kidney formed of nephric tubules arising in the anterior region of the nephric ridge; usually forms only as a transient embryonic structure. Compare with mesonephros, metanephros, and opisthonephros.

proprioceptor A specialized interoceptor that responds to limb position, joint angle, and state of muscle contraction. Compare with exteroceptor and interoceptor.

protandry Reproduction in which the same individual produces sperm and then later in life produces eggs, but does not do both concurrently.

protostome An animal whose mouth forms from or near the embryonic blastopore.

proximal Toward the base of an attached part where it joins the body. Compare with distal.

pterylae Feather tracks.

punctuated equilibrium A description of phylogenetic patterns in which long periods of little or no change are punctuated by short periods of prolific change before returning to a period of little change. Compare with quantum evolution.

quadrupedal Walking or running by means of four legs. Compare with bipedal.

quantum evolution Adaptive evolutionary change within a lineage characterized by long periods of little change that are suddenly interrupted by short bursts of rapid change. Compare with punctuated equilibrium.

radial symmetry A regular arrangement of the body about a central axis.

radiation receptor A sensory organ that responds to light and other forms of electromagnetic radiation.

raptors Predatory birds that use talons, including hawks, eagles, falcons, and owls.

rate modulation The proportionate increase in the contractile force of a muscle as the rate of nerve impulses increases. Compare with motor unit.

receptor The end of a nerve fiber that responds to stimuli. Compare with effector.

recrudescence Renewal of reproductive interest and readiness of reproductive tracts, usually on a seasonal basis.

recruitment The process of initiating contraction of additional muscle cells within a muscle organ during its activity.

reflex Involuntary action of effectors mediated by the nervous system.

refractive index A measure of the light-bending properties of an object.

refractory Not responsive; usually time dependent.

relative wind The apparent direction of the airstream across an airfoil; depends on the angle of attack, speed of the airfoil, and so forth.

release-inhibiting hormone A hormone that depresses the responsiveness of target tissues.

releasing hormone A hormone that initiates target tissue activity. Compare with release-inhibiting hormone.

renal capsule The expanded end of the uriniferous tubule that surrounds the vascular glomerulus; ultrafiltrate first forms in the renal capsule; also known as Bowman's capsule.

respiratory bladder A gas bladder enriched with capillaries that allow it to function primarily in gas exchange.

rete A compact, dense network of capillaries.

reticulum The second of four chambers in the complex ruminant stomach; a specialized region of the esophagus. Compare with abomasum, omasum, and rumen.

rhachitomous vertebra An aspidospondylous vertebra characteristic of some crossopterygians and early amphibians.

roe Fish eggs still within the ovary.

rumen The first of four chambers in the complex ruminant stomach; an expanded specialization of the esophagus. Compare with abomasum, omasum, and reticulum.

ruminant A placental mammal with a rumen, a specialized expansion of the digestive tract that processes plant material; Ruminantia.

rumination Remastication together with microbial fermentation in ruminants.

sagittal plane Any plane parallel with the long axis of the animal's body and oriented dorsoventrally.

sarcolemma The plasma membrane of a muscle cell.

sarcomere A repeating unit of overlapping myofilaments that composes the contractile myofibril of a muscle cell.

sarcoplasm The cytoplasm of a muscle cell.

scaling Compensatory adjustments in proportion to maintain performance with changes in size.

scansorial locomotion Climbing of trees with claws. Compare with brachiation.

schizocoelom The body cavity formed by splitting of the mesoderm. Compare with enterocoelom.

scotopic vision Sensitivity to dim light. Compare with photopic vision.

secondary cartilage Cartilage that forms after initial bone ossification is complete; formed usually in response to mechanical stress, especially on the margins of intramembranous bone.

secondary neurulation Formation of the embryonic neural tube via cavitation, the later neurocoel, within a solid rod of cells, the neural keel. Compare with primary neurulation.

secondary plant compounds Chemicals produced by plants that are toxic or unpalatable to herbivores.

sectorial teeth Teeth with opposing sharp ridges specialized for cutting.

segmentation A body made up of repeating sections or parts; metamerism.

seismic Relating to an earth vibration.

selection force or pressure The biological or physical demands arising from the environment that affect the survival of the individual living there.

selenodont Teeth with crescent-shaped cusps, as in artiodactyls. Compare with bunodont and lophodont.

sensory neuron A nerve cell carrying responses from a sensory organ to the brain or spinal cord. Compare with motor neuron.

serial homology Similarity between successively repeated features in the same individual.

serial theory The hypothesis that jaws evolved from one of the anterior branchial arches. Compare with composite theory.

serous gland An organ secreting a thin, watery fluid.

sesamoid bone A bone that develops directly in a tendon, for example, the patella (kneecap).

sessile Pertaining to an animal attached to a fixed substrate in its environment.

sheer force The direction of an applied force that tends to slide sections of an object across each other.

signaling *See* cell signaling.

sinusoids Tiny vascular channels that are slightly larger than capillaries and lined or partially lined only by endothelium.

sister group In taxonomy, the particular outgroup most closely related to the ingroup. Compare with ingroup, outgroup.

solenocyte A single excretory cell with a projecting circle of microvilli around a central flagellum.

solute Molecules dissolved in solution.

somatic Pertaining to the body, usually to the skeleton, muscles, and skin but not to the viscera.

somatosensory system All proprioceptive neurons and neurons receiving stimulation from the skin.

special cranial nerve A bundle of fibers that detect stimuli from the local senses: sight, smell, hearing, balance, and lateral line. Compare with general cranial nerve.

spermatophore A package of sperm ready for delivery or presentation to the female.

sphincter A band of muscle around a tube or opening that functions to constrict or close it.

spinal nerve Any nerve entering or departing from the spinal cord. Compare with cranial nerve.

spiracle A reduced gill slit that is first in series.

splanchnocranium That part of the skull arising first to support the pharyngeal slits and later contributing to the jaws and other structures of the head; branchial arches and derivatives; visceral cranium.

spontaneous generation The concept that fully formed organisms arise directly and naturally from inanimate matter.

stall Loss of lift due to the onset of turbulent flow across an airfoil.

stapes One of the three middle ear bones in mammals, phylogenetically derived from the columella (hyomandibula).

stem group A paraphyletic assemblage of extinct taxa related to but not part of the crown group. Compare with crown group, total group.

stenohaline Having a narrow tolerance to salinity differences.

stereocilia Very long microvilli.

stereoscopic vision Denoting the ability to see images in three dimensions.

stereospondylous vertebra A monospondylous vertebra in which the single centrum (an intercentrum) is separate (aspidospondylous). Compare with embolomerous vertebra.

stigmata An extensively subdivided pharyngeal slit.

stolon A rootlike process of ascidians and other invertebrates that may fragment into pieces that asexually grow into more individuals.

stomodeum The embryonic invagination of surface ectoderm that contributes to the mouth.

stratified Formed of layers.

stratigraphy The geology of the origin, composition, and relative chronology of strata.

stratum (pl., strata) The geological term for a layer of rocks deposited during about the same geological time.

strength The load a structure withstands before failing or breaking.

streptostyly The condition in which the quadrate bone is movable relative to the braincase.

stress A measure of internal forces acting within a body, resulting from a load, often expressed as average force per unit area.

Stylophora A fossil group of basal echinoderms.

subcortical region That portion of the telencephalon exclusive of the cerebral cortex.

subterranean Pertaining to life lived underground.

suction feeding The capture of prey by means of a sudden muscular expansion of the buccal cavity that creates a vacuum to draw in water carrying the prey.

sulcus A furrow on the surface of the brain. Compare with gyrus.

surfactant A soluble compound that reduces surface tension, as in the lungs.

suspension feeding Feeding based on filtering suspended food particles from water; usually involves cilia and secreted mucus; filter feeding or ciliary-mucus feeding.

swim bladder A gas bladder functioning primarily in buoyancy control.

synaphomorphy A trait shared by two or more taxa and their immediate, last common ancestor. Compare with pleslomorphy.

synapse The region of contact between two neurons or between a neuron and an effector organ.

synapticules Cross-linking connections between pharyngeal bars in amphioxus.

synarthrosis A joint through which little or no movement is permitted between articulated skeletal elements.

syncytium Multinucleated cytoplasm; an aggregation of cells without cell boundaries.

synergist Two or more muscles cooperating to produce motion in the same direction. Compare with antagonist and fixator.

synovial joint A diarthrosis.

tachycardia Abnormally fast heart rate. Compare with bradycardia.

tactile Pertaining to touch.

talons Specialized bird claws used in striking or catching live prey.

taphonomy The study of how organisms decay and become fossils.

target tissue A group of related cells that respond to a particular hormone.

telolecithal Pertaining to eggs in which yolk stores are concentrated at one pole.

temporal region An area of the skull behind the orbit, completing the posterior wall of the braincase.

tendon A noncontractile, fibrous connective tissue band joining a muscle organ to a bone or cartilage. Compare with aponeurosis.

tensile force 1. The direction of an applied force that tends to pull apart an object. 2. The force produced by muscle contraction.

thermoreceptor A radiation receptor sensitive to infrared energy.

thermoregulation The process by which body temperature is established and maintained.

tidal volume The total quantity of air inhaled and exhaled in one breath. Compare with dead space.

tissue fluid A clear liquid outside of blood or lymphatic vessels that bathes cells.

tonic fibers Slow contracting muscle fibers that produce prolonged, sustained contractions with low force. Compare with twitch fibers.

tonus Partial muscle contraction with low force when a muscle is in a relaxation state.

total group The monophyletic clade formed of the stem group plus the crown group. Compare with crown group, stem group.

tract A collection of nerve fibers coursing together in the brain or spinal cord. Compare with nerve.

transverse plane A plane passing from one side of an organism to the other so as to divide the body into anterior and posterior parts.

transverse process A general term for any bony or cartilaginous projection from the centrum or neural arch.

trophoblast The outer cellular layer of the mammalian blastocyst.

trot A gait characterized by diagonally opposite feet coming in contact with the ground simultaneously.

turbulent flow The movement of fluid particles in irregular paths. Compare with laminar flow.

tusks Specialized, long teeth protruding from the mouth; elongate incisors (elephants), left upper incisor (narwhal), canines (walruses).

twitch fibers Fast-contracting muscle fibers whose force may or may not fatigue quickly; phasic fibers. Compare with tonic fibers.

tympanum The eardrum or tympanic membrane.

ungulate A hoofed placental mammal belonging to the orders Perissodactyla (horses) and Artiodactyla (cattle, deer, pigs).

unguligrade A foot posture in which the weight is carried on the tips of the toes (as in horses).

uniform extinctions Gradual loss of species over long geological time. Syn., background extinction.

up-regulation Increased food intake prompts increased anatomical and metabolic prominence of the alimentary canal, termed *up-regulation*. Reduced food intake reverses this response, termed *down-regulation*.

ureotelism Excretion of nitrogen in the form of urea.

ureter The metanephric duct arising as a uteteric diverticulum and draining the metanephros.

uricotelism Excretion of nitrogen in the form of uric acid.

uriniferous tubule The functional unit of the kidney composed of the nephron and collecting tubule.

vasoconstriction The narrowing of a blood vessel; usually resulting from smooth muscle contraction. Compare with vasodilation.

vasodilation The widening of a blood vessel; may be active or passive enlargement. Compare with vasoconstriction.

vasoreceptor Monitors pressure and gas levels in blood passing through the heart and systemic arches.

vein A blood vessel carrying blood toward the heart; blood carried may be low or high in oxygen tension. Compare with artery.

velocity The rate of change of displacement; how fast a body is traveling in a particular direction.

ventilation The active movement of water or air across respiratory exchange surfaces. Compare with perfusion.

ventral Toward the belly or bottom of an animal; opposite of dorsal.

vertebra One of several bone or cartilage blocks firmly joined into a backbone that defines the major body axis of vertebrates.

vestibular apparatus A sensory organ of the inner ear composed of semicircular canals and associated compartments, such as the sacculus, utriculus, and cochlea (lagena).

vestigial Evolutionary decline; reduction of a trait in descendants. Compare with atavism.

villus A fingerlike projection of a tissue layer, as in the small intestine. Compare with microvillus.

viscosity The resistance of a fluid to flow.

viviparity The reproductive pattern of live birth; birth of young not encased in a shell.

vomeronasal organ A chemosensory organ present in the nasal chamber or roof of the mouth of some tetrapods.

vomodor A chemical detected by sensory cells of the vomeronasal organ. Compare with odor.

warren An underground maze of excavated passageways used by animals, usually rabbits.

wolffian duct Mesonephric duct.

zygapophysis The projection of a neural arch that articulates with the adjacent neural arch.

Credits

Chapter 1

Figure 1.18: Redrawn from W.J. Bock, "The Role of Adaptive Mechanisms in the Origin of Higher Levels of Organization, "*Systematic Zoology*, 14: 272–287, 165. Reprinted with permission.; **1.23:** From Plough, Harold Henry, Sea Squirts of the Atlantic Continental Shelf from Maine to Texas; **1.43:** Redrawn from C.R. Longwell and R.F. Flint, *Introduction to Physical Geology*, 2nd edition, 1969, John Wiley & Sons, NY.; **1.47:** From F.A. Jenkins and W.A. Weijs, Journal of Zoology, 188: 379–410, London. Reprinted with permission of Wiley-Blackwell.

Chapter 2

Figure 2.9: Based on E.E. Ruppert and E.J. Balser, "Nephridia in the Larvae of Hemichordates and Echinoderms," Biol. Bull. 171: 188–196.; **2.10:** (a,b) Source: G. Stiasny, 1910, "Zur Kenntnis der lebenseise von Balanoglossus clavigerus," Zoolisches Anzeiger 35, Gustav Fischer Verlag; © After E.W. Knight-Jones, 1956, "On the nervous system of Saccoglossus cambrensis (Enteropneusta)," Proc. Trans. Roy, Soc., B, 236: 315–354.; **2.11:** From C. Burden-Jones, 1956, "Observations on the Enteropneust, Protoglossus kohleri (Caullery and Mesnil)," Proceedings of the Zoological Society of London. Reprinted with permission of Wiley-Blackwell.; **2.15b:** After R. P. S. Jeffries, 1973, "The Ordovician fossil Lagynocystis pyramidalis (Barrande) and the ancestry of amphioxus" Philosophical Transactions of the Royal Society of London The Royal Society of London. Reprinted by permission.; **2.19:** Based on J. Brandenburg and G. Kummel, 1961, "Die Feinstruktur der Solenocyten," *Journal of Ultrastruc. Research*, 5: 437–452; **2.20a:** From H.E. Lehman, Chordate Development, 1977, The Bermuda Biological Station and the Department of Zoology, University of North Carolina at Chapel Hill. Image courtesy of Hunter Texbooks, Inc. Reprinted with permission.; **2.22b:** With kind permission of Springer Science + Business Media. Redrawn from S.A. Torrence, 1986, "Sensory endings of the ascidian static organ (Chordata, Ascidacea)" Zoomorphology 106:62. Copyright © 1986.; **2.22c:** With kind permission of Springer Science + Business Media: S.A. Torrence and R.A. Cloney, "Nervous system of Ascidian larvae: Caudal primary sensory neurons," Zoomorphology, 00:106, fig. 3. Copyright © 1982 Springer-Verlag, Heidelberg.; **2.25:** Plough, Harold Henry. *Sea Squirts of the Atlantic Continental Shelf from Maine to Texas*. Pp. 2, Figure 2, 106, Figure 52. © 1978 The Johns Hopkins University Press. Reprinted with permission of The Johns Hopkins University Press.; **2.26:** (a) Modified with permission from P. R. Flood, "Filter characteristics of appendicularin food catching nets," *Experientia 34:* 173–175, Birkhaeuser Verlag, Basel, Swizerland. (b,c) Modified from A. Alldredge, 1976," Appendicularians," *Scientific American*, 235 (1): 95–102.

Chapter 3

Figure 3.4: From P.C.J. Donoghue, P.L. Forey, and R.J. Aldridge, "Condont Affinity and chordate Phylogeny" in Biological Reviews of the Cambridge Philosophical Society, Vol. 75. (2000): 191–251. Reprinted with permission of Wiley-Blackwell.; **3.5:** (a,b) After Romar and Parson, 1986, The Vertebrate Body, Saunders College publishing, from Dean; (c) After Dean; (d) after Jensen, 1966.; **3.6a:** From Mallatt, J., and J.-Y. Chen. 2003. Fossil sister group of craniates: predicted and found. Journal of Morphology. 258:1–31. © 2003. Reprinted with permission of Wiley-Liss, Inc., a subsidiary of John Wiley & Sons, Inc.; **3.9:** Redrawn from J.A. Moy-Thomas and R.S. Miles, 1971, Paleozoic Fishes, W. B. Saunders Company, (a) after Gross, (b) after White and (c) after Elliott.; **3.10:** (a) After J. Pan and S. Wang, 1980: 1, "New finding of Galeaspirotmes in South China" *Acta Paleotologia Sinica*, Beijing. (b,c) Redrawn from J. A. Moy-Thomas and R.S. Miles, 1971 Paleozoic Fishes, W.B. Saunders Company, (a) after Gross, (b) after White, and (c) after Elliot.; **3.12:** Redrawn from E.A. Steinsi, 1969, in J. Pivetau, ed. *Traite de Paleontologie* 4(2): 71692, Masono S.A., Paris.; **Box 3.3, Fig. 1:** *Traite de Zoologie*, 1948, after Blanc, D'Auberton, and Plessis; (c) based on Newton Parker.; **3.18:** Redrawn from A.S. Romer and T.S. Parsons, 1986, *The Vertebrate Body*, Saunders College Publishing, (a) and (b) after Traquair, (c) after Millot.; **3.20 a-b:** Material adapted from Hans-Peter Schultz and Linda Trueb (eds.), Origins of the Higher Group of Tetrapods: Controversy and Consensus. Copyright © 1991 by Cornell University. Used by permission of the publisher, Cornell University Press.; **3.21:** Based on Coates, Ruta, and Friedman; **3.22:** From J.I. Coates. Reproduced by permission of the Royal Society of Edinburgh from *Transactions of the Royal Society of Edinburgh: Earth Sciences*, Vol. 87 (1996), pp. 363–421.; **3.24 and 3.26:** After R.L. Carroll, *Vertebrate Paleontology and Evolution*, W.H. Freeman & Company, from Beerbower, 1963.; **3.30:** Modified from A.S. Romer, 1966, *Vertebrate Paleontology*, 3rd edition, University of Chicago Press, after Jaekel.; **3.31:** After R. L. Carroll, *Vertebrate Paleontology and Evolution*, W.H. Freeman & Company, from Heaton and Reisz.; **3.34:** After A.S. Romer, 1996, *Vertebrate Paleontology*, 3rd edition, University of Chicago Press, (a) from Williston: (b) from Eaton.; **3.35:** After

D. Lambert, 1983, *A Field Guide to Dinosaurs,* The Diagram Group, Avon.; **3.38:** After R.T. Peterson, 1963, *The Birds,* Life Nature Library, Time Inc.

Chapter 4

Figure 4.1: Based on T. McMahon and J.T. Bonner, 1983, *On Size and Life,* 1983, Scientific American Books, adapted from H.G. Wells, et al., *The Science of Life,* 1931, The Literary Guild, NY.; **4.9:** After T. McMahon and J.T. Bonner, 1983, *On Size and Life,* Scientific American Books, adapted from P.B. Medawar, 1945, "Size, Shape, and Age" in *Essays on Growth and Form Presented to D'Archy Thompson,* pp. 157–187, Clarendo Press, Oxford.; Figure **4.11:** Adapted with permission from On Size and Life by T. McMahon and J.T. Bonner, Copyright 1983 by Thomas McMahon and John Tyler Bonner. Reprinted by permission of Henry Holt and Company, LLC.; **4.12:** Bonner, John Tyler: Size and Cycle. © 1965 Princeton University Press, 1993 renewed PUP Reprinted by permission of Princeton University Press.; **4.14:** From B. Kummer, 1951, "Zur Einstehung der menschlichen Schadelform," Verh. Anat. Ges. 49 (suppl. To Anat. Anz.98) 140–146. Reprinted with kind permission of Springer Science + Business Media. **4.15:** Modified from D.A. W. Thompson, *On Growth and Form,* 2nd edition, 1942 Cambridge University Press, Cambridge.; **4.16:** Based on V. Geist, 1971, *Mountain Sheep,* 2nd edition, University of Chicago Press.; **4.26:** Source: M. Hildebrand, "How Animals Run," *Scientific American,* May 1960.; **4.30:** (a,b,d) Based on J.Z. Young, 1981, *The Life of Vertebrates,* after Storer from photographs of the National Advisory Committee for Aeronautics; and from Jones. (c) Modified from T. A. McMahon and H.T. Bonner, 1983. *On Size and Life.*

Chapter 5

Figure 5.8: Redrawn from H.E. Leman, Chordate Development, p. 30, © Hunter Publishing Co. Image coutesy of Hunter Textbooks, Inc. Reprinted with permission, (f, g) based on Pflugfelder 1962; (i, j) adapted from Grasse, 1958 after Parker and Hawell, 1930. **5.12:** Adapted from W.E. Duellman and L. Trueb, 1986, *Biology of Amphibians,* based on K.L. Gosner, 1960, *Herpetologica,* 16: 183–190.; **5.25:** After R.V. Krstic´, 1984, *General Histology of the Mammal,* Springer-Verlag.; **5.27:** (a-d) After R.V. Krstic´, 1984, *General Histology of the Mammal,* Springer-Verlag. (e) After Halsted and Middleton, 1972, Bare Bones.; **5.29:** From L.B. Arey, Developmental Anatomy, 1966.; **5.31:** After B.M. Patten and B.M. Carlson, 1974, *Foundations of Embryology,* The McGraw-Hill Companies.; **5.32:** After A. McLaren, "The Embryo" in *Reproduction in Mammals,* 2: Embryonic and Fetal Development, p. 32, ed. by C.R. Austin and R.V. Short, 1972, Cambridge University Press.; **5.34:** After H.E. Evans, 1979, Ch. 2, Reproduction and Prenatal Development in Evans and Christensen, *Miller's Anatomy of the Dog,* pp. 18 and 20. **5.37:** From McNamara, Kenneth J., Shapes of Time: The Evolution of Growth and Development. Figure on pp. 41–43. © 1997 The Johns Hopkins University Press. Reprinted with permission from The Johns Hopkins University Press.

Chapter 6

Figure 6.3: Redrawn from M.H. Wake, ed., Hyman's Comparative Vertebrate Anatomy, 3rd edition, © 1979 by The University of Chicago. All rights reserved. Reprinted by permission.; **6.4:** With kind permission of Springer Science + Business Media: Adapted from R. Olson, 1961, "The skin of amphioxus," Zeitschrift Fur Zellforschung Und Mikroskopische Anatomie, Bd. 54, 90–104. Copyright © 1961.; **6.8:** Based on J.A. Moy-Thomas, 1971, *Paleozoic Fishes,* 2nd edition, W.B. Saunders Company, after Kiaer.; **6.9:** (a) Based on H.M. Smith, 1960, *Evolution of Chordate Structure,* Holt, Rinehart, and Winston, Inc. (b) Redrawn from J. Bereieter-Han, et al., *Biology of the Integument,* Vol. 2, 1986 Springer-Verlag, Berlin.; **6.18:** After P.J. Regal, 1975, "The evolutionary origin of feathers," *Quarterly Review of Biology,* 50: 35–66.; **6.22:** Modified from A.G. Lyne, 1952, "Notes on external characters of the pouch young of four species of bandicoot," *Proc. Zool. Soc. London* 122: 625–650.; **6.25:** From William J. Banks, Applied Veterinary Histology, 2nd Edition. Copyright © 1986. Reprinted by permission of Lippincott Williams & Wilkins.; **6.30:** Adapted from A. Pivorunas, 1979, "The feeding mechanisms of baleen whales," *American Scientist,* July/August.; **6.33:** After H.M. Smith, *Evolution of Chordate Structure,* Holt, Rinehart, and Winston.

Chapter 7

Figure 7.4: After P.F.A. Maderson (ed.), 1987, *Developmental and Evolutionary Aspects of the Neural Crest,* John Wiley & Sons.; **7.6:** After G.R. deBeer, 1985, *The Development of the Vertebrate Skull,* plate 13, University of Chicago Press.; **7.9:** After E.S. Goodrich, 1958, *Studies on the Structure and Development of Vertebrates,* Vol. 1, Dover Publishing, Inc., **Box 7.1, Fig. 1:** (a) After M. Jollie, 1971, "A theory concerning the early evolution of the visceral arches" in *Acta Zool.* 52: 85–96. (b,c) After J. Reader, 1986, *The Rise of Life,* Alfred A. Knopf. (d) After Jollie.; **7.13:** After Jarvik, 1980, *Basic Structure and Evolution of Vertebrates,* Vol. 1, Academic Press.; **7.14:** After E. Stensi, 1964, "Les cyclostomes folliles ou ostracodermes" in *Traite de Paleontologie,* Vol. 1. Mason et Cie, Paris, p. 161.; **7.15:** After E. Stensi, 1964, in Traite de Paleontologie, Vol. 1, fig. 115, p. 335, Mason et Cie, Paris.; **7.16:** (a) After E. Stensi, 1969, "Elasmorbranchiomorphi Placodermata Arthrodires," in J. Piveteau (ed.), Traite de Paleontologie, Vol. 2, Mason et Cie, Paris p. 165. (b) After E. Stensi, 1969, "Elasmorbranchiuomorphi Placodermata Arthrodires," in J. Piveteau (ed.) *Traite de Paleontologie,* Vol 2, Mason et Cie, Paris, p. 448.; **7.21:** After R.L. Carroll, 1988, *Vertebrate Paleontology and Evolution,* W.H. Freeman & Company.; **7.23:** After G. Lauder, 1980, "Evolution of the feeding mechanism in primitive actinopterygian fishes: A functional anatomical analysis of Polyperus, Lepsosteus and Amia" in *Jrn. Morph.* 163: 283–317.; **7.24:** After L.B. Radinsky, 1987, *The Evolution of Vertebrate Design,* University of Chicago Press, Chicago-London.; **7.25:** Based on K.F. Liem, 1978, "Modulatory multiplicity in the functional repertoire of the feeding mechanism in cichlid fishes," in J. Morph. 158: 323–360.; **7.29:** After E. Jarvik, 1980, *Basic Structure and Evolution of Vertebrates,* Vol. 1, Academic Press.; **7.31:** After J.Z. Young, 1981, *The Life of Vertebrates,* Clarendon Press, Oxford.; **7.32:** Based on G.V. Lauder, 1985, "Aquatic feeding in lower vertebrates: in *Functional Vertebrate Morphology* edited by M. Hildebrand et al., Harvard University Press.; **7.33:** After R.L. Carroll, 1988, *Vertebrate Paleontolgoy and Evolution,* W.H. Freeman and Company.; **7.35:** Based on H.M. Smith, 1960, *Evolution of Chordate Structure,* Holt, Rinehart, & Winston.; **7.37:** (a,b,c,d) After R.L. Carroll, 1988, *Vertebrate Paleontology and Evolution,* W.H. Freeman and Company. (e)

After A.S. Romer, 1966, *Vertebrate Paleontology,* University of Chicago Press, Chicago-London.; **7.39:** After M. Jollie, 1962, *Chordate Morphology,* Reinhold Publishing Corp.; **7.41:** After M. Hildebrand, et al., (eds.) 1985, *Functional Vertebrate Morphology,* Belknap Press of Harvard University Press.; **7.51:** After R.L. Carroll, 1988, *Vertebrate Paleontology and Evolution,* W.H. Freeman and Company.; **7.52:** After L.H. Hyman, 1942, *Comparative Vertebrate Anatomy,* University of Chicago Press.; **7.54:** Based on H.E. Evans, 1979 "Reproduction and Prenatal Development (Chapter 21)" from *Miller's Anatomy of the Dog,* Saunders Company.; **7.58:** After H.M. Smith, 1960, *Evolution of Chordate Structure,* Holt, Rinehart, & Winston.; **7.59:** After P.E. Wheeler, 1978, "Elaborate CNS cooling structures in large dinosaurs" in *Nature,* 275: 441–442.; **7.60:** After G. Kooyman, *The Weddell Seal. Vertebrates: Adaptation (Rdgs. From Scien. Amer.)* August 1969, p. 235.; **Box 7.4, Fig. 1:** After A.S. Romer, 1966, *Vertebrate Paleontology,* 3rd edition, University of Chicago Press; R.L. Carroll, 1988, *Vertebrate Paleontology and Evolution,* W.H. Freeman and Company; and H.F. Osborn, 1924, *Andrewsarchus in Amer. Mus. Novitates* 146: 1–5.; **7.64:** Source: McGinnis, W., and R. Krumlauf. 1992. "Homeobox genes and azial patterning." *Cell* 68: 283–302.; **7.65:** After R.L. Carroll, 1988, *Vertebrate Paleontology and Evolution,* W.H. Freeman and Company.

Chapter 8

Figure 8.8: (d,e) After H.M. Smith, 1960, *Evolution of Chordate Structure,* Holt, Rinehart and Winston. (f) After M. Hildebrand, *Analysis of Vertebrate Structure,* 2nd edition, John Wiley & Sons.; **8.9:** Based on A.S. Romer and T.S. Parsons, 1986, *The Vertebrate Body.*; **8.13:** After M.J. Wake (ed.), 1979 Hyman's Comparative Vertebrate Anatomy, *University of Chicago Press,* p. 203.; **8.15:** After J.A. Moy-Thomas, 1971, Paleozoic Fishes.; **8.18:** Based on J. Laerm, 1976, "The development, function, and design of amphicoelous vertebrae in teleost fishes," in *Zool. J. Linn. Soc.* 58: 237–54, figure 4 & 5.; **8.21:** After S.M. Andrews, et al., 1977, "Problems in vertebrate evolution: publ. for the Linnean Soc. Of London by Academic Press; and E Jarvik, 1980, *Basic Structure and Evolution of Vertebrates.*; **8.22:** (a) After J. A. Moy-Thomas, 1971, *Paleozoic Fishes.* (c,d) After E. Jarvik, 1980, *Basic Structure and Evolution of Vertebrates,* Vol. 1 Academic Press.; **8.25:** After F. R. Parrington, 1967, "The vertebrae of early tetrapods: in *Colloq. Int. Cent. Nat. Rech. Sci.* 163: 267–279.; **8.26:** (a,b) After R.L. Carroll, 1988, *Vertebrate Paleontology and Evolution,* W.H. Freeman and Company. (c) After A.S. Romer, 1956, *Osteology of the Reptiles,* University of Chicago Press. (d,e,f,g) After R.L. Carroll, 1988, *Vertebrate Paleontology and Evolution,* W.H. Freeman and Company.; **8.27:** (a) M.J. Wake (ed.), 1979, *Hyman's Comparative Vertebrate Anatomy,* University of Chicago Press. (c) After G.H. Dalrymple, 1979, "Packaging problems of head retraction in trionychid turtles" in *Amer. Soc. Of Ichth. & Herpt.* 4: 655–66, fig, 1.; **Box 8.1, Fig. 1:** After G.S. Kranz, 1981, *The process of Human Evolution,* Schenkman Publishing Co., Cambridge, MA.; **8.34:** After R.L. Carroll, 1988, *Vertebrate Paleontology and Evolution,* W.H. Freeman and Company.; **8.37:** (a) After R.L. Carroll, 1988, *Vertebrate Paleontology and Evolution,* W.H. Freeman and Company. (b) After A.S. Romer, 1966, *Vertebrate Paleontology,* University of Chicago Press.; **8.39:** After George C. Kent and Larry Miller, *Comparative Anatomy of the Vertebrates,* 8th edition, 1997. The McGraw-Hill Companies.

Chapter 9

Figure 9.8: Based on research of Shubin and Alberch.; **9.12:** (a,b) After R. L. Carroll, *Vertebrate Paleontology and Evolution,* W. H. Freeman and Company. (c) After E. Jarvik, 1980, *Basic Structure and Evolution of Vertebrates,* Academic Press.; **9.14:** (a) After E. Jarvik, 1980, *Basic Structure and Evolution of Vertebrates, Academic Press.* (b) After M. Jollie, 1962, *Chordate Morphology,* Reinhold Books in the Biological Sciences; and two additional sources.; **9.18:** After R.L. Carroll, 1988, *Vertebrate Paleontology and Evolution,* W.H. Freeman & Company.; **9.20:** After A.S. Romer and T.S. Parsons, 1986, *The Vertebrate Body,* 6th edition, Saunders College Publishing.; **9.22:** After A. S. Romer and T.S. Parsons, 1986, *The Vertebrate Body,* 6th edition, Saunders College Publishing.; **9.23:** After E. Jarvik, 1980, *Basic Structure and Evolution of Vertebrates,* Academic Press.; **9.25:** After H.M. Smith, 1960, *Evolution of Chordate Structure,* Holt, Rinehart, & Winston.; **9.26:** After H.M. Smith, 1960, *Evolution of Chordate Structure,* Holt, Rinehart & Winston.; **9.30:** (c) After F.H. Pough, et al., 1989, *Vertebrate Life,* MacMillan.; **9.37:** After M. Hildebrand, *Analysis of Vertebrate Structure,* 2nd edition, John Wiley & Sons.; **9.39:** (a) After M. Hildebreand, 1968, "How Animals Run" in readings from *Scientific American,* May 1960. (b) After A.S. Romer, 1956, *Osteology of the Reptiles,* University of Chicago Press. (c) After F.H. Pough, et al., 1989, *Vertebrate Life,* MacMillan.; **9.42:** After M. Hildebrand, 1968, "How Animals Run" in readings from *Scientific American,* May 1960.; **9.44:** After M. Hildebrand, 1988, *Analysis of Vertebrate Structure,* 3rd edition, John Wiley & Sons.; **9.49:** (c) After R.T. Peterson, 1963, *The Birds, Life Nature Library,* Time, Inc. (d) After F.H. Pough, et al., 1989, *Vertebrate Life,* MacMillan.; **9.50:** After F.H. Pough, et al., 1989, *Vertebrate Life,* MacMillan.; **9.54:** After M. Hildebrand, et al. (ed.), 1985, *Functional Vertebrate Morphology,* Belknap Press, Harvard University.; **9.56:** After C. Gans, 1974, *Biomechanics,* J.B. Lippincott Company.; **9.57:** From Gasc, J.P., Jouffroy, F. K., Renous, S. and Blottnitz, F. V. (1986). Morphofunctional study of the digging system of the Namib Desert Golden Mole (Eremitalpa granti namibensis): cineflurographical and anatomical analysis. J. Zool., Lond. 208, 9–35. Reprinted with permission from John Wiley & Sons.; **9.58:** After M. Hildebrand, 1983, *Analysis of Vertebrate Structure,* 3rd edition, John Wiley & Sons.

Chapter 10

Figure 10.3: Based on R.V. Krstic´, 1984, *General Histology of the Mammal,* Springer-Verlag.; **10.16:** From J.V. Basmajian, Muscles Alive, p. 313, 1974.; **10.27:** Modified from Romer and Parsons, 1986, *The Vertebrate Body,* 6th edition. After Mauer and Furbinger.; **Box 10.3, Fig. 1:** After P. W. Webb, 1984, "Form & Function in Fish Swimming," *Scientific American,* July, pp. 72–82.; **10.33:** After P. Goody, 1976, *Horse Anatomy,* J.A. Allen, London.; **10.34:** Modified from H.E. Evans, 1982, "Anatomy of the Budgerigar" in *Diseases of Cage and Aviary Birds,* 2nd edition, ed. By Margaret L. Petrak, Lea & Febiger, Philadelphia.

Chapter 11

Figure 11.6: Modified from M.E. Feder and W.W. Burggren, 1985, "Skin breathing in vertebrates" *Scientific American,* 235(5): 126–142.; **11.16:** After K.F. Liem, 1977, "Respiration and Circulation," in A. Kluge (ed.), *Chordate Structure and Function,* MacMillan Publishing, from Jensen, 1966.; **11.17:** (a) After Mallatt. (b) After Johansen, figure 7–5, in Kluge, *Chordate*

Structure & Function, MacMillan Publishing, from Jensen, 1966.; **11.18:** Adapted from G.M. Hughe and C.M. Ballintjin, "The muscular basis of the respiratory pumps in the dogfish (Scyliorhinus canicula)," *Journal of Experimental Biology,* 43:363–383. Used with permission of company of Biologists Ltd.; **11.20:** Redrawn with permission from D.J. Randall, et al., The Evolution of Air Breathing in Vertebrates, copyright © 1981, Cambridge University Press. Reprinted with permission of Cambridge University Press.; **11.23:** Reprinted from *Biol. Journal of Linnean Society,* Vol. 12, R.J. Wassersug and K. Hoff, "A comparative study of the buccal pumping mechanism of tadpoles," pp. 225–259, 1979, by permission of the publisher Academic Press Limited London.; **11.24:** Based on N. Gradwell, 1971, "ascaphus tadpole: experiments on the suction and gill irrigation mechanisms" *Can. J. Zool.* 49:307–332.; **11.31:** Redrawn from M.E. Miller, 1964, *Anatomy of the Dog,* W.B. Saunders Company.; **11.33:** Source: M. Hildebrand, et al., (eds.) 1985, *Functional Vertebrate Morphology.* Belknap Press of Harvard University Press.

Chapter 12

Figure 12.22: Adapted from B.I. Balinsky, 1981, *An Introduction to Embryology,* 5th edition, W.B. Saunders College Publishing.; **Box 12.1, Fig. 1:** After Burggren and Johansen.; **12.30:** After M.H. Wake, ed., Hyman's Comparative Vertebrate Anatomy, 3rd edition, © 1979 by The University of Chicago. Reprinted by permission.; **12.45:** Redrawn from M.A. Baker, "A brain-cooling system in mammals," *Vertebrates: Physiology,* 1979, W.H. Freeman and Company.; **12.47:** From J.V. Basmajian, Primary Anatomy, 7th ed, 1976, Williams and Wilkins, Baltimore.; **12.48:** After C. Gans, *Biology of the Reptilia,* Vol. 6, figure 1, p. 318, Academic Press.; **12.49:** After H.M. Smith, 1960, *Evolution of Chordate Structure,* Holt, Rinehart, & Winston.; **12.50:** After C. Gans, *Biology of the Reptilia,* Vol. 6, Academic Press.

Chapter 13

Figure 13.5: After A.S. Romer and T.S. Parsons, 1985, *The Vertebrate Body,* Saunders College Publishing.; **13.10:** After H. M. Smith, 1960, Evolution of Chordate Structure, Holt, Rinehart, & Winston.; **13.16:** After H.M. Smith, 1960, *Evolution of Chordate Structure,* Holt, Rinehart, & Winston.; **13.17:** Modified from A.S. Romer and T.S. Parsons, 1985, *The Vertebrate Body,* Saunders College Publishing.; **13.18:** After H.M. Smith, 1960, *Evolution of Chordate Structure,* Holt, Rinehart, & Winston.; **13.23:** After H. Tuchmann-Duplessis and P. Haegel, 1974,*Illustrated Human Embryology,* Vol. II, Organogenesis, Springer-Verlag.; **13.27:** Platypus and Echidna, after Hume; Kangaroo and Dog after Stevens; Koala after Degabriele.; **13.28:** After C.E. Stevens, 1980, and C.J.F. Harrop and I. D. Hume, 1978, in *Comparative Physiology: Primitive Mammals,* ed. By K. Schmidt-Nielson, et al., Cambridge University Press.; **13.32:** (a,b,d) After Romer and Parsons; (c) After Pernkoff; (e) After A. Bellairs, 1970, *The Life of Reptiles,* Universe Books, NY.; **13.33:** After H.E. Evans, 1982, "Anatomy of the Budgerigar" in *Diseases of Cage and Aviary Birds,* ed. Margaret L. Petrak, 2nd edition, Lea & Febiger, Philadelphia.; **13.35:** Modified from E. Kochva, 1978.; **13.37:** Modified from M.E. Miller, et al., 1964, *Anatomy of the Dog,* W.B. Saunders Company.; **13.40:** Modified from D.D. Davis, 1964, "The Giant Panda, A morphological study of evolutionary mechanisms," *Fieldiana: Zoology Memoirs,* Vol. 3, Chicago

Natural History Museum.; **13.41:** After M. Hildebrand, et al., eds., 1985, *Functional Vertebrate Morphology,* Belknap Press of Harvard University Press.; **13.42:** (a) After T.J. Dawson, 1977, "Kangaroos: Vertebrates Adaptation" *Scientific American,* W.H. Freeman and Company. (b) Modified from E. Pernkoff, 1929.

Chapter 14

Figure 14.5: After F.H. Pough et al., 1989, *Vertebrate Life,* MacMillan.; **14.8:** Modified from J.A. Willett, 1965, "The male urogenital system in the Sirenidae," *Jrnl. Tennessee Academy of Science,* 40:1.; **14.9:** Modified after E.J. Braun and W.H. Dantzler, 1972, *Amer. Jrnl. Physiol.* 222:4.; **14.11:** Modified from Schmidt-Nielson.; **14.15:** Redrawn from J.J. Berridge and J.L. Oschman, 1972, Transporting Epithelia, Academic Press, London. Reprinted by permission of the author.; **14.23:** (a) After M.S. Hardisty, 1979, *Biology of the Cyclostomes,* Chapman and Hall Ltd., redrawn from micrograph of Patzner, 1974. (b) After J.G.D. Lambert, 1970, "The ovary of the guppy, Poeccilia reticulate," *Gen. Comp. Endocrin.* 15:464–476.; **14.25:** Redrawn from Romer and Parsons; (a) After Borcea; (b) after Parker, Kerr; (c) after Hyrtl, Goodrich.; **14.28:** Modified and redrawn from Romer and Parsons, (a) After McEwen; (c) after Ozawa; (d) after Parker.; **14.31:** From R. van den Hurk, Morphological and functional Aspects of the Testis of the Black Molly, mollienisia (=Poecillia) latipinna, Thesis, University of Turecht, The Netherlands, 1975. Figure a (left) also appeared in the teste of oviparous and vivparous teleosts" in Proc. Kon. Ned. Akad. Wetensch C76, 270–279, 1973. Reprinted with kind permission of Springer Science + Business Media.; **14.32:** Redrawn from Romer and Parsons after (a) Conel; (b) Borcea; (c) Kerr, Parker; (d) Edwards.; **14.34:** (a) After C.L. Baker and W.W. Taylor, Jr., 1964, "The urogenital system of the male Ambystoma" *Jrnl. Tenn. Acad. Sci.* 41:1 (b) After P. Strickland, 1966, "The male urogenital system of Gyrinophilus Danielsi dunni," *Jrnl. Tenn. Acad. Sci.* 41:1.; **14.36:** Redrawn from Romer and Parsons (a) and (b) after McEwen; (c) after van den Broek; (d) after Oseler and Lasmprecht.; **14.37:** After P.E. Lake, 1981, "Male genital organs," *Form and Function in Birds,* Vol. 2, ed. By A.S. King and J. McLelland, Academic Press.; **14.38:** Material adapted from Ari Van Tienhoven, Reproductive Physiology of Vertebrates, 2nd Edition. Copyright © 1983 by Cornell University Press. Used by permission of the publisher, Cornell University Press.; **14.39:** Material adapted from Ari Van Tienhoven, Reproductive Physiology of Vertebrates, 2nd Edition. Copyright © 1983 by Cornell University Press. Used by permission of the publisher, Cornell University Press.; **14.42:** This figure was published in H.E. Evans, "Introduction and Anatomy." in Zoo and Wild Animal Medicine, edited by M.E. Fowler. Copyright © 1979 Elsevier Inc. Reprinted with permission from Elsevier.; **14.44:** After A.S. King, 1981, *Form and Function in Birds,* Vol. 2, (c) redrawn from Liebe, 1914; (d) based on Liebe and Rauterfeld, et al., 1974.; **14.45:** After M.E. Miller, et al., 1964, *Anatomy of the Dog,* W.B. Saunders and Company.; **14.46:** Redrawn from Romer and Parsons after Dean.; **14.47:** (a) After P.E. Lake, 1981, "Male genital organs," *Form and Function in Birds,* Vol. 2, ed. By A.S. King and J. McLelland, Academic Press, from Gadow 1887.; **14.48:** After M. Gabe and H. Saint-Girons, 1965, *Mem. Mus. Nat. Hist. Nat.,* Paris, Serie A, t. XXXIII, fasc. 4, pp. 149–292.; **14.54:** After M.P. Hardy and J.N. Dent, 1986, "Transport of sperm within the cloaca of the female red-spotted newt," *J. Morph.* 190:259–270.; **14.55:** Modified from C.R. Austin and

R. V. Short, 1973, eds., *Reproduction in Mammals,* Book 1, Germ Cells and Fertilization, Cambridge University Press.

Chapter 15

Figure 15.1: (b) From R.V. Krstic´, *General Histology of the Mammal,* Springer-Verlag, Verlin-Heode; berg-NY-Tokyo, p. 89. (c) From R.V. Krstic´, *General Histology of the Mammal,* Springer-Verlag, Verlin-Heidelberg-NY-Tokyo, p. 93.; **15.6:** (a,b) After L.S. Stone and H. Steinitz, 1953, *Exp. Zool.,* 124:469. (c) After E. Witschi, 1949, Z. Naturforsch, 4:230.; **15.9:** (a,c,d,f) After P.J. Bentley, 1976, *Comparative Vertebrate Endocrinology,* Figs. 2.9 (a); 2.10 (c,d); 2.11(f) Cambridge University Press, Cambridge. (e,f) After Hartman and Brownell, 1949.; **15.11:** (a) From R.V. Krsti´c, *General Histology of the Mammal,* Springer-Verlag, Verlin-Heidelberg-NY-Tokyo, p. 89. (b) From R.V. Krstic´, *General Histology of the Mammal,* Springer-Verlag, Verlin-Heidelberg-NY-Tokyo, p. 91.; **15.12:** After Bentley.

Chapter 16

Figure 16.3: From R.V. Krstic´. General Histology of the Mammal © Springer-Verlag Berlin Heidelberg 1984. With kind permission of Springer Science+Business Media.; **16.12:** (a) After Stensio; (b) after Lindstrom 1949 via Jollie, 1962, *Chordate Morphology,* Reinhold Publishing Corp., NY, p. 401; (c) After Marinelli and Strenger, 1954 via M. Jollie, 1952 *Chordate Morphology,* Reinhold Publishing Corp., NY, p. 401; (d) After Norris and Hughes, 1920, via M. Jollie, 1962, *Chordate Morphology,* Reinhold Publishing Corp., NY, p. 401.; **16.19:** After M. Hildebrand, 1988, *Analysis of Vertebrate Structure,* John Wiley & Sons, NY.; **16.26:** After H. Smith, 1960, *Evolution of Chordate Structure,* Holt, Rinehart, & Winston, NY.; **16.32:** After W.J.H. Nauta and M. Feirtag, 1979, "The organization of the brain," *Scientific American,* 1979(9):102.; **16.34:** After Ebbessona and Northcutt, 1976; Roberts, 1981; and Kremers, 1981.; **16.35:** (a,b) Redrawn from A.S. Romer and T.S. Parsons, 1985, *The Vertebrate Body,* Saunders College Publishing after Ahlborn. (c) After R.E. Davis and R.G. Northcutt, eds., 1983, *Fish Neurobiology,* Vol. 2, University of Michigan Press, Ann Arbor. (d,e,f,g,h) Redrawn from A.S. Romer and T.S. Parsons, 1985, *The Vertebrate Body,* Saunders College Publishing.; **16.36:** Redrawn from A.S. Romer and T.S. Parsons, 1985, *The Vertebrate Body,* Saunders College Publishing, after Butschli, Weinstein, Sisson.; **16.37:** (a,d,e) Redrawn from A.S. Ropmer and T.S. Parsons, 1985, *The Vertebrate Body,* Saunders College publishing after (a) Haller, Burcjhardt; (d) Loo. (b,c) Modified from M. Jollie, 1962, *Chordate Morphology,* Reinhold Publishing Corp., NY.

Chapter 17

Figure 17.6: After H. Tuchmann-Duplessis, et al., 1974, *Illustrated Human Embryology,* Vol. III Nervous System and Endocrine Glands, Springer-Verlag, NY.; **17.9:** After M.H. Wake (ed.), 1979, *Hyman's Comparative Vertebrate Anatomy,* University of Chicago Press.; **17.13:** After M.A. Miller, et al, 1964, *Anatomy of the Dog,* W.B. Saunders, Philadelphia.; **17.14:** After M. Halpern and J.L. Kubie, 1984. "The role of the ophidian vomeronasal system in species-typical behavior: in *Trends Neurosci,* 7(12):472–477.; **17.15:** After F.E. Bloom, et al., 1985, *Brain, Mind, and Behavior,* W.H. Freeman and Company.; **17.18:** After H. Tuchmann-Duplessis, et al., 1974, *Illustrated Human Embryology,* Vol. III Nervous System and Endocrine Glands, Springer-Verlag, NY.; **17.19:** After H.M. Smith, 1960, Evolution of Chordate Structure, Holt, Rinehart,

& Winston.; **17.22:** After Q. Bone and N.B. Marshall, 1982, *Biology of fishes,* Blackie & Sons, Ltd., Glasgow & London.; **17.28:** From M.H. Wake (ed.), Hyman's Comparative Vertebrate Anatomy, Copyright © 1979, University of Chicago Press. Reprinted by permission.; **17.29:** After H.M. Smith, 1960, *Revolution of Chordate Structure,* Holt, Rinehart, & Winston, after Neal and Rand.; **17.30:** After deCock Buning, 1983.; **17.32:** After M.J. Lannoo, 1987, "Neuromast topography in anuran amphibians" in *Scientific American,* November.; **17.33:** After D.E.Parker, 1980, "The vestibular apparatus: in *Scientific American,* November.; **17.37 (top):** Adapted from R.G. Northcutt and R.E. Davis, eds., Fish Neurobiology, Vol. 1. Copyright © 1983, The University of Michigan Press, Ann Arbor. Reprinted by permission.; **17.38:** After Q. Bone and N.B. Marshall, Biology of Fishes. From A.N. Popper and Sheryl Coombs, "Auditory mechanisms in Teleost fishes" in American Scientist, 1980, 68:420–440. Copyright © 1980 Sigma Xi, The Scientific Research Society. Reprinted by permission.; **17.39:** (a) After A.S. Romer and T.S. Parsons, 1986, *The Vertebrate Body,* Saunders College Publishing. (b) After E.G. Weaver, 1985, *The Amphian Ear,* Princeton University Press.; **17.41:** After A.S. Romer and T.S. Parsons, 1986, *The Vertebrate Body,* Saunders College Publishing.; **17.42:** After H.E. Evans, 1982, "Diseases of Cage and Aviary Birds" in *Anatomy of the Budgerigar,* Margaret L. Petrak, (ed.), Lea & Febiger, Philadelphia.; **17.43:** (a) After Knudson; (b,c) after Norberg. From "The Hearing of the Barn Owl" edited by E.I. Knudson, drawn by Tom Prentiss in Scientific American, December 1981. Reprinted by permission of Nelson H. Prentiss.; **17.44:** (a) After H.M. Smith, 1960, *Evolution of Chordate Structure,* Holt, Rinehart, Winston. (b) After A.S. Romer and T.S. Parsons, 1986, *The Vertebrate Body,* Saunders College Publishing. (c) After E. van Beneden and C. van Bambeke, 1965, "Zure funktionsmorphologie des akustishen organs." *Extract des Archives des Biologie.;* **17.45:** After E. van Beneden and C. van Bambeke, 1965, "Zure funktionsmorphologie des akustishen organs." *Extract des Archives des Biologie.;* **17.46:** After W. Raschi, 1986, "A morphological analysis of the ampullaie of Lorenzini in selected skates (Pisces, Rajoidei)" in *J. Morph.* 189: p. 231.; **17.47:** After H.W. Lissmann, 1963, "Electric location by fishes" in *Scientific American,* March, p. 163.; **17.48:** After R.G. Northcutt and P.E. Davis (eds.), *Fish Neurobiology,* Vol.1, University of Michigan Press, Ann Arbor.

Photo Credits

Chapter 1

Figure 1.2: Neg. no. 326697, courtesy of Department of Library Services, American Museum of Natural History; **1.3:** Courtesy of the National Library of Medicine (USNLM); **1.4:** © Museum of Comparative Zoology Harvard University, © President and Fellows of Harvard College; **1.5b:** Photo courtesy of Bibliotheque Centrale du Museum National d' Histoire Naturelle Paris, 2008; **1.6:** © National Portrait Gallery, London; **1.7 & 1.8 & 1.10a:** © Stock Montage; **1.32:** Neg. no. 336671, courtesy of Department of Library Services, American Museum of Natural History; **1.33:** Courtesy of Dr. R. Wild Staatl, Museum fur Naturkunde, Stuttgart; **1.34:** © John Cunningham/ Visuals Unlimited; **1.35a:** Neg. #330491, courtesy of Department of Library Services, American Museum of Natural History;

1.35b: Neg. #35608, courtesy of Department of Library Services, American Museum of Natural History; **1.36a:** Neg. #324393, courtesy of Department of Library Services, American Museum of Natural History; **1.37a:** Photo courtesy National Park Service; **1.37b:** Photos courtesy of Dr. David Taylor, Executive Director, Northwest Museum of Natural History, Portland State University, Portland, Oregon; **1.39:** Chasmosaurus by Eleanor Kish. Reproduced with permission of the Canadian Museum of Nature, Ottawa, Canada.; **1.42b:** © Grand Canyon National Park; **1.42c:** Courtesy of Zion National Park; **1.42d-f:** Courtesy of National Park Services; **1.46a:** From Bruce A. Young, University of Massachusetts and Kenneth V. Kardong, Washington State University, "Naja haje" (On-line), Digital Morpholoty, at http://digimorph.org/specimens/Naja_haje. **1.46b:** Kindly supplied by Bruce A. Young; p. 46(left & right): © East London Museum.

Chapter 2

Figure 2.15a: From Bryan Wilbur, Ashley Gosselin-Ildari, 2001, "Pisaster sp. Digital Morphology at digimorph.org/specimens/ pisaster_sp

Chapter 4

Figures 4.18a & b: © Kenneth Kardong; **4.39a & b:** Reprinted from Cowin, Hart, Blaser, Kohn, "Functional Adaptation in Long Bones: Establishing in Vivo Values for Surface Remodeling Rate Coefficients, *J. Biomechanics*, Vol. 18, No. 9, 1985. Copyright © 1985 with permission from Elsevier; **4.41b:** Reprinted by permission of the publisher from "Functional Vertebrate Morphology," **fig. 1.10a,** p. 13. Edited by Milton Hildebrand, et al., Cambridge, Mass.: The Belknap Press of Harvard University Press, © 1985 by the Presidents and Fellows of Harvard College.

Chapter 5

Figure 5.38: Photo courtesy of M. Richardson

Chapter 6

Page 218: © Punchstock RF; **6.12a:** From *Comparative Vertebrate Histology*, 1969. © Donald I. Patt & Gail R. Patt. Reprinted by permission; p. 221(background & insert): Courtesy of Deborah A. Hutchinson and Alan H. Savitzky, part of the research team including A. Mori, J. Meinwald, F.C. Schroeder, G.M.

Burghardt; p. 228(background & insert): Photos courtesy of and based on the research of John P. Dumbacher.

Chapter 10

Figures 10.7a & b: Modified from and thanks to Young, Magnon, and Goslow, 1990.

Chapter 11

Figures 11.27a & b: From D.L. Uchtel and K.V. Kardong in *Journal of Morphology*, Vol. 169, Issue 1, pp. 29-47, 1981 "Ultrastructure of the Lung of the Rattlesnake, Crotalus viridis oreganos," Reprinted by permission of Wiley-Liss, a division of John Wiley & Sons Inc.

Chapter 12

Figure 12.10: From *Early Embryology of the Chick* by Bradley M. Patten, 1995. Reprinted by permission of the McGraw-Hill Companies; p. 485 **a-c:** Waterman, Allyn J: Olsen, Ingrith D; Johansen, Hermana; Noback, Charles R; Liem, Karel; Kluge, Arnold; Frye, B.E, *Chordate Structure & Function*, 2nd edition, © 1977. Reprinted by permission of Pearson Education, Inc., Upper Saddle River, NJ.

Chapter 14

Figure 14.40: © Dr. Edward Zalisko.

Chapter 15

Page 619: © The Trustees of the British Museum, Department of Prints and Drawings. All Rights Reserved.

Chapter 17

Figure 17.27: From *Brain, Mind & Behavior* by Floyd E. Bloom, Arlyne Lazerson, and Laura Hofstadter, © 1985 by Educational Broadcasting Corp. Used with permission of W.H. Freeman and Co.; **17.37:** From Northcutt & Davis *Fish Neurobiology*, Vol. 1, pg. 147, © The University of Michigan Press.; **17.44a:** Ruf, 1. (2007): Having a Closer Look: micro (Tomalysis of the bony Cabyrinth in rodents. ICVM8 Abstracts, 107.

Chapter 18

Figure 18.1b: © Crown Copyright. NMR

Index

Aardvark, 125
Abdominal air sacs, 438
Abdominal muscles, 434
Abdominal ribs. *See* Gastralia
Abducens nerve, 634, 638*f*, 639*t*
Abduction, 390*f*
Abductors, 390
Abomasum, 530
Absorption, 535
Abundance phylogeny, 25*f*
Acanthodes
    appendicular skeleton, 335*f*
    axial skeleton, 305*f*
    skull phylogeny, 256, 258*f*
Acanthodians, 94, 97–98
    appendicular skeleton
        phylogeny, 334, 335*f*
    skull phylogeny of, 256, 258*f*
Acanthostega, 105
    axial skeleton, 312
    gastralia, 299
    manus and pes, 342, 345*f*
    ribs, 299
Acceleration, 139–40, 144, 199, 200*f*
Accelerator muscle, 269
Accessory air-breathing organs, 417–18, 420*f*, 491
Accessory digestive glands, 504
Accessory "hearts," 475
Accessory urinary ducts, 573
Accommodation, 159
A cells, 604
Acellular bone, 182
Acetabulum, 327
Acetylcholine, 642
Acid-base regulation, 444–46
    cardiac shunt and, 487
Acidophils, 607
Acini, 602
Acinonyx, digits of, 356*f*
Acipenser
    gill ventilation in, 429*b*
    limb development, 332*f*

oviducts of, 568*f*
    pectoral girdle of, 341*f*
Acipenserids, 100
Acipenseriformes, 100
Acoelous centra, 296, 297*f*
Acorn worms, 56*f*, 57–60, 57*f*
Acquired characteristics, 6, 7
Acrochordus javanicus, cutaneous respiration of, 419*f*
Acrodonts, teeth of, 512
Acromegaly, 609
ACTH. *See* Adrenocorticotropic hormone (ACTH)
Actinistia, 102
Actinopterygians, 99–101
    aortic arches of, 462
    appendicular skeleton, 334–36
    buccal cavity, 505
    cloaca, 582
    gas bladders on, 416
    hydrostatic function of, 428
    pancreatic islets, 603*f*
    pectoral girdle of, 341*f*
    skull phylogeny of, 260–61, 261*f*, 264*f*
    suspension feeding, 286
    telencephalon, 661
    trunk vertebrae, 307*f*
    venous vessels, 469
Actinotrichia, 326
Actiocetus, skull of, 290*b*
Action potential, 626, 628
Action/reaction, third law of, 141
Active component, 379–80
Active muscle, 376
Active state, 376
Active tension, 380
Adaptation, 5, 13*f*, 14
Adduction, 390*f*
Adductor femoris, 399*t*, 401
Adductor mandibulae, 406
Adductors, 390, 406

Adenohypophyseal placodes, 189
Adenohypophysis, 593*t*, 604, 605, 609–10
ADH. *See* Antidiuretic hormone (ADH)
Adipose tissue, 232
Adrenal gland, 593*t*, 598–602
Adrenergic control, 642, 642*f*
Adrenocortical tissue, 598, 599*f*, 600*f*
Adrenocorticotropic hormone (ACTH), 601, 610
Advanced groups, 722–23
Adventitia, 520
Aepyornis, 115, 130*f*
Aerial locomotion, 349, 358–67, 367*f*, 404–5
Aerodynamic circulation, 363, 365*f*
Aerodynamics, 225, 361–62, 364*f*, 365*f*
Afferent artery, 461
Afferent neurons, 628, 629*f*
Afferent spinal nerves, 629*f*, 631*f*
Afferent spiracular artery, 461
African elephant, 130*f*
African lungfishes. *See* *Protopterus* (African lungfishes)
Afrotheria, 124*f*, 126
Agassiz, Louis, 6, 6*f*
Aging, 163, 538*b*
Aglomerular kidneys, 558
Agnathans, 85
    appendicular skeleton phylogeny, 331
    axial skeleton phylogeny, 304
    fossils, 89–93
    jaws and mouths, 509–10*b*
    living, 86–89
    origin of, 85
    ovary, species with single, 569*t*

overview, 93–94
pallium, 662
pancreas, 604*t*
phylogeny, 87*f*
pituitary gland, 607*t*
respiratory system, 424–26
skull phylogeny, 255, 256*f*, 257*f*
splanchnocranium, 293*f*
Air
    breathing in, 444
    feeding in, 287–88
    vision in, 684–86, 687*f*
Air-breathing fishes, 422–23
Air-breathing mechanisms, 448–49
Air-breathing organs, 446–47, 491. *See also* Respiratory organs
Air capillaries, 437–38
Airfoils, 362, 363–64, 365*f*
Air passages, 520*f*. *See also* Buccal cavity
Air sacs, 438, 449–50
Air ventilation
    aspiration pump, 423–24
    buccal pump, 422–23
Akinesis, 253*f*, 282, 291–92
Akinetic skulls, 254
Albatross, 130*f*, 363*f*
Aldosterone, 601
Algebra, vector, 731–32
Alimentary canal (digestive tract), 504, 520–21, 520*f*
    cloaca, 525
    endocrine system and, 611, 612*f*
    esophagus, 521
    fishes, 527–28
    intestines, 523–25, 526*f*
    in lampreys, 528*f*
    specializations of, 525, 526*f*, 527
    stomach, 521, 523

tetrapods, 528–31
in tetrapods, 529f
vascularization of, 527
Alisphenoid bone, 253, 275
Alkaline blood, 487
Allantoic cavity, 191
Allantoic placenta, 192
Allantoic vessels, 191
Allantois, 191, 191t
Alligators, 115. *See also*
    Crocodilians
alimentary canal of, 529
axial skeleton of, 315f
brain, 656f, 658f
gizzards, 536
heart, 483
muscle homology, 391f, 408t
nitrogen elimination, 554
pectoral girdle, 340f, 341f
pelvic girdle of, 343f
penis, 581f
skulls of, 275f
stomach of, 528f
teeth of, 512f
Allin, Edgar F., 283f
Allometry, 134, 135f, 136f, 138f
All-or-none muscle
    excitation, 381
Allosaurus, 115, 356f
Alpha-keratins, 216
Alpha motor neurons, 673
Alula, 361
Alveolar ducts, 232
Alveolar exocrine glands, 180
Alveolar lungs, 440
Alveoli, 437
Amacrine cells, 682
Ambiens, 399t, 400
Amble, 356
Ambystoma, 575f
Ambystoma gracile, 198, 513f
Ambystoma macrodactylum, 580f
Ambystoma texanum, 580f
Ambystoma tigrinum, 199,
    419f, 423f
Ameloblasts, 508
Amia (bowfin), 45b, 526f
axial skeleton of, 308f
genital ducts (female),
    568f, 570f
oviducts of, 568f
pectoral girdle, 335f, 341f
skull phylogeny of, 260, 261f
structural analysis, 718
Amia calva, axial skeleton
    of, 307f
Ammocoete larva, 89
Ammocoetes, 88f, 89, 286, 424,
    446, 594
Ammonia, 445, 554, 555f
Ammonotelism, 554, 555f

Ammotragus, 138f
Amnion, 86, 190, 191t
Amniotes, 86, 108–10, 267f
aortic arches, 465f
appendicular musculature,
    392–93
aspiration pump, 424
axial skeleton phylogeny,
    312–14
cranial nerves, 631, 634t
digits of, 356f
double circulation, 455–56
embryonic development, 194
extraembryonic membranes,
    190, 195f
eyes, 683
forebrain, 652–54
genital ducts (female),
    569, 570f
genital ducts (male), 573
kidneys, 549f, 551, 553
opthamalic nerve, 633–34
pectoral girdle, 338, 340f
pelvic girdle, 342f
pineal gland, 658
ribs, 298f
sauropsids, 110–19
skull, 264–68
spinal accessory nerve, 634
spinal reflexes, 641
stem-amniotes, 110
synapsida, 119–26
tectum, 660f
testis, 572
vertebral column, 322f
Amniotic cavity, 191
Amniotic folds, 191
Amoebocytes, 69
Amphibians, 104, 106–8
acid-base regulation, 446
adrenal tissues, 599f, 600f
air-breathing
    mechanisms, 449
alimentary canal, 526f,
    528–29
aortic arches, 462f, 463–65
autonomic nervous
    system, 645f
axial musculature, 395–96
axial skeleton, 309–10
body cavities, 196f
branchial arches, 247t
buccal pump, 423
CCK cells, 623–24
cerebral hemispheres, 664f
chondrocranium, 244t
circulation in, 456
cleavage, 165
cloaca, 582, 584f
copulatory organs, 576–77
cranial musculature, 410f

cranial nerves, 631, 633
cutaneous respiration,
    417, 419f
dehydration, 555b
ear functions, 702–4
embryonic development, 196f
eyes, 683, 685f
fertilization in, 588–89
fossorial locomotion, 369
gastrulation and neurulation,
    171, 173f, 174f
genital ducts (female),
    569, 571f
genital ducts (male),
    573, 575f
gliding and parachuting, 360f
hearts, 474–75, 478–80, 492
hypobranchial nerve, 636
integument of, 219–20
kidneys, 551
manus, 345f
midbrain, 658
nasal passages, 680f
nasal sac, 676
nephrons, 554f
nitrogen elimination, 554
olfactory bulb and tract, 676f
pallium, 663
pancreas, 604t
parathyroid glands, 597f
pectoral girdle, 337–38,
    340f, 341f
pelvic girdle, 342f, 343f
pes, 346f
photoreception, 686
pineal complex, 690
pituitary gland, 607t
respiratory systems, 430,
    432–33
salt balance, 561f
skulls, 263–64, 265f, 266f
somatic and visceral circuits
    in, 641f
spermatophores, 580f
stomach, 528f
suction feeding, 286–87
thyroid gland, 596, 597f
urinary bladder, 588
venous vessels, 469,
    471f, 472
vocalization, 435b
vomeronasal area, 676, 677
water balance, 558
Amphicoelous centra, 296, 297f
Amphicoelous vertebrae, 307f
Amphioxus, 62f, 63–66,
    63f–64f, 65b, 66f, 594
cleavage, 164–65
gastrulation and neurulation,
    169, 170f
integument of, 215f

larval, 66f
specialized notochord of, 66f
suspension feeding, 286
Amphiplatyan, 296
Amphisbaena, 113f
Amphistylic jaw suspension, 248
AmphiTbx4/5 gene, 330
Amphiuma tridactylum, buccal
    pump in, 423f
Amplexus, 576
Ampullae, 696
Ampullae of Lorenzini,
    709, 712f
Ampullary receptor, 709, 711f
Amygdala, 662, 664
Amylase, 532
Anadromous fishes, 558b
Analogy, 15–16, 15f
Anal regions, of fishes, 583f
Anamniotes, 86, 190
aortic arches, 462f
breathing and embryos,
    418–20
cranial nerves, 631
embryonic development, 194
somatic and visceral circuits
    in, 641f
spinal reflexes, 641
testis, 572, 574f
Anapsids, 93, 111
dermatocranium
    evolution, 267f
skull, 109–10, 255
temporal fenestrae in, 268f
Anatomy, 15
autonomic nervous system,
    643–45
ear, 698–700
small intestine, 524f
spinal nerve, 630f
stomach, 523f
tooth, 507
Anatosaurus, fossilized, 32f
Anaximander, 4
Anchiornis, 116
Andrewsarchus, skull of, 290b
Androgens, 572, 598
Angelfish, swimming-related
    musculature of, 400b
Angiogenesis, 456
Angiogenic cells, 458f
Angiotensin I, 612f, 613
Angiotensin II, 612f, 613
Angiotensinogen, 612–13
Angle of attack, 361
Anguilla anguilla, cutaneous
    respiration of, 419f
Angular bone, 250
Animal pole, 163
Animal-to-fossil process. *See*
    Fossilization

Ankles, of archosaurs, 347*f*
Ankylosaurs, 115
Ankylosis, 189
Anlagen, 411
Annelids, 17, 19*f,* 75
Anomodontia, skulls of, 278*f*
Antagonists, 388
Anteaters, teeth of, 515*f*
Antelopes, 125
Anterior cardinal veins, 63,
      460, 466
Anterior chamber, 682
Anterior lobes, 605
Anterior mesenteric
      artery, 460
Anterior region, 16
Anterior thoracic air sacs, 438
Anterior vena cava, 460–61
Anthracosaurs, 105, 105*f,* 110,
      310*f,* 311–12, 313*f*
Anthropoids, 125
Antidiuretic hormone
      (ADH), 608
Antlers, 234–35, 234*f*
Antorbital fenestra, 114
Antrum, 613
Ants, 130*f*
Anurans
   aortic arches, 468*f,* 479*f*
   appendicular
         musculature, 402
   development, external view
         of, 174*f*
   genital ducts (male), 576*f*
   pancreas, 604*t*
   sternum, 300*f*
   systemic arches, 468*f*
   venous vessels, 471*f*
Anus, 504
Aorta, 457, 459
Aortic arches, 457, 461–66,
      467*f,* 468*f,* 479*f,* 481
   evolution overview, 466
Apatosaurus, 115
Apes, 125, 687, 725, 727*f*
Apical ectodermal ridge (AER),
      206–7
Apical foramen, 507
Apical surface, 179
Apnea, 415, 487
Apocrine glands, 231
Apoda. *See* Gymnophiona
      (Apoda)
Aponeuroses, 376
Appendicularia. *See* Larvaceans
Appendicular musculature
      comparative
         anatomy, 397–405,
         406*f,* 407*f*
      embryonic development,
         392–93

Appendicular skeleton, 325
   components, 326–27
   evolution of, 346–48
   fish phylogeny, 331, 334–36,
         337*f*
   form and function, 348–71
   modularity and, 725, 728*f*
   overview, 370–71
   paired fins, origin of,
         327–31, 332*f*
   tetrapod phylogeny, 336–46
Appendix, 530*b*
Appositional growth, 184
Apteryx, pelvic girdle of, 343*f*
Aquatic environment. *See* Fluid
      environment
Aqueous humor, 682
Arachnoid, 646
Arboreal locomotion, 349, 368*f*
Arboreal theory, 367, 368*f*
Arborize, 672
Archaeopteryx, 116, 119*b,*
      123*b,* 407*f*
   aerial locomotion, 361
   epigenomics, 207–8
   feathers, 228, 229*f*
   fossilized, 32*f*
   mosaic evolution and, 725
   structural analysis, 718
   supracoracoideus
         muscles, 407*f*
Arch centers, 302
Archean eon, 40, 40*f*
Archenteron, 167
Archetypes, 12, 13*f,* 14*f*
Archimedes, 147
Archinephric duct, 566, 566*f*
Archipterygial fin, 327
Architheuthis, 130*f*
Archosauromorpha, 112,
      114–15
Archosaurs, 114
   ankles of, 347*f*
   appendicluar skeleton
         of, 353
   digits of, 356*f*
   supracoracoideus muscles
         of, 406*f*
Arctodus simus, reconstruction
      of, 34*f*
Arcualia, 301–2
Arduino, Giovanni, 37
Area, 129, 132–33
Areola, 232
Argali sheep, 138*f*
Aricularian hypothesis, 76–77
Aristotle, 7
Arithmetic growth, 137
Armadillos
   adrenal tissues of, 600*f*
   cervical vertebrae of, 314

forelimbs of, 386*f*
   skull (embryonic
         development), 248*f*
Armor. *See* Dermal armor
Arrector pili, 230
Arterial vessels, 461–66, 467*f,*
      468*f*
Arteries, 452–55, 459–60
Arterioles, 453
Arthrodire, 130*f*
Arthropods, evolution of, 75, 75*f*
Articular bones, 253, 700
Articular cartilage, 187
Articulations. *See* Joints
Artificial taxon, 24
Artiodactyls, 125, 143–44
   appendicular musculature
         of, 403
   gait of, 356
   manus of, 345*f*
   teeth, 537*f*
Ascaphus
   copulatory organs in, 576–77
   fertilization in, 588
   pectoral girdle of, 341*f*
Ascaphus truei tadpoles,
         respiratory system
         of, 432
Ascending nerve tracts, 650,
      651*f,* 652*t*
Ascidiacea, 68–70, 70*f–71f*
Ascidian tadpole, 68
Askeptosaurus, 112
Aspidospondylous vertebra, 296
Aspiration effect, 474
Aspiration filling, 474*f*
Aspiration pump, 423–24, 449
Aspondyly, 295
Associated nerve endings, 673.
         See also Associated
         sensory receptors
Associated sensory receptors,
      673, 674*f,* 675*f*
Association neurons, 638
Astragalus, 343
Astrocytes, 625, 626*f*
Astylosternus robustus, cutaneous
      respiration in, 420*f*
Asymmetrical gaits, 357
Atavisms, 209*f*
Atherinomorphs, 101
Atlas, 275, 313
Atmosphere (atm), 147
Atretic follicles, 613, 614*f*
Atretic oocytes, 569
Atrial chambers, 477
Atriopore, 63
Atrioventricular (AV) node, 474
Atrioventricular (AV) plug, 477
Atrioventricular (AV)
         valves, 474

Atrium, 69, 456, 473
Atrophy, 151, 151*f,* 153–54
Attachments, 376
Auditory bulla, 275
Auditory nerve, 634, 638*f*
Auditory papilla, 704
Auditory system, 696–9
Auks, 349*f,* 721
Auricle, 654
Auricularia larva, 56
Auricular operculum, 107
Australian lungfish. *See*
         Neoceratodus
         (Australian lungfish)
Autonomic nervous system
         (ANS), 629, 641–46,
         645*f*
Autopalatine, 256
Autopodium, 326
Avascular bone, 182
Aves, 115–19, 118. *See also*
         Birds
Axial column, 296
Axial complex, 68
Axial musculature
      appendicular musculature
         and, 397–98
      comparative anatomy,
         394–97
      embryonic development, 393
Axial skeleton, 295
      components, 295–300, 301*f*
      embryonic development,
         301–4
      form and function, 314–21,
         324*f*
      modularity and, 725, 728*f*
      overview, 321–24
      phylogeny, 304–14
Axis, 313, 327
Axon, 626, 672
Azygos veins, 472

Background extinction, 41
Baculum, 582
Bakker, Robert, 117*b*
Balaena, axial skeletons in, 322*f*
Balaenoptera, axial skeletons
         in, 322*f*
Balance
   bipedals, 318*f*
   salt, 554–62, 560–62
Baleen, 235–36, 236*f,*
         289*b,* 290*b*
Baleen whales, 125
Balfour, F. M., 328, 329
Balinsky, Boris I., 172*f*
Baluchitherium, 130*f*
Bandicoots, 193, 195*f*
Banks, William J., 234
Barbary sheep, 138*f*

Barbs, 223
Barbules, 223
Basal cells, 675
Basal ganglia, 663
Basal lamina, 179
Basal layer, 177
Basal plate, 242
Basals, 326
Basapophyses, 297
Base, 507
Basement membrane, 177, 212
Basibranchial element, 246
Basilar membrane, 700
Basilosaurus, axial skeletons
    in, 322*f*
Basioccipital bone, 250
Basisphenoid, 244*t*
Basmajian, J. V., 388*f,* 497*f*
Basophils, 607
Bateson, William, 204
Batoidea, 97
Bats, 115, 125
    cutaneous respiration, 419*f*
    diversity and patterns in, 724*f*
    forelimb modifications for
        flight, 360*f*
    sternum, 300*f*
    suspension feeding, 285
B cells, 602
Bdellostoma, 550, 574*f*
Beaks, 274
Beaumont, William, 525*b*
Beavers
    scales of, 236, 237*f*
    teeth of, 511*f*
Belly, of muscle organs, 376
Benchley, Robert, 382*b*
Bends, 447*b,* 448*b*
Benthic protochordates, 55
Bermuda parrot fish thyroid
    glands, 595*f*
Beta-keratins, 216
Bias, 214, 215*f*
Biceps brachii, 398, 399*t*
Biceps femoris, 399*t,* 401
Bichirs, 100, 568*f*
    axial skeleton of, 308*f*
    brain, 654*f*
    genital ducts (male), 575*f*
    oviducts of, 568*f*
Bicipital ribs, 298
Bicornuate uteri, 585
Bicuspids, 512–13
Bi-directional food flow, 264
Bi-directional ventilation,
    415–16
Big brown bat, cutaneous
    respiration of, 419*f*
Bilaminar blastoderm, 166
Bilateral symmetry, 16, 18*f*
Bilateria, 48, 50*f,* 50*t*

Bile, 533, 534*f*
Binocular vision, 158, 688
Bioengineering, 137
Biogenetic laws, 201–4
Biological design, 128–29
    biomechanics, 137–55
    biophysics and other
        physical processes,
        155–59
    overview, 159–60
    shape, 128–29, 134–37, 137*f*
    size, 129–34, 137
Biological failure, 149–51, 151*f*
Biological role, 19–20, 19*f,*
    46–47
Biology, evolutionary, 2, 3–14
Bioluminescence, 710*b*
Biomechanics, 137–55, 146*f*
    force laws, 141–42, 142*f*
    free bodies and forces, 142
    fundamental principles,
        138–41
    land and fluid, 144–48
    machines, 148, 149*f*
    materials, strength of,
        148–51, 151*t*
    mechanical stress, tissue
        response to, 151–55
    torques and levers, 142–44
Biophysics, 155–59
Biosynthesis, cardiac shunts
    and, 487
Bipartite uteri, 585
Bipedal balance, 318*f*
Bipedalism, 114
Bipedal posture, 320*b,* 321*b*
Bipolar cells, 682
Bipolar neurons, 626, 627*f*
Birds, 115–19, 118, 119*b,*
    120*f,* 191*t*
    adrenal tissues, 600*f*
    aerial locomotion, 349,
        358–67
    air sacs, 449–50
    alimentary canal, 526*f,*
        529–30
    aortic arches, 465, 468*f*
    appendicular musculature,
        403–4
    appendicular skeleton, 349*f*
    aspiration pump, 424
    axial musculature, 397
    body cavities, 197*f*
    bones, 715*f*
    branchial arches, 247*t*
    breathing and embryos, 418,
        420, 421*f*
    cardiovascular system design,
        491–92
    castration of, 618*b*
    CCK cells, 623–24

cerebral hemispheres, 664*f*
chondrocranium, 244*t*
cleavage, 165–66
cloaca, 582, 584*f,* 585*f*
copulatory organs, 577*f*
cranial kinesis, 268
cursorial locomotion, 355
depth perception, 688–89
dermatocranium
    evolution, 267*f*
digits, 356*f*
diversity and patterns
    in, 724*f*
ear functions, 704–5,
    706*f,* 707*f*
embryonic development,
    192*f,* 194
epigenomics, 205–9
extraembryonic membranes,
    190–91, 191*f,* 191*t*
eyes, 683, 685*f*
feather color, 238
feather regeneration,
    224–25, 226*f*
feathers, 20
feces, 536
flight, 19–20, 19*f,* 20*f,*
    367–69, 368*f,* 718–19
foregut fermentation, 539
forelimb modifications for
    flight, 360*f*
functional coupling of
    feathers, 720
functional matrices in, 721
gas exchange, 443*f*
gastrulation and neurulation,
    171–72, 175*f,* 176*f*
genital ducts (female), 571*f*
genital ducts (male), 576*f*
gizzards, 536
hearts, 474, 489–90
heat blocks, 495*f*
integument, 214*f,* 221–28,
    229*f,* 230*b*
intromittent organs,
    579, 582*f*
jaw articulations, 291
kidneys, 553, 553*f*
long bones of, 363*f*
lungs, 440, 441*f,* 442*f,*
    449–50
magnetic fields and, 712–13
manus and pes, 345*f,* 346
midbrain, 658
modularity, 725, 728*f*
muscle fiber colors, 377–78
nasal passages, 677
nephrons, 554*f*
nitrogen elimination, 554
olfactory bulb and tract, 676*f*
oral glands, 532

ovary, species with single,
    569*t*
pallium, 663
pancreas, 604*t*
pancreatic islets, 602
pectoral girdle, 339, 340*f,*
    341*f*
pelvic girdle, 342*f,* 343*f*
penis, 578
phylogeny of, 24*f*
pineal complex, 690,
    691, 692*f*
pituitary gland, 607
poisonous, 230*b*
preadaptation, 20, 20*f*
prolactin, 609
respiratory system, 437–40,
    440, 441*f*–443*f*
salt balance, 561–62
scales, 222*f,* 236
size and, 130*f*
skull phylogeny, 272–74,
    274, 276*f,* 277*f*
spinal accessory nerve, 634
spinal cords, 650*f*
sternum, 300*f*
stomach, 528*f*
supracoracoideus
    muscles, 407*f*
systemic arches, 468*f*
telencephalon, 661
testis and epididymis, 578*f*
thyroid glands, 595*f,* 596
venous vessels, 472
ventilation, 442*f*
vertebral column, 320–21,
    323*f*
visual accommodation, 687*f*
Birds of prey
    feeding by, 288
    flight of, 361
Birth, cardiovascular changes at,
    492–94
Birth canal, in humans,
    320*b,* 354*f*
Bison, 125, 319*f*
Blackbelly salamanders,
    courtship in, 580*f*
Blackburn, Daniel, 233
Black-tailed godwit, 135*f*
Blastocoel, 164, 167*f*
Blastocyst, 166, 167*f,* 589
Blastoderm, 165
Blastodisc, 165, 166
Blastomeres, 164–166
Blastopore, 49, 169
Blastula, 163, 165*f*
Blood, 452, 527, 534*f*
    alkaline, 487
Blood cell formation,
    embryonic, 457*f*

Blood clot, 452
Blood islands, 456
Blood pressure, 453–54, 455*f*
Blood vessels, 452*f. See also*
    Arterial vessels;
    Arteries; Capillaries
Blowholes, 431*b*
Blue whales, 130*f,* 289*b*
BMP (bone morphogenetic
    protein), 78
Boa constrictors
    cutaneous respiration of, 419*f*
    infrared receptors in, 693
Body
    free, 142
    neuron, 626
    vertebra, 295
Body cavities, 196*f*
Body design segmentation,
    16–19. *See also* Design
    similarities, 15–16, 15*f*
    symmetry, 16, 18*f*
Body mass. *See* Mass
Body parts, *Hox* genes and,
    204, 205*f*
Body plans, of chordates, 54, 76*f*
Body size and shape.
    See Shape; Size
Body support, in fishes and
    dinosaurs, 317*f*
Body veins, general, 468, 469*f*
Bolitoglossa occidentalis, 199
Bolus, 503
Bombesin, 623
Bone fractures, 149–50, 151*f,*
    153*b. See also* Bone
    remodeling and repair
Bone histology
    comparative, 186
    dinosaurs, 117*b,* 123*b*
Bone-muscle lever systems,
    385–88
Bone remodeling and repair,
    156*f,* 186–87, 188*f*
Bones. *See also* Appendicular
    skeleton; Axial
    skeleton; Skull
    architecture of, 184*f*
    as connective tissue, 182
    dermal, 213
    development and growth,
    182–89
    evolution of, 186*b*
    responsiveness of, 151–55
Bone spicules, 183
Bonner, J. T., 135*f,* 136*f*
Bony fishes
    aortic arches, 462
    appendicular skeleton
    phylogeny,
    334–36, 337*f*

axial skeleton phylogeny, 302
bombesin, 623
cloaca, 582
eyes, 683
genital ducts (female), 569
gill ventilation, 429*b*
hearts, 476
hyoid arch, 407
integument, 218
kidneys, 553
nephrons, 554*f*
pancreas, 535
pancreatic islets, 602
pineal complex, 690
pyloric ceca, 528
respiratory system, 427, 428*f*
scale types of, 220*f*
vertebrae, 302
vertebral column, 322*f*
Boswell, James, 560*b*
Bothriolepis, 258*f,* 333*f*
Bound, 357
Boundaries, of buccal cavity,
    504–5, 506*f*
Boundary layer, 145
Bovine stomach, foregut
    fermentation in, 540*f*
Bowfins, 101
    axial skeleton phylogeny, 305
    brain, 655*f*
    genital ducts (female), 570*f*
    genital ducts (male), 575*f*
    holoblastic cleavage, 165*f*
    oviducts, 568*f*
    skull, 261, 261*f*
Bowman's capsule, 551
Braces, 150*f*
Brachial vein, 466
Brachiation, 349, 354*f*
Brachiocephalic artery, 465
Brachiosaurus, 115, 147*f*
Brachydont teeth, 514
Bradycardia, 473
Brain, 651–66
    form and function, 654–66
    functional associations
    of central nervous
    systems parts and, 666
    functional matrices in, 721
    neurogenesis, 627*b*
    phylogeny, 651–54
    regions of, 646, 647*f,* 648*f,*
    653*f*
    structural analysis, 718
Braincase, 242, 248–250, 253
Brain stem, 651
Bramble, D., 517*f*
Branchial arches, 245, 247*t,*
    408, 639*t*
Branchial basket, 53
Branchial grooves, 518

Branchial heart, 475, 476
Branchial pore, 58
Branchial pouch, 58
Branchiomeric musculature,
    393, 397, 399*t,*
    405–8, 409*f,* 410*f*
Branchiostoma, 62*f,* 65*b*
Breasts, 232
Breathing
    air, 444
    embryos and, 418, 420, 421*f*
    medulla and, 660*f*
    water, 444
    whales, 431*b*
Broca, P. P., 657*b*
Broca, Paul P., 666
Broca's motor speech area, 657*b*
Broca's speech area, 666
Brockmann bodies, 602
Bronchi, 416
Bronchioles, 416
Brood patch, 221, 609
Brown hydra, 130*f*
Brown trout, cutaneous
    respiration of, 419*f*
Brunner's glands, 525
Buccal cavity, 504–20
    boundaries, 504–5, 506*f*
    evolution of, 509–10*b*
    human speech, 511*b*
    palate, 505–6, 507*f*
    teeth, 506–8, 511*f,* 512–14
    tongue, 516–17
Buccal cirri, 62
Buccal pump, 422–23, 428*f,*
    433, 449
Buccinator, 408
Buccopharyngeal
    membrane, 504
Budding, 70
Buffon, George Louis LeClerc
    de, 31*b*
Bufo, pancreatic islets in, 602
Bufo japonicus, 222*b*
Bufonidae, 108
Bulbar valves, 476
Bulbourethral gland, 563
Bulbus arteriosus, 473
Bulbus cordis, 456, 473
Bullfrogs
    cutaneous respiration
    of, 419*f*
    hearts of, 479*f*
    thyroid and parathyroid
    glands in, 597*f*
Bunodont teeth, 514
Buoyancy, 147–48
Burghardt, G. M., 222*b*
Burrowing. *See* Fossorial
    locomotion
Burrowing protochordates, 55

Bustard, aortic arches of, 465*f*
Butterfly fish, swimming
    specializations in,
    400*b,* 401*b*

Caecilians, 106, 108, 263–64,
    676. *See also*
    Gymnophiona
    (Apoda)
Caiman crocodilius, 499*f*
Caimans, 115, 434, 499*f,* 526*f*
Caisson disease, 447*b*
Calamoichthys calabaricus,
    cutaneous respiration
    of, 419*f*
Calamus, 223
Calcaneum, 343
Calcification, 181
Calcitonin, 596
Calcium, 152*f,* 153, 186*b*
Calcium regulation, 598
Callus, 216
Calorie, 140
Cambered wings, 362
Camels, 125, 539*f*
Cancellous bone, 182
Canines, 514
Caninus, 408
Canis dirus, skull of, 281*f*
Canis latrans, teeth of, 514
Cannon bone, 343
Canter, 357, 357*f*
Capillaries, 452–55
Capillary bed, 453
Capitulum, 298
Capons, 618*b*
Captorhinidae, 112
Captorhinomorph, 112*f*
Captorhinus, digits of, 356*f*
Capybara, adrenal tissues of, 600*f*
Carapace, 111, 236
Carassius carassius, cutaneous
    respiration of, 419*f*
Carbohydrates, 537
Carbon dioxide, cardiac shunt
    and, 487
Carbon dioxide expulsion,
    415, 443, 445–46.
    See also Circulatory
    system; Respiratory
    system
Carcharhinus acronotus, teeth
    of, 513*f*
Cardia, 521
Cardiac glands, 521
Cardiac muscles, 373, 375
Cardiac shunt, 481, 489–90
Cardinal hearts, 476
Cardinal veins, 63, 460, 466
Cardiovascular function,
    ontogeny of, 492

Cardiovascular system, 452
    arteries, veins, and
        capillaries, 452–55
    blood, 452
    circulation, single and
        double, 455–56
    embryonic development,
        456–57, 458f, 459f
    environmental influences on,
        490–95, 496f
    hearts, 473–90
    phylogeny, 457–73, 459–90
Carina, 299
Carnassials, 515, 537f
Carnivora, 125
Carotid arteries, 457, 463
Carotid body, 463
Carotid duct, 463
Carotid rete, 495
Carp, muscle fibers of, 388, 389f
Carpals, 339
Cartilage, 181, 183, 244f
Cassowary, 118
Castrati, 618–19b
Castration, 618–19b
Catadromous fishes, 558b
Catarrhine monkeys, 125
Catastrophic extinction, 41
Catecholamines, 598
Catfish
    accessory air-breathing
        organs of, 417–18
    chemoreceptors in, 675f
    electric organs in, 373f
Cats
    adrenal tissues, 600f
    aortic arches, 465f
    appendicular
        musculature, 402f
    biomechanics, 321b
    cranial and shoulder
        musculature, 402f
    facial muscles, 411f
    muscle homologies, 408t
    pelvic girdle, 343f
    pelvic musculature, 402f
    pes, 346f
    saber-tooth, 519b
    salivary glands, 533f
    sternum, 300f
CAT scans, 42
Cat shark, cutaneous respiration
    of, 419f
Cattle, 125
Caudal artery, 459
Caudal hearts, 476
Caudal lobes, 607
Caudal region, 16, 18f, 295, 319
Caudal skeleton and fins, 305,
    307–8
Caudate nucleus, 663

Caudofemoralis, 399t, 400
Cavefish, midbrain of, 653
Cavernous urethra, 582
Cavum arteriosum, 481
Cavum pulmonale, 481
Cavum venosum, 481
C cells, 597
CCK. See Cholecystokinin
    (CCK)
CCK-PZ. See Cholecystokinin-
    pancreozymin
    (CCK-PZ)
Cecal fermenters. See Intestinal
    (hindgut) fermenters
Cecum, 525–26
Celiac artery, 460
Cell division, 164
Cells. See specific types of cells
Cell signaling, 210
Cellular bone, 182
Cellulases, 537
Cellulose, 537
Cemental annuli, 507
Cementocytes, 507
Cementum, 507, 508
Cenozoic era, 41
Center of mass, 140, 141f
Center of pressure, 227
Centra, 295–97
Central blood sinus, 59
Centrales, 343
Central nervous system, 625, 646
    brain, 652–66
    development, 647f
    embryology, 646, 649, 653f
    overview, 669–70
    spinal cord, 649–51, 651f,
        653f, 653t
Central tendon, 197, 435
Centrum, 83, 84f
Cephalaspidomorphs, skull
    phylogeny of,
    255, 256f
Cephalic clasper, 97
Cephalic lobes, 607
Cephalization, 84
Cephalochordata, 50, 51,
    61–66, 62f
Cephalochordates, 594
Ceratobranchial element, 245
Ceratopsians, 115
Ceratotrichia, 326
Cerebellum, 651, 654–55,
    654–58, 660f
        of electric fish, 712
Cerebral cortex, 661–663
Cerebral ganglion, 68, 69
Cerebral hemispheres, 660–61,
    664f
Cerebrospinal fluid (CSF),
    646, 648f

Cerebrum, 651, 662
Cervical air sacs, 438
Cervical ganglia, 630
Cervical region, 295, 319
Cervical vertebrae, 313,
    313f, 318f
Cetacea, 125
Chameleon, alimentary canal
    of, 526f
Chaucer, Geoffrey, 560b
Cheek, 505
Cheek pouches, 505
Cheek teeth, 514
Cheese mite, 130f
Cheetahs
    cursorial locomotion of,
        357–58, 359f
    digits of, 356f
    gait of, 357–58, 359f
Cheirolepis, skull phylogeny
    of, 260
Chelonians
    adrenal tissues of, 600f
    hearts of, 481–83, 487
    pectoral girdle of, 341f
Chemical breakdown, of food,
    536–40, 539–40, 541f
Chemoreceptors, 674–79
    mouth, 678–79, 682f
    nasal passages, 675–77,
        678f–680f
    vomeronasal area, 677–78,
        681f
Chewing. See Mastication
Chicken embryos
    blood cell formation in, 457f
    cardiovascular system, 459f
    heart formation, 458f
    heart growth, 459f
    respiration in, 421f
Chickens
    pes, 346f
    skeleton of, 316f
    vestibular apparatus of, 706f
Chief cells, 521, 597
Chimaeras, 95, 97
    alimentary canal of, 527f
    anal regions of, 583f
    cloaca, 582, 583f
    copulatory organs, 576
    intromittent organs in, 578f
Chimpanzees, 125, 321b, 511b,
    600f
Chiridium, 104, 326
Chiroptera, 125
Chlamydoselachus, intromittent
    organs in, 578f
Choanae, 102, 262, 505, 676
Choanate fishes, olfactory
    organs, 680f
Choanichthyes, 102

Cholecystokinin (CCK), 612,
    623–24
Cholecystokinin-pancreozymin
    (CCK-PZ), 612
Cholinergic control, 642, 642f
Chonate fishes, nasal passages
    in, 676
Chondrichthyans, 94, 95–97
    adrenal tissues, 599f
    appendicular musculature,
        393
    appendicular skeleton
        phylogeny, 333–334,
        333f
    axial skeleton phylogeny,
        304, 305f
    fins and squalene, 430
    hearts, 476
    integument, 214f, 217–18
    intromittent organs, 578f
    ovary, species with
        single, 569t
    pancreas, 604t
    pancreatic islets, 602, 603f
    pituitary gland, 606, 607t
    respiratory system, 426
    skull phylogeny, 257–58,
        259f
    telencephalon, 661
    water balance, 557
Chondrocranium, 242, 243f,
    244f, 244t, 247, 250f
Chondrocytes, 181
Chondrosteans, 99, 332f
Chordal centers, 302
Chordamesoderm, 169, 170f
Chordates, 48
    body plan, 54
    characteristics, 52f
    characteristics of, 50–54
    Cladistic classification
        of, 743
    dorsoventral inversion, 78–80
    Linnaean classification
        of, 740–42
    origins, 74–81
    overview, 80–81
    phylogeny, 48–50
    prechordate inversion, 78–80
    protochordates, 54–61
    tadpoles, 77
Chordin, 78
Choresine beetle, 230f
Chorioallantoic membrane,
    167, 191
Chorion, 191, 191t
Chorionic gonadotropin (CG),
    616, 622b
Choriovitelline membrane, 167
Choroid, 681–82
Choroid plexus, 646

Chromaffin bodies, 598
Chromaffin hormones, 598
Chromaffin tissue (bodies),
    598, 600
Chromatophores, 213
    amphibians, 220
    birds, 225
    fishes, 217, 218
    mammals, 228–29
    skin color and, 238
Chromophobes, 607
Chronic overload, 382b
Chrysopelea, 367, 367f
Chuckwalla, cutaneous
    respiration of, 419f
Chyme, 503, 611
Cilia, 53, 131f, 421, 693
Ciliary body, 682
Ciliary muscle, 682
Ciliary pumps, 446
Ciliated funnel, 69
Cingulated gyrus, 661
Cingulate gyrus, 666
Cingulum, 516
Circadian rhythms, 604, 611
Circulating blood, 452
Circulation
    fetal (placental mammals),
        492, 493f, 494f
    hemodynamics of, 453–54
    lymphatic, 497f
    microcirculation, 454–55
    pituitary gland, 608f
    single and double, 455–56
Circulatory system, 451–52. *See
    also* Cardiovascular
    system
    evolution of, 501f
    lymphatic system, 496–99,
        500
    overview, 499–502
    respiratory system and,
        414, 415f
Circuli, 218
Cistern, 232
Clades, 24–29
Cladistia, 100
Cladistic chordate classification,
    743
Cladistics, 26, 28f
Cladogram, 26
Cladoselache, 96
    axial skeleton phylogeny,
        305f, 306f
    skull phylogeny of, 259f
Claspers, 331, 576
Classification, 27f
    muscles, 373–75
    vertebrates, 86
Clavicle, 333f, 334, 341f
Claws, 233–34

Cleavage, 49, 49f, 164–67
Cleidoic eggs, 194
Cleithrum, 334, 341f
Climate, dinosaurs and, 117b
Climatius, appendicular
    skeleton, 335f
Climbing, 349, 368f
Clitoris, 572
Cloaca, 504, 525, 545, 582–86,
    586f, 587f
Cloacal bursa, 585
Cloacal membrane, 504
Cloacal opening, 504
Cloney, R. A., 69f
Closing plate, 518
Clot, 452
Club cell, 216
Clumps, 180
Clupeomorphs, 101
Cobra, 42f
Coccosteus, axial skeleton of, 305f
Cochlea, 696, 709f
Coelacanth, 45b
Coelacanths, 45b, 102
    air-breathing organs, 447
    appendicular skeleton,
        335, 336f
    brain, 654f
    cloaca, 582
    cranial kinesis, 264f
    cranial nerves, 633
    pancreatic islets, 602
    vestibular apparatus, 698f
    water balance, 557, 560
Coeliac ganglia, 630
Coelom, 48, 51f, 167, 168f, 169
Coelom development, 196–97
Coelomic fold, 196
Coelurosaurus, 112
Coitus, 576
Coleonyx, cloaca, 585f
Collagen, 217
Collagen fibrils, 181
Collar, 57
Collar cord, 56, 59
Collar ganglion, 60
Collateral branches, 628
Collateral ganglia, 630
Collecting loop, 461
Collecting tubule, 551
Colloid, 593
Colon, 523
Colonial protochordates, 56
Color
    muscle fibers, 377–78
    skin, 238
Color changes, 238f
Color vision, 686–87
Colubrid snake
    cranial nerves of, 635f
    dentition of, 512f

Colugo, gliding and
    parachuting by, 360f
Columba
    copulatory organs in, 577f
    genital ducts (female)
        of, 571f
Columella, 248
Columnar cells, 179
Commissures, 662
Common cardinal veins,
    460, 466
Common carotid artery, 463
Compact bone, 154, 182, 184f
Comparative anatomy, for
    muscles, 394–410
Comparative bone
    histology, 186
Comparative morphology, 1–4
Comparative vertebrate
    morphology, 1–4
Competition, 5
Composite bones, 280f
Composite skull, 243f, 292
Composite theory, 246, 247f
Compound bone, 271
Compression, 422
Compression teeth, 537f
Compressive forces, 148, 149f
Compressive strength,
    148, 149f
Compsognathus,
    supracoracoideus
    muscles of, 406f
Compund glands, 180
Comuboniscus, pectoral girdle
    of, 341f
Conal valves, 473
Conchae, 676
Concurrent exchange, 156–58
Cones, 516, 682
Connective tissue, 177, 180–82
    categories of, 180f
Conodonts, 89–90, 93, 506
Constrictor colli, 407
Constrictor muscles, 390, 406
Contia, 113f
Contour feathers, 223, 223f, 359
Contraceptives, oral, 614b, 615b
Contraction
    heart, 474
    muscle, 376–77
Conus arteriosus, 473
Convergence, 628, 645
Convergent design, 16, 17f, 722f
Cooling, 496f
Co-option, 725, 727
Coots, appendicular skeleton
    of, 349f
Coprodeum, 529, 582, 584
Coprolites, 527
Coprophagy, 535–36, 541

Coprourodeal fold, 582
Copulation, 576
Copulatory organs,
    576–82, 583f
Coracobrachialis, 398, 399t
Coracoid, 338–39, 341f
Cords, 180
Cormorants, appendicular
    skeleton of, 349f
Cornea, 159, 681
Corneal muscle, 683
Cornified layer of
    epidermis, 215
Cornua, 253
Corona radiata, 164, 167f
Coronary ligament, 196
Coronary vessels, 473, 642
Coronoid process, 253
Coronoids, 250
Corpora cavernosa, 578
Corpus, 253, 654
Corpus albicans, 613
Corpus callosum, 662
Corpuscle of Ruffini, 673
Corpus luteum, 610, 613
Corpus spongiosum, 582
Cortex, 163, 563, 600, 601f
Cortical bone, 182
Cortical granules, 163
Cortical region, 661
Corticospinal tracts, 650
Corticosteroid hormones, 598
Corticotropin. *See*
    Adrenocorticotropic
    hormone (ACTH)
Corticotropin-releasing
    hormone (CRH),
    616, 620f
Corvus, testis and epididymis
    in, 578f
Cosmine, 102
Cosmoid scale, 102, 214f, 218
Costa, O. G., 65b
Costal (diaphragm), 435
Costal rib, 299
Cotylosauria, 109f, 110
Countercurrent exchange,
    156–58
Countercurrent gas
    transfer, 443f
Courtenay-Latimer, M., 45b
Cousteau, Jacques, 447b
Covering, epithelium, 179–80
Covering cells, 709
Coyotes
    cursorial locomotion of, 358t
    teeth of, 514
Cranford, Ted, 431b
Cranial autonomic system, 644
Cranial kinesis, 253–55, 264f,
    268, 271f, 277f

Cranial musculature
  comparative anatomy,
    405–410
  embryonic development,
    392f, 393–94, 394f
Cranial nerves, 189, 190f,
    190t, 629, 630–37,
    633–37, 638f, 639f,
    639t, 640f
Cranial neural crest, 245f,
    288–89, 291f
Cranial region, 16, 18f
Cranial shield, 256
Cranial skeleton, 241
Craniates, 84, 87f
Craniosacral outflow, 642
Craniostylic jaw
    suspension, 248
Cranium, 84, 242
Creodonts, 519b
Cretaceous birds, 118
Cretinism, 596
CRH. See Corticotropin-
    releasing hormone
    (CRH)
Cribriform plate, 277, 675
Cristae, 696, 697f
Crocodiles, 45b
Crocodilians, 115, 130f
  adrenal tissues, 600f
  alimentary canal, 529
  cloaca, 584f
  crurotarsal joints, 347f
  gastralia, 299
  gizzards, 536
  heart, 483, 487, 488f,
    489f, 492
  integument, 221
  integument of, 220
  lymphatic system, 499f
  pancreas, 604t
  pectoral girdle, 339
  penis, 578, 579, 581f
  pituitary gland, 607
  respiratory system, 434
  skulls, 265, 267f, 272
  sniffing by, 680f
  spinal cords, 650f
  venous vessels, 472f
  ventilation, 437f
  vestibular apparatus, 698f
Crocodylus palustris, penis, 581f
Cromwell, Oliver, 560b
Crop, 525
Crop milk, 609
Crosscurrent exchange, 156–58
Crosscurrent gas exchange, 443f
Crossopterygians, 102
Cross-sectional area, 383
Crotalus atrox, hemipenes, 581f
Crown, 507, 508

Crown group, 28
Crown height, 514f
Crows, cranial kinesis in, 277f
Cruising (swimming), 400b
Crura, 197
Crural (diaphragm), 435
Crurotarsal joint, 346
Cryptobranchus alleganiensis,
    cutaneous respiration
    of, 419f
Cryptodira, 112
Cryptorchidism, 576
CSF. See Cerebrospinal fluid
    (CSF)
Ctenacanthus, axial skeleton
    phylogeny, 306f
Ctenoid scale, 214f, 218
Ctenosaura, adrenal tissues
    of, 600f
Cubes, 129, 132f
Cuboidal cells, 179
Culmann, Karl, 154
Cupula, 694
Cursorial locomotion, 349,
    352f, 353–58, 359f,
    402–3
Cursorial theory, 368f, 369
Cusps, 507
Cutaneous artery, 464
Cutaneous respiration, 219,
    417, 419f, 420f
Cutaneous respiratory organs,
    417, 419f, 420f
Cuticle, 229
Cuticular plate, 693
Cuvier, Georges, 11–12, 11f,
    14, 75, 290
Cyanosis, 493
Cycle (in gaits), 350
Cycloid scale, 214f, 218
Cyclostomes, 86, 93, 596, 598
  cranial nerves, 631
  genital ducts (male), 572–73,
    574f
  kidneys, 550f
  nasal passages, 676
  ovaries, 568
  pancreas, 534–35
  pancreatic islets, 602, 603f
  skull phylogeny, 255
  thyroid glands, 595f
Cynodonts, 120–21, 278f,
    291, 292f
Cynognathus, pelvic girdle
    of, 343f
Cyprinids, 101
Cytotrophoblast, 192

Dale, Henry, 3b
Dalton, John, 686b
Daphnia, 130f

Darwin, Charles, 4–5, 4f,
    9–10
  Archaeopteryx and, 119b
  critics and controversy,
    10–11
  evolutionary change, 4–5
  morphology and, 47
  natural selection, 7
Darwinism, 10. See also
    Evolution
Darwin's rhea, alimentary canal
    of, 526f
Dasyatis, intromittent organs
    in, 578f
Da Vinci, Leonardo, 579b
D cells, 604
Dead space, 416
De Beer, Gavin, 725
Deciduous dentition, 507, 508f
Deer, 125, 143–44, 357f
Deer mice, 46
Deer mouse, stomach of, 528f
Deglutition, 518
Dehydration, 555b, 556f
Deinonychus, ventilation in,
    434–35
Delamination, 168
Delayed development, 589
Delayed fertilization, 589
Delayed implantation, 192,
    589, 590f
Deltoideus, 398, 399t
Dendograms, 28
Dendrites, 626, 672
Dendrograms, 21–22, 23f, 25f
Density, 140, 146, 147
Dental formula, 514
Dentary bone, 248, 250
Dentin, 217, 507
Depression, 390f
Depressor labii
    mandibularis, 408
Depressor mandibulae, 407
Depressors, 390
Depth perception, 158, 688–89
Depulsation, 476
Derived condition, 24
Derived groups, 723f
Derived quantities, 139–40
Dermal armor, 236–37
Dermal bones, 184, 213,
    248–50
Dermal bone series, 249–50
Dermal fin rays, 326
Dermal melanophore, 238
Dermal papillae, 229, 508
Dermal scale, 216
Dermatocranium, 242, 243f,
    248–50, 267f
Dermatome, 168f, 169, 213,
    392, 630

Dermis, 212, 213–15, 215f,
    229, 239f
Dermomyotome, 213
Dermoptera, 125
Descending nerve tracts, 650,
    651f, 653t
Descriptive morphology,
    717–18
Design, 46f. See also
    Biological design;
    Form; Function;
    Morphology;
    Structure
  analysis of, 41–42, 714–15,
    716f
  body design, 15–19
  cardiovascular system,
    490–95, 496f
  convergent, 722f
  form and function of, 3–4,
    42–46, 44f
  grand, 4
  optimum, 721
  skull, 284–88
  vertebrae, 317–18
  vertebral column, 317f
Desmognathus
    quadramaculatus, 580f
Deuterostomes, 48–50, 50f, 50t,
    79f. See also Chordates
Diabetes insipidus, 608
Diabetes mellitus, 602
Diadectes, 110
Diadectomorphs, 110
Diagonal-sequence gait,
    350, 351f
Dial, K., 368f, 369
Diaphragm, 196, 435, 436
Diaphragmatic muscles,
    197, 434
Diaphysis, 182f, 183
Diapophysis, 297, 298
Diapsids, 111, 112, 267f
  skull, 110
Diarthrognathus, 290–91
Diarthrosis, 187
Diastolic pressure, 453
Diatryma, 115, 118f
Didelphis, 341f
Didelphis
  extraembryonic membranes
    in, 195f
  pelvic girdle of, 343f
  skulls of, 280f
  supracoracoideus muscles
    of, 406f
Diencephalon, 604, 651, 658
Differential scattering, 238
Differentiation, 210
Diffusion, 156, 413–14
Digastric muscle, 407

Digesta, 503
Digestion
    chemical breakdown of,
        536–40, 539–40
    ectothermic, 487
    endocrine control of, 612*f*
    mechanical breakdown,
        536, 537*f*
    toxins, 542
Digestive enzymes, 536, 537
Digestive glands, 504, 531–35
Digestive system, 503–4
    absorption, 535
    alimentary canal, 520–31
    buccal cavity, 504–520
    chemical breakdown of food,
        536–40, 539–40
    embryonic development
        of, 505*f*
    feces, 535–36
    fermenters, gastric versus
        intestinal, 540–42
    function and evolution of,
        535–42
    glands of digestion, 531–35
    mechanical breakdown of
        food, 536, 537*f*
    overview, 543–44
    pharynx, 517–20
    toxin digestion, 542
Digestive tract, 503
Digging. *See* Fossorial
        locomotion
Digital cushion, 233
Digitigrade posture, 355
Digits, 339, 356*f*
Dikinetic, 268
Dilators, 390
Dimetrodon, 122*f*, 341*f*, 343*f*
Dinesen, Isak, 560*b*
Dinocephalia, skulls of, 278*f*
Dinornis, 115
Dinosaurs, 114–15, 123*b*. See
        also Paleontology
    air passage of, 284, 285*f*
    body support, 317*f*
    feathered, 227*b*
    heresies and hearsay, 117*b*
    locomotion, 352*f*
    mesotarsal joints, 347*f*
    modularity, 725, 728*f*
    pelvic girdle of, 343*f*
    ventilation and locomotion,
        434–35
Dinosaur tracks, 33*f*
Dioecious protochordates, 56
Diphycercal, 102
Diphycercal tail, 305
Diphyodont dentition, 507
Diphyodonty, 232, 282
Diploceraspis, 108*f*

Diplodocus, 115, 130*f*, 144, 145*f*
Diploid, 164
Diplospondyly, 295
Dipnoans, 102–103, 469,
        606, 607*t*
Dipterus, 102*f*
Direct development, 198
Discoidal cleavage, 164, 166*f*
Diseases, bones and, 153
Disraeli, Benjamin, 10, 161
Dissociation, 725
Distal convoluted tubule, 552
Distal region, 16
Distance, of shortening, 385
Distance advantage, 143
Distaplia, 69*f*
Divergence, 628, 645, 725
Diversity
    birds, 118
    terrestrial vertebrates,
        124*f*, 126*f*
Diving, cardiac shunts and, 487
Diving birds, cardiovascular
        system design in,
        491–92
DNA, 14
Dogfish
    brain of, 654*f*
    jaws and mouths of, 510*b*
    thyroid glands, 595*f*
Dogs
    cranial nerve
        development, 637*f*
    heat transfer in, 495, 496*f*
    jumping mechanics, 718*f*
    salivary glands of, 533*f*
    skull of, 282*f*
    teeth in, 508*f*
    ventilation in, 439*f*
    vomeronasal organs of, 681*f*
Dolphins, 447*b*, 448*b*, 495, 526*f*
Dorsal, 18*f*
Dorsal aorta, 64, 64*f*, 457, 459
Dorsal blood vessel, 59
Dorsal branchial muscles, 406
Dorsal hollow nerve chord, 51,
        54, 54*f*
Dorsal horns, 649
Dorsalis trunci, 395
Dorsal lamina, 69
Dorsal lip of the blastopore, 169
Dorsal mesentery, 168*f*
Dorsal muscles, 399*t*, 400
Dorsal pallium, 662, 663
Dorsal pancreatic diverticulum,
        533–34
Dorsal ramus, 394
Dorsal region, 16, 319
Dorsal ribs, 298
Dorsal roots, 629, 641
Dorsal sac, 690

Dorsal thalamus, 658
Dorsal ventricular ridge (DVR),
        663, 664, 666
Dorso-bronchi, 438
Dorsolateral placodes, 190,
        190*f*, 190*t*, 636, 637*f*
Dorsomedial musculature, 397
Dorsoventral inversion,
        prechordates, 78–80
Dorsoventral rotation, 297*f*
Double circulation, 455–56
Double jaw articulation, 292*f*
Down feathers, 223*f*, 224
Down-regulation, 542
Downstroke, 360, 362*f*
Downstroke-upstroke
        transition, 360
Draco, gliding and parachuting
        by, 360*f*
Drag, 145–48, 361, 362,
        364–65, 365*f*, 366*f*
Drosophila melanogaster, Hox
        genes in, 205*f*
Dual pump, 421–22
Duck-billed dinosaurs, air
        passage of, 284, 285*f*
Ducks
    appendicular skeleton
        of, 349*f*
    penis, 579, 582
Ductus arteriosus, 462, 492
Ductus deferens, 563, 566*f*
Ductus venosus, 492
Dullemeijer, P., 155*f*
Dumbacher, John P., 230*f*
Duodenal glands, 525
Duodenum, 523, 593*t*
Duplex uterus, 585
Duplication, 725
Dura mater, 646
Duty factor, 350
Duvernoy's gland, 532
DVR. *See* Dorsal ventricular
        ridge (DVR)
Dwarfism, 609
Dynamic fluids, 145–48, 147*f*
Dynamics, 142

Eagles, 118, 361
Ears
    anatomy of, 698–700
    functions of, 701–9
    middle ear bones, 277, 281,
        283*f*, 285*f*
Eastlick, Herbert, 3*b*
Eccrine glands, 231
Ecdysis, 221
Echidna
    adrenal tissues of, 600*f*
    alimentary canal of, 526*f*
    cerebral cortex, 661

pectoral girdle of, 341*f*
    stomach of, 528*f*
Echinoderms, 61, 61*f*, 76–77
Ecological analysis, 722
Ecomorphology, 46
Ectoderm, 49, 163, 168*f*, 178*f*
Ectodermal placodes, 68,
        189, 190*f*
Ectomesenchyme, 169
Ectopterygoid bone, 250
Ectothermic digestion, 487
Ectotherms, 117*b*, 123*b*, 225
Edaphosaurus, 120, 122*f*
Edema, 496
Eels
    accessory air-breathing
        organs of, 418
    air-breathing by, 420*f*
    cutaneous respiration, 419*f*
    electric organs in, 373*f*
    locomotion of, 311*f*
    mucus of, 237
Effectors, 625
Efferent artery, 461
Efferent ductules, 563
Efferent neurons, 628, 629*f*
Efferent spinal nerves,
        629*f*, 631*f*
Efferent spiracular artery, 461
Egg fossils, 31*f*
Eggs, 163, 204, 545, 563.
        See also Cleavage
Egyptian cobra, 42*f*
Einstein, Albert, 141, 560*b*
Ejaculation, 572
Elands, cooling in, 496*f*
Elaphe, swallowing by, 288*f*
Elasmobranchs, 96–97
    autonomic nervous
        system, 644*f*
    fins and squalene, 430
    genital ducts (male), 574*f*
    ovaries, 568–69
    pallium, 663
    pancreas, 535, 604*t*
    pineal complex, 690
    pituitary gland, 606
    respiratory system, 426–27
    salt balance, 561*f*
    ventral lobe, 606
    vertebrae, 301, 302, 306*f*
    water balance, 560
Elastic cartilage, 181
Elastic component, 380
Elastic fibers, 181
Electric catfish, 373*f*
Electric eels, 373*f*, 418, 420*f*
Electric fishes, 709, 711, 712*f*
    cerebellum of, 712
Electric organs, 373*f*
Electric rays, 97

Electromagnetic spectrum, 679–81, 683f

Electromyogram (EMG), 43–44, 44f

Electrophorus
accessory air-breathing organs of, 418
air-breathing by, 420f
electric organs in, 373f

Electroplaques, 709

Electroreceptors, 709–12, 713

Elephant birds, 118

Elephant nose, electric organs in, 373f

Elephants, 125, 130f
facial muscles, 411f
teeth, 508, 511f, 512, 515f

Elephant trunk snake, cutaneous respiration of, 419f

Elevation, 390f

Elk antlers, 235f

Emarginations, 265, 267, 267f, 269f

Embolomerous vertebra, 296

Embryonic area, 163

Embryonic development, 162, 163
adaptive changes, 729–30
adrenal glands, 599f
axial skeleton, 301–4
breathing, 418, 420, 421f
cardiovascular function, 492
cardiovascular system, 456–57, 458f, 459f
central nervous system, 646, 649, 653f
chondrocranium, 242, 244, 244f
cleavage, 164–67
cranial nerves, 637f
digestive system, 505f
evolution and, 201f
eyes, 682, 684f
fertilization, 163–64
gastrulation and neurulation, 167–75, 176f, 177f
gut (mammalian), 522f
human skulls, 280f
integument, 213, 214f
life history and, 201–9
liver, 534f
metamorphosis and, 197–98
middle ear, 699f
muscles, 391–93, 394f
nasal passages, 675–76, 678f
olfactory system, 678f
ontogeny and phylogeny, 201–9
overview of, 193–95
pharyngeal pouches, 596f

pituitary gland, 605f
principles of, 202f
protostomes and deuterostomes, 51, 51f
reproductive system, 563–67
spinal nerves, 631f
splanchnocranium, 244–48
telencephalon, 662f
tetrapod limbs, 331, 332f
thyroid gland, 595f
tranformation grids in, 136f
urinary system, 546–53
venous vessels, 470f

Embryonic diapause, 589, 616

Embryonic disc, 165

Embryonic implantation. *See* Implantation

Embryonic induction, 240

Embryonic shield, 170

Emergency life history state, 601

Emotions, 666, 667

Empedocles, 4

Emulsify, 533

Emus, 118, 274

Emys (European pond turtle), 269f

Enamel, 217, 507

Enamel organ, 508

Encapsulated nerve endings, 673

Encapsulated sensory receptors, 673

End-bulb of Krause, 673

Endocardial tubes, 456

Endocardium, 456

Endochondral bones, 182, 183–84, 185f, 244t

Endocrine coordination
environment and endocrine system, 621–22
hormonal control, 620–21
metamorphosis (frogs), 616, 620
reproduction (mammalian), 613–16

Endocrine glands, 180, 592
adrenal gland, 598–602
gonads, 610
pancreatic islets, 602–4
parathyroid gland, 596, 597
pineal gland, 610–11
pituitary gland, 604–610
secondary organs, 611–13
thyroid gland, 592–96
ultimobranchial body, 596–97

Endocrine portion of pancreas, 602

Endocrine system, 592
adrenal gland, 598–602
environment and, 621–22
evolution, 622–24

gonads, 610
hormonal control, 620–21
metamorphosis (frogs), 616, 620
overview, 624
pancreatic islets, 602–4
parathyroid gland, 596, 597
pineal gland, 610–11
pituitary gland, 604–610
reproduction (mammalian), 613–16
secondary endocrine organs, 611–13
thyroid gland, 592–96
ultimobranchial body, 596–97

Endocytosis, 179

Endoderm, 49, 163, 168f, 178f

Endolymph, 695–96

Endomesoderm, 169, 170

Endomysium, 376

Endoskeleton, 89, 98, 99, 241

Endostylar artery, 63

Endostyle, 51, 53, 69

Endothelium, 179

Endotherms, 117b, 123b, 225

Energetics, gait selection and, 358f

English system of measurement, 139

Engrailed-1 gene expression, 330

Ensatina eschscholtzii, cutaneous respiration of, 419f

Enteric autonomic system, 644

Enterocoelom, 49

Enterocoely, 169

Enterocrinin, 612

Enterogastrone, 611–12

Enteropneusta, 57–60

Entoplastron, 338

Environment. *See also* Fluid environment; Terrestrial environment
bones and, 153
cardiovascular system and, 490–95, 496f
endocrine system and, 621–22

Enzymes, digestive, 536, 537

Eocaptorhinus, 112f

Eocene carnivore, skull of, 290b

Eocene cetacean, skull of, 290b

Eogyrinus, 104, 336

EOP. *See* Erythropoietin (EOP)

Eosuchia, 113

Epaxial musculature, 393

Epaxial myotomes, 410

Ependymal cells, 68, 625, 626f

Epiblast, 165–66, 166

Epiboly, 168

gonads, 610
hormonal control, 620–21
metamorphosis (frogs), 616, 620
overview, 624
pancreatic islets, 602–4
parathyroid gland, 596, 597
pineal gland, 610–11
pituitary gland, 604–610
reproduction (mammalian), 613–16
secondary endocrine organs, 611–13
thyroid gland, 592–96
ultimobranchial body, 596–97

Epibranchial element, 245

Epibranchial groove, 58

Epibranchial placodes, 190, 190f, 190t

Epidermal cells, 216

Epidermal melanophore, 238

Epidermal placodes, 84

Epidermal scale, 216

Epidermis, 212, 213, 215–16, 228–29, 232f, 239f

Epididymis, 563, 578f

Epigenomics, 205–9

Epiglottis, 519

Epihyoidean, 407

Epimeres, 168f, 169, 213

Epimyocardium, 456

Epimysium, 376

Epinephrine, 598, 642

Epiphyseal plate, 183

Epiphysis, 182f, 183, 610. *See also* Pineal gland

Epiphysis cerebri, 690

Epipterygoid bone, 247t, 253, 265

Episternal cartilage, 299

Epithalamus, 658, 690

Epithelium, 177–80, 179–80

Epoöphoron, 565

Eptesicus fucus, cutaneous respiration of, 419f

Equus, digits of, 356f

Erpetoichthys, 100

Erupt, 508

Eryops
appendicular skeleton, 336, 337f, 339f
axial skeleton of, 312
pectoral girdle of, 341f
pelvic girdle of, 343f

Erythrocytes, 452

Erythrophore, 238

Erythropoietin (EOP), 613

Esophagus, 504, 506f, 521, 524f

Essay on the Principle of Population, An (Malthus), 8

Estemmenosuchus, 231

Estivation, 101b

Estrogens, 598

Ethmoid plate, 242

Ethmoid region, 244t, 257

Ethmoid unit, 262

Ethmoturbinate bone, 277

Euagnathans, phylogeny of, 91f

Euarchontoglires, 124f

Euautostylic jaw suspension, 248

Eunuchs, 618–19b

Eupodotis, aortic arches of, 465f

Eureptilia, 109, 111, 112

Eureptilian radiation, 112

European eel, cutaneous respiration of, 419f
European starling, wing-beat cycle of, 362f
Euryapsids, 109f, 112, 267f
Euryhaline animals, 558b, 560
Eustachian tube, 698
Eusthenopteron, 103, 335
 appendicular skeleton, 337f
 axial skeleton of, 308, 309
 limb development, 332f
 skull of, 243f
Eutherian mammals
 appendicular skeleton phylogeny, 331, 333
 copulatory organs, 576
Eutherian mammals (placental mammals), 122, 124, 124f, 192
 cerebral hemispheres, 664f
 cleavage, 166–67
 cloaca, 582
 commissure, 661–62
 embryonic development, 194
 extraembryonic membranes, 191t
 fetal circulation, 492, 493f, 494f
 gastrulation and neurulation, 173–75, 177f
 genital ducts (male), 574–75
 glands, 232
 molar patterns, 516f
 reproductive organs, 587f
 skull phylogeny, 275, 277, 280f, 282f
 urinary bladder, 588f
 urogenital sinus, 586f
Eutherian placenta, 192–93
Evans, H., 706f
Evolution
 agnathan, 93–94
 annelid/arthropod to chordate, 75–76
 aortic arches, 467f
 appendicular skeleton, 346–48
 Archaeopteryx and, 119b
 bone, 186b
 brains, 654f
 cerebral hemispheres, 664f
 circulatory system, 501f
 digestive system, 535–42
 embryology and, 201f
 endocrine system, 622–24
 facial muscles, 411f
 feathers, 225
 flight feathers, 715, 717f
 hair, 231
 hearing, 706–9, 709

historical development of, 3–14
 historical predecessors, 4–14
 Hox genes and, 204–5, 206f
 jaws and mouths, 509–10b
 kidneys, 562
 mode and tempo of, 727–30
 morphology and, 2
 mosaic, 725, 726f, 727f
 mouth, 509–10b
 patterns of, 26f
 pectoral girdle, 340f
 pelvic girdle, 342f
 peripheral nervous system, 637–38
 quantum, 24
 remodeling through, 21
 reproductive system, 588–89
 respiratory organs, 444–50
 steps in, 24f
 supracoracoideus muscles, 406f
 tetrapod vertebrae, 310f
 urinary bladder, 588f
 ventilatory mechanisms, 449f
 vertebrate, 84
 water to land transition, 347–348, 348f
 whales, 289b, 290b
Evolutionary analysis, 722–27, 728f
Evolutionary biology, 2
Evolutionary morphology, 2, 19–21
Evolution of Man, The (Haeckel), 21
Excavations. See Paleontology
Exchange, 156–58
 systems of, 157f
Excretion, 553–54, 554, 554f
Exercise, of muscles, 382b
Exhalation, 416
Exhalation phase, 422, 423f
Exoccipitals, 250
Exocoelomic membrane, 174
Exocrine glands, 180
Exocrine portion of pancreas, 602
Exocytosis, 179
Exoskeleton, 92, 98, 241
Exposures, 33f, 34
Expulsion, 422
Extant groups, 29f, 30f
Extension, 390f
Extensors, 390
External auditory meatus, 698
External carotids, 457
External ear, 698, 700f
External elastic membranes (EEM), 452f

External fertilization, 588–89, 590f
External gills, 416
External jugular veins, 471f, 472
External naris, 262, 676
External orifice, 698
External respiration, 415
Exteroceptors, 645, 672
Extinct groups, 29f, 30f
Extinctions, 41
Extinct mammals, 123
Extracellular matrix, 177, 180
Extracolumella, 281, 700
Extraembryonic area, 163
Extraembryonic membranes, 190–93
Extrafusal muscle cell, 673
Extrapolation, 2
Extrascapulars, 260
Extrinsic eye muscles, 393–94, 394f
Eye muscles, extrinsic, 393–94, 394f
Eye photoreceptors, 681–91
Eyes. *See also* Optics
 embryonic development, 682, 684f
 form and function, 684–90, 692f
 phylogeny, 682–83, 685f
 structure of, 681–82, 683f

Facet, 298
Facial expression, muscles for, 407–8
Facial nerve, 634, 638f, 639t, 678
Facial pits, 693
Facial series, 249–50
Falcons, 118
False ribs, 298
Fangs, 514
Farmer, Colleen, 487
Farmer, D. S., 644f, 645f
Fascia, 376
Fascicle, 376
Fasciculus cuneatus, 650
Fasciculus gracilis, 650
Fasting, 542
Fast twitch fibers, 378–79
Fatigue fracture, 149–50
Faveoli, 430
Faviform, 440
Feather color, 238
Feather follicles, 224, 224f
Feathers, 214f, 223–28, 229f
 aerial locomotion and, 359, 361f
 birds, 118
 dinosaurs and, 227b

evolution, 225, 228f, 715, 717f
 evolutionary remodeling, 728–29
 flight, 715, 717f
 flight funciton of asymmetrical feather vane, 228f
 functional coupling of, 720
 growth, 224f
 morphology of, 223f
 origin of, 369
 pennaceous, 223
 plumulaceous, 224
 regeneration, 224–25, 226f
Feces, 535–36
Feeding. *See also* Digestion; Digestive system; Lingual feeding; Suction feeding; Suspension feeding
 air, 287–88
 fasting and, 542
 sharks, 259f
 vertebrate origins and, 84, 85–86
 water, 285–87
 whales, 289b, 290b
Felids, 519b
Felis, pelvic girdle of, 343f
Female genital systems, embryonic development of, 566f
Female reproduction, in mammals, 613–16, 617f
Female reproductive system, 564f, 567–72
Femoral glands, 221
Fenestrae, 250
Fenestra ovalis, 700
Fenestra rotunda, 700
Fermentation, 537
 gastric, 537, 539–41
 intestinal, 540–541
 size and, 541–42
 stomach and, 539f
Fermenters, gastric versus intestinal, 531f, 537, 540–42
Fertility, 588
Fertilization, 163–64, 613
 delayed, 589
 external and internal, 588–89, 590f
Fetal circulation, in placental mammals, 492, 493f, 494f
Fetal extraembryonic membranes, 195f
Fetal zone, 601

Fibers. *See* Muscle fibers
Fibroblast, 181
Fibrocartilage, 181
"Fight-or-flee" readiness, 601, 642
Filiform papillae, 516
Filoplumes, 223*f*, 225
Filters, 71
Filtration kidneys, 557
Fimbria, 566
Fin-fold theory, 328–30
Fin rays, 326
Fins, 326. *See also* Appendicular skeleton
    caudal, 305, 307–8
    paired, 326, 327–31, 332*f*
    as stabilizers, 327*f*
Fin whales, 289*b*
First law of inertia, 141
Fish
    homocercal and heterocercal, 1–2, 2*f*
    male urogenital system, 574*f*
    water-breathing (dual pump), 422*f*
Fisher, J., 24*f*
Fishes, 86
    acid-base regulation, 445–46
    adrenal tissues, 599*f*
    air-breathing, 420*f*, 422–23, 428*f*
    air-breathing organs, 417–18, 446–47, 491
    alimentary canal, 527–28
    anal regions, 583*f*
    aortic arches, 461–63, 467*f*
    appendicular musculature, 392–93, 397
    appendicular skeleton phylogeny, 331, 334–36, 337*f*
    autonomic nervous system, 644
    axial musculature, 394–95
    axial skeleton, 301–2, 304–9
    body cavities, 196*f*
    body support, 317*f*
    brains, 654*f*
    branchial arches, 247*t*
    branchiomeric musculature, 409*f*
    buccal pump, 422–23
    calcium regulation, 598
    cerebral hemispheres, 664*f*
    cerebrospinal fluid and meninges, 648*f*
    chemoreceptors, 674, 675*f*
    chondrocranium, 244*t*
    cleavage, 165, 166*f*
    cloaca, 582, 583*f*
    copulatory organs, 576

cranial nerves, 631, 633*f*
cutaneous respiration, 416–17, 419*f*, 420*f*
dual pumps, 421–22
ear functions, 701, 702*f*
electric, 709, 711, 712*f*
electroreceptors, 709–12
embryonic development, 193
eyes, 683, 684
fluid environment and, 314–15
fluid mechanics, 718, 719*f*
forebrain, 651
gaits, 350
gas bladders, 416–17
gas transfer, 443–44
gastrulation and neurulation, 169–70, 171*f*, 172*f*
genital ducts (female), 567–69, 570*f*
genital ducts (male), 572–76, 574*f*, 578*f*
gliding and parachuting, 360*f*
hearts, 473, 475–78
hyoid arch, 407
hypobranchial nerve, 636
integument, 214*f*, 216–18
jaws and mouths of, 509–10*b*
kidneys, 550–51, 553
land, movement on, 448
lateral line canals, 695
locomotion, 311*f*
midbrain, 652–54, 658
mucus, 237
muscle fibers, 388, 389*f*
nasal passages, 676
olfactory bulb and tract, 676*f*
olfactory organs, 680*f*
pancreas, 535, 604*t*
pectoral girdle, 336, 340*f*
photophores, 710*b*
pituitary gland, 606–7
pulmonary system, 469
ram ventilation, 421
respiratory system, 427–30
ribs, 298*f*
salt balance, 561
scales, 236
single circulation, 455
skull phylogeny, 256–63, 264*f*
spinal reflexes, 641
stomach, 528*f*
swimming specializations, 400*b*, 401*b*
telencephalon, 661
thyroid glands, 595*f*
urinary bladder, 586, 588*f*
venous vessels, 469, 471*f*
vertebrae, 322*f*
visual accommodation, 687*f*

water balance, 556–57, 558
water-breathing, 444
Fissiped, 125
Fissure, 661
Fixators, 390
Flagella, 131*f*
Flailing, 369
Flamingo, 118
Flatfishes, 97
Fleas, 130*f*
Fleming, Alexander, 3*b*
Flexion, 390*f*
Flexor hallucis longus, 399*t*, 401
Flexors, 390
Flies, 130*f*
Flight, 115, 118
    aerial locomotion analysis of, 359
    appendicular musculature and, 404–5
    functional analysis of, 718–19
    origin of, 367–69, 368*f*
    types of, 361, 363*f*, 364*f*
Flight feathers, 223, 223*f*, 225, 229*f*, 715, 717*f*
Floating ribs, 298
Flocculus, 654
Flow separation, 145
Fluid environment
    axial skeletons in, 314–15
    breathing in, 444
    dehydration in, 555*b*
    feeding in, 285–87
    locomotion in, 311*f*, 312
    swimming, 349–50
    vision in, 685, 687*f*
    water balance, 556*f*
Fluid mechanics, 718, 719*f*
Fluids, life in, 145–48, 146*f*
Flying fish, gliding and parachuting by, 360*f*
Flying frogs, gliding and para-chuting by, 360*f*, 367*f*
Flying geckos, gliding and parachuting by, 360*f*
Flying lemurs, 125
Flying squirrels, 125, 360*f*, 367*f*
Follicle, 180, 563, 593
Follicle cells, 164, 563, 567
Follicle-stimulating hormone (FSH), 588, 609, 613, 614, 614*b*, 616
Food digestion. *See* Digestion
Food passages, 520*f*. *See also* Buccal cavity
Foot, 326
Footfalls, 350. *See also* Gait
Footplate, 700
Foot postures, 355
Foramen magnum, 250

Foramen of Panizza, 487
Foramen ovale, 492
Force
    as derived quantity, 140
    direction of, 148, 149*f*
    free bodies and, 142
    graded, 380–82, 383*f*
    vectors and, 141*f*
Force advantage, 143
Force generation, whole muscle, 379–80
Force laws, basic, 141–42, 142*f*
Force phase, 422
Forebrain, 646, 647*f*, 651–54, 658–60
Foregut, 504, 531*f*
Foregut fermentation, 537–540
Foregut fermenters. *See* Gastric (foregut) fermenters
Forelimbs, 344*f*. *See also* Appendicular skeleton
    flight-related modifications in, 360*f*
    pectoral girdle and, 397–98
    strength versus speed in, 386*f*
Forest chameleon, alimentary canal of, 526*f*
Form, 15. *See also* Morphology
    appendicular skeleton, 348–71
    axial skeleton, 314–21, 324*f*
    brain, 654–66
    electroreceptors, 709–12
    eyes, 684–90, 692*f*
    lymphatic system, 498–99
    nasal passages, 677, 680*f*
    parathyroid, 598
    pineal complex, 691
    respiratory system, 438, 440–44, 443–44, 445*f*
    vertebrate design, 3–4, 15
Formed elements, 452
Form-function models, 719
Fornix, 666
Fossil forgeries, 119*b*
Fossilization, 30–32, 35, 35*f*
Fossils, 28
    agnathans, 89–93
    Archaeopteryx, 119*b*
    chronology of, 38*f*
    dating, 37–41
    fossilization and, 30–32, 35, 35*f*
    living, 45*b*
    recovery and restoration, 32–35
    saber-tooth mammals, 519*b*
Fossorial locomotion, 349, 369–70, 370*f*
    skeletal adaptations for, 371*f*

Four-stroke buccal pump, 422, 423*f*
Fovea, 682
Foxes, cursorial locomotion of, 358*t*
Fracture propagation, 150*f*
Fractures. *See* Bone fractures; Bone remodeling and repair
Frank-Starling reflex, 474, 475
Frazzetta, T. H., 259*f,* 271*f,* 513*f,* 727
Free bodies, 142
Free-body diagram, 142, 142*f*
Free nerve endings. *See* Free sensory receptors
Free sensory receptors, 672–73
Free surface, 179
Frequencies, discrimination of, 705–6, 709*f*
Friction drag, 145, 146
Frigate birds, wing design of, 363*f*
Frog (hoof), 233
Frogs, 106–7, 130*f*
   aerial locomotion, 358
   aortic arches, 462*f,* 464
   appendicular musculature, 403*f*
   axial musculature, 395–96
   brain, 655*f,* 659*f*
   copulatory organs in, 576–77
   cutaneous respiration, 419*f,* 420*f*
   dehydration, 555*b*
   diversity and patterns in, 724*f*
   ears, 703–4
   fertilization, 588
   genital ducts (female), 571*f*
   gliding and parachuting by, 360*f,* 367*f*
   hearts, 479
   integument, 219–20
   life cycle, 162*f*
   melanophores, 610*f*
   metamorphosis, 616, 620
   muscle homology, 391*f*
   pectoral girdle, 341*f*
   pelvic girdle, 343*f*
   pineal complex, 690, 692*f*
   poison dart, 218*b*
   respiratory system, 432–33
   skulls, 264, 266*f*
   spinal cords, 650*f*
   thyroid glands, 595*f*
   vestibular apparatus, 698*f*
   vocalization, 435*b*
Frontal bones, 250
Frontal cilia, 59
Frontal plane, 16

Frontoparietal bones, 264
Fruit fly, *Hox* genes in, 205*f*
FSH. *See* Follicle-stimulating hormone (FSH)
Fulcrum, 142
Function, 15, 19–20. *See also* Morphology
   adrenal gland, 601–2
   appendicular skeleton, 348–71
   autonomic nervous system, 642
   axial skeleton, 314–21, 324*f*
   biological role and, 19–20
   bird feathers, 225
   brain, 654–66
   cardiovascular, 492
   centra, 297*f*
   design, 3–4, 15, 42–46
   digestive system, 535–42
   electroreceptors, 709–12
   eyes, 684–90, 692*f*
   hormonal control, 620, 621*f*
   kidneys, 553–62
   lymphatic system, 498–99
   multiple, 720–21
   muscles, 377–91
   nasal passages, 677, 680*f*
   pancreatic islets, 602–4
   parathyroid gland, 598
   peripheral nervous system, 638–46
   phrenology and, 665*b*
   pineal complex, 691
   pituitary gland, 607–10
   reproductive system, 588–89, 590*f*
   respiratory system, 438, 440–44, 443–44, 445*f*
   skull, 284–88
   thyroid gland, 594–96
Functional analysis, 718–22
Functional compromise, 719–20, 721*f*
Functional coupling, 719–20
Functional matrix, 720
Functional morphology, 2
Functional similarity, of muscles, 390
Functional unit, 719
Fundic glands, 521
Fundulus, copulatory organs in, 576
Fundus, 521
Funnel, 566
Fur, 230
Furcula, 338
Fusiform, 93
Fusion, 411

Gadow, Hans, 301, 302, 582
Gage, Fred, 627*b*
Gagnan, Emile, 447*b*
Gait, 350, 351*f*
   cursorial locomotion and, 356–57, 358*f*
   humans, 387*b*
Galaeaspids, axial skeleton of, 304
Galago, extraembryonic membranes in, 195*f*
Galápagos Islands, 9
Gallbladder, 533
Gallop, 357, 358*f*
Game birds, muscle fiber colors in, 377–78
Gametes, 163, 563–65
Gamma motor neurons, 673
Ganglia, 68, 69, 626, 629–30
Ganglion cells, 682
Ganoid scale, 100, 214*f,* 218
Gap junctions, 68
Gars, 101
Garstang, W., 76–78, 81
Gaskell, W. H., 75
Gaster, 376
Gastralia, 220, 295*f,* 299
Gas transfer
   patterns of, 438, 440, 443, 443*f*
   rates of, 443–44, 445*f*
Gastric (foregut) fermentation, 537, 539–40, 544*f*
Gastric (foregut) fermenters, 531*f,* 540–42, 544*f*
Gastric glands, 521
Gastric juice, 521
Gastric pits, 521
Gastric secretion, 525*b*
Gastrin, 611
Gastrocnemius, 401
Gastrocnemius externus, 399*t,* 401
Gastrocnemius internus, 399*t,* 401
Gastrocnemius lateralis, 399*t,* 401
Gastrocnemius medialis, 399*t,* 401
Gastrocoel, 167, 168*f*
Gastrointestinal (GI) tract. *See* Alimentary canal (digestive tract)
Gastrula, 163
Gastrulation, 49, 167–75, 176*f,* 177*f*
Gavials, 115

Geckos, gliding and parachuting by, 360*f*
Gegenbaur, C., 327–28, 330–31
Gelatinous matrix, 71
Geldings, 618*b*
Gemelli muscles, 401
General cranial nerves, 631
General sensory organs
   associated sensory receptors, 673, 674*f,* 675*f*
   components of, 672
   encapsulated sensory receptors, 673
   free sensory receptors, 672–73
   perception mechanisms, 674
Genes. *See Hox* genes
Genetic mutations, 10
Genital arteries, 459
Genital ducts
   female, 567–69, 570*f,* 571*f*
   male, 572–76, 577*f,* 578*f,* 582
Genital ridge, 563
Genital systems, embryonic development of, 566*f*
Geological ages, 39–41
Geological time, 30*f,* 83*f*
Geometric growth, 137
Gephyrostegus, 313*f,* 406*f*
Germ cells, 563
Gestation, delays in, 589, 590*f*
Gestation period, 164
GH. *See* Growth hormone (GH)
Giant flying squirrels, gliding and parachuting by, 360*f*
Giant panda, pancreatic ducts of, 535*f*
Gigantism, 609
Gill arches, 245
Gill-arch theory, 327–28
Gill coverings, 416*f*
Gill curtain, 415
Gill filaments, 424
Gill lamellae, 255
Gill rays, 426
Gills, 85, 416
   air breathing and, 418
   bony fishes, 427
   gas transfer and, 445*f*
   respiration through, 427
   sharks, 426*f*
Gill slits, 53
Gill ventilation, 429*b*
   amphibian tadpoles, 432*f*
   frogs, 433*f*
   sharks, 426*f*
   sturgeons, 429*b*
   teleosts, 427*f*
Gingiva, 507

Giraffes, 125, 130f
  cervical vertebrae of, 314
  horns of, 236f
  limb tendons of, 376f
Girdles, 326. *See also* Pectoral
    girdle; Pelvic girdle
Gizzard, 529, 536
Glands, 180
  adrenal, 598–602
  bulbourethral, 563
  digestive, 531–35
  duodenal, 525
  endocrine, 592
  intestinal, 523
  mammals, 230–33
  parathyroid, 596, 597
  pineal, 610–11
  pituitary, 604–10
  rectal, 528
  salivary, 504
  salt, 561
  shell, 567
  stomach and, 521
Glandular epithelium, 179f,
    180, 521
Glans penis, 582
Glenoid, 341f
Glenoid fossa, 253, 327
Glia, 625
Gliders, 21
Gliding, 358–59, 360f, 367,
    367f, 368f
Globus pallidus, 663
Glomerular filtrate, 557
Glomerulus, 59, 546, 551
Glomus, 550
Glossopharyngeal nerve, 634,
    635f, 638f, 639t, 678
Glottis, 416, 519
Glucagons, 534, 604
Glucocorticoids, 598
Gluteus complex, 399t, 400
Glyptodon, 229, 236, 237f
Gnathostomes, 84, 86, 91f,
    94–97, 427
  alimentary canal, 528
  axial skeleton, 295f, 304–5
  cerebral hemispheres, 664f
  cloaca, 582
  dual pumps, 421–22
  jaws and mouths of, 509–10b
  phylogeny, 91f
  skull phylogeny, 255–56
  spinal nerves, 630
  splanchnocranium, 293f
  suspension feeding, 286
  thyroid, 593
GnRH. *See* Gonadotropin-
    releasing hormone
    (GnRH)
Goats, 125

Goblet cell, 216
Godwit, 135f
Goethe, Johann Wolfgang
    von, 13
Goethe, W., 251b
Golden mole, extraembryonic
    membranes in, 195f
Goldfish, cutaneous respiration
    of, 419f
Golgi tendon organs, 673, 674f
Gonadotropin-releasing
    hormone
    (GnRH), 616
Gonadotropins, 609
Gonads, 563–65, 610
  castration of, 618b
Gonopodium, 576
Goose, brain of, 656f
Gordon, J. E., 149t
Gorillas, 125
Goslow, G. E., 381f
Gosselin-Ildari, Ashley, 61f
Gothic cathedral, 149f
Gould, John, 9
Gracilis, 399t, 401
Graded force, 380–81, 383f
Graded potential, 626
Grades, 24–29
Grand design, 4
Granular cell, 216
Granulosa, 613
Granulosa lutein cells, 613
Graves' disease, 596
Gravity, 128, 132f, 144–45, 145f
Gray matter, 649, 653f
Great auks, appendicular
    skeleton of, 349f
Greek combining forms,
    736–39
Green heron, aortic arches
    of, 465f
Green iguana, alimentary canal
    of, 526f
Green lizard, cutaneous
    respiration of, 419f
Grinding teeth, 537f
Ground substance, 180, 181f
Grouse, alimentary canal
    of, 526f
Growth, thyroid gland and, 596
Growth hormone (GH), 609
Guard hairs, 230
Gulls, adrenal tissues of, 600f
Gustatory cells, 678–79
Gut, 504, 544f
Gut tube cavity, 168f
Gymnarchus, 373f, 709, 712f
Gymnophiona (Apoda),
    107f, 108
Gyri, 661
Gyrinophilus, 575f

Habenular nucleus, 658
Habenular region of brain, 691
Hadean eon, 40, 40f
Haeckel, Ernst, 21–22, 201–2
Hagfishes, 86, 88f, 89, 93, 93b
  aortic arches, 461
  autonomic nervous system,
    644
  axial skeleton, 294, 304
  circulation, 475f
  cranial nerves, 636f
  forebrain, 651
  genital ducts (male), 574f
  heart, 475–76
  integument, 217
  kidneys, 550, 553
  nephrons, 554f
  ovaries, 568f
  pancreas, 535, 604t
  pancreatic islets, 602
  pituitary gland, 606, 607t
  respiratory system, 425
  skull phylogeny, 255
  vertebrate origins and,
    562–63
  vestibular apparatus, 698f
  water balance, 560
Haikouella, 89, 93, 255, 294
Haikouichthys, 89, 93, 255, 294,
    328, 331
Hair, 214f, 229–31
Hair bundle, 693
Hair cells, 693, 694, 696f
Hair cortex, 229
Hair follicle, 229
Hair medulla, 229
Hair papillae, 230
Hair tracts, 231f
Hairy frog, cutaneous
    respiration in, 420f
Haldane, J. B. S., 128, 133, 160
Half bound, 357, 357f
Hampé, Armand, 208
Hamsters, facial muscles
    of, 411f
Hamstrings, 388f, 401
Hands, 326
Haploids, 163
Harderian glands, 531
Hard palate, 282, 505
Hares
  cranial kinesis in, 254b
  alimentary canal of, 526f
  flight of, 361
  vision in, 688f
  wing shape of, 364f

Head, 84, 251b
Head (muscle origin), 390
Head shield, 92
Hearing, evolution of, 706–9,
    709f. *See also* Auditory
    system; Ears
Heart flow, 492
Heart formation,
    embryonic, 458f
Heart growth, embryonic, 459f
Hearts, 473–90
  amphibians, 478–80
  aspiration filling, 474f
  basic vertebrate heart,
    473–75
  birds and mammals, 489–90
  fishes, 475–78
  reptiles, 480–88, 489f
  structure of, 474f
Heart vesicle, 59
Heat blocks, 158, 495
Heat distribution, 455
Heat transfer, 494–95, 496f
Hellbender, cutaneous
    respiration of, 419f
Heloderma, adrenal tissues
    of, 600f
Hemal arches, 83, 84f, 295
Hemal spine, 83, 84f
Hemibranch, 426
Hemichordata, 54, 56–57
Hemichordate larva,
    metamorphosis
    of, 59f
Hemicyclaspis, skull phylogeny
    of, 255, 256f
Hemipenes, 579, 581f
Hemodynamics, 453–54,
    484–86b
Hemoglobin, 452
Hemopoiesis, 456
Hemopoietic tissues, 181
Hemotumescence, 578
Henslow, John, 9
Hepatic diverticulum, 196, 533
Hepatic portal vein, 460,
    466–67, 469f
Hepatic sinusoids, 466, 533
Hepatic veins, 63, 466
Hepatocytes, 533
Herons, aortic arches of, 465f
Herpes zoster virus, 633b
Herring gulls, 349f, 600f
Hesperornis, 118
Heterocercal tail, 1–2, 2f, 305
Heterochrony, 198–201
Heterocoelous centra, 296, 297f
Heterodonts, 123, 507, 512f
Heterostracans, 255, 304
Higher primates, 125
High gear muscles, 386

High-resolution digital tomography, 42
Hildebrand, Milton, 350
Hindbrain, 646, 647f, 654
Hindgut, 504, 531f
Hindgut fermenters. See Intestinal (hindgut) fermenters
Hindlimbs. See also Appendicular skeleton
  pectoral girdle and, 344f, 398–405, 406f, 407f
  of reptile, primitive bird, and modern bird, 208f
Hinge, 220
Hip girdle. See Pelvic girdle
Hippocampus, 627b, 662
Hippocampus, 574f
Hippopotamuses
  facial muscles of, 411f
  gait of, 356
  stomach of, 528f
Hips, of dinosaurs, 114f
Histochemical profiles, of muscle fibers, 378f
Histogenesis, 177
Historical constraints, on evolutionary analysis, 722
Historical predecessors
  evolution, 4–14
  morphology, 11–14
Holoblastic cleavage, 164, 165f
Holobranch, 426
Holocephalans, 96f, 97, 427, 578f
Holonephric concept, 547
Holonephros, 547, 549
Holospondylous vertebra, 296
Holosteans, 99, 101
Homeostasis, 553, 598f
Homeotic genes, 204
Homeotic mutants, 204
Hominidae, 125
Hominids, 125, 725
Homocercal tail, 1–2, 2f, 305
Homodont teeth, 507
Homo erectus, 725
Homo habilis, 725
Homology, 15–16, 15f
  muscle, 390–91
Homoplasy, 15–16, 15f, 16
Homo sapiens neanderthalensis, 35
Hooded pitohui, 230b
Hoof wall, 233
Hooves, 233–34
Hoplerythrinus, 491f
Hoplosternum, 417–18, 420f, 491f
Hopson, James A., 283f

Horizontal cells, 682
Horizontal septum, 298, 393, 394
Hormonal control, 617f, 620–21
Hormonal receptor sites, 621f
Hormones, 592, 593t
  ACTH, 601, 610
  ADH, 608
  bones and, 153
  chromaffin, 598
  corticosteroid, 598
  CRH, 616, 620f
  evolution and, 622–24
  FSH, 588, 609, 613, 614, 614b, 616
  gastrointestinal, 611–12
  GnRH, 616
  growth, 609
  ICSH, 610
  LH, 610, 613, 614, 616, 621
  MSH, 610
  neurohormones, 607
  parathyroid, 596
  pars nervosa and, 609f
  releasing-inhibiting, 607
  thyroid, 592–96
  TSH, 594, 595f, 609, 616, 620f
Horns, 234–35, 234f
Horses, 125, 130f
  brain, 656f, 658f
  castration, 618b
  cursorial locomotion, 357–58, 358f, 359f
  digits, 356f
  diversity and patterns in, 724f
  gait, 356, 357–58, 359f
  hindgut fermentation, 542
  hooves, 233f, 234f
  limb muscles, 356f
  mosaic evolution, 726f
  pes, 346f
  stomach, 528f
  superficial musculature, 404f
  teeth, 538b
  ventilation and locomotion, 440f
Housefly, 130f
Hovering flight, 361, 364f
Hox genes, 204–5, 206f, 208f, 314b
  axial patterning by, 314b
  cranial neural crest and, 289, 291f
  digestive system and, 543f
  embryonic changes and, 729–30
  jaws and, 247
  turtles, 313

Humans (*homo sapiens*), 35, 125, 130f
  adrenal tissues, 600f
  alimentary canal, 526f
  allometry, 134f
  aortic arches of, 465f
  appendicular skeleton, 354b
  appendix, 530b
  autonomic nervous system, 644–45
  biomechanics, 321b
  bipedal posture, 320b, 321b
  brain, 659f
  castration, 618–19b
  cranial nerves, 639f
  cursorial locomotion, 358t
  cutaneous respiration, 419f
  diversity and patterns in, 724f
  embryonic development, 193f, 202f
  facial muscles, 411f
  food and air passages, 520f
  functional matrices in, 721f
  gray and white matter, 653f
  missing-link fallacy, 725, 727f
  ovarian and uterine cycles, 615f
  pes, 346f
  pharynx, 519–20
  photoreception, 687
  reproductive systems, 564f
  running gait, 387b
  skulls, 280f
  speech, 511b
  stomach, 528f
  thyroid glands, 595f
Hummingbirds
  wing beats of, 361, 363f
  wing design of, 363f
Humpback whales, 289b, 290b
Hutchinson, Deborah A., 222b
Huxley, J. S., 130f
Huxley, T. H., 11, 11f, 13, 15, 115, 251b, 723
Hyaline cartilage, 181
Hybodus, axial skeleton phylogeny, 306f
Hydra, 130f
Hydrogen, 445
Hydrostatic functions, 428
Hydrostatic pressure, 496
Hydrostatic skeleton, 18, 18f
Hyoid apparatus (hyoid), 248, 253, 516
Hyoid arch, 246, 407–8
Hyomandibula, 246, 698
Hyostylic jaw suspension, 248
Hypaxial musculature, 393
Hypermorphosis, 199, 200f

Hyperoodon, axial skeletons in, 322f
Hyperopia, 159
Hyperosmotic fish, 556–57
Hyperplasia, 151
Hyperthyroidism, 596
Hypertrophy, 151, 151f, 153–54
Hypoblast, 166
Hypobranchial element, 245
Hypobranchial musculature, 393, 408, 410–11
Hypobranchial nerves, 633
Hypobranchial ridge, 58
Hypocercal tail, 305
Hypodermis, 212
Hypoglossal nerve, 634, 635f, 639t
Hypomere, 168f, 169, 391
Hypophyseal pouch, 505
Hypophysis, 604. See also Pituitary gland
Hyposmotic fish, 556f, 557
Hypotensive shock, 455
Hypothalamus, 658, 661f
Hypotheses, descriptive, 718
Hypothesis testing, 3b
Hypothyroidism, 596
Hypoxia, 416
Hypsodont teeth, 514
Hypurals, 305
Hyracoidea (hyraxes), 125
Hystricomorphs, 125

Ichthyopterygia, 112
Ichthyornis, 118
Ichthyosaurs, 112
  convergent design, 722f
  eyes of, 684
  fossilized, 31, 31f
  manus of, 345f
Ichthyostega, 105, 106f
  acid-base regulation, 446
  appendicular skeleton, 336, 338f
  axial skeleton, 312
  manus and pes, 342, 345f
  pectoral girdle, 341f
  pelvic girdle, 343f
  respiratory system of, 430
  ribs, 299
Ichthystegalians, 107f
ICSH. See Interstitial cell-stimulating hormone (ICSH)
Iguana
  alimentary canal of, 526f
  pelvic girdle of, 343f
Ileocolic valve, 523
Ileum, 523
Iliac arteries, 459
Iliacus, 399t, 400

Iliac vein, 466
Iliocostalis, 397
Iliofemoralis, 399t, 400
Iliotibialis, 399t, 400
Ilium, 336, 339
Immature bone, 182
Impedance matching, 701
Implantation, 563
    delayed, 192, 589, 590f
Impotence, 588
Incipient structures, 20
Incisors, 514
Incremental lines of von
        Ebner, 507
Incubation, 164
Incus, 248, 281, 700
Index fossils, 38, 38f
Indifferent gonad, 563
Induced drag, 366, 366f
Induction, 206, 240
Inertia, 141
Infectious diseases, bones
        and, 151
Inferior (caudal), 18f
Inferior colliculi, 658
Inferior jugular veins, 469
Inferior region, 16
Infertility, 588
Information, visual,
        689–90, 691f
Information transmission,
        in nervous system,
        626, 628
Infralabial glands, 531
Infrared radiation, 679
Infrared receptors, 691–93
Infraspinatus, 398, 399t
Infundibulum, 567, 604, 605t
Ingression, 166, 168
Ingroup, 26
Inguinal canal, 573
Inhalation, 416
Inhalation phase, 422, 423f
Inheritance of acquired
        characteristics, 6, 7
Inhibition, 628
Inner cell mass, 166, 167f
Inner ear, 698, 700, 701f,
        702f, 704
Innervation, 213
Innominate bone, 313, 339
Innovations, in vertebrates,
        83–84
Insectivores, 656f
Insect-net theory, 368f, 369
Insertion, 390
Insulation, of dinosaurs, 123b
Insulin, 534, 602
Integument, 212
    embryonic development
        of, 213, 214f

general features of, 213–16
overview, 239–40
phylogeny, 216–33
specializations of, 213f,
        233–40
Interatrial septum, 477
Interbranchial septa, 255, 416
Intercalated disks, 375
Intercardinal anastomosis, 468
Intercentrum, 295, 310f, 311f,
        312–13
Interclavicle, 336, 341f
Interclavicular sac, 438
Interhemal arches, 295
Interhyoideus, 407
Intermandibularis, 407
Intermediate mesoderm, 169
Intermediate tubule, 552
Intermedium, 343
Intermittent stress, 154
Internal carotids, 457
Internal elastic membranes
        (IEM), 452f
Internal fertilization,
        588–89, 590f
Internal gills, 416
Internal jugular veins, 472
Internal nares, 505
Internal narial valve, 432
Internal naris, 676
Internal respiration, 415
International (SI) units, 733–35
Interneural arches, 295
Interneurons, 638
Internuncial neurons, 638
Interoceptors, 645, 672
Interparietal bone, 274–75
Interpharyngobranchials, 411
Interstitial cells, 572, 610
Interstitial cell-stimulating
        hormone (ICSH), 610
Interstitial space
        (interstitium), 177
Intertemporal bone, 250
Interventricular canal, 481
Interventricular septum, 477
Intervertebral cartilage, 297
Intervertebral disks, 83,
        84f, 297
Intervertebral ligament, 297
Intestinal glands, 523
Intestinal (hindgut)
        fermentation, 540,
        541f, 544f
Intestinal (hindgut) fermenters,
        531f, 540–42, 544f
Intestinal juice, 525
Intestines, 504, 523, 524f,
        525, 526f
In-torque, 142
Intrafusal muscle cells, 673

Intramembranous bone, 182
    development, 184, 187f
Intraoral transport, 516, 517f
Intromittent organs, 576, 578f,
        579, 582f, 584
Invagination, 54, 168
Inversion, prechordates,
        78–80
Involuntary muscles, 373
Involution, 168
Iridescent colors, 238
Iridophore, 238
Iris, 682
Irreducible design, 12f
Ischemia, 455
Ischiotrochantericus, 401
Ischium, 336, 339
Ischnacanthus, 97f
Islets of Langerhans. See
        Pancreatic islets
Isolecithal, 163
Isometry, 134, 136f
Isosmotic animals, 556f, 557
Isthmus, 568, 569
Isurid shark, brain of, 654f

Jackdaw, testis and epididymis
        in, 578f
Jackrabbits, cranial kinesis
        in, 254b
Jacobson's organ, 676
Japanese toad, 222b
Jarvik, Erik, 103, 246, 328
Jaw articulations, 281,
        284f, 285f
Jaw attachments, types of,
        247–48
Jaw musculature, 393
Jaws, 246
    actinopterygians, 261f
    as machines, 149f
    morphology, 253
    origin of, 246–47
Jaw suspension, 248f
Jefferson, Thomas, 31b
Jeju, respiratory system of, 427
Jejunum, 523
Jenkins, F. A., 43f, 362f
Joey, 616, 617f
Johansen, K., 43f
Johansen, Kjell, 484b
Joint capsule, 187
Joints, 187, 189
Joule, 140
Joule, James, 140
Jugal, 248
Jugular lymphatics, 496
Jugular veins, 469, 472
Jumping mechanics, of
        dogs, 718f
Juxtaglomerular cells, 612

Kangaroo rats, 314, 555b
Kangaroos, 124
    adrenal tissues, 600f
    alimentary canal, 526f
    cerebral cortex, 661
    reproductive cycle, 616, 617f
    reproductive organs,
        585, 587f
    stomach, 528f, 539f
Kardong, Kenneth V., 42f
Kelvin, Lord, 10, 11
Keratinization, 215, 216, 228
Keratinized layer of
        epidermis, 215
Keratinizing system, 216
Keratinocytes, 216, 228
Keratins, 215–16
Kiaeraspis, cranial nerves of, 636f
Kidney organization, tripartite
        concept of, 547–50
Kidneys
    acid-base regulation by, 445
    avian, 553f
    endocrine system and,
        612–13
    filtration, 557
    fish, 84
    function and structure,
        553–62
    homeostatic problems and
        adaption, 557f
    mammalian structure, 552f
    phylogeny, 550–53
    renal arteries and, 459–60
    renal portal veins and, 460
    secretion, 557
    structure (mammals), 545–47
    three-part, 549f
Killfish, copulatory organs
        in, 576
Kinematic chain, 148, 148f
Kinetic skulls, 253–55
King, A. S., 581f, 584f
Kinocilium, 693
Kiwi, 118
Knudson, E. I., 707f
Koalas, 192
    alimentary canal of, 526f
    feces, 536
    toxin digestion, 542
Komodo dragon, 130f

Labial pits, 693
Labyrinthodonts, 104–6
    axial skeleton, 310
    pectoral girdle of, 341f
    teeth, 103, 262, 512, 513f
    vertebral column, 322f
Lacerta viridis, cutaneous
        respiration of, 419f
Lacrimal bone, 249

Lacrimal glands, 531
Lactation, 232
Lacteals, 499, 527
Lactic acid, 445
Lacunae, 181
Laerm, J., 307f
Lagena, 696, 698f
Lagomorphs, teeth in, 508
Lake Titicaca frog, cutaneous
        respiration of, 420f
Lamarck, Jean-Baptiste de, 6–8,
        7f, 11
Lamella, 182
Lamellar, 217
Lamellar bone, 182, 183f
Lampreys, 88–89, 93, 93b
    alimentary canal,
        527–28, 528f
    aortic arches, 461
    autonomic nervous
        system, 644
    axial muscles, 395f
    axial skeleton, 294, 304
    brain, 654f, 655f
    cardiovascular system, 461f
    cerebral hemispheres, 664f
    cloacal and anal regions
        of, 583f
    cranial nerves, 636f
    embryonic development, 171f
    endostyle, 594
    eyes, 683, 685f
    gills, 427
    heart, 476
    integument, 216f, 217
    kidneys, 550–51, 553
    muscular pumps, 446
    nephrons, 554f
    pallium, 662
    pancreas, 535
    pineal complex, 690, 692f
    pituitary gland, 606, 607t
    respiratory system, 424–26
    skull phylogeny, 255
    somatic and visceral circuits
        in, 641f
    spinal cords, 650f
    suspension feeding, 286
    venous vessels, 471f
    vestibular apparatus, 698f
    visual accommodation, 687f
Land. See Gravity; Terrestrial
        environment
Langerhans cells, 228
Larch Mountain salamanders,
        head of, 680f
Large intestine, 504, 523, 524f
Larsen, J. H., 680f
Larus, adrenal tissues of, 600f
Larvacea (Appendicularia),
        70–73

Larvaceans, 286
Larynx, 435b, 511b, 519
Lateral abdominal veins,
        460, 466
Lateral cilia, 58–59
Lateral cranial fissure, 256
Lateral geniculate nuclei, 689
Lateral line, 190, 190t
Lateral line canals, 694, 695
Lateral line nerves, 634t,
        636–37
Lateral line system, 104, 263,
        694–95, 697f
Lateral musculature, 397
Lateral pallium, 662–63
Lateral plate mesoderm, 169
Lateral rotation, 297f
Lateral-sequence gait, 350, 351f
Lateral undulatory
        locomotion, 311f
Latero-bronchi, 438
Laterosphenoid, 244t
Lateroventral spinothalamic
        tract, 650
Latimeria, 102
    air-breathing organs, 447
    appendicular skeleton,
        335, 336f
    axial skeleton, 309f
    brain, 654f
    cranial nerves, 633
    pancreatic islets, 602
    pituitary gland, 607t
    water balance, 557, 560
Latimeria (coelacanth), 45b
Latin combining forms, 736–39
Latissimus dorsi, 398, 399t
Laurasiatheria, 124f
Law of the general to the
        specific, 202
Leading foot, 357
Leaf-eating monkeys, stomach
        of, 528f
Leakey, Mary, 32
Leaping, 367, 368f
Lecithotrophic embryos, 164
Left atrial chamber, 477
Left systemic arch, 464
Left systemic trunk, 481
Lemurs, 125, 600f
Length, 129, 132–33, 138–39
Lens, 159, 682–86
Lens primordium, 682
Lepidosauria, 113, 113f, 604t
Lepidosauromorpha, 112
Lepidosiren, 103f, 447, 477
Lepidotrichia, 99, 326
Lepisosteus, pectoral girdle
        of, 341f
Lepospondyls, 104, 105f, 107f,
        108, 296, 309–10

Letdown, 232
Leucocytes, 452
Levation, 390f
Levator hyomandibulae, 407
Levator labii superioiris, 408
Levator nasolabialis, 408
Levator operculi, 407
Levator palatoquadrati, 407
Levators, 390, 397, 399t
Levator scapulae, 397
Lever advantage, 386
Lever arm, 142
Levers, 142–44
Lever systems
    bone-muscle, 385–88
    principles of, 143f
Leydig cells, 219, 220
LH. See Luteinizing hormone
        (LH)
Life cycle, of frogs, 162f
Life history, 161–63
    coelom development, 196–97
    embryology, 163–75, 176f,
        177f, 194
    extraembryonic membranes,
        190–95
    maturation, 197–201
    ontogeny and phylogeny,
        201–9
    organogenesis, 176–90
    overview, 210–11
Lift, 358, 359, 364, 366f
Ligamentum arteriosum, 492
Light, 680, 683–85, 710b
Lightfoot, Dr. John, 10
Light organs, 710b
Limbic system, 660, 666–69
Limb loss, in snakes, 208f
Limb muscles, in lizards
        and horses, 356f
Limb posture, changes in,
        352–53
Limbs, 326–27. See also
        Appendicular skeleton
    lizards, 206–7
    pelycosaurs, 134f
    strength versus speed in,
        385f, 386f
Limiting membrane, 434
Lines of arrested growth
        (LAGs), 182, 183f
Lingual feeding, 269, 272f, 287,
        516
Lingual glands, 531
Lingual process, 269
Lingual vein, 472
Lining epithelium, 179–80
Linnaean chordate
        classification, 740–42
Linnaeus, Carolus, 5–6, 5f, 11
Lions, 357, 358t, 724f

Lipases, 536
Lips, 504–5
Lissamphibia, 105f, 106–8
Liver, 504, 533, 534f
Living fossils, 45b
Livingstone, Dr. David,
        187, 188f
Lizards, 113, 113f
    adrenal tissues, 600f
    alimentary canal, 526f, 529
    autonomic nervous
        system, 645f
    axial musculature, 396f
    blood flow through heart
        of, 485b
    body cavities, 196f
    brain, 659f
    cloaca, 584f, 585f
    copulatory organs, 577f
    cranial kinesis, 268, 271f
    cutaneous respiration, 419f
    dentition, 512f
    dermatocranium
        evolution, 267f
    diversity and patterns
        in, 724f
    epigenetics in, 206–7
    extraembryonic
        membranes, 195f
    eyes, 685f
    facial muscles, 411f
    fossorial locomotion, 369
    functional coupling of
        locomotion and
        respiration, 720
    gastralia, 299
    gliding and parachuting
        by, 360f
    heart, 480f, 484–86b,
        487, 492
    hemipenes, 579, 581f
    hemodynamics, 484–86b
    Hox genes, 729f, 730
    integument, 220–21, 225
    limb formation in, 207f
    limb muscles, 356f
    lingual feeding, 272f
    lungs, 433, 434f
    manus, 345f
    muscle development
        (embryonic), 392f
    muscle homology, 391f, 408t
    pectoral girdle, 339
    pelvic girdle, 342f
    pes, 346f
    pineal complex, 690, 692f
    placenta, 193, 194
    skulls, 268–72, 270f
    snake evolution and, 694b
    sternum, 300f
    thyroid glands, 595f, 596

Llamas, stomach of, 528f
Load, 148, 149, 150f, 376
Load fracture, 149–50, 151f
Lobe fins, 347
Lobsters, allometry of, 138f
Lobules, 232
Locomotion. See also
    Appendicular
      skeleton; Flight
  aerial, 349, 358–67, 367f
  appendicular musculature
     and, 401–5, 406f, 407f
  arboreal, 349, 368f
  axial skeleton and, 310,
     311f, 312
  body size and, 131f
  cursorial, 349, 352f,
     353–58, 359f
  digit orientation, 352f
  fossorial, 349, 369–70
  gaits, 350, 351f
  lateral undulatory, 311f
  mucus and, 237
  saltatorial, 349
  scansorial, 249
  shoulder girdle and, 353f
  swimming, 349–50
  terrestrial, 349, 350–58
  ventilation and, 434–35,
     438f, 440f
Locomotor modules, 725, 728f
Loewi, Otto, 3b
Longissimus, 397
Loop of Henle, 552
Lophodont teeth, 514
Lorises, 125
Lower jaw, 253
Lower primates, 125
Lower temporal bar, 110
Low gear muscles, 386
Lumbar lymphatics, 496
Lumbar region, 319
Lumen, 179
Luminal glands, 504
Lunar valves, 483
Lungfishes, 43f, 100, 101b,
    102–3
  accessory air-breathing
     organs, 491
  air-breathing by, 420f
  alimentary canal, 527f
  ammonia and, 554, 562
  anal region, 583f
  aortic arches, 462–63
  appendicular skeleton,
     335, 336f
  buccal pump, 423
  circulation, 456
  cloaca, 582, 583f
  cranial musculature, 410f
  diversity and patterns in, 724f

fossorial locomotion, 369
genital ducts (female), 570f
genital ducts (male), 574f
gill-arch theory and, 327–28
hearts, 477–78
limb development, 332f
lobe fins, 347
lungs, 417, 428
nephrons, 554f
pallium, 663
pituitary gland, 606
respiratory system, 433
skull phylogeny, 261–62
venous vessels, 469, 471f
Lungless salamanders
  cutaneous respiration
     of, 419f
  hearts of, 478, 479
Lungs, 416, 447–48
  alveolar, 440
  birds, 440, 441f, 442f,
     449–50
  fishes, 428–30
  frogs, 432, 433f
  gas bladders and, 416
  mammals, 440f
  parabronchial, 440
  reptiles, 433–35
Luteinizing hormone (LH),
    610, 613, 614, 616
Lyell, Charles, 37
Lymph, 451–52, 496
  formation, 497f
Lymphatic circulation, 497f
Lymphatic cisterns, 498
Lymphatic sacs, 498
Lymphatic system, 451–52,
    496–99, 500
Lymphatic tissue, 498
Lymphatic vessels, 496–98, 499f
Lymph "hearts," 496, 498
Lymph nodes, 497f, 498
Lymphocytes, 181
Lymphoid tissue, 181
Lymph return, 498

Mabuya, extraembryonic
     membranes in, 195f
Mabuya lizards, 193
Machines, 148, 149f
Macrolecithal, 163
Macula, 696, 697f
Macula neglecta, 701
Magnetic fields, 713
Main beam, 234
Major calyx, 551
Malapterus, electric organs
     in, 373f
Male genital systems,
     embryonic
     development of, 566f

Male reproduction, in
     mammals, 613
Male reproductive system, 564f,
    572–82, 583f
Mallatt, Jon, 509b, 510b
Malleus, 248, 281, 700
Malthus, Thomas, 8, 9
Mammals, 121–26, 122–26
  acid-base regulation, 445
  adrenal glands, 599f, 601f
  adrenal tissues, 600f
  akinesis, 253f, 282, 291–92
  alimentary canals, 524f,
     526f, 530–31
  aortic arches, 465–66, 468f
  appendicular
     musculature, 402–3
  aspiration pump, 424
  autonomic nervous
     system, 642, 643f,
     644–45, 645f
  axial musculature, 397
  body cavities, 196f
  bombesin-secreting cells,
     623–24
  branchial arches, 247t
  buccal cavity, 505
  cardiovascular system design,
     491–92
  castration of, 618b
  CCK cells, 623–24
  cerebral cortex, 661
  cerebral hemispheres, 664f
  cerebrospinal fluid and
     meninges,
     648f–650f
  cervical vertebrae, 313f
  chondrocranium, 244t
  cleavage, 165f, 166–67
  cloaca, 582
  cooling, 496f
  cranial nerves, 635f
  cutaneous respiration, 419f
  dehydration, 555b
  depth perception, 688f, 689f
  dermatocranium
     evolution, 267f
  digits of, 356f
  diversity and patterns, 724f
  ear functions, 705, 707, 708f
  embryonic development, 194
  emergence of, 289–91
  endocrine tissues and
     secretions, 593t
  extraembryonic membranes,
     191–93
  fetal circulation, 492,
     493f, 494f
  gaits, 357f
  gas exchange, 437, 440f
  gastrocnemius, 401

gastrointestinal
    hormones, 611f
gastrulation and neurulation,
    173–75, 176f
genital ducts (female), 571f
genital ducts (male), 573,
    575, 576f
genital systems, embryonic
    development of, 566f
gliding and parachuting, 360f
gonadotropic hormones, 613f
gut (embryonic
    development), 522f
hearts, 474, 489–90
heat blocks, 495f
heat distribution in, 495f
hyoid arch, 407–8
integument of, 214f, 228–33
kidneys, 545–47, 551–53,
    553, 557–58, 559f
locomotion and
    ventilation, 440f
lungs, 440f
mastication, 537f
maturation of ovarian
    follicles, 614f
medulla and breathing, 660f
midbrain, 658
molar patterns of, 516f
muscle homology, 391f, 399t
muscular "sling" of, 397, 397f
nasal passages, 677, 680f
nephrons, 554f
nitrogen elimination, 554
olfactory bulb and tract, 676f
oral glands, 532
ovary, species with
    single, 569t
pallium, 663
pancreas, 604t
pancreatic islets, 603f
pars nervosa, 609f
pectoral girdle, 340f
pelvic girdle, 342f, 353f
penis, 578, 582
pineal complex, 690, 691
pituitary gland, 607t, 608–9
prolactin, 609
reflexes, 640t
reproduction, 613–16
reproductive organs, 587f
reproductive system,
    563–89, 564f
rib cage movements, 439f
saber-tooth, 519b
scales, 236
skulls, 265, 274–75, 279f,
    280f, 281, 284f, 285f
spinal accessory nerve, 634
spinal reflexes, 640f, 640t
splanchnocranium, 293f

Mammals—*Cont.*
stomach, 528*f*
supracoracoideus
muscles, 406*f*
systemic arches, 468*f*
teeth, 507–8, 511*f*, 512,
514–16, 538*b*
telencephalon, 661
temporal fenestrae, 268*f*
thyroid gland, 594*f*, 595*f*,
596
trapezius muscle group, 397
urinary bladder, 588*f*
urogenital sinus, 586*f*
venous vessels, 472, 473*f*
ventilation, 435–37,
438*f*–440*f*
vertebrae development,
203*f*, 303*f*
vertebral column, 313–14
visual accommodation, 687*f*
Mammary glands, 214*f*,
232–33
Mammillary bodies, 658
Mammoths, 32
Manatees, 125
Mandible, 253
Mandibular, 633
Mandibular arch, 246, 406–7
Mandibular bone, 256
Mandibular condyle, 253
Mandibular glands, 532
Mandibular ramus, 639*t*
Mandibular series, 249*t*, 250
Mandibular symphysis, 250
Maneuvering (swimming), 400*b*
Manubrium, 299
Manus, 326, 339, 342–43,
346, 347*f*
Marsh, O. C., 119*b*
Marsupials, 123, 124*f*, 582
adrenal tissues, 600*f*
alimentary canal, 526*f*
cardiovascular changes at
birth, 493–94
cleavage, 167*f*
cloaca, 584*f*, 585
embryonic development, 194
extraembryonic membranes,
191*t*, 195*f*
gastrulation and neurulation
in, 174, 177*f*
genital ducts (male), 575
glands, 232
pectoral girdle, 341*f*
pelvic girdle, 343*f*
placenta, 193
reproductive organs,
585, 587*f*
saber-teeth, 519*b*
skull phylogeny, 275, 280*f*

Mass, 129, 132*f*, 133–34,
138–39, 140, 141*f*
Masseter, 406
Mass extinction, 41
Massospondylus, 123*b*
Mastication, 282, 288, 292,
536, 537*f*
Mastodons, 31*b*
Mastoid muscle group, 397
Mastoid process, 277
Materials, strength of,
148–51, 151*f*
Matrix cells, 230
Matrotrophic embryos, 164
Maturation, 197–201, 563,
613, 614*f*
Mature bone, 182
Maxilla, 249
Maxillary, 633
Maxillary ramus, 639*t*
Maxilloturbinate bone, 277
McMahon, T., 135*f*, 136*f*
Measurement, units of, 139.
See also Acceleration;
Force; Length; Mass;
Time; Velocity
Mechanical advantage, 143
Mechanical breakdown, of food,
536, 537*f*
Mechanical effectors, 625
Mechanical stress
bones responsiveness
to, 151, 152–53,
153*b*, 155*f*
tissue response to, 151–55
Mechanism, 148
Mechanoreceptors,
693–709, 713
auditory system, 696–700
ear functions, 701–9
lateral line system,
694–95, 697*f*
vestibular apparatus, 695–96,
697*f*, 698*f*
Meckel's cartilage, 246
Medial musculature, 397
Medial pallium, 662,
663, 666
Medial region, 16
Median eminence, 604
Medulla, 551, 563, 600, 601*f*
Medulla oblongata, 651, 654
Medullary bone, 182
Megaleia rufa (red kangaroo),
616, 617*f*
Meibomian glands, 231
Meinwald, J., 222*b*
Meiotic cell division, 567
Meissner's corpuscle, 673
Melanin, 610
Melanophore, 238, 610

Melanophore-stimulating
hormone (MSH), 610
Melanosomes, 238
Melatonin, 611
Melville, Herman, 431*b*
Membranes, extraembryonic,
190–95
Memory, 667
Meninges, 646, 648*f*
Menopause, 588
Mental bone, 253
Merkel cells, 228
Meroblastic cleavage, 164
Merycism, 540
Mesencephalon, 646
Mesenchyme, 169, 391
Mesenteric ganglia, 630
Mesentery, 197
Mesethmoid, 244*t*, 277
Mesocoel, 59
Mesoderm, 49, 163, 168*f*,
169, 178*f*
Mesokinesis, 268
Mesolecithal, 163
Mesomere, 168*f*, 169
Mesonephric duct, 547
Mesonephric tubules, 547
Mesonephros, 547, 548*f*
Mesopterygium, 334
Mesorchium, 572
Mesosaurs, 110–11
Mesotarsal joint, 346
Mesothelium, 179
Mesovarium, 567
Mesozoic era, 40*f*, 41, *41*
Metabolism
body size and, 133*f*
nitrogen, 553–54, 554
thyroid gland, 594, 596
Metacarpals, 339
Metacoel, 59
Metakinesis, 268
Metamere, 16
Metamerism. *See* Segmentation
Metamorphic climax, 616, 620*f*
Metamorphosis, 162
ascidian, 68, 69*f*
echinoderms to chordates, 77
frogs, 162*f*, 616, 620
life history and, 197–98
Saltentia, 108
thyroid gland and, 596
Metanephric duct, 566, 566*f*
Metanephric tubules, 547
Metanephros, 547, 548*f*
Metaphysis, 182*f*, 183
Metaplasia, 152
Metapodial element, 339
Metapterygial fin, 327
Metapterygial stem, 327
Metapterygium, 334

Metapterygoid, 256
Metatarsals, 339
Metatherians, 122
Metautostylic jaw
suspension, 248
Metric system, 139
Meyer, Hermann von, 154
Mice
Hox genes in, 205*f*, 314*b*
thyroid glands in, 595*f*
Microcirculation, 454–55
Microglia, 625, 626*f*
Microlecithal, 163
Microridges, 216
Microsaurs, 108
Microvilli, 163–64, 523, 693
Midbrain, 646, 647*f*, 651, 658
Middle ear, 698, 699*f*, 700*f*
Middle ear bones, 277, 281,
283*f*–285*f*
Middle ear cavity, 698
Middle ear ossicles, 698
Midsaggital plane, 16
Migration, 411
Migratory waterfowl, flight
of, 361
Milk, 232
Mineralization, 181
Mineralocorticoids, 598
Mink
alimentary canal of, 526*f*
muscle homologies of, 408*t*
Minor calyx, 551
Missing-link fallacy, 725, 727*f*
Missing links, 31*b*
Mites, 130*f*
Mitotic cell division, 567
Mitral cell, 675
Mixed nerves, 630
Mnierpes, accessory
air-breathing organs
of, 418
Moas, 115, 118
Modularity, 725, 727, 728*f*
Molariform teeth, 514
Molars, 514, 516*f*
Molecular mechanisms, of
muscle contraction,
376–77
Moles, 124
extraembryonic membranes
in, 195*f*
fossorial locomotion of, 369,
370*f*
Molting, 221, 596
Moment, 142
Moment arm, 142
Monkeys, 125
alimentary canal of, 526*f*
buccal cavity of, 505
stomach of, 528*f*

Monocular vision, 158, 688
Monoecious protochordates, 56
Monokinesis, 268
Monophyletic groups, 26
Monopterus albus, cutaneous respiration in, 417, 420*f*
Monospondyly, 295
Monotremes, 122, 123, 124*f*
   adrenal tissues, 600*f*
   cleavage, 166, 167*f*
   cloaca, 582, 584*f*, 585
   extraembryonic membranes, 191*t*
   gastrulation and neurulation in, 173, 177*f*
   glands, 232
   pectoral girdle, 340*f*, 341*f*
   skulls, 275, 279*f*
Morganucodon, 341*f*
Mori, A., 221*b*
Morphogenesis, 210
Morphological cross section, 383
Morphology, 1, 714–15, 715*f*, 716*f*
   comparative, 1–4
   concepts, 15–19
   ecological analysis, 722
   evolution, mode and tempo of, 727–30
   evolutionary, 2, 19–21
   evolutionary analysis, 722–27, 728*f*
   functional, 2
   functional analysis, 42–46, 718–22
   historical predecessors (evolution), 4–14
   historical predecessors (morphology), 11–14
   idealistic, 16
   overview, 47
   paleontology, 29–42
   phylogeny, 21–29, 21*f*
   promise of, 730
   skull, 250–53
   structural analysis, 717–18
   tools, 41–47
Morula, 163, 165*f*, 167*f*
Mosaic evolution, 725, 726*f*, 727*f*
Moschops, 122*f*
Moss, Melvin, 720
Motion, second law of, 141, 142*f*, 144
Motor circuits, 646*f*
Motor end plate, 375
Motor neurons. *See* Efferent neurons
Motor pattern, 388

Motor unit, 381
Mountain sheep, horns of, 235*f*
Mouse. *See* Mice
Mouth, 678–79, 682*f*. *See also* Buccal cavity
Movement. *See* Locomotion
MSH. *See* Melanophore-stimulating hormone (MSH)
Mucoid coat, 166
Mucosa, 520, 524*f*
Mucous cuticle, 216
Mucous glands, 220
Mucus, 237
Mudpuppies, 199, 419*f*
Mudskippers
   cutaneous respiration of, 419*f*
   land, movement on, 448
   mucus of, 237
Mule deer, teeth of, 514
Müller, Gerd, 208
Müllerian duct, 565, 566*f*
Multicellular gland, 180
Multipolar neurons, 626, 627*f*
Mummification, 32*f*
Muscle actions
   sequencing of, 387–88, 389*f*
   types of, 388–90, 390
Muscle-bone lever systems, 385–88
Muscle cells, 376
Muscle classifications, 373–75
Muscle contraction, 376–77
Muscle fibers, 376, 377–79
   muscle organs and, 379–85
   orientation, 383–84
Muscle force, 379–81, 383*f*
Muscle homologies, 390–91
Muscle mechanics, overview of, 388
Muscle organs, 376, 379–85
Muscles
   actions, 388–390, 388*f*
   skeletal, 374*f*
Muscles, lever principles in, 144*f*
Muscle spindles, 673, 674*f*
Muscle tissue, 177
Muscularis externa, 520
Muscular mechanisms, for ventilation, 421–24
Muscular pumps, 446
Muscular ridge, 481
Muscular "sling," 397, 397*f*
Muscular system, 372–73
   comparative anatomy, 394–410
   embryonic development of, 391–93, 394*f*
   function, 377–91

   organization, 373–77
   overview, 410–11
Musculature. *See* Cranial musculature; Postcranial musculature
Mus musculus, Hox genes in, 205*f*
Mutation, 10
Mycoel, 168*f*
Myelin, 626
Myelinated nerves, 626, 627*f*
Myelocytes, 181
Myeloid tissue, 181
Myenteric plexuses, 644
Myllokunmingia, 89, 328, 331
Mylohyoid, 407
Myocardium, 456
Myocoel, 169
Myoepithelial cells, 180, 608
Myofibrils, 374
Myofilaments, 374
Myoglobin, 377
Myomeres, 54, 394
Myometrium, 608
Myomorphs, 125
Myomuscular junctions, 68
Myopia, 159
Myosepta, 298, 394
Myosin, 379
Myotomes, 168*f*, 169, 213, 392, 393, 410, 630
Mysticetes, 125, 290*b*, 322*f*
Myxine, cranial nerves of, 636*f*
Myxinoidea, 86, 88, 88*f*

Nail matrix, 233
Nails, 233–34
Namib Golden mole, fossorial locomotion of, 369, 370*f*
Narwhals, teeth of, 515*f*
Nasal bones, 264
Nasal capsules, 242, 262–63, 264*f*
Nasal chamber, 676
Nasal glands, 531
Nasal passages, 675–77, 678*f*–680*f*
Nasal placode, 505
Nasal sacs, 93, 262, 264*f*, 676, 679*f*
Nasal turbinates, 225
Nasohypophyseal opening, 89, 93, 255
   canal, 506*f*, 636*f*
Nasolabial grooves, 676
Nasolacrimal duct, 263
Nasopalatine duct, 677
Nasopharyngeal duct, 677
Nasopharynx, 511*b*

Nasoturbinate bone, 277
Naturalists, 6
Natural selection, 5, 725
   critics and controversy, 10–11
   Darwin, Charles, 5, 9–10
   DNA and, 14
   Huxley, T. H., 11, 11*f*
   vertebrate design, 3–4, 15
   Wallace, A. R., 8–9, 8*f*, 10
Natural taxon, 24
Natural Theology (Paley), 6
Neandertals, 34, 35
Nectridean lepospondyls, 108
Necturus
   adrenal tissues, 599*f*
   aortic arches, 463, 464*f*
   appendicular musculature, 402*f*
   cranial and shoulder musculature, 402*f*
   genital ducts (male), 573
   manus, 345*f*
   muscle homologies, 408*t*
   pelvic musculature, 402*f*
   pes, 346*f*
   pineal complex, 692*f*
   skull, 266*f*
   venous vessels, 472
Necturus maculosus, 199, 419*f*
Neoceratodus (Australian lungfish), 101*b*, 103*f*
   air-breathing organs, 447
   aortic arches, 462
   appendicular skeleton, 327–28, 335, 336*f*
   axial skeleton, 309*f*
   cranial musculature, 410*f*
   gas bladder, 417
   hearts, 477
   limb development, 332*f*
Neognathae, 118
Neomorphs, 331
Neopterygii, 100–101
Neoteny, 199, 200*f*
Nepetalactone, 675
Nephric duct, 546
Nephric ridge, 546
Nephric tubules. *See* Nephrons
Nephridial pore, 59
Nephridial tubule, 64
Nephridium, 56*f*, 59, 65*f*
Nephrocoel, 168*f*, 169, 546
Nephrogenic region, 547, 549–50
Nephrons, 546–47, 551, 554*f*
Nephrotome, 546–47
Nerve endings. *See* Sensory receptors
Nerve fibers, 626, 627*f*, 672
Nerve impulses, 626

Nerves
cranial, 189, 190f, 190t, 629, 630–37, 633–37, 638f, 639f, 639t, 640f
evolution, 637–38
myelinated, 626, 627f
olfactory, 675
phrenic, 197
somatic, 628, 629f, 634t
spinal, 629–30, 631f, 632f, 633b
unmyelinated, 626, 627f
visceral, 628, 629f, 634t
Nerve tracts, 68, 626, 650–51, 653t
Nervous system, 625. See also Central nervous system; Peripheral nervous system
information transmission, 626, 628, 628f
neurogenesis, 627b
neuroglia, 625, 626f
neurons, 625–26, 627f
neurosecretory cells, 628
overview, 669–70
Nervous tissue, 177
Nervus terminalis, 633, 634t
Neural arches, 83, 84f, 295, 305f
Neural control, 617f
Neural crest, 68, 168f, 189–90, 189b. See also Cranial neural crest
derivatives, 189t
Neural crest cells, 84, 168
Neural folds, 168, 168f
Neural keel, 168
Neural plate, 54, 168, 168f
Neural spine, 83, 84f, 317–18, 319f
Neural tube, 167, 168f
Neuroanatomy, 665b
Neurocoel, 54, 167, 168f
Neurocranium, 242
Neurogenesis, 627b
Neuroglia, 625, 626f
Neurohormones, 607
Neurohypophysis, 593t, 604, 605t, 608–9
Neurology, 657b
Neuromast organ, 693–94
Neurons, 625–26, 627b, 627f, 628, 629f, 638, 640
structure of, 626f
Neurosecretory cells, 628
Neurosecretory neurons, 607
Neurotransmitters, 628, 642f
Neurula, 163
Neurulation, 167–75, 176f, 177f
Newton, Isaac, 141
Nilsson, S., 644f, 645f

Nimravids, 519b
Nipples, 232
Nitrogen, 447b, 448b
Nitrogen narcosis, 447b
Nitrogenous wastes, excretion of, 554, 555f, 562
Nodal, 315
Nodes of Ranvier, 626
Nonglandular epithelium, 521
Nonlamellar bone, 182, 183f
Norepinephrine, 598, 642
Norris, Kenneth, 431b
Northcutt, Glenn, 718
Northwestern salamanders, 198
Noses, of dinosaurs, 123b. See also Nasal passages
Nostril. See External naris
Nothosaurs, 113
Notochords, 19, 51–52, 53f, 168f, 203f, 294, 304, 306f, 308
Notophthalmus, sperm storage in, 590f
Nuchal crest, 275
Nuclear bag intrafusal fiber, 673
Nuclear chain intrafusal fiber, 673
Nucleus, 626
Nucleus pulposus, 52, 84, 297
Nuptial pads, 220
Nutrition, bones and, 153

Obliquus abdominis, 434
Obturator externus, 399t, 400
Obturator internus, 401
Occipital arch, 242
Occipital bones, 244t, 250
Occipital cartilages, 242
Occipital condyle, 250
Occipital nerves, 631
Occipitospinal nerves, 631
Occlusal surfaces, 507, 514f, 515
Ocellus, 68
Octaval nerve, 634. See also Auditory nerve
Octavolateralis system, 636
Oculomotor nerve, 633, 634t, 638f, 639t
Odocoileus hemionus, teeth of, 514
Odontoblasts, 189, 508
Odontocetes, 125, 290b, 322f
Odor detection, 678
Oikopleura, 72f, 73f
Oken, Lorenz, 13, 251b
Olfaction, 675, 678, 713. See also Nasal passages
Olfactory bulbs, 661, 675, 676f
Olfactory epithelium, 675, 677f
Olfactory nerve, 633, 634t, 638f, 675

Olfactory placodes, 189, 190f, 190t, 675
Olfactory sensory cells, 675
Olfactory system. See Chemoreceptors; Nasal passages; Vomeronasal area
Olfactory tract, 675, 676f
Oligocene whale, skull of, 290b
Oligodendroglia, 625, 626f
Omasum, 530
Omosternum, 299
One-way valves, in veins, 454f
Ontogeny, 197. See also Embryonic development
Oocytes, 567, 569
Oogenesis, 567, 567f
Oogonia, 567
Opercular bones, 260
Opercularis, 703
Operculum, 97, 256, 416, 703
Ophiacodon, cervical vertebrae of, 313f
Ophthalmic nerve, 633, 639t
Opisthocoelous centra, 296, 297f
Opisthonephric duct, 566f
Opisthonephros, 547, 548f
Opisthoproctus, 710b
Opossum, 43f
Opossums
adrenal tissues, 600f
alimentary canal, 526f
brain, 659f
cerebral cortex, 661
extraembryonic membranes, 195f
food and air passages, 520f
muscle homology, 391f
pectoral girdle, 341f
pelvic girdle, 343f
reproductive organs, 585, 587f
skulls, 280f
supracoracoideus muscles, 406f
Opsins, 686
Optical illusion, 713f
Optic capsules, 242, 244t
Optic chiasma, 158, 689
Optic cup, 682
Optic nerve, 633, 634t, 638f, 682
Optic placodes, 189, 190t
Optics, 158–59
Optic stalk, 682
Optic vesicles, 682
Optimum design, 721
Oral contraceptives, 614b
Oral glands, 531–32

Oral hood, 62, 62f
Oral opening, 504
Oral valve, 432
Orangutans, 125, 600f
Orbicularis oris, 408
Orbital region, 257
Orbital series, 249
Orbitosphenoid bones, 244t, 264
Ordovician era, 40f, 41
Organization
kidneys, 547–50
muscles, 373–77
Organ of Corti, 696
Organogenesis, 163, 176–77, 178f
bone development and growth, 182–90
connective tissues, 180–82
epithelium, 177, 179–80
histogenesis, 177
neural crest and ectodermal placodes, 189–90
Organs. See Air-breathing organs; Copulatory organs; Muscle organs; Respiratory organs; Sensory organs; specific organs
Origin of Species, The (Darwin), 4f, 10, 11
Origins. See also Evolution
bird flight, 367–69
birds, 116
chordates, 74–81
jaws, 246–47
muscles, 390
paired fins, 327–31, 332f
pectoral girdle, 340f
vertebrates, 84–86, 562–63
Ornithischians, 114, 114f
cervical vertebrae, 313f
pelvic girdle, 341f, 342f
Ornithomimus, digits of, 356f
Ornithopods, 115
Ornithurine, 118
Oropharyngeal cavity, 505
Osmoconformers, 556f, 557, 559–60
Osmoregulation, 553, 554–62
Osmoregulators, 556f, 557
Osmotic pressure, 496
Ossification, 181
Osteichthyans, 98–99
ovary, species with single, 569t
pancreas, 604t
pectoral girdle, 340f
pituitary gland, 606, 607t
Osteoblasts, 182, 187f
Osteoclasts, 182, 187f

Osteocytes, 182
Osteoderms, 220
Osteolepiforms, 103
Osteolepis, 102*f*, 308
Osteon, 182, 184*f*, 187*f*
Osteostracans, 93, 304
Ostium, 566
Ostracoderms, 86, 90, 91*f*,
    92, 92*f*, 93
  appendicular skeletons, 331
  axial skeleton phylogeny, 304
  cranial nerves, 636*f*
  nasal passages, 676
  pineal complex, 690
  skull phylogeny, 255,
    256*f*, 257*f*
Ostriches, 115, 118, 130*f*
  cloaca, 584–85
  cursorial locomotion of, 355
  digits of, 356*f*
  intromittent organs, 579
  skulls of, 274, 277*f*
Otic capsules, 242, 244*t*
Oticoocciptal region, 257
Otic placode, 190, 190*f*,
  190*t*, 696
Otic vesicle, 190*f*
Otoconia, 696
Otolith, 68
Otolith receptor, 696
Ouranosaurus, 318*f*
Outgroup, 26
Out-torque, 142
Ova, 563
Oval, 430
Ovarian arteries, 459
Ovarian cells, 164
Ovarian cycles, in humans, 615*f*
Ovarian ducts, 569
Ovarian follicles, 567, 614*f*
Ovaries, 567, 568*f*, 569*t*, 593*t*
  castration of, 618*b*
  single, species with, 569*t*
Oviducts, 563, 566*f*, 568*f*,
  569–70, 571*f*, 572*f*
Oviparity, 589
Oviparous vertebrates, 164
Oviposition, 164
Ovis ammon ammon, 138*f*
Ovulation, 164, 563, 613
Ovum, 163. *See also* Eggs
Owen, Richard, 12–14, 13*f*,
  16, 251*b*
Owls, 118, 704–5, 707*f*
Oxygen, SCUBA diving
  and, 447*b*
Oxygen consumption, 133*f*
Oxygen delivery, 414–15,
  443, 444. *See also*
    Circulatory system;
    Respiratory system

Oxyphil cells, 597
Oxytocin, 232, 608–9

Pace, 356, 357*f*
Pacemaker, 474
Pachycephalosaurs, 116*f*
Pachystomias, 710*b*
Pacinian corpuscles, 673
Paddlefish, 100, 305
Padian, Kevin, 369
Pads, 233
Paedocypris progenetica, 130*f*
Paedomorphosis, 77, 198, 199,
  200*f*, 201
Paired fins, 94, 326,
  327–31, 332*f*
Palaeonisciformes, 99, 100
Palaeoniscoids, 100, 305
Palatal folds, 505
Palatal series, 249*t*, 250
Palate, 505–6, 507*f*
Palatine bone, 250
Palatine glands, 531
Palatoquadrate, 246, 256
Paleognathae, 118
Paleognathous birds, skulls
  of, 274, 277*f*
Paleontology, 29–41
  animal-to-fossil process, 35,
    35*f*, 36*f*, 37
  fossil dating, 37–41, 38*f*
  fossilization and fossils,
    30–32
  recovery and restoration,
    32–35
Paleostylic stage, 247
Paleothyris, 313*f*
Paleozoic era, 41
Paley, William, 6
Pallas, P. S., 65*b*
Pallidum, 663
Pallium, 662–63, 666
Palmate, 235
Pampiniform plexus, 575
Pancreas, 504, 533–35,
  602, 604*t*
Pancreatic diverticulum,
  533–34, 602
Pancreatic ducts, 535*f*
Pancreatic islets, 534, 593*t*,
  602–4
Pancreatic juice, 534
Pancreatic polypeptide (PP), 604
Panda, pancreatic ducts of, 535*f*
Panda's thumb, 209*b*
Panderichthyids, 103
Panderichthys, 103
Pangolin, alimentary canal
  of, 526*f*
Papez, James, 666
Papez circuit, 660, 666

Papilla amphibiorum, 702
Papilla basilaris, 702
Papillary cone, 683
Papillary layer, 229
Parabranchial
    compartments, 426
Parabronchi, 437
Parabronchial lungs, 440
Parachordals, 242
Parachuting, 358–59, 360*f*, 368*f*
Paradise snake, 367*f*
Parafollicular cells, 597
Parakeets
  alimentary canal of, 530*f*
  superficial musculature
    of, 405*f*
Parallax, 158, 689
Parallel muscle, 383–84
Paraphyletic groups, 28
Paraphysis, 690
Parapineal organ, 690
Parapophyses, 297, 298
Parareptilia, 109, 111
Parasitic drag, 364
Parasphenoid bone, 250
Parasympathetic nervous
    system, 642, 643*f*
Parathormone (parathyroid
    hormone), 596, 598
Parathyroid glands, 593*t*,
  596, 597
Paraxial mesoderm, 169, 392
Parexus, 97*f*
Parietal arteries, 459
Parietal cells, 521
Parietal eye, 690
Parietal foramen, 250
Parietal organ, 690
Parition, 164
Paroöphoron, 565
Parotid glands, 532
Parotoid glands, 108
Parr, 596
Parrots, 360
Pars distalis, 604, 606
Pars intermedia, 604, 606
Pars nervosa, 604, 609*f*
Pars tuberalis, 604
Pars ventralis, 606
Partial pressures, 156
Parturition, 164
Pascal (Pa), 147
Passeres, aortic arches of, 465*f*
Passive diffusion, 413–14
Passive tension, 380
Patagialis muscle, 404
Patagium, 404
Patten, William, 75, 81
Pattern
  gas transfer, 438, 440,
    443, 443*f*

phylogeny, 23–24
  unity of, 723, 724*f*
Patterning, 55, 78, 210
  feather regeneration,
    225–28, 226*f*
Pattern theory of sensation, 674
Pecten, 683
Pectineus, 399*t*, 400
Pectoantebrachialis, 398, 402*f*
Pectoral fins, 94
Pectoral girdle, 326
  Amia, 335*f*
  dual origin of, 346–47
  locomotion and, 352*f*
  phylogeny, 336–39, 340*f*,
    341*f*, 349*f*
Pectoralis, 361, 398, 399*t*
Pectoralis group, 399*t*
Pectoralis major, 398
Pectoralis minor, 398
Pectoral region, 16
Pedicel, 513
Pedicellate teeth, 107
Pedicels, 64
Peduncle, 400*b*
Pelage, 122, 230
Pelagic protochordates, 55
Pelagic sea snake, cutaneous
    respiration of, 419*f*
Pelamis platurus, cutaneous
    respiration of, 419*f*
Pelicans, 600*f*
Pelvic, 96
Pelvic claspers, 95
Pelvic fins, 94
Pelvic girdle, 326, 339, 342*f*,
  343*f*, 349*f*, 353*f*
Pelvic region, 16
Pelvis, in humans, 320*b*
Pelycosaurs, 119–20, 134*f*
  cervical vertebrae, 313*f*
  pectoral girdle, 341*f*
  pelvic girdle, 343*f*
  skeleton, 319*f*
  skulls, 265, 278*f*, 281, 284*f*
Penguins, 115, 349*f*
Penile erections, in dogs, 583*f*
Penis, 578–79, 581*f*, 582
Pennaceous feathers, 223
Pentadactylous limbs, 342
Peramorphosis, 198, 199, 200*f*
Perception, 671. *See also*
    Sensory organs
Perception mechanisms, 674
Perch
  alimentary canal of, 527*f*
  muscle fiber colors in, 378
Percomorphs, 101
Perfection, concept of, 8
Performance, 721–22
Perfusion, 415, 443, 444*f*

Periblast, 165
Pericardial cavity, 69, 196, 474
Pericardium, 474
Perichondral bones, 184
Perichondrium, 181
Perichordal rings, 302
Perichordal tube, 302
Periderm, 213
Perikaryon, 626
Perilymph, 695–96
Perimysium, 376
Periodontal ligament, 508
Periodontal membrane, 507
Periophthalmus cantonensis,
    cutaneous respiration
    of, 419*f*
Periosteal bones, 184
Periosteum, 181, 184*f*
Peripheral blood, 452
Peripheral nervous system, 625,
    628–46
    cranial nerves, 630–37, 638*f*
    evolution, 637–38
    functions of, 638–46
    overview, 669–70
    spinal nerves, 629–30
Perissodactyls, 125, 345*f,* 402
Peristalsis, 503
Peristaltic waves, 644
Peritoneal funnel, 546
Perivitelline space, 164
Permanent dentition, 507–8
Permissive hormones, 592
Personality, 665*b*
Pes, 326, 339, 342–43, 346, 347*f*
Petrolacosaurus, 112
Petromyzontida, 88–89
Petrosal, 277
Phalanges, 339
Phalarope, appendicular
    skeleton of, 349*f*
Phanerozoic eon, 40, 41
Pharyngeal bars, 58
Pharyngeal basket, 53
Pharyngeal furrows, 433
Pharyngeal musculature, 393
Pharyngeal pouches,
    517–18, 518*t*
Pharyngeal ramus, 631, 633*f*
Pharyngeal slits, 51, 52–53,
    416, 433
Pharyngobranchial element, 245
Pharynx, 52, 58, 58*f,* 504, 506*f,*
    509*b,* 511*b,* 517–18
Phasic fibers. *See* Twitch fibers
pH balance, 445, 446*f*
Pheasants, 118, 364*f*
Pheromones, 675
Philosophie Zoologique
    (Lamarck), 6
Phonemes, 511*b*

Phorusrhacus, 115, 118*f*
Phosphate, 186*b*
Photophores, 710*b*
Photopic photoreception, 686
Photoreception, 686–88
Photoreceptors, 681–91
Phrenic nerve, 197
Phrenology, 665*b*
Phylogenetic systematics, 26
Phylogeny, 21–29, 21*f*
    actinopterygian, 99*f*
    adrenal gland, 598–600
    agnathans, 87*f*
    allometric trends in, 136*f*
    amniotes, 110*f*
    appendicular skeleton, 331,
        334–36
    axial skeleton, 304–14, 316*f*
    brain, 651–54
    cardiovascular system,
        457–73, 457–90
    chordates, 48–50, 78–80
    cranial nerves, 640*f*
    dendrograms, 21–22, 23*f*
    deuterostomes, 79*f*
    dinosaurs, 116
    electroreceptors, 709,
        711*f,* 712*f*
    epigenomics and, 206–9
    euagnathans, 91*f*
    eyes, 682–83, 685*f*
    gnathostomes, 91*f,* 94*f*
    grades and clades, 24–29
    hemichordates, 60–61
    integument, 216–33
    kidneys, 550–53
    life history and, 201–9
    mammals, 124*f*
    nasal passages, 676–77,
        679*f,* 680*f*
    olfactory organs, 680*f*
    pancreatic islets, 602, 603*f*
    patterns of, 23–24
    pineal complex, 690–91, 692*f*
    pituitary gland, 606–7
    respiratory system, 424–40,
        439*f–443f*
    saber-tooth mammals, 519*b*
    simplification, 23
    skull, 255–84, 284*f,* 285*f*
    Synapsida, 120*f*
    Teleostomi, 98*f,* 104
    tetrapods, 105*f*
    thyroid gland, 592–96, 597*f*
    transformation grids in, 137*f*
    of vertibrate pituitary, 606
Physeter, axial skeletons in, 322*f*
Physical processes, 155–59
Physiological cross section, 383
Physiological properties, of
    muscle fibers, 379*t*

Physoclistous swim bladder,
    428, 429*f,* 430
Phystostomous swim bladder,
    428, 429*f*
Pia mater, 646
Pibranchial groove, 62
Picus, tongue protrusion in, 517*f*
Piezoelectricity, 155
Pigeons
    form-function models, 719
    prolactin in, 609
    synsacrum and innominate
        of, 316*f*
    vision in, 688*f*
Pigment, 238
Pigs, 125
Pike
    muscle fiber colors in, 378
    swimming specializations
        in, 401*b*
Pineal complex, 690–91, 692*f*
Pineal foramen, 610
Pineal gland, 610–11, 658, 690
Pinealocytes, 691
Pineal organ, 690
Pink muscles, 388, 389*f*
Pinna, 698
Pinnate muscle, 383–84
Pinniped, 125
Piriform lobe, 675
Pisiform, 343
Pitch, 327
Pit membrane, 693
Pituicytes, 607
Pituitary, phylogeny of, 606
Pituitary dwarfism, 609
Pituitary gigantism, 609
Pituitary gland, 604–10
Pit vipers
    infrared receptors in, 693
    striking features of, 274*b*
Placenta, 124, 192–93,
    572, 593*t*
    development, 193–95
Placental mammals. *See*
    Eutherian mammals
Placentome, 193
Placoderms, 94–95, 256
Placoid scales, 214*f,* 217
Plaice
    cutaneous respiration
        of, 419*f*
    mucus of, 237
Plankton, 289*b*
Planktonic protochordates, 55
Plantaris, 399*t,* 401
Plantigrade posture, 355
Plasma, 452
Plasma membrane, 163
Plastron, 111, 237, 299
Platelets, 452

Platypus
    alimentary canal of, 526*f*
    cerebral cortex, 661
    mosaic evolution and, 725,
        727*f*
    pectoral girdle of, 341*f*
Platyrrhine monkeys, 125
Platysma, 407
Pleisochelys, skull of, 269*f*
Plesiosaurs, 113
Pleural cavity, 196, 435
Pleurapophyses, 299
Pleurocentrum, 295, 310*f,* 311*f,*
    312–13
Pleurodires, 112
Pleurodonts, teeth of, 512
Pleuronectes platessa, cutaneous
    respiration of, 419*f*
Pleuroperitoneal cavity, 196
Pleuroperitoneal membrane,
    196
Pleurosphenoid, 244*t*
Plexuses, 644
Plies, 214
Plug cells, 709
Plumulaceous feathers, 224
Pneumatic duct, 416
Podocytes, 60, 64
Poecilia reticulata, ovaries
    of, 568*f*
Points, 234
Poison dart frogs, 218*b*
Poison glands, 220
Poisonous birds, 230*b*
Polar body, 567
Polar cartilages, 242
Polarity, 163, 204
Poliomyelitis, 657*b*
Polydactyly, 339
Polyphalangy, 339
Polyphyletic groups, 28
Polyphyodont dentition, 507
Polyphyodonty, 282
Polypteriformes, 100
Polypterus, 100
    axial skeleton, 308*f*
    brain, 654*f*
    cerebral hemispheres, 664*f*
    genital ducts (male), 575*f*
    oviducts, 568*f*
    pituitary gland, 607*t*
Polyspondyly, 295–96
Pongids, 125
Pons, 651, 654
Porolepiforms, 103
Porpoises, 125
    convergent design, 722*f*
    teeth of, 515*f*
    vertebrae of, 322*f*
Portal heart, 476
Portal system, 460

Port Jackson sharks, 308f, 512f
Positional information, 204
Positions, *Hox* genes and, 204, 205f
Postanal tail, 51, 52f, 54
Postaxial sides of fins, 327
Postcardinal vein, 460
Postcava, 460–61
Postcranial musculature
    comparative anatomy, 394–405, 406f, 407f
    embryonic development, 391f, 392–93
Postcranial skeleton, 241
Postdisplacement, 199, 200f
Posterior cardinal veins, 63, 460, 466
Posterior chamber, 682
Posterior lobes, 605
Posterior mesenteric artery, 460
Posterior region, 16
Posterior thoracic air sacs, 438
Posterior vena cava, 460–61
Postfrontal bone, 249
Postganglionic neuron, 640
Posthepatic diaphragms, 197
Posthepatic septum, 434
Postorbital bone, 249
Postorbital ligament, 274
Postparietal bones, 250
Postsacral region, 295
Posttemporal, 334
Posttrematic branches, 461
Posttrematic ramus, 631, 633f
Posture
    bipedal, 320b, 321b
    cursorial locomotion and, 355
    limbs, 352–53
Potency, 588
Pouches, 416
Power, 140
PP cells, 604
Prairie dogs, 369
Prairie falcon, 118
Preadaptation, 20–21, 562, 728
Prearticular bone, 250
Preaxial sides of fins, 327
Precambrian eons, 40, 40f
Precapillary sphincters, 454
Precardinal vein, 460
Precava, 460–61
Predator-to-prey ratios, among dinosaurs, 117b
Predisplacement, 199, 200f
Prefrontal bone, 249
Preganglionic neuron, 640
Pregnancy tests, 622b
Pregnathostome jaws, 509b, 510b
Prehension, 288

Prehepatic diaphragm, 197
Premaxilla, 249
Premaxillary glands, 531
Premetamorphosis, 616, 620f
Premolars, 514
Preorbitalis, 406
Preotic myotomes, 393
Preservationism, 202f, 203
Pressure, 140, 156, 227
Pressure drag, 145, 146, 366
Pretectal area, 689
Pretrematic branches, 461
Pretrematic ramus, 631, 633f
Prevertebrate, 84–85
Prey capture, 285–88
Primary afferent sensory nerve, 673
Primary bronchi, 438
Primary egg envelope, 163
Primary feathers, 359
Primary lamellae, 424
Primary neurulation, 168
Primary oocytes, 567
Primary palate, 250, 505
Primary pharyngeal bars, 58
Primary sclerotomes, 302
Primary spermatocyte, 572
Primates, 125
    cerebral cortex of, 661
    embryonic development, 193f
    extraembryonic membranes in, 195f
    eye structure, 683f
    mastication by, 537f
    olfactory bulb and tract, 676f
    visual perception, 688f, 690f
Prime motion, 388
Primitive amniotes. *See also* Amniotes
    digits, 356f
    pectoral girdle, 338, 340f
    pelvic girdle, 342f
    skulls, 264–68, 272f, 276f, 279f
    vertebral column, 322f
Primitive condition, 24
Primitive groove, 171
Primitive groups, 722–23, 723f
Primitive meninx, 649
Primitive node, 171
Primitive streak, 171
Primitive synapsids, skulls of, 274
Primitive terapods, 104–6. *See also* Tetrapods
Principal cells, 593
Principal islet, 602
Principles of Geology (Lyell), 37
Probainognathus, 291, 292f

Proboscidea, 124f, 125
Proboscis, 57
Procoelous centra, 296, 297f
Procoracoid bone, 338, 339, 341f
Procoracoid plate, 299
Proctodeum, 504, 584
Profile drag, 146, 366
Proganochelys, 112
Progenesis, 199, 200f
Progesterone, 614b, 615f
Progestogens, 598, 610
Prokinesis, 268
Prolactin (PRL), 609
Prometamorphosis, 616, 620f
Pronators, 390
Pronephric duct, 547, 566f
Pronephric tubules, 547
Pronephros, 547, 548f
Pronghorn, 46
Pronghorns, cursorial locomotion of, 358t
Pronghorns, horns of, 236f
Pronk, 356, 357f
Prootic bones, 264
Proportional relationships, 132
Proprioception, 673
Proprioceptors, 645, 672
Propterygium, 334
Propulsion, 351f
Prosauropods, 115
Prosencephalon, 646
Prosimians, 125
Prostate, 563
Protandrous, 72
Proteases, 536
Proterozoic eon, 40, 40f
Protoceratops, cervical vertebrae of, 313f
Protochordates, 55–56, 594
    cephalochordata, 61–66, 62f, 594
    chordates and, 78f
    hemichordata, 56–57
    integument of, 215f
    overview, 73–74
    urochordata, 66–74, 67f, 74f, 594
Protocoel, 59
Protoderm, 166, 167f
Protopterus (African lungfishes), 101b, 103f
    air-breathing by, 420f
    air-breathing organs, 447
    aortic arches, 462
    buccal pump, 423
    genital ducts (female), 570f
    genital ducts (male), 574f
    hearts, 477, 478
    lungs, 428
    venous vessels, 471f

Protorohippus, epigenomics in, 208
Protostomes, 48, 49, 50f, 50t, 51f
Protraction, 390f
Protractor lentis muscle, 683
Protractors, 390
Proventriculus, 529
Proximal convoluted tubule, 552
Proximal pars distalis, 606
Proximal region, 16
Prozeuglodon, skull of, 290b
Pseudemoia, extraembryonic membranes in, 195f
Pseudemys
    adrenal tissues of, 600f
    cloaca, 584f
    manus of, 345f
    penis, 581f
Pseudemys scripta
    cutaneous respiration of, 419f
    lymphatic vessels in, 498f, 499f
Pseudopods, 610
Pseudostratified epithelium, 179
Psoas, 399t, 400
Psoas minor, 398, 399t, 400
Pteranodon, 114f
Pteraspidomorpha, 92, 93
Pteraspis, skull phylogeny of, 257f
Pterobranchia, 57, 60, 60f
Pterolepis, skull phylogeny of, 257f
Pterosaurs, 114, 123b, 130f, 360f, 361
Pterygiophores, 326
Pterygoid bone, 250
Pterygoideus muscles, 407
Pterylae, 223
Pubioischiac bars, 334
Pubis, 336, 339
Puboischiofemoralis externus, 399t, 400
Puboischiofemoralis internus, 399t, 400
Puboischiotibialis, 399t, 401
Pudu, digits of, 356f
Pulmocutaneous artery, 464
Pulmonary arch, 464
Pulmonary artery, 462
Pulmonary fold, 196
Pulmonary system, 469
Pulmonary trunk, 481
Pulmonary veins, 469
Pulp, 507
Pulp cavity, 507
Pulsatile vesicle, 60

Pulse pump, 422, 423
Pulvinar nucleus, 689
Pumps
    aspiration, 423–24, 449
    buccal, 422–23, 428f, 449
    ciliary, 446
    dual, 421–22
    muscular, 446
    pulse, 423
Punctuated equilibrium, 24
Pupil, 682
Purkinje fibers, 474, 489
Putamen, 662, 663
Pyloric ceca, 528
Pyloric glands, 521
Pylorus, 521
Pyramidal tract, 665f
Pyriformis, 399t, 400
Pyriform recess, 519
Pythons, 130f, 693

Quadrate bone, 253, 700
Quadratojugal bone, 250
Quadratus femoris, 399t, 400
Quadrupeds, 86
Quail, adrenal tissues of, 600f
Quantities
    basic, 138–39
    derived, 139–40
Quantum evolution, 24
Quetzalcoatlus, 114, 130f
Quills, 231

Rabbits
    alimentary canal of, 526f
    coprophagy, 541
    cranial kinesis in, 254b
    fossorial locomotion of, 369
    olfactory bulb and
        tract, 676f
    pregnancy tests and, 622b
Raccoons, adrenal tissues
    of, 600f
Rachis, 223
Radial cleavage, 49, 51f
Radiale, 343
Radials, 326
Radial symmetry, 16, 18f
Radiation receptors,
    679–93, 713
    infrared receptors, 691–93
    photoreceptors, 681–91
Radiography, 42–43
Radiometric dating, 39, 39f
Radiopaque, 484b
Raja, electric organs in, 373f
Raja laevis, electroreceptors
    in, 711f
Ramus, 253
Ramus communicans, 629
Ram ventilation, 421, 427

Rana
    adrenal tissues of, 599f
    copulatory organs, 577f
    genital ducts (female), 571f
    pelvic girdle of, 343f
Rana cancrivora, dehydration
    in, 555b
Rana catesbeiana
    cutaneous respiration
        of, 419f
    hearts of, 479f
    thyroid and parathyroid
        glands in, 597f
Raptors, 118
Rate, 381
Rate modulation, 381
Ratfish, 97, 427
Rathke's pouch, 604, 605t
Ratites, 118
Rat snakes, swallowing by, 288f
Rattlesnakes
    alimentary canal of, 526f
    hemipenes, 581f
    lungs in, 436f
    striking features of, 274b
Ray, John, 6
Rays, 95–97
    copulatory organs, 576
    electric organs in, 373f
    forebrain, 651
    genital ducts (female)
        of, 570f
    intromittent organs in, 578f
    telencephalon, 661
Razorbills, appendicular
    skeleton of, 349f
Reaction. See Action/reaction,
    third law of
Recapitulation, 199, 201
Receptors, 620–21, 625
Recovery, of fossils, 32–35
Recrudescence, 588, 622
Rectal gland, 528
Rectangular Cartesian reference
    system, 140
Rectocoprodeal fold, 582
Rectum, 504, 523
Rectus femoris, 399t, 400
Red blood cells, 452
Red-eared turtle, cutaneous
    respiration of, 419f
Red kangaroos, reproductive
    cycle of, 616, 617f
Red muscles, 373, 388, 389f
Redstart, adrenal tissues
    of, 600f
Red-tailed hawk, alimentary
    canal of, 526f
Reedfish, cutaneous respiration
    of, 419f
Refection, 535

Reference, units of, 139. See also
    Acceleration; Force;
    Length; Mass; Time;
    Velocity
Reference systems, 140
Reflex, in mammals, 640t. See
    also Spinal reflexes
Refraction, 159
Refractive index, 159, 684
Regulatory genes, 204
Reissner's membrane, 700
Relativity, theories of, 141
Release-inhibiting
    hormones, 607
Releasing hormones, 607
Remodeling, 728–29
Renal arteries, 459–60
Renal capsule, 551
Renal corpuscle, 551
Renal pelvis, 551
Renal portal system,
    467–68, 469f
Renal portal veins, 460
Renin, 601, 612
Replacement bones, 183
Reproduction
    mammals, 613–16
    thyroid gland and, 596
Reproductive kidney, 566
Reproductive potential, 5
Reproductive system, 545,
    563–89
    cloaca, 582–86, 586f, 587f
    embryoninc development,
        563–67
    female, 567–72
    function and evolution,
        588–89, 590f
    gestation delays, 589, 590f
    male, 572–82, 583f
    overview, 589, 591
    structure (mammals),
        563, 564f
    urinary bladder, 586, 588
Reproductive tracts, 565, 566f
Reptiles, 110, 111–19, 680f
    air-breathing
        mechanisms, 449
    alimentary canal, 526f, 529
    aortic arches, 464–65, 468f
    aspiration pump, 424
    autonomic nervous
        system, 645f
    axial musculature, 396–97
    body cavities, 196f
    branchial arches, 247t
    breathing and embryos,
        418–20
    CCK cells, 623–24
    cerebral hemispheres, 664f
    cervical vertebrae, 313f

chondrocranium, 244t
circulation, 456
cleavage, 165–66
cloaca, 582, 584f
cranial kinesis, 268
cutaneous respiration,
    417, 419f
ear functions, 704, 705f
embryonic development, 194
epigenomics, 206–8
extraembryonic membranes,
    190–91, 191t
eyes, 683
gait, 356
gastrulation and neurulation,
    171, 172
genital ducts (female), 571f
genital ducts (male), 576f
gliding and parachuting, 360f
hearts, 480–89, 480f, 482f,
    483f, 484b–486b,
    488f, 489f
integument, 220–21, 225
intromittent organs, 579
lymphatic system, 499f
mesotarsal and crurotarsal
    joints, 347f
midbrain, 658
muscle development
    (embryonic), 392f
muscle homology, 391f, 399t
nasal sac, 676–77
nephrons, 554f
nitrogen elimination, 554
olfactory bulb and tract, 676f
oral glands, 532f
ovary, species with
    single, 569t
pallium, 663
pancreas, 604t
pectoral girdle, 339, 340f
pelvic girdle, 342f, 343f, 353f
penises, 578–79, 581f, 582
pineal complex, 690
pituitary gland, 607
placenta, 193
respiratory system, 433–35,
    436f–438f
salt balance, 561
scales, 236
secondary palate, 284f
skull phylogeny, 268–72
splanchnocranium, 293f
stomach, 528f
supracoracoideus
    muscles, 406f
systemic arches, 468f
teeth, 513–14
telencephalon, 661
thyroid glands, 595f
urinary bladder, 588f

venous vessels, 472
water balance, 556
Respiration, 414–15
Respiratory gas bladders, 417
Respiratory membrane, 191t
Respiratory organs, 415
  evolution of, 444–50
  types of, 416–21, 421f
Respiratory system, 413–16
  form and function, 438,
    440–44, 443–44, 445f
  overview, 450
  phylogeny, 424–40,
    439f–443f
  respiratory organs, types of,
    416–21, 421f, 444–50
  ventilatory mechanisms,
    421–24
Respiratory tree, 437
Respiratory unit, 426
Resting muscle, 376
Resting state, 376
Restoration, of fossils, 32–35
Rete, 158
Rete mirabile, 429f, 430
Rete testis, 563
Reticular formation, 667
Reticular groove, 540
Reticular lamina, 179
Reticular layer, 229
Reticular systems, 660
Reticulated python, 130f
Reticulum, 530
Retina, 159, 682, 683f
  retinal connections, 688f
Retractors, 390, 683
Reverse recapitulation, 201
Reynolds number, 146–47
Rhachitomous vertebra,
    296, 310f
Rhamphorhynchoids, 114
Rhamphorhynchus, 114f
Rheas, 118, 356f
Rhesus, buccal cavity of, 506f
Rhinoceroses, 125
  alimentary canal of, 526f
  facial muscles of, 409f
  horns of, 236f
  hyoid arch, 408
Rhipidistians, 102–103
  appendicular skeleton,
    335–36, 337f
  axial skeleton, 308, 309
  gaits, 350
  land, movement on, 448
  limb development, 332f
  lobe fins, 347
  skulls, 263, 265f
Rhombencephalon, 646
Rhomboideus complex muscles,
    397, 399t

Rhynchocephalian, skull of, 270f
Rhynchokinesis, 277f
Rhynchosaurs, 113
Rib cage, 299
Rib cage movements, in
    mammals, 439f
Ribs, 297–99
Richardson, M., 201f
Ridge soaring, 364f
Right atrial chamber, 477
Right systemic arch, 464
Right systemic trunk, 481
Right-to-left shunt. See Cardiac
    shunt
Right whales, 289b, 515f
RNA, 205
Roadrunners, digits of, 356f
Rockskippers, accessory
    airbreathing organs
    of, 418
Rodents, 125
  cerebral cortex, 661
  cloaca, 585
  coprophagy, 541
  fossorial locomotion, 369
  teeth, 508
Rods, 682, 686
Roe, 100
Role, 19–20, 19f
Roll, 327
Romer, Alfred, 29, 104
"Romer's gap," 104
Roofing bones, 250
Root, 229, 507, 508
Root canal, 507
Rorquals, 289b
Rostral, 606
Rostrum, 97
Rotators, 390
Rough-legged hawks, wing
    shape of, 364f
Rubrospinal tract, 651
Ruf, Irina, 708f
Ruffed grouse, alimentary canal
    of, 526f
Rugae, 521
Rumen, 530
Ruminants, 125
  alimentary canal of, 530
  gastric fermentation in, 539,
    541–42
  hindgut fermentation, 541f
  stomach of, 528f
Ruminating, 539
Rumination, 539, 540
Running gait, of humans, 387f

Saber-teeth, 519b
Sacciform cell, 216
Sacculus, 696
Saccus vasculosus, 606

Sacral region, 295, 319
Sagittal plane, 16
Sahelanthropus tchadensis, 725
Saint-Hillaire, Geoffroy, 75,
    79, 81
Salamanders, 106–7
  alimentary canal, 526f
  aortic arches, 462f, 463,
    464f, 479f
  appendicular
    musculature, 402f
  axial musculature, 395
  axial skeleton, 309–10
  copulatory organs, 577
  courtship, 580f
  cranial and shoulder
    musculature, 402f
  cutaneous respiration, 419f
  diversity and patterns in, 724f
  ears, 704
  facial muscles, 411f
  fertilization, 589
  functional coupling of
    respiration and
    feeding, 719–20
  gait, 356
  genital ducts (female), 571f
  genital ducts (male), 573,
    575f
  hearts, 478–79, 480
  integument, 219–20
  locomotion, 311f, 352f
  lymphatic vessels, 498f
  maturation, 198, 199
  muscle homologies, 399t
  nasolabial grooves, 676
  pectoral girdle, 337–338, 341f
  pelvic girdle, 343f
  pelvic musculature, 402f
  respiratory systems (larvae),
    430, 432
  skull, 245f, 264, 265f
  sperm storage, 590f
  stomach, 528f
  suction feeding by, 265f, 266f
  suspension feeding, 287
  teeth, 512–13
  thyroid and parathyroid
    glands, 597f
  urogenital organs, 551f
  venous vessels, 472
  vestibular apparatus, 698f
Salientia (Anura), 107–8
Saliva, 532
Salivary glands, 504, 531, 532
Salmo, pectoral girdle of, 341f
Salmon
  brain of, 654
  cloacal and anal regions
    of, 583f
  spinal cords of, 650f

Salmonids, 101
  genital ducts (female)
    of, 569
  genital ducts (male), 573
Salmo trutta, cutaneous
    respiration of, 419f
Saltatorial locomotion, 349
Salt balance, 554–62, 560–62
Salt gland, 222, 561
Saltwater crocodile, 130f
Sarcolemma, 373
Sarcomeres, 374
Sarcoplasmic reticulum, 374
Sarcopterygians, 99, 101–4
  appendicular skeleton,
    335–36, 337f
  axial skeleton phylogeny,
    308–9
  pelvic girdle, 342f
  skull phylogeny, 261–62
Sartorius, 399t, 400
Saurischians
  digits of, 356f
  pelvic girdle of, 342f, 343f
Sauromalus obesus, cutaneous
    respiration of, 419f
Sauropodomorphs, 115, 116f
Sauropods, 115
Sauropsids, 109, 111–19,
    338–39, 340f,
    342f, 487
Sauropterygians, 113
Savitzky, Alan H., 222b
Scala media, 700
Scalar quantity, 140
Scala tympani, 700
Scala vestibuli, 700
"Scale of nature" (scala
    naturae), 7
Scales, 214f, 218, 220, 225, 236
Scaling, 129
Scaly-tailed squirrels, 125
Scansorial locomotion, 349
Scapula, 338, 341f
Scapulocoracoid, 334, 338
Sceloporus, 113f
Scent glands, 221, 232
Schizocoelom, 49, 51f
Schizocoely, 169
Schroeder, F. C., 222b
Schwann cells, 625
Schwenk, K., 517f
Scientific method, 3b
Sciuromorphs, 125
Sclera, 681
Scleral ossicles, 249, 681
Sclerotome, 168f, 169, 213,
    302–3, 392
Scotopic photoreception, 686
Screens, 71
Scroll valve, 527

Scrotum, 573, 575
SCUBA diving, 447*b*, 448*b*
Sculpin, swimming
    specializations
    in, 401*b*
Scute, 220
Scyliorhinus canicula, cutaneous
    respiration of, 419*f*
Scyllium, jaws of, 246*f*
Seabirds, pharynx of, 518
Sea cows, 125
Seals, 125, 447*b*, 448*b*
Sea snakes, cutaneous respiration
    of, 417, 419*f*
Sea squirts. *See* Ascidiacea
Sea turtles, magnetic fields
    and, 712–13
Sebaceous glands, 231
Sebum, 231
Secondary afferent sensory
    nerve, 673
Secondary bronchi, 438
Secondary cartilage, 184
Secondary egg envelope, 164
Secondary endocrine organs,
    611–13
Secondary lamellae, 424
Secondary meninx, 649
Secondary neurulation, 168
Secondary oocytes, 567
Secondary palate, 282,
    284*f*, 505
Secondary pharyngeal bars, 58
Secondary plant
    compounds, 542
Secondary sclerotomes, 303
Secondary spermatocyte, 572
Second law of motion, 141,
    142*f*, 144
Secretary birds, digits of, 356*f*
Secretin, 611
Secretion kidneys, 557
Sectorial teeth, 515
Segment, 16
Segmentation, 16–19
Seismic sound waves, 703
Selenodont teeth, 514
Semen (seminal fluid), 563
Semicircular canals, 696
Semilunar valves, 476
Semimembranosus, 399*t*, 401
Seminal fluid, 563
Seminal vesicle, 563
Seminiferous tubules, 563
Semitendinosus, 399*t*, 401
Senescence, 163
Sensation, pattern theory
    of, 674
Sensory circuits, 646*f*
Sensory neurons, 628. *See also*
    Afferent neurons

Sensory organs, 671–72
    associated sensory receptors,
        673, 674*f*, 675*f*
    chemoreceptors, 674–79,
        680*f*–682*f*
    components of, 672
    electroreceptors, 709–12
    encapsulated sensory
        receptors, 673
    free sensory receptors,
        672–73
    magnetic fields and, 712–13
    mechanoreceptors, 693–709
    overview, 712–13
    perception mechanisms, 674
    radiation receptors, 679–93
Sensory receptors, 671, 672*f*
    associated, 673, 674*f*, 675*f*
    encapsulated, 673
    free, 672–73
Sensory vesicle, 68
Septa, 18, 662, 675. *See also*
        Primary pharyngeal
        bars
Septal, 430
Septomaxilla, 249
Serial homology, 16
Serial theory, 246, 247*f*
Serosa, 196, 521
Serratus muscles, 397, 399*t*
Sertoli cells, 572
Sesamoid bones, 184
Sessile protochordates, 56
Seymouria, 106*f*
    locomotion of, 312
    pectoral girdle of, 341*f*
Seymouromorpha, 110
Shaft, 229
Shape, 128–29, 134–37, 137*f*
    dynamic fluids and, 146
    Hox genes, 204–5, 206*f*
    wings, 364*f*
Sharks, 95–97, 333*f*
    adrenal tissues, 599*f*
    alimentary canal of, 527*f*
    anal regions of, 583*f*
    aortic arches, 463*f*
    appendicular
        musculature, 402*f*
    appendicular skeleton, 327,
        332*f*, 333
    autonomic nervous
        system, 644*f*
    axial skeleton phylogeny,
        305–7
    brain, 654*f*, 655*f*, 658*f*, 659*f*
    branchial arches, 247*t*
    branchiomeric
        musculature, 409*f*
    buccal cavity, 505, 506*f*, 510*b*
    cerebral hemispheres, 664*f*

    circulatory pattern, 460*f*
    cloaca, 582, 583*f*
    copulatory organs, 576
    cranial musculature,
        402*f*, 410*f*
    cranial nerves, 636*f*, 638*f*
    cutaneous respiration, 419*f*
    diversity and patterns
        in, 724*f*
    extrinsic eye
        musculature, 394*f*
    eyes, 685*f*
    feeding in, 259*f*
    fins and squalene, 430
    forebrain, 651
    genital ducts (male), 575*f*
    gills, 426*f*
    gill ventilation, 426*f*
    hyoid arch, 407
    intromittent organs, 578*f*
    jaws, 246, 246*f*, 247*t*, 248
    locomotion, 311*f*
    mandibular arch, 406
    medulla oblongata, 654
    muscle development
        (embryonic), 392*f*
    muscle fibers, 389*f*
    muscle homology, 391*f*, 408*t*
    nasal sacs, 679*f*
    nephrons, 554*f*
    oviducts, 568*f*
    pancreas, 535
    pelvic musculature, 402*f*
    pineal complex, 692*f*
    ram ventilation, 421
    respiratory system, 426–27
    salt balance, 561*f*
    scales, 236
    shoulder musculature, 402*f*
    skin, 217*f*
    skull phylogeny,
        257–58, 259*f*
    spinal cords, 650*f*
    stomach, 528*f*
    swimming forces in, 395*f*
    teeth, 239*f*, 512*f*, 513*f*
    telencephalon, 661
    venous vessels, 469*f*, 471*f*
    vestibular apparatus, 698*f*
Sharman, G. G., 617*f*
Shear forces, 148, 149*t*
Shear strength, 148, 149*t*
Shearwater, wing shape of, 364*f*
Sheep, 125, 130*f*, 138*f*,
        236*f*, 688*f*
Shell, of turtles, 299, 301*f*
Shell glands, 567
Shell membrane, 166, 167*f*
Shell-secreting regions, 567
Shh expression, 330
Shingles, 633*b*

Shock. *See* Hypotensive shock
Shortening, 385
Shoulder girdle, 326. *See also*
        Pectoral girdle
Shrews, 600*f*
Shunts, 454
Siberian argali, 138*f*
Sight. *See* Optics
Silurian era, 40*f*, 41
Siluroids, 101
Similarities, 15–16, 15*f*
Simons, J. R., 308*f*
Simple epithelia, 179
Simple glands, 180
Simplex uterus, 585
Simplification, 23
Simpson, G. G., 24, 29
Single circulation, 455–56
Sinoatrial (SA) node, 474
Sinoatrial (SA) valves, 474
Sinus, 460
Sinus venosus, 63, 456, 473
Sirenia, 125
Siren salamander (male),
        urogenital organs
        of, 551*f*
Sister group, 24
Size, 129–34, 137, 146, 541–42
Skates, 97
    adrenal tissues, 599*f*
    electric organs in, 373*f*
    electroreceptors in, 711*f*
    telencephalon, 661
    thyroid glands, 595*f*
Skeletal muscles, 373, 374*f*
    classification, 374–75
    structure of, 375–76
Skeletal system, of
        pteraspidomorpha,
        93. *See also*
        Appendicular
        skeleton; Axial
        skeleton; Skull
Skeletal tissues, organization
        of, 242*f*
Skeleton
    rigid, 18, 19
Skeletons
    chitinous outer, 18
    hydrostatic, 18, 18*f*
Skimmers, 291
Skin. *See also* Integument
    bird, 222*f*
    bony-fish, 220*f*
    color, 237*b*
    mammalian, 229*f*
    shark, 217*f*
Skin color, 238
Skin derivatives, 214*f*
Skinks, extraembryonic
        membranes in, 195*f*

Skull, 241–42
  chondrocranium, 242, 244, 244*f*, 244*t*
  composite, 243*f*
  cranial kinesis, 253–55
  dermatocranium, 248–50
  function and design, 284–88
  lizard, 270*f*
  modularity and, 725, 728*f*
  morphology overview, 250–53
  overview, 288–93
  phylogeny of, 255–84, 284*f*, 285*f*
  splanchnocranium, 244–48
  theories related to, 13
Skull bones, mobility of, 253*f*
Skull fenestrae, 265, 267, 268*f*
Skull fenestration, 109–10
Slider-crank mechanism, 274
Slime glands, 217
Slip, 390
Sloth, alimentary canal of, 526*f*
Slow twitch fibers, 378–79
Small intestine, 504, 523, 524*f*
Smilodon, teeth of, 519*b*
Smith, A. B., 29*f*
Smith, H. M., 648*f*, 708*f*
Smith, Homer W., 84, 560*b*, 562
Smith, J. L. B., 45*b*
Smolt, 596
Smooth muscles, 373, 375
Snails, 130*f*
Snakes, 113–14, 130*f*
  adrenal tissues, 600*f*
  alimentary canal, 526*f*
  axial skeleton, 313
  cloaca, 584*f*
  cranial kinesis, 268, 273*f*
  cutaneous respiration, 419*f*
  dermatocranium evolution, 267*f*
  ears, 704
  facial muscles, 411*f*
  fossorial locomotion, 369, 370, 370*f*
  functional matrices in, 720–21
  gliding, 367, 367*f*
  hemipenes, 579, 581*f*
  Hox genes, 729*f*, 730
  hypaxial musculature, 397
  infrared receptors, 693
  limb loss, 208*f*
  liver, 533
  lungs, 436*f*
  mosaic evolution, 726*f*
  pharynx, 518

placenta, 193
  sensory organs and origins, 694*b*, 695*b*
  skulls, 271, 273*f*
  stomach, 528*f*
  striking features, 274*b*
  swallowing, 518*f*
  teeth, 512*f*, 513–14
  thyroid glands, 595*f*
  tongue flicking, 681*f*
  toxins in prey, 222*b*
  venom glands, 532
  vertebrae, 316*f*
  vestibular apparatus, 698*f*
Sniffing, 677, 680*f*
Soaring flight, 361, 363*f*, 364*f*
Socket, 507
Soft palate, 282, 506
Sole, 233
Soleus, 399*t*, 401
Solitary protochordates, 56
Soma, 626
Somatic mesoderm, 168*f*, 169
Somatic muscles, 373
Somatic nerves, 628, 629*f*, 634*t*
Somatic reflex arcs, 638, 640, 641*f*
Somatic sensory organs, 672
Somatopleure, 168*f*, 169
Somatosensory system, 645
Somatostatin, 604
Somites, 168*f*, 169, 392
Somitomeres, 169, 392, 393
Songbirds, aortic arches of, 465*f*
Sonic hedgehog, 207, 330
Sound detection, by swim bladders, 430
Sounds, speech and, 511*b*. *See also* Auditory system; Ears
Sound vibrations, distribution of, 705–6, 709*f*
South American lungfish. *See* Lepidosiren
Southern musk turtle, cutaneous respiration of, 419*f*
Spathe, 223
Special cranial nerves, 630
Special sensory organs, 672, 674
  chemoreceptors, 674–79, 680*f*–682*f*
  electroreceptors, 709–12
  magnetic fields and, 712–13
  mechanoreceptors, 693–709
  radiation receptors, 679–93
Special stains, 602
Specific nerve energies, theory of, 674
Spectacle, 694*b*
Speech, of humans, 511*b*

Speed, 144
  strength versus, 143*f*, 385*f*, 386*f*
Speed advantage, 143
Sperm, 163, 545
Spermatheca, 577, 590*f*
Spermatic arteries, 459
Spermatid, 572
Spermatocyst, 572
Spermatocytes, 572
Spermatogenesis, 572, 573*f*
Spermatogonia, 572
Spermatophore, 107, 576, 580*f*
Spermatozoa, 572
Sperm production, in teleosts, 574*f*
Sperm storage, in salamanders, 590*f*
Sperm whales, 431*b*
Sphenethmoid, 244*t*
Sphenodon, 110, 113, 113*f*
  axial musculature, 396*f*
  cloaca, 584*f*
  dentition, 512*f*
  gastralia, 299
  genital ducts (female), 571*f*
  intromittent organs, 578
  pectoral girdle, 341*f*
  pineal complex, 690
  skulls, 265, 267*f*, 268, 270*f*, 272
  vestibular apparatus, 698*f*
Sphenoid bones, 244*t*, 250
Sphincter muscles, 390
Spinal accessory nerve, 634, 635*f*, 639*t*
Spinal autonomic system, 644
Spinal cord, 649–51, 651*f*, 652*f*, 653*t*
Spinal nerves, 629–30, 631*f*, 632*f*, 633*b*
Spinal reflexes, 638, 640–41, 649–50, 651*f*, 652*f*
Spinal tracts, 650–51, 652*f*, 653*t*
Spine. *See* Axial skeleton; Neural spine; Vertebrae
Spinocerebellar tracts, 650
Spinocortical associations, 667–69
Spinothalamic tract, 650, 668*f*
Spiracle, 96, 256, 257, 426, 461
Spiracular pseudobranch, 426
Spiral cleavage, 49, 51*f*
Spiral valve, 478, 525
Splanchnic mesoderm, 168*f*, 169
Splanchnocranium, 242, 243*f*, 244–48, 245*f*, 250*f*, 292, 293*f*

Splanchnopleure, 168*f*, 169
Splenials, 250
Splitting, 411
Spongy bone, 154, 182, 184*f*
Squalene, 430
Squalus
  appendicular skeleton, 334*f*
  axial skeleton phylogeny, 306*f*
  cranial musculature, 410*f*
  cranial nerves, 636*f*, 638*f*
  eyes, 685*f*
  skull phylogeny, 259*f*
Squamates, 113. *See also* Lizards; Snakes
  blood flow through heart of, 482*f*
  cranial kinesis in, 268*f*
  hearts of, 481–83
  systemic circulation, 483*f*
Squamosal bone, 248, 250
Squamosal-stapedial ligament, 704
Squamous cells, 179
Squirrels, 125. *See also* Flying squirrels
Stagnation lines, 363
Stall, 361
Stance phase, 388
Stapedial muscle, 703
Stapedius, 407
Stapes, 248, 281, 698, 700, 702
Stargazers, electric organs in, 373*f*
Static fluids, 147
Statics, 142
Stegocephali, 104
Stegosaurs, 115
Stem-amniotes, 110
Stem group, 28
Stenohaline animals, 558*b*, 560
Stereocilia, 179, 693
Stereoscopic vision, 158, 688
Stereospondylous vertebra, 296
Sterility, 588
Sternal (diaphragm), 435
Sternal plate, 299
Sternal rib, 299
Sternebrae, 299
Sternotherus minor, cutaneous respiration of, 419*f*
Sternum, 299, 300*f*
Steroids, 598
Stigmata, 68
Stize, T. E., 667*f*
Stolons, 70
Stomach, 504, 521, 523, 524*f*, 526*f*, 528*f*, 539*f*, 540*f*
Stomochord, 59
Stomodeum, 504, 604
Strata, 35

Stratified epithelia, 179
Stratigraphy, 37–38, 37f
Stratum basale, 213
Stratum compactum, 213
Stratum corneum, 215, 216f, 228
Stratum externa, 233
Stratum germanitivum, 213
Stratum granulosum, 228
Stratum intermedium, 221
Stratum internum, 233
Stratum lucidum, 228
Stratum medium, 233
Stratum spinosum, 228
Streamlining, 145, 146f, 315
Strength, 148, 149t
    speed versus, 143f, 385f, 386f
Strepsirrhini, 125
Streptostyly, 268
Stress
    intermittent, 154
    mechanical, 151–55
Stress trajectories, 155f
Stretch reflex, 673, 675f
Striatum, 662, 663
Stride length, 355
Stride rate, 355, 356f
Structural analysis, 717–18
Structural coupling, 720f
Structural genes, 204
Structure
    adrenal gland, 598–600
    electroreceptors, 709, 711f, 712f
    eyes, 681–82, 683f
    hormonal control, 620, 621f
    kidneys, 545–47, 553–62
    pancreatic islets, 602, 603f
    pineal complex, 690, 691f
    pituitary gland, 604–6, 607t
    reproductive system (mammals), 563, 564f
    skeletal muscles, 375–76
    thyroid gland, 592–94, 595f
Sturgeons, 45b, 100
    alimentary canal of, 527f
    axial skeleton phylogeny, 305, 307f
    cloacal and anal regions of, 583f
    genital ducts (male), 575f
    gill ventilation in, 429b
    oviducts of, 568f
    vestibular apparatus, 698f
Sturnus, pectoral girdle of, 341f
Styloichthys, 102
Styloid process, 253
Stylophora, 61
Stylopodium, 326
Suarischia, 114–15, 116f
Subcardinal anastomosis, 468

Subcardinal veins, 468
Subclavian arteries, 459
Subclavian lymphatics, 496
Subclavian vein, 460, 466
Subcortical region, 661
Subgerminal space, 165f, 166
Subintestinal vein, 466
Sublingual glands, 531, 532
Submucosa, 520
Submucosal plexuses, 644
Subneural gland, 69
Subpallium, 662, 663–64, 667
Subscapularis, 398, 399t
Substantia nigra, 663
Subterranean animals, 369
Subunguis, 233f
Subungulates, 125
Suckling, muscles for, 407
Suction feeding, 255, 260, 263f, 266f, 285, 286–87
Suction phase, 422
Sugars, 537
Sulci, 661
Summation, 628
Superficial fascia, 212
Superficial musculature, 404f, 405f
Superior colliculi, 658, 689
Superior (cranial), 18f
Superior region, 16
Supinators, 390
Supportive rods, 62
Supracardinal anastomosis, 468
Supracleithrum, 334, 341f
Supracoracoideus muscles, 398, 399t, 406f, 407f
Supralabial glands, 531
Supraoccipital bone, 250
Suprarenal, 600
Supraspinatus, 398, 399t
Supratemporal bone, 250
Surangular bone, 250
Surface area, 129, 133
Surface tension, 132f
Surfactant, 421
Survival of the few, 5
Suspension feeding, 285, 287f
Suspension-feeding organisms, 53
Suspension mucous feeding, 58f
Suspension phase, 350, 357, 359f
Suspensorium, 247, 260–61
Suspensory ligament, 682
Sustentacular cells, 675
Sutures, 189
Swallowing, 288
Swallows, 361, 364f
Sweet, Samuel S., 136f
Swifts, 361
Swim bladders, 99, 416–17, 428, 429f, 430

Swimming, 349–50
    functional analysis of, 718
    musculature and, 401b
    sharks, 395f
Swimming birds, appendicular skeleton of, 349f
Swing phase, 388
Symbranchus, air-breathing by, 420f
Symmetrical gaits, 357
Symmetry, 16, 18f
Sympathetic chain, 629
Sympathetic nervous system, 642, 642t, 643f
Symphyses, 189
Symplectic bone, 248
Synapses, 628
Synapsids, 109, 119–26
    cervical vertebrae of, 313f
    dermatocranium evolution, 267f
    pectoral girdle, 340f
    pelvic girdle, 342f
    skull, 110, 274–84, 279f, 280f, 281, 284f, 285f
Synapticules, 62
Synarthrosis, 187
Synchondrosis, 187
Syncytiotrophoblast, 192
Syndesmosis, 187
Synergists, 388
Synostosis, 187
Synovial capsule, 187
Synovial fluid, 187
Synovial joint, 187
Synovial membrane, 187
Synsacrum, 313, 316f
Système Internationale (SI), 139
Systemic arches, 468f
Systemic system, 466–69, 469f, 470f
Systems of exchange, 157f
Systolic pressure, 453

Tabular bone, 250
Tachycardia, 473
Tachyglossus, 341f
Tadpoles, gill ventilation by, 432f
Tailed frogs, 432, 588
Talons, 118, 233
Tapetum lucidum, 682
Tapeworms, 17, 18f, 130f
Target tissues, 592, 620–21
Tarpon, 130f
Tarsals, 339
Tarsiers, 515f
Tarsometatarsus, 343
Tassle-finned fish, 102
Taste buds, 678–79, 682f
Taste pore, 678
Taxa, 24, 26, 27f, 28

Taxonomists, 2
Taxonomy, 42
Taylor, David, 33f
T-box gene, 330
Tbx4, 330
Tbx5, 330
Teat, 232
Tectorial membrane, 700
Tectospinal tract, 650
Tectum, 651, 658, 660f
Teeth, 214f, 506–16, 511f, 512–16, 519b, 537f, 538b
    in dogs, 508f
    in snakes, 513
    tooth structure, 507f
Tegmentum, 651, 658
Teiid lizard, dentition of, 512f
Telencephalon, 651, 660–61, 662t, 666
Teleosts, 94, 97–104, 604t
    adrenal tissues, 599f
    alimentary canal, 528
    aortic arches, 462f
    appendicular musculature, 393
    autonomic nervous system, 644f
    axial musculature of, 396f
    axial skeleton phylogeny, 305, 307f
    branchial arches, 247t
    branchiomeric musculature, 409f
    calcium regulation, 598
    CCK cells, 623f
    cloaca, 582
    contraction cycle, 477f
    copulatory organs, 576
    cutaneous respiration, 417
    diversity and patterns, 724f
    embryonic development, 190, 193
    eyes, 685f
    female urogenital system of, 570f
    forebrain, 651
    genital ducts (female), 568f, 569
    genital ducts (male), 573, 574f, 575f
    gill ventilation, 427f
    heart, 473, 476f, 477f
    kidneys, 554, 558
    land, movement on, 448
    midbrain, 652
    muscle fibers, 388
    nephrons, 554f
    ovaries, 568f
    pallium, 663
    pectoral girdle, 340f
    pineal complex, 692f

pituitary gland, 606–7
skulls, 261, 262*f*, 263*f*
spawning in, 579
sperm production, 574*f*
stomach, 528*f*
suction feeding, 263*f*
thyroid gland, 594, 595*f*
urinary bladder, 588*f*
vertebrae, 302*f*, 322*f*
vestibular apparatus, 698*f*
visual accommodation, 687*f*
Teleost scale, 218
Telmatobius coleus, cutaneous
respiration in, 420*f*
Telolecithal, 163
Temnospondyli, 105, 107*f*
lineage of, 310–11, 311–12
pectoral girdle of, 341*f*
pelvic girdle of, 343*f*
Temporal arches (bars), 109
Temporal bone, 277, 281*f*
Temporal fenestrae, 109,
267, 268*f*
Temporalis, 407
Temporal notch, 107, 250
Temporal region, 250
Temporal series, 249*t*, 250
Tendons, 376
Tennyson, Alfred Lord, 503
Tensile force, 148, 149*f*, 376
Tensile strength, 148, 149*f*
Tension-length curves, 377,
380, 381*f*
Tensor fascia latae, 399*t*, 400
Teres major, 398
Teres minor, 398
Terminal nerve, 633, 634*t*
Terminal web, 693
Terrestrial environment
appendicular skeleton and,
347–48
axial skeletons in,
315–17, 318*f*
breathing in, 444
dehydration in, 555*b*
locomotion of, 311*f*, 312
vision in, 685–86, 687*f*
water balance, 556*f*
Terrestrial feeding, 285–87
Terrestrial locomotion, 347–48,
349, 350–58. *See also*
Locomotion
Tertiary egg envelope, 164
Test. *See* Tunic
Testes, 572, 573*f*, 574*f*,
578*f*, 593*t*
castration of, 618–19*b*
Testicular duct, 573
Testosterone, 572, 610, 613
Test (tunic), 68
Testudines, 111–12, 267*f*
Testudo, shell of, 301*f*

Tetraiodothyronine, 592–93
Tetrapods, 86, 104–8, 346*f*
alimentary canal,
528–31, 529*f*
appendicular musculature,
397–405, 406*f*, 407*f*
appendicular skeleton,
336–46, 349*f*
autonomic nervous system,
645, 645*f*
axial musculature, 395–97
axial skeleton, 302–4, 309–14
cerebrospinal fluid and
meninges, 648*f*
cloaca, 582, 584*f*
ears, 701–2, 706–7
forebrain, 651–54
gaits, 352
genital ducts (female),
569, 571*f*
genital ducts (male), 573,
575–76, 577*f*, 578*f*
gut, 544*f*
hyoid arch, 407
integument, 219–33
kidneys, 551–53
limbs, 330–31, 332*f*, 345*f*
locomotion, 311*f*
muscle homology, 391*f*
nasal passages, 676, 677
pancreas, 535
pectoral girdle, 336–39,
340*f*, 341*f*, 349*f*
pelvic girdle, 339, 342*f*,
343*f*, 349*f*
pituitary gland, 606*f*, 607
postcranial muscle
development, 391*f*
respiratory system, 430
skull phylogeny,
263–64, 266*f*
specializations among, 401
splanchnocranium, 293*f*
sterna, 300*f*
swimming, 349–50
systemic arches, 468*f*
terrestrial environment and,
315–17, 318*f*
tongue, 516, 517*f*
urinary bladder, 586, 588
vertebra (embryonic
development), 302*f*
vertebral column, 295,
318–19
Thacher, J. K., 328, 329
Thalamus, 651, 658, 660
Thaliacea, 73, 74*f*
Thamnophis, adrenal tissues
of, 600*f*
Theca, 613
Theca lutein cells, 613
Thecodonts, 114, 512

Thelodonts, 93
Theory of specific nerve
energies, 674
Therapsids, 119–21
cervical vertebrae, 313*f*
locomotion, 353
pectoral girdle, 340*f*
pelvic girdle, 342*f*, 343*f*
secondary palates, 285*f*
skulls, 264, 274, 278*f*
supracoracoideus
muscles, 406*f*
temporal fenestrae, 268*f*
vertebral column, 322*f*
Therian mammals, 122, 123
ear anatomy of, 700*f*
pectoral girdle of, 340*f*
urinary bladder of, 588
Therian radiation, 125*f*
Theriodontia, skulls of, 278*f*
Thermals, 361, 364*f*
Thermoreceptors, 693. *See also*
Infrared receptors
Theropods, 114–15, 116*f*
Thescelosaurus, pelvic girdle
of, 343*f*
Thick filaments, 374
Thin filaments, 374
Third eye, 690
Third law of action/
reaction, 141
Thompson, D'Arcy, 134
Thoracic lymphatics, 496
Thoracolumbar outflow, 642
Thoracolumbar region, 319
Thorax region, 319
Thread cells, 217
Thrinaxodon
cervical vertebrae of, 313*f*
supracoracoideus muscles
of, 406*f*
Thrombus, 452
Thylacosmilidae, 519*b*
Thymus, 596*f*
Thyreophora, 116*f*
Thyroid gland, 51, 53, 597*f*
Thyroid hormones, 592–96
Thyroid secretion and
mobilization, 595*f*
Thyroid-stimulating hormone,
594, 595*f*, 609,
616, 620*f*
Thyrotropin, 594, 595*f*, 609.
See also Thyroid-
stimulating hormone
Thyrotropin-releasing hormone
(TRH), 620*f*
Thyroxine, 592
Tibialis anterior, 399*t*, 400
Tidal ventilation, 415–16
Tidal volume, 416
Tiger keelback snake, 222*b*

Tiger salamander, 199
alimentary canal of, 526*f*
cutaneous respiration
of, 419*f*
Tiktaalik, 103, 336
Time, 138–39
Tinamous, 118
Tines, 234
Tissue fluid, 496
Tissue response, to mechanical
stress, 151–55
Tissues
adrenocortical, 598,
599*f*, 600
chromaffin, 598
connective, 177, 180–82
endocrine, 593*t*
lymphatic, 498
nervous, 177
target, 592, 620–21
Titanophoneus, 122*f*
Titanotheres, 134, 136*f*
Toads, 108
alimentary canal, 526*f*
pancreatic islets, 602
toxic glands in, 222*b*
vocalization by, 435*b*
Tongue, 516–17
Tongue bar, 58
Tongue flicking, in snakes, 681*f*
Tonic fibers, 378–79
Tooth anatomy, 507
Tooth attachment, types
of, 512*f*
Tooth development, 508,
511*f*, 512
Toothed whales, 125
Tooth fractures, 151
Tooth structure, 507*f*
Tornaria larva, 56, 56*f*, 59
Torpedo
electric organs in, 373*f*
genital ducts (female), 570*f*
genital ducts (male), 574*f*
Torques, 142–44
Torsion, 305
Tortoise, 112
Total group, 28
Total tension, 380
Toucans, 360
Toxins
birds, 230*b*
digestion of, 542
snakes and toads, 222*b*
Trabeculae, 154, 183, 242, 473
Trabeculate, 473
Trachea, 416, 438
Traditional systematics, 25
Trailing foot, 357
Transcranial joint, 268
Transducer, 672
Transfer, 422

Transformation grids, 134, 136, 137f
Transitional epithelium, 180
Transverse plane, 16
Transverse process, 297
Transverse septum, 196
Transverse tubules, 375
Transversospinalis, 397
Transversus abdominis, 434
Trapezius muscle group, 397, 399t
Triceps, 398, 399t
Triceratops, 33f, 115
Trigeminal nerve (trigeminus), 633–34, 638f
Triiodothyronine, 593
Trimming, 307
Trionyx
  axial skeleton of, 315f
  skull of, 269f
Tripartite concept, of kidney organization, 547–50
Tripodal gait, 351f
Triturus viridescens, thyroid and parathyroid glands in, 597f
Trochlear nerve, 633, 634t, 638f, 639t
Trophoblast, 166, 167
Tropical catfish, accessory airbreathing organs of, 417–18
Tropical frogs, 130f
Tropical salamanders, 199
Tropidonotus, cloaca, 584f
Trot, 350, 351f, 356, 357f
Trout
  cutaneous respiration, 419f
  differentiation in, 172f
  eyes, 686, 688f
  fluid mechanics, 719f
  gastrulation in, 172f
  swimming specializations in, 401b
  visual accommodation in, 688f
True horns, 235
True ribs, 298
Truncus arteriosus, 473
Trunk, 57
Trunk region, 295
Trunk vertebra, 298f, 306f, 318f
Trypsin, 534
TSH. See Thyroid-stimulating hormone
Tuberculum, 298
Tuberous receptor, 709, 711f
Tubotympanic recess, 518
Tubular exocrine glands, 180
Tubular nerve cord, 54
Tubulidentata, 125

Tubuli recti, 563
Tulerpeton, 105, 342
Tuna
  ram ventilation in, 421
  swimming specializations in, 401b
  thyroid glands in, 595f
Tunic, 67
Tunica adventitia, 452
Tunica albuginea, 563
Tunica intima, 452
Tunica media, 452
Tunicates, 66
Tunicin, 68
Turbinals, 676
Turbinate bones, 495
Turbinates, 277
Turkey, muscle fiber colors in, 377–78
Turtles, 111–12
  alimentary canal, 526f
  axial skeleton, 313, 315f
  blood flow through heart of, 482f
  cloaca, 584f
  cross section of body, 315f
  cutaneous respiration, 417, 419f
  fossorial locomotion, 369
  heart, 481–483
  integument, 220–21
  lymphatic vessels, 498f, 499f
  magnetic fields and, 712–13
  manus, 345f
  mucus, 237
  nitrogen elimination, 554
  pectoral girdle, 340f, 341f
  penis, 578–79, 581f
  plastron, 299, 301f
  skulls, 268, 269f
  stomach, 528f
  thyroid glands, 595f
  venous vessels, 472f
  ventilation, 434, 437f
  visual accommodation, 687f
Turtle shell, 237, 237f
Tusks, 515
Twitch fibers, 378–79
Two-stroke buccal pump, 422, 423f
Tympanic bulla, 277
Tympanic ear, 707
Tympanum, 108, 698, 702
Type, diversity of, 723, 724f
Typhlosole, 525
Tyrannosaurus, 115, 130f, 343f

Ulnare, 343
Ultimobranchial body, 596–97
Ultraviolet (UV) radiation, 679, 687–88

Umbilical arteries, 492
Umbilical vein, 492
Umbilical vessels, 191
Uncinate process, 299
Underfur, 230
Undulatory locomotion, lateral, 311f
Unguis, 233f
Unguligrade posture, 355
Unicellular gland, 180, 216
Unidirectional food flow, 264
Unidirectional ventilation, 415f
Uniform extinctions, 41
Uniform pool, 440, 443f
Unilaminar protoderm, 166
Unipolar neurons, 626, 627f
Units, 139
Unmyelinated nerves, 626, 627f
Upper jaw, 253
Upper temporal bar, 110
Up-regulation, 542
Upstroke, 360, 362f
Upstroke-downstroke transition, 360
Uranoscopus, electric organs in, 373f
Ureotelism, 554, 555f
Ureter, 551, 566f
Ureteric diverticulum, 547
Urethra, 551
Uricotelism, 554, 555f
Urinary bladder, 551, 586, 588
Urinary ducts, accessory, 573
Urinary system, 545
  embryonic development, 546–63
  evolution, 562–63
  kidney function and structure, 553–62
  kidney phylogeny, 550–53
  kidney structure (mammals), 545–47
  overview, 589, 591
Urine, 545, 560b
Uriniferous kidney, 566
Uriniferous tubule, 551, 552f
Urochordata, 50, 66–74, 67f, 74f
Urochordates, 594
Urodela, 107–8, 604t
Urodeles
  genital ducts (male), 576f
  sternum, 300f
  venous channels, 471f
Urodeum, 529, 582, 584
Urogenital papilla, 582
Urogenital ridge, 565f
Urogenital sinus, 545, 582, 586f
Urogenital system. See also
  Reproductive system;
  Urinary system
  ducts, 566f

  of male fishes, 574f
  of teleost female, 570
Uroproctodeal fold, 582, 584
Uropygial gland, 222
Ussher, James, 10
Uterine cycles, in humans, 615f
Uterus, 193f, 563, 570–72, 572, 585
Utriculus, 696
Uvea, 681–82

Vagina, 572
Vagus nerve, 634, 635f, 639t, 678
Vallecula, 519
Vane, 223
Varanid lizards, hemodynamics of, 484–86b
Varanus, copulatory organs in, 577f
Varanus exanthematicus, heart of, 484b, 486b
Varanus niloticus, heart of, 484–486, 484b, 485b
Variation, 10–11
Vasa recta, 559
Vasa vasorum, 452f
Vascular bone, 182
Vascularization, 213, 527
Vascular supply, within pituitary gland, 608f
Vas deferens, 563, 566f
Vasoconstriction, 453
Vasodilation, 453
Vasopressin, 608, 609f
Vastus, 400
Vault series, 249t, 250
Vector algebra, 731–32
Vectors, 140–41
Vegetal pole, 163
Veins, 452–55, 460
Velar folds (velum), 424
Velar tentacles, 63
Velociraptors, 115
Velocity, 139–40, 146, 385
Velocity ratio, 143
Velum, 63
Venom gland, 532
Venous vessels, 466–73, 473f
Vent (cloacal opening), 504
Ventilation, 415. See also
  Respiratory system
  birds, 442f
  ciliary pumps, 446
  crocodilians, 437f
  dogs, 439f
  frogs, 433f
  hagfishes, 425f
  lampreys, 425f
  lizards, 434f
  locomotion and, 434, 438f, 440f

mammals, 435–37, 438, 439f, 440f
muscular pumps, 446
perfusion and, 443, 444f
sharks, 426f
sturgeons, 429b
tadpoles, 432f
teleost, 427f
turtles, 434, 437f
Ventilatory mechanisms, 421–24, 449f
Ventral, 18f
Ventral abdominal vein, 466, 472
Ventral aorta, 457
Ventral arches, 295
Ventral blood vessel, 59
Ventral branchial muscles, 406
Ventral horns, 649
Ventral lobe, 606
Ventral mesentery, 168f
Ventral muscles, 397, 399t, 400
Ventral pancreatic diverticulum, 534
Ventral ramus, 394
Ventral region, 16
Ventral ribs, 298
Ventral roots, 629, 641
Ventral thalamus, 658, 660
Ventricles, 456, 473, 646, 647f
Ventro-bronchi, 438
Venules, 452f, 453
Vermiform appendix, 530
Verne, Jules, 447b
Verrell, P., 580f
Vertebrae, 83. See also Axial skeleton
archetype, 13, 14f
components of, 295–300, 301f
design of, 317–18
embryonic development and, 203f
Vertebral column, 83–84, 294, 313–14
design of, 317f
form and function, 314–21, 322f
regionalization of, 318–21, 322f
regions of, 295
Vertebral rib, 299
Vertebral segment, 295
Vertebrate design. See Design; Form; Function; Morphology
Vertebrates, 82
agnathans, 86–94
amniotes, 108–26
classification, 86

diversity of, 83f
gnathostomes, 94–97
innovations, 83–84
origin of, 84–86, 562–63
teleosts, 97–104
tetrapods, 104–8
Vessels. See also Arteries; Capillaries; Veins
arterial, 461–66, 467f, 468f
blood, 452f
lymphatic, 496–98, 499f
venous, 463f, 466–73
Vestibular apparatus, 695–96, 697f, 698f, 706f
Vestibular epithelium, 696
Vestibule, 676
V-formation flying, 364, 366f
Viagra, 579b
Vibrissae, 34, 122, 231
Villi, 523
Viperid snake, venom gland in, 532f
Vipers, striking features of, 274b
Visceral cranium. See Splanchnocranium
Visceral ganglia, 68, 630
Visceral muscles, 373
Visceral nerves, 628, 629f, 634t
Visceral reflex arc, 640, 645
Visceral sensory organs, 672
Visceral skeleton, 393
Viscosity, 146
Visual accommodation, 684–86, 687f, 688f
Visual perception, 689–90, 691f
Visual receptors, 713
Vitelline membrane, 163
Vitelline veins, 466
Vitelline vessels, 191
Vitellogenin, 163
Vitreal chamber, 682
Vitreous body, 682
Vitreous humor, 682
Viviparity, 164, 589
Viviparous vertebrates, 164
Vocalization
amphibians, 435b
humans, 511b
Volkmann's canal, 184f
Volume, 129, 132–33, 148
Voluntary muscles, 373
Vomer bone, 249f, 249t, 250
Vomerolfaction, 678, 713
Vomeronasal area, 677–78, 681f
Vomeronasal organ, 516, 676, 678, 681f
Vomodors, 678
Von Baer, Karl Ernst, 202
Von Baer's law, 202
Vortices, 364, 366f

Voyage of the Beagle, The (Darwin), 4f
Vultures, flight of, 361

Wake, D., 517f
Walk, 358f
Wallace, A. R., 8–9, 8f, 10, 11
Walls, G. A., 685f, 687f, 694b
Walruses, 125, 515f
Warblers, adrenal tissues of, 600f
Warrens, 369
Water. See also Fluid environment
breathing in, 444
feeding in, 285–87
vision in, 684–86, 687f, 688f
Water balance, 554–62
Waterbirds, lifestyles of, 118f
Water conservation, 558–59
Water conservers, 556
Water elimination, 557–58
Water flea, 130f
Water fluxes, 556
Waterfowl, flight of, 361
Water moccasin, skull of, 273f
Water pressure, 147f
Water-to-land transition, 446–449
Water ventilation, through dual pump, 421–22
Watt, 140
Watt, James, 140
Wax glands, 231
Weasels, fossorial locomotion, 367
Weberian ossicles, 701
Weddell seals, jaws of, 284, 286f
Weight, 140
Weight distribution, 145f
Weijs, W. A., 43f
Wells, G. P., 130f
Wells, H. G., 130f
Whalebone. See Baleen
Whales, 125, 130f
axial skeletons, 322f
baleen, 235–36
blowholes, 431b
diversity and patterns in, 724f
evolution of, 289b, 290b
heat transfer, 495
stomach, 528f
teeth, 515f
Whale shark, 130f
Wheel organ, 62
White matter, 649, 653f
White muscles, 373, 388, 389f
Whole muscle force generation, 379–80
Whole muscle tension-length curves, 380, 381f
Wilbur, Bryan, 61f

Wing-assisted inclined running (WAIR), 369
Wing design, 363f
Wing shape, 364f
Wisdom of God Manifested in the Works of the Creation, The (Ray), 6
Wolffian duct, 565, 566f
Wolff's Law, 154–55
Woodburne, R. T., 648f
"Woodpecker" finch, 46
Woodpeckers
adrenal tissues, 600f
tongue protrusion in, 517f
Woolly mammoths, 32
Woolly monkey, alimentary canal of, 526f
Work, 140
Worms
acorn, 56f, 57–60, 57f
segmentation of, 17
Woven bone, 182
Wulst, 663

Xenarthra, 124f, 126
Xiphihumeralis, 398
Xiphisternum, 299
Xiphoid cartilage, 299
X-rays, 42

Yarrell, William, 65b
Yaw, 327
Yolk, 166, 167f, 504
Yolk accumulation, 163t
Yolk platelets, 163
Yolk sac, 170, 190, 191, 191t
Yolk sac placenta, 192
Yolk stalk, 504
Young, Bruce A., 42f
Younginiformes, skulls of, 268
Yunnanozoon, 89

Zebra fish, discoidal cleavage in, 166f
Zebras, alimentary canal of, 526f
Zeugopodium, 326
Zona fasciculata, 601
Zona glomerulosa, 601
Zonal bone, 182
Zona pellucida, 163, 167f
Zona reticularis, 601
Zooid, 60
Zygantrum, 313, 316f
Zygapophyses, 297
Zygomatic arch, 274
Zygomatic gland, 532
Zygomaticus, 408
Zygorthiza, axial skeletons in, 322f
Zygosphene, 313, 316f
Zygotes, 163, 165f